Intermediate Algebra

Intermediate Algebra

Sixth Edition

Elayn Martin-Gay

University of New Orleans

PEARSON

Boston   Columbus   Indianapolis   New York   San Francisco   Upper Saddle River
Amsterdam   Cape Town   Dubai   London   Madrid   Milan   Munich   Paris   Montréal   Toronto
Delhi   Mexico City   São Paulo   Sydney   Hong Kong   Seoul   Singapore   Taipei   Tokyo

Editorial Director, Mathematics: *Christine Hoag*
Acquisitions Editor: *Mary Beckwith*
Executive Content Editor: *Kari Heen*
Associate Content Editor: *Christine Whitlock*
Editorial Assistant: *Matthew Summers*
Executive Director, Development: *Carol Trueheart*
Senior Development Editor: *Dawn Nuttall*
Senior Managing Editor: *Karen Wernholm*
Production Project Manager: *Patty Bergin*
Senior Design Specialist: *Heather Scott*
Associate Director of Design, USHE North and West: *Andrea Nix*
Digital Assets Manager: *Marianne Groth*
Supplements Production Project Manager: *Katherine Roz*
Executive Manager, Course Production: *Peter Silvia*
Media Producer: *Audra Walsh*
Director of Content Development: *Rebecca Williams*
Content Project Supervisor: *Janet Szykowny*
Executive Marketing Manager: *Michelle Renda*
Marketing Assistant: *Susan Mai*
Senior Author Support/Technology Specialist: *Joe Vetere*
Senior Media Buyer: *Ginny Michaud*
Permissions Project Supervisor: *Michael Joyce*
Procurement Specialist: *Linda Cox*
Production Management, Interior Design, Composition, and Answer Art: *Integra*
Text Art: *Scientific Illustrators*
Cover Design and Image: *Tamara Newnam*

Many of the designations used by manufacturers and sellers to distinguish their products are claimed as trademarks. Where those designations appear in this book, and Pearson Education was aware of a trademark claim, the designations have been printed in initial caps or all caps.

Library of Congress Cataloging-in-Publication Data

Martin-Gay, K. Elayn
   Intermediate algebra / Elayn Martin-Gay.—6th ed.
     p. cm.
   ISBN-13: 978-0-321-78504-6
   ISBN-10: 0-321-78504-5
  1. Algebra—Textbooks.   I. Title.
   QA152.3.M36 2013
   512.9—dc23
                              2011013318

2 3 4 5 6 7 8 9 10—CKV—15   14   13   12

www.pearsonhighered.com

ISBN-10: 0-321-78504-5
ISBN-13: 978-0-321-78504-6

This book is dedicated to students everywhere—and we should all be students. After all, is there anyone among us who really knows too much? Take that hint and continue to learn something new every day of your life.

Best of wishes from a fellow student: Elayn Martin-Gay

Contents

APPENDICES

Student Resources

These resources, located in the back of the text, give you a variety of tools conveniently located in one place to help you succeed in math.

Study Skills Builders

Attitude and Study Tips:

1. Have You Decided to Complete This Course Successfully?
2. Tips for Studying for an Exam
3. What to Do the Day of an Exam
4. Are You Satisfied with Your Performance on a Particular Quiz or Exam?
5. How Are You Doing?
6. Are You Preparing for Your Final Exam?

Organizing Your Work:

7. Learning New Terms
8. Are You Organized?
9. Organizing a Notebook
10. How Are Your Homework Assignments Going?

MyMathLab and MathXL:

11. Tips for Turning in Your Homework on Time
12. Tips for Doing Your Homework Online
13. Organizing Your Work
14. Getting Help with Your Homework Assignments
15. Tips for Preparing for an Exam
16. How Well Do You Know the Resources Available to You in MyMathLab?

Additional Help Inside and Outside Your Textbook:

17. How Well Do You Know Your Textbook?
18. Are You Familiar with Your Textbook Supplements?
19. Are You Getting All the Mathematics Help That You Need?

Bigger Picture–Study Guide Outline

Practice Final Exam

Answers to Selected Exercises

A New Tool to Help You Succeed

Introducing Martin-Gay's New Student Organizer

The new **Student Organizer** guides you through three important parts of studying effectively—note-taking, practice, and homework.

It is designed to help you organize your learning materials and develop the study habits you need to be successful. The Student Organizer includes:

- How to prepare for class
- Space to take class notes
- Step-by-step worked examples
- Your Turn exercises (modeled after the examples)
- Answers to the Your Turn exercises as well as worked-out solutions via references to the Martin-Gay text and videos
- Helpful hints and directions for completing homework assignments

A flexible design allows instructors to assign any or all parts of the Student Organizer.

The Student Organizer is available in a loose-leaf, notebook-ready format. It is also available for download in MyMathLab.

For more information, please go to

www.pearsonhighered.com/martingay

www.mypearsonstore.com
   (search Martin-Gay, Intermediate Algebra, Sixth Edition)
your Martin-Gay MyMathLab® course

Martin-Gay Video Resources to Help You Succeed

Interactive DVD Lecture Series

Active Learning at Your Pace

Designed for use on your computer or DVD player, these interactive videos include a 15–20 minute lecture for every section in the text as well as Concept Checks, Study Skills Builders, and a Practice Final Exam.

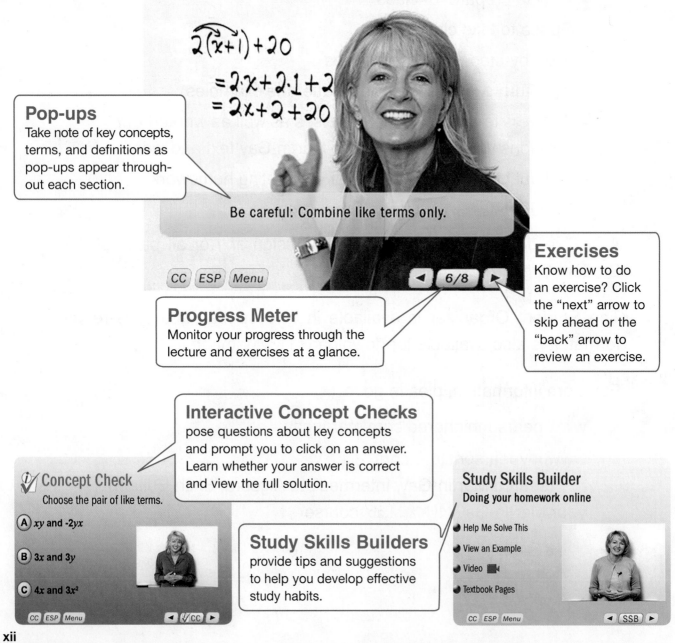

Pop-ups
Take note of key concepts, terms, and definitions as pop-ups appear throughout each section.

$$2(x+1) + 20$$
$$= 2 \cdot x + 2 \cdot 1 + 2$$
$$= 2x + 2 + 20$$

Be careful: Combine like terms only.

CC ESP Menu   ◄ 6/8 ►

Exercises
Know how to do an exercise? Click the "next" arrow to skip ahead or the "back" arrow to review an exercise.

Progress Meter
Monitor your progress through the lecture and exercises at a glance.

Interactive Concept Checks
pose questions about key concepts and prompt you to click on an answer. Learn whether your answer is correct and view the full solution.

✓ Concept Check
Choose the pair of like terms.

Ⓐ xy and $-2yx$

Ⓑ $3x$ and $3y$

Ⓒ $4x$ and $3x^2$

CC ESP Menu   ◄ ✓/CC ►

Study Skills Builders
provide tips and suggestions to help you develop effective study habits.

Study Skills Builder
Doing your homework online

● Help Me Solve This
● View an Example
● Video
● Textbook Pages

CC ESP Menu   ◄ SSB ►

Chapter Test Prep Videos

Step-by-step solutions on video for all chapter test exercises from the text. Available via:

- Interactive DVD Lecture Series
- MyMathLab®
- You Tube™

English and Spanish Subtitles Available

AlgebraPrep Apps for the iPhone™ and iPod Touch®

Your 24/7 Algebra Tutor—Anytime, Anywhere!

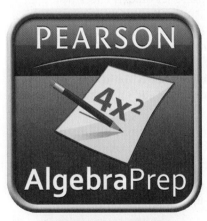

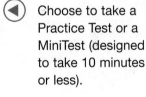

Choose to take a Practice Test or a MiniTest (designed to take 10 minutes or less).

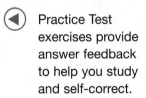

Practice Test exercises provide answer feedback to help you study and self-correct.

Step-by-step video solutions give you the guidance of an expert tutor whenever you need help.

Preface

Intermediate Algebra, **Sixth Edition,** was written to provide a solid foundation in algebra for students who might not have previous experience in algebra. Specific care was taken to make sure students have the most up-to-date, relevant text preparation for their next mathematics course or for nonmathematical courses that require an understanding of algebraic fundamentals. I have tried to achieve this by writing a user-friendly text that is keyed to objectives and contains many worked-out examples. As suggested by AMATYC and the NCTM Standards (plus Addenda), real-life and real-data applications, data interpretation, conceptual understanding, problem solving, writing, cooperative learning, appropriate use of technology, mental mathematics, number sense, estimation, critical thinking, and geometric concepts are emphasized and integrated throughout the book.

The many factors that contributed to the success of the previous editions have been retained. In preparing the Sixth Edition, I considered comments and suggestions of colleagues, students, and many users of the prior edition throughout the country.

What's New in the Sixth Edition?

- **The Martin-Gay Program** has been revised and enhanced with a new design in the text and MyMathLab to actively encourage students to use the text, video program, and Student Organizer as an integrated learning system.

- **The Student Organizer** is designed by me to help students develop the study habits they need to be successful. This Organizer guides students through the three main components of studying effectively—note-taking, practice, and homework—and helps them develop the habits that will enable them to succeed in future courses. The Student Organizer can be packaged with the text in loose-leaf, notebook-ready format and is also available for download in MyMathLab.

- **New Vocabulary, Readiness & Video Check** questions have been added prior to every section exercise set. These exercises quickly check a student's understanding of new vocabulary words. The **readiness** exercises center on a student's understanding of a concept that is necessary in order to continue to the exercise set. **New video check questions for the Martin-Gay Interactive Lecture videos** are now included in every section for each learning objective. **These exercises are all available for assignment in MyMathLab** and are a great way to assess whether students have viewed and understood the key concepts presented in the videos.

- **The Interactive DVD Lecture Series,** featuring your text author (Elayn Martin-Gay), provides students with active learning at their own pace. The videos offer the following resources and more:

  A complete lecture for each section of the text highlights key examples and exercises from the text. New "pop-ups" reinforce key terms, definitions, and concepts.

  An interface with menu navigation features allows students to quickly find and focus on the examples and exercises they need to review.

  Interactive Concept Check exercises measure students' understanding of key concepts and common trouble spots.

  The Interactive DVD Lecture Series also includes the following resources for test prep:

  The Practice Final Exam helps students prepare for an end-of-course final. Students can watch full video solutions to each exercise.

The Chapter Test Prep Videos help students during their most teachable moment—when they are preparing for a test. This innovation provides step-by-step solutions for the Chapter Test exercises found at the end of each chapter in the text. The videos are captioned in English and Spanish. For the Sixth Edition, the chapter test prep videos are also available on YouTube™.

- **The Martin-Gay MyMathLab course** has been updated and revised to provide more exercise coverage, including assignable video check questions, and an expanded video program. There are section lecture videos for every section, students can also access at the specific objective level, and there are an increased number of watch clips at the exercise level to help students while doing homework in MathXL. Suggested homework assignments have been premade for assignment at the instructor's discretion.

- **New MyMathLab Ready to Go courses** (access code required) provide students with all the same great MyMathLab features that you're used to, but make it easier for instructors to get started. Each course includes preassigned homework and quizzes to make creating your course even simpler. Ask your Pearson representative about the details for this particular course or to see a copy of this course.

- **A new section** (9.4) devoted specifically to exponential growth and decay and applications has been added. This section includes the definition and examples of half-life.

- **The new Student Resources** section, located in the back of the text, gives students a variety of tools that are conveniently located in one place to help them achieve success in math.

  — **Study Skills Builders** give students tips and suggestions on successful study habits and help them take responsibility for their learning. Assignable exercises check students' progress in improving their skills.

  — The **Bigger Picture—Study Guide Outline** covers key concepts of the course—simplifying expressions and solving equations and inequalities—to help students transition from thinking section-by-section to thinking about how the material they are learning fits into mathematics as a whole. This outline provides a model for students on how to organize and develop their own study guide.

  — The **Practice Final Exam** helps students prepare for the end-of-the-course exam. Students can also watch the step-by-step solutions to all the Practice Final Exam exercises on the new Interactive DVD Lecture Series and in MyMathLab.

  — The **Answers to Selected Exercises** section allows students to check their answers for all Practice exercises; odd-numbered Vocabulary, Readiness & Video Check exercises; odd-numbered section exercises; Chapter Review and Cumulative Review exercises; and all Integrated Review and Chapter Test exercises.

- **New guided application exercises** appear in many sections throughout the text, beginning with Section 2.2. These applications prompt students on how to set up the application and get started with the solution process. These guided exercises will help students prepare to solve application exercises on their own.

- **Enhanced emphasis on Study Skills** helps students develop good study habits and makes it more convenient for instructors to incorporate or assign study skills in their courses. The following changes have been made in the Sixth Edition:

  Section 1.1, Tips for Success in Mathematics, has been updated to include helpful hints for doing homework online in MyMathLab. Exercises pertaining to doing homework online in MyMathLab are now included in the exercise set for 1.1.

The Study Skills Builders, formerly located at the end of select exercise sets, are now included in the new **Student Resources** section at the back of the book and organized by topic for ease of assignment. This section now also includes new Study Skills Builders on doing homework online in MyMathLab.

- All exercise sets have been reviewed and updated to ensure that even- and odd-numbered exercises are paired.

Key Pedagogical Features

The following key features have been retained and/or updated for the Sixth Edition of the text:

Problem-Solving Process   This is formally introduced in Chapter 2 with a four-step process that is integrated throughout the text. The four steps are **Understand, Translate, Solve,** and **Interpret.** The repeated use of these steps in a variety of examples shows their wide applicability. Reinforcing the steps can increase students' comfort level and confidence in tackling problems.

Exercise Sets Revised and Updated   The exercise sets have been carefully examined and extensively revised. Special focus was placed on making sure that even- and odd-numbered exercises are paired.

Examples   Detailed, step-by-step examples were added, deleted, replaced, or updated as needed. Many of these reflect real life. Additional instructional support is provided in the annotated examples.

Practice Exercises   Throughout the text, each worked-out example has a parallel Practice Exercise. These invite students to be actively involved in the learning process. Students should try each Practice Exercise after finishing the corresponding example. Learning by doing will help students grasp ideas before moving on to other concepts. Answers to the Practice Exercises are provided in the back of the text.

Helpful Hints   Helpful Hints contain practical advice on applying mathematical concepts. Strategically placed where students are most likely to need immediate reinforcement, Helpful Hints help students avoid common trouble areas and mistakes.

Concept Checks   This feature allows students to gauge their grasp of an idea as it is being presented in the text. Concept Checks stress conceptual understanding at the point-of-use and help suppress misconceived notions before they start. Answers appear at the bottom of the page. Exercises related to Concept Checks are included in the exercise sets.

Mixed Practice Exercises   Found in the section exercise sets, each requires students to determine the problem type and strategy needed to solve it just as they would need to do on a test.

Integrated Reviews   A unique, mid-chapter exercise set that helps students assimilate new skills and concepts that they have learned separately over several sections. These reviews provide yet another opportunity for students to work with "mixed" exercises as they master the topics.

Vocabulary Check   Provides an opportunity for students to become more familiar with the use of mathematical terms as they strengthen their verbal skills. These appear at the end of each chapter before the Chapter Highlights. Vocabulary, Readiness & Video Check exercises also provide vocabulary practice at the section level.

Chapter Highlights Found at the end of every chapter, these contain key definitions and concepts with examples to help students understand and retain what they have learned and help them organize their notes and study for tests.

Chapter Review The end of every chapter contains a comprehensive review of topics introduced in the chapter. The Chapter Review offers exercises keyed to every section in the chapter, as well as Mixed Review exercises that are not keyed to sections.

Chapter Test and Chapter Test Prep Video The Chapter Test is structured to include those problems that involve common student errors. The **Chapter Test Prep Videos** give students instant access to a step-by-step video solution of each exercise in the Chapter Test.

Cumulative Review Follows every chapter in the text (except Chapter 1). Each odd-numbered exercise contained in the Cumulative Review is an earlier worked example in the text that is referenced in the back of the book along with the answer.

Writing Exercises ＼ These exercises occur in almost every exercise set and require students to provide a written response to explain concepts or justify their thinking.

Applications Real-world and real-data applications have been thoroughly updated and many new applications are included. These exercises occur in almost every exercise set and show the relevance of mathematics and help students gradually, and continuously, develop their problem-solving skills.

Review and Preview Exercises These exercises occur in each exercise set (except in Chapter 1) and are keyed to earlier sections. They review concepts learned earlier in the text that will be needed in the next section or chapter.

Exercise Set Resource Icons Located at the opening of each exercise set, these icons remind students of the resources available for extra practice and support:

See Student Resources descriptions on page xviii for details on the individual resources available.

Exercise Icons These icons facilitate the assignment of specialized exercises and let students know what resources can support them.

    ●   Video icon: exercise worked on the Interactive DVD Lecture Series and in MyMathLab.

    △   Triangle icon: identifies exercises involving geometric concepts.

    ＼   Pencil icon: indicates a written response is needed.

    ▦   Calculator icon: optional exercises intended to be solved using a scientific or graphing calculator.

Optional: Calculator Exploration Boxes and Calculator Exercises The optional Calculator Explorations provide key strokes and exercises at appropriate points to give an opportunity for students to become familiar with these tools. Section exercises that are best completed by using a calculator are identified by ▦ for ease of assignment.

Student and Instructor Resources

STUDENT RESOURCES

Student Organizer

Guides students through the 3 main components of studying effectively–note-taking, practice, and homework.

The organizer includes before-class preparation exercises, note-taking pages in a 2-column format for use in class, and examples paired with exercises for practice for each section. It is 3-hole-punched. Also available in MyMathLab.

Student Solutions Manual

Provides complete worked-out solutions to

- the odd-numbered section exercises; all Practice Exercises; all exercises in the Integrated Reviews, Chapter Reviews, Chapter Tests, and Cumulative Reviews

Interactive DVD Lecture Series

Provides students with active learning at their pace. The videos offer:

- A complete lecture for each text section. The interface allows easy navigation to examples and exercises students need to review.
- Interactive Concept Check exercises
- Study Skills Builders
- Practice Final Exam
- Chapter Test Prep Videos

Chapter Test Prep Videos

- Step-by-step solutions to every exercise in each Chapter Practice Test.
- Available in MyMathLab® and on YouTube, and in the Interactive DVD Lecture Series.

INSTRUCTOR RESOURCES

Annotated Instructor's Edition

Contains all the content found in the student edition, plus the following:

- Answers to exercises on the same text page
- Answers to graphing exercises and all video exercises
- Teaching Tips throughout the text placed at key points.
- Classroom Examples in the margin paired to each example in the text.

Instructor's Resource Manual with Tests and Mini-Lectures

- Mini-lectures for each text section
- Additional Practice worksheets for each section
- Several forms of test per chapter–free response and multiple choice
- Group activities
- Video key to the example number in the video questions and section exercises worked in the videos
- Answers to all items

Instructor's Solutions Manual
TestGen® (Available for download from the IRC)

Online Resources
MyMathLab® (access code required)

MathXL® (access code required)

Acknowledgments

There are many people who helped me develop this text, and I will attempt to thank some of them here. Cindy Trimble and Carrie Green were *invaluable* for contributing to the overall accuracy of the text. Dawn Nuttall, Courtney Slade, and JoAnne Thomasson were *invaluable* for their many suggestions and contributions during the development and writing of this Sixth Edition. Debbie Meyer and Amanda Zagnoli of Integra-Chicago provided guidance throughout the production process.

A very special thank you goes to my editor, Mary Beckwith, for being there 24/7/365, as my students say. Last, my thanks to the staff at Pearson for all their support: Patty Bergin, Heather Scott, Michelle Renda, Chris Hoag, and Greg Tobin.

I would like to thank the following reviewers for their input and suggestions:

Sandi Athanassiou, *University of Missouri-Columbia*
Michelle Beerman, *Pasco-Hernandez Community College*
Monika Bender, *Central Texas College*
Bob Hervey, *Hillsborough Community College*
Michael Maltenfort, *Truman College*
Jorge Romero, *Hillsborough Community College*
Joseph Wakim, *Brevard Community College*
Flo Wilson, *Central Texas College*
Marie Caruso and students, *Middlesex Community College*

I would also like to thank the following dedicated group of instructors who participated in our focus groups, Martin-Gay Summits, and our design review for the series. Their feedback and insights have helped to strengthen this edition of the text. These instructors include:

Billie Anderson, *Tyler Junior College*
Joey Anderson, *Central Piedmont Community College*
Cedric Atkins, *Mott Community College*
Teri Barnes, *McLennan Community College*
Andrea Barnett, *Tri-County Technical College*
Lois Beardon, *Schoolcraft College*
Michelle Beerman, *Pasco-Hernandez Community College*
Laurel Berry, *Bryant & Stratton College*
John Beyers, *University of Maryland*
Jennifer Brahier, *Pensacola Junior College*
Bob Brown, *Community College of Baltimore County–Essex*
Lisa Brown, *Community College of Baltimore County–Essex*
NeKeith Brown, *Richland College*
Sue Brown, *Guilford Technical Community College*
Gail Burkett, *Palm Beach State College*
Cheryl Cantwell, *Seminole Community College*
Janie Chapman, *Spartanburg Community College*
Jackie Cohen, *Augusta State College*
Julie Dewan, *Mohawk Valley Community College*
Janice Ervin, *Central Piedmont Community College*
Karen Estes, *St. Petersburg College*
Richard Fielding, *Southwestern College*
Sonia Ford, *Midland College*
Julie Francavilla, *State College of Florida*
Cindy Gaddis, *Tyler Junior College*
Nita Graham, *St. Louis Community College*
Pauline Hall, *Iowa State College*
Elizabeth Hamman, *Cypress College*
Kathy Hoffmaster, *Thomas Nelson Community College*
Pat Hussey, *Triton College*
Dorothy Johnson, *Lorain County Community College*
Sonya Johnson, *Central Piedmont Community College*

Irene Jones, *Fullerton College*
Paul Jones, *University of Cincinnati*
Mike Kirby, *Tidewater Community College*
Kathy Kopelousos, *Lewis and Clark Community College*
Nancy Lange, *Inver Hills Community College*
Judy Langer, *Westchester Community College*
Lisa Lindloff, *McLennan Community College*
Sandy Lofstock, *St. Petersburg College*
Kathy Lovelle, *Westchester Community College*
Jamie Malek, *Florida State College*
Jean McArthur, *Joliet Junior College*
Kevin McCandless, *Evergreen Valley College*
Daniel Miller, *Niagara County Community College*
Marcia Molle, *Metropolitan Community College*
Carol Murphy, *San Diego Miramar College*
Charlotte Newsom, *Tidewater Community College*
Greg Nguyen, *Fullerton College*
Eric Ollila, *Jackson Community College*
Linda Padilla, *Joliet Junior College*
Rena Petrello, *Moorpark College*
Davidson Pierre, *State College of Florida*
Marilyn Platt, *Gaston College*
Susan Poss, *Spartanburg Community College*
Natalie Rivera, *Estrella Mountain Community College*
Judy Roane, *Pearl River Community College*
Claudinna Rowley, *Montgomery Community College, Rockville*
Ena Salter, *State College of Florida*
Carole Shapero, *Oakton Community College*
Janet Sibol, *Hillsborough Community College*
Anne Smallen, *Mohawk Valley Community College*
Mike Stack, *South Suburban College*
Barbara Stoner, *Reading Area Community College*
Jennifer Strehler, *Oakton Community College*
Ellen Stutes, *Louisiana State University Eunice*
Tanomo Taguchi, *Fullerton College*
Sam Tinsley, *Richland College*
Linda Tucker, *Rose State College*
MaryAnn Tuerk, *Elgin Community College*
Gwen Turbeville, *J. Sargeant Reynolds Community College*
Walter Wang, *Baruch College*
Leigh Ann Wheeler, *Greenville Technical Community College*
Jenny Wilson, *Tyler Junior College*
Valerie Wright, *Central Piedmont Community College*

A special thank you to those students who participated in our design review: Katherine Browne, Mike Bulfin, Nancy Canipe, Ashley Carpenter, Jeff Chojnachi, Roxanne Davis, Mike Dieter, Amy Dombrowski, Kay Herring, Todd Jaycox, Kaleena Levan, Matt Montgomery, Tony Plese, Abigail Polkinghorn, Harley Price, Eli Robinson, Avery Rosen, Robyn Schott, Cynthia Thomas, and Sherry Ward.

Elayn Martin-Gay

About the Author

Elayn Martin-Gay has taught mathematics at the University of New Orleans for more than 25 years. Her numerous teaching awards include the local University Alumni Association's Award for Excellence in Teaching, and Outstanding Developmental Educator at University of New Orleans, presented by the Louisiana Association of Developmental Educators.

Prior to writing textbooks, Elayn Martin-Gay developed an acclaimed series of lecture videos to support developmental mathematics students in their quest for success. These highly successful videos originally served as the foundation material for her texts. Today, the videos are specific to each book in the Martin-Gay series. The author has also created Chapter Test Prep Videos to help students during their most "teachable moment"—as they prepare for a test—along with Instructor-to-Instructor videos that provide teaching tips, hints, and suggestions for each developmental mathematics course, including basic mathematics, prealgebra, beginning algebra, and intermediate algebra. Her most recent innovations are the AlgebraPrep Apps for the iPhone and iPod Touch. These Apps embrace the different learning styles, schedules, and paces of students and provide them with quality math tutoring.

Elayn is the author of 12 published textbooks as well as interactive multimedia mathematics, all specializing in developmental mathematics courses. She has participated as an author across the broadest range of educational materials: textbooks, videos, tutorial software, and courseware. This provides the opportunity of various combinations for an integrated teaching and learning package that offers great consistency for the student.

Applications Index

Real Numbers and Algebraic Expressions

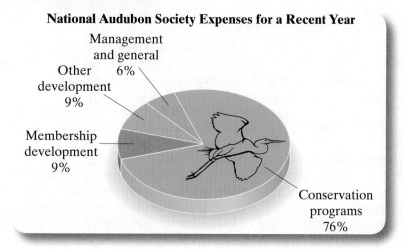

National Audubon Society Expenses for a Recent Year

Management and general 6%

Other development 9%

Membership development 9%

Conservation programs 76%

The National Audubon Society is a U.S. nonprofit organization dedicated to conservation. It is named in honor of John James Audubon, a Franco-American naturalist who painted and described the birds of North America in his famous book *Birds of America* published in sections between 1827 and 1838.

The Audubon Society is over a century old and funds conservation programs focusing on birds.

The bar graph below shows the differing wingbeats per second for selected birds. In the Chapter 1 Review, Exercises 3 and 4, we study the hummingbird wingbeats per second further.

In arithmetic, we add, subtract, multiply, divide, raise to powers, and take roots of numbers. In algebra, we add, subtract, multiply, divide, raise to powers, and take roots of variables. Letters, such as x, that represent numbers are called **variables.** Understanding these algebraic expressions depends on your understanding of arithmetic expressions. This chapter reviews the arithmetic operations on real numbers and the corresponding algebraic expressions.

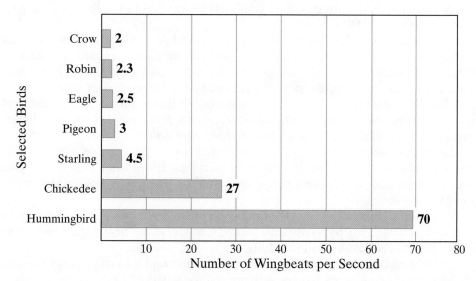

Source: National Audubon Society website

1.1 Tips for Success in Mathematics

OBJECTIVES

1 Get Ready for This Course.

2 Understand Some General Tips for Success.

3 Understand How to Use This Text.

4 Get Help as Soon as You Need It.

5 Learn How to Prepare for and Take an Exam.

6 Develop Good Time Management.

Before reading this section, remember that your instructor is your best source of information. Please see your instructor for any additional help or information.

OBJECTIVE

1 Getting Ready for This Course

Now that you have decided to take this course, remember that a *positive attitude* will make all the difference in the world. Your belief that you can succeed is just as important as your commitment to this course. Make sure you are ready for this course by having the time and positive attitude that it takes to succeed.

Next, make sure that you have scheduled your math course at a time that will give you the best chance for success. For example, if you are also working, you may want to check with your employer to make sure that your work hours will not conflict with your course schedule.

On the day of your first class period, double-check your schedule and allow yourself extra time to arrive on time in case of traffic problems or difficulty locating your classroom. Make sure that you bring at least your textbook, paper, and a writing instrument. Are you required to have a lab manual, graph paper, calculator, or some other supplies besides this text? If so, also bring this material with you.

OBJECTIVE

2 General Tips for Success

Below are some general tips that will increase your chance for success in a mathematics class. Many of these tips will also help you in other courses you may be taking.

Exchange names and phone numbers or email addresses with at least one other person in class. This contact person can be a great help if you miss an assignment or want to discuss math concepts or exercises that you find difficult.

Choose to attend all class periods. If possible, sit near the front of the classroom. This way, you will see and hear the presentation better. It may also be easier for you to participate in classroom activities.

Do your homework. You've probably heard the phrase "practice makes perfect" in relation to music and sports. It also applies to mathematics. You will find that the more time you spend solving mathematics exercises, the easier the process becomes. Be sure to schedule enough time to complete your assignments before the due date assigned by your instructor.

Check your work. Review the steps you made while working a problem. Learn to check your answers in the original problems. You may also compare your answers with the "Answers to Selected Exercises" section in the back of the book. If you have made a mistake, try to figure out what went wrong. Then correct your mistake. If you can't find what went wrong, don't erase your work or throw it away. Bring your work to your instructor, a tutor in a math lab, or a classmate. It is easier for someone to find where you had trouble if he or she looks at your original work.

Learn from your mistakes and be patient with yourself. Everyone, even your instructor, makes mistakes. (That definitely includes me—Elayn Martin-Gay.) Use your errors to learn and to become a better math student. The key is finding and understanding your errors.

Was your mistake a careless one, or did you make it because you can't read your own math writing? If so, try to work more slowly or write more neatly and make a conscious effort to check your work carefully.

Did you make a mistake because you don't understand a concept? Take the time to review the concept or ask questions to understand it better.

Did you skip too many steps? Skipping steps or trying to do too many steps mentally may lead to preventable mistakes.

Know how to get help if you need it. It's all right to ask for help. In fact, it's a good idea to ask for help whenever there is something that you don't understand.

Make sure you know when your instructor has office hours and how to find his or her office. Find out whether math tutoring services are available on your campus. Check on the hours, location, and requirements of the tutoring service.

▶ **Helpful Hint**

MyMathLab® and **MathXL®** If you are doing your homework online, you can work and rework those exercises that you struggle with until you master them. Try working through all the assigned exercises twice before the due date.

▶ **Helpful Hint**

MyMathLab® and **MathXL®** If you are completing your homework online, it's important to work each exercise on paper before submitting the answer. That way, you can check your work and follow your steps to find and correct any mistakes.

> **Helpful Hint**
>
> **MyMathLab®** and **MathXL®** When assignments are turned in online, keep a hard copy of your complete written work. You will need to refer to your written work to be able to ask questions and to study for tests later.

> **Helpful Hint**
>
> **MyMathLab®** and **MathXL®** Be aware of assignments and due dates set by your instructor. Don't wait until the last minute to submit work online. Allow 6–8 hours before the deadline in case you have technology trouble.

> **Helpful Hint**
>
> **MyMathLab®** In MyMathLab, you have access to the following video resources:
> - Lecture Videos for each section
> - Chapter Test Prep Videos
>
> Use these videos provided by the author to prepare for class, review, and study for tests.

Organize your class materials, including homework assignments, graded quizzes and tests, and notes from your class or lab. All of these items will make valuable references throughout your course and when studying for upcoming tests and the final exam. Make sure that you can locate these materials when you need them.

Read your textbook before class. Reading a mathematics textbook is unlike reading a novel or a newspaper. Your pace will be much slower. It is helpful to have paper and a pencil with you when you read. Try to work out examples on your own as you encounter them in your text. You should also write down any questions that you want to ask in class. When you read a mathematics textbook, sometimes some of the information in a section will be unclear. But after you hear a lecture or watch a lecture video on that section, you will understand it much more easily than if you had not read your text beforehand.

Don't be afraid to ask questions. You are not the only person in class with questions. Other students are normally grateful that someone has spoken up.

Turn in assignments on time. This way you can be sure that you will not lose points for being late. Show every step of a problem and be neat and organized. Also be sure that you understand which problems are assigned for homework. If allowed, you can always double-check the assignment with another student in your class.

OBJECTIVE

3   Using This Text

Many helpful resources are available to you. It is important to become familiar with and use these resources. They should increase your chances for success in this course.

- *Practice Exercises.* Each example in every section has a parallel Practice exercise. As you read a section, try each Practice exercise after you've finished the corresponding example. This "learn-by-doing" approach will help you grasp ideas before you move on to other concepts. Answers are at the back of the text.
- *Chapter Test Prep Videos.* These videos provide solutions to all of the Chapter Test exercises worked out by the author. This supplement is very helpful before a test or exam.
- *Interactive DVD Lecture Series.* Exercises marked with a ⊙ are fully worked out by the author on the DVDs. The lecture series provides approximately 20 minutes of instruction per section.
- *Symbols at the Beginning of an Exercise Set.* If you need help with a particular section, the symbols listed at the beginning of each exercise set will remind you of the numerous supplements available.
- *Examples.* The main section of exercises in each exercise set is referenced by an example(s). There is also often a section of exercises entitled "Mixed Practice," which combines exercises from multiple objectives or sections. These are mixed exercises written to prepare you for your next exam. Use all of this referencing if you have trouble completing an assignment from the exercise set.
- *Icons (Symbols).* Make sure that you understand the meaning of the icons that are beside many exercises. ⊙ tells you that the corresponding exercise may be viewed on the video segment that corresponds to that section. ⬎ tells you that this exercise is a writing exercise in which you should answer in complete sentences. △ tells you that the exercise involves geometry.
- *Integrated Reviews.* Found in the middle of each chapter, these reviews offer you a chance to practice–in one place–the many concepts that you have learned separately over several sections.
- *End-of-Chapter Opportunities.* There are many opportunities at the end of each chapter to help you understand the concepts of the chapter.

    Vocabulary Checks contain key vocabulary terms introduced in the chapter.

    Chapter Highlights contain chapter summaries and examples.

    Chapter Reviews contain review problems. The first part is organized section by section and the second part contains a set of mixed exercises.

    Chapter Tests are sample tests to help you prepare for an exam. The Chapter Test Prep Videos, found in this text, contain all the Chapter Test exercises worked by the author.

Cumulative Reviews are reviews consisting of material from the beginning of the book to the end of that particular chapter.

- *Student Resources in Your Textbook.* You will find a **Student Resources** section at the back of this textbook. It contains the following to help you study and prepare for tests:

  Study Skill Builders contain study skills advice. To increase your chance for success in the course, read these study tips and answer the questions.

  Bigger Picture–Study Guide Outline provides you with a study guide outline of the course, with examples.

  Practice Final provides you with a Practice Final Exam to help you prepare for a final. The video solutions to each question are provided in the Interactive DVD Lecture Series and within MyMathLab®.

- *Resources to Check Your Work.* The **Answers to Selected Exercises** section provides answers to all odd-numbered section exercises and all chapter test exercises.

OBJECTIVE

4 Getting Help

If you have trouble completing assignments or understanding the mathematics, get help as soon as you need it! This tip is presented as an objective on its own because it is so important. In mathematics, usually the material presented in one section builds on your understanding of the previous section. This means that if you don't understand the concepts covered during a class period, there is a good chance that you will not understand the concepts covered during the next class period. If this happens to you, get help as soon as you can.

Where can you get help? Many suggestions have been made in this section on where to get help, and now it is up to you to get it. Try your instructor, a tutoring center, or a math lab, or you may want to form a study group with fellow classmates. If you do decide to see your instructor or go to a tutoring center, make sure that you have a neat notebook and are ready with your questions.

OBJECTIVE

5 Preparing for and Taking an Exam

Make sure that you allow yourself plenty of time to prepare for a test. If you think that you are a little "math anxious," it may be that you are not preparing for a test in a way that will ensure success. The way that you prepare for a test in mathematics is important. To prepare for a test:

1. Review your previous homework assignments.

2. Review any notes from class and section-level quizzes you have taken. (If this is a final exam, also review chapter tests you have taken.)

3. Review concepts and definitions by reading the Chapter Highlights at the end of each chapter.

4. Practice working out exercises by completing the Chapter Review found at the end of each chapter. (If this is a final exam, go through a Cumulative Review. There is one found at the end of each chapter except Chapter 1. Choose the review found at the end of the latest chapter that you have covered in your course.) *Don't stop here!*

5. It is important to place yourself in conditions similar to test conditions to find out how you will perform. In other words, as soon as you feel that you know the material, get a few blank sheets of paper and take a sample test. A Chapter Test is available at the end of each chapter, or you can work selected problems from the Chapter Review. Your instructor may also provide you with a review sheet. During this sample test, do not use your notes or your textbook. Then check your sample test. If you are not satisfied with the results, study the areas that you are weak in and try again.

6. On the day of the test, allow yourself plenty of time to arrive at where you will be taking your exam.

▶ Helpful Hint

MyMathLab® and **MathXL®**

- Use the **Help Me Solve This** button to get step-by-step help for the exercise you are working. You will need to work an additional exercise of the same type before you can get credit for having worked it correctly.

- Use the **Video** button to view a video clip of the author working a similar exercise.

▶ Helpful Hint

MyMathLab® and **MathXL®** Review your written work for previous assignments. Then, go back and rework previous assignments. Open a previous assignment, and click **Similar Exercise** to generate new exercises. Rework the exercises until you fully understand them and can work them without help features.

When taking your test:

1. Read the directions on the test carefully.

2. Read each problem carefully as you take the test. Make sure that you answer the question asked.

3. Watch your time and pace yourself so that you can attempt each problem on your test.

4. If you have time, check your work and answers.

5. Do not turn your test in early. If you have extra time, spend it double-checking your work.

OBJECTIVE

6 Managing Your Time

As a college student, you know the demands that classes, homework, work, and family place on your time. Some days you probably wonder how you'll ever get everything done. One key to managing your time is developing a schedule. Here are some hints for making a schedule:

1. Make a list of all of your weekly commitments for the term. Include classes, work, regular meetings, extracurricular activities, etc. You may also find it helpful to list such things as laundry, regular workouts, grocery shopping, etc.

2. Next, estimate the time needed for each item on the list. Also make a note of how often you will need to do each item. Don't forget to include time estimates for the reading, studying, and homework you do outside of your classes. You may want to ask your instructor for help estimating the time needed.

3. In the exercise set that follows, you are asked to block out a typical week on the schedule grid given. Start with items with fixed time slots like classes and work.

4. Next, include the items on your list with flexible time slots. Think carefully about how best to schedule items such as study time.

5. Don't fill up every time slot on the schedule. Remember that you need to allow time for eating, sleeping, and relaxing! You should also allow a little extra time in case some items take longer than planned.

6. If you find that your weekly schedule is too full for you to handle, you may need to make some changes in your workload, classload, or other areas of your life. You may want to talk to your advisor, manager or supervisor at work, or someone in your college's academic counseling center for help with such decisions.

1.1 Exercise Set MyMathLab®

1. What is your instructor's name?

2. What are your instructor's office location and office hours?

3. What is the best way to contact your instructor?

4. Do you have the name and contact information of at least one other student in class?

5. Will your instructor allow you to use a calculator in this class?

6. Why is it important that you write step-by-step solutions to homework exercises and keep a hard copy of all work submitted?

7. Is a tutoring service available on campus? If so, what are its hours? What services are available?

8. Have you attempted this course before? If so, write down ways that you might improve your chances of success during this second attempt.

9. List some steps that you can take if you begin having trouble understanding the material or completing an assignment. If you are completing your homework in MyMathLab® and MathXL®, list the resources you can use for help.

10. How many hours of studying does your instructor advise for each hour of instruction?

11. What does the ╲ icon in this text mean?

12. What does the ▶ icon in this text mean?

13. What does the △ icon in this text mean?

14. What are Practice exercises?

15. When might be the best time to work a Practice exercise?

16. Where are the answers to Practice exercises?

17. What answers are contained in this text and where are they?

18. What and where are the study skills builders?

19. What and where are Integrated Reviews?

20. How many times is it suggested that you work through the homework exercises in MathXL® before the submission deadline?

21. How far in advance of the assigned due date is it suggested that homework be submitted online? Why?

22. Chapter Highlights are found at the end of each chapter. Find the Chapter 1 Highlights and explain how you might use it and how it might be helpful.

23. Chapter Reviews are found at the end of each chapter. Find the Chapter 1 Review and explain how you might use it and how it might be useful.

24. Chapter Tests are found at the end of each chapter. Find the Chapter 1 Test and explain how you might use it and how it might be helpful when preparing for an exam on Chapter 1. Include how the Chapter Test Prep Videos may help. If you are working in MyMathLab® and MathXL®, how can you use previous homework assignments to study?

25. Read or reread objective 6 and fill out the schedule grid below.

	Monday	*Tuesday*	*Wednesday*	*Thursday*	*Friday*	*Saturday*	*Sunday*
1:00 a.m.							
2:00 a.m.							
3:00 a.m.							
4:00 a.m.							
5:00 a.m.							
6:00 a.m.							
7:00 a.m.							
8:00 a.m.							
9:00 a.m.							
10:00 a.m.							
11:00 a.m.							
12:00 p.m.							
1:00 p.m.							
2:00 p.m.							
3:00 p.m.							
4:00 p.m.							
5:00 p.m.							
6:00 p.m.							
7:00 p.m.							
8:00 p.m.							
9:00 p.m.							
10:00 p.m.							
11:00 p.m.							
Midnight							

1.2 | Algebraic Expressions and Sets of Numbers

OBJECTIVES

1 Identify and Evaluate Algebraic Expressions.

2 Identify Natural Numbers, Whole Numbers, Integers, and Rational and Irrational Real Numbers.

3 Find the Absolute Value of a Number.

4 Find the Opposite of a Number.

5 Write Phrases as Algebraic Expressions.

OBJECTIVE

1   Evaluating Algebraic Expressions

Recall that letters that represent numbers are called **variables.** An **algebraic expression** (or simply **expression**) is formed by numbers and variables connected by the operations of addition, subtraction, multiplication, division, raising to powers, or taking roots. For example,

$$2x, \qquad \frac{x+5}{6}, \qquad \sqrt{y}-1.6, \qquad \text{and} \qquad z^3$$

are algebraic expressions or, more simply, expressions. (Recall that the expression $2x$ means $2 \cdot x$.)

Algebraic expressions occur often during problem solving. For example, the average cost to own and operate a car in the United States for 2009 was $0.453 per mile. The expression $0.453m$ gives the total cost to operate a car annually for m miles. (*Source:* AAA)

To find the cost of driving a car for 12,000 miles, for example, we replace the variable m with 12,000 and perform the indicated operation. This process is called **evaluating** an expression, and the result is called the **value** of the expression for the given replacement value.

In our example, when $m = 12,000$,

$$0.453m = 0.453(12,000) = 5436$$

▶ **Helpful Hint**

Recall that $0.453m$ means $0.453 \times m$

Thus, it costs $5436 to own and operate a car for 12,000 miles of driving.

EXAMPLE 1   **Finding the Area of a Tile**

The research department of a flooring company is considering a new flooring design that contains parallelograms. The area of a parallelogram with base b and height h is bh. Find the area of a parallelogram with base 10 centimeters and height 8.2 centimeters.

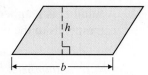

Solution   We replace b with 10 and h with 8.2 in the algebraic expression bh.

$$bh = 10 \cdot 8.2 = 82$$

The area is 82 square centimeters □

PRACTICE

1   The tile edging for a bathroom is in the shape of a triangle. The area of a triangle with base b and height h is $A = \frac{1}{2}bh$. Find the area of the tile if the base measures 3.5 cm and the height measures 8 cm.

Algebraic expressions simplify to different values depending on replacement values. (Order of operations is needed for simplifying many expressions. We fully review this in Section 1.3.)

EXAMPLE 2 Evaluate: $3x - y$ when $x = 15$ and $y = 4$.

Solution We replace x with 15 and y with 4 in the expression.

$$3x - y = 3 \cdot 15 - 4 = 45 - 4 = 41 \qquad \square$$

PRACTICE
2 Evaluate $2p - q$ when $p = 17$ and $q = 3$.

When evaluating an expression to solve a problem, we often need to think about the kind of number that is appropriate for the solution. For example, if we are asked to determine the maximum number of parking spaces for a parking lot to be constructed, an answer of $98\frac{1}{10}$ is not appropriate because $\frac{1}{10}$ of a parking space is not realistic.

OBJECTIVE
2 Identifying Common Sets of Numbers

Let's review some common sets of numbers and their graphs on a number line. To construct a number line, we draw a line and label a point 0 with which we associate the number 0. This point is called the **origin.** Choose a point to the right of 0 and label it 1. The distance from 0 to 1 is called the **unit distance** and can be used to locate more points. The **positive numbers** lie to the right of the origin, and the **negative numbers** lie to the left of the origin. The number 0 is neither positive nor negative.

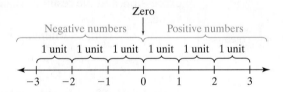

> ▶ **Helpful Hint**
> 0 is neither a positive number nor a negative number.

✓**CONCEPT CHECK**
Use the definitions of positive numbers, negative numbers, and zero to describe the meaning of *nonnegative numbers*.

A number is **graphed** on a number line by shading the point on the number line that corresponds to the number. Some common sets of numbers and their graphs include:

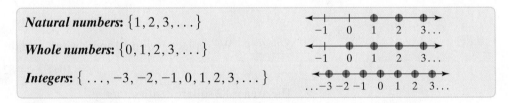

Natural numbers: $\{1, 2, 3, \ldots\}$

Whole numbers: $\{0, 1, 2, 3, \ldots\}$

Integers: $\{\ldots, -3, -2, -1, 0, 1, 2, 3, \ldots\}$

Each listing of three dots,…, is called an **ellipsis** and means to continue in the same pattern.

A **set** is a collection of objects. The objects of a set are called its **members or elements.** When the elements of a set are listed, such as those displayed in the box above, the set is written in **roster** form.

Answer to Concept Check:
a number that is 0 or positive

A set can also be written in **set builder notation,** which describes the members of a set but does not list them. The following set is written in set builder notation.

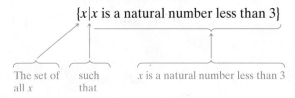

$$\{x | x \text{ is a natural number less than 3}\}$$

The set of all x — such that — x is a natural number less than 3

This same set written in roster form is $\{1, 2\}$.

A set that contains *no* elements is called the **empty set** (or **null set**), symbolized by $\{\ \}$ or $\varnothing$. The set

$$\{x | x \text{ is a month with 32 days}\} \text{ is } \varnothing \text{ or } \{\ \}$$

because no month has 32 days. The set has no elements.

> **▶ Helpful Hint**
>
> Use $\{\ \}$ or $\varnothing$ to write the empty set. $\{\varnothing\}$ is **not** the empty set because it has one element: $\varnothing$.

EXAMPLE 3   Write each set in roster form. (List the elements of each set.)

a. $\{x | x \text{ is a natural number greater than } 100\}$

b. $\{x | x \text{ is a whole number between 1 and 6}\}$

Solution

a. $\{101, 102, 103, \dots\}$ **b.** $\{2, 3, 4, 5\}$ ☐

PRACTICE
3   Write each set in roster form. (List the elements of each set.)

a. $\{x | x \text{ is a whole number between 5 and 10}\}$

b. $\{x | x \text{ is a natural number greater than } 40\}$

The symbol $\in$ denotes that an element is in a particular set. The symbol $\in$ is read as "is an element of." For example, the true statement

$$3 \text{ is an element of } \{1, 2, 3, 4, 5\}$$

can be written in symbols as

$$3 \in \{1, 2, 3, 4, 5\}$$

The symbol $\notin$ is read as "is not an element of." In symbols, we write the true statement "p is not an element of $\{a, 5, g, j, q\}$" as

$$p \notin \{a, 5, g, j, q\}$$

EXAMPLE 4   Determine whether each statement is true or false.

a. $3 \in \{x | x \text{ is a natural number}\}$ **b.** $7 \notin \{1, 2, 3\}$

Solution

a. True, since 3 is a natural number and therefore an element of the set.

b. True, since 7 is not an element of the set $\{1, 2, 3\}$. ☐

PRACTICE
4   Determine whether each statement is true or false.

a. $7 \in \{x | x \text{ is a natural number}\}$ **b.** $6 \notin \{1, 3, 5, 7\}$

We can use set builder notation to describe three other common sets of numbers.

> **Identifying Numbers**
>
> **Real Numbers:** $\{x \mid x \text{ corresponds to a point on the number line}\}$
>
>
>
> **Rational numbers:** $\left\{\dfrac{a}{b} \middle| a \text{ and } b \text{ are integers and } b \neq 0\right\}$
>
> **Irrational numbers:** $\{x \mid x \text{ is a real number and } x \text{ is not a rational number}\}$

> ▶ **Helpful Hint**
> Notice from the definition that all real numbers are either rational or irrational.

Every rational number can be written as a decimal that either repeats in a pattern or terminates. For example,

Rational Numbers

$$\frac{1}{2} = 0.5 \qquad\qquad \frac{5}{4} = 1.25$$

$$\frac{2}{3} = 0.6666666\ldots = 0.\overline{6} \qquad \frac{1}{11} = 0.090909\ldots = 0.\overline{09}$$

An irrational number written as a decimal neither terminates nor repeats. For example, π and $\sqrt{2}$ are irrational numbers. Their decimal form neither terminates nor repeats. Decimal approximations of each are below:

Irrational Numbers

$$\pi \approx 3.141592\ldots \qquad \sqrt{2} \approx 1.414213\ldots$$

Notice that every integer is also a rational number since each integer can be written as the quotient of itself and 1:

$$3 = \frac{3}{1}, \quad 0 = \frac{0}{1}, \quad -8 = \frac{-8}{1}$$

Not every rational number, however, is an integer. The rational number $\dfrac{2}{3}$, for example, is not an integer. Some square roots are rational numbers and some are irrational numbers. For example, $\sqrt{2}$, $\sqrt{3}$, and $\sqrt{7}$ are irrational numbers while $\sqrt{25}$ is a rational number because $\sqrt{25} = 5 = \dfrac{5}{1}$. The set of rational numbers together with the set of irrational numbers make up the set of real numbers. To help you make the distinction between rational and irrational numbers, here are a few examples of each.

Rational Numbers			*Irrational Numbers*
Number	*Equivalent Quotient of Integers,* $\dfrac{a}{b}$		
$-\dfrac{2}{3}$	=	$\dfrac{-2}{3}$ or $\dfrac{2}{-3}$	$\sqrt{5}$
$\sqrt{36}$	=	$\dfrac{6}{1}$	$\dfrac{\sqrt{6}}{7}$
5	=	$\dfrac{5}{1}$	$-\sqrt{3}$
0	=	$\dfrac{0}{1}$	π
1.2	=	$\dfrac{12}{10}$	$\dfrac{2}{\sqrt{3}}$
$3\dfrac{7}{8}$	=	$\dfrac{31}{8}$	

Some rational and irrational numbers are graphed below.

Earlier, we mentioned that every integer is also a rational number. In other words, all the elements of the set of integers are also elements of the set of rational numbers. When this happens, we say that the set of integers, set Z, is a subset of the set of rational numbers, set Q. In symbols,

$$\underbrace{Z \subseteq Q}_{\text{is a subset of}}$$

The natural numbers, whole numbers, integers, rational numbers, and irrational numbers are each a subset of the set of real numbers. The relationships among these sets of numbers are shown in the following diagram.

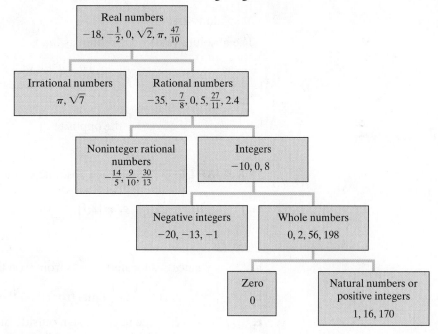

EXAMPLE 5   Determine whether the following statements are true or false.

a. 3 is a real number.

b. $\dfrac{1}{5}$ is an irrational number.

c. Every rational number is an integer.

d. $\{1, 5\} \subseteq \{2, 3, 4, 5\}$

Solution

a. True. Every whole number is a real number.

b. False. The number $\dfrac{1}{5}$ is a rational number since it is in the form $\dfrac{a}{b}$ with a and b integers and $b \neq 0$.

c. False. The number $\dfrac{2}{3}$, for example, is a rational number, but it is not an integer.

d. False. Since the element 1 in the first set is not an element of the second set.   □

PRACTICE

5   Determine whether the following statements are true or false.

a. -5 is a real number.

b. $\sqrt{8}$ is a rational number.

c. Every whole number is a rational number.

d. $\{2, 4\} \subset \{1, 3, 4, 7\}$

OBJECTIVE

3 Finding the Absolute Value of a Number

The number line can also be used to visualize distance, which leads to the concept of absolute value. The **absolute value** of a real number a, written as $|a|$, is the distance between a and 0 on the number line. Since distance is always positive or zero, $|a|$ is always positive or zero.

Using the number line, we see that

$$|4| = 4 \quad \text{and also} \quad |-4| = 4$$

Why? Because both 4 and -4 are a distance of 4 units from 0.

An equivalent definition of the absolute value of a real number a is given next.

Absolute Value

The absolute value of a, written as $|a|$, is

$$|a| = \begin{cases} a \text{ if } a \text{ is 0 or a positive number} \\ -a \text{ if } a \text{ is a negative number} \end{cases}$$

$\underbrace{}$

the opposite of

EXAMPLE 6 Find each absolute value.

a. $|3|$ **b.** $\left|-\dfrac{1}{7}\right|$ **c.** $-|2.7|$ **d.** $-|-8|$ **e.** $|0|$

Solution

a. $|3| = 3$ since 3 is located 3 units from 0 on the number line.

b. $\left|-\dfrac{1}{7}\right| = \dfrac{1}{7}$ since $-\dfrac{1}{7}$ is $\dfrac{1}{7}$ units from 0 on the number line.

c. $-|2.7| = -2.7$. The negative sign outside the absolute value bars means to take the opposite of the absolute value of 2.7.

d. $-|-8| = -8$. Since $|-8|$ is 8, $-|-8| = -8$.

e. $|0| = 0$ since 0 is located 0 units from 0 on the number line. ☐

PRACTICE

6 Find each absolute value.

a. $|4|$ **b.** $\left|-\dfrac{1}{2}\right|$ **c.** $|1|$ **d.** $-|6.8|$ **e.** $-|-4|$

··

✓CONCEPT CHECK
Explain how you know that $|14| = -14$ is a false statement.

Answer to Concept Check:
$|14| = 14$ since the absolute value of a number is the distance between the number and 0, and distance cannot be negative.

OBJECTIVE

4 Finding the Opposite of a Number

The number line can also help us visualize opposites. Two numbers that are the same distance from 0 on the number line but are on opposite sides of 0 are called **opposites.**

See the definition illustrated on the number lines below.

The opposite of 6.5 is -6.5

The opposite of $\dfrac{2}{3}$ is $-\dfrac{2}{3}$.

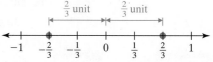

The opposite of -4 is 4.

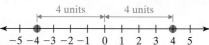

> ▶ **Helpful Hint**
> The opposite of 0 is 0.

> **Opposite**
> The opposite of a number a is the number $-a$.

Above, we state that the opposite of a number a is $-a$. This means that the opposite of -4 is $-(-4)$. But from the number line above, the opposite of -4 is 4. This means that $-(-4) = 4$, and in general, we have the following property.

> **Double Negative Property**
> For every real number a, $-(-a) = a$.

EXAMPLE 7   Write the opposite of each number.

a. 8   **b.** $\dfrac{1}{5}$ **c.** -9.6

Solution

a. The opposite of 8 is -8.

b. The opposite of $\dfrac{1}{5}$ is $-\dfrac{1}{5}$.

c. The opposite of -9.6 is $-(-9.6) = 9.6$.   □

PRACTICE
7   Write the opposite of each number.

a. 5.4   **b.** $-\dfrac{3}{5}$ **c.** 18

OBJECTIVE
5   Writing Phrases as Algebraic Expressions

Often, solving problems involves translating a phrase to an algebraic expression. The following is a partial list of key words and phrases and their usual direct translations.

Selected Key Words/Phrases and Their Translations			
Addition	*Subtraction*	*Multiplication*	*Division*
sum	difference of	product	quotient
plus	minus	times	divide
added to	subtracted from	multiply	into
more than	less than	twice	ratio
increased by	decreased by	of	
total	less		

EXAMPLE 8   Translate each phrase to an algebraic expression. Use the variable x to represent each unknown number.

a. Eight times a number

b. Three more than eight times a number

c. The quotient of a number and -7

d. One and six-tenths subtracted from twice a number

e. Six less than a number

f. Twice the sum of four and a number

Solution

a. $8 \cdot x$ or $8x$

b. $8x + 3$

c. $x \div -7$ or $\dfrac{x}{-7}$

d. $2x - 1.6$ or $2x - 1\dfrac{6}{10}$

e. $x - 6$

f. $2(4 + x)$ □

PRACTICE

8   Translate each phrase to an algebraic expression. Use the variable x to represent the unknown number.

a. The product of 3 and a number

b. Five less than twice a number

c. Three and five-eighths more than a number

d. The quotient of a number and 2

e. Fourteen subtracted from a number

f. Five times the sum of a number and ten

Vocabulary, Readiness & Video Check

Use the choices below to fill in each blank. Not all choices will be used.

whole numbers	integers	rational number	a
natural numbers	value	irrational number	$-a$
absolute value	expression	variables	

1. Letters that represent numbers are called _____.
2. Finding the _____ of an expression means evaluating the expression.
3. The _____ of a number is that number's distance from 0 on the number line.
4. A(n) _____ is formed by numbers and variables connected by operations such as addition, subtraction, multiplication, division, raising to powers, and/or taking roots.
5. The _____ are $\{1, 2, 3, \ldots\}$.
6. The _____ are $\{0, 1, 2, 3, \ldots\}$.
7. The _____ are $\{\ldots -3, -2, -1, 0, 1, 2, 3, \ldots\}$.
8. The number $\sqrt{5}$ is a(n) _____.
9. The number $\dfrac{5}{7}$ is a(n) _____.
10. The opposite of a is _____.

Martin-Gay Interactive Videos

See Video 1.2

Watch the section lecture video and answer the following questions.

OBJECTIVE 1

11. When the algebraic expression in Example 2 is evaluated for the variable, why is 10,612.8 not the final answer for this application?

OBJECTIVE 2

12. Based on the lecture before Example 3, what is the relationship among real numbers, rational numbers, and irrational numbers?

OBJECTIVE 3

13. Based on the lecture before ⊞ Example 7, complete the following statements. The absolute value of a number is that number's _____ from zero on a number line. Or, more formally, $|a| = a$ if a is _____ or a _____ number. Also, $|a| = -a$ if a is a _____ number.

OBJECTIVE 4

14. Based on the lecture before ⊞ Example 10, the description of opposite using a number line is given. Explain the difference in the absolute value of a number and the opposite of a number using the number line descriptions.

OBJECTIVE 5

15. From ⊞ Example 12, why must we be careful when translating phrases dealing with subtraction?

1.2 Exercise Set   MyMathLab®

Find the value of each algebraic expression at the given replacement values. See Examples 1 and 2.

1. $5x$ when $x = 7$

2. $3y$ when $y = 45$

3. $9.8z$ when $z = 3.1$

4. $7.1a$ when $a = 1.5$

▶ **5.** ab when $a = \dfrac{1}{2}$ and $b = \dfrac{3}{4}$

6. yz when $y = \dfrac{2}{3}$ and $z = \dfrac{1}{5}$

7. $3x + y$ when $x = 6$ and $y = 4$

8. $2a - b$ when $a = 12$ and $b = 7$

9. The B737-400 aircraft flies an average speed of 400 miles per hour.

The expression $400t$ gives the distance traveled by the aircraft in t hours. Find the distance traveled by the B737-400 in 5 hours.

10. The algebraic expression $1.5x$ gives the total length of shelf space needed in inches for x encyclopedias. Find the length of shelf space needed for a set of 30 encyclopedias.

△ **11.** Employees at Walmart constantly reorganize and reshelve merchandise. In doing so, they calculate floor space needed for displays. The algebraic expression $l \cdot w$ gives the floor space needed in square units for a display that measures length l units and width w units. Calculate the floor space needed for a display whose length is 5.1 feet and whose width is 4 feet.

12. The algebraic expression $\dfrac{x}{5}$ can be used to calculate the distance in miles that you are from a flash of lightning, where x is the number of seconds between the time you see a flash of lightning and the time you hear the thunder. Calculate the distance that you are from the flash of lightning if you hear the thunder 2 seconds after you see the lightning.

▶ **13.** The B737-400 aircraft costs $2948 dollars per hour to operate. The algebraic expression $2948t$ gives the total cost to operate the aircraft for t hours. Find the total cost to operate the B737-400 for 3.6 hours.

14. Flying the SR-71A jet, Capt. Elden W. Joersz, USAF, set a record speed of 2193.16 miles per hour. At this speed, the algebraic expression $2193.16t$ gives the total distance flown in t hours. Find the distance flown by the SR-71A in 1.7 hours.

Write each set in roster form. (List the elements of each set.) See Example 3.

15. $\{x \mid x \text{ is a natural number less than } 6\}$

16. $\{x \mid x \text{ is a natural number greater than } 6\}$

17. $\{x \mid x \text{ is a natural number between } 10 \text{ and } 17\}$

18. $\{x \mid x \text{ is an odd natural number}\}$

19. $\{x \mid x \text{ is a whole number that is not a natural number}\}$

20. $\{x \mid x \text{ is a natural number less than } 1\}$

21. $\{x \mid x \text{ is an even whole number less than } 9\}$

22. $\{x \mid x \text{ is an odd whole number less than } 9\}$

Graph each set on a number line.

23. $\{0, 2, 4, 6\}$

24. $\{1, 3, 5, 7\}$

25. $\left\{\dfrac{1}{2}, \dfrac{2}{3}\right\}$

26. $\left\{\dfrac{1}{4}, \dfrac{1}{3}\right\}$

27. $\{-2, -6, -10\}$

28. $\{-1, -2, -3\}$

29. $\left\{-\dfrac{1}{3}, -1\dfrac{1}{3}\right\}$

30. $\left\{-\dfrac{1}{2}, -1\dfrac{1}{2}\right\}$

List the elements of the set $\left\{3, 0, \sqrt{7}, \sqrt{36}, \dfrac{2}{5}, -134\right\}$ *that are also elements of the given set. See Example 4.*

31. Whole numbers

32. Integers

33. Natural numbers

34. Rational numbers

35. Irrational numbers

36. Real numbers

Place $\in$ *or* $\notin$ *in the space provided to make each statement true. See Example 4.*

37. -11 ____ $\{x \mid x \text{ is an integer}\}$

38. 0 ____ $\{x \mid x \text{ is a positive integer}\}$

39. -6 ____ $\{2, 4, 6, \ldots\}$　　**40.** 12 ____ $\{1, 2, 3, \ldots\}$

41. 12 ____ $\{1, 3, 5, \ldots\}$　　**42.** 0 ____ $\{1, 2, 3, \ldots\}$

43. $\dfrac{1}{2}$ ____ $\{x \mid x \text{ is an irrational number}\}$

44. 0 ____ $\{x \mid x \text{ is a natural number}\}$

Determine whether each statement is true or false. See Examples 4 and 5. Use the following sets of numbers.

$$N = \text{set of natural numbers}$$
$$Z = \text{set of integers}$$
$$I = \text{set of irrational numbers}$$
$$Q = \text{set of rational numbers}$$
$$\mathbb{R} = \text{set of real numbers}$$

45. $Z \subseteq \mathbb{R}$　　　　　　**46.** $\mathbb{R} \subseteq N$

47. $-1 \in Z$　　　　　　**48.** $\dfrac{1}{2} \in Q$

49. $0 \in N$　　　　　　**50.** $Z \subseteq Q$

51. $\sqrt{5} \notin I$　　　　　　**52.** $\pi \notin \mathbb{R}$

53. $N \subseteq Z$　　　　　　**54.** $I \subseteq N$

55. $\mathbb{R} \subseteq Q$　　　　　　**56.** $N \subseteq Q$

Find each absolute value. See Example 6.

57. $-|2|$　　　　　　**58.** $|8|$

59. $|-4|$　　　　　　**60.** $|-6|$

61. $|0|$　　　　　　**62.** $|-1|$

63. $-|-3|$　　　　　　**64.** $-|-11|$

Write the opposite of each number. See Example 7.

65. -6.2　　　　　　**66.** -7.8

67. $\dfrac{4}{7}$　　　　　　**68.** $\dfrac{9}{5}$

69. $-\dfrac{2}{3}$　　　　　　**70.** $-\dfrac{14}{3}$

71. 0　　　　　　**72.** 10.3

Translating *Write each phrase as an algebraic expression. Use the variable x to represent each unknown number. See Example 8.*

73. Twice a number.　　**74.** Six times a number.

75. Five more than twice a number.

76. One more than six times a number.

77. Ten less than a number.

78. A number minus seven.

79. The sum of a number and two.

80. The difference of twenty-five and a number.

81. A number divided by eleven.

82. The quotient of a number and thirteen.

83. Twelve, minus three times a number.

84. Four, subtracted from three times a number.

85. A number plus two and three-tenths

86. Fifteen and seven-tenths plus a number

87. A number less than one and one-third.

88. Two and three-fourths less than a number.

89. The quotient of five and the difference of four and a number.

90. The quotient of four and the sum of a number and one.

91. Twice the sum of a number and three.

92. Eight times the difference of a number and nine.

CONCEPT EXTENSIONS

Use the bar graph below to complete the given table by estimating the millions of tourists predicted for each country. (Use whole numbers.)

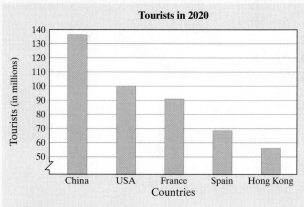

Source: Dan Smith, *The State of the World Atlas.*

93. China	
94. France	
95. Spain	
96. Hong Kong	

97. Explain why $-(-2)$ and $-|-2|$ simplify to different numbers.

98. The boxed definition of absolute value states that $|a| = -a$ if a is a negative number. Explain why $|a|$ is always non-negative, even though $|a| = -a$ for negative values of a.

99. In your own words, explain why every natural number is also a rational number but not every rational number is a natural number.

100. In your own words, explain why every irrational number is a real number but not every real number is an irrational number.

101. In your own words, explain why the empty set is a subset of every set.

102. In your own words, explain why every set is a subset of itself.

1.3 Operations on Real Numbers and Order of Operations

OBJECTIVES

1. Add and Subtract Real Numbers.

2. Multiply and Divide Real Numbers.

3. Evaluate Expressions Containing Exponents.

4. Find Roots of Numbers.

5. Use the Order of Operations.

6. Evaluate Algebraic Expressions.

OBJECTIVE

1   Adding and Subtracting Real Numbers

When solving problems, we often have to add real numbers. For example, if the New Orleans Saints lose 5 yards in one play, then lose another 7 yards in the next play, their total loss may be described by $-5 + (-7)$.

The addition of two real numbers may be summarized by the following.

> **Adding Real Numbers**
>
> 1. To add two numbers with the *same sign*, add their absolute values and attach their common sign.
>
> 2. To add two numbers with *different signs*, subtract the smaller absolute value from the larger absolute value and attach the sign of the number with the larger absolute value.

For example, to add $-5 + (-7)$, first add their absolute values.

$$|-5| = 5, |-7| = 7, \quad \text{and} \quad 5 + 7 = 12$$

Next, attach their common negative sign.

$$-5 + (-7) = -12$$

(This represents a total loss of 12 yards for the New Orleans Saints in the example above.)

To find $(-4) + 3$, first subtract their absolute values.

$$|-4| = 4, \quad |3| = 3, \quad \text{and} \quad 4 - 3 = 1$$

Next, attach the sign of the number with the larger absolute value.

$$(-4) + 3 = -1$$

EXAMPLE 1   Add.

a. $-3 + (-11)$ **b.** $3 + (-7)$ **c.** $-10 + 15$

d. $-8.3 + (-1.9)$ **e.** $-\dfrac{1}{4} + \dfrac{1}{2}$ **f.** $-\dfrac{2}{3} + \dfrac{3}{7}$

Solution

a. $-3 + (-11) = -(3 + 11) = -14$

b. $3 + (-7) = -4$

c. $-10 + 15 = 5$

d. $-8.3 + (-1.9) = -10.2$

e. $-\dfrac{1}{4} + \dfrac{1}{2} = -\dfrac{1}{4} + \dfrac{1 \cdot 2}{2 \cdot 2} = -\dfrac{1}{4} + \dfrac{2}{4} = \dfrac{1}{4}$

f. $-\dfrac{2}{3} + \dfrac{3}{7} = -\dfrac{14}{21} + \dfrac{9}{21} = -\dfrac{5}{21}$

PRACTICE

1   Add.

a. $-6 + (-2)$ **b.** $5 + (-8)$ **c.** $-4 + 9$

d. $(-3.2) + (-4.9)$ **e.** $-\dfrac{3}{5} + \dfrac{2}{3}$ **f.** $-\dfrac{5}{11} + \dfrac{3}{22}$

Subtraction of two real numbers may be defined in terms of addition.

> **Subtracting Real Numbers**
>
> If a and b are real numbers,
> $$a - b = a + (-b)$$

In other words, to subtract a real number, we add its opposite.

EXAMPLE 2 Subtract.

a. $2 - 8$ **b.** $-8 - (-1)$ **c.** $-11 - 5$ **d.** $10.7 - (-9.8)$

e. $-\dfrac{2}{3} - \dfrac{1}{2}$ **f.** $1 - 0.06$ **g.** Subtract 7 from 4.

Solution   Add the opposite   Add the opposite

a. $2 - 8 = 2 + (-8) = -6$ **b.** $-8 - (-1) = -8 + (1) = -7$

c. $-11 - 5 = -11 + (-5) = -16$ **d.** $10.7 - (-9.8) = 10.7 + 9.8 = 20.5$

e. $-\dfrac{2}{3} - \dfrac{1}{2} = -\dfrac{2 \cdot 2}{3 \cdot 2} - \dfrac{1 \cdot 3}{2 \cdot 3} = -\dfrac{4}{6} + \left(-\dfrac{3}{6}\right) = -\dfrac{7}{6}$

f. $1 - 0.06 = 1 + (-0.06) = 0.94$ **g.** $4 - 7 = 4 + (-7) = -3$ □

PRACTICE
2   Subtract.

a. $3 - 11$ **b.** $-6 - (-3)$ **c.** $-7 - 5$ **d.** $4.2 - (-3.5)$

e. $-\dfrac{5}{7} - \dfrac{1}{3}$ **f.** $3 - 1.2$ **g.** Subtract 9 from 2.

To add or subtract three or more real numbers, add or subtract from left to right.

EXAMPLE 3 Simplify the following expressions.

a. $11 + 2 - 7$ **b.** $-5 - 4 + 2$

Solution

a. $11 + 2 - 7 = 13 - 7 = 6$ **b.** $-5 - 4 + 2 = -9 + 2 = -7$ □

PRACTICE
3   Simplify the following expressions.

a. $13 + 5 - 6$ **b.** $-6 - 2 + 4$

OBJECTIVE
2   **Multiplying and Dividing Real Numbers**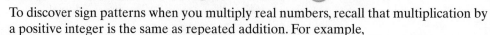

To discover sign patterns when you multiply real numbers, recall that multiplication by a positive integer is the same as repeated addition. For example,

$$3(2) = 2 + 2 + 2 = 6$$
$$3(-2) = (-2) + (-2) + (-2) = -6$$

Notice here that $3(-2) = -6$. This illustrates that the product of two numbers with different signs is negative. We summarize sign patterns for multiplying any two real numbers as follows.

> **Multiplying Two Real Numbers**
>
> The product of two numbers with the *same* sign is positive.
> The product of two numbers with *different* signs is negative.

Also recall that the product of zero and any real number is zero.

$$0 \cdot a = 0$$

> **Product Property of 0**
> $0 \cdot a = 0$ Also $a \cdot 0 = 0$

EXAMPLE 4 Multiply.

a. $(-8)(-1)$ **b.** $-2\left(\dfrac{1}{6}\right)$ **c.** $-1.2(0.3)$ **d.** $0(-11)$

e. $\dfrac{1}{5}\left(-\dfrac{10}{11}\right)$ **f.** $(7)(1)(-2)(-3)$ **g.** $8(-2)(0)$

Solution

a. Since the signs of the two numbers are the same, the product is positive. Thus $(-8)(-1) = +8$, or 8.

b. Since the signs of the two numbers are different or unlike, the product is negative. Thus $-2\left(\dfrac{1}{6}\right) = -\dfrac{2}{6} = -\dfrac{1}{3}$.

c. $-1.2(0.3) = -0.36$

d. $0(-11) = 0$

e. $\dfrac{1}{5}\left(-\dfrac{10}{11}\right) = -\dfrac{10}{55} = -\dfrac{2}{11}$

f. To multiply three or more real numbers, you may multiply from left to right.

$$
\begin{aligned}
(7)(1)(-2)(-3) &= 7(-2)(-3) \\
&= -14(-3) \\
&= 42
\end{aligned}
$$

g. Since zero is a factor, the product is zero.

$$(8)(-2)(0) = 0$$ □

PRACTICE
4 Multiply.

a. $(-5)(3)$ **b.** $(-7)\left(-\dfrac{1}{14}\right)$ **c.** $5.1(-2)$ **d.** $14(0)$

e. $\left(-\dfrac{1}{4}\right)\left(\dfrac{8}{13}\right)$ **f.** $6(-1)(-2)(3)$ **g.** $5(-2.3)$

> ▶ **Helpful Hint**
> The following sign patterns may be helpful when we are multiplying.
> **1.** An odd number of negative factors gives a negative product.
> **2.** An even number of negative factors gives a positive product.

Recall that $\dfrac{8}{4} = 2$ because $2 \cdot 4 = 8$. Likewise, $\dfrac{8}{-4} = -2$ because $(-2)(-4) = 8$.

Also, $\dfrac{-8}{4} = -2$ because $(-2)4 = -8$, and $\dfrac{-8}{-4} = 2$ because $2(-4) = -8$. From these examples, we can see that the sign patterns for division are the same as for multiplication.

> **Dividing Two Real Numbers**
>
> The quotient of two numbers with the *same* sign is positive.
> The quotient of two numbers with *different* signs is negative.

Recall from your knowledge of fractions that division by a nonzero real number b is the same as multiplication by $\frac{1}{b}$. In other words,

$$\frac{a}{b} = a \div b = a \cdot \frac{1}{b}$$

This means that to simplify $\frac{a}{b}$, we can divide by b or multiply by $\frac{1}{b}$. The nonzero numbers b and $\frac{1}{b}$ are called **reciprocals.**

Notice that b *must* be a nonzero number. We do not define division by 0. For example, $5 \div 0$, or $\frac{5}{0}$, is undefined. To see why, recall that if $5 \div 0 = n$, a number, then $n \cdot 0 = 5$. This is not possible since $n \cdot 0 = 0$ for any number n and is never 5. Thus far, we have learned that we cannot divide 5 or any other nonzero number by 0.

Can we divide 0 by 0? By the same reasoning, if $0 \div 0 = n$, a number, then $n \cdot 0 = 0$. This is true for any number n so that the quotient $0 \div 0$ would not be a single number. To avoid this, we say that

> Division by 0 is undefined.

EXAMPLE 5 Divide.

a. $\dfrac{20}{-4}$　　**b.** $\dfrac{-9}{-3}$　　**c.** $-\dfrac{3}{8} \div 3$　　**d.** $\dfrac{-40}{10}$　　**e.** $\dfrac{-1}{10} \div \dfrac{-2}{5}$　　**f.** $\dfrac{8}{0}$

Solution

a. Since the signs are different or unlike, the quotient is negative and $\dfrac{20}{-4} = -5$.

b. Since the signs are the same, the quotient is positive and $\dfrac{-9}{-3} = 3$.

c. $-\dfrac{3}{8} \div 3 = -\dfrac{3}{8} \cdot \dfrac{1}{3} = -\dfrac{1}{8}$　　　**d.** $\dfrac{-40}{10} = -4$

e. $\dfrac{-1}{10} \div \dfrac{-2}{5} = -\dfrac{1}{10} \cdot -\dfrac{5}{2} = \dfrac{1}{4}$　　**f.** $\dfrac{8}{0}$ is undefined.　　□

PRACTICE

5 Divide.

a. $\dfrac{-16}{8}$　　　　　**b.** $\dfrac{-15}{-3}$　　　　　**c.** $-\dfrac{2}{3} \div 4$

d. $\dfrac{54}{-9}$　　　　**e.** $-\dfrac{1}{12} \div \left(-\dfrac{3}{4}\right)$　　　**f.** $\dfrac{0}{-7}$

With sign rules for division, we can understand why the positioning of the negative sign in a fraction does not change the value of the fraction. For example,

$$\frac{-12}{3} = -4, \quad \frac{12}{-3} = -4, \quad \text{and} \quad -\frac{12}{3} = -4$$

Since all the fractions equal -4, we can say that

$$\frac{-12}{3} = \frac{12}{-3} = -\frac{12}{3}$$

In general, the following holds true.

> If a and b are real numbers and $b \neq 0$, then $\dfrac{a}{-b} = \dfrac{-a}{b} = -\dfrac{a}{b}$.

OBJECTIVE

3 Evaluating Expressions Containing Exponents

Recall that when two numbers are multiplied, they are called **factors.** For example, in $3 \cdot 5 = 15$, the 3 and 5 are called factors.

A natural number *exponent* is a shorthand notation for repeated multiplication of the same factor. This repeated factor is called the **base,** and the number of times it is used as a factor is indicated by the **exponent.** For example,

$$\overset{\text{exponent}}{4^{\scriptsize 3}} = \underbrace{4 \cdot 4 \cdot 4}_{4 \text{ is a factor 3 times.}} = 64$$
base

> **Exponents**
>
> If a is a real number and n is a natural number, then the **nth power of a,** or **a raised to the nth power,** written as a^n, is the product of n factors, each of which is a.
>
> $$\overset{\text{exponent}}{a^{\scriptsize n}} = \underbrace{a \cdot a \cdot a \cdot a \cdot \,\cdots\, \cdot a}_{a \text{ is a factor } n \text{ times.}}$$
> base

It is not necessary to write an exponent of 1. For Example, 3 is assumed to be 3^1.

EXAMPLE 6 Evaluate each expression.

a. 3^2 **b.** $\left(\dfrac{1}{2}\right)^4$ **c.** -5^2

d. $(-5)^2$ **e.** -5^3 **f.** $(-5)^3$

Solution

a. $3^2 = 3 \cdot 3 = 9$ **b.** $\left(\dfrac{1}{2}\right)^4 = \left(\dfrac{1}{2}\right)\left(\dfrac{1}{2}\right)\left(\dfrac{1}{2}\right)\left(\dfrac{1}{2}\right) = \dfrac{1}{16}$

c. $-5^2 = -(5 \cdot 5) = -25$ **d.** $(-5)^2 = (-5)(-5) = 25$

e. $-5^3 = -(5 \cdot 5 \cdot 5) = -125$ **f.** $(-5)^3 = (-5)(-5)(-5) = -125$ □

PRACTICE

6 Evaluate each expression.

a. 2^3 **b.** $\left(\dfrac{1}{3}\right)^2$ **c.** -12^2

d. $(-12)^2$ **e.** -4^3 **f.** $(-4)^3$

✓CONCEPT CHECK

When $(-8.2)^7$ is evaluated, will the value be positive or negative? How can you tell without making any calculations?

Answer to Concept Check
negative; the exponent is an odd
number

> ▶ **Helpful Hint**
> Be very careful when simplifying expressions such as -5^2 and $(-5)^2$.
>
> $$-5^2 = -(5 \cdot 5) = -25 \quad \text{and} \quad (-5)^2 = (-5)(-5) = 25$$
>
> Without parentheses, the base to square is 5, not -5.

OBJECTIVE
4 Finding Roots of Numbers

The opposite of squaring a number is taking the **square root** of a number. For example, since the square of 4, or 4^2, is 16, we say that a square root of 16 is 4. The notation $\sqrt{a}$ denotes the **positive**, or **principal**, **square root** of a nonnegative number a. We then have in symbols that $\sqrt{16} = 4$. The negative square root of 16 is written $-\sqrt{16} = -4$. The square root of a negative number such as $\sqrt{-16}$ is not a real number. Why? There is no real number that, when squared, gives a negative number.

EXAMPLE 7 Find the square roots.

a. $\sqrt{9}$ **b.** $\sqrt{25}$ **c.** $\sqrt{\dfrac{1}{4}}$ **d.** $-\sqrt{36}$ **e.** $\sqrt{-36}$

Solution

a. $\sqrt{9} = 3$ since 3 is positive and $3^2 = 9$. **b.** $\sqrt{25} = 5$ since $5^2 = 25$.

c. $\sqrt{\dfrac{1}{4}} = \dfrac{1}{2}$ since $\left(\dfrac{1}{2}\right)^2 = \dfrac{1}{4}$. **d.** $-\sqrt{36} = -6$

e. $\sqrt{-36}$ is not a real number.

PRACTICE
7 Find the square roots.

a. $\sqrt{49}$ **b.** $\sqrt{\dfrac{1}{16}}$ **c.** $-\sqrt{64}$ **d.** $\sqrt{-64}$ **e.** $\sqrt{100}$

We can find roots other than square roots. Since $(-2)^3$, is -8, we say that the **cube root** of -8 is -2. This is written as

$$\sqrt[3]{-8} = -2.$$

Also, since $3^4 = 81$ and 3 is positive,

$$\sqrt[4]{81} = 3.$$

EXAMPLE 8 Find the roots.
a. $\sqrt[3]{-27}$ **b.** $\sqrt[5]{1}$ **c.** $\sqrt[4]{16}$
Solution

a. $\sqrt[3]{-27} = -3$ since $(-3)^3 = -27$.
b. $\sqrt[5]{1} = 1$ since $1^5 = 1$.
c. $\sqrt[4]{16} = 2$ since 2 is positive and $2^4 = 16$.

PRACTICE
8 Find the roots.

a. $\sqrt[3]{64}$ **b.** $\sqrt[5]{-1}$ **c.** $\sqrt[4]{10,000}$

Of course, as mentioned in Section 1.2, not all roots simplify to rational numbers. We study radicals further in Chapter 7.

OBJECTIVE

5 Using the Order of Operations

Expressions containing more than one operation are written to follow a particular agreed-upon **order of operations.** For example, when we write $3 + 2 \cdot 10$, we mean to multiply first and then add.

Order of Operations

Simplify expressions using the order that follows.

If grouping symbols such as parentheses are present, simplify expressions within those first, starting with the innermost set. If fraction bars are present, simplify the numerator and denominator separately.

1. Evaluate exponential expressions, roots, or absolute values in order from left to right.
2. Multiply or divide in order from left to right.
3. Add or subtract in order from left to right.

▶ **Helpful Hint**

Fraction bars, radical signs, and absolute value bars can sometimes be used as grouping symbols. For example,

	Fraction Bar	*Radical Sign*	*Absolute Value Bars*
Grouping Symbol	$\dfrac{-1-7}{6-11}$	$\sqrt{15+1}$	$\lvert -7.2 - \sqrt{4} \rvert$
Not Grouping Symbol	$-\dfrac{8}{9}$	$\sqrt{9}$	$\lvert -3.2 \rvert$

EXAMPLE 9   Simplify.

a. $20 \div 2 \cdot 10$ **b.** $1 + 2(1-4)^2$ **c.** $\dfrac{\lvert -2 \rvert^3 + 1}{-7 - \sqrt{4}}$

Solution

a. Be careful! Here, we multiply or divide in order from left to right. Thus, divide, then multiply.

$$20 \div 2 \cdot 10 = 10 \cdot 10 = 100$$

b. Remember order of operations so that you are *not* tempted to add 1 and 2 first.

$$
\begin{aligned}
1 + 2(1-4)^2 &= 1 + 2(-3)^2 && \text{Simplify inside grouping symbols first.}\\
&= 1 + 2(9) && \text{Write } (-3)^2 \text{ as 9.}\\
&= 1 + 18 && \text{Multiply.}\\
&= 19 && \text{Add.}
\end{aligned}
$$

c. Simplify the numerator and the denominator separately; then divide.

$$
\begin{aligned}
\frac{\lvert -2 \rvert^3 + 1}{-7 - \sqrt{4}} &= \frac{2^3 + 1}{-7 - 2} && \text{Write } \lvert -2 \rvert \text{ as 2 and } \sqrt{4} \text{ as 2.}\\
&= \frac{8 + 1}{-9} && \text{Write } 2^3 \text{ as 8.}\\
&= \frac{9}{-9} = -1 && \text{Simplify the numerator, then divide.}
\end{aligned}
$$

9 Simplify.

a. $14 - 3 \cdot 4$ **b.** $3(5 - 8)^2$ **c.** $\dfrac{|-5|^2 + 4}{\sqrt{4} - 3}$

Besides parentheses, other symbols used for grouping expressions are brackets [ ] and braces { }. These other grouping symbols are commonly used when we group expressions that already contain parentheses.

EXAMPLE 10 Simplify: $3 - [(4 - 6) + 2(5 - 9)]$

Solution $3 - [(4 - 6) + 2(5 - 9)] = 3 - [-2 + 2(-4)]$ Simplify within the innermost sets of parentheses.

$$= 3 - [-2 + (-8)]$$
$$= 3 - [-10]$$
$$= 13$$

> ▶ Helpful Hint
> When grouping symbols occur within grouping symbols, remember to perform operations on the innermost set first.

PRACTICE
10 Simplify: $5 - [(3 - 5) + 6(2 - 4)]$.

EXAMPLE 11 Simplify: $\dfrac{-5\sqrt{30 - 5} + (-2)^2}{4^2 + |7 - 10|}$

Solution Here, the fraction bar, radical sign, and absolute value bars serve as grouping symbols. Thus, we simplify within the radical sign and absolute value bars first, remembering to calculate above and below the fraction bar separately.

$$\frac{-5\sqrt{30 - 5} + (-2)^2}{4^2 + |7 - 10|} = \frac{-5\sqrt{25} + (-2)^2}{4^2 + |-3|} = \frac{-5 \cdot 5 + 4}{16 + 3} = \frac{-25 + 4}{16 + 3}$$

$$= \frac{-21}{19} \text{ or } -\frac{21}{19}$$

PRACTICE
11 Simplify: $\dfrac{-2\sqrt{12 + 4} - (-3)^2}{6^2 + |1 - 9|}$

✓CONCEPT CHECK
True or false? If two people use the order of operations to simplify a numerical expression and neither makes a calculation error, it is not possible that they each obtain a different result. Explain.

OBJECTIVE
6 **Evaluating Algebraic Expressions**

Recall from Section 1.2 that an algebraic expression is formed by numbers and variables connected by the operations of addition, subtraction, multiplication, division, raising to powers, and/or taking roots. Also, if numbers are substituted for the variables in an algebraic expression and the operations performed, the result is called **the value of the expression** for the given replacement values. This entire process is called **evaluating an expression.**

Answer to Concept Check:
true; answers may vary

EXAMPLE 12 Evaluate each expression when $x = 4$ and $y = -3$.

a. $3x - 7y$ **b.** $-2y^2$ **c.** $\dfrac{\sqrt{x}}{y} - \dfrac{y}{x}$

Solution   For each expression, replace x with 4 and y with -3.

a. $\begin{aligned} 3x - 7y &= 3 \cdot 4 - 7(-3) && \text{Let } x = 4 \text{ and } y = -3. \\ &= 12 - (-21) && \text{Multiply.} \\ &= 12 + 21 && \text{Write as an addition.} \\ &= 33 && \text{Add.} \end{aligned}$

b. $\begin{aligned} -2y^2 &= -2(-3)^2 && \text{Let } y = -3. \\ &= -2(9) && \text{Write } (-3)^2 \text{ as 9.} \\ &= -18 && \text{Multiply.} \end{aligned}$

> ▶ **Helpful Hint**
> In $-2(-3)^2$, the exponent 2 goes with the base of -3 only.

c. $\begin{aligned} \frac{\sqrt{x}}{y} - \frac{y}{x} &= \frac{\sqrt{4}}{-3} - \frac{-3}{4} \\ &= -\frac{2}{3} + \frac{3}{4} && \text{Write } \sqrt{4} \text{ as 2.} \\ &= -\frac{2}{3} \cdot \frac{4}{4} + \frac{3}{4} \cdot \frac{3}{3} && \text{The LCD is 12.} \\ &= -\frac{8}{12} + \frac{9}{12} && \text{Write each fraction with a denominator of 12.} \\ &= \frac{1}{12} && \text{Add.} \end{aligned}$

PRACTICE
12   Evaluate each expression when $x = 16$ and $y = -5$

a. $2x - 7y$ **b.** $-4y^2$ **c.** $\dfrac{\sqrt{x}}{y} - \dfrac{y}{x}$

Sometimes variables such as x_1 and x_2 will be used in this book. The small 1 and 2 are called **subscripts**. The variable x_1 can be read as "x sub 1," and the variable x_2 can be read as "x sub 2." The important thing to remember is that they are two different variables. For example, if $x_1 = -5$ and $x_2 = 7$, then

$$x_1 - x_2 = -5 - 7 = -12.$$

EXAMPLE 13 The algebraic expression $\dfrac{5(x - 32)}{9}$ represents the equivalent temperature in degrees Celsius when x is the temperature in degrees Fahrenheit. Complete the following table by evaluating this expression at the given values of x.

Degrees Fahrenheit	x	-4	10	32
Degrees Celsius	$\dfrac{5(x - 32)}{9}$			

Solution To complete the table, evaluate $\dfrac{5(x - 32)}{9}$ at each given replacement value.

When $x = -4$,

$$\frac{5(x - 32)}{9} = \frac{5(-4 - 32)}{9} = \frac{5(-36)}{9} = -20$$

When $x = 10$,

$$\frac{5(x - 32)}{9} = \frac{5(10 - 32)}{9} = \frac{5(-22)}{9} = -\frac{110}{9}$$

When $x = 32$,

$$\frac{5(x - 32)}{9} = \frac{5(32 - 32)}{9} = \frac{5 \cdot 0}{9} = 0$$

The completed table is

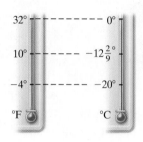

Degrees Fahrenheit	x	-4	10	32
Degrees Celsius	$\dfrac{5(x - 32)}{9}$	-20	$-\dfrac{110}{9}$	0

Thus, $-4°$F is equivalent to $-20°$C, $10°$F is equivalent to $-\dfrac{110°}{9}$C, and $32°$F is equivalent to $0°$C.   □

PRACTICE
13   The algebraic expression $\dfrac{9}{5}x + 32$ represents the equivalent temperature in degrees Fahrenheit when x is the temperature in degrees Celsius. Complete the following table by evaluating this expression at the given values of x.

Degrees Celsius	x	-5	10	25
Degrees Fahrenheit	$\dfrac{9}{5}x + 32$			

Vocabulary, Readiness & Video Check

Use the choices below to fill in each blank. Some choices may be used more than once and some used not at all.

exponent	undefined	base	1	$\dfrac{-a}{-b}$	$\dfrac{a}{b}$
square root	reciprocal	0	9	$\dfrac{-a}{b}$	$\dfrac{a}{-b}$

1. $0 \cdot a =$ _____ .

2. $\dfrac{0}{4}$ simplifies to _____ while $\dfrac{4}{0}$ is _____ .

3. The _____ of the nonzero number b is $\dfrac{1}{b}$.

4. The fraction $-\dfrac{a}{b} =$ _____ $=$ _____ .

5. A(n) _____ is a shorthand notation for repeated multiplication of the same number.

6. In $(-5)^2$, the 2 is the _____ and the -5 is the _____ .

7. The opposite of squaring a number is taking the _____ of a number.

8. Using order of operations, $9 \div 3 \cdot 3 =$ _____ .

Martin-Gay Interactive Videos

See Video 1.3

Watch the section lecture video and answer the following questions.

9. From ▣ Example 2, we learn that subtraction of two real numbers may be defined in terms of what operation?

10. From ▣ Examples 3 and 4, explain the significance of the signs of numbers when multiplying and dividing two real numbers.

11. Explain the significance of the use of parentheses when comparing ▣ Examples 7 and 8.

12. From ▣ Example 9, what does the $\sqrt{}$ notation mean?

13. In the lecture before ▣ Example 12, what reason is given for needing the order of operations?

14. In ▣ Example 15, why can you not first add the 3 and 2 in the numerator?

1.3 Exercise Set   MyMathLab®

Add or subtract as indicated. See Examples 1 through 3

1. $-3 + 8$

2. $12 + (-7)$

3. $-14 + (-10)$

4. $-5 + (-9)$

5. $-4.3 - 6.7$

6. $-8.2 - (-6.6)$

7. $13 - 17$

8. $15 - (-1) \quad 16$

9. $\dfrac{11}{15} - \left(-\dfrac{3}{5}\right)$

10. $\dfrac{7}{10} - \dfrac{4}{5}$

11. $19 - 10 - 11$

12. $-13 - 4 + 9$

13. $-\dfrac{4}{5} - \left(-\dfrac{3}{10}\right)$

14. $-\dfrac{5}{2} - \left(-\dfrac{2}{3}\right)$

15. Subtract 14 from 8.

16. Subtract 9 from -3.

Multiply or divide as indicated. See Examples 4 and 5.

17. $-5 \cdot 12$

18. $-3 \cdot 8$

19. $-17 \cdot 0$

20. $-5 \cdot 0$

21. $\dfrac{0}{-2}$

22. $\dfrac{-2}{0}$

23. $\dfrac{-9}{3}$

24. $\dfrac{-20}{5}$

25. $\dfrac{-12}{-4}$

26. $\dfrac{-36}{-6}$

27. $3\left(-\dfrac{1}{18}\right)$

28. $5\left(-\dfrac{1}{50}\right)$

29. $(-0.7)(-0.8)$

30. $(-0.9)(-0.5)$

31. $9.1 \div (-1.3)$

32. $22.5 \div (-2.5)$

33. $-4(-2)(-1)$

34. $-5(-3)(-2)$

Evaluate each expression. See Example 6.

35. -7^2

36. $(-7)^2$

37. $(-6)^2$

38. -6^2

39. $(-2)^3$

40. -2^3

41. $\left(-\dfrac{1}{3}\right)^3$

42. $\left(-\dfrac{1}{2}\right)^4$

Find the following roots. See Examples 7 and 8.

43. $\sqrt{49}$

44. $\sqrt{81}$

45. $-\sqrt{\dfrac{4}{9}}$

46. $-\sqrt{\dfrac{4}{25}}$

47. $\sqrt[3]{64}$

48. $\sqrt[5]{32}$

49. $\sqrt[4]{81}$

50. $\sqrt[3]{1}$

51. $\sqrt{-100}$

52. $\sqrt{-25}$

MIXED PRACTICE

Simplify each expression. See Examples 1 through 11.

53. $3(5 - 7)^4$

54. $7(3 - 8)^2$

55. $-3^2 + 2^3$

56. $-5^2 - 2^4$

57. $\dfrac{3.1 - (-1.4)}{-0.5}$

58. $\dfrac{4.2 - (-8.2)}{-0.4}$

59. $(-3)^2 + 2^3$

60. $(-15)^2 - 2^4$

61. $-8 \div 4 \cdot 2$

62. $-20 \div 5 \cdot 4$

63. $-8\left(-\dfrac{3}{4}\right) - 8$

64. $-10\left(-\dfrac{2}{5}\right) - 10$

65. $2 - [(7 - 6) + (9 - 19)]$

66. $8 - [(4 - 7) + (8 - 1)]$

67. $\dfrac{(-9 + 6)(-1^2)}{-2 - 2}$

68. $\dfrac{(-1 - 2)(-3^2)}{-6 - 3}$

69. $(\sqrt[3]{8})(-4) - (\sqrt{9})(-5)$

70. $(\sqrt[3]{27})(-5) - (\sqrt{25})(-3)$

71. $25 - [(3 - 5) + (14 - 18)]^2$

72. $10 - [(4 - 5)^2 + (12 - 14)]^4$

▶ **73.** $\dfrac{(3 - \sqrt{9}) - (-5 - 1.3)}{-3}$ **74.** $\dfrac{-\sqrt{16} - (6 - 2.4)}{-2}$

75. $\dfrac{|3 - 9| - |-5|}{-3}$ **76.** $\dfrac{|-14| - |2 - 7|}{-15}$

77. $\dfrac{3(-2 + 1)}{5} - \dfrac{-7(2 - 4)}{1 - (-2)}$ **78.** $\dfrac{-1 - 2}{2(-3) + 10} - \dfrac{2(-5)}{-1(8) + 1}$

79. $\dfrac{\frac{1}{3} \cdot 9 - 7}{3 + \frac{1}{2} \cdot 4}$ **80.** $\dfrac{\frac{1}{5} \cdot 20 - 6}{10 + \frac{1}{4} \cdot 12}$

81. $3\{-2 + 5[1 - 2(-2 + 5)]\}$ **82.** $2\{-1 + 3[7 - 4(-10 + 12)]\}$

83. $\dfrac{-4\sqrt{80 + 1} + (-4)^2}{3^3 + |-2(3)|}$ **84.** $\dfrac{(-2)^4 + 3\sqrt{120 - 20}}{4^3 + |5(-1)|}$

Evaluate each expression when $x = 9$ and $y = -2$. See Example 12.

85. $9x - 6y$ **86.** $4x - 10y$

87. $-3y^2$ **88.** $-7y^2$

89. $\dfrac{\sqrt{x}}{y} - \dfrac{y}{x}$ **90.** $\dfrac{y}{2x} - \dfrac{\sqrt{x}}{3y}$

▶ **91.** $\dfrac{3 + 2|x - y|}{x + 2y}$ **92.** $\dfrac{5 + 2|y - x|}{x + 6y}$

93. $\dfrac{y^3 + \sqrt{x - 5}}{|4x - y|}$ **94.** $\dfrac{y^2 + \sqrt{x + 7}}{|3x - y|}$

See Example 13.

△ **95.** The algebraic expression $8 + 2y$ represents the perimeter of a rectangle with width 4 and length y.

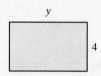

a. Complete the table that follows by evaluating this expression at the given values of y.

Length	y	5	7	10	100
Perimeter	$8 + 2y$				

b. Use the results of the table in part **a** to answer the following question. As the width of a rectangle remains the same and the length increases, does the perimeter increase or decrease? Explain how you arrived at your answer.

△ **96.** The algebraic expression πr^2 represents the area of a circle with radius r.

a. Complete the table below by evaluating this expression at given values of r. (Use 3.14 for π.)

Radius	r	2	3	7	10
Area	πr^2				

b. As the radius of a circle increases (see the table), does its area increase or decrease? Explain your answer.

97. The algebraic expression $\dfrac{100x + 5000}{x}$ represents the cost per bookshelf (in dollars) of producing x bookshelves.

a. Complete the table below.

Number of Bookshelves	x	10	100	1000
Cost per Bookshelf	$\dfrac{100x + 5000}{x}$			

b. As the number of bookshelves manufactured increases (see the table), does the cost per bookshelf increase or decrease? Why do you think that this is so?

98. If c is degrees Celsius, the algebraic expression $1.8c + 32$ represents the equivalent temperature in degrees Fahrenheit.

a. Complete the table below.

Degrees Celsius	c	-10	0	50
Degrees Fahrenheit	$1.8c + 32$			

b. As degrees Celsius increase (see the table), do degrees Fahrenheit increase or decrease?

CONCEPT EXTENSIONS

Choose the fraction(s) equivalent to the given fraction. (There may be more than one correct choice.)

99. $-\dfrac{1}{7}$ **a.** $\dfrac{-1}{-7}$ **b.** $\dfrac{-1}{7}$ **c.** $\dfrac{1}{-7}$ **d.** $\dfrac{1}{7}$

100. $\dfrac{-x}{y}$ **a.** $\dfrac{x}{-y}$ **b.** $-\dfrac{x}{y}$ **c.** $\dfrac{x}{y}$ **d.** $\dfrac{-x}{-y}$

101. $\dfrac{5}{-(x + y)}$ **a.** $\dfrac{5}{(x + y)}$ **b.** $\dfrac{-5}{(x + y)}$

  c. $\dfrac{-5}{-(x + y)}$ **d.** $-\dfrac{5}{(x + y)}$

102. $-\dfrac{(y + z)}{3y}$ **a.** $\dfrac{-(y + z)}{3y}$ **b.** $\dfrac{-(y + z)}{-3y}$

  c. $\dfrac{(y + z)}{3y}$ **d.** $\dfrac{(y + z)}{-3y}$

103. $\dfrac{-9x}{-2y}$ **a.** $\dfrac{-9x}{2y}$ **b.** $\dfrac{9x}{2y}$ **c.** $\dfrac{9x}{-2y}$ **d.** $-\dfrac{9x}{2y}$

104. $\dfrac{-a}{-b}$ **a.** $\dfrac{a}{b}$ **b.** $\dfrac{a}{-b}$ **c.** $\dfrac{-a}{b}$ **d.** $-\dfrac{a}{b}$

Find the value of the expression when $x_1 = 2$, $x_2 = 4$, $y_1 = -3$, $y_2 = 2$.

105. $\dfrac{y_2 - y_1}{x_2 - x_1}$ **106.** $\sqrt{(x_2 - x_1)^2 + (y_2 - y_1)^2}$

Each circle below represents a whole, or 1. Determine the unknown fractional part of each circle.

107.

108.

109. Most of Mauna Kea, a volcano on Hawaii, lies below sea level. If this volcano begins at 5998 meters below sea level and then rises 10,203 meters, find the height of the volcano above sea level.

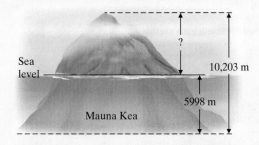

110. The highest point on land on Earth is the top of Mt. Everest in the Himalayas, at an elevation of 29,028 feet above sea level. The lowest point on land is the Dead Sea, between Israel and Jordan, at 1319 feet below sea level. Find the difference in elevations.

Insert parentheses so that each expression simplifies to the given number.

111. $2 + 7 \cdot 1 + 3; 36$

112. $6 - 5 \cdot 2 + 2; -6$

The following graph is called a broken-line graph, or simply a line graph. This particular graph shows the past, present, and future predicted U.S. population over 65. Just as with a bar graph, to find the population over 65 for a particular year, read the height of the corresponding point. To read the height, follow the point horizontally to the left until you reach the vertical axis. Use this graph to answer Exercises 113 through 118.

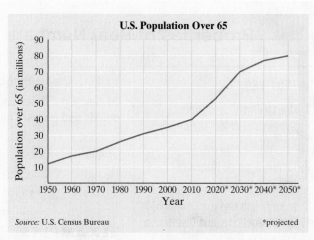

U.S. Population Over 65

Source: U.S. Census Bureau   *projected

113. Estimate the population over 65 in the year 1970.

114. Estimate the predicted population over 65 in the year 2050.

115. Estimate the predicted population over 65 in the year 2030.

116. Estimate the population over 65 in the year 2000.

117. Is the population over 65 increasing as time passes or decreasing? Explain how you arrived at your answer.

118. The percent of Americans over 65 in 1950 was 8.1%. The percent of Americans over 65 in 2050 is expected to be 2.5 times the percent over 65 in 1950. Estimate the percent of Americans expected to be over age 65 in 2050.

119. Explain why -3^2 and $(-3)^2$ simplify to different numbers.

120. Explain why -3^3 and $(-3)^3$ simplify to the same number.

Use a calculator to approximate each square root. For Exercises 125 and 126, simplify the expression. Round answers to four decimal places.

121. $\sqrt{10}$

122. $\sqrt{273}$

123. $\sqrt{7.9}$

124. $\sqrt{19.6}$

125. $\dfrac{-1.682 - 17.895}{(-7.102)(-4.691)}$

126. $\dfrac{(-5.161)(3.222)}{7.955 - 19.676}$

Integrated Review   ALGEBRAIC EXPRESSIONS AND OPERATIONS ON WHOLE NUMBERS

Sections 1.2–1.3

Find the value of each expression when $x = -1$, $y = 3$, and $z = -4$.

1. z^2 **2.** $-z^2$ **3.** $\dfrac{4x - z}{2y}$ **4.** $x(y - 2z)$

Perform the indicated operations.

5. $-7 - (-2)$

6. $\dfrac{9}{10} - \dfrac{11}{12}$

7. $\dfrac{-13}{2 - 2}$

8. $(1.2)^2 - (2.1)^2$

9. $\sqrt{64} - \sqrt[3]{64}$

10. $-5^2 - (-5)^2$

11. $9 + 2[(8 - 10)^2 + (-3)^2]$

12. $8 - 6[\sqrt[3]{8}(-2) + \sqrt{4}(-5)]$

Translating *For Exercises 13 and 14, write each phrase as an algebraic expression. Use x to represent each unknown number.*

13. Subtract twice a number from -15.

14. Five more than three times a number.

15. Name the whole number that is not a natural number.

16. True or false: A real number is either a rational number or an irrational number but never both.

1.4 Properties of Real Numbers and Algebraic Expressions

OBJECTIVES

1 Use Operation and Order Symbols to Write Mathematical Sentences.

2 Identify Identity Numbers and Inverses.

3 Identify and Use the Commutative, Associative, and Distributive Properties.

4 Write Algebraic Expressions.

5 Simplify Algebraic Expressions.

OBJECTIVE

1 Using Symbols to Write Mathematical Sentences

In Section 1.2, we used the symbol $=$ to mean "is equal to." All of the following key words and phrases also imply equality.

Equality

equals	is/was	represents	is the same as
gives	yields	amounts to	is equal to

EXAMPLES   Write each sentence as an equation.

1. The sum of x and 5 is 20.

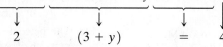

$$x + 5 = 20$$

2. Two times the sum of 3 and y amounts to 4.

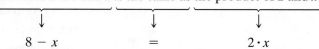

$$2 \qquad (3 + y) \qquad = \qquad 4$$

3. The difference of 8 and x is the same as the product of 2 and x.

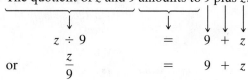

$$8 - x \qquad = \qquad 2 \cdot x$$

4. The quotient of z and 9 amounts to 9 plus z.

$$z \div 9 \qquad = \qquad 9 + z$$

or $\qquad \dfrac{z}{9} \qquad\qquad = \qquad 9 + z$

PRACTICE

1–4   Write each sentence using mathematical symbols.

1. The product of -4 and x is 20.

2. Three times the difference of z and 3 equals 9.

3. The sum of x and 5 is the same as 3 less than twice x.

4. The sum of y and 2 is 4 more than the quotient of z and 8.

If we want to write in symbols that two numbers are not equal, we can use the symbol $\neq$, which means "**is not equal to.**" For example,

$$3 \neq 2$$

Graphing two numbers on a number line gives us a way to compare two numbers. For two real numbers a and b, we say a **is less than** b if on the number line a lies to the left of b. Also, if b is to the right of a on the number line, then b **is greater than** a. The symbol $<$ means "**is less than.**" Since a is less than b, we write

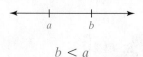

$$b < a$$

The symbol $>$ means "**is greater than.**" Since b is greater than a, we write

$$b > a$$

> **Helpful Hint**
>
> Notice that if $a < b$, then $b > a$. For example, since $-1 < 7$, then $7 > -1$.

EXAMPLE 5   Insert $<$, $>$, or $=$ between each pair of numbers to form a true statement.

a. -1 -2 **b.** $\dfrac{12}{4}$ 3   **c.** -5 0   **d.** -3.5 -3.05

e. $\dfrac{5}{8}$ $\dfrac{3}{8}$ **f.** $\dfrac{2}{3}$ $\dfrac{3}{4}$

Solution

a. $-1 > -2$ since -1 lies to the right of -2 on the number line.

b. $\dfrac{12}{4} = 3$.

c. $-5 < 0$ since -5 lies to the left of 0 on the number line.

d. $-3.5 < -3.05$ since -3.5 lies to the left of -3.05 on the number line.

e. $\dfrac{5}{8} > \dfrac{3}{8}$ The denominators are the same, so $\dfrac{5}{8} > \dfrac{3}{8}$ since $5 > 3$.

f. $\dfrac{2}{3} < \dfrac{3}{4}$ By dividing, we see that $\dfrac{3}{4} = 0.75$ and $\dfrac{2}{3} = 0.666\ldots$.

Thus $\dfrac{2}{3} < \dfrac{3}{4}$ since $0.666\ldots < 0.75$.   □

PRACTICE

5   Insert $<$, $>$, or $=$ between each pair of numbers to form a true statement.

a. -6 -5 **b.** $\dfrac{24}{3}$ 8   **c.** 0 -7 **d.** -2.76 -2.67

e. $\dfrac{9}{10}$ $\dfrac{7}{10}$ **f.** $\dfrac{2}{3}$ $\dfrac{7}{9}$

> ▶ Helpful Hint
> When inserting the $>$ or $<$ symbol, think of the symbols as arrowheads that point toward the smaller number when the statement is true.

In addition to $<$ and $>$, there are the inequality symbols $\leq$ and $\geq$. The symbol
$\leq$ means "**is less than or equal to**" and the symbol
$\geq$ means "**is greater than or equal to**"

For example, the following are true statements.

$$10 \leq 10 \quad \text{since} \quad 10 = 10$$
$$-8 \leq 13 \quad \text{since} \quad -8 < 13$$
$$-5 \geq -5 \quad \text{since} \quad -5 = -5$$
$$-7 \geq -9 \quad \text{since} \quad -7 > -9$$

EXAMPLE 6   Write each sentence using mathematical symbols.

a. The sum of 5 and y is greater than or equal to 7.

b. 11 is not equal to z.

c. 20 is less than the difference of 5 and twice x.

Solution

a. $5 + y \geq 7$ **b.** $11 \neq z$ **c.** $20 < 5 - 2x$ □

PRACTICE

6   Write each sentence using mathematical symbols.

a. The difference of x and 3 is less than or equal to 5.

b. y is not equal to -4.

c. Two is less than the sum of 4 and one-half z.

OBJECTIVE

2   **Identifying Identities and Inverses**

Of all the real numbers, two of them stand out as extraordinary: 0 and 1. **Zero** is the only number that, when *added* to any real number, results in the same real number. Zero is thus called the **additive identity.** Also, **one** is the only number that, when *multiplied* by any real number, results in the same real number. One is thus called the **multiplicative identity.**

	Addition	*Multiplication*
Identity Properties	The additive identity is 0. $a + 0 = 0 + a = a$	The multiplicative identity is 1. $a \cdot 1 = 1 \cdot a = a$

In Section 1.2, we learned that a and $-a$ are opposites.

Another name for opposite is **additive inverse.** For example, the additive inverse of 3 is -3. Notice that the sum of a number and its opposite is always 0.

In Section 1.3, we learned that, for a nonzero number, b and $\dfrac{1}{b}$ are reciprocals.

Another name for reciprocal is **multiplicative inverse.** For example, the multiplicative inverse of $-\dfrac{2}{3}$ is $-\dfrac{3}{2}$. Notice that the product of a number and its reciprocal is always 1.

	Opposite or Additive Inverse	*Reciprocal or Multiplicative Inverse*
Inverse Properties	For each number a, there is a unique number $-a$ called the **additive inverse** or **opposite** of a such that $a + (-a) = (-a) + a = 0$	For each nonzero a, there is a unique number $\dfrac{1}{a}$ called the **multiplicative inverse** or **reciprocal** of a such that $a \cdot \dfrac{1}{a} = \dfrac{1}{a} \cdot a = 1$

EXAMPLE 7   Write the additive inverse, or opposite, of each.

a. 4   **b.** $\dfrac{3}{7}$ **c.** -11.2

Solution

a. The opposite of 4 is -4.

b. The opposite of $\dfrac{3}{7}$ is $-\dfrac{3}{7}$.

c. The opposite of -11.2 is $-(-11.2) = 11.2$.   □

PRACTICE

7   Write the additive inverse, or opposite, of each.

a. -7 **b.** 4.7   **c.** $-\dfrac{3}{8}$

EXAMPLE 8   Write the multiplicative inverse, or reciprocal, of each.

a. 11   **b.** -9 **c.** $\dfrac{7}{4}$

Solution

a. The reciprocal of 11 is $\dfrac{1}{11}$.

b. The reciprocal of -9 is $-\dfrac{1}{9}$.

c. The reciprocal of $\dfrac{7}{4}$ is $\dfrac{4}{7}$ because $\dfrac{7}{4} \cdot \dfrac{4}{7} = 1$.   □

PRACTICE
8   Write the multiplicative inverse, or reciprocal, of each.

a. $-\dfrac{5}{3}$ **b.** 14   **c.** -2

▶ **Helpful Hint**
The number 0 has no reciprocal. Why? There is no number that when multiplied by 0 gives a product of 1.

✓CONCEPT CHECK
Can a number's additive inverse and multiplicative inverse ever be the same? Explain.

OBJECTIVE
3   **Using the Commutative, Associative, and Distributive Properties**

In addition to these special real numbers, all real numbers have certain properties that allow us to write equivalent expressions—that is, expressions that have the same value. These properties will be especially useful in Chapter 2 when we solve equations.

The **commutative properties** state that the order in which two real numbers are added or multiplied does not affect their sum or product.

Commutative Properties
For real numbers *a* and *b*,

$$\textit{Addition: } \quad a + b = b + a$$
$$\textit{Multiplication: } \quad a \cdot b = b \cdot a$$

The **associative properties** state that regrouping numbers that are added or multiplied does not affect their sum or product.

Associative Properties
For real numbers *a*, *b*, and *c*,

$$\textit{Addition: } \quad (a + b) + c = a + (b + c)$$
$$\textit{Multiplication: } \quad (a \cdot b) \cdot c = a \cdot (b \cdot c)$$

Answer to Concept Check:
no; answers may vary

EXAMPLE 9 Use the commutative property of addition to write an expression equivalent to $7x + 5$.

Solution $$7x + 5 = 5 + 7x.$$ □

PRACTICE
9 Use the commutative property of addition to write an expression equivalent to $8 + 13x$.

EXAMPLE 10 Use the associative property of multiplication to write an expression equivalent to $4 \cdot (9y)$. Then simplify this equivalent expression.

Solution $$4 \cdot (9y) = (4 \cdot 9)y = 36y.$$ □

PRACTICE
10 Use the associative property of multiplication to write an expression equivalent to $3 \cdot (11b)$. Then simplify the equivalent expression.

The **distributive property** states that multiplication distributes over addition.

Distributive Property

For real numbers a, b, and c,

$$a(b + c) = ab + ac$$

Also,

$$a(b - c) = ab - ac$$

EXAMPLE 11 Use the distributive property to multiply.

a. $3(2x + y)$ **b.** $-(3x - 1)$ **c.** $0.7a(b - 2)$

Solution

a. $3(2x + y) = 3 \cdot 2x + 3 \cdot y$ Apply the distributive property.
$\qquad\qquad = 6x + 3y$ Apply the associative property of multiplication.

b. Recall that $-(3x - 1)$ means $-1(3x - 1)$.

$$-1(3x - 1) = -1(3x) + (-1)(-1)$$
$$= -3x + 1$$

c. $0.7a(b - 2) = 0.7a \cdot b - 0.7a \cdot 2 = 0.7ab - 1.4a$ □

PRACTICE
11 Use the distributive property to multiply.

a. $4(x + 5y)$ **b.** $-(3 - 2z)$ **c.** $0.3x(y - 3)$

Answer to Concept Check:
no; $6(2a)(3b) = 12a(3b) = 36ab$

✓CONCEPT CHECK
Is the statement below true? Why or why not?

$$6(2a)(3b) = 6(2a) \cdot 6(3b)$$

OBJECTIVE

4   Writing Algebraic Expressions

As mentioned earlier, an important step in problem solving is to be able to write algebraic expressions from word phrases. Sometimes this involves a direct translation, but often an indicated operation is not directly stated but rather implied.

EXAMPLE 12   Write each as an algebraic expression.

a. A vending machine contains x quarters. Write an expression for the *value* of the quarters.

b. The number of grams of fat in x pieces of bread if each piece of bread contains 2 grams of fat.

c. The cost of x desks if each desk costs $156.

d. Sales tax on a purchase of x dollars if the tax rate is 9%.

Each of these examples implies finding a product.

Solution

a. The value of the quarters is found by multiplying the value of a quarter (0.25 dollar) by the number of quarters.

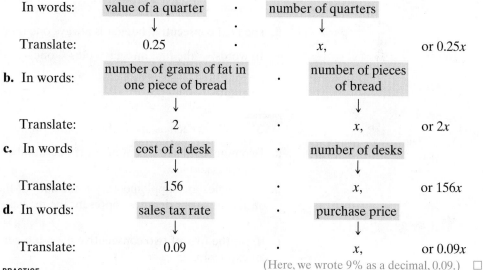

| In words: | value of a quarter | · | number of quarters | |
| Translate: | 0.25 | · | x, | or $0.25x$ |

b. In words: number of grams of fat in one piece of bread · number of pieces of bread

| Translate: | 2 | · | x, | or $2x$ |

c. In words: cost of a desk · number of desks

| Translate: | 156 | · | x, | or $156x$ |

d. In words: sales tax rate · purchase price

| Translate: | 0.09 | · | x, | or $0.09x$ |

(Here, we wrote 9% as a decimal, 0.09.)   □

PRACTICE

12   Write each as an algebraic expression.

a. A parking meter contains x dimes. Write an expression for the value of the dimes.

b. The grams of carbohydrates in y cookies if each cookie has 26 g of carbohydrates.

c. The cost of z birthday cards if each birthday card costs $1.75.

d. The amount of money you save on a new cell phone costing t dollars if it has a 15% discount.

Let's continue writing phrases as algebraic expressions. Two or more unknown numbers in a problem may sometimes be related. If so, try letting a variable represent one unknown number and then represent the other unknown number or numbers as expressions containing the same variable.

EXAMPLE 13   Write each as an algebraic expression.

a. Two numbers have a sum of 20. If one number is x, represent the other number as an expression in x.

b. The older sister is 8 years older than her younger sister. If the age of the younger sister is x, represent the age of the older sister as an expression in x.

△**c.** Two angles are complementary if the sum of their measures is 90°. If the measure of one angle is x degrees, represent the measure of the other angle as an expression in x.

d. If x is the first of two consecutive integers, represent the second integer as an expression in x.

Solution

a. If two numbers have a sum of 20 and one number is x, the other number is "the rest of 20."

In words: twenty minus x
↓   ↓   ↓
Translate: 20   − x

b. The older sister's age is

In words: eight years added to younger sister's age
↓   ↓   ↓
Translate: 8   + x

c. In words: ninety minus x
↓   ↓   ↓
Translate: 90   − x

d. The next consecutive integer is always one more than the previous integer.

In words: the first integer plus one
↓   ↓   ↓
Translate: x +   1

☐

PRACTICE
13   Write each as an algebraic expression.

a. Two numbers have a sum of 16. If one number is x, represent the other number as an expression in x.

b. Two angles are supplementary if the sum of their measures is 180°. If the measure of one angle is x degrees, represent the measure of the other angle as an expression in x.

c. If x is the first of two consecutive even integers, represent the next even integer as an expression in x.

d. One brother is 9 years younger than another brother. If the age of the younger brother is x, represent the age of the older brother as an expression in x.

OBJECTIVE
5 Simplifying Algebraic Expressions

Often, an expression may be **simplified** by removing grouping symbols and combining any like terms. The **terms** of an expression are the addends of the expression. For example, in the expression $3x^2 + 4x$, the terms are $3x^2$ and $4x$.

Expression	*Terms*
$-2x + y$	$-2x, y$
$3x^2 - \dfrac{y}{5} + 7$	$3x^2, -\dfrac{y}{5}, 7$

Terms with the same variable(s) raised to the same power are called **like terms**. We can add or subtract like terms by using the distributive property. This process is called **combining like terms**.

EXAMPLE 14   Use the distributive property to simplify each expression.

a. $3x - 5x + 4$ **b.** $7yz + yz$ **c.** $4z + 6.1$

Solution

a. $3x - 5x + 4 = (3 - 5)x + 4$ Apply the distributive property.

$$= -2x + 4$$

b. $7yz + yz = (7 + 1)yz = 8yz$

c. $4z + 6.1$ cannot be simplified further since $4z$ and 6.1 are not like terms.   □

PRACTICE
14   Use the distributive property to simplify.

a. $6ab - ab$ **b.** $4x - 5 + 6x$ **c.** $17p - 9$

Let's continue to use properties of real numbers to simplify expressions. Recall that the distributive property can also be used to multiply. For example,

$$-2(x + 3) = -2(x) + (-2)(3) = -2x - 6$$

The associative and commutative properties may sometimes be needed to rearrange and group like terms when we simplify expressions.

$$-7x^2 + 5 + 3x^2 - 2 = -7x^2 + 3x^2 + 5 - 2$$
$$= (-7 + 3)x^2 + (5 - 2)$$
$$= -4x^2 + 3$$

EXAMPLE 15   Simplify each expression.

a. $3xy - 2xy + 5 - 7 + xy$ **b.** $7x^2 + 3 - 5(x^2 - 4)$

c. $(2.1x - 5.6) - (-x - 5.3)$ **d.** $\frac{1}{2}(4a - 6b) - \frac{1}{3}(9a + 12b - 1) + \frac{1}{4}$

Solution

a. $3xy - 2xy + 5 - 7 + xy = 3xy - 2xy + xy + 5 - 7$ Apply the commutative property.

$$= (3 - 2 + 1)xy + (5 - 7)$$ Apply the distributive property.

$$= 2xy - 2$$ Simplify.

b. $7x^2 + 3 - 5(x^2 - 4) = 7x^2 + 3 - 5x^2 + 20$ Apply the distributive property.

$$= 2x^2 + 23$$ Simplify.

c. Think of $-(-x - 5.3)$ as $-1(-x - 5.3)$ and use the distributive property.

$(2.1x - 5.6) - 1(-x - 5.3) = 2.1x - 5.6 + 1x + 5.3$

$$= 3.1x - 0.3.$$ Combine like terms

d. $\frac{1}{2}(4a - 6b) - \frac{1}{3}(9a + 12b - 1) + \frac{1}{4}$

$$= 2a - 3b - 3a - 4b + \frac{1}{3} + \frac{1}{4}$$ Use the distributive property.

$$= -a - 7b + \frac{7}{12}$$ Combine like terms.   □

PRACTICE
15 Simplify each expression.

a. $5pq - 2pq - 11 - 4pq + 18$

b. $3x^2 + 7 - 2(x^2 - 6)$

c. $(3.7x + 2.5) - (-2.1x - 1.3)$

d. $\frac{1}{5}(15c - 25d) - \frac{1}{2}(8c + 6d + 1) + \frac{3}{4}$

Answer to Concept Check:
$x - 4(x - 5) = x - 4x + 20$
$\qquad\qquad\qquad = -3x + 20$

✓CONCEPT CHECK

Find and correct the error in the following.

$$x - 4(x - 5) = x - 4x - 20$$
$$= -3x - 20$$

Vocabulary, Readiness & Video Check

Complete the table by filling in the symbols.

	Symbol	Meaning		Symbol	Meaning
1.		is less than	**2.**		is greater than
3.		is not equal to	**4.**		is equal to
5.		is greater than or equal to	**6.**		is less than or equal to

Use the choices below to fill in each blank. Not all choices will be used.

like   terms   distributive $-a$ commutative

unlike   combining   associative $\frac{1}{a}$

7. The opposite of nonzero number a is _____.

8. The reciprocal of nonzero number a is _____.

9. The _____ property has to do with "order."

10. The _____ property has to do with "grouping."

11. $a(b + c) = ab + ac$ illustrates the _____ property.

12. The _____ of an expression are the addends of the expression.

Martin-Gay Interactive Videos

See Video 1.4

Watch the section lecture video and answer the following questions.

OBJECTIVE
1
13. In the lecture before ▱ Example 1, what 6 symbols are discussed that can be used to compare two numbers?

OBJECTIVE
2
14. Complete these statements based on the lecture given before ▱ Example 4. *Reciprocal* is the same as _____ inverse and *opposite* is the same as _____ inverse.

OBJECTIVE
3
15. The commutative and associative properties are discussed in ▱ Examples 7 and 8 and the lecture before. What's the one word used again and again to describe the commutative property? The associative property?

OBJECTIVE
4
16. In ▱ Example 10, what important point are you told to keep in mind when working with applications that have to do with money?

OBJECTIVE
5
17. From ▱ Examples 12–14, how do we simplify algebraic expressions? If the expression contains parentheses, what property might we apply first?

1.4 Exercise Set   MyMathLab®

MIXED PRACTICE

***Translating.** Write each sentence using mathematical symbols. See Examples 1 through 4 and 6 through 8.*

1. The sum of 10 and x is -12.
2. The difference of y and 3 amounts to 12.
3. Twice x, plus 5, is the same as -14.
4. Three more than the product of 4 and c is 7.
5. The quotient of n and 5 is 4 times n.
6. The quotient of 8 and y is 3 more than y.
7. The difference of z and one-half is the same as the product of z and one-half.
8. Five added to one-fourth q is the same as 4 more than q.
9. The product of 7 and x is less than or equal to -21.
10. 10 subtracted from the reciprocal of x is greater than 0.
11. Twice the difference of x and 6 is greater than the reciprocal of 11.
12. Four times the sum of 5 and x is not equal to the opposite of 15.
13. Twice the difference of x and 6 is -27.
14. 5 times the sum of 6 and y is -35.

Insert $<$, $>$, or $=$ between each pair of numbers to form a true statement. See Example 5.

15. -16 -17
16. -14 -24
17. 7.4 7.40
18. $\dfrac{7}{2}$ $\dfrac{35}{10}$
19. $\dfrac{7}{11}$ $\dfrac{9}{11}$
20. $\dfrac{9}{20}$ $\dfrac{3}{20}$
21. $\dfrac{1}{2}$ $\dfrac{5}{8}$
22. $\dfrac{3}{4}$ $\dfrac{7}{8}$
23. -7.9 -7.09
24. -13.07 -13.7

Fill in the chart. See Examples 7 and 8.

	Number	Opposite	Reciprocal
25.	5		
26.	7		
27.		8	
28.			$-\dfrac{1}{4}$
29.	$-\dfrac{1}{7}$		
30.	$\dfrac{1}{11}$		
31.	0		
32.	1		
33.			$\dfrac{8}{7}$
34.		$\dfrac{23}{5}$	

Use a commutative property to write an equivalent expression. See Example 9.

35. $7x + y$
36. $3a + 2b$
37. $z \cdot w$
38. $r \cdot s$
39. $\dfrac{1}{3} \cdot \dfrac{x}{5}$
40. $\dfrac{x}{2} \cdot \dfrac{9}{10}$

Use an associative property to write an equivalent expression. See Example 10.

41. $5 \cdot (7x)$
42. $3 \cdot (10z)$
43. $(x + 1.2) + y$
44. $5q + (2r + s)$
45. $(14z) \cdot y$
46. $(9.2x) \cdot y$

Use the distributive property to find the product. See Example 11.

47. $3(x + 5)$
48. $7(y + 2)$
49. $-(2a + b)$
50. $-(c + 7d)$
51. $2(6x + 5y + 2z)$
52. $5(3a + b + 9c)$
53. $-4(x - 2y + 7)$
54. $-10(2a - 3b - 4)$
55. $0.5x(6y - 3)$
56. $1.2m(9n - 4)$

Complete the statement to illustrate the given property.

57. $3x + 6 = $ _____ Commutative property of addition
58. $8 + 0 = $ ____ Additive identity property
59. $\dfrac{2}{3} + \left(-\dfrac{2}{3}\right) = $ ____ Additive inverse property
60. $4(x + 3) = $ _____ Distributive property
61. $7 \cdot 1 = $ ____ Multiplicative identity property
62. $0 \cdot (-5.4) = $ ____ Multiplication property of zero
63. $10(2y) = $ _____ Associative property
64. $9y + (x + 3z) = $ _____ Associative property

***Translating** Write each of the following as an algebraic expression. See Examples 12 and 13.*

65. Write an expression for the amount of money (in dollars) in d dimes.
66. Write an expression for the amount of money (in dollars) in n nickels.
67. Two numbers have a sum of 112. If one number is x, represent the other number as an expression in x.
68. Two numbers have a sum of 25. If one number is x, represent the other number as an expression in x.
△ 69. Two angles are supplementary if the sum of their measures is $180°$. If the measure of one angle is x degrees, represent the measure of the other angle as an expression in x.
△ 70. If the measure of an angle is $5x$ degrees, represent the measure of its complement as an expression in x.
71. The cost of x compact discs if each compact disc costs $6.49.

72. The cost of y books if each book costs $35.61.

73. If x is an odd integer, represent the next odd integer as an expression in x.

74. If $2x$ is an even integer, represent the next even integer as an expression in x.

MIXED PRACTICE

Simplify each expression. See Examples 11, 14, and 15.

▶ **75.** $-9 + 4x + 18 - 10x$

76. $5y - 14 + 7y - 20y$

77. $5k - (3k - 10)$

78. $-11c - (4 - 2c)$

79. $(3x + 4) - (6x - 1)$

80. $(8 - 5y) - (4 + 3y)$

▶ **81.** $3(xy - 2) + xy + 15 - x^2$

82. $-4(yz + 3) - 7yz + 1 + y^2$

83. $-(n + 5) + (5n - 3)$

84. $-(8 - t) + (2t - 6)$

85. $4(6n^2 - 3) - 3(8n^2 + 4)$

86. $5(2z^3 - 6) + 10(3 - z^3)$

87. $3x - 2(x - 5) + x$

88. $7n + 3(2n - 6) - 2$

89. $1.5x + 2.3 - 0.7x - 5.9$

90. $6.3y - 9.7 + 2.2y - 11.1$

91. $\dfrac{3}{4}b - \dfrac{1}{2} + \dfrac{1}{6}b - \dfrac{2}{3}$

92. $\dfrac{7}{8}a - \dfrac{11}{12} - \dfrac{1}{2}a + \dfrac{5}{6}$

93. $2(3x + 7)$

94. $4(5y + 12)$

▶ **95.** $\dfrac{1}{2}(10x - 2) - \dfrac{1}{6}(60x - 5y)$

96. $\dfrac{1}{4}(8x - 4) - \dfrac{1}{5}(20x - 6y)$

97. $\dfrac{1}{6}(24a - 18b) - \dfrac{1}{7}(7a - 21b - 2) - \dfrac{1}{5}$

98. $\dfrac{1}{3}(6x - 33y) - \dfrac{1}{8}(24x - 40y + 1) - \dfrac{1}{3}$

CONCEPT EXTENSIONS

In each statement, a property of real numbers has been incorrectly applied. Correct the right-hand side of each statement. See the second Concept Check in this section.

99. $3(x + 4) = 3x + 4$

100. $4 + 8y = 4y + 8$

101. $5(7y) = (5 \cdot 7)(5 \cdot y)$

102. $5(3a)(7b) = 5(3a) \cdot 5(7b)$

To demonstrate the distributive property geometrically, represent the area of the larger rectangle in two ways: first as width times length and second as the sum of the areas of the smaller rectangles.

Example:

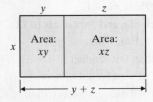

Area of larger rectangle: $x(y + z)$
Area of larger rectangle: $xy + xz$
Thus: $x(y + z) = xy + xz$

103.

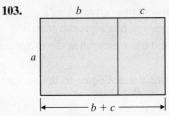

104.

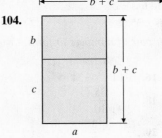

105. Name the only real number that is its own opposite, and explain why this is so.

106. Name the only real number that has no reciprocal, and explain why this is so.

107. Is division commutative? Explain why or why not.

108. Is subtraction commutative? Explain why or why not.

109. Evaluate $24 \div (6 \div 3)$ and $(24 \div 6) \div 3$. Use these two expressions and discuss whether division is associative.

110. Evaluate $12 - (5 - 3)$ and $(12 - 5) - 3$. Use these two expressions and discuss whether subtraction is associative.

Simplify each expression.

111. $8.1z + 7.3(z + 5.2) - 6.85$

112. $6.5y - 4.4(1.8x - 3.3) + 10.95$

Chapter 1 Vocabulary Check

Fill in each blank with one of the words or phrases listed below.

distributive   real   reciprocals   absolute value   opposite   associative

inequality   commutative   whole   algebraic expression   exponent   variable

1. A(n) _____ is formed by numbers and variables connected by the operations of addition, subtraction, multiplication, division, raising to powers, and/or taking roots.

2. The _____ of a number a is $-a$.

3. $3(x - 6) = 3x - 18$ by the _____ property.

4. The _____ of a number is the distance between that number and 0 on the number line.

5. A(n) _____ is a shorthand notation for repeated multiplication of the same factor.

6. A letter that represents a number is called a(n) _____.

7. The symbols $<$ and $>$ are called _____ symbols.

8. If a is not 0, then a and $\dfrac{1}{a}$ are called _____.

9. $A + B = B + A$ by the _____ property.

10. $(A + B) + C = A + (B + C)$ by the _____ property.

11. The numbers $0, 1, 2, 3, \ldots$ are called _____ numbers.

12. If a number corresponds to a point on the number line, we know that number is a _____ number.

Chapter 1 Highlights

DEFINITIONS AND CONCEPTS	EXAMPLES

Section 1.2   Algebraic Expressions and Sets of Numbers

Letters that represent numbers are called **variables.**

An **algebraic expression** is formed by numbers and variables connected by the operations of addition, subtraction, multiplication, division, raising to powers, and/or taking roots.

To **evaluate** an algebraic expression containing variables, substitute the given numbers for the variables and simplify. The result is called the **value** of the expression.

Natural numbers: $\{1, 2, 3, \ldots\}$

Whole numbers: $\{0, 1, 2, 3, \ldots\}$

Integers: $\{\ldots, -3, -2, -1, 0, 1, 2, 3, \ldots\}$

Each listing of three dots above is called an **ellipsis,** which means the pattern continues.

The members of a set are called its **elements.**

Set builder notation describes the elements of a set but does not list them.

Real numbers: $\{x \mid x$ corresponds to a point on the number line$\}$.

Rational numbers: $\left\{\dfrac{a}{b} \,\middle|\, a \text{ and } b \text{ are integers and } b \neq 0\right\}$.

Examples of variables are

$$x, a, m, y$$

Examples of algebraic expressions are

$$7y, \quad -3, \quad \frac{x^2 - 9}{-2} + 14x, \quad \sqrt{3} + \sqrt{m}$$

Evaluate $2.7x$ if $x = 3$.

$$2.7x = 2.7(3)$$
$$= 8.1$$

Given the set $\left\{-9.6, -5, -\sqrt{2}, 0, \dfrac{2}{5}, 101\right\}$ list the elements that belong to the set of

Natural numbers: 101

Whole numbers: 0, 101

Integers: $-5, 0, 101$

Real numbers: $-9.6, -5, -\sqrt{2}, 0, \dfrac{2}{5}, 101$

Rational numbers: $-9.6, -5, 0, \dfrac{2}{5}, 101$

Irrational numbers: $-\sqrt{2}$

(continued)

DEFINITIONS AND CONCEPTS	EXAMPLES

Section 1.2 Algebraic Expressions and Sets of Numbers (continued)

Irrational numbers: $\{x \mid x$ is a real number and x is not a rational number$\}$.

If 3 is an element of set A, we write $3 \in A$.

If all the elements of set A are also in set B, we say that set A is a **subset** of set B, and we write $A \subseteq B$.

Absolute value:

$$|a| = \begin{cases} a \text{ if } a \text{ is } 0 \text{ or a positive number} \\ -a \text{ if } a \text{ is a negative number} \end{cases}$$

The opposite of a number a is the number $-a$.

List the elements in the set
$\{x \mid x$ is an integer between -2 and $5\}$.

$$\{-1, 0, 1, 2, 3, 4\}$$
$$\{1, 2, 4\} \subseteq \{1, 2, 3, 4\}.$$

$$|3| = 3, |0| = 0, |-7.2| = 7.2$$

The opposite of 5 is -5. The opposite of -11 is 11.

Section 1.3 Operations on Real Numbers and Order of Operations

Adding real numbers

1. To add two numbers with the same sign, add their absolute values and attach their common sign.

2. To add two numbers with different signs, subtract the smaller absolute value from the larger absolute value and attach the sign of the number with the larger absolute value.

Subtracting real numbers:

$$a - b = a + (-b)$$

Multiplying and dividing real numbers:
The product or quotient of two numbers with the same sign is positive.

The product or quotient of two numbers with different signs is negative.

$$\frac{2}{7} + \frac{1}{7} = \frac{3}{7}$$
$$-5 + (-2.6) = -7.6$$
$$-18 + 6 = -12$$
$$20.8 + (-10.2) = 10.6$$
$$18 - 21 = 18 + (-21) = -3$$

$$(-8)(-4) = 32 \qquad \frac{-8}{-4} = 2$$
$$8 \cdot 4 = 32 \qquad \frac{8}{4} = 2$$
$$-17 \cdot 2 = -34 \qquad \frac{-14}{2} = -7$$
$$4(-1.6) = -6.4 \qquad \frac{22}{-2} = -11$$

A natural number **exponent** is a shorthand notation for repeated multiplication of the same factor.

The notation $\sqrt{a}$ denotes the **positive, or principal, square root** of a nonnegative number a.

$$\sqrt{a} = b \text{ if } b^2 = a \text{ and } b \text{ is positive}.$$

Also,

$$\sqrt[3]{a} = b \text{ if } b^3 = a$$
$$\sqrt[4]{a} = b \text{ if } b^4 = a \text{ and } b \text{ is positive}$$

$$3^4 = 3 \cdot 3 \cdot 3 \cdot 3 = 81$$

$$\sqrt{49} = 7$$
$$\sqrt[3]{64} = 4$$
$$\sqrt[4]{16} = 2$$

Order of Operations

Simplify expressions using the order that follows. If grouping symbols such as parentheses are present, simplify expressions within those first, starting with the innermost set. If fraction bars are present, simplify the numerator and denominator separately.

1. Raise to powers or take roots in order from left to right.

2. Multiply or divide in order from left to right.

3. Add or subtract in order from left to right.

Simplify $\dfrac{42 - 2(3^2 - \sqrt{16})}{-8}$.

$$\frac{42 - 2(3^2 - \sqrt{16})}{-8} = \frac{42 - 2(9 - 4)}{-8}$$
$$= \frac{42 - 2(5)}{-8}$$
$$= \frac{42 - 10}{-8}$$
$$= \frac{32}{-8} = -4$$

DEFINITIONS AND CONCEPTS	EXAMPLES

Section 1.4 Properties of Real Numbers and Algebraic Expressions

Symbols: $=$ is equal to

$\neq$ is not equal to

$>$ is greater than

$<$ is less than

$\geq$ is greater than or equal to

$\leq$ is less than or equal to

$$-5 = -5$$
$$-5 \neq -3$$
$$1.7 > 1.2$$
$$-1.7 < -1.2$$
$$\frac{5}{3} \geq \frac{5}{3}$$
$$-\frac{1}{2} \leq \frac{1}{2}$$

Identity:

$$a + 0 = a \qquad 0 + a = a$$
$$a \cdot 1 = a \qquad 1 \cdot a = a$$

$$3 + 0 = 3 \qquad 0 + 3 = 3$$
$$-1.8 \cdot 1 = -1.8 \qquad 1 \cdot -1.8 = -1.8$$

Inverse:

$$a + (-a) = 0 \qquad -a + a = 0$$
$$a \cdot \frac{1}{a} = 1 \qquad \frac{1}{a} \cdot a = 1, a \neq 0$$

$$7 + (-7) = 0 \qquad -7 + 7 = 0$$
$$5 \cdot \frac{1}{5} = 1 \qquad \frac{1}{5} \cdot 5 = 1$$

Commutative:

$$a + b = b + a$$
$$a \cdot b = b \cdot a$$

$$x + 7 = 7 + x$$
$$9 \cdot y = y \cdot 9$$

Associative:

$$(a + b) + c = a + (b + c)$$
$$(a \cdot b) \cdot c = a \cdot (b \cdot c)$$

$$(3 + 1) + 10 = 3 + (1 + 10)$$
$$(3 \cdot 1) \cdot 10 = 3 \cdot (1 \cdot 10)$$

Distributive:

$$a(b + c) = ab + ac$$

$$6(x + 5) = 6 \cdot x + 6 \cdot 5$$
$$= 6x + 30$$

Chapter 1 Review

(1.2) *Find the value of each algebraic expression at the given replacement values.*

1. $7x$ when $x = 3$
2. st when $s = 1.6$ and $t = 5$

The hummingbird has an average wing speed of 70 beats per second. The expression 70t gives the number of wingbeats in t seconds.

3. Calculate the number of wingbeats in *1 minute*. (*Hint:* How many seconds are in 1 minute?)
4. Calculate the number of wingbeats in *1 hour* for the hummingbird. (See the Hint for Exercise 3.)

List the elements in each set.

5. $\{x \mid x$ is an odd integer between -2 and $4\}$
6. $\{x \mid x$ is an even integer between -3 and $7\}$
7. $\{x \mid x$ is a negative whole number$\}$
8. $\{x \mid x$ is a natural number that is not a rational number$\}$
9. $\{x \mid x$ is a whole number greater than $5\}$
10. $\{x \mid x$ is an integer less than $3\}$

Determine whether each statement is true or false if $A = \{6, 10, 12\}$, $B = \{5, 9, 11\}$, $C = \{\ldots, -3, -2, -1, 0, 1, 2, 3, \ldots\}$, $D = \{2, 4, 6, \ldots, 16\}$ $E = \{x \mid x$ is a rational number$\}$, $F = \{\ \}$, $G = \{x \mid x$ is an irrational number$\}$, *and* $H = \{x \mid x$ is a real number$\}$.

11. $10 \in D$
12. $59 \in B$
13. $\sqrt{169} \notin G$
14. $0 \notin F$
15. $\pi \in E$
16. $\pi \in H$
17. $\sqrt{4} \in G$
18. $-9 \in E$
19. $A \subseteq D$
20. $C \nsubseteq B$
21. $C \nsubseteq E$
22. $F \subseteq H$

List the elements of the set $\left\{5, -\dfrac{2}{3}, \dfrac{8}{2}, \sqrt{9}, 0.3, \sqrt{7}, 1\dfrac{5}{8}, -1, \pi\right\}$
that are also elements of each given set.

23. Whole numbers

24. Natural numbers

25. Rational numbers

26. Irrational numbers

27. Real numbers

28. Integers

Find the opposite.

29. $-\dfrac{3}{4}$ **30.** 0.6

31. 0 **32.** 1

(1.3) Find the reciprocal.

33. $-\dfrac{3}{4}$ **34.** 0.6

35. 0 **36.** 1

Simplify.

37. $-7 + 3$ **38.** $-10 + (-25)$

39. $5(-0.4)$ **40.** $(-3.1)(-0.1)$

41. $-7 - (-15)$ **42.** $9 - (-4.3)$

43. $(-6)(-4)(0)(-3)$ **44.** $(-12)(0)(-1)(-5)$

45. $(-24) \div 0$ **46.** $0 \div (-45)$

47. $(-36) \div (-9)$ **48.** $60 \div (-12)$

49. $\left(-\dfrac{4}{5}\right) - \left(-\dfrac{2}{3}\right)$ **50.** $\left(\dfrac{5}{4}\right) - \left(-2\dfrac{3}{4}\right)$

51. Determine the unknown fractional part.

52. The Bertha Rogers gas well in Washita County, Oklahoma, is the deepest well in the United States. From the surface, this now-capped well extends 31,441 feet into the earth. The elevation of the nearby Cordell Municipal Airport is 1589 feet above sea level. Assuming that the surface elevation of the well is the same as at the Cordell Municipal Airport, find the elevation relative to sea level of the *bottom* of the Bertha Rogers gas well. *(Sources: U.S. Geological Survey, Oklahoma Department of Transportation)*

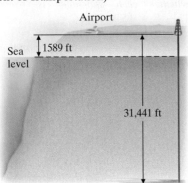

Simplify.

53. $-5 + 7 - 3 - (-10)$ **54.** $8 - (-3) + (-4) + 6$

55. $3(4 - 5)^4$ **56.** $6(7 - 10)^2$

57. $\left(-\dfrac{8}{15}\right) \cdot \left(-\dfrac{2}{3}\right)^2$ **58.** $\left(-\dfrac{3}{4}\right)^2 \cdot \left(-\dfrac{10}{21}\right)$

59. $-\dfrac{6}{15} \div \dfrac{8}{25}$ **60.** $\dfrac{4}{9} \div -\dfrac{8}{45}$

61. $-\dfrac{3}{8} + 3(2) \div 6$ **62.** $5(-2) - (-3) - \dfrac{1}{6} + \dfrac{2}{3}$

63. $|2^3 - 3^2| - |5 - 7|$ **64.** $|5^2 - 2^2| + |9 \div (-3)|$

65. $(2^3 - 3^2) - (5 - 7)$ **66.** $(5^2 - 2^4) + [9 \div (-3)]$

67. $\dfrac{(8 - 10)^3 - (-4)^2}{2 + 8(2) \div 4}$ **68.** $\dfrac{(2 + 4)^2 + (-1)^5}{12 \div 2 \cdot 3 - 3}$

69. $\dfrac{(4 - 9) + 4 - 9}{10 - 12 \div 4 \cdot 8}$ **70.** $\dfrac{3 - 7 - (7 - 3)}{15 + 30 \div 6 \cdot 2}$

71. $\dfrac{\sqrt{25}}{4 + 3 \cdot 7}$ **72.** $\dfrac{\sqrt{64}}{24 - 8 \cdot 2}$

Find the value of each expression when $x = 0, y = 3,$ *and* $z = -2.$

73. $x^2 - y^2 + z^2$ **74.** $\dfrac{5x + z}{2y}$

75. $\dfrac{-7y - 3z}{-3}$ **76.** $(x - y + z)^2$

The algebraic expression $2\pi r$ *represents the circumference of (distance around) a circle of radius r.*

77. Complete the table below by evaluating the expression at the given values of r. (Use 3.14 for π.)

Radius	r	1	10	100
Circumference	$2\pi r$			

78. As the radius of a circle increases, does the circumference of the circle increase or decrease?

(1.4) Simplify each expression.

79. $5xy - 7xy + 3 - 2 + xy$

80. $4x + 10x - 19x + 10 - 19$

81. $6x^2 + 2 - 4(x^2 + 1)$

82. $-7(2x^2 - 1) - x^2 - 1$

83. $(3.2x - 1.5) - (4.3x - 1.2)$

84. $(7.6x + 4.7) - (1.9x + 3.6)$

Translating Write each statement using mathematical symbols.

85. Twelve is the product of x and negative 4.

86. The sum of n and twice n is negative fifteen.

87. Four times the sum of y and three is -1.

88. The difference of t and five, multiplied by six, is four.

89. Seven subtracted from z is six.

90. Ten less than the product of x and nine is five.

91. The difference of x and 5 is at least 12.

92. The opposite of four is less than the product of y and seven.

93. Two-thirds is not equal to twice the sum of n and one-fourth.

94. The sum of t and six is not more than negative twelve.

Name the property illustrated.

95. $(M + 5) + P = M + (5 + P)$

96. $5(3x - 4) = 15x - 20$

97. $(-4) + 4 = 0$

98. $(3 + x) + 7 = 7 + (3 + x)$

99. $(XY)Z = (YZ)X$

100. $\left(-\dfrac{3}{5}\right) \cdot \left(-\dfrac{5}{3}\right) = 1$

101. $T \cdot 0 = 0$

102. $(ab)c = a(bc)$

103. $A + 0 = A$

104. $8 \cdot 1 = 8$

Complete the equation using the given property.

105. $5x - 15z = $ _____ Distributive property

106. $(7 + y) + (3 + x) = $ _____ Commutative property

107. $0 = $ _____ Additive inverse property

108. $1 = $ _____ Multiplicative inverse property

109. $[(3.4)(0.7)]5 = $ _____ Associative property

110. $7 = $ _____ Additive identity property

Insert $<$, $>$, or $=$ to make each statement true.

111. $-9 \quad -12$

112. $0 \quad -6$

113. $-3 \quad -1$

114. $7 \quad |-7|$

115. $-5 \quad -(-5)$

116. $-(-2) \quad -2$

MIXED REVIEW

Complete the table.

	Number	Opposite of Number	Reciprocal of Number
117.	$-\dfrac{3}{4}$		
118.		-5	

Simplify each expression.

119. $-2\left(5x + \dfrac{1}{2}\right) + 7.1$

120. $\sqrt{36} \div 2 \cdot 3$

121. $-\dfrac{7}{11} - \left(-\dfrac{1}{11}\right)$

122. $10 - (-1) + (-2) + 6$

123. $\left(-\dfrac{2}{3}\right)^3 \div \dfrac{10}{9}$

124. $\dfrac{(3 - 5)^2 + (-1)^3}{1 + 2(3 - (-1))^2}$

125. $\dfrac{1}{3}(9x - 3y) - (4x - 1) + 4y$

126. The average price for an ounce of gold in the United States during a month in 2011 was \$1536. The algebraic expression $1536z$ gives the average cost of z ounces of gold during this period. Find the average cost if 7.5 ounces of gold was purchased during this time. (*Source: Business News America*)

The bar graph shows the U.S. life expectancy at birth for females born in the years shown. Use the graph to calculate the increase in life expectancy over each ten-year period shown. (The first row has been completed for you.)

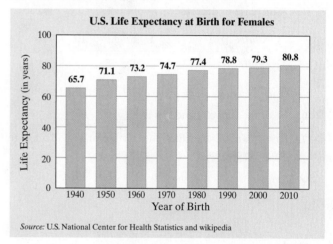

U.S. Life Expectancy at Birth for Females

Source: U.S. National Center for Health Statistics and wikipedia

	Year	Increase in Life Expectancy (in years) from 10 Years Earlier
	1950	5.4
127.	1960	
128.	1970	
129.	1980	
130.	1990	
131.	2000	
132.	2010	

Determine whether each statement is true or false.

1. $-2.3 > -2.33$ **2.** $-6^2 = (-6)^2$

3. $-5 - 8 = -(5 - 8)$

4. $(-2)(-3)(0) = \dfrac{-4}{0}$

5. All natural numbers are integers.

6. All rational numbers are integers.

Simplify.

7. $5 - 12 \div 3(2)$ **8.** $5^2 - 3^4$

9. $(4 - 9)^3 - |-4 - 6|^2$ **10.** $12 + \{6 - [5 - 2(-5)]\}$

11. $\dfrac{6(7 - 9)^3 + (-2)}{(-2)(-5)(-5)}$ **12.** $\dfrac{(4 - \sqrt{16}) - (-7 - 20)}{-2(1 - 4)^2}$

Evaluate each expression when $q = 4$, $r = -2$, and $t = 1$.

13. $q^2 - r^2$ **14.** $\dfrac{5t - 3q}{3r - 1}$

15. The algebraic expression $8.75x$ represents the total cost for x adults to attend the theater.

  a. Complete the table that follows.

  b. As the number of adults increases, does the total cost increase or decrease?

Adults	x	1	3	10	20
Total Cost	$8.75x$				

Write each statement using mathematical symbols.

16. Twice the sum of x and five is 30.

17. The square of the difference of six and y, divided by seven, is less than -2.

18. The product of nine and z, divided by the absolute value of -12, is not equal to 10.

19. Three times the quotient of n and five is the opposite of n.

20. Twenty is equal to 6 subtracted from twice x.

21. Negative two is equal to x divided by the sum of x and five.

Name each property illustrated.

22. $6(x - 4) = 6x - 24$

23. $(4 + x) + z = 4 + (x + z)$

24. $(-7) + 7 = 0$

25. $(-18)(0) = 0$

Solve.

26. Write an expression for the total amount of money (in dollars) in n nickels and d dimes.

27. Find the reciprocal and opposite of $-\dfrac{7}{11}$.

Simplify each expression.

28. $\dfrac{1}{3}a - \dfrac{3}{8} + \dfrac{1}{6}a - \dfrac{3}{4}$

29. $4y + 10 - 2(y + 10)$

30. $(8.3x - 2.9) - (9.6x - 4.8)$

CHAPTER 2

Equations, Inequalities, and Problem Solving

Today, it seems that most people in the world want to stay connected most of the time. In fact, 86% of U.S. citizens own cell phones. Also, computers with Internet access are just as important in our lives. Thus, the merging of these two into Wi-Fi-enabled cell phones might be the next big technological explosion. In Section 2.1, Objective 1, and Section 2.2, Exercises 35 and 36, you will find the projected increase in the number of Wi-Fi-enabled cell phones in the United States as well as the percent increase. (*Source:* Techcrunchies.com)

Mathematics is a tool for solving problems in such diverse fields as transportation, engineering, economics, medicine, business, and biology. We solve problems using mathematics by modeling real-world phenomena with mathematical equations or inequalities. Our ability to solve problems using mathematics, then, depends in part on our ability to solve equations and inequalities. In this chapter, we solve linear equations and inequalities in one variable and graph their solutions on number lines.

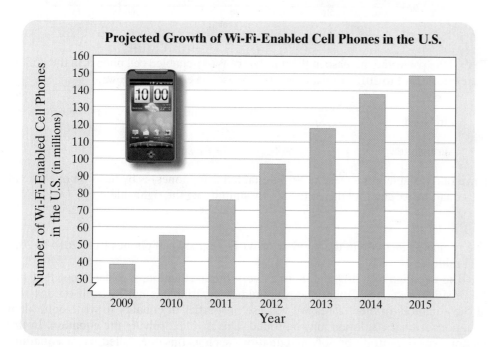

2.1 Linear Equations in One Variable

OBJECTIVES

1 Solve Linear Equations Using Properties of Equality.

2 Solve Linear Equations That Can Be Simplified by Combining Like Terms.

3 Solve Linear Equations Containing Fractions or Decimals.

4 Recognize Identities and Equations with No Solution.

OBJECTIVE

1 Solving Linear Equations Using Properties of Equality

Linear equations model many real-life problems. For example, we can use a linear equation to calculate the increase in the number (in millions) of Wi-Fi-enabled cell phones.

Wi-Fi-enabled cell phones let you carry your Internet access with you. There are already several of these smart phones available, and this technology will continue to expand. Predicted numbers of Wi-Fi-enabled cell phones in the United States for various years are shown below.

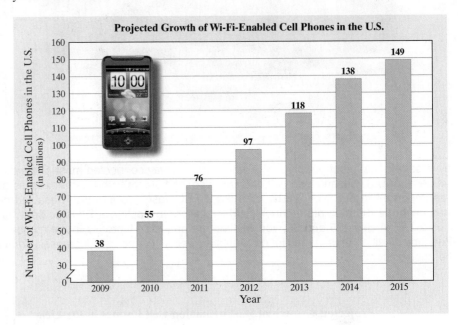

To find the projected increase in the number of Wi-Fi-enabled cell phones in the United States from 2014 to 2015, for example, we can use the equation below.

In words:	Increase in cell phones	is	cell phones in 2015	minus	cell phones in 2014
Translate:	x	$=$	149	$-$	138

Since our variable x (increase in Wi-Fi-enabled cell phones) is by itself on one side of the equation, we can find the value of x by simplifying the right side.

$$x = 11$$

The projected increase in the number of Wi-Fi-enabled cell phones from 2014 to 2015 is 11 million.

The **equation** $x = 149 - 138$, like every other equation, is a statement that two expressions are equal. Oftentimes, the unknown variable is not by itself on one side of the equation. In these cases, we will use properties of equality to write equivalent equations so that a solution may be found. This is called **solving the equation.** In this section, we concentrate on solving equations such as this one, called **linear equations** in one variable. Linear equations are also called **first-degree equations** since the exponent on the variable is 1.

Linear Equations in One Variable

$$3x = -15 \qquad 7 - y = 3y \qquad 4n - 9n + 6 = 0 \qquad z = -2$$

> **Linear Equations in One Variable**
>
> A linear equation in one variable is an equation that can be written in the form
> $$ax + b = c$$
> where $a, b,$ and c are real numbers and $a \neq 0$.

When a variable in an equation is replaced by a number and the resulting equation is true, then that number is called a **solution** of the equation. For example, 1 is a solution of the equation $3x + 4 = 7$, since $3(1) + 4 = 7$ is a true statement. But 2 is not a solution of this equation, since $3(2) + 4 = 7$ is not a true statement. The **solution set** of an equation is the set of solutions of the equation. For example, the solution set of $3x + 4 = 7$ is $\{1\}$.

To **solve an equation** is to find the solution set of an equation. Equations with the same solution set are called **equivalent equations.** For example,

$$3x + 4 = 7 \qquad 3x = 3 \qquad x = 1$$

are equivalent equations because they all have the same solution set, namely $\{1\}$. To solve an equation in x, we start with the given equation and write a series of simpler equivalent equations until we obtain an equation of the form

$$x = \textbf{number}$$

Two important properties are used to write equivalent equations.

> **The Addition and Multiplication Properties of Equality**
>
> If $a, b,$ and $c,$ are real numbers, then
>
> $$a = b \text{ and } a + c = b + c \text{ are equivalent equations.}$$
> Also, $a = b$ and $\quad ac = bc$ are equivalent equations as long as $c \neq 0$.

The **addition property of equality** guarantees that the same number may be added to both sides of an equation, and the result is an equivalent equation. The **multiplication property of equality** guarantees that both sides of an equation may be multiplied by the same nonzero number, and the result is an equivalent equation. Because we define subtraction in terms of addition $(a - b = a + (-b))$, and division in terms of multiplication $\left(\dfrac{a}{b} = a \cdot \dfrac{1}{b} \right)$, these properties also guarantee that we may *subtract* the same number from both sides of an equation, or *divide* both sides of an equation by the same nonzero number and the result is an equivalent equation.

For example, to solve $2x + 5 = 9$, use the addition and multiplication properties of equality to isolate x—that is, to write an equivalent equation of the form

$$x = \textbf{number}$$

We will do this in the next example.

EXAMPLE 1   Solve for x: $2x + 5 = 9$.

Solution   First, use the addition property of equality and subtract 5 from both sides. We do this so that our only variable term, $2x$, is by itself on one side of the equation.

$$2x + 5 = 9$$
$$2x + 5 - 5 = 9 - 5 \qquad \text{Subtract 5 from both sides.}$$
$$2x = 4 \qquad \text{Simplify.}$$

Now that the variable term is isolated, we can finish solving for x by using the multiplication property of equality and dividing both sides by 2.

$$\frac{2x}{2} = \frac{4}{2} \qquad \text{Divide both sides by 2.}$$
$$x = 2 \qquad \text{Simplify.}$$

Check: To see that 2 is the solution, replace x in the original equation with 2.

$$2x + 5 = 9 \quad \text{Original equation}$$
$$2(2) + 5 \overset{?}{=} 9 \quad \text{Let } x = 2.$$
$$4 + 5 \overset{?}{=} 9$$
$$9 = 9 \quad \text{True}$$

Since we arrive at a true statement, 2 is the solution or the solution set is $\{2\}$. □

PRACTICE
1   Solve for x: $3x + 7 = 22$.

■

EXAMPLE 2   Solve: $0.6 = 2 - 3.5c$.

Solution We use both the addition property and the multiplication property of equality.

$$0.6 = 2 - 3.5c$$
$$0.6 - 2 = 2 - 3.5c - 2 \quad \text{Subtract 2 from both sides.}$$
$$-1.4 = -3.5c \quad \text{Simplify. The variable term is now isolated.}$$
$$\frac{-1.4}{-3.5} = \frac{-3.5c}{-3.5} \quad \text{Divide both sides by } -3.5.$$
$$0.4 = c \quad \text{Simplify } \frac{-1.4}{-3.5}.$$

> ▶ **Helpful Hint**
> Don't forget that
> $$0.4 = c \text{ and } c = 0.4 \text{ are}$$
> equivalent equations.
>
> We may solve an equation so that the variable is alone on either side of the equation.

Check:

$$0.6 = 2 - 3.5c$$
$$0.6 \overset{?}{=} 2 - 3.5(0.4) \quad \text{Replace } c \text{ with } 0.4.$$
$$0.6 \overset{?}{=} 2 - 1.4 \quad \text{Multiply.}$$
$$0.6 = 0.6 \quad \text{True}$$

The solution is 0.4, or the solution set is $\{0.4\}$. □

PRACTICE
2   Solve: $2.5 = 3 - 2.5t$.

■

OBJECTIVE
2   **Solving Linear Equations That Can Be Simplified by Combining Like Terms** ▶

Often, an equation can be simplified by removing any grouping symbols and combining any like terms.

EXAMPLE 3   Solve: $-4x - 1 + 5x = 9x + 3 - 7x$.

Solution First we simplify both sides of this equation by combining like terms. Then, let's get variable terms on the same side of the equation by using the addition property of equality to subtract $2x$ from both sides. Next, we use this same property to add 1 to both sides of the equation.

$$-4x - 1 + 5x = 9x + 3 - 7x$$
$$x - 1 = 2x + 3 \quad \text{Combine like terms.}$$
$$x - 1 - 2x = 2x + 3 - 2x \quad \text{Subtract } 2x \text{ from both sides.}$$
$$-x - 1 = 3 \quad \text{Simplify.}$$
$$-x - 1 + 1 = 3 + 1 \quad \text{Add 1 to both sides.}$$
$$-x = 4 \quad \text{Simplify.}$$

Notice that this equation is not solved for x since we have $-x$ or $-1x$, not x. To solve for x, we divide both sides by -1.

$$\frac{-x}{-1} = \frac{4}{-1} \quad \text{Divide both sides by } -1.$$

$$x = -4 \quad \text{Simplify.}$$

Check to see that the solution is −4, or the solution set is $\{-4\}$. □

PRACTICE
3 Solve: $-8x - 4 + 6x = 5x + 11 - 4x$.

If an equation contains parentheses, use the distributive property to remove them.

EXAMPLE 4 Solve: $2(x - 3) = 5x - 9$.

Solution First, use the distributive property.

$$2(x - 3) = 5x - 9$$
$$2x - 6 = 5x - 9 \quad \text{Use the distributive property.}$$

Next, get variable terms on the same side of the equation by subtracting $5x$ from both sides.

$$2x - 6 - 5x = 5x - 9 - 5x \quad \text{Subtract } 5x \text{ from both sides.}$$
$$-3x - 6 = -9 \quad \text{Simplify.}$$
$$-3x - 6 + 6 = -9 + 6 \quad \text{Add 6 to both sides.}$$
$$-3x = -3 \quad \text{Simplify.}$$
$$\frac{-3x}{-3} = \frac{-3}{-3} \quad \text{Divide both sides by } -3.$$
$$x = 1$$

Let $x = 1$ in the original equation to see that 1 is the solution. □

PRACTICE
4 Solve: $3(x - 5) = 6x - 3$.

OBJECTIVE
3 **Solving Linear Equations Containing Fractions or Decimals**

If an equation contains fractions, we first clear the equation of fractions by multiplying both sides of the equation by the *least common denominator* (LCD) of all fractions in the equation.

EXAMPLE 5 Solve for y: $\dfrac{y}{3} - \dfrac{y}{4} = \dfrac{1}{6}$.

Solution First, clear the equation of fractions by multiplying both sides of the equation by 12, the LCD of denominators 3, 4, and 6.

$$\frac{y}{3} - \frac{y}{4} = \frac{1}{6}$$

$$12\left(\frac{y}{3} - \frac{y}{4}\right) = 12\left(\frac{1}{6}\right) \quad \text{Multiply both sides by the LCD 12.}$$

$$12\left(\frac{y}{3}\right) - 12\left(\frac{y}{4}\right) = 2 \quad \text{Apply the distributive property.}$$

$$4y - 3y = 2 \quad \text{Simplify.}$$

$$y = 2 \quad \text{Simplify.}$$

Check: To check, let $y = 2$ in the original equation.

$$\frac{y}{3} - \frac{y}{4} = \frac{1}{6} \quad \text{Original equation.}$$

$$\frac{2}{3} - \frac{2}{4} \stackrel{?}{=} \frac{1}{6} \quad \text{Let } y = 2.$$

$$\frac{8}{12} - \frac{6}{12} \stackrel{?}{=} \frac{1}{6} \qquad \text{Write fractions with the LCD.}$$

$$\frac{2}{12} \stackrel{?}{=} \frac{1}{6} \qquad \text{Subtract.}$$

$$\frac{1}{6} = \frac{1}{6} \qquad \text{Simplify.}$$

This is a true statement, so the solution is 2. □

PRACTICE
5 Solve for y: $\dfrac{y}{2} - \dfrac{y}{5} = \dfrac{1}{4}$.

As a general guideline, the following steps may be used to solve a linear equation in one variable.

> **Solving a Linear Equation in One Variable**
>
> **Step 1.** Clear the equation of fractions by multiplying both sides of the equation by the least common denominator (LCD) of all denominators in the equation.
> **Step 2.** Use the distributive property to remove grouping symbols such as parentheses.
> **Step 3.** Combine like terms on each side of the equation.
> **Step 4.** Use the addition property of equality to rewrite the equation as an equivalent equation with variable terms on one side and numbers on the other side.
> **Step 5.** Use the multiplication property of equality to isolate the variable.
> **Step 6.** Check the proposed solution in the original equation.

EXAMPLE 6 Solve for x: $\dfrac{x + 5}{2} + \dfrac{1}{2} = 2x - \dfrac{x - 3}{8}$.

Solution Multiply both sides of the equation by 8, the LCD of 2 and 8.

> ▶ **Helpful Hint**
> When we multiply both sides of an equation by a number, the distributive property tells us that each term of the equation is multiplied by the number.

$$8\left(\frac{x + 5}{2} + \frac{1}{2}\right) = 8\left(2x - \frac{x - 3}{8}\right) \qquad \text{Multiply both sides by 8.}$$

$$8\left(\frac{x + 5}{2}\right) + 8 \cdot \frac{1}{2} = 8 \cdot 2x - 8\left(\frac{x - 3}{8}\right) \qquad \text{Apply the distributive property.}$$

$$4(x + 5) + 4 = 16x - (x - 3) \qquad \text{Simplify.}$$

$$4x + 20 + 4 = 16x - x + 3 \qquad \text{Use the distributive property to remove parentheses.}$$

$$4x + 24 = 15x + 3 \qquad \text{Combine like terms.}$$

$$-11x + 24 = 3 \qquad \text{Subtract } 15x \text{ from both sides.}$$

$$-11x = -21 \qquad \text{Subtract 24 from both sides.}$$

$$\frac{-11x}{-11} = \frac{-21}{-11} \qquad \text{Divide both sides by } -11.$$

$$x = \frac{21}{11} \qquad \text{Simplify.}$$

Check: To check, verify that replacing x with $\dfrac{21}{11}$ makes the original equation true. The solution is $\dfrac{21}{11}$. □

PRACTICE
6 Solve for x: $x - \dfrac{x - 2}{12} = \dfrac{x + 3}{4} + \dfrac{1}{4}$.

If an equation contains decimals, you may want to first clear the equation of decimals.

EXAMPLE 7   Solve: $0.3x + 0.1 = 0.27x - 0.02$.

Solution   To clear this equation of decimals, we multiply both sides of the equation by 100. Recall that multiplying a number by 100 moves its decimal point two places to the right.

$$100(0.3x + 0.1) = 100(0.27x - 0.02)$$

$$100(0.3x) + 100(0.1) = 100(0.27x) - 100(0.02)$$ Use the distributive property.

$$30x + 10 = 27x - 2$$ Multiply.

$$30x - 27x = -2 - 10$$ Subtract $27x$ and 10 from both sides.

$$3x = -12$$ Simplify.

$$\frac{3x}{3} = \frac{-12}{3}$$ Divide both sides by 3.

$$x = -4$$ Simplify.

Check to see that the solution is -4.   □

PRACTICE
7   Solve: $0.15x - 0.03 = 0.2x + 0.12$.

✓**CONCEPT CHECK**

Explain what is wrong with the following:

$$3x - 5 = 16$$
$$3x = 11$$
$$\frac{3x}{3} = \frac{11}{3}$$
$$x = \frac{11}{3}$$

OBJECTIVE
4   Recognizing Identities and Equations with No Solution ▶

So far, each linear equation that we have solved has had a single solution. A linear equation in one variable that has exactly one solution is called a **conditional equation.** We will now look at two other types of equations: contradictions and identities.

An equation in one variable that has no solution is called a **contradiction,** and an equation in one variable that has every number (for which the equation is defined) as a solution is called an **identity.** For review: A linear equation in one variable with

No solution	Is a	Contradiction
Every real number as a solution (as long as the equation is defined)	Is an	Identity

The next examples show how to recognize contradictions and identities.

EXAMPLE 8   Solve for x: $3x + 5 = 3(x + 2)$.

Solution   First, use the distributive property and remove parentheses.

$$3x + 5 = 3(x + 2)$$
$$3x + 5 = 3x + 6$$ Apply the distributive property.
$$3x + 5 - 3x = 3x + 6 - 3x$$ Subtract $3x$ from both sides.
$$5 = 6$$

Answer to Concept Check:
Add 5 on the right side instead of subtracting 5.
$$3x - 5 = 16$$
$$3x = 21$$
$$x = 7$$
Therefore, the correct solution is 7.

▶ Helpful Hint
A solution set of $\{0\}$ and a solution set of $\{\,\}$ are not the same. The solution set $\{0\}$ means 1 solution, 0. The solution set $\{\,\}$ means *no* solution.

The equation $5 = 6$ is a false statement no matter what value the variable x might have. Thus, the original equation has no solution. Its solution set is written either as $\{\,\}$ or $\varnothing$. This equation is a contradiction.

PRACTICE
8 Solve for x: $4x - 3 = 4(x + 5)$.

EXAMPLE 9 Solve for x: $6x - 4 = 2 + 6(x - 1)$.

Solution First, use the distributive property and remove parentheses.

$$6x - 4 = 2 + 6(x - 1)$$
$$6x - 4 = 2 + 6x - 6 \qquad \text{Apply the distributive property.}$$
$$6x - 4 = 6x - 4 \qquad \text{Combine like terms.}$$

At this point, we might notice that both sides of the equation are the same, so replacing x by any real number gives a true statement. Thus the solution set of this equation is the set of real numbers, and the equation is an identity. Continuing to "solve" $6x - 4 = 6x - 4$, we eventually arrive at the same conclusion.

$$6x - 4 + 4 = 6x - 4 + 4 \quad \text{Add 4 to both sides.}$$
$$6x = 6x \quad \text{Simplify.}$$
$$6x - 6x = 6x - 6x \quad \text{Subtract } 6x \text{ from both sides.}$$
$$0 = 0 \quad \text{Simplify.}$$

Since $0 = 0$ is a true statement for every value of x, all real numbers are solutions. The solution set is the set of all real numbers or $\mathbb{R}$, $\{x \mid x \text{ is a real number}\}$, and the equation is called an identity.

PRACTICE
9 Solve for x: $5x - 2 = 3 + 5(x - 1)$.

▶ Helpful Hint
For linear equations, *any* false statement such as $5 = 6, 0 = 1$, or $-2 = 2$ informs us that the original equation has no solution. Also, *any* true statement such as $0 = 0, 2 = 2$, or $-5 = -5$ informs us that the original equation is an identity.

Vocabulary, Readiness & Video Check

Use the choices below to fill in the blanks. Not all choices will be used.

| multiplication | value | like |
| addition | solution | equivalent |

1. Equations with the same solution set are called _____ equations.

2. A value for the variable in an equation that makes the equation a true statement is called a(n) _____ of the equation.

3. By the _____ property of equality, $y = -3$ and $y - 7 = -3 - 7$ are equivalent equations.

4. By the _____ property of equality, $2y = -3$ and $\dfrac{2y}{2} = \dfrac{-3}{2}$ are equivalent equations.

Identify each as an equation or an expression.

5. $\dfrac{1}{3}x - 5$ _____

6. $2(x - 3) = 7$ _____

7. $\dfrac{5}{9}x + \dfrac{1}{3} = \dfrac{2}{9} - x$ _____

8. $\dfrac{5}{9}x + \dfrac{1}{3} - \dfrac{2}{9} - x$ _____

Martin-Gay Interactive Videos

See Video 2.1

Watch the section lecture video and answer the following questions.

OBJECTIVE
1

9. Complete these statements based on the lecture given before ▯ Example 1. The addition property of equality allows us to add the same number to (or subtract the same number from) _____ of an equation and have an equivalent equation. The multiplication property of equality allows us to multiply (or divide) both sides of an equation by the _____ nonzero number and have an equivalent equation.

OBJECTIVE
2

10. From ▯ Example 2, if an equation is simplified by removing parentheses before the properties of equality are applied, what property is used?

OBJECTIVE
3

11. In ▯ Example 3, what is the main reason given for first removing fractions from the equation?

OBJECTIVE
4

12. Complete this statement based on ▯ Example 4. When solving a linear equation and all variable terms subtract out and:
a. you have a _____ statement, then the equation has all real numbers for which the equation is defined as solutions.
b. you have a _____ statement, then the equation has no solution.

2.1 Exercise Set   MyMathLab®

Solve each equation and check. See Examples 1 and 2.

1. $-5x = -30$

2. $-2x = 18$

3. $-10 = x + 12$

4. $-25 = y + 30$

5. $x - 2.8 = 1.9$

6. $y - 8.6 = -6.3$

7. $5x - 4 = 26 + 2x$

8. $5y - 3 = 11 + 3y$

9. $-4.1 - 7z = 3.6$

10. $10.3 - 6x = -2.3$

11. $5y + 12 = 2y - 3$

12. $4x + 14 = 6x + 8$

Solve each equation and check. See Examples 3 and 4.

13. $3x - 4 - 5x = x + 4 + x$

14. $13x - 15x + 8 = 4x + 2 - 24$

15. $8x - 5x + 3 = x - 7 + 10$

16. $6 + 3x + x = -x + 8 - 26 + 24$

17. $5x + 12 = 2(2x + 7)$

18. $2(4x + 3) = 7x + 5$

19. $3(x - 6) = 5x$

20. $6x = 4(x - 5)$

21. $-2(5y - 1) - y = -4(y - 3)$

22. $-4(3n - 2) - n = -11(n - 1)$

Solve each equation and check. See Examples 5 through 7.

23. $\dfrac{x}{2} + \dfrac{x}{3} = \dfrac{3}{4}$

24. $\dfrac{x}{2} + \dfrac{x}{5} = \dfrac{5}{4}$

25. $\dfrac{3t}{4} - \dfrac{t}{2} = 1$

26. $\dfrac{4r}{5} - \dfrac{r}{10} = 7$

27. $\dfrac{n - 3}{4} + \dfrac{n + 5}{7} = \dfrac{5}{14}$

28. $\dfrac{2 + h}{9} + \dfrac{h - 1}{3} = \dfrac{1}{3}$

29. $0.6x - 10 = 1.4x - 14$

30. $0.3x + 2.4 = 0.1x + 4$

31. $\dfrac{3x - 1}{9} + x = \dfrac{3x + 1}{3} + 4$

32. $\dfrac{2z + 7}{8} - 2 = z + \dfrac{z - 1}{2}$

33. $1.5(4 - x) = 1.3(2 - x)$

34. $2.4(2x + 3) = -0.1(2x + 3)$

Solve each equation. See Examples 8 and 9.

35. $4(n + 3) = 2(6 + 2n)$

36. $6(4n + 4) = 8(3 + 3n)$

37. $3(x + 1) + 5 = 3x + 2$

38. $4(x + 2) + 4 = 4x - 8$

39. $2(x - 8) + x = 3(x - 6) + 2$

40. $5(x - 4) + x = 6(x - 2) - 8$

41. $4(x + 5) = 3(x - 4) + x$

42. $9(x - 2) = 8(x - 3) + x$

MIXED PRACTICE

Solve each equation. See Examples 1 through 9.

43. $\dfrac{3}{8} + \dfrac{b}{3} = \dfrac{5}{12}$

44. $\dfrac{a}{2} + \dfrac{7}{4} = 5$

45. $x - 10 = -6x - 10$

46. $4x - 7 = 2x - 7$

47. $5(x - 2) + 2x = 7(x + 4) - 38$

48. $3x + 2(x + 4) = 5(x + 1) + 3$

49. $y + 0.2 = 0.6(y + 3)$

50. $-(w + 0.2) = 0.3(4 - w)$

51. $\dfrac{1}{4}(a + 2) = \dfrac{1}{6}(5 - a)$

52. $\dfrac{1}{3}(8 + 2c) = \dfrac{1}{5}(3c - 5)$

53. $2y + 5(y - 4) = 4y - 2(y - 10)$

54. $9c - 3(6 - 5c) = c - 2(3c + 9)$

55. $6x - 2(x - 3) = 4(x + 1) + 4$

56. $10x - 2(x + 4) = 8(x - 2) + 6$

57. $\dfrac{m - 4}{3} - \dfrac{3m - 1}{5} = 1$ **58.** $\dfrac{n + 1}{8} - \dfrac{2 - n}{3} = \dfrac{5}{6}$

59. $8x - 12 - 3x = 9x - 7$

60. $10y - 18 - 4y = 12y - 13$

61. $-(3x - 5) - (2x - 6) + 1 = -5(x - 1) - (3x + 2) + 3$

62. $-4(2x - 3) - (10x + 7) - 2 = -(12x - 5) - (4x + 9) - 1$

63. $\dfrac{1}{3}(y + 4) + 6 = \dfrac{1}{4}(3y - 1) - 2$

64. $\dfrac{1}{5}(2y - 1) - 2 = \dfrac{1}{2}(3y - 5) + 3$

65. $2[7 - 5(1 - n)] + 8n = -16 + 3[6(n + 1) - 3n]$

66. $3[8 - 4(n - 2)] + 5n = -20 + 2[5(1 - n) - 6n]$

REVIEW AND PREVIEW

Translating. Translate each phrase into an expression. Use the variable x to represent each unknown number. See Section 1.2.

67. The quotient of 8 and a number

68. The sum of 8 and a number

69. The product of 8 and a number

70. The difference of 8 and a number

71. Five subtracted from twice a number

72. Two more than three times a number

CONCEPT EXTENSIONS

Find the error for each proposed solution. Then correct the proposed solution. See the Concept Check in this section.

73. $2x + 19 = 13$

    $2x = 32$

    $\dfrac{2x}{2} = \dfrac{32}{2}$

    $x = 16$

74. $-3(x - 4) = 10$

    $-3x - 12 = 10$

    $-3x = 22$

    $\dfrac{-3x}{-3} = \dfrac{22}{-3}$

    $x = -\dfrac{22}{3}$

75. $9x + 1.6 = 4x + 0.4$

    $5x = 1.2$

    $\dfrac{5x}{5} = \dfrac{1.2}{5}$

    $x = 0.24$

76. $\dfrac{x}{3} + 7 = \dfrac{5x}{3}$

    $x + 7 = 5x$

    $7 = 4x$

    $\dfrac{7}{4} = \dfrac{4x}{4}$

    $\dfrac{7}{2} = x$

By inspection, decide which equations have no solution and which equations have all real numbers as solutions.

77. $2x + 3 = 2x + 3$

78. $5x - 3 = 5x - 3$

79. $2x + 1 = 2x + 3$

80. $5x - 2 = 5x - 7$

81. a. Simplify the expression $4(x + 1) + 1$.

   b. Solve the equation $4(x + 1) + 1 = -7$.

   c. Explain the difference between solving an equation for a variable and simplifying an expression.

82. Explain why the multiplication property of equality does not include multiplying both sides of an equation by 0. (*Hint:* Write down a false statement and then multiply both sides by 0. Is the result true or false? What does this mean?)

83. In your own words, explain why the equation $x + 7 = x + 6$ has no solution, while the solution set of the equation $x + 7 = x + 7$ contains all real numbers.

84. In your own words, explain why the equation $x = -x$ has one solution—namely, 0—while the solution set of the equation $x = x$ is all real numbers.

Find the value of K such that the equations are equivalent.

85. $3.2x + 4 = 5.4x - 7$

    $3.2x = 5.4x + K$

86. $-7.6y - 10 = -1.1y + 12$

    $-7.6y = -1.1y + K$

87. $\dfrac{7}{11}x + 9 = \dfrac{3}{11}x - 14$

    $\dfrac{7}{11}x = \dfrac{3}{11}x + K$

88. $\dfrac{x}{6} + 4 = \dfrac{x}{3}$

    $x + K = 2x$

89. Write a linear equation in x whose only solution is 5.

90. Write an equation in x that has no solution.

Solve the following.

91. $x(x - 6) + 7 = x(x + 1)$

92. $7x^2 + 2x - 3 = 6x(x + 4) + x^2$

93. $3x(x + 5) - 12 = 3x^2 + 10x + 3$

94. $x(x + 1) + 16 = x(x + 5)$

Solve and check.

95. $2.569x = -12.48534$

96. $-9.112y = -47.537304$

97. $2.86z - 8.1258 = -3.75$

98. $1.25x - 20.175 = -8.15$

2.2 | An Introduction to Problem Solving

OBJECTIVES

1 Write Algebraic Expressions That Can Be Simplified.

2 Apply the Steps for Problem Solving.

OBJECTIVE

1   Writing and Simplifying Algebraic Expressions

To prepare for problem solving, we practice writing algebraic expressions that can be simplified.

Our first example involves consecutive integers and perimeter. Recall that *consecutive integers* are integers that follow one another in order. Study the examples of consecutive, even, and odd integers and their representations.

Consecutive Integers: **Consecutive Even Integers:**   **Consecutive Odd Integers:**

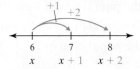   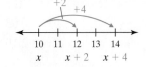   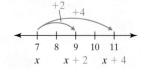

> ▶ **Helpful Hint**
>
> You may want to begin this section by studying key words and phrases and their translations in Sections 1.2 Objective 5 and 1.4 Objective 4.

EXAMPLE 1   Write the following as algebraic expressions. Then simplify.

a. The sum of three consecutive integers if x is the first consecutive integer.

 b. The perimeter of a triangle with sides of length x, $5x$, and $6x - 3$.

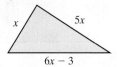

Solution

a. Recall that if x is the first integer, then the next consecutive integer is 1 more, or $x + 1$ and the next consecutive integer is 1 more than $x + 1$, or $x + 2$.

In words:	first integer	plus	next consecutive integer	plus	next consecutive integer
Translate:	x	$+$	$(x + 1)$	$+$	$(x + 2)$

Then $x + (x + 1) + (x + 2) = x + x + 1 + x + 2$

                         $= 3x + 3$ Simplify by combining like terms.

b. The perimeter of a triangle is the sum of the lengths of the sides.

In words:	side	$+$	side	$+$	side
Translate:	x	$+$	$5x$	$+$	$(6x - 3)$

Then $x + 5x + (6x - 3) = x + 5x + 6x - 3$

                         $= 12x - 3$ Simplify.   □

PRACTICE

1   Write the following algebraic expressions. Then simplify.

a. The sum of three consecutive odd integers if x is the first consecutive odd integer

b. The perimeter of a trapezoid with bases x and $2x$ and sides of $x + 2$ and $2x - 3$

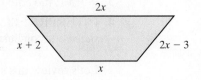

EXAMPLE 2 **Writing Algebraic Expressions Representing Metropolitan Regions**

The most populous metropolitan region in the United States is New York City, although it is only the sixth most populous metropolis in the world. Tokyo is the most populous metropolitan region. Mexico City is the fifth most populous metropolis in the world. Mexico City's population is 0.03 million more than New York's, and Tokyo's is twice that of New York, decreased by 2.19 million. Write the sum of the populations of these three metropolitan regions as an algebraic expression. Let x be the population of New York (in millions). (*Source:* United Nations, Department of Economic and Social Affairs)

Solution:

If x = the population of New York (in millions), then

$x + 0.03$ = the population of Mexico City (in millions) and

$2x - 2.19$ = the population of Tokyo (in millions)

In words:

population of New York	population of Mexico City	population of Tokyo

Translate: x + $(x + 0.03)$ + $(2x - 2.19)$

Then $x + (x + 0.03) + (2x - 2.19) = x + x + 2x + 0.03 - 2.19$

$= 4x - 2.16$ Combine like terms.

In Exercise 57, we will find the actual populations of these cities. □

PRACTICE

2 The three busiest airports in Europe are in London, England; Paris, France; and Frankfurt, Germany. The airport in London has 12.9 million more arrivals and departures than the Frankfurt airport. The Paris airport has 5.2 million more arrivals and departures than the Frankfurt airport. Write the sum of the arrivals and departures from these three cities as a simplified algebraic expression. Let x be the number of arrivals and departures at the Frankfurt airport. (*Source:* Association of European Airlines)

OBJECTIVE

2 **Applying Steps for Problem Solving**

Our main purpose for studying algebra is to solve problems. The following problem-solving strategy will be used throughout this text and may also be used to solve real-life problems that occur outside the mathematics classroom.

> **General Strategy for Problem Solving**
>
> **1.** UNDERSTAND the problem. During this step, become comfortable with the problem. Some ways of doing this are to:
>
> Read and reread the problem.
>
> Propose a solution and check. Pay careful attention to how you check your proposed solution. This will help when writing an equation to model the problem.
>
> Construct a drawing.
>
> **Choose a variable to represent the unknown.** (Very important part)
>
> **2.** TRANSLATE the problem into an equation.
>
> **3.** SOLVE the equation.
>
> **4.** INTERPRET the results: *Check* the proposed solution in the stated problem and *state* your conclusion.

Let's review this strategy by solving a problem involving unknown numbers.

EXAMPLE 3   **Finding Unknown Numbers**

Find three numbers such that the second number is 3 more than twice the first number, and the third number is four times the first number. The sum of the three numbers is 164.

Solution

> ▶ **Helpful Hint**
>
> The purpose of guessing a solution is not to guess correctly but to gain confidence and to help understand the problem and how to model it.

1. **UNDERSTAND** the problem. First let's read and reread the problem and then propose a solution. For example, if the first number is 25, then the second number is 3 more than twice 25, or 53. The third number is four times 25, or 100. The sum of 25, 53, and 100 is 178, not the required sum, but we have gained some valuable information about the problem. First, we know that the first number is less than 25 since our guess led to a sum greater than the required sum. Also, we have gained some information as to how to model the problem.

   Next let's assign a variable and use this variable to represent any other unknown quantities. If we let

$$x = \text{the first number, then}$$
$$2x + 3 = \text{the second number}$$

   ↑   ↑
   | 3 more than
   twice the second number

$$4x = \text{the third number}$$

2. **TRANSLATE** the problem into an equation. To do so, we use the fact that the sum of the numbers is 164. First let's write this relationship in words and then translate to an equation.

In words:	first number	added to	second number	added to	third number	is	164
	↓	↓	↓	↓	↓	↓	↓
Translate:	x	$+$	$(2x + 3)$	$+$	$4x$	$=$	164

3. **SOLVE** the equation.

$$x + (2x + 3) + 4x = 164$$
$$x + 2x + 4x + 3 = 164 \quad \text{Remove parentheses.}$$
$$7x + 3 = 164 \quad \text{Combine like terms.}$$
$$7x = 161 \quad \text{Subtract 3 from both sides.}$$
$$x = 23 \quad \text{Divide both sides by 7.}$$

4. **INTERPRET.** Here, we *check* our work and *state* the solution. Recall that if the first number $x = 23$, then the second number $2x + 3 = 2 \cdot 23 + 3 = 49$ and the third number $4x = 4 \cdot 23 = 92$.

   Check: Is the second number 3 more than twice the first number? Yes, since 3 more than twice 23 is $46 + 3$, or 49. Also, their sum, $23 + 49 + 92 = 164$, is the required sum.

   State: The three numbers are 23, 49, and 92.   □

PRACTICE

3   Find three numbers such that the second number is 8 less than triple the first number, the third number is five times the first number, and the sum of the three numbers is 118.

Many of today's rates and statistics are given as percents. Interest rates, tax rates, nutrition labeling, and percent of households in a given category are just a few examples. Before we practice solving problems containing percents, let's briefly review the meaning of percent and how to find a percent of a number.

The word *percent* means "per hundred," and the symbol % denotes percent. This means that 23% is 23 per hundred, or $\frac{23}{100}$. Also,

$$41\% = \frac{41}{100} = 0.41$$

To find a percent of a number, we multiply.

$$16\% \text{ of } 25 = 16\% \cdot 25 = 0.16 \cdot 25 = 4$$

Thus, 16% of 25 is 4.

Study the table below. It will help you become more familiar with finding percents.

Percent	*Meaning/Shortcut*	*Example*
50%	$\frac{1}{2}$ or half of a number	50% of 60 is 30.
25%	$\frac{1}{4}$ or a quarter of a number	25% of 60 is 15.
10%	0.1 or $\frac{1}{10}$ of a number (move the decimal point 1 place to the left)	10% of 60 is 6.0 or 6.
1%	0.01 or $\frac{1}{100}$ of a number (move the decimal point 2 places to the left)	1% of 60 is 0.60 or 0.6.
100%	1 or all of a number	100% of 60 is 60.
200%	2 or double a number	200% of 60 is 120.

✓CONCEPT CHECK

Suppose you are finding 112% of a number x. Which of the following is a correct description of the result? Explain.

a. The result is less than x.   **b.** The result is equal to x.   **c.** The result is greater than x.

Next, we solve a problem containing a percent.

EXAMPLE 4 **Finding the Original Price of a Computer**

Suppose that a computer store just announced an 8% decrease in the price of a particular computer model. If this computer sells for $2162 after the decrease, find the original price of this computer.

Solution

1. UNDERSTAND. Read and reread the problem. Recall that a percent decrease means a percent of the original price. Let's guess that the original price of the computer is $2500. The amount of decrease is then 8% of $2500, or $(0.08)(\$2500) = \$200$. This means that the new price of the computer is the original price minus the decrease, or $\$2500 - \$200 = \$2300$. Our guess is incorrect, but we now have an idea of how to model this problem. In our model, we will let $x = $ the original price of the computer.

2. TRANSLATE.

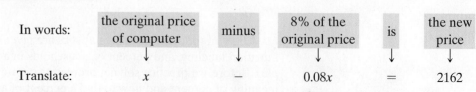

In words:	the original price of computer	minus	8% of the original price	is	the new price
Translate:	x	$-$	$0.08x$	$=$	2162

3. SOLVE the equation.

$$x - 0.08x = 2162$$

$$0.92x = 2162 \qquad \text{Combine like terms.}$$

$$x = \frac{2162}{0.92} = 2350 \quad \text{Divide both sides by 0.92.}$$

4. INTERPRET.

Check: If the original price of the computer was $2350, the new price is

$$\$2350 - (0.08)(\$2350) = \$2350 - \$188$$

$$= \$2162 \qquad \text{The given new price}$$

State: The original price of the computer was $2350.   □

PRACTICE

4   At the end of the season, the cost of a snowboard was reduced by 40%. If the snowboard sells for $270 after the decrease, find the original price of the board.

 EXAMPLE 5   **Finding the Lengths of a Triangle's Sides**

A pennant in the shape of an isosceles triangle is to be constructed for the Slidell High School Athletic Club and sold at a fund-raiser. The company manufacturing the pennant charges according to perimeter, and the athletic club has determined that a perimeter of 149 centimeters should make a nice profit. If each equal side of the triangle is twice the length of the third side, increased by 12 centimeters, find the lengths of the sides of the triangular pennant.

Solution

1. UNDERSTAND. Read and reread the problem. Recall that the perimeter of a triangle is the distance around. Let's guess that the third side of the triangular pennant is 20 centimeters. This means that each equal side is twice 20 centimeters, increased by 12 centimeters, or $2(20) + 12 = 52$ centimeters.

This gives a perimeter of $20 + 52 + 52 = 124$ centimeters. Our guess is incorrect, but we now have a better understanding of how to model this problem.
Now we let the third side of the triangle $= x$.

the first side $=$	twice	the third side	increased by 12
↓	↓	↓	↓
$=$	2	x	$+$ 12

or $2x + 12$

the second side $= 2x + 12$

2. TRANSLATE.

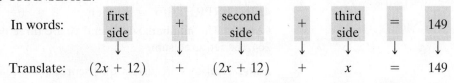

In words:	first side	$+$	second side	$+$	third side	$=$	149
	↓	↓	↓	↓	↓	↓	↓
Translate:	$(2x + 12)$	$+$	$(2x + 12)$	$+$	x	$=$	149

3. SOLVE the equation.

$$(2x + 12) + (2x + 12) + x = 149$$

$2x + 12 + 2x + 12 + x = 149$ Remove parentheses.

$5x + 24 = 149$ Combine like terms.

$5x = 125$ Subtract 24 from both sides.

$x = 25$ Divide both sides by 5.

4. INTERPRET. If the third side is 25 centimeters, then the first side is $2(25) + 12 = 62$ centimeters, and the second side is 62 centimeters.

Check: The first and second sides are each twice 25 centimeters increased by 12 centimeters or 62 centimeters. Also, the perimeter is $25 + 62 + 62 = 149$ centimeters, the required perimeter.

State: The lengths of the sides of the triangle are 25 centimeters, 62 centimeters, and 62 centimeters. □

PRACTICE

5 For its 40th anniversary, North Campus Community College is placing rectangular banners on all the light poles on campus. The perimeter of these banners is 160 inches. If the longer side of each banner is 16" less than double the width, find the dimensions of the banners.

EXAMPLE 6 **Finding Consecutive Integers**

Kelsey Ohleger was helping her friend Benji Burnstine study for an algebra exam. Kelsey told Benji that her three latest art history quiz scores are three consecutive even integers whose sum is 264. Help Benji find the scores.

Solution

1. UNDERSTAND. Read and reread the problem. Since we are looking for consecutive even integers, let

x = the first integer. Then

$x + 2$ = the second consecutive even integer

$x + 4$ = the third consecutive even integer.

2. TRANSLATE.

In words:	first integer	+	second even integer	+	third even integer	=	264
Translate:	x	+	$(x + 2)$	+	$(x + 4)$	=	264

3. SOLVE.

$$x + (x + 2) + (x + 4) = 264$$

$3x + 6 = 264$ Combine like terms.

$3x = 258$ Subtract 6 from both sides.

$x = 86$ Divide both sides by 3.

4. INTERPRET. If $x = 86$, then $x + 2 = 86 + 2$ or 88, and $x + 4 = 86 + 4$ or 90.

Check: The numbers 86, 88, and 90 are three consecutive even integers. Their sum is 264, the required sum.

State: Kelsey's art history quiz scores are 86, 88, and 90. □

PRACTICE

6 Find three consecutive odd integers whose sum is 81.

Vocabulary, Readiness & Video Check

Fill in each blank with $<$, $>$, *or* $=$. *(Assume that the unknown number is a positive number.)*

1. 130% of a number ___ the number.

2. 70% of a number ___ the number.

3. 100% of a number ___ the number.

4. 200% of a number ___ the number.

Complete the table. The first row has been completed for you.

	First Integer	All Described Integers
Three consecutive integers	18	18, 19, 20
5. Four consecutive integers	31	
6. Three consecutive odd integers	31	
7. Three consecutive even integers	18	
8. Four consecutive even integers	92	
9. Three consecutive integers	y	
10. Three consecutive even integers	z (z is even)	
11. Four consecutive integers	p	
12. Three consecutive odd integers	s (s is odd)	

Martin-Gay Interactive Videos

See Video 2.2

Watch the section lecture video and answer the following questions.

OBJECTIVE
1

13. In Example 2, once the phrase is translated, what property is used to remove parentheses?

OBJECTIVE
2

14. The equation for Example 3 is solved and the result is $x = 45$. Why is 45 not the complete solution to the application?

2.2 Exercise Set   MyMathLab®

Write the following as algebraic expressions. Then simplify. See Examples 1 and 2.

△ **1.** The perimeter of a square with side length y.

y

△ **2.** The perimeter of a rectangle with length x and width $x - 5$.

x

$x - 5$

▶ **3.** The sum of three consecutive integers if the first is z.

4. The sum of three consecutive odd integers if the first integer is x.

▶ **5.** The total amount of money (in cents) in x nickels, $(x + 3)$ dimes, and $2x$ quarters. (*Hint:* The value of a nickel is 5 cents, the value of a dime is 10 cents, and the value of a quarter is 25 cents.)

6. The total amount of money (in cents) in y quarters, $7y$ dimes, and $(2y - 1)$ nickels. (Use the hint for Exercise 5.)

△ **7.** A piece of land along Bayou Liberty is to be fenced and subdivided as shown so that each rectangle has the same dimensions. Express the total amount of fencing needed as an algebraic expression in x.

x

$\leftarrow 2x + 1 \rightarrow$ $\leftarrow ? \rightarrow$ $\leftarrow ? \rightarrow$

8. A flooded piece of land near the Mississippi River in New Orleans is to be surveyed and divided into 4 rectangles of equal dimension. Express the total amount of fencing needed as an algebraic expression in x.

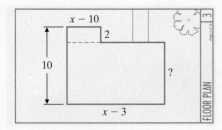

△ **9.** Write the perimeter of the floor plan shown as an algebraic expression in x.

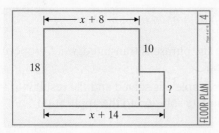

10. Write the perimeter of the floor plan shown as an algebraic expression in x.

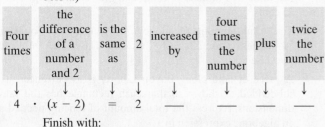

Solve. For Exercises 11 and 12, the solutions have been started for you. See Example 3.

11. Four times the difference of a number and 2 is the same as 2, increased by four times the number, plus twice the number. Find the number.

Start the solution:

1. UNDERSTAND the problem. Reread it as many times as needed.

2. TRANSLATE into an equation. (Fill in the blanks below.)

Four times	the difference of a number and 2	is the same as	2	increased by	four times the number	plus	twice the number
↓	↓	↓	↓	↓	↓	↓	↓
4 ·	$(x-2)$	=	2	___	___	___	___

Finish with:

3. SOLVE and 4. INTERPRET

12. Twice the sum of a number and 3 is the same as five times the number, minus 1, minus four times the number. Find the number.

Start the solution:

1. UNDERSTAND the problem. Reread it as many times as needed.

2. TRANSLATE into an equation. (Fill in the blanks below.)

Twice	the sum of a number and 3	is the same as	five times the number	minus	1	minus	four times the number
↓	↓	↓	↓	↓	↓	↓	↓
2	$(x+3)$	=	___	___	1	___	___

Finish with:

3. SOLVE and 4. INTERPRET

▶ **13.** A second number is five times a first number. A third number is 100 more than the first number. If the sum of the three numbers is 415, find the numbers.

14. A second number is 6 less than a first number. A third number is twice the first number. If the sum of the three numbers is 306, find the numbers.

Solve. See Example 4.

15. The United States consists of 2271 million acres of land. Approximately 29% of this land is federally owned. Find the number of acres that are not federally owned. (*Source:* U.S. General Services Administration)

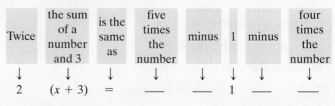

16. The state of Nevada contains the most federally owned acres of land in the United States. If 90% of the state's 70 million acres of land is federally owned, find the number of acres that are not federally owned. (*Source:* U.S. General Services Administration)

17. In 2010, 8476 earthquakes occurred in the United States. Of these, 91.4% were minor tremors with magnitudes of 3.9 or less on the Richter scale. How many minor earthquakes occurred in the United States in 2010? Round to the nearest whole. (*Source:* U.S. Geological Survey National Earthquake Information Center)

18. Of the 1543 tornadoes that occurred in the United States during 2010, 27.7% occurred during the month of June. How many tornadoes occurred in the United States during June 2010? Round to the nearest whole. (*Source:* Storm Prediction Center)

19. In a recent survey, 15% of online shoppers in the United States say that they prefer to do business only with large, well-known retailers. In a group of 1500 online shoppers, how many are willing to do business with any size retailers? (*Source:* Inc.com)

20. In 2010, the restaurant industry employed 9% of the U.S. workforce. If there are estimated to be 141 million Americans in the workforce, how many people are employed by the restaurant industry? Round to the nearest tenth of a million. (*Source:* National Restaurant Association, U.S. Bureau of Labor Statistics)

The following graph is called a circle graph or a pie chart. The circle represents a whole, or in this case, 100%. This particular graph shows the number of minutes per day that people use email at work. Use this graph to answer Exercises 21 through 24.

Time Spent on Email at Work

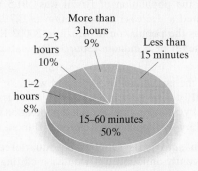

Source: Pew Internet & American Life Project

21. What percent of email users at work spend less than 15 minutes on email per day?

22. Among email users at work, what is the most common time spent on email per day?

23. If it were estimated that a large company has 4633 employees, how many of these would you expect to be using email more than 3 hours per day?

24. If it were estimated that a medium-size company has 250 employees, how many of these would you expect to be using email between 2 and 3 hours per day?

Use the diagrams to find the unknown measures of angles or lengths of sides. Recall that the sum of the angle measures of a triangle is 180°. See Example 5.

25.

26.

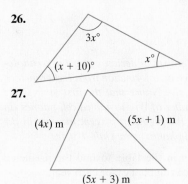

27.

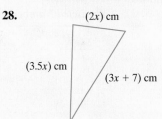

Perimeter is 102 meters.

28.

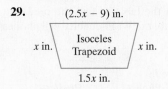

Perimeter is 75 centimeters.

29.

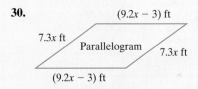

Perimeter is 99 inches.

30.

(9.2x − 3) ft
7.3x ft   Parallelogram   7.3x ft
(9.2x − 3) ft

Perimeter is 324 feet.

Solve. See Example 6.

31. The sum of three consecutive integers is 228. Find the integers.

32. The sum of three consecutive odd integers is 327. Find the integers.

33. The ZIP codes of three Nevada locations—Fallon, Fernley, and Gardnerville Ranchos—are three consecutive even integers. If twice the first integer added to the third is 268,222, find each ZIP code.

34. During a recent year, the average SAT scores in math for the states of Alabama, Louisiana, and Michigan were 3 consecutive integers. If the sum of the first integer, second integer, and three times the third integer is 2637, find each score.

MIXED PRACTICE

Solve. See Examples 1 through 6.

Many companies predict the growth or decline of various technologies. The following data is based on information from Techcrunchies, a technological information site. Notice that the first table is the predicted increase in the number of Wi-Fi-enabled cell phones (in millions), and the second is the predicted percent increase in the number of Wi-Fi-enabled cell phones in the United States.

35. Use the middle column in the table to find the predicted number of Wi-Fi-enabled cell phones for each year.

Year	Increase in Wi-Fi-Enabled Cell Phones	Predicted Number
2010	$2x - 21$	
2012	$\frac{5}{2}x + 2$	
2014	$3x + 24$	
Total	290 million	

36. Use the middle column in the table to find the predicted percent increase in the number of Wi-Fi-enabled cell phones for each year.

Year	Percent Increase in Wi-Fi-Enabled Cell Phones since 2009	Predicted Percent Increase
2010	x	
2011	$2x + 10$	
2012	$4x - 25$	
	300%	

Solve.

37. The occupations of biomedical engineers, skin care specialists, and physician assistants are among the 10 with the largest growth from 2008 to 2018. The number of physician assistant jobs will grow 7 thousand less than three times the number of biomedical engineer jobs. The number of skin care specialist jobs will grow 9 thousand more than half the number of biomedical engineer jobs. If the total growth of these three jobs is predicted to be 56 thousand, find the predicted growth of each job. (*Source:* U.S. Department of Labor, Bureau of Labor Statistics)

38. The occupations of farmer or rancher, file clerk, and telemarketer are among the 10 jobs with the largest decline from 2008 to 2018. The number of file clerk jobs is predicted to decline 11 thousand more than the number of telemarketer jobs. The number of farmer or rancher jobs is predicted to decline 3 thousand more than twice the number of telemarketer jobs. If the total decline of these three jobs is predicted to be 166 thousand, find the predicted decline of each job. (*Source:* U.S. Department of Labor, Bureau of Labor Statistics)

39. The B767-300ER aircraft has 88 more seats than the B737-200 aircraft. The F-100 has 32 fewer seats than the B737-200 aircraft. If their total number of seats is 413, find the number of seats for each aircraft. (*Source:* Air Transport Association of America)

40. Cowboy Stadium, home of the Dallas Cowboys of the NFL, seats approximately 9800 more fans than does Candlestick Park, home of the San Francisco 49ers. Soldier Field, home of the Chicago Bears, seats 8700 fewer fans than Candlestick Park. If the total seats in these three stadiums is 211,700, how many seats are in each of the three stadiums?

41. A new fax machine was recently purchased for an office in Hopedale for $464.40 including tax. If the tax rate in Hopedale is 8%, find the price of the fax machine before tax.

42. A premedical student at a local university was complaining that she had just paid $158.60 for her human anatomy book, including tax. Find the price of the book before taxes if the tax rate at this university is 9%.

43. The median compensation for a U.S. university president was $436,000 for the 2008–2009 academic year. Calculate the salary of a university president who received a 2.3% raise.

44. In 2009, the population of Brazil was 191.5 million. This represented a decrease in population of 3.7% from 2000. What was the population of Brazil in 2000? Round to the nearest tenth of a million. (*Source:* Population Reference Bureau)

45. In 2010, the population of Swaziland was 1,200,000 people. From 2010 to 2050, Swaziland's population is expected to increase by 50%. Find the expected population of Swaziland in 2050. (*Source:* Population Reference Bureau)

46. Dana, an auto parts supplier headquartered in Toledo, Ohio, recently announced it would be cutting 11,000 jobs worldwide. This is equivalent to 15% of Dana's workforce. Find the size of Dana's workforce prior to this round of job layoffs. Round to the nearest whole. (*Source:* Dana Corporation)

Recall that two angles are complements of each other if their sum is 90°. Two angles are supplements of each other if their sum is 180°. Find the measure of each angle.

47. One angle is three times its supplement increased by 20°. Find the measures of the two supplementary angles.

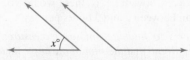

48. One angle is twice its complement increased by 30°. Find the measure of the two complementary angles.

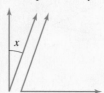

Recall that the sum of the angle measures of a triangle is 180°.

△ **49.** Find the measures of the angles of a triangle if the measure of one angle is twice the measure of a second angle and the third angle measures 3 times the second angle decreased by 12.

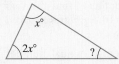

△ **50.** Find the angles of an isoceles triangle whose two base angles are equal and whose third angle is 10° less than three times a base angle.

▶ **51.** Two frames are needed with the same perimeter: one frame in the shape of a square and one in the shape of an equilateral triangle. Each side of the triangle is 6 centimeters longer than each side of the square. Find the side lengths of each frame. (An equilateral triangle has sides that are the same length.)

52. Two frames are needed with the same perimeter: one frame in the shape of a square and one in the shape of a regular pentagon. Each side of the square is 7 inches longer than each side of the pentagon. Find the side lengths of each frame. (A regular polygon has sides that are the same length.)

53. The sum of the first and third of three consecutive even integers is 156. Find the three even integers.

54. The sum of the second and fourth of four consecutive integers is 110. Find the four integers.

55. Daytona International Speedway in Florida has 37,000 more grandstand seats than twice the number of grandstand seats at Darlington Motor Raceway in South Carolina. Together, these two race tracks seat 220,000 NASCAR fans. How many seats does each race track have? (*Source: NASCAR*)

56. For the 2010–2011 National Hockey League season, the payroll for the San Jose Sharks was $5,986,667 more than that of the Montreal Canadiens. The total payroll for these two teams was $113,103,333. What was the payroll for these two teams for the 2010–2011 NHL season?

57. The sum of the populations of the metropolitan regions of New York, Tokyo, and Mexico City is 75.56 million. Use this information and Example 2 in this section to find the population of each metropolitan region. (*Source:* United Nations Department of Economic and Social Affairs)

58. The airports in London, Paris, and Frankfurt have a total of 177.1 million annual arrivals and departures. Use this information and Practice 2 in this section to find the number at each airport.

△ **59.** Suppose the perimeter of the triangle in Example 1b in this section is 483 feet. Find the length of each side.

△ **60.** Suppose the perimeter of the trapezoid in Practice 1b in this section is 110 meters. Find the lengths of its sides and bases.

61. Incandescent, fluorescent, and halogen bulbs are lasting longer today than ever before. On average, the number of bulb hours for a fluorescent bulb is 25 times the number of bulb hours for a halogen bulb. The number of bulb hours for an incandescent bulb is 2500 less than the halogen bulb. If the total number of bulb hours for the three types of bulbs is 105,500, find the number of bulb hours for each type. (*Source: Popular Science* magazine)

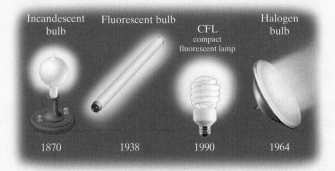

62. Falkland Islands, Iceland, and Norway are the top three countries that have the greatest Internet penetration rate (percent of population) in the world. Falkland Islands has a 6.8 percent greater penetration rate than Iceland. Norway has a 2.3 percent less penetration rate than Iceland. If the sum of the penetration rates is 284.1, find the Internet penetration rate in each of these countries. (*Source*: Internet World Stats)

63. During the 2010 Major League Baseball season, the number of wins for the Milwaukee Brewers, Houston Astros, and Chicago Cubs was three consecutive integers. Of these three teams, the Milwaukee Brewers had the most wins. The Chicago Cubs had the least wins. The total number of wins by these three teams was 228. How many wins did each team have in the 2010 season?

64. In the 2010 Winter Olympics, Austria won more medals than the Russian Federation, which won more medals than South Korea. If the numbers of medals won by these three countries are three consecutive integers whose sum is 45, find the number of medals won by each. (*Source*: Vancouver 2010)

65. The three tallest hospitals in the world are Guy's Tower in London, Queen Mary Hospital in Hong Kong, and Galter Pavilion in Chicago. These buildings have a total height of 1320 feet. Guy's Tower is 67 feet taller than Galter Pavilion, and the Queen Mary Hospital is 47 feet taller than Galter Pavilion. Find the heights of the three hospitals.

△ **66.** The official manual for traffic signs is the *Manual on Uniform Traffic Control Devices* published by the Government Printing Office. The rectangular sign below has a length 12 inches more than twice its height. If the perimeter of the sign is 312 inches, find its dimensions.

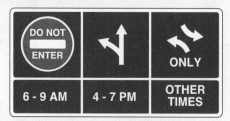

REVIEW AND PREVIEW

Find the value of each expression for the given values. See Section 1.3.

67. $4ab - 3bc$; $a = -5$, $b = -8$, and $c = 2$

68. $ab + 6bc$; $a = 0$, $b = -1$, and $c = 9$

69. $n^2 - m^2$; $n = -3$ and $m = -8$

70. $2n^2 + 3m^2$; $n = -2$ and $m = 7$

71. $P + PRT$; $P = 3000$, $R = 0.0325$, and $T = 2$

72. $\frac{1}{3}lwh$; $l = 37.8$, $w = 5.6$, and $h = 7.9$

CONCEPT EXTENSIONS

73. For Exercise 36, the percents have a sum of 300%. Is this possible? Why or why not?

74. In your own words, explain the differences in the tables for Exercises 35 and 36.

75. Find an angle such that its supplement is equal to 10 times its complement.

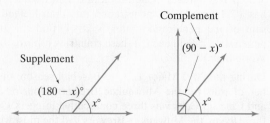

76. Find an angle such that its supplement is equal to twice its complement increased by 50°.

The average annual number of cigarettes smoked by an American adult continues to decline. For the years 2000–2009, the equation $y = -94.8x + 2049$ approximates this data. Here, x is the number of years after 2000 and y is the average annual number of cigarettes smoked. (Source: Centers for Disease Control)

77. If this trend continues, find the year in which the average annual number of cigarettes smoked is zero. To do this, let $y = 0$ and solve for x.

78. Predict the average annual number of cigarettes smoked by an American adult in 2015. To do so, let $x = 15$ (since $2015 - 2000 = 15$) and find y.

79. Predict the average annual number of cigarettes smoked by an American adult in 2020. To do so, let $x = 20$ (since $2020 - 2000 = 20$) and find y.

80. Use the result of Exercise 78 to predict the average *daily* number of cigarettes smoked by an American adult in 2015. Round to the nearest whole. Do you think this number represents the average daily number of cigarettes smoked by an adult? Why or why not?

81. Determine whether there are three consecutive integers such that their sum is three times the second integer.

82. Determine whether there are two consecutive odd integers such that 7 times the first exceeds 5 times the second by 54.

To break even in a manufacturing business, income or revenue R must equal the cost of production C. Use this information to answer Exercises 83 through 86.

83. The cost C to produce x skateboards is $C = 100 + 20x$. The skateboards are sold wholesale for $24 each, so revenue R is given by $R = 24x$. Find how many skateboards the manufacturer needs to produce and sell to break even. (*Hint:* Set the cost expression equal to the revenue expression and solve for x.)

84. The revenue R from selling x computer boards is given by $R = 60x$, and the cost C of producing them is given by $C = 50x + 5000$. Find how many boards must be sold to break even. Find how much money is needed to produce the break-even number of boards.

85. In your own words, explain what happens if a company makes and sells fewer products than the break-even number.

86. In your own words, explain what happens if more products than the break-even number are made and sold.

2.3 | Formulas and Problem Solving

OBJECTIVES

1 Solve a Formula for a Specified Variable.

2 Use Formulas to Solve Problems.

OBJECTIVE

1   Solving a Formula for a Specified Variable

Solving problems that we encounter in the real world sometimes requires us to express relationships among measured quantities. An equation that describes a known relationship among quantities such as distance, time, volume, weight, money, and gravity is called a **formula.** Some examples of formulas are

Formula	*Meaning*
$I = PRT$	Interest = principal $\cdot$ rate $\cdot$ time
$A = lw$	Area of a rectangle = length $\cdot$ width
$d = rt$	Distance = rate $\cdot$ time
$C = 2\pi r$	Circumference of a circle = $2 \cdot \pi \cdot$ radius
$V = lwh$	Volume of a rectangular solid = length $\cdot$ width $\cdot$ height

Other formulas are listed in the front cover of this text. Notice that the formula for the volume of a rectangular solid $V = lwh$ is solved for V since V is by itself on one side of the equation with no V's on the other side of the equation. Suppose that the volume of a rectangular solid is known as well as its width and its length, and we wish to find its height. One way to find its height is to begin by solving the **formula** $V = lwh$ for h.

EXAMPLE 1   Solve: $V = lwh$ for h.

Solution   To solve $V = lwh$ for h, isolate h on one side of the equation. To do so, divide both sides of the equation by lw.

$$V = lwh$$

$$\frac{V}{lw} = \frac{lw\,h}{lw} \qquad \text{Divide both sides by } lw.$$

$$\frac{V}{lw} = h \text{ or } h = \frac{V}{lw} \qquad \text{Simplify.}$$

Thus we see that to find the height of a rectangular solid, we divide its volume by the product of its length and its width. $\square$

PRACTICE

1   Solve: $I = PRT$ for T.

The following steps may be used to solve formulas and equations in general for a specified variable.

Solving Equations for a Specified Variable

Step 1. Clear the equation of fractions by multiplying each side of the equation by the least common denominator.

Step 2. Use the distributive property to remove grouping symbols such as parentheses.

Step 3. Combine like terms on each side of the equation.

Step 4. Use the addition property of equality to rewrite the equation as an equivalent equation with terms containing the specified variable on one side and all other terms on the other side.

Step 5. Use the distributive property and the multiplication property of equality to isolate the specified variable.

EXAMPLE 2 Solve: $3y - 2x = 7$ for y.

Solution This is a linear equation in two variables. Often an equation such as this is solved for y to reveal some properties about the graph of this equation, which we will learn more about in Chapter 3. Since there are no fractions or grouping symbols, we begin with Step 4 and isolate the term containing the specified variable y by adding $2x$ to both sides of the equation.

$$3y - 2x = 7$$
$$3y - 2x + 2x = 7 + 2x \quad \text{Add } 2x \text{ to both sides.}$$
$$3y = 7 + 2x$$

To solve for y, divide both sides by 3.

$$\frac{3y}{3} = \frac{7 + 2x}{3} \quad \text{Divide both sides by 3.}$$

$$y = \frac{2x + 7}{3} \quad \text{or} \quad y = \frac{2x}{3} + \frac{7}{3}$$

PRACTICE
2 Solve: $7x - 2y = 5$ for y.

△ **EXAMPLE 3** Solve: $A = \frac{1}{2}(B + b)h$ for b.

Solution Since this formula for finding the area of a trapezoid contains fractions, we begin by multiplying both sides of the equation by the LCD 2.

$$A = \frac{1}{2}(B + b)h$$

$$2 \cdot A = 2 \cdot \frac{1}{2}(B + b)h \quad \text{Multiply both sides by 2.}$$

$$2A = (B + b)h \quad \text{Simplify.}$$

Next, use the distributive property and remove parentheses.

$$2A = (B + b)h$$
$$2A = Bh + bh \qquad\qquad \text{Apply the distributive property.}$$
$$2A - Bh = bh \qquad\qquad \text{Isolate the term containing } b \text{ by}$$
$$\frac{2A - Bh}{h} = \frac{bh}{h} \qquad\qquad \text{subtracting } Bh \text{ from both sides.}$$
$$\qquad\qquad\qquad \text{Divide both sides by } h.$$

$$\frac{2A - Bh}{h} = b \quad \text{or} \quad b = \frac{2A - Bh}{h}$$

> ▶ **Helpful Hint**
> Remember that we may get the specified variable alone on either side of the equation.

PRACTICE
3 Solve: $A = P + Prt$ for r.

OBJECTIVE
2 Using Formulas to Solve Problems ▶

In this section, we also solve problems that can be modeled by known formulas. We use the same problem-solving steps that were introduced in the previous section.

Formulas are very useful in problem solving. For example, the compound interest formula

$$A = P\left(1 + \frac{r}{n}\right)^{nt}$$

is used by banks to compute the amount A in an account that pays compound interest. The variable P represents the principal or amount invested in the account, r is the annual rate of interest, t is the time in years, and n is the number of times compounded per year.

EXAMPLE 4 **Finding the Amount in a Savings Account**

Karen Estes just received an inheritance of $10,000 and plans to place all the money in a savings account that pays 5% compounded quarterly to help her son go to college in 3 years. How much money will be in the account in 3 years?

Solution

1. UNDERSTAND. Read and reread the problem. The appropriate formula to solve this problem is the compound interest formula

$$A = P\left(1 + \frac{r}{n}\right)^{nt}$$

Make sure that you understand the meaning of all the variables in this formula.

A = amount in the account after t years

P = principal or amount invested

t = time in years

r = annual rate of interest

n = number of times compounded per year

2. TRANSLATE. Use the compound interest formula and let $P = \$10,000$, $r = 5\% = 0.05$, $t = 3$ years, and $n = 4$ since the account is compounded quarterly, or 4 times a year.

Formula: $\quad A = P\left(1 + \dfrac{r}{n}\right)^{nt}$

Substitute: $\quad A = 10,000\left(1 + \dfrac{0.05}{4}\right)^{4\cdot3}$

3. SOLVE. We simplify the right side of the equation.

$$A = 10,000\left(1 + \frac{0.05}{4}\right)^{4\cdot3}$$

$A = 10,000(1.0125)^{12}$ Simplify $1 + \dfrac{0.05}{4}$ and write $4 \cdot 3$ as 12.

$A \approx 10,000(1.160754518)$ Approximate $(1.0125)^{12}$.

$A \approx 11,607.55$ Multiply and round to two decimal places.

4. INTERPRET.

Check: Repeat your calculations to make sure that no error was made. Notice that $11,607.55 is a reasonable amount to have in the account after 3 years.

State: In 3 years, the account will contain $11,607.55.   □

PRACTICE

4 Russ placed $8000 into his credit union account paying 6% compounded semiannually (twice a year). How much will be in Russ's account in 4 years?

Graphing Calculator Explorations

To solve Example 4, we approximated the expression

$$10{,}000\left(1 + \frac{0.05}{4}\right)^{4 \cdot 3}$$

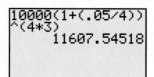

Use the keystrokes shown in the accompanying calculator screen to evaluate this expression using a graphing calculator. Notice the use of parentheses.

EXAMPLE 5 **Finding Cycling Time**

The fastest average speed by a cyclist across the continental United States is 15.4 mph, by Pete Penseyres. If he traveled a total distance of about 3107.5 miles at this speed, find his time cycling. Write the time in days, hours, and minutes. (*Source: The Guinness Book of World Records*)

Solution

1. **UNDERSTAND.** Read and reread the problem. The appropriate formula is the distance formula

$$d = rt \quad \text{where}$$
$$d = \text{distance traveled} \quad r = \text{rate} \quad \text{and} \quad t = \text{time}$$

2. **TRANSLATE.** Use the distance formula and let $d = 3107.5$ miles and $r = 15.4$ mph.

$$d = rt$$
$$3107.5 = 15.4t$$

3. **SOLVE.**

$$\frac{3107.5}{15.4} = \frac{15.4t}{15.4} \quad \text{Divide both sides by 15.4.}$$
$$201.79 \approx t$$

The time is approximately 201.79 hours. Since there are 24 hours in a day, we divide 201.79 by 24 and find that the time is approximately 8.41 days. Now, let's convert the decimal part of 8.41 days back to hours. To do this, multiply 0.41 by 24 and the result is 9.84 hours. Next, we convert the decimal part of 9.84 hours to minutes by multiplying by 60 since there are 60 minutes in an hour. We have $0.84 \cdot 60 \approx 50$ minutes rounded to the nearest whole. The time is then approximately

8 days, 9 hours, 50 minutes.

4. **INTERPRET.**

Check: Repeat your calculations to make sure that an error was not made.

State: Pete Penseyres's cycling time was approximately 8 days, 9 hours, 50 minutes. □

PRACTICE

5 Nearly 4800 cyclists from 36 U.S. states and 6 countries rode in the Pan-Massachusetts Challenge recently to raise money for cancer research and treatment. If the riders of a certain team traveled their 192-mile route at an average speed of 7.5 miles per hour, find the time they spent cycling. Write the answer in hours and minutes.

Vocabulary, Readiness & Video Check

Solve each equation for the specified variable. See Examples 1 through 3.

1. $2x + y = 5$ for y

2. $7x - y = 3$ for y

3. $a - 5b = 8$ for a

4. $7r + s = 10$ for s

5. $5j + k - h = 6$ for k

6. $w - 4y + z = 0$ for z

Martin-Gay Interactive Videos

See Video 2.3

Watch the section lecture video and answer the following questions.

OBJECTIVE 1

7. Based on the lecture before Example 1, what two things does solving an equation for a specific variable mean?

OBJECTIVE 2

8. As the solution is checked at the end of Example 3, why do you think it is mentioned to be especially careful that you use the correct formula when solving problems?

2.3 Exercise Set   MyMathLab®

Solve each equation for the specified variable. See Examples 1–3.

1. $d = rt$; for t

2. $W = gh$; for g

3. $I = PRT$; for R

△ **4.** $V = lwh$; for l

▶ **5.** $9x - 4y = 16$; for y

6. $2x + 3y = 17$; for y

△ **7.** $P = 2L + 2W$; for W

8. $A = 3M - 2N$; for N

9. $J = AC - 3$; for A

10. $y = mx + b$; for x

11. $W = gh - 3gt^2$; for g

12. $A = Prt + P$; for P

13. $T = C(2 + AB)$; for B

14. $A = 5H(b + B)$; for b

△ **15.** $C = 2\pi r$; for r

△ **16.** $S = 2\pi r^2 + 2\pi rh$; for h

17. $E = I(r + R)$; for r

18. $A = P(1 + rt)$; for t

▶ **19.** $s = \dfrac{n}{2}(a + L)$; for L

20. $C = \dfrac{5}{9}(F - 32)$; for F

21. $N = 3st^4 - 5sv$; for v

22. $L = a + (n - 1)d$; for d

△ **23.** $S = 2LW + 2LH + 2WH$; for H

24. $T = 3vs - 4ws + 5vw$; for v

In this exercise set, round all dollar amounts to two decimal places. Solve. See Example 4.

25. Complete the table and find the balance A if \$3500 is invested at an annual percentage rate of 3% for 10 years and compounded n times a year.

n	1	2	4	12	365
A					

26. Complete the table and find the balance A if \$5000 is invested at an annual percentage rate of 6% for 15 years and compounded n times a year.

n	1	2	4	12	365
A					

27. A principal of \$6000 is invested in an account paying an annual percentage rate of 4%. Find the amount in the account after 5 years if the account is compounded

a. semiannually   **b.** quarterly

c. monthly

28. A principal of \$25,000 is invested in an account paying an annual percentage rate of 5%. Find the amount in the account after 2 years if the account is compounded

a. semiannually   **b.** quarterly

c. monthly

MIXED PRACTICE

Solve. For Exercises 29 and 30, the solutions have been started for you. Round all dollar amounts to two decimal places. See Examples 4 and 5.

29. Omaha, Nebraska, is about 90 miles from Lincoln, Nebraska. Irania Schmidt must go to the law library in Lincoln to get a document for the law firm she works for. Find how long it takes her to drive *round-trip* if she averages 50 mph.

Start the solution:

1. UNDERSTAND the problem. Reread it as many times as needed.

2. TRANSLATE into an equation. (Fill in the blanks below.) Here, we simply use the formula $d = r \cdot t$.

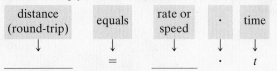

Finish with:

3. SOLVE and 4. INTERPRET

30. It took the Selby family $5\frac{1}{2}$ hours round-trip to drive from their house to their beach house 154 miles away. Find their average speed.

Start the solution:

1. UNDERSTAND the problem. Reread it as many times as needed.

2. TRANSLATE into an equation. (Fill in the blanks below.) Here, we simply use the formula $d = r \cdot t$.

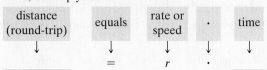

Finish with:

3. SOLVE and 4. INTERPRET

31. The day's high temperature in Phoenix, Arizona, was recorded as 104°F. Write 104°F as degrees Celsius. [Use the formula $C = \frac{5}{9}(F - 32)$.]

32. The annual low temperature in Nome, Alaska, was recorded as −15°C. Write −15°C as degrees Fahrenheit. [Use the formula $F = \frac{9}{5}C + 32$.]

△ **33.** A package of floor tiles contains 24 one-foot-square tiles. Find how many packages should be bought to cover a square ballroom floor whose side measures 64 feet.

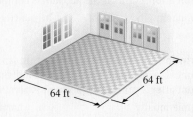

△ **34.** One-foot-square ceiling tiles are sold in packages of 50. Find how many packages must be bought for a rectangular ceiling 18 feet by 12 feet.

▶ **35.** If the area of a triangular kite is 18 square feet and its base is △ 4 feet, find the height of the kite.

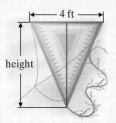

36. Bailey, Ethan, Avery, Mia, and Madison would like to go to Disneyland in 3 years. Their total cost should be $4500. If each invests $800 in a savings account paying 5.5% interest compounded semiannually, will they have enough in 3 years?

△ **37.** A gallon of latex paint can cover 500 square feet. Find how many gallon containers of paint should be bought to paint two coats on each wall of a rectangular room whose dimensions are 14 feet by 16 feet (assume 8-foot ceilings).

△ **38.** A gallon of enamel paint can cover 300 square feet. Find how many gallon containers of paint should be bought to paint three coats on a wall measuring 21 feet by 8 feet.

To prepare for Exercises 43 and 44, use the volume formulas below to solve Exercises 39–42. Remember, volume is measured in cubic units.

$$\text{Cylinder: } V = \pi r^2 h \qquad \text{Sphere: } V = \frac{4}{3}\pi r^3$$

39. The cylinder below has an exact volume of 980π cubic meters. Find its height.

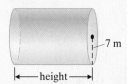

40. The battery below is in the shape of a cylinder and has an exact volume of 825π cubic millimeters. Find its height.

41. The steel ball below is in the shape of a sphere and has a diameter of 12 millimeters.

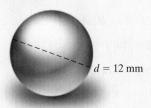

a. Find the exact volume of the sphere.

b. Find a 2-decimal-place approximation for the volume.

42. The spherical ball below has a diameter of 18 centimeters.

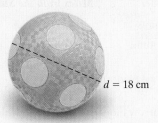

$d = 18$ cm

    a. Find the exact volume of the ball.

    b. Find a 2-decimal-place approximation of the volume.

△ **43.** A portion of the external tank of the Space Shuttle *Endeavour* was a liquid hydrogen tank. If the ends of the tank are hemispheres, find the volume of the tank. To do so, answer parts **a** through **c**. (*Note: Endeavour* completed its last mission in 2011.)

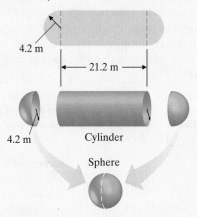

4.2 m

21.2 m

4.2 m   Cylinder

Sphere

    a. Find the volume of the cylinder shown. Round to 2 decimal places.

    b. Find the volume of the sphere shown. Round to 2 decimal places.

    c. Add the results of parts **a** and **b**. This sum is the approximate volume of the tank.

44. A vitamin is in the shape of a cylinder with a hemisphere at each end, as shown. Use the art in Exercise 43 to help find the volume of the vitamin.

4 mm

15 mm

    a. Find the volume of the cylinder part. Round to two decimal places.

    b. Find the volume of the sphere formed by joining the two hemispherical ends. Round to two decimal places.

    c. Add the results of parts **a** and **b** to find the approximate volume of the vitamin.

45. Amelia Earhart was the first woman to fly solo nonstop coast to coast, setting the women's nonstop transcontinental speed record. She traveled 2447.8 miles in 19 hours 5 minutes. Find the average speed of her flight in miles per hour. (Change 19 hours 5 minutes into hours and use the formula $d = rt$.) Round to the nearest tenth of a mile per hour.

△ **46.** In 1945, Arthur C. Clarke, a scientist and science-fiction writer, predicted that an artificial satellite placed at a height of 22,248 miles directly above the equator would orbit the globe at the same speed with which the earth was rotating. This belt along the equator is known as the Clarke belt. Use the formula for circumference of a circle and find the "length" of the Clarke belt. (*Hint:* Recall that the radius of the earth is approximately 4000 miles. Round to the nearest whole mile.)

22,248 mi

△ **47.** The deepest hole in the ocean floor is beneath the Pacific Ocean and is called Hole 504B. It is located off the coast of Ecuador. Scientists are drilling it to learn more about the earth's history. Currently, the hole is in the shape of a cylinder whose volume is approximately 3800 cubic feet and whose length is 1.3 miles. Find the radius of the hole to the nearest hundredth of a foot. (*Hint:* Make sure the same units of measurement are used.)

48. The deepest man-made hole is called the Kola Superdeep Borehole. It is approximately 8 miles deep and is located near a small Russian town in the Arctic Circle. If it takes 7.5 hours to remove the drill from the bottom of the hole, find the rate that the drill can be retrieved in feet per second. Round to the nearest tenth. (*Hint:* Write 8 miles as feet, 7.5 hours as seconds, then use the formula $d = rt$.)

△ **49.** Eartha is the world's largest globe. It is located at the headquarters of DeLorme, a mapmaking company in Yarmouth, Maine. Eartha is 41.125 feet in diameter. Find its exact circumference (distance around) and then approximate its circumference using 3.14 for π. (*Source:* DeLorme)

△ **50.** Eartha is in the shape of a sphere. Its radius is about 20.6 feet. Approximate its volume to the nearest cubic foot. (*Source:* DeLorme)

51. Find *how much interest* $10,000 earns in 2 years in a certificate of deposit paying 8.5% interest compounded quarterly.

52. Find how long it takes Mark to drive 135 miles on I-10 if he merges onto I-10 at 10 a.m. and drives nonstop with his cruise control set on 60 mph.

The calorie count of a serving of food can be computed based on its composition of carbohydrate, fat, and protein. The calorie count C for a serving of food can be computed using the formula $C = 4h + 9f + 4p$, where h is the number of grams of carbohydrate contained in the serving, f is the number of grams of fat contained in the serving, and p is the number of grams of protein contained in the serving.

53. Solve this formula for f, the number of grams of fat contained in a serving of food.

54. Solve this formula for h, the number of grams of carbohydrate contained in a serving of food.

55. A serving of cashews contains 14 grams of fat, 7 grams of carbohydrate, and 6 grams of protein. How many calories are in this serving of cashews?

56. A serving of chocolate candies contains 9 grams of fat, 30 grams of carbohydrate, and 2 grams of protein. How many calories are in this serving of chocolate candies?

57. A serving of raisins contains 130 calories and 31 grams of carbohydrate. If raisins are a fat-free food, how much protein is provided by this serving of raisins?

58. A serving of yogurt contains 120 calories, 21 grams of carbohydrate, and 5 grams of protein. How much fat is provided by this serving of yogurt? Round to the nearest tenth of a gram.

REVIEW AND PREVIEW

Determine which numbers in the set $\{-3, -2, -1, 0, 1, 2, 3\}$ are solutions of each inequality. See Sections 1.4 and 2.1.

59. $x < 0$ **60.** $x > 1$

61. $x + 5 \le 6$

62. $x - 3 \ge -7$

63. In your own words, explain what real numbers are solutions of $x < 0$.

64. In your own words, explain what real numbers are solutions of $x > 1$.

CONCEPT EXTENSIONS

Solar system distances are so great that units other than miles or kilometers are often used. For example, the astronomical unit (AU) is the average distance between Earth and the sun, or 92,900,000 miles. Use this information to convert each planet's distance in miles from the sun to astronomical units. Round to three decimal places. The planet Mercury's AU from the sun has been completed for you. (Source: National Space Science Data Center)

	Planet	Miles from the Sun	AU from the Sun
	Mercury	36 million	0.388
65.	Venus	67.2 million	
66.	Earth	92.9 million	
67.	Mars	141.5 million	
68.	Jupiter	483.3 million	
69.	Saturn	886.1 million	

	Planet	Miles from the Sun	AU from the Sun
70.	Uranus	1783 million	
71.	Neptune	2793 million	
72.	Pluto (dwarf planet)	3670 million	

73. To borrow money at a rate of r, which loan plan should you choose—one compounding 4 times a year or 12 times a year? Explain your choice.

74. If you are investing money in a savings account paying a rate of r, which account should you choose—an account compounded 4 times a year or 12 times a year? Explain your choice.

75. To solve the formula $W = gh - 3gt^2$ for g, explain why it is a good idea to factor g first from the terms on the right side of the equation. Then perform this step and solve for g.

76. An orbit such as Clarke's belt in Exercise 46 is called a geostationary orbit. In your own words, why do you think that communications satellites are placed in geostationary orbits?

*The measure of the chance or likelihood of an event occurring is its **probability**. A formula basic to the study of probability is the formula for the probability of an event when all the outcomes are equally likely. This formula is*

$$\text{Probability of an event} = \frac{\text{number of ways that the event can occur}}{\text{number of possible outcomes}}$$

For example, to find the probability that a single spin on the spinner below will result in red, notice first that the spinner is divided into 8 parts, so there are 8 possible outcomes. Next, notice that there is only one sector of the spinner colored red, so the number of ways that the spinner can land on red is 1. Then this probability denoted by P(red) is

$$P(\text{red}) = \tfrac{1}{8}$$

Find each probability in simplest form.

77. $P(\text{green})$ **78.** $P(\text{yellow})$

79. $P(\text{black})$ **80.** $P(\text{blue})$

81. $P(\text{green or blue})$ **82.** $P(\text{black or yellow})$

83. $P(\text{red, green, or black})$ **84.** $P(\text{yellow, blue, or black})$

85. $P(\text{white})$

86. $P(\text{red, yellow, green, blue, or black})$

87. From the previous probability formula, what do you think is always the probability of an event that is impossible to occur?

88. What do you think is always the probability of an event that is sure to occur?

2.4 | Linear Inequalities and Problem Solving

Relationships among measurable quantities are not always described by equations. For example, suppose that a salesperson earns a base of $600 per month plus a commission of 20% of sales. Suppose we want to find the minimum amount of sales needed to receive a total income of *at least* $1500 per month. Here, the phrase "at least" implies that an income of $1500 *or more* is acceptable. In symbols, we can write

$$\text{income} \geq 1500$$

This is an example of an inequality, and we will solve this problem in Example 8.

A **linear inequality** is similar to a linear equation except that the equality symbol is replaced with an inequality symbol, such as $<$, $>$, $\leq$, or $\geq$.

Linear Inequalities in One Variable

$3x + 5 \geq 4$	$2y < 0$	$3(x - 4) > 5x$	$\dfrac{x}{3} \leq 5$
↑	↑	↑	↑
is greater than or equal to	is less than	is greater than	is less than or equal to

> **Linear Inequality in One Variable**
>
> A linear inequality in one variable is an inequality that can be written in the form
>
> $$ax + b < c$$
>
> where a, b, and c are real numbers and $a \neq 0$.

In this section, when we make definitions, state properties, or list steps about an inequality containing the symbol $<$, we mean that the definition, property, or steps also apply to inequalities containing the symbols $>$, $\leq$, and $\geq$.

OBJECTIVE
1 Using Interval Notation

A **solution** of an inequality is a value of the variable that makes the inequality a true statement. The **solution set** of an inequality is the set of all solutions. Notice that the solution set of the inequality $x > 2$, for example, contains all numbers greater than 2. Its graph is an interval on the number line since an infinite number of values satisfy the variable. If we use open/closed-circle notation, the graph of $\{x \mid x > 2\}$ looks like the following.

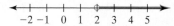

In this text **interval notation** will be used to write solution sets of inequalities. To help us understand this notation, a different graphing notation will be used. Instead of an open circle, we use a parenthesis. With this new notation, the graph of $\{x \mid x > 2\}$ now looks like

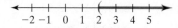

and can be represented in interval notation as $(2, \infty)$. The symbol ∞ is read "infinity" and indicates that the interval includes *all* numbers greater than 2. The left parenthesis indicates that 2 *is not* included in the interval.

When 2 *is* included in the interval, we use a bracket. The graph of $\{x \mid x \geq 2\}$ is below

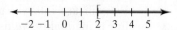

and can be represented as $[2, \infty)$.

The following table shows three equivalent ways to describe an interval: in set notation, as a graph, and in interval notation.

Set Notation	Graph	Interval Notation
$\{x\|x < a\}$	![number line with parenthesis at a, shaded left] a	$(-\infty, a)$
$\{x\|x > a\}$	![number line with parenthesis at a, shaded right] a	(a, ∞)
$\{x\|x \le a\}$	![number line with bracket at a, shaded left] a	$(-\infty, a]$
$\{x\|x \ge a\}$	![number line with bracket at a, shaded right] a	$[a, \infty)$
$\{x\|a < x < b\}$	![number line with parentheses at a and b] $a \quad b$	(a, b)
$\{x\|a \le x \le b\}$	![number line with brackets at a and b] $a \quad b$	$[a, b]$
$\{x\|a < x \le b\}$	![number line with parenthesis at a, bracket at b] $a \quad b$	$(a, b]$
$\{x\|a \le x < b\}$	![number line with bracket at a, parenthesis at b] $a \quad b$	$[a, b)$

▶ **Helpful Hint**

Notice that a parenthesis is always used to enclose ∞ and $-\infty$.

✓**CONCEPT CHECK**

Explain what is wrong with writing the interval $(5, \infty]$.

EXAMPLE 1 Graph each set on a number line and then write in interval notation.

a. $\{x\|x \ge 2\}$ **b.** $\{x\|x < -1\}$ **c.** $\{x\|0.5 < x \le 3\}$

Solution

a. ![number line from -2 to 4, bracket at 2 shaded right] $-2\ -1\quad 0\quad 1\quad 2\quad 3\quad 4$ $[2, \infty)$

b. ![number line from -3 to 3, parenthesis at -1 shaded left] $-3\ -2\ -1\quad 0\quad 1\quad 2\quad 3$ $(-\infty, -1)$

c. ![number line from -1 to 5, parenthesis at 0.5 bracket at 3] 0.5 $-1\quad 0\quad 1\quad 2\quad 3\quad 4\quad 5$ $(0.5, 3]$

PRACTICE

1 Graph each set on a number line and then write in interval notation.

a. $\{x \mid x < 3.5\}$
b. $\{x \mid x \ge -3\}$
c. $\{x \mid -1 \le x < 4\}$

OBJECTIVE

2 Solving Linear Inequalities Using the Addition Property

We will use interval notation to write solutions of linear inequalities. To solve a linear inequality, we use a process similar to the one used to solve a linear equation. We use properties of inequalities to write equivalent inequalities until the variable is isolated.

Answer to Concept Check:
should be $(5, \infty)$ since a parenthesis is always used to enclose ∞

> **Addition Property of Inequality**
>
> If a, b, and c are real numbers, then
>
> $$a < b \quad \text{and} \quad a + c < b + c$$
>
> are equivalent inequalities.

In other words, we may add the same real number to both sides of an inequality, and the resulting inequality will have the same solution set. This property also allows us to subtract the same real number from both sides.

EXAMPLE 2 Solve: $x - 2 < 5$. Graph the solution set and write it in interval notation.

Solution

$$x - 2 < 5$$
$$x - 2 + 2 < 5 + 2 \quad \text{Add 2 to both sides.}$$
$$x < 7 \quad \text{Simplify.}$$

The solution set is $\{x \mid x < 7\}$, which in interval notation is $(-\infty, 7)$. The graph of the solution set is

$$\xleftarrow{\hspace{1.5cm}} \begin{array}{ccccccc} & & &) & & & \\ 4 & 5 & 6 & 7 & 8 & 9 & 10 \end{array} \xrightarrow{\hspace{0.5cm}}$$

PRACTICE

2 Solve: $x + 5 > 9$. Graph the solution set and write it in interval notation.

> ▶ **Helpful Hint**
>
> In Example 2, the solution set is $\{x \mid x < 7\}$. This means that *all* numbers less than 7 are solutions. For example, $6.9, 0, -\pi, 1$, and -56.7 are solutions, just to name a few. To see this, replace x in $x - 2 < 5$ with each of these numbers and see that the result is a true inequality.

EXAMPLE 3 Solve: $3x + 4 \geq 2x - 6$. Graph the solution set and write it in interval notation.

Solution

$$3x + 4 \geq 2x - 6$$
$$3x + 4 - 2x \geq 2x - 6 - 2x \quad \text{Subtract } 2x \text{ from both sides.}$$
$$x + 4 \geq -6 \quad \text{Combine like terms.}$$
$$x + 4 - 4 \geq -6 - 4 \quad \text{Subtract 4 from both sides.}$$
$$x \geq -10 \quad \text{Simplify.}$$

The solution set is $\{x \mid x \geq -10\}$, which in interval notation is $[-10, \infty)$. The graph of the solution set is

$$\xleftarrow{\hspace{1cm}} \begin{array}{cccccc} [& & & & & \\ -11 & -10 & -9 & -8 & -7 & -6 \end{array} \xrightarrow{\hspace{0.5cm}}$$

PRACTICE

3 Solve: $3x + 1 \leq 2x - 3$. Graph the solution set and write it in interval notation.

OBJECTIVE

3 Solving Linear Inequalities Using the Multiplication and Addition Properties ▶

Next, we introduce and use the multiplication property of inequality to solve linear inequalities. To understand this property, let's start with the true statement $-3 < 7$ and multiply both sides by 2.

$$-3 < 7$$
$$-3(2) < 7(2) \quad \text{Multiply by 2.}$$
$$-6 < 14 \quad \text{True}$$

The statement remains true.

Notice what happens if both sides of $-3 < 7$ are multiplied by -2.

$$-3 < 7$$
$$-3(-2) < 7(-2) \quad \text{Multiply by } -2.$$
$$6 < -14 \quad \text{False}$$

The inequality $6 < -14$ is a false statement. However, **if the direction of the inequality sign is reversed,** the result is true.

$$6 > -14 \quad \text{True}$$

These examples suggest the following property.

Multiplication Property of Inequality

If $a, b,$ and c are real numbers and c is **positive,** then

$$a < b \text{ and } ac < bc$$

are equivalent inequalities.

If $a, b,$ and c are real numbers and c is **negative,** then

$$a < b \text{ and } ac > bc$$

are equivalent inequalities.

In other words, we may multiply both sides of an inequality by the same positive real number and the result is an equivalent inequality.

We may also multiply both sides of an inequality by the same **negative number** and **reverse the direction of the inequality symbol,** and the result is an equivalent inequality. The multiplication property holds for division also, since division is defined in terms of multiplication.

▶ Helpful Hint

Whenever both sides of an inequality are multiplied or divided by a negative number, the direction of the inequality symbol **must be** reversed to form an equivalent inequality.

EXAMPLE 4 Solve and graph the solution set. Write the solution set in interval notation.

a. $\dfrac{1}{4}x \le \dfrac{3}{8}$ **b.** $-2.3x < 6.9$

Solution

a.
$$\frac{1}{4}x \le \frac{3}{8}$$

▶ Helpful Hint

The inequality symbol is the same since we are multiplying by a *positive* number.

$$4 \cdot \frac{1}{4}x \le 4 \cdot \frac{3}{8} \quad \text{Multiply both sides by 4.}$$

$$x \le \frac{3}{2} \quad \text{Simplify.}$$

The solution set is $\left\{ x \mid x \le \dfrac{3}{2} \right\}$, which in interval notation is $\left(-\infty, \dfrac{3}{2} \right]$. The graph of the solution set is

b.
$$-2.3x < 6.9$$

$$\frac{-2.3x}{-2.3} > \frac{6.9}{-2.3} \qquad \text{Divide both sides by } -2.3 \text{ and reverse the inequality symbol.}$$
$$x > -3 \qquad \text{Simplify.}$$

The solution set is $\{x \mid x > -3\}$, which is $(-3, \infty)$ in interval notation. The graph of the solution set is

PRACTICE
4   Solve and graph the solution set. Write the solution set in interval notation.

a. $\dfrac{2}{5}x \geq \dfrac{4}{15}$

b. $-2.4x < 9.6$

✓CONCEPT CHECK

In which of the following inequalities must the inequality symbol be reversed during the solution process?

a. $-2x > 7$ **b.** $2x - 3 > 10$
c. $-x + 4 + 3x < 7$ **d.** $-x + 4 < 5$

To solve linear inequalities in general, we follow steps similar to those for solving linear equations.

Solving a Linear Inequality in One Variable

Step 1. Clear the inequality of fractions by multiplying both sides of the inequality by the least common denominator (LCD) of all fractions in the inequality.

Step 2. Use the distributive property to remove grouping symbols such as parentheses.

Step 3. Combine like terms on each side of the inequality.

Step 4. Use the addition property of inequality to write the inequality as an equivalent inequality with variable terms on one side and numbers on the other side.

Step 5. Use the multiplication property of inequality to isolate the variable.

EXAMPLE 5   Solve: $-(x - 3) + 2 \leq 3(2x - 5) + x$.

Solution
$$-(x - 3) + 2 \leq 3(2x - 5) + x$$
$$-x + 3 + 2 \leq 6x - 15 + x \qquad \text{Apply the distributive property.}$$
$$5 - x \leq 7x - 15 \qquad \text{Combine like terms.}$$
$$5 - x + x \leq 7x - 15 + x \qquad \text{Add } x \text{ to both sides.}$$
$$5 \leq 8x - 15 \qquad \text{Combine like terms.}$$
$$5 + 15 \leq 8x - 15 + 15 \qquad \text{Add 15 to both sides.}$$
$$20 \leq 8x \qquad \text{Combine like terms.}$$
$$\frac{20}{8} \leq \frac{8x}{8} \qquad \text{Divide both sides by 8.}$$
$$\frac{5}{2} \leq x, \quad \text{or} \quad x \geq \frac{5}{2} \qquad \text{Simplify.}$$

Answer to Concept Check: a, d

The solution set written in interval notation is $\left[\dfrac{5}{2}, \infty\right)$ and its graph is

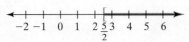

□

PRACTICE
5 Solve: $-(4x + 6) \leq 2(5x + 9) + 2x$. Graph and write the solution set in interval notation.

■

EXAMPLE 6 Solve: $\dfrac{2}{5}(x - 6) \geq x - 1$.

Solution
$$\dfrac{2}{5}(x - 6) \geq x - 1$$

$$5\left[\dfrac{2}{5}(x - 6)\right] \geq 5(x - 1) \qquad \text{Multiply both sides by 5 to eliminate fractions.}$$

$$2(x - 6) \geq 5(x - 1)$$

$$2x - 12 \geq 5x - 5 \qquad \text{Apply the distributive property.}$$

$$-3x - 12 \geq -5 \qquad \text{Subtract } 5x \text{ from both sides.}$$

$$-3x \geq 7 \qquad \text{Add 12 to both sides.}$$

$$\dfrac{-3x}{-3} \leq \dfrac{7}{-3} \qquad \text{Divide both sides by } -3 \text{ and reverse the inequality symbol.}$$

$$x \leq -\dfrac{7}{3} \qquad \text{Simplify.}$$

The solution set written in interval notation is $\left(-\infty, -\dfrac{7}{3}\right]$ and its graph is

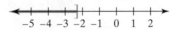

□

PRACTICE
6 Solve: $\dfrac{3}{5}(x - 3) \geq x - 7$. Graph and write the solution set in interval notation.

■

EXAMPLE 7 Solve: $2(x + 3) > 2x + 1$.

Solution
$$2(x + 3) > 2x + 1$$

$$2x + 6 > 2x + 1 \qquad \text{Distribute on the left side.}$$

$$2x + 6 - 2x > 2x + 1 - 2x \qquad \text{Subtract } 2x \text{ from both sides.}$$

$$6 > 1 \qquad \text{Simplify.}$$

$6 > 1$ is a true statement for all values of x, so this inequality and the original inequality are true for all numbers. The solution set is $\{x \mid x \text{ is a real number}\}$, or $(-\infty, \infty)$ in interval notation, and its graph is

□

PRACTICE
7 Solve: $4(x - 2) < 4x + 5$. Graph and write the solution set in interval notation.

■

OBJECTIVE

4   Solving Problems Modeled by Linear Inequalities

Application problems containing words such as "at least," "at most," "between," "no more than," and "no less than" usually indicate that an inequality is to be solved instead of an equation. In solving applications involving linear inequalities, we use the same procedure as when we solved applications involving linear equations.

EXAMPLE 8   **Calculating Income with Commission**

A salesperson earns $600 per month plus a commission of 20% of sales. Find the minimum amount of sales needed to receive a total income of at least $1500 per month.

Solution

1. UNDERSTAND. Read and reread the problem. Let x = amount of sales.

2. TRANSLATE. As stated in the beginning of this section, we want the income to be greater than or equal to $1500. To write an inequality, notice that the salesperson's income consists of $600 plus a commission (20% of sales).

In words:	600	+	commission (20% of sales)	$\geq$	1500
	$\downarrow$		$\downarrow$		$\downarrow$
Translate:	600	+	0.20x	$\geq$	1500

3. SOLVE the inequality for x.

$$600 + 0.20x \geq 1500$$
$$600 + 0.20x - 600 \geq 1500 - 600$$
$$0.20x \geq 900$$
$$x \geq 4500$$

4. INTERPRET.

Check: The income for sales of $4500 is

$$600 + 0.20(4500), \text{ or } 1500.$$

Thus, if sales are greater than or equal to $4500, income is greater than or equal to $1500.

State: The minimum amount of sales needed for the salesperson to earn at least $1500 per month is $4500 per month.   □

PRACTICE

8   A salesperson earns $900 a month plus a commission of 15% of sales. Find the minimum amount of sales needed to receive a total income of at least $2400 per month.

EXAMPLE 9   **Finding the Annual Consumption**

In the United States, the annual consumption of cigarettes is declining. The consumption c in billions of cigarettes per year since the year 2000 can be approximated by the formula

$$c = -9.4t + 431$$

where t is the number of years after 2000. Use this formula to predict the years that the consumption of cigarettes will be less than 200 billion per year.

Solution

1. UNDERSTAND. Read and reread the problem. To become familiar with the given formula, let's find the cigarette consumption after 20 years, which would be the year $2000 + 20$, or 2020. To do so, we substitute 20 for t in the given formula.

$$c = -9.4(20) + 431 = 243$$

Thus, in 2020, we predict cigarette consumption to be about 243 billion.

Variables have already been assigned in the given formula. For review, they are c = the annual consumption of cigarettes in the United States in billions of cigarettes and

$$t = \text{the number of years after 2000}$$

2. TRANSLATE. We are looking for the years that the consumption of cigarettes c is less than 200. Since we are finding years t, we substitute the expression in the formula given for c, or

$$-9.4t + 431 < 200$$

3. SOLVE the inequality.

$$-9.4t + 431 < 200$$
$$-9.4t < -231$$
$$\frac{-9.4t}{-9.4} > \frac{-231}{-9.4}$$
$$t > \text{approximately } 24.6$$

4. INTERPRET.

Check: Substitute a number greater than 24.6 and see that c is less than 200.

State: The annual consumption of cigarettes will be less than 200 billion more than 24.6 years after 2000, or approximately for $25 + 2000 = 2025$ and all years after.   □

PRACTICE

9   Use the formula given in Example 9 to predict when the consumption of cigarettes will be less than 275 billion per year.

Vocabulary, Readiness & Video Check

Match each graph with the interval notation that describes it.

1.

-7 -6 -5 -4 -3 -2 -1 0

a. $(-5, \infty)$ b. $(-5, -\infty)$
c. $(\infty, -5)$ d. $(-\infty, -5)$

2.

-13 -12 -11 -10 -9 -8

a. $(-\infty, -11]$ b. $(-11, \infty)$
c. $[-11, \infty)$ d. $(-\infty, -11)$

3.

-2.5 $\frac{7}{4}$

-3 -2 -1 0 1 2 3 4

a. $\left[\frac{7}{4}, -2.5\right)$ b. $\left(-2.5, \frac{7}{4}\right]$
c. $\left[-2.5, \frac{7}{4}\right)$ d. $\left(\frac{7}{4}, -2.5\right)$

4.

$-\frac{10}{3}$ 0.2

-4 -3 -2 -1 0 1 2 3

a. $\left[-\frac{10}{3}, 0.2\right)$ b. $\left(0.2, -\frac{10}{3}\right]$
c. $\left(-\frac{10}{3}, 0.2\right]$ d. $\left[0.2, -\frac{10}{3}\right)$

Use the choices below to fill in each blank.

$(-\infty, -0.4)$ $(-\infty, -0.4]$ $[-0.4, \infty)$ $(-0.4, \infty)$ $(\infty, -0.4]$

5. The set $\{x \mid x \geq -0.4\}$ written in interval notation is _____.

6. The set $\{x \mid x < -0.4\}$ written in interval notation is _____.

7. The set $\{x \mid x \leq -0.4\}$ written in interval notation is _____.

8. The set $\{x \mid x > -0.4\}$ written in interval notation is _____.

Martin-Gay Interactive Videos

See Video 2.4

Watch the section lecture video and answer the following questions.

OBJECTIVE 1

9. Using Example 1 as a reference, explain how the graph of the solution set of an inequality can help you write the solution set in interval notation.

OBJECTIVE 2

10. From the lecture before Example 3, explain the addition property of inequality in your own words. What equality property does it closely resemble?

OBJECTIVE 3

11. Based on the lecture before Example 4, complete the following statement. If you multiply or divide both sides of an inequality by the _____ nonzero negative number, you must _____ the direction of the inequality symbol.

OBJECTIVE 4

12. What words or phrases in the Example 7 statement tell you this is an inequality application (besides the last line telling you to use an inequality)?

2.4 Exercise Set   MyMathLab®

Graph the solution set of each inequality and write it in interval notation. See Example 1.

1. $\{x \mid x < -3\}$

2. $\{x \mid x > 5\}$

3. $\{x \mid x \geq 0.3\}$

4. $\{x \mid x < -0.2\}$

5. $\{x \mid -7 \leq x\}$

6. $\{x \mid -7 \geq x\}$

7. $\{x \mid -2 < x < 5\}$

8. $\{x \mid -5 \leq x \leq -1\}$

9. $\{x \mid 5 \geq x > -1\}$

10. $\{x \mid -3 > x \geq -7\}$

Solve. Graph the solution set and write it in interval notation. See Examples 2 through 4.

11. $x - 7 \geq -9$

12. $x + 2 \leq -1$

13. $7x < 6x + 1$

14. $11x < 10x + 5$

15. $8x - 7 \leq 7x - 5$

16. $7x - 1 \geq 6x - 1$

17. $\dfrac{3}{4}x \geq 6$

18. $\dfrac{5}{6}x \geq 5$

19. $5x < -23.5$

20. $4x > -11.2$

21. $-3x \geq 9$

22. $-4x \geq 8$

Solve. Write the solution set using interval notation. See Examples 5 through 7.

23. $-2x + 7 \geq 9$

24. $8 - 5x \leq 23$

25. $15 + 2x \geq 4x - 7$

26. $20 + x < 6x - 15$

27. $4(2x + 1) > 4$

28. $6(2 - 3x) \geq 12$

29. $3(x - 5) < 2(2x - 1)$

30. $5(x + 4) \leq 4(2x + 3)$

31. $\dfrac{5x + 1}{7} - \dfrac{2x - 6}{4} \geq -4$

32. $\dfrac{1 - 2x}{3} + \dfrac{3x + 7}{7} > 1$

33. $-3(2x - 1) < -4[2 + 3(x + 2)]$

34. $-2(4x + 2) > -5[1 + 2(x - 1)]$

MIXED PRACTICE

Solve. Write the solution set using interval notation. See Examples 1 through 7.

35. $x + 9 < 3$

36. $x - 9 < -12$

37. $-x < -4$

38. $-x > -2$

39. $-7x \leq 3.5$

40. $-6x \leq 4.2$

41. $\dfrac{1}{2} + \dfrac{2}{3} \geq \dfrac{x}{6}$

42. $\dfrac{3}{4} - \dfrac{2}{3} \geq \dfrac{x}{6}$

43. $-5x + 4 \leq -4(x - 1)$

44. $-6x + 2 < -3(x + 4)$

45. $\dfrac{3}{4}(x - 7) \geq x + 2$

46. $\dfrac{4}{5}(x + 1) \leq x + 1$

47. $0.8x + 0.6x \geq 4.2$

48. $0.7x - x > 0.45$

49. $4(x - 6) + 2x - 4 \geq 3(x - 7) + 10x$

50. $7(2x + 3) + 4x \leq 7 + 5(3x - 4) + x$

51. $14 - (5x - 6) \geq -6(x + 1) - 5$

52. $13y - (9y + 2) \leq 5(y - 6) + 10$

53. $4(x - 1) \geq 4x - 8$

54. $8(x + 3) \leq 7(x + 5) + x$

55. $3x + 1 < 3(x - 2)$

56. $7x < 7(x - 2)$

57. $0.4(4x - 3) < 1.2(x + 2)$

58. $0.2(8x - 2) < 1.2(x - 3)$

59. $\frac{2}{5}x - \frac{1}{4} \le \frac{3}{10}x - \frac{4}{5}$

60. $\frac{7}{12}x - \frac{1}{3} \le \frac{3}{8}x - \frac{5}{6}$

61. $\frac{1}{2}(3x - 4) \le \frac{3}{4}(x - 6) + 1$

62. $\frac{2}{3}(x + 3) < \frac{1}{6}(2x - 8) + 2$

63. $\frac{-x + 2}{2} - \frac{1 - 5x}{8} < -1$

64. $\frac{3 - 4x}{6} - \frac{1 - 2x}{12} \le -2$

65. $\frac{x + 5}{5} - \frac{3 + x}{8} \ge -\frac{3}{10}$

66. $\frac{x - 4}{2} - \frac{x - 2}{3} > \frac{5}{6}$

67. $\frac{x + 3}{12} + \frac{x - 5}{15} < \frac{2}{3}$

68. $\frac{3x + 2}{18} - \frac{1 + 2x}{6} \le -\frac{1}{2}$

Solve. See Examples 8 and 9. For Exercises 69 through 76, ***a.*** *answer with an inequality and* ***b.****, in your own words, explain the meaning of your answer to part (a).*

Exercises 69 and 70 are written to help you get started. Exercises 71 and 72 are written to help you write and solve the inequality.

69. Shureka Washburn has scores of 72, 67, 82, and 79 on her algebra tests.

    a. Use an inequality to find the scores she must make on the final exam to pass the course with an average of 77 or higher, given that the final exam counts as two tests.

    b. In your own words, explain the meaning of your answer to part (a).

70. In a Winter Olympics 5000-meter speed-skating event, Hans Holden scored times of 6.85, 7.04, and 6.92 minutes on his first three trials.

    a. Use an inequality to find the times he can score on his last trial so that his average time is under 7.0 minutes.

    b. In your own words, explain the meaning of your answer to part (a).

71. A clerk must use the elevator to move boxes of paper. The elevator's maximum weight limit is 1500 pounds. If each box of paper weighs 66 pounds and the clerk weighs 147 pounds, use an inequality to find the number of whole boxes she can move on the elevator at one time.

    Start the solution:

    1. UNDERSTAND the problem. Reread it as many times as needed.

    2. TRANSLATE into an inequality. (Fill in the blanks below.)

    Let x = number of boxes

Clerk's weight		number of boxes	times	weight of each box	$\le$	elevator maximum weight
$\downarrow$		$\downarrow$	$\downarrow$	$\downarrow$	$\downarrow$	$\downarrow$
___	+	x	$\cdot$	___	$\le$	___

Finish with:

3. SOLVE and

4. INTERPRET

72. To mail a large envelope first class, the U.S. Post Office charges 88 cents for the first ounce and 17 cents per ounce for each additional ounce. Use an inequality to find the number of whole ounces that can be mailed for no more than $2.00.

Start the solution:

1. UNDERSTAND the problem. Reread it as many times as needed.

2. TRANSLATE into an inequality. (Fill in the blanks below.)

Let x = number of additional ounces (after first ounce)

Price of first ounce		number of additional ounces	times	price per additional ounce	$\le$	200 maximum cents
$\downarrow$		$\downarrow$	$\downarrow$	$\downarrow$	$\downarrow$	$\downarrow$
___	+	x	$\cdot$	___	$\le$	___

Finish with:

3. SOLVE and

4. INTERPRET

73. A small plane's maximum takeoff weight is 2000 pounds or less. Six passengers weigh an average of 160 pounds each. Use an inequality to find the luggage and cargo weights the plane can carry.

74. A shopping mall parking garage charges $1 for the first half-hour and 60 cents for each additional half-hour. Use an inequality to find how long you can park if you have only $4.00 in cash.

75. A car rental company offers two subcompact rental plans.

Plan A: $36 per day and unlimited mileage

Plan B: $24 per day plus $0.15 per mile

Use an inequality to find the number of daily miles for which plan A is more economical than plan B.

76. Northeast Telephone Company offers two billing plans for local calls.

Plan 1: $25 per month for unlimited calls

Plan 2: $13 per month plus $0.06 per call

Use an inequality to find the number of monthly calls for which plan 1 is more economical than plan 2.

77. At room temperature, glass used in windows actually has some properties of a liquid. It has a very slow, viscous flow. (Viscosity is the property of a fluid that resists internal flow. For example, lemonade flows more easily than fudge syrup. Fudge syrup has a higher viscosity than lemonade.) Glass does not become a true liquid until temperatures are greater than or equal to 500°C. Find the Fahrenheit temperatures for which glass is a liquid. (Use the formula $F = \frac{9}{5}C + 32$.)

78. Stibnite is a silvery white mineral with a metallic luster. It is one of the few minerals that melts easily in a match flame or at temperatures of approximately 977°F or greater. Find the Celsius temperatures at which stibnite melts. [Use the formula $C = \frac{5}{9}(F - 32)$.]

79. Although beginning salaries vary greatly according to your field of study, the equation

$$s = 1284.7t + 48,133$$

can be used to approximate and to predict average beginning salaries for candidates with bachelor's degrees. The variable s is the starting salary and t is the number of years after 2000. (*Source: Statistical Abstract of the U.S.*)

a. Approximate the year in which beginning salaries for candidates will be greater than $70,000. (Round your answer up and use it to calculate the year.)

b. Determine the year you plan to graduate from college. Use this year to find the corresponding value of t and approximate a beginning salary for a bachelor's degree candidate.

80. a. Use the formula in Example 9 to estimate the years that the consumption of cigarettes will be less than 50 billion per year.

b. Use your answer to part (a) to describe the limitations of your answer.

The average consumption per person per year of whole milk *w can be approximated by the equation*

$$w = -2.11t + 69.2$$

where t is the number of years after 2000 and w is measured in pounds. The average consumption of skim milk *s per person per year can be approximated by the equation*

$$s = -0.42t + 29.9$$

where t is the number of years after 2000 and s is measured in pounds. The consumption of whole milk is shown on the graph in blue and the consumption of skim milk is shown on the graph in red. Use this information to answer Exercises 81 through 90.

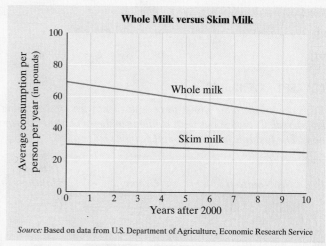

Source: Based on data from U.S. Department of Agriculture, Economic Research Service

81. Is the consumption of whole milk increasing or decreasing over time? Explain how you arrived at your answer.

82. Is the consumption of skim milk increasing or decreasing over time? Explain how you arrived at your answer.

83. Predict the consumption of whole milk in 2015. (*Hint:* Find the value of t that corresponds to 2015.)

84. Predict the consumption of skim milk in 2015. (*Hint:* Find the value of t that corresponds to 2015.)

85. Determine when the consumption of whole milk will be less than 45 pounds per person per year.

86. For 2000 through 2010, the consumption of whole milk was greater than the consumption of skim milk. Explain how this can be determined from the graph.

87. Both lines have a negative slope, that is, the amount of each type of milk consumed per person per year is decreasing as time goes on. However, the amount of whole milk being consumed is decreasing faster than the amount of skim milk being consumed. Explain how this could be.

88. Do you think it is possible that the consumption of whole milk will eventually be the same as the consumption of skim milk? Explain your answer.

89. The consumption of skim milk will be greater than the consumption of whole milk when $s > w$.

a. Find when this will occur by substituting the given equivalent expression for w and the given equivalent expression for s and solving for t.

b. Estimate to the nearest whole the first year when this will occur.

90. How will the two lines in the graph appear if the consumption of whole milk is the same as the consumption of skim milk?

REVIEW AND PREVIEW

List or describe the integers that make both inequalities true. See Section 1.4.

91. $x < 5$ and $x > 1$

92. $x \geq 0$ and $x \leq 7$

93. $x \geq -2$ and $x \geq 2$

94. $x < 6$ and $x < -5$

Solve each equation for x. See Section 2.1.

95. $2x - 6 = 4$

96. $3x - 12 = 3$

97. $-x + 7 = 5x - 6$

98. $-5x - 4 = -x - 4$

CONCEPT EXTENSIONS

Each row of the table shows three equivalent ways of describing an interval. Complete this table by filling in the equivalent descriptions. The first row has been completed for you.

	Set Notation	Graph	Interval Notation
	$\{x \mid x < -3\}$	← ──)── → at -3	$(-\infty, -3)$
99.		← ──[── → at 2	
100.		← ──(── → at -4	
101.	$\{x \mid x < 0\}$		
102.	$\{x \mid x \le 5\}$		
103.			$(-2, 1.5]$
104.			$[-3.7, 4)$

Each inequality below (Exercises 105–108) is solved by dividing both sides by the coefficient of x. Determine whether the inequality symbol will be reversed during this solution process.

105. $-3x \le 14$

106. $3x > -14$

107. $-x \ge -23$

108. $-x \le 9$

109. Solve: $2x - 3 = 5$

110. Solve: $2x - 3 > 5$

111. Solve: $2x - 3 < 5$

112. Read the equations and inequalities for Exercises 109, 110, and 111 and their solutions. In your own words, write down your thoughts.

113. When graphing the solution set of an inequality, explain how you know whether to use a parenthesis or a bracket.

114. Explain what is wrong with the interval notation $(-6, -\infty)$.

115. Explain how solving a linear inequality is similar to solving a linear equation.

116. Explain how solving a linear inequality is different from solving a linear equation.

Integrated Review) LINEAR EQUATIONS AND INEQUALITIES

Sections 2.1– 2.4

Solve each equation or inequality. For inequalities, write the solution set in interval notation.

1. $-4x = 20$

2. $-4x < 20$

3. $\dfrac{3x}{4} \ge 2$

4. $5x + 3 \ge 2 + 4x$

5. $6(y - 4) = 3(y - 8)$

6. $-4x \le \dfrac{2}{5}$

7. $-3x \ge \dfrac{1}{2}$

8. $5(y + 4) = 4(y + 5)$

9. $7x < 7(x - 2)$

10. $\dfrac{-5x + 11}{2} \le 7$

11. $-5x + 1.5 = -19.5$

12. $-5x + 4 = -26$

13. $5 + 2x - x = -x + 3 - 14$

14. $12x + 14 < 11x - 2$

15. $\dfrac{x}{5} - \dfrac{x}{4} = \dfrac{x - 2}{2}$

16. $12x - 12 = 8(x - 1)$

17. $2(x - 3) > 70$

18. $-3x - 4.7 = 11.8$

19. $-2(b - 4) - (3b - 1) = 5b + 3$

20. $8(x + 3) < 7(x + 5) + x$

21. $\dfrac{3t + 1}{8} = \dfrac{5 + 2t}{7} + 2$

22. $4(x - 6) - x = 8(x - 3) - 5x$

23. $\dfrac{x}{6} + \dfrac{3x - 2}{2} < \dfrac{2}{3}$

24. $\dfrac{y}{3} + \dfrac{y}{5} = \dfrac{y + 3}{10}$

25. $5(x - 6) + 2x > 3(2x - 1) - 4$

26. $14(x - 1) - 7x \le 2(3x - 6) + 4$

27. $\dfrac{1}{4}(3x + 2) - x \ge \dfrac{3}{8}(x - 5) + 2$

28. $\dfrac{1}{3}(x - 10) - 4x > \dfrac{5}{6}(2x + 1) - 1$

2.5 Compound Inequalities

OBJECTIVES

1 Find the Intersection of Two Sets.

2 Solve Compound Inequalities Containing **and**.

3 Find the Union of Two Sets.

4 Solve Compound Inequalities Containing **or**.

Two inequalities joined by the words **and** or **or** are called **compound inequalities.**

Compound Inequalities

$$x + 3 < 8 \quad \text{and} \quad x > 2$$

$$\frac{2x}{3} \geq 5 \quad \text{or} \quad -x + 10 < 7$$

OBJECTIVE

1 Finding the Intersection of Two Sets

The solution set of a compound inequality formed by the word **and** is the **intersection** of the solution sets of the two inequalities. We use the symbol ∩ to represent "intersection."

Intersection of Two Sets

The intersection of two sets, A and B, is the set of all elements common to both sets. A intersect B is denoted by $A \cap B$.

$A \cap B$

EXAMPLE 1   If $A = \{x \mid x$ is an even number greater than 0 and less than 10$\}$ and $B = \{3, 4, 5, 6\}$, find $A \cap B$.

Solution   Let's list the elements in set A.

$$A = \{2, 4, 6, 8\}$$

The numbers 4 and 6 are in sets A and B. The intersection is $\{4, 6\}$.   □

PRACTICE

1   If $A = \{x \mid x$ is an odd number greater than 0 and less than 10$\}$ and $B = \{1, 2, 3, 4\}$, find $A \cap B$.

OBJECTIVE

2 Solving Compound Inequalities Containing "and"

A value is a solution of a compound inequality formed by the word **and** if it is a solution of *both* inequalities. For example, the solution set of the compound inequality $x \leq 5$ and $x \geq 3$ contains all values of x that make the inequality $x \leq 5$ a true statement **and** the inequality $x \geq 3$ a true statement. The first graph shown below is the graph of $x \leq 5$, the second graph is the graph of $x \geq 3$, and the third graph shows the intersection of the two graphs. The third graph is the graph of $x \leq 5$ **and** $x \geq 3$.

$\{x \mid x \leq 5\}$ $(-\infty, 5]$

$\{x \mid x \geq 3\}$ $[3, \infty)$

$\{x \mid x \leq 5 \text{ and } x \geq 3\}$ $[3, 5]$
also $\{x \mid 3 \leq x \leq 5\}$
(see below)

Since $x \geq 3$ is the same as $3 \leq x$, the compound inequality $3 \leq x$ and $x \leq 5$ can be written in a more compact form as $3 \leq x \leq 5$. The solution set $\{x \mid 3 \leq x \leq 5\}$ includes all numbers that are greater than or equal to 3 and at the same time less than or equal to 5.

In interval notation, the set $\{x \mid x \leq 5 \text{ and } x \geq 3\}$ or the set $\{x \mid 3 \leq x \leq 5\}$ is written as $[3, 5]$.

> ▶ **Helpful Hint**
> Don't forget that some compound inequalities containing "and" can be written in a more compact form.

Compound Inequality	**Compact Form**	**Interval Notation**
$2 \leq x$ and $x \leq 6$	$2 \leq x \leq 6$	$[2, 6]$

Graph:

EXAMPLE 2 Solve: $x - 7 < 2$ and $2x + 1 < 9$

Solution First we solve each inequality separately.

$$x - 7 < 2 \quad \text{and} \quad 2x + 1 < 9$$
$$x < 9 \quad \text{and} \quad 2x < 8$$
$$x < 9 \quad \text{and} \quad x < 4$$

Now we can graph the two intervals on two number lines and find their intersection. Their intersection is shown on the third number line.

$\{x \mid x < 9\}$ $(-\infty, 9)$

$\{x \mid x < 4\}$ $(-\infty, 4)$

$\{x \mid x < 9 \text{ and } x < 4\} = \{x \mid x < 4\}$ $(-\infty, 4)$

The solution set is $(-\infty, 4)$.

PRACTICE
2   Solve: $x + 3 < 8$ and $2x - 1 < 3$. Write the solution set in interval notation.

EXAMPLE 3 Solve: $2x \geq 0$ and $4x - 1 \leq -9$.

Solution First we solve each inequality separately.

$$2x \geq 0 \quad \text{and} \quad 4x - 1 \leq -9$$
$$x \geq 0 \quad \text{and} \quad 4x \leq -8$$
$$x \geq 0 \quad \text{and} \quad x \leq -2$$

Now we can graph the two intervals and find their intersection.

$\{x \mid x \geq 0\}$ $[0, \infty)$

$\{x \mid x \leq -2\}$ $(-\infty, -2]$

$\{x \mid x \geq 0 \text{ and } x \leq -2\} = \varnothing$ $\varnothing$

There is no number that is greater than or equal to 0 *and* less than or equal to -2. The solution set is $\varnothing$.

PRACTICE
3   Solve: $4x \leq 0$ and $3x + 2 > 8$. Write the solution set in interval notation.

> ▶ **Helpful Hint**
> Example 3 shows that some compound inequalities have no solution. Also, some have all real numbers as solutions.

To solve a compound inequality written in a compact form, such as $2 < 4 - x < 7$, we get x alone in the "middle part." Since a compound inequality is really two inequalities in one statement, we must perform the same operations on all three parts of the inequality. For example:

$$2 < 4 - x < 7 \text{ means } 2 < 4 - x \quad and \quad 4 - x < 7,$$

EXAMPLE 4   Solve: $2 < 4 - x < 7$

Solution   To get x alone, we first subtract 4 from all three parts.

$$2 < 4 - x < 7$$
$$2 - 4 < 4 - x - 4 < 7 - 4 \quad \text{Subtract 4 from all three parts.}$$
$$-2 < -x < 3 \quad \text{Simplify.}$$
$$\frac{-2}{-1} > \frac{-x}{-1} > \frac{3}{-1} \quad \text{Divide all three parts by } -1 \text{ and reverse the inequality symbols.}$$
$$2 > x > -3$$

> ▶ **Helpful Hint**
> Don't forget to reverse both inequality symbols.

This is equivalent to $-3 < x < 2$.

The solution set in interval notation is $(-3, 2)$, and its graph is shown.

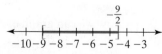

PRACTICE
4   Solve: $3 < 5 - x < 9$. Write the solution set in interval notation.

EXAMPLE 5   Solve: $-1 \le \dfrac{2x}{3} + 5 \le 2$.

Solution   First, clear the inequality of fractions by multiplying all three parts by the LCD 3.

$$-1 \le \frac{2x}{3} + 5 \le 2$$
$$3(-1) \le 3\left(\frac{2x}{3} + 5\right) \le 3(2) \quad \text{Multiply all three parts by the LCD 3.}$$
$$-3 \le 2x + 15 \le 6 \quad \text{Use the distributive property and multiply.}$$
$$-3 - 15 \le 2x + 15 - 15 \le 6 - 15 \quad \text{Subtract 15 from all three parts.}$$
$$-18 \le 2x \le -9 \quad \text{Simplify.}$$
$$\frac{-18}{2} \le \frac{2x}{2} \le \frac{-9}{2} \quad \text{Divide all three parts by 2.}$$
$$-9 \le x \le -\frac{9}{2} \quad \text{Simplify.}$$

The graph of the solution is shown.

The solution set in interval notation is $\left[-9, -\dfrac{9}{2}\right]$.

PRACTICE
5   Solve: $-4 \le \dfrac{x}{2} - 1 \le 3$. Write the solution set in interval notation.

OBJECTIVE

3   Finding the Union of Two Sets

The solution set of a compound inequality formed by the word **or** is the **union** of the solution sets of the two inequalities. We use the symbol ∪ to denote "union."

▶ **Helpful Hint**

The word *either* in this definition means "one or the other or both."

Union of Two Sets

The **union** of two sets, A and B, is the set of elements that belong to *either* of the sets. A union B is denoted by $A \cup B$.

$$A \quad B$$
$$A \cap B$$

EXAMPLE 6   If $A = \{x \mid x \text{ is an even number greater than 0 and less than 10}\}$ and $B = \{3, 4, 5, 6\}$, find $A \cup B$.

Solution   Recall from Example 1 that $A = \{2, 4, 6, 8\}$. The numbers that are in either set or both sets are $\{2, 3, 4, 5, 6, 8\}$. This set is the union.   □

PRACTICE

6   If $A = \{x \mid x \text{ is an odd number greater than 0 and less than 10}\}$ and $B = \{2, 3, 4, 5, 6\}$, find $A \cup B$.

OBJECTIVE

4   Solving Compound Inequalities Containing "or"

A value is a solution of a compound inequality formed by the word **or** if it is a solution of **either** inequality. For example, the solution set of the compound inequality $x \leq 1$ **or** $x \geq 3$ contains all numbers that make the inequality $x \leq 1$ a true statement **or** the inequality $x \geq 3$ a true statement.

$\{x \mid x \leq 1\}$![number line -1 to 6, ray to left from 1] $(-\infty, 1]$

$\{x \mid x \geq 3\}$![number line -1 to 6, ray to right from 3] $[3, \infty)$

$\{x \mid x \leq 1 \text{ or } x \geq 3\}$![number line -1 to 6, both rays] $(-\infty, 1] \cup [3, \infty)$

In interval notation, the set $\{x \mid x \leq 1 \text{ or } x \geq 3\}$ is written as $(-\infty, 1] \cup [3, \infty)$.

EXAMPLE 7   Solve: $5x - 3 \leq 10 \text{ or } x + 1 \geq 5$.

Solution   First we solve each inequality separately.

$$5x - 3 \leq 10 \quad or \quad x + 1 \geq 5$$
$$5x \leq 13 \quad or \quad x \geq 4$$
$$x \leq \frac{13}{5} \quad or \quad x \geq 4$$

Now we can graph each interval and find their union.

$\left\{x \mid x \leq \dfrac{13}{5}\right\}$![number line with 13/5 marked, ray to left] $\left(-\infty, \dfrac{13}{5}\right]$

$\{x \mid x \geq 4\}$![number line -1 to 6, ray to right from 4] $[4, \infty)$

$\left\{x \mid x \leq \dfrac{13}{5} \text{ or } x \geq 4\right\}$![number line with both rays] $\left(-\infty, \dfrac{13}{5}\right] \cup [4, \infty)$

The solution set is $\left(-\infty, \dfrac{13}{5}\right] \cup [4, \infty)$.

PRACTICE
7 Solve: $8x + 5 \leq 8$ or $x - 1 \geq 2$. Write the solution set in interval notation.

EXAMPLE 8   Solve: $-2x - 5 < -3$ *or* $6x < 0$.

Solution   First we solve each inequality separately.

$$-2x - 5 < -3 \quad or \quad 6x < 0$$
$$-2x < 2 \quad or \quad x < 0$$
$$x > -1 \quad or \quad x < 0$$

Now we can graph each interval and find their union.

$\{x \mid x > -1\}$ $(-1, \infty)$

$\{x \mid x < 0\}$ $(-\infty, 0)$

$\{x \mid x > -1 \ or \ x < 0\}$ $(-\infty, \infty)$

$=$ all real numbers

The solution set is $(-\infty, \infty)$.

PRACTICE
8 Solve: $-3x - 2 > -8$ *or* $5x > 0$. Write the solution set in interval notation.

Answer to Concept Check:
b is not correct

✓CONCEPT CHECK
Which of the following is *not* a correct way to represent the set of all numbers between -3 and 5?
a. $\{x \mid -3 < x < 5\}$ **b.** $-3 < x$ or $x < 5$
c. $(-3, 5)$ **d.** $x > -3$ and $x < 5$

Vocabulary, Readiness & Video Check

Use the choices below to fill in each blank.

or	∪	∅
and	∩	compound

1. Two inequalities joined by the words "and" or "or" are called _____ inequalities.
2. The word _____ means intersection.
3. The word _____ means union.
4. The symbol _____ represents intersection.
5. The symbol _____ represents union.
6. The symbol _____ is the empty set.

Martin-Gay Interactive Videos

See Video 2.5 🍐

Watch the section lecture video and answer the following questions.

OBJECTIVE
1
7. Based on Example 1 and the lecture before, complete the following statement. For an element to be in the intersection of sets A and B, the element must be in set A _____ in set B.

OBJECTIVE
2
8. In Example 2, how can using three number lines help us find the solution to this "and" compound inequality?

OBJECTIVE
3
9. Based on Example 4 and the lecture before, complete the following statement. For an element to be in the union of sets A and B, the element must be in set A _____ in set B.

OBJECTIVE
4
10. In Example 5, how can using three number lines help us find the solution to this "or" compound inequality?

2.5 Exercise Set

MyMathLab®

MIXED PRACTICE

If $A = \{x | x \text{ is an even integer}\}$, $B = \{x | x \text{ is an odd integer}\}$, $C = \{2, 3, 4, 5\}$, and $D = \{4, 5, 6, 7\}$, list the elements of each set. See Examples 1 and 6.

1. $C \cup D$
2. $C \cap D$
3. $A \cap D$
4. $A \cup D$
5. $A \cup B$
6. $A \cap B$
7. $B \cap D$
8. $B \cup D$
9. $B \cup C$
10. $B \cap C$
11. $A \cap C$
12. $A \cup C$

Solve each compound inequality. Graph the solution set and write it in interval notation. See Examples 2 and 3.

13. $x < 1 \text{ and } x > -3$
14. $x \leq 0 \text{ and } x \geq -2$
15. $x \leq -3 \text{ and } x \geq -2$
16. $x < 2 \text{ and } x > 4$
17. $x < -1 \text{ and } x < 1$
18. $x \geq -4 \text{ and } x > 1$

Solve each compound inequality. Write solutions in interval notation. See Examples 2 and 3.

19. $x + 1 \geq 7 \text{ and } 3x - 1 \geq 5$
20. $x + 2 \geq 3 \text{ and } 5x - 1 \geq 9$
21. $4x + 2 \leq -10 \text{ and } 2x \leq 0$
22. $2x + 4 > 0 \text{ and } 4x > 0$
23. $-2x < -8 \text{ and } x - 5 < 5$
24. $-7x \leq -21 \text{ and } x - 20 \leq -15$

Solve each compound inequality. See Examples 4 and 5.

25. $5 < x - 6 < 11$
26. $-2 \leq x + 3 \leq 0$
27. $-2 \leq 3x - 5 \leq 7$
28. $1 < 4 + 2x < 7$
29. $1 \leq \frac{2}{3}x + 3 \leq 4$
30. $-2 < \frac{1}{2}x - 5 < 1$
31. $-5 \leq \frac{-3x + 1}{4} \leq 2$
32. $-4 \leq \frac{-2x + 5}{3} \leq 1$

Solve each compound inequality. Graph the solution set and write it in interval notation. See Examples 7 and 8.

33. $x < 4 \text{ or } x < 5$
34. $x \geq -2 \text{ or } x \leq 2$
35. $x \leq -4 \text{ or } x \geq 1$
36. $x < 0 \text{ or } x < 1$
37. $x > 0 \text{ or } x < 3$
38. $x \geq -3 \text{ or } x \leq -4$

Solve each compound inequality. Write solutions in interval notation. See Examples 7 and 8.

39. $-2x \leq -4 \text{ or } 5x - 20 \geq 5$
40. $-5x \leq 10 \text{ or } 3x - 5 \geq 1$
41. $x + 4 < 0 \text{ or } 6x > -12$
42. $x + 9 < 0 \text{ or } 4x > -12$
43. $3(x - 1) < 12 \text{ or } x + 7 > 10$
44. $5(x - 1) \geq -5 \text{ or } 5 + x \leq 11$

MIXED PRACTICE

Solve each compound inequality. Write solutions in interval notation. See Examples 1 through 8.

45. $x < \frac{2}{3} \text{ and } x > -\frac{1}{2}$
46. $x < \frac{5}{7} \text{ and } x < 1$
47. $x < \frac{2}{3} \text{ or } x > -\frac{1}{2}$
48. $x < \frac{5}{7} \text{ or } x < 1$
49. $0 \leq 2x - 3 \leq 9$
50. $3 < 5x + 1 < 11$

51. $\dfrac{1}{2} < x - \dfrac{3}{4} < 2$

52. $\dfrac{2}{3} < x + \dfrac{1}{2} < 4$

53. $x + 3 \geq 3 \; and \; x + 3 \leq 2$

54. $2x - 1 \geq 3 \; and \; -x > 2$

55. $3x \geq 5 \; or \; -\dfrac{5}{8}x - 6 > 1$

56. $\dfrac{3}{8}x + 1 \leq 0 \; or \; -2x < -4$

57. $0 < \dfrac{5 - 2x}{3} < 5$

58. $-2 < \dfrac{-2x - 1}{3} < 2$

59. $-6 < 3(x - 2) \leq 8$

60. $-5 < 2(x + 4) < 8$

61. $-x + 5 > 6 \; and \; 1 + 2x \leq -5$

62. $5x \leq 0 \; and \; -x + 5 < 8$

▶ **63.** $3x + 2 \leq 5 \; or \; 7x > 29$

64. $-x < 7 \; or \; 3x + 1 < -20$

65. $5 - x > 7 \; and \; 2x + 3 \geq 13$

66. $-2x < -6 \; or \; 1 - x > -2$

67. $-\dfrac{1}{2} \leq \dfrac{4x - 1}{6} < \dfrac{5}{6}$

68. $-\dfrac{1}{2} \leq \dfrac{3x - 1}{10} < \dfrac{1}{2}$

69. $\dfrac{1}{15} < \dfrac{8 - 3x}{15} < \dfrac{4}{5}$

70. $-\dfrac{1}{4} < \dfrac{6 - x}{12} < -\dfrac{1}{6}$

71. $0.3 < 0.2x - 0.9 < 1.5$

72. $-0.7 \leq 0.4x + 0.8 < 0.5$

REVIEW AND PREVIEW

Evaluate the following. See Sections 1.2 and 1.3.

73. $|-7| - |19|$

74. $|-7 - 19|$

75. $-(-6) - |-10|$

76. $|-4| - (-4) + |-20|$

Find by inspection all values for x that make each equation true.

77. $|x| = 7$

78. $|x| = 5$

79. $|x| = 0$

80. $|x| = -2$

CONCEPT EXTENSIONS

Use the graph to answer Exercises 81 and 82.

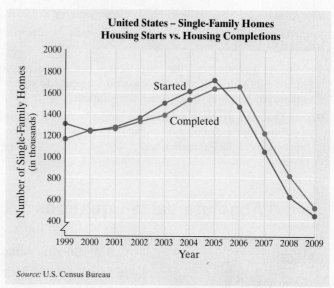

Source: U.S. Census Bureau

81. For which years were the number of single-family housing starts greater than 1500 and the number of single-family home completions greater than 1500?

82. For which years were the number of single-family housing starts less than 1000 or the number of single-family housing completions greater than 1500?

83. In your own words, describe how to find the union of two sets.

84. In your own words, describe how to find the intersection of two sets.

*Solve each compound inequality for x. See the example below. To solve $x - 6 < 3x < 2x + 5$, notice that this inequality contains a variable not only in the middle but also on the left and the right. When this occurs, we solve by rewriting the inequality using the word **and**.*

$$x - 6 < 3x \quad and \quad 3x < 2x + 5$$
$$-6 < 2x \quad and \quad x < 5$$
$$-3 < x$$
$$x > -3 \quad and \quad x < 5$$

$x > -3$

$x < 5$

$-3 < x < 5 \; or \; (-3, 5)$

85. $2x - 3 < 3x + 1 < 4x - 5$

86. $x + 3 < 2x + 1 < 4x + 6$

87. $-3(x - 2) \leq 3 - 2x \leq 10 - 3x$

88. $7x - 1 \leq 7 + 5x \leq 3(1 + 2x)$

89. $5x - 8 < 2(2 + x) < -2(1 + 2x)$

90. $1 + 2x < 3(2 + x) < 1 + 4x$

The formula for converting Fahrenheit temperatures to Celsius temperatures is $C = \dfrac{5}{9}(F - 32)$. Use this formula for Exercises 91 and 92.

91. During a recent year, the temperatures in Chicago ranged from $-29°C$ to $35°C$. Use a compound inequality to convert these temperatures to Fahrenheit temperatures.

92. In Oslo, the average temperature ranges from $-10°$ to $18°$ Celsius. Use a compound inequality to convert these temperatures to the Fahrenheit scale.

Solve.

93. Christian D'Angelo has scores of 68, 65, 75, and 78 on his algebra tests. Use a compound inequality to find the scores he can make on his final exam to receive a C in the course. The final exam counts as two tests, and a C is received if the final course average is from 70 to 79.

94. Wendy Wood has scores of 80, 90, 82, and 75 on her chemistry tests. Use a compound inequality to find the range of scores she can make on her final exam to receive a B in the course. The final exam counts as two tests, and a B is received if the final course average is from 80 to 89.

2.6 Absolute Value Equations

OBJECTIVE

1 Solve Absolute Value Equations.

OBJECTIVE

1 Solving Absolute Value Equations

In Chapter 1, we defined the absolute value of a number as its distance from 0 on a number line.

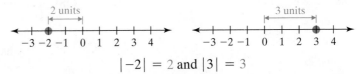

$$|-2| = 2 \text{ and } |3| = 3$$

In this section, we concentrate on solving equations containing the absolute value of a variable or a variable expression. Examples of absolute value equations are

$$|x| = 3 \qquad -5 = |2y + 7| \qquad |z - 6.7| = |3z + 1.2|$$

Since distance and absolute value are so closely related, absolute value equations and inequalities (see Section 2.7) are extremely useful in solving distance-type problems such as calculating the possible error in a measurement.

For the absolute value equation $|x| = 3$, its solution set will contain all numbers whose distance from 0 is 3 units. Two numbers are 3 units away from 0 on the number line: 3 and -3.

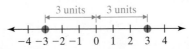

Thus, the solution set of the equation $|x| = 3$ is $\{3, -3\}$. This suggests the following:

> **Solving Equations of the Form $|X| = a$**
>
> If a is a positive number, then $|X| = a$ is equivalent to $X = a$ or $X = -a$.

EXAMPLE 1 Solve: $|p| = 2$.

Solution Since 2 is positive, $|p| = 2$ is equivalent to $p = 2$ or $p = -2$.

To check, let $p = 2$ and then $p = -2$ in the original equation.

$|p| = 2$	Original equation	$|p| = 2$	Original equation
$|2| = 2$	Let $p = 2$.	$|-2| = 2$	Let $p = -2$.
$2 = 2$	True	$2 = 2$	True

The solutions are 2 and -2 or the solution set is $\{2, -2\}$. □

PRACTICE

1 Solve: $|q| = 3$.

If the expression inside the absolute value bars is more complicated than a single variable, we can still apply the absolute value property.

> ▶ **Helpful Hint**
> For the equation $|X| = a$ in the box on the previous page, X can be a single variable or a variable expression.

EXAMPLE 2 Solve: $|5w + 3| = 7$.

Solution Here the expression inside the absolute value bars is $5w + 3$. If we think of the expression $5w + 3$ as X in the absolute value property, we see that $|X| = 7$ is equivalent to

$$X = 7 \quad \text{or} \quad X = -7$$

Then substitute $5w + 3$ for X, and we have

$$5w + 3 = 7 \quad \text{or} \quad 5w + 3 = -7$$

Solve these two equations for w.

$$
\begin{aligned}
5w + 3 &= 7 \quad &\text{or} \quad 5w + 3 &= -7 \\
5w &= 4 \quad &\text{or} \quad 5w &= -10 \\
w &= \frac{4}{5} \quad &\text{or} \quad w &= -2
\end{aligned}
$$

Check: To check, let $w = -2$ and then $w = \frac{4}{5}$ in the original equation.

Let $w = -2$	Let $w = \frac{4}{5}$
$|5(-2) + 3| = 7$	$\left|5\left(\frac{4}{5}\right) + 3\right| = 7$
$|-10 + 3| = 7$	$|4 + 3| = 7$
$|-7| = 7$	$|7| = 7$
$7 = 7 \quad$ True	$7 = 7 \quad$ True

Both solutions check, and the solutions are -2 and $\frac{4}{5}$ or the solution set is $\left\{-2, \frac{4}{5}\right\}$. ☐

PRACTICE
2 Solve: $|2x - 3| = 5$.

EXAMPLE 3 Solve: $\left|\frac{x}{2} - 1\right| = 11$.

Solution $\left|\frac{x}{2} - 1\right| = 11$ is equivalent to

$$
\begin{aligned}
\frac{x}{2} - 1 &= 11 \quad &\text{or} \quad & \frac{x}{2} - 1 = -11 \\
2\left(\frac{x}{2} - 1\right) &= 2(11) \quad &\text{or} \quad & 2\left(\frac{x}{2} - 1\right) = 2(-11) \quad \text{Clear fractions.} \\
x - 2 &= 22 \quad &\text{or} \quad & x - 2 = -22 \quad \text{Apply the distributive property.} \\
x &= 24 \quad &\text{or} \quad & x = -20
\end{aligned}
$$

The solutions are 24 and -20. ☐

PRACTICE
3 Solve: $\left|\frac{x}{5} + 1\right| = 15$.

To apply the absolute value property, first make sure that the absolute value expression is isolated.

> ▶ **Helpful Hint**
>
> If the equation has a single absolute value expression containing variables, isolate the absolute value expression first.

EXAMPLE 4 Solve: $|2x| + 5 = 7$.

Solution We want the absolute value expression alone on one side of the equation, so begin by subtracting 5 from both sides. Then apply the absolute value property.

$$|2x| + 5 = 7$$
$$|2x| = 2 \qquad \text{Subtract 5 from both sides.}$$
$$2x = 2 \quad \text{or} \quad 2x = -2$$
$$x = 1 \quad \text{or} \quad x = -1$$

The solutions are -1 and 1.

PRACTICE
4 Solve: $|3x| + 8 = 14$.

EXAMPLE 5 Solve: $|y| = 0$.

Solution We are looking for all numbers whose distance from 0 is zero units. The only number is 0. The solution is 0.

PRACTICE
5 Solve: $|z| = 0$.

The next two examples illustrate a special case for absolute value equations. This special case occurs when an isolated absolute value is equal to a negative number.

EXAMPLE 6 Solve: $2|x| + 25 = 23$.

Solution First, isolate the absolute value.

$$2|x| + 25 = 23$$
$$2|x| = -2 \quad \text{Subtract 25 from both sides.}$$
$$|x| = -1 \quad \text{Divide both sides by 2.}$$

The absolute value of a number is never negative, so this equation has no solution. The solution set is $\{\ \}$ or $\varnothing$.

PRACTICE
6 Solve: $3|z| + 9 = 7$.

EXAMPLE 7 Solve: $\left|\dfrac{3x + 1}{2}\right| = -2$.

Solution Again, the absolute value of any expression is never negative, so no solution exists. The solution set is $\{\ \}$ or $\varnothing$.

PRACTICE
7 Solve: $\left|\dfrac{5x + 3}{4}\right| = -8$.

Given two absolute value expressions, we might ask, when are the absolute values of two expressions equal? To see the answer, notice that

$$|2| = |2|, \quad |-2| = |-2|, \quad |-2| = |2|, \quad \text{and} \quad |2| = |-2|$$

same   same   opposites   opposites

Two absolute value expressions are equal when the expressions inside the absolute value bars are equal to or are opposites of each other.

EXAMPLE 8   Solve: $|3x + 2| = |5x - 8|$.

Solution   This equation is true if the expressions inside the absolute value bars are equal to or are opposites of each other.

$$3x + 2 = 5x - 8 \quad \text{or} \quad 3x + 2 = -(5x - 8)$$

Next, solve each equation.

$$
\begin{aligned}
3x + 2 &= 5x - 8 \quad &\text{or} \quad 3x + 2 &= -5x + 8 \\
-2x + 2 &= -8 \quad &\text{or} \quad 8x + 2 &= 8 \\
-2x &= -10 \quad &\text{or} \quad 8x &= 6 \\
x &= 5 \quad &\text{or} \quad x &= \frac{3}{4}
\end{aligned}
$$

The solutions are $\frac{3}{4}$ and 5.

PRACTICE
8   Solve: $|2x + 4| = |3x - 1|$.

EXAMPLE 9   Solve: $|x - 3| = |5 - x|$.

Solution

$$
\begin{aligned}
x - 3 &= 5 - x \quad &\text{or} \quad x - 3 &= -(5 - x) \\
2x - 3 &= 5 \quad &\text{or} \quad x - 3 &= -5 + x \\
2x &= 8 \quad &\text{or} \quad x - 3 - x &= -5 + x - x \\
x &= 4 \quad &\text{or} \quad -3 &= -5 \quad \text{False}
\end{aligned}
$$

Recall from Section 2.1 that when an equation simplifies to a false statement, the equation has no solution. Thus, the only solution for the original absolute value equation is 4.

PRACTICE
9   Solve: $|x - 2| = |8 - x|$.

✓CONCEPT CHECK
True or false? Absolute value equations always have two solutions. Explain your answer.

The following box summarizes the methods shown for solving absolute value equations.

Absolute Value Equations

$|X| = a$ $\begin{cases} \text{If } a \text{ is positive, then solve } X = a \text{ or } X = -a. \\ \text{If } a \text{ is } 0, \text{ solve } X = 0. \\ \text{If } a \text{ is negative, the equation } |X| = a \text{ has no solution.} \end{cases}$

$|X| = |Y|$ Solve $X = Y$ or $X = -Y$.

Answer to Concept Check:
false; answers may vary

Vocabulary, Readiness & Video Check

Match each absolute value equation with an equivalent statement.

1. $|x - 2| = 5$
2. $|x - 2| = 0$
3. $|x - 2| = |x + 3|$
4. $|x + 3| = 5$
5. $|x + 3| = -5$

A. $x - 2 = 0$
B. $x - 2 = x + 3$ or $x - 2 = -(x + 3)$
C. $x - 2 = 5$ or $x - 2 = -5$
D. $\varnothing$
E. $x + 3 = 5$ or $x + 3 = -5$

Martin-Gay Interactive Videos

See Video 2.6

Watch the section lecture videos and answer the following question.

OBJECTIVE 1

6. As explained in Example 3, why is *a* positive in the rule "$|X| = a$ is equivalent to $X = a$ or $X = -a$"?

2.6 Exercise Set   MyMathLab®

Solve each absolute value equation. See Examples 1 through 7.

1. $|x| = 7$
2. $|y| = 15$
3. $|3x| = 12.6$
4. $|6n| = 12.6$
5. $|2x - 5| = 9$
6. $|6 + 2n| = 4$
7. $\left|\dfrac{x}{2} - 3\right| = 1$
8. $\left|\dfrac{n}{3} + 2\right| = 4$
9. $|z| + 4 = 9$
10. $|x| + 1 = 3$
11. $|3x| + 5 = 14$
12. $|2x| - 6 = 4$
13. $|2x| = 0$
14. $|7z| = 0$
15. $|4n + 1| + 10 = 4$
16. $|3z - 2| + 8 = 1$
17. $|5x - 1| = 0$
18. $|3y + 2| = 0$

Solve. See Examples 8 and 9.

19. $|5x - 7| = |3x + 11|$
20. $|9y + 1| = |6y + 4|$
21. $|z + 8| = |z - 3|$
22. $|2x - 5| = |2x + 5|$

MIXED PRACTICE

Solve each absolute value equation. See Examples 1 through 9.

23. $|x| = 4$
24. $|x| = 1$
25. $|y| = 0$
26. $|y| = 8$
27. $|z| = -2$
28. $|y| = -9$
29. $|7 - 3x| = 7$
30. $|4m + 5| = 5$
31. $|6x| - 1 = 11$
32. $|7z| + 1 = 22$
33. $|4p| = -8$
34. $|5m| = -10$

35. $|x - 3| + 3 = 7$
36. $|x + 4| - 4 = 1$
37. $\left|\dfrac{z}{4} + 5\right| = -7$
38. $\left|\dfrac{c}{5} - 1\right| = -2$
39. $|9v - 3| = -8$
40. $|1 - 3b| = -7$
41. $|8n + 1| = 0$
42. $|5x - 2| = 0$
43. $|1 + 6c| - 7 = -3$
44. $|2 + 3m| - 9 = -7$
45. $|5x + 1| = 11$
46. $|8 - 6c| = 1$
47. $|4x - 2| = |-10|$
48. $|3x + 5| = |-4|$
49. $|5x + 1| = |4x - 7|$
50. $|3 + 6n| = |4n + 11|$
51. $|6 + 2x| = -|-7|$
52. $|4 - 5y| = -|-3|$
53. $|2x - 6| = |10 - 2x|$
54. $|4n + 5| = |4n + 3|$
55. $\left|\dfrac{2x - 5}{3}\right| = 7$
56. $\left|\dfrac{1 + 3n}{4}\right| = 4$
57. $2 + |5n| = 17$
58. $8 + |4m| = 24$
59. $\left|\dfrac{2x - 1}{3}\right| = |-5|$
60. $\left|\dfrac{5x + 2}{2}\right| = |-6|$
61. $|2y - 3| = |9 - 4y|$
62. $|5z - 1| = |7 - z|$
63. $\left|\dfrac{3n + 2}{8}\right| = |-1|$
64. $\left|\dfrac{2r - 6}{5}\right| = |-2|$
65. $|x + 4| = |7 - x|$
66. $|8 - y| = |y + 2|$
67. $\left|\dfrac{8c - 7}{3}\right| = -|-5|$
68. $\left|\dfrac{5d + 1}{6}\right| = -|-9|$

REVIEW AND PREVIEW

The circle graph shows the types of cheese produced in the United States in 2010. Use this graph to answer Exercises 69 through 72. See Section 2.2.

U.S. Cheese¹ Production by Variety, 2010

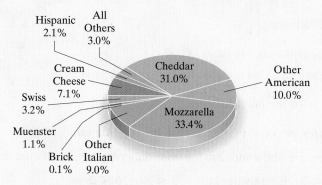

Hispanic 2.1%
All Others 3.0%
Cream Cheese 7.1%
Swiss 3.2%
Muenster 1.1%
Brick 0.1%
Other Italian 9.0%
Cheddar 31.0%
Other American 10.0%
Mozzarella 33.4%

¹Excludes Cottage Cheese

Source: USDA, Dairy Products Annual Survey

69. In 2010, cheddar cheese made up what percent of U.S. cheese production?

70. Which cheese had the highest U.S. production in 2010?

71. A circle contains 360°. Find the number of degrees found in the 9% sector for Other Italian Cheese.

72. In 2010, the total production of cheese in the United States was 10,109,293,000 pounds. Find the amount of cream cheese produced during that year.

List five integer solutions of each inequality. See Sections 1.2 through 1.4.

73. $|x| \leq 3$

74. $|x| \geq -2$

75. $|y| > -10$

76. $|y| < 0$

CONCEPT EXTENSIONS

Without going through a solution procedure, determine the solution of each absolute value equation or inequality.

77. $|x - 7| = -4$ **78.** $|x - 7| < -4$

79. Write an absolute value equation representing all numbers x whose distance from 0 is 5 units.

80. Write an absolute value equation representing all numbers x whose distance from 0 is 2 units.

81. Explain why some absolute value equations have two solutions.

82. Explain why some absolute value equations have one solution.

83. Write an absolute value equation representing all numbers x whose distance from 1 is 5 units.

84. Write an absolute value equation representing all numbers x whose distance from 7 is 2 units.

85. Describe how solving an absolute value equation such as $|2x - 1| = 3$ is similar to solving an absolute value equation such as $|2x - 1| = |x - 5|$.

86. Describe how solving an absolute value equation such as $|2x - 1| = 3$ is different from solving an absolute value equation such as $|2x - 1| = |x - 5|$.

Write each as an equivalent absolute value equation.

87. $x = 6$ or $x = -6$

88. $2x - 1 = 4$ or $2x - 1 = -4$

89. $x - 2 = 3x - 4$ or $x - 2 = -(3x - 4)$

90. For what value(s) of c will an absolute value equation of the form $|ax + b| = c$ have

a. one solution?

b. no solution?

c. two solutions?

2.7 Absolute Value Inequalities

OBJECTIVES

1 Solve Absolute Value Inequalities of the Form $|X| < a$.

2 Solve Absolute Value Inequalities of the Form $|X| > a$.

OBJECTIVE

1   Solving Absolute Value Inequalities of the Form $|X| < a$

The solution set of an absolute value inequality such as $|x| < 2$ contains all numbers whose distance from 0 is less than 2 units, as shown below.

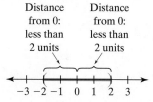

Distance from 0: less than 2 units   Distance from 0: less than 2 units

-3 -2 -1 0 1 2 3

The solution set is $\{x | -2 < x < 2\}$, or $(-2, 2)$ in interval notation.

EXAMPLE 1   Solve: $|x| \leq 3$ and graph the solution set.

Solution   The solution set of this inequality contains all numbers whose distance from 0 is less than or equal to 3. Thus 3, -3, and all numbers between 3 and -3 are in the solution set.

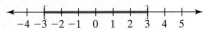

The solution set is $[-3, 3]$.

1 Solve: $|x| < 5$ and graph the solution set.

In general, we have the following.

> **Solving Absolute Value Inequalities of the Form $|X| < a$**
>
> If a is a positive number, then $|X| < a$ is equivalent to $-a < X < a$.

This property also holds true for the inequality symbol $\leq$.

EXAMPLE 2 Solve for m: $|m - 6| < 2$. Graph the solution set.

Solution Replace X with $m - 6$ and a with 2 in the preceding property, and we see that

$$|m - 6| < 2 \quad \text{is equivalent to} \quad -2 < m - 6 < 2$$

Solve this compound inequality for m by adding 6 to all three parts.

$$-2 < m - 6 < 2$$
$$-2 + 6 < m - 6 + 6 < 2 + 6 \qquad \text{Add 6 to all three parts.}$$
$$4 < m < 8 \qquad \text{Simplify.}$$

The solution set is $(4, 8)$, and its graph is shown.

2 Solve for b: $|b + 1| < 3$. Graph the solution set.

> ▶ **Helpful Hint**
>
> Before using an absolute value inequality property, isolate the absolute value expression on one side of the inequality.

EXAMPLE 3 Solve for x: $|5x + 1| + 1 \leq 10$. Graph the solution set.

Solution First, isolate the absolute value expression by subtracting 1 from both sides.

$$|5x + 1| + 1 \leq 10$$
$$|5x + 1| \leq 10 - 1 \qquad \text{Subtract 1 from both sides.}$$
$$|5x + 1| \leq 9 \qquad \text{Simplify.}$$

Since 9 is positive, we apply the absolute value property for $|X| \leq a$.

$$-9 \leq 5x + 1 \leq 9$$
$$-9 - 1 \leq 5x + 1 - 1 \leq 9 - 1 \qquad \text{Subtract 1 from all three parts.}$$
$$-10 \leq 5x \leq 8 \qquad \text{Simplify.}$$
$$-2 \leq x \leq \frac{8}{5} \qquad \text{Divide all three parts by 5.}$$

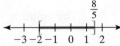

The solution set is $\left[-2, \dfrac{8}{5}\right]$, and the graph is shown above.

3 Solve for x: $|3x - 2| + 5 \leq 9$. Graph the solution set.

EXAMPLE 4   Solve for x: $\left|2x - \dfrac{1}{10}\right| < -13$.

Solution   The absolute value of a number is always nonnegative and can never be less than -13. Thus this absolute value inequality has no solution. The solution set is $\{\ \}$ or $\varnothing$. ☐

PRACTICE
4   Solve for x: $\left|3x + \dfrac{5}{8}\right| < -4$.

EXAMPLE 5   Solve for x: $\left|\dfrac{2(x + 1)}{3}\right| \le 0$.

Solution   Recall that "$\le$" means "is less than or equal to." The absolute value of any expression will never be less than 0, but it may be equal to 0. Thus, to solve $\left|\dfrac{2(x + 1)}{3}\right| \le 0$, we solve $\left|\dfrac{2(x + 1)}{3}\right| = 0$

$$\frac{2(x + 1)}{3} = 0$$

$$3\left[\frac{2(x + 1)}{3}\right] = 3(0) \quad \text{Clear the equation of fractions.}$$

$$2x + 2 = 0 \quad \text{Apply the distributive property.}$$

$$2x = -2 \quad \text{Subtract 2 from both sides.}$$

$$x = -1 \quad \text{Divide both sides by 2.}$$

The solution set is $\{-1\}$. ☐

PRACTICE
5   Solve for x: $\left|\dfrac{3(x - 2)}{5}\right| \le 0$.

OBJECTIVE
2   **Solving Absolute Value Inequalities of the Form $|X| > a$**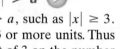

Let us now solve an absolute value inequality of the form $|X| > a$, such as $|x| \ge 3$. The solution set contains all numbers whose distance from 0 is 3 or more units. Thus the graph of the solution set contains 3 and all points to the right of 3 on the number line or -3 and all points to the left of -3 on the number line.

Distance from 0:   Distance from 0:
greater than or   greater than or
equal to 3 units   equal to 3 units

$$-3\ -2\ -1\ \ 0\ \ 1\ \ 2\ \ 3$$

This solution set is written as $\{x | x \le -3 \text{ or } x \ge 3\}$. In interval notation, the solution is $(-\infty, -3] \cup [3, \infty)$, since "or" means "union." In general, we have the following.

> **Solving Absolute Value Inequalities of the Form $|X| > a$**
>
> If a is a positive number, then $|X| > a$ is equivalent to $X < -a$ or $X > a$.

This property also holds true for the inequality symbol $\ge$.

EXAMPLE 6   Solve for y: $|y - 3| > 7$.

Solution   Since 7 is positive, we apply the property for $|X| > a$.

$$|y - 3| > 7 \text{ is equivalent to } y - 3 < -7 \text{ or } y - 3 > 7$$

Next, solve the compound inequality.

$$y - 3 < -7 \qquad \text{or} \qquad y - 3 > 7$$

$$y - 3 + 3 < -7 + 3 \quad \text{or} \quad y - 3 + 3 > 7 + 3 \qquad \text{Add 3 to both sides.}$$

$$y < -4 \qquad \text{or} \qquad y > 10 \qquad \text{Simplify.}$$

The solution set is $(-\infty, -4) \cup (10, \infty)$, and its graph is shown.

PRACTICE

6   Solve for y: $|y + 4| \geq 6$.

Example 7 illustrates another special case of absolute value inequalities when an isolated absolute value expression is less than, less than or equal to, greater than, or greater than or equal to a negative number or 0.

EXAMPLE 7   Solve: $|2x + 9| + 5 > 3$.

Solution   First isolate the absolute value expression by subtracting 5 from both sides.

$$|2x + 9| + 5 > 3$$

$$|2x + 9| + 5 - 5 > 3 - 5 \qquad \text{Subtract 5 from both sides.}$$

$$|2x + 9| > -2 \qquad \text{Simplify.}$$

The absolute value of any number is always nonnegative and thus is always greater than -2. This inequality and the original inequality are true for all values of x. The solution set is $\{x \mid x \text{ is a real number}\}$ or $(-\infty, \infty)$, and its graph is shown.

PRACTICE

7   Solve: $|4x + 3| + 5 > 3$. Graph the solution set.

> **✓CONCEPT CHECK**
> Without taking any solution steps, how do you know that the absolute value inequality $|3x - 2| > -9$ has a solution? What is its solution?

EXAMPLE 8   Solve: $\left|\dfrac{x}{3} - 1\right| - 7 \geq -5$.

Solution   First, isolate the absolute value expression by adding 7 to both sides.

$$\left|\frac{x}{3} - 1\right| - 7 \geq -5$$

$$\left|\frac{x}{3} - 1\right| - 7 + 7 \geq -5 + 7 \qquad \text{Add 7 to both sides.}$$

$$\left|\frac{x}{3} - 1\right| \geq 2 \qquad \text{Simplify.}$$

Next, write the absolute value inequality as an equivalent compound inequality and solve.

$$\frac{x}{3} - 1 \leq -2 \qquad \text{or} \qquad \frac{x}{3} - 1 \geq 2$$

$$3\left(\frac{x}{3} - 1\right) \leq 3(-2) \quad \text{or} \quad 3\left(\frac{x}{3} - 1\right) \geq 3(2) \qquad \text{Clear the inequalities of fractions.}$$

$$x - 3 \leq -6 \qquad \text{or} \qquad x - 3 \geq 6 \qquad \text{Apply the distributive property.}$$

$$x \leq -3 \qquad \text{or} \qquad x \geq 9 \qquad \text{Add 3 to both sides.}$$

Answer to Concept Check:
$(-\infty, \infty)$ since the absolute value is always nonnegative

The solution set is $(-\infty, -3] \cup [9, \infty)$, and its graph is shown.

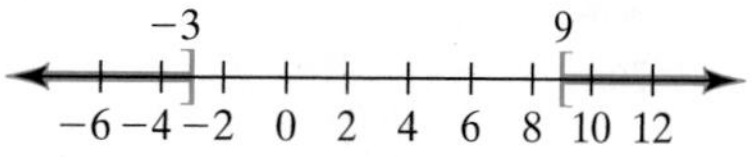

PRACTICE 8 Solve: $\left|\frac{x}{2} - 3\right| - 5 > -2$. Graph the solution set.

The following box summarizes the types of absolute value equations and inequalities.

Solving Absolute Value Equations and Inequalities with $a > 0$

Algebraic Solution

$|X| = a$ is equivalent to $X = a$ or $X = -a$.

$|X| < a$ is equivalent to $-a < X < a$.

$|X| > a$ is equivalent to $X < -a$ or $X > a$.

Solution Graph

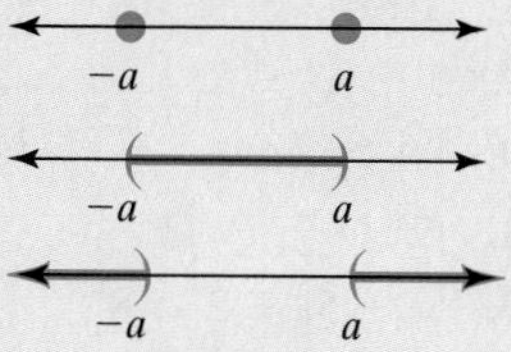

Vocabulary, Readiness & Video Check

Match each absolute value statement with an equivalent statement.

1. $|2x + 1| = 3$
2. $|2x + 1| \le 3$
3. $|2x + 1| < 3$
4. $|2x + 1| \ge 3$
5. $|2x + 1| > 3$

A. $2x + 1 > 3$ or $2x + 1 < -3$
B. $2x + 1 \ge 3$ or $2x + 1 \le -3$
C. $-3 < 2x + 1 < 3$
D. $2x + 1 = 3$ or $2x + 1 = -3$
E. $-3 \le 2x + 1 \le 3$

Martin-Gay Interactive Videos

See Video 2.7

Watch the section lecture video and answer the following questions.

OBJECTIVE 1

6. In Example 3, how can you reason that the inequality has no solution even if you don't know the rule?

OBJECTIVE 2

7. In Example 4, why is the union symbol used when the solution is written in interval notation?

2.7 Exercise Set

MyMathLab®

Solve each inequality. Then graph the solution set and write it in interval notation. See Examples 1 through 4.

1. $|x| \le 4$
2. $|x| < 6$
3. $|x - 3| < 2$
4. $|y - 7| \le 5$
5. $|x + 3| < 2$
6. $|x + 4| < 6$
7. $|2x + 7| \le 13$
8. $|5x - 3| \le 18$
9. $|x| + 7 \le 12$
10. $|x| + 6 \le 7$
11. $|3x - 1| < -5$
12. $|8x - 3| < -2$
13. $|x - 6| - 7 \le -1$
14. $|z + 2| - 7 < -3$

Solve each inequality. Graph the solution set and write it in interval notation. See Examples 6 through 8.

15. $|x| > 3$
16. $|y| \ge 4$
17. $|x + 10| \ge 14$
18. $|x - 9| \ge 2$
19. $|x| + 2 > 6$
20. $|x| - 1 > 3$
21. $|5x| > -4$
22. $|4x - 11| > -1$
23. $|6x - 8| + 3 > 7$
24. $|10 + 3x| + 1 > 2$

Solve each inequality. Graph the solution set and write it in interval notation. See Example 5.

25. $|x| \le 0$ **26.** $|x| \ge 0$

27. $|8x + 3| > 0$

28. $|5x - 6| < 0$

MIXED PRACTICE

Solve each inequality. Graph the solution set and write it in interval notation. See Examples 1 through 8.

29. $|x| \le 2$ **30.** $|z| < 8$

31. $|y| > 1$ **32.** $|x| \ge 10$

33. $|x - 3| < 8$ **34.** $|-3 + x| \le 10$

35. $|0.6x - 3| > 0.6$ **36.** $|1 + 0.3x| \ge 0.1$

37. $5 + |x| \le 2$ **38.** $8 + |x| < 1$

39. $|x| > -4$ **40.** $|x| \le -7$

41. $|2x - 7| \le 11$ **42.** $|5x + 2| < 8$

43. $|x + 5| + 2 \ge 8$ **44.** $|-1 + x| - 6 > 2$

45. $|x| > 0$ **46.** $|x| < 0$

47. $9 + |x| > 7$ **48.** $5 + |x| \ge 4$

49. $6 + |4x - 1| \le 9$ **50.** $-3 + |5x - 2| \le 4$

51. $\left|\dfrac{2}{3}x + 1\right| > 1$ **52.** $\left|\dfrac{3}{4}x - 1\right| \ge 2$

53. $|5x + 3| < -6$ **54.** $|4 + 9x| \ge -6$

55. $\left|\dfrac{8x - 3}{4}\right| \le 0$ **56.** $\left|\dfrac{5x + 6}{2}\right| \le 0$

57. $|1 + 3x| + 4 < 5$ **58.** $|7x - 3| - 1 \le 10$

59. $\left|\dfrac{x + 6}{3}\right| > 2$

60. $\left|\dfrac{7 + x}{2}\right| \ge 4$

61. $-15 + |2x - 7| \le -6$

62. $-9 + |3 + 4x| < -4$

63. $\left|2x + \dfrac{3}{4}\right| - 7 \le -2$

64. $\left|\dfrac{3}{5} + 4x\right| - 6 < -1$

MIXED PRACTICE

Solve each equation or inequality for x. (Sections 2.6, 2.7)

65. $|2x - 3| < 7$ **66.** $|2x - 3| > 7$

67. $|2x - 3| = 7$ **68.** $|5 - 6x| = 29$

69. $|x - 5| \ge 12$ **70.** $|x + 4| \ge 20$

71. $|9 + 4x| = 0$ **72.** $|9 + 4x| \ge 0$

73. $|2x + 1| + 4 < 7$ **74.** $8 + |5x - 3| \ge 11$

75. $|3x - 5| + 4 = 5$ **76.** $|5x - 3| + 2 = 4$

77. $|x + 11| = -1$ **78.** $|4x - 4| = -3$

79. $\left|\dfrac{2x - 1}{3}\right| = 6$ **80.** $\left|\dfrac{6 - x}{4}\right| = 5$

81. $\left|\dfrac{3x - 5}{6}\right| > 5$

82. $\left|\dfrac{4x - 7}{5}\right| < 2$

REVIEW AND PREVIEW

Recall the formula:

$$\text{Probability of an event} = \frac{\text{number of ways that the event can occur}}{\text{number of possible outcomes}}$$

Find the probability of rolling each number on a single toss of a die. (Recall that a die is a cube with each of its six sides containing 1, 2, 3, 4, 5, and 6 black dots, respectively.) See Section 2.3.

83. $P(\text{rolling a 2})$ **84.** $P(\text{rolling a 5})$

85. $P(\text{rolling a 7})$ **86.** $P(\text{rolling a 0})$

87. $P(\text{rolling a 1 or 3})$

88. $P(\text{rolling a 1, 2, 3, 4, 5, or 6})$

Consider the equation $3x - 4y = 12$. For each value of x or y given, find the corresponding value of the other variable that makes the statement true. See Section 2.3.

89. If $x = 2$, find y.   **90.** If $y = -1$, find x.

91. If $y = -3$, find x.   **92.** If $x = 4$, find y.

CONCEPT EXTENSIONS

93. Write an absolute value inequality representing all numbers x whose distance from 0 is less than 7 units.

94. Write an absolute value inequality representing all numbers x whose distance from 0 is greater than 4 units.

95. Write $-5 \le x \le 5$ as an equivalent inequality containing an absolute value.

96. Write $x > 1$ or $x < -1$ as an equivalent inequality containing an absolute value.

97. Describe how solving $|x - 3| = 5$ is different from solving $|x - 3| < 5$.

98. Describe how solving $|x + 4| = 0$ is similar to solving $|x + 4| \le 0$.

The expression $|x_T - x|$ is defined to be the absolute error in x, where x_T is the true value of a quantity and x is the measured value or value as stored in a computer.

99. If the true value of a quantity is 3.5 and the absolute error must be less than 0.05, find the acceptable measured values.

100. If the true value of a quantity is 0.2 and the approximate value stored in a computer is $\dfrac{51}{256}$, find the absolute error.

Chapter 2 Vocabulary Check

Fill in each blank with one of the words or phrases listed below.

contradiction	linear inequality in one variable	compound inequality	solution
absolute value	consecutive integers	identity	union
formula	linear equation in one variable	intersection	

1. The statement "$x < 5$ or $x > 7$" is called a(n) _____ .

2. An equation in one variable that has no solution is called a(n) _____ .

3. The _____ of two sets is the set of all elements common to both sets.

4. The _____ of two sets is the set of all elements that belong to either of the sets.

5. An equation in one variable that has every number (for which the equation is defined) as a solution is called a(n) _____ .

6. The equation $d = rt$ is also called a(n) _____ .

7. A number's distance from 0 is called its _____ .

8. When a variable in an equation is replaced by a number and the resulting equation is true, then that number is called a(n) _____ of the equation.

9. The integers $17, 18, 19$ are examples of _____ .

10. The statement $5x - 0.2 < 7$ is an example of a(n) _____ .

11. The statement $5x - 0.2 = 7$ is an example of a(n) _____ .

Chapter 2 Highlights

DEFINITIONS AND CONCEPTS	EXAMPLES
Section 2.1 Linear Equations in One Variable	

An **equation** is a statement that two expressions are equal.	Equations: $$5 = 5 \qquad 7x + 2 = -14 \qquad 3(x-1)^2 = 9x^2 - 6$$
A **linear equation in one variable** is an equation that can be written in the form $ax + b = c$, where a, b, and c are real numbers and a is not 0.	Linear equations: $$7x + 2 = -14 \qquad x = -3$$ $$5(2y - 7) = -2(8y - 1)$$
A **solution** of an equation is a value for the variable that makes the equation a true statement.	Check to see that -1 is a solution of $3(x - 1) = 4x - 2$. $$3(-1 - 1) = 4(-1) - 2$$ $$3(-2) = -4 - 2$$ $$-6 = -6 \qquad \text{True}$$ Thus, -1 is a solution.
Equivalent equations have the same solution.	$x - 12 = 14$ and $x = 26$ are equivalent equations.
	(continued)

DEFINITIONS AND CONCEPTS	EXAMPLES

Section 2.1 Linear Equations in One Variable (continued)

The **addition property of equality** guarantees that the same number may be added to (or subtracted from) both sides of an equation, and the result is an equivalent equation.

Solve for x: $-3x - 2 = 10$.

$$-3x - 2 + 2 = 10 + 2 \quad \text{Add 2 to both sides.}$$
$$-3x = 12$$
$$\frac{-3x}{-3} = \frac{12}{-3} \quad \text{Divide both sides by } -3.$$
$$x = -4$$

The **multiplication property of equality** guarantees that both sides of an equation may be multiplied by (or divided by) the same nonzero number, and the result is an equivalent equation.

To solve linear equations in one variable:

Solve for x:

$$x - \frac{x-2}{6} = \frac{x-7}{3} + \frac{2}{3}$$

1. Clear the equation of fractions.

1. $6\left(x - \frac{x-2}{6}\right) = 6\left(\frac{x-7}{3} + \frac{2}{3}\right)$ Multiply both sides by 6.

$6x - (x-2) = 2(x-7) + 2(2)$ Remove grouping symbols.

2. Remove grouping symbols such as parentheses.

2. $6x - x + 2 = 2x - 14 + 4$

3. Simplify by combining like terms.

3. $5x + 2 = 2x - 10$

4. Write variable terms on one side and numbers on the other side using the addition property of equality.

4. $5x + 2 - 2 = 2x - 10 - 2$ Subtract 2.

$5x = 2x - 12$

$5x - 2x = 2x - 12 - 2x$ Subtract $2x$.

$3x = -12$

5. Isolate the variable using the multiplication property of equality.

5. $\dfrac{3x}{3} = \dfrac{-12}{3}$ Divide by 3.

$x = -4$

6. Check the proposed solution in the original equation.

6. $-4 - \dfrac{-4-2}{6} \overset{?}{=} \dfrac{-4-7}{3} + \dfrac{2}{3}$ Replace x with -4 in the original equation.

$-4 - \dfrac{-6}{6} \overset{?}{=} \dfrac{-11}{3} + \dfrac{2}{3}$

$-4 - (-1) \overset{?}{=} \dfrac{-9}{3}$

$-3 = -3$ True

Section 2.2 An Introduction to Problem Solving

Problem-Solving Strategy

Colorado is shaped like a rectangle whose length is about 1.3 times its width. If the perimeter of Colorado is 2070 kilometers, find its dimensions.

1. UNDERSTAND the problem.

1. Read and reread the problem. Guess a solution and check your guess.

Let x = width of Colorado in kilometers. Then
$1.3x$ = length of Colorado in kilometers

$1.3x$

DEFINITIONS AND CONCEPTS	**EXAMPLES**

Section 2.2 An Introduction to Problem Solving (continued)

2. TRANSLATE the problem.

2. In words:

twice the length	+	twice the width	=	perimeter
↓		↓		↓

Translate: $2(1.3x)$ + $2x$ = 2070

3. SOLVE the equation.

3. $2.6x + 2x = 2070$

$4.6x = 2070$

$x = 450$

4. INTERPRET the results.

4. If $x = 450$ kilometers, then $1.3x = 1.3(450) = 585$ kilometers. *Check:* The perimeter of a rectangle whose width is 450 kilometers and length is 585 kilometers is $2(450) + 2(585) = 2070$ kilometers, the required perimeter. *State:* The dimensions of Colorado are 450 kilometers by 585 kilometers

Section 2.3 Formulas and Problem Solving

An equation that describes a known relationship among quantities is called a **formula.**

Formulas:

$A = \pi r^2$ (area of a circle)

$I = PRT$ (interest = principal · rate · time)

To solve a formula for a specified variable, use the steps for solving an equation. Treat the specified variable as the only variable of the equation.

Solve $A = 2HW + 2LW + 2LH$ for H.

$A - 2LW = 2HW + 2LH$ Subtract $2LW$.

$A - 2LW = H(2W + 2L)$ Factor out H.

$\dfrac{A - 2LW}{2W + 2L} = \dfrac{H(2W + 2L)}{2W + 2L}$ Divide by $2W + 2L$.

$\dfrac{A - 2LW}{2W + 2L} = H$ Simplify.

Section 2.4 Linear Inequalities and Problem Solving

A **linear inequality in one variable** is an inequality that can be written in the form $ax + b < c$, where $a, b,$ and c are real numbers and $a \neq 0$. (The inequality symbols $\leq$, $>$, and $\geq$ also apply here.)

The **addition property of inequality** guarantees that the same number may be added to (or subtracted from) both sides of an inequality, and the resulting inequality will have the same solution set.

The **multiplication property of inequality** guarantees that both sides of an inequality may be multiplied by (or divided by) the same **positive** number, and the resulting inequality will have the same solution set. We may also multiply (or divide) both sides of an inequality by the same **negative** number and **reverse the direction of the inequality symbol**, and the result is an inequality with the same solution set.

To solve a linear inequality in one variable:

Linear inequalities:

$5x - 2 \leq -7$ $3y > 1$ $\dfrac{z}{7} < -9(z - 3)$

$x - 9 \leq -16$

$x - 9 + 9 \leq -16 + 9$ Add 9.

$x \leq -7$

Solve.

$6x < -66$

$\dfrac{6x}{6} < \dfrac{-66}{6}$ Divide by 6. Do not reverse direction of inequality symbol.

$x < -11$

Solve.

$-6x < -66$

$\dfrac{-6x}{-6} > \dfrac{-66}{-6}$ Divide by -6. Reverse direction of inequality symbol.

$x > 11$

Solve for x:

$\dfrac{3}{7}(x - 4) \geq x + 2$

(continued)

DEFINITIONS AND CONCEPTS	EXAMPLES

Section 2.4 Linear Inequalities and Problem Solving (continued)

Steps:

1. Clear the equation of fractions.

2. Remove grouping symbols such as parentheses.

3. Simplify by combining like terms.

4. Write variable terms on one side and numbers on the other side using the addition property of inequality.

5. Isolate the variable using the multiplication property of inequality.

Step 1. $7\left[\dfrac{3}{7}(x-4)\right] \geq 7(x+2)$ Multiply by 7.

$3(x-4) \geq 7(x+2)$

Step 2. $3x - 12 \geq 7x + 14$ Apply the distributive property.

Step 3. No like terms on each side of the inequality.

Step 4. $-4x - 12 \geq 14$ Subtract $7x$.

$-4x \geq 26$ Add 12.

Step 5. $\dfrac{-4x}{-4} \leq \dfrac{26}{-4}$ Divide by -4. Reverse direction of inequality

$x \leq -\dfrac{13}{2}$ symbol.

Section 2.5 Compound Inequalities

Two inequalities joined by the words **and** or **or** are called **compound inequalities.**

The solution set of a compound inequality formed by the word **and** is the **intersection** $\cap$ of the solution sets of the two inequalities.

Compound inequalities:

$x - 7 \leq 4$ and $x \geq -21$

$2x + 7 > x - 3$ or $5x + 2 > -3$

Solve for x:

$x < 5$ and $x < 3$

$\{x \mid x < 5\}$ $(-\infty, 5)$

$\{x \mid x < 3\}$ $(-\infty, 3)$

$\{x \mid x < 3$ and $x < 5\}$ $(-\infty, 3)$

The solution set of a compound inequality formed by the word **or** is the **union**, $\cup$, of the solution sets of the two inequalities.

Solve for x:

$x - 2 \geq -3$ or $2x \leq -4$

$x \geq -1$ or $x \leq -2$

$\{x \mid x \geq -1\}$ $[-1, \infty)$

$\{x \mid x \leq -2\}$ $(-\infty, -2]$

$\{x \mid x \leq -2$ or $x \geq -1\}$ $(-\infty, -2]$ $\cup [-1, \infty)$

Section 2.6 Absolute Value Equations

If a is a positive number, then $|x| = a$ is equivalent to $x = a$ or $x = -a$.

Solve for y:

$$|5y - 1| - 7 = 4$$

$|5y - 1| = 11$

$5y - 1 = 11$ or $5y - 1 = -11$

$5y = 12$ or $5y = -10$ Add 1.

$y = \dfrac{12}{5}$ or $y = -2$ Divide by 5.

The solutions are -2 and $\dfrac{12}{5}$.

DEFINITIONS AND CONCEPTS	EXAMPLES

Section 2.6   Absolute Value Equations (continued)

If a is negative, then $|x| = a$ has no solution.

Solve for x:

$$\left|\frac{x}{2} - 7\right| = -1$$

The solution set is $\{\ \}$ or $\varnothing$.

If an absolute value equation is of the form $|x| = |y|$, solve $x = y$ or $x = -y$.

Solve for x:

$$|x - 7| = |2x + 1|$$

$x - 7 = 2x + 1$ or $x - 7 = -(2x + 1)$

$ x = 2x + 8$ $x - 7 = -2x - 1$

$ -x = 8$ $x = -2x + 6$

$ x = -8$ or $3x = 6$

$\phantom{x - 7 x = -8 \text{ or } 3} x = 2$

The solutions are -8 and 2.

Section 2.7   Absolute Value Inequalities

If a is a positive number, then $|x| < a$ is equivalent to $-a < x < a$.

Solve for y:

$$|y - 5| \le 3$$

$-3 \le y - 5 \le 3$

$-3 + 5 \le y - 5 + 5 \le 3 + 5$ Add 5.

$2 \le y \le 8$

The solution set is $[2, 8]$.

If a is a positive number, then $|x| > a$ is equivalent to $x < -a$ or $x > a$.

Solve for x:

$$\left|\frac{x}{2} - 3\right| > 7$$

$\dfrac{x}{2} - 3 < -7$ or $\dfrac{x}{2} - 3 > 7$

$x - 6 < -14$ or $x - 6 > 14$ Multiply by 2.

$x < -8$ or $x > 20$ Add 6.

The solution set is $(-\infty, -8) \cup (20, \infty)$.

Chapter 2 **Review**

(2.1) *Solve each linear equation.*

1. $4(x - 5) = 2x - 14$

2. $x + 7 = -2(x + 8)$

3. $3(2y - 1) = -8(6 + y)$

4. $-(z + 12) = 5(2z - 1)$

5. $n - (8 + 4n) = 2(3n - 4)$

6. $4(9v + 2) = 6(1 + 6v) - 10$

7. $0.3(x - 2) = 1.2$

8. $1.5 = 0.2(c - 0.3)$

9. $-4(2 - 3x) = 2(3x - 4) + 6x$

10. $6(m - 1) + 3(2 - m) = 0$

11. $6 - 3(2g + 4) - 4g = 5(1 - 2g)$

12. $20 - 5(p + 1) + 3p = -(2p - 15)$

13. $\dfrac{x}{3} - 4 = x - 2$

14. $\dfrac{9}{4}y = \dfrac{2}{3}y$

15. $\dfrac{3n}{8} - 1 = 3 + \dfrac{n}{6}$

16. $\dfrac{z}{6} + 1 = \dfrac{z}{2} + 2$

17. $\dfrac{y}{4} - \dfrac{y}{2} = -8$

18. $\dfrac{2x}{3} - \dfrac{8}{3} = x$

19. $\dfrac{b - 2}{3} = \dfrac{b + 2}{5}$

20. $\dfrac{2t - 1}{3} = \dfrac{3t + 2}{15}$

21. $\dfrac{2(t + 1)}{3} = \dfrac{2(t - 1)}{3}$

22. $\dfrac{3a - 3}{6} = \dfrac{4a + 1}{15} + 2$

(2.2) Solve.

23. Twice the difference of a number and 3 is the same as 1 added to three times the number. Find the number.

24. One number is 5 more than another number. If the sum of the numbers is 285, find the numbers.

25. Find 40% of 130.

26. Find 1.5% of 8.

27. The length of a rectangular playing field is 5 meters less than twice its width. If 230 meters of fencing goes around the field, find the dimensions of the field.

28. In 2008, the median earnings of young adults with bachelor's degrees were $46,000. This represents a 28% increase over the median earnings of young adults with associate's degrees. Find the median earnings of young adults with associate's degrees in 2008. Round to the nearest whole dollar. (*Source: National Center for Educational Statistics*)

29. Find four consecutive integers such that twice the first subtracted from the sum of the other three integers is 16.

30. Determine whether there are two consecutive odd integers such that 5 times the first exceeds 3 times the second by 54.

31. A car rental company charges $19.95 per day for a compact car plus 12 cents per mile for every mile over 100 miles driven per day. If Mr. Woo's bill for 2 days' use is $46.86, find how many miles he drove.

32. The cost C of producing x number of scientific calculators is given by $C = 4.50x + 3000$, and the revenue R from selling them is given by $R = 16.50x$. Find the number of calculators that must be sold to break even. (Recall that to break even, revenue = cost.)

(2.3) Solve each equation for the specified variable.

△ **33.** $V = lwh$ for w

△ **34.** $C = 2\pi r$ for r

35. $5x - 4y = -12$ for y

36. $5x - 4y = -12$ for x

37. $y - y_1 = m(x - x_1)$ for m

38. $y - y_1 = m(x - x_1)$ for x

39. $E = I(R + r)$ for r

40. $S = vt + gt^2$ for g

41. $T = gr + gvt$ for g

42. $I = Prt + P$ for P

43. A principal of $3000 is invested in an account paying an annual percentage rate of 3%. Find the amount (to the nearest cent) in the account after 7 years if the amount is compounded

   a. semiannually.

   b. weekly.

44. The high temperature in Slidell, Louisiana, one day was 90° Fahrenheit. Convert this temperature to degrees Celsius.

△ **45.** Angie Applegate has a photograph for which the length is 2 inches longer than the width. If she increases each dimension by 4 inches, the area is increased by 88 square inches. Find the original dimensions.

△ **46.** One-square-foot floor tiles come 24 to a package. Find how many packages are needed to cover a rectangular floor 18 feet by 21 feet.

(2.4) Solve each linear inequality. Write your answers in interval notation.

47. $3(x - 5) > -(x + 3)$

48. $-2(x + 7) \geq 3(x + 2)$

49. $4x - (5 + 2x) < 3x - 1$

50. $3(x - 8) < 7x + 2(5 - x)$

51. $24 \geq 6x - 2(3x - 5) + 2x$

52. $\dfrac{x}{3} + \dfrac{1}{2} > \dfrac{2}{3}$

53. $x + \dfrac{3}{4} < -\dfrac{x}{2} + \dfrac{9}{4}$

54. $\dfrac{x - 5}{2} \leq \dfrac{3}{8}(2x + 6)$

Solve.

55. George Boros can pay his housekeeper $15 per week to do his laundry, or he can have the laundromat do it at a cost of 50 cents per pound for the first 10 pounds and 40 cents for each additional pound. Use an inequality to find the weight at which it is more economical to use the housekeeper than the laundromat.

56. In the Olympic gymnastics competition, Nana must average a score of 9.65 to win the silver medal. Seven of the eight judges have reported scores of 9.5, 9.7, 9.9, 9.7, 9.7, 9.6, and 9.5. Use an inequality to find the minimum score that Nana must receive from the last judge to win the silver medal.

(2.5) Solve each inequality. Write your answers in interval notation.

57. $1 \leq 4x - 7 \leq 3$

58. $-2 \leq 8 + 5x < -1$

59. $-3 < 4(2x - 1) < 12$

60. $-6 < x - (3 - 4x) < -3$

61. $\dfrac{1}{6} < \dfrac{4x - 3}{3} \leq \dfrac{4}{5}$

62. $x \leq 2$ and $x > -5$

63. $3x - 5 > 6$ or $-x < -5$

64. Ceramic firing temperatures usually range from 500° to 1000° Fahrenheit. Use a compound inequality to convert this range to the Celsius scale. Round to the nearest degree.

65. Carol would like to pay cash for a car when she graduates from college and estimates that she can afford a car that costs between $4000 and $8000. She has saved $500 so far and plans to earn the rest of the money by working the next two summers. If Carol plans to save the same amount each summer, use a compound inequality to find the range of money she must save each summer to buy the car.

(2.6) Solve each absolute value equation.

66. $|x - 7| = 9$

67. $|8 - x| = 3$

68. $|2x + 9| = 9$

69. $|-3x + 4| = 7$

70. $|3x - 2| + 6 = 10$

71. $5 + |6x + 1| = 5$

72. $-5 = |4x - 3|$

73. $|5 - 6x| + 8 = 3$

74. $-8 = |x - 3| - 10$

75. $\left|\dfrac{3x - 7}{4}\right| = 2$

76. $|6x + 1| = |15 + 4x|$

(2.7) Solve each absolute value inequality. Graph the solution set and write it in interval notation.

77. $|5x - 1| < 9$

78. $|6 + 4x| \geq 10$

79. $|3x| - 8 > 1$

80. $9 + |5x| < 24$

81. $|6x - 5| \leq -1$

82. $\left|3x + \dfrac{2}{5}\right| \geq 4$

83. $\left|\dfrac{x}{3} + 6\right| - 8 > -5$

84. $\left|\dfrac{4(x - 1)}{7}\right| + 10 < 2$

MIXED REVIEW

Solve.

85. $\dfrac{x - 2}{5} + \dfrac{x + 2}{2} = \dfrac{x + 4}{3}$

86. $\dfrac{2z - 3}{4} - \dfrac{4 - z}{2} = \dfrac{z + 1}{3}$

△ **87.** $A = \dfrac{h}{2}(B + b)$ for B

△ **88.** $V = \dfrac{1}{3}\pi r^2 h$ for h

89. China, the United States, and France are predicted to be the top tourist destinations by 2020. In this year, the United States is predicted to have 9 million more tourists than France, and China is predicted to have 44 million more tourists than France. If the total number of tourists predicted for these three countries is 332 million, find the number predicted for each country in 2020.

90. Erasmos Gonzalez left Los Angeles at 11 a.m. and drove nonstop to San Diego, 130 miles away. If he arrived at 1:15 p.m., find his average speed, rounded to the nearest mile per hour.

△ **91.** Determine which container holds more ice cream, an 8 inch × 5 inch × 3 inch box or a cylinder with radius of 3 inches and height of 6 inches.

Solve. If an inequality, write your solutions in interval notation.

92. $48 + x \geq 5(2x + 4) - 2x$

93. $\dfrac{3(x - 2)}{5} > \dfrac{-5(x - 2)}{3}$

94. $0 \leq \dfrac{2(3x + 4)}{5} \leq 3$

95. $x \leq 2$ or $x > -5$

96. $-2x \leq 6$ and $-2x + 3 < -7$

97. $|7x| - 26 = -5$

98. $\left|\dfrac{9 - 2x}{5}\right| = -3$

99. $|x - 3| = |7 + 2x|$

100. $|6x - 5| \geq -1$

101. $\left|\dfrac{4x - 3}{5}\right| < 1$

Chapter 2 Test   MyMathLab®    Test Prep VIDEOS   You Tube™

Solve each equation.

▶ **1.** $8x + 14 = 5x + 44$

▶ **2.** $9(x + 2) = 5[11 - 2(2 - x) + 3]$

▶ **3.** $3(y - 4) + y = 2(6 + 2y)$

▶ **4.** $7n - 6 + n = 2(4n - 3)$

▶ **5.** $\dfrac{7w}{4} + 5 = \dfrac{3w}{10} + 1$

▶ **6.** $\dfrac{z + 7}{9} + 1 = \dfrac{2z + 1}{6}$

▶ **7.** $|6x - 5| - 3 = -2$

▶ **8.** $|8 - 2t| = -6$

▶ **9.** $|2x - 3| = |4x + 5|$

▶ **10.** $|x - 5| = |x + 2|$

Solve each equation for the specified variable.

▶ **11.** $3x - 4y = 8$ for y

▶ **12.** $S = gt^2 + gvt$ for g

▶ **13.** $F = \dfrac{9}{5}C + 32$ for C

Solve each inequality. Write your solutions in interval notation.

▶ **14.** $3(2x - 7) - 4x > -(x + 6)$

▶ **15.** $\dfrac{3x - 2}{3} - \dfrac{5x + 1}{4} \geq 0$

16. $-3 < 2(x - 3) \le 4$

17. $|3x + 1| > 5$

18. $|x - 5| - 4 < -2$

19. $x \ge 5$ and $x \ge 4$

20. $x \ge 5$ or $x \ge 4$

21. $-1 \le \dfrac{2x - 5}{3} < 2$

22. $6x + 1 > 5x + 4$ or $1 - x > -4$

23. Find 12% of 80.

Solve.

24. In 2009, Ford sold 4,817,000 new vehicles worldwide. This represents a 28.32% decrease over the number of new vehicles sold by Ford in 2003. Use this information to find the number of new vehicles sold by Ford in 2003. Round to the nearest thousand. (*Source:* Ford Motor Company)

25. A circular dog pen has a circumference of 78.5 feet. Approximate π by 3.14 and estimate how many hunting dogs could be safely kept in the pen if each dog needs at least 60 square feet of room.

26. In 2018, the number of people employed as registered nurses in the United States is expected to be 3,200,000. This represents a 22% increase over the number of people employed as registered nurses in 2008. Find the number of registered nurses employed in 2008. Round to the nearest thousand. (*Source:* U.S. Bureau of Labor Statistics)

27. Find the amount of money in an account after 10 years if a principal of $2500 is invested at 3.5% interest compounded quarterly. (Round to the nearest cent.)

28. The three states where international travelers spend the most money are Florida, California, and New York. International travelers spend $4 billion more money in California than New York, and in Florida they spend $1 billion less than twice the amount spent in New York. If total international spending in these three states is $39 billion, find the amount spent in each state. (*Source:* Travel Industry Asso. of America)

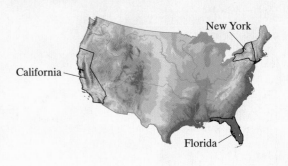

New York

California

Florida

Chapter 2 **Cumulative Review**

List the elements in each set.

1. a. $\{x \mid x \text{ is a natural number greater than } 100\}$
   b. $\{x \mid x \text{ is a whole number between 1 and 6}\}$

2. a. $\{x \mid x \text{ is an integer between } -3 \text{ and } 5\}$
   b. $\{x \mid x \text{ is a whole number between 3 and 5}\}$

3. Find each absolute value.

   a. $|3|$ **b.** $\left|-\dfrac{1}{7}\right|$

   c. $-|2.7|$ **d.** $-|-8|$

   e. $|0|$

4. Find the opposite of each number.

   a. $\dfrac{2}{3}$ **b.** -9 **c.** 1.5

5. Add.

   a. $-3 + (-11)$ **b.** $3 + (-7)$
   c. $-10 + 15$ **d.** $-8.3 + (-1.9)$
   e. $-\dfrac{1}{4} + \dfrac{1}{2}$ **f.** $-\dfrac{2}{3} + \dfrac{3}{7}$

6. Subtract.

   a. $-2 - (-10)$ **b.** $1.7 - 8.9$
   c. $-\dfrac{1}{2} - \dfrac{1}{4}$

7. Find the square roots.

   a. $\sqrt{9}$ **b.** $\sqrt{25}$

   c. $\sqrt{\dfrac{1}{4}}$ **d.** $-\sqrt{36}$

   e. $\sqrt{-36}$

8. Multiply or divide.

   a. $-3(-2)$ **b.** $-\dfrac{3}{4}\left(-\dfrac{4}{7}\right)$

   c. $\dfrac{0}{-2}$ **d.** $\dfrac{-20}{-2}$

9. Evaluate each algebraic expression when $x = 4$ and $y = -3$.
   a. $3x - 7y$ **b.** $-2y^2$
   c. $\dfrac{\sqrt{x}}{y} - \dfrac{y}{x}$

10. Find the roots.

   a. $\sqrt[4]{1}$ **b.** $\sqrt[3]{8}$ **c.** $\sqrt[4]{81}$

11. Write each sentence as an equation.

   a. The sum of x and 5 is 20.
   b. Two times the sum of 3 and y amounts to 4.
   c. The difference of 8 and x is the same as the product of 2 and x.
   d. The quotient of z and 9 amounts to 9 plus z.

12. Insert $<$, $>$, or $=$ between each pair of numbers to form a true statement.
    a. $-3 \quad -5$
    b. $\dfrac{-12}{-4} \quad 3$
    c. $0 \quad -2$

13. Use the commutative property of addition to write an expression equivalent to $7x + 5$.

14. Use the associative property of multiplication to write an expression equivalent to $5 \cdot (7x)$. Then simplify the expression.

Solve for x.

15. $2x + 5 = 9$

16. $11.2 = 1.2 - 5x$

17. $6x - 4 = 2 + 6(x - 1)$

18. $2x + 1.5 = -0.2 + 1.6x$

19. Write the following as algebraic expressions. Then simplify.
    a. The sum of three consecutive integers if x is the first consecutive integer
    b. The perimeter of a triangle with sides of length x, $5x$, and $6x - 3$

20. Write the following as algebraic expressions. Then simplify.
    a. The sum of three consecutive even integers if x is the first consecutive integers
    b. The perimeter of a square with side length $3x + 1$

21. Find three numbers such that the second number is 3 more than twice the first number, and the third number is four times the first number. The sum of the three numbers is 164.

22. Find two numbers such that the second number is 2 more than three times the first number, and the difference of the two numbers is 24.

23. Solve $3y - 2x = 7$ for y.

24. Solve $7x - 4y = 10$ for x.

25. Solve $A = \dfrac{1}{2}(B + b)h$ for b.

26. Solve $P = 2l + 2w$ for l.

27. Graph each set on a number line and then write in interval notation.
    a. $\{x \mid x \geq 2\}$
    b. $\{x \mid x < -1\}$
    c. $\{x \mid 0.5 < x \leq 3\}$

28. Graph each set on a number line and then write in interval notation.
    a. $\{x \mid x \leq -3\}$
    b. $\{x \mid -2 \leq x < 0.1\}$

Solve.

29. $-(x - 3) + 2 \leq 3(2x - 5) + x$

30. $2(7x - 1) - 5x > -(-7x) + 4$

31. $2(x + 3) > 2x + 1$

32. $4(x + 1) - 3 < 4x + 1$

33. If $A = \{x \mid x$ is an even number greater than 0 and less than 10$\}$ and $B = \{3, 4, 5, 6\}$, find $A \cap B$.

34. Find the union: $\{-2, 0, 2, 4\} \cup \{-1, 1, 3, 5\}$

35. Solve: $x - 7 < 2 \; and \; 2x + 1 < 9$

36. Solve: $x + 3 \leq 1 \; or \; 3x - 1 < 8$

37. If $A = \{x \mid x$ is an even number greater than 0 and less than 10$\}$ and $B = \{3, 4, 5, 6\}$, find $A \cup B$.

38. Find the intersection: $\{-2, 0, 2, 4\} \cap \{-1, 1, 3, 5\}$

39. Solve: $-2x - 5 < -3 \; or \; 6x < 0$

40. Solve: $-2x - 5 < -3 \; and \; 6x < 0$

Solve.

41. $|p| = 2$

42. $|x| = 5$

43. $\left| \dfrac{x}{2} - 1 \right| = 11$

44. $\left| \dfrac{y}{3} + 2 \right| = 10$

45. $|x - 3| = |5 - x|$

46. $|x + 3| = |7 - x|$

47. $|x| \leq 3$

48. $|x| > 1$

49. $|2x + 9| + 5 > 3$

50. $|3x + 1| + 9 < 1$

Graphs and Functions

The linear equations and inequalities we explored in Chapter 2 are statements about a single variable. This chapter examines statements about two variables: linear equations and inequalities in two variables. We focus particularly on graphs of those equations and inequalities that lead to the notion of relation and to the notion of function, perhaps the single most important and useful concept in all of mathematics.

We define online courses as courses in which at least 80% of the content is delivered online. Although there are many types of course delivery used by instructors, the bar graph below shows the increase in percent of students taking at least one online course. Notice that the two functions, $f(x)$ and $g(x)$, both approximate the percent of students taking at least one online course. Also, for both functions, x is the number of years since 2000. In Section 3.2, Exercises 81–86, we use these functions to predict the growth of online courses.

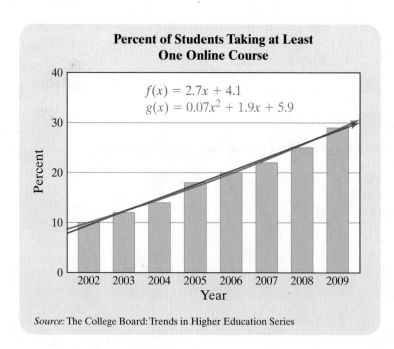

Percent of Students Taking at Least One Online Course

$f(x) = 2.7x + 4.1$
$g(x) = 0.07x^2 + 1.9x + 5.9$

Source: The College Board: Trends in Higher Education Series

3.1 | Graphing Equations

OBJECTIVES

1 Plot Ordered Pairs.

2 Determine Whether an Ordered Pair of Numbers Is a Solution to an Equation in Two Variables.

3 Graph Linear Equations.

4 Graph Nonlinear Equations.

Graphs are widely used today in newspapers, magazines, and all forms of newsletters. A few examples of graphs are shown here.

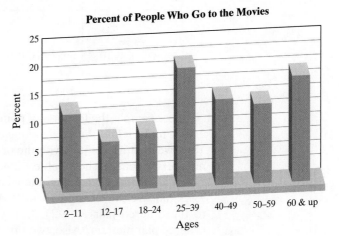

Percent of People Who Go to the Movies

Source: Motion Picture Association of America

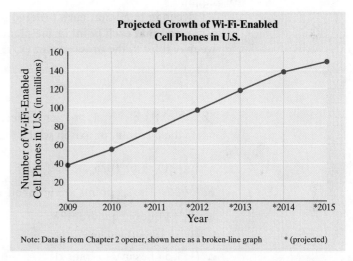

Projected Growth of Wi-Fi-Enabled Cell Phones in U.S.

Note: Data is from Chapter 2 opener, shown here as a broken-line graph   * (projected)

To help us understand how to read these graphs, we will review their basis—the rectangular coordinate system.

OBJECTIVE

1   Plotting Ordered Pairs on a Rectangular Coordinate System

One way to locate points on a plane is by using a **rectangular coordinate system,** which is also called a **Cartesian coordinate system** after its inventor, René Descartes (1596–1650).

A rectangular coordinate system consists of two number lines that intersect at right angles at their 0 coordinates. We position these axes on paper such that one number line is horizontal and the other number line is then vertical. The horizontal number line is called the ***x*-axis** (or the axis of the **abscissa**), and the vertical number line is called the ***y*-axis** (or the axis of the **ordinate**). The point of intersection of these axes is named the **origin.**

Notice in the left figure on the next page that the axes divide the plane into four regions. These regions are called **quadrants.** The top-right region is quadrant I. Quadrants II, III, and IV are numbered counterclockwise from the first quadrant as shown. The *x*-axis and the *y*-axis are not in any quadrant.

Each point in the plane can be located, or **plotted,** or graphed by describing its position in terms of distances along each axis from the origin. An **ordered pair,** represented by the notation (x, y), records these distances.

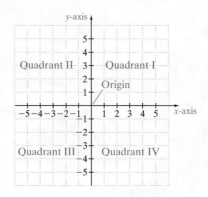

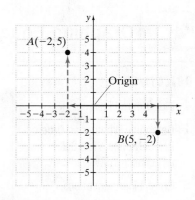

For example, the location of point A in the above figure on the right is described as 2 units to the left of the origin along the x-axis and 5 units upward parallel to the y-axis. Thus, we identify point A with the ordered pair $(-2, 5)$. Notice that the order of these numbers is *critical*. The x-value -2 is called the **x-coordinate** and is associated with the x-axis. The y-value 5 is called the **y-coordinate** and is associated with the y-axis.

Compare the location of point A with the location of point B, which corresponds to the ordered pair $(5, -2)$. Can you see that the order of the coordinates of an ordered pair matters? Also, two ordered pairs are considered equal and correspond to the same point if and only if their x-coordinates are equal and their y-coordinates are equal.

Keep in mind that **each ordered pair corresponds to exactly one point in the real plane and that each point in the plane corresponds to exactly one ordered pair.** Thus, we may refer to the ordered pair (x, y) as the point (x, y).

EXAMPLE 1 Plot each ordered pair on a Cartesian coordinate system and name the quadrant or axis in which the point is located.

a. $(2, -1)$ **b.** $(0, 5)$ **c.** $(-3, 5)$ **d.** $(-2, 0)$ **e.** $\left(-\dfrac{1}{2}, -4\right)$ **f.** $(1.5, 1.5)$

Solution The six points are graphed as shown.

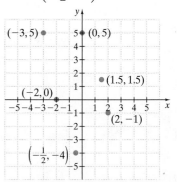

a. $(2, -1)$ lies in quadrant IV.

b. $(0, 5)$ is on the y-axis.

c. $(-3, 5)$ lies in quadrant II.

d. $(-2, 0)$ is on the x-axis.

e. $\left(-\dfrac{1}{2}, -4\right)$ is in quadrant III.

f. $(1.5, 1.5)$ is in quadrant I.

PRACTICE

1 Plot each ordered pair on a Cartesian coordinate system and name the quadrant or axis in which the point is located.

a. $(3, -4)$ **b.** $(0, -2)$ **c.** $(-2, 4)$ **d.** $(4, 0)$ **e.** $\left(-1\dfrac{1}{2}, -2\right)$ **f.** $(2.5, 3.5)$

Notice that the y-coordinate of any point on the x-axis is 0. For example, the point with coordinates $(-2, 0)$ lies on the x-axis. Also, the x-coordinate of any point on the y-axis is 0. For example, the point with coordinates $(0, 5)$ lies on the y-axis. These points that lie on the axes do not lie in any quadrants.

✓ CONCEPT CHECK
Which of the following correctly describes the location of the point $(3, -6)$ in a rectangular coordinate system?

a. 3 units to the left of the y-axis and 6 units above the x-axis
b. 3 units above the x-axis and 6 units to the left of the y-axis
c. 3 units to the right of the y-axis and 6 units below the x-axis
d. 3 units below the x-axis and 6 units to the right of the y-axis

Many types of real-world data occur in pairs. Study the graph below and notice the paired data $(2013, 57)$ and the corresponding plotted point, both in blue.

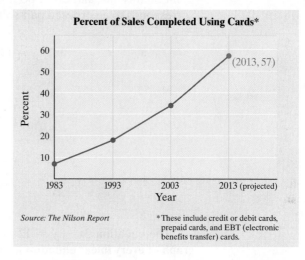

Percent of Sales Completed Using Cards*

Source: The Nilson Report

*These include credit or debit cards, prepaid cards, and EBT (electronic benefits transfer) cards.

This paired data point, $(2013, 57)$, means that in the year 2013, it is predicted that 57% of sales will be completed using some type of card (credit, debit, etc.).

OBJECTIVE

2   Determining Whether an Ordered Pair Is a Solution

Solutions of equations in two variables consist of two numbers that form a true statement when substituted into the equation. A convenient notation for writing these numbers is as ordered pairs. A solution of an equation containing the variables x and y is written as a pair of numbers in the order (x, y). If the equation contains other variables, we will write ordered pair solutions in alphabetical order.

EXAMPLE 2   Determine whether $(0, -12)$, $(1, 9)$, and $(2, -6)$ are solutions of the equation $3x - y = 12$.

Solution   To check each ordered pair, replace x with the x-coordinate and y with the y-coordinate and see whether a true statement results.

Let $x = 0$ and $y = -12$.

$$3x - y = 12$$
$$3(0) - (-12) \stackrel{?}{=} 12$$
$$0 + 12 \stackrel{?}{=} 12$$
$$12 = 12 \quad \text{True}$$

Let $x = 1$ and $y = 9$.

$$3x - y = 12$$
$$3(1) - 9 \stackrel{?}{=} 12$$
$$3 - 9 \stackrel{?}{=} 12$$
$$-6 = 12 \quad \text{False}$$

Let $x = 2$ and $y = -6$.

$$3x - y = 12$$
$$3(2) - (-6) \stackrel{?}{=} 12$$
$$6 + 6 \stackrel{?}{=} 12$$
$$12 = 12 \quad \text{True}$$

Thus, $(1, 9)$ is not a solution of $3x - y = 12$, but both $(0, -12)$ and $(2, -6)$ are solutions.   □

PRACTICE

2   Determine whether $(1, 4)$, $(0, 6)$, and $(3, -4)$ are solutions of the equation $4x + y = 8$.

Answer to Concept Check: **c**

OBJECTIVE

3 **Graphing Linear Equations**

The equation $3x - y = 12$, from Example 2, actually has an infinite number of ordered pair solutions. Since it is impossible to list all solutions, we visualize them by graphing.

A few more ordered pairs that satisfy $3x - y = 12$ are $(4, 0)$, $(3, -3)$, $(5, 3)$, and $(1, -9)$. These ordered pair solutions along with the ordered pair solutions from Example 2 are plotted on the following graph.

The graph of $3x - y = 12$ is the single line containing these points. Every ordered pair solution of the equation corresponds to a point on this line, and every point on this line corresponds to an ordered pair solution.

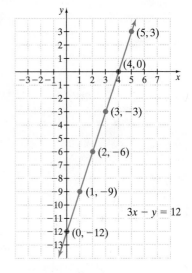

x	y	$3x - y = 12$
5	3	$3 \cdot 5 - 3 = 12$
4	0	$3 \cdot 4 - 0 = 12$
3	-3	$3 \cdot 3 - (-3) = 12$
2	-6	$3 \cdot 2 - (-6) = 12$
1	-9	$3 \cdot 1 - (-9) = 12$
0	-12	$3 \cdot 0 - (-12) = 12$

The equation $3x - y = 12$ is called a linear equation in two variables, and **the graph of every linear equation in two variables is a line.**

Linear Equation in Two Variables

A **linear equation in two variables** is an equation that can be written in the form

$$Ax + By = C$$

where A and B are not both 0. This form is called **standard form.**

Some examples of equations in standard form:

$$3x - y = 12$$
$$-2.1x + 5.6y = 0$$

▶ Helpful Hint

Remember: A linear equation is written in standard form when all of the variable terms are on one side of the equation and the constant is on the other side.

Many real-life applications are modeled by linear equations. Suppose you have a part-time job at a store that sells office products.

Your pay is $3000 plus 20% or $\frac{1}{5}$ of the price of the products you sell. If we let x represent products sold and y represent monthly salary, the linear equation that models your salary is

$$y = 3000 + \frac{1}{5}x$$

(Although this equation is not written in standard form, it is a linear equation. To see this, subtract $\frac{1}{5}x$ from both sides.)

Some ordered pair solutions of this equation are below.

Products Sold	x	0	1000	2000	3000	4000	10,000
Monthly Salary	y	3000	3200	3400	3600	3800	5000

For example, we say that the ordered pair $(1000, 3200)$ is a solution of the equation $y = 3000 + \frac{1}{5}x$ because when x is replaced with 1000 and y is replaced with 3200, a true statement results.

$$y = 3000 + \frac{1}{5}x$$

$$3200 \overset{?}{=} 3000 + \frac{1}{5}(1000) \quad \text{Let } x = 1000 \text{ and } y = 3200.$$

$$3200 \overset{?}{=} 3000 + 200$$

$$3200 = 3200 \qquad \text{True}$$

A portion of the graph of $y = 3000 + \frac{1}{5}x$ is shown in the next example.

Since we assume that the smallest amount of product sold is none, or 0, then x must be greater than or equal to 0. Therefore, only the part of the graph that lies in quadrant I is shown. Notice that the graph gives a visual picture of the correspondence between products sold and salary.

> ▶ Helpful Hint
> A line contains an infinite number of points and each point corresponds to an ordered pair that is a solution of its corresponding equation.

EXAMPLE 3   Use the graph of $y = 3000 + \frac{1}{5}x$ to answer the following questions.

a. If the salesperson sells \$8000 of products in a particular month, what is the salary for that month?

b. If the salesperson wants to make more than \$5000 per month, what must be the total amount of products sold?

Solution

a. Since x is products sold, find 8000 along the x-axis and move vertically up until you reach a point on the line. From this point on the line, move horizontally to the left until you reach the y-axis. Its value on the y-axis is 4600, which means if \$8000 worth of products is sold, the salary for the month is \$4600.

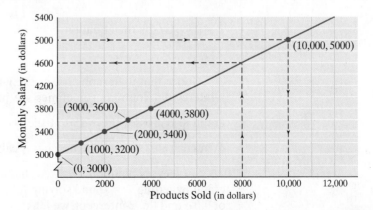

b. Since y is monthly salary, find 5000 along the y-axis and move horizontally to the right until you reach a point on the line. Either read the corresponding x-value from the labeled ordered pair or move vertically downward until you reach the x-axis. The corresponding x-value is 10,000. This means that $10,000 worth of products sold gives a salary of $5000 for the month. For the salary to be greater than $5000, products sold must be greater that $10,000. ☐

PRACTICE
3   Use the graph in Example 3 to answer the following questions.

a. If the salesperson sells $6000 of products in a particular month, what is the salary for that month?

b. If the salesperson wants to make more than $4800 per month, what must be the total amount of products sold?

· ■

Recall from geometry that a line is determined by two points. This means that to graph a linear equation in two variables, just two solutions are needed. We will find a third solution, just to check our work. To find ordered pair solutions of linear equations in two variables, we can choose an x-value and find its corresponding y-value, or we can choose a y-value and find its corresponding x-value. The number 0 is often a convenient value to choose for x and for y.

EXAMPLE 4   Graph the equation $y = -2x + 3$.

Solution   This is a linear equation. (In standard form it is $2x + y = 3$.) Find three ordered pair solutions, and plot the ordered pairs. The line through the plotted points is the graph. Since the equation is solved for y, let's choose three x-values. We'll choose $0, 2$, and then -1 for x to find our three ordered pair solutions.

Let $x = 0$	Let $x = 2$	Let $x = -1$
$y = -2x + 3$	$y = -2x + 3$	$y = -2x + 3$
$y = -2 \cdot 0 + 3$	$y = -2 \cdot 2 + 3$	$y = -2(-1) + 3$
$y = 3$ Simplify.	$y = -1$ Simplify.	$y = 5$ Simplify.

The three ordered pairs $(0, 3)$, $(2, -1)$, and $(-1, 5)$ are listed in the table, and the graph is shown.

x	y
0	3
2	-1
-1	5

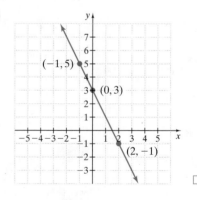

PRACTICE
4   Graph the equation $y = -3x - 2$.

· ■

Notice that the graph crosses the y-axis at the point $(0, 3)$. This point is called the **y-intercept.** (You may sometimes see just the number 3 called the y-intercept.) This graph also crosses the x-axis at the point $\left(\frac{3}{2}, 0\right)$. This point is called the **x-intercept.** (You may also see just the number $\frac{3}{2}$ called the x-intercept.)

Since every point on the y-axis has an x-value of 0, we can find the y-intercept of a graph by letting $x = 0$ and solving for y. Also, every point on the x-axis has a y-value of 0. To find the x-intercept, we let $y = 0$ and solve for x.

> **Finding *x*- and *y*-Intercepts**
>
> To find an *x*-intercept, let $y = 0$ and solve for *x*.
> To find a *y*-intercept, let $x = 0$ and solve for *y*.

We will study intercepts further in Section 3.3.

EXAMPLE 5 Graph the linear equation $y = \dfrac{1}{3}x$.

Solution To graph, we find ordered pair solutions, plot the ordered pairs, and draw a line through the plotted points. We will choose *x*-values and substitute in the equation. To avoid fractions, we choose *x*-values that are multiples of 3. To find the *y*-intercept, we let $x = 0$.

If $x = 0$, then $y = \dfrac{1}{3}(0)$, or 0.

If $x = 6$, then $y = \dfrac{1}{3}(6)$, or 2.

If $x = -3$, then $y = \dfrac{1}{3}(-3)$, or -1.

$y = \dfrac{1}{3}x$

x	*y*
0	0
6	2
−3	−1

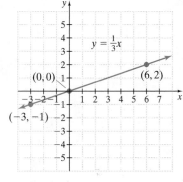

> **Helpful Hint**
>
> Notice that by using multiples of 3 for *x*, we avoid fractions.

> **Helpful Hint**
>
> Since the equation $y = \dfrac{1}{3}x$ is solved for *y*, we choose *x*-values for finding points. This way, we simply need to evaluate an expression to find the *x*-value, as shown.

This graph crosses the *x*-axis at $(0, 0)$ and the *y*-axis at $(0, 0)$. This means that the *x*-intercept is $(0, 0)$ and that the *y*-intercept is $(0, 0)$.

PRACTICE
5 Graph the linear equation $y = -\dfrac{1}{2}x$.

OBJECTIVE
4 Graphing Nonlinear Equations

Not all equations in two variables are linear equations, and not all graphs of equations in two variables are lines.

EXAMPLE 6 Graph $y = x^2$.

Solution This equation is not linear because the x^2 term does not allow us to write it in the form $Ax + By = C$. Its graph is not a line. We begin by finding ordered pair solutions. Because this graph is solved for *y*, we choose *x*-values and find corresponding *y*-values.

If $x = -3$, then $y = (-3)^2$, or 9.

If $x = -2$, then $y = (-2)^2$, or 4.

If $x = -1$, then $y = (-1)^2$, or 1.

If $x = 0$, then $y = 0^2$, or 0.

If $x = 1$, then $y = 1^2$, or 1.

If $x = 2$, then $y = 2^2$, or 4.

If $x = 3$, then $y = 3^2$, or 9.

x	*y*
−3	9
−2	4
−1	1
0	0
1	1
2	4
3	9

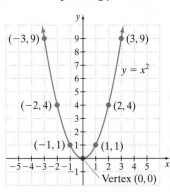

Study the table a moment and look for patterns. Notice that the ordered pair solution $(0, 0)$ contains the smallest *y*-value because any other *x*-value squared will give a positive result. This means that the point $(0, 0)$ will be the lowest point on the graph. Also notice that all other *y*-values correspond to two different *x*-values. For example, $3^2 = 9$, and also $(-3)^2 = 9$. This means that the graph will be a mirror image of itself across the *y*-axis. Connect the plotted points with a smooth curve to sketch the graph.

This curve is given a special name, **a parabola.** We will study more about parabolas in later chapters. ☐

PRACTICE
6   Graph $y = 2x^2$. ■

EXAMPLE 7   Graph the equation $y = |x|$.

Solution   This is not a linear equation since it cannot be written in the form $Ax + By = C$. Its graph is not a line. Because we do not know the shape of this graph, we find many ordered pair solutions. We will choose x-values and substitute to find corresponding y-values.

If $x = -3$, then $y = |-3|$, or 3.

If $x = -2$, then $y = |-2|$, or 2.

If $x = -1$, then $y = |-1|$, or 1.

If $x = 0$, then $y = |0|$, or 0.

If $x = 1$, then $y = |1|$, or 1.

If $x = 2$, then $y = |2|$, or 2.

If $x = 3$, then $y = |3|$, or 3.

x	y
−3	3
−2	2
−1	1
0	0
1	1
2	2
3	3

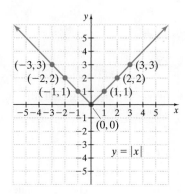

Again, study the table of values for a moment and notice any patterns.

From the plotted ordered pairs, we see that the graph of this absolute value equation is V-shaped. ☐

PRACTICE
7   Graph $y = -|x|$. ■

Graphing Calculator Explorations

In this section, we begin a study of graphing calculators and graphing software packages for computers. These graphers use the same point plotting technique that we introduced in this section. The advantage of this graphing technology is, of course, that graphing calculators and computers can find and plot ordered pair solutions much faster than we can. Note, however, that the features described in these boxes may not be available on all graphing calculators.

The rectangular screen where a portion of the rectangular coordinate system is displayed is called a **window.** We call it a **standard window** for graphing when both the x- and y-axes display coordinates between −10 and 10. This information is often displayed in the window menu on a graphing calculator as

Xmin = −10

Xmax = 10

Xscl = 1   The scale on the x-axis is one unit per tick mark.

Ymin = −10

Ymax = 10

Yscl = 1   The scale on the y-axis is one unit per tick mark.

To use a graphing calculator to graph the equation $y = -5x + 4$, press the $\boxed{Y =}$ key and enter the keystrokes

.

↑

(Check your owner's manual to make sure the "negative" key is pressed here and not the "subtraction" key.)

The top row should now read $Y_1 = -5x + 4$. Next, press the GRAPH key, and the display should look like this:

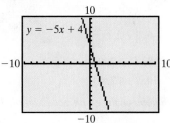

Use a standard window and graph the following equations. (Unless otherwise stated, we will use a standard window when graphing.)

1. $y = -3.2x + 7.9$ **2.** $y = -x + 5.85$

3. $y = \dfrac{1}{4}x - \dfrac{2}{3}$ **4.** $y = \dfrac{2}{3}x - \dfrac{1}{5}$

5. $y = |x - 3| + 2$ **6.** $y = |x + 1| - 1$

7. $y = x^2 + 3$ **8.** $y = (x + 3)^2$

Vocabulary, Readiness & Video Check

Use the choices below to fill in each blank. Some choices may not be used.

line	parabola	1	3
origin	V-shaped	2	4

1. The intersection of the *x*-axis and *y*-axis is a point called the _____.

2. The rectangular coordinate system has _____ quadrants and _____ axes.

3. The graph of a single ordered pair of numbers is how many points? _____

4. The graph of $Ax + By = C$, where A and B are not both 0, is a(n) _____.

5. The graph of $y = |x|$ looks _____.

6. The graph of $y = x^2$ is a _____.

Martin-Gay Interactive Videos

See Video 3.1

Watch the section lecture video and answer the following questions.

OBJECTIVE 1

7. Several points are plotted in ⊞ Examples 1–3. Where do you start when using this method to plot a point? How does the 1st coordinate tell you to move? How does the 2nd coordinate tell you to move? Include the role of signs in your answer.

OBJECTIVE 2

8. Based on ⊞ Examples 4 and 5, complete the following statement. An ordered pair is a solution of an equation in _____ variables if, when the variables are replaced with their ordered pair values, a _____ statement results.

OBJECTIVE 3

9. From ⊞ Example 6 and the lecture before, what is the graph of an equation? If you need only two points to determine a line, why are three ordered pair solutions or points found for a linear equation?

OBJECTIVE 4

10. Based on ⊞ Examples 7 and 8, complete the following statements. When graphing a nonlinear equation, first recognize it as a nonlinear equation and know that the graph is _____ a line. If you don't know the _____ of the graph, plot enough points until you see a pattern.

3.1 Exercise Set   MyMathLab®

Plot each point and name the quadrant or axis in which the point lies. See Example 1.

1. $(3, 2)$ **2.** $(2, -1)$

3. $(-5, 3)$ **4.** $(-3, -1)$

5. $\left(5\frac{1}{2}, -4\right)$ **6.** $\left(-2, 6\frac{1}{3}\right)$

7. $(0, 3.5)$ **8.** $(-5.2, 0)$

9. $(-2, -4)$ **10.** $(-4.2, 0)$

Determine the coordinates of each point on the graph. See Example 1.

11. Point C

12. Point D

13. Point E

14. Point F

15. Point B

16. Point A

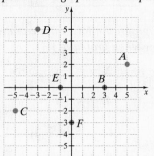

Determine whether each ordered pair is a solution of the given equation. See Example 2.

17. $y = 3x - 5$; $(0, 5), (-1, -8)$

18. $y = -2x + 7$; $(1, 5), (-2, 3)$

19. $-6x + 5y = -6$; $(1, 0), \left(2, \frac{6}{5}\right)$

20. $5x - 3y = 9$; $(0, 3), \left(\frac{12}{5}, -1\right)$

21. $y = 2x^2$; $(1, 2), (3, 18)$

22. $y = 2|x|$; $(-1, 2), (0, 2)$

23. $y = x^3$; $(2, 8), (3, 9)$

24. $y = x^4$; $(-1, 1), (2, 16)$

25. $y = \sqrt{x} + 2$; $(1, 3), (4, 4)$

26. $y = \sqrt[3]{x} - 4$; $(1, -3), (8, 6)$

MIXED PRACTICE

Determine whether each equation is linear or not. Then graph the equation by finding and plotting ordered pair solutions. See Examples 3 through 7.

27. $x + y = 3$ **28.** $y - x = 8$

29. $y = 4x$ **30.** $y = 6x$

31. $y = 4x - 2$ **32.** $y = 6x - 5$

33. $y = |x| + 3$ **34.** $y = |x| + 2$

35. $2x - y = 5$ **36.** $4x - y = 7$

37. $y = 2x^2$ **38.** $y = 3x^2$

39. $y = x^2 - 3$ **40.** $y = x^2 + 3$

41. $y = -2x$ **42.** $y = -3x$

43. $y = -2x + 3$ **44.** $y = -3x + 2$

45. $y = |x + 2|$ **46.** $y = |x - 1|$

47. $y = x^3$
     (*Hint*: Let $x = -3, -2, -1, 0, 1, 2$.)

48. $y = x^3 - 2$
     (*Hint*: Let $x = -3, -2, -1, 0, 1, 2$.)

49. $y = -|x|$ **50.** $y = -x^2$

51. $y = \frac{1}{3}x - 1$ **52.** $y = \frac{1}{2}x - 3$

53. $y = -\frac{3}{2}x + 1$ **54.** $y = -\frac{2}{3}x + 1$

REVIEW AND PREVIEW

Solve the following equations. See Section 2.1.

55. $3(x - 2) + 5x = 6x - 16$

56. $5 + 7(x + 1) = 12 + 10x$

57. $3x + \frac{2}{5} = \frac{1}{10}$ **58.** $\frac{1}{6} + 2x = \frac{2}{3}$

Solve the following inequalities. See Section 2.4.

59. $3x \leq -15$

60. $-3x > 18$

61. $2x - 5 > 4x + 3$

62. $9x + 8 \leq 6x - 4$

CONCEPT EXTENSIONS

Without graphing, visualize the location of each point. Then give its location by quadrant or x- or y-axis.

63. $(4, -2)$ **64.** $(-42, 17)$

65. $(0, -100)$ **66.** $(-87, 0)$

67. $(-10, -30)$ **68.** $(0, 0)$

Given that x is a positive number and that y is a positive number, determine the quadrant or axis in which each point lies.

69. $(x, -y)$ **70.** $(-x, y)$

71. $(x, 0)$ **72.** $(0, -y)$

73. $(-x, -y)$ **74.** $(0, 0)$

Solve. See the Concept Check in this section.

75. Which correctly describes the location of the point $(-1, 5.3)$ in a rectangular coordinate system?
   a. 1 unit to the right of the y-axis and 5.3 units above the x-axis
   b. 1 unit to the left of the y-axis and 5.3 units above the x-axis
   c. 1 unit to the left of the y-axis and 5.3 units below the x-axis
   d. 1 unit to the right of the y-axis and 5.3 units below the x-axis

76. Which correctly describes the location of the point $\left(0, -\frac{3}{4}\right)$ in a rectangular coordinate system?
   a. on the x-axis and $\frac{3}{4}$ unit to the left of the y-axis
   b. on the x-axis and $\frac{3}{4}$ unit to the right of the y-axis
   c. on the y-axis and $\frac{3}{4}$ unit above the x-axis
   d. on the y-axis and $\frac{3}{4}$ unit below the x-axis

For Exercises 77 through 80, match each description with the graph that best illustrates it.

77. Moe worked 40 hours per week until the fall semester started. He quit and didn't work again until he worked 60 hours a week during the holiday season starting mid-December.

78. Kawana worked 40 hours a week for her father during the summer. She slowly cut back her hours to not working at all during the fall semester. During the holiday season in December, she started working again and increased her hours to 60 hours per week.

79. Wendy worked from July through February, never quitting. She worked between 10 and 30 hours per week.

80. Bartholomew worked from July through February. During the holiday season between mid-November and the beginning of January, he worked 40 hours per week. The rest of the time, he worked between 10 and 40 hours per week.

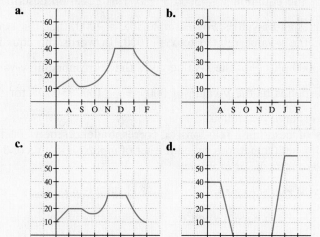

This broken-line graph shows the hourly minimum wage and the years it increased. Use this graph for Exercises 81 through 84.

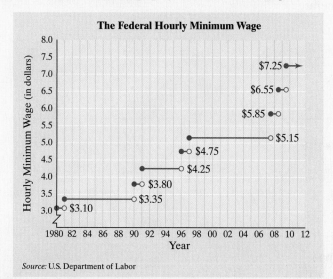

Source: U.S. Department of Labor

81. What was the first year that the minimum hourly wage rose above $5.00?

82. What was the first year that the minimum hourly wage rose above $6.00?

83. Why do you think that this graph is shaped the way it is?

84. The federal hourly minimum wage started in 1938 at $0.25. How much will it have increased by 2011?

85. Graph $y = x^2 - 4x + 7$. Let $x = 0, 1, 2, 3, 4$ to generate ordered pair solutions.

86. Graph $y = x^2 + 2x + 3$. Let $x = -3, -2, -1, 0, 1$ to generate ordered pair solutions.

△ **87.** The perimeter y of a rectangle whose width is a constant 3 inches and whose length is x inches is given by the equation

$$y = 2x + 6$$

   a. Draw a graph of this equation.

   b. Read from the graph the perimeter y of a rectangle whose length x is 4 inches.

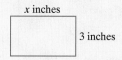

88. The distance y traveled in a train moving at a constant speed of 50 miles per hour is given by the equation

$$y = 50x$$

   where x is the time in hours traveled.

   a. Draw a graph of this equation.

   b. Read from the graph the distance y traveled after 6 hours.

*For income tax purposes, the owner of Copy Services uses a method called **straight-line depreciation** to show the loss in value of a copy machine he recently purchased. He assumes that he can use the machine for 7 years. The following graph shows the value of the machine over the years. Use this graph to answer Exercises 89 through 94.*

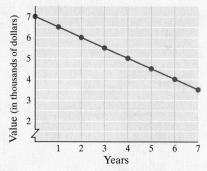

89. What was the purchase price of the copy machine?

90. What is the depreciated value of the machine in 7 years?

91. What loss in value occurred during the first year?

92. What loss in value occurred during the second year?

93. Why do you think that this method of depreciating is called straight-line depreciation?

94. Why is the line tilted downward?

95. On the same set of axes, graph $y = 2x, y = 2x - 5$, and $y = 2x + 5$. What patterns do you see in these graphs?

96. On the same set of axes, graph $y = 2x, y = x$, and $y = -2x$. Describe the differences and similarities in these graphs.

Write each statement as an equation in two variables. Then graph each equation.

97. The *y*-value is 5 more than three times the *x*-value.

98. The *y*-value is −3 decreased by twice the *x*-value.

99. The *y*-value is 2 more than the square of the *x*-value.

100. The *y*-value is 5 decreased by the square of the *x*-value.

Use a graphing calculator to verify the graphs of the following exercises.

101. Exercise 39

102. Exercise 40

103. Exercise 47

104. Exercise 48

3.2 Introduction to Functions

OBJECTIVES

1 Define Relation, Domain, and Range.

2 Identify Functions.

3 Use the Vertical Line Test for Functions.

4 Find the Domain and Range of a Function.

5 Use Function Notation.

OBJECTIVE

1 Defining Relation, Domain, and Range

Recall our example from the last section about products sold and monthly salary. We modeled the data given by the equation $y = 3000 + \frac{1}{5}x$. This equation describes a relationship between *x*-values and *y*-values. For example, if $x = 1000$, then this equation describes how to find the *y*-value related to $x = 1000$. In words, the equation $y = 3000 + \frac{1}{5}x$ says that 3000 plus $\frac{1}{5}$ of the *x*-value gives the corresponding *y*-value. The *x*-value of 1000 corresponds to the *y*-value of $3000 + \frac{1}{5} \cdot 1000 = 3200$ for this equation, and we have the ordered pair (1000, 3200).

There are other ways of describing relations or correspondences between two numbers or, in general, a first set (sometimes called the set of *inputs*) and a second set (sometimes called the set of *outputs*). For example,

First Set: Input $\longrightarrow$	*Correspondence* $\longrightarrow$	*Second Set: Output*
People in a certain city	Each person's age, to the nearest year	The set of nonnegative integers

A few examples of ordered pairs from this relation might be (Ana, 4), (Bob, 36), (Trey, 21), and so on.

Below are just a few other ways of describing relations between two sets and the ordered pairs that they generate.

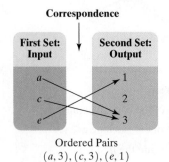

Ordered Pairs
$(a, 3), (c, 3), (e, 1)$

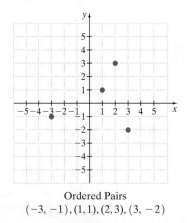

Ordered Pairs
$(-3, -1), (1, 1), (2, 3), (3, -2)$

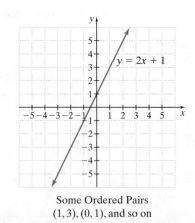

Some Ordered Pairs
$(1, 3), (0, 1),$ and so on

Relation, Domain, and Range

A **relation** is a set of ordered pairs.

The **domain** of the relation is the set of all first components of the ordered pairs.

The **range** of the relation is the set of all second components of the ordered pairs.

For example, the domain for our relation on the left above is $\{a, c, e\}$ and the range is $\{1, 3\}$. Notice that the range does not include the element 2 of the second set.

This is because no element of the first set is assigned to this element. If a relation is defined in terms of *x*- and *y*-values, we will agree that the domain corresponds to *x*-values and that the range corresponds to *y*-values that have *x*-values assigned to them.

> ▶ **Helpful Hint**
>
> Remember that the range includes only elements that are paired with domain values. For the correspondence to the right, the range is $\{a\}$.

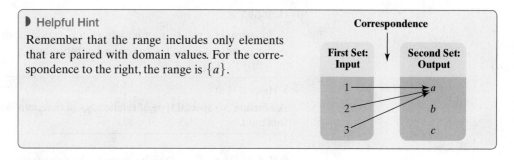

EXAMPLE 1   Determine the domain and range of each relation.

a. $\{(2,3),(2,4),(0,-1),(3,-1)\}$

b. 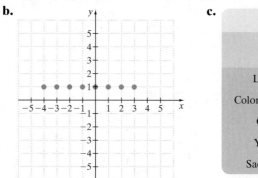   **c.**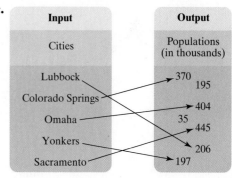

Solution

a. The domain is the set of all first coordinates of the ordered pairs, $\{2,0,3\}$. The range is the set of all second coordinates, $\{3,4,-1\}$.

b. Ordered pairs are not listed here but are given in graph form. The relation is $\{(-4,1),(-3,1),(-2,1),(-1,1),(0,1),(1,1),(2,1),(3,1)\}$. The domain is $\{-4,-3,-2,-1,0,1,2,3\}$. The range is $\{1\}$.

c. The domain is the set of inputs, {Lubbock, Colorado Springs, Omaha, Yonkers, Sacramento}. The range is the numbers in the set of outputs that correspond to elements in the set of inputs {370, 404, 445, 206, 197}.

> ▶ **Helpful Hint**
>
> Domain or range elements that occur more than once need to be listed only once.

PRACTICE
1   Determine the domain and range of each relation.

a. $\{(4,1),(4,-3),(5,-2),(5,6)\}$

b.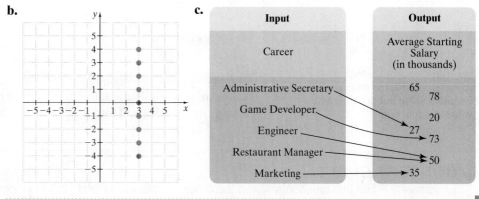

OBJECTIVE

2 **Identifying Functions**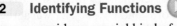

Now we consider a special kind of relation called a function.

> **Function**
>
> A **function** is a relation in which each first component in the ordered pairs corresponds to *exactly* one second component.

> ▶ **Helpful Hint**
>
> A function is a special type of relation, so all functions are relations, but not all relations are functions.

EXAMPLE 2 Determine whether the following relations are also functions.

a. $\{(-2,5),(2,7),(-3,5),(9,9)\}$

b.

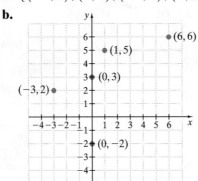

c.

Input	Correspondence	Output
People in a certain city	Each person's age	The set of nonnegative integers

Solution

a. Although the ordered pairs $(-2,5)$ and $(-3,5)$ have the same *y*-value, each *x*-value is assigned to only one *y*-value, so this set of ordered pairs is a function.

b. The *x*-value 0 is assigned to two *y*-values, -2 and 3, in this graph, so this relation does not define a function.

c. This relation is a function because although two people may have the same age, each person has only one age. This means that each element in the first set is assigned to only one element in the second set. □

PRACTICE

2 Determine whether the following relations are also functions.

a. $\{(3,1),(-3,-4),(8,5),(9,1)\}$ **b.**

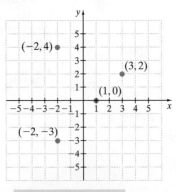

c.

Input	Correspondence	Output
People in a certain city	Birth date (day of month)	Set of nonnegative integers

✓**CONCEPT CHECK**

Explain why a function can contain both the ordered pairs $(1,3)$ and $(2,3)$ but not both $(3,1)$ and $(3,2)$.

We will call an equation such as $y = 2x + 1$ a **relation** since this equation defines a set of ordered pair solutions.

EXAMPLE 3 Is the relation $y = 2x + 1$ also a function?[*]

Solution The relation $y = 2x + 1$ is a function if each x-value corresponds to just one y-value. For each x-value substituted in the equation $y = 2x + 1$, the multiplication and addition performed on each gives a single result, so only one y-value will be associated with each x-value. Thus, $y = 2x + 1$ is a function.

[*]For further discussion including the graph, see Objective 3. ☐

PRACTICE
3 Is the relation $y = -3x + 5$ also a function?

■

EXAMPLE 4 Is the relation $x = y^2$ also a function?[*]

Solution In $x = y^2$, if $y = 3$, then $x = 9$. Also, if $y = -3$, then $x = 9$. In other words, we have the ordered pairs $(9, 3)$ and $(9, -3)$. Since the x-value 9 corresponds to two y-values, 3 and -3, $x = y^2$ is not a function.

[*]For further discussion including the graph, see Objective 3. ☐

PRACTICE
4 Is the relation $y = -x^2$ also a function?

■

OBJECTIVE
3 Using the Vertical Line Test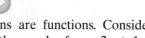

As we have seen so far, not all relations are functions. Consider the graphs of $y = 2x + 1$ and $x = y^2$ shown next. For the graph of $y = 2x + 1$, notice that each x-value corresponds to only one y-value. Recall from Example 3 that $y = 2x + 1$ is a function.

Graph of Example 3:
$y = 2x + 1$

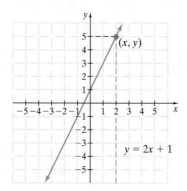

Graph of Example 4:
$x = y^2$

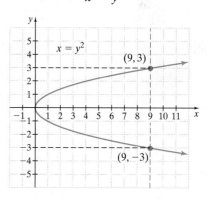

For the graph of $x = y^2$ the x-value 9, for example, corresponds to two y-values, 3 and -3, as shown by the vertical line. Recall from Example 4 that $x = y^2$ is not a function.

Graphs can be used to help determine whether a relation is also a function by the following **vertical line test.**

Vertical Line Test

If no vertical line can be drawn so that it intersects a graph more than once, the graph is the graph of a function.

EXAMPLE 5 Determine whether the following graphs are graphs of functions.

a.
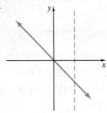

b.
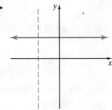

c.
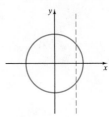

Solution

Yes, this is the graph of a function since no vertical line will intersect this graph more than once.

Yes, this is the graph of a function.

No, this is not the graph of a function. Note that vertical lines can be drawn that intersect the graph in two points.

d.
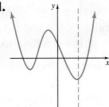

e.
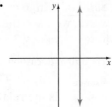

Solution

Yes, this is the graph of a function.

No, this is not the graph of a function. A vertical line can be drawn that intersects this line at every point.

PRACTICE
5 Determine whether the following graphs are graphs of functions.

a.
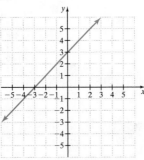

b.
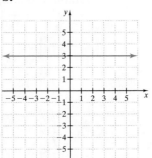

c.
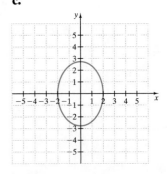

d.
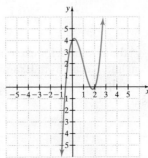

e.
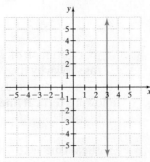

Recall that the graph of a linear equation in two variables is a line, and a line that is not vertical will pass the vertical line test. Thus, **all linear equations are functions except those whose graph is a vertical line.**

✓CONCEPT CHECK

Determine which equations represent functions. Explain your answer.

a. $y = |x|$ 　　　　 **b.** $y = x^2$ 　　　　 **c.** $x + y = 6$

OBJECTIVE

4 **Finding the Domain and Range of a Function**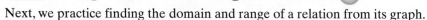

Next, we practice finding the domain and range of a relation from its graph.

EXAMPLE 6 Find the domain and range of each relation. Determine whether the relation is also a function.

a.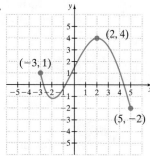

b.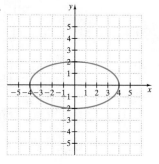

c.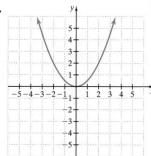

d.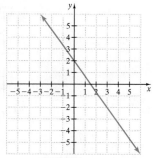

Solution By the vertical line test, graphs **a**, **c**, and **d** are graphs of functions. The domain is the set of values of x and the range is the set of values of y. We read these values from each graph.

a.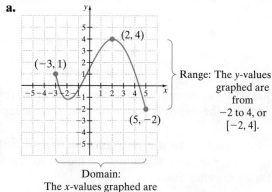

Range: The y-values graphed are from -2 to 4, or $[-2, 4]$.

Domain: The x-values graphed are from -3 to 5, or $[-3, 5]$.

b.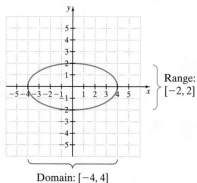

Range: $[-2, 2]$

Domain: $[-4, 4]$

▶ Helpful Hint

In Example 6, Part **a,** notice that the graph contains the end points $(-3, 1)$ and $(5, -2)$ whereas the graphs in Parts **c** and **d** contain arrows that indicate that they continue forever.

c.

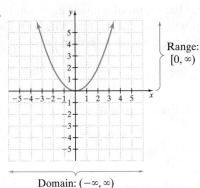

Range: $[0, \infty)$

Domain: $(-\infty, \infty)$

d.

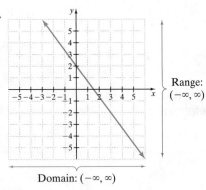

Range: $(-\infty, \infty)$

Domain: $(-\infty, \infty)$

PRACTICE

6 Find the domain and range of each relation. Determine whether each relation is also a function.

a.

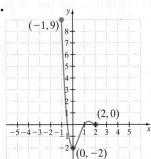

b.

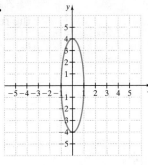

c.

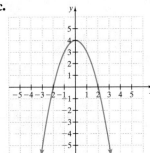

d.

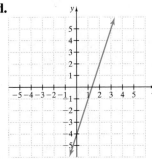

OBJECTIVE

5 Using Function Notation

Many times letters such as f, g, and h are used to name functions.

Function Notation

To denote that y is a function of x, we can write

$$y = \underbrace{f(x)}_{\text{Function Notation}} \quad (\text{Read "}f\text{ of }x\text{."})$$

This notation means that **y is a function of x** or that y *depends on x*. For this reason, y is called the **dependent variable** and x the **independent variable.**

For example, to use function notation with the function $y = 4x + 3$, we write $f(x) = 4x + 3$. The notation $f(1)$ means to replace x with 1 and find the resulting y or function value. Since

$$f(x) = 4x + 3$$

then

$$f(1) = 4(1) + 3 = 7$$

This means that when $x = 1$, y or $f(x) = 7$. The corresponding ordered pair is $(1, 7)$. Here, the input is 1 and the output is $f(1)$ or 7. Now let's find $f(2), f(0),$ and $f(-1)$.

$$f(x) = 4x + 3 \qquad\qquad f(x) = 4x + 3 \qquad\qquad f(x) = 4(x) + 3$$
$$f(2) = 4(2) + 3 \qquad\qquad f(0) = 4(0) + 3 \qquad\qquad f(-1) = 4(-1) + 3$$
$$= 8 + 3 \qquad\qquad\qquad = 0 + 3 \qquad\qquad\qquad = -4 + 3$$
$$= 11 \qquad\qquad\qquad\qquad = 3 \qquad\qquad\qquad\qquad = -1$$

Ordered Pairs:

$(2, 11) \qquad\qquad\qquad\qquad (0, 3) \qquad\qquad\qquad\qquad (-1, -1)$

> ▶ **Helpful Hint**
>
> Make sure you remember that $f(2) = 11$ corresponds to the ordered pair $(2, 11)$.

> ▶ **Helpful Hint**
>
> Note that $f(x)$ is a special symbol in mathematics used to denote a function. The symbol $f(x)$ is read "f of x." It does *not* mean $f \cdot x$ (f times x).

EXAMPLE 7 If $f(x) = 7x^2 - 3x + 1$ and $g(x) = 8x - 2$, find the following.

a. $f(1)$ **b.** $g(1)$ **c.** $f(-2)$ **d.** $g(0)$

Solution

a. Substitute 1 for x in $f(x) = 7x^2 - 3x + 1$ and simplify.

$$f(x) = 7x^2 - 3x + 1$$
$$f(1) = 7(1)^2 - 3(1) + 1 = 5$$

b. $g(x) = 8x - 2$
$$g(1) = 8(1) - 2 = 6$$

c. $f(x) = 7x^2 - 3x + 1$
$$f(-2) = 7(-2)^2 - 3(-2) + 1 = 35$$

d. $g(x) = 8x - 2$
$$g(0) = 8(0) - 2 = -2 \qquad\qquad\qquad\qquad\qquad \square$$

PRACTICE

7 If $f(x) = 3x - 2$ and $g(x) = 5x^2 + 2x - 1$, find the following.

a. $f(1)$ **b.** $g(1)$ **c.** $f(0)$ **d.** $g(-2)$

✓ **CONCEPT CHECK**

Suppose $y = f(x)$ and we are told that $f(3) = 9$. Which is not true?

a. When $x = 3$, $y = 9$.
b. A possible function is $f(x) = x^2$.
c. A point on the graph of the function is $(3, 9)$.
d. A possible function is $f(x) = 2x + 4$.

If it helps, think of a function, f, as a machine that has been programmed with a certain correspondence or rule. An input value (a member of the domain) is then fed into the machine, the machine does the correspondence or rule, and the result is the output (a member of the range).

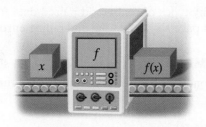

EXAMPLE 8 Given the graphs of the functions f and g, find each function value by inspecting the graphs.

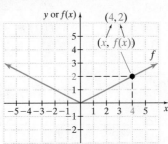
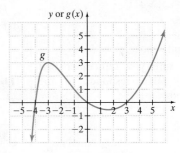

a. $f(4)$ **b.** $f(-2)$ **c.** $g(5)$ **d.** $g(0)$

e. Find all x-values such that $f(x) = 1$.

f. Find all x-values such that $g(x) = 0$.

Solution

a. To find $f(4)$, find the y-value when $x = 4$. We see from the graph that when $x = 4$, y or $f(x) = 2$. Thus, $f(4) = 2$.

b. $f(-2) = 1$ from the ordered pair $(-2, 1)$.

c. $g(5) = 3$ from the ordered pair $(5, 3)$.

d. $g(0) = 0$ from the ordered pair $(0, 0)$.

e. To find x-values such that $f(x) = 1$, we are looking for any ordered pairs on the graph of f whose $f(x)$ or y-value is 1. They are $(2, 1)$ and $(-2, 1)$. Thus $f(2) = 1$ and $f(-2) = 1$. The x-values are 2 and -2.

f. Find ordered pairs on the graph of g whose $g(x)$ or y-value is 0. They are $(3, 0)$, $(0, 0)$, and $(-4, 0)$. Thus $g(3) = 0$, $g(0) = 0$, and $g(-4) = 0$. The x-values are $3, 0$, and -4. □

PRACTICE

8 Given the graphs of the functions f and g, find each function value by inspecting the graphs.

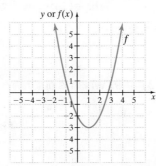
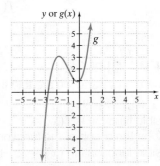

a. $f(1)$ **b.** $f(0)$ **c.** $g(-2)$ **d.** $g(0)$

e. Find all x-values such that $f(x) = 1$.

f. Find all x-values such that $g(x) = -2$.

Many types of real-world paired data form functions. The broken-line graphs on the next page show the total and online enrollment in postsecondary institutions.

EXAMPLE 9 The following graph shows the total and online enrollments in postsecondary institutions as functions of time.

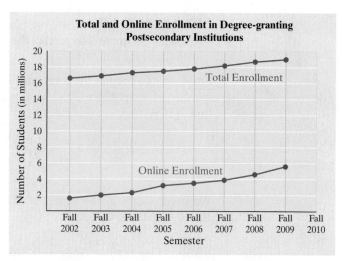

Source: Projections of Education Statistics to 2018, National Center for Education Statistics

a. Approximate the total enrollment in fall 2009.

b. In fall 2002, the total enrollment was 16.6 million students. Find the increase in total enrollment from fall 2002 to fall 2009.

Solution

a. Find the semester Fall 2009 and move upward until you reach the top broken-line graph. From the point on the graph, move horizontally to the left until the vertical axis is reached. In fall 2009, approximately 19 million students, or 19,000,000 students, were enrolled in degree-granting postsecondary institutions.

b. The increase from fall 2002 to fall 2009 is 19 million − 16.6 million = 2.4 million or 2,400,000 students.   ☐

PRACTICE

9   Use the graph in Example 9 and approximate the online enrollment in fall 2009.

Notice that each graph separately in Example 9 is the graph of a function since for each year there is only one total enrollment and only one online enrollment. Also notice that each graph resembles the graph of a line. Often, businesses depend on equations that closely fit data-defined functions like this one to model the data and predict future trends. For example, by a method called **least squares,** the function $f(x) = 0.34x + 16$ approximates the data for the red graph, and the function $f(x) = 0.55x + 0.3$ approximates the data for the blue graph. For each function, x is the number of years since 2000, and $f(x)$ is the number of students (in millions). The graphs and the data functions are shown next.

▶ **Helpful Hint**

Each function graphed is the graph of a function and passes the vertical line test.

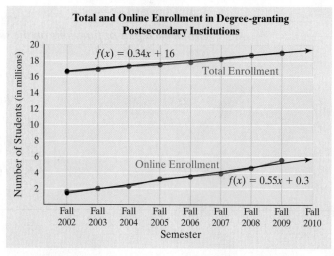

Source: Projections of Education Statistics to 2018, National Center for Education Statistics

EXAMPLE 10 Use the function $f(x) = 0.34x + 16$ and the discussion following Example 9 to predict the total enrollment in degree-granting postsecondary institutions for fall 2010.

Solution To predict the total enrollment in fall 2010, remember that x represents the number of years since 2000, so $x = 2010 - 2000 = 10$. Use $f(x) = 0.34x + 16$ and find $f(10)$.

$$f(x) = 0.34x + 16$$
$$f(10) = 0.34(10) + 16$$
$$= 19.4$$

We predict that in the semester fall 2010, the total enrollment was 19.4 million, or 19,400,000 students. □

PRACTICE
10 Use $f(x) = 0.55x + 0.3$ to approximate the online enrollment in fall 2010. ∎

Graphing Calculator Explorations

It is possible to use a graphing calculator to sketch the graph of more than one equation on the same set of axes. For example, graph the functions $f(x) = x^2$ and $g(x) = x^2 + 4$ on the same set of axes.

To graph on the same set of axes, press the ⬚Y = key and enter the equations on the first two lines.

$$Y_1 = x^2$$
$$Y_2 = x^2 + 4$$

Then press the ⬚GRAPH key as usual. The screen should look like this.

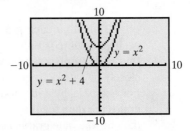

Notice that the graph of y or $g(x) = x^2 + 4$ is the graph of $y = x^2$ moved 4 units upward.

Graph each pair of functions on the same set of axes. Describe the similarities and differences in their graphs.

1. $f(x) = |x|$
  $g(x) = |x| + 1$

2. $f(x) = x^2$
  $h(x) = x^2 - 5$

3. $f(x) = x$
  $H(x) = x - 6$

4. $f(x) = |x|$
  $G(x) = |x| + 3$

5. $f(x) = -x^2$
  $F(x) = -x^2 + 7$

6. $f(x) = x$
  $F(x) = x + 2$

Vocabulary, Readiness & Video Check

Use the choices below to fill in each blank. Some choices may not be used.

domain	vertical	relation	$(1.7, -2)$
range	horizontal	function	$(-2, 1.7)$

1. A _____ is a set of ordered pairs.

2. The _____ of a relation is the set of all second components of the ordered pairs.

3. The _____ of a relation is the set of all first components of the ordered pairs.

4. A _____ is a relation in which each first component in the ordered pairs corresponds to *exactly* one second component.

5. By the vertical line test, all linear equations are functions except those whose graphs are _____ lines.

6. If $f(-2) = 1.7$, the corresponding ordered pair is _____.

Martin-Gay Interactive Videos

See Video 3.2

Watch the section lecture video and answer the following questions.

OBJECTIVE 1

7. Based on the lecture before Example 1, why can an equation in two variables define a relation?

OBJECTIVE 2

8. Based on the lecture before Example 2, can a relation in which a second component corresponds to more than one first component be a function?

OBJECTIVE 3

9. Based on Example 3 and the lecture before, explain why the vertical line test works.

OBJECTIVE 4

10. From Example 8, do all linear equations in two variables define functions? Explain.

OBJECTIVE 5

11. From Examples 9 and 10, what is the connection between function notation to evaluate a function at certain values and ordered pair solutions of the function?

3.2   Exercise Set   MyMathLab®

Find the domain and the range of each relation. Also determine whether the relation is a function. See Examples 1 and 2.

1. $\{(-1, 7), (0, 6), (-2, 2), (5, 6)\}$

2. $\{(4, 9), (-4, 9), (2, 3), (10, -5)\}$

3. $\{(-2, 4), (6, 4), (-2, -3), (-7, -8)\}$

4. $\{(6, 6), (5, 6), (5, -2), (7, 6)\}$

5. $\{(1, 1), (1, 2), (1, 3), (1, 4)\}$

6. $\{(1, 1), (2, 1), (3, 1), (4, 1)\}$

7. $\left\{\left(\frac{3}{2}, \frac{1}{2}\right), \left(1\frac{1}{2}, -7\right), \left(0, \frac{4}{5}\right)\right\}$

8. $\{(\pi, 0), (0, \pi), (-2, 4), (4, -2)\}$

9. $\{(-3, -3), (0, 0), (3, 3)\}$

10. $\left\{\left(\frac{1}{2}, \frac{1}{4}\right), \left(0, \frac{7}{8}\right), (0.5, \pi)\right\}$

11.

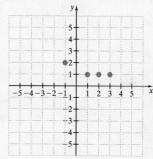

12.

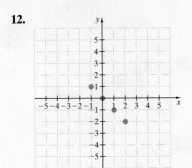

13.

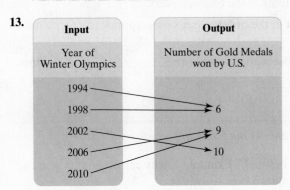

Input	Output
Year of Winter Olympics	Number of Gold Medals won by U.S.

1994
1998 → 6
2002 → 9
2006 → 10
2010

14.

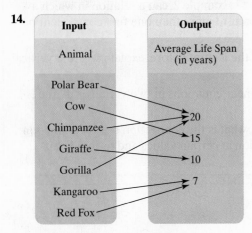

Input	Output
Animal	Average Life Span (in years)

Polar Bear
Cow → 20
Chimpanzee → 15
Giraffe → 10
Gorilla
Kangaroo → 7
Red Fox

15.

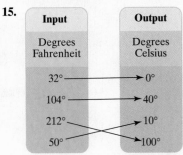

Input	Output
Degrees Fahrenheit	Degrees Celsius

32° → 0°
104° → 40°
212° → 10°
50° → 100°

16.

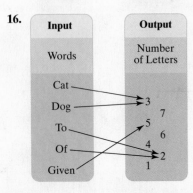

Input	Output
Words	Number of Letters

Cat
Dog → 3
7
To → 5
6
Of → 4
2
Given → 1

17.

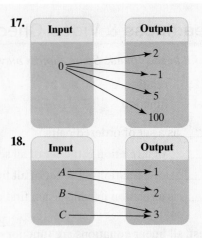

Input	Output
0	2, −1, 5, 100

18.

Input	Output
A	1
B	2
C	3

In Exercises 19 through 22, determine whether the relation is a function. See Example 2.

	First Set: Input	Correspondence	Second Set: Output
19.	Class of algebra students	Final grade average	nonnegative numbers
20.	People who live in Cincinnati, Ohio	Birth date	days of the year
21.	blue, green, brown	Eye color	People who live in Cincinnati, Ohio
22.	Whole numbers from 0 to 4	Number of children	50 women in a water aerobics class

Use the vertical line test to determine whether each graph is the graph of a function. See Example 5.

23.   **24.**

25.   **26.**

27.   **28.**

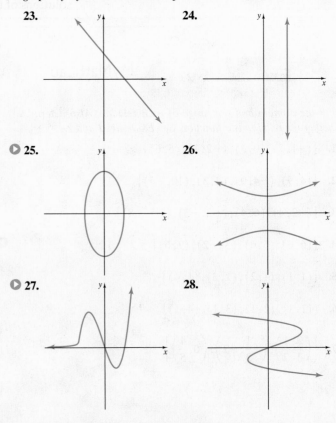

Find the domain and the range of each relation. Use the vertical line test to determine whether each graph is the graph of a function. See Example 6.

29.

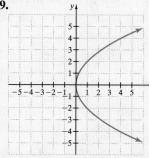

30.

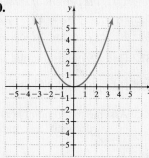

31.

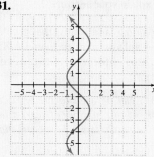

32.

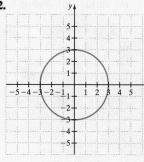

33.

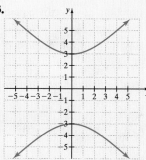

34.

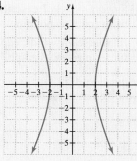

35.

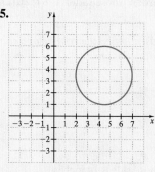

36.

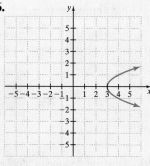

37.

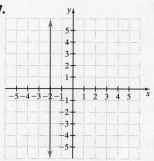

38.

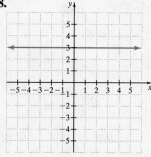

39.

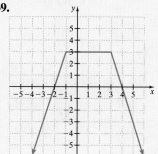

40.

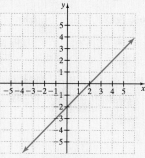

MIXED PRACTICE

Decide whether each is a function. See Examples 3 through 6.

41. $y = x + 1$

42. $y = x - 1$

43. $x = 2y^2$

44. $y = x^2$

45. $y - x = 7$

46. $2x - 3y = 9$

47. $y = \dfrac{1}{x}$

48. $y = \dfrac{1}{x - 3}$

49. $y = 5x - 12$

50. $y = \dfrac{1}{2}x + 4$

51. $x = y^2$

52. $x = |y|$

If $f(x) = 3x + 3, g(x) = 4x^2 - 6x + 3$, and $h(x) = 5x^2 - 7$, find the following. See Example 7.

53. $f(4)$

54. $f(-1)$

55. $h(-3)$

56. $h(0)$

57. $g(2)$

58. $g(1)$

59. $g(0)$

60. $h(-2)$

Given the following functions, find the indicated values. See Example 7.

61. $f(x) = \dfrac{1}{2}x;$

    a. $f(0)$ **b.** $f(2)$ **c.** $f(-2)$

62. $g(x) = -\dfrac{1}{3}x;$

    a. $g(0)$ **b.** $g(-1)$ **c.** $g(3)$

63. $g(x) = 2x^2 + 4;$

    a. $g(-11)$ **b.** $g(-1)$ **c.** $g\left(\dfrac{1}{2}\right)$

64. $h(x) = -x^2$;

   a. $h(-5)$ **b.** $h\left(-\dfrac{1}{3}\right)$ **c.** $h\left(\dfrac{1}{3}\right)$

65. $f(x) = -5$;

   a. $f(2)$ **b.** $f(0)$ **c.** $f(606)$

66. $h(x) = 7$;

   a. $h(7)$ **b.** $h(542)$ **c.** $h\left(-\dfrac{3}{4}\right)$

67. $f(x) = 1.3x^2 - 2.6x + 5.1$

   a. $f(2)$ **b.** $f(-2)$ **c.** $f(3.1)$

68. $g(x) = 2.7x^2 + 6.8x - 10.2$

   a. $g(1)$ **b.** $g(-5)$ **c.** $g(7.2)$

Use the graph of the functions below to answer Exercises 69 through 80. See Example 8.

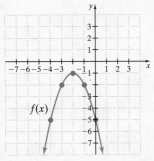

69. If $f(1) = -10$, write the corresponding ordered pair.

70. If $f(-5) = -10$, write the corresponding ordered pair.

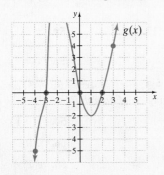

71. If $g(4) = 56$, write the corresponding ordered pair.

72. If $g(-2) = 8$, write the corresponding ordered pair.

73. Find $f(-1)$.

74. Find $f(-2)$.

75. Find $g(2)$.

76. Find $g(-4)$.

77. Find all values of x such that $f(x) = -5$.

78. Find all values of x such that $f(x) = -2$.

79. Find all positive values of x such that $g(x) = 4$.

80. Find all values of x such that $g(x) = 0$.

From the Chapter 3 opener, we have two functions to describe the percent of college students taking at least one online course. For both functions, x is the number of years since 2000 and y (or f(x) or g(x)) is the percent of students taking at least one online course.

$$f(x) = 2.7x + 4.1 \quad \text{or} \quad g(x) = 0.07x^2 + 1.9x + 5.9$$

Use this for Exercises 81–86. See Examples 9 and 10.

81. Find $f(9)$ and describe in words what this means.

82. Find $g(9)$ and describe in words what this means.

83. Assume the trend of $g(x)$ continues. Find $g(16)$ and describe in words what this means.

84. Assume the trend of $f(x)$ continues. Find $f(16)$ and describe in words what this means.

85. Use Exercises 81–84 and compare $f(9)$ and $g(9)$, then $f(16)$ and $g(16)$. As x increases, are the function values staying about the same or not? Explain your answer.

86. Use the Chapter 3 opener graph and study the graphs of $f(x)$ and $g(x)$. Use these graphs to answer Exercise 85. Explain your answer. Use this information to answer Exercises 87 and 88.

The function $f(x) = 0.42x + 10.5$, can be used to predict diamond production. For this function, x is the number of years after 2000, and f(x) is the value (in billions of dollars) of the years diamond production.

87. Use the function to predict diamond production in 2012.

88. Use the function to predict diamond production in 2015.

89. Since $y = x + 7$ describes a function, rewrite the equation using function notation.

90. In your own words, explain how to find the domain of a function given its graph.

The function $A(r) = \pi r^2$ may be used to find the area of a circle if we are given its radius.

△ **91.** Find the area of a circle whose radius is 5 centimeters. (Do not approximate π.)

△ **92.** Find the area of a circular garden whose radius is 8 feet. (Do not approximate π.)

The function $V(x) = x^3$ may be used to find the volume of a cube if we are given the length x of a side.

93. Find the volume of a cube whose side is 14 inches.

94. Find the volume of a die whose side is 1.7 centimeters.

Forensic scientists use the following functions to find the height of a woman if they are given the length of her femur bone f or her tibia bone t in centimeters.

$$H(f) = 2.59f + 47.24.$$

$$H(t) = 2.72t + 61.28$$

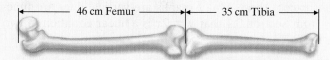

← 46 cm Femur → ← 35 cm Tibia →

95. Find the height of a woman whose femur measures 46 centimeters.

96. Find the height of a woman whose tibia measures 35 centimeters.

The dosage in milligrams D of Ivermectin, a heartworm preventive, for a dog who weighs x pounds is given by

$$D(x) = \frac{136}{25}x$$

97. Find the proper dosage for a dog that weighs 30 pounds.

98. Find the proper dosage for a dog that weighs 50 pounds.

99. The per capita consumption (in pounds) of all beef in the United States is given by the function $C(x) = -0.74x + 67.1$, where x is the number of years since 2000. (*Source:* U.S. Department of Agriculture and Cattle Network)

  a. Find and interpret $C(4)$.

  b. Estimate the per capita consumption of beef in the United States in 2014.

100. The amount of money (in billions of dollars) spent by the Boeing Company and subsidiaries on research and development annually is represented by the function $R(x) = 0.382x + 2.21$, where x is the number of years after 2005. (*Source*: Boeing Corporation)

  a. Find and interpret $R(2)$.

  b. Estimate the amount of money spent on research and development by Boeing in 2014.

REVIEW AND PREVIEW

Complete the given table and use the table to graph the linear equation. See Section 3.1.

101. $x - y = -5$

x	0		1
y		0	

102. $2x + 3y = 10$

x	0		
y		0	2

103. $7x + 4y = 8$

x	0		
y		0	-1

104. $5y - x = -15$

x	0		-2
y		0	

105. $y = 6x$

x	0		-1
y		0	

106. $y = -2x$

x	0		-2
y		0	

△ **107.** Is it possible to find the perimeter of the following geometric figure? If so, find the perimeter.

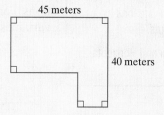

45 meters

40 meters

108. Is it possible to find the area of the figure in Exercise 107? If so, find the area.

CONCEPT EXTENSIONS

For Exercises 109 through 112, suppose that $y = f(x)$ and it is true that $f(7) = 50$. Determine whether each is true or false. See the third Concept Check in this section.

109. An ordered pair solution of the function is $(7, 50)$.

110. When x is 50, y is 7.

111. A possible function is $f(x) = x^2 + 1$

112. A possible function is $f(x) = 10x - 20$.

Given the following functions, find the indicated values.

113. $h(x) = x^2 + 7$;

  a. $h(3)$ **b.** $h(a)$

114. $f(x) = x^2 - 12$;

  a. $f(12)$ **b.** $f(a)$

115. $f(x) = 3x - 12$;

  a. $f(4)$
  b. $f(a)$
  c. $f(-x)$
  d. $f(x + h)$

116. $f(x) = 2x + 7$

  a. $f(2)$
  b. $f(a)$
  c. $f(-x)$
  d. $f(x + h)$

117. What is the greatest number of x-intercepts that a function may have? Explain your answer.

118. What is the greatest number of y-intercepts that a function may have? Explain your answer.

119. In your own words, explain how to find the domain of a function given its graph.

120. Explain the vertical line test and how it is used.

121. Describe a function whose domain is the set of people in your hometown.

122. Describe a function whose domain is the set of people in your algebra class.

3.3 Graphing Linear Functions

OBJECTIVE

1   Graphing Linear Functions

In this section, we identify and graph linear functions. By the vertical line test, we know that all linear equations except those whose graphs are vertical lines are functions. For example, we know from Section 3.1 that $y = 2x$ is a linear equation in two variables. Its graph is shown.

x	$y = 2x$
1	2
0	0
−1	−2

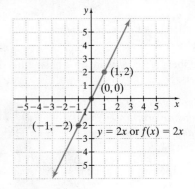

Because this graph passes the vertical line test, we know that $y = 2x$ is a function. If we want to emphasize that this equation describes a function, we may write $y = 2x$ as $f(x) = 2x$.

EXAMPLE 1   Graph $g(x) = 2x + 1$. Compare this graph with the graph of $f(x) = 2x$.

Solution   To graph $g(x) = 2x + 1$, find three ordered pair solutions.

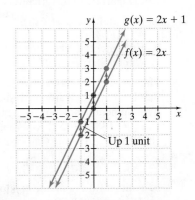

x	$f(x) = 2x$	$g(x) = 2x + 1$
0	0	1
−1	−2	−1
1	2	3

add 1 (over $f(x)$ and $g(x)$ columns)

Notice that y-values for the graph of $g(x) = 2x + 1$ are obtained by adding 1 to each y-value of each corresponding point of the graph of $f(x) = 2x$. The graph of $g(x) = 2x + 1$ is the same as the graph of $f(x) = 2x$ shifted upward 1 unit.

PRACTICE

1   Graph $g(x) = 4x - 3$ and $f(x) = 4x$ on the same axes.

In general, a **linear function** is a function that can be written in the form $f(x) = mx + b$. For example, $g(x) = 2x + 1$ is in this form, with $m = 2$ and $b = 1$.

EXAMPLE 2   Graph the linear functions $f(x) = -3x$ and $g(x) = -3x - 6$ on the same set of axes.

Solution   To graph $f(x)$ and $g(x)$, find ordered pair solutions.

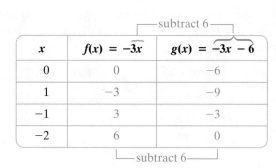

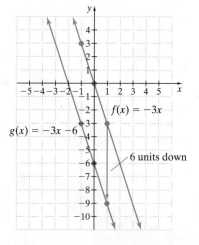

Each y-value for the graph of $g(x) = -3x - 6$ is obtained by subtracting 6 from the y-value of the corresponding point of the graph of $f(x) = -3x$. The graph of $g(x) = -3x - 6$ is the same as the graph of $f(x) = -3x$ shifted down 6 units.   □

PRACTICE
2   Graph the linear functions $f(x) = -2x$ and $g(x) = -2x + 5$ on the same set of axes.

OBJECTIVE
2   **Graphing Linear Functions by Using Intercepts**

Notice that the y-intercept of the graph of $g(x) = -3x - 6$ in the preceding figure is $(0, -6)$. In general, if *a linear function is written in the form* $f(x) = mx + b$ *or* $y = mx + b$, *the y-intercept is* $(0, b)$. This is because if x is 0, then $f(x) = mx + b$ becomes $f(0) = m \cdot 0 + b = b$, and we have the ordered pair solution $(0, b)$. We will study this form more in the next section.

EXAMPLE 3   Find the y-intercept of the graph of each equation.

a. $f(x) = \dfrac{1}{2}x + \dfrac{3}{7}$
b. $y = -2.5x - 3.2$

Solution

a. The y-intercept of $f(x) = \dfrac{1}{2}x + \dfrac{3}{7}$ is $\left(0, \dfrac{3}{7}\right)$.

b. The y-intercept of $y = -2.5x - 3.2$ is $(0, -3.2)$.   □

PRACTICE
3   Find the y-intercept of the graph of each equation.

a. $f(x) = \dfrac{3}{4}x - \dfrac{2}{5}$
b. $y = 2.6x + 4.1$

In general, to find the y-intercept of the graph of an equation not in the form $y = mx + b$, let $x = 0$ since any point on the y-axis has an x-coordinate of 0. To find the x-intercept of a line, let $y = 0$ or $f(x) = 0$ since any point on the x-axis has a y-coordinate of 0.

> **Finding x- and y-Intercepts**
>
> To find an x-intercept, let $y = 0$ or $f(x) = 0$ and solve for x.
> To find a y-intercept, let $x = 0$ and solve for y.

Intercepts are usually easy to find and plot since one coordinate is 0.

EXAMPLE 4 Find the intercepts and graph: $3x + 4y = -12$.

Solution To find the y-intercept, we let $x = 0$ and solve for y. To find the x-intercept, we let $y = 0$ and solve for x. Let's let $x = 0$, then $y = 0$; and then let $x = 2$ to find a third point as a check.

Let $x = 0$.	Let $y = 0$.	Let $x = 2$.
$3x + 4y = -12$	$3x + 4y = -12$	$3x + 4y = -12$
$3 \cdot 0 + 4y = -12$	$3x + 4 \cdot 0 = -12$	$3 \cdot 2 + 4y = -12$
$4y = -12$	$3x = -12$	$6 + 4y = -12$
$y = -3$	$x = -4$	$4y = -18$
		$y = -\dfrac{18}{4} = -4\dfrac{1}{2}$
$(0, -3)$	$(-4, 0)$	$\left(2, -4\dfrac{1}{2}\right)$

The ordered pairs are on the table below. The graph of $3x + 4y = -12$ is the line drawn through these points, as shown.

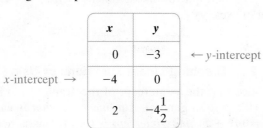

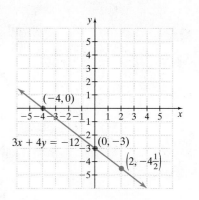

x	y	
0	-3	← y-intercept
-4	0	x-intercept →
2	$-4\dfrac{1}{2}$	

PRACTICE
4 Find the intercepts and graph: $4x - 5y = -20$.

Notice that the equation $3x + 4y = -12$ describes a linear function—"linear" because its graph is a line and "function" because the graph passes the vertical line test.

If we want to emphasize that the equation $3x + 4y = -12$ from Example 4 describes a function, first solve the equation for y.

$$3x + 4y = -12$$
$$4y = -3x - 12 \quad \text{Subtract } 3x \text{ from both sides.}$$
$$\frac{4y}{4} = \frac{-3x}{4} - \frac{12}{4} \quad \text{Divide both sides by 4.}$$
$$y = -\frac{3}{4}x - 3 \quad \text{Simplify.}$$

Next, since $y = f(x)$, replace y with $f(x)$.

$$f(x) = -\frac{3}{4}x - 3$$

> ▶ **Helpful Hint**
>
> Any linear equation that describes a function can be written using function notation. To do so,
>
> **1.** solve the equation for y and then
> **2.** replace y with $f(x)$, as we did above.

EXAMPLE 5   Graph $x = -2y$ by plotting intercepts.

Solution   Let $y = 0$ to find the x-intercept and $x = 0$ to find the y-intercept.

$$\text{If } y = 0 \quad \text{then} \qquad \text{If } x = 0 \quad \text{then}$$
$$x = -2(0) \quad \text{or} \qquad 0 = -2y \quad \text{or}$$
$$x = 0 \qquad\qquad\qquad 0 = y$$
$$(0, 0) \qquad\qquad\qquad (0, 0)$$

Ordered pairs   Both the x-intercept and y-intercept are $(0, 0)$. This happens when the graph passes through the origin. Since two points are needed to determine a line, we must find at least one more ordered pair that satisfies $x = -2y$. Let $y = -1$ to find a second ordered pair solution and let $y = 1$ as a check point.

$$\text{If } y = -1 \quad \text{then} \qquad \text{If } y = 1 \quad \text{then}$$
$$x = -2(-1) \quad \text{or} \qquad x = -2(1) \quad \text{or}$$
$$x = 2 \qquad\qquad\qquad x = -2$$

The ordered pairs are $(0, 0)$, $(2, -1)$, and $(-2, 1)$. Plot these points to graph $x = -2y$.

x	y
0	0
2	−1
−2	1

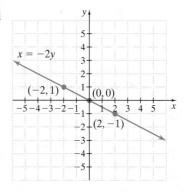

PRACTICE
5   Graph $y = -3x$ by plotting intercepts.

OBJECTIVE
3   **Graphing Vertical and Horizontal Lines**

The equations $x = c$ and $y = c$, where c is a real number constant, are both linear equations in two variables. Why? Because $x = c$ can be written as $x + 0y = c$ and $y = c$ can be written as $0x + y = c$. We graph these two special linear equations below.

EXAMPLE 6   Graph $x = 2$.

Solution   The equation $x = 2$ can be written as $x + 0y = 2$. For any y-value chosen, notice that x is 2. No other value for x satisfies $x + 0y = 2$. Any ordered pair whose x-coordinate is 2 is a solution to $x + 0y = 2$ because 2 added to 0 times any value of y is $2 + 0$, or 2. We will use the ordered pairs $(2, 3)$, $(2, 0)$, and $(2, -3)$ to graph $x = 2$.

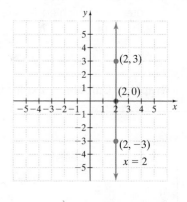

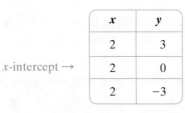

The graph is a vertical line with x-intercept $(2, 0)$. Notice that this graph **is not the graph of a function,** and it has no y-intercept because x is never 0.

PRACTICE
6   Graph $x = -4$.

EXAMPLE 7 Graph $y = -3$.

**Solution** The equation $y = -3$ can be written as $0x + y = -3$. For any x-value chosen, y is -3. If we choose 4, 0, and -2 as x-values, the ordered pair solutions are $(4, -3)$, $(0, -3)$, and $(-2, -3)$. We will use these ordered pairs to graph $y = -3$.

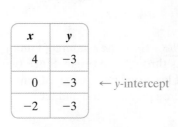

x	y
4	-3
0	-3	← y-intercept
-2	-3

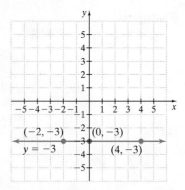

The graph is a horizontal line with y-intercept $(0, -3)$ and no x-intercept. Notice that this graph **is the graph of a function**. □

PRACTICE

7 Graph $y = 4$.

From Examples 6 and 7, we have the following generalization.

Graphing Vertical and Horizontal Lines

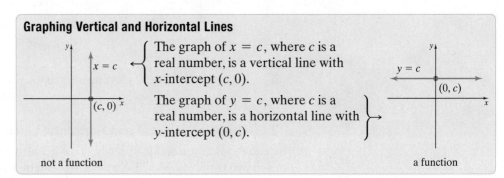

The graph of $x = c$, where c is a real number, is a vertical line with x-intercept $(c, 0)$.

The graph of $y = c$, where c is a real number, is a horizontal line with y-intercept $(0, c)$.

not a function

a function

Concept Check Answer
a; answers may vary

✓**CONCEPT CHECK**
Determine which equations represent functions. Explain your answer.
a. $y = 14$ **b.** $x = -5$

Graphing Calculator Explorations

You may have noticed by now that to use the $\boxed{Y =}$ key on a graphing calculator to graph an equation, the equation must be solved for y.

Graph each function by first solving the function for y.

1. $x = 3.5y$

2. $-2.7y = x$

3. $5.78x + 2.31y = 10.98$

4. $-7.22x + 3.89y = 12.57$

5. $y - |x| = 3.78$

6. $3y - 5x^2 = 6x - 4$

7. $y - 5.6x^2 = 7.7x + 1.5$

8. $y + 2.6|x| = -3.2$

Vocabulary, Readiness & Video Check

Use the choices below to fill in each blank. Some choices may be used more than once and some not at all.

horizontal	y	$(c, 0)$	$(b, 0)$	$(m, 0)$	linear
vertical	x	$(0, c)$	$(0, b)$	$(0, m)$	$f(x)$

1. A(n) _____ function can be written in the form $f(x) = mx + b$.

2. In the form $f(x) = mx + b$, the *y*-intercept is _____.

3. The graph of $x = c$ is a(n) _____ line with *x*-intercept _____.

4. The graph of $y = c$ is a(n) _____ line with *y*-intercept _____.

5. To find an *x*-intercept, let _____ = 0 or _____ = 0 and solve for _____.

6. To find a *y*-intercept, let _____ = 0 and solve for _____.

Martin-Gay Interactive Videos

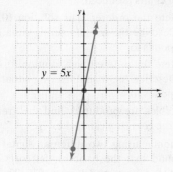

See Video 3.3

Watch the section lecture video and answer the following questions.

OBJECTIVE 1

7. Based on the lecture before Example 1, in what form can a linear function be written?

OBJECTIVE 2

8. In Example 2, the goal is to use the *x*- and *y*-intercepts to graph a line. Yet once two intercepts are found, a third point is also found before the line is graphed. Why?

OBJECTIVE 3

9. From 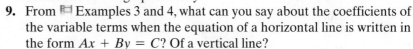 Examples 3 and 4, what can you say about the coefficients of the variable terms when the equation of a horizontal line is written in the form $Ax + By = C$? Of a vertical line?

3.3 Exercise Set MyMathLab®

Graph each linear function. See Examples 1 and 2.

1. $f(x) = -2x$
2. $f(x) = 2x$

▶ 3. $f(x) = -2x + 3$
4. $f(x) = 2x + 6$

5. $f(x) = \dfrac{1}{2}x$
6. $f(x) = \dfrac{1}{3}x$

7. $f(x) = \dfrac{1}{2}x - 4$
8. $f(x) = \dfrac{1}{3}x - 2$

The graph of $f(x) = 5x$ follows. Use this graph to match each linear function with its graph. See Examples 1 through 3.

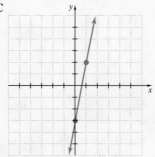

A

B

C

D
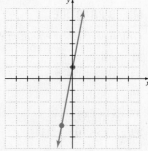

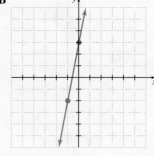

$y = 5x$

9. $f(x) = 5x - 3$
10. $f(x) = 5x - 2$

11. $f(x) = 5x + 1$
12. $f(x) = 5x + 3$

Graph each linear function by finding x- and y-intercepts. Then write each equation using function notation. See Examples 4 and 5.

13. $x - y = 3$

14. $x - y = -4$

15. $x = 5y$

16. $2x = y$

17. $-x + 2y = 6$

18. $x - 2y = -8$

19. $2x - 4y = 8$

20. $4x + 6y = 12$

Graph each linear equation. See Examples 6 and 7.

21. $x = -1$ **22.** $y = 5$

23. $y = 0$ **24.** $x = 0$

25. $y + 7 = 0$ **26.** $x - 3 = 0$

Match each equation below with its graph.

A 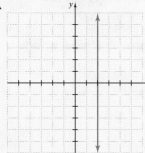   **B**

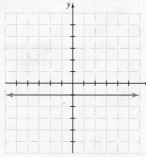

C 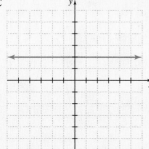   **D**

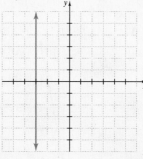

27. $y = 2$ **28.** $x = -3$

29. $x - 2 = 0$ **30.** $y + 1 = 0$

MIXED PRACTICE

Graph each linear equation. See Examples 1 through 7.

31. $x + 2y = 8$ **32.** $x - 3y = 3$

33. $3x + 5y = 7$ **34.** $3x - 2y = 5$

35. $x + 8y = 8$ **36.** $x - 3y = 9$

37. $5 = 6x - y$ **38.** $4 = x - 3y$

39. $-x + 10y = 11$ **40.** $-x + 9 = -y$

41. $y = \dfrac{3}{2}$ **42.** $x = \dfrac{3}{2}$

43. $2x + 3y = 6$ **44.** $4x + y = 5$

45. $x + 3 = 0$ **46.** $y - 6 = 0$

47. $f(x) = \dfrac{3}{4}x + 2$ **48.** $f(x) = \dfrac{4}{3}x + 2$

49. $f(x) = x$ **50.** $f(x) = -x$

51. $f(x) = \dfrac{1}{2}x$ **52.** $f(x) = -2x$

53. $f(x) = 4x - \dfrac{1}{3}$ **54.** $f(x) = -3x + \dfrac{3}{4}$

55. $x = -3$ **56.** $f(x) = 3$

REVIEW AND PREVIEW

Solve the following. See Sections 2.6 and 2.7.

57. $|x - 3| = 6$

58. $|x + 2| < 4$

59. $|2x + 5| > 3$

60. $|5x| = 10$

61. $|3x - 4| \le 2$

62. $|7x - 2| \ge 5$

Simplify. See Section 1.3.

63. $\dfrac{-6 - 3}{2 - 8}$ **64.** $\dfrac{4 - 5}{-1 - 0}$ **65.** $\dfrac{-8 - (-2)}{-3 - (-2)}$

66. $\dfrac{-12 - (-3)}{-10 - (-9)}$ **67.** $\dfrac{0 - 6}{5 - 0}$ **68.** $\dfrac{2 - 2}{3 - 5}$

CONCEPT EXTENSIONS

Think about the appearance of each graph. Without graphing, determine which equations represent functions. Explain each answer. See the Concept Check in this section.

69. $x = -1$

70. $y = 5$

71. $y = 2x$

72. $x + y = -5$

Solve.

73. Broyhill Furniture found that it takes 2 hours to manufacture each table for one of its special dining room sets. Each chair takes 3 hours to manufacture. A total of 1500 hours is available to produce tables and chairs of this style. The linear equation that models this situation is $2x + 3y = 1500$, where x represents the number of tables produced and y the number of chairs produced.

  a. Complete the ordered pair solution $(0, \ )$ of this equation. Describe the manufacturing situation this solution corresponds to.

  b. Complete the ordered pair solution $(\ , 0)$ for this equation. Describe the manufacturing situation this solution corresponds to.

  c. If 50 tables are produced, find the greatest number of chairs the company can make.

74. While manufacturing two different digital camera models, Kodak found that the basic model costs $55 to produce, whereas the deluxe model costs $75. The weekly budget for these two models is limited to $33,000 in production costs. The linear equation that models this situation is

$55x + 75y = 33,000$, where x represents the number of basic models and y the number of deluxe models.

a. Complete the ordered pair solution $(0, \ )$ of this equation. Describe the manufacturing situation this solution corresponds to.

b. Complete the ordered pair solution $(\ , 0)$ of this equation. Describe the manufacturing situation this solution corresponds to.

c. If 350 deluxe models are produced, find the greatest number of basic models that can be made in one week.

75. The cost of renting a car for a day is given by the linear function $C(x) = 0.2x + 24$, where $C(x)$ is in dollars and x is the number of miles driven.

a. Find the cost of driving the car 200 miles.

b. Graph $C(x) = 0.2x + 24$.

c. How can you tell from the graph of $C(x)$ that as the number of miles driven increases, the total cost increases also?

76. The cost of renting a piece of machinery is given by the linear function $C(x) = 4x + 10$, where $C(x)$ is in dollars and x is given in hours.

a. Find the cost of renting the piece of machinery for 8 hours.

b. Graph $C(x) = 4x + 10$.

c. How can you tell from the graph of $C(x)$ that as the number of hours increases, the total cost increases also?

77. The yearly cost of tuition (in-state) and required fees for attending a public two-year college full time can be estimated by the linear function $f(x) = 64x + 2083$, where x is the number of years after 2000 and $f(x)$ is the total cost. (*Source*: The College Board)

a. Use this function to approximate the yearly cost of attending a two-year college in the year 2016. [*Hint*: Find $f(16)$.]

b. Use the given function to predict in what year the yearly cost of tuition and required fees will exceed $3200. [*Hint*: Let $f(x) = 3200$, solve for x, then round your solution up to the next whole year.]

c. Use this function to approximate the yearly cost of attending a two-year college in the present year. If you attend a two-year college, is this amount greater than or less than the amount that is currently charged by the college you attend?

78. The yearly cost of tuition (in-state) and required fees for attending a public four-year college full time can be estimated by the linear function $f(x) = 318x + 4467$, where x is the number of years after 2000 and $f(x)$ is the total cost in dollars. (*Source*: The College Board)

a. Use this function to approximate the yearly cost of attending a four-year college in the year 2016. [*Hint*: Find $f(16)$.]

b. Use the given function to predict in what year the yearly cost of tuition and required fees will exceed $10,000. [*Hint*: Let $f(x) = 10,000$, solve for x, then round your solution up to the next whole year.]

c. Use this function to approximate the yearly cost of attending a four-year college in the present year. If you attend a four-year college, is this amount greater than or less than the amount that is currently charged by the college you attend?

79. In your own words, explain how to find x- and y-intercepts.

80. Explain why it is a good idea to use three points to graph a linear equation.

81. Discuss whether a vertical line ever has a y-intercept.

82. Discuss whether a horizontal line ever has an x-intercept.

The graph of $f(x)$ or $y = -4x$ is given below. Without actually graphing, describe the shape and location of

83. $y = -4x + 2$

84. $y = -4x - 5$

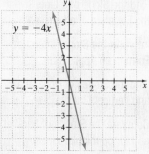

It is true that for any function $f(x)$, the graph of $f(x) + k$ is the same as the graph of $f(x)$ shifted k units up if k is positive and $|k|$ units down if k is negative. (We study this further in Section 3.7.)

The graph of $y = |x|$ is

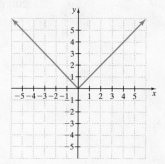

Without actually graphing, match each equation with its graph.

a. $y = |x| - 1$ b. $y = |x| + 1$
c. $y = |x| - 3$ d. $y = |x| + 3$

85.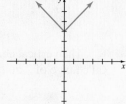

86.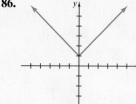

87.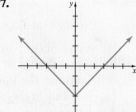

88.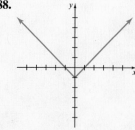

 Use a graphing calculator to verify the results of each exercise.

89. Exercise 9   90. Exercise 10

91. Exercise 17   92. Exercise 18

3.4 The Slope of a Line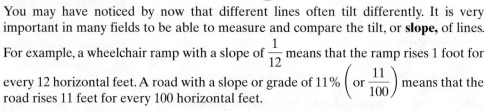

OBJECTIVE
1 Finding Slope Given Two Points

You may have noticed by now that different lines often tilt differently. It is very important in many fields to be able to measure and compare the tilt, or **slope,** of lines.

For example, a wheelchair ramp with a slope of $\frac{1}{12}$ means that the ramp rises 1 foot for every 12 horizontal feet. A road with a slope or grade of 11% $\left(\text{or } \frac{11}{100}\right)$ means that the road rises 11 feet for every 100 horizontal feet.

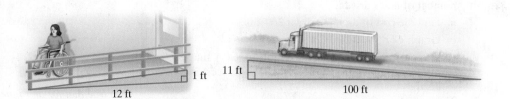

We measure the slope of a line as a ratio of **vertical change** to **horizontal change.** Slope is usually designated by the letter m.

Suppose that we want to measure the slope of the following line.

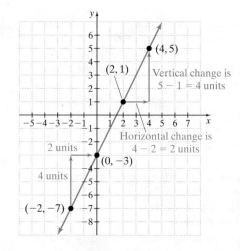

The vertical change between *both* pairs of points on the line is 4 units per horizontal change of 2 units. Then

$$\text{slope } m = \frac{\text{change in } y \text{ (vertical change)}}{\text{change in } x \text{ (horizontal change)}} = \frac{4}{2} = 2$$

Notice that slope is a **rate of change** between points. A slope of 2 or $\frac{2}{1}$ means that between pairs of points on the line, the rate of change is a vertical change of 2 units per horizontal change of 1 unit.

Consider the line in the box on the next page, which passes through the points (x_1, y_1) and (x_2, y_2). (The notation x_1 is read "x-sub-one.") The vertical change, or *rise*, between these points is the difference of the y-coordinates: $y_2 - y_1$. The horizontal change, or *run*, between the points is the difference of the x-coordinates: $x_2 - x_1$.

Slope of a Line

Given a line passing through points (x_1, y_1) and (x_2, y_2), the **slope** m of the line is

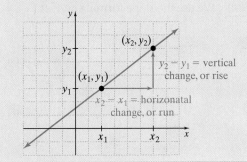

$y_2 - y_1$ = vertical change, or rise

$x_2 - x_1$ = horizonatal change, or run

$$m = \frac{\text{rise}}{\text{run}} = \frac{y_2 - y_1}{x_2 - x_1}, \text{ as long as}$$
$$x_2 \neq x_1.$$

✓**CONCEPT CHECK**

In the definition of slope, we state that $x_2 \neq x_1$. Explain why.

EXAMPLE 1   Find the slope of the line containing the points $(0, 3)$ and $(2, 5)$. Graph the line.

Solution   We use the slope formula. It does not matter which point we call (x_1, y_1) and which point we call (x_2, y_2). We'll let $(x_1, y_1) = (0, 3)$ and $(x_2, y_2) = (2, 5)$.

$$m = \frac{y_2 - y_1}{x_2 - x_1}$$

$$= \frac{5 - 3}{2 - 0} = \frac{2}{2} = 1$$

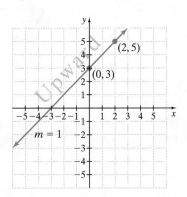

Notice in this example that the slope is *positive* and that the graph of the line containing $(0, 3)$ and $(2, 5)$ moves *upward*—that is, the y-values increase—as we go from left to right.   □

PRACTICE

1   Find the slope of the line containing the points $(4, 0)$ and $(-2, 3)$. Graph the line.

▶ **Helpful Hint**

The slope of a line is the same no matter which 2 points of a line you choose to calculate slope. The line in Example 1 also contains the point $(-3, 0)$. Below, we calculate the slope of the line using $(0, 3)$ as (x_1, y_1) and $(-3, 0)$ as (x_2, y_2).

$$m = \frac{y_2 - y_1}{x_2 - x_1} = \frac{0 - 3}{-3 - 0} = \frac{-3}{-3} = 1 \quad \text{Same slope as found in Example 1.}$$

Answer to Concept Check:
So that the denominator is not 0

EXAMPLE 2 Find the slope of the line containing the points $(5, -4)$ and $(-3, 3)$. Graph the line.

Solution We use the slope formula and let $(x_1, y_1) = (5, -4)$ and $(x_2, y_2) = (-3, 3)$.

$$m = \frac{y_2 - y_1}{x_2 - x_1}$$

$$= \frac{3 - (-4)}{-3 - 5} = \frac{7}{-8} = -\frac{7}{8}$$

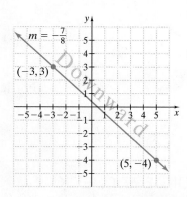

Notice in this example that the slope is negative and that the graph of the line through $(5, -4)$ and $(-3, 3)$ moves downward—that is, the y-values decrease—as we go from left to right. □

PRACTICE

2 Find the slope of the line containing the points $(-5, -4)$ and $(5, 2)$. Graph the line.

> ▶ **Helpful Hint**
>
> When we are trying to find the slope of a line through two given points, it makes no difference which given point is called (x_1, y_1) and which is called (x_2, y_2). Once an x-coordinate is called x_1, however, make sure its corresponding y-coordinate is called y_1.

✓**CONCEPT CHECK**
Find and correct the error in the following calculation of slope of the line containing the points $(12, 2)$ and $(4, 7)$.

$$m = \frac{12 - 4}{2 - 7} = \frac{8}{-5} = -\frac{8}{5}$$

OBJECTIVE

2 Finding Slope Given an Equation

As we have seen, the slope of a line is defined by two points on the line. Thus, if we know the equation of a line, we can find its slope.

EXAMPLE 3 Find the slope of the line whose equation is $f(x) = \frac{2}{3}x + 4$.

Solution Two points are needed on the line defined by $f(x) = \frac{2}{3}x + 4$ or $y = \frac{2}{3}x + 4$ to find its slope. We will use intercepts as our two points.

If $x = 0$, then	If $y = 0$, then
$y = \frac{2}{3} \cdot 0 + 4$	$0 = \frac{2}{3}x + 4$
$y = 4$	$-4 = \frac{2}{3}x$ Subtract 4.
	$\frac{3}{2}(-4) = \frac{3}{2} \cdot \frac{2}{3}x$ Multiply by $\frac{3}{2}$.
	$-6 = x$

Answer to Concept Check:
$m = \dfrac{2 - 7}{12 - 4} = \dfrac{-5}{8} = -\dfrac{5}{8}$

Use the points $(0, 4)$ and $(-6, 0)$ to find the slope. Let (x_1, y_1) be $(0, 4)$ and (x_2, y_2) be $(-6, 0)$. Then

$$m = \frac{y_2 - y_1}{x_2 - x_1} = \frac{0 - 4}{-6 - 0} = \frac{-4}{-6} = \frac{2}{3}$$

PRACTICE
3   Find the slope of the line whose equation is $f(x) = -4x + 6$.

Analyzing the results of Example 3, you may notice a striking pattern:

The slope of $y = \frac{2}{3}x + 4$ is $\frac{2}{3}$, the same as the coefficient of x.

Also, the y-intercept is $(0, 4)$, as expected.

When a linear equation is written in the form $f(x) = mx + b$ or $y = mx + b$, m is the slope of the line and $(0, b)$ is its y-intercept. The form $y = mx + b$ is appropriately called the **slope–intercept form.**

> **Slope–Intercept Form**
>
> When a linear equation in two variables is written in slope–intercept form,
>
> $$\underset{\uparrow}{\text{slope}} \quad \underset{\uparrow}{y\text{-intercept is } (0, b)}$$
> $$y = mx + b$$
>
> then m is the slope of the line and $(0, b)$ is the y-intercept of the line.

EXAMPLE 4   Find the slope and the y-intercept of the line $3x - 4y = 4$.

Solution   We write the equation in slope–intercept form by solving for y.

$$3x - 4y = 4$$
$$-4y = -3x + 4 \qquad \text{Subtract } 3x \text{ from both sides.}$$
$$\frac{-4y}{-4} = \frac{-3x}{-4} + \frac{4}{-4} \qquad \text{Divide both sides by } -4.$$
$$y = \frac{3}{4}x - 1 \qquad \text{Simplify.}$$

The coefficient of x, $\frac{3}{4}$, is the slope, and the y-intercept is $(0, -1)$.

PRACTICE
4   Find the slope and the y-intercept of the line $2x - 3y = 9$.

OBJECTIVE
3   Interpreting Slope–Intercept Form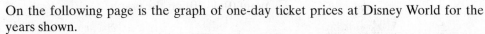

On the following page is the graph of one-day ticket prices at Disney World for the years shown.

Notice that the graph resembles the graph of a line. Recall that businesses often depend on equations that closely fit graphs like this one to model the data and to predict future trends. By the **least squares** method, the linear function $f(x) = 3.32x + 44.10$ approximates the data shown, where x is the number of years since 2000 and y is the ticket price for that year.

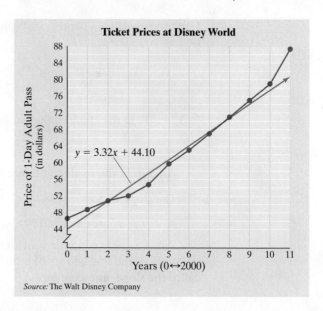

Source: The Walt Disney Company

> ▶ **Helpful Hint**
>
> The notation 0 ↔ 2000 means that the number 0 corresponds to the year 2000, 1 corresponds to the year 2001, and so on.

EXAMPLE 5 **Predicting Future Prices**

The adult one-day pass price y for Disney World is given by

$$y = 3.32x + 44.10$$

where x is the number of years since 2000.

a. Use this equation to predict the ticket prices for 2015.

b. What does the slope of this equation mean?

c. What does the y-intercept of this equation mean?

Solution:

a. To predict the price of a pass in 2015, we need to find y when x is 15. (Since the year 2000 corresponds to $x = 0$, the year 2015 corresponds to the year $2015 - 2000 = 15$.)

$$y = 3.32x + 44.10$$
$$= 3.32(15) + 44.10 \quad \text{Let } x = 15$$
$$= 93.90$$

We predict that in the year 2015, the price of an adult one-day pass to Disney World will be about $93.90.

b. The slope of $y = 3.32x + 41.10$ is 3.32. We can think of this number as $\dfrac{\text{rise}}{\text{run}}$, or $\dfrac{3.32}{1}$.

This means that the ticket price increases on average by $3.32 each year.

c. The y-intercept of $y = 3.32x + 44.10$ is 44.10. Notice that it corresponds to the point $(0, 44.10)$ on the graph.
    ↑   ↑
    Year price

That means that at year $x = 0$, or 2000, the ticket price was about $44.10.   □

PRACTICE

5 For the period 1980 through 2020, the number of people y age 85 or older living in the United States is given by the equation $y = 110{,}520x + 2{,}127{,}400$, where x is the number of years since 1980. (*Source:* Based on data and estimates from the U.S. Bureau of the Census)

a. Estimate the number of people age 85 or older living in the United States in 2010.

b. What does the slope of this equation mean?

c. What does the *y*-intercept of this equation mean?

OBJECTIVE
4   Finding Slopes of Horizontal and Vertical Lines

Next we find the slopes of two special types of lines: vertical lines and horizontal lines.

> **EXAMPLE 6**   Find the slope of the line $x = -5$.

Solution   Recall that the graph of $x = -5$ is a vertical line with *x*-intercept $(-5, 0)$. To find the slope, we find two ordered pair solutions of $x = -5$. Of course, solutions of $x = -5$ must have an *x*-value of -5. We will let $(x_1, y_1) = (-5, 0)$ and $(x_2, y_2) = (-5, 4)$. Then

$$
\begin{aligned}
m &= \frac{y_2 - y_1}{x_2 - x_1} \\[2mm]
&= \frac{4 - 0}{-5 - (-5)} \\[2mm]
&= \frac{4}{0}
\end{aligned}
$$

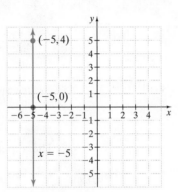

Since $\dfrac{4}{0}$ is undefined, we say that the slope of the vertical line $x = -5$ is undefined. □

PRACTICE
6   Find the slope of the line $x = 4$.

> **EXAMPLE 7**   Find the slope of the line $y = 2$.

Solution   Recall that the graph of $y = 2$ is a horizontal line with *y*-intercept $(0, 2)$. To find the slope, we find two points on the line, such as $(0, 2)$ and $(1, 2)$, and use these points to find the slope.

$$
\begin{aligned}
m &= \frac{2 - 2}{1 - 0} \\[2mm]
&= \frac{0}{1} \\[2mm]
&= 0
\end{aligned}
$$

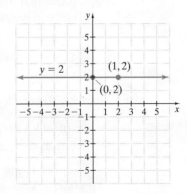

The slope of the horizontal line $y = 2$ is 0. □

PRACTICE
7   Find the slope of the line $y = -3$.

From the previous two examples, we have the following generalization.

> The slope of any vertical line is undefined.
> The slope of any horizontal line is 0.

▶ Helpful Hint

Slope of 0 and undefined slope are not the same. Vertical lines have undefined slope, whereas horizontal lines have slope of 0.

The following four graphs summarize the overall appearance of lines with positive, negative, zero, or undefined slopes.

Appearance of Lines with Given Slopes

Increasing line, positive slope	Decreasing line, negative slope	Horizontal line, zero slope	Vertical line, undefined slope

The appearance of a line can give us further information about its slope.

The graphs of $y = \frac{1}{2}x + 1$ and $y = 5x + 1$ are shown to the right. Recall that the graph of $y = \frac{1}{2}x + 1$ has a slope of $\frac{1}{2}$ and that the graph of $y = 5x + 1$ has a slope of 5.

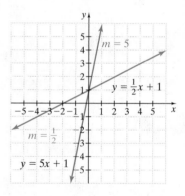

Notice that the line with the slope of 5 is steeper than the line with the slope of $\frac{1}{2}$. This is true in general for positive slopes.

> For a line with positive slope m, as m increases, the line becomes steeper.

To see why this is so, compare the slopes from above.
$\frac{1}{2}$ means a vertical change of 1 unit per horizontal change of 2 units; 5 or $\frac{10}{2}$ means a vertical change of 10 units per horizontal change of 2 units.

For larger positive slopes, the vertical change is greater for the same horizontal change. Thus, larger positive slopes mean steeper lines.

OBJECTIVE

5 Comparing Slopes of Parallel and Perpendicular Lines

Slopes of lines can help us determine whether lines are parallel. Parallel lines are distinct lines with the same steepness, so it follows that they have the same slope.

> ### Parallel Lines
> Two nonvertical lines are parallel if they have the same slope and different *y*-intercepts.

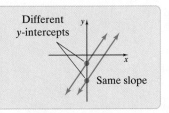

How do the slopes of perpendicular lines compare? (Two lines intersecting at right angles are called **perpendicular lines**.) Suppose that a line has a slope of $\frac{a}{b}$. If the line is rotated 90°, the rise and run are now switched, except that the run is now negative. This means that the new slope is $-\frac{b}{a}$. Notice that

$$\left(\frac{a}{b}\right) \cdot \left(-\frac{b}{a}\right) = -1$$

This is how we tell whether two lines are perpendicular.

> ### Perpendicular Lines
> Two nonvertical lines are perpendicular if the product of their slopes is -1.

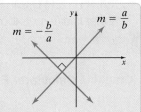

In other words, two nonvertical lines are perpendicular if the slope of one is the negative reciprocal of the slope of the other.

EXAMPLE 8   Are the following pairs of lines parallel, perpendicular, or neither?

a. $3x + 7y = 4$ **b.** $-x + 3y = 2$
 $6x + 14y = 7$ $2x + 6y = 5$

Solution   Find the slope of each line by solving each equation for *y*.

a. $3x + 7y = 4$ $\qquad\qquad$ $6x + 14y = 7$

 $7y = -3x + 4$ $\qquad\qquad$ $14y = -6x + 7$

 $\dfrac{7y}{7} = \dfrac{-3x}{7} + \dfrac{4}{7}$ $\qquad$ $\dfrac{14y}{14} = \dfrac{-6x}{14} + \dfrac{7}{14}$

 $y = -\dfrac{3}{7}x + \dfrac{4}{7}$ $\qquad\quad$ $y = -\dfrac{3}{7}x + \dfrac{1}{2}$

 ↑ $\qquad$ ↖ $\qquad\qquad\qquad$ ↑ $\qquad$ ↖

 slope $\quad$ *y*-intercept $\qquad\quad$ slope $\quad$ *y*-intercept

 $\left(0, \dfrac{4}{7}\right)$ $\qquad\qquad\qquad$ $\left(0, \dfrac{1}{2}\right)$

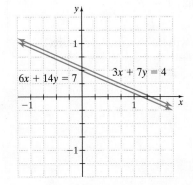

The slopes of both lines are $-\dfrac{3}{7}$.

The *y*-intercepts are different, so the lines are not the same.

Therefore, the lines are parallel. (Their graphs are shown in the margin.)

b. $-x + 3y = 2$ $2x + 6y = 5$

$3y = x + 2$ $6y = -2x + 5$

$\dfrac{3y}{3} = \dfrac{x}{3} + \dfrac{2}{3}$ $\dfrac{6y}{6} = \dfrac{-2x}{6} + \dfrac{5}{6}$

$y = \dfrac{1}{3}x + \dfrac{2}{3}$ $y = -\dfrac{1}{3}x + \dfrac{5}{6}$

     ↑   ↖   ↑   ↖

  slope y-intercept   slope y-intercept

$\left(0, \dfrac{2}{3}\right)$ $\left(0, \dfrac{5}{6}\right)$

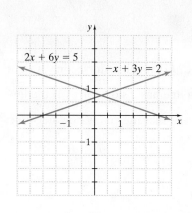

The slopes are not the same and their product is not -1. $\left[\left(\dfrac{1}{3}\right)\cdot\left(-\dfrac{1}{3}\right) = -\dfrac{1}{9}\right]$

Therefore, the lines are neither parallel nor perpendicular. (Their graphs are shown in the margin.)

PRACTICE

8   Are the following pairs of lines parallel, perpendicular, or neither?

a. $x - 2y = 3$ **b.** $4x - 3y = 2$

$2x + y = 3$ $-8x + 6y = -6$

✓CONCEPT CHECK

What is *different* about the equations of two parallel lines?

Graphing Calculator Explorations

Many graphing calculators have a TRACE feature. This feature allows you to trace along a graph and see the corresponding x- and y-coordinates appear on the screen. Use this feature for the following exercises.

Graph each function and then use the TRACE feature to complete each ordered pair solution. (Many times, the tracer will not show an exact x- or y-value asked for. In each case, trace as closely as you can to the given x- or y-coordinate and approximate the other, unknown coordinate to one decimal place.)

1. $y = 2.3x + 6.7$ **2.** $y = -4.8x + 2.9$

$x = 5.1, y = ?$ $x = -1.8, y = ?$

3. $y = -5.9x - 1.6$ **4.** $y = 0.4x - 8.6$

$x = ?, y = 7.2$ $x = ?, y = -4.4$

5. $y = x^2 + 5.2x - 3.3$ **6.** $y = 5x^2 - 6.2x - 8.3$

$x = 2.3, y = ?$ $x = 3.2, y = ?$

$x = ?, y = 36$ $x = ?, y = 12$

(There will be two answers here.)   (There will be two answers here.)

Answer to Concept Check:
y-intercepts are different

Vocabulary, Readiness & Video Check

Use the choices below to fill in each blank. Some choices may be used more than once and some not at all.

horizontal	the same	-1	y-intercepts	$(0, b)$	slope
vertical	different	m	x-intercepts	$(b, 0)$	slope–intercept

1. The measure of the steepness or tilt of a line is called _____.
2. The slope of a line through two points is measured by the ratio of _____ change to _____ change.
3. If a linear equation is in the form $y = mx + b$, or $f(x) = mx + b$, the slope of the line is _____ and the y-intercept is _____.
4. The form $y = mx + b$ or $f(x) = mx + b$ is the _____ form.
5. The slope of a _____ line is 0.
6. The slope of a _____ line is undefined.
7. Two non-vertical perpendicular lines have slopes whose product is _____.
8. Two non-vertical lines are parallel if they have _____ slope and different _____.

Martin-Gay Interactive Videos

See Video 3.4

Watch the section lecture video and answer the following questions.

OBJECTIVE 1

9. Based on ▭ Examples 1 and 2, complete the following statements. A positive slope means the line _____ from left to right. A negative slope means the line _____ from left to right.

OBJECTIVE 2

10. From ▭ Example 3, how do you write an equation in slope–intercept form? Once the equation is in slope–intercept form, how do you determine the slope?

OBJECTIVE 3

11. ▭ Example 4 gives a linear equation that models a real-life application. The equation is rewritten in slope–intercept form. How does this help us better understand how the equation relates to the application? What specific information is discovered?

OBJECTIVE 4

12. In the lecture after ▭ Example 6, different slopes are summarized. What's the difference between zero slope and undefined slope? What does "no slope" mean?

OBJECTIVE 5

13. From the lecture before ▭ Example 7, why do the slope rules for parallel or perpendicular lines indicate nonvertical lines only?

3.4 Exercise Set   MyMathLab®

Find the slope of the line that goes through the given points. See Examples 1 and 2.

1. $(3, 2), (8, 11)$
2. $(1, 6), (7, 11)$
3. $(3, 1), (1, 8)$
4. $(2, 9), (6, 4)$
5. $(-2, 8), (4, 3)$
6. $(3, 7), (-2, 11)$
▶ 7. $(-2, -6), (4, -4)$
8. $(-3, -4), (-1, 6)$
9. $(-3, -1), (-12, 11)$
10. $(3, -1), (-6, 5)$
11. $(-2, 5), (3, 5)$
12. $(4, 2), (4, 0)$
13. $(-1, 1), (-1, -5)$
14. $(-2, -5), (3, -5)$

15. $(0, 6), (-3, 0)$
16. $(5, 2), (0, 5)$
▶ 17. $(-1, 2), (-3, 4)$
18. $(3, -2), (-1, -6)$

Decide whether a line with the given slope slants upward or downward from left to right or is horizontal or vertical.

19. $m = \dfrac{7}{6}$
20. $m = -3$
21. $m = 0$
22. m is undefined

Two lines are graphed on each set of axes. Decide whether l_1 or l_2 has the greater slope. See the boxed material on page 158.

23.

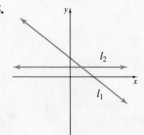

24.

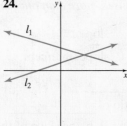

25.

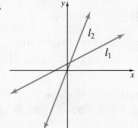

26.

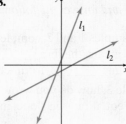

27.

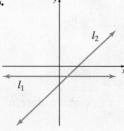

28.

C

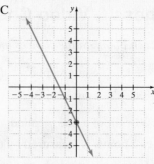

D

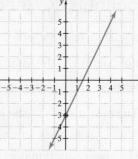

37. $f(x) = 2x + 3$ **38.** $f(x) = 2x - 3$
39. $f(x) = -2x + 3$ **40.** $f(x) = -2x - 3$

Find the slope of each line. See Examples 6 and 7.

▶ **41.** $x = 1$ **42.** $y = -2$
▶ **43.** $y = -3$ **44.** $x = 4$
  45. $x + 2 = 0$ **46.** $y - 7 = 0$

MIXED PRACTICE

Find the slope and the y-intercept of each line. See Examples 3 through 7.

47. $f(x) = -x + 5$
48. $f(x) = x + 2$
49. $-6x + 5y = 30$
50. $4x - 7y = 28$
51. $3x + 9 = y$
52. $2y - 7 = x$
53. $y = 4$ **54.** $x = 7$
55. $f(x) = 7x$ **56.** $f(x) = \frac{1}{7}x$
57. $6 + y = 0$ **58.** $x - 7 = 0$
59. $2 - x = 3$ **60.** $2y + 4 = -7$

Find the slope and the y-intercept of each line. See Examples 3 and 4.

29. $f(x) = 5x - 2$
30. $f(x) = -2x + 6$
31. $2x + y = 7$
32. $-5x + y = 10$
▶ **33.** $2x - 3y = 10$
34. $-3x - 4y = 6$
35. $f(x) = \frac{1}{2}x$
36. $f(x) = -\frac{1}{4}x$

Match each graph with its equation. See Examples 1 and 2.

A

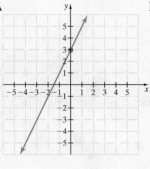

B

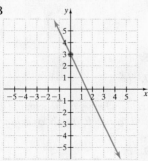

Decide whether the lines are parallel, perpendicular, or neither. See Example 8.

61. $y = 12x + 6$ **62.** $y = 5x + 8$
       $y = 12x - 2$ $y = 5x - 8$
▶ **63.** $y = -9x + 3$ **64.** $y = 2x - 12$
       $y = \frac{3}{2}x - 7$ $y = \frac{1}{2}x - 6$
△ **65.** $f(x) = -3x + 6$ △ **66.** $f(x) = 7x - 6$
       $g(x) = \frac{1}{3}x + 5$ $g(x) = -\frac{1}{7}x + 2$
△ **67.** $-4x + 2y = 5$ △ **68.** $-8x + 20y = 7$
       $2x - y = 7$ $2x - 5y = 0$
△ **69.** $-2x + 3y = 1$ △ **70.** $2x - y = -10$
       $3x + 2y = 12$ $2x + 4y = 2$

Use the points shown on the graphs to determine the slope of each line. See Examples 1 and 2.

71.

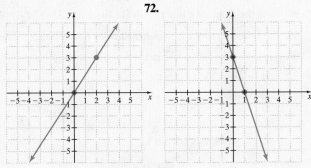

72.

73.

74.

Find each slope. See Examples 1 and 2.

75. Find the pitch, or slope, of the roof shown.

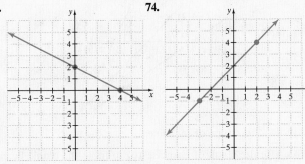
8 ft
12 ft

76. Upon takeoff, a Delta Airlines jet climbs to 3 miles as it passes over 25 miles of land below it. Find the slope of its climb.

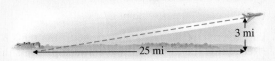
3 mi
25 mi

77. Driving down Bald Mountain in Wyoming, Bob Dean finds that he descends 1600 feet in elevation by the time he is 2.5 miles (horizontally) away from the high point on the mountain road. Find the slope of his descent rounded to two decimal places (1 mile = 5280 feet).

78. Find the grade, or slope, of the road shown.

15 ft
100 ft

Solve. See Example 5.

79. The life expectancy y for females born in the United States is given by the equation $y = 0.16x + 71.6$, where x is the number of years after 1950. (*Source:* U.S. National Center for Health Statistics)

   a. Find the life expectancy of an American female born in 1980.

   b. Find and interpret the slope of the equation.

   c. Find and interpret the y-intercept of the equation.

80. The average annual income y of an American woman with a bachelor's degree is given by the equation $y = 4207.4x + 38{,}957$, where x is the number of years after 2005. (*Source:* Based on data from U.S. Bureau of the Census, 2005–2009)

   a. Find the average income of an American woman with a bachelor's degree in 2008.

   b. Find and interpret the slope of the equation.

   c. Find and interpret the y-intercept of the equation.

81. One of the top 10 occupations in terms of job growth in the next few years is expected to be physician assistants. The number of people, y, in thousands, employed as physician assistants in the United States can be estimated by the linear equation $29x - 10y = -750$, where x is the number of years after 2008. (*Source:* Based on projections from the U.S. Bureau of Labor Statistics, 2008–2018)

   a. Find the slope and the y-intercept of the linear equation.

   b. What does the slope mean in this context?

   c. What does the y-intercept mean in this context?

82. One of the fastest growing occupations over the next few years is expected to be network system and data communications analysts. The number of people, y, in thousands, employed as network system and data communications analysts in the United States can be estimated by the linear equation $78x - 5y = -1460$, where x is the number of years after 2008. (*Source:* Based on projections from the U.S. Bureau of Labor Statistics 2008–2018)

   a. Find the slope and the y-intercept of the linear equation.

   b. What does the slope mean in this context?

   c. What does the y-intercept mean in this context?

83. The number of U.S. admissions y (in billions) to movie theaters can be estimated by the linear equation $y = -0.04x + 1.62$, where x is the number of years after 2002. (*Source:* Motion Picture Association of America)

a. Use this equation to estimate the number of movie admissions in the United States in 2007.

b. Use this equation to predict in what year the number of movie admissions in the United States will be below 1 billion. (*Hint:* Let $y = 1$ and solve for x.)

c. Use this equation to estimate the number of movie admissions in the present year. Do you go to the movies? Do your friends?

84. The amount of restaurant sales y (in billions of dollars) in the United States can be estimated by the linear equation $y = 13.34x + 5.36$, where x is the number of years after 1970. (*Source:* Based on data from the National Restaurant Association)

a. Use this equation to approximate the amount of restaurant sales in 2005.

b. Use this equation to approximate the year in which the amount of restaurant sales will exceed $700 billion.

c. Use this equation to approximate the amount of restaurant sales in the current year. Do you go out to eat often? Do your friends?

REVIEW AND PREVIEW

Simplify and solve for y. See Section 2.3.

85. $y - 2 = 5(x + 6)$

86. $y - 0 = -3[x - (-10)]$

87. $y - (-1) = 2(x - 0)$

88. $y - 9 = -8[x - (-4)]$

CONCEPT EXTENSIONS

Each slope calculation is incorrect. Find the error and correct the calculation. See the second Concept Check in this section.

89. $(-2, 6)$ and $(7, -14)$

$$m = \frac{-14 - 6}{7 - 2} = \frac{-20}{5} = -4$$

90. $(-1, 4)$ and $(-3, 9)$

$$m = \frac{9 - 4}{-3 - 1} = \frac{5}{-4} \text{ or } -\frac{5}{4}$$

91. $(-8, -10)$ and $(-11, -5)$

$$m = \frac{-10 - 5}{-8 - 11} = \frac{-15}{-19} = \frac{15}{19}$$

92. $(0, -4)$ and $(-6, -6)$

$$m = \frac{0 - (-6)}{-4 - (-6)} = \frac{6}{2} = 3$$

△ **93.** Find the slope of a line parallel to the line $f(x) = -\frac{7}{2}x - 6$.

△ **94.** Find the slope of a line parallel to the line $f(x) = x$.

△ **95.** Find the slope of a line perpendicular to the line

$$f(x) = -\frac{7}{2}x - 6.$$

△ **96.** Find the slope of a line perpendicular to the line $f(x) = x$.

△ **97.** Find the slope of a line parallel to the line $5x - 2y = 6$.

△ **98.** Find the slope of a line parallel to the line $-3x + 4y = 10$.

△ **99.** Find the slope of a line perpendicular to the line $5x - 2y = 6$.

△ **100.** Find the slope of a line perpendicular to the line

$$-3x + 4y = 10.$$

Each line below has negative slope.

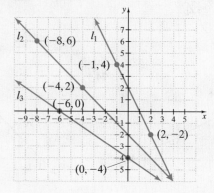

101. Find the slope of each line.

102. Use the result of Exercise 101 to fill in the blank. For lines with negative slopes, the steeper line has the _____ (greater/lesser) slope.

The following graph shows the altitude of a seagull in flight over a time period of 30 seconds.

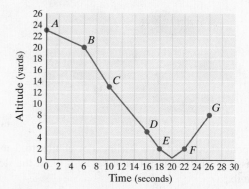

103. Find the coordinates of point B.

104. Find the coordinates of point C.

105. Find the rate of change of altitude between points B and C. (Recall that the rate of change between points is the slope between points. This rate of change will be in yards per second.)

106. Find the rate of change of altitude (in yards per second) between points F and G.

107. Explain how merely looking at a line can tell us whether its slope is negative, positive, undefined, or zero.

108. Explain why the graph of $y = b$ is a horizontal line.

109. Explain whether two lines, both with positive slopes, can be perpendicular.

110. Explain why it is reasonable that nonvertical parallel lines have the same slope.

111. Professional plumbers suggest that a sewer pipe should be sloped 0.25 inch for every foot. Find the recommended slope for a sewer pipe. (*Source: Rules of Thumb* by Tom Parker, Houghton Mifflin Company)

112. a. On a single screen, graph $y = \frac{1}{2}x + 1$, $y = x + 1$, and $y = 2x + 1$. Notice the change in slope for each graph.

  b. On a single screen, graph $y = -\frac{1}{2}x + 1$, $y = -x + 1$, and $y = -2x + 1$. Notice the change in slope for each graph.

c. Determine whether the following statement is true or false for slope m of a given line. As $|m|$ becomes greater, the line becomes steeper.

113. Support the result of Exercise 67 by graphing the pair of equations on a graphing calculator.

114. Support the result of Exercise 70 by graphing the pair of equations on a graphing calculator. (*Hint:* Use the window showing $[-15, 15]$ on the x-axis and $[-10, 10]$ on the y-axis.)

3.5   Equations of Lines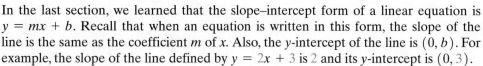

OBJECTIVES

1 Graph a Line Using Its Slope and y-Intercept.

2 Use the Slope–Intercept Form to Write the Equation of a Line.

3 Use the Point–Slope Form to Write the Equation of a Line.

4 Write Equations of Vertical and Horizontal Lines.

5 Find Equations of Parallel and Perpendicular Lines.

OBJECTIVE

1 **Graphing a Line Using Its Slope and y-Intercept**

In the last section, we learned that the slope–intercept form of a linear equation is $y = mx + b$. Recall that when an equation is written in this form, the slope of the line is the same as the coefficient m of x. Also, the y-intercept of the line is $(0, b)$. For example, the slope of the line defined by $y = 2x + 3$ is 2 and its y-intercept is $(0, 3)$.

We may also use the slope–intercept form to graph a linear equation.

EXAMPLE 1   Graph $y = \frac{1}{4}x - 3$.

Solution   Recall that the slope of the graph of $y = \frac{1}{4}x - 3$ is $\frac{1}{4}$ and the y-intercept is $(0, -3)$. To graph the line, we first plot the y-intercept $(0, -3)$. To find another point on the line, we recall that slope is $\dfrac{\text{rise}}{\text{run}} = \dfrac{1}{4}$. Another point may then be plotted by starting at $(0, -3)$, rising 1 unit up, and then running 4 units to the right. We are now at the point $(4, -2)$. The graph is the line through these two points.

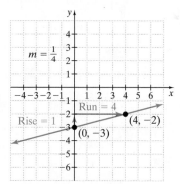

Notice that the line does have a y-intercept of $(0, -3)$ and a slope of $\frac{1}{4}$. □

PRACTICE

1   Graph $y = \frac{3}{4}x + 2$.

EXAMPLE 2   Graph $2x + 3y = 12$.

Solution   First, we solve the equation for y to write it in slope–intercept form. In slope–intercept form, the equation is $y = -\frac{2}{3}x + 4$. Next we plot the y-intercept $(0, 4)$.

To find another point on the line, we use the slope $-\frac{2}{3}$, which can be written as $\dfrac{\text{rise}}{\text{run}} = \dfrac{-2}{3}$. We start at $(0, 4)$ and move down 2 units since the numerator of the slope is -2; then we move 3 units to the right since the denominator of the slope is 3. We arrive at the point $(3, 2)$. The line through these points is the graph, shown below to the left.

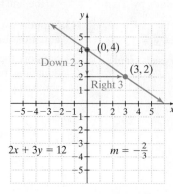

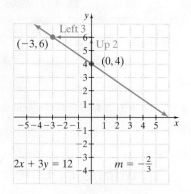

The slope $\dfrac{-2}{3}$ can also be written as $\dfrac{2}{-3}$, so to find another point, we could start at $(0, 4)$ and move up 2 units and then 3 units to the left. We would arrive at the point $(-3, 6)$. The line through $(-3, 6)$ and $(0, 4)$ is the same line as shown previously through $(3, 2)$ and $(0, 4)$. See the graph above to the right.   □

PRACTICE
2   Graph $x + 2y = 6$.

OBJECTIVE
2   Using Slope–Intercept Form to Write Equations of Lines

We may also use the slope–intercept form to write the equation of a line given its slope and y-intercept. The equation of a line is a linear equation in 2 variables that, if graphed, would produce the line described.

EXAMPLE 3   Write an equation of the line with y-intercept $(0, -3)$ and slope of $\dfrac{1}{4}$.

Solution   We want to write a linear equation in 2 variables that describes the line with y-intercept $(0, -3)$ and has a slope of $\dfrac{1}{4}$. We are given the slope and the y-intercept. Let $m = \dfrac{1}{4}$ and $b = -3$ and write the equation in slope–intercept form, $y = mx + b$.

$$y = mx + b$$
$$y = \frac{1}{4}x + (-3) \quad \text{Let } m = \frac{1}{4} \text{ and } b = -3.$$
$$y = \frac{1}{4}x - 3 \quad \text{Simplify.} \qquad □$$

PRACTICE
3   Write an equation of the line with y-intercept $(0, 4)$ and slope of $-\dfrac{3}{4}$.

✓CONCEPT CHECK

What is wrong with the following equation of a line with y-intercept $(0, 4)$ and slope 2?

$$y = 4x + 2$$

Answer to Concept Check:
y-intercept and slope were switched,
should be $y = 2x + 4$

OBJECTIVE
3   Using Point–Slope Form to Write Equations of Lines

When the slope of a line and a point on the line are known, the equation of the line can also be found. To do this, use the slope formula to write the slope of a line that passes through points (x_1, y_1) and (x, y). We have

$$m = \frac{y - y_1}{x - x_1}$$

Multiply both sides of this equation by $x - x_1$ to obtain

$$y - y_1 = m(x - x_1)$$

This form is called the **point–slope form** of the equation of a line.

Point–Slope Form of the Equation of a Line

The **point–slope form** of the equation of a line is

$$\overset{\text{slope}}{\underset{\text{point}}{y - y_1 = m(x - x_1)}}$$

where m is the slope of the line and (x_1, y_1) is a point on the line.

EXAMPLE 4   Find an equation of the line with slope -3 containing the point $(1, -5)$. Write the equation in slope–intercept form $y = mx + b$.

Solution   Because we know the slope and a point of the line, we use the point–slope form with $m = -3$ and $(x_1, y_1) = (1, -5)$.

$$
\begin{aligned}
y - y_1 &= m(x - x_1) &&\text{Point–slope form} \\
y - (-5) &= -3(x - 1) &&\text{Let } m = -3 \text{ and } (x_1, y_1) = (1, -5). \\
y + 5 &= -3x + 3 &&\text{Apply the distributive property.} \\
y &= -3x - 2 &&\text{Write in slope–intercept form.}
\end{aligned}
$$

In slope–intercept form, the equation is $y = -3x - 2$.   □

PRACTICE
4   Find an equation of the line with slope -4 containing the point $(-2, 5)$. Write the equation in slope–intercept form $y = mx + b$.

▶ Helpful Hint
Remember, "slope–intercept form" means the equation is "solved for y."

EXAMPLE 5   Find an equation of the line through points $(4, 0)$ and $(-4, -5)$. Write the equation using function notation.

Solution   First, find the slope of the line.

$$m = \frac{-5 - 0}{-4 - 4} = \frac{-5}{-8} = \frac{5}{8}$$

Next, use the point–slope form. Replace (x_1, y_1) by either $(4, 0)$ or $(-4, -5)$ in the point–slope equation. We will choose the point $(4, 0)$. The line through $(4, 0)$ with slope $\frac{5}{8}$ is

$$
\begin{aligned}
y - y_1 &= m(x - x_1) &&\text{Point–slope form.} \\
y - 0 &= \frac{5}{8}(x - 4) &&\text{Let } m = \frac{5}{8} \text{ and } (x_1, y_1) = (4, 0). \\
8y &= 5(x - 4) &&\text{Multiply both sides by 8.} \\
8y &= 5x - 20 &&\text{Apply the distributive property.}
\end{aligned}
$$

To write the equation using function notation, we solve for y, then replace y with $f(x)$.

$$8y = 5x - 20$$

$$y = \frac{5}{8}x - \frac{20}{8} \quad \text{Divide both sides by 8.}$$

$$f(x) = \frac{5}{8}x - \frac{5}{2} \quad \text{Write using function notation.} \qquad \square$$

PRACTICE

5   Find an equation of the line through points $(-1, 2)$ and $(2, 0)$. Write the equation using function notation.

> ▶ **Helpful Hint**
>
> If two points of a line are given, either one may be used with the point–slope form to write an equation of the line.

EXAMPLE 6   Find an equation of the line graphed. Write the equation in standard form.

Solution   First, find the slope of the line by identifying the coordinates of the noted points on the graph.

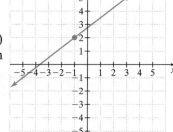

The points have coordinates $(-1, 2)$ and $(3, 5)$.

$$m = \frac{5 - 2}{3 - (-1)} = \frac{3}{4}$$

Next, use the point–slope form. We will choose $(3, 5)$ for (x_1, y_1), although it makes no difference which point we choose. The line through $(3, 5)$ with slope $\frac{3}{4}$ is

$$y - y_1 = m(x - x_1) \quad \text{Point–slope form}$$

$$y - 5 = \frac{3}{4}(x - 3) \quad \text{Let } m = \frac{3}{4} \text{ and } (x_1, y_1) = (3, 5).$$

$$4(y - 5) = 3(x - 3) \quad \text{Multiply both sides by 4.}$$

$$4y - 20 = 3x - 9 \quad \text{Apply the distributive property.}$$

To write the equation in standard form, move x- and y-terms to one side of the equation and any numbers (constants) to the other side.

$$4y - 20 = 3x - 9$$

$$-3x + 4y = 11 \quad \text{Subtract } 3x \text{ from both sides and add 20 to both sides.}$$

The equation of the graphed line is $-3x + 4y = 11$. $\qquad \square$

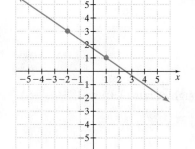

PRACTICE

6   Find an equation of the line graphed in the margin. Write the equation in standard form.

The point–slope form of an equation is very useful for solving real-world problems.

EXAMPLE 7   **Predicting Sales**

Southern Star Realty is an established real estate company that has enjoyed constant growth in sales since 2000. In 2002, the company sold 200 houses, and in 2007, the company sold 275 houses. Use these figures to predict the number of houses this company will sell in the year 2016.

Solution

1. UNDERSTAND. Read and reread the problem. Then let

$x =$ the number of years after 2000 and

$y =$ the number of houses sold in the year corresponding to x.

The information provided then gives the ordered pairs $(2, 200)$ and $(7, 275)$. To better visualize the sales of Southern Star Realty, we graph the linear equation that passes through the points $(2, 200)$ and $(7, 275)$.

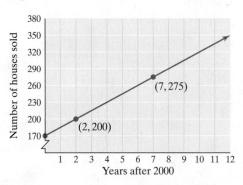

2. TRANSLATE. We write a linear equation that passes through the points $(2, 200)$ and $(7, 275)$. To do so, we first find the slope of the line.

$$m = \frac{275 - 200}{7 - 2} = \frac{75}{5} = 15$$

Then, using the point–slope form and the point $(2, 200)$ to write the equation, we have

$$y - y_1 = m(x - x_1)$$
$$y - 200 = 15(x - 2) \quad \text{Let } m = 15 \text{ and } (x_1, y_1) = (2, 200).$$
$$y - 200 = 15x - 30 \quad \text{Multiply.}$$
$$y = 15x + 170 \quad \text{Add 200 to both sides.}$$

3. SOLVE. To predict the number of houses sold in the year 2016, we use $y = 15x + 170$ and complete the ordered pair $(16, \ )$, since $2016 - 2000 = 16$.

$$y = 15(16) + 170 \quad \text{Let } x = 16.$$
$$y = 410$$

4. INTERPRET.

Check: Verify that the point $(16, 410)$ is a point on the line graphed in Step 1.

State: Southern Star Realty should expect to sell 410 houses in the year 2016.   □

PRACTICE

7   Southwest Florida, including Fort Myers and Cape Coral, has been a growing real estate market in past years. In 2002, there were 7513 house sales in the area, and in 2006, there were 9198 house sales. Use these figures to predict the number of house sales there will be in 2014.

OBJECTIVE

4   **Writing Equations of Vertical and Horizontal Lines**

Two special types of linear equations are linear equations whose graphs are vertical and horizontal lines.

EXAMPLE 8 Find an equation of the horizontal line containing the point $(2, 3)$.

Solution Recall that a horizontal line has an equation of the form $y = b$. Since the line contains the point $(2, 3)$, the equation is $y = 3$, as shown to the right.

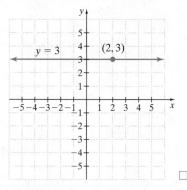

PRACTICE
8 Find the equation of the horizontal line containing the point $(6, -2)$.

EXAMPLE 9 Find an equation of the line containing the point $(2, 3)$ with undefined slope.

Solution Since the line has undefined slope, the line must be vertical. A vertical line has an equation of the form $x = c$. Since the line contains the point $(2, 3)$, the equation is $x = 2$, as shown to the right.

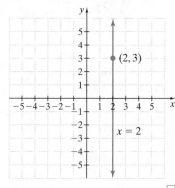

PRACTICE
9 Find an equation of the line containing the point $(6, -2)$ with undefined slope.

OBJECTIVE
5 Finding Equations of Parallel and Perpendicular Lines

Next, we find equations of parallel and perpendicular lines.

⚠ **EXAMPLE 10** Find an equation of the line containing the point $(4, 4)$ and parallel to the line $2x + 3y = -6$. Write the equation in standard form.

Solution Because the line we want to find is *parallel* to the line $2x + 3y = -6$, the two lines must have equal slopes. Find the slope of $2x + 3y = -6$ by writing it in the form $y = mx + b$. In other words, solve the equation for y.

$$2x + 3y = -6$$

$$3y = -2x - 6 \qquad \text{Subtract } 2x \text{ from both sides.}$$

$$y = \frac{-2x}{3} - \frac{6}{3} \qquad \text{Divide by 3.}$$

$$y = -\frac{2}{3}x - 2 \qquad \text{Write in slope–intercept form.}$$

The slope of this line is $-\frac{2}{3}$. Thus, a line parallel to this line will also have a slope of $-\frac{2}{3}$. The equation we are asked to find describes a line containing the point $(4, 4)$ with a slope of $-\frac{2}{3}$. We use the point–slope form.

> **Helpful Hint**
> Multiply both sides of the equation $2x + 3y = 20$ by -1 and it becomes $-2x - 3y = -20$. Both equations are in standard form, and their graphs are the same line.

$$y - y_1 = m(x - x_1)$$

$$y - 4 = -\frac{2}{3}(x - 4) \quad \text{Let } m = -\frac{2}{3}, x_1 = 4, \text{ and } y_1 = 4.$$

$$3(y - 4) = -2(x - 4) \quad \text{Multiply both sides by 3.}$$

$$3y - 12 = -2x + 8 \quad \text{Apply the distributive property.}$$

$$2x + 3y = 20 \quad \text{Write in standard form.} \qquad \square$$

PRACTICE

10   Find an equation of the line containing the point $(8, -3)$ and parallel to the line $3x + 4y = 1$. Write the equation in standard form.

.... ■

EXAMPLE 11   Write a function that describes the line containing the point $(4, 4)$ that is perpendicular to the line $2x + 3y = -6$.

Solution   In the previous example, we found that the slope of the line $2x + 3y = -6$ is $-\frac{2}{3}$. A line perpendicular to this line will have a slope that is the negative reciprocal of $-\frac{2}{3}$, or $\frac{3}{2}$. From the point–slope equation, we have

$$y - y_1 = m(x - x_1)$$

$$y - 4 = \frac{3}{2}(x - 4) \quad \text{Let } x_1 = 4, y_1 = 4, \text{ and } m = \frac{3}{2}.$$

$$2(y - 4) = 3(x - 4) \quad \text{Multiply both sides by 2.}$$

$$2y - 8 = 3x - 12 \quad \text{Apply the distributive property.}$$

$$2y = 3x - 4 \quad \text{Add 8 to both sides.}$$

$$y = \frac{3}{2}x - 2 \quad \text{Divide both sides by 2.}$$

$$f(x) = \frac{3}{2}x - 2 \quad \text{Write using function notation.} \qquad \square$$

PRACTICE

11   Write a function that describes the line containing the point $(8, -3)$ that is perpendicular to the line $3x + 4y = 1$.

.... ■

Forms of Linear Equations

$Ax + By = C$	**Standard form** of a linear equation A and B are not both 0.
$y = mx + b$	**Slope–intercept form** of a linear equation The slope is m, and the y-intercept is $(0, b)$.
$y - y_1 = m(x - x_1)$	**Point–slope form** of a linear equation The slope is m, and (x_1, y_1) is a point on the line.
$y = c$	**Horizontal line** The slope is 0, and the y-intercept is $(0, c)$.
$x = c$	**Vertical line** The slope is undefined, and the x-intercept is $(c, 0)$.

Parallel and Perpendicular Lines

Nonvertical parallel lines have the same slope. The product of the slopes of two nonvertical perpendicular lines is -1.

Vocabulary, Readiness & Video Check

State the slope and the y-intercept of each line with the given equation.

1. $y = -4x + 12$

2. $y = \frac{2}{3}x - \frac{7}{2}$

3. $y = 5x$

4. $y = -x$

5. $y = \frac{1}{2}x + 6$

6. $y = -\frac{2}{3}x + 5$

Decide whether the lines are parallel, perpendicular, or neither.

7. $y = 12x + 6$
$y = 12x - 2$

8. $y = -5x + 8$
$y = -5x - 8$

9. $y = -9x + 3$
$y = \frac{3}{2}x - 7$

10. $y = 2x - 12$
$y = \frac{1}{2}x - 6$

Martin-Gay Interactive Videos

See Video 3.5

Watch the section lecture video and answer the following questions.

OBJECTIVE 1

11. Complete these statements based on ▢ Example 1. To graph a line using its slope and *y*-intercept, first write the equation in _____ form. Graph the one point you now know, the _____. Use the _____ to find a second point.

OBJECTIVE 2

12. From ▢ Example 2, given a *y*-intercept point, how do you know which value to use for *b* in the slope–intercept form?

OBJECTIVE 3

13. ▢ Example 4 discusses how to find an equation of a line given two points. Under what circumstances might the slope–intercept form be chosen over the point–slope form to find an equation?

OBJECTIVE 4

14. Solve ▢ Examples 5 and 6 again, this time using the point $(-1, 3)$ in each exercise.

OBJECTIVE 5

15. Solve ▢ Example 7 again, this time write the equation of the line in function notation, *parallel* to the given line through the given point.

3.5 Exercise Set   MyMathLab®

Graph each linear equation. See Examples 1 and 2.

1. $y = 5x - 2$

2. $y = 2x + 1$

▶ **3.** $4x + y = 7$

4. $3x + y = 9$

5. $-3x + 2y = 3$

6. $-2x + 5y = -16$

Use the slope–intercept form of the linear equation to write the equation of each line with the given slope and y-intercept. See Example 3.

▶ **7.** Slope -1; *y*-intercept $(0, 1)$

8. Slope $\frac{1}{2}$; *y*-intercept $(0, -6)$

9. Slope 2; *y*-intercept $\left(0, \frac{3}{4}\right)$

10. Slope -3; *y*-intercept $\left(0, -\frac{1}{5}\right)$

11. Slope $\frac{2}{7}$; *y*-intercept $(0, 0)$

12. Slope $-\frac{4}{5}$; *y*-intercept $(0, 0)$

Find an equation of the line with the given slope and containing the given point. Write the equation in slope–intercept form. See Example 4.

▶ **13.** Slope 3; through $(1, 2)$

14. Slope 4; through $(5, 1)$

15. Slope -2; through $(1, -3)$

16. Slope -4; through $(2, -4)$

17. Slope $\frac{1}{2}$; through $(-6, 2)$

18. Slope $\frac{2}{3}$; through $(-9, 4)$

19. Slope $-\frac{9}{10}$; through $(-3, 0)$

20. Slope $-\frac{1}{5}$; through $(4, -6)$

Find an equation of the line passing through the given points. Use function notation to write the equation. See Example 5.

21. $(2, 0), (4, 6)$

22. $(3, 0), (7, 8)$

23. $(-2, 5), (-6, 13)$

24. $(7, -4), (2, 6)$

25. $(-2, -4), (-4, -3)$

26. $(-9, -2), (-3, 10)$

27. $(-3, -8), (-6, -9)$

28. $(8, -3), (4, -8)$

29. $\left(\frac{3}{5}, \frac{4}{10}\right)$ and $\left(-\frac{1}{5}, \frac{7}{10}\right)$

30. $\left(\frac{1}{2}, -\frac{1}{4}\right)$ and $\left(\frac{3}{2}, \frac{3}{4}\right)$

Find an equation of each line graphed. Write the equation in standard form. See Example 6.

31.
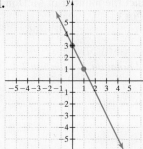

32.
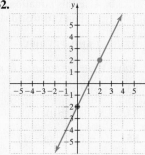

33.
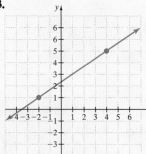

34.
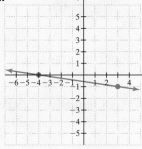

Use the graph of the following function f(x) to find each value.

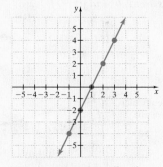

35. $f(0)$

36. $f(-1)$

37. $f(2)$

38. $f(1)$

39. Find x such that $f(x) = -6$.

40. Find x such that $f(x) = 4$.

Write an equation of each line. See Examples 8 and 9.

41. Slope 0; through $(-2, -4)$

42. Horizontal; through $(-3, 1)$

43. Vertical; through $(4, 7)$

44. Vertical; through $(2, 6)$

45. Horizontal; through $(0, 5)$

46. Undefined slope; through $(0, 5)$

Find an equation of each line. Write the equation using function notation. See Examples 10 and 11.

47. Through $(3, 8)$; parallel to $f(x) = 4x - 2$

48. Through $(1, 5)$; parallel to $f(x) = 3x - 4$

49. Through $(2, -5)$; perpendicular to $3y = x - 6$

50. Through $(-4, 8)$; perpendicular to $2x - 3y = 1$

51. Through $(-2, -3)$; parallel to $3x + 2y = 5$

52. Through $(-2, -3)$; perpendicular to $3x + 2y = 5$

MIXED PRACTICE

Find the equation of each line. Write the equation in standard form unless indicated otherwise. See Examples 3 through 5, and 8 through 11.

53. Slope 2; through $(-2, 3)$

54. Slope 3; through $(-4, 2)$

55. Through $(1, 6)$ and $(5, 2)$; use function notation.

56. Through $(2, 9)$ and $(8, 6)$

57. With slope $-\frac{1}{2}$; y-intercept 11

58. With slope -4; y-intercept $\frac{2}{9}$; use function notation.

59. Through $(-7, -4)$ and $(0, -6)$

60. Through $(2, -8)$ and $(-4, -3)$

61. Slope $-\frac{4}{3}$; through $(-5, 0)$

62. Slope $-\frac{3}{5}$; through $(4, -1)$

63. Vertical line; through $(-2, -10)$

64. Horizontal line; through $(1, 0)$

65. Through $(6, -2)$; parallel to the line $2x + 4y = 9$

66. Through $(8, -3)$; parallel to the line $6x + 2y = 5$

67. Slope 0; through $(-9, 12)$

68. Undefined slope; through $(10, -8)$

69. Through $(6, 1)$; parallel to the line $8x - y = 9$

70. Through $(3, 5)$; perpendicular to the line $2x - y = 8$

71. Through $(5, -6)$; perpendicular to $y = 9$

72. Through $(-3, -5)$; parallel to $y = 9$

73. Through $(2, -8)$ and $(-6, -5)$; use function notation.

74. Through $(-4, -2)$ and $(-6, 5)$; use function notation.

Solve. See Example 7.

75. A rock is dropped from the top of a 400-foot building. After 1 second, the rock is traveling 32 feet per second. After 3 seconds, the rock is traveling 96 feet per second. Let y be the rate of descent and x be the number of seconds since the rock was dropped.

    a. Write a linear equation that relates time x to rate y. [*Hint:* Use the ordered pairs $(1, 32)$ and $(3, 96)$.]

    b. Use this equation to determine the rate of travel of the rock 4 seconds after it was dropped.

76. A fruit company recently released a new applesauce. By the end of its first year, profits on this product amounted to $30,000. The anticipated profit for the end of the fourth year is $66,000. The ratio of change in time to change in profit is constant. Let x be years and y be profit.

    a. Write a linear equation that relates profit and time. [*Hint:* Use the ordered pairs $(1, 30{,}000)$ and $(4, 66{,}000)$.]

    b. Use this equation to predict the company's profit at the end of the seventh year.

    c. Predict when the profit should reach $126,000.

77. The Whammo Company has learned that by pricing a newly released Frisbee at $6, sales will reach 2000 per day. Raising the price to $8 will cause the sales to fall to 1500 per day. Assume that the ratio of change in price to change in daily sales is constant and let x be the price of the Frisbee and y be number of sales.

    a. Find the linear equation that models the price–sales relationship for this Frisbee. [*Hint:* The line must pass through $(6, 2000)$ and $(8, 1500)$.]

    b. Use this equation to predict the daily sales of Frisbees if the price is set at $7.50.

78. The Pool Fun Company has learned that, by pricing a newly released Fun Noodle at $3, sales will reach 10,000 Fun Noodles per day during the summer. Raising the price to $5 will cause the sales to fall to 8000 Fun Noodles per day. Let x be price and y be the number sold.

    a. Assume that the relationship between sales price and number of Fun Noodles sold is linear and write an equation describing this relationship. [*Hint:* The line must pass through $(3, 10{,}000)$ and $(5, 8000)$.]

    b. Use this equation to predict the daily sales of Fun Noodles if the price is $3.50.

79. The number of people employed in the United States as registered nurses was 2619 thousand in 2008. By 2018, this number is expected to rise to 3200 thousand. Let y be the number of registered nurses (in thousands) employed in the United States in the year x, where $x = 0$ represents 2008. (*Source:* U.S. Bureau of Labor Statistics)

    a. Write a linear equation that models the number of people (in thousands) employed as registered nurses in year x.

    b. Use this equation to estimate the number of people employed as registered nurses in 2012.

80. In 2008, IBM had 398,500 employees worldwide. By 2010, this number had increased to 426,751. Let y be the number of IBM employees worldwide in the year x, where $x = 0$ represents 2008. (*Source:* IBM Corporation)

    a. Write a linear equation that models the growth in the number of IBM employees worldwide, in terms of the year x.

    b. Use this equation to predict the number of IBM employees worldwide in 2013.

81. In 2010, the average price of a new home sold in the United States was $272,900. In 2005, the average price of a new home in the United States was $297,000. Let y be the average price of a new home in the year x, where $x = 0$ represents the year 2005. (*Source:* Based on data from U.S. census)

    a. Write a linear equation that models the average price of a new home in terms of the year x. [*Hint:* The line must pass through the points $(0, 297{,}000)$ and $(5, 272{,}900)$.]

    b. Use this equation to predict the average price of a new home in 2013.

82. The number of McDonald's restaurants worldwide in 2010 was 32,737. In 2005, there were 31,046 McDonald's restaurants worldwide. Let y be the number of McDonald's restaurants in the year x, where $x = 0$ represents the year 2005. (*Source:* McDonald's Corporation)

    a. Write a linear equation that models the growth in the number of McDonald's restaurants worldwide in terms of the year x. [*Hint:* The line must pass through the points $(0, 31{,}046)$ and $(5, 32{,}737)$.]

    b. Use this equation to predict the number of McDonald's restaurants worldwide in 2013.

REVIEW AND PREVIEW

Solve. Write the solution in interval notation. See Section 2.4.

83. $2x - 7 \le 21$

84. $-3x + 1 > 0$

85. $5(x - 2) \ge 3(x - 1)$

86. $-2(x + 1) \le -x + 10$

87. $\dfrac{x}{2} + \dfrac{1}{4} < \dfrac{1}{8}$ **88.** $\dfrac{x}{5} - \dfrac{3}{10} \ge \dfrac{x}{2} - 1$

CONCEPT EXTENSIONS

Answer true or false.

89. A vertical line is always perpendicular to a horizontal line.

90. A vertical line is always parallel to a vertical line.

Example:

Find an equation of the perpendicular bisector of the line segment whose endpoints are (2, 6) and (0, −2).

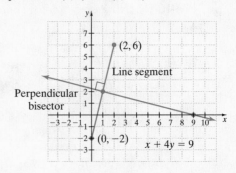

Solution:

A perpendicular bisector is a line that contains the midpoint of the given segment and is perpendicular to the segment.

Step 1. The midpoint of the segment with endpoints $(2, 6)$ and $(0, -2)$ is $(1, 2)$.

Step 2. The slope of the segment containing points $(2, 6)$ and $(0, -2)$ is 4.

Step 3. A line perpendicular to this line segment will have slope of $-\dfrac{1}{4}$.

Step 4. The equation of the line through the midpoint $(1, 2)$ with a slope of $-\dfrac{1}{4}$ will be the equation of the perpendicular bisector. This equation in standard form is $x + 4y = 9$.

Find an equation of the perpendicular bisector of the line segment whose endpoints are given. See the previous example.

△ **91.** $(3, -1); (-5, 1)$

△ **92.** $(-6, -3); (-8, -1)$

△ **93.** $(-2, 6); (-22, -4)$

△ **94.** $(5, 8); (7, 2)$

△ **95.** $(2, 3); (-4, 7)$

△ **96.** $(-6, 8); (-4, -2)$

97. Describe how to see if the graph of $2x - 4y = 7$ passes through the points $(1.4, -1.05)$ and $(0, -1.75)$. Then follow your directions and check these points.

Use a graphing calculator with a TRACE feature to see the results of each exercise.

98. Exercise 56; graph the equation and verify that it passes through $(2, 9)$ and $(8, 6)$.

99. Exercise 55; graph the function and verify that it passes through $(1, 6)$ and $(5, 2)$.

100. Exercise 62; graph the equation. See that it has a negative slope and passes through $(4, -1)$.

101. Exercise 61; graph the equation. See that it has a negative slope and passes through $(-5, 0)$.

102. Exercise 48: Graph the equation and verify that it passes through $(1, 5)$ and is parallel to $y = 3x - 4$.

103. Exercise 47: Graph the equation and verify that it passes through $(3, 8)$ and is parallel to $y = 4x - 2$.

Integrated Review   LINEAR EQUATIONS IN TWO VARIABLES

Sections 3.1–3.5

Below is a review of equations of lines.

Forms of Linear Equations

$Ax + By = C$	**Standard form** of a linear equation. A and B are not both 0.
$y = mx + b$	**Slope–intercept form** of a linear equation. The slope is m, and the y-intercept is $(0, b)$.
$y - y_1 = m(x - x_1)$	**Point–slope form** of a linear equation. The slope is m, and (x_1, y_1) is a point on the line.
$y = c$	**Horizontal line** The slope is 0, and the y-intercept is $(0, c)$.
$x = c$	**Vertical line** The slope is undefined and the x-intercept is $(c, 0)$.

Parallel and Perpendicular Lines

Nonvertical parallel lines have the same slope. The product of the slopes of two nonvertical perpendicular lines is -1.

Graph each linear equation.

1. $y = -2x$

2. $3x - 2y = 6$

3. $x = -3$

4. $y = 1.5$

Find the slope of the line containing each pair of points.

5. $(-2, -5), (3, -5)$

6. $(5, 2), (0, 5)$

Find the slope and y-intercept of each line.

7. $y = 3x - 5$

8. $5x - 2y = 7$

Determine whether each pair of lines is parallel, perpendicular, or neither.

9. $y = 8x - 6$
$\quad y = 8x + 6$

10. $y = \dfrac{2}{3}x + 1$
$\quad 2y + 3x = 1$

Find the equation of each line. Write the equation in the form $x = a$, $y = b$, or $y = mx + b$. For Exercises 14 through 17, write the equation in the form $f(x) = mx + b$.

11. Through $(1, 6)$ and $(5, 2)$

12. Vertical line; through $(-2, -10)$

13. Horizontal line; through $(1, 0)$

14. Through $(2, -9)$ and $(-6, -5)$

15. Through $(-2, 4)$ with slope -5

16. Slope -4; y-intercept $\left(0, \dfrac{1}{3}\right)$

17. Slope $\dfrac{1}{2}$; y-intercept $(0, -1)$

18. Through $\left(\dfrac{1}{2}, 0\right)$ with slope 3

19. Through $(-1, -5)$; parallel to $3x - y = 5$

20. Through $(0, 4)$; perpendicular to $4x - 5y = 10$

21. Through $(2, -3)$; perpendicular to $4x + y = \dfrac{2}{3}$

22. Through $(-1, 0)$; parallel to $5x + 2y = 2$

23. Undefined slope; through $(-1, 3)$

24. $m = 0$; through $(-1, 3)$

3.6 Graphing Piecewise-Defined Functions and Shifting and Reflecting Graphs of Functions

OBJECTIVES

1 Graph Piecewise-Defined Functions.

2 Vertical and Horizontal Shifts.

3 Reflect Graphs.

OBJECTIVE

1 Graphing Piecewise-Defined Functions

Throughout Chapter 3, we have graphed functions. There are many special functions. In this objective, we study functions defined by two or more expressions. The expression used to complete the function varies with and depends upon the value of x. Before we actually graph these piecewise-defined functions, let's practice finding function values.

EXAMPLE 1 Evaluate $f(2)$, $f(-6)$, and $f(0)$ for the function

$$f(x) = \begin{cases} 2x + 3 & \text{if } x \le 0 \\ -x - 1 & \text{if } x > 0 \end{cases}$$

Then write your results in ordered pair form.

Solution Take a moment and study this function. It is a single function defined by two expressions depending on the value of x. From above, if $x \le 0$, use $f(x) = 2x + 3$. If $x > 0$, use $f(x) = -x - 1$. Thus

$f(2) = -(2) - 1$	$f(-6) = 2(-6) + 3$	$f(0) = 2(0) + 3$
$\quad = -3 \;$ since $2 > 0$	$\quad = -9 \;$ since $-6 \le 0$	$\quad = 3 \;$ since $0 \le 0$
$f(2) = -3$	$f(-6) = -9$	$f(0) = 3$
Ordered pairs: $(2, -3)$	$(-6, -9)$	$(0, 3)$ □

PRACTICE

1 Evaluate $f(4)$, $f(-2)$, and $f(0)$ for the function

$$f(x) = \begin{cases} -4x - 2 & \text{if } x \le 0 \\ x + 1 & \text{if } x > 0 \end{cases}$$

Now, let's graph a piecewise-defined function.

EXAMPLE 2   Graph $f(x) = \begin{cases} 2x + 3 & \text{if } x \le 0 \\ -x - 1 & \text{if } x > 0 \end{cases}$

Solution   Let's graph each piece.

$$\text{If } x \le 0, \qquad\qquad\qquad\qquad \text{If } x > 0,$$
$$f(x) = 2x + 3 \qquad\qquad\qquad\qquad f(x) = -x - 1$$

Values ≤ 0
x	$f(x) = 2x + 3$
0	3 Closed circle
-1	1
-2	-1

Values > 0
x	$f(x) = -x - 1$
1	-2
2	-3
3	-4

The graph of the first part of $f(x)$ listed will look like a ray with a closed-circle end point at $(0, 3)$. The graph of the second part of $f(x)$ listed will look like a ray with an open-circle end point. To find the exact location of the open-circle end point, use $f(x) = -x - 1$ and find $f(0)$. Since $f(0) = -0 - 1 = -1$, we graph the values from the second table and place an open circle at $(0, -1)$.

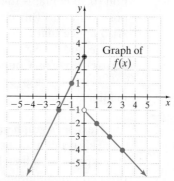

Graph of $f(x)$

Notice that this graph is the graph of a function because it passes the vertical line test. The domain of this function is $(-\infty, \infty)$ and the range is $(-\infty, 3]$.

PRACTICE
2   Graph

$$f(x) = \begin{cases} -4x - 2 & \text{if } x \le 0 \\ x + 1 & \text{if } x > 0 \end{cases}$$

OBJECTIVE
2   Vertical and Horizontal Shifting

Review of Common Graphs

We now take common graphs and learn how more complicated graphs are actually formed by shifting and reflecting these common graphs. These shifts and reflections are called transformations, and it is possible to combine transformations. A knowledge of these transformations will help you simplify future graphs.

Let's begin with a review of the graphs of four common functions. Many of these functions we graphed in earlier sections.

First, **let's graph the linear function $f(x) = x$, or $y = x$.** Ordered pair solutions of this graph consist of ordered pairs whose x- and y-values are the same.

x	y or $f(x) = x$
-3	-3
0	0
1	1
4	4

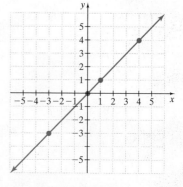

Next, **let's graph the nonlinear function** $f(x) = x^2$ **or** $y = x^2$.

This equation is not linear because the x^2 term does not allow us to write it in the form $Ax + By = C$. Its graph is not a line. We begin by finding ordered pair solutions. Because this graph is solved for $f(x)$, or y, we choose x-values and find corresponding $f(x)$, or y-values.

If $x = -3$, then $y = (-3)^2$, or 9.

If $x = -2$, then $y = (-2)^2$, or 4.

If $x = -1$, then $y = (-1)^2$, or 1.

If $x = 0$, then $y = 0^2$, or 0.

If $x = 1$, then $y = 1^2$, or 1.

If $x = 2$, then $y = 2^2$, or 4.

If $x = 3$, then $y = 3^2$, or 9.

x	$f(x)$ or y
-3	9
-2	4
-1	1
0	0
1	1
2	4
3	9

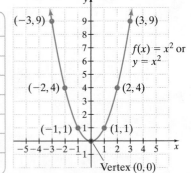

Study the table for a moment and look for patterns. Notice that the ordered pair solution $(0, 0)$ contains the smallest y-value because any other x-value squared will give a positive result. This means that the point $(0, 0)$ will be the lowest point on the graph. Also notice that all other y-values correspond to two different x-values, for example, $3^2 = 9$ and $(-3)^2 = 9$. This means that the graph will be a mirror image of itself across the y-axis. Connect the plotted points with a smooth curve to sketch its graph.

This curve is given a special name, a **parabola.** We will study more about parabolas in later chapters.

Next, **let's graph another nonlinear function,** $f(x) = |x|$ **or** $y = |x|$.

This is not a linear equation since it cannot be written in the form $Ax + By = C$. Its graph is not a line. Because we do not know the shape of this graph, we find many ordered pair solutions. We will choose x-values and substitute to find corresponding y-values.

If $x = -3$, then $y = |-3|$, or 3.

If $x = -2$, then $y = |-2|$, or 2.

If $x = -1$, then $y = |-1|$, or 1.

If $x = 0$, then $y = |0|$, or 0.

If $x = 1$, then $y = |1|$, or 1.

If $x = 2$, then $y = |2|$, or 2.

If $x = 3$, then $y = |3|$, or 3.

x	y
-3	3
-2	2
-1	1
0	0
1	1
2	2
3	3

Again, study the table of values for a moment and notice any patterns.

From the plotted ordered pairs, we see that the graph of this absolute value equation is V-shaped.

Finally, a fourth common function, $f(x) = \sqrt{x}$ or $y = \sqrt{x}$. For this graph, you need to recall basic facts about square roots and use your calculator to approximate some square roots to help locate points. Recall also that the square root of a negative number is not a real number, so be careful when finding your domain.

Now **let's graph the square root function** $f(x) = \sqrt{x}$, **or** $y = \sqrt{x}$.

To graph, we identify the domain, evaluate the function for several values of x, plot the resulting points, and connect the points with a smooth curve. Since $\sqrt{x}$ represents the nonnegative square root of x, the domain of this function is the set of all nonnegative numbers, $\{x | x \geq 0\}$, or $[0, \infty)$. We have approximated $\sqrt{3}$ on the next page to help us locate the point corresponding to $(3, \sqrt{3})$.

If $x = 0$, then $y = \sqrt{0}$, or 0.

If $x = 1$, then $y = \sqrt{1}$, or 1.

If $x = 3$, then $y = \sqrt{3}$, or 1.7.

If $x = 4$, then $y = \sqrt{4}$, or 2.

If $x = 9$, then $y = \sqrt{9}$, or 3.

x	$f(x) = \sqrt{x}$
0	0
1	1
3	$\sqrt{3} \approx 1.7$
4	2
9	3

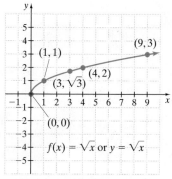

Notice that the graph of this function passes the vertical line test, as expected.

Below is a summary of our four common graphs. Take a moment and study these graphs. Your success in the rest of this section depends on your knowledge of these graphs.

Common Graphs

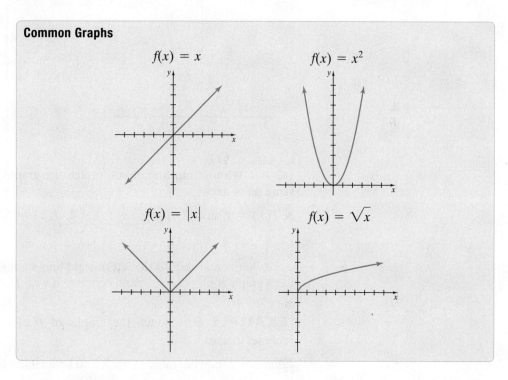

Your knowledge of the slope–intercept form, $f(x) = mx + b$, will help you understand simple shifting of transformations such as vertical shifts. For example, what is the difference between the graphs of $f(x) = x$ and $g(x) = x + 3$?

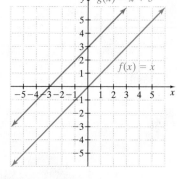

$f(x) = x$ $g(x) = x + 3$
slope, $m = 1$ slope, $m = 1$
y-intercept is $(0, 0)$ y-intercept is $(0, 3)$

Notice that the graph of $g(x) = x + 3$ is the same as the graph of $f(x) = x$, but moved upward 3 units. This is an example of a **vertical shift** and is true for graphs in general.

Vertical Shifts (Upward and Downward)
Let *k* be a Positive Number

Graph of	Same As	Moved
$g(x) = f(x) + k$	$f(x)$	k units upward
$g(x) = f(x) - k$	$f(x)$	k units downward

EXAMPLES Without plotting points, sketch the graph of each pair of functions on the same set of axes.

3. $f(x) = x^2$ and $g(x) = x^2 + 2$ **4.** $f(x) = \sqrt{x}$ and $g(x) = \sqrt{x} - 3$

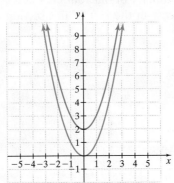
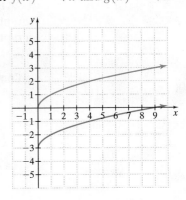

PRACTICES
3–4 Without plotting points, sketch the graphs of each pair of functions on the same set of axes.

3. $f(x) = x^2$ and $g(x) = x^2 - 3$ **4.** $f(x) = \sqrt{x}$ and $g(x) = \sqrt{x} + 1$

A horizontal shift to the left or right may be slightly more difficult to understand. Let's graph $g(x) = |x - 2|$ and compare it with $f(x) = |x|$.

EXAMPLE 5 Sketch the graphs of $f(x) = |x|$ and $g(x) = |x - 2|$ on the same set of axes.

Solution Study the table to the left to understand the placement of both graphs.

x	$f(x) = |x|$	$g(x) = |x - 2|$
−3	3	5
−2	2	4
−1	1	3
0	0	2
1	1	1
2	2	0
3	3	1

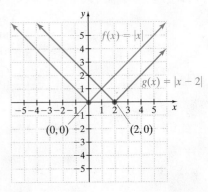

PRACTICE
5 Sketch the graphs of $f(x) = |x|$ and $g(x) = |x - 3|$ on the same set of axes.

The graph of $g(x) = |x - 2|$ is the same as the graph of $f(x) = |x|$, but moved 2 units to the right. This is an example of a **horizontal shift** and is true for graphs in general.

Horizontal Shift (To the Left or Right)
Let *h* be a Positive Number

Graph of	Same as	Moved
$g(x) = f(x - h)$	$f(x)$	h units to the right
$g(x) = f(x + h)$	$f(x)$	h units to the left

▶ Helpful Hint

Notice that $f(x - h)$ corresponds to a shift to the right and $f(x + h)$ corresponds to a shift to the left.

Vertical and horizontal shifts can be combined.

EXAMPLE 6   Sketch the graphs of $f(x) = x^2$ and $g(x) = (x - 2)^2 + 1$ on the same set of axes.

Solution   The graph of $g(x)$ is the same as the graph of $f(x)$ shifted 2 units to the right and 1 unit up.

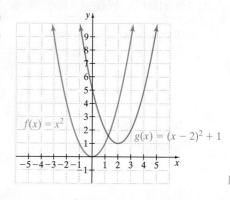

PRACTICE

6   Sketch the graphs of $f(x) = |x|$ and $g(x) = |x - 2| + 3$ on the same set of axes.

OBJECTIVE

3   Reflecting Graphs ▶

Another type of transformation is called a **reflection.** In this section, we will study reflections (mirror images) about the *x*-axis only. For example, take a moment and study these two graphs. The graph of $g(x) = -x^2$ can be verified, as usual, by plotting points.

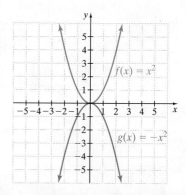

Reflection about the *x*-axis

The graph of $g(x) = -f(x)$ is the graph of $f(x)$ reflected about the *x*-axis.

EXAMPLE 7 Sketch the graph of $h(x) = -|x - 3| + 2$.

Solution The graph of $h(x) = -|x - 3| + 2$ is the same as the graph of $f(x) = |x|$ reflected about the x-axis, then moved three units to the right and two units upward.

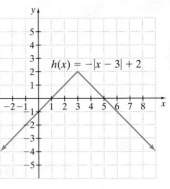

PRACTICE
7 Sketch the graph of $h(x) = -(x + 2)^2 - 1$.

There are other transformations, such as stretching, that won't be covered in this section. For a review of this transformation, see the Appendix.

Vocabulary, Readiness & Video Check

Match each equation with its graph.

1. $y = \sqrt{x}$

A
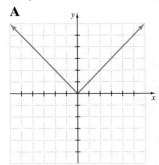

2. $y = x^2$

B
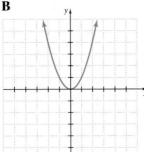

3. $y = x$

C
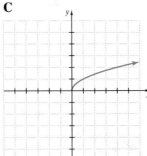

4. $y = |x|$

D
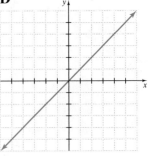

Martin-Gay Interactive Videos

See Video 3.6

Watch the section lecture video and answer the following questions.

OBJECTIVE 1
5. In Example 1, only one piece of the function is defined for the value $x = -1$. Why do we find $f(-1)$ for $f(x) = x + 3$?

OBJECTIVE 2
6. For Examples 2–8, why is it helpful to be familiar with common graphs and their basic shapes?

OBJECTIVE 3
7. Based on the lecture before Example 9, complete the following statement. The graph of $f(x) = -\sqrt{x + 6}$ has the same shape as the graph of $f(x) = \sqrt{x + 6}$ but it is reflected about the _____.

3.6 Exercise Set   MyMathLab®

Graph each piecewise-defined function. See Examples 1 and 2.

1. $f(x) = \begin{cases} 2x & \text{if } x < 0 \\ x + 1 & \text{if } x \geq 0 \end{cases}$

2. $f(x) = \begin{cases} 3x & \text{if } x < 0 \\ x + 2 & \text{if } x \geq 0 \end{cases}$

3. $f(x) = \begin{cases} 4x + 5 & \text{if } x \leq 0 \\ \dfrac{1}{4}x + 2 & \text{if } x > 0 \end{cases}$

4. $f(x) = \begin{cases} 5x + 4 & \text{if } x \leq 0 \\ \dfrac{1}{3}x - 1 & \text{if } x > 0 \end{cases}$

5. $g(x) = \begin{cases} -x & \text{if } x \le 1 \\ 2x + 1 & \text{if } x > 1 \end{cases}$

6. $g(x) = \begin{cases} 3x - 1 & \text{if } x \le 2 \\ -x & \text{if } x > 2 \end{cases}$

7. $f(x) = \begin{cases} 5 & \text{if } x < -2 \\ 3 & \text{if } x \ge -2 \end{cases}$ 8. $f(x) = \begin{cases} 4 & \text{if } x < -3 \\ -2 & \text{if } x \ge -3 \end{cases}$

MIXED PRACTICE

(Sections 3.2, 3.6) Graph each piecewise-defined function. Use the graph to determine the domain and range of the function. See Examples 1 and 2.

9. $f(x) = \begin{cases} -2x & \text{if } x \le 0 \\ 2x + 1 & \text{if } x > 0 \end{cases}$

10. $g(x) = \begin{cases} -3x & \text{if } x \le 0 \\ 3x + 2 & \text{if } x > 0 \end{cases}$

11. $h(x) = \begin{cases} 5x - 5 & \text{if } x < 2 \\ -x + 3 & \text{if } x \ge 2 \end{cases}$

12. $f(x) = \begin{cases} 4x - 4 & \text{if } x < 2 \\ -x + 1 & \text{if } x \ge 2 \end{cases}$

▶ 13. $f(x) = \begin{cases} x + 3 & \text{if } x < -1 \\ -2x + 4 & \text{if } x \ge -1 \end{cases}$

14. $h(x) = \begin{cases} x + 2 & \text{if } x < 1 \\ 2x + 1 & \text{if } x \ge 1 \end{cases}$

15. $g(x) = \begin{cases} -2 & \text{if } x \le 0 \\ -4 & \text{if } x \ge 1 \end{cases}$

16. $f(x) = \begin{cases} -1 & \text{if } x \le 0 \\ -3 & \text{if } x \ge 2 \end{cases}$

MIXED PRACTICE

Sketch the graph of function. See Examples 3 through 6.

▶ 17. $f(x) = |x| + 3$ 18. $f(x) = |x| - 2$

▶ 19. $f(x) = \sqrt{x} - 2$ 20. $f(x) = \sqrt{x} + 3$

▶ 21. $f(x) = |x - 4|$ 22. $f(x) = |x + 3|$

23. $f(x) = \sqrt{x + 2}$ 24. $f(x) = \sqrt{x - 2}$

▶ 25. $y = (x - 4)^2$ 26. $y = (x + 4)^2$

27. $f(x) = x^2 + 4$ 28. $f(x) = x^2 - 4$

29. $f(x) = \sqrt{x - 2} + 3$ 30. $f(x) = \sqrt{x - 1} + 3$

31. $f(x) = |x - 1| + 5$ 32. $f(x) = |x - 3| + 2$

▶ 33. $f(x) = \sqrt{x + 1} + 1$ 34. $f(x) = \sqrt{x + 3} + 2$

▶ 35. $f(x) = |x + 3| - 1$ 36. $f(x) = |x + 1| - 4$

37. $g(x) = (x - 1)^2 - 1$ 38. $h(x) = (x + 2)^2 + 2$

39. $f(x) = (x + 3)^2 - 2$ 40. $f(x) = (x + 2)^2 + 4$

Sketch the graph of each function. See Examples 3 through 7.

▶ 41. $f(x) = -(x - 1)^2$ 42. $g(x) = -(x + 2)^2$

43. $h(x) = -\sqrt{x} + 3$ 44. $f(x) = -\sqrt{x + 3}$

▶ 45. $h(x) = -|x + 2| + 3$ 46. $g(x) = -|x + 1| + 1$

47. $f(x) = (x - 3) + 2$ 48. $f(x) = (x - 1) + 4$

REVIEW AND PREVIEW

Match each equation with its graph. See Section 3.3.

49. $y = -1$ 50. $x = -1$

51. $x = 3$ 52. $y = 3$

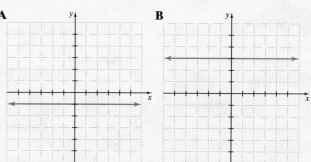

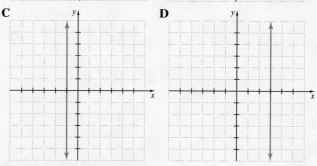

CONCEPT EXTENSIONS

53. Draw a graph whose domain is $(-\infty, 5]$ and whose range is $[2, \infty)$.

54. In your own words, describe how to graph a piecewise-defined function.

55. Graph: $f(x) = \begin{cases} -\dfrac{1}{2}x & \text{if } x \le 0 \\ x + 1 & \text{if } 0 < x \le 2 \\ 2x - 1 & \text{if } x > 2 \end{cases}$

56. Graph: $f(x) = \begin{cases} -\dfrac{1}{3}x & \text{if } x \le 0 \\ x + 2 & \text{if } 0 < x \le 4 \\ 3x - 4 & \text{if } x > 4 \end{cases}$

Write the domain and range of the following exercises.

57. Exercise 29   58. Exercise 30

59. Exercise 45   60. Exercise 46

Without graphing, find the domain of each function.

61. $f(x) = 5\sqrt{x - 20} + 1$

62. $g(x) = -3\sqrt{x + 5}$

63. $h(x) = 5|x - 20| + 1$

64. $f(x) = -3|x + 5.7|$

65. $g(x) = 9 - \sqrt{x + 103}$

66. $h(x) = \sqrt{x - 17} - 3$

Sketch the graph of each piecewise-defined function. Write the domain and range of each function.

67. $f(x) = \begin{cases} |x| & \text{if } x \le 0 \\ x^2 & \text{if } x > 0 \end{cases}$ 68. $f(x) = \begin{cases} x^2 & \text{if } x < 0 \\ \sqrt{x} & \text{if } x \ge 0 \end{cases}$

69. $g(x) = \begin{cases} |x - 2| & \text{if } x < 0 \\ -x^2 & \text{if } x \ge 0 \end{cases}$

70. $g(x) = \begin{cases} -|x + 1| - 1 & \text{if } x < -2 \\ \sqrt{x + 2} - 4 & \text{if } x \ge -2 \end{cases}$

3.7 | Graphing Linear Inequalities

OBJECTIVES

1 Graph Linear Inequalities.

2 Graph the Intersection or Union of Two Linear Inequalities.

OBJECTIVE

1   Graphing Linear Inequalities

Recall that the graph of a linear equation in two variables is the graph of all ordered pairs that satisfy the equation, and we determined that the graph is a line. Here we graph **linear inequalities** in two variables; that is, we graph all the ordered pairs that satisfy the inequality.

If the equal sign in a linear equation in two variables is replaced with an inequality symbol, the result is a linear inequality in two variables.

Examples of Linear Inequalities in Two Variables

$$3x + 5y \geq 6 \qquad 2x - 4y < -3$$
$$4x > 2 \qquad y \leq 5$$

To graph the linear inequality $x + y < 3$, for example, we first graph the related **boundary** equation $x + y = 3$. The resulting boundary line contains all ordered pairs the sum of whose coordinates is 3. This line separates the plane into two **half-planes.** All points "above" the boundary line $x + y = 3$ have coordinates that satisfy the inequality $x + y > 3$, and all points "below" the line have coordinates that satisfy the inequality $x + y < 3$.

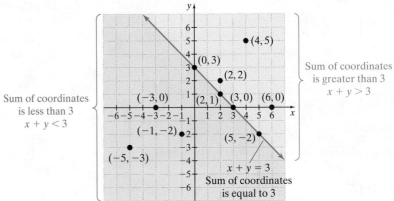

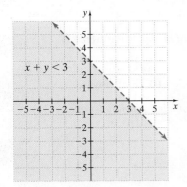

The graph, or **solution region,** for $x + y < 3$, then, is the half-plane below the boundary line and is shown shaded in the graph on the left. The boundary line is shown dashed since it is not a part of the solution region. These ordered pairs on this line satisfy $x + y = 3$ and not $x + y < 3$.

The following steps may be used to graph linear inequalities in two variables.

Graphing a Linear Inequality in Two Variables

Step 1. Graph the boundary line found by replacing the inequality sign with an equal sign. If the inequality sign is $<$ or $>$, graph a dashed line indicating that points on the line are not solutions of the inequality. If the inequality sign is $\leq$ or $\geq$, graph a solid line indicating that points on the line are solutions of the inequality.

Step 2. Choose a **test point not on the boundary line** and substitute the coordinates of this test point into the **original inequality.**

Step 3. If a true statement is obtained in Step 2, shade the half-plane that contains the test point. If a false statement is obtained, shade the half-plane that does not contain the test point.

EXAMPLE 1 Graph $2x - y < 6$.

Solution First, the boundary line for this inequality is the graph of $2x - y = 6$. Graph a dashed boundary line because the inequality symbol is $<$. Next, choose a test point on either side of the boundary line. The point $(0, 0)$ is not on the boundary line, so we use this point. Replacing x with 0 and y with 0 in the *original inequality* $2x - y < 6$ leads to the following:

$$2x - y < 6$$
$$2(0) - 0 < 6 \quad \text{Let } x = 0 \text{ and } y = 0.$$
$$0 < 6 \quad \text{True}$$

Because $(0, 0)$ satisfies the inequality, so does every point on the same side of the boundary line as $(0, 0)$. Shade the half-plane that contains $(0, 0)$. The half-plane graph of the inequality is shown at the right.

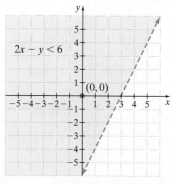

Every point in the shaded half-plane satisfies the original inequality. Notice that the inequality $2x - y < 6$ does not describe a function since its graph does not pass the vertical line test. □

PRACTICE

1 Graph $3x + y < 8$.

In general, linear inequalities of the form $Ax + By \le C$, where A and B are not both 0, do not describe functions.

EXAMPLE 2 Graph $3x \ge y$.

Solution First, graph the boundary line $3x = y$. Graph a solid boundary line because the inequality symbol is $\ge$. Test a point not on the boundary line to determine which half-plane contains points that satisfy the inequality. We choose $(0, 1)$ as our test point.

$$3x \ge y$$
$$3(0) \ge 1 \quad \text{Let } x = 0 \text{ and } y = 1.$$
$$0 \ge 1 \quad \text{False}$$

This point does not satisfy the inequality, so the correct half-plane is on the opposite side of the boundary line from $(0, 1)$. The graph of $3x \ge y$ is the boundary line together with the shaded region shown.

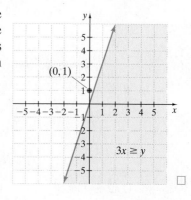

PRACTICE

2 Graph $x \ge 3y$.

> **✓CONCEPT CHECK**
> If a point on the boundary line is included in the solution of an inequality in two variables, should the graph of the boundary line be solid or dashed?

OBJECTIVE

2 Graphing Intersections or Unions of Linear Inequalities

The intersection and the union of linear inequalities can also be graphed, as shown in the next two examples. Recall from Section 2.5 that the graph of two inequalities joined by *and* is the intersection of the graphs of the two inequalities. Also, the graph of two inequalities joined by *or* is the union of the graphs of the two inequalities.

EXAMPLE 3 Graph $x \geq 1$ and $y \geq 2x - 1$.

Solution Graph each inequality. The word *and* means the intersection. The intersection of the two graphs is all points common to both regions, as shown by the dark pink shading in the third graph.

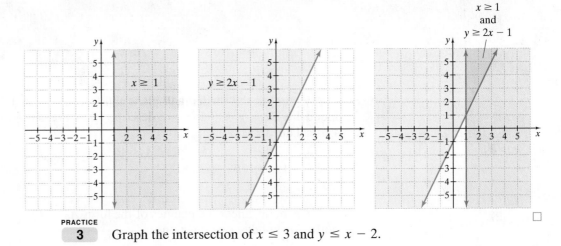

PRACTICE
3 Graph the intersection of $x \leq 3$ and $y \leq x - 2$.

EXAMPLE 4 Graph $x + \dfrac{1}{2}y \geq -4$ or $y \leq -2$.

Solution Graph each inequality. The word *or* means the union. The union of the two inequalities is both shaded regions, including the solid boundary lines shown in the third graph.

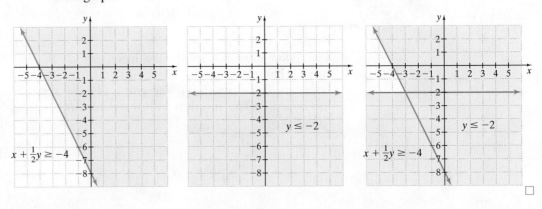

PRACTICE
4 Graph $2x - 3y \leq -2$ or $y \geq 1$.

Answer to Concept Check:
Solid

Vocabulary, Readiness & Video Check

Martin-Gay Interactive Videos

Watch the section lecture video and answer the following questions.

OBJECTIVE 1

1. From Example 1 and the lecture before, how do you find the equation of the boundary line? How do you determine if the points on the boundary line are solutions to the inequality?

OBJECTIVE 2

2. Based on Example 2, describe how you find the intersection of two linear inequalities in two variables.

See Video 3.7

3.7   Exercise Set   MyMathLab®

Graph each inequality. See Examples 1 and 2.

1. $x < 2$
2. $x > -3$
3. $x - y \geq 7$
4. $3x + y \leq 1$
▶ **5.** $3x + y > 6$
6. $2x + y > 2$
7. $y \leq -2x$
8. $y \leq 3x$
9. $2x + 4y \geq 8$
10. $2x + 6y \leq 12$
11. $5x + 3y > -15$
12. $2x + 5y < -20$

Graph each union or intersection. See Examples 3 and 4.

13. $x \geq 3$ and $y \leq -2$
14. $x \geq 3$ or $y \leq -2$
15. $x \leq -2$ or $y \geq 4$
16. $x \leq -2$ and $y \geq 4$
17. $x - y < 3$ and $x > 4$
18. $2x > y$ and $y > x + 2$
19. $x + y \leq 3$ or $x - y \geq 5$
20. $x - y \leq 3$ or $x + y > -1$

MIXED PRACTICE

Graph each inequality.

21. $y \geq -2$
22. $y \leq 4$
23. $x - 6y < 12$
24. $x - 4y < 8$
25. $x > 5$
26. $y \geq -2$
27. $-2x + y \leq 4$
28. $-3x + y \leq 9$
29. $x - 3y < 0$
30. $x + 2y > 0$
31. $3x - 2y \leq 12$
32. $2x - 3y \leq 9$
33. $x - y > 2$ or $y < 5$
34. $x - y < 3$ or $x > 4$
▶ **35.** $x + y \leq 1$ and $y \leq -1$

36. $y \geq x$ and $2x - 4y \geq 6$
37. $2x + y > 4$ or $x \geq 1$
38. $3x + y < 9$ or $y \leq 2$
39. $x \geq -2$ and $x \leq 1$
40. $x \geq -4$ and $x \leq 3$
41. $x + y \leq 0$ or $3x - 6y \geq 12$
42. $x + y \leq 0$ and $3x - 6y \geq 12$
43. $2x - y > 3$ and $x > 0$
44. $2x - y > 3$ or $x > 0$

Match each inequality with its graph.

45. $y \leq 2x + 3$
46. $y < 2x + 3$
47. $y > 2x + 3$
48. $y \geq 2x + 3$

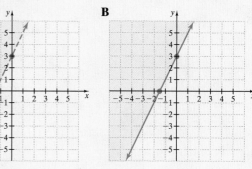

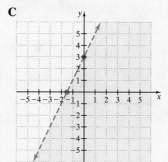
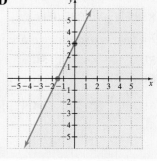

Write the inequality whose graph is given.

49.

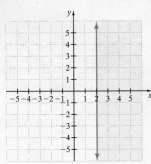

50.

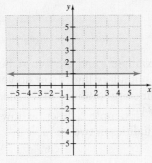

51.

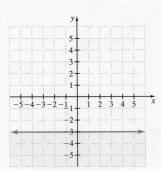

52.

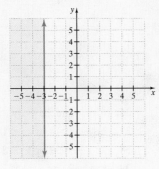

53.

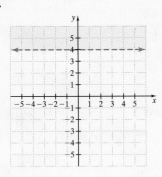

54.

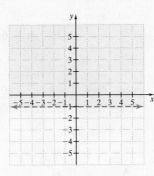

55.

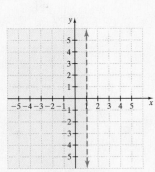

56.

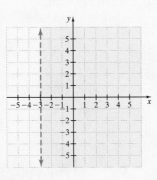

REVIEW AND PREVIEW

Evaluate each expression. See Sections 1.3 and 1.4.

57. 2^3

58. 3^2

59. -5^2

60. $(-5)^2$

61. $(-2)^4$

62. -2^4

63. $\left(\dfrac{3}{5}\right)^3$

64. $\left(\dfrac{2}{7}\right)^2$

Find the domain and the range of each relation. Determine whether the relation is also a function. See Section 3.2.

65.

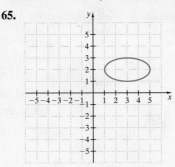

66.

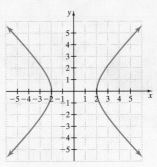

CONCEPT EXTENSIONS

67. Explain when a dashed boundary line should be used in the graph of an inequality.

68. Explain why, after the boundary line is sketched, we test a point on either side of this boundary in the original inequality.

Solve.

69. Chris-Craft manufactures boats out of Fiberglas and wood. Fiberglas hulls require 2 hours of work, whereas wood hulls require 4 hours of work. Employees work at most 40 hours a week. The following inequalities model these restrictions, where x represents the number of Fiberglas hulls produced and y represents the number of wood hulls produced.

$$\begin{cases} x \geq 0 \\ y \geq 0 \\ 2x + 4y \leq 40 \end{cases}$$

Graph the intersection of these inequalities.

70. Rheem Abo-Zahrah decides that she will study at most 20 hours every week and that she must work at least 10 hours every week. Let x represent the hours studying and y represent the hours working. Write two inequalities that model this situation and graph their intersection.

Chapter 3 Vocabulary Check

Fill in each blank with one of the words or phrases listed below.

relation	standard	slope–intercept	range	point–slope
line	slope	*x*	parallel	perpendicular
function	domain	*y*	linear function	linear inequality

1. A(n) _____ is a set of ordered pairs.

2. The graph of every linear equation in two variables is a(n) _____.

3. The statement $-x + 2y > 0$ is called a(n) _____ in two variables.

4. _____ form of linear equation in two variables is $Ax + By = C$.

5. The _____ of a relation is the set of all second components of the ordered pairs of the relation.

6. _____ lines have the same slope and different y-intercepts.

7. _____ form of a linear equation in two variables is $y = mx + b$.

8. A(n) _____ is a relation in which each first component in the ordered pairs corresponds to exactly one second component.

9. In the equation $y = 4x - 2$, the coefficient of x is the _____ of its corresponding graph.

10. Two lines are _____ if the product of their slopes is -1.

11. To find the x-intercept of a linear equation, let _____ $= 0$ and solve for the other variable.

12. The _____ of a relation is the set of all first components of the ordered pairs of the relation.

13. A(n) _____ is a function that can be written in the form $f(x) = mx + b$.

14. To find the y-intercept of a linear equation, let _____ $= 0$ and solve for the other variable.

15. The equation $y - 8 = -5(x + 1)$ is written in _____ form.

Chapter 3 Highlights

DEFINITIONS AND CONCEPTS	EXAMPLES

Section 3.1   Graphing Equations

The **rectangular coordinate system,** or **Cartesian coordinate system,** consists of a vertical and a horizontal number line intersecting at their 0 coordinate. The vertical number line is called the **y-axis,** and the horizontal number line is called the **x-axis.** The point of intersection of the axes is called the **origin.** The axes divide the plane into four regions called **quadrants.**

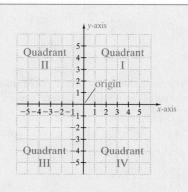

(continued)

DEFINITIONS AND CONCEPTS	EXAMPLES

Section 3.1 Graphing Equations (continued)

To **plot** or **graph** an ordered pair means to find its corresponding point on a rectangular coordinate system.

To plot or graph the ordered pair $(-2, 5)$, start at the origin. Move 2 units to the left along the x-axis, then 5 units upward parallel to the y-axis.

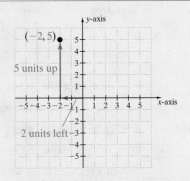

An ordered pair is a **solution** of an equation in two variables if replacing the variables by the corresponding coordinates results in a true statement.

Determine whether $(-2, 3)$ is a solution of

$$3x + 2y = 0$$
$$3(-2) + 2(3) = 0$$
$$-6 + 6 = 0$$
$$0 = 0 \quad \text{True}$$

$(-2, 3)$ is a solution.

A **linear equation in two variables** is an equation that can be written in the form $Ax + By = C$, where $A, B,$ and C are real numbers and A and B are not both 0. The form $Ax + By = C$ is called **standard form.**

Linear Equations in Two Variables

$$y = -2x + 5, \quad x = 7$$
$$y - 3 = 0, \quad 6x - 4y = 10$$

$6x - 4y = 10$ is in standard form.

The graph of a linear equation in two variables is a line. To graph a linear equation in two variables, find three ordered pair solutions. (Use the third ordered pair to check.) Plot the solution points and draw the line connecting the points.

Graph $3x + y = -6$.

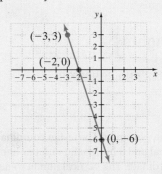

x	y
0	-6
-2	0
-3	3

To graph an equation that is not linear, find a sufficient number of ordered pair solutions so that a pattern may be discovered.

Graph $y = x^3 + 2$.

x	y
-2	-6
-1	1
0	2
1	3
2	10

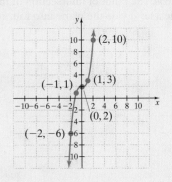

DEFINITIONS AND CONCEPTS	EXAMPLES

Section 3.2   Introduction to Functions

A **relation** is a set of ordered pairs. The **domain** of the relation is the set of all first components of the ordered pairs. The **range** of the relation is the set of all second components of the ordered pairs.

Relation

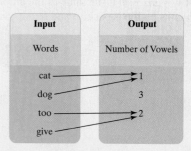

Domain: $\{\text{cat, dog, too, give}\}$
Range: $\{1, 2\}$

A **function** is a relation in which each element of the first set corresponds to exactly one element of the second set.

The previous relation is a function. Each word contains exactly one number of vowels.

Vertical Line Test

If no vertical line can be drawn so that it intersects a graph more than once, the graph is the graph of a function.

Find the domain and the range of the relation. Also determine whether the relation is a function.

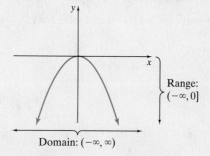

By the vertical line test, this graph is the graph of a function.

The symbol $f(x)$ means **function of x** and is called **function notation**.

If $f(x) = 2x^2 - 5$, find $f(-3)$.

$$f(-3) = 2(-3)^2 - 5 = 2(9) - 5 = 13$$

Section 3.3   Graphing Linear Functions

A **linear function** is a function that can be written in the form $f(x) = mx + b$.

Linear Functions

$$f(x) = -3, g(x) = 5x, h(x) = -\frac{1}{3}x - 7$$

To graph a linear function, find three ordered pair solutions. (Use the third ordered pair to check.) Graph the solutions and draw a line through the plotted points.

Graph $f(x) = -2x$.

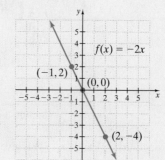

x	y or $f(x)$
-1	2
0	0
2	-4

(continued)

DEFINITIONS AND CONCEPTS	EXAMPLES

Section 3.3 Graphing Linear Functions (continued)

The graph of $y = mx + b$ is the same as the graph of $y = mx$ but shifted b units up if b is positive and b units down if b is negative.

Graph $g(x) = -2x + 3$.

This is the same as the graph of $f(x) = -2x$ shifted 3 units up.

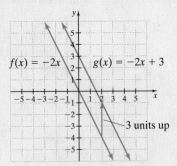

The x-coordinate of a point where a graph crosses the x-axis is called an **x-intercept.** The y-coordinate of a point where a graph crosses the y-axis is called a **y-intercept.**

To find an x-intercept, let $y = 0$ or $f(x) = 0$ and solve for x.

To find a y-intercept, let $x = 0$ and solve for y.

Graph $5x - y = -5$ by finding intercepts.

$$
\begin{array}{ll}
\text{If } x = 0, \text{ then} & \text{If } y = 0, \text{ then} \\
5x - y = -5 & 5x - y = -5 \\
5 \cdot 0 - y = -5 & 5x - 0 = -5 \\
-y = -5 & 5x = -5 \\
y = 5 & x = -1 \\
(0, 5) & (-1, 0)
\end{array}
$$

Ordered pairs are $(0, 5)$ and $(-1, 0)$.

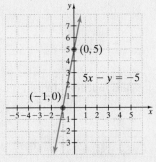

The graph of $x = c$ is a vertical line with x-intercept $(c, 0)$.

The graph of $y = c$ is a horizontal line with y-intercept $(0, c)$.

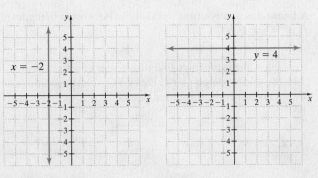

Section 3.4 The Slope of a Line

The **slope** m of the line through (x_1, y_1) and (x_2, y_2) is given by

$$m = \frac{y_2 - y_1}{x_2 - x_1} \text{ as long } x_2 \neq x_1$$

Find the slope of the line through $(-1, 7)$ and $(-2, -3)$.

$$m = \frac{y_2 - y_1}{x_2 - x_1} = \frac{-3 - 7}{-2 - (-1)} = \frac{-10}{-1} = 10$$

DEFINITIONS AND CONCEPTS	**EXAMPLES**

Section 3.4   The Slope of a Line (continued)

The **slope–intercept form** of a linear equation is $y = mx + b$, where m is the slope of the line and $(0, b)$ is the y-intercept.

Find the slope and y-intercept of $-3x + 2y = -8$.

$$2y = 3x - 8$$

$$\frac{2y}{2} = \frac{3x}{2} - \frac{8}{2}$$

$$y = \frac{3}{2}x - 4$$

The slope of the line is $\dfrac{3}{2}$, and the y-intercept is $(0, -4)$.

Nonvertical parallel lines have the same slope.

If the product of the slopes of two lines is -1, the lines are perpendicular.

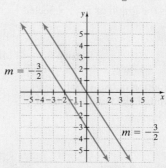 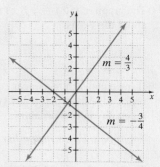

The slope of a horizontal line is 0.

The slope of a vertical line is undefined.

The slope of $y = -2$ is 0.

The slope of $x = 5$ is undefined.

Section 3.5   Equations of Lines

We can use the slope–intercept form to write an equation of a line given its slope and y-intercept.

Write an equation of the line with y-intercept $(0, -1)$ and slope $\dfrac{2}{3}$.

$$y = mx + b$$

$$y = \frac{2}{3}x - 1$$

The point–slope form of the equation of a line is $y - y_1 = m(x - x_1)$, where m is the slope of the line and (x_1, y_1) is a point on the line.

Find an equation of the line with slope 2 containing the point $(1, -4)$. Write the equation in standard form: $Ax + By = C$.

$$y - y_1 = m(x - x_1)$$

$$y - (-4) = 2(x - 1)$$

$$y + 4 = 2x - 2$$

$$-2x + y = -6 \qquad \text{Standard form}$$

Section 3.6   Graphing Piecewise-Defined Functions and Shifting and Reflecting Graphs of Functions

Vertical shifts (upward and downward) let k be a positive number.

Graph of	*Same as*	*Moved*
$g(x) = f(x) + k$	$f(x)$	k units upward
$g(x) = f(x) - k$	$f(x)$	k units downward

The graph of $h(x) = -|x - 3| + 1$ is the same as the graph of $f(x) = |x|$, reflected about the x-axis, shifted 3 units right, and 1 unit up.

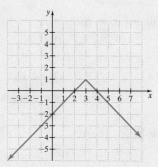

(continued)

DEFINITIONS AND CONCEPTS	EXAMPLES

Section 3.6 Graphing Piecewise-Defined Functions and Shifting and Reflecting Graphs of Functions (continued)

Horizontal shift (to the left or right)
let h be a positive number.

Graph of	*Same as*	*Moved*
$g(x) = f(x - h)$	$f(x)$	h units to the right
$g(x) = f(x + h)$	$f(x)$	h units to the left

Reflection about the x-axis

The graph of $g(x) = -f(x)$ is the graph of $f(x)$ reflected about the x-axis.

Section 3.7 Graphing Linear Inequalities

If the equal sign in a linear equation in two variables is replaced with an inequality symbol, the result is a **linear inequality in two variables.**

To graph a linear inequality

1. Graph the boundary line by graphing the related equation. Draw the line solid if the inequality symbol is $\leq$ or $\geq$. Draw the line dashed if the inequality symbol is $<$ or $>$.

2. Choose a test point not on the line. Substitute its coordinates into the original inequality.

3. If the resulting inequality is true, shade the **half-plane** that contains the test point. If the inequality is not true, shade the half-plane that does not contain the test point.

Linear Inequalities in Two Variables

$$x \leq -5 \quad y \geq 2$$
$$3x - 2y > 7 \quad x < -5$$

Graph $2x - 4y > 4$.

1. Graph $2x - 4y = 4$. Draw a dashed line because the inequality symbol is $>$.

2. Check the test point $(0, 0)$ in the inequality $2x - 4y > 4$.

$$2 \cdot 0 - 4 \cdot 0 > 4 \quad \text{Let } x = 0 \text{ and } y = 0.$$
$$0 > 4 \quad \text{False}$$

3. The inequality is false, so we shade the half-plane that does not contain $(0, 0)$.

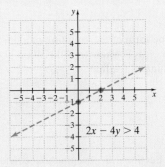

$2x - 4y > 4$

Chapter 3 Review

(3.1) *Plot the points and name the quadrant or axis in which each point lies.*

1. $A(2, -1), B(-2, 1), C(0, 3), D(-3, -5)$

2. $A(-3, 4), B(4, -3), C(-2, 0), D(-4, 1)$

Determine whether each ordered pair is a solution of the given equation.

3. $7x - 8y = 56; (0, 56), (8, 0)$

4. $-2x + 5y = 10; (-5, 0), (1, 1)$

5. $x = 13; (13, 5), (13, 13)$

6. $y = 2; (7, 2), (2, 7)$

Determine whether each equation is linear or not. Then graph the equation.

7. $y = 3x$ **8.** $y = 5x$

9. $3x - y = 4$ **10.** $x - 3y = 2$

11. $y = |x| + 4$ **12.** $y = x^2 + 4$

13. $y = -\dfrac{1}{2}x + 2$ **14.** $y = -x + 5$

15. $y = 2x - 1$ **16.** $y = \dfrac{1}{3}x + 1$

17. $y = -1.36x$ **18.** $y = 2.1x + 5.9$

(3.2) Find the domain and range of each relation. Also determine whether the relation is a function.

19. $\left\{ \left(-\dfrac{1}{2}, \dfrac{3}{4} \right), (6, 0.75), (0, -12), (25, 25) \right\}$

20. $\left\{ \left(\dfrac{3}{4}, -\dfrac{1}{2} \right), (0.75, 6), (-12, 0), (25, 25) \right\}$

21.

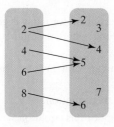

22.

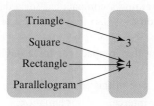

23.

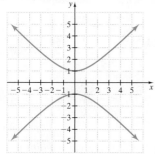

24.

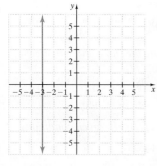

25.

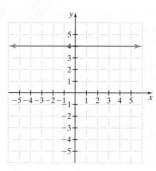

26.

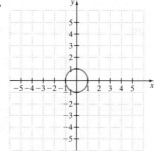

If $f(x) = x - 5$, $g(x) = -3x$, and $h(x) = 2x^2 - 6x + 1$, find the following.

27. $f(2)$ **28.** $g(0)$

29. $g(-6)$ **30.** $h(-1)$

31. $h(1)$ **32.** $f(5)$

The function $J(x) = 2.54x$ may be used to calculate the weight of an object on Jupiter J given its weight on Earth x.

33. If a person weighs 150 pounds on Earth, find the equivalent weight on Jupiter.

34. A 2000-pound probe on Earth weighs how many pounds on Jupiter?

Use the graph of the function below to answer Exercises 35 through 38.

35. Find $f(-1)$.   **36.** Find $f(1)$.

37. Find all values of x such that $f(x) = 1$.

38. Find all values of x such that $f(x) = -1$.

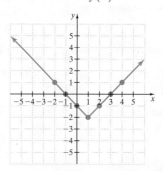

(3.3) Graph each linear function.

39. $f(x) = \dfrac{1}{5}x$ **40.** $f(x) = -\dfrac{1}{3}x$

41. $g(x) = -2x + 3$ **42.** $g(x) = 4x - 1$

The graph of $f(x) = 3x$ is sketched below. Use this graph to match each linear function with its graph.

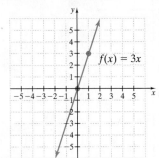

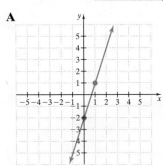 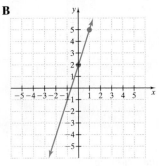

A   **B**

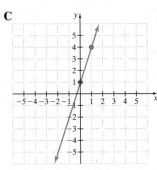 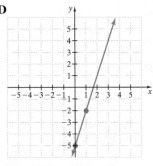

C   **D**

43. $f(x) = 3x + 1$ **44.** $f(x) = 3x - 2$

45. $f(x) = 3x + 2$ **46.** $f(x) = 3x - 5$

Graph each linear equation by finding intercepts if possible.

47. $4x + 5y = 20$ **48.** $3x - 2y = -9$

49. $4x - y = 3$ **50.** $2x + 6y = 9$

51. $y = 5$ **52.** $x = -2$

Graph each linear equation.

53. $x - 2 = 0$ **54.** $y + 3 = 0$

(3.4) Find the slope of the line through each pair of points.

55. $(2, 8)$ and $(6, -4)$ **56.** $(-3, 9)$ and $(5, 13)$

57. $(-7, -4)$ and $(-3, 6)$ **58.** $(7, -2)$ and $(-5, 7)$

Find the slope and y-intercept of each line.

59. $f(x) = -3x + \dfrac{1}{2}$

60. $g(x) = 2x + 4$

61. $6x - 15y = 20$

62. $4x + 14y = 21$

Find the slope of each line.

63. $y - 3 = 0$ **64.** $x = -5$

Two lines are graphed on each set of axes. Decide whether l_1 or l_2 has the greater slope.

65. 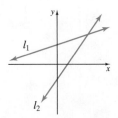   **66.**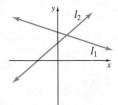

67.   **68.**

69. The cost C, in dollars, of renting a minivan for a day is given by the linear equation $C = 0.3x + 42$, where x is number of miles driven.

   a. Find the cost of renting the minivan for a day and driving it 150 miles.

   b. Find and interpret the slope of this equation.

   c. Find and interpret the y-intercept of this equation.

70. The cost C, in dollars, of renting a compact car for a day is given by the linear equation $C = 0.4x + 19$, where x is the number of miles driven.

   a. Find the cost of renting the compact car for a day and driving it 325 miles.

   b. Find and interpret the slope of this equation.

   c. Find and interpret the y-intercept of this equation.

Decide whether the lines are parallel, perpendicular, or neither.

△ **71.** $f(x) = -2x + 6$ **72.** $y = \dfrac{3}{4}x + 1$
     $g(x) = 2x - 1$
     $y = -\dfrac{4}{3}x + 1$

△ **73.** $-x + 3y = 2$ **74.** $x - 2y = 6$
     $6x - 18y = 3$ $4x + y = 8$

(3.5) Graph each line passing through the given point with the given slope.

75. Through $(2, -3)$ with slope $\dfrac{2}{3}$

76. Through $(-2, 0)$ with slope -3

Graph each linear equation using the slope and y-intercept.

77. $y = -x + 1$ **78.** $y = 4x - 3$

79. $3x - y = 6$ **80.** $y = -5x$

Find an equation of the line satisfying the given conditions.

81. Horizontal; through $(3, -1)$

82. Vertical; through $(-2, -4)$

△ **83.** Parallel to the line $x = 6$; through $(-4, -3)$

84. Slope 0; through $(2, 5)$

Find the standard form equation of each line satisfying the given conditions.

85. Through $(-3, 5)$; slope 3

86. Slope 2; through $(5, -2)$

87. Through $(-6, -1)$ and $(-4, -2)$

88. Through $(-5, 3)$ and $(-4, -8)$

△ **89.** Through $(-2, 3)$; perpendicular to $x = 4$

△ **90.** Through $(-2, -5)$; parallel to $y = 8$

Find an equation of each line satisfying the given conditions. Write each equation using function notation.

91. Slope $-\dfrac{2}{3}$; y-intercept $(0, 4)$

92. Slope -1; y-intercept $(0, -2)$

△ **93.** Through $(2, -6)$; parallel to $6x + 3y = 5$

△ **94.** Through $(-4, -2)$; parallel to $3x + 2y = 8$

△ **95.** Through $(-6, -1)$; perpendicular to $4x + 3y = 5$

96. Through $(-4, 5)$; perpendicular to $2x - 3y = 6$

△ **97.** The value of a building bought in 2000 continues to increase as time passes. Seven years after the building was bought, it was worth \$210,000; 12 years after it was bought, it was worth \$270,000.

   a. Assuming that this relationship between the number of years past 2000 and the value of the building is linear, write an equation describing this relationship. [*Hint:* Use ordered pairs of the form (years past 2000, value of the building).]

   b. Use this equation to estimate the value of the building in 2018.

98. The value of an automobile bought in 2006 continues to decrease as time passes. Two years after the car was bought, it was worth \$17,500; four years after it was bought, it was worth \$14,300.

   a. Assuming that this relationship between the number of years past 2006 and the value of the car is linear, write an equation describing this relationship. [*Hint:* Use ordered pairs of the form (years past 2006, value of the automobile).]

   b. Use this equation to estimate the value of the automobile in 2012.

(3.6) Graph each function.

99. $f(x) = \begin{cases} -3x & \text{if } x < 0 \\ x - 3 & \text{if } x \geq 0 \end{cases}$

100. $g(x) = \begin{cases} -\dfrac{1}{5}x & \text{if } x \leq -1 \\ -4x + 2 & \text{if } x > -1 \end{cases}$

Graph each function.

101. $y = \sqrt{x} - 4$ **102.** $f(x) = \sqrt{x - 4}$

103. $g(x) = |x - 2| - 2$ **104.** $h(x) = -(x + 3)^2 - 1$

(3.7) Graph each linear inequality.

105. $3x + y > 4$ **106.** $\dfrac{1}{2}x - y < 2$

107. $5x - 2y \leq 9$ **108.** $2x \leq 6y$

109. $y < 1$ **110.** $x > -2$

111. Graph $y > 2x + 3$ or $x \leq -3$.

112. Graph $2x < 3y + 8$ and $y \geq -2$.

MIXED REVIEW

Graph each linear equation or inequality.

113. $3x - 2y = -9$ **114.** $5x - 3y < 10$

115. $3y \geq x$ **116.** $x = -4y$

Write an equation of the line satisfying each set of conditions. If possible, write the equation in the form $y = mx + b$.

117. Vertical; through $\left(-7, -\dfrac{1}{2}\right)$

118. Slope 0; through $\left(-4, \dfrac{9}{2}\right)$

119. Slope $\dfrac{3}{4}$; through $(-8, -4)$

120. Through $(-3, 8)$ and $(-2, 3)$

121. Through $(-6, 1)$; parallel to $y = -\dfrac{3}{2}x + 11$

122. Through $(-5, 7)$; perpendicular to $5x - 4y = 10$

Graph each piecewise-defined function.

123. $f(x) = \begin{cases} x - 2 & \text{if } x \leq 0 \\ -\dfrac{x}{3} & \text{if } x \geq 3 \end{cases}$

124. $g(x) = \begin{cases} 4x - 3 & \text{if } x \leq 1 \\ 2x & \text{if } x > 1 \end{cases}$

Graph each function.

125. $f(x) = \sqrt{x - 2}$ **126.** $f(x) = |x + 1| - 3$

Chapter 3 **Test**

1. Plot the points and name the quadrant or axis in which each is located: $A(6, -2), B(4, 0), C(-1, 6)$.

Graph each line.

2. $2x - 3y = -6$

3. $4x + 6y = 7$

4. $f(x) = \dfrac{2}{3}x$

5. $y = -3$

6. Find the slope of the line that passes through $(5, -8)$ and $(-7, 10)$.

7. Find the slope and the y-intercept of the line $3x + 12y = 8$.

Graph each nonlinear function. Suggested x-values have been given for ordered pair solutions.

8. $f(x) = (x - 1)^2$ Let $x = -2, -1, 0, 1, 2, 3, 4$

9. $g(x) = |x| + 2$ Let $x = -3, -2, -1, 0, 1, 2, 3$

Find an equation of each line satisfying the given conditions. Write Exercises 10–14 in standard form. Write Exercises 15–17 using function notation.

10. Horizontal; through $(2, -8)$

11. Vertical; through $(-4, -3)$

12. Perpendicular to $x = 5$; through $(3, -2)$

13. Through $(4, -1)$; slope -3

14. Through $(0, -2)$; slope 5

15. Through $(4, -2)$ and $(6, -3)$

16. Through $(-1, 2)$; perpendicular to $3x - y = 4$

17. Parallel to $2y + x = 3$; through $(3, -2)$

18. Line L_1 has the equation $2x - 5y = 8$. Line L_2 passes through the points $(1, 4)$ and $(-1, -1)$. Determine whether these lines are parallel lines, perpendicular lines, or neither.

Graph each inequality.

19. $x \le -4$

20. $2x - y > 5$

21. The intersection of $2x + 4y < 6$ and $y \le -4$

Find the domain and range of each relation. Also determine whether the relation is a function.

22.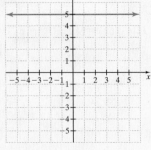

23.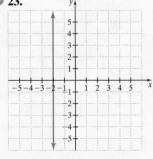

24.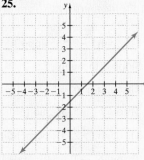

25.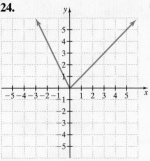

26. For the 2009 Major League Baseball season, the following linear equation describes the relationship between a team's payroll x (in millions of dollars) and the number of games y that team won during the regular season.

$$y = 0.096x + 72.81$$

Round to the nearest whole. (*Sources:* Based on data from Major League Baseball and *USA Today*)

a. According to this equation, how many games would have been won during the 2009 season by a team with a payroll of \$90 million?

b. The Baltimore Orioles had a payroll of \$67 million in 2009. According to this equation, how many games would they have won during the season?

c. According to this equation, what payroll would have been necessary in 2009 to have won 95 games during the season?

d. Find and interpret the slope of the equation.

Graph each function. For Exercises 27 and 29, state the domain and the range of the function.

27. $f(x) = \begin{cases} -\dfrac{1}{2}x & \text{if } x \le 0 \\ 2x - 3 & \text{if } x > 0 \end{cases}$

28. $f(x) = (x - 4)^2$

29. $g(x) = -|x + 2| - 1$

30. $h(x) = \sqrt{x} - 1$

Chapter 3 **Cumulative Review**

1. Evaluate: $3x - y$ when $x = 15$ and $y = 4$.

2. Add.
   a. $-4 + (-3)$
   b. $\dfrac{1}{2} - \left(-\dfrac{1}{3}\right)$
   c. $7 - 20$

3. Determine whether the following statements are true or false.
   a. 3 is a real number.
   b. $\frac{1}{5}$ is an irrational number.
   c. Every rational number is an integer.
   d. $\{1, 5\} \subseteq \{2, 3, 4, 5\}$

4. Write the opposite of each number.
   a. -7
   b. 0
   c. $\dfrac{1}{4}$

5. Subtract.
   a. $2 - 8$
   b. $-8 - (-1)$
   c. $-11 - 5$
   d. $10.7 - (-9.8)$
   e. $-\dfrac{2}{3} - \dfrac{1}{2}$
   f. $1 - 0.06$
   g. Subtract 7 from 4.

6. Multiply or divide.
   a. $\dfrac{-42}{-6}$
   b. $\dfrac{0}{14}$
   c. $-1(-5)(-2)$

7. Evaluate each expression.
   a. 3^2
   b. $\left(\dfrac{1}{2}\right)^4$
   c. -5^2
   d. $(-5)^2$
   e. -5^3
   f. $(-5)^3$

8. Which property is illustrated?
   a. $5(x + 7) = 5 \cdot x + 5 \cdot 7$
   b. $5(x + 7) = 5(7 + x)$

9. Insert $<, >,$ or $=$ between each pair of numbers to form a true statement.
   a. $-1 \quad -2$
   b. $\dfrac{12}{4} \quad 3$
   c. $-5 \quad 0$
   d. $-3.5 \quad -3.05$
   e. $\dfrac{5}{8} \quad \dfrac{3}{8}$
   f. $\dfrac{2}{3} \quad \dfrac{3}{4}$

10. Evaluate $2x^2$ for
    a. $x = 7$
    b. $x = -7$

11. Write the multiplicative inverse, or reciprocal, of each.
    a. 11
    b. -9
    c. $\dfrac{7}{4}$

12. Simplify $-2 + 3[5 - (7 - 10)]$.

13. Solve: $0.6 = 2 - 3.5c$

14. Solve: $2(x - 3) = -40$.

15. Solve for x: $3x + 5 = 3(x + 2)$

16. Solve: $5(x - 7) = 4x - 35 + x$.

17. Write the following as algebraic expressions. Then simplify.
    a. The sum of three consecutive integers if x is the first consecutive integer
    b. The perimeter of a triangle with sides of length $x, 5x,$ and $6x - 3$.

18. Find 25% of 16.

19. Kelsey Ohleger was helping her friend Benji Burnstine study for an algebra exam. Kelsey told Benji that her three latest art history quiz scores are three consecutive even integers whose sum is 264. Help Benji find the scores.

20. Find 3 consecutive odd integers whose sum is 213.

21. Solve $V = lwh$ for h.

22. Solve $7x + 3y = 21$ for y

23. Solve: $x - 2 < 5$.

24. Solve: $-x - 17 \geq 9$.

25. Solve: $\dfrac{2}{5}(x - 6) \geq x - 1$

26. $3x + 10 > \dfrac{5}{2}(x - 1)$.

27. Solve: $2x \geq 0$ *and* $4x - 1 \leq -9$

28. Solve: $x - 2 < 6$ *and* $3x + 1 > 1$.

29. Solve: $5x - 3 \leq 10$ *or* $x + 1 \geq 5$

30. Solve: $x - 2 < 6$ *or* $3x + 1 > 1$.

31. Solve: $|5w + 3| = 7$

32. Solve: $|5x - 2| = 3$.

33. Solve: $|3x + 2| = |5x - 8|$

34. Solve: $|7x - 2| = |7x + 4|$

35. Solve: $|5x + 1| + 1 \leq 10$

36. Solve: $|-x + 8| - 2 \leq 8$

37. Solve for y: $|y - 3| > 7$

38. Solve for x: $|x + 3| > 1$.

39. Determine whether $(0, -12)$, $(1, 9)$, and $(2, -6)$ are solutions of the equation $3x - y = 12$.

40. Find the slope and y-intercept of $7x + 2y = 10$.

41. Is the relation $y = 2x + 1$ also a function?

42. Determine whether the graph below is the graph of a function.

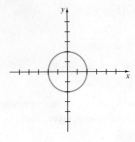

43. Find the y-intercept of the graph of each equation.
 a. $f(x) = \dfrac{1}{2}x + \dfrac{3}{7}$
 b. $y = -2.5x - 3.2$

44. Find the slope of the line through $(-1, 6)$ and $(0, 9)$.

45. Find the slope of the line whose equation is $f(x) = \dfrac{2}{3}x + 4$.

46. Find an equation of the vertical line through $\left(-2, -\dfrac{3}{4}\right)$.

47. Write an equation of the line with y-intercept $(0, -3)$ and slope of $\dfrac{1}{4}$.

48. Find an equation of the horizontal line through $\left(-2, -\dfrac{3}{4}\right)$.

49. Graph: $2x - y < 6$.

50. Write an equation of the line through $(-2, 5)$ and $(-4, 7)$.

Systems of Equations

Although the number of cell phone calls is slowly decreasing, for the past few years, the number of text messages has increased significantly. In fact, teenagers age 13–17 send about 3300 messages per month while those age 18–24 send "only" about 1600 per month. In Section 4.3, Exercises 57 and 58, we will use the functions graphed below to solve the system of equations.

In this chapter, two or more equations in two or more variables are solved simultaneously. Such a collection of equations is called a **system of equations.** Systems of equations are good mathematical models for many real-world problems because these problems may involve several related patterns.

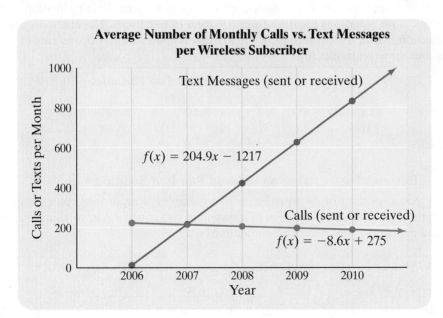

Average Number of Monthly Calls vs. Text Messages per Wireless Subscriber

Text Messages (sent or received)

$f(x) = 204.9x - 1217$

Calls (sent or received)

$f(x) = -8.6x + 275$

Calls or Texts per Month

Year

4.1 Solving Systems of Linear Equations in Two Variables

OBJECTIVES

1 Determine Whether an Ordered Pair Is a Solution of a System of Two Linear Equations.

2 Solve a System by Graphing.

3 Solve a System by Substitution.

4 Solve a System by Elimination.

An important problem that often occurs in the fields of business and economics concerns the concepts of revenue and cost. For example, suppose that a small manufacturing company begins to manufacture and sell compact disc storage units. The revenue of a company is the company's income from selling these units, and the cost is the amount of money that a company spends to manufacture these units. The following coordinate system shows the graphs of revenue and cost for the storage units.

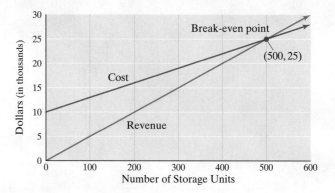

These lines intersect at the point $(500, 25)$. This means that when 500 storage units are manufactured and sold, both cost and revenue are \$25,000. In business, this point of intersection is called the **break-even point.** Notice that for x-values (units sold) less than 500, the cost graph is above the revenue graph, meaning that cost of manufacturing is greater than revenue, and so the company is losing money. For x-values (units sold) greater than 500, the revenue graph is above the cost graph, meaning that revenue is greater than cost, and so the company is making money.

Recall from Chapter 3 that each line is a graph of some linear equation in two variables. Both equations together form a **system of equations.** The common point of intersection is called the **solution of the system.** Some examples of systems of linear equations in two variables are

Systems of Linear Equations in Two Variables

$$\begin{cases} x - 2y = -7 \\ 3x + y = 0 \end{cases} \qquad \begin{cases} x = 5 \\ x + \dfrac{y}{2} = 9 \end{cases} \qquad \begin{cases} x - 3 = 2y + 6 \\ y = 1 \end{cases}$$

OBJECTIVE
1 Determining Whether an Ordered Pair Is a Solution

Recall that a solution of an equation in two variables is an ordered pair (x, y) that makes the equation true. A **solution of a system** of two equations in two variables is an ordered pair (x, y) that makes both equations true.

EXAMPLE 1 Determine whether the given ordered pair is a solution of the system.

a. $\begin{cases} -x + y = 2 \\ 2x - y = -3 \end{cases} \quad (-1, 1)$ **b.** $\begin{cases} 5x + 3y = -1 \\ x - y = 1 \end{cases} \quad (-2, 3)$

Solution

a. We replace x with -1 and y with 1 in each equation.

$-x + y = 2$	First equation	$2x - y = -3$	Second equation
$-(-1) + (1) \stackrel{?}{=} 2$	Let $x = -1$ and $y = 1$.	$2(-1) - (1) \stackrel{?}{=} -3$	Let $x = -1$ and $y = 1$.
$1 + 1 \stackrel{?}{=} 2$		$-2 - 1 \stackrel{?}{=} -3$	
$2 = 2$	True	$-3 = -3$	True

Since $(-1, 1)$ makes *both* equations true, it is a solution. Using set notation, the solution set is $\{(-1, 1)\}$.

b. We replace x with -2 and y with 3 in each equation.

$$5x + 3y = -1 \quad \text{First equation} \qquad\qquad x - y = 1 \quad \text{Second equation}$$
$$5(-2) + 3(3) \overset{?}{=} -1 \quad \text{Let } x = -2 \text{ and } y = 3. \qquad (-2) - (3) \overset{?}{=} 1 \quad \text{Let } x = -2 \text{ and } y = 3.$$
$$-10 + 9 \overset{?}{=} -1 \qquad\qquad\qquad\qquad\qquad -5 = 1 \quad \text{False}$$
$$-1 = -1 \quad \text{True}$$

Since the ordered pair $(-2, 3)$ does not make *both* equations true, it is not a solution of the system. $\square$

PRACTICE
1   Determine whether the given ordered pair is a solution of the system.

a. $\begin{cases} -x - 4y = 1 \\ 2x + y = 5 \end{cases} (3, -1)$ 　　　　 **b.** $\begin{cases} 4x + y = -4 \\ -x + 3y = 8 \end{cases} (-2, 4)$

OBJECTIVE
2   Solving a System by Graphing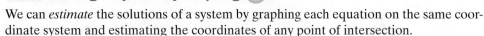

We can *estimate* the solutions of a system by graphing each equation on the same coordinate system and estimating the coordinates of any point of intersection.

EXAMPLE 2   Solve the system by graphing.

$$\begin{cases} x + y = 2 \\ 3x - y = -2 \end{cases}$$

Solution: First we graph the linear equations on the same rectangular coordinate system. These lines intersect at one point as shown. The coordinates of the point of intersection appear to be $(0, 2)$. We check this estimated solution by replacing x with 0 and y with 2 in *both* equations.

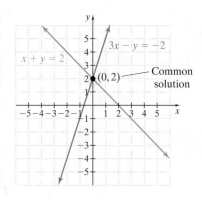

$$x + y = 2 \quad \text{First equation} \qquad\qquad 3x - y = -2 \quad \text{Second equation}$$
$$0 + 2 \overset{?}{=} 2 \quad \text{Let } x = 0 \text{ and } y = 2. \qquad\qquad 3(0) - 2 \overset{?}{=} -2 \quad \text{Let } x = 0 \text{ and } y = 2.$$
$$2 = 2 \quad \text{True} \qquad\qquad\qquad\qquad\qquad -2 = -2 \quad \text{True}$$

The ordered pair $(0, 2)$ is the solution of the system. A system that has at least one solution, such as this one, is said to be **consistent.** $\square$

PRACTICE
2   Solve the system by graphing. $\begin{cases} y = 5x \\ 2x + y = 7 \end{cases}$

▶ **Helpful Hint**
Reading values from graphs may not be accurate. Until a proposed solution is checked in both equations of the system, we can only assume that we have *estimated* a solution.

In Example 2, we have a **consistent system.** To review, a system that has at least one solution is said to be consistent.

Later, we will talk about **dependent equations.** For now, we define an **independent equation** to be an equation in a system of equations that cannot be algebraically derived from any other equation in the system.

Thus for:

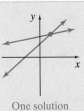

Consistent System: The system has at least one solution.

Independent Equations: Each equation in the system cannot be algebraically derived from the other.

One solution

EXAMPLE 3 Solve the system by graphing.

$$\begin{cases} x - 2y = 4 \\ x = 2y \end{cases}$$

Solution: We graph each linear equation.

The lines appear to be parallel. To be sure, let's write each equation in slope–intercept form, $y = mx + b$. To do so, we solve for y.

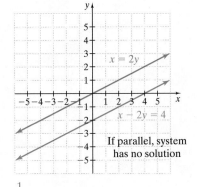

$x - 2y = 4$ First equation

$-2y = -x + 4$ Subtract x from both sides.

$y = \dfrac{1}{2}x - 2$ Divide both sides by -2

$x = 2y$ Second equation

$\dfrac{1}{2}x = y$ Divide both sides by 2.

$y = \dfrac{1}{2}x$

The graphs of these equations have the same slope, $\dfrac{1}{2}$, but different y-intercepts, so these lines are parallel. Therefore, the system has no solution since the equations have no common solution (there are no intersection points). A system that has no solution is said to be **inconsistent.**   □

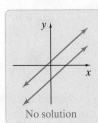

PRACTICE 3   Solve the system by graphing. $\begin{cases} y = \dfrac{3}{4}x + 1 \\ 3x - 4y = 12 \end{cases}$

In Example 3, we have an **inconsistent system.** To review, a system that has no solution is said to be inconsistent.

Let's now talk about the equations in this system. Each equation in this system cannot be algebraically derived from the other, so each equation is independent of the other.

Thus:

Inconsistent System: The system has no solution.

Independent Equations: Each equation in the system cannot be algebraically derived from the other.

No solution

> ▶ **Helpful Hint**
> - If a system of equations has *at least one solution*, the system is *consistent*.
> - If a system of equations has *no solution*, the system is *inconsistent*.

The pairs of equations in Examples 2 and 3 are called independent because their graphs differ. In Example 4, we see an example of dependent equations.

EXAMPLE 4   Solve the system by graphing.

$$\begin{cases} 2x + 4y = 10 \\ x + 2y = 5 \end{cases}$$

Solution: We graph each linear equation. We see that the graphs of the equations are the same line. To confirm this, notice that if both sides of the second equation are multiplied by 2, the result is the first equation. This means that the equations have identical solutions. Any ordered pair solution of one equation satisfies the other equation also. These equations are said to be **dependent equations.** The solution set of the system is $\{(x, y) \mid x + 2y = 5\}$ or, equivalently, $\{(x, y) \mid 2x + 4y = 10\}$ since the lines describe identical ordered pairs. Written the second way, the solution set is read "the set of all ordered pairs (x, y), such that $2x + 4y = 10$." There is an infinite number of solutions to this system.

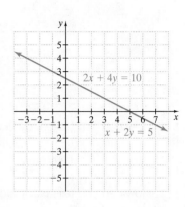

PRACTICE
4   Solve the system by graphing. $\begin{cases} 3x - 2y = 4 \\ -9x + 6y = -12 \end{cases}$

In Example 4, we have a **consistent system** since the system has at least one solution. In fact, the system in Example 4 has an infinite number of solutions.

Let's now define **dependent equations.** We define a **dependent equation** to be an equation in a system of equations that can be algebraically derived from another equation in the system.

Thus:

y graph with line through origin	**Consistent System:** The system has at least one solution.
Infinite number of solutions	**Dependent Equations:** An equation in a system of equations can be algebraically derived from another.

> ▶ **Helpful Hint**
> - If the graphs of two equations *differ,* they are *independent* equations.
> - If the graphs of two equations are the *same,* they are *dependent* equations.

Answer to Concept Check: b, c

✓CONCEPT CHECK

The equations in the system are dependent and the system has an infinite number of solutions. Which ordered pairs below are solutions?

$$\begin{cases} -x + 3y = 4 \\ 2x + 8 = 6y \end{cases}$$

a. $(4, 0)$ **b.** $(-4, 0)$ **c.** $(-1, 1)$

We can summarize the information discovered in Examples 2 through 4 as follows.

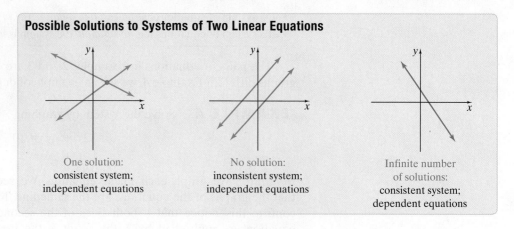

Possible Solutions to Systems of Two Linear Equations

One solution:
consistent system;
independent equations

No solution:
inconsistent system;
independent equations

Infinite number
of solutions:
consistent system;
dependent equations

✓**CONCEPT CHECK**

How can you tell just by looking at the following system that it has no solution?

$$\begin{cases} y = 3x + 5 \\ y = 3x - 7 \end{cases}$$

How can you tell just by looking at the following system that it has infinitely many solutions?

$$\begin{cases} x + y = 5 \\ 2x + 2y = 10 \end{cases}$$

OBJECTIVE

3   Solving a System by Substitution

Graphing the equations of a system by hand is often a good method of finding approximate solutions of a system, but it is not a reliable method of finding exact solutions of a system. We turn instead to two algebraic methods of solving systems. We use the first method, the **substitution method,** to solve the system

$$\begin{cases} 2x + 4y = -6 & \text{First equation} \\ x = 2y - 5 & \text{Second equation} \end{cases}$$

EXAMPLE 5   Use the substitution method to solve the system.

$$\begin{cases} 2x + 4y = -6 & \text{First equation} \\ x = 2y - 5 & \text{Second equation} \end{cases}$$

Solution   In the second equation, we are told that x is equal to $2y - 5$. Since they are equal, we can *substitute* $2y - 5$ for x in the first equation. This will give us an equation in one variable, which we can solve for y.

$$2x + 4y = -6 \qquad \text{First equation}$$

$$2(\overbrace{2y - 5}) + 4y = -6 \qquad \text{Substitute } 2y - 5 \text{ for } x.$$

$$4y - 10 + 4y = -6$$

$$8y = 4$$

$$y = \frac{4}{8} = \frac{1}{2} \qquad \text{Solve for } y.$$

Answer to Concept Check:
answers may vary

The y-coordinate of the solution is $\dfrac{1}{2}$. To find the x-coordinate, we replace y with $\dfrac{1}{2}$ in the second equation, $x = 2y - 5$.

$$x = 2y - 5$$
$$x = 2\left(\frac{1}{2}\right) - 5 = 1 - 5 = -4$$

The ordered pair solution is $\left(-4, \dfrac{1}{2}\right)$. Check to see that $\left(-4, \dfrac{1}{2}\right)$ satisfies both equations of the system.   □

PRACTICE
5   Use the substitution method to solve the system.

$$\begin{cases} y = 4x + 7 \\ 2x + y = 4 \end{cases}$$

Solving a System of Two Equations Using the Substitution Method

Step 1. Solve one of the equations for one of its variables.

Step 2. Substitute the expression for the variable found in Step 1 into the other equation.

Step 3. Find the value of one variable by solving the equation from Step 2.

Step 4. Find the value of the other variable by substituting the value found in Step 3 into the equation from Step 1.

Step 5. Check the ordered pair solution in *both* original equations.

▶ **Helpful Hint**
If a system of equations contains equations with fractions, the first step is to clear the equations of fractions.

EXAMPLE 6   Use the substitution method to solve the system.

$$\begin{cases} -\dfrac{x}{6} + \dfrac{y}{2} = \dfrac{1}{2} \\ \dfrac{x}{3} - \dfrac{y}{6} = -\dfrac{3}{4} \end{cases}$$

Solution   First we multiply each equation by its least common denominator to clear the system of fractions. We multiply the first equation by 6 and the second equation by 12.

$$\begin{cases} 6\left(-\dfrac{x}{6} + \dfrac{y}{2}\right) = 6\left(\dfrac{1}{2}\right) \\ 12\left(\dfrac{x}{3} - \dfrac{y}{6}\right) = 12\left(-\dfrac{3}{4}\right) \end{cases} \quad \text{simplifies to} \quad \begin{cases} -x + 3y = 3 \quad \text{First equation} \\ 4x - 2y = -9 \quad \text{Second equation} \end{cases}$$

▶ **Helpful Hint**
To avoid tedious fractions, solve for a variable whose coefficient is 1 or -1 if possible.

To use the substitution method, we now solve the first equation for x.

$$-x + 3y = 3 \quad \text{First equation}$$
$$3y - 3 = x \quad \text{Solve for } x.$$

Next we replace x with $3y - 3$ in the second equation.

$$4x - 2y = -9 \quad \text{Second equation}$$
$$4(\overbrace{3y - 3}) - 2y = -9$$
$$12y - 12 - 2y = -9$$
$$10y = 3$$
$$y = \frac{3}{10} \quad \text{Solve for } y.$$

To find the corresponding x-coordinate, we replace y with $\dfrac{3}{10}$ in the equation $x = 3y - 3$. Then

$$x = 3\left(\frac{3}{10}\right) - 3 = \frac{9}{10} - 3 = \frac{9}{10} - \frac{30}{10} = -\frac{21}{10}$$

The ordered pair solution is $\left(-\dfrac{21}{10}, \dfrac{3}{10}\right)$. Check to see that this solution satisfies both original equations. ☐

PRACTICE

6 Use the substitution method to solve the system.

$$\begin{cases} -\dfrac{x}{3} + \dfrac{y}{4} = \dfrac{1}{2} \\ \dfrac{x}{4} - \dfrac{y}{2} = -\dfrac{1}{4} \end{cases}$$

OBJECTIVE

4 **Solving a System by Elimination**

The **elimination method,** or **addition method,** is a second algebraic technique for solving systems of equations. For this method, we rely on a version of the addition property of equality, which states that "equals added to equals are equal."

If $A = B$ and $C = D$ then $A + C = B + D$.

EXAMPLE 7 Use the elimination method to solve the system.

$$\begin{cases} x - 5y = -12 & \text{First equation} \\ -x + y = 4 & \text{Second equation} \end{cases}$$

Solution Since the left side of each equation is equal to the right side, we add equal quantities by adding the left sides of the equations and the right sides of the equations. This sum gives us an equation in one variable, y, which we can solve for y.

$$\begin{array}{ll} x - 5y = -12 & \text{First equation} \\ \underline{-x + y = 4} & \text{Second equation} \\ -4y = -8 & \text{Add} \\ y = 2 & \text{Solve for } y. \end{array}$$

The y-coordinate of the solution is 2. To find the corresponding x-coordinate, we replace y with 2 in either original equation of the system. Let's use the second equation.

$$\begin{array}{ll} -x + y = 4 & \text{Second equation} \\ -x + 2 = 4 & \text{Let } y = 2. \\ -x = 2 \\ x = -2 \end{array}$$

The ordered pair solution is $(-2, 2)$. Check to see that $(-2, 2)$ satisfies both equations of the system. ☐

PRACTICE

7 Use the elimination method to solve the system.

$$\begin{cases} 3x - y = 5 \\ 5x + y = 11 \end{cases}$$

The steps below summarize the elimination method.

Solving a System of Two Linear Equations Using the Elimination Method

Step 1. Rewrite each equation in standard form, $Ax + By = C$.

Step 2. If necessary, multiply one or both equations by some nonzero number so that the coefficients of a variable are opposites of each other.

Step 3. Add the equations.

Step 4. Find the value of one variable by solving the equation from Step 3.

Step 5. Find the value of the second variable by substituting the value found in Step 4 into either original equation.

Step 6. Check the proposed ordered pair solution in *both* original equations.

EXAMPLE 8   Use the elimination method to solve the system.

$$\begin{cases} 3x - 2y = 10 \\ 4x - 3y = 15 \end{cases}$$

Solution   If we add the two equations, the sum will still be an equation in two variables. Notice, however, that we can eliminate y when the equations are added if we multiply both sides of the first equation by 3 and both sides of the second equation by -2. Then

$$\begin{cases} 3(3x - 2y) = 3(10) \\ -2(4x - 3y) = -2(15) \end{cases} \quad \text{simplifies to} \quad \begin{cases} 9x - 6y = 30 \\ -8x + 6y = -30 \end{cases}$$

Next we add the left sides and add the right sides.

$$\begin{array}{r} 9x - 6y = 30 \\ -8x + 6y = -30 \\ \hline x = 0 \end{array}$$

To find y, we let $x = 0$ in either equation of the system.

$$\begin{aligned} 3x - 2y &= 10 & \text{First equation} \\ 3(0) - 2y &= 10 & \text{Let } x = 0. \\ -2y &= 10 \\ y &= -5 \end{aligned}$$

The ordered pair solution is $(0, -5)$. Check to see that $(0, -5)$ satisfies both equations of the system. ☐

PRACTICE
8   Use the elimination method to solve the system.

$$\begin{cases} 3x - 2y = -6 \\ 4x + 5y = -8 \end{cases}$$

EXAMPLE 9   Use the elimination method to solve the system.

$$\begin{cases} 3x + \dfrac{y}{2} = 2 \\ 6x + y = 5 \end{cases}$$

Solution   If we multiply both sides of the first equation by -2, the coefficients of x in the two equations will be opposites. Then

$$\begin{cases} -2\left(3x + \dfrac{y}{2}\right) = -2(2) \\ 6x + y = 5 \end{cases} \quad \text{simplifies to} \quad \begin{cases} -6x - y = -4 \\ 6x + y = 5 \end{cases}$$

Now we can add the left sides and add the right sides.

$$
\begin{array}{r}
-6x - y = -4 \\
6x + y = 5 \\
\hline
0 = 1 \quad \text{False}
\end{array}
$$

The resulting equation, $0 = 1$, is false for all values of y or x. Thus, the system has no solution. The solution set is $\{\ \}$ or $\varnothing$. This system is inconsistent, and the graphs of the equations are parallel lines. □

PRACTICE
9   Use the elimination method to solve the system.

$$
\begin{cases}
8x + y = 6 \\
2x + \dfrac{y}{4} = -2
\end{cases}
$$

EXAMPLE 10   Use the elimination method to solve the system.

$$
\begin{cases}
-5x - 3y = 9 \\
10x + 6y = -18
\end{cases}
$$

Solution   To eliminate x when the equations are added, we multiply both sides of the first equation by 2. Then

$$
\begin{cases}
2(-5x - 3y) = 2(9) \\
10x + 6y = -18
\end{cases}
\quad \text{simplifies to} \quad
\begin{cases}
-10x - 6y = 18 \\
10x + 6y = -18
\end{cases}
$$

Next we add the equations.

$$
\begin{array}{r}
-10x - 6y = 18 \\
10x + 6y = -18 \\
\hline
0 = 0
\end{array}
$$

The resulting equation, $0 = 0$, is true for all possible values of y or x. Notice in the original system that if both sides of the first equation are multiplied by -2, the result is the second equation. This means that the two equations are equivalent. They have the same solution set and there is an infinite number of solutions. Thus, the equations of this system are dependent, and the solution set of the system is

$$\{(x, y) \mid -5x - 3y = 9\} \quad \text{or, equivalently,} \quad \{(x, y) \mid 10x + 6y = -18\}. \quad □$$

PRACTICE
10   Use the elimination method to solve the system.

$$
\begin{cases}
-3x + 2y = -1 \\
9x - 6y = 3
\end{cases}
$$

> ▶ **Helpful Hint**
> Remember that not all ordered pairs are solutions of the system in Example 10, only the infinite number of ordered pairs that satisfy $-5x - 3y = 9$ or equivalently $10x + 6y = -18$.

Graphing Calculator Explorations

A graphing calculator may be used to approximate solutions of systems of equations by graphing each equation on the same set of axes and approximating any points of intersection. For example, approximate the solution of the system

$$\begin{cases} y = -2.6x + 5.6 \\ y = 4.3x - 4.9 \end{cases}$$

First use a standard window and graph both equations on a single screen.

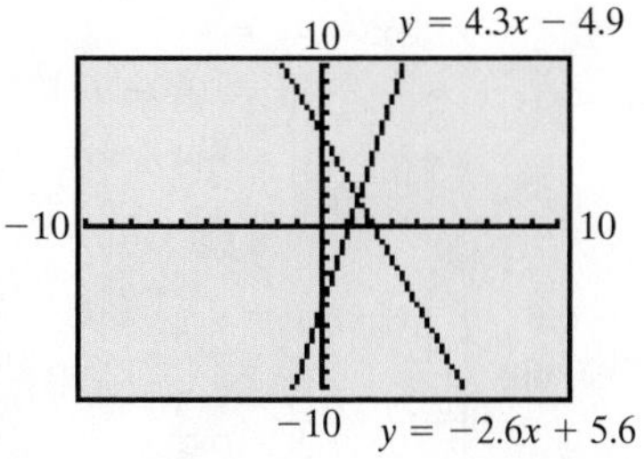

The two lines intersect. To approximate the point of intersection, trace to the point of intersection and use an Intersect feature of the graphing calculator or a Zoom In feature.

Using either method, we find that the approximate point of intersection is (1.52, 1.64).

Solve each system of equations. Approximate the solutions to two decimal places.

1. $y = -1.65x + 3.65$
$y = 4.56x - 9.44$

2. $y = 7.61x + 3.48$
$y = -1.26x - 6.43$

3. $2.33x - 4.72y = 10.61$
$5.86x + 6.22y = -8.89$

4. $-7.89x - 5.68y = 3.26$
$-3.65x + 4.98y = 11.77$

Vocabulary, Readiness & Video Check

Match each graph with the solution of the corresponding system.

A

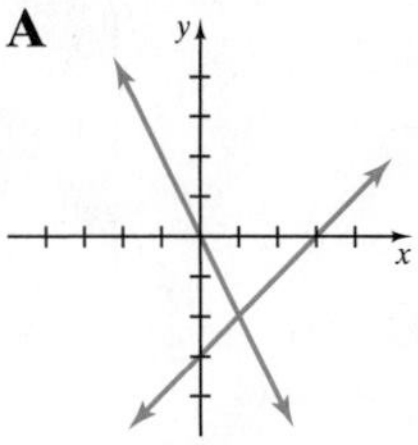

B

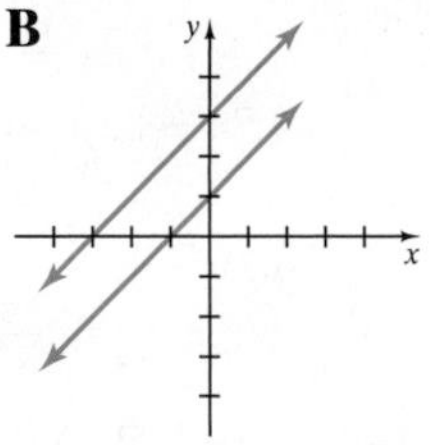

C

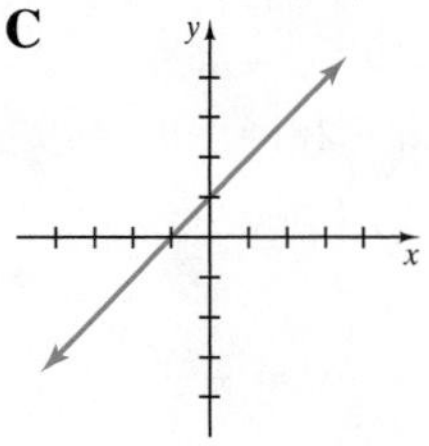

D

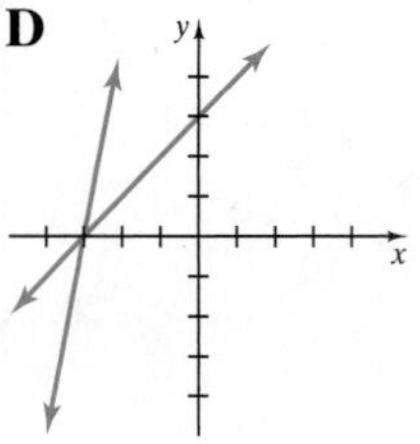

1. no solution

2. Infinite number of solutions

3. $(1, -2)$

4. $(-3, 0)$

Martin-Gay Interactive Videos

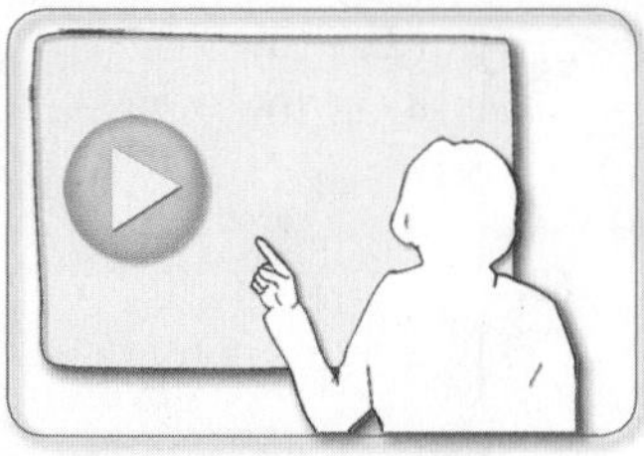

See Video 4.1

Watch the section lecture video and answer the following questions.

OBJECTIVE 1

5. In Example 1, the ordered pair is a solution of the first equation of the system. Why is this not enough to determine if the ordered pair is a solution of the system?

OBJECTIVE 2

6. From Example 2, what potential drawbacks does the graphing method have?

OBJECTIVE 3

7. The system in Example 3 needs one of its equations solved for a variable as a first step. What important point is then emphasized?

OBJECTIVE 4

8. Why is Step 2 in Solving a System by Elimination skipped in Example 5?

4.1 Exercise Set

MyMathLab®

Determine whether each given ordered pair is a solution of each system. See Example 1.

1. $\begin{cases} x - y = 3 \\ 2x - 4y = 8 \end{cases}$ $(2, -1)$

2. $\begin{cases} x - y = -4 \\ 2x + 10y = 4 \end{cases}$ $(-3, 1)$

3. $\begin{cases} 2x - 3y = -9 \\ 4x + 2y = -2 \end{cases}$ $(3, 5)$

4. $\begin{cases} 2x - 5y = -2 \\ 3x + 4y = 4 \end{cases}$ $(4, 2)$

5. $\begin{cases} y = -5x \\ x = -2 \end{cases}$ $(-2, 10)$

6. $\begin{cases} y = 6 \\ x = -2y \end{cases}$ $(-12, 6)$

7. $\begin{cases} 3x + 7y = -19 \\ -6x = 5y + 8 \end{cases}$ $\left(\frac{2}{3}, -3\right)$

8. $\begin{cases} 4x + 5y = -7 \\ -8x = 3y - 1 \end{cases}$ $\left(\frac{3}{4}, -2\right)$

Solve each system by graphing. See Examples 2, through 4.

9. $\begin{cases} x + y = 1 \\ x - 2y = 4 \end{cases}$

10. $\begin{cases} 2x - y = 8 \\ x + 3y = 11 \end{cases}$

11. $\begin{cases} 2y - 4x = 0 \\ x + 2y = 5 \end{cases}$

12. $\begin{cases} 4x - y = 6 \\ x - y = 0 \end{cases}$

13. $\begin{cases} 3x - y = 4 \\ 6x - 2y = 4 \end{cases}$

14. $\begin{cases} -x + 3y = 6 \\ 3x - 9y = 9 \end{cases}$

Solve each system of equations by the substitution method. See Examples 5 and 6.

15. $\begin{cases} x + y = 10 \\ y = 4x \end{cases}$

16. $\begin{cases} 5x + 2y = -17 \\ x = 3y \end{cases}$

17. $\begin{cases} 4x - y = 9 \\ 2x + 3y = -27 \end{cases}$

18. $\begin{cases} 3x - y = 6 \\ -4x + 2y = -8 \end{cases}$

19. $\begin{cases} \frac{1}{2}x + \frac{3}{4}y = -\frac{1}{4} \\ \frac{3}{4}x - \frac{1}{4}y = 1 \end{cases}$

20. $\begin{cases} \frac{2}{5}x + \frac{1}{5}y = -1 \\ x + \frac{2}{5}y = -\frac{8}{5} \end{cases}$

21. $\begin{cases} \frac{x}{3} + y = \frac{4}{3} \\ -x + 2y = 11 \end{cases}$

22. $\begin{cases} \frac{x}{8} - \frac{y}{2} = 1 \\ \frac{x}{3} - y = 2 \end{cases}$

Solve each system of equations by the elimination method. See Examples 7 through 10.

23. $\begin{cases} -x + 2y = 0 \\ x + 2y = 5 \end{cases}$

24. $\begin{cases} -2x + 3y = 0 \\ 2x + 6y = 3 \end{cases}$

25. $\begin{cases} 5x + 2y = 1 \\ x - 3y = 7 \end{cases}$

26. $\begin{cases} 6x - y = -5 \\ 4x - 2y = 6 \end{cases}$

27. $\begin{cases} \frac{3}{4}x + \frac{5}{2}y = 11 \\ \frac{1}{16}x - \frac{3}{4}y = -1 \end{cases}$

28. $\begin{cases} \frac{2}{3}x + \frac{1}{4}y = -\frac{3}{2} \\ \frac{1}{2}x - \frac{1}{4}y = -2 \end{cases}$

29. $\begin{cases} 3x - 5y = 11 \\ 2x - 6y = 2 \end{cases}$

30. $\begin{cases} 6x - 3y = -3 \\ 4x + 5y = -9 \end{cases}$

31. $\begin{cases} x - 2y = 4 \\ 2x - 4y = 4 \end{cases}$

32. $\begin{cases} -x + 3y = 6 \\ 3x - 9y = 9 \end{cases}$

33. $\begin{cases} 3x + y = 1 \\ 2y = 2 - 6x \end{cases}$

34. $\begin{cases} y = 2x - 5 \\ 8x - 4y = 20 \end{cases}$

MIXED PRACTICE

Solve each system of equations.

35. $\begin{cases} 2x + 5y = 8 \\ 6x + y = 10 \end{cases}$

36. $\begin{cases} x - 4y = -5 \\ -3x - 8y = 0 \end{cases}$

37. $\begin{cases} 2x + 3y = 1 \\ x - 2y = 4 \end{cases}$

38. $\begin{cases} -2x + y = -8 \\ x + 3y = 11 \end{cases}$

39. $\begin{cases} \frac{1}{3}x + y = \frac{4}{3} \\ -\frac{1}{4}x - \frac{1}{2}y = -\frac{1}{4} \end{cases}$

40. $\begin{cases} \frac{3}{4}x - \frac{1}{2}y = -\frac{1}{2} \\ x + y = -\frac{3}{2} \end{cases}$

41. $\begin{cases} 2x + 6y = 8 \\ 3x + 9y = 12 \end{cases}$

42. $\begin{cases} x = 3y - 1 \\ 2x - 6y = -2 \end{cases}$

43. $\begin{cases} 4x + 2y = 5 \\ 2x + y = -1 \end{cases}$

44. $\begin{cases} 3x + 6y = 15 \\ 2x + 4y = 3 \end{cases}$

45. $\begin{cases} 10y - 2x = 1 \\ 5y = 4 - 6x \end{cases}$

46. $\begin{cases} 3x + 4y = 0 \\ 7x = 3y \end{cases}$

47. $\begin{cases} 5x - 2y = 27 \\ -3x + 5y = 18 \end{cases}$

48. $\begin{cases} 3x + 4y = 2 \\ 2x + 5y = -1 \end{cases}$

49. $\begin{cases} x = 3y + 2 \\ 5x - 15y = 10 \end{cases}$

50. $\begin{cases} y = \frac{1}{7}x + 3 \\ x - 7y = -21 \end{cases}$

51. $\begin{cases} 2x - y = -1 \\ y = -2x \end{cases}$

52. $\begin{cases} x = \frac{1}{5}y \\ x - y = -4 \end{cases}$

53. $\begin{cases} 2x = 6 \\ y = 5 - x \end{cases}$

54. $\begin{cases} x = 3y + 4 \\ -y = 5 \end{cases}$

55. $\begin{cases} \dfrac{x + 5}{2} = \dfrac{6 - 4y}{3} \\ \dfrac{3x}{5} = \dfrac{21 - 7y}{10} \end{cases}$

56. $\begin{cases} \dfrac{y}{5} = \dfrac{8 - x}{2} \\ x = \dfrac{2y - 8}{3} \end{cases}$

57. $\begin{cases} 4x - 7y = 7 \\ 12x - 21y = 24 \end{cases}$

58. $\begin{cases} 2x - 5y = 12 \\ -4x + 10y = 20 \end{cases}$

59. $\begin{cases} \frac{2}{3}x - \frac{3}{4}y = -1 \\ -\frac{1}{6}x + \frac{3}{8}y = 1 \end{cases}$

60. $\begin{cases} \frac{1}{2}x - \frac{1}{3}y = -3 \\ \frac{1}{8}x + \frac{1}{6}y = 0 \end{cases}$

61. $\begin{cases} 0.7x - 0.2y = -1.6 \\ 0.2x - y = -1.4 \end{cases}$

62. $\begin{cases} -0.7x + 0.6y = 1.3 \\ 0.5x - 0.3y = -0.8 \end{cases}$

63. $\begin{cases} 4x - 1.5y = 10.2 \\ 2x + 7.8y = -25.68 \end{cases}$

64. $\begin{cases} x - 3y = -5.3 \\ 6.3x + 6y = 3.96 \end{cases}$

REVIEW AND PREVIEW

Determine whether the given replacement values make each equation true or false. See Section 1.3.

65. $3x - 4y + 2z = 5$; $x = 1, y = 2,$ and $z = 5$

66. $x + 2y - z = 7$; $x = 2, y = -3,$ and $z = 3$

67. $-x - 5y + 3z = 15$; $x = 0, y = -1,$ and $z = 5$

68. $-4x + y - 8z = 4$; $x = 1, y = 0,$ and $z = -1$

Add the equations. See Section 4.1.

69. $\begin{array}{l} 3x + 2y - 5z = 10 \\ -3x + 4y + z = 15 \end{array}$

70. $\begin{array}{l} x + 4y - 5z = 20 \\ 2x - 4y - 2z = -17 \end{array}$

71. $\begin{array}{l} 10x + 5y + 6z = 14 \\ -9x + 5y - 6z = -12 \end{array}$

72. $\begin{array}{l} -9x - 8y - z = 31 \\ 9x + 4y - z = 12 \end{array}$

CONCEPT EXTENSIONS

Without graphing, determine whether each system has one solution, no solution, or an infinite number of solutions. See the second Concept Check in this section.

73. $\begin{cases} y = 2x - 5 \\ y = 2x + 1 \end{cases}$

74. $\begin{cases} y = 3x - \dfrac{1}{2} \\ y = -2x + \dfrac{1}{5} \end{cases}$

75. $\begin{cases} x + y = 3 \\ 5x + 5y = 15 \end{cases}$

76. $\begin{cases} y = 5x - 2 \\ y = -\dfrac{1}{5}x - 2 \end{cases}$

77. Can a system consisting of two linear equations have exactly two solutions? Explain why or why not.

78. Suppose the graph of the equations in a system of two equations in two variables consists of a circle and a line. Discuss the possible number of solutions for this system.

The concept of supply and demand is used often in business. In general, as the unit price of a commodity increases, the demand for that commodity decreases. Also, as a commodity's unit price increases, the manufacturer normally increases the supply. The point where supply is equal to demand is called the equilibrium point. The following shows the graph of a demand equation and the graph of a supply equation for previously rented DVDs. The x-axis represents the number of DVDs in thousands, and the y-axis represents the cost of a DVD. Use this graph to answer Exercises 79 through 82.

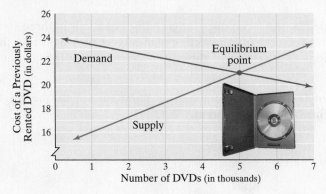

79. Find the number of DVDs and the price per DVD when supply equals demand.

80. When x is between 3 and 4, is supply greater than demand or is demand greater than supply?

81. When x is greater than 6, is supply greater than demand or is demand greater than supply?

82. For what x-values are the y-values corresponding to the supply equation greater than the y-values corresponding to the demand equation?

The revenue equation for a certain brand of toothpaste is $y = 2.5x$, where x is the number of tubes of toothpaste sold and y is the total income for selling x tubes. The cost equation is $y = 0.9x + 3000$, where x is the number of tubes of toothpaste manufactured and y is the cost of producing x tubes. The following set of axes shows the graph of the cost and revenue equations. Use this graph for Exercises 83 through 88.

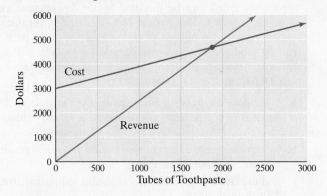

83. Find the coordinates of the point of intersection, or break-even point, by solving the system

$$\begin{cases} y = 2.5x \\ y = 0.9x + 3000 \end{cases}$$

84. Explain the meaning of the x-value of the point of intersection.

85. If the company sells 2000 tubes of toothpaste, does the company make money or lose money?

86. If the company sells 1000 tubes of toothpaste, does the company make money or lose money?

87. For what x-values will the company make a profit? (*Hint:* For what x-values is the revenue graph "higher" than the cost graph?)

88. For what x-values will the company lose money? (*Hint:* For what x-values is the revenue graph "lower" than the cost graph?)

89. Write a system of two linear equations in x and y that has the ordered pair solution $(2, 5)$.

90. Which method would you use to solve the system?

$$\begin{cases} 5x - 2y = 6 \\ 2x + 3y = 5 \end{cases}$$

Explain your choice.

91. The amount y of bottled water consumed per person in the United States (in gallons) in the year x can be modeled by the linear equation $y = 1.47x + 9.26$. The amount y of carbonated diet soft drinks consumed per person in the United States (in gallons) in the year x can be modeled by the linear equation $y = 0.13x + 13.55$. In both models, $x = 0$ represents the year 1995. (*Source:* Based on data from the Economic Research Service, U.S. Department of Agriculture)

   a. What does the slope of each equation tell you about the patterns of bottled water and carbonated diet soft drink consumption in the United States?

   b. Solve this system of equations. (Round your final results to the nearest whole numbers.)

   c. Explain the meaning of your answer to part (b).

92. The amount of U.S. federal government income y (in billions of dollars) for fiscal year x, from 2006 through 2009 ($x = 0$ represents 2006), can be modeled by the linear equation $y = -95x + 2406$. The amount of U.S. federal government expenditures y (in billions of dollars) for the same period can be modeled by the linear equation $y = 285x + 2655$. (*Source:* Based on data from Financial Management Service, U.S. Department of the Treasury, 2006–2009)

   a. What does the slope of each equation tell you about the patterns of U.S. federal government income and expenditures?

   b. Solve this system of equations. (Round your final results to the nearest whole numbers.)

   c. Did expenses ever equal income during the period from 2006 through 2009?

Solve each system. To do so, you may want to let $a = \dfrac{1}{x}$ (if x is in the denominator) and let $b = \dfrac{1}{y}$ (if y is in the denominator.)

93.
$$\begin{cases} \dfrac{1}{x} + y = 12 \\ \dfrac{3}{x} - y = 4 \end{cases}$$

94.
$$\begin{cases} x + \dfrac{2}{y} = 7 \\ 3x + \dfrac{3}{y} = 6 \end{cases}$$

95.
$$\begin{cases} \dfrac{1}{x} + \dfrac{1}{y} = 5 \\ \dfrac{1}{x} - \dfrac{1}{y} = 1 \end{cases}$$

96.
$$\begin{cases} \dfrac{2}{x} + \dfrac{3}{y} = 5 \\ \dfrac{5}{x} - \dfrac{3}{y} = 2 \end{cases}$$

97.
$$\begin{cases} \dfrac{2}{x} - \dfrac{4}{y} = 5 \\ \dfrac{1}{x} - \dfrac{2}{y} = \dfrac{3}{2} \end{cases}$$

98.
$$\begin{cases} \dfrac{2}{x} + \dfrac{3}{y} = -1 \\ \dfrac{3}{x} - \dfrac{2}{y} = 18 \end{cases}$$

99.
$$\begin{cases} \dfrac{3}{x} - \dfrac{2}{y} = -18 \\ \dfrac{2}{x} + \dfrac{3}{y} = 1 \end{cases}$$

100.
$$\begin{cases} \dfrac{5}{x} + \dfrac{7}{y} = 1 \\ -\dfrac{10}{x} - \dfrac{14}{y} = 0 \end{cases}$$

4.2   Solving Systems of Linear Equations in Three Variables

OBJECTIVE

1 Solve a System of Three Linear Equations in Three Variables.

In this section, the algebraic methods of solving systems of two linear equations in two variables are extended to systems of three linear equations in three variables. We call the equation $3x - y + z = -15$, for example, a **linear equation in three variables** since there are three variables and each variable is raised only to the power 1. A solution of this equation is an **ordered triple (x, y, z)** that makes the equation a true statement. For example, the ordered triple $(2, 0, -21)$ is a solution of $3x - y + z = -15$ since replacing x with 2, y with 0, and z with -21 yields the true statement $3(2) - 0 + (-21) = -15$. The graph of this equation is a plane in three-dimensional space, just as the graph of a linear equation in two variables is a line in two-dimensional space.

Although we will not discuss the techniques for graphing equations in three variables, visualizing the possible patterns of intersecting planes gives us insight into the possible patterns of solutions of a system of three three-variable linear equations. There are four possible patterns.

1. Three planes have a single point in common. This point represents the single solution of the system. This system is **consistent.**

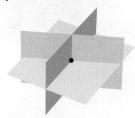

2. Three planes intersect at no point common to all three. This system has no solution. A few ways that this can occur are shown. This system is **inconsistent.**

3. Three planes intersect at all the points of a single line. The system has infinitely many solutions. This system is **consistent.**

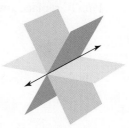

4. Three planes coincide at all points on the plane. The system is consistent, and the equations are **dependent.**

OBJECTIVE

1 Solving a System of Three Linear Equations in Three Variables

Just as with systems of two equations in two variables, we can use the elimination or substitution method to solve a system of three equations in three variables. To use the elimination method, we eliminate a variable and obtain a system of two equations in two variables. Then we use the methods we learned in the previous section to solve the system of two equations.

EXAMPLE 1 Solve the system.

$$\begin{cases} 3x - y + z = -15 & \text{Equation (1)} \\ x + 2y - z = 1 & \text{Equation (2)} \\ 2x + 3y - 2z = 0 & \text{Equation (3)} \end{cases}$$

Solution Add equations (1) and (2) to eliminate z.

$$\begin{array}{r} 3x - y + z = -15 \\ \underline{x + 2y - z = 1} \\ 4x + y = -14 \quad \text{Equation (4)} \end{array}$$

Next, add two *other* equations and *eliminate z again*. To do so, multiply both sides of equation (1) by 2 and add this resulting equation to equation (3). Then

$$\begin{cases} 2(3x - y + z) = 2(-15) \\ 2x + 3y - 2z = 0 \end{cases} \text{ simplifies to } \begin{cases} 6x - 2y + 2z = -30 \\ \underline{2x + 3y - 2z = 0} \\ 8x + y = -30 \quad \text{Equation (5)} \end{cases}$$

> ▶ Helpful Hint
> Don't forget to add two other equations and to **eliminate the same variable.**

Now solve equations (4) and (5) for x and y. To solve by elimination, multiply both sides of equation (4) by -1 and add this resulting equation to equation (5). Then

$$\begin{cases} -1(4x + y) = -1(-14) \\ 8x + y = -30 \end{cases} \text{ simplifies to } \begin{cases} -4x - y = 14 \\ \underline{8x + y = -30} \\ 4x = -16 \end{cases}$$

Add the equations.

$$x = -4 \quad \text{Solve for } x.$$

Replace x with -4 in equation (4) or (5).

$$4x + y = -14 \quad \text{Equation (4)}$$
$$4(-4) + y = -14 \quad \text{Let } x = -4.$$
$$y = 2 \quad \text{Solve for } y.$$

Finally, replace x with -4 and y with 2 in equation (1), (2), or (3).

$$x + 2y - z = 1 \quad \text{Equation (2)}$$
$$-4 + 2(2) - z = 1 \quad \text{Let } x = -4 \text{ and } y = 2.$$
$$-4 + 4 - z = 1$$
$$-z = 1$$
$$z = -1$$

The solution is $(-4, 2, -1)$. To check, let $x = -4$, $y = 2$, and $z = -1$ in all three original equations of the system.

Equation (1)	***Equation (2)***	***Equation (3)***
$3x - y + z = -15$	$x + 2y - z = 1$	$2x + 3y - 2z = 0$
$3(-4) - 2 + (-1) \stackrel{?}{=} -15$	$-4 + 2(2) - (-1) \stackrel{?}{=} 1$	$2(-4) + 3(2) - 2(-1) \stackrel{?}{=} 0$
$-12 - 2 - 1 \stackrel{?}{=} -15$	$-4 + 4 + 1 \stackrel{?}{=} 1$	$-8 + 6 + 2 \stackrel{?}{=} 0$
$-15 = -15$ True	$1 = 1$ True	$0 = 0$ True

All three statements are true, so the solution is $(-4, 2, -1)$. ☐

PRACTICE 1 Solve the system. $\begin{cases} 3x + 2y - z = 0 \\ x - y + 5z = 2 \\ 2x + 3y + 3z = 7 \end{cases}$

EXAMPLE 2 Solve the system.

$$\begin{cases} 2x - 4y + 8z = 2 & (1) \\ -x - 3y + z = 11 & (2) \\ x - 2y + 4z = 0 & (3) \end{cases}$$

Solution Add equations (2) and (3) to eliminate x, and the new equation is

$$-5y + 5z = 11 \quad (4)$$

To eliminate x again, multiply both sides of equation (2) by 2 and add the resulting equation to equation (1). Then

$$\begin{cases} 2x - 4y + 8z = 2 \\ 2(-x - 3y + z) = 2(11) \end{cases} \begin{matrix} \text{simplifies} \\ \text{to} \end{matrix} \begin{cases} 2x - 4y + 8z = 2 \\ \underline{-2x - 6y + 2z = 22} \\ -10y + 10z = 24 \quad (5) \end{cases}$$

Next, solve for y and z using equations (4) and (5). Multiply both sides of equation (4) by -2 and add the resulting equation to equation (5).

$$\begin{cases} -2(-5y + 5z) = -2(11) \\ -10y + 10z = 24 \end{cases} \quad \begin{array}{c} \text{simplifies} \\ \text{to} \end{array} \quad \begin{cases} 10y - 10z = -22 \\ \underline{-10y + 10z = 24} \\ \qquad\qquad 0 = 2 \quad \text{False} \end{cases}$$

Since the statement is false, this system is inconsistent and has no solution. The solution set is the empty set $\{\ \}$ or $\varnothing$. $\qquad\qquad\square$

PRACTICE
2   Solve the system. $\begin{cases} 6x - 3y + 12z = 4 \\ -6x + 4y - 2z = 7 \\ -2x + y - 4z = 3 \end{cases}$

The elimination method is summarized next.

Solving a System of Three Linear Equations by the Elimination Method

Step 1.   Write each equation in standard form $Ax + By + Cz = D$.

Step 2.   Choose a pair of equations and use the equations to eliminate a variable.

Step 3.   Choose any **other** pair of equations and eliminate the **same variable** as in Step 2.

Step 4.   Two equations in two variables should be obtained from Step 2 and Step 3. Use methods from Section 4.1 to solve this system for both variables.

Step 5.   To solve for the third variable, substitute the values of the variables found in Step 4 into any of the original equations containing the third variable.

Step 6.   Check the ordered triple solution in *all three* original equations.

> ▶ **Helpful Hint**
> Make sure you read closely and follow Step 3.

✓**CONCEPT CHECK**
In the system

$$\begin{cases} x + y + z = 6 & \text{Equation (1)} \\ 2x - y + z = 3 & \text{Equation (2)} \\ x + 2y + 3z = 14 & \text{Equation (3)} \end{cases}$$

equations (1) and (2) are used to eliminate y. Which action could be used to finish solving best? Why?
a. Use (1) and (2) to eliminate z.   **b.** Use (2) and (3) to eliminate y.
c. Use (1) and (3) to eliminate x.

EXAMPLE 3   Solve the system.

$$\begin{cases} 2x + 4y \qquad = 1 & (1) \\ 4x \qquad - 4z = -1 & (2) \\ \qquad y - 4z = -3 & (3) \end{cases}$$

Solution   Notice that equation (2) has no term containing the variable y. Let us eliminate y using equations (1) and (3). Multiply both sides of equation (3) by -4 and add the resulting equation to equation (1). Then

$$\begin{cases} 2x + 4y \qquad = 1 \\ -4(y - 4z) = -4(-3) \end{cases} \quad \text{simplifies to} \quad \begin{cases} 2x + 4y \qquad = 1 \\ \underline{\qquad -4y + 16z = 12} \\ 2x \qquad + 16z = 13 \quad (4) \end{cases}$$

Next, solve for z using equations (4) and (2). Multiply both sides of equation (4) by -2 and add the resulting equation to equation (2).

$$\begin{cases} -2(2x + 16z) = -2(13) \\ 4x - 4z = -1 \end{cases} \quad \text{simplifies to} \quad \begin{cases} -4x - 32z = -26 \\ \underline{4x - 4z = -1} \\ -36z = -27 \\ z = \dfrac{3}{4} \end{cases}$$

Replace z with $\dfrac{3}{4}$ in equation (3) and solve for y.

$$y - 4\left(\dfrac{3}{4}\right) = -3 \quad \text{Let } z = \dfrac{3}{4} \text{ in equation (3).}$$
$$y - 3 = -3$$
$$y = 0$$

Replace y with 0 in equation (1) and solve for x.

$$2x + 4(0) = 1$$
$$2x = 1$$
$$x = \dfrac{1}{2}$$

The solution is $\left(\dfrac{1}{2}, 0, \dfrac{3}{4}\right)$. Check to see that this solution satisfies all three equations of the system. □

PRACTICE
3 Solve the system. $\begin{cases} 3x + 4y = 0 \\ 9x - 4z = 6 \\ -2y + 7z = 1 \end{cases}$

EXAMPLE 4 Solve the system.

$$\begin{cases} x - 5y - 2z = 6 \quad (1) \\ -2x + 10y + 4z = -12 \quad (2) \\ \dfrac{1}{2}x - \dfrac{5}{2}y - z = 3 \quad (3) \end{cases}$$

Solution Multiply both sides of equation (3) by 2 to eliminate fractions and multiply both sides of equation (2) by $-\dfrac{1}{2}$ so that the coefficient of x is 1. The resulting system is then

$$\begin{cases} x - 5y - 2z = 6 \quad (1) \\ x - 5y - 2z = 6 \quad \text{Multiply (2) by } -\dfrac{1}{2}. \\ x - 5y - 2z = 6 \quad \text{Multiply (3) by 2.} \end{cases}$$

All three equations are identical, and therefore equations (1), (2), and (3) are all equivalent. There are infinitely many solutions of this system. The equations are dependent. The solution set can be written as $\{(x, y, z) \mid x - 5y - 2z = 6\}$. □

PRACTICE
4 Solve the system. $\begin{cases} 2x + y - 3z = 6 \\ x + \dfrac{1}{2}y - \dfrac{3}{2}z = 3 \\ -4x - 2y + 6z = -12 \end{cases}$

As mentioned earlier, we can also use the substitution method to solve a system of linear equations in three variables.

EXAMPLE 5 Solve the system:

$$\begin{cases} x - 4y - 5z = 35 & (1) \\ x - 3y = 0 & (2) \\ -y + z = -55 & (3) \end{cases}$$

Solution Notice in equations (2) and (3) that a variable is missing. Also notice that both equations contain the variable y. Let's use the substitution method by solving equation (2) for x and equation (3) for z and substituting the results in equation (1).

$$x - 3y = 0 \qquad (2)$$
$$x = 3y \qquad \text{Solve equation (2) for } x. \cdot$$
$$-y + z = -55 \qquad (3)$$
$$z = y - 55 \quad \text{Solve equation (3) for } z.$$

Now substitute $3y$ for x and $y - 55$ for z in equation (1).

$$x - 4y - 5z = 35 \qquad (1)$$

> ▶ **Helpful Hint**
> Do not forget to distribute.

$$3y - 4y - 5(y - 55) = 35 \qquad \text{Let } x = 3y \text{ and } z = y - 55.$$
$$3y - 4y - 5y + 275 = 35 \qquad \text{Use the distributive law and multiply.}$$
$$-6y + 275 = 35 \qquad \text{Combine like terms.}$$
$$-6y = -240 \quad \text{Subtract 275 from both sides.}$$
$$y = 40 \qquad \text{Solve.}$$

To find x, recall that $x = 3y$ and substitute 40 for y. Then $x = 3y$ becomes $x = 3 \cdot 40 = 120$. To find z, recall that $z = y - 55$ and substitute 40 for y, also. Then $z = y - 55$ becomes $z = 40 - 55 = -15$. The solution is $(120, 40, -15)$. ☐

PRACTICE
5 Solve the system. $\begin{cases} x + 2y + 4z = 16 \\ x + 2z = -4 \\ y - 3z = 30 \end{cases}$

Vocabulary, Readiness & Video Check

Solve.

1. Choose the equation(s) that has $(-1, 3, 1)$ as a solution.
   a. $x + y + z = 3$ **b.** $-x + y + z = 5$ **c.** $-x + y + 2z = 0$ **d.** $x + 2y - 3z = 2$

2. Choose the equation(s) that has $(2, 1, -4)$ as a solution.
   a. $x + y + z = -1$ **b.** $x - y - z = -3$ **c.** $2x - y + z = -1$ **d.** $-x - 3y - z = -1$

3. Use the result of Exercise 1 to determine whether $(-1, 3, 1)$ is a solution of the system below. Explain your answer.
   $\begin{cases} x + y + z = 3 \\ -x + y + z = 5 \\ x + 2y - 3z = 2 \end{cases}$

4. Use the result of Exercise 2 to determine whether $(2, 1, -4)$ is a solution of the system below. Explain your answer.
   $\begin{cases} x + y + z = -1 \\ x - y - z = -3 \\ 2x - y + z = -1 \end{cases}$

Martin-Gay Interactive Videos

Watch the section lecture video and answer the following question.

OBJECTIVE 1

5. From Example 1 and the lecture before, why does Step 3 stress that the same variable be eliminated from two other equations?

See Video 4.2

4.2 Exercise Set

MyMathLab®

Solve each system. See Examples 1 through 5.

1. $\begin{cases} x - y + z = -4 \\ 3x + 2y - z = 5 \\ -2x + 3y - z = 15 \end{cases}$

2. $\begin{cases} x + y - z = -1 \\ -4x - y + 2z = -7 \\ 2x - 2y - 5z = 7 \end{cases}$

3. $\begin{cases} x + y = 3 \\ 2y = 10 \\ 3x + 2y - 3z = 1 \end{cases}$

4. $\begin{cases} 5x = 5 \\ 2x + y = 4 \\ 3x + y - 4z = -15 \end{cases}$

5. $\begin{cases} 2x + 2y + z = 1 \\ -x + y + 2z = 3 \\ x + 2y + 4z = 0 \end{cases}$

6. $\begin{cases} 2x - 3y + z = 5 \\ x + y + z = 0 \\ 4x + 2y + 4z = 4 \end{cases}$

7. $\begin{cases} x - 2y + z = -5 \\ -3x + 6y - 3z = 15 \\ 2x - 4y + 2z = -10 \end{cases}$

8. $\begin{cases} 3x + y - 2z = 2 \\ -6x - 2y + 4z = -4 \\ 9x + 3y - 6z = 6 \end{cases}$

9. $\begin{cases} 4x - y + 2z = 5 \\ 2y + z = 4 \\ 4x + y + 3z = 10 \end{cases}$

10. $\begin{cases} 5y - 7z = 14 \\ 2x + y + 4z = 10 \\ 2x + 6y - 3z = 30 \end{cases}$

11. $\begin{cases} x + 5z = 0 \\ 5x + y = 0 \\ y - 3z = 0 \end{cases}$

12. $\begin{cases} x - 5y = 0 \\ x - z = 0 \\ -x + 5z = 0 \end{cases}$

13. $\begin{cases} 6x - 5z = 17 \\ 5x - y + 3z = -1 \\ 2x + y = -41 \end{cases}$

14. $\begin{cases} x + 2y = 6 \\ 7x + 3y + z = -33 \\ x - z = 16 \end{cases}$

15. $\begin{cases} x + y + z = 8 \\ 2x - y - z = 10 \\ x - 2y - 3z = 22 \end{cases}$

16. $\begin{cases} 5x + y + 3z = 1 \\ x - y + 3z = -7 \\ -x + y = 1 \end{cases}$

17. $\begin{cases} x + 2y - z = 5 \\ 6x + y + z = 7 \\ 2x + 4y - 2z = 5 \end{cases}$

18. $\begin{cases} 4x - y + 3z = 10 \\ x + y - z = 5 \\ 8x - 2y + 6z = 10 \end{cases}$

19. $\begin{cases} 2x - 3y + z = 2 \\ x - 5y + 5z = 3 \\ 3x + y - 3z = 5 \end{cases}$

20. $\begin{cases} 4x + y - z = 8 \\ x - y + 2z = 3 \\ 3x - y + z = 6 \end{cases}$

21. $\begin{cases} -2x - 4y + 6z = -8 \\ x + 2y - 3z = 4 \\ 4x + 8y - 12z = 16 \end{cases}$

22. $\begin{cases} -6x + 12y + 3z = -6 \\ 2x - 4y - z = 2 \\ -x + 2y + \frac{z}{2} = -1 \end{cases}$

23. $\begin{cases} 2x + 2y - 3z = 1 \\ y + 2z = -14 \\ 3x - 2y = -1 \end{cases}$

24. $\begin{cases} 7x + 4y = 10 \\ x - 4y + 2z = 6 \\ y - 2z = -1 \end{cases}$

25. $\begin{cases} x + 2y - z = 5 \\ -3x - 2y - 3z = 11 \\ 4x + 4y + 5z = -18 \end{cases}$

26. $\begin{cases} 3x - 3y + z = -1 \\ 3x - y - z = 3 \\ -6x + 4y + 3z = -8 \end{cases}$

27. $\begin{cases} \frac{3}{4}x - \frac{1}{3}y + \frac{1}{2}z = 9 \\ \frac{1}{6}x + \frac{1}{3}y - \frac{1}{2}z = 2 \\ \frac{1}{2}x - y + \frac{1}{2}z = 2 \end{cases}$

28. $\begin{cases} \frac{1}{3}x - \frac{1}{4}y + z = -9 \\ \frac{1}{2}x - \frac{1}{3}y - \frac{1}{4}z = -6 \\ x - \frac{1}{2}y - z = -8 \end{cases}$

REVIEW AND PREVIEW

Translating Solve. See Section 2.2.

29. The sum of two numbers is 45 and one number is twice the other. Find the numbers.

30. The difference of two numbers is 5. Twice the smaller number added to five times the larger number is 53. Find the numbers.

Solve. See Section 2.1.

31. $2(x - 1) - 3x = x - 12$

32. $7(2x - 1) + 4 = 11(3x - 2)$

33. $-y - 5(y + 5) = 3y - 10$

34. $z - 3(z + 7) = 6(2z + 1)$

CONCEPT EXTENSIONS

35. Write a single linear equation in three variables that has $(-1, 2, -4)$ as a solution. (There are many possibilities.) Explain the process you used to write an equation.

36. Write a system of three linear equations in three variables that has $(2, 1, 5)$ as a solution. (There are many possibilities.) Explain the process you used to write an equation.

37. Write a system of linear equations in three variables that has the solution $(-1, 2, -4)$. Explain the process you used to write your system.

38. When solving a system of three equation in three unknowns, explain how to determine that a system has no solution.

39. The fraction $\dfrac{1}{24}$ can be written as the following sum:

$$\frac{1}{24} = \frac{x}{8} + \frac{y}{4} + \frac{z}{3}$$

where the numbers x, y, and z are solutions of

$$\begin{cases} x + y + z = 1 \\ 2x - y + z = 0 \\ -x + 2y + 2z = -1 \end{cases}$$

Solve the system and see that the sum of the fractions is $\dfrac{1}{24}$.

40. The fraction $\dfrac{1}{18}$ can be written as the following sum:

$$\frac{1}{18} = \frac{x}{2} + \frac{y}{3} + \frac{z}{9}$$

where the numbers x, y, and z are solutions of

$$\begin{cases} x + 3y + z = -3 \\ -x + y + 2z = -14 \\ 3x + 2y - z = 12 \end{cases}$$

Solve the system and see that the sum of the fractions is $\dfrac{1}{18}$.

Solving systems involving more than three variables can be accomplished with methods similar to those encountered in this section. Apply what you already know to solve each system of equations in four variables.

41. $\begin{cases} x + y \quad\;\; - w = 0 \\ \quad\;\; y + 2z + w = 3 \\ x \quad\;\;\; - z \quad\;\;\;\; = 1 \\ 2x - y \quad\;\;\; - w = -1 \end{cases}$

42. $\begin{cases} 5x + 4y \quad\qquad = 29 \\ \quad\;\; y + z \; - w = -2 \\ 5x \qquad + z \quad\;\; = 23 \\ \quad\;\; y - z \; + w = 4 \end{cases}$

43. $\begin{cases} x + y + z + w = 5 \\ 2x + y + z + w = 6 \\ x + y + z \quad\;\; = 2 \\ x + y \qquad\quad = 0 \end{cases}$

44. $\begin{cases} 2x \qquad - z \quad\;\; = -1 \\ \quad\;\; y + z + \; w = 9 \\ \quad\;\; y \quad\;\; - 2w = -6 \\ x + y \qquad\quad = 3 \end{cases}$

45. Write a system of three linear equations in three variables that are dependent equations.

46. What is the solution to the system in Exercise 45?

4.3   Systems of Linear Equations and Problem Solving

OBJECTIVES

1 Solve Problems That Can Be Modeled by a System of Two Linear Equations.

2 Solve Problems with Cost and Revenue Functions.

3 Solve Problems That Can Be Modeled by a System of Three Linear Equations.

OBJECTIVE

1 Solving Problem Modeled by Systems of Two Equations

Thus far, we have solved problems by writing one-variable equations and solving for the variable. Some of these problems can be solved, perhaps more easily, by writing a system of equations, as illustrated in this section.

EXAMPLE 1   **Predicting Equal Consumption of Red Meat and Poultry**

America's consumption of red meat has decreased most years since 2000, while consumption of poultry has increased. The function $y = -0.56x + 113.6$ approximates the annual pounds of red meat consumed per capita, where x is the number of years since 2000. The function $y = 0.76x + 68.57$ approximates the annual pounds of poultry consumed per capita, where x is also the number of years since 2000. If this trend continues, determine the year when the annual consumption of red meat and poultry will be equal. (*Source:* USDA: Economic Research Service)

(Continued on next page)

Annual U.S. per Capita Consumption of Red Meat and Poultry

Solution:

1. UNDERSTAND. Read and reread the problem and guess a year. Let's guess the year 2020. This year is 20 years since 2000, so $x = 20$. Now let $x = 20$ in each given function.

$$\text{Red meat: } y = -0.56x + 113.6 = -0.56(20) + 113.6 = 102.4 \text{ pounds}$$
$$\text{Poultry: } \quad y = 0.76x + 68.57 = 0.76(20) + 68.57 = 83.77 \text{ pounds}$$

Since the projected pounds in 2020 for red meat and poultry are not the same, we guessed incorrectly, but we do have a better understanding of the problem. We know that the year will be later than 2020.

2. TRANSLATE. We are already given the system of equations.

3. SOLVE. We want to know the year x in which pounds y are the same, so we solve the system:

$$\begin{cases} y = -0.56x + 113.6 \\ y = 0.76x + 68.57 \end{cases}$$

Since both equations are solved for y, one way to solve is to use the substitution method.

$$y = 0.76x + 68.57 \qquad \text{Second equation}$$

$$\overbrace{-0.56x + 113.6} = 0.76x + 68.57 \qquad \text{Let } y = -0.56x + 113.6$$
$$-1.32x = -45.03$$
$$x = \frac{-45.03}{-1.32} \approx 34.11$$

4. INTERPRET. Since we are only asked to find the year, we need only solve for x.

Check: To check, see whether $x \approx 34.11$ gives approximately the same number of pounds of red meat and poultry.

$$\text{Red meat: } y = -0.56x + 113.6 = -0.56(34.11) + 113.6 \approx 94.4984 \text{ pounds}$$
$$\text{Poultry: } \quad y = 0.76x + 68.57 = 0.76(34.11) + 68.57 \approx 94.4936 \text{ pounds}$$

Since we rounded the number of years, the numbers of pounds do differ slightly. They differ only by 0.0048, so we can assume we solved correctly.

State: The consumption of red meat and poultry will be the same about 34.11 years after 2000, or 2034.11. Thus, in the year 2034, we predict the consumption will be the same. ☐

PRACTICE

1 Read Example 1. If we use the years 2005, 2006, 2007, and 2008 only to write functions approximating the consumption of red meat and poultry, we have the following:

$$\text{Red Meat: } y = -0.54x + 110.6$$
$$\text{Poultry: } \quad y = -0.36x + 74.1$$

where x is the number of years since 2005 and y is pounds per year consumed.

a. Assuming this trend continues, predict the year when consumption of red meat and poultry will be the same. Round to the nearest year.

b. Does your answer differ from the answer to Example 1? Why or why not?

EXAMPLE 2 **Finding Unknown Numbers**

A first number is 4 less than a second number. Four times the first number is 6 more than twice the second. Find the numbers.

Solution

1. UNDERSTAND. Read and reread the problem and guess a solution. If a first number is 10 and this is 4 less than a second number, the second number is 14. Four times the first number is $4(10)$, or 40. This is not equal to 6 more than twice the second number, which is $2(14) + 6$ or 34. Although we guessed incorrectly, we now have a better understanding of the problem.

Since we are looking for two numbers, we will let

$$x = \text{first number}$$
$$y = \text{second number}$$

2. TRANSLATE. Since we have assigned two variables to this problem, we will translate the given facts into two equations. For the first statement we have

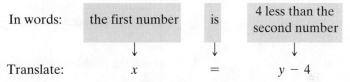

In words:	the first number	is	4 less than the second number
Translate:	x	$=$	$y - 4$

Next we translate the second statement into an equation.

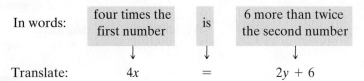

In words:	four times the first number	is	6 more than twice the second number
Translate:	$4x$	$=$	$2y + 6$

3. SOLVE. Here we solve the system

$$\begin{cases} x = y - 4 \\ 4x = 2y + 6 \end{cases}$$

Since the first equation expresses x in terms of y, we will use substitution. We substitute $y - 4$ for x in the second equation and solve for y.

$$4x = 2y + 6 \quad \text{Second equation}$$
$$4(y - 4) = 2y + 6$$
$$4y - 16 = 2y + 6 \quad \text{Let } x = y - 4.$$
$$2y = 22$$
$$y = 11$$

Now we replace y with 11 in the equation $x = y - 4$ and solve for x. Then $x = y - 4$ becomes $x = 11 - 4 = 7$. The ordered pair solution of the system is $(7, 11)$.

(Continued on next page)

4. INTERPRET. Since the solution of the system is $(7, 11)$, then the first number we are looking for is 7 and the second number is 11.

Check: Notice that 7 *is* 4 less than 11, and 4 times 7 *is* 6 more than twice 11. The proposed numbers, 7 and 11, are correct.

State: The numbers are 7 and 11. ☐

PRACTICE

2 A first number is 5 more than a second number. Twice the first number is 2 less than 3 times the second number. Find the numbers.

···■

EXAMPLE 3 **Finding the Rate of Speed**

Two cars leave Indianapolis, one traveling east and the other west. After 3 hours, they are 297 miles apart. If one car is traveling 5 mph faster than the other, what is the speed of each?

Solution

1. UNDERSTAND. Read and reread the problem. Let's guess a solution and use the formula $d = rt$ (distance = rate · time) to check. Suppose that one car is traveling at a rate of 55 miles per hour. This means that the other car is traveling at a rate of 50 miles per hour since we are told that one car is traveling 5 mph faster than the other. To find the distance apart after 3 hours, we will first find the distance traveled by each car. One car's distance is rate · time = $55(3) = 165$ miles. The other car's distance is rate · time = $50(3) = 150$ miles. Since one car is traveling east and the other west, their distance apart is the sum of their distances, or 165 miles + 150 miles = 315 miles. Although this distance apart is not the required distance of 297 miles, we now have a better understanding of the problem.

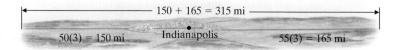

Let's model the problem with a system of equations. We will let

$$x = \text{speed of one car}$$
$$y = \text{speed of the other car}$$

We summarize the information on the following chart. Both cars have traveled 3 hours. Since distance = rate · time, their distances are $3x$ and $3y$ miles, respectively.

	Rate •	*Time* =	*Distance*
One Car	x	3	$3x$
Other Car	y	3	$3y$

2. TRANSLATE. We can now translate the stated conditions into two equations.

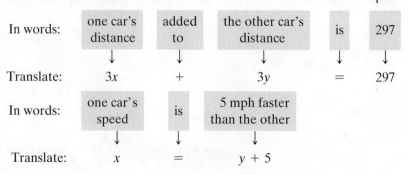

3. SOLVE. Here we solve the system

$$\begin{cases} 3x + 3y = 297 \\ x = y + 5 \end{cases}$$

Again, the substitution method is appropriate. We replace x with $y + 5$ in the first equation and solve for y.

$$3x + 3y = 297 \quad \text{First equation}$$
$$3\overparen{(y + 5)} + 3y = 297 \quad \text{Let } x = y + 5.$$
$$3y + 15 + 3y = 297$$
$$6y = 282$$
$$y = 47$$

To find x, we replace y with 47 in the equation $x = y + 5$. Then $x = 47 + 5 = 52$. The ordered pair solution of the system is $(52, 47)$.

4. INTERPRET. The solution $(52, 47)$ means that the cars are traveling at 52 mph and 47 mph, respectively.

Check: Notice that one car is traveling 5 mph faster than the other. Also, if one car travels 52 mph for 3 hours, the distance is $3(52) = 156$ miles. The other car traveling for 3 hours at 47 mph travels a distance of $3(47) = 141$ miles. The sum of the distances $156 + 141$ is 297 miles, the required distance.

State: The cars are traveling at 52 mph and 47 mph.   ☐

> ▶ **Helpful Hint**
> Don't forget to attach units if appropriate.

PRACTICE

3   In 2007, the French train TGV V150 became the fastest conventional rail train in the world. It broke the 1990 record of the next fastest conventional rail train, the French TGV Atlantique. Assume the V150 and the Atlantique left the same station in Paris, with one heading west and one heading east. After 2 hours, they were 2150 kilometers apart. If the V150 is 75 kph faster than the Atlantique, what is the speed of each?

··

EXAMPLE 4   **Mixing Solutions**

Lynn Pike, a pharmacist, needs 70 liters of a 50% alcohol solution. She has available a 30% alcohol solution and an 80% alcohol solution. How many liters of each solution should she mix to obtain 70 liters of a 50% alcohol solution?

Solution

1. UNDERSTAND. Read and reread the problem. Next, guess the solution. Suppose that we need 20 liters of the 30% solution. Then we need $70 - 20 = 50$ liters of the 80% solution. To see if this gives us 70 liters of a 50% alcohol solution, let's find the amount of pure alcohol in each solution.

number of liters	×	alcohol strength	=	amount of pure alcohol
20 liters	×	0.30	=	6 liters
50 liters	×	0.80	=	40 liters
70 liters	×	0.50	=	35 liters

Since 6 liters + 40 liters = 46 liters and not 35 liters, our guess is incorrect, but we have gained some insight as to how to model and check this problem.

We will let

$$x = \text{amount of 30\% solution, in liters}$$
$$y = \text{amount of 80\% solution, in liters}$$

(Continued on next page)

and use a table to organize the given data.

	Number of Liters	*Alcohol Strength*	*Amount of Pure Alcohol*
30% Solution	x	30%	$0.30x$
80% Solution	y	80%	$0.80y$
50% Solution Needed	70	50%	$(0.50)(70)$

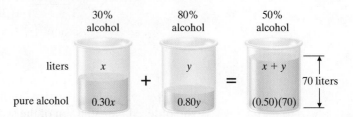

2. **TRANSLATE.** We translate the stated conditions into two equations.

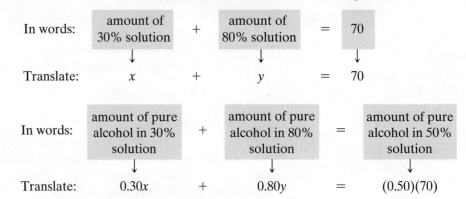

3. **SOLVE.** Here we solve the system

$$\begin{cases} x + y = 70 \\ 0.30x + 0.80y = (0.50)(70) \end{cases}$$

To solve this system, we use the elimination method. We multiply both sides of the first equation by -3 and both sides of the second equation by 10. Then

$$\begin{cases} -3(x + y) = -3(70) \\ 10(0.30x + 0.80y) = 10(0.50)(70) \end{cases} \quad \begin{matrix} \text{simplifies} \\ \text{to} \end{matrix} \quad \begin{cases} -3x - 3y = -210 \\ \underline{3x + 8y = 350} \\ \quad\quad 5y = 140 \\ \quad\quad\quad y = 28 \end{cases}$$

Now we replace y with 28 in the equation $x + y = 70$ and find that $x + 28 = 70$, or $x = 42$.

The ordered pair solution of the system is $(42, 28)$.

4. **INTERPRET.**

Check: Check the solution in the same way that we checked our guess.

State: The pharmacist needs to mix 42 liters of 30% solution and 28 liters of 80% solution to obtain 70 liters of 50% solution. □

PRACTICE

4 Keith Robinson is a chemistry teacher who needs 1 liter of a solution of 5% hydrochloric acid to carry out an experiment. If he only has a stock solution of 99% hydrochloric acid, how much water (0% acid) and how much stock solution (99%) of HCL must he mix to get 1 liter of 5% solution? Round answers to the nearest hundredth of a liter.

✓CONCEPT CHECK

Suppose you mix an amount of 25% acid solution with an amount of 60% acid solution. You then calculate the acid strength of the resulting acid mixture. For which of the following results should you suspect an error in your calculation? Why?

a. 14%   **b.** 32%   **c.** 55%

OBJECTIVE

2 **Solving Problems with Cost and Revenue Functions**

Recall that businesses are often computing cost and revenue functions or equations to predict sales, to determine whether prices need to be adjusted, and to see whether the company is making or losing money. Recall also that the value at which revenue equals cost is called the break-even point. When revenue is less than cost, the company is losing money; when revenue is greater than cost, the company is making money.

EXAMPLE 5   **Finding a Break-Even Point**

A manufacturing company recently purchased $3000 worth of new equipment to offer new personalized stationery to its customers. The cost of producing a package of personalized stationery is $3.00, and it is sold for $5.50. Find the number of packages that must be sold for the company to break even.

Solution

1. UNDERSTAND. Read and reread the problem. Notice that the cost to the company will include a one-time cost of $3000 for the equipment and then $3.00 per package produced. The revenue will be $5.50 per package sold.

   To model this problem, we will let

$$x = \text{number of packages of personalized stationery}$$
$$C(x) = \text{total cost of producing } x \text{ packages of stationery}$$
$$R(x) = \text{total revenue from selling } x \text{ packages of stationery}$$

2. TRANSLATE. The revenue equation is

In words:	revenue for selling x packages of stationery	=	price per package	·	number of packages
Translate:	$R(x)$	=	5.5	·	x

The cost equation is

In words:	cost for producing x packages of stationery	=	cost per package	·	number of packages	+	cost for equipment
Translate:	$C(x)$	=	3	·	x	+	3000

Since the break-even point is when $R(x) = C(x)$, we solve the equation

$$5.5x = 3x + 3000$$

(Continued on next page)

3. SOLVE.

$$5.5x = 3x + 3000$$

$$2.5x = 3000 \qquad \text{Subtract } 3x \text{ from both sides.}$$

$$x = 1200 \qquad \text{Divide both sides by 2.5.}$$

4. INTERPRET.

Check: To see whether the break-even point occurs when 1200 packages are produced and sold, see if revenue equals cost when $x = 1200$. When $x = 1200$, $R(x) = 5.5x = 5.5(1200) = 6600$ and $C(x) = 3x + 3000 = 3(1200) + 3000 = 6600$. Since $R(1200) = C(1200) = 6600$, the break-even point is 1200.

State: The company must sell 1200 packages of stationery to break even. The graph of this system is shown.

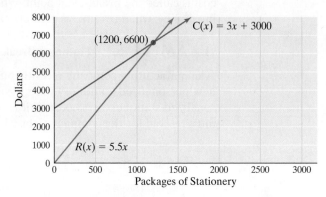

PRACTICE

5   An online-only electronics firm recently purchased $3000 worth of new equipment to create shock-proof packaging for its products. The cost of producing one shock-proof package is $2.50, and the firm charges the customer $4.50 for the packaging. Find the number of packages that must be sold for the company to break even.

OBJECTIVE

3   Solving Problems Modeled by Systems of Three Equations

To introduce problem solving by writing a system of three linear equations in three variables, we solve a problem about triangles.

EXAMPLE 6   Finding Angle Measures

The measure of the largest angle of a triangle is 80° more than the measure of the smallest angle, and the measure of the remaining angle is 10° more than the measure of the smallest angle. Find the measure of each angle.

Solution

1. UNDERSTAND. Read and reread the problem. Recall that the sum of the measures of the angles of a triangle is 180°. Then guess a solution. If the smallest angle measures 20°, the measure of the largest angle is 80° more, or $20° + 80° = 100°$. The measure of the remaining angle is 10° more than the measure of the smallest angle, or $20° + 10° = 30°$. The sum of these three angles is $20° + 100° + 30° = 150°$, not the required 180°. We now know that the measure of the smallest angle is greater than 20°.

To model this problem, we will let

x = degree measure of the smallest angle

y = degree measure of the largest angle

z = degree measure of the remaining angle

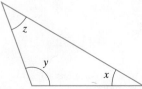

2. TRANSLATE. We translate the given information into three equations.

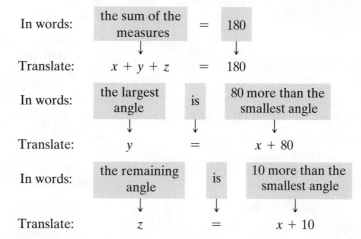

In words: the sum of the measures   =   180

Translate: $x + y + z$ =   180

In words: the largest angle   is   80 more than the smallest angle

Translate: y = $x + 80$

In words: the remaining angle   is   10 more than the smallest angle

Translate: z = $x + 10$

3. SOLVE. We solve the system

$$\begin{cases} x + y + z = 180 \\ y = x + 80 \\ z = x + 10 \end{cases}$$

Since y and z are both expressed in terms of x, we will solve using the substitution method. We substitute $y = x + 80$ and $z = x + 10$ in the first equation. Then

$$x + y + z = 180$$

$$x + (x + 80) + (x + 10) = 180 \quad \text{First equation}$$
$$3x + 90 = 180 \quad \text{Let } y = x + 80 \text{ and } z = x + 10.$$
$$3x = 90$$
$$x = 30$$

Then $y = x + 80 = 30 + 80 = 110$,   and $z = x + 10 = 30 + 10 = 40$.   The ordered triple solution is $(30, 110, 40)$.

4. INTERPRET.

Check: Notice that $30° + 40° + 110° = 180°$. Also, the measure of the largest angle, $110°$, is $80°$ more than the measure of the smallest angle, $30°$. The measure of the remaining angle, $40°$, is $10°$ more than the measure of the smallest angle, $30°$.   □

PRACTICE
6   The measure of the largest angle of a triangle is $40°$ more than the measure of the smallest angle, and the measure of the remaining angle is $20°$ more than the measure of the smallest angle. Find the measure of each angle.

.... ■

Vocabulary, Readiness & Video Check

Martin-Gay Interactive Videos

See Video 4.3

Watch the section lecture video and answer the following questions.

OBJECTIVE
1

1. In ▭ Example 1 and the lecture before, the problem-solving steps for solving applications are mentioned. What is the difference here from when we've used these steps in the past?

OBJECTIVE
2

2. Based on ▭ Example 2, explain the meaning of a break-even point. How do you find the break-even point algebraically?

OBJECTIVE
3

3. In ▭ Example 3, why is the ordered triple not the final stated solution to the application?

4.3 Exercise Set — MyMathLab®

MIXED PRACTICE

Solve. See Examples 1 through 4. For Exercises 1 and 2, the solutions have been started for you.

1. One number is two more than a second number. Twice the first is 4 less than 3 times the second. Find the numbers.

Start the solution:

1. UNDERSTAND the problem. Since we are looking for two numbers, let

$$x = \text{one number}$$
$$y = \text{second number}$$

2. TRANSLATE. Since we have assigned two variables, we will translate the facts into two equations. (Fill in the blanks.)

First equation:

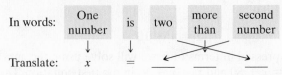

In words: | One number | is | two | more than | second number |

Translate: x = ___ ___ ___

Second equation:

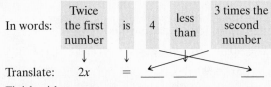

In words: | Twice the first number | is | 4 | less than | 3 times the second number |

Translate: $2x$ = ___ ___ ___

Finish with:

3. SOLVE the system and

4. INTERPRET the results.

2. Three times one number minus a second is 8, and the sum of the numbers is 12. Find the numbers.

Start the solution:

1. UNDERSTAND the problem. Since we are looking for two numbers, let

$$x = \text{one number}$$
$$y = \text{second number}$$

2. TRANSLATE. Since we have assigned two variables, we will translate the facts into two equations. (Fill in the blanks.)

First equation:

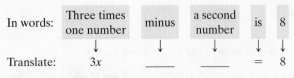

In words: | Three times one number | minus | a second number | is | 8 |

Translate: $3x$ ___ ___ = 8

Second equation:

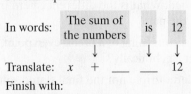

In words: | The sum of the numbers | is | 12 |

Translate: x + ___ ___ 12

Finish with:

3. SOLVE the system and

4. INTERPRET the results.

3. The United States has the world's only large-deck aircraft carriers that can hold up to 72 aircraft. The Enterprise class carrier is longest in length, while the Nimitz class carrier is the second longest. The total length of these two carriers is 2193 feet; the difference of their lengths is only 9 feet. (*Source: USA Today*)

a. Find the length of each class carrier.

b. If a football field has a length of 100 yards, determine the length of the Enterprise class carrier in terms of number of football fields.

4. The rate of growth of participation of women in sports has been increasing since Title IX was enacted in the 1970s. In 2009, the number of women participating in swimming was 2.6 million less than twice the number participating in running. If the total number of women participating in these two sports was 40.9 million, find the number of participants in each sport. (*Source:* Sporting Goods Association of America, 2010 report)

5. A Delta 727 traveled 2520 miles with the wind in 4.5 hours and 2160 miles against the wind in the same amount of time. Find the speed of the plane in still air and the speed of the wind.

6. Terry Watkins can row about 10.6 kilometers in 1 hour downstream and 6.8 kilometers upstream in 1 hour. Find how fast he can row in still water and find the speed of the current.

7. Find how many quarts of 4% butterfat milk and 1% butterfat milk should be mixed to yield 60 quarts of 2% butterfat milk.

8. A pharmacist needs 500 milliliters of a 20% phenobarbital solution but has only 5% and 25% phenobarbital solutions available. Find how many milliliters of each she should mix to get the desired solution.

9. In recent years, the United Kingdom was the most popular host country for U.S. students traveling abroad to study. Italy was the second most popular destination. A total of 58,704 students visited one of the two countries. If 3980 more U.S. students studied in the United Kingdom than Italy, how many students studied abroad in each country? (*Source:* Institute of International Education, Open Doors 2010)

10. The enrollment at both the University of Texas at El Paso (UTEP) and the University of New Hampshire at Durham (UNH) decreased for the 2010–2011 school year. The enrollment at UTEP is 1977 less than twice the enrollment at UNH. Together, these two schools enrolled 29,796 students. Find the number of students enrolled at each school. (*Source:* UTEP and UNH)

11. Karen Karlin bought some large frames for $15 each and some small frames for $8 each at a closeout sale. If she bought 22 frames for $239, find how many of each type she bought.

12. Hilton University Drama Club sold 311 tickets for a play. Student tickets cost 50 cents each; nonstudent tickets cost $1.50. If total receipts were $385.50, find how many tickets of each type were sold.

13. One number is two less than a second number. Twice the first is 4 more than 3 times the second. Find the numbers.

14. Twice one number plus a second number is 42, and the first number minus the second number is −6. Find the numbers.

15. In the United States, the percent of adult blogging has changed within the various age ranges. From 2007 to 2009, the function $y = -4.5x + 24$ can be used to estimate the percent of adults under 30 who blogged, and the function $y = 2x + 7$ can be used to estimate the percent of adults over 30 who blogged. For both functions, x is the number of years after 2007. (*Source:* Pew Internet & American Life Project)

    a. If this trend continues, predict the year in which the percent of adults under 30 and the percent of adults over 30 who blog was the same.

    b. Use these equations to predict the percent of adults under 30 who blog and the percent of adults over 30 who blog for the current year.

16. The rate of fatalities per 100 million vehicle-miles has been decreasing for both automobiles and light trucks (pickups, sport-utility vehicles, and minivans). For the years 2001 through 2009, the function $y = -0.06x + 1.7$ can be used to estimate the rate of fatalities per 100 million vehicle-miles for automobiles during this period, and the function $y = -0.08x + 2.1$ can be used to estimate the rate of fatalities per 100 million vehicle-miles for light trucks during this period. For both functions, x is the number of years since 2000. (*Source:* Bureau of Statistics, U.S. Department of Transportation)

    a. If this trend continues, predict the year in which the fatality rate for automobiles equals the fatality rate for light trucks.

    b. Use these equations to predict the fatality rate per million vehicle-miles for automobiles and light trucks for the current year.

17. An office supply store in San Diego sells 7 writing tablets and 4 pens for $6.40. Also, 2 tablets and 19 pens cost $5.40. Find the price of each.

18. A Candy Barrel shop manager mixes M&M's worth $2.00 per pound with trail mix worth $1.50 per pound. Find how many pounds of each she should use to get 50 pounds of a party mix worth $1.80 per pound.

19. An airplane takes 3 hours to travel a distance of 2160 miles with the wind. The return trip takes 4 hours against the wind. Find the speed of the plane in still air and the speed of the wind.

20. Two cyclists start at the same point and travel in opposite directions. One travels 4 mph faster than the other. In 4 hours, they are 112 miles apart. Find how fast each is traveling.

21. The annual U.S. per capita consumption of cheddar cheese has remained about the same since the millennium, while the consumption of mozzarella cheese has increased. For the years 2000–2010, the function $y = 0.06x + 9.7$ approximates the annual U.S. per capita consumption of cheddar cheese in pounds, and the function $y = 0.21x + 9.3$ approximates the annual U.S. per capita consumption of mozzarella cheese in pounds. For both functions, x is the number of years after 2000.

    a. Explain how the given function verifies that the consumption of cheddar cheese has remained the same, while the given function verifies that the consumption of mozzarella cheese has increased.

    b. Based on this information, determine the year in which the pounds of cheddar cheese consumed equaled the pounds of mozzarella cheese consumed. (*Source:* Based on data from the U.S. Department of Agriculture)

22. Two of the major job categories defined by the U.S. Department of Labor are manufacturing jobs and jobs in the service sector. Jobs in the manufacturing sector have decreased nearly every year since the 1960s. During the same time period, service sector jobs have been steadily increasing. For the years from 1988 through 2009, the function $y = -0.225x + 16.1$ approximates the percent of jobs in the U.S. economy that are manufacturing jobs, while the function $y = 0.45x + 21.7$ approximates the percent of jobs that are service sector jobs. (*Source:* Based on data from the U.S. Department of Labor)

a. Explain how the decrease in manufacturing jobs can be verified by the given function, and the increase of service sector jobs can be verified by the given function.

b. Based on this information, determine the year when the percent of manufacturing jobs and the percent of service sector jobs were the same.

△ **23.** The perimeter of a triangle is 93 centimeters. If two sides are equally long and the third side is 9 centimeters longer than the others, find the lengths of the three sides.

24. Jack Reinholt, a car salesman, has a choice of two pay arrangements: a weekly salary of $200 plus 5% commission on sales or a straight 15% commission. Find the amount of weekly sales for which Jack's earnings are the same regardless of the pay arrangement.

25. Hertz car rental agency charges $25 daily plus 10 cents per mile. Budget charges $20 daily plus 25 cents per mile. Find the daily mileage for which the Budget charge for the day is twice that of the Hertz charge for the day.

26. Carroll Blakemore, a drafting student, bought three templates and a pencil one day for $6.45. Another day, he bought two pads of paper and four pencils for $7.50. If the price of a pad of paper is three times the price of a pencil, find the price of each type of item.

△ **27.** In the figure, line *l* and line *m* are parallel lines cut by transversal *t*. Find the values of *x* and *y*.

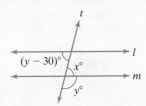

△ **28.** Find the values of *x* and *y* in the following isosceles triangle.

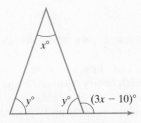

Given the cost function C(x) and the revenue function R(x), find the number of units x that must be sold to break even. See Example 5.

29. $C(x) = 30x + 10,000 \quad R(x) = 46x$

30. $C(x) = 12x + 15,000 \quad R(x) = 32x$

31. $C(x) = 1.2x + 1500 \quad R(x) = 1.7x$

32. $C(x) = 0.8x + 900 \quad R(x) = 2x$

33. $C(x) = 75x + 160,000 \quad R(x) = 200x$

34. $C(x) = 105x + 70,000 \quad R(x) = 245x$

▶ **35.** The planning department of Abstract Office Supplies has been asked to determine whether the company should introduce a new computer desk next year. The department estimates that $6000 of new manufacturing equipment will need to be purchased and that the cost of constructing each desk

will be $200. The department also estimates that the revenue from each desk will be $450.

a. Determine the revenue function $R(x)$ from the sale of *x* desks.

b. Determine the cost function $C(x)$ for manufacturing *x* desks.

c. Find the break-even point.

36. Baskets, Inc., is planning to introduce a new woven basket. The company estimates that $500 worth of new equipment will be needed to manufacture this new type of basket and that it will cost $15 per basket to manufacture. The company also estimates that the revenue from each basket will be $31.

a. Determine the revenue function $R(x)$ from the sale of *x* baskets.

b. Determine the cost function $C(x)$ for manufacturing *x* baskets.

c. Find the break-even point. Round up to the nearest whole basket.

Solve. See Example 6.

37. Rabbits in a lab are to be kept on a strict daily diet that includes 30 grams of protein, 16 grams of fat, and 24 grams of carbohydrates. The scientist has only three food mixes available with the following grams of nutrients per unit.

	Protein	*Fat*	*Carbohydrate*
Mix A	4	6	3
Mix B	6	1	2
Mix C	4	1	12

Find how many units of each mix are needed daily to meet each rabbit's dietary need.

38. Gerry Gundersen mixes different solutions with concentrations of 25%, 40%, and 50% to get 200 liters of a 32% solution. If he uses twice as much of the 25% solution as of the 40% solution, find how many liters of each kind he uses.

△ **39.** The perimeter of a quadrilateral (four-sided polygon) is 29 inches. The longest side is twice as long as the shortest side. The other two sides are equally long and are 2 inches longer than the shortest side. Find the lengths of all four sides.

△ **40.** The measure of the largest angle of a triangle is 90° more than the measure of the smallest angle, and the measure of the remaining angle is 30° more than the measure of the smallest angle. Find the measure of each angle.

41. The sum of three numbers is 40. The first number is five more than the second number. It is also twice the third. Find the numbers.

42. The sum of the digits of a three-digit number is 15. The tens-place digit is twice the hundreds-place digit, and the ones-place digit is 1 less than the hundreds-place digit. Find the three-digit number.

43. During the 2010–2011 regular NBA season, the top-scoring player was Kevin Durant of the Oklahoma City Thunder.

Durant scored a total of 2161 points during the regular season. The number of free throws (each worth one point) he made was 14 more than four times the number of three-point field goals he made. The number of two-point field goals that Durant made was 28 less than the number of free throws he made. How many free throws, two-point field goals, and three-point field goals did Kevin Durant make during the 2010–2011 NBA season? (*Source*: National Basketball Association)

44. For 2010, the WNBA's top scorer was Diana Taurasi of the Phoenix Mercury. She scored a total of 745 points during the regular season. The number of two-point field goals Taurasi made was 36 fewer than two times the number of three-point field goals she made. The number of free throws (each worth one point) she made was 61 more than the number of two-point field goals she made. Find how many free throws, two-point field goals, and three-point field goals Diana Taurasi made during the 2010 regular season. (*Source*: Women's National Basketball Association)

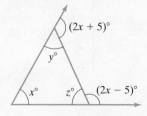

△ **45.** Find the values of x, y, and z in the following triangle.

△ **46.** The sum of the measures of the angles of a quadrilateral is 360°. Find the values of x, y, and z in the following quadrilateral.

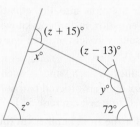

REVIEW AND PREVIEW

Multiply both sides of equation (1) by 2 and add the resulting equation to equation (2). See Section 4.2.

47. $3x - y + z = 2$ (1)
$\quad\;\, -x + 2y + 3z = 6$ (2)

48. $2x + y + 3z = 7$ (1)
$\quad\;\, -4x + y + 2z = 4$ (2)

Multiply both sides of equation (1) by −3 and add the resulting equation to equation (2). See Section 4.2.

49. $x + 2y - z = 0$ (1)
$\quad\;\, 3x + y - z = 2$ (2)

50. $2x - 3y + 2z = 5$ (1)
$\quad\; x - 9y + z = -1$ (2)

CONCEPT EXTENSIONS

51. The number of personal bankruptcy petitions field in the United States was consistently on the rise until there was a major change in the bankruptcy law. The year 2007 was the year in which the fewest personal bankruptcy petitions were filed in 15 years, but the rate soon began to rise. In 2010, the number of petitions filed was 30,000 less than twice the number of petitions filed in 2007. This is equivalent to an increase of 750,000 petitions filed from 2007 to 2010. Find how many personal bankruptcy petitions were filed in each year. (*Source*: Based on data from the Administrative Office of the United States Courts)

52. Two major job categories defined by the U.S. Department of Labor are service occupations and sales occupations. In 2010, the median weekly earnings for males in the service occupations were $120 more than the median weekly earnings for females in the service occupations. Also, the median weekly earnings for females in the sales occupations were $289 less than the median weekly earnings for males in the sales occupations. The median weekly earnings for males in the service occupations were $303 less than twice the median weekly earnings for females in the same category. The median weekly earning for males in the sales occupations were $227 less than twice the median weekly earnings for females in the same category. (*Source*: Based on data from the Bureau of Labor Statistics, U.S. Department of Labor)

a. Find the median weekly earnings for females in the service occupations in the United States for 2010.

b. Find the median weekly earnings for females in the sales occupations in the United States for 2010.

c. Of the four groups of workers described in the problem, which group makes the greatest weekly earnings? Which group makes the least earnings?

53. Find the values of $a, b,$ and c such that the equation $y = ax^2 + bx + c$ has ordered pair solutions $(1, 6), (-1, -2),$ and $(0, -1)$. To do so, substitute each ordered pair solution into the equation. Each time, the result is an equation in three unknowns: $a, b,$ and c. Then solve the resulting system of three linear equations in three unknowns, $a, b,$ and c.

54. Find the values of $a, b,$ and c such that the equation $y = ax^2 + bx + c$ has ordered pair solutions $(1, 2), (2, 3),$ and $(-1, 6)$. (*Hint:* See Exercise 53.)

55. Data (x, y) for the total number (in thousands) of college-bound students who took the ACT assessment in the year x are approximately $(3, 927), (11, 1179),$ and $(19, 1495),$ where $x = 3$ represents 1993 and $x = 11$ represents 2001. Find the values $a, b,$ and c such that the equation $y = ax^2 + bx + c$ models these data. According to your model, how many students will take the ACT in 2015? (*Source:* ACT, Inc.)

56. Monthly normal rainfall data (x, y) for Portland, Oregon, are $(4, 2.47), (7, 0.58), (8, 1.07),$ where x represents time in months (with $x = 1$ representing January) and y represents rainfall in inches. Find the values of $a, b,$ and c rounded to 2 decimal places such that the equation $y = ax^2 + bx + c$ models this data. According to your model, how much rain should Portland expect during September? (*Source:* National Climatic Data Center)

The function $f(x) = -8.6x + 275$ represents the U.S. average number of monthly calls (sent or received) per wireless subscriber and the function $f(x) = 204.9x - 1217$ represents the average number of text messages (sent or received) per wireless subscriber. For both functions, x is the number of years since 2000, and these functions are good for the years 2006–2010.

57. Solve the system formed by these functions. Round each coordinate to the nearest whole number.

58. Use your answer to Exercise 57 to predict the year in which the monthly calls and text messages are/were the same.

Integrated Review · SYSTEMS OF LINEAR EQUATIONS

Sections 4.1–4.3

The graphs of various systems of equations are shown. Match each graph with the solution of its corresponding system.

A
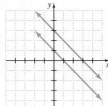

B
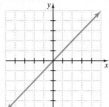

C
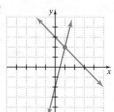

D
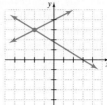

1. Solution: $(1, 2)$ **2.** Solution: $(-2, 3)$ **3.** No solution   **4.** Infinite number of solutions

Solve each system by elimination or substitution.

5. $\begin{cases} x + y = 4 \\ y = 3x \end{cases}$

6. $\begin{cases} x - y = -4 \\ y = 4x \end{cases}$

7. $\begin{cases} x + y = 1 \\ x - 2y = 4 \end{cases}$

8. $\begin{cases} 2x - y = 8 \\ x + 3y = 11 \end{cases}$

9. $\begin{cases} 2x + 5y = 8 \\ 6x + y = 10 \end{cases}$

10. $\begin{cases} \dfrac{1}{8}x - \dfrac{1}{2}y = -\dfrac{5}{8} \\ -3x - 8y = 0 \end{cases}$

11. $\begin{cases} 4x - 7y = 7 \\ 12x - 21y = 24 \end{cases}$

12. $\begin{cases} 2x - 5y = 3 \\ -4x + 10y = -6 \end{cases}$

13. $\begin{cases} y = \dfrac{1}{3}x \\ 5x - 3y = 4 \end{cases}$

14. $\begin{cases} y = \dfrac{1}{4}x \\ 2x - 4y = 3 \end{cases}$

15. $\begin{cases} x + y = 2 \\ -3y + z = -7 \\ 2x + y - z = -1 \end{cases}$

16. $\begin{cases} y + 2z = -3 \\ x - 2y = 7 \\ 2x - y + z = 5 \end{cases}$

17. $\begin{cases} 2x + 4y - 6z = 3 \\ -x + y - z = 6 \\ x + 2y - 3z = 1 \end{cases}$

18. $\begin{cases} x - y + 3z = 2 \\ -2x + 2y - 6z = -4 \\ 3x - 3y + 9z = 6 \end{cases}$

19. $\begin{cases} x + y - 4z = 5 \\ x - y + 2z = -2 \\ 3x + 2y + 4z = 18 \end{cases}$

20. $\begin{cases} 2x - y + 3z = 2 \\ x + y - 6z = 0 \\ 3x + 4y - 3z = 6 \end{cases}$

21. A first number is 8 less than a second number. Twice the first number is 11 more than the second number. Find the numbers.

△ **22.** The sum of the measures of the angles of a quadrilateral is 360°. The two smallest angles of the quadrilateral have the same measure. The third angle measures 30° more than the measure of one of the smallest angles and the fourth angle measures 50° more than the measure of one of the smallest angles. Find the measure of each angle.

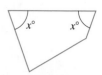

4.4 Solving Systems of Equations by Matrices

OBJECTIVES

1 Use Matrices to Solve a System of Two Equations.

2 Use Matrices to Solve a System of Three Equations.

By now, you may have noticed that the solution of a system of equations depends on the coefficients of the equations in the system and not on the variables. In this section, we introduce solving a system of equations by a **matrix.**

OBJECTIVE

1 Using Matrices to Solve a System of Two Equations

A matrix (plural: **matrices**) is a rectangular array of numbers. The following are examples of matrices.

$$\begin{bmatrix} 1 & 0 \\ 0 & 1 \end{bmatrix} \quad \begin{bmatrix} 2 & 1 & 3 & -1 \\ 0 & -1 & 4 & 5 \\ -6 & 2 & 1 & 0 \end{bmatrix} \quad \begin{bmatrix} a & b & c \\ d & e & f \end{bmatrix}$$

The numbers aligned horizontally in a matrix are in the same **row.** The numbers aligned vertically are in the same **column.**

$$\begin{array}{l} \text{row 1} \rightarrow \\ \text{row 2} \rightarrow \end{array} \begin{bmatrix} 2 & 1 & 0 \\ -1 & 6 & 2 \end{bmatrix}$$

column 1 column 2 column 3

This matrix has 2 rows and 3 columns. It is called a 2 × 3 (read "two by three") matrix.

> **Helpful Hint**
>
> Before writing the corresponding matrix associated with a system of equations, make sure that the equations are written in standard form.

To see the relationship between systems of equations and matrices, study the example below.

System of Equations
(in standard form)

$$\begin{cases} 2x - 3y = 6 & \text{Equation 1} \\ x + y = 0 & \text{Equation 2} \end{cases}$$

Corresponding Matrix

$$\begin{bmatrix} 2 & -3 & \vdots & 6 \\ 1 & 1 & \vdots & 0 \end{bmatrix} \begin{array}{l} \text{Row 1} \\ \text{Row 2} \end{array}$$

Notice that the rows of the matrix correspond to the equations in the system. The coefficients of each variable are placed to the left of a vertical dashed line. The constants are placed to the right. Each of the numbers in the matrix is called an **element.**

The method of solving systems by matrices is to write this matrix as an equivalent matrix from which we easily identify the solution. Two matrices are equivalent if they represent systems that have the same solution set. The following **row operations** can be performed on matrices, and the result is an equivalent matrix.

Elementary Row Operations

1. Any two rows in a matrix may be interchanged.

2. The elements of any row may be multiplied (or divided) by the same nonzero number.

3. The elements of any row may be multiplied (or divided) by a nonzero number and added to their corresponding elements in any other row.

> **Helpful Hint**
>
> Notice that these *row* operations are the same operations that we can perform on *equations* in a system.

To solve a system of two equations in x and y by matrices, write the corresponding matrix associated with the system. Then use elementary row operations to write equivalent matrices until you have a matrix of the form

$$\begin{bmatrix} 1 & a & | & b \\ 0 & 1 & | & c \end{bmatrix},$$

where a, b, and c are constants. Why? If a matrix associated with a system of equations is in this form, we can easily solve for x and y. For example,

Matrix		*System of Equations*

$$\begin{bmatrix} 1 & 2 & | & -3 \\ 0 & 1 & | & 5 \end{bmatrix} \quad \text{corresponds to} \quad \begin{cases} 1x + 2y = -3 \\ 0x + 1y = 5 \end{cases} \quad \text{or} \quad \begin{cases} x + 2y = -3 \\ y = 5 \end{cases}$$

In the second equation, we have $y = 5$. Substituting this in the first equation, we have $x + 2(5) = -3$ or $x = -13$. The solution of the system is the ordered pair $(-13, 5)$.

EXAMPLE 1   Use matrices to solve the system.

$$\begin{cases} x + 3y = 5 \\ 2x - y = -4 \end{cases}$$

Solution   The corresponding matrix is $\begin{bmatrix} 1 & 3 & | & 5 \\ 2 & -1 & | & -4 \end{bmatrix}$. We use elementary row operations to write an equivalent matrix that looks like $\begin{bmatrix} 1 & a & | & b \\ 0 & 1 & | & c \end{bmatrix}$.

For the matrix given, the element in the first row, first column is already 1, as desired. Next we write an equivalent matrix with a 0 below the 1. To do this, we multiply row 1 by -2 and add to row 2. *We will change only row 2.*

$$\begin{bmatrix} 1 & 3 & | & 5 \\ -2(1) + 2 & -2(3) + (-1) & | & -2(5) + (-4) \end{bmatrix} \quad \text{simplifies to} \quad \begin{bmatrix} 1 & 3 & | & 5 \\ 0 & -7 & | & -14 \end{bmatrix}$$

   ↑   ↑   ↑   ↑   ↑   ↑

row 1   row 2   row 1   row 2   row 1   row 2
element element element element element element

Now we change the -7 to a 1 by use of an elementary row operation. We divide row 2 by -7, then

$$\begin{bmatrix} 1 & 3 & | & 5 \\ \frac{0}{-7} & \frac{-7}{-7} & | & \frac{-14}{-7} \end{bmatrix} \quad \text{simplifies to} \quad \begin{bmatrix} 1 & 3 & | & 5 \\ 0 & 1 & | & 2 \end{bmatrix}$$

This last matrix corresponds to the system

$$\begin{cases} x + 3y = 5 \\ y = 2 \end{cases}$$

To find x, we let $y = 2$ in the first equation, $x + 3y = 5$.

$$x + 3y = 5 \qquad \text{First equation}$$
$$x + 3(2) = 5 \qquad \text{Let } y = 2.$$
$$x = -1$$

The ordered pair solution is $(-1, 2)$. Check to see that this ordered pair satisfies both equations. □

PRACTICE

1   Use matrices to solve the system.

$$\begin{cases} x + 4y = -2 \\ 3x - y = 7 \end{cases}$$

EXAMPLE 2   Use matrices to solve the system.

$$\begin{cases} 2x - y = 3 \\ 4x - 2y = 5 \end{cases}$$

**Solution**   The corresponding matrix is $\begin{bmatrix} 2 & -1 & \vdots & 3 \\ 4 & -2 & \vdots & 5 \end{bmatrix}$. To get 1 in the row 1, column 1 position, we divide the elements of row 1 by 2.

$$\begin{bmatrix} \dfrac{2}{2} & \dfrac{-1}{2} & \vdots & \dfrac{3}{2} \\ 4 & -2 & \vdots & 5 \end{bmatrix} \quad \text{simplifies to} \quad \begin{bmatrix} 1 & -\dfrac{1}{2} & \vdots & \dfrac{3}{2} \\ 4 & -2 & \vdots & 5 \end{bmatrix}$$

To get 0 under the 1, we multiply the elements of row 1 by -4 and add the new elements to the elements of row 2.

$$\begin{bmatrix} 1 & -\dfrac{1}{2} & \vdots & \dfrac{3}{2} \\ -4(1) + 4 & -4\left(-\dfrac{1}{2}\right) - 2 & \vdots & -4\left(\dfrac{3}{2}\right) + 5 \end{bmatrix} \quad \text{simplifies to} \quad \begin{bmatrix} 1 & -\dfrac{1}{2} & \vdots & \dfrac{3}{2} \\ 0 & 0 & \vdots & -1 \end{bmatrix}$$

The corresponding system is $\begin{cases} x - \dfrac{1}{2}y = \dfrac{3}{2} \\ 0 = -1 \end{cases}$. The equation $0 = -1$ is false for all y or x values; hence, the system is inconsistent and has no solution.   □

PRACTICE
2   Use matrices to solve the system.

$$\begin{cases} x - 3y = 3 \\ -2x + 6y = 4 \end{cases}$$

✔CONCEPT CHECK
Consider the system

$$\begin{cases} 2x - 3y = 8 \\ x + 5y = -3 \end{cases}$$

What is wrong with its corresponding matrix shown below?

$$\begin{bmatrix} 2 & -3 & \vdots & 8 \\ 5 & 8 & \vdots & -3 \end{bmatrix}$$

OBJECTIVE
2   Using Matrices to Solve a System of Three Equations

To solve a system of three equations in three variables using matrices, we will write the corresponding matrix in the form

$$\begin{bmatrix} 1 & a & b & \vdots & d \\ 0 & 1 & c & \vdots & e \\ 0 & 0 & 1 & \vdots & f \end{bmatrix}$$

EXAMPLE 3   Use matrices to solve the system.

$$\begin{cases} x + 2y + z = 2 \\ -2x - y + 2z = 5 \\ x + 3y - 2z = -8 \end{cases}$$

Answer to Concept Check:

matrix should be $\begin{bmatrix} 2 & -3 & \vdots & 8 \\ 1 & 5 & \vdots & -3 \end{bmatrix}$

(Continued on next page)

Solution The corresponding matrix is $\begin{bmatrix} 1 & 2 & 1 & \vdots & 2 \\ -2 & -1 & 2 & \vdots & 5 \\ 1 & 3 & -2 & \vdots & -8 \end{bmatrix}$. Our goal is to write

an equivalent matrix with 1's along the diagonal (see the numbers in red) and 0's below the 1's. The element in row 1, column 1 is already 1. Next we get 0's for each element in the rest of column 1. To do this, first we multiply the elements of row 1 by 2 and add the new elements to row 2. Also, we multiply the elements of row 1 by -1 and add the new elements to the elements of row 3. We *do not change row 1*. Then

$$\begin{bmatrix} 1 & 2 & 1 & \vdots & 2 \\ 2(1)-2 & 2(2)-1 & 2(1)+2 & \vdots & 2(2)+5 \\ -1(1)+1 & -1(2)+3 & -1(1)-2 & \vdots & -1(2)-8 \end{bmatrix} \text{ simplifies to } \begin{bmatrix} 1 & 2 & 1 & \vdots & 2 \\ 0 & 3 & 4 & \vdots & 9 \\ 0 & 1 & -3 & \vdots & -10 \end{bmatrix}$$

We continue down the diagonal and use elementary row operations to get 1 where the element 3 is now. To do this, we interchange rows 2 and 3.

$$\begin{bmatrix} 1 & 2 & 1 & \vdots & 2 \\ 0 & 3 & 4 & \vdots & 9 \\ 0 & 1 & -3 & \vdots & -10 \end{bmatrix} \text{ is equivalent to } \begin{bmatrix} 1 & 2 & 1 & \vdots & 2 \\ 0 & 1 & -3 & \vdots & -10 \\ 0 & 3 & 4 & \vdots & 9 \end{bmatrix}$$

Next we want the new row 3, column 2 element to be 0. We multiply the elements of row 2 by -3 and add the result to the elements of row 3.

$$\begin{bmatrix} 1 & 2 & 1 & \vdots & 2 \\ 0 & 1 & -3 & \vdots & -10 \\ -3(0)+0 & -3(1)+3 & -3(-3)+4 & \vdots & -3(-10)+9 \end{bmatrix} \text{ simplifies to}$$

$$\begin{bmatrix} 1 & 2 & 1 & \vdots & 2 \\ 0 & 1 & -3 & \vdots & -10 \\ 0 & 0 & 13 & \vdots & 39 \end{bmatrix}$$

Finally, we divide the elements of row 3 by 13 so that the final diagonal element is 1.

$$\begin{bmatrix} 1 & 2 & 1 & \vdots & 2 \\ 0 & 1 & -3 & \vdots & -10 \\ \dfrac{0}{13} & \dfrac{0}{13} & \dfrac{13}{13} & \vdots & \dfrac{39}{13} \end{bmatrix} \text{ simplifies to } \begin{bmatrix} 1 & 2 & 1 & \vdots & 2 \\ 0 & 1 & -3 & \vdots & -10 \\ 0 & 0 & 1 & \vdots & 3 \end{bmatrix}$$

This matrix corresponds to the system

$$\begin{cases} x + 2y + z = 2 \\ y - 3z = -10 \\ z = 3 \end{cases}$$

We identify the z-coordinate of the solution as 3. Next, we replace z with 3 in the second equation and solve for y.

$$y - 3z = -10 \quad \text{Second equation}$$
$$y - 3(3) = -10 \quad \text{Let } z = 3.$$
$$y = -1$$

To find x, we let $z = 3$ and $y = -1$ in the first equation.

$$x + 2y + z = 2 \quad \text{First equation}$$
$$x + 2(-1) + 3 = 2 \quad \text{Let } z = 3 \text{ and } y = -1.$$
$$x = 1$$

The ordered triple solution is $(1, -1, 3)$. Check to see that it satisfies all three equations in the original system.

PRACTICE
3   Use matrices to solve the system.

$$\begin{cases} x + 3y - z = 0 \\ 2x + y + 3z = 5 \\ -x - 2y + 4z = 7 \end{cases}$$

Vocabulary, Readiness & Video Check

Use the choices below to fill in each blank.

column   element   row   matrix

1. A(n) _____ is a rectangular array of numbers.
2. Each of the numbers in a matrix is called a(n) _____.
3. The numbers aligned horizontally in a matrix are in the same _____.
4. The numbers aligned vertically in a matrix are in the same _____.

Answer true or false for each statement about operations within a matrix forming an equivalent matrix.

5. Any two columns may be interchanged. _____
6. Any two rows may be interchanged. _____
7. The elements in a row may be added to their corresponding elements in another row. _____
8. The elements of a column may be multiplied by any nonzero number. _____

Martin-Gay Interactive Videos

See Video 4.4

Watch the section lecture video and answer the following questions.

OBJECTIVE 1
9. From the lecture before Example 1, what elementary row operations can be performed on matrices? Which operation is not performed during Example 1?

OBJECTIVE 2
10. In Example 2, why do you think the suggestion is made to write neatly when using matrices to solve systems?

4.4   Exercise Set   MyMathLab®

Solve each system of linear equations using matrices. See Example 1.

1. $\begin{cases} x + y = 1 \\ x - 2y = 4 \end{cases}$

2. $\begin{cases} 2x - y = 8 \\ x + 3y = 11 \end{cases}$

3. $\begin{cases} x + 3y = 2 \\ x + 2y = 0 \end{cases}$

4. $\begin{cases} 4x - y = 5 \\ 3x + 3y = 0 \end{cases}$

Solve each system of linear equations using matrices. See Example 2.

5. $\begin{cases} x - 2y = 4 \\ 2x - 4y = 4 \end{cases}$

6. $\begin{cases} -x + 3y = 6 \\ 3x - 9y = 9 \end{cases}$

7. $\begin{cases} 3x - 3y = 9 \\ 2x - 2y = 6 \end{cases}$

8. $\begin{cases} 9x - 3y = 6 \\ -18x + 6y = -12 \end{cases}$

Solve each system of linear equations using matrices. See Example 3.

9. $\begin{cases} x + y = 3 \\ 2y = 10 \\ 3x + 2y - 4z = 12 \end{cases}$

10. $\begin{cases} 5x = 5 \\ 2x + y = 4 \\ 3x + y - 5z = -15 \end{cases}$

11. $\begin{cases} 2y - z = -7 \\ x + 4y + z = -4 \\ 5x - y + 2z = 13 \end{cases}$

12. $\begin{cases} 4y + 3z = -2 \\ 5x - 4y = 1 \\ -5x + 4y + z = -3 \end{cases}$

MIXED PRACTICE

Solve each system of linear equations using matrices. See Examples 1 through 3.

13. $\begin{cases} x - 4 = 0 \\ x + y = 1 \end{cases}$

14. $\begin{cases} 3y = 6 \\ x + y = 7 \end{cases}$

15. $\begin{cases} x + y + z = 2 \\ 2x \quad - z = 5 \\ \quad 3y + z = 2 \end{cases}$

16. $\begin{cases} x + 2y + z = 5 \\ x - y - z = 3 \\ \quad y + z = 2 \end{cases}$

17. $\begin{cases} 5x - 2y = 27 \\ -3x + 5y = 18 \end{cases}$

18. $\begin{cases} 4x - y = 9 \\ 2x + 3y = -27 \end{cases}$

19. $\begin{cases} 4x - 7y = 7 \\ 12x - 21y = 24 \end{cases}$

20. $\begin{cases} 2x - 5y = 12 \\ -4x + 10y = 20 \end{cases}$

21. $\begin{cases} 4x - y + 2z = 5 \\ \quad 2y + z = 4 \\ 4x + y + 3z = 10 \end{cases}$

22. $\begin{cases} 5y - 7z = 14 \\ 2x + y + 4z = 10 \\ 2x + 6y - 3z = 30 \end{cases}$

▶ 23. $\begin{cases} 4x + y + z = 3 \\ -x + y - 2z = -11 \\ x + 2y + 2z = -1 \end{cases}$

24. $\begin{cases} x + y + z = 9 \\ 3x - y + z = -1 \\ -2x + 2y - 3z = -2 \end{cases}$

REVIEW AND PREVIEW

Determine whether each graph is the graph of a function. See Section 3.2.

25.

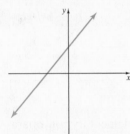

26.

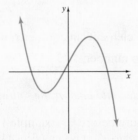

27.

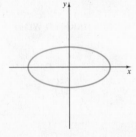

28.

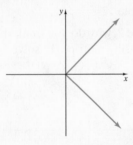

Evaluate. See Section 1.3.

29. $(-1)(-5) - (6)(3)$

30. $(2)(-8) - (-4)(1)$

31. $(4)(-10) - (2)(-2)$

32. $(-7)(3) - (-2)(-6)$

33. $(-3)(-3) - (-1)(-9)$

34. $(5)(6) - (10)(10)$

CONCEPT EXTENSIONS

Solve. See the Concept Check in the section.

35. For the system $\begin{cases} x \quad + z = 7 \\ \quad y + 2z = -6 \\ 3x - y \quad = 0 \end{cases}$, which is the correct corresponding matrix?

a. $\begin{bmatrix} 1 & 1 & \vdots & 7 \\ 1 & 2 & \vdots & -6 \\ 3 & -1 & \vdots & 0 \end{bmatrix}$

b. $\begin{bmatrix} 1 & 0 & 1 & \vdots & 7 \\ 1 & 2 & 0 & \vdots & -6 \\ 3 & -1 & 0 & \vdots & 0 \end{bmatrix}$

c. $\begin{bmatrix} 1 & 0 & 1 & \vdots & 7 \\ 0 & 1 & 2 & \vdots & -6 \\ 3 & -1 & 0 & \vdots & 0 \end{bmatrix}$

36. For the system $\begin{cases} x - 6 = 0 \\ 2x - 3y = 1 \end{cases}$, which is the correct corresponding matrix?

a. $\begin{bmatrix} 1 & -6 & \vdots & 0 \\ 2 & -3 & \vdots & 1 \end{bmatrix}$ **b.** $\begin{bmatrix} 1 & 0 & \vdots & 6 \\ 2 & -3 & \vdots & 1 \end{bmatrix}$ **c.** $\begin{bmatrix} 1 & 0 & \vdots & -6 \\ 2 & -3 & \vdots & 1 \end{bmatrix}$

37. The amount of electricity y generated by geothermal sources (in billions of kilowatts) from 2000 to 2009 can be modeled by the linear equation $y - 0.11x = 14.05$, where x represents the number of years after 2000. Similarly, the amount of electricity y generated by wind power (in billions of kilowatts) during the same time period can be modeled by the linear equation $5.13x - y = 0.65$. (*Source:* Based on data from Energy Information Administration, U.S. Department of Energy)

a. The data used to form these two models were incomplete. It is impossible to tell from the data the year in which the electricity generated by geothermal sources was the same as the electricity generated by wind power. Use matrix methods to estimate the year in which this occurred.

b. The earliest data for wind power was in 1989, where 2.1 billion kilowatts of electricity was generated. Can this data be determined from the given equation? Why do you think that is?

c. According to these models, will the percent of electricity generated by geothermal ever go to zero? Why?

d. Can you think of an explanation why the amount of electricity generated by wind power is increasing so much faster than the amount of electricity generated by geothermal power?

38. The most popular amusement park in the world (according to attendance) is Walt Disney World's Magic Kingdom, whose yearly attendance in millions can be approximated by the equation $y = 0.25x + 16.2$ where x is the number of years after 2005. In second place is Walt Disney World's Disneyland, whose yearly attendance in millions can be approximated by the equation $y = 0.35x + 14.5$ where x is the number of years after 2005. Find the year when attendance in Disneyland is equal to the attendance in Magic Kingdom. (*Source:* Themed Entertainment Association, Economics Research Associates)

39. For the system $\begin{cases} 2x - 3y = 8 \\ x + 5y = -3 \end{cases}$, explain what is wrong with writing the corresponding matrix as $\begin{bmatrix} 2 & 3 & \vdots & 8 \\ 0 & 5 & \vdots & -3 \end{bmatrix}$.

40. For the system $\begin{cases} 5x + 2y = 0 \\ -y = 2 \end{cases}$, explain what is wrong with writing the corresponding matrix as $\begin{bmatrix} 5 & 2 & \vdots & 0 \\ -1 & 0 & \vdots & 2 \end{bmatrix}$.

4.5 | Systems of Linear Inequalities

OBJECTIVE

1 Graph a System of Linear Inequalities.

OBJECTIVE

1   Graphing Systems of Linear Inequalities

In Section 3.7, we solved linear inequalities in two variables as well as their union and intersection. Just as two or more linear equations make a system of linear equations, two or more linear inequalities make a **system of linear inequalities.** Systems of inequalities are very important in a process called linear programming. Many businesses use linear programming to find the most profitable way to use limited resources such as employees, machines, or buildings.

A **solution of a system of linear inequalities** is an ordered pair that satisfies each inequality in the system. The set of all such ordered pairs is the solution set of the system. Graphing this set gives us a picture of the solution set. We can graph a system of inequalities by graphing each inequality in the system and identifying the region of overlap.

Graphing the Solutions of a System of Linear Inequalities

Step 1. Graph each inequality in the system on the same set of axes.

Step 2. The solutions of the system are the points common to the graphs of all the inequalities in the system.

EXAMPLE 1   Graph the solutions of the system: $\begin{cases} 3x \geq y \\ x + 2y \leq 8 \end{cases}$

Solution   We begin by graphing each inequality on the *same* set of axes. The graph of the solutions of the system is the region contained in the graphs of both inequalities. In other words, it is their intersection.

First let's graph $3x \geq y$. The boundary line is the graph of $3x = y$. We sketch a solid boundary line since the inequality $3x \geq y$ means $3x > y$ or $3x = y$. The test point $(1, 0)$ satisfies the inequality, so we shade the half-plane that includes $(1, 0)$.

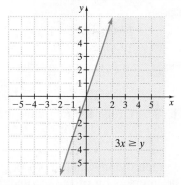

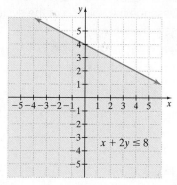

Next we sketch a solid boundary line $x + 2y = 8$ on the same set of axes. The test point $(0, 0)$ satisfies the inequality $x + 2y \leq 8$, so we shade the half-plane that includes $(0, 0)$. (For clarity, the graph of $x + 2y \leq 8$ is shown here on a separate set of axes.) An ordered pair solution of the system must satisfy both inequalities. These solutions are points that lie in both shaded regions. The solution of the system is the darkest shaded region. This solution includes parts of both boundary lines.

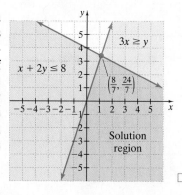

PRACTICE

1   Graph the solutions of the system: $\begin{cases} 4x \geq y \\ x + 3y \geq 6 \end{cases}$

In linear programming, it is sometimes necessary to find the coordinates of the **corner point:** the point at which the two boundary lines intersect. To find the corner point for the system of Example 1, we solve the related linear system

$$\begin{cases} 3x = y \\ x + 2y = 8 \end{cases}$$

using either the substitution or the elimination method. The lines intersect at $\left(\dfrac{8}{7}, \dfrac{24}{7}\right)$, the corner point of the graph.

EXAMPLE 2 Graph the solutions of the system: $\begin{cases} x - y < 2 \\ x + 2y > -1 \\ y < 2 \end{cases}$

Solution First we graph all three inequalities on the same set of axes. All boundary lines are dashed lines since the inequality symbols are $<$ and $>$. The solution of the system is the region shown by the shading. In this example, the boundary lines are *not* part of the solution.

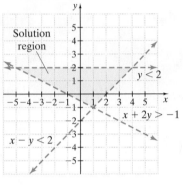

PRACTICE
2 Graph the solutions of the system: $\begin{cases} x - y < 1 \\ y < 4 \\ 3x + y > -3 \end{cases}$.

✓CONCEPT CHECK
Describe the solution of the system of inequalities:

$$\begin{cases} x \leq 2 \\ x \geq 2 \end{cases}$$

EXAMPLE 3 Graph the solutions of the system: $\begin{cases} -3x + 4y \leq 12 \\ x \leq 3 \\ x \geq 0 \\ y \geq 0 \end{cases}$

Solution We graph the inequalities on the same set of axes. The intersection of the inequalities is the solution region. It is the only region shaded in this graph and includes the portions of all four boundary lines that border the shaded region.

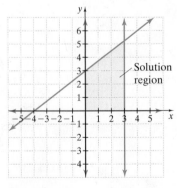

PRACTICE
3 Graph the solutions of the system: $\begin{cases} -2x + 5y \leq 10 \\ x \leq 4 \\ x \geq 0 \\ y \geq 0 \end{cases}$.

Answer to Concept Check:
the line $x = 2$

Vocabulary, Readiness & Video Check

Use the choices below to fill in each blank. Not all choices will be used.

solution   union   system

corner   intersection

1. Two or more linear inequalities form a(n) _____ of linear inequalities.

2. An ordered pair that satisfies each inequality in a system is a(n) _____ of the system.

3. The point where two boundary lines intersect is a(n) _____ point.

4. The solution region of a system of inequalities consists of the _____ of the solution regions of the inequalities in the system.

Martin-Gay Interactive Videos

See Video 4.5

Watch the section lecture video and answer the following question.

OBJECTIVE
1

5. In Example 1, do the solutions of the first inequality of the system limit where we can choose the test point for the second inequality? Why or why not?

4.5 Exercise Set   MyMathLab®

MIXED PRACTICE

Graph the solutions of each system of linear inequalities. See Examples 1 through 3.

1. $\begin{cases} y \geq x + 1 \\ y \geq 3 - x \end{cases}$

2. $\begin{cases} y \geq x - 3 \\ y \geq -1 - x \end{cases}$

3. $\begin{cases} y < 3x - 4 \\ y \leq x + 2 \end{cases}$

4. $\begin{cases} y \leq 2x + 1 \\ y > x + 2 \end{cases}$

5. $\begin{cases} y < -2x - 2 \\ y > x + 4 \end{cases}$

6. $\begin{cases} y \leq 2x + 4 \\ y \geq -x - 5 \end{cases}$

7. $\begin{cases} y \geq -x + 2 \\ y \leq 2x + 5 \end{cases}$

8. $\begin{cases} y \geq x - 5 \\ y \geq -3x + 3 \end{cases}$

9. $\begin{cases} x \geq 3y \\ x + 3y \leq 6 \end{cases}$

10. $\begin{cases} -2x > y \\ x + 2y < 3 \end{cases}$

11. $\begin{cases} x \leq 2 \\ y \geq -3 \end{cases}$

12. $\begin{cases} x \geq -3 \\ y \geq -2 \end{cases}$

13. $\begin{cases} y \geq 1 \\ x < -3 \end{cases}$

14. $\begin{cases} y > 2 \\ x \geq -1 \end{cases}$

15. $\begin{cases} y + 2x \geq 0 \\ 5x - 3y \leq 12 \\ y \leq 2 \end{cases}$

16. $\begin{cases} y + 2x \leq 0 \\ 5x + 3y \geq -2 \\ y \leq 4 \end{cases}$

17. $\begin{cases} 3x - 4y \geq -6 \\ 2x + y \leq 7 \\ y \geq -3 \end{cases}$

18. $\begin{cases} 4x - y \geq -2 \\ 2x + 3y \leq -8 \\ y \geq -5 \end{cases}$

19. $\begin{cases} 2x + y \leq 5 \\ x \leq 3 \\ x \geq 0 \\ y \geq 0 \end{cases}$

20. $\begin{cases} 3x + y \leq 4 \\ x \leq 4 \\ x \geq 0 \\ y \geq 0 \end{cases}$

Match each system of inequalities to the corresponding graph.

21. $\begin{cases} y < 5 \\ x > 3 \end{cases}$ **22.** $\begin{cases} y > 5 \\ x < 3 \end{cases}$

23. $\begin{cases} y \le 5 \\ x < 3 \end{cases}$ **24.** $\begin{cases} y > 5 \\ x \ge 3 \end{cases}$

A 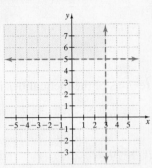   **B**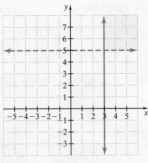

C 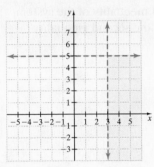   **D**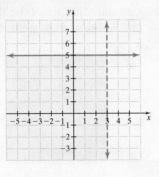

REVIEW

Evaluate each expression. See Section 1.3.

25. $(-3)^2$ **26.** $(-5)^3$ **27.** $\left(\dfrac{2}{3}\right)^2$

28. $\left(\dfrac{3}{4}\right)^3$

Perform each indicated operation. See Section 1.3.

29. $(-2)^2 - (-3) + 2(-1)$

30. $5^2 - 11 + 3(-5)$

31. $8^2 + (-13) - 4(-2)$

32. $(-12)^2 + (-1)(2) - 6$

CONCEPT EXTENSIONS

Solve. See the Concept Check in this section.

33. Describe the solution of the system: $\begin{cases} y \le 3 \\ y \ge 3 \end{cases}$.

34. Describe the solution of the system: $\begin{cases} x \le 5 \\ x \le 3 \end{cases}$.

35. Explain how to decide which region to shade to show the solution region of the following system.

$$\begin{cases} x \ge 3 \\ y \ge -2 \end{cases}$$

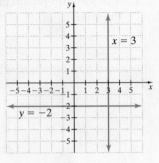

36. Tony Noellert budgets his time at work today. Part of the day he can write bills; the rest of the day he can use to write purchase orders. The total time available is at most 8 hours. Less than 3 hours is to be spent writing bills.

    a. Write a system of inequalities to describe the situation. (Let $x = $ hours available for writing bills and $y = $ hours available for writing purchase orders.)

    b. Graph the solutions of the system.

Chapter 4 Vocabulary Check

Fill in each blank with one of the words or phrases listed below.

matrix	consistent	system of equations	triple	row
solution	inconsistent	element	column	

1. Two or more linear equations in two variables form a(n) _____.

2. A(n) _____ of a system of two equations in two variables is an ordered pair that makes both equations true.

3. A(n) _____ system of equations has at least one solution.

4. A solution of a system of three equations in three variables is an ordered _____ that makes all three equations true.

5. A(n) _____ system of equations has no solution.

6. A(n) _____ is a rectangular array of numbers.

7. Each of the numbers in a matrix is called a(n) _____.

8. The numbers aligned horizontally in a matrix are in the same _____.

9. The numbers aligned vertically in a matrix are in the same _____.

Chapter 4 Highlights

DEFINITIONS AND CONCEPTS	EXAMPLES

Section 4.1 Solving Systems of Linear Equations in Two Variables

A **system of linear equations** consists of two or more linear equations.

$$\begin{cases} x - 3y = 6 \\ \quad\quad y = \dfrac{1}{2}x \end{cases} \quad \begin{cases} x + 2y - \;\;z = 1 \\ 3x - \;\;y + 4z = 0 \\ \quad\quad 5y + \;\;z = 6 \end{cases}$$

A **solution** of a system of two equations in two variables is an ordered pair (x, y) that makes both equations true.

Determine whether $(2, -5)$ is a solution of the system.

$$\begin{cases} x + \;\;y = -3 \\ 2x - 3y = 19 \end{cases}$$

Replace x with 2 and y with -5 in both equations.

$$x + y = -3 \qquad\qquad 2x - 3y = 19$$
$$2 + (-5) \overset{?}{=} -3 \qquad 2(2) - 3(-5) \overset{?}{=} 19$$
$$-3 = -3 \;\; \text{True} \qquad\qquad 4 + 15 \overset{?}{=} 19$$
$$19 = 19 \;\; \text{True}$$

$(2, -5)$ is a solution of the system.

Geometrically, a solution of a system in two variables is a point of intersection of the graphs of the equations.

Solve by graphing: $\begin{cases} y = 2x - 1 \\ x + 2y = 13 \end{cases}$

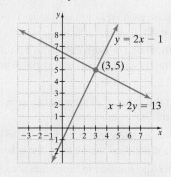

(continued)

DEFINITIONS AND CONCEPTS	EXAMPLES

Section 4.1 Solving Systems of Linear Equations in Two Variables (continued)

A system of equations with at least one solution is a **consistent system**. A system that has no solution is an **inconsistent system**.

If the graphs of two linear equations are identical, the equations are **dependent**.

If their graphs are different, the equations are **independent**.

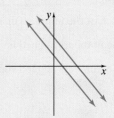

One solution:
Independent equations
Consistent system

No solution:
Independent equations
Inconsistent system

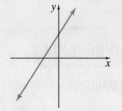

Infinite number of solutions:
Dependent equations
Consistent system

Solving a System of Linear Equations by the Substitution Method

Step 1. Solve one equation for a variable.

Step 2. Substitute the expression for the variable into the other equation.

Step 3. Solve the equation from Step 2 to find the value of one variable.

Step 4. Substitute the value from Step 3 in either original equation to find the value of the other variable.

Step 5. Check the solution in both equations.

Solve by substitution:

$$\begin{cases} y = x + 2 \\ 3x - 2y = -5 \end{cases}$$

Since the first equation is solved for y, substitute $x + 2$ for y in the second equation.

$$\begin{aligned} 3x - 2y &= -5 \quad \text{Second equation} \\ 3x - 2(x + 2) &= -5 \quad \text{Let } y = x + 2. \\ 3x - 2x - 4 &= -5 \\ x - 4 &= -5 \quad \text{Simplify.} \\ x &= -1 \quad \text{Add 4.} \end{aligned}$$

To find y, let $x = -1$ in $y = x + 2$, so $y = -1 + 2 = 1$. The solution $(-1, 1)$ checks.

Solving a System of Linear Equations by the Elimination Method

Step 1. Rewrite each equation in standard form, $Ax + By = C$.

Step 2. Multiply one or both equations by a nonzero number so that the coefficients of a variable are opposites.

Step 3. Add the equations.

Step 4. Find the value of the remaining variable by solving the resulting equation.

Step 5. Substitute the value from Step 4 into either original equation to find the value of the other variable.

Step 6. Check the solution in both equations.

Solve by elimination:

$$\begin{cases} x - 3y = -3 \\ -2x + y = 6 \end{cases}$$

Multiply both sides of the first equation by 2.

$$\begin{aligned} 2x - 6y &= -6 \\ \underline{-2x + y = 6} \\ -5y &= 0 \quad \text{Add.} \\ y &= 0 \quad \text{Divide by } -5. \end{aligned}$$

To find x, let $y = 0$ in an original equation.

$$\begin{aligned} x - 3y &= -3 \\ x - 3 \cdot 0 &= -3 \\ x &= -3 \end{aligned}$$

The solution $(-3, 0)$ checks.

DEFINITIONS AND CONCEPTS	EXAMPLES

Section 4.2 Solving Systems of Linear Equations in Three Variables

A **solution** of an equation in three variables x, y, and z is an **ordered triple** (x, y, z) that makes the equation a true statement.

Verify that $(-2, 1, 3)$ is a solution of $2x + 3y - 2z = -7$. Replace x with -2, y with 1, and z with 3.

$$2(-2) + 3(1) - 2(3) \overset{?}{=} -7$$

$$-4 + 3 - 6 \overset{?}{=} -7$$

$$-7 = -7 \quad \text{True}$$

$(-2, 1, 3)$ is a solution.

Solving a System of Three Linear Equations by the Elimination Method

Step 1. Write each equation in standard form, $Ax + By + Cz = D$.

Step 2. Choose a pair of equations and use them to eliminate a variable.

Step 3. Choose any other pair of equations and eliminate the same variable.

Step 4. Solve the system of two equations in two variables from Steps 2 and 3.

Step 5. Solve for the third variable by substituting the values of the variables from Step 4 into any of the original equations.

Step 6. Check the solution in all three original equations.

Solve:

$$\begin{cases} 2x + y - z = 0 & (1) \\ x - y - 2z = -6 & (2) \\ -3x - 2y + 3z = -22 & (3) \end{cases}$$

1. Each equation is written in standard form.

2.
$$\begin{array}{ll} 2x + y - z = 0 & (1) \\ \underline{x - y - 2z = -6} & (2) \\ 3x \quad\quad - 3z = -6 & (4) \quad \text{Add.} \end{array}$$

3. Eliminate y from equations (1) and (3) also.

$$\begin{array}{ll} 4x + 2y - 2z = 0 & \text{Multiply equation} \\ \underline{-3x - 2y + 3z = -22} \quad (3) & \text{(1) by 2.} \\ x \quad\quad + z = -22 \quad (5) & \text{Add.} \end{array}$$

4. Solve.

$$\begin{cases} 3x - 3z = -6 & (4) \\ x + z = -22 & (5) \end{cases}$$

$$\begin{array}{ll} x - z = -2 & \\ \underline{x + z = -22} & \text{Divide equation (4) by 3.} \\ 2x \quad\quad = -24 & (5) \\ x = -12 & \end{array}$$

To find z, use equation (5).

$$x + z = -22$$
$$-12 + z = -22$$
$$z = -10$$

5. To find y, use equation (1).

$$2x + y - z = 0$$
$$2(-12) + y - (-10) = 0$$
$$-24 + y + 10 = 0$$
$$y = 14$$

6. The solution $(-12, 14, -10)$ checks.

DEFINITIONS AND CONCEPTS	EXAMPLES

Section 4.3 Systems of Linear Equations and Problem Solving

Two numbers have a sum of 11. Twice one number is 3 less than 3 times the other. Find the numbers.

1. UNDERSTAND the problem.

1. Read and reread.

$$x = \text{one number}$$
$$y = \text{other number}$$

2. TRANSLATE.

2. In words: | sum of numbers | is | 11 |

Translate: $\quad x + y \quad = \quad 11$

In words: | twice one number | is | 3 less than 3 times the other number |

Translate: $\quad 2x \quad = \quad 3y - 3$

3. SOLVE.

3. Solve the system: $\begin{cases} x + y = 11 \\ 2x = 3y - 3 \end{cases}$

In the first equation, $x = 11 - y$. Substitute into the other equation.

$$2x = 3y - 3$$
$$2(11 - y) = 3y - 3$$
$$22 - 2y = 3y - 3$$
$$-5y = -25$$
$$y = 5$$

Replace y with 5 in the equation $x = 11 - y$. Then $x = 11 - 5 = 6$. The solution is $(6, 5)$.

4. INTERPRET.

4. *Check:* See that $6 + 5 = 11$ is the required sum and that twice 6 is 3 times 5 less 3. *State:* The numbers are 6 and 5.

Section 4.4 Solving Systems of Equations by Matrices

A **matrix** is a rectangular array of numbers.

$$\begin{bmatrix} -7 & 0 & 3 \\ 1 & 2 & 4 \end{bmatrix} \quad \begin{bmatrix} a & b & c \\ d & e & f \\ g & h & i \end{bmatrix}$$

The **matrix** corresponding to a system is composed of the coefficients of the variables and the constants of the system.

The matrix corresponding to the system

$$\begin{cases} x - y = 1 \\ 2x + y = 11 \end{cases} \text{ is } \begin{bmatrix} 1 & -1 & \vdots & 1 \\ 2 & 1 & \vdots & 11 \end{bmatrix}$$

The following **row operations** can be performed on matrices, and the result is an equivalent matrix.

Elementary row operations:

1. Interchange any two rows.

2. Multiply (or divide) the elements of one row by the same nonzero number.

3. Multiply (or divide) the elements of one row by the same nonzero number and add them to their corresponding elements in any other row.

Use matrices to solve: $\begin{cases} x - y = 1 \\ 2x + y = 11 \end{cases}$.

The corresponding matrix is

$$\begin{bmatrix} 1 & -1 & \vdots & 1 \\ 2 & 1 & \vdots & 11 \end{bmatrix}$$

Use row operations to write an equivalent matrix with 1's along the diagonal and 0's below each 1 in the diagonal. Multiply row 1 by -2 and add to row 2. Change row 2 only.

$$\begin{bmatrix} 1 & -1 & \vdots & 1 \\ -2(1) + 2 & -2(-1) + 1 & \vdots & -2(1) + 11 \end{bmatrix}$$

simplifies to $\begin{bmatrix} 1 & -1 & \vdots & 1 \\ 0 & 3 & \vdots & 9 \end{bmatrix}$

DEFINITIONS AND CONCEPTS	EXAMPLES
Section 4.5 Solving Systems of Equations by Matrices (continued)	

Divide row 2 by 3.

$$\begin{bmatrix} 1 & -1 & | & 1 \\ \dfrac{0}{3} & \dfrac{3}{3} & | & \dfrac{9}{3} \end{bmatrix} \text{ simplifies to } \begin{bmatrix} 1 & -1 & | & 1 \\ 0 & 1 & | & 3 \end{bmatrix}$$

This matrix corresponds to the system

$$\begin{cases} x - y = 1 \\ \quad\ \ y = 3 \end{cases}$$

Let $y = 3$ in the first equation.

$$x - 3 = 1$$
$$x = 4$$

The ordered pair solution is $(4, 3)$.

Section 4.5 Systems of Linear Inequalities	

A **system of linear inequalities** consists of two or more linear inequalities.

To graph a system of inequalities, graph each inequality in the system. The overlapping region is the solution of the system.

$$\begin{cases} x - y \geq 3 \\ \quad\ \ y \leq -2x \end{cases}$$

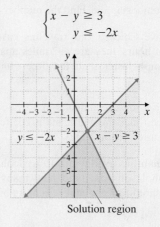

Solution region

Chapter 4 Review

(4.1) *Solve each system of equations in two variables by each method: (a) graphing, (b) substitution, and (c) elimination.*

1. $\begin{cases} 3x + 10y = 1 \\ x + 2y = -1 \end{cases}$

2. $\begin{cases} y = \dfrac{1}{2}x + \dfrac{2}{3} \\ 4x + 6y = 4 \end{cases}$

3. $\begin{cases} 2x - 4y = 22 \\ 5x - 10y = 15 \end{cases}$

4. $\begin{cases} 3x - 6y = 12 \\ 2y = x - 4 \end{cases}$

5. $\begin{cases} \dfrac{1}{2}x - \dfrac{3}{4}y = -\dfrac{1}{2} \\ \dfrac{1}{8}x + \dfrac{3}{4}y = \dfrac{19}{8} \end{cases}$

6. The revenue equation for a certain style of backpack is $y = 32x$, where x is the number of backpacks sold and y is the income in dollars for selling x backpacks. The cost equation for these units is $y = 15x + 25{,}500$, where x is the number of backpacks manufactured and y is the cost in dollars for manufacturing x backpacks. Find the number of units to be sold for the company to break even. (*Hint:* Solve the system of equations formed by the two given equations.)

(4.2) *Solve each system of equations in three variables.*

7. $\begin{cases} x \quad\ + z = 4 \\ 2x - y \quad\ = 4 \\ x + y - z = 0 \end{cases}$

8. $\begin{cases} 2x + 5y \quad\ = 4 \\ x - 5y + z = -1 \\ 4x \quad\ - z = 11 \end{cases}$

9. $\begin{cases} 4y + 2z = 5 \\ 2x + 8y \quad\ = 5 \\ 6x + \quad\ 4z = 1 \end{cases}$

10. $\begin{cases} 5x + 7y \quad\ = 9 \\ 14y - z = 28 \\ 4x \quad\ + 2z = -4 \end{cases}$

11. $\begin{cases} 3x - 2y + 2z = 5 \\ -x + 6y + z = 4 \\ 3x + 14y + 7z = 20 \end{cases}$

12. $\begin{cases} x + 2y + 3z = 11 \\ y + 2z = 3 \\ 2x + 2z = 10 \end{cases}$

13. $\begin{cases} 7x - 3y + 2z = 0 \\ 4x - 4y - z = 2 \\ 5x + 2y + 3z = 1 \end{cases}$

14. $\begin{cases} x - 3y - 5z = -5 \\ 4x - 2y + 3z = 13 \\ 5x + 3y + 4z = 22 \end{cases}$

(4.3) Use systems of equations to solve.

15. The sum of three numbers is 98. The sum of the first and second is two more than the third number, and the second is four times the first. Find the numbers.

16. One number is three times a second number, and twice the sum of the numbers is 168. Find the numbers.

17. Two cars leave Chicago, one traveling east and the other west. After 4 hours, they are 492 miles apart. If one car is traveling 7 mph faster than the other, find the speed of each.

18. The foundation for a rectangular Hardware Warehouse has a length three times the width and is 296 feet around. Find the dimensions of the building.

19. James Callahan has available a 10% alcohol solution and a 60% alcohol solution. Find how many liters of each solution he should mix to make 50 liters of a 40% alcohol solution.

20. An employee at See's Candy Store needs a special mixture of candy. She has creme-filled chocolates that sell for $3.00 per pound, chocolate-covered nuts that sell for $2.70 per pound, and chocolate-covered raisins that sell for $2.25 per pound. She wants to have twice as many raisins as nuts in the mixture. Find how many pounds of each she should use to make 45 pounds worth $2.80 per pound.

21. Chris Kringler has $2.77 in her coin jar—all in pennies, nickels, and dimes. If she has 53 coins in all and four more nickels than dimes, find how many of each type of coin she has.

22. If $10,000 and $4000 are invested such that $1250 in interest is earned in one year, and if the rate of interest on the larger investment is 2% more than that on the smaller investment, find the rates of interest.

23. The perimeter of an isosceles (two sides equal) triangle is 73 centimeters. If the unequal side is 7 centimeters longer than the two equal sides, find the lengths of the three sides.

24. The sum of three numbers is 295. One number is five more than the second and twice the third. Find the numbers.

(4.4) Use matrices to solve each system.

25. $\begin{cases} 3x + 10y = 1 \\ x + 2y = -1 \end{cases}$

26. $\begin{cases} 3x - 6y = 12 \\ 2y = x - 4 \end{cases}$

27. $\begin{cases} 3x - 2y = -8 \\ 6x + 5y = 11 \end{cases}$

28. $\begin{cases} 6x - 6y = -5 \\ 10x - 2y = 1 \end{cases}$

29. $\begin{cases} 3x - 6y = 0 \\ 2x + 4y = 5 \end{cases}$

30. $\begin{cases} 5x - 3y = 10 \\ -2x + y = -1 \end{cases}$

31. $\begin{cases} 0.2x - 0.3y = -0.7 \\ 0.5x + 0.3y = 1.4 \end{cases}$

32. $\begin{cases} 3x + 2y = 8 \\ 3x - y = 5 \end{cases}$

33. $\begin{cases} x + z = 4 \\ 2x - y = 0 \\ x + y - z = 0 \end{cases}$

34. $\begin{cases} 2x + 5y = 4 \\ x - 5y + z = -1 \\ 4x - z = 11 \end{cases}$

35. $\begin{cases} 3x - y = 11 \\ x + 2z = 13 \\ y - z = -7 \end{cases}$

36. $\begin{cases} 5x + 7y + 3z = 9 \\ 14y - z = 28 \\ 4x + 2z = -4 \end{cases}$

37. $\begin{cases} 7x - 3y + 2z = 0 \\ 4x - 4y - z = 2 \\ 5x + 2y + 3z = 1 \end{cases}$

38. $\begin{cases} x + 2y + 3z = 14 \\ y + 2z = 3 \\ 2x - 2z = 10 \end{cases}$

(4.5) Graph the solution of each system of linear inequalities.

39. $\begin{cases} y \geq 2x - 3 \\ y \leq -2x + 1 \end{cases}$

40. $\begin{cases} y \leq -3x - 3 \\ y \leq 2x + 7 \end{cases}$

41. $\begin{cases} x + 2y > 0 \\ x - y \leq 6 \end{cases}$

42. $\begin{cases} x - 2y \geq 7 \\ x + y \leq -5 \end{cases}$

43. $\begin{cases} 3x - 2y \leq 4 \\ 2x + y \geq 5 \\ y \leq 4 \end{cases}$

44. $\begin{cases} 4x - y \leq 0 \\ 3x - 2y \geq -5 \\ y \geq -4 \end{cases}$

45. $\begin{cases} x + 2y \leq 5 \\ x \leq 2 \\ x \geq 0 \\ y \geq 0 \end{cases}$

46. $\begin{cases} x + 3y \leq 7 \\ y \leq 5 \\ x \geq 0 \\ y \geq 0 \end{cases}$

MIXED REVIEW

Solve each system.

47. $\begin{cases} y = x - 5 \\ y = -2x + 2 \end{cases}$

48. $\begin{cases} \dfrac{2}{5}x + \dfrac{3}{4}y = 1 \\ x + 3y = -2 \end{cases}$

49. $\begin{cases} 5x - 2y = 10 \\ x = \dfrac{2}{5}y + 2 \end{cases}$

50. $\begin{cases} x - 4y = 4 \\ \dfrac{1}{8}x - \dfrac{1}{2}y = 3 \end{cases}$

51. $\begin{cases} x - 3y + 2z = 0 \\ 9y - z = 22 \\ 5x + 3z = 10 \end{cases}$

52. One number is five less than three times a second number. If the sum of the numbers is 127, find the numbers.

53. The perimeter of a triangle is 126 units. The length of one side is twice the length of the shortest side. The length of the third side is fourteen more than the length of the shortest side. Find the lengths of the sides of the triangles.

54. Graph the solution of the system: $\begin{cases} y \le 3x - \dfrac{1}{2} \\ 3x + 4y \ge 6 \end{cases}$.

55. In recent years, the number of newspapers printed as morning editions has been increasing and the number of newspapers printed as evening editions has been decreasing. The number y of daily morning newspapers in existence is approximated by the equation $y = 12.6x + 716.1$, where x is the number of years since 1997. The number y of daily evening newspapers in existence is approximated by $y = -23.8x + 806.3$, where x is the number of years since 1997. Use these equations to determine the year in which the number of newspapers printed as morning editions was the same as the number of newspapers printed as evening editions. (Round to the nearest whole number.) (*Source:* Based on data from Newspaper Association of America)

Chapter 4 Test   MyMathLab®    CHAPTER Test Prep VIDEOS   You Tube™

Solve each system of equations graphically and then solve by the elimination method or the substitution method.

1. $\begin{cases} 2x - y = -1 \\ 5x + 4y = 17 \end{cases}$

2. $\begin{cases} 7x - 14y = 5 \\ x = 2y \end{cases}$

Solve each system.

3. $\begin{cases} 4x - 7y = 29 \\ 2x + 5y = -11 \end{cases}$

4. $\begin{cases} 15x + 6y = 15 \\ 10x + 4y = 10 \end{cases}$

5. $\begin{cases} 2x - 3y = 4 \\ 3y + 2z = 2 \\ x - z = -5 \end{cases}$

6. $\begin{cases} 3x - 2y - z = -1 \\ 2x - 2y = 4 \\ 2x - 2z = -12 \end{cases}$

7. $\begin{cases} \dfrac{x}{2} + \dfrac{y}{4} = -\dfrac{3}{4} \\ x + \dfrac{3}{4}y = -4 \end{cases}$

Use matrices to solve each system.

8. $\begin{cases} x - y = -2 \\ 3x - 3y = -6 \end{cases}$

9. $\begin{cases} x + 2y = -1 \\ 2x + 5y = -5 \end{cases}$

10. $\begin{cases} x - y - z = 0 \\ 3x - y - 5z = -2 \\ 2x + 3y = -5 \end{cases}$

11. A motel in New Orleans charges $90 per day for double occupancy and $80 per day for single occupancy. If 80 rooms are occupied for a total of $6930, how many rooms of each kind are occupied?

12. The research department of a company that manufactures children's fruit drinks is experimenting with a new flavor. A 17.5% fructose solution is needed, but only 10% and 20% solutions are available. How many gallons of the 10% fructose solution should be mixed with the 20% fructose solution to obtain 20 gallons of a 17.5% fructose solution?

13. A company that manufactures boxes recently purchased $2000 worth of new equipment to offer gift boxes to its customers. The cost of producing a package of gift boxes is $1.50 and it is sold for $4.00. Find the number of packages that must be sold for the company to break even.

14. The measure of the largest angle of a triangle is 3 less than five times the measure of the smallest angle. The measure of the remaining angle is 1 less than twice the measure of the smallest angle. Find the measure of each angle.

Graph the solutions of each system of linear inequalities.

15. $\begin{cases} 2y - x \ge 1 \\ x + y \ge -4 \\ y \le 2 \end{cases}$

Chapter 4 Cumulative Review

1. Determine whether each statement is true or false.

    a. $3 \in \{x \mid x \text{ is a natural number}\}$

    b. $7 \notin \{1, 2, 3\}$

2. Determine whether each statement is true or false.

    a. $\{0, 7\} \subseteq \{0, 2, 4, 6, 8\}$

    b. $\{1, 3, 5\} \subseteq \{1, 3, 5, 7\}$

3. Simplify the following expressions.

    a. $11 + 2 - 7$

    b. $-5 - 4 + 2$

4. Subtract.

    a. $-7 - (-2)$

    b. $14 - 38$

5. Write the additive inverse, or opposite, of each.

    a. 4

    b. $\dfrac{3}{7}$

    c. -11.2

6. Write the reciprocal of each.

    a. 5

    b. $-\dfrac{2}{3}$

7. Use the distributive property to multiply.

    a. $3(2x + y)$

    b. $-(3x - 1)$

    c. $0.7a(b - 2)$

8. Use the distributive property to multiply.

    a. $7(3x - 2y + 4)$

    b. $-(-2s - 3t)$

9. Use the distributive property to simplify each expression.

    a. $3x - 5x + 4$

    b. $7yz + yz$

    c. $4z + 6.1$

10. Use the distributive property to simplify each expression.

    a. $5y^2 - 1 + 2(y^2 + 2)$

    b. $(7.8x - 1.2) - (5.6x - 2.4)$

Solve.

11. $-4x - 1 + 5x = 9x + 3 - 7x$

12. $8y - 14 = 6y - 14$

13. $0.3x + 0.1 = 0.27x - 0.02$

14. $2(m - 6) - m = 4(m - 3) - 3m$

15. A pennant in the shape of an isosceles triangle is to be constructed for the Slidell High School Athletic Club and sold at a fund-raiser. The company manufacturing the pennant charges according to perimeter, and the athletic club has determined that a perimeter of 149 centimeters should make a nice profit. If each equal side of the triangle is twice the length of the third side, increased by 12 centimeters, find the lengths of the sides of the triangular pennant.

16. A quadrilateral has 4 angles whose sum is $360°$. In a particular quadrilateral, two angles have the same measure. A third

angle is $10°$ more than the measure of one of the equal angles, and the fourth angle is half the measure of one of the equal angles. Find the measures of the angles.

17. Solve: $3x + 4 \geq 2x - 6$. Graph the solution set.

18. Solve: $5(2x - 1) > -5$

19. Solve: $2 < 4 - x < 7$

20. Solve: $-1 < \dfrac{-2x - 1}{3} < 1$

21. Solve: $|2x| + 5 = 7$

22. Solve: $|x - 5| = 4$

23. Solve for m: $|m - 6| < 2$

24. $|2x + 1| > 5$

25. Plot each ordered pair on a Cartesian coordinate system and name the quadrant or axis in which the point is located.

    a. $(2, -1)$ **b.** $(0, 5)$

    c. $(-3, 5)$ **d.** $(-2, 0)$

    e. $\left(-\dfrac{1}{2}, -4\right)$ **f.** $(1.5, 1.5)$

26. Name the quadrant or axis each point is located.

    a. $(-1, -5)$

    b. $(4, -2)$

    c. $(0, 2)$

27. Is the relation $x = y^2$ also a function?

28. Graph: $-2x + \dfrac{1}{2}y = -2$

29. If $f(x) = 7x^2 - 3x + 1$ and $g(x) = 3x - 2$, find the following.

    a. $f(1)$ **b.** $g(1)$

    c. $f(-2)$ **d.** $g(0)$

30. If $f(x) = 3x^2$, find the following.

    a. $f(5)$ **b.** $f(-2)$

31. Graph $g(x) = 2x + 1$. Compare this graph with the graph of $f(x) = 2x$.

32. Find the slope of the line containing $(-2, 6)$ and $(0, 9)$.

33. Find the slope and the y-intercept of the line $3x - 4y = 4$.

34. Find the slope and y-intercept of the line defined by $y = 2$.

35. Are the following pairs of lines parallel, perpendicular, or neither?

    a. $3x + 7y = 4$

        $6x + 14y = 7$

    b. $-x + 3y = 2$

        $2x + 6y = 5$

36. Find an equation of the line through $(0, -9)$ with slope $\dfrac{1}{5}$.

37. Find an equation of the line through points $(4, 0)$ and $(-4, -5)$. Write the equation using function notation.

38. Find an equation of the line through $(-2, 6)$ perpendicular to $f(x) = \dfrac{1}{2}x - \dfrac{1}{3}$.

39. Graph $3x \geq y$.

40. Graph: $x \geq 1$.

41. Determine whether the given ordered pair is a solution of the system.

a. $\begin{cases} -x + y = 2 \\ 2x - y = -3 \end{cases}$ $(-1, 1)$

b. $\begin{cases} 5x + 3y = -1 \\ x - y = 1 \end{cases}$ $(-2, 3)$

42. Solve the system.
$$\begin{cases} 5x + y = -2 \\ 4x - 2y = -10 \end{cases}$$

43. Solve the system.
$$\begin{cases} 3x - y + z = -15 \\ x + 2y - z = 1 \\ 2x + 3y - 2z = 0 \end{cases}$$

44. Solve the system.
$$\begin{cases} x - 2y + z = 0 \\ 3x - y - 2z = -15 \\ 2x - 3y + 3z = 7 \end{cases}$$

45. Use matrices to solve the system.
$$\begin{cases} x + 3y = 5 \\ 2x - y = -4 \end{cases}$$

46. Use matrices to solve the system.
$$\begin{cases} -6x + 8y = 0 \\ 9x - 12y = 2 \end{cases}$$

Exponents, Polynomials, and Polynomial Functions

Linear equations are important for solving problems. They are not sufficient, however, to solve all problems. Many real-world phenomena are modeled by polynomials. We begin this chapter by reviewing exponents. We will then study operations on polynomials and how polynomials can be used in problem solving.

A hybrid vehicle is one with a gasoline engine and an electric motor, each of which is used to propel the vehicle.

With the arrival of electric-drive cars, it is very difficult to predict future sales of hybrids based on past sales, but we will attempt to do so.

In Section 5.8, Exercises 103 and 104, we attempt to predict sales of hybrids in future years.

Hybrid Sales for a Month in 2010

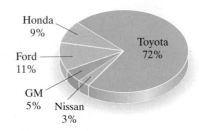

Honda 9%

Ford 11%

GM 5%

Nissan 3%

Toyota 72%

5.1   Exponents and Scientific Notation

OBJECTIVE

1   Using the Product Rule

Recall that exponents may be used to write repeated factors in a more compact form. As we have seen in the previous chapters, exponents can be used when the repeated factor is a number or a variable. For example,

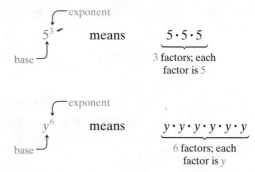

Expressions such as 5^3 and y^6 that contain exponents are called **exponential expressions.**

Exponential expressions can be multiplied, divided, added, subtracted, and themselves raised to powers. In this section, we review operations on exponential expressions.

We review multiplication first. To multiply x^2 by x^3, use the definition of an exponent.

$$x^2 \cdot x^3 = \underbrace{(x \cdot x)(x \cdot x \cdot x)}_{x \text{ is a factor 5 times}}$$

$$= x^5$$

Notice that the result is exactly the same if we add the exponents.

$$x^2 \cdot x^3 = x^{2+3} = x^5$$

This suggests the following.

Product Rule for Exponents

If m and n are positive integers and a is a real number, then

$$a^m \cdot a^n = a^{m+n}$$

In other words, the *product* of exponential expressions with a common base is the common base raised to a power equal to the *sum* of the exponents of the factors.

EXAMPLE 1   Use the product rule to simplify.

a. $2^2 \cdot 2^5$ 　　　　**b.** $x^7 x^3$ 　　　　**c.** $y \cdot y^2 \cdot y^4$

Solution

a. $2^2 \cdot 2^5 = 2^{2+5} = 2^7$

b. $x^7 x^3 = x^{7+3} = x^{10}$

c. $y \cdot y^2 \cdot y^4 = (y^1 \cdot y^2) \cdot y^4$

$$= y^3 \cdot y^4$$

$$= y^7$$

PRACTICE

1   Use the product rule to simplify.

a. $3^4 \cdot 3^2$ 　　　　**b.** $x^5 \cdot x^2$ 　　　　**c.** $y \cdot y^3 \cdot y^5$

EXAMPLE 2 Use the product rule to simplify.

a. $(3x^6)(5x)$ **b.** $(-2.4x^3p^2)(4xp^{10})$

Solution Here, we use properties of multiplication to group together like bases.

a. $(3x^6)(5x) = 3(5)x^6x^1 = 15x^7$

b. $(-2.4x^3p^2)(4xp^{10}) = -2.4(4)x^3x^1p^2p^{10} = -9.6x^4p^{12}$

PRACTICE

2 Use the product rule to simplify.

a. $(5z^3)(7z)$ **b.** $(-4.1t^5q^3)(5tq^5)$

OBJECTIVE

2 Evaluating Expressions Raised to the 0 Power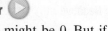

The definition of a^n does not include the possibility that n might be 0. But if it did, then, by the product rule,

$$\underbrace{a^0 \cdot a^n}_{} = a^{0+n} = a^n = \underbrace{1 \cdot a^n}_{}$$

From this, we reasonably define that $a^0 = 1$ as long as a does not equal 0.

> **Zero Exponent**
>
> If a does not equal 0, then $a^0 = 1$.

EXAMPLE 3 Evaluate the following.

a. 7^0 **b.** -7^0 **c.** $(2x + 5)^0$ **d.** $2x^0$

Solution

a. $7^0 = 1$

b. Without parentheses, only 7 is raised to the 0 power.

$$-7^0 = -(7^0) = -(1) = -1$$

c. $(2x + 5)^0 = 1$

d. $2x^0 = 2(1) = 2$

PRACTICE

3 Evaluate the following.

a. 5^0 **b.** -5^0 **c.** $(3x - 8)^0$ **d.** $3x^0$

OBJECTIVE

3 Using the Quotient Rule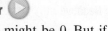

To find quotients of exponential expressions, we again begin with the definition of a^n to simplify $\dfrac{x^9}{x^2}$. For example,

$$\frac{x^9}{x^2} = \frac{x \cdot x \cdot x \cdot x \cdot x \cdot x \cdot x \cdot x \cdot x}{x \cdot x} = x^7$$

(Assume for the next two sections that denominators containing variables are not 0.)

Notice that the result is exactly the same if we subtract the exponents.

$$\frac{x^9}{x^2} = x^{9-2} = x^7$$

This suggests the following.

Quotient Rule for Exponents

If a is a nonzero real number and n and m are integers, then

$$\frac{a^m}{a^n} = a^{m-n}$$

In other words, the *quotient* of exponential expressions with a common base is the common base raised to a power equal to the *difference* of the exponents.

EXAMPLE 4   Use the quotient rule to simplify.

a. $\dfrac{x^7}{x^4}$ **b.** $\dfrac{5^8}{5^2}$ **c.** $\dfrac{20x^6}{4x^5}$ **d.** $\dfrac{12y^{10}z^7}{14y^8z^7}$

Solution

a. $\dfrac{x^7}{x^4} = x^{7-4} = x^3$

b. $\dfrac{5^8}{5^2} = 5^{8-2} = 5^6$

c. $\dfrac{20x^6}{4x^5} = 5x^{6-5} = 5x^1,\ \text{or}\ 5x$

d. $\dfrac{12y^{10}z^7}{14y^8z^7} = \dfrac{6}{7}y^{10-8} \cdot z^{7-7} = \dfrac{6}{7}y^2z^0 = \dfrac{6}{7}y^2,\quad \text{or}\quad \dfrac{6y^2}{7}$

PRACTICE

4   Use the quotient rule to simplify.

a. $\dfrac{z^8}{z^3}$ **b.** $\dfrac{3^9}{3^3}$ **c.** $\dfrac{45x^7}{5x^3}$ **d.** $\dfrac{24a^{14}b^6}{18a^7b^6}$

OBJECTIVE

4   Evaluating Exponents Raised to the Negative *n*th Power

When the exponent of the denominator is larger than the exponent of the numerator, applying the quotient rule yields a negative exponent. For example,

$$\frac{x^3}{x^5} = x^{3-5} = x^{-2}$$

Using the definition of a^n, though, gives us

$$\frac{x^3}{x^5} = \frac{x \cdot x \cdot x}{x \cdot x \cdot x \cdot x \cdot x} = \frac{1}{x^2}$$

From this, we reasonably define $x^{-2} = \dfrac{1}{x^2}$ or, in general, $a^{-n} = \dfrac{1}{a^n}$.

Negative Exponents

If a is a real number other than 0 and n is a positive integer, then

$$a^{-n} = \frac{1}{a^n}$$

EXAMPLE 5 Simplify and write with positive exponents only.

a. 5^{-2} **b.** $(-4)^{-4}$ **c.** $2x^{-3}$ **d.** $(3x)^{-1}$

e. $\dfrac{m^5}{m^{15}}$ **f.** $\dfrac{3^3}{3^6}$ **g.** $2^{-1} + 3^{-2}$ **h.** $\dfrac{1}{t^{-5}}$

Solution

a. $5^{-2} = \dfrac{1}{5^2} = \dfrac{1}{25}$

b. $(-4)^{-4} = \dfrac{1}{(-4)^4} = \dfrac{1}{256}$

> ▶ Helpful Hint
>
> Study Example 5c. Make sure you understand that
> $2x^{-3} = \dfrac{2}{x^3}$

c. $2x^{-3} = 2 \cdot \dfrac{1}{x^3} = \dfrac{2}{x^3}$ Without parentheses, only x is raised to the -3 power.

d. $(3x)^{-1} = \dfrac{1}{(3x)^1} = \dfrac{1}{3x}$ With parentheses, both 3 and x are raised to the -1 power.

e. $\dfrac{m^5}{m^{15}} = m^{5-15} = m^{-10} = \dfrac{1}{m^{10}}$

f. $\dfrac{3^3}{3^6} = 3^{3-6} = 3^{-3} = \dfrac{1}{3^3} = \dfrac{1}{27}$

g. $2^{-1} + 3^{-2} = \dfrac{1}{2^1} + \dfrac{1}{3^2} = \dfrac{1}{2} + \dfrac{1}{9} = \dfrac{9}{18} + \dfrac{2}{18} = \dfrac{11}{18}$

h. $\dfrac{1}{t^{-5}} = \dfrac{1}{\frac{1}{t^5}} = 1 \div \dfrac{1}{t^5} = 1 \cdot \dfrac{t^5}{1} = t^5$

PRACTICE
5   Simplify and write with positive exponents only.

a. 6^{-2} **b.** $(-2)^{-6}$ **c.** $3x^{-5}$ **d.** $(5y)^{-1}$ **e.** $\dfrac{k^4}{k^{11}}$

f. $\dfrac{5^3}{5^5}$ **g.** $5^{-1} + 2^{-2}$ **h.** $\dfrac{1}{z^{-8}}$

> ▶ Helpful Hint
>
> Notice that when a factor containing an exponent is moved from the numerator to the denominator or from the denominator to the numerator, the sign of its exponent changes.
>
> $$x^{-3} = \dfrac{1}{x^3}, \qquad 5^{-2} = \dfrac{1}{5^2} = \dfrac{1}{25}$$
>
> $$\dfrac{1}{y^{-4}} = y^4, \qquad \dfrac{1}{2^{-3}} = 2^3 = 8$$

EXAMPLE 6 Simplify and write with positive exponents only.

a. $\dfrac{x^{-9}}{x^2}$ **b.** $\dfrac{5p^4}{p^{-3}}$ **c.** $\dfrac{2^{-3}}{2^{-1}}$ **d.** $\dfrac{2x^{-7}y^2}{10xy^{-5}}$ **e.** $\dfrac{(3x^{-3})(x^2)}{x^6}$

Solution

a. $\dfrac{x^{-9}}{x^2} = x^{-9-2} = x^{-11} = \dfrac{1}{x^{11}}$

b. $\dfrac{5p^4}{p^{-3}} = 5 \cdot p^{4-(-3)} = 5p^7$

c. $\dfrac{2^{-3}}{2^{-1}} = 2^{-3-(-1)} = 2^{-2} = \dfrac{1}{2^2} = \dfrac{1}{4}$

d. $\dfrac{2x^{-7}y^2}{10xy^{-5}} = \dfrac{x^{-7-1} \cdot y^{2-(-5)}}{5} = \dfrac{x^{-8}y^7}{5} = \dfrac{y^7}{5x^8}$

e. Simplify the numerator first.

$$\dfrac{(3x^{-3})(x^2)}{x^6} = \dfrac{3x^{-3+2}}{x^6} = \dfrac{3x^{-1}}{x^6} = 3x^{-1-6} = 3x^{-7} = \dfrac{3}{x^7}$$ □

PRACTICE
6 Simplify and write with positive exponents only.

a. $\dfrac{z^{-8}}{z^3}$ **b.** $\dfrac{7t^3}{t^{-5}}$ **c.** $\dfrac{3^{-2}}{3^{-4}}$ **d.** $\dfrac{5a^{-5}b^3}{15a^2b^{-4}}$ **e.** $\dfrac{(2x^{-5})(x^6)}{x^5}$

✓**CONCEPT CHECK**
Find and correct the error in the following:

$$\dfrac{y^{-6}}{y^{-2}} = y^{-6-2} = y^{-8} = \dfrac{1}{y^8}$$

EXAMPLE 7 Simplify. Assume that a and t are nonzero integers and that x is not 0.

a. $x^{2a} \cdot x^3$ **b.** $\dfrac{x^{2t-1}}{x^{t-5}}$

Solution

a. $x^{2a} \cdot x^3 = x^{2a+3}$ Use the product rule.

b. $\dfrac{x^{2t-1}}{x^{t-5}} = x^{(2t-1)-(t-5)}$ Use the quotient rule.

$= x^{2t-1-t+5} = x^{t+4}$ □

PRACTICE
7 Simplify. Assume that a and t are nonzero integers and that x is not 0.

a. $x^{3a} \cdot x^4$ **b.** $\dfrac{x^{3t-2}}{x^{t-3}}$

OBJECTIVE
5 **Converting Between Scientific Notation and Standard Notation**

Very large and very small numbers occur frequently in nature. For example, the distance between the earth and the sun is approximately 150,000,000 kilometers. A helium atom has a diameter of 0.000 000 022 centimeter. It can be tedious to write these very large and very small numbers in standard notation like this. **Scientific notation** is a convenient shorthand notation for writing very large and very small numbers.

Helium atom

0.000000022
centimeter

150,000,000 km

Scientific Notation

A positive number is written in **scientific notation** if it is written as the product of a number a, where $1 \leq a < 10$, and an integer power r of 10:

$$a \times 10^r$$

The following are examples of numbers written in scientific notation.

diameter of helium atom: 2.2×10^{-8} cm

approximate distance between Earth and the sun: 1.5×10^8 km

Writing a Number in Scientific Notation

Step 1. Move the decimal point in the original number until the new number has a value between 1 and 10.

Step 2. Count the number of decimal places the decimal point was moved in Step 1. If the original number is 10 or greater, the count is positive. If the original number is less than 1, the count is negative.

Step 3. Write the product of the new number in Step 1 by 10 raised to an exponent equal to the count found in Step 2.

EXAMPLE 8   Write each number in scientific notation.

a. 730,000   **b.** 0.00000104

Solution

a. Step 1. Move the decimal point until the number is between 1 and 10.

$$730{,}000.$$

Step 2. The decimal point is moved 5 places and the original number is 10 or greater, so the count is positive 5.

Step 3. $730{,}000 = 7.3 \times 10^5$.

b. Step 1. Move the decimal point until the number is between 1 and 10.

$$0.00000104$$

Step 2. The decimal point is moved 6 places and the original number is less then 1, so the count is -6.

Step 3. $0.00000104 = 1.04 \times 10^{-6}$.   □

PRACTICE

8   Write each number in scientific notation.

a. 65,000   **b.** 0.000038

To write a scientific notation number in standard form, we reverse the preceding steps.

Writing a Scientific Notation Number in Standard Notation

Move the decimal point in the number the same number of places as the exponent on 10. If the exponent is positive, move the decimal point to the right. If the exponent is negative, move the decimal point to the left.

EXAMPLE 9   Write each number in standard notation.

a. 7.7×10^8 **b.** 1.025×10^{-3}

Solution

a. $7.7 \times 10^8 = 770,000,000$ Since the exponent is positive, move the decimal point 8 places to the right. Add zeros as needed.

b. $1.025 \times 10^{-3} = 0.001025$ Since the exponent is negative, move the decimal point 3 places to the left. Add zeros as needed.   □

PRACTICE
9   Write each number in standard notation.

Answers to Concept Check:
a, c, d

a. 6.2×10^5 **b.** 3.109×10^{-2}

✓CONCEPT CHECK
Which of the following numbers have values that are less than 1?

a. 3.5×10^{-5} **b.** 3.5×10^5 **c.** -3.5×10^5 **d.** -3.5×10^{-5}

Scientific Calculator Explorations

Multiply 5,000,000 by 700,000 on your calculator. The display should read $\boxed{3.5 \quad 12}$ or $\boxed{3.5 \text{ E } 12}$, which is the product written in scientific notation. Both these notations mean 3.5×10^{12}.

To enter a number written in scientific notation on a calculator, find the key marked $\boxed{\text{EE}}$. (On some calculators, this key may be marked $\boxed{\text{EXP}}$.)
To enter 7.26×10^{13}, press the keys

$$\boxed{7.26} \quad \boxed{\text{EE}} \quad \boxed{13}$$

The display will read $\boxed{7.26 \quad 13}$ or $\boxed{7.26 \text{ E } 13}$.

Use your calculator to perform each operation indicated.

1. Multiply 3×10^{11} and 2×10^{32}.   **2.** Divide 6×10^{14} by 3×10^9.

3. Multiply 5.2×10^{23} and 7.3×10^4.   **4.** Divide 4.38×10^{41} by 3×10^{17}.

Vocabulary, Readiness & Video Check

State the base of the exponent 5 in each expression.
1. $9x^5$ **2.** yz^5 **3.** -3^5 **4.** $(-3)^5$ **5.** $(y^7)^5$ **6.** $9 \cdot 2^5$

Martin-Gay Interactive Videos

See Video 5.1

Watch the section lecture video and answer the following questions.

OBJECTIVE
1

7. Why are we reminded that multiplication is commutative and associative during the simplifying of ⊞ Example 2?

OBJECTIVE
2

8. ⊞ Example 3 does not contain parentheses yet a discussion of parentheses is an important part of evaluating. Explain.

OBJECTIVE
3

9. When applying the quotient rule in ⊞ Example 5, how do you know which exponents to subtract?

OBJECTIVE
4

10. What important reminder is made at the beginning of ⊞ Example 6?

OBJECTIVE
5

11. From ⊞ Examples 10 and 11, explain how the direction of the movement of the decimal point in Step 1 might suggest another way to determine the sign of the exponent on 10.

5.1 Exercise Set

MyMathLab®

Write each expression with positive exponents. See Examples 5 and 6.

1. $5x^{-1}y^{-2}$

2. $7xy^{-4}$

3. $a^2b^{-1}c^{-5}$

4. $a^{-4}b^2c^{-6}$

5. $\dfrac{y^{-2}}{x^{-4}}$

6. $\dfrac{x^{-7}}{z^{-3}}$

Use the product rule to simplify each expression. See Examples 1 and 2.

7. $4^2 \cdot 4^3$

8. $3^3 \cdot 3^5$

9. $x^5 \cdot x^3$

10. $a^2 \cdot a^9$

11. $m \cdot m^7 \cdot m^6$

12. $n \cdot n^{10} \cdot n^{12}$

13. $(4xy)(-5x)$

14. $(-7xy)(7y)$

15. $(-4x^3p^2)(4y^3x^3)$

16. $(-6a^2b^3)(-3ab^3)$

Evaluate each expression. See Example 3.

17. -8^0

18. $(-9)^0$

19. $(4x + 5)^0$

20. $(3x - 1)^0$

21. $-x^0$

22. $-5x^0$

23. $4x^0 + 5$

24. $8x^0 + 1$

Use the quotient rule to simplify. See Example 4.

25. $\dfrac{a^5}{a^2}$

26. $\dfrac{x^9}{x^4}$

27. $-\dfrac{26z^{11}}{2z^7}$

28. $-\dfrac{16x^5}{8x}$

29. $\dfrac{x^9y^6}{x^8y^6}$

30. $\dfrac{a^{12}b^2}{a^9b}$

31. $\dfrac{12x^4y^7}{9xy^5}$

32. $\dfrac{24a^{10}b^{11}}{10ab^3}$

33. $\dfrac{-36a^5b^7c^{10}}{6ab^3c^4}$

34. $\dfrac{49a^3bc^{14}}{-7abc^8}$

Simplify and write using positive exponents only. See Examples 5 and 6.

35. 4^{-2}

36. 2^{-3}

37. $(-3)^{-3}$

38. $(-6)^{-2}$

39. $\dfrac{x^7}{x^{15}}$

40. $\dfrac{z}{z^3}$

41. $5a^{-4}$

42. $10b^{-1}$

43. $\dfrac{x^{-7}}{y^{-2}}$

44. $\dfrac{p^{-13}}{q^{-3}}$

45. $\dfrac{x^{-2}}{x^5}$

46. $\dfrac{z^{-12}}{z^{10}}$

47. $\dfrac{8r^4}{2r^{-4}}$

48. $\dfrac{3s^3}{15s^{-3}}$

49. $\dfrac{x^{-9}x^4}{x^{-5}}$

50. $\dfrac{y^{-7}y}{y^8}$

51. $\dfrac{2a^{-6}b^2}{18ab^{-5}}$

52. $\dfrac{18ab^{-6}}{3a^{-3}b^6}$

53. $\dfrac{(24x^8)(x)}{20x^{-7}}$

54. $\dfrac{(30z^2)(z^5)}{55z^{-4}}$

MIXED PRACTICE

Simplify and write using positive exponents only. See Examples 1 through 6.

55. $-7x^3 \cdot 20x^9$

56. $-3y \cdot -9y^4$

57. $x^7 \cdot x^8 \cdot x$

58. $y^6 \cdot y \cdot y^9$

59. $2x^3 \cdot 5x^7$

60. $-3z^4 \cdot 10z^7$

61. $(5x)^0 + 5x^0$

62. $4y^0 - (4y)^0$

63. $\dfrac{z^{12}}{z^{15}}$

64. $\dfrac{x^{11}}{x^{20}}$

65. $3^0 - 3t^0$

66. $4^0 + 4x^0$

67. $\dfrac{y^{-3}}{y^{-7}}$

68. $\dfrac{y^{-6}}{y^{-9}}$

69. $4^{-1} + 3^{-2}$

70. $1^{-3} - 4^{-2}$

71. $3x^{-1}$

72. $(4x)^{-1}$

73. $\dfrac{r^4}{r^{-4}}$

74. $\dfrac{x^{-5}}{x^3}$

75. $\dfrac{x^{-7}y^{-2}}{x^2y^2}$

76. $\dfrac{a^{-5}b^7}{a^{-2}b^{-3}}$

77. $(-4x^2y)(3x^4)(-2xy^5)$

78. $(-6a^4b)(2b^3)(-3ab^6)$

79. $2^{-4} \cdot x$

80. $5^{-2} \cdot y$

81. $\dfrac{5^{17}}{5^{13}}$

82. $\dfrac{10^{25}}{10^{23}}$

83. $\dfrac{8^{-7}}{8^{-6}}$

84. $\dfrac{13^{-10}}{13^{-9}}$

85. $\dfrac{9^{-5}a^4}{9^{-3}a^{-1}}$

86. $\dfrac{11^{-9}b^3}{11^{-7}b^{-4}}$

87. $\dfrac{14x^{-2}yz^{-4}}{2xyz}$

88. $\dfrac{30x^{-7}yz^{-14}}{3xyz}$

Simplify. Assume that variables in the exponents represent nonzero integers and that x, y, and z are not 0. See Example 7.

89. $x^5 \cdot x^{7a}$

90. $x^{4a} \cdot x^7$

91. $\dfrac{x^{3t-1}}{x^t}$

92. $\dfrac{y^{4p-2}}{y^{3p}}$

93. $y^{2p} \cdot y^{9p}$

94. $x^{9y} \cdot x^{-7y}$

95. $\dfrac{z^{6x}}{z^7}$

96. $\dfrac{y^6}{y^{4z}}$

97. $\dfrac{x^{3t} \cdot x^{4t-1}}{x^t}$

98. $\dfrac{z^{5x} \cdot z^{x-7}}{z^x}$

Write each number in scientific notation. See Example 8.

99. 31,250,000

100. 678,000

101. 0.016

102. 0.007613

103. 67,413

104. 36,800,000

105. 0.0125

106. 0.00084

107. 0.000053

108. 98,700,000,000

Write each number in scientific notation.

109. The approximate distance between Jupiter and the sun is 778,300,000 kilometers. (*Source:* National Space Data Center)

110. For the 2010–2011 National Football League season, the payroll of the Super Bowl champion Green Bay Packers was approximately $72,245,500. (*Source: USA Today*)

111. The gross domestic product (GDP) of a region is the value of all final goods and services produced within a given year. The GDP of the European Union for 2010 was estimated to be equivalent to U.S. $16,228,000,000,000. (*Source:* Wikipedia)

112. By the end of 2010, Nintendo had sold 34,550,000 Wii units. (*Source:* Nintendo)

113. Lake Mead, created from the Colorado River by the Hoover Dam, has a capacity of 124,000,000,000 cubic feet of water. (*Source:* U.S. Bureau of Reclamation)

114. The temperature of the core of the sun is about 27,000,000°F.

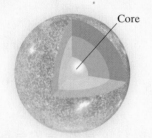

Core

115. A pulsar is a rotating neutron star that gives off sharp, regular pulses of radio waves. For one particular pulsar, the rate of pulses is every 0.001 second.

116. To convert from cubic inches to cubic meters, multiply by 0.0000164.

Write each number in standard notation, without exponents. See Example 9.

▶ **117.** 3.6×10^{-9} **118.** 2.7×10^{-5}

▶ **119.** 9.3×10^{7} **120.** 6.378×10^{8}

121. 1.278×10^{6} **122.** 7.6×10^{4}

123. 7.35×10^{12} **124.** 1.66×10^{-5}

125. 4.03×10^{-7} **126.** 8.007×10^{8}

Write each number in standard notation.

127. The estimated world population in 1 C.E. was 3.0×10^{8}. (*Source: World Almanac and Book of Facts*)

128. There are 3.949×10^{6} miles of highways, roads, and streets in the United States. (*Source:* Bureau of Transportation Statistics)

129. Chiricahua National Monument in Arizona contains 9.5×10^{-3} square kilometers of privately owned land.

130. Hagerman Fossil Beds National Monument in Idaho contains 6.68×10^{-2} square kilometers of privately owned land.

131. In 2010, the Coca-Cola Company sold 2.55×10^{10} unit cases of beverages worldwide. (*Source:* the Coca-Cola Company)

132. In 2010, the revenue for McDonald's was 2.27×10^{10}. (*Source:* McDonald's Corporation)

REVIEW AND PREVIEW

Evaluate. See Sections 1.3 and 5.1.

133. $(5 \cdot 2)^{2}$ **134.** $5^{2} \cdot 2^{2}$

135. $\left(\dfrac{3}{4}\right)^{3}$ **136.** $\dfrac{3^{3}}{4^{3}}$

137. $(2^{3})^{2}$ **138.** $(2^{2})^{3}$

CONCEPT EXTENSIONS

139. Explain how to convert a number from standard notation to scientific notation.

140. Explain how to convert a number from scientific notation to standard notation.

141. Explain why $(-5)^{0}$ simplifies to 1 but -5^{0} simplifies to -1.

142. Explain why both $4x^{0} - 3y^{0}$ and $(4x - 3y)^{0}$ simplify to 1.

143. Simplify where possible.

     a. $x^{a} \cdot x^{a}$ **b.** $x^{a} + x^{a}$

     c. $\dfrac{x^{a}}{x^{b}}$ **d.** $x^{a} \cdot x^{b}$

     e. $x^{a} + x^{b}$

144. Which numbers are equal to 36,000? Of these, which is written in scientific notation?

     a. 36×10^{3} **b.** 360×10^{2}

     c. 0.36×10^{5} **d.** 3.6×10^{4}

Without calculating, determine which number is larger.

145. 7^{11} or 7^{13} **146.** 5^{10} or 5^{9}

147. 7^{-11} or 7^{-13} **148.** 5^{-10} or 5^{-9}

5.2 More Work with Exponents and Scientific Notation

OBJECTIVES

1 Use the Power Rules for Exponents.

2 Use Exponent Rules and Definitions to Simplify Exponential Expressions.

3 Compute Using Scientific Notation.

OBJECTIVE

1 Using the Power Rules

The volume of the cube shown whose side measures x^2 units is $(x^2)^3$ cubic units. To simplify an expression such as $(x^2)^3$, we use the definition of a^n. Then

$$(x^2)^3 = \underbrace{(x^2)(x^2)(x^2)}_{x^2 \text{ is a factor 3 times}} = x^{2+2+2} = x^6$$

x^2 units

Notice that the result is exactly the same if the exponents are multiplied.

$$(x^2)^3 = x^{2\cdot3} = x^6$$

This suggests that the power of an exponential expression raised to a power is the product of the exponents. Two additional rules for exponents are given in the following box.

The Power Rule and Power of a Product or Quotient Rules for Exponents

If a and b are real numbers and m and n are integers, then

$$(a^m)^n = a^{m\cdot n} \qquad \text{Power rule}$$
$$(ab)^m = a^m b^m \qquad \text{Power of a product}$$
$$\left(\frac{a}{b}\right)^n = \frac{a^n}{b^n} \ (b \neq 0) \quad \text{Power of a quotient}$$

EXAMPLE 1 Use the power rule to simplify the following expressions. Use positive exponents to write all results.

a. $(x^5)^7$ **b.** $(2^2)^3$ **c.** $(5^{-1})^2$ **d.** $(y^{-3})^{-4}$

Solution

a. $(x^5)^7 = x^{5\cdot7} = x^{35}$ **b.** $(2^2)^3 = 2^{2\cdot3} = 2^6 = 64$

c. $(5^{-1})^2 = 5^{-1\cdot2} = 5^{-2} = \frac{1}{5^2} = \frac{1}{25}$ **d.** $(y^{-3})^{-4} = y^{-3(-4)} = y^{12}$ □

PRACTICE

1 Use the power rule to simplify the following expressions. Use positive exponents to write all results.

a. $(z^3)^5$ **b.** $(5^2)^2$ **c.** $(3^{-1})^3$ **d.** $(x^{-4})^{-6}$

EXAMPLE 2 Use the power rules to simplify the following. Use positive exponents to write all results.

a. $(5x^2)^3$ **b.** $\left(\frac{2}{3}\right)^3$ **c.** $\left(\frac{3p^4}{q^5}\right)^2$ **d.** $\left(\frac{2^{-3}}{y}\right)^{-2}$ **e.** $(x^{-5}y^2z^{-1})^7$

Solution

a. $(5x^2)^3 = 5^3 \cdot (x^2)^3 = 5^3 \cdot x^{2\cdot3} = 125x^6$

b. $\left(\frac{2}{3}\right)^3 = \frac{2^3}{3^3} = \frac{8}{27}$

c. $\left(\frac{3p^4}{q^5}\right)^2 = \frac{(3p^4)^2}{(q^5)^2} = \frac{3^2 \cdot (p^4)^2}{(q^5)^2} = \frac{9p^8}{q^{10}}$

d. $\left(\dfrac{2^{-3}}{y}\right)^{-2} = \dfrac{(2^{-3})^{-2}}{y^{-2}}$

$= \dfrac{2^6}{y^{-2}} = 64y^2$ Use the negative exponent rule.

e. $(x^{-5}y^2z^{-1})^7 = (x^{-5})^7 \cdot (y^2)^7 \cdot (z^{-1})^7$

$= x^{-35}y^{14}z^{-7} = \dfrac{y^{14}}{x^{35}z^7}$

☐

PRACTICE

2   Use the power rules to simplify the following. Use positive exponents to write all results.

a. $(2x^3)^5$ **b.** $\left(\dfrac{3}{5}\right)^2$ **c.** $\left(\dfrac{2a^5}{b^7}\right)^4$ **d.** $\left(\dfrac{3^{-2}}{x}\right)^{-1}$ **e.** $(a^{-2}b^{-5}c^4)^{-2}$

OBJECTIVE

2   Using Exponent Rules and Definitions to Simplify Expressions

In the next few examples, we practice the use of several of the rules and definitions for exponents. The following is a summary of these rules and definitions.

Summary of Rules for Exponents

If a and b are real numbers and m and n are integers, then

Product rule	$a^m \cdot a^n = a^{m+n}$	
Zero exponent	$a^0 = 1$	$(a \neq 0)$
Negative exponent	$a^{-n} = \dfrac{1}{a^n}$	$(a \neq 0)$
Quotient rule	$\dfrac{a^m}{a^n} = a^{m-n}$	$(a \neq 0)$
Power rule	$(a^m)^n = a^{m \cdot n}$	
Power of a product	$(ab)^m = a^m \cdot b^m$	
Power of a quotient	$\left(\dfrac{a}{b}\right)^m = \dfrac{a^m}{b^m}$	$(b \neq 0)$

EXAMPLE 3   Simplify each expression. Use positive exponents to write the answers.

a. $(2x^0y^{-3})^{-2}$ **b.** $\left(\dfrac{x^{-5}}{x^{-2}}\right)^{-3}$ **c.** $\left(\dfrac{2}{7}\right)^{-2}$ **d.** $\dfrac{5^{-2}x^{-3}y^{11}}{x^2y^{-5}}$

Solution

a. $(2x^0y^{-3})^{-2} = 2^{-2}(x^0)^{-2}(y^{-3})^{-2}$

$= 2^{-2}x^0y^6$

$= \dfrac{1(y^6)}{2^2}$ Write x^0 as 1.

$= \dfrac{y^6}{4}$

b. $\left(\dfrac{x^{-5}}{x^{-2}}\right)^{-3} = \dfrac{(x^{-5})^{-3}}{(x^{-2})^{-3}} = \dfrac{x^{15}}{x^6} = x^{15-6} = x^9$

c. $\left(\dfrac{2}{7}\right)^{-2} = \dfrac{2^{-2}}{7^{-2}} = \dfrac{7^2}{2^2} = \dfrac{49}{4}$

d. $\dfrac{5^{-2}x^{-3}y^{11}}{x^2y^{-5}} = (5^{-2})\left(\dfrac{x^{-3}}{x^2}\right)\left(\dfrac{y^{11}}{y^{-5}}\right) = 5^{-2}x^{-3-2}y^{11-(-5)} = 5^{-2}x^{-5}y^{16}$

$$= \dfrac{y^{16}}{5^2x^5} = \dfrac{y^{16}}{25x^5}$$

PRACTICE

3 Simplify each expression. Use positive exponents to write the answers.

a. $(3ab^{-5})^{-3}$ **b.** $\left(\dfrac{y^{-7}}{y^{-4}}\right)^{-5}$ **c.** $\left(\dfrac{3}{8}\right)^{-2}$ **d.** $\dfrac{9^{-2}a^{-4}b^3}{a^2b^{-5}}$

EXAMPLE 4 Simplify each expression. Use positive exponents to write the answers.

a. $\left(\dfrac{3x^2y}{y^{-9}z}\right)^{-2}$ **b.** $\left(\dfrac{3a^2}{2x^{-1}}\right)^3\left(\dfrac{x^{-3}}{4a^{-2}}\right)^{-1}$

Solution There is often more than one way to simplify exponential expressions. Here, we will simplify inside the parentheses if possible before we apply the power rules for exponents.

a. $\left(\dfrac{3x^2y}{y^{-9}z}\right)^{-2} = \left(\dfrac{3x^2y^{10}}{z}\right)^{-2} = \dfrac{3^{-2}x^{-4}y^{-20}}{z^{-2}} = \dfrac{z^2}{3^2x^4y^{20}} = \dfrac{z^2}{9x^4y^{20}}$

b. $\left(\dfrac{3a^2}{2x^{-1}}\right)^3\left(\dfrac{x^{-3}}{4a^{-2}}\right)^{-1} = \dfrac{27a^6}{8x^{-3}} \cdot \dfrac{x^3}{4^{-1}a^2} = \dfrac{27 \cdot 4 \cdot a^6x^3x^3}{8 \cdot a^2} = \dfrac{27a^4x^6}{2}$

PRACTICE

4 Simplify each expression. Use positive exponents to write the answers.

a. $\left(\dfrac{5a^4b}{a^{-8}c}\right)^{-3}$ **b.** $\left(\dfrac{2x^4}{5y^{-2}}\right)^3\left(\dfrac{x^{-4}}{10y^{-2}}\right)^{-1}$

EXAMPLE 5 Simplify each expression. Assume that a and b are integers and that x and y are not 0.

a. $x^{-b}(2x^b)^2$ **b.** $\dfrac{(y^{3a})^2}{y^{a-6}}$

Solution

a. $x^{-b}(2x^b)^2 = x^{-b}2^2x^{2b} = 4x^{-b+2b} = 4x^b$

b. $\dfrac{(y^{3a})^2}{y^{a-6}} = \dfrac{y^{6a}}{y^{a-6}} = y^{6a-(a-6)} = y^{6a-a+6} = y^{5a+6}$

PRACTICE

5 Simplify each expression. Assume that a and b are integers and that x and y are not 0.

a. $x^{-2a}(3x^a)^3$ **b.** $\dfrac{(y^{3b})^3}{y^{4b-3}}$

OBJECTIVE

3 Computing Using Scientific Notation

To perform operations on numbers written in scientific notation, we use properties of exponents.

EXAMPLE 6   Perform the indicated operations. Write each result in scientific notation.

a. $(8.1 \times 10^5)(5 \times 10^{-7})$ **b.** $\dfrac{1.2 \times 10^4}{3 \times 10^{-2}}$

Solution

a. $(8.1 \times 10^5)(5 \times 10^{-7}) = 8.1 \times 5 \times 10^5 \times 10^{-7}$ Not in scientific notation because
$= 40.5 \times 10^{-2}$ 40.5 is not between 1 and 10.
$= (4.05 \times 10^1) \times 10^{-2}$
$= 4.05 \times 10^{-1}$

b. $\dfrac{1.2 \times 10^4}{3 \times 10^{-2}} = \left(\dfrac{1.2}{3}\right)\left(\dfrac{10^4}{10^{-2}}\right) = 0.4 \times 10^{4-(-2)}$
$= 0.4 \times 10^6 = (4 \times 10^{-1}) \times 10^6 = 4 \times 10^5$ ☐

PRACTICE
6   Perform the indicated operations. Write each result in scientific notation.

a. $(3.4 \times 10^4)(5 \times 10^{-7})$ **b.** $\dfrac{1.6 \times 10^8}{4 \times 10^{-2}}$

EXAMPLE 7   Use scientific notation to simplify $\dfrac{2000 \times 0.000021}{700}$. Write the result in scientific notation.

Solution

$$\dfrac{2000 \times 0.000021}{700} = \dfrac{(2 \times 10^3)(2.1 \times 10^{-5})}{7 \times 10^2} = \dfrac{2(2.1)}{7} \cdot \dfrac{10^3 \cdot 10^{-5}}{10^2}$$
$$= 0.6 \times 10^{-4}$$
$$= (6 \times 10^{-1}) \times 10^{-4}$$
$$= 6 \times 10^{-5}$$ ☐

PRACTICE
7   Use scientific notation to simplify $\dfrac{2400 \times 0.0000014}{800}$. Write the result in scientific notation.

Vocabulary, Readiness & Video Check

Simplify. See Examples 1 through 4.

1. $(x^4)^5$ **2.** $(5^6)^2$ **3.** $x^4 \cdot x^5$ **4.** $x^7 \cdot x^8$ **5.** $(y^6)^7$
6. $(x^3)^4$ **7.** $(z^4)^9$ **8.** $(z^3)^7$ **9.** $(z^{-6})^{-3}$ **10.** $(y^{-4})^{-2}$

Martin-Gay Interactive Videos

See Video 5.2 🍐

Watch the section lecture video and answer the following questions

OBJECTIVE 1
11. Based on ▣ Example 1 and the lecture before, what are the differences between the power rule and product rule for exponents?

OBJECTIVE 2
12. Name all the rules and definitions used to simplify ▣ Example 5.

OBJECTIVE 3
13. In the ▣ Example 6 multiplication problem, why is the first product not the final answer? How can we tell that this first product is not written in scientific notation?

5.2 Exercise Set MyMathLab®

Simplify. Write each answer using positive exponents only. See Examples 1 and 2.

1. $(3^{-1})^2$

2. $(2^{-2})^2$

▷ **3.** $(x^4)^{-9}$

4. $(y^7)^{-3}$

5. $(3x^2y^3)^2$

6. $(4x^3yz)^2$

7. $\left(\dfrac{2x^5}{y^{-3}}\right)^4$

8. $\left(\dfrac{3a^{-4}}{b^7}\right)^3$

9. $(2a^2bc^{-3})^{-6}$

10. $(6x^{-6}y^7z^0)^{-2}$

▷ **11.** $\left(\dfrac{x^7y^{-3}}{z^{-4}}\right)^{-5}$

12. $\left(\dfrac{a^{-2}b^{-5}}{c^{-11}}\right)^{-6}$

13. $(-2^{-2}y^{-1})^{-3}$

14. $(-4^{-6}y^{-6})^{-4}$

Simplify. Write each answer using positive exponents only. See Examples 3 and 4.

15. $\left(\dfrac{a^{-4}}{a^{-5}}\right)^{-2}$

16. $\left(\dfrac{x^{-9}}{x^{-4}}\right)^{-3}$

17. $\left(\dfrac{6p^6}{p^{12}}\right)^2$

18. $\left(\dfrac{4p^6}{p^9}\right)^3$

19. $(-8y^3xa^{-2})^{-3}$

20. $(-5y^0x^2a^3)^{-3}$

21. $\left(\dfrac{3}{4}\right)^{-3}$

22. $\left(\dfrac{5}{8}\right)^{-2}$

▷ **23.** $\left(\dfrac{2a^{-2}b^5}{4a^2b^7}\right)^{-2}$

24. $\left(\dfrac{5x^7y^4}{10x^3y^{-2}}\right)^{-3}$

25. $\left(\dfrac{x^{-2}y^{-2}}{a^{-3}}\right)^{-7}$

26. $\left(\dfrac{x^{-1}y^{-2}}{z^{-3}}\right)^{-5}$

MIXED PRACTICE

Simplify. Write each answer using positive exponents only. See Examples 1 through 4.

27. $(y^{-5})^2$

28. $(z^{-2})^{13}$

29. $(5^{-1})^3$

30. $(8^2)^{-1}$

31. $(x^7)^{-9}$

32. $(y^{-4})^5$

33. $\left(\dfrac{x^4}{y^{-3}}\right)^{-5}$

34. $\left(\dfrac{a^{-3}}{b^{-6}}\right)^{-4}$

▷ **35.** $(4x^2)^2$

36. $(-8x^3)^2$

37. $\left(\dfrac{4^{-4}}{y^3x}\right)^{-2}$

38. $\left(\dfrac{7^{-3}}{ab^2}\right)^{-2}$

39. $\left(\dfrac{2x^{-3}}{y^{-1}}\right)^{-3}$

40. $\left(\dfrac{n^5}{2m^{-2}}\right)^{-4}$

41. $\dfrac{4^{-1}x^2yz}{x^{-2}yz^3}$

42. $\dfrac{8^{-2}x^{-3}y^{11}}{x^2y^{-5}}$

43. $\left(\dfrac{3x^5}{6x^4}\right)^4$

44. $\left(\dfrac{8^{-3}}{y^2}\right)^{-2}$

▷ **45.** $\dfrac{(y^3)^{-4}}{y^3}$

46. $\dfrac{2(y^3)^{-3}}{y^{-3}}$

47. $\dfrac{3^{-2}a^{-5}b^6}{4^{-2}a^{-7}b^{-3}}$

48. $\dfrac{2^{-3}m^{-4}n^{-5}}{5^{-2}m^{-5}n}$

49. $(4x^6y^5)^{-2}(6x^4y^3)$

50. $(5x^2y^4)^{-2}(3x^9y^4)$

51. $x^6(x^6bc)^{-6}$

52. $y^2(y^2bx)^{-4}$

53. $\dfrac{2^{-3}x^2y^{-5}}{5^{-2}x^7y^{-1}}$

54. $\dfrac{7^{-1}a^{-3}b^5}{a^2b^{-2}}$

55. $\left(\dfrac{2x^2}{y^4}\right)^3\left(\dfrac{2x^5}{y}\right)^{-2}$

56. $\left(\dfrac{3z^{-2}}{y}\right)^2\left(\dfrac{9y^{-4}}{z^{-3}}\right)^{-1}$

Simplify the following. Assume that variables in the exponents represent integers and that all other variables are not 0. See Example 5.

57. $(x^{3a+6})^3$

58. $(x^{2b+7})^2$

59. $\dfrac{x^{4a}(x^{4a})^3}{x^{4a-2}}$

60. $\dfrac{x^{-5y+2}x^{2y}}{x}$

61. $(b^{5x-2})^2$

62. $(c^{2a+3})^3$

63. $\dfrac{(y^{2a})^8}{y^{a-3}}$

64. $\dfrac{(y^{4a})^7}{y^{2a-1}}$

65. $\left(\dfrac{2x^{3t}}{x^{2t-1}}\right)^4$

66. $\left(\dfrac{3y^{5a}}{y^{-a+1}}\right)^2$

67. $\dfrac{25x^{2a+1}y^{a-1}}{5x^{3a+1}y^{2a-3}}$

68. $\dfrac{16x^{-5-3a}y^{-2a-b}}{2x^{-5+3b}y^{-2b-a}}$

Perform each indicated operation. Write each answer in scientific notation. See Examples 6 and 7.

▷ **69.** $(5 \times 10^{11})(2.9 \times 10^{-3})$

70. $(3.6 \times 10^{-12})(6 \times 10^9)$

71. $(2 \times 10^5)^3$

72. $(3 \times 10^{-7})^3$

73. $\dfrac{3.6 \times 10^{-4}}{9 \times 10^2}$

74. $\dfrac{1.2 \times 10^9}{2 \times 10^{-5}}$

75. $\dfrac{0.0069}{0.023}$

76. $\dfrac{0.00048}{0.0016}$

77. $\dfrac{18{,}200 \times 100}{91{,}000}$

78. $\dfrac{0.0003 \times 0.0024}{0.0006 \times 20}$

79. $\dfrac{6000 \times 0.006}{0.009 \times 400}$

80. $\dfrac{0.00016 \times 300}{0.064 \times 100}$

81. $\dfrac{0.00064 \times 2000}{16{,}000}$

82. $\dfrac{0.00072 \times 0.003}{0.00024}$

▷ **83.** $\dfrac{66{,}000 \times 0.001}{0.002 \times 0.003}$

84. $\dfrac{0.0007 \times 11{,}000}{0.001 \times 0.0001}$

85. $\dfrac{9.24 \times 10^{15}}{(2.2 \times 10^{-2})(1.2 \times 10^{-5})}$

86. $\dfrac{(2.6 \times 10^{-3})(4.8 \times 10^{-4})}{1.3 \times 10^{-12}}$

Solve.

87. To convert from square inches to square meters, multiply by 6.452×10^{-4}. The area of the following square is 4×10^{-2} square inches. Convert this area to square meters.

4×10^{-2} sq in.

88. To convert from cubic inches to cubic meters, multiply by 1.64×10^{-5}. A grain of salt is in the shape of a cube. If an average size of a grain of salt is 3.8×10^{-6} cubic inches, convert this volume to cubic meters.

3.8×10^{-6} cu in.

REVIEW AND PREVIEW

Simplify each expression. See Section 1.4.

89. $-5y + 4y - 18 - y$

90. $12m - 14 - 15m - 1$

91. $-3x - (4x - 2)$

92. $-9y - (5 - 6y)$

93. $3(z - 4) - 2(3z + 1)$

94. $5(x - 3) - 4(2x - 5)$

CONCEPT EXTENSIONS

95. Each side of the cube shown is $\dfrac{2x^{-2}}{y}$ meters. Find its volume.

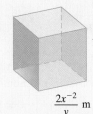

$\dfrac{2x^{-2}}{y}$ m

96. The lot shown is in the shape of a parallelogram with base $\dfrac{3x^{-1}}{y^{-3}}$ feet and height $5x^{-7}$ feet. Find its area.

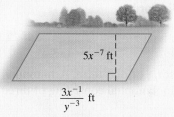

$5x^{-7}$ ft

$\dfrac{3x^{-1}}{y^{-3}}$ ft

97. The density D of an object is equivalent to the quotient of its mass M and volume V. Thus $D = \dfrac{M}{V}$. Express in scientific notation the density of an object whose mass is 500,000 pounds and whose volume is 250 cubic feet.

98. The density of ordinary water is 3.12×10^{-2} tons per cubic foot. The volume of water in the largest of the Great Lakes,

Lake Superior, is 4.269×10^{14} cubic feet. Use the formula $D = \dfrac{M}{V}$ (see Exercise 97) to find the mass (in tons) of the water in Lake Superior. Express your answer in scientific notation. (*Source:* National Ocean Service)

99. Is there a number a such that $a^{-1} = a^{1}$? If so, give the value of a.

100. Is there a number a such that a^{-2} is a negative number? If so, give the value of a.

101. Explain whether 0.4×10^{-5} is written in scientific notation.

102. In your own words, explain why $a^0 = 1$ (as long as a is not 0).

103. The estimated population of the United States in April 2010 was 3.08×10^8 people. The land area of the United States is 3.536×10^6 square miles. Find the population density (number of people per square mile) for the United States in 2010. Round to the nearest whole. (*Source:* U.S. Census Bureau)

104. The estimated population of China at the end of 2010 was 1.34×10^9 people. The land area of China is 3.705×10^6 square miles. Find the population density (number of people per square mile) for China at the end of 2010. Round to the nearest whole. (*Source:* U.S. Bureau of the Census)

105. In March 2011, the value of goods and services imported into the United States was $\$2.21 \times 10^{11}$. The estimated population was 3.11×10^8 people. Find the average value of imports per person in the United States for March 2011. Round to the nearest dollar. (*Source:* U.S. Bureau of the Census, Bureau of Economic Analysis)

106. In 2010, the population of the European Union was 4.92×10^8 people. At the same time, the population of the United States was 3.10×10^8 people. How many times greater was the population of the European Union than the population of the United States? Round to the nearest tenth. (*Source:* Population Reference Bureau)

5.3 Polynomials and Polynomial Functions

OBJECTIVES

1   Identify Term, Constant, Polynomial, Monomial, Binomial, Trinomial, and the Degree of a Term and of a Polynomial.

2   Define Polynomial Functions.

3   Review Combining Like Terms.

4   Add Polynomials.

5   Subtract Polynomials.

6   Recognize the Graph of a Polynomial Function from the Degree of the Polynomial.

OBJECTIVE

1   Identifying Polynomial Terms and Degrees of Terms and Polynomials

A **term** is a number or the product of a number and one or more variables raised to powers. The **numerical coefficient,** or simply the **coefficient,** is the numerical factor of a term.

Term	Numerical Coefficient of Term
$-1.2x^5$	-1.2
x^3y	1
$-z$	-1
2	2
$\dfrac{x^9}{7}$ $\left(\text{or } \dfrac{1}{7}x^9\right)$	$\dfrac{1}{7}$

If a term contains only a number, it is called a **constant term,** or simply a **constant.**

A **polynomial** is a finite sum of terms in which all variables are raised to nonnegative integer powers and no variables appear in any denominator.

Polynomials	Not Polynomials	
$4x^5y + 7xz$	$5x^{-3} + 2x$	Negative integer exponent
$-5x^3 + 2x + \dfrac{2}{3}$	$\dfrac{6}{x^2} - 5x + 1$	Variable in denominator

A polynomial that contains only one variable is called a **polynomial in one variable.** For example, $3x^2 - 2x + 7$ is a **polynomial in x.** This polynomial in x is written in *descending order* since the terms are listed in descending order of the variable's exponents. (The term 7 can be thought of as $7x^0$.) The following examples are polynomials in one variable written in **descending order.**

$$4x^3 - \frac{7}{8}x^2 + 5 \qquad y^2 - 4.7 \qquad \frac{8a^4}{11} - 7a^2 + \frac{4a}{3}$$

A **monomial** is a polynomial consisting of one term. A **binomial** is a polynomial consisting of two terms. A **trinomial** is a polynomial consisting of three terms.

Monomials	Binomials	Trinomials
ax^2	$x + y$	$x^2 + 4xy + y^2$
$-3x$	$6y^2 - 2.9$	$-x^4 + 3x^3 + 0.1$
4	$\dfrac{5}{7}z^3 - 2z$	$8y^2 - 2y - \dfrac{10}{17}$

By definition, all monomials, binomials, and trinomials are also polynomials.

Each term of a polynomial has a **degree.**

> **Helpful Hint**
> We usually write answers that are polynomials in one variable in descending order.

Degree of a Term

The **degree of a term** is the sum of the exponents on the *variables* contained in the term.

EXAMPLE 1 Find the degree of each term.

a. $3x^2$ **b.** -2^3x^5 **c.** y **d.** $12x^2yz^3$ **e.** 5.27

Solution

a. The exponent on x is 2, so the degree of the term is 2.

b. The exponent on x is 5, so the degree of the term is 5. (Recall that the degree is the sum of the exponents on only the *variables*.)

c. The degree of y, or y^1, is 1.

d. The degree is the sum of the exponents on the variables, or $2 + 1 + 3 = 6$.

e. The degree of 5.27, which can be written as $5.27x^0$, is 0.   ☐

PRACTICE

1 Find the degree of each term.

a. $4x^5$ **b.** -4^3y^3 **c.** z **d.** $65a^3b^7c$ **e.** 36

From the preceding example, we can say that the degree of a constant is 0. Also, the term 0 has no degree.

Each polynomial also has a degree.

Degree of a Polynomial

The **degree of a polynomial** is the largest degree of all its terms.

EXAMPLE 2 Find the degree of each polynomial and indicate whether the polynomial is also a monomial, binomial, or trinomial.

	Polynomial	Degree	Classification
a.	$7x^3 - \dfrac{3}{4}x + 2$	3	Trinomial
b.	$-xyz$	$1 + 1 + 1 = 3$	Monomial
c.	$x^4 - 16.5$	4	Binomial

☐

PRACTICE

2 Find the degree of each polynomial and indicate whether the polynomial is also a monomial, binomial, or trinomial.

	Polynomial	Degree	Classification
a.	$3x^4 + 2x^2 - 3$		
b.	$9abc^3$		
c.	$8x^5 + 5x^3$		

EXAMPLE 3 Find the degree of the polynomial

$$3xy + x^2y^2 - 5x^2 - 6.7$$

Solution The degree of each term is

$$3xy + x^2y^2 - 5x^2 - 6.7$$
$$\downarrow \qquad \downarrow \qquad \downarrow \qquad \downarrow$$

Degree: 2   4   2   0

The largest degree of any term is 4, so the degree of this polynomial is 4. ☐

PRACTICE
3 Find the degree of the polynomial $2x^3y - 3x^3y^2 - 9y^5 + 9.6$.

OBJECTIVE
2 **Defining Polynomial Functions** ▶

At times, it is convenient to use function notation to represent polynomials. For example, we may write $P(x)$ to represent the polynomial $3x^2 - 2x - 5$. In symbols, this is

$$P(x) = 3x^2 - 2x - 5$$

This function is called a **polynomial function** because the expression $3x^2 - 2x - 5$ is a polynomial.

> ▶ Helpful Hint
>
> Recall that the symbol $P(x)$ **does not mean** P times x. It is a special symbol used to denote a function.

EXAMPLE 4 If $P(x) = 3x^2 - 2x - 5$, find the following.

a. $P(1)$ **b.** $P(-2)$

Solution

a. Substitute 1 for x in $P(x) = 3x^2 - 2x - 5$ and simplify.

$$P(x) = 3x^2 - 2x - 5$$
$$P(1) = 3(1)^2 - 2(1) - 5 = -4$$

b. Substitute -2 for x in $P(x) = 3x^2 - 2x - 5$ and simplify.

$$P(x) = 3x^2 - 2x - 5$$
$$P(-2) = 3(-2)^2 - 2(-2) - 5 = 11$$ ☐

PRACTICE
4 If $P(x) = -5x^2 + 2x - 8$, find the following.

a. $P(-1)$ **b.** $P(3)$

Many real-world phenomena are modeled by polynomial functions. If the polynomial function model is given, we can often find the solution of a problem by evaluating the function at a certain value.

EXAMPLE 5 **Finding the Height of an Object**

The world's highest bridge, the Millau Viaduct in France, is 1125 feet above the River Tarn. An object is dropped from the top of this bridge. Neglecting air resistance, the height of the object at time t seconds is given by the polynomial function $P(t) = -16t^2 + 1125$. Find the height of the object when $t = 1$ second and when $t = 8$ seconds.

Solution   To find the height of the object at 1 second, we find $P(1)$.

$$P(t) = -16t^2 + 1125$$
$$P(1) = -16(1)^2 + 1125$$
$$P(1) = 1109$$

When $t = 1$ second, the height of the object is 1109 feet.
To find the height of the object at 8 seconds, we find $P(8)$.

$$P(t) = -16t^2 + 1125$$
$$P(8) = -16(8)^2 + 1125$$
$$P(8) = 101$$

When $t = 8$ seconds, the height of the object is 101 feet. Notice that as time t increases, the height of the object decreases. □

PRACTICE
5   The largest natural bridge is in the canyons at the base of Navajo Mountain, Utah. From the base to the top of the arch, it measures 290 feet. Neglecting air resistance, the height of an object dropped off the bridge is given by the polynomial function $P(t) = -16t^2 + 290$ at time t seconds. Find the height of the object at time $t = 0$ second and $t = 2$ seconds.

OBJECTIVE
3   Combining Like Terms Review

Before we add polynomials, recall that terms are considered to be **like terms** if they contain exactly the same variables raised to exactly the same powers.

Like Terms	*Unlike Terms*
$-5x^2, -x^2$	$4x^2, 3x$
$7xy^3z, -2xzy^3$	$12x^2y^3, -2xy^3$

To simplify a polynomial, we **combine like terms** by using the distributive property. For example, by the distributive property,

$$5x + 7x = (5 + 7)x = 12x$$

EXAMPLE 6   Simplify by combining like terms.

a. $-12x^2 + 7x^2 - 6x$ **b.** $3xy - 2x + 5xy - x$

Solution   By the distributive property,

a. $-12x^2 + 7x^2 - 6x = (-12 + 7)x^2 - 6x = -5x^2 - 6x$

b. Use the associative and commutative properties to group together like terms; then combine.

$$3xy - 2x + 5xy - x = 3xy + 5xy - 2x - x$$
$$= (3 + 5)xy + (-2 - 1)x$$
$$= 8xy - 3x$$ □

▶ **Helpful Hint**
These two terms are unlike terms. They cannot be combined.

PRACTICE
6   Simplify by combining like terms.

a. $8x^4 - 5x^4 - 5x$ **b.** $4ab - 5b + 3ab + 2b$

OBJECTIVE
4   Adding Polynomials

Now we have reviewed the necessary skills to add polynomials.

> **Adding Polynomials**
> To add polynomials, combine all like terms.

EXAMPLE 7 Add $11x^3 - 12x^2 + x - 3$ and $x^3 - 10x + 5$.

Solution $(11x^3 - 12x^2 + x - 3) + (x^3 - 10x + 5)$

$\quad = 11x^3 + x^3 - 12x^2 + x - 10x - 3 + 5$ Group like terms.

$\quad = 12x^3 - 12x^2 - 9x + 2$ Combine like terms.   □

PRACTICE
7   Add $5x^3 - 3x^2 - 9x - 8$ and $x^3 + 9x^2 + 2x$.

EXAMPLE 8 Add.

a. $(7x^3y - xy^3 + 11) + (6x^3y - 4)$ **b.** $(3a^3 - b + 2a - 5) + (a + b + 5)$

Solution

a. To add, remove the parentheses and group like terms.

$$(7x^3y - xy^3 + 11) + (6x^3y - 4)$$
$$= 7x^3y - xy^3 + 11 + 6x^3y - 4$$
$$= 7x^3y + 6x^3y - xy^3 + 11 - 4 \quad \text{Group like terms.}$$
$$= 13x^3y - xy^3 + 7 \quad \text{Combine like terms.}$$

b.
$$(3a^3 - b + 2a - 5) + (a + b + 5)$$
$$= 3a^3 - b + 2a - 5 + a + b + 5$$
$$= 3a^3 - b + b + 2a + a - 5 + 5 \quad \text{Group like terms.}$$
$$= 3a^3 + 3a \quad \text{Combine like terms.} \quad □$$

PRACTICE
8   Add.

a. $(3a^4b - 5ab^2 + 7) + (9ab^2 - 12)$ **b.** $(2x^5 - 3y + x - 6) + (4y - 2x - 3)$

Sometimes it is more convenient to add polynomials vertically. To do this, line up like terms beneath one another and add like terms. An example is shown later in this section.

OBJECTIVE
5   **Subtracting Polynomials**

The definition of subtraction of real numbers can be extended to apply to polynomials. To subtract a number, we add its opposite.

$$a - b = a + (-b)$$

Likewise, to subtract a polynomial, we add its opposite. In other words, if P and Q are polynomials, then

$$P - Q = P + (-Q)$$

The polynomial $-Q$ is the **opposite,** or **additive inverse,** of the polynomial Q. We can find $-Q$ by writing the opposite of each term of Q.

✓**CONCEPT CHECK**
Which polynomial is the opposite of $16x^3 - 5x + 7$?

a. $-16x^3 - 5x + 7$ **b.** $-16x^3 + 5x - 7$ **c.** $16x^3 + 5x + 7$ **d.** $-16x^3 + 5x + 7$

Subtracting Polynomials

To subtract a polynomial, add its opposite.

For example,

To subtract, change the signs; then add.

$$(3x^2 + 4x - 7) - (3x^2 - 2x - 5) = (3x^2 + 4x - 7) + (-3x^2 + 2x + 5)$$

$$= 3x^2 + 4x - 7 - 3x^2 + 2x + 5$$

$$= 6x - 2 \qquad \text{Combine like terms.}$$

EXAMPLE 9   Subtract: $(12z^5 - 12z^3 + z) - (-3z^4 + z^3 + 12z)$

Solution   To subtract, add the opposite of the second polynomial to the first polynomial.

$$(12z^5 - 12z^3 + z) - (-3z^4 + z^3 + 12z)$$

$$= 12z^5 - 12z^3 + z + 3z^4 - z^3 - 12z) \quad \text{Add the opposite of the polynomial being subtracted.}$$

$$= 12z^5 + 3z^4 - 12z^3 - z^3 + z - 12z \quad \text{Group like terms.}$$

$$= 12z^5 + 3z^4 - 13z^3 - 11z \quad \text{Combine like terms.} \qquad \square$$

PRACTICE
9   Subtract: $(13a^4 - 7a^3 - 9) - (-2a^4 + 8a^3 - 12)$.

✓**CONCEPT CHECK**
Why is the following subtraction incorrect?

$$\require{cancel}\cancel{\begin{array}{l}(7z - 5) - (3z - 4) \\ = 7z - 5 - 3z - 4 \\ = 4z - 9\end{array}}$$

EXAMPLE 10   Subtract $4x^3y^2 - 3x^2y^2 + 2y^2$ *from* $10x^3y^2 - 7x^2y^2$.

Solution   If we subtract 2 from 8, the difference is $8 - 2 = 6$. Notice the order of the numbers and then write "Subtract $4x^3y^2 - 3x^2y^2 + 2y^2$ from $10x^3y^2 - 7x^2y^2$" as a mathematical expression.

$$(10x^3y^2 - 7x^2y^2) - (4x^3y^2 - 3x^2y^2 + 2y^2)$$

$$= 10x^3y^2 - 7x^2y^2 - 4x^3y^2 + 3x^2y^2 - 2y^2 \quad \text{Remove parentheses.}$$

$$= 6x^3y^2 - 4x^2y^2 - 2y^2 \quad \text{Combine like terms.} \qquad \square$$

PRACTICE
10   Subtract $5x^2y^2 - 3xy^2 + 5y^3$ from $11x^2y^2 - 7xy^2$.

Answers to Concept Checks:
b;

With parentheses removed, the expression should be

$$7z - 5 - 3z + 4 = 4z - 1$$

To add or subtract polynomials vertically, just remember to line up like terms. For example, perform the subtraction, from Example 10,

$$(10x^3y^2 - 7x^2y^2) - (4x^3y^2 - 3x^2y^2 + 2y^2)$$

vertically.

Add the opposite of the second polynomial.

$$
\begin{array}{ll}
10x^3y^2 - 7x^2y^2 & \\
-(4x^3y^2 - 3x^2y^2 + 2y^2) & \text{is equivalent to}
\end{array}
\qquad
\begin{array}{l}
10x^3y^2 - 7x^2y^2 \\
\underline{-4x^3y^2 + 3x^2y^2 - 2y^2} \\
6x^3y^2 - 4x^2y^2 - 2y^2
\end{array}
$$

Polynomial functions, like polynomials, can be added, subtracted, multiplied, and divided. For example, if

$$P(x) = x^2 + x + 1$$

then

$$2P(x) = 2(x^2 + x + 1) = 2x^2 + 2x + 2 \quad \text{Use the distributive property.}$$

Also, if $Q(x) = 5x^2 - 1$, then

$$P(x) + Q(x) = (x^2 + x + 1) + (5x^2 - 1) = 6x^2 + x.$$

A useful business and economics application of subtracting polynomial functions is finding the profit function $P(x)$ when given a revenue function $R(x)$ and a cost function $C(x)$. In business, it is true that

$$\text{profit} = \text{revenue} - \text{cost, or}$$
$$P(x) = R(x) - C(x)$$

For example, if the revenue function is $R(x) = 7x$ and the cost function is $C(x) = 2x + 5000$, then the profit function is

$$P(x) = R(x) - C(x)$$

or

$$
\begin{aligned}
P(x) &= 7x - (2x + 5000) \quad &\text{Substitute } R(x) = 7x \\
P(x) &= 5x - 5000 \quad &\text{and } C(x) = 2x + 5000.
\end{aligned}
$$

Problem-solving exercises involving profit are in the exercise set.

OBJECTIVE

6   Recognizing Graphs of Polynomial Functions from Their Degree

In this section, we reviewed how to find the degree of a polynomial. Knowing the degree of a polynomial can help us recognize the graph of the related polynomial function. For example, we know from Section 3.1 that the graph of the polynomial function $f(x) = x^2$ is a parabola as shown to the left.

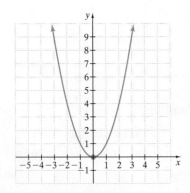

The polynomial x^2 has degree 2. The graphs of all polynomial functions of degree 2 will have this same general shape—opening upward, as shown, or downward. Graphs of polynomial functions of degree 2 or 3 will, in general, resemble one of the graphs shown next.

General Shapes of Graphs of Polynomial Functions

Degree 2

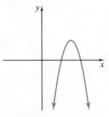

Coefficient of x^2
is a positive number.

Coefficient of x^2
is a negative number.

Degree 3

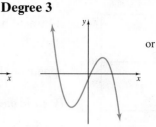

or

Coefficient of x^3
is a positive number.

or

Coefficient of x^3
is a negative number.

EXAMPLE 11 Determine which of the following graphs most closely resembles the graph of $f(x) = 5x^3 - 6x^2 + 2x + 3$.

A 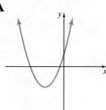 **B**  **C** 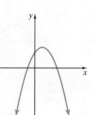 **D**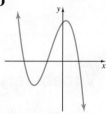

Solution The degree of $f(x)$ is 3, which means that its graph has the shape of B or D. The coefficient of x^3 is 5, a positive number, so the graph has the shape of B. ☐

PRACTICE
11 Determine which of the following graphs most closely resembles the graph of $f(x) = x^3 - 3$.

A 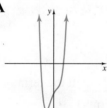 **B** 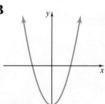 **C** 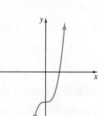 **D**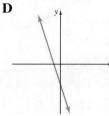

Graphing Calculator Explorations

A graphing calculator may be used to visualize addition and subtraction of polynomials in one variable. For example, to visualize the following polynomial subtraction statement

$$(3x^2 - 6x + 9) - (x^2 - 5x + 6) = 2x^2 - x + 3$$

graph both

$$Y_1 = (3x^2 - 6x + 9) - (x^2 - 5x + 6) \quad \text{Left side of equation}$$

and

$$Y_2 = 2x^2 - x + 3 \quad \text{Right side of equation}$$

on the same screen and see that their graphs coincide. (*Note:* If the graphs do not coincide, we can be sure that a mistake has been made in combining polynomials or in calculator keystrokes. If the graphs appear to coincide, we cannot be sure that our work is correct. This is because it is possible for the graphs to differ so slightly that we do not notice it.)

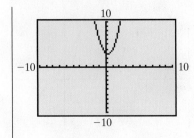

The graphs of Y_1 and Y_2 are shown. The graphs appear to coincide, so the subtraction statement

$$(3x^2 - 6x + 9) - (x^2 - 5x + 6) = 2x^2 - x + 3$$

appears to be correct.

Perform the indicated operations. Then visualize by using the procedure described above.

1. $(2x^2 + 7x + 6) + (x^3 - 6x^2 - 14)$

2. $(-14x^3 - x + 2) + (-x^3 + 3x^2 + 4x)$

3. $(1.8x^2 - 6.8x - 1.7) - (3.9x^2 - 3.6x)$

4. $(-4.8x^2 + 12.5x - 7.8) - (3.1x^2 - 7.8x)$

5. $(1.29x - 5.68) + (7.69x^2 - 2.55x + 10.98)$

6. $(-0.98x^2 - 1.56x + 5.57) + (4.36x - 3.71)$

Vocabulary, Readiness & Video Check

Use the choices below to fill in each blank. Not all choices will be used.

monomial	trinomial	like	degree	coefficient
binomial	polynomial	unlike	variables	term

1. The numerical factor of a term is the _____.

2. A(n) _____ is a finite sum of terms in which all variables are raised to nonnegative integer powers and no variables appear in any denominator.

3. A(n) _____ is a polynomial with 2 terms.

4. A(n) _____ is a polynomial with 1 term.

5. A(n) _____ is a polynomial with 3 terms.

6. The degree of a term is the sum of the exponents on the _____ in the term.

7. The _____ of a polynomial is the largest degree of all its terms.

8. _____ terms contain the same variables raised to the same powers.

Martin-Gay Interactive Videos

See Video 5.3

Watch the section lecture video and answer the following questions.

OBJECTIVE 1

9. In ▭ Example 3, why is the degree of each term found when this is not asked for?

OBJECTIVE 2

10. From ▭ Example 4, finding the value of a polynomial function at a given replacement value is similar to what?

OBJECTIVE 3

11. How is the polynomial in ▭ Example 5 simplified?

OBJECTIVE 4

12. From ▭ Example 6, how do you add polynomials?

OBJECTIVE 5

13. From ▭ Examples 7 and 8, how do you change a polynomial subtraction problem into an equivalent addition problem?

OBJECTIVE 6

14. Which of the basic graph shapes, A, B, C, or D, in ▭ Examples 9 and 10 most closely resembles the graph of $f(x) = -2x^2 - 4x + 5$? Explain.

5.3 Exercise Set MyMathLab®

Find the degree of each term. See Example 1.

1. 4

2. 7

3. $5x^2$

4. $-z^3$

5. $-3xy^2$

6. $12x^3z$

7. -8^7y^3

8. $-9^{11}y^5$

9. $3.78ab^3c^5$

10. $9.11r^2st^{12}$

Find the degree of each polynomial and indicate whether the polynomial is a monomial, binomial, trinomial, or none of these. See Examples 2 and 3.

11. $6x + 0.3$

12. $7x - 0.8$

13. $3x^2 - 2x + 5$

14. $5x^2 - 3x - 2$

15. -3^4xy^2

16. -7^5abc

17. $x^2y - 4xy^2 + 5x + y^4$

18. $-2x^2y - 3y^2 + 4x + y^5$

If $P(x) = x^2 + x + 1$ and $Q(x) = 5x^2 - 1$, find the following. See Example 4.

19. $P(7)$

20. $Q(4)$

21. $Q(-10)$

22. $P(-4)$

23. $Q\left(\dfrac{1}{4}\right)$

24. $P\left(\dfrac{1}{2}\right)$

Refer to Example 5 for Exercises 25 through 28.

25. Find the height of the object at $t = 2$ seconds.

26. Find the height of the object at $t = 4$ seconds.

27. Find the height of the object at $t = 6$ seconds.

28. Approximate (to the nearest second) how long it takes before the object hits the ground. (*Hint:* The object hits the ground when $P(x) = 0$.)

Simplify each polynomial by combining like terms. See Example 6.

29. $5y + y - 6y^2 - y^2$

30. $-x + 3x - 4x^2 - 9x^2$

31. $4x^2y + 2x - 3x^2y - \dfrac{1}{2} - 7x$

32. $-8xy^2 + 4x - x + 2xy^2 - \dfrac{11}{15}$

33. $7x^2 - 2xy + 5y^2 - x^2 + xy + 11y^2$

34. $-a^2 + 18ab - 2b^2 + 14a^2 - 12ab - b^2$

Add. See Examples 7 and 8.

35. $(9y^2 + y - 8) + (9y^2 - y - 9)$

36. $(x^2 + 4x - 7) + (8x^2 + 9x - 7)$

37. $(x^2 + xy - y^2)$ and $(2x^2 - 4xy + 7y^2)$

38. $(6x^2 + 5x + 7)$ and $(x^2 + 6x - 3)$

39. $\begin{array}{r} x^2 - 6x + 3 \\ + \quad (2x + 5) \\ \hline \end{array}$

40. $\begin{array}{r} -2x^2 + 3x - 9 \\ + \quad (2x - 3) \\ \hline \end{array}$

41. $(7x^3y - 4xy + 8) + (5x^3y + 4xy + 8x)$

42. $(9xyz + 4x - y) + (-9xyz - 3x + y + 2)$

43. $(0.6x^3 + 1.2x^2 - 4.5x + 9.1) + (3.9x^3 - x^2 + 0.7x)$

44. $(9.3y^2 - y + 12.8) + (2.6y^2 + 4.4y - 8.9)$

Subtract. See Examples 9 and 10.

45. $(9y^2 - 7y + 5) - (8y^2 - 7y + 2)$

46. $(2x^2 + 3x + 12) - (20x^2 - 5x - 7)$

47. Subtract $(6x^2 - 3x)$ from $(4x^2 + 2x)$.

48. Subtract $(8y^2 + 4x)$ from $(y^2 + x)$.

49. $\begin{array}{r} 6y^2 - 6y + 4 \\ -(-y^2 + 6y + 7) \\ \hline \end{array}$

50. $\begin{array}{r} -4x^3 + 4x^2 - 4x \\ -(2x^3 - 2x^2 + 3x) \\ \hline \end{array}$

51. $(9x^3 - 2x^2 + 4x - 7) - (2x^3 - 6x^2 - 4x + 3)$

52. $(3x^2 + 6xy + 3y^2) - (8x^2 - 6xy - y^2)$

53. Subtract $\left(y^2 + 4yx + \dfrac{1}{7}\right)$ from $\left(-19y^2 + 7yx + \dfrac{1}{7}\right)$.

54. Subtract $\left(13x^2 + x^2y - \dfrac{1}{4}\right)$ from $\left(3x^2 - 4x^2y - \dfrac{1}{4}\right)$.

MIXED PRACTICE

Perform indicated operations and simplify. See Examples 6 through 10.

55. $(-3x + 8) + (-3x^2 + 3x - 5)$

56. $(-5y^2 - 2y + 4) + (3y + 7)$

57. $(5y^4 - 7y^2 + x^2 - 3) + (-3y^4 + 2y^2 + 4)$

58. $(8x^4 - 14x^2 + x + 6) + (-12x^6 - 21x^4 - 9x^2)$

59. $(4x^2 - 6x + 2) - (-x^2 + 3x + 5)$

60. $(7x^2 + x + 1) - (6x^2 + x - 1)$

61. $(5x^2 + x + 9) - (2x^2 - 9)$

62. $(4x - 4) - (-x - 4)$

63. $(5x - 11) + (-x - 2)$

64. $(3x^2 - 2x) + (-5x^2 - 9x)$

65. $(3x^3 - b + 2a - 6) + (-4x^3 + b + 6a - 6)$

66. $(9y^3 - a + 7b - 3) + (-2y^3 + a + 6b - 8)$

67. $(14ab - 10a^2b + 6b^2) - (18a^2 - 20a^2b - 6b^2)$

68. $(13x^2 - 26x^2y^2 + 4) - (19x^2 + x^2y^2 - 11)$

69. $3x^2 + 15x + 8$
  $+(2x^2 + 7x + 8)$

70. $9x^2 + 9x - 4$
  $+(7x^2 - 3x - 4)$

71. $(7x^2 - 5) + (-3x^2 - 2) - (4x^2 - 7)$

72. $(9y^2 - 3) + (-4y^2 + 1) - (5y^2 - 2)$

73. $(-3 + 4x^2 + 7xy^2) + (2x^3 - x^2 + xy^2)$

74. $(-3x^2y + 4) + (-7x^2y - 8y)$

75. $3x^2 - 4x + 8$
  $-\quad (5x - 7)$

76. $-3x^2 - 5x + 1$
  $-\quad (3x + 12)$

77. Subtract $(3x + 7)$ from the sum of $(7x^2 + 4x + 9)$ and $(8x^2 + 7x - 8)$.

78. Subtract $(9x + 8)$ from the sum of $(3x^2 - 2x - x^3 + 2)$ and $(5x^2 - 8x - x^3 + 4)$.

79. $\left(\dfrac{2}{3}x^2 - \dfrac{1}{6}x + \dfrac{5}{6}\right) - \left(\dfrac{1}{3}x^2 + \dfrac{5}{6}x - \dfrac{1}{6}\right)$

80. $\left(\dfrac{3}{16}x^2 + \dfrac{5}{8}x - \dfrac{1}{4}\right) - \left(\dfrac{5}{16}x^2 - \dfrac{3}{8}x + \dfrac{3}{4}\right)$

Use the information below to solve Exercises 81 and 82.
The surface area of a rectangular box is given by the polynomial function

$$SA = 2HL + 2LW + 2HW$$

and is measured in square units. In business, surface area is often calculated to help determine cost of materials.

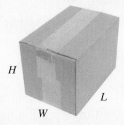

81. A rectangular box is to be constructed to hold a new camcorder. The box is to have dimensions 5 inches by 4 inches by 9 inches. Find the surface area of the box.

82. Suppose it has been determined that a box of dimensions 4 inches by 4 inches by 8.5 inches can be used to contain the camcorder in Exercise 81. Find the surface area of this box and calculate the square inches of material saved by using this box instead of the box in Exercise 81.

83. A projectile is fired upward from the ground with an initial velocity of 300 feet per second. Neglecting air resistance, the height of the projectile at any time t can be described by the polynomial function $P(t) = -16t^2 + 300t$. Find the height of the projectile at each given time.
  a. $t = 1$ second
  b. $t = 2$ seconds
  c. $t = 10$ seconds
  d. $t = 14$ seconds

  e. Explain why the height increases and then decreases as time passes.
  f. Approximate (to the nearest second) how long before the object hits the ground.

84. A worker at the Hoover Dam bypass bridge was spending his lunch break tossing a football. He threw the ball upward with an initial velocity of 10 feet per second but missed the ball, and it went over his head and down toward the river. The height of the football above the Colorado River at any time t can be described by the polynomial function $P(t) = -16t^2 + 10t + 910$. Find the height of the football at each given time.

  a. $t = 0$ seconds
  b. $t = \dfrac{1}{2}$ second
  c. $t = 3$ seconds
  d. $t = 5$ seconds
  e. Explain why the height increases and then decreases as time passes.
  f. Approximate (to the nearest second) how long before the football lands in the river.

85. The polynomial function $P(x) = 45x - 100,000$ models the relationship between the number of computer briefcases x that a company sells and the profit the company makes, $P(x)$. Find $P(4000)$, the profit from selling 4000 computer briefcases.

86. The total cost (in dollars) for MCD, Inc., Manufacturing Company to produce x blank CDs per week is given by the polynomial function $C(x) = 0.2x + 5000$. Find the total cost of producing 20,000 CDs per week.

87. The total revenues (in dollars) for an art supply company to sell x boxes of colored pencils per week over the Internet is given by the polynomial function $R(x) = 11x$. Find the total revenue from selling 1500 boxes of colored pencils.

88. The total revenues (in dollars) for MCD, Inc., Manufacturing Company to sell x blank CDs per week is given by the polynomial function $R(x) = 0.9x$. Find the total revenue from selling 20,000 CDs per week.

Match each equation with its graph. See Example 11.

89. $f(x) = 3x^2 - 2$

90. $h(x) = 5x^3 - 6x + 2$

91. $g(x) = -2x^3 - 3x^2 + 3x - 2$

92. $F(x) = -2x^2 - 6x + 2$

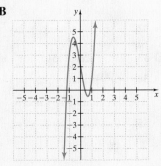

C

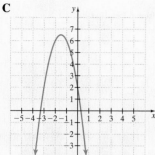

D

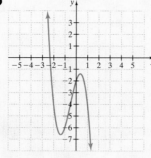

REVIEW AND PREVIEW

Multiply. See Section 1.4.

93. $5(3x - 2)$

94. $-7(2z - 6y)$

95. $-2(x^2 - 5x + 6)$

96. $5(-3y^2 - 2y + 7)$

CONCEPT EXTENSIONS

Solve. See the Concept Checks in this section.

97. Which polynomial(s) is the opposite of $8x - 6$?

   a. $-(8x - 6)$ **b.** $8x + 6$

   c. $-8x + 6$ **d.** $-8x - 6$

98. Which polynomial(s) is the opposite of $-y^5 + 10y^3 - 2.3$?

   a. $y^5 + 10y^3 + 2.3$ **b.** $-y^5 - 10y^3 - 2.3$

   c. $y^5 + 10y^3 - 2.3$ **d.** $y^5 - 10y^3 + 2.3$

99. Correct the subtraction.

$$(12x - 1.7) - (15x + 6.2) = 12x - 1.7 - 15x + 6.2$$
$$= -3x + 4.5$$

100. Correct the addition.

$$(12x - 1.7) + (15x + 6.2) = 12x - 1.7 + 15x + 6.2$$
$$= 27x + 7.9$$

101. Write a function, $P(x)$, so that $P(0) = 7$.

102. Write a function, $R(x)$, so that $R(1) = 2$.

103. In your own words, describe how to find the degree of a term.

104. In your own words, describe how to find the degree of a polynomial.

Perform the indicated operations.

105. $(4x^{2a} - 3x^a + 0.5) - (x^{2a} - 5x^a - 0.2)$

106. $(9y^{5a} - 4y^{3a} + 1.5y) - (6y^{5a} - y^{3a} + 4.7y)$

107. $(8x^{2y} - 7x^y + 3) + (-4x^{2y} + 9x^y - 14)$

108. $(14z^{5x} + 3z^{2x} + z) - (2z^{5x} - 10z^{2x} + 3z)$

Find each perimeter.

△ **109.**

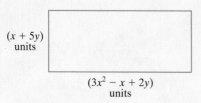

$(x + 5y)$ units

$(3x^2 - x + 2y)$ units

△ **110.**

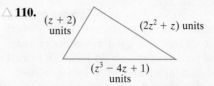

$(z + 2)$ units $(2z^2 + z)$ units

$(z^3 - 4z + 1)$ units

If $P(x) = 3x + 3$, $Q(x) = 4x^2 - 6x + 3$, and $R(x) = 5x^2 - 7$, find the following.

111. $P(x) + Q(x)$

112. $R(x) + P(x)$

113. $Q(x) - R(x)$

114. $P(x) - Q(x)$

115. $2[Q(x)] - R(x)$

116. $-5[P(x)] - Q(x)$

117. $3[R(x)] + 4[P(x)]$

118. $2[Q(x)] + 7[R(x)]$

*If $P(x)$ is the polynomial given, find **a.** $P(a)$, **b.** $P(-x)$, and **c.** $P(x + h)$.*

119. $P(x) = 2x - 3$

120. $P(x) = 8x + 3$

121. $P(x) = 4x$

122. $P(x) = -4x$

123. $P(x) = 4x - 1$

124. $P(x) = 3x - 2$

125. The function $f(x) = 0.19x^2 + 5.67x + 43.7$ can be used to approximate the amount of restaurant food-and-drink sales, where x is the number of years since 1970 and $f(x)$ or y is the sales (in billions of dollars.)

   a. Approximate the restaurant food-and-drink sales in 2005.

   b. Approximate the restaurant food-and-drink sales in 2010.

   c. Use this function to estimate the restaurant food-and-drink sales in 2015.

   d. From parts (a), (b), and (c), determine whether the restaurant food-and-drink sales is increasing at a steady rate. Explain why or why not.

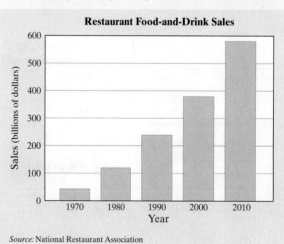

Restaurant Food-and-Drink Sales

Source: National Restaurant Association

126. The function $f(x) = 0.0007x^2 + 0.24x + 7.98$ can be used to approximate the total cheese production in the United States from 2000 to 2009, where x is the number of years after 2000 and y is pounds of cheese (in billions). Round answers to the nearest hundredth of a billion. (*Source*: National Agricultural Statistics Service, USDA)

a. Approximate the number of pounds of cheese produced in the United States in 2000.

b. Approximate the number of pounds of cheese produced in the United States in 2005.

c. Use this function to estimate the pounds of cheese produced in the United States in 2015.

d. From parts (a), (b), and (c), determine whether the number of pounds of cheese produced in the United States is increasing at a steady rate. Explain why or why not.

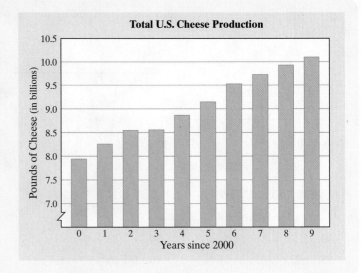

Total U.S. Cheese Production

5.4 | Multiplying Polynomials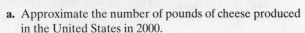

OBJECTIVES

1 Multiply Two Polynomials.

2 Multiply Binomials.

3 Square Binomials.

4 Multiply the Sum and Difference of Two Terms.

5 Multiply Three or More Polynomials.

6 Evaluate Polynomial Functions.

> ▶ **Helpful Hint**
> See Sections 5.1 and 5.2 to review exponential expressions further.

OBJECTIVE

1 Multiplying Two Polynomials

Properties of real numbers and exponents are used continually in the process of multiplying polynomials. To multiply monomials, for example, we apply the commutative and associative properties of real numbers and the product rule for *exponents*.

EXAMPLE 1 Multiply.

a. $(2x^3)(5x^6)$ **b.** $(7y^4z^4)(-xy^{11}z^5)$

Solution Group like bases and apply the product rule for exponents.

a. $(2x^3)(5x^6) = 2(5)(x^3)(x^6) = 10x^9$

b. $(7y^4z^4)(-xy^{11}z^5) = 7(-1)x(y^4y^{11})(z^4z^5) = -7xy^{15}z^9$ □

PRACTICE

1 Multiply.

a. $(3x^4)(2x^2)$ **b.** $(-5m^4np^3)(-8mnp^5)$

To multiply a monomial by a polynomial other than a monomial, we use an expanded form of the distributive property.

$$a(b + c + d + \cdots + z) = ab + ac + ad + \cdots + az$$

Notice that the monomial a is multiplied by each term of the polynomial.

EXAMPLE 2 Multiply.

a. $2x(5x - 4)$ **b.** $-3x^2(4x^2 - 6x + 1)$ **c.** $-xy(7x^2y + 3xy - 11)$

Solution Apply the distributive property.

a. $2x(5x - 4) = 2x(5x) + 2x(-4)$ Use the distributive property.

              $= 10x^2 - 8x$ Multiply.

b. $-3x^2(4x^2 - 6x + 1) = -3x^2(4x^2) + (-3x^2)(-6x) + (-3x^2)(1)$
$$= -12x^4 + 18x^3 - 3x^2$$

c. $-xy(7x^2y + 3xy - 11) = -xy(7x^2y) + (-xy)(3xy) + (-xy)(-11)$
$$= -7x^3y^2 - 3x^2y^2 + 11xy$$ □

PRACTICE
2   Multiply.

a. $3x(7x - 1)$ **b.** $-5a^2(3a^2 - 6a + 5)$ **c.** $-mn^3(5m^2n^2 + 2mn - 5m)$

To multiply any two polynomials, we can use the following.

> **Multiplying Two Polynomials**
>
> To multiply any two polynomials, use the distributive property and multiply each term of one polynomial by each term of the other polynomial. Then combine any like terms.

✓ CONCEPT CHECK
Find the error:

$$4x(x - 5) + 2x$$
$$= 4x(x) + 4x(-5) + 4x(2x)$$
$$= 4x^2 - 20x + 8x^2$$
$$= 12x^2 - 20x$$

EXAMPLE 3   Multiply and simplify the product if possible.

a. $(x + 3)(2x + 5)$ **b.** $(2x - 3)(5x^2 - 6x + 7)$

Solution

a. Multiply each term of $(x + 3)$ by $(2x + 5)$.

$(x + 3)(2x + 5) = x(2x + 5) + 3(2x + 5)$ Apply the distributive property.
$$= 2x^2 + 5x + 6x + 15$$ Apply the distributive property again.
$$= 2x^2 + 11x + 15$$ Combine like terms.

b. Multiply each term of $(2x - 3)$ by each term of $(5x^2 - 6x + 7)$.

$(2x - 3)(5x^2 - 6x + 7) = 2x(5x^2 - 6x + 7) + (-3)(5x^2 - 6x + 7)$

$$= 10x^3 - 12x^2 + 14x - 15x^2 + 18x - 21$$
$$= 10x^3 - 27x^2 + 32x - 21$$ Combine like terms. □

PRACTICE
3   Multiply and simplify the product if possible.

a. $(x + 5)(2x + 3)$ **b.** $(3x - 1)(x^2 - 6x + 2)$

Answer to Concept Check:
$4x(x - 5) + 2x$
$= 4x(x) + 4x(-5) + 2x$
$= 4x^2 - 20x + 2x$
$= 4x^2 - 18x$

Sometimes polynomials are easier to multiply vertically, in the same way we multiply real numbers. When multiplying vertically, we line up like terms in the **partial products** vertically. This makes combining like terms easier.

EXAMPLE 4 Multiply vertically $(4x^2 + 7)(x^2 + 2x + 8)$.

Solution

$$
\begin{array}{r}
x^2 + 2x + 8 \\
4x^2 + 7 \\
\hline
7x^2 + 14x + 56 \\
4x^4 + 8x^3 + 32x^2 \\
\hline
4x^4 + 8x^3 + 39x^2 + 14x + 56
\end{array}
$$

$7(x^2 + 2x + 8)$

$4x^2(x^2 + 2x + 8)$

Combine like terms.   □

PRACTICE
4 Multiply vertically: $(3x^2 + 2)(x^2 - 4x - 5)$. ■

OBJECTIVE
2 **Multiplying Binomials** ▶

When multiplying a binomial by a binomial, we can use a special order of multiplying terms, called the **FOIL** order. The letters of FOIL stand for "**F**irst-**O**uter-**I**nner-**L**ast." To illustrate this method, let's multiply $(2x - 3)$ by $(3x + 1)$.

Multiply the **F**irst terms of each binomial. $(2x - 3)(3x + 1)$ $\mathbf{F}$ $2x(3x) = 6x^2$

Multiply the **O**uter terms of each binomial. $(2x - 3)(3x + 1)$ $\mathbf{O}$ $2x(1) = 2x$

Multiply the **I**nner terms of each binomial. $(2x - 3)(3x + 1)$ $\mathbf{I}$ $-3(3x) = -9x$

Multiply the **L**ast terms of each binomial. $(2x - 3)(3x + 1)$ $\mathbf{L}$ $-3(1) = -3$
  Combine like terms.

$$6x^2 + 2x - 9x - 3 = 6x^2 - 7x - 3$$

EXAMPLE 5 Use the FOIL order to multiply $(x - 1)(x + 2)$.

Solution

$$
\begin{array}{ccccccc}
\text{First} & & \text{Outer} & & \text{Inner} & & \text{Last} \\
\downarrow & & \downarrow & & \downarrow & & \downarrow
\end{array}
$$

$$
\begin{aligned}
(x - 1)(x + 2) &= x \cdot x + 2 \cdot x + (-1)x + (-1)(2) \\
&= x^2 + 2x - x - 2 \\
&= x^2 + x - 2 \quad \text{Combine like terms.} \quad \square
\end{aligned}
$$

PRACTICE
5 Use the FOIL order to multiply $(x - 5)(x + 3)$. ■

EXAMPLE 6 Multiply.

a. $(2x - 7)(3x - 4)$ **b.** $(3x^2 + y)(5x^2 - 2y)$

Solution

$$
\begin{array}{cccc}
\text{First} & \text{Outer} & \text{Inner} & \text{Last} \\
\downarrow & \downarrow & \downarrow & \downarrow
\end{array}
$$

$$
\begin{aligned}
\textbf{a. } (2x - 7)(3x - 4) &= 2x(3x) + 2x(-4) + (-7)(3x) + (-7)(-4) \\
&= 6x^2 - 8x - 21x + 28 \\
&= 6x^2 - 29x + 28
\end{aligned}
$$

$$
\begin{array}{cccc}
\text{F} & \text{O} & \text{I} & \text{L} \\
\downarrow & \downarrow & \downarrow & \downarrow
\end{array}
$$

$$
\begin{aligned}
\textbf{b. } (3x^2 + y)(5x^2 - 2y) &= 15x^4 - 6x^2y + 5x^2y - 2y^2 \\
&= 15x^4 - x^2y - 2y^2 \quad \square
\end{aligned}
$$

PRACTICE

6 Multiply.

a. $(3x - 5)(2x - 7)$

b. $(2x^2 - 3y)(4x^2 + y)$

OBJECTIVE

3 Squaring Binomials

The **square of a binomial** is a special case of the product of two binomials. By the FOIL order for multiplying two binomials, we have

$$(a + b)^2 = (a + b)(a + b)$$

$$\begin{array}{cccc} F & O & I & L \\ \downarrow & \downarrow & \downarrow & \downarrow \end{array}$$

$$= a^2 + ab + ba + b^2$$

$$= a^2 + 2ab + b^2$$

This product can be visualized geometrically by analyzing areas.

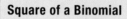

Area of larger square: $(a + b)^2$

Sum of areas of smaller rectangles: $a^2 + 2ab + b^2$

Thus, $(a + b)^2 = a^2 + 2ab + b^2$

The same pattern occurs for the square of a difference. In general,

---

Square of a Binomial

$$(a + b)^2 = a^2 + 2ab + b^2 \qquad (a - b)^2 = a^2 - 2ab + b^2$$

In other words, a binomial squared is the sum of the first term squared, twice the product of both terms, and the second term squared.

EXAMPLE 7 Multiply.

a. $(x + 5)^2$ **b.** $(x - 9)^2$ **c.** $(3x + 2z)^2$ **d.** $(4m^2 - 3n)^2$

Solution

a.
$$(a + b)^2 = a^2 + 2 \cdot a \cdot b + b^2$$
$$\downarrow \quad \downarrow \quad\quad \downarrow \quad \downarrow\downarrow\downarrow \quad\quad \downarrow$$
$$(x + 5)^2 = x^2 + 2 \cdot x \cdot 5 + 5^2 = x^2 + 10x + 25$$

b. $(x - 9)^2 = x^2 - 2 \cdot x \cdot 9 + 9^2 = x^2 - 18x + 81$

c. $(3x + 2z)^2 = (3x)^2 + 2(3x)(2z) + (2z)^2 = 9x^2 + 12xz + 4z^2$

d. $(4m^2 - 3n)^2 = (4m^2)^2 - 2(4m^2)(3n) + (3n)^2 = 16m^4 - 24m^2n + 9n^2$ □

PRACTICE

7 Multiply.

a. $(x + 6)^2$ **b.** $(x - 2)^2$ **c.** $(3x + 5y)^2$ **d.** $(3x^2 - 8b)^2$

▶ Helpful Hint

Note that $(a + b)^2 = a^2 + 2ab + b^2$, **not** $a^2 + b^2$. Also,

$$(a - b)^2 = a^2 - 2ab + b^2, \quad \textbf{not} \quad a^2 - b^2.$$

4 Multiplying the Sum and Difference of Two Terms

Another special product applies to the sum and difference of the same two terms. Multiply $(a + b)(a - b)$ to see a pattern.

$$(a + b)(a - b) = a^2 - ab + ba - b^2$$
$$= a^2 - b^2$$

Product of the Sum and Difference of Two Terms

$$(a + b)(a - b) = a^2 - b^2$$

The product of the sum and difference of the same two terms is the difference of the first term squared and the second term squared.

EXAMPLE 8 Multiply.

a. $(x - 3)(x + 3)$ **b.** $(4y + 1)(4y - 1)$

c. $(x^2 + 2y)(x^2 - 2y)$ **d.** $\left(3m^2 - \dfrac{1}{2}\right)\left(3m^2 + \dfrac{1}{2}\right)$

Solution

a. $(a + b)(a - b) = a^2 - b^2$

$(x + 3)(x - 3) = x^2 - 3^2 = x^2 - 9$

b. $(4y + 1)(4y - 1) = (4y)^2 - 1^2 = 16y^2 - 1$

c. $(x^2 + 2y)(x^2 - 2y) = (x^2)^2 - (2y)^2 = x^4 - 4y^2$

d. $\left(3m^2 - \dfrac{1}{2}\right)\left(3m^2 + \dfrac{1}{2}\right) = (3m^2)^2 - \left(\dfrac{1}{2}\right)^2 = 9m^4 - \dfrac{1}{4}$ □

PRACTICE

8 Multiply.

a. $(x - 7)(x + 7)$ **b.** $(2a + 5)(2a - 5)$

c. $\left(5x^2 + \dfrac{1}{4}\right)\left(5x^2 - \dfrac{1}{4}\right)$ **d.** $(a^3 - 4b^2)(a^3 + 4b^2)$

EXAMPLE 9 Multiply $[3 + (2a + b)]^2$.

Solution Think of 3 as the first term and $(2a + b)$ as the second term, and apply the method for squaring a binomial.

$$[a \quad + \quad b \quad]^2 = a^2 + 2(a) \cdot \quad b \quad + \quad b^2$$
$$[3 + (2a + b)]^2 = 3^2 + 2(3)(2a + b) + (2a + b)^2$$
$$= 9 + 6(2a + b) + (2a + b)^2$$
$$= 9 + 12a + 6b + (2a)^2 + 2(2a)(b) + b^2 \quad \text{Square } (2a + b).$$
$$= 9 + 12a + 6b + 4a^2 + 4ab + b^2 \quad □$$

PRACTICE

9 Multiply $[2 + (3x - y)]^2$.

EXAMPLE 10   Multiply $[(5x - 2y) - 1][(5x - 2y) + 1]$.

Solution   Think of $(5x - 2y)$ as the first term and 1 as the second term and apply the method for the product of the sum and difference of two terms.

$$\overbrace{(a \quad\quad - b)}\ \overbrace{(a \quad\quad + b)} = \overbrace{a^2} \quad - b^2$$
$$[\overbrace{(5x - 2y)} - 1][\overbrace{(5x - 2y)} + 1] = \overbrace{(5x - 2y)^2} - 1^2$$
$$= (5x)^2 - 2(5x)(2y) + (2y)^2 - 1 \quad \text{Square}$$
$$= 25x^2 - 20xy + 4y^2 - 1 \quad\quad\quad (5x - 2y). \ \square$$

PRACTICE
10   Multiply $[(3x - y) - 5][(3x - y) + 5]$.

OBJECTIVE
5   Multiplying Three or More Polynomials

To multiply three or more polynomials, more than one method may be needed.

EXAMPLE 11   Multiply: $(x - 3)(x + 3)(x^2 - 9)$

Solution   We multiply the first two binomials, the sum and difference of two terms. Then we multiply the resulting two binomials, the square of a binomial.

$$(x - 3)(x + 3)(x^2 - 9) = (x^2 - 9)(x^2 - 9) \quad \text{Multiply } (x - 3)(x + 3).$$
$$= (x^2 - 9)^2$$
$$= x^4 - 18x^2 + 81 \quad \text{Square } (x^2 - 9). \quad \square$$

PRACTICE
11   Multiply $(x + 4)(x - 4)(x^2 - 16)$.

OBJECTIVE
6   Evaluating Polynomial Functions

Our work in multiplying polynomials is often useful in evaluating polynomial functions.

EXAMPLE 12   If $f(x) = x^2 + 5x - 2$, find $f(a + 1)$.

Solution   To find $f(a + 1)$, replace x with the expression $a + 1$ in the polynomial function $f(x)$.

$$f(x) = x^2 + 5x - 2$$
$$f(a + 1) = (a + 1)^2 + 5(a + 1) - 2$$
$$= a^2 + 2a + 1 + 5a + 5 - 2$$
$$= a^2 + 7a + 4 \quad \square$$

PRACTICE
12   If $f(x) = x^2 - 3x + 5$, find $f(h + 1)$.

Graphing Calculator Explorations

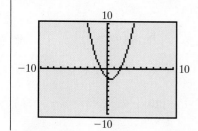

In the previous section, we used a graphing calculator to visualize addition and subtraction of polynomials in one variable. In this section, the same method is used to visualize multiplication of polynomials in one variable. For example, to see that

$$(x - 2)(x + 1) = x^2 - x - 2,$$

graph both $Y_1 = (x - 2)(x + 1)$ and $Y_2 = x^2 - x - 2$ on the same screen and see whether their graphs coincide.

By tracing along both graphs, we see that the graphs of Y_1 and Y_2 appear to coincide, and thus $(x - 2)(x + 1) = x^2 - x - 2$ appears to be correct.

Multiply. Then use a graphing calculator to visualize the results.

1. $(x + 4)(x - 4)$ **2.** $(x + 3)(x + 3)$

3. $(3x - 7)^2$ **4.** $(5x - 2)^2$

5. $(5x + 1)(x^2 - 3x - 2)$ **6.** $(7x + 4)(2x^2 + 3x - 5)$

Vocabulary, Readiness & Video Check

Use the choices to fill in each blank.

1. $(6x^3)\left(\dfrac{1}{2}x^3\right) = $ _____

    a. $3x^3$ **b.** $3x^6$ **c.** $10x^6$ **d.** $\dfrac{13}{2}x^6$

2. $(x + 7)^2 = $ _____

    a. $x^2 + 49$ **b.** $x^2 - 49$ **c.** $x^2 + 14x + 49$ **d.** $x^2 + 7x + 49$

3. $(x + 7)(x - 7) = $ _____

    a. $x^2 + 49$ **b.** $x^2 - 49$ **c.** $x^2 + 14x - 49$ **d.** $x^2 + 7x - 49$

4. The product of $(3x - 1)(4x^2 - 2x + 1)$ is a polynomial of degree _____.

    a. 3   **b.** 12   **c.** $12x^3$ **d.** 2

5. If $f(x) = x^2 + 1$ then $f(a + 1) = $ _____

    a. $(a + 1)^2$ **b.** $a + 1$ **c.** $(a + 1)^2 + (a + 1)$ **d.** $(a + 1)^2 + 1$

6. $[x + (2y + 1)]^2 = $ _____

    a. $[x + (2y + 1)][x - (2y + 1)]$ **b.** $[x + (2y + 1)][x + (2y + 1)]$ **c.** $[x + (2y + 1)][x + (2y - 1)]$

Martin-Gay Interactive Videos

See Video 5.4

Watch the section lecture video and answer the following questions.

OBJECTIVE 1

7. From Examples 1 and 2, what property is key when multiplying polynomials? What exponent rule is often used when applying this property?

OBJECTIVE 2

8. From Examples 3 and 4, for what type of multiplication of polynomials is the FOIL method used?

OBJECTIVE 3

9. Name at least one other method you can use to multiply Example 5.

OBJECTIVE 4

10. From Example 6, why does multiplying the sum and difference of the same two terms always give you a binomial answer?

OBJECTIVE 5

11. From Example 9, how do you multiply three or more polynomials?

OBJECTIVE 6

12. From Example 10, can a polynomial function be evaluated only for a numerical value?

5.4 Exercise Set

MyMathLab®

Multiply. See Examples 1 through 4.

1. $(-4x^3)(3x^2)$ **2.** $(-6a)(4a)$

3. $(8.6a^4b^5c)(10ab^3c^2)$ **4.** $(7.1xy^2z^{11})(10xy^7z)$

5. $3x(4x + 7)$ **6.** $5x(6x - 4)$

7. $-6xy(4x + y)$ **8.** $-8y(6xy + 4x)$

9. $-4ab(xa^2 + ya^2 - 3)$ **10.** $-6b^2z(z^2a + baz - 3b)$

11. $(x - 3)(2x + 4)$ **12.** $(y + 5)(3y - 2)$

▶ **13.** $(2x + 3)(x^3 - x + 2)$ **14.** $(a + 2)(3a^2 - a + 5)$

15. $\begin{array}{r} 3x - 2 \\ \times\ 5x + 1 \\ \hline \end{array}$ **16.** $\begin{array}{r} 2z - 4 \\ \times\ 6z - 2 \\ \hline \end{array}$

17. $\begin{array}{r} 3m^2 + 2m - 1 \\ \times\qquad 5m + 2 \\ \hline \end{array}$ **18.** $\begin{array}{r} 2x^2 - 3x - 4 \\ \times\qquad 4x + 5 \\ \hline \end{array}$

19. $-6a^2b^2(5a^2b^2 - 6a - 6b)$ **20.** $7x^2y^3(-3ax - 4xy + z)$

Multiply the binomials. See Examples 5 and 6.

▶ **21.** $(x - 3)(x + 4)$ **22.** $(c - 3)(c + 1)$

23. $(5x - 8y)(2x - y)$ **24.** $(2n - 9m)(n - 7m)$

25. $\left(4x + \dfrac{1}{3}\right)\left(4x - \dfrac{1}{2}\right)$ **26.** $\left(4y - \dfrac{1}{3}\right)\left(3y - \dfrac{1}{8}\right)$

27. $(5x^2 - 2y^2)(x^2 - 3y^2)$ **28.** $(4x^2 - 5y^2)(x^2 - 2y^2)$

Use special products to multiply. See Examples 7 and 8.

29. $(x + 4)^2$ **30.** $(x - 5)^2$

▶ **31.** $(6y - 1)(6y + 1)$ **32.** $(7x - 9)(7x + 9)$

33. $(3x - y)^2$ **34.** $(4x + z)^2$

▶ **35.** $(7ab + 3c)(7ab - 3c)$ **36.** $(3xy - 2b)(3xy + 2b)$

37. $\left(3x + \dfrac{1}{2}\right)\left(3x - \dfrac{1}{2}\right)$ **38.** $\left(2x - \dfrac{1}{3}\right)\left(2x + \dfrac{1}{3}\right)$

Use special products to multiply. See Examples 9 and 10.

39. $[3 + (4b + 1)]^2$ **40.** $[5 - (3b - 3)]^2$

▶ **41.** $[(2s - 3) - 1][(2s - 3) + 1]$

42. $[(2y + 5) + 6][(2y + 5) - 6]$

43. $[(xy + 4) - 6]^2$ **44.** $[(2a^2 + 4a) + 1]^2$

Multiply. See Example 11.

▶ **45.** $(x + y)(x - y)(x^2 - y^2)$ **46.** $(z - y)(z + y)(z^2 - y^2)$

47. $(x - 2)^4$ **48.** $(x - 1)^4$

49. $(x - 5)(x + 5)(x^2 + 25)$ **50.** $(x + 3)(x - 3)(x^2 + 9)$

MIXED PRACTICE

Multiply. See Examples 1 through 11.

51. $-8a^2b(3b^2 - 5b + 20)$ **52.** $-9xy^2(3x^2 - 2x + 10)$

▶ **53.** $(6x + 1)^2$ **54.** $(4x + 7)^2$

55. $(5x^3 + 2y)(5x^3 - 2y)$ **56.** $(3x^4 + 2y)(3x^4 - 2y)$

57. $(2x^3 + 5)(5x^2 + 4x + 1)$ **58.** $(3y^3 - 1)(3y^3 - 6y + 1)$

59. $(3x^2 + 2x - 1)^2$ **60.** $(4x^2 + 4x - 4)^2$

▶ **61.** $(3x - 1)(x + 3)$ **62.** $(5d - 3)(d + 6)$

63. $(3x^4 + 1)(3x^2 + 5)$ **64.** $(4x^3 - 5)(5x^2 + 6)$

65. $(3x + 1)^2$ **66.** $(4x + 6)^2$

67. $(3b - 6y)(3b + 6y)$ **68.** $(2x - 4y)(2x + 4y)$

69. $(7x - 3)(7x + 3)$ **70.** $(4x + 1)(4x - 1)$

71. $\begin{array}{r} 3x^2 + 4x - 4 \\ \times\qquad 3x + 6 \\ \hline \end{array}$ **72.** $\begin{array}{r} 6x^2 + 2x - 1 \\ \times\qquad 3x - 6 \\ \hline \end{array}$

73. $(4x^2 - 2x + 5)(3x + 1)$ **74.** $(5x^2 - x - 2)(2x - 1)$

75. $[(xy + 4) - 6]^2$ **76.** $[(2a^2 + 4) - 1]^2$

77. $(11a^2 + 1)(2a + 1)$ **78.** $(13x^2 + 1)(3x + 1)$

79. $\left(\dfrac{2}{3}n - 2\right)\left(\dfrac{1}{2}n - 9\right)$ **80.** $\left(\dfrac{3}{5}y - 6\right)\left(\dfrac{1}{3}y - 10\right)$

81. $(3x + 1)(3x - 1)(2y + 5x)$

82. $(2a + 1)(2a - 1)(6a + 7b)$

If $f(x) = x^2 - 3x$, find the following. See Example 12.

83. $f(a)$ **84.** $f(c)$

▶ **85.** $f(a + h)$ **86.** $f(a + 5)$

87. $f(b - 2)$ **88.** $f(a - b)$

REVIEW AND PREVIEW

Use the slope-intercept form of a line, $y = mx + b$, to find the slope of each line. See Section 3.4.

89. $y = -2x + 7$ **90.** $y = \dfrac{3}{2}x - 1$

91. $3x - 5y = 14$ **92.** $x + 7y = 2$

Use the vertical line test to determine whether the following are graphs of functions. See Section 3.2.

93.

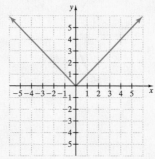

94.

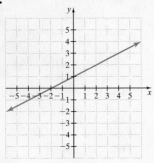

CONCEPT EXTENSIONS

Solve. See the Concept Check in this section.

95. Find the error: $7y(3z - 2) + 1$
$$= 21yz - 14y + 7y$$
$$= 21yz - 7y$$

96. Find the error: $2x + 3x(12 - x)$
$$= 5x(2 - x)$$
$$= 60x - 5x^2$$

97. Explain how to multiply a polynomial by a polynomial.

98. Explain why $(3x + 2)^2$ does not equal $9x^2 + 4$.

99. If $F(x) = x^2 + 3x + 2$, find
  a. $F(a + h)$
  b. $F(a)$
  c. $F(a + h) - F(a)$

100. If $g(x) = x^2 + 2x + 1$, find
  a. $g(a + h)$
  b. $g(a)$
  c. $g(a + h) - g(a)$

Multiply. Assume that variables represent positive integers.

101. $5x^2y^n(6y^{n+1} - 2)$

102. $-3yz^n(2y^3z^{2n} - 1)$

103. $(x^a + 5)(x^{2a} - 3)$

104. $(x^a + y^{2b})(x^a - y^{2b})$

For Exercises 105 through 108, write the result as a simplified polynomial.

△ **105.** Find the area of the circle. Do not approximate π.

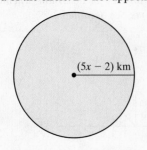

(5x − 2) km

△ **106.** Find the volume of the cylinder. Do not approximate π.

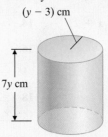

(y − 3) cm

7y cm

Find the area of each shaded region.

△ **107.**

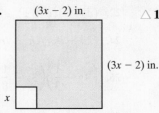

(3x − 2) in.

(3x − 2) in.

x

x

△ **108.**

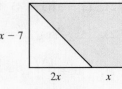

x − 7

2x

x

109. Perform each indicated operation. Explain the difference between the two problems.
  a. $(3x + 5) + (3x + 7)$
  b. $(3x + 5)(3x + 7)$

110. Explain when the FOIL method can be used to multiply polynomials.

If $R(x) = x + 5$, $Q(x) = x^2 - 2$, and $P(x) = 5x$, find the following.

111. $P(x) \cdot R(x)$

112. $P(x) \cdot Q(x)$

113. $[Q(x)]^2$

114. $[R(x)]^2$

115. $R(x) \cdot Q(x)$

116. $P(x) \cdot R(x) \cdot Q(x)$

5.5 | The Greatest Common Factor and Factoring by Grouping

OBJECTIVE

1   Identifying the GCF

Factoring is the reverse process of multiplying. It is the process of writing a polynomial as a product.

$$6x^2 + 13x - 5 = (3x - 1)(2x + 5)$$

(factoring / multiplying)

In the next few sections, we review techniques for factoring polynomials. These techniques are used at the end of this chapter to solve polynomial equations.

To factor a polynomial, we first factor out the **greatest common factor** (GCF) of its terms, using the distributive property. The GCF of a list of terms or monomials is the product of the GCF of the numerical coefficients and each GCF of the powers of a common variable.

Finding the GCF of a List of Monomials

Step 1. Find the GCF of the numerical coefficients.

Step 2. Find the GCF of the variable factors.

Step 3. The product of the factors found in Steps 1 and 2 is the GCF of the monomials.

EXAMPLE 1   Find the GCF of $20x^3y$, $10x^2y^2$, and $35x^3$.

Solution   The GCF of the numerical coefficients 20, 10, and 35 is 5, the largest integer that is a factor of each integer. The GCF of the variable factors x^3, x^2, and x^3 is x^2 because x^2 is the largest factor common to all three powers of x. The variable y is not a common factor because it does not appear in all three monomials. The GCF is thus

$$5 \cdot x^2, \quad \text{or} \quad 5x^2$$

To see this in factored form,

$$20x^3y = 2 \cdot 2 \cdot 5 \cdot x^2 \cdot x \cdot y$$
$$10x^2y^2 = 2 \cdot 5 \cdot x^2 \cdot y^2$$
$$35x^3 = 7 \cdot 5 \cdot x^2 \cdot x$$
$$\textbf{GCF} = 5 \cdot x^2$$

PRACTICE

1   Find the GCF of $32x^4y^2$, $48x^3y$, $24y^2$.

OBJECTIVE

2   Factoring Out the GCF of a Polynomial's Terms

A first step in factoring polynomials is to use the distributive property and write the polynomial as a product of the GCF of its monomial terms and a simpler polynomial. This is called **factoring out** the GCF.

EXAMPLE 2   Factor.

a. $8x^2 + 4$ **b.** $5y - 2z^4$ **c.** $6x^2 - 3x^3 + 12x^4$

Solution

a. The GCF of terms $8x^2$ and 4 is 4.

$$8x^2 + 4 = 4 \cdot 2x^2 + 4 \cdot 1 \quad \text{Factor out 4 from each term.}$$
$$= 4(2x^2 + 1) \quad \text{Apply the distributive property.}$$

The factored form of $8x^2 + 4$ is $4(2x^2 + 1)$. To check, multiply $4(2x^2 + 1)$ to see that the product is $8x^2 + 4$.

b. There is no common factor of the terms $5y$ and $-2z^4$ other than 1 (or -1).

c. The greatest common factor of $6x^2$, $-3x^3$, and $12x^4$ is $3x^2$. Thus,

$$6x^2 - 3x^3 + 12x^4 = 3x^2 \cdot 2 - 3x^2 \cdot x + 3x^2 \cdot 4x^2$$
$$= 3x^2(2 - x + 4x^2) \qquad \square$$

PRACTICE

2 Factor.

a. $6x^2 + 9 + 15x$ **b.** $3x - 8y^3$ **c.** $8a^4 - 2a^3$

> ▶ **Helpful Hint**
> To verify that the GCF has been factored out correctly, multiply the factors together and see that their product is the original polynomial.

EXAMPLE 3 Factor $17x^3y^2 - 34x^4y^2$.

Solution The GCF of the two terms is $17x^3y^2$, which we factor out of each term.

$$17x^3y^2 - 34x^4y^2 = 17x^3y^2 \cdot 1 - 17x^3y^2 \cdot 2x$$
$$= 17x^3y^2(1 - 2x) \qquad \square$$

PRACTICE

3 Factor $64x^5y^2 - 8x^3y^2$.

> ▶ **Helpful Hint**
> If the GCF happens to be one of the terms in the polynomial, a factor of 1 will remain for this term when the GCF is factored out.
> For example, in the polynomial $21x^2 + 7x$, the GCF of $21x^2$ and $7x$ is $7x$, so
> $$21x^2 + 7x = 7x \cdot 3x + 7x \cdot 1 = 7x(3x + 1)$$

> ✔**CONCEPT CHECK**
> Which factorization of $12x^2 + 9x - 3$ is correct?
> **a.** $3(4x^2 + 3x + 1)$ **b.** $3(4x^2 + 3x - 1)$ **c.** $3(4x^2 + 3x - 3)$ **d.** $3(4x^2 + 3x)$

EXAMPLE 4 Factor $-3x^3y + 2x^2y - 5xy$.

Solution Two possibilities are shown for factoring this polynomial. First, the common factor xy is factored out.

$$-3x^3y + 2x^2y - 5xy = xy(-3x^2 + 2x - 5)$$

Also, the common factor $-xy$ can be factored out as shown.

$$-3x^3y + 2x^2y - 5xy = -xy(3x^2) + (-xy)(-2x) + (-xy)(5)$$
$$= -xy(3x^2 - 2x + 5)$$

Both of these alternatives are correct. $\qquad \square$

PRACTICE

4 Factor $-9x^4y^2 + 5x^2y^2 + 7xy^2$.

EXAMPLE 5   Factor $2(x - 5) + 3a(x - 5)$.

Solution   The greatest common factor is the binomial factor $(x - 5)$.

$$2(x - 5) + 3a(x - 5) = (x - 5)(2 + 3a)$$ □

PRACTICE
5   Factor $3(x + 4) + 5b(x + 4)$.

EXAMPLE 6   Factor $7x(x^2 + 5y) - (x^2 + 5y)$.

Solution $7x(x^2 + 5y) - (x^2 + 5y) = 7x(x^2 + 5y) - 1(x^2 + 5y)$

$$= (x^2 + 5y)(7x - 1)$$ □

▶ **Helpful Hint**
Notice that we wrote $-(x^2 + 5y)$ as $-1(x^2 + 5y)$ to aid in factoring.

PRACTICE
6   Factor $8b(a^3 + 2y) - (a^3 + 2y)$.

OBJECTIVE
3   Factoring Polynomials by Grouping ▶

Sometimes it is possible to factor a polynomial by grouping the terms of the polynomial and looking for common factors in each group. This method of factoring is called **factoring by grouping**.

EXAMPLE 7   Factor $ab - 6a + 2b - 12$.

Solution   First look for the GCF of all four terms. The GCF of all four terms is 1. Next, group the first two terms and the last two terms and factor out common factors from each group.

$$ab - 6a + 2b - 12 = (ab - 6a) + (2b - 12)$$

Factor a from the first group and 2 from the second group.

$$= a(b - 6) + 2(b - 6)$$

Now we see a GCF of $(b - 6)$. Factor out $(b - 6)$ to get

$$a(b - 6) + 2(b - 6) = (b - 6)(a + 2)$$

Check: To check, multiply $(b - 6)$ and $(a + 2)$ to see that the product is $ab - 6a + 2b - 12$.   □

PRACTICE
7   Factor $xy - 5x - 10 + 2y$.

▶ **Helpful Hint**
Notice that the polynomial $a(b - 6) + 2(b - 6)$ is *not* in factored form. It is a *sum*, not a *product*. The factored form is $(b - 6)(a + 2)$.

EXAMPLE 8   Factor $x^3 + 5x^2 + 3x + 15$.

Solution $x^3 + 5x^2 + 3x + 15 = (x^3 + 5x^2) + (3x + 15)$ Group pairs of terms.

$$= x^2(x + 5) + 3(x + 5)$$ Factor each binomial.

$$= (x + 5)(x^2 + 3)$$ Factor out the common factor, $(x + 5)$.   □

PRACTICE
8   Factor $a^3 + 2a^2 + 5a + 10$.

EXAMPLE 9 Factor $m^2n^2 + m^2 - 2n^2 - 2$.

Solution $m^2n^2 + m^2 - 2n^2 - 2 = (m^2n^2 + m^2) + (-2n^2 - 2)$ Group pairs of terms.

$$= m^2(n^2 + 1) - 2(n^2 + 1)$$ Factor each binomial.

$$= (n^2 + 1)(m^2 - 2)$$ Factor out the common factor, $(n^2 + 1)$.   □

PRACTICE
9   Factor $x^2y^2 + 3y^2 - 5x^2 - 15$.

EXAMPLE 10   Factor $xy + 2x - y - 2$.

Solution $xy + 2x - y - 2 = (xy + 2x) + (-y - 2)$ Group pairs of terms.

$$= x(y + 2) - 1(y + 2)$$ Factor each binomial.

$$= (y + 2)(x - 1)$$ Factor out the common factor $(y + 2)$.   □

PRACTICE
10   Factor $pq + 3p - q - 3$.

Vocabulary, Readiness & Video Check

Use the choices below to fill in each blank. Some choices will be used more than once and some not at all.

least   greatest   sum   product   factoring x^3 x^7 true   false

1. The reverse process of multiplying is _____.
2. The greatest common factor (GCF) of x^7, x^3, x^5 is _____.
3. In general, the GCF of a list of common variables raised to powers is the _____ exponent in the list.
4. Factoring means writing as a _____.
5. True or false: A factored form of $2xy^3 + 10xy$ is $2xy \cdot y^2 + 2xy \cdot 5$. _____
6. True or false: A factored form of $x^3 - 6x^2 + x$ is $x(x^2 - 6x)$. _____
7. True or false: A factored form of $5x - 5y + x^3 - x^2y$ is $5(x - y) + x^2(x - y)$. _____
8. True or false: A factored form of $5x - 5y + x^3 - x^2y$ is $(x - y)(5 + x^2)$. _____

Martin-Gay Interactive Videos

See Video 5.5

Watch the section lecture video and answer the following questions.

OBJECTIVE
1
9. From Example 1, how do you find the GCF of a list of common variables raised to powers?

OBJECTIVE
2
10. From Example 2, when factoring out a GCF from a polynomial, what two operations may help? How?

OBJECTIVE
3
11. In Examples 4 and 5, what are you reminded to do first when factoring a polynomial? How do you know to check to see if factoring by grouping might work?

5.5 Exercise Set   MyMathLab®

Find the GCF of each list of monomials. See Example 1.

1. a^8, a^5, a^3
2. b^9, b^2, b^5
3. $x^2y^3z^3, y^2z^3, xy^2z^2$
4. $xy^2z^3, x^2y^2z^2, x^2y^3$
5. $6x^3y, 9x^2y^2, 12x^2y$
6. $4xy^2, 16xy^3, 8x^2y^2$

 7. $10x^3yz^3, 20x^2z^5, 45xz^3$
8. $12y^2z^4, 9xy^3z^4, 15x^2y^2z^3$

Factor out the GCF in each polynomial. See Examples 2 through 6.

9. $18x - 12$
10. $21x + 14$
11. $4y^2 - 16xy^3$
12. $3z - 21xz^4$

13. $6x^5 - 8x^4 + 2x^3$

14. $9x + 3x^2 - 6x^3$

15. $8a^3b^3 - 4a^2b^2 + 4ab + 16ab^2$

16. $12a^3b - 6ab + 18ab^2 - 18a^2b$

17. $6(x + 3) + 5a(x + 3)$

18. $2(x - 4) + 3y(x - 4)$

19. $2x(z + 7) + (z + 7)$

20. $9x(y - 2) + (y - 2)$

21. $3x(6x^2 + 5) - 2(6x^2 + 5)$

22. $4x(2y^2 + 3) - 5(2y^2 + 3)$

23. $20a^3 + 5a^2 + 1$

24. $12b^4 - 3b^2 + 1$

25. $39x^3y^3 - 26x^2y^3$

26. $38x^2y^4 - 57xy^4$

Factor each polynomial by grouping. See Examples 7 through 10.

27. $ab + 3a + 2b + 6$

28. $ab + 2a + 5b + 10$

29. $ac + 4a - 2c - 8$

30. $bc + 8b - 3c - 24$

31. $2xy - 3x - 4y + 6$

32. $12xy - 18x - 10y + 15$

33. $12xy - 8x - 3y + 2$

34. $20xy - 15x - 4y + 3$

MIXED PRACTICE

Factor each polynomial. See Examples 1 through 10.

35. $6x^3 + 9$

36. $6x^2 - 8$

37. $x^3 + 3x^2$

38. $x^4 - 4x^3$

39. $8a^3 - 4a$

40. $12b^4 + 3b^2$

41. $8m^3 + 4m^2 + 1$

42. $30n^4 + 15n^2 + 1$

43. $-20x^2y + 16xy^3$

44. $-18xy^3 + 27x^4y$

45. $10a^2b^3 + 5ab^2 - 15ab^3$

46. $10ef - 20e^2f^3 + 30e^3f$

47. $9abc^2 + 6a^2bc - 6ab + 3bc$

48. $4a^2b^2c - 6ab^2c - 4ac + 8a$

49. $4x(y - 2) - 3(y - 2)$

50. $8y(z + 8) - 3(z + 8)$

51. $6xy + 10x + 9y + 15$

52. $15xy + 20x + 6y + 8$

53. $xy + 3y - 5x - 15$

54. $xy + 4y - 3x - 12$

55. $6ab - 2a - 9b + 3$

56. $16ab - 8a - 6b + 3$

57. $12xy + 18x + 2y + 3$

58. $20xy + 8x + 5y + 2$

59. $2m(n - 8) - (n - 8)$

60. $3a(b - 4) - (b - 4)$

61. $15x^3y^2 - 18x^2y^2$

62. $12x^4y^2 - 16x^3y^3$

63. $2x^2 + 3xy + 4x + 6y$

64. $3x^2 + 12x + 4xy + 16y$

65. $5x^2 + 5xy - 3x - 3y$

66. $4x^2 + 2xy - 10x - 5y$

67. $x^3 + 3x^2 + 4x + 12$

68. $x^3 + 4x^2 + 3x + 12$

69. $x^3 - x^2 - 2x + 2$

70. $x^3 - 2x^2 - 3x + 6$

REVIEW AND PREVIEW

Simplify the following. See Section 5.1.

71. $(5x^2)(11x^5)$ **72.** $(7y)(-2y^3)$

73. $(5x^2)^3$ **74.** $(-2y^3)^4$

Find each product by using the FOIL order of multiplying binomials. See Section 5.4.

75. $(x + 2)(x - 5)$

76. $(x - 7)(x - 1)$

77. $(x + 3)(x + 2)$

78. $(x - 4)(x + 2)$

79. $(y - 3)(y - 1)$

80. $(s + 8)(s + 10)$

CONCEPT EXTENSIONS

Solve. See the Concept Check in this section.

81. Which factorization of $10x^2 - 2x - 2$ is correct?

  a. $2(5x^2 - x + 1)$ **b.** $2(5x^2 - x)$

  c. $2(5x^2 - x - 2)$ **d.** $2(5x^2 - x - 1)$

82. Which factorization of $x^4 + 5x^3 - x^2$ is correct?

  a. $-1(x^4 + 5x^3 + x^2)$ **b.** $x^2(x^2 + 5x^3 - x^2)$

  c. $x^2(x^2 + 5x - 1)$ **d.** $5x^2(x^2 + 5x - 5)$

Solve.

△ **83.** The area of the material needed to manufacture a tin can is given by the polynomial $2\pi r^2 + 2\pi rh$, where the radius is r and height is h. Factor this expression.

84. To estimate the cost of a new product, one expression used by the production department is $4\pi r^2 + \frac{4}{3}\pi r^3$. Write an equivalent expression by factoring $4\pi r^2$ from both terms.

85. At the end of T years, the amount of money A in a savings account earning simple interest from an initial investment of $5600 at rate r is given by the formula $A = 5600 + 5600rt$. Write an equivalent equation by factoring the expression $5600 + 5600rt$.

△ **86.** An open-topped box has a square base and a height of 10 inches. If each of the bottom edges of the box has length x inches, find the amount of material needed to construct the box. Write the answer in factored form.

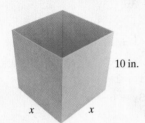

87. When $3x^2 - 9x + 3$ is factored, the result is $3(x^2 - 3x + 1)$. Explain why it is necessary to include the term 1 in this factored form.

88. Construct a trinomial whose greatest common factor is $5x^2y^3$.

89. A factored polynomial can be in many forms. For example, a factored form of $xy - 3x - 2y + 6$ is $(x - 2)(y - 3)$. Which of the following is not a factored form of $xy - 3x - 2y + 6$?

  a. $(2 - x)(3 - y)$ **b.** $(-2 + x)(-3 + y)$
  c. $(y - 3)(x - 2)$ **d.** $(-x + 2)(-y + 3)$

90. Which factorization of $12x^2 + 9x + 3$ is correct?

  a. $3(4x^2 + 3x + 1)$
  b. $3(4x^2 + 3x - 1)$
  c. $3(4x^2 + 3x - 3)$
  d. $3(4x^2 + 3x)$

91. Explain why $9(5 - x) + y(5 - x)$ is not a factored form of $45 - 9x + 5y - xy$.

92. Consider the following sequence of algebraic steps:

$$x^3 - 6x^2 + 2x - 10 = (x^3 - 6x^2) + (2x - 10)$$
$$= x^2(x - 6) + 2(x - 5)$$

Explain whether the final result is the factored form of the original polynomial.

93. The amount E of voltage in an electrical circuit is given by the formula

$$IR_1 + IR_2 = E$$

Write an equivalent equation by factoring the expression $IR_1 + IR_2$.

94. At the end of T years, the amount of money A in a savings account earning simple interest from an initial investment of P dollars at rate R is given by the formula

$$A = P + PRT$$

Write an equivalent equation by factoring the expression $P + PRT$.

Factor out the greatest common factor. Assume that variables used as exponents represent positive integers.

95. $x^{3n} - 2x^{2n} + 5x^n$

96. $3y^n + 3y^{2n} + 5y^{8n}$

97. $6x^{8a} - 2x^{5a} - 4x^{3a}$

98. $3x^{5a} - 6x^{3a} + 9x^{2a}$

99. An object is thrown upward from the ground with an initial velocity of 64 feet per second. The height $h(t)$ in feet of the object after t seconds is given by the polynomial function

$$h(t) = -16t^2 + 64t$$

  a. Write an equivalent factored expression for the function $h(t)$ by factoring $-16t^2 + 64t$.

  b. Find $h(1)$ by using

$$h(t) = -16t^2 + 64t$$

  and then by using the factored form of $h(t)$.

  c. Explain why the values found in part (b) are the same.

100. An object is dropped from the gondola of a hot-air balloon at a height of 224 feet. The height $h(t)$ of the object after t seconds is given by the polynomial function

$$h(t) = -16t^2 + 224$$

  a. Write an equivalent factored expression for the function $h(t)$ by factoring $-16t^2 + 224$.

  b. Find $h(2)$ by using $h(t) = -16t^2 + 224$ and then by using the factored form of the function.

  c. Explain why the values found in part **b** are the same.

101. The polynomial function $f(x) = 201x^3 - 2517x^2 + 6975x + 83,634$ models the number of patents granted by the United States Patent Office for the years 2000–2010, where x represents the number of years since 2000 and $f(x)$ is the number of patents granted. Write an equivalent expression for $f(x)$ by factoring the greatest common factor from the terms of $f(x)$. (*Source:* United States Patent Office)

102. The number of college students in the United States is increasing. The polynomial function $f(x) = 22x^2 + 274x + 15,628$ models the number of students enrolled in American colleges for the years 2000–2010, where x is the years after 2000 and $f(x)$ is the number of college students (in thousands). Write an equivalent expression for $f(x)$ by factoring the greatest common factor from the terms of $f(x)$. (*Source:* U.S. Department of Education)

5.6 Factoring Trinomials ▷

OBJECTIVE

1 Factoring Trinomials of the Form $x^2 + bx + c$ ▷

In the previous section, we used factoring by grouping to factor four-term polynomials. In this section, we present techniques for factoring trinomials. Since $(x - 2)(x + 5) = x^2 + 3x - 10$, we say that $(x - 2)(x + 5)$ is a factored form of $x^2 + 3x - 10$. Taking a close look at how $(x - 2)$ and $(x + 5)$ are multiplied suggests a pattern for factoring trinomials of the form

$$x^2 + bx + c$$

The pattern for factoring is summarized next.

Factoring a Trinomial of the Form $x^2 + bx + c$

Find two numbers whose product is c and whose sum is b. The factored form of $x^2 + bx + c$ is

$$(x + \text{one number})(x + \text{other number})$$

EXAMPLE 1 Factor $x^2 + 10x + 16$.

Solution We look for two integers whose product is 16 and whose sum is 10. Since our integers must have a positive product and a positive sum, we look at only positive factors of 16.

Positive Factors of 16	Sum of Factors
1, 16	$1 + 16 = 17$
4, 4	$4 + 4 = 8$
2, 8	$2 + 8 = 10$ Correct pair

The correct pair of numbers is 2 and 8 because their product is 16 and their sum is 10. Thus,

$$x^2 + 10x + 16 = (x + 2)(x + 8)$$

Check: To check, see that $(x + 2)(x + 8) = x^2 + 10x + 16$. □

PRACTICE

1 Factor $x^2 + 5x + 6$

EXAMPLE 2 Factor $x^2 - 12x + 35$.

Solution We need to find two integers whose product is 35 and whose sum is -12. Since our integers must have a positive product and a negative sum, we consider only negative factors of 35.

Negative Factors of 35	Sum of Factors
$-1, -35$	$-1 + (-35) = -36$
$-5, -7$	$-5 + (-7) = -12$ Correct pair

The numbers are -5 and -7.

$$x^2 - 12x + 35 = [x + (-5)][x + (-7)]$$
$$= (x - 5)(x - 7)$$

Check: To check, see that $(x - 5)(x - 7) = x^2 - 12x + 35$. □

PRACTICE
2 Factor $x^2 - 11x + 24$.

EXAMPLE 3 Factor $5x^3 - 30x^2 - 35x$.

Solution First we factor out the greatest common factor, $5x$.

$$5x^3 - 30x^2 - 35x = 5x(x^2 - 6x - 7)$$

Next we try to factor $x^2 - 6x - 7$ by finding two numbers whose product is -7 and whose sum is -6. The numbers are 1 and -7.

$$5x^3 - 30x^2 - 35x = 5x(x^2 - 6x - 7)$$
$$= 5x(x + 1)(x - 7) \quad □$$

PRACTICE
3 Factor $3x^3 - 9x^2 - 30x$.

> ▶ Helpful Hint
>
> If the polynomial to be factored contains a common factor that is factored out, don't forget to include that common factor in the final factored form of the original polynomial.

EXAMPLE 4 Factor $2n^2 - 38n + 80$.

Solution The terms of this polynomial have a greatest common factor of 2, which we factor out first.

$$2n^2 - 38n + 80 = 2(n^2 - 19n + 40)$$

Next we factor $n^2 - 19n + 40$ by finding two numbers whose product is 40 and whose sum is -19. Both numbers must be negative since their sum is -19. Possibilities are

$$-1 \text{ and } -40, \quad -2 \text{ and } -20, \quad -4 \text{ and } -10, \quad -5 \text{ and } -8$$

None of the pairs has a sum of -19, so no further factoring with integers is possible. The factored form of $2n^2 - 38n + 80$ is

$$2n^2 - 38n + 80 = 2(n^2 - 19n + 40) \quad □$$

PRACTICE
4 Factor $2b^2 - 18b - 22$.

We call a polynomial such as $n^2 - 19n + 40$, which cannot be factored further, a **prime polynomial.**

OBJECTIVE
2 **Factoring Trinomials of the Form $ax^2 + bx + c$** ▶

Next, we factor trinomials of the form $ax^2 + bx + c$, where the coefficient a of x^2 is not 1. Don't forget that the first step in factoring any polynomial is to factor out the greatest common factor of its terms. We will review two methods here. The first method we'll call trial and check.

Method 1—Factoring $ax^2 + bx + c$ by Trial and Check

EXAMPLE 5

Factor $2x^2 + 11x + 15$.

Solution Factors of $2x^2$ are $2x$ and x. Let's try these factors as first terms of the binomials.

$$2x^2 + 11x + 15 = (2x + \quad)(x + \quad)$$

Next we try combinations of factors of 15 until the correct middle term, $11x$, is obtained. We will try only positive factors of 15 since the coefficient of the middle term, 11, is positive. Positive factors of 15 are 1 and 15 and 3 and 5.

$(2x + 1)(x + 15)$

$1x$

$\dfrac{30x}{31x}$ Incorrect middle term

$(2x + 15)(x + 1)$

$15x$

$\dfrac{2x}{17x}$ Incorrect middle term

$(2x + 3)(x + 5)$

$3x$

$\dfrac{10x}{13x}$ Incorrect middle term

$(2x + 5)(x + 3)$

$5x$

$\dfrac{6x}{11x}$ Correct middle term

Thus, the factored form of $2x^2 + 11x + 15$ is $(2x + 5)(x + 3)$.   □

PRACTICE

5   Factor $2x^2 + 13x + 6$.

Factoring a Trinomial of the Form $ax^2 + bx + c$

Step 1. Write all pairs of factors of ax^2.

Step 2. Write all pairs of factors of c, the constant term.

Step 3. Try various combinations of these factors until the correct middle term bx is found.

Step 4. If no combination exists, the polynomial is **prime.**

EXAMPLE 6   Factor $3x^2 - x - 4$.

Solution   Factors of $3x^2$: $3x \cdot x$

Factors of -4: $-1 \cdot 4$, $1 \cdot -4$, $-2 \cdot 2$, $2 \cdot -2$

Let's try possible combinations of these factors.

$(3x - 1)(x + 4)$

$-1x$

$\dfrac{12x}{11x}$ Incorrect middle term

$(3x + 4)(x - 1)$

$4x$

$\dfrac{-3x}{1x}$ Incorrect middle term

$(3x - 4)(x + 1)$

$-4x$

$\dfrac{3x}{-1x}$ Correct middle term

Thus, $3x^2 - x - 4 = (3x - 4)(x + 1)$.   □

PRACTICE

6   Factor $4x^2 + 5x - 6$.

> **Helpful Hint—Sign Patterns**
>
> A positive constant in a trinomial tells us to look for two numbers with the same sign. The sign of the coefficient of the middle term tells us whether the signs are both positive or both negative.
>
>
>
> both   same   both   same
> positive   sign   negative   sign
> ↓   ↓   ↓   ↓
>
>
>
> $$2x^2 + 7x + 3 = (2x + 1)(x + 3) \qquad 2x^2 - 7x + 3 = (2x - 1)(x - 3)$$
>
> A negative constant in a trinomial tells us to look for two numbers with opposite signs.
>
>
>
> opposite   opposite
> signs   signs
> ↓   ↓
>
>
>
> $$2x^2 - 5x - 3 = (2x + 1)(x - 3) \qquad 2x^2 + 5x - 3 = (2x - 1)(x + 3)$$

EXAMPLE 7   Factor $12x^3y - 22x^2y + 8xy$.

Solution   First we factor out the greatest common factor of the terms of this trinomial, $2xy$.

$$12x^3y - 22x^2y + 8xy = 2xy(6x^2 - 11x + 4)$$

Now we try to factor the trinomial $6x^2 - 11x + 4$.
Factors of $6x^2$: $2x \cdot 3x$, $6x \cdot x$

Let's try $2x$ and $3x$.

$$2xy(6x^2 - 11x + 4) = 2xy(2x + (3x +)$$

The constant term, 4, is positive and the coefficient of the middle term, -11, is negative, so we factor 4 into negative factors only.
Negative factors of 4: $-4(-1)$, $-2(-2)$

Let's try -4 and -1.

$2xy(2x - 4)(3x - 1)$
$-12x$
$\underline{-2x}$
$-14x$ Incorrect middle term

This combination cannot be correct because one of the factors, $(2x - 4)$, has a common factor of 2. This cannot happen if the polynomial $6x^2 - 11x + 4$ has no common factors.

Now let's try -1 and -4.

$2xy(2x - 1)(3x - 4)$
$-3x$
$\underline{-8x}$
$-11x$ Correct middle term

Thus,

$$12x^3y - 22x^2y + 8xy = 2xy(2x - 1)(3x - 4)$$

If this combination had not worked, we would have tried -2 and -2 as factors of 4 and then $6x$ and x as factors of $6x^2$.   □

PRACTICE
7   Factor $18b^4 - 57b^3 + 30b^2$.

▶ Helpful Hint

If a trinomial has no common factor (other than 1), then none of its binomial factors will contain a common factor (other than 1).

EXAMPLE 8   Factor $16x^2 + 24xy + 9y^2$.

Solution   No greatest common factor can be factored out of this trinomial.

Factors of $16x^2$: $16x \cdot x$, $8x \cdot 2x$, $4x \cdot 4x$
Factors of $9y^2$: $y \cdot 9y$, $3y \cdot 3y$

We try possible combinations until the correct factorization is found.

$$16x^2 + 24xy + 9y^2 = (4x + 3y)(4x + 3y) \quad \text{or} \quad (4x + 3y)^2 \qquad \square$$

PRACTICE
8   Factor $25x^2 + 20xy + 4y^2$.

The trinomial $16x^2 + 24xy + 9y^2$ in Example 8 is an example of a **perfect square trinomial** since its factors are two identical binomials. In the next section, we examine a special method for factoring perfect square trinomials.

Method 2—Factoring $ax^2 + bx + c$ by Grouping

There is another method we can use when factoring trinomials of the form $ax^2 + bx + c$: Write the trinomial as a four-term polynomial and then factor by grouping.

> **Factoring a Trinomial of the Form $ax^2 + bx + c$ by Grouping**
>
> **Step 1.** Find two numbers whose product is $a \cdot c$ and whose sum is b.
> **Step 2.** Write the term bx as a sum by using the factors found in Step 1.
> **Step 3.** Factor by grouping.

EXAMPLE 9   Factor $6x^2 + 13x + 6$.

Solution   In this trinomial, $a = 6$, $b = 13$, and $c = 6$.

Step 1. Find two numbers whose product is $a \cdot c$, or $6 \cdot 6 = 36$, and whose sum is b, 13. The two numbers are 4 and 9.

Step 2. Write the middle term, $13x$, as the sum $4x + 9x$.

$$6x^2 + 13x + 6 = 6x^2 + 4x + 9x + 6$$

Step 3. Factor $6x^2 + 4x + 9x + 6$ by grouping.

$$(6x^2 + 4x) + (9x + 6) = 2x(3x + 2) + 3(3x + 2)$$
$$= (3x + 2)(2x + 3) \qquad \square$$

PRACTICE
9   Factor $20x^2 + 23x + 6$.

✓CONCEPT CHECK
Name one way that a factorization can be checked.

Answer to Concept Check:
Answers may vary. A sample is: by multiplying the factors to see that the product is the original polynomial.

EXAMPLE 10   Factor $18x^2 - 9x - 2$.

Solution   In this trinomial, $a = 18$, $b = -9$, and $c = -2$.

Step 1. Find two numbers whose product is $a \cdot c$ or $18(-2) = -36$ and whose sum is b, -9. The two numbers are -12 and 3.

Step 2. Write the middle term, $-9x$, as the sum $-12x + 3x$.

$$18x^2 - 9x - 2 = 18x^2 - 12x + 3x - 2$$

Step 3. Factor by grouping.

$$(18x^2 - 12x) + (3x - 2) = 6x(3x - 2) + 1(3x - 2)$$
$$= (3x - 2)(6x + 1). \qquad \square$$

PRACTICE
10 Factor $15x^2 + 4x - 3$.

OBJECTIVE

3 Factoring by Substitution

A complicated-looking polynomial may be a simpler trinomial in disguise. Revealing the simpler trinomial is possible by substitution.

EXAMPLE 11 Factor $2(a + 3)^2 - 5(a + 3) - 7$.

Solution The quantity $(a + 3)$ is in two of the terms of this polynomial. **Substitute x** for $(a + 3)$, and the result is the following simpler trinomial.

$$2(a + 3)^2 - 5(a + 3) - 7 \quad \text{Original trinomial.}$$
$$= 2(x)^2 - 5(x) - 7 \quad \text{Substitute } x \text{ for } (a + 3).$$

Now factor $2x^2 - 5x - 7$.

$$2x^2 - 5x - 7 = (2x - 7)(x + 1)$$

But the quantity in the original polynomial was $(a + 3)$, not x. Thus, we need to reverse the substitution and replace x with $(a + 3)$.

$$(2x - 7)(x + 1) \qquad \text{Factored expression.}$$
$$= [2(a + 3) - 7][(a + 3) + 1] \quad \text{Substitute } (a + 3) \text{ for } x.$$
$$= (2a + 6 - 7)(a + 3 + 1) \qquad \text{Remove inside parentheses.}$$
$$= (2a - 1)(a + 4) \qquad \text{Simplify.}$$

Thus, $2(a + 3)^2 - 5(a + 3) - 7 = (2a - 1)(a + 4)$. $\qquad \square$

PRACTICE
11 Factor $3(x + 1)^2 - 7(x + 1) - 20$.

EXAMPLE 12 Factor $5x^4 + 29x^2 - 42$.

Solution Again, substitution may help us factor this polynomial more easily. Since this polynomial contains the variable x, we will choose a different substitution variable. Let $y = x^2$, so $y^2 = (x^2)^2$, or x^4. Then

$$5x^4 + 29x^2 - 42$$

becomes

$$5y^2 + 29y - 42$$

which factors as

$$5y^2 + 29y - 42 = (5y - 6)(y + 7)$$

Next, replace y with x^2 to get

$$(5x^2 - 6)(x^2 + 7) \qquad \square$$

PRACTICE
12 Factor $6x^4 - 11x^2 - 10$.

Vocabulary, Readiness & Video Check

1. Find two numbers whose product is 10 and whose sum is 7.
2. Find two numbers whose product is 12 and whose sum is 8.
3. Find two numbers whose product is 24 and whose sum is 11.
4. Find two numbers whose product is 30 and whose sum is 13.

Martin-Gay Interactive Videos

See Video 5.6

Watch the section lecture video and answer the following questions.

OBJECTIVE 1
5. In Example 2, we know one of the factors of -24 is negative. How do we determine which one?

OBJECTIVE 1
6. In Example 3, once the GCF is factored out how do we know both factors of $+8$ are negative?

OBJECTIVE 2a
7. From Example 4, explain in your own words how to factor a trinomial with a first-term coefficient $\neq 1$ by trial and error.

OBJECTIVE 2b
8. From Example 6 and the lecture before, why does writing the middle term of your trinomial as the sum of two terms suggest we'd need to factor by grouping?

OBJECTIVE 3
9. In Example 8, the goal is to factor the polynomial, so why are we not done once we have substituted, then factored?

5.6 Exercise Set   MyMathLab®

Factor each trinomial. See Examples 1 through 4.

1. $x^2 + 9x + 18$
2. $x^2 + 9x + 20$
3. $x^2 - 12x + 32$
4. $x^2 - 12x + 27$
5. $x^2 + 10x - 24$
6. $x^2 + 3x - 54$
7. $x^2 - 2x - 24$
8. $x^2 - 9x - 36$
9. $3x^2 - 18x + 24$
10. $5x^2 - 45x + 70$
11. $4x^2z + 28xz + 40z$
12. $x^2y^2 + 4xy^2 + 3y^2$
13. $2x^2 - 24x - 64$
14. $3n^2 - 6n - 51$

Factor each trinomial. See Examples 5 through 10.

15. $5x^2 + 16x + 3$
16. $3x^2 + 8x + 4$
17. $2x^2 - 11x + 12$
18. $3x^2 - 19x + 20$
19. $2x^2 + 25x - 20$
20. $6x^2 + 13x + 8$
21. $4x^2 - 12x + 9$
22. $25x^2 - 30x + 9$
23. $12x^2 + 10x - 50$
24. $12y^2 - 48y + 45$
25. $3y^4 - y^3 - 10y^2$
26. $2x^2z + 5xz - 12z$
27. $6x^3 + 8x^2 + 24x$
28. $18y^3 + 6y^2 + 2y$
29. $2x^2 - 5xy - 3y^2$
30. $6x^2 + 11xy + 4y^2$
31. $28y^2 + 22y + 4$
32. $24y^3 - 2y^2 - y$
33. $2x^2 + 15x - 27$
34. $3x^2 + 14x + 15$

Use substitution to factor each polynomial completely. See Examples 11 and 12.

35. $x^4 + x^2 - 6$
36. $x^4 - x^2 - 20$
37. $(5x + 1)^2 + 8(5x + 1) + 7$
38. $(3x - 1)^2 + 5(3x - 1) + 6$
39. $x^6 - 7x^3 + 12$
40. $x^6 - 4x^3 - 12$
41. $(a + 5)^2 - 5(a + 5) - 24$
42. $(3c + 6)^2 + 12(3c + 6) - 28$

MIXED PRACTICE

Factor each polynomial completely. See Examples 1 through 12.

43. $x^2 - 24x - 81$
44. $x^2 - 48x - 100$
45. $x^2 - 15x - 54$
46. $x^2 - 15x + 54$
47. $3x^2 - 6x + 3$
48. $8x^2 - 8x + 2$
49. $3x^2 - 5x - 2$
50. $5x^2 - 14x - 3$
51. $8x^2 - 26x + 15$
52. $12x^2 - 17x + 6$
53. $18x^4 + 21x^3 + 6x^2$
54. $20x^5 + 54x^4 + 10x^3$
55. $x^2 + 8xz + 7z^2$
56. $a^2 - 2ab - 15b^2$
57. $x^2 - x - 12$
58. $x^2 + 4x - 5$
59. $3a^2 + 12ab + 12b^2$
60. $2x^2 + 16xy + 32y^2$
61. $x^2 + 4x + 5$
62. $x^2 + 5x + 8$
63. $2(x + 4)^2 + 3(x + 4) - 5$
64. $3(x + 3)^2 + 2(x + 3) - 5$

65. $6x^2 - 49x + 30$

66. $4x^2 - 39x + 27$

67. $x^4 - 5x^2 - 6$

68. $x^4 - 5x^2 + 6$

69. $6x^3 - x^2 - x$

70. $12x^3 + x^2 - x$

71. $12a^2 - 29ab + 15b^2$

72. $16y^2 + 6yx - 27x^2$

73. $9x^2 + 30x + 25$

74. $4x^2 + 12x + 9$

75. $3x^2y - 11xy + 8y$

76. $5xy^2 - 9xy + 4x$

77. $2x^2 + 2x - 12$

78. $3x^2 + 6x - 45$

▶ **79.** $(x - 4)^2 + 3(x - 4) - 18$

80. $(x - 3)^2 - 2(x - 3) - 8$

81. $2x^6 + 3x^3 - 9$

82. $3x^6 - 14x^3 + 8$

83. $72xy^4 - 24xy^2z + 2xz^2$

84. $36xy^2 - 48xyz^2 + 16xz^4$

85. $2x^3y + 2x^2y - 12xy$

86. $3x^2y^3 + 6x^2y^2 - 45x^2y$

87. $x^2 + 6xy + 5y^2$

88. $x^2 + 6xy + 8y^2$

REVIEW AND PREVIEW

Multiply. See Section 5.4.

89. $(x - 3)(x + 3)$

90. $(x - 4)(x + 4)$

91. $(2x + 1)^2$

92. $(3x + 5)^2$

93. $(x - 2)(x^2 + 2x + 4)$

94. $(y + 1)(y^2 - y + 1)$

CONCEPT EXTENSIONS

95. Find all positive and negative integers b such that $x^2 + bx + 6$ is factorable.

96. Find all positive and negative integers b such that $x^2 + bx - 10$ is factorable.

△ **97.** The volume $V(x)$ of a box in terms of its height x is given by the function $V(x) = x^3 + 2x^2 - 8x$. Factor this expression for $V(x)$.

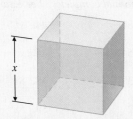

△ **98.** Based on your results from Exercise 97, find the length and width of the box if the height is 5 inches and the dimensions of the box are whole numbers.

99. Suppose that a movie is being filmed in New York City. An action shot requires an object to be thrown upward with an initial velocity of 80 feet per second off the top of 1 Madison Square Plaza, a height of 576 feet. The height $h(t)$ in feet of the object after t seconds is

given by the function $h(t) = -16t^2 + 80t + 576$. (*Source: The World Almanac*)

a. Find the height of the object at $t = 0$ seconds, $t = 2$ seconds, $t = 4$ seconds, and $t = 6$ seconds.

\ **b.** Explain why the height of the object increases and then decreases as time passes.

c. Factor the polynomial $-16t^2 + 80t + 576$.

576 ft

100. Suppose that an object is thrown upward with an initial velocity of 64 feet per second off the edge of a 960-foot cliff. The height $h(t)$ in feet of the object after t seconds is given by the function

$$h(t) = -16t^2 + 64t + 960$$

a. Find the height of the object at $t = 0$ seconds, $t = 3$ seconds, $t = 6$ seconds, and $t = 9$ seconds.

\ **b.** Explain why the height of the object increases and then decreases as time passes.

c. Factor the polynomial $-16t^2 + 64t + 960$.

Factor. Assume that variables used as exponents represent positive integers.

101. $x^{2n} + 10x^n + 16$

102. $x^{2n} - 7x^n + 12$

103. $x^{2n} - 3x^n - 18$

104. $x^{2n} + 7x^n - 18$

105. $2x^{2n} + 11x^n + 5$

106. $3x^{2n} - 8x^n + 4$

107. $4x^{2n} - 12x^n + 9$

108. $9x^{2n} + 24x^n + 16$

Recall that a graphing calculator may be used to check addition, subtraction, and multiplication of polynomials. In the same manner, a graphing calculator may be used to check factoring of polynomials in one variable. For example, to see that

$$2x^3 - 9x^2 - 5x = x(2x + 1)(x - 5)$$

graph $Y_1 = 2x^3 - 9x^2 - 5x$ *and* $Y_2 = x(2x + 1)(x - 5)$. *Then trace along both graphs to see that they coincide. Factor the following and use this method to check your results.*

109. $x^4 + 6x^3 + 5x^2$

110. $x^3 + 6x^2 + 8x$

111. $30x^3 + 9x^2 - 3x$

112. $-6x^4 + 10x^3 - 4x^2$

5.7 | Factoring by Special Products

OBJECTIVE

1   Factoring a Perfect Square Trinomial

In the previous section, we considered a variety of ways to factor trinomials of the form $ax^2 + bx + c$. In Example 8, we factored $16x^2 + 24xy + 9y^2$ as

$$16x^2 + 24xy + 9y^2 = (4x + 3y)^2$$

Recall that $16x^2 + 24xy + 9y^2$ is a perfect square trinomial because its factors are two identical binomials. A perfect square trinomial can be factored quickly if you recognize the trinomial as a perfect square.

A trinomial is a perfect square trinomial if it can be written so that its first term is the square of some quantity a, its last term is the square of some quantity b, and its middle term is twice the product of the quantities a and b.

The following special formulas can be used to factor perfect square trinomials.

> **Perfect Square Trinomials**
>
> $$a^2 + 2ab + b^2 = (a + b)^2$$
> $$a^2 - 2ab + b^2 = (a - b)^2$$

Notice that these formulas above are the same special products from Section 5.4 for the square of a binomial.

From

$$a^2 + 2ab + b^2 = (a + b)^2,$$

we see that

$$16x^2 + 24xy + 9y^2 = (4x)^2 + 2(4x)(3y) + (3y)^2 = (4x + 3y)^2$$

EXAMPLE 1   Factor $m^2 + 10m + 25$.

Solution   Notice that the first term is a square: $m^2 = (m)^2$, the last term is a square: $25 = 5^2$; and $10m = 2 \cdot 5 \cdot m$.

Thus,

$$m^2 + 10m + 25 = m^2 + 2(m)(5) + 5^2 = (m + 5)^2 \qquad \square$$

PRACTICE

1   Factor $b^2 + 16b + 64$.

EXAMPLE 2   Factor $12a^2x - 12abx + 3b^2x$.

Solution   The terms of this trinomial have a GCF of $3x$, which we factor out first.

$$12a^2x - 12abx + 3b^2x = 3x(4a^2 - 4ab + b^2)$$

Now, the polynomial $4a^2 - 4ab + b^2$ is a perfect square trinomial. Notice that the first term is a square: $4a^2 = (2a)^2$; the last term is a square: $b^2 = (b)^2$; and $4ab = 2(2a)(b)$. The factoring can now be completed as

$$3x(4a^2 - 4ab + b^2) = 3x(2a - b)^2 \qquad \square$$

PRACTICE

2   Factor $45x^2b - 30xb + 5b$.

> ▶ **Helpful Hint**
>
> If you recognize a trinomial as a perfect square trinomial, use the special formulas to factor. However, methods for factoring trinomials in general from Section 5.6 will also result in the correct factored form.

OBJECTIVE

2 Factoring the Difference of Two Squares ▶

We now factor special types of binomials, beginning with the **difference of two squares.** The special product pattern presented in Section 5.4 for the product of a sum and a difference of two terms is used again here. However, the emphasis is now on factoring rather than on multiplying.

Difference of Two Squares

$$a^2 - b^2 = (a + b)(a - b)$$

Notice that a binomial is a difference of two squares when it is the difference of the square of some quantity a and the square of some quantity b.

EXAMPLE 3 Factor the following.

a. $x^2 - 9$ **b.** $16y^2 - 9$ **c.** $50 - 8y^2$ **d.** $x^2 - \dfrac{1}{4}$

Solution

a. $x^2 - 9 = x^2 - 3^2$
$= (x + 3)(x - 3)$

b. $16y^2 - 9 = (4y)^2 - 3^2$
$= (4y + 3)(4y - 3)$

c. First factor out the common factor of 2.
$50 - 8y^2 = 2(25 - 4y^2)$
$= 2(5 + 2y)(5 - 2y)$

d. $x^2 - \dfrac{1}{4} = x^2 - \left(\dfrac{1}{2}\right)^2 = \left(x + \dfrac{1}{2}\right)\left(x - \dfrac{1}{2}\right)$ ▢

PRACTICE

3 Factor the following.

a. $x^2 - 16$

b. $25b^2 - 49$

c. $45 - 20x^2$

d. $y^2 - \dfrac{1}{81}$

The binomial $x^2 + 9$ is a **sum of two squares** and cannot be factored by using real numbers. **In general, except for factoring out a GCF, the sum of two squares usually cannot be factored by using real numbers.**

▶ Helpful Hint

The sum of two squares whose GCF is 1 usually cannot be factored by using real numbers. For example, $x^2 + 9$ is a prime polynomial.

EXAMPLE 4 Factor the following.

a. $p^4 - 16$ **b.** $(x + 3)^2 - 36$

Solution

a. $p^4 - 16 = (p^2)^2 - 4^2$
$= (p^2 + 4)(p^2 - 4)$

The binomial factor $p^2 + 4$ cannot be factored by using real numbers, but the binomial factor $p^2 - 4$ is a difference of squares.

$$(p^2 + 4)\overbrace{(p^2 - 4)} = (p^2 + 4)(p + 2)(p - 2)$$

b. Factor $(x + 3)^2 - 36$ as the difference of squares.

$$(x + 3)^2 - 36 = (x + 3)^2 - 6^2$$
$$= [(x + 3) + 6][(x + 3) - 6] \quad \text{Factor.}$$
$$= [x + 3 + 6][x + 3 - 6] \quad \text{Remove parentheses.}$$
$$= (x + 9)(x - 3) \quad \text{Simplify.} \quad \square$$

PRACTICE
4   Factor the following.

a. $x^4 - 10{,}000$

b. $(x + 2)^2 - 49$

✓ **CONCEPT CHECK**
Is $(x - 4)(y^2 - 9)$ completely factored? Why or why not?

EXAMPLE 5   Factor $x^2 + 4x + 4 - y^2$.

Solution   Factoring by grouping comes to mind since the sum of the first three terms of this polynomial is a perfect square trinomial.

$$x^2 + 4x + 4 - y^2 = (x^2 + 4x + 4) - y^2 \quad \text{Group the first three terms.}$$
$$= (x + 2)^2 - y^2 \quad \text{Factor the perfect square trinomial.}$$

This is not factored yet since we have a *difference*, not a *product*. Since $(x + 2)^2 - y^2$ is a difference of squares, we have

$$(x + 2)^2 - y^2 = [(x + 2) + y][(x + 2) - y]$$
$$= (x + 2 + y)(x + 2 - y) \quad \square$$

PRACTICE
5   Factor $m^2 + 6m + 9 - n^2$.

OBJECTIVE
3   **Factoring the Sum or Difference of Two Cubes**

Although the sum of two squares usually cannot be factored, the sum of two cubes, as well as the difference of two cubes, can be factored as follows.

Sum and Difference of Two Cubes

$$a^3 + b^3 = (a + b)(a^2 - ab + b^2)$$
$$a^3 - b^3 = (a - b)(a^2 + ab + b^2)$$

To check the first pattern, let's find the product of $(a + b)$ and $(a^2 - ab + b^2)$.

$$(a + b)(a^2 - ab + b^2) = a(a^2 - ab + b^2) + b(a^2 - ab + b^2)$$
$$= a^3 - a^2b + ab^2 + a^2b - ab^2 + b^3$$
$$= a^3 + b^3$$

EXAMPLE 6   Factor $x^3 + 8$.

Solution   First we write the binomial in the form $a^3 + b^3$. Then we use the formula

$$a^3 + b^3 = (a + b)(a^2 - a \cdot b + b^2), \quad \text{where } a \text{ is } x \text{ and } b \text{ is } 2.$$
$$\downarrow \quad \downarrow \qquad \downarrow \quad \downarrow \quad \downarrow \quad \downarrow \quad \downarrow \qquad \downarrow$$
$$x^3 + 8 = x^3 + 2^3 = (x + 2)(x^2 - x \cdot 2 + 2^2)$$

Thus, $x^3 + 8 = (x + 2)(x^2 - 2x + 4)$ $\quad \square$

PRACTICE
6   Factor $x^3 + 64$.

EXAMPLE 7 Factor $p^3 + 27q^3$.

Solution
$$p^3 + 27q^3 = p^3 + (3q)^3$$
$$= (p + 3q)[p^2 - (p)(3q) + (3q)^2]$$
$$= (p + 3q)(p^2 - 3pq + 9q^2)$$

PRACTICE
7 Factor $a^3 + 8b^3$.

EXAMPLE 8 Factor $y^3 - 64$.

Solution This is a difference of cubes since $y^3 - 64 = y^3 - 4^3$.

From

$$a^3 - b^3 = (a - b)(a^2 + a \cdot b + b^2)$$
$$\downarrow \quad \downarrow \quad \downarrow \quad \downarrow \downarrow \quad \downarrow \downarrow \quad \downarrow$$
$$y^3 - 4^3 = (y - 4)(y^2 + y \cdot 4 + 4^2)$$
$$= (y - 4)(y^2 + 4y + 16)$$

PRACTICE
8 Factor $27 - y^3$.

▶ **Helpful Hint**

When factoring sums or differences of cubes, be sure to notice the sign patterns.

Same sign

$$x^3 + y^3 = (x + y)(x^2 - xy + y^2)$$

Opposite sign

Always positive

Same sign

$$x^3 - y^3 = (x - y)(x^2 + xy + y^2)$$

Opposite sign

EXAMPLE 9 Factor $125q^2 - n^3q^2$.

Solution First we factor out a common factor of q^2.

$$125q^2 - n^3q^2 = q^2(125 - n^3)$$
$$= q^2(5^3 - n^3)$$

Opposite sign   Positive

$$= q^2(5 - n)[5^2 + (5)(n) + (n^2)]$$
$$= q^2(5 - n)(25 + 5n + n^2)$$

Thus, $125q^2 - n^3q^2 = q^2(5 - n)(25 + 5n + n^2)$. The trinomial $25 + 5n + n^2$ cannot be factored further.

PRACTICE
9 Factor $b^3x^2 - 8x^2$.

Vocabulary, Readiness & Video Check

Write each term as a square. For example, $25x^2$ as a square is $(5x)^2$.
1. $81y^2$ **2.** $4z^2$ **3.** $64x^6$ **4.** $49y^6$

Write each term as a cube.
5. $8x^3$ **6.** $27y^3$ **7.** $64x^6$ **8.** x^3y^6

Martin-Gay Interactive Videos

See Video 5.7

Watch the section lecture video and answer the following questions.

OBJECTIVE 1
9. From Example 1, what is the first step to see if you have a perfect square trinomial? How do you then finish determining that you do indeed have a perfect square trinomial?

OBJECTIVE 2
10. In Example 2, the original binomial is rewritten to write each term as a square. Give two reasons why this is helpful.

OBJECTIVE 3
11. In Examples 4 and 5, what tips are given to remember how to factor the sum or difference of two cubes rather than memorizing the formulas?

5.7 Exercise Set   MyMathLab®

Factor. See Examples 1 and 2.

1. $x^2 + 6x + 9$ **2.** $x^2 - 10x + 25$

3. $4x^2 - 12x + 9$ **4.** $9a^2 - 30a + 25$

5. $25x^2 + 10x + 1$ **6.** $4a^2 + 12a + 9$

7. $3x^2 - 24x + 48$ **8.** $2x^2 + 28x + 98$

9. $9y^2x^2 + 12yx^2 + 4x^2$ **10.** $4x^2y^3 - 4xy^3 + y^3$

11. $16x^2 - 56xy + 49y^2$ **12.** $81x^2 + 36xy + 4y^2$

Factor. See Examples 3 through 5.

13. $x^2 - 25$ **14.** $y^2 - 100$

15. $\dfrac{1}{9} - 4z^2$ **16.** $\dfrac{1}{16} - y^2$

17. $(y + 2)^2 - 49$ **18.** $(x - 1)^2 - z^2$

19. $64x^2 - 100$ **20.** $4x^2 - 36$

21. $(x + 2y)^2 - 9$ **22.** $(3x + y)^2 - 25$

23. $x^2 + 6x + 9 - y^2$ **24.** $x^2 + 12x + 36 - y^2$

25. $x^2 + 16x + 64 - x^4$ **26.** $x^2 + 20x + 100 - x^4$

Factor. See Examples 6 through 9.

27. $x^3 + 27$ **28.** $y^3 + 1$

29. $z^3 - 1$ **30.** $x^3 - 8$

31. $m^3 + n^3$ **32.** $p^3 + 125q^3$

33. $27y^2 - x^3y^2$ **34.** $64q^2 - q^2p^3$

35. $8ab^3 + 27a^4$ **36.** $a^3b + 8b^4$

37. $250y^3 - 16x^3$ **38.** $54y^3 - 128$

MIXED PRACTICE

Factor completely. See Examples 1 through 9.

39. $x^2 - 12x + 36$ **40.** $x^2 - 18x + 81$

41. $18x^2y - 2y$ **42.** $12xy^2 - 108x$

43. $9x^2 - 49$ **44.** $25x^2 - 4$

45. $x^4 - 1$ **46.** $x^4 - 256$

47. $x^6 - y^3$ **48.** $x^3 - y^6$

49. $8x^3 + 27y^3$ **50.** $125x^3 + 8y^3$

51. $4x^2 + 4x + 1 - z^2$ **52.** $9y^2 + 12y + 4 - x^2$

53. $3x^6y^2 + 81y^2$ **54.** $x^2y^9 + x^2y^3$

55. $n^3 - \dfrac{1}{27}$ **56.** $p^3 + \dfrac{1}{125}$

57. $-16y^2 + 64$ **58.** $-12y^2 + 108$

59. $x^2 - 10x + 25 - y^2$

60. $x^2 - 18x + 81 - y^2$

61. $a^3b^3 + 125$

62. $x^3y^3 + 216$

63. $\dfrac{x^2}{25} - \dfrac{y^2}{9}$

64. $\dfrac{a^2}{4} - \dfrac{b^2}{49}$

65. $(x + y)^3 + 125$

66. $(r + s)^3 + 27$

REVIEW AND PREVIEW

Solve the following equations. See Section 2.1.

67. $x - 5 = 0$ **68.** $x + 7 = 0$

69. $3x + 1 = 0$ **70.** $5x - 15 = 0$

71. $-2x = 0$ **72.** $3x = 0$

73. $-5x + 25 = 0$ **74.** $-4x - 16 = 0$

CONCEPT EXTENSIONS

Determine whether each polynomial is factored completely. See the Concept Check in this section.

75. $5x(x^2 - 4)$

76. $x^2y^2(x^3 - y^3)$

77. $7y(a^2 + a + 1)$ **78.** $9z(x^2 + 4)$

△ **79.** A manufacturer of metal washers needs to determine the cross-sectional area of each washer. If the outer radius of the washer is R and the radius of the hole is r, express the area of the washer as a polynomial. Factor this polynomial completely.

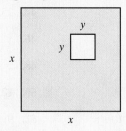

△ **80.** Express the area of the shaded region as a polynomial. Factor the polynomial completely.

Express the volume of each solid as a polynomial. To do so, subtract the volume of the "hole" from the volume of the larger solid. Then factor the resulting polynomial.

△ **81.** 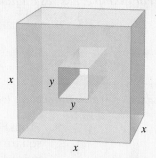   △ **82.**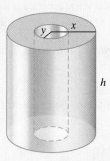

Find the value of c that makes each trinomial a perfect square trinomial.

83. $x^2 + 6x + c$ **84.** $y^2 + 10y + c$

85. $m^2 - 14m + c$ **86.** $n^2 - 2n + c$

87. $x^2 + cx + 16$ **88.** $x^2 + cx + 36$

89. Factor $x^6 - 1$ completely, using the following methods from this chapter.

  a. Factor the expression by treating it as the difference of two squares, $(x^3)^2 - 1^2$.

  b. Factor the expression, treating it as the difference of two cubes, $(x^2)^3 - 1^3$.

  c. Are the answers to parts (a) and (b) the same? Why or why not?

90. Factor $x^{12} - 1$ completely, using the following methods from this chapter:

  a. Factor the expression by treating it as the difference of two squares, $(x^3)^4 - 1^4$.

  b. Factor the expression by treating it as the difference of two cubes, $(x^4)^3 - 1^3$.

  c. Are the answers to parts (a) and (b) the same? Why or why not?

Factor. Assume that variables used as exponents represent positive integers.

91. $x^{2n} - 25$

92. $x^{2n} - 36$

93. $36x^{2n} - 49$

94. $25x^{2n} - 81$

95. $x^{4n} - 16$

96. $x^{4n} - 625$

Integrated Review OPERATIONS ON POLYNOMIALS AND FACTORING STRATEGIES

Sections 5.1–5.7

Operations on Polynomials

Perform each indicated operation.

1. $(-y^2 + 6y - 1) + (3y^2 - 4y - 10)$

2. $(5z^4 - 6z^2 + z + 1) - (7z^4 - 2z + 1)$

3. Subtract $(x - 5)$ from $(x^2 - 6x + 2)$.

4. $(2x^2 + 6x - 5) + (5x^2 - 10x)$

5. $(5x - 3)^2$

6. $(5x^2 - 14x - 3) - (5x + 1)$

7. $\dfrac{2x^4}{x} - \dfrac{3x^2}{x} + \dfrac{5x}{x} - \dfrac{2}{2}$

8. $(4x - 1)(x^2 - 3x - 2)$

Factoring Strategies

The key to proficiency in factoring polynomials is to practice until you are comfortable with each technique. A strategy for factoring polynomials completely is given next.

Factoring a Polynomial

Step 1. Are there any common factors? If so, factor out the greatest common factor.

Step 2. How many terms are in the polynomial?

    a. If there are *two* terms, decide if one of the following formulas may be applied:

        i. Difference of two squares: $a^2 - b^2 = (a - b)(a + b)$

        ii. Difference of two cubes: $a^3 - b^3 = (a - b)(a^2 + ab + b^2)$

        iii. Sum of two cubes: $a^3 + b^3 = (a + b)(a^2 - ab + b^2)$

    b. If there are *three* terms, try one of the following:

        i. Perfect square trinomial: $a^2 + 2ab + b^2 = (a + b)^2$
$$a^2 - 2ab + b^2 = (a - b)^2$$

        ii. If not a perfect square trinomial, factor by using the methods presented in Sections 5.5 and 5.6.

    c. If there are *four* or more terms, try factoring by grouping.

Step 3. See whether any factors in the factored polynomial can be factored further.

A few examples are worked for you below.

EXAMPLE 1   Factor each polynomial completely.

a. $8a^2b - 4ab$ **b.** $36x^2 - 9$ **c.** $2x^2 - 5x - 7$

d. $5p^2 + 5 + qp^2 + q$ **e.** $9x^2 + 24x + 16$ **f.** $y^2 + 25$

Solution

a. Step 1. The terms have a common factor of *4ab*, which we factor out.

$$8a^2b - 4ab = 4ab(2a - 1)$$

   Step 2. There are two terms, but the binomial $2a - 1$ is not the difference of two squares or the sum or difference of two cubes.

   Step 3. The factor $2a - 1$ cannot be factored further.

b. Step 1. Factor out a common factor of 9.

$$36x^2 - 9 = 9(4x^2 - 1)$$

   Step 2. The factor $4x^2 - 1$ has two terms, and it is the difference of two squares.

$$9(4x^2 - 1) = 9(2x + 1)(2x - 1)$$

   Step 3. No factor with more than one term can be factored further.

c. Step 1. The terms of $2x^2 - 5x - 7$ contain no common factor other than 1 or -1.

   Step 2. There are three terms. The trinomial is not a perfect square, so we factor by methods from Section 5.6.

$$2x^2 - 5x - 7 = (2x - 7)(x + 1)$$

   Step 3. No factor with more than one term can be factored further.

d. Step 1. There is no common factor of all terms of $5p^2 + 5 + qp^2 + q$.

   Step 2. The polynomial has four terms, so try factoring by grouping.

$$\begin{aligned}
5p^2 + 5 + qp^2 + q &= (5p^2 + 5) + (qp^2 + q) \quad \text{Group the terms.} \\
&= 5(p^2 + 1) + q(p^2 + 1) \\
&= (p^2 + 1)(5 + q)
\end{aligned}$$

Step 3. No factor can be factored further.

e. Step 1. The terms of $9x^2 + 24x + 16$ contain no common factor other than 1 or -1.

Step 2. The trinomial $9x^2 + 24x + 16$ is a perfect square trinomial, and $9x^2 + 24x + 16 = (3x + 4)^2$.

Step 3. No factor can be factored further.

f. Step 1. There is no common factor of $y^2 + 25$ other than 1.

Step 2. This binomial is the sum of two squares and is prime.

Step 3. The binomial $y^2 + 25$ cannot be factored further. ☐

PRACTICE

1 Factor each polynomial completely.

a. $12x^2y - 3xy$ **b.** $49x^2 - 4$

c. $5x^2 + 2x - 3$ **d.** $3x^2 + 6 + x^3 + 2x$

e. $4x^2 + 20x + 25$ **f.** $b^2 + 100$

EXAMPLE 2 Factor each polynomial completely.

a. $27a^3 - b^3$ **b.** $3n^2m^4 - 48m^6$ **c.** $2x^2 - 12x + 18 - 2z^2$

d. $8x^4y^2 + 125xy^2$ **e.** $(x - 5)^2 - 49y^2$

Solution

a. This binomial is the difference of two cubes.

$$27a^3 - b^3 = (3a)^3 - b^3$$
$$= (3a - b)[(3a)^2 + (3a)(b) + b^2]$$
$$= (3a - b)(9a^2 + 3ab + b^2)$$

b. $3n^2m^4 - 48m^6 = 3m^4(n^2 - 16m^2)$ Factor out the GCF, $3m^4$.

$\qquad\qquad\quad = 3m^4(n + 4m)(n - 4m)$ Factor the difference of squares.

c. $2x^2 - 12x + 18 - 2z^2 = 2(x^2 - 6x + 9 - z^2)$ The GCF is 2.

$\qquad\qquad\qquad\qquad = 2[(x^2 - 6x + 9) - z^2]$ Group the first three terms together.

$\qquad\qquad\qquad\qquad = 2[(x - 3)^2 - z^2]$ Factor the perfect square trinomial.

$\qquad\qquad\qquad\qquad = 2[(x - 3) + z][(x - 3) - z]$ Factor the difference of squares.

$\qquad\qquad\qquad\qquad = 2(x - 3 + z)(x - 3 - z)$

d. $8x^4y^2 + 125xy^2 = xy^2(8x^3 + 125)$ The GCF is xy^2.

$\qquad\qquad\qquad = xy^2[(2x)^3 + 5^3]$

$\qquad\qquad\qquad = xy^2(2x + 5)[(2x)^2 - (2x)(5) + 5^2]$ Factor the sum of cubes.

$\qquad\qquad\qquad = xy^2(2x + 5)(4x^2 - 10x + 25)$

e. This binomial is the difference of squares.

$$(x - 5)^2 - 49y^2 = (x - 5)^2 - (7y)^2$$
$$= [(x - 5) + 7y][(x - 5) - 7y]$$
$$= (x - 5 + 7y)(x - 5 - 7y)$$ ☐

PRACTICE
2   Factor each polynomial completely.

a. $64x^3 + y^3$

b. $7x^2y^2 - 63y^4$

c. $3x^2 + 12x + 12 - 3b^2$

d. $x^5y^4 + 27x^2y$

e. $(x + 7)^2 - 81y^2$

Factor completely.

9. $x^2 - 8x + 16 - y^2$

10. $12x^2 - 22x - 20$

11. $x^4 - x$

12. $(2x + 1)^2 - 3(2x + 1) + 2$

13. $14x^2y - 2xy$

14. $24ab^2 - 6ab$

15. $4x^2 - 16$

16. $9x^2 - 81$

17. $3x^2 - 8x - 11$

18. $5x^2 - 2x - 3$

19. $4x^2 + 8x - 12$

20. $6x^2 - 6x - 12$

21. $4x^2 + 36x + 81$

22. $25x^2 + 40x + 16$

23. $8x^3 + 125y^3$

24. $27x^3 - 64y^3$

25. $64x^2y^3 - 8x^2$

26. $27x^5y^4 - 216x^2y$

27. $(x + 5)^3 + y^3$

28. $(y - 1)^3 + 27x^3$

29. $(5a - 3)^2 - 6(5a - 3) + 9$

30. $(4r + 1)^2 + 8(4r + 1) + 16$

31. $7x^2 - 63x$

32. $20x^2 + 23x + 6$

33. $ab - 6a + 7b - 42$

34. $20x^2 - 220x + 600$

35. $x^4 - 1$

36. $15x^2 - 20x$

37. $10x^2 - 7x - 33$

38. $45m^3n^3 - 27m^2n^2$

39. $5a^3b^3 - 50a^3b$

40. $x^4 + x$

41. $16x^2 + 25$

42. $20x^3 + 20y^3$

43. $10x^3 - 210x^2 + 1100x$

44. $9y^2 - 42y + 49$

45. $64a^3b^4 - 27a^3b$

46. $y^4 - 16$

47. $2x^3 - 54$

48. $2sr + 10s - r - 5$

49. $3y^5 - 5y^4 + 6y - 10$

50. $64a^2 + b^2$

51. $100z^3 + 100$

52. $250x^4 - 16x$

53. $4b^2 - 36b + 81$

54. $2a^5 - a^4 + 6a - 3$

55. $(y - 6)^2 + 3(y - 6) + 2$

56. $(c + 2)^2 - 6(c + 2) + 5$

△ **57.** Express the area of the shaded region as a polynomial. Factor the polynomial completely.

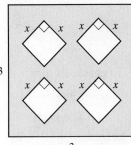

5.8 Solving Equations by Factoring and Problem Solving

OBJECTIVES

1 Solve Polynomial Equations by Factoring.

2 Solve Problems That Can Be Modeled by Polynomial Equations.

3 Find the *x*-Intercepts of a Polynomial Function.

OBJECTIVE

1 Solving Polynomial Equations by Factoring

In this section, your efforts to learn factoring start to pay off. We use factoring to solve polynomial equations, which in turn helps us solve problems that can be modeled by polynomial equations and helps us sketch the graph of polynomial functions.

A **polynomial equation** is the result of setting two polynomials equal to each other. Examples of polynomial equations are

$$3x^3 - 2x^2 = x^2 + 2x - 1 \qquad 2.6x + 7 = -1.3 \qquad -5x^2 - 5 = -9x^2 - 2x + 1$$

A polynomial equation is in **standard form** if one side of the equation is 0. In standard form, the polynomial equations above are

$$3x^3 - 3x^2 - 2x + 1 = 0 \qquad 2.6x + 8.3 = 0 \qquad 4x^2 + 2x - 6 = 0$$

The degree of a simplified polynomial equation in standard form is the same as the highest degree of any of its terms. A polynomial equation of degree 2 is also called a **quadratic equation.**

A solution of a polynomial equation in one variable is a value of the variable that makes the equation true. The method presented in this section for solving polynomial equations is called the **factoring method.** This method is based on the **zero factor property.**

Zero Factor Property

If a and b are real numbers and $a \cdot b = 0$, then $a = 0$ or $b = 0$. This property is true for three or more factors also.

In other words, if the product of two or more real numbers is zero, then at least one number must be zero.

EXAMPLE 1   Solve: $(x + 2)(x - 6) = 0$.

Solution By the zero factor property, $(x + 2)(x - 6) = 0$ only if $x + 2 = 0$ or $x - 6 = 0$.

$$x + 2 = 0 \quad \text{or} \quad x - 6 = 0 \qquad \text{Apply the zero factor property.}$$
$$x = -2 \quad \text{or} \qquad x = 6 \qquad \text{Solve each linear equation.}$$

To check, let $x = -2$ and then let $x = 6$ in the original equation.

Let $x = -2$.   Let $x = 6$.

Then $(x + 2)(x - 6) = 0$ Then $(x + 2)(x - 6) = 0$

becomes $(-2 + 2)(-2 - 6) \stackrel{?}{=} 0$ becomes $(6 + 2)(6 - 6) \stackrel{?}{=} 0$

$(0)(-8) \stackrel{?}{=} 0$ $(8)(0) \stackrel{?}{=} 0$

$0 = 0$ True $0 = 0$ True

Both -2 and 6 check, so they are both solutions. The solution set is $\{-2, 6\}$. ☐

PRACTICE

1 Solve: $(x + 8)(x - 5) = 0$.

EXAMPLE 2   Solve: $2x^2 + 9x - 5 = 0$.

Solution   To use the zero factor property, one side of the equation must be 0, and the other side must be in factored form.

$$2x^2 + 9x - 5 = 0$$
$$(2x - 1)(x + 5) = 0 \qquad \text{Factor.}$$
$$2x - 1 = 0 \quad \text{or} \quad x + 5 = 0 \qquad \text{Set each factor equal to zero.}$$
$$2x = 1$$
$$x = \frac{1}{2} \quad \text{or} \quad x = -5 \qquad \text{Solve each linear equation.}$$

The solutions are -5 and $\frac{1}{2}$. To check, let $x = \frac{1}{2}$ in the original equation; then let $x = -5$ in the original equation. The solution set is $\left\{ -5, \frac{1}{2} \right\}$. □

PRACTICE
2   Solve: $3x^2 + 10x - 8 = 0$.

Solving Polynomial Equations by Factoring

Step 1.   Write the equation in standard form so that one side of the equation is 0.
Step 2.   Factor the polynomial completely.
Step 3.   Set each factor containing a variable equal to 0.
Step 4.   Solve the resulting equations.
Step 5.   Check each solution in the original equation.

Since it is not always possible to factor a polynomial, not all polynomial equations can be solved by factoring. Other methods of solving polynomial equations are presented in Chapter 8.

EXAMPLE 3   Solve: $x(2x - 7) = 4$.

Solution   First, write the equation in standard form; then, factor.

$$x(2x - 7) = 4$$
$$2x^2 - 7x = 4 \qquad \text{Multiply.}$$
$$2x^2 - 7x - 4 = 0 \qquad \text{Write in standard form.}$$
$$(2x + 1)(x - 4) = 0 \qquad \text{Factor.}$$
$$2x + 1 = 0 \quad \text{or} \quad x - 4 = 0 \qquad \text{Set each factor equal to zero.}$$
$$2x = -1 \qquad \text{Solve.}$$
$$x = -\frac{1}{2} \quad \text{or} \quad x = 4$$

The solutions are $-\frac{1}{2}$ and 4. Check both solutions in the original equation. □

PRACTICE
3   Solve: $x(3x + 14) = -8$.

▶ **Helpful Hint**

To apply the zero factor property, one side of the equation must be 0, and the other side of the equation must be factored. To solve the equation $x(2x - 7) = 4$, for example, you may **not** set each factor equal to 4.

EXAMPLE 4 Solve: $3(x^2 + 4) + 5 = -6(x^2 + 2x) + 13$.

Solution Rewrite the equation so that one side is 0.

$$3(x^2 + 4) + 5 = -6(x^2 + 2x) + 13.$$
$$3x^2 + 12 + 5 = -6x^2 - 12x + 13 \qquad \text{Apply the distributive property.}$$
$$9x^2 + 12x + 4 = 0 \qquad \text{Rewrite the equation so that one side is 0.}$$
$$(3x + 2)(3x + 2) = 0 \qquad \text{Factor.}$$
$$3x + 2 = 0 \quad \text{or} \quad 3x + 2 = 0 \qquad \text{Set each factor equal to 0.}$$
$$3x = -2 \quad \text{or} \quad 3x = -2$$
$$x = -\frac{2}{3} \quad \text{or} \quad x = -\frac{2}{3} \qquad \text{Solve each equation.}$$

The solution is $-\dfrac{2}{3}$. Check by substituting $-\dfrac{2}{3}$ into the original equation. $\square$

PRACTICE
4 Solve: $8(x^2 + 3) + 4 = -8x(x + 3) + 19$.

If the equation contains fractions, we clear the equation of fractions as a first step.

EXAMPLE 5 Solve: $2x^2 = \dfrac{17}{3}x + 1$.

Solution
$$2x^2 = \frac{17}{3}x + 1$$
$$3(2x^2) = 3\left(\frac{17}{3}x + 1\right) \qquad \text{Clear the equation of fractions.}$$
$$6x^2 = 17x + 3 \qquad \text{Apply the distributive property.}$$
$$6x^2 - 17x - 3 = 0 \qquad \text{Rewrite the equation in standard form.}$$
$$(6x + 1)(x - 3) = 0 \qquad \text{Factor.}$$
$$6x + 1 = 0 \quad \text{or} \quad x - 3 = 0 \qquad \text{Set each factor equal to zero.}$$
$$6x = -1$$
$$x = -\frac{1}{6} \quad \text{or} \quad x = 3 \qquad \text{Solve each equation.}$$

The solutions are $-\dfrac{1}{6}$ and 3. $\square$

PRACTICE
5 Solve: $4x^2 = \dfrac{15}{2}x + 1$.

EXAMPLE 6 Solve: $x^3 = 4x$.

Solution
$$x^3 = 4x$$
$$x^3 - 4x = 0 \qquad \text{Rewrite the equation so that one side is 0.}$$
$$x(x^2 - 4) = 0 \qquad \text{Factor out the GCF, } x.$$
$$x(x + 2)(x - 2) = 0 \qquad \text{Factor the difference of squares.}$$
$$x = 0 \quad \text{or} \quad x + 2 = 0 \quad \text{or} \quad x - 2 = 0 \qquad \text{Set each factor equal to 0.}$$
$$x = 0 \quad \text{or} \quad x = -2 \quad \text{or} \quad x = 2 \qquad \text{Solve each equation.}$$

The solutions are $-2, 0,$ and 2. Check by substituting into the original equation. $\square$

PRACTICE
6 Solve: $x^3 = 2x^2 + 3x$.

Notice that the *third*-degree equation of Example 6 yielded *three* solutions.

EXAMPLE 7   Solve: $x^3 + 5x^2 = x + 5$.

Solution   First, write the equation so that one side is 0.

$$x^3 + 5x^2 - x - 5 = 0$$
$$(x^3 - x) + (5x^2 - 5) = 0 \qquad \text{Factor by grouping.}$$
$$x(x^2 - 1) + 5(x^2 - 1) = 0$$
$$(x^2 - 1)(x + 5) = 0$$
$$(x + 1)(x - 1)(x + 5) = 0 \qquad \text{Factor the difference of squares.}$$
$$x + 1 = 0 \quad \text{or} \quad x - 1 = 0 \quad \text{or} \quad x + 5 = 0 \qquad \text{Set each factor equal to 0.}$$
$$x = -1 \quad \text{or} \qquad x = 1 \quad \text{or} \qquad x = -5 \qquad \text{Solve each equation.}$$

The solutions are -5, -1, and 1. Check in the original equation.   □

PRACTICE
7   Solve: $x^3 - 9x = 18 - 2x^2$.

✓CONCEPT CHECK

Which solution strategies are incorrect? Why?

  a. Solve $(y - 2)(y + 2) = 4$ by setting each factor equal to 4.
  b. Solve $(x + 1)(x + 3) = 0$ by setting each factor equal to 0.
  c. Solve $z^2 + 5z + 6 = 0$ by factoring $z^2 + 5z + 6$ and setting each factor equal to 0.
  d. Solve $x^2 + 6x + 8 = 10$ by factoring $x^2 + 6x + 8$ and setting each factor equal to 0.

OBJECTIVE
2   Solving Problems Modeled by Polynomial Equations

Some problems may be modeled by polynomial equations. To solve these problems, we use the same problem-solving steps that were introduced in Section 2.2. When solving these problems, keep in mind that a solution of an equation that models a problem is not always a solution to the problem. For example, a person's weight or the length of a side of a geometric figure is always a positive number. Discard solutions that do not make sense as solutions of the problem.

EXAMPLE 8   **Finding the Return Time of a Rocket**

An Alpha III model rocket is launched from the ground with an A8–3 engine. Without a parachute, the height of the rocket h at time t seconds is approximated by the equation

$$h = -16t^2 + 144t$$

Find how long it takes the rocket to return to the ground.

Solution

1. UNDERSTAND. Read and reread the problem. The equation $h = -16t^2 + 144t$ models the height of the rocket. Familiarize yourself with this equation by finding a few values.

When $t = 1$ second, the height of the rocket is

$$h = -16(1)^2 + 144(1) = 128 \text{ feet}$$

When $t = 2$ seconds, the height of the rocket is

$$h = -16(2)^2 + 144(2) = 224 \text{ feet}$$

2. TRANSLATE. To find how long it takes the rocket to return to the ground, we want to know what value of t makes the height h equal to 0. That is, we want to solve $h = 0$.

$$-16t^2 + 144t = 0$$

3. SOLVE the quadratic equation by factoring.

$$-16t^2 + 144t = 0$$
$$-16t(t - 9) = 0$$
$$-16t = 0 \quad \text{or} \quad t - 9 = 0$$
$$t = 0 \quad \text{or} \quad t = 9$$

4. INTERPRET. The height h is 0 feet at time 0 seconds (when the rocket is launched) and at time 9 seconds.

Check: See that the height of the rocket at 9 seconds equals 0.

$$h = -16(9)^2 + 144(9) = -1296 + 1296 = 0$$

State: The rocket returns to the ground 9 seconds after it is launched. □

PRACTICE

8 A model rocket is launched from the ground. Its height h in feet at time t seconds is approximated by the equation $h = -16t^2 + 96t$. Find how long it takes the rocket to return to the ground.

Some of the exercises at the end of this section use the **Pythagorean theorem.** Before we review this theorem, recall that a **right triangle** is a triangle that contains a 90° angle, or right angle. The **hypotenuse** of a right triangle is the side opposite the right angle and is the longest side of the triangle. The **legs** of a right triangle are the other sides of the triangle.

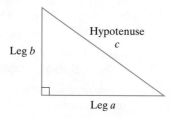

Pythagorean Theorem

In a right triangle, the sum of the squares of the lengths of the two legs is equal to the square of the length of the hypotenuse.

$$(\text{leg})^2 + (\text{leg})^2 = (\text{hypotenuse})^2 \quad \text{or} \quad a^2 + b^2 = c^2$$

△ **EXAMPLE 9** Using the Pythagorean Theorem

While framing an addition to an existing home, a carpenter used the Pythagorean theorem to determine whether a wall was "square"—that is, whether the wall formed a right angle with the floor. He used a triangle whose sides are three consecutive integers. Find a right triangle whose sides are three consecutive integers.

Solution

1. UNDERSTAND. Read and reread the problem.

Let $x, x + 1$, and $x + 2$ be three consecutive integers. Since these integers represent lengths of the sides of a right triangle, we have

$$x = \text{one leg}$$
$$x + 1 = \text{other leg}$$
$$x + 2 = \text{hypotenuse (longest side)}$$

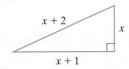

2. TRANSLATE. By the Pythagorean theorem, we have

In words:	$(\text{leg})^2$	$+$	$(\text{leg})^2$	$=$	$(\text{hypotenuse})^2$
	$\downarrow$		$\downarrow$		$\downarrow$
Traslate:	$(x)^2$	$+$	$(x + 1)^2$	$=$	$(x + 2)^2$

3. SOLVE the equation.

$$x^2 + (x + 1)^2 = (x + 2)^2$$
$$x^2 + x^2 + 2x + 1 = x^2 + 4x + 4 \qquad \text{Multiply.}$$
$$2x^2 + 2x + 1 = x^2 + 4x + 4$$
$$x^2 - 2x - 3 = 0 \qquad\qquad \text{Write in standard form.}$$
$$(x - 3)(x + 1) = 0$$
$$x - 3 = 0 \quad \text{or} \quad x + 1 = 0$$
$$x = 3 \qquad\qquad x = -1$$

4. INTERPRET. Discard $x = -1$ since length cannot be negative. If $x = 3$, then $x + 1 = 4$ and $x + 2 = 5$.

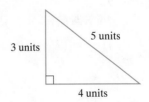

Check: To check, see that $(\text{leg})^2 + (\text{leg})^2 = (\text{hypotenuse})^2$

$$3^2 + 4^2 = 5^2$$
$$9 + 16 = 25 \quad \text{True}$$

State: The lengths of the sides of the right triangle are 3, 4, and 5 units. The carpenter used this information, for example, by marking off lengths of 3 and 4 feet on the floor and framing respectively. If the diagonal length between these marks was 5 feet, the wall was "square." If not, adjustments were made.

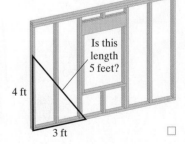

PRACTICE

9   Find a right triangle whose sides are consecutive even integers.

OBJECTIVE

3   Finding the x-Intercepts of Polynomial Functions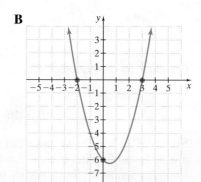

Recall that to find the x-intercepts of the graph of a function, let $f(x) = 0$ or $y = 0$ and solve for x. This fact gives us a visual interpretation of the results of this section.

From Example 1, we know that the solutions of the equation $(x + 2)(x - 6) = 0$ are -2 and 6. These solutions give us important information about the related polynomial function $p(x) = (x + 2)(x - 6)$. We know that when x is -2 or when x is 6, the value of $p(x)$ is 0.

$$p(x) = (x + 2)(x - 6)$$
$$p(-2) = (-2 + 2)(-2 - 6) = (0)(-8) = 0$$
$$p(6) = (6 + 2)(6 - 6) = (8)(0) = 0$$

Thus, we know that $(-2, 0)$ and $(6, 0)$ are the x-intercepts of the graph of $p(x)$.

We also know that the graph of $p(x)$ does not cross the x-axis at any other point. For this reason, and the fact that $p(x) = (x + 2)(x - 6) = x^2 - 4x - 12$ has degree 2, we conclude that the graph of p must look something like one of these two graphs:

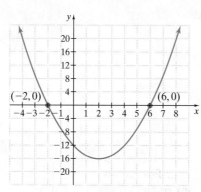

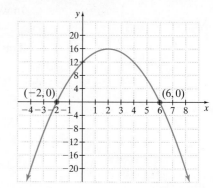

> **Helpful Hint**
>
> From our discussion on the bottom of page 276, the graph of $p(x) = (x + 2)(x - 6) = x^2 - 4x - 12$ must look like the graph on the left since the coefficient of x^2 is positive.

In a later chapter, we explore these graphs more fully. For the moment, know that the solutions of a polynomial equation are the x-intercepts of the graph of the related function and that the x-intercepts of the graph of a polynomial function are the solutions of the related polynomial equation. These values are also called **roots**, or **zeros**, of a polynomial function.

EXAMPLE 10   **Match Each Function with Its Graph**

$$f(x) = (x - 3)(x + 2) \quad g(x) = x(x + 2)(x - 2) \quad h(x) = (x - 2)(x + 2)(x - 1)$$

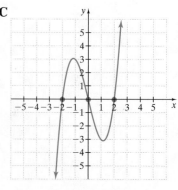

Solution   The graph of the function $f(x) = (x - 3)(x + 2)$ has two x-intercepts, $(3, 0)$ and $(-2, 0)$, because the equation $0 = (x - 3)(x + 2)$ has two solutions, 3 and -2.

The graph of $f(x)$ is graph B.

The graph of the function $g(x) = x(x + 2)(x - 2)$ has three x-intercepts, $(0, 0)$, $(-2, 0)$, and $(2, 0)$, because the equation $0 = x(x + 2)(x - 2)$ has three solutions, $0, -2,$ and 2.

The graph of $g(x)$ is graph C.

The graph of the function $h(x) = (x - 2)(x + 2)(x - 1)$ has three x-intercepts, $(2, 0)$, $(-2, 0)$, and $(1, 0)$, because the equation $0 = (x - 2)(x + 2)(x - 1)$ has three solutions, 2, -2, and 1.

The graph of $h(x)$ is graph A.

PRACTICE
10   Match each function with its graph.

$$f(x) = (x - 1)(x + 3) \quad g(x) = x(x + 3)(x - 2) \quad h(x) = (x - 3)(x + 2)(x - 2)$$

A 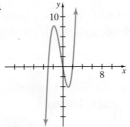   **B** 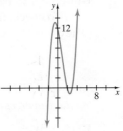   **C**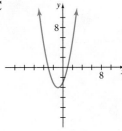

Graphing Calculator Explorations

We can use a graphing calculator to approximate real-number solutions of any quadratic equation in standard form, whether the associated polynomial is factorable or not. For example, let's solve the quadratic equation $x^2 - 2x - 4 = 0$. The solutions of this equation will be the x-intercepts of the graph of the function $f(x) = x^2 - 2x - 4$. (Recall that to find x-intercepts, we let $f(x) = 0$, or $y = 0$.) When we use a standard window, the graph of this function looks like this.

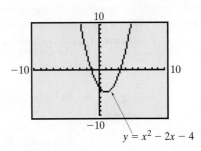

$y = x^2 - 2x - 4$

The graph appears to have one x-intercept between -2 and -1 and one between 3 and 4. To find the x-intercept between 3 and 4 to the nearest hundredth, we can use a zero feature, a Zoom feature, which magnifies a portion of the graph around the cursor, or we can redefine our window. If we redefine our window to

Xmin $= 2$	Ymin $= -1$
Xmax $= 5$	Ymax $= 1$
Xscl $= 1$	Yscl $= 1$

the resulting screen is

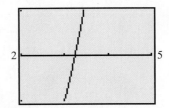

By using the Trace feature, we can now see that one of the intercepts is between 3.21 and 3.25. To approximate to the nearest hundredth, Zoom again or redefine the window to

Xmin = 3.2	Ymin = −0.1
Xmax = 3.3	Ymax = 0.1
Xscl = 1	Yscl = 1

If we use the Trace feature again, we see that, to the nearest hundredth, the x-intercept is 3.24. By repeating this process, we can approximate the other x-intercept to be −1.24.

To check, find $f(3.24)$ and $f(-1.24)$. Both of these values should be close to 0. (They will not be exactly 0 because we approximated these solutions.)

$$f(3.24) = 0.0176 \quad \text{and} \quad f(-1.24) = 0.0176$$

Solve each of these quadratic equations by graphing the related function and approximating the x-intercepts to the nearest thousandth.

1. $x^2 + 3x - 2 = 0$

2. $5x^2 - 7x + 1 = 0$

3. $2.3x^2 - 4.4x - 5.6 = 0$

4. $0.2x^2 + 6.2x + 2.1 = 0$

5. $0.09x^2 - 0.13x - 0.08 = 0$

6. $x^2 + 0.08x - 0.01 = 0$

Vocabulary, Readiness & Video Check

Solve each equation for the variable. See Example 1.

1. $(x - 3)(x + 5) = 0$

2. $(y + 5)(y + 3) = 0$

3. $(z - 3)(z + 7) = 0$

4. $(c - 2)(c - 4) = 0$

5. $x(x - 9) = 0$

6. $w(w + 7) = 0$

Martin-Gay Interactive Videos

See Video 5.8

Watch the section lecture video and answer the following questions.

OBJECTIVE 1

7. As shown in ▦ Examples 1–3, what two things have to be true in order to use the zero factor property?

OBJECTIVE 2

8. In ▦ Example 5, why aren't both solutions to the equation accepted?

OBJECTIVE 3

9. From ▦ Example 6, how can finding the x-intercepts of the graph of a quadratic equation in two variables lead to solving a quadratic equation by factoring?

5.8   Exercise Set

MyMathLab®

Solve each equation. See Example 1.

▶ **1.** $(x + 3)(3x - 4) = 0$

2. $(5x + 1)(x - 2) = 0$

3. $3(2x - 5)(4x + 3) = 0$

4. $8(3x - 4)(2x - 7) = 0$

Solve each equation. See Examples 2 through 5.

5. $x^2 + 11x + 24 = 0$

6. $y^2 - 10y + 24 = 0$

7. $12x^2 + 5x - 2 = 0$

8. $3y^2 - y - 14 = 0$

9. $z^2 + 9 = 10z$

10. $n^2 + n = 72$

▶ **11.** $x(5x + 2) = 3$

12. $n(2n - 3) = 2$

13. $x^2 - 6x = x(8 + x)$

14. $n(3 + n) = n^2 + 4n$

15. $\dfrac{z^2}{6} - \dfrac{z}{2} - 3 = 0$

16. $\dfrac{c^2}{20} - \dfrac{c}{4} + \dfrac{1}{5} = 0$

17. $\dfrac{x^2}{2} + \dfrac{x}{20} = \dfrac{1}{10}$

18. $\dfrac{y^2}{30} = \dfrac{y}{15} + \dfrac{1}{2}$

19. $\dfrac{4t^2}{5} = \dfrac{t}{5} + \dfrac{3}{10}$

20. $\dfrac{5x^2}{6} - \dfrac{7x}{2} + \dfrac{2}{3} = 0$

Solve each equation. See Examples 6 and 7.

21. $(x + 2)(x - 7)(3x - 8) = 0$

22. $(4x + 9)(x - 4)(x + 1) = 0$

23. $(x^2 - 1)(6x + 1) = 0$

24. $(x^2 - 25)(3x + 1) = 0$

25. $y^3 = 9y$ **26.** $n^3 = 16n$

▶ **27.** $x^3 - x = 2x^2 - 2$ **28.** $m^3 = m^2 + 12m$

MIXED PRACTICE

Solve each equation. See Examples 1 through 7.

29. $(2x + 7)(x - 10) = 0$ **30.** $(x + 4)(5x - 1) = 0$

31. $3x(x - 5) = 0$ **32.** $4x(2x + 3) = 0$

33. $x^2 - 2x - 15 = 0$ **34.** $x^2 + 6x - 7 = 0$

35. $12x^2 + 2x - 2 = 0$ **36.** $8x^2 + 13x + 5 = 0$

37. $w^2 - 5w = 36$ **38.** $x^2 + 32 = 12x$

39. $25x^2 - 40x + 16 = 0$ **40.** $9n^2 + 30n + 25 = 0$

41. $2r^3 + 6r^2 = 20r$ **42.** $-2t^3 = 108t - 30t^2$

43. $z(5z - 4)(z + 3) = 0$ **44.** $2r(r + 3)(5r - 4) = 0$

▶ **45.** $2z(z + 6) = 2z^2 + 12z - 8$ **46.** $3c^2 - 8c + 2 = c(3c - 8)$

47. $(x - 1)(x + 4) = 24$ **48.** $(2x - 1)(x + 2) = -3$

49. $\dfrac{x^2}{4} - \dfrac{5}{2}x + 6 = 0$ **50.** $\dfrac{x^2}{18} + \dfrac{x}{2} + 1 = 0$

51. $y^2 + \dfrac{1}{4} = -y$ **52.** $\dfrac{x^2}{10} + \dfrac{5}{2} = x$

53. $y^3 + 4y^2 = 9y + 36$ **54.** $x^3 + 5x^2 = x + 5$

55. $2x^3 = 50x$ **56.** $m^5 = 36m^3$

57. $x^2 + (x + 1)^2 = 61$ **58.** $y^2 + (y + 2)^2 = 34$

59. $m^2(3m - 2) = m$ **60.** $x^2(5x + 3) = 26x$

61. $3x^2 = -x$ **62.** $y^2 = -5y$

63. $x(x - 3) = x^2 + 5x + 7$ **64.** $z^2 - 4z + 10 = z(z - 5)$

65. $3(t - 8) + 2t = 7 + t$ **66.** $7c - 2(3c + 1) = 5(4 - 2c)$

67. $-3(x - 4) + x = 5(3 - x)$ **68.** $-4(a + 1) - 3a = -7(2a - 3)$

Solve. See Examples 8 and 9.

69. One number exceeds another by five, and their product is 66. Find the numbers.

70. If the sum of two numbers is 4 and their product is $\dfrac{15}{4}$, find the numbers.

△ **71.** An electrician needs to run a cable from the top of a 60-foot tower to a transmitter box located 45 feet away from the base of the tower. Find how long he should make the cable.

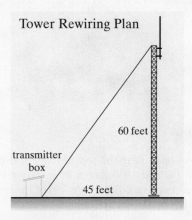

Tower Rewiring Plan

60 feet

transmitter box

45 feet

△ **72.** A stereo system installer needs to run speaker wire along the two diagonals of a rectangular room whose dimensions are 40 feet by 75 feet. Find how much speaker wire she needs.

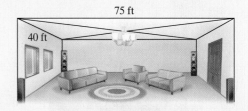

75 ft

40 ft

73. If the cost, $C(x)$, for manufacturing x units of a certain product is given by $C(x) = x^2 - 15x + 50$, find the number of units manufactured at a cost of $9500.

△ **74.** Determine whether any three consecutive integers represent the lengths of the sides of a right triangle.

△▶ **75.** The shorter leg of a right triangle is 3 centimeters less than the other leg. Find the length of the two legs if the hypotenuse is 15 centimeters.

△ **76.** The longer leg of a right triangle is 4 feet longer than the other leg. Find the length of the two legs if the hypotenuse is 20 feet.

△ **77.** Marie Mulroney has a rectangular board 12 inches by 16 inches around which she wants to put a uniform border of shells. If she has enough shells for a border whose area is 128 square inches, determine the width of the border.

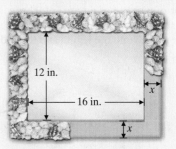

12 in.

16 in.

x

x

△ **78.** A gardener has a rose garden that measures 30 feet by 20 feet. He wants to put a uniform border of pine bark around the outside of the garden. Find how wide the border should be if he has enough pine bark to cover 336 square feet.

79. While hovering near the top of Ribbon Falls in Yosemite National Park at 1600 feet, a helicopter pilot accidentally drops his sunglasses. The height $h(t)$ of the sunglasses after t seconds is given by the polynomial function

$$h(t) = -16t^2 + 1600$$

When will the sunglasses hit the ground? (*Hint:* Let the height, $h(t)$, be 0 and solve for x.)

80. After t seconds, the height $h(t)$ of a model rocket launched from the ground into the air is given by the function

$$h(t) = -16t^2 + 80t$$

Find how long it takes the rocket to reach a height of 96 feet.

△ **81.** The floor of a shed has an area of 90 square feet. The floor is in the shape of a rectangle whose length is 3 feet less than twice the width. Find the length and the width of the floor of the shed.

△ **82.** A vegetable garden with an area of 200 square feet is to be fertilized. If the length of the garden is 1 foot less than three times the width, find the dimensions of the garden.

83. The function $W(x) = 0.5x^2$ gives the number of servings of wedding cake that can be obtained from a two-layer x-inch square wedding cake tier. What size square wedding cake tier is needed to serve 50 people? (*Source:* Based on data from the *Wilton 2000 Yearbook of Cake Decorating*)

84. Use the function in Exercise 83 to determine what size square wedding cake tier is needed to serve 200 people.

85. Suppose that a movie is being filmed in New York City. An action shot requires an object to be thrown upward with an initial velocity of 80 feet per second off the top of 1 Madison Square Plaza, a height of 576 feet. The height $h(t)$ in feet of the object after t seconds is given by the function

$$h(t) = -16t^2 + 80t + 576.$$

Determine how long before the object strikes the ground. (*Hint:* Let the height, $h(t)$, be 0 and solve for x.) (*Source: The World Almanac*)

576 ft

86. Suppose that an object is thrown upward with an initial velocity of 64 feet per second off the edge of a 960-foot-cliff. The height $h(t)$ in feet of the object after t seconds is given by the function

$$h(t) = -16t^2 + 64t + 960$$

Determine how long before the object strikes the ground. (See the hint for Exercise 85 and note Exercise 100, Section 5.6.)

Match each polynomial function with its graph. See Example 10.

87. $f(x) = (x - 2)(x + 5)$

88. $g(x) = (x + 1)(x - 6)$

89. $h(x) = x(x + 3)(x - 3)$

90. $F(x) = (x + 1)(x - 2)(x + 5)$

91. $G(x) = 2x^2 + 9x + 4$

92. $H(x) = 2x^2 - 7x - 4$

A
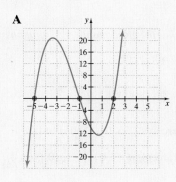

B
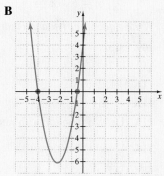

C
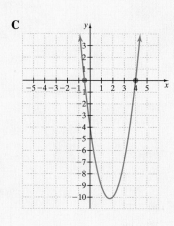

D
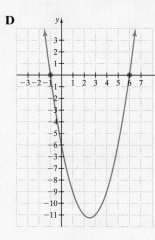

E
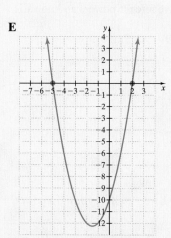

F
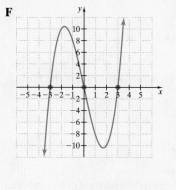

REVIEW AND PREVIEW

Write the x- and y-intercepts for each graph and determine whether the graph is the graph of a function. See Sections 3.1 and 3.2.

93.
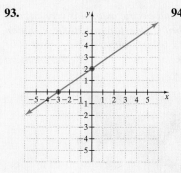

94.
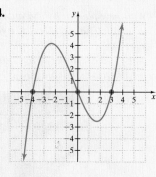

95
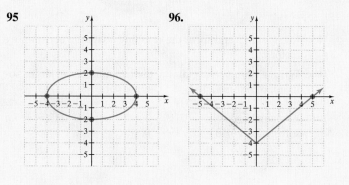

96.

CONCEPT EXTENSIONS

Each exercise contains an error. Find and correct the error. See the Concept Check in the section.

97. $(x - 5)(x + 2) = 0$

$x - 5 = 0$ or $x + 2 = 0$

$x = -5$ or $x = -2$

98. $(4x - 5)(x + 7) = 0$

$4x - 5 = 0$ or $x + 7 = 0$

$x = \dfrac{4}{5}$ or $x = -7$

99. $y(y - 5) = -6$

$y = -6$ or $y - 5 = -5$

$y = -6$ or $y = 0$

100. $3x^2 - 19x = 14$

$-16x = 14$

$x = -\dfrac{14}{16}$

$x = -\dfrac{7}{8}$

Solve.

101. $(x^2 + x - 6)(3x^2 - 14x - 5) = 0$

102. $(x^2 - 9)(x^2 + 8x + 16) = 0$

With new plug-in hybrid autos and electric autos, it is very difficult to predict future sales of hybrids. One forecast is the equation $y = 3x^2 - 25x + 345$, where x is the number of years past 2007 and y is the number of hybrid sales in thousands. Use this equation for Exercises 103 and 104.

103. Let $y = 317$ and solve the resulting quadratic equation by factoring.

104. Write a sentence explaining the meaning of the larger solution of Exercise 103.

105. Explain how solving $2(x - 3)(x - 1) = 0$ differs from solving $2x(x - 3)(x - 1) = 0$.

106. Explain why the zero factor property works for more than two numbers whose product is 0.

107. Is the following step correct? Why or why not?

$$x(x - 3) = 5$$
$$x = 5 \quad \text{or} \quad x - 3 = 5$$

108. Are the following steps correct? Why or why not?

$$x^2 + x = 6$$
$$x(x + 1) = 6$$
$$x = 6 \quad \text{or} \quad x + 1 = 6$$

Write a quadratic equation that has the given numbers as solutions.

109. $5, 3$

110. $6, 7$

111. $-1, 2$

112. $4, -3$

113. Draw a function with intercepts $(-3, 0)$, $(5, 0)$, and $(0, 4)$.

114. Draw a function with intercepts $(-7, 0)$, $\left(-\dfrac{1}{2}, 0\right)$, $(4, 0)$, and $(0, -1)$.

Chapter 5 Vocabulary Check

Fill in each blank with one of the words or phrases listed below.

| quadratic equation | scientific notation | polynomial | exponents | 1 | monomial |
| binomial | trinomial | degree of a polynomial | degree of a term | 0 | factoring |

1. A(n) _____ is a finite sum of terms in which all variables are raised to nonnegative integer powers and no variables appear in any denominator.

2. _____ is the process of writing a polynomial as a product.

3. _____ are used to write repeated factors in a more compact form.

4. The _____ is the sum of the exponents on the variables contained in the term.

5. A(n) _____ is a polynomial with one term.

6. If a is not 0, $a^0 =$ _____.

7. A(n) _____ is a polynomial with three terms.

8. A polynomial equation of degree 2 is also called a(n) _____.

9. A positive number is written in _____ if it is written as the product of a number a, where $1 \le a < 10$, and a power of 10.

10. The _____ is the largest degree of all its terms.

11. A(n) _____ is a polynomial with two terms.

12. If a and b are real numbers and $a \cdot b =$ _____, then $a = 0$ or $b = 0$.

Chapter 5 Highlights

DEFINITIONS AND CONCEPTS	EXAMPLES

Section 5.1   Exponents and Scientific Notation

Product rule: $a^m \cdot a^n = a^{m+n}$

Zero exponent: $a^0 = 1, a \neq 0$

Quotient rule: $\dfrac{a^m}{a^n} = a^{m-n}, a \neq 0$

Negative exponent: $a^{-n} = \dfrac{1}{a^n}, a \neq 0$

A positive number is written in **scientific notation** if it is written as the product of a number a, where $1 \leq a < 10$, and an integer power of 10: $a \times 10^r$.

$$x^2 \cdot x^3 = x^5$$

$$7^0 = 1, (-10)^0 = 1$$

$$\frac{y^{10}}{y^4} = y^{10-4} = y^6$$

$$3^{-2} = \frac{1}{3^2} = \frac{1}{9}, \frac{x^{-5}}{x^{-7}} = x^{-5-(-7)} = x^2$$

Numbers written in scientific notation

$$568{,}000 = 5.68 \times 10^5$$

$$0.0002117 = 2.117 \times 10^{-4}$$

Section 5.2   More Work with Exponents and Scientific Notation

Power rules:

$$(a^m)^n = a^{m \cdot n}$$
$$(ab)^m = a^m b^m$$
$$\left(\frac{a}{b}\right)^n = \frac{a^n}{b^n}, b \neq 0$$

$$(7^8)^2 = 7^{16}$$
$$(2y)^3 = 2^3 y^3 = 8y^3$$
$$\left(\frac{5x^{-3}}{x^2}\right)^{-2} = \frac{5^{-2} x^6}{x^{-4}}$$
$$= 5^{-2} \cdot x^{6-(-4)}$$
$$= \frac{x^{10}}{5^2}, \text{ or } \frac{x^{10}}{25}$$

Section 5.3   Polynomials and Polynomial Functions

A **polynomial** is a finite sum of terms in which all variables have exponents raised to nonnegative integer powers and no variables appear in a denominator.

Polynomials

$$1.3x^2 \quad \text{(monomial)}$$
$$-\frac{1}{3}y + 5 \quad \text{(binomial)}$$
$$6z^2 - 5z + 7 \quad \text{(trinomial)}$$

A function P is a **polynomial function** if $P(x)$ is a polynomial.

For the polynomial function
$$P(x) = -x^2 + 6x - 12, \text{ find } P(-2)$$
$$P(-2) = -(-2)^2 + 6(-2) - 12 = -28.$$

To add polynomials, combine all like terms.

Add

$$(3y^2x - 2yx + 11) + (-5y^2x - 7)$$
$$= -2y^2x - 2yx + 4$$

To subtract polynomials, change the signs of the terms of the polynomial being subtracted, then add.

Subtract

$$(-2z^3 - z + 1) - (3z^3 + z - 6)$$
$$= -2z^3 - z + 1 - 3z^3 - z + 6$$
$$= -5z^3 - 2z + 7$$

DEFINITIONS AND CONCEPTS	EXAMPLES

Section 5.4 Multiplying Polynomials

To multiply two polynomials, use the distributive property and multiply each term of one polynomial by each term of the other polynomial; then combine like terms.

Special products

$$(a + b)^2 = a^2 + 2ab + b^2$$
$$(a - b)^2 = a^2 - 2ab + b^2$$
$$(a + b)(a - b) = a^2 - b^2$$

The FOIL method may be used when multiplying two binomials.

Multiply

$$(x^2 - 2x)(3x^2 - 5x + 1)$$
$$= 3x^4 - 5x^3 + x^2 - 6x^3 + 10x^2 - 2x$$
$$= 3x^4 - 11x^3 + 11x^2 - 2x$$
$$(3m + 2n)^2 = 9m^2 + 12mn + 4n^2$$
$$(z^2 - 5)^2 = z^4 - 10z^2 + 25$$
$$(7y + 1)(7y - 1) = 49y^2 - 1$$

Multiply

$$(x^2 + 5)(2x^2 - 9)$$
$$\text{F} \qquad \text{O} \qquad \text{I} \qquad \text{L}$$
$$= x^2(2x^2) + x^2(-9) + 5(2x^2) + 5(-9)$$
$$= 2x^4 - 9x^2 + 10x^2 - 45$$
$$= 2x^4 + x^2 - 45$$

Section 5.5 The Greatest Common Factor and Factoring by Grouping

The greatest common factor (GCF) of the terms of a polynomial is the product of the GCF of the numerical coefficients and the GCF of the variable factors.

To factor a polynomial by grouping, group the terms so that each group has a common factor. Factor out these common factors. Then see if the new groups have a common factor.

Factor: $14xy^3 - 2xy^2 = 2 \cdot 7 \cdot x \cdot y^3 - 2 \cdot x \cdot y^2$.
The GCF is $2 \cdot x \cdot y^2$, or $2xy^2$.

$$14xy^3 - 2xy^2 = 2xy^2(7y - 1)$$

Factor $x^4y - 5x^3 + 2xy - 10$.

$$x^4y - 5x^3 + 2xy - 10 = x^3(xy - 5) + 2(xy - 5)$$
$$= (xy - 5)(x^3 + 2)$$

Section 5.6 Factoring Trinomials

To factor $ax^2 + bx + c$ by trial and check:

Step 1. Write all pairs of factors of ax^2.

Step 2. Write all pairs of factors of c.

Step 3. Try combinations of these factors until the middle term bx is found.

To factor $ax^2 + bx + c$ by grouping:

Find two numbers whose product is $a \cdot c$ and whose sum is b, write the term bx using the two numbers found, and then factor by grouping.

Factor: $28x^2 - 27x - 10$.

Factors of $28x^2$: $28x$ and x, $2x$ and $14x$, $4x$ and $7x$.

Factors of -10: -2 and 5, 2 and -5, -10 and 1, 10 and -1.

$$28x^2 - 27x - 10 = (7x + 2)(4x - 5)$$

$$2x^2 - 3x - 5 = 2x^2 - 5x + 2x - 5$$
$$= x(2x - 5) + 1(2x - 5)$$
$$= (2x - 5)(x + 1)$$

Section 5.7 Factoring by Special Products

Perfect square trinomial

$$a^2 + 2ab + b^2 = (a + b)^2$$
$$a^2 - 2ab + b^2 = (a - b)^2$$

Difference of two squares

$$a^2 - b^2 = (a + b)(a - b)$$

Sum and difference of two cubes

$$a^3 + b^3 = (a + b)(a^2 - ab + b^2)$$
$$a^3 - b^3 = (a - b)(a^2 + ab + b^2)$$

Factor

$$25x^2 + 30x + 9 = (5x + 3)^2$$
$$49z^2 - 28z + 4 = (7z - 2)^2$$

$$36x^2 - y^2 = (6x + y)(6x - y)$$

$$8y^3 + 1 = (2y + 1)(4y^2 - 2y + 1)$$
$$27p^3 - 64q^3 = (3p - 4q)(9p^2 + 12pq + 16q^2)$$

DEFINITIONS AND CONCEPTS	EXAMPLES

Section 5.7   Factoring by Special Products (continued)

To factor a polynomial:

Step 1. Factor out the GCF.

Step 2. If the polynomial is a binomial, see if it is a difference of two squares or a sum or difference of two cubes. If it is a trinomial, see if it is a perfect square trinomial. If not, try factoring by methods of Section 5.6. If it is a polynomial with 4 or more terms, try factoring by grouping.

Step 3. See if any factors can be factored further.

Factor $10x^4y + 5x^2y - 15y$.

$$10x^4y + 5x^2y - 15y = 5y(2x^4 + x^2 - 3)$$
$$= 5y(2x^2 + 3)(x^2 - 1)$$
$$= 5y(2x^2 + 3)(x + 1)(x - 1)$$

Section 5.8   Solving Equations by Factoring and Problem Solving

To solve polynomial equations by factoring:

Step 1. Write the equation so that one side is 0.

Step 2. Factor the polynomial completely.

Step 3. Set each factor equal to 0.

Step 4. Solve the resulting equations.

Step 5. Check each solution.

Solve.
$$2x^3 - 5x^2 = 3x$$
$$2x^3 - 5x^2 - 3x = 0$$
$$x(2x + 1)(x - 3) = 0$$
$$x = 0 \quad \text{or} \quad 2x + 1 = 0 \quad \text{or} \quad x - 3 = 0$$
$$x = 0 \quad \text{or} \quad x = -\frac{1}{2} \quad \text{or} \quad x = 3$$

The solutions are $0, -\frac{1}{2}$, and 3.

Chapter 5 **Review**

(5.1) Evaluate.

1. $(-2)^2$

2. $(-3)^4$

3. -2^2

4. -3^4

5. 8^0

6. -9^0

7. -4^{-2}

8. $(-4)^{-2}$

Simplify each expression. Write each answer with positive exponents only. Assume that variables in the exponents represent integers and that all other variables are not 0.

9. $-xy^2 \cdot y^3 \cdot xy^2z$

10. $(-4xy)(-3xy^2b)$

11. $a^{-14} \cdot a^5$

12. $\dfrac{a^{16}}{a^{17}}$

13. $\dfrac{x^{-7}}{x^4}$

14. $\dfrac{9a(a^{-3})}{18a^{15}}$

15. $\dfrac{y^{6p-3}}{y^{6p+2}}$

16. $(3x^{2a+b}y^{-3b})^2$

Write in scientific notation.

17. 36,890,000

18. 0.000362

Write each number in standard notation, without exponents.

19. 1.678×10^{-6}

20. 4.1×10^5

(5.2) Simplify. Use only positive exponents.

21. $(8^5)^3$

22. $\left(\dfrac{a}{4}\right)^2$

23. $(3x)^3$

24. $(-4x)^{-2}$

25. $\left(\dfrac{6x}{5}\right)^2$

26. $(8^6)^{-3}$

27. $\left(\dfrac{4}{3}\right)^{-2}$

28. $(-2x^3)^{-3}$

29. $\left(\dfrac{8p^6}{4p^4}\right)^{-2}$

30. $(-3x^{-2}y^2)^3$

31. $\left(\dfrac{x^{-5}y^{-3}}{z^3}\right)^{-5}$

32. $\dfrac{4^{-1}x^3yz}{x^{-2}yx^4}$

33. $(5xyz)^{-4}(x^{-2})^{-3}$

34. $\dfrac{2(3yz)^{-3}}{y^{-3}}$

Simplify each expression.

35. $x^{4a}(3x^{5a})^3$

36. $\dfrac{4y^{3x-3}}{2y^{2x+4}}$

Use scientific notation to find the quotient. Express each quotient in scientific notation.

37. $\dfrac{(0.00012)(144,000)}{0.0003}$

38. $\dfrac{(-0.00017)(0.00039)}{3000}$

Simplify. Use only positive exponents.

39. $\dfrac{27x^{-5}y^5}{18x^{-6}y^2} \cdot \dfrac{x^4y^{-2}}{x^{-2}y^3}$

40. $\dfrac{3x^5}{y^{-4}} \cdot \dfrac{(3xy^{-3})^{-2}}{(z^{-3})^{-4}}$

(5.3) Find the degree of each polynomial.

41. $x^2y - 3xy^3z + 5x + 7y$

42. $3x + 2$

Simplify by combining like terms.

43. $4x + 8x - 6x^2 - 6x^2y$

44. $-8xy^3 + 4xy^3 - 3x^3y$

Add or subtract as indicated.

45. $(3x + 7y) + (4x^2 - 3x + 7) + (y - 1)$

46. $(4x^2 - 6xy + 9y^2) - (8x^2 - 6xy - y^2)$

47. $(3x^2 - 4b + 28) + (9x^2 - 30) - (4x^2 - 6b + 20)$

48. Add $(9xy + 4x^2 + 18)$ and $(7xy - 4x^3 - 9x)$.

49. Substract $(3x - 5)$ from $2x$.

50. Subtract $(x - 7)$ from the sum of $(3x^2y - 7xy - 4)$ and $(9x^2y + x)$.

51. $\begin{array}{r} x^2 - 5x + 7 \\ -(x + 4) \\ \hline \end{array}$

52. $\begin{array}{r} x^3 \quad\ + 2xy^2 - y \\ + (x - 4xy^2 - 7) \\ \hline \end{array}$

If $P(x) = 9x^2 - 7x + 8$, find the following.

53. $P(6)$ **54.** $P(0)$

55. $P(-2)$ **56.** $P(-3)$

If $P(x) = 2x - 1$ and $Q(x) = x^2 + 2x - 5$, find the following.

57. $P(x) + Q(x)$

58. $2 \cdot P(x) - Q(x)$

△ **59.** Find the perimeter of the rectangle.

$x^2y + 5$ cm

$2x^2y - 6x + 1$ cm

△ **60.** Find the perimeter of the triangle.

$(3x + 11)$ in. $(3x + 11)$ in.

$3x$ in.

(5.4) Multiply.

61. $-6x(4x^2 - 6x + 1)$

62. $-4ab^2(3ab^3 + 7ab + 1)$

63. $(x - 4)(2x + 9)$

64. $(-3xa + 4b)^2$

65. $(9x^2 + 4x + 1)(4x - 3)$

66. $(2x - 1)(x^2 + 2x - 5)$

67. $(5x - 9y)(3x + 9y)$

68. $\left(x - \dfrac{1}{3}\right)\left(x + \dfrac{2}{3}\right)$

69. $(x^2 + 9x + 1)^2$

70. $(m^2 + m - 2)^2$

Multiply, using special products.

71. $(3x - y)^2$

72. $(4x + 9)^2$

73. $(x + 3y)(x - 3y)$

74. $[4 + (3a - b)][4 - (3a - b)]$

△ **75.** Find the area of the rectangle.

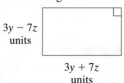

$3y - 7z$ units

$3y + 7z$ units

△ **76.** Find the area of the square.

$(11y + 9)$ units

Multiply. Assume that all variable exponents represent integers.

77. $4a^b(3a^{b+2} - 7)$

78. $(3x^a - 4)(3x^a + 4)$

(5.5) Factor out the greatest common factor.

79. $16x^3 - 24x^2$

80. $36y - 24y^2$

81. $6ab^2 + 8ab - 4a^2b^2$

82. $14a^2b^2 - 21ab^2 + 7ab$

83. $6a(a + 3b) - 5(a + 3b)$

84. $4x(x - 2y) - 5(x - 2y)$

Factor by grouping.

85. $xy - 6y + 3x - 18$

86. $ab - 8b + 4a - 32$

87. $pq - 3p - 5q + 15$

88. $x^3 - x^2 - 2x + 2$

For Exercises 89 and 90, smaller squares are cut from larger rectangles. Write the area of each shaded region as a factored polynomial.

△ **89.**

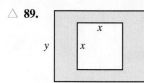

△ **90.**

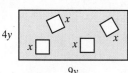

(5.6) *Completely factor each polynomial.*

91. $x^2 - 14x - 72$

92. $x^2 + 16x - 80$

93. $2x^2 - 18x + 28$

94. $3x^2 + 33x + 54$

95. $2x^3 - 7x^2 - 9x$

96. $3x^3 + 2x^2 - 16x$

97. $6x^2 + 17x + 10$

98. $15x^2 - 91x + 6$

99. $4x^2 + 2x - 12$

100. $9x^2 - 12x - 12$

101. $y^2(x + 6)^2 - 2y(x + 6)^2 - 3(x + 6)^2$

102. $(x + 5)^2 + 6(x + 5) + 8$

103. $x^4 - 6x^2 - 16$

104. $x^4 + 8x^2 - 20$

(5.7) *Factor each polynomial completely.*

105. $x^2 - 100$

106. $x^2 - 81$

107. $2x^2 - 32$

108. $6x^2 - 54$

109. $81 - x^4$

110. $16 - y^4$

111. $(y + 2)^2 - 25$

112. $(x - 3)^2 - 16$

113. $x^3 + 216$

114. $y^3 + 512$

115. $8 - 27y^3$

116. $1 - 64y^3$

117. $6x^4y + 48xy$

118. $2x^5 + 16x^2y^3$

119. $x^2 - 2x + 1 - y^2$

120. $x^2 - 6x + 9 - 4y^2$

121. $4x^2 + 12x + 9$

122. $16a^2 - 40ab + 25b^2$

△ **123.** The volume of the cylindrical shell is $(\pi R^2 h - \pi r^2 h)$ cubic units. Write this volume as a factored expression.

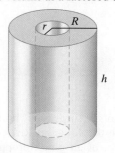

△ **124.** The surface area of the cylindrical shell (except for its top and bottom as drawn) is $(2\pi Rh + 2\pi rh)$ sq units. Write this as a factored expression.

(5.8) *Solve each polynomial equation for the variable.*

125. $(3x - 1)(x + 7) = 0$

126. $3(x + 5)(8x - 3) = 0$

127. $5x(x - 4)(2x - 9) = 0$

128. $6(x + 3)(x - 4)(5x + 1) = 0$

129. $2x^2 = 12x$

130. $4x^3 - 36x = 0$

131. $(1 - x)(3x + 2) = -4x$

132. $2x(x - 12) = -40$

133. $3x^2 + 2x = 12 - 7x$

134. $2x^2 + 3x = 35$

135. $x^3 - 18x = 3x^2$

136. $19x^2 - 42x = -x^3$

137. $12x = 6x^3 + 6x^2$

138. $8x^3 + 10x^2 = 3x$

139. The sum of a number and twice its square is 105. Find the number.

△ **140.** The length of a rectangular piece of carpet is 5 meters less than twice its width. Find the dimensions of the carpet if its area is 33 square meters.

141. A scene from an adventure film calls for a stunt dummy to be dropped from above the second-story platform of the Eiffel Tower, a distance of 400 feet. Its height $h(t)$ at time t seconds is given by

$$h(t) = -16t^2 + 400$$

Determine when the stunt dummy will reach the ground.

400 ft

142. The Royal Gorge suspension bridge in Colorado is 1053 feet above the Arkansas River. Neglecting air resistance, the height of an object dropped off the bridge is given by the polynomial function $P(t) = -16t^2 + 1053$ after time t seconds. Find the height of the object when $t = 1$ second and when $t = 8$ seconds.

MIXED REVIEW

Perform the indicated operation.

143. $(x + 5)(3x^2 - 2x + 1)$

144. $(3x^2 + 4x - 1.2) - (5x^2 - x + 5.7)$

145. $(3x^2 + 4x - 1.2) + (5x^2 - x + 5.7)$

146. $\left(7ab - \dfrac{1}{2}\right)^2$

If $P(x) = -x^2 + x - 4$, find

147. $P(5)$ **148.** $P(-2)$

Factor each polynomial completely.

149. $12y^5 - 6y^4$

150. $x^2y + 4x^2 - 3y - 12$

151. $6x^2 - 34x - 12$

152. $y^2(4x + 3)^2 - 19y(4x + 3)^2 - 20(4x + 3)^2$

153. $4z^7 - 49z^5$

154. $5x^4 + 4x^2 - 9$

Solve each equation.

155. $8x^2 = 24x$ **156.** $x(x - 11) = 26$

Chapter 5 Test   MyMathLab®   CHAPTER Test Prep VIDEOS   ▶   You Tube™

Simplify. Use positive exponents to write the answers.

1. $(-9x)^{-2}$

2. $-3xy^{-2}(4xy^2)z$

3. $\dfrac{6^{-1}a^2b^{-3}}{3^{-2}a^{-5}b^2}$

4. $\left(\dfrac{-xy^{-5}z}{xy^3}\right)^{-5}$

Write Exercises 5 and 6 in scientific notation.

5. 630,000,000

6. 0.01200

7. Write 5×10^{-6} without exponents.

8. Use scientific notation to find the quotient.

$$\dfrac{(0.0024)(0.00012)}{0.00032}$$

Perform the indicated operations.

9. $(4x^3y - 3x - 4) - (9x^3y + 8x + 5)$

10. $-3xy(4x + y)$

11. $(3x + 4)(4x - 7)$

12. $(5a - 2b)(5a + 2b)$

13. $(6m + n)^2$

14. $(2x - 1)(x^2 - 6x + 4)$

Factor each polynomial completely.

15. $16x^3y - 12x^2y^4$

16. $x^2 - 13x - 30$

17. $4y^2 + 20y + 25$

18. $6x^2 - 15x - 9$

19. $4x^2 - 25$

20. $x^3 + 64$

21. $3x^2y - 27y^3$

22. $6x^2 + 24$

23. $16y^3 - 2$

24. $x^2y - 9y - 3x^2 + 27$

Solve the equation for the variable.

25. $3n(7n - 20) = 96$

26. $(x + 2)(x - 2) = 5(x + 4)$

27. $2x^3 + 5x^2 = 8x + 20$

28. Write the area of the shaded region as a factored polynomial.

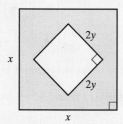

29. A pebble is hurled upward from the top of the Canada Trust Tower, which is 880 feet tall, with an initial velocity of 96 feet per second. Neglecting air resistance, the height $h(t)$ of the pebble after t seconds is given by the polynomial function

$$h(t) = -16t^2 + 96t + 880$$

a. Find the height of the pebble when $t = 1$.

b. Find the height of the pebble when $t = 5.1$.

c. Factor the polynomial $-16t^2 + 96t + 880$.

d. When will the pebble hit the ground?

Chapter 5 Cumulative Review

1. Find the roots.
 a. $\sqrt[3]{27}$
 b. $\sqrt[5]{1}$
 c. $\sqrt[4]{16}$

2. Find the roots.
 a. $\sqrt[3]{64}$
 b. $\sqrt[4]{81}$
 c. $\sqrt[5]{32}$

3. Solve: $2(x - 3) = 5x - 9$.

4. Solve: $0.3y + 2.4 = 0.1y + 4$

5. Karen Estes just received an inheritance of $10,000 and plans to place all the money in a savings account that pays 5% compounded quarterly to help her son go to college in 3 years. How much money will be in the account in 3 years?

6. A gallon of latex paint can cover 400 square feet. How many gallon containers of paint should be bought to paint two coats on each wall of a rectangular room whose dimensions are 14 feet by 18 feet? (Assume 8-foot ceilings.)

7. Solve and graph the solution set. Write the solution set in interval notation.
 a. $\frac{1}{4}x \le \frac{3}{8}$
 b. $-2.3x < 6.9$

8. Solve. Graph the solution set. Write the solution set in interval notation. $x + 2 \le \frac{1}{4}(x - 7)$

Solve.

9. $-1 \le \frac{2x}{3} + 5 \le 2$

10. $-\frac{1}{3} < \frac{3x + 1}{6} \le \frac{1}{3}$

11. $|y| = 0$

12. $8 + |4c| = 24$

13. $\left|2x - \frac{1}{10}\right| < -13$

14. Solve: $|5x - 1| + 9 > 5$

15. Graph the linear equation $y = \frac{1}{3}x$.

16. Graph the linear equation $y = 3x$.

17. Evaluate $f(2), f(-6),$ and $f(0)$ for the function
$$f(x) = \begin{cases} 2x + 3 & \text{if } x \le 0 \\ -x - 1 & \text{if } x > 0 \end{cases}$$
Write your results in ordered pair form.

18. If $f(x) = 3x^2 + 2x + 3$, find $f(-3)$.

19. Graph $x = 2$.

20. Graph $y - 5 = 0$

21. Find the slope of the line $y = 2$.

22. Find the slope of the line $f(x) = -2x - 3$.

23. Find an equation of the horizontal line containing the point $(2, 3)$.

24. Find the equation of the vertical line containing the point $(-3, 2)$.

25. Graph the union of $x + \frac{1}{2}y \ge -4$ or $y \le -2$.

26. Find the equation of the line containing the point $(-2, 3)$ and slope of 0.

27. Use the substitution method to solve the system.
$$\begin{cases} 2x + 4y = -6 \\ x = 2y - 5 \end{cases}$$

28. Use the substitution method to solve the system.
$$\begin{cases} 4x - 2y = 8 \\ y = 3x - 6 \end{cases}$$

29. Solve the system. $\begin{cases} 2x + 4y = 1 \\ 4x - 4z = -1 \\ y - 4z = -3 \end{cases}$

30. Solve the system. $\begin{cases} x + y - \frac{3}{2}z = \frac{1}{2} \\ -y - 2z = 14 \\ x - \frac{2}{3}y = -\frac{1}{3} \end{cases}$

31. A first number is 4 less than a second number. Four times the first number is 6 more than twice the second. Find the numbers.

32. One solution contains 20% acid and a second solution contains 60% acid. How many ounces of each solution should be mixed to have 50 ounces of a 30% acid solution?

33. Use matrices to solve the system. $\begin{cases} 2x - y = 3 \\ 4x - 2y = 5 \end{cases}$

34. Use matrices to solve the system. $\begin{cases} 4y = 8 \\ x + y = 7 \end{cases}$

35. The measure of the largest angle of a triangle is 80° more than the measure of the smallest angle, and the measure of the remaining angle is 10° more than the measure of the smallest angle. Find the measure of each angle.

36. Find an equation of the line with slope $\dfrac{1}{2}$ through the point $(0, 5)$. Write the equation using function notation.

37. Write each number in scientific notation.

  a. 730,000

  b. 0.00000104

38. Write each number in scientific notation.

  a. 8,250,000

  b. 0.0000346

39. Simplify each expression. Use positive exponents to write the answers.

  a. $(2x^0y^{-3})^{-2}$ **b.** $\left(\dfrac{x^{-5}}{x^{-2}}\right)^{-3}$

  c. $\left(\dfrac{2}{7}\right)^{-2}$ **d.** $\dfrac{5^{-2}x^{-3}y^{11}}{x^2y^{-5}}$

40. Simplify each expression. Use positive exponents to write the answers.

  a. $(4a^{-1}b^0)^{-3}$ **b.** $\left(\dfrac{a^{-6}}{a^{-8}}\right)^{-2}$

  c. $\left(\dfrac{2}{3}\right)^{-3}$ **d.** $\dfrac{3^{-2}a^{-2}b^{12}}{a^4b^{-5}}$

41. Find the degree of the polynomial $3xy + x^2y^2 - 5x^2 - 6.7$.

42. Subtract $(5x^2 + 3x)$ from $(3x^2 - 2x)$.

43. Multiply.

  a. $(2x^3)(5x^6)$

  b. $(7y^4z^4)(-xy^{11}z^5)$

44. Multiply.

  a. $(3y^6)(4y^2)$

  b. $(6a^3b^2)(-a^2bc^4)$

Factor.

45. $17x^3y^2 - 34x^4y^2$

46. $12x^3y - 3xy^3$

47. $x^2 + 10x + 16$

48. $5a^2 + 14a - 3$

Solve.

49. $2x^2 + 9x - 5 = 0$

50. $3x^2 - 10x - 8 = 0$

CHAPTER 6 Rational Expressions

During the past 10 years, music has changed from a physical album, tape, or CD to a digital form. This has enabled consumers to download music and move it among different players. In the year 2012, it is predicted that consumer spending on CDs will fall below consumer spending on digital music. In the Chapter Review, Exercises 107 and 108, we will review the future of digital sales.

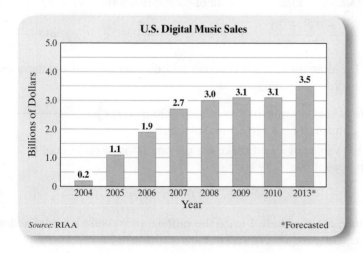

Polynomials are to algebra what integers are to arithmetic. We have added, subtracted, multiplied, and raised polynomials to powers, each operation yielding another polynomial, just as these operations on integers yield another integer. But when we divide one integer by another, the result may or may not be another integer. Likewise, when we divide one polynomial by another, we may or may not get a polynomial in return. The quotient $x \div (x + 1)$ is not a polynomial; it is a *rational expression* that can be written as $\dfrac{x}{x + 1}$.

In this chapter, we study these new algebraic forms known as rational expressions and the *rational functions* they generate.

6.1 | Rational Functions and Multiplying and Dividing Rational Expressions

Recall that a *rational number,* or *fraction,* is a number that can be written as the quotient $\frac{p}{q}$ of two integers p and q as long as q is not 0. A **rational expression** is an expression that can be written as the quotient $\frac{P}{Q}$ of two polynomials P and Q as long as Q is not 0.

Examples of Rational Expressions

$$\frac{3x + 7}{2} \qquad \frac{5x^2 - 3}{x - 1} \qquad \frac{7x - 2}{2x^2 + 7x + 6}$$

Rational expressions are sometimes used to describe functions. For example, we call the function $f(x) = \frac{x^2 + 2}{x - 3}$ a **rational function** since $\frac{x^2 + 2}{x - 3}$ is a rational expression.

OBJECTIVE

1 Finding the Domain of a Rational Expression

As with fractions, a rational expression is **undefined** if the denominator is 0. If a variable in a rational expression is replaced with a number that makes the denominator 0, we say that the rational expression is **undefined** for this value of the variable. For example, the rational expression $\frac{x^2 + 2}{x - 3}$ is undefined when x is 3 because replacing x with 3 results in a denominator of 0. For this reason, we must exclude 3 from the domain of the function $f(x) = \frac{x^2 + 2}{x - 3}$.

The domain of f is then

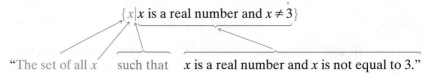

$$\{x \mid x \text{ is a real number and } x \neq 3\}$$

"The set of all x such that x is a real number and x is not equal to 3."

In this section, we will use this set builder notation to write domains. Unless told otherwise, we assume that the domain of a function described by an equation is the set of all real numbers for which the equation is defined.

EXAMPLE 1   Find the domain of each rational function.

a. $f(x) = \dfrac{8x^3 + 7x^2 + 20}{2}$ **b.** $g(x) = \dfrac{5x^2 - 3}{x - 1}$ **c.** $f(x) = \dfrac{7x - 2}{x^2 - 2x - 15}$

Solution   The domain of each function will contain all real numbers except those values that make the denominator 0.

a. No matter what the value of x, the denominator of $f(x) = \dfrac{8x^3 + 7x^2 + 20}{2}$ is never 0, so the domain of f is $\{x \mid x \text{ is a real number}\}$.

b. To find the values of x that make the denominator of $g(x)$ equal to 0, we solve the equation "denominator = 0":

$$x - 1 = 0 \quad \text{or} \quad x = 1$$

The domain must exclude 1 since the rational expression is undefined when x is 1. The domain of g is $\{x \mid x \text{ is a real number and } x \neq 1\}$.

c. We find the domain by setting the denominator equal to 0.

$$x^2 - 2x - 15 = 0 \quad \text{Set the denominator equal to 0 and solve.}$$
$$(x - 5)(x + 3) = 0$$

$$x - 5 = 0 \quad \text{or} \quad x + 3 = 0$$
$$x = 5 \quad \text{or} \quad x = -3$$

If x is replaced with 5 or with -3, the rational expression is undefined.

The domain of f is $\{x \mid x \text{ is a real number and } x \neq 5, x \neq -3\}$.

PRACTICE
1   Find the domain of each rational function.

a. $f(x) = \dfrac{4x^5 - 3x^2 + 2}{-6}$ **b.** $g(x) = \dfrac{6x^2 + 1}{x + 3}$ **c.** $h(x) = \dfrac{8x - 3}{x^2 - 5x + 6}$

✓CONCEPT CHECK

For which of these values (if any) is the rational expression $\dfrac{x - 3}{x^2 + 2}$ undefined?

a. 2   **b.** 3   **c.** -2 **d.** 0   **e.** None of these

OBJECTIVE
2   Simplifying Rational Expressions

Recall that a fraction is in lowest terms or simplest form if the numerator and denominator have no common factors other than 1 (or -1). For example, $\dfrac{3}{13}$ is in lowest terms since 3 and 13 have no common factors other than 1 (or -1).

To **simplify** a rational expression, or to write it in lowest terms, we use a method similar to simplifying a fraction.

Recall that to simplify a fraction, we essentially "remove factors of 1." Our ability to do this comes from these facts:

- If $c \neq 0$, then $\dfrac{c}{c} = 1$. For example, $\dfrac{7}{7} = 1$ and $\dfrac{-8.65}{-8.65} = 1$.

- $n \cdot 1 = n$. For example, $-5 \cdot 1 = -5$, $126.8 \cdot 1 = 126.8$, and $\dfrac{a}{b} \cdot 1 = \dfrac{a}{b}, b \neq 0$.

In other words, we have the following:

$$\dfrac{a \cdot c}{b \cdot c} = \underbrace{\dfrac{a}{b} \cdot \dfrac{c}{c}}_{} = \dfrac{a}{b} \uparrow$$

$$\text{Since } \tfrac{a}{b} \cdot 1 = \tfrac{a}{b}$$

Let's practice simplifying a fraction by simplifying $\dfrac{15}{65}$.

$$\dfrac{15}{65} = \dfrac{3 \cdot 5}{13 \cdot 5} = \dfrac{3}{13} \cdot \dfrac{5}{5} = \dfrac{3}{13} \cdot 1 = \dfrac{3}{13}$$

Let's use the same technique and simplify the rational expression $\dfrac{(x + 2)^2}{x^2 - 4}$.

$$\dfrac{(x + 2)^2}{x^2 - 4} = \dfrac{(x + 2)\,(x + 2)}{(x - 2)\,(x + 2)}$$

$$= \dfrac{(x + 2)}{(x - 2)} \cdot \dfrac{x + 2}{x + 2}$$

$$= \dfrac{x + 2}{x - 2} \cdot 1$$

$$= \dfrac{x + 2}{x - 2}$$

This means that the rational expression $\dfrac{(x+2)^2}{x^2-4}$ has the same value as the rational expression $\dfrac{x+2}{x-2}$ for all values of x except 2 and -2. (Remember that when x is 2, the denominators of both rational expressions are 0 and that when x is -2, the original rational expression has a denominator of 0.)

As we simplify rational expressions, we will assume that the simplified rational expression is equivalent to the original rational expression for all real numbers except those for which either denominator is 0.

Just as for numerical fractions, we can use a shortcut notation. Remember that as long as exact factors in both the numerator and denominator are divided out, we are "removing a factor of 1." We can use the following notation:

$$\frac{(x+2)^2}{x^2-4} = \frac{(x+2)\,(x+2)}{(x-2)\,(x+2)} \qquad \text{A factor of 1 is identified by the shading.}$$

$$= \frac{x+2}{x-2} \qquad \text{"Remove" the factor of 1.}$$

This "removing a factor of 1" is stated in the principle below:

Fundamental Principle of Rational Expressions

For any rational expression $\dfrac{P}{Q}$ and any polynomial R, where $R \neq 0$,

$$\frac{PR}{QR} = \frac{P}{Q} \cdot \frac{R}{R} = \frac{P}{Q} \cdot 1 = \frac{P}{Q}$$

or, simply,

$$\frac{PR}{QR} = \frac{P}{Q}$$

In general, the following steps may be used to simplify rational expressions or to write a rational expression in lowest terms.

Simplifying or Writing a Rational Expression in Lowest Terms

Step 1. Completely factor the numerator and denominator of the rational expression.

Step 2. Divide out factors common to the numerator and denominator. (This is the same as "removing a factor of 1.")

For now, we assume that variables in a rational expression do not represent values that make the denominator 0.

EXAMPLE 2 Simplify each rational expression.

a. $\dfrac{2x^2}{10x^3 - 2x^2}$ **b.** $\dfrac{9x^2 + 13x + 4}{8x^2 + x - 7}$

Solution

a. $\dfrac{2x^2}{10x^3 - 2x^2} = \dfrac{2x^2 \cdot 1}{2x^2\,(5x-1)} = 1 \cdot \dfrac{1}{5x-1} = \dfrac{1}{5x-1}$

b. $\dfrac{9x^2 + 13x + 4}{8x^2 + x - 7} = \dfrac{(9x + 4)\,(x + 1)}{(8x - 7)\,(x + 1)}$ Factor the numerator and denominator.

$\qquad\qquad\qquad = \dfrac{9x + 4}{8x - 7} \cdot 1$ Since $\dfrac{x + 1}{x + 1} = 1$

$\qquad\qquad\qquad = \dfrac{9x + 4}{8x - 7}$ Simplest form   □

PRACTICE
2   Simplify each rational expressions.

a. $\dfrac{5z^4}{10z^5 - 5z^4}$ **b.** $\dfrac{5x^2 + 13x + 6}{6x^2 + 7x - 10}$

EXAMPLE 3   Simplify each rational expression.

a. $\dfrac{2 + x}{x + 2}$ **b.** $\dfrac{2 - x}{x - 2}$

Solution

a. $\dfrac{2 + x}{x + 2} = \dfrac{x + 2}{x + 2} = 1$ By the commutative property of addition, $2 + x = x + 2$.

b. $\dfrac{2 - x}{x - 2}$

The terms in the numerator of $\dfrac{2 - x}{x - 2}$ differ by sign from the terms of the denominator, so the polynomials are opposites of each other and the expression simplifies to -1. To see this, we factor out -1 from the numerator or the denominator. If -1 is factored from the numerator, then

$$\dfrac{2 - x}{x - 2} = \dfrac{-1(-2 + x)}{x - 2} = \dfrac{-1\,(x - 2)}{x - 2} = \dfrac{-1}{1} = -1$$

If -1 is factored from the denominator, the result is the same.

> **Helpful Hint**
> When the numerator and the denominator of a rational expression are opposites of each other, the expression simplifies to -1.

$$\dfrac{2 - x}{x - 2} = \dfrac{2 - x}{-1(-x + 2)} = \dfrac{2 - x}{-1\,(2 - x)} = \dfrac{1}{-1} = -1 \qquad □$$

PRACTICE
3   Simplify each rational expression.

a. $\dfrac{x + 3}{3 + x}$ **b.** $\dfrac{3 - x}{x - 3}$

EXAMPLE 4   Simplify $\dfrac{18 - 2x^2}{x^2 - 2x - 3}$.

Solution $\dfrac{18 - 2x^2}{x^2 - 2x - 3} = \dfrac{2(9 - x^2)}{(x + 1)(x - 3)}$ Factor.

$\qquad\qquad\qquad\qquad = \dfrac{2(3 + x)(3 - x)}{(x + 1)(x - 3)}$ Factor completely.

$\qquad\qquad\qquad\qquad = \dfrac{2(3 + x) \cdot -1\,(x - 3)}{(x + 1)\,(x - 3)}$ Notice the opposites $3 - x$ and $x - 3$. Write $3 - x$ as $-1(x - 3)$ and simplify.

$\qquad\qquad\qquad\qquad = -\dfrac{2(3 + x)}{x + 1}$ □

PRACTICE
4   Simplify $\dfrac{20 - 5x^2}{x^2 + x - 6}$.

▶ Helpful Hint

Recall that for a fraction,

$$\frac{a}{-b} = \frac{-a}{b} = -\frac{a}{b}$$

For example

$$\frac{-(x+1)}{(x+2)} = \frac{(x+1)}{-(x+2)} = -\frac{x+1}{x+2}$$

✓CONCEPT CHECK

Which of the following expressions are equivalent to $\dfrac{x}{8-x}$?

a. $\dfrac{-x}{x-8}$ **b.** $\dfrac{-x}{8-x}$ **c.** $\dfrac{x}{x-8}$ **d.** $\dfrac{-x}{-8+x}$

EXAMPLE 5   Simplify each rational expression.

a. $\dfrac{x^3 + 8}{2 + x}$ **b.** $\dfrac{2y^2 + 2}{y^3 - 5y^2 + y - 5}$

Solution

a. $\dfrac{x^3 + 8}{2 + x} = \dfrac{(x+2)(x^2 - 2x + 4)}{x + 2}$ Factor the sum of the two cubes.

$= x^2 - 2x + 4$ Divide out common factors.

b. $\dfrac{2y^2 + 2}{y^3 - 5y^2 + y - 5} = \dfrac{2(y^2 + 1)}{(y^3 - 5y^2) + (y - 5)}$ Factor the numerator.

$= \dfrac{2(y^2 + 1)}{y^2(y - 5) + 1(y - 5)}$ Factor the denominator by grouping.

$= \dfrac{2(y^2 + 1)}{(y - 5)(y^2 + 1)}$

$= \dfrac{2}{y - 5}$ Divide out common factors.   □

PRACTICE
5   Simplify each rational expression.

a. $\dfrac{x^3 + 64}{4 + x}$ **b.** $\dfrac{5z^2 + 10}{z^3 - 3z^2 + 2z - 6}$

✓CONCEPT CHECK

Does $\dfrac{n}{n + 2}$ simplify to $\dfrac{1}{2}$? Why or why not?

Answers to Concept Checks:
a and d
no; answers may vary.

OBJECTIVE
3   Multiplying Rational Expressions

Arithmetic operations on rational expressions are performed in the same way as they are on rational numbers.

Multiplying Rational Expressions

The rule for multiplying rational expressions is

$$\frac{P}{Q} \cdot \frac{R}{S} = \frac{PR}{QS} \quad \text{as long as } Q \neq 0 \text{ and } S \neq 0.$$

To multiply rational expressions, you may use these steps:

Step 1. Completely factor each numerator and denominator.

Step 2. Use the rule above and multiply the numerators and the denominators.

Step 3. Simplify the product by dividing the numerator and denominator by their common factors.

When we multiply rational expressions, notice that we factor each numerator and denominator first. This helps when we apply the fundamental principle to write the product in simplest form.

EXAMPLE 6   Multiply.

a. $\dfrac{1 + 3n}{2n} \cdot \dfrac{2n - 4}{3n^2 - 2n - 1}$ **b.** $\dfrac{x^3 - 1}{-3x + 3} \cdot \dfrac{15x^2}{x^2 + x + 1}$

Solution

a. $\dfrac{1 + 3n}{2n} \cdot \dfrac{2n - 4}{3n^2 - 2n - 1} = \dfrac{1 + 3n}{2n} \cdot \dfrac{2(n - 2)}{(3n + 1)(n - 1)}$ Factor.

$$= \frac{(1 + 3n) \cdot 2(n - 2)}{2n(3n + 1)(n - 1)}$$ Multiply.

$$= \frac{n - 2}{n(n - 1)}$$ Divide out common factors.

b. $\dfrac{x^3 - 1}{-3x + 3} \cdot \dfrac{15x^2}{x^2 + x + 1} = \dfrac{(x - 1)(x^2 + x + 1)}{-3(x - 1)} \cdot \dfrac{3 \cdot 5x^2}{x^2 + x + 1}$ Factor.

$$= \frac{(x - 1)(x^2 + x + 1) \cdot 3 \cdot 5x^2}{-1 \cdot 3(x - 1)(x^2 + x + 1)}$$ Multiply.

$$= \frac{5x^2}{-1} = -5x^2$$ Simplest form   □

PRACTICE

6   Multiply.

a. $\dfrac{2 + 5n}{3n} \cdot \dfrac{6n + 3}{5n^2 - 3n - 2}$ **b.** $\dfrac{x^3 - 8}{-6x + 12} \cdot \dfrac{6x^2}{x^2 + 2x + 4}$

OBJECTIVE

4 **Dividing Rational Expressions**

Recall that two numbers are reciprocals of each other if their product is 1. Similarly, if $\dfrac{P}{Q}$ is a rational expression, then $\dfrac{Q}{P}$ is its **reciprocal,** since

$$\frac{P}{Q} \cdot \frac{Q}{P} = \frac{P \cdot Q}{Q \cdot P} = 1$$

The following are examples of expressions and their reciprocals.

Expression		Reciprocal
$\dfrac{3}{x}$	$\longrightarrow$	$\dfrac{x}{3}$
$\dfrac{2 + x^2}{4x - 3}$	$\longrightarrow$	$\dfrac{4x - 3}{2 + x^2}$
x^3	$\longrightarrow$	$\dfrac{1}{x^3}$
0	$\longrightarrow$	no reciprocal

Dividing Rational Expressions

The rule for dividing rational expressions is

$$\frac{P}{Q} \div \frac{R}{S} = \frac{P}{Q} \cdot \frac{S}{R} = \frac{PS}{QR} \quad \text{as long as } Q \neq 0, S \neq 0, \text{ and } R \neq 0.$$

To divide by a rational expression, use the rule above and multiply by its reciprocal. Then simplify if possible.

Notice that division of rational expressions is the same as for rational numbers.

EXAMPLE 7 Divide.

a. $\dfrac{8m^2}{3m^2 - 12} \div \dfrac{40}{2 - m}$ **b.** $\dfrac{18y^2 + 9y - 2}{24y^2 - 10y + 1} \div \dfrac{3y^2 + 17y + 10}{8y^2 + 18y - 5}$

Solution

a. $\dfrac{8m^2}{3m^2 - 12} \div \dfrac{40}{2 - m} = \dfrac{8m^2}{3m^2 - 12} \cdot \dfrac{2 - m}{40}$ Multiply by the reciprocal of the divisor.

$= \dfrac{8m^2(2 - m)}{3(m + 2)(m - 2) \cdot 40}$ Factor and multiply.

$= \dfrac{8m^2 \cdot -1\,(m - 2)}{3(m + 2)\,(m - 2) \cdot 8 \cdot 5}$ Write $(2 - m)$ as $-1(m - 2)$.

$= -\dfrac{m^2}{15(m + 2)}$ Simplify.

b. $\dfrac{18y^2 + 9y - 2}{24y^2 - 10y + 1} \div \dfrac{3y^2 + 17y + 10}{8y^2 + 18y - 5}$

$= \dfrac{18y^2 + 9y - 2}{24y^2 - 10y + 1} \cdot \dfrac{8y^2 + 18y - 5}{3y^2 + 17y + 10}$ Multiply by the reciprocal.

$= \dfrac{(6y - 1)\,(3y + 2)}{(6y - 1)\,(4y - 1)} \cdot \dfrac{(4y - 1)\,(2y + 5)}{(3y + 2)\,(y + 5)}$ Factor.

$= \dfrac{2y + 5}{y + 5}$ Simplest form □

PRACTICE
7 Divide.

a. $\dfrac{6y^3}{3y^2 - 27} \div \dfrac{42}{3 - y}$ **b.** $\dfrac{10x^2 + 23x - 5}{5x^2 - 51x + 10} \div \dfrac{2x^2 + 9x + 10}{7x^2 - 68x - 20}$

> ▶ **Helpful Hint**
> When dividing rational expressions, do not divide out common factors until the division problem is rewritten as a multiplication problem.

EXAMPLE 8   Perform each indicated operation.

$$\frac{x^2 - 25}{(x + 5)^2} \cdot \frac{3x + 15}{4x} \div \frac{x^2 - 3x - 10}{x}$$

Solution $\dfrac{x^2 - 25}{(x + 5)^2} \cdot \dfrac{3x + 15}{4x} \div \dfrac{x^2 - 3x - 10}{x}$

$$= \frac{x^2 - 25}{(x + 5)^2} \cdot \frac{3x + 15}{4x} \cdot \frac{x}{x^2 - 3x - 10} \qquad \text{To divide, multiply by the reciprocal}$$

$$= \frac{(x + 5)(x - 5)}{(x + 5)(x + 5)} \cdot \frac{3(x + 5)}{4x} \cdot \frac{x}{(x - 5)(x + 2)}$$

$$= \frac{3}{4(x + 2)} \qquad \qquad \square$$

PRACTICE
8   Perform each indicated operation.

$$\frac{x^2 - 16}{(x - 4)^2} \cdot \frac{5x - 20}{3x} \div \frac{x^2 + x - 12}{x}$$

OBJECTIVE
5   Using Rational Functions in Applications

Rational functions occur often in real-life situations.

EXAMPLE 9   **Cost for Pressing Compact Discs**

For the ICL Production Company, the rational function $C(x) = \dfrac{2.6x + 10,000}{x}$ describes the company's cost per disc for pressing x compact discs. Find the cost per disc for pressing:

a. 100 compact discs
b. 1000 compact discs

Solution

a. $C(100) = \dfrac{2.6(100) + 10,000}{100} = \dfrac{10,260}{100} = 102.6$

The cost per disc for pressing 100 compact discs is $102.60.

b. $C(1000) = \dfrac{2.6(1000) + 10,000}{1000} = \dfrac{12,600}{1000} = 12.6$

The cost per disc for pressing 1000 compact discs is $12.60. Notice that as more compact discs are produced, the cost per disc decreases. $\qquad \square$

PRACTICE
9   A company's cost per tee shirt for silk screening x tee shirts is given by the rational function $C(x) = \dfrac{3.2x + 400}{x}$. Find the cost per tee shirt for printing:

a. 100 tee shirts   **b.** 1000 tee shirts

Graphing Calculator Explorations

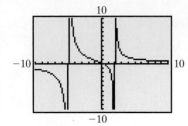

Recall that since the rational expression $\dfrac{7x - 2}{(x - 2)(x + 5)}$ is not defined when $x = 2$ or when $x = -5$, we say that the domain of the rational function $f(x) = \dfrac{7x - 2}{(x - 2)(x + 5)}$ is all real numbers except 2 and -5. This domain can be written as $\{x \mid x$ is a real number and $x \neq 2, x \neq -5\}$. This means that the graph of $f(x)$ should not cross the vertical lines $x = 2$ and $x = -5$. The graph of $f(x)$ in *connected* mode is to the left. In connected mode, the graphing calculator tries to connect all dots of the graph so that the result is a smooth curve. This is what has happened in the graph. Notice that the graph appears to contain vertical lines at $x = 2$ and at $x = -5$. We know that this cannot happen because the function is not defined at $x = 2$ and at $x = -5$. We also know that this cannot happen because the graph of this function would not pass the vertical line test.

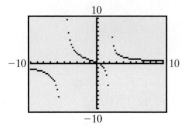

   The graph of $f(x)$ in *dot* mode is to the left. In dot mode, the graphing calculator will not connect dots with a smooth curve. Notice that the vertical lines have disappeared, and we have a better picture of the graph. The graph, however, actually appears more like the hand-drawn graph below. By using a Table feature, a Calculate Value feature, or by tracing, we can see that the function is not defined at $x = 2$ and at $x = -5$.

Find the domain of each rational function. Then graph each rational function and use the graph to confirm the domain.

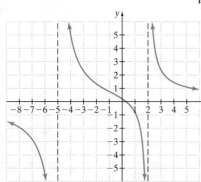

1. $f(x) = \dfrac{x + 1}{x^2 - 4}$

2. $g(x) = \dfrac{5x}{x^2 - 9}$

3. $h(x) = \dfrac{x^2}{2x^2 + 7x - 4}$

4. $f(x) = \dfrac{3x + 2}{4x^2 - 19x - 5}$

Vocabulary, Readiness & Video Check

Use the choices below to fill in each blank. Some choices may not be used.

1	true	rational	simplified	$\dfrac{-a}{-b}$	$\dfrac{-a}{b}$	$\dfrac{a}{-b}$
-1	false	domain	0			

1. A _____ expression is an expression that can be written as the quotient $\dfrac{P}{Q}$ of two polynomials P and Q as long as $Q \neq 0$.

2. A rational expression is undefined if the denominator is _____.

3. The _____ of the rational function $f(x) = \dfrac{2}{x}$ is $\{x \mid x$ is a real number and $x \neq 0\}$.

4. A rational expression is _____ if the numerator and denominator have no common factors other than 1 or -1.

5. The expression $\dfrac{x^2 + 2}{2 + x^2}$ simplifies to _____.

6. The expression $\dfrac{y - z}{z - y}$ simplifies to _____.

7. For a rational expression, $-\dfrac{a}{b} =$ _____ $=$ _____.

8. True or false: $\dfrac{a - 6}{a + 2} = \dfrac{-(a - 6)}{-(a + 2)} = \dfrac{-a + 6}{-a - 2}.$ _____

Martin-Gay Interactive Videos

See Video 6.1 🍐

Watch the section lecture video and answer the following questions.

OBJECTIVE 1

9. From Examples 1 and 2, why can't the denominators of rational expressions be zero? How does this relate to the domain of a rational function?

OBJECTIVE 2

10. What conclusion does Example 4 lead us to?

OBJECTIVE 3

11. Does a person need to be comfortable with factoring polynomials in order to be successful with multiplying rational expressions? Explain, referencing ▭ Example 6 in your answer.

OBJECTIVE 4

12. Based on ▭ Example 7, complete the following statements. Division of rational expressions is similar to division of _____. Therefore, to divide rational expressions, it's the first rational expression times the _____ of the second rational expression.

OBJECTIVE 5

13. In ▭ Example 8d, we find the domain of the function. Do you think the domain should be restricted in any way given the context of this application? Explain.

6.1 Exercise Set   MyMathLab®

Find the domain of each rational function. See Example 1.

1. $f(x) = \dfrac{5x - 7}{4}$

2. $g(x) = \dfrac{4 - 3x}{2}$

3. $s(t) = \dfrac{t^2 + 1}{2t}$

4. $v(t) = -\dfrac{5t + t^2}{3t}$

▶ **5.** $f(x) = \dfrac{3x}{7 - x}$

6. $f(x) = \dfrac{-4x}{-2 + x}$

7. $f(x) = \dfrac{x}{3x - 1}$

8. $g(x) = \dfrac{-2}{2x + 5}$

9. $R(x) = \dfrac{3 + 2x}{x^3 + x^2 - 2x}$

10. $h(x) = \dfrac{5 - 3x}{2x^2 - 14x + 20}$

▶ **11.** $C(x) = \dfrac{x + 3}{x^2 - 4}$

12. $R(x) = \dfrac{5}{x^2 - 7x}$

Simplify each rational expression. See Examples 2 through 5.

▶ **13.** $\dfrac{8x - 16x^2}{8x}$

14. $\dfrac{3x - 6x^2}{3x}$

15. $\dfrac{x^2 - 9}{3 + x}$

16. $\dfrac{x^2 - 25}{5 + x}$

17. $\dfrac{9y - 18}{7y - 14}$

18. $\dfrac{6y - 18}{2y - 6}$

▶ **19.** $\dfrac{x^2 + 6x - 40}{x + 10}$

20. $\dfrac{x^2 - 8x + 16}{x - 4}$

▶ **21.** $\dfrac{x - 9}{9 - x}$

22. $\dfrac{x - 4}{4 - x}$

23. $\dfrac{x^2 - 49}{7 - x}$

24. $\dfrac{x^2 - y^2}{y - x}$

25. $\dfrac{2x^2 - 7x - 4}{x^2 - 5x + 4}$

26. $\dfrac{3x^2 - 11x + 10}{x^2 - 7x + 10}$

27. $\dfrac{x^3 - 125}{2x - 10}$

28. $\dfrac{4x + 4}{x^3 + 1}$

29. $\dfrac{3x^2 - 5x - 2}{6x^3 + 2x^2 + 3x + 1}$

30. $\dfrac{2x^2 - x - 3}{2x^3 - 3x^2 + 2x - 3}$

31. $\dfrac{9x^2 - 15x + 25}{27x^3 + 125}$

32. $\dfrac{8x^3 - 27}{4x^2 + 6x + 9}$

Multiply and simplify. See Example 6.

33. $\dfrac{2x - 4}{15} \cdot \dfrac{6}{2 - x}$

34. $\dfrac{10 - 2x}{7} \cdot \dfrac{14}{5x - 25}$

35. $\dfrac{18a - 12a^2}{4a^2 + 4a + 1} \cdot \dfrac{4a^2 + 8a + 3}{4a^2 - 9}$

36. $\dfrac{a - 5b}{a^2 + ab} \cdot \dfrac{b^2 - a^2}{10b - 2a}$

▶ **37.** $\dfrac{9x + 9}{4x + 8} \cdot \dfrac{2x + 4}{3x^2 - 3}$

38. $\dfrac{2x^2 - 2}{10x + 30} \cdot \dfrac{12x + 36}{3x - 3}$

39. $\dfrac{2x^3 - 16}{6x^2 + 6x - 36} \cdot \dfrac{9x + 18}{3x^2 + 6x + 12}$

40. $\dfrac{x^2 - 3x + 9}{5x^2 - 20x - 105} \cdot \dfrac{x^2 - 49}{x^3 + 27}$

41. $\dfrac{a^3 + a^2b + a + b}{5a^3 + 5a} \cdot \dfrac{6a^2}{2a^2 - 2b^2}$

42. $\dfrac{4a^2 - 8a}{ab - 2b + 3a - 6} \cdot \dfrac{8b + 24}{3a + 6}$

43. $\dfrac{x^2 - 6x - 16}{2x^2 - 128} \cdot \dfrac{x^2 + 16x + 64}{3x^2 + 30x + 48}$

44. $\dfrac{2x^2 + 12x - 32}{x^2 + 16x + 64} \cdot \dfrac{x^2 + 10x + 16}{x^2 - 3x - 10}$

Divide and simplify. See Example 7.

45. $\dfrac{2x}{5} \div \dfrac{6x + 12}{5x + 10}$

46. $\dfrac{7}{3x} \div \dfrac{14 - 7x}{18 - 9x}$

47. $\dfrac{a + b}{ab} \div \dfrac{a^2 - b^2}{4a^3b}$

48. $\dfrac{6a^2b^2}{a^2 - 4} \div \dfrac{3ab^2}{a - 2}$

49. $\dfrac{x^2 - 6x + 9}{x^2 - x - 6} \div \dfrac{x^2 - 9}{4}$

50. $\dfrac{x^2 - 4}{3x + 6} \div \dfrac{2x^2 - 8x + 8}{x^2 + 4x + 4}$

51. $\dfrac{x^2 - 6x - 16}{2x^2 - 128} \div \dfrac{x^2 + 10x + 16}{x^2 + 16x + 64}$

52. $\dfrac{a^2 - a - 6}{a^2 - 81} \div \dfrac{a^2 - 7a - 18}{4a + 36}$

▶ **53.** $\dfrac{3x - x^2}{x^3 - 27} \div \dfrac{x}{x^2 + 3x + 9}$

54. $\dfrac{x^2 - 3x}{x^3 - 27} \div \dfrac{2x}{2x^2 + 6x + 18}$

55. $\dfrac{8b + 24}{3a + 6} \div \dfrac{ab - 2b + 3a - 6}{a^2 - 4a + 4}$

56. $\dfrac{2a^2 - 2b^2}{a^3 + a^2b + a + b} \div \dfrac{6a^2}{a^3 + a}$

MIXED PRACTICE

Perform each indicated operation. See Examples 2 through 8.

57. $\dfrac{x^2 - 9}{4} \cdot \dfrac{x^2 - x - 6}{x^2 - 6x + 9}$

58. $\dfrac{x^2 - 4}{9} \cdot \dfrac{x^2 - 6x + 9}{x^2 - 5x + 6}$

59. $\dfrac{2x^2 - 4x - 30}{5x^2 - 40x - 75} \div \dfrac{x^2 - 8x + 15}{x^2 - 6x + 9}$

60. $\dfrac{4a + 36}{a^2 - 7a - 18} \div \dfrac{a^2 - a - 6}{a^2 - 81}$

61. Simplify: $\dfrac{r^3 + s^3}{r + s}$

62. Simplify: $\dfrac{m^3 - n^3}{m - n}$

63. $\dfrac{4}{x} \div \dfrac{3xy}{x^2} \cdot \dfrac{6x^2}{x^4}$

64. $\dfrac{4}{x} \cdot \dfrac{3xy}{x^2} \div \dfrac{6x^2}{x^4}$

65. $\dfrac{3x^2 - 5x - 2}{y^2 + y - 2} \cdot \dfrac{y^2 + 4y - 5}{12x^2 + 7x + 1} \div \dfrac{5x^2 - 9x - 2}{8x^2 - 2x - 1}$

66. $\dfrac{x^2 + x - 2}{3y^2 - 5y - 2} \cdot \dfrac{12y^2 + y - 1}{x^2 + 4x - 5} \div \dfrac{8y^2 - 6y + 1}{5y^2 - 9y - 2}$

67. $\dfrac{5a^2 - 20}{3a^2 - 12a} \div \dfrac{a^3 + 2a^2}{2a^2 - 8a} \cdot \dfrac{9a^3 + 6a^2}{2a^2 - 4a}$

68. $\dfrac{5a^2 - 20}{3a^2 - 12a} \div \left(\dfrac{a^3 + 2a^2}{2a^2 - 8a} \cdot \dfrac{9a^3 + 6a^2}{2a^2 - 4a} \right)$

69. $\dfrac{5x^4 + 3x^2 - 2}{x - 1} \cdot \dfrac{x + 1}{x^4 - 1}$

70. $\dfrac{3x^4 - 10x^2 - 8}{x - 2} \cdot \dfrac{3x + 6}{15x^2 + 10}$

Find each function value. See Example 9.

71. If $f(x) = \dfrac{x + 8}{2x - 1}$, find $f(2), f(0),$ and $f(-1)$.

72. If $f(x) = \dfrac{x - 2}{-5 + x}$, find $f(-5), f(0),$ and $f(10)$.

73. If $g(x) = \dfrac{x^2 + 8}{x^3 - 25x}$, find $g(3), g(-2),$ and $g(1)$.

74. If $s(t) = \dfrac{t^3 + 1}{t^2 + 1}$, find $s(-1), s(1),$ and $s(2)$.

▶ **75.** The total revenue from the sale of a popular book is approximated by the rational function $R(x) = \dfrac{1000x^2}{x^2 + 4}$, where x is the number of years since publication and $R(x)$ is the total revenue in millions of dollars.

   a. Find the total revenue at the end of the first year.

   b. Find the total revenue at the end of the second year.

   c. Find the revenue during the second year only.

   d. Find the domain of function R.

76. The function $f(x) = \dfrac{100,000x}{100 - x}$ models the cost in dollars for removing x percent of the pollutants from a bayou in which a nearby company dumped creosol.

   a. Find the cost of removing 20% of the pollutants from the bayou. [*Hint:* Find $f(20)$.]

   b. Find the cost of removing 60% of the pollutants and then 80% of the pollutants.

   c. Find $f(90)$, then $f(95)$, and then $f(99)$. What happens to the cost as x approaches 100%?

   d. Find the domain of function f.

REVIEW AND PREVIEW

Perform each indicated operation. See Section 1.3.

77. $\dfrac{4}{5} + \dfrac{3}{5}$

78. $\dfrac{4}{10} - \dfrac{7}{10}$

79. $\dfrac{5}{28} - \dfrac{2}{21}$

80. $\dfrac{5}{13} + \dfrac{2}{7}$

81. $\dfrac{3}{8} + \dfrac{1}{2} - \dfrac{3}{16}$

82. $\dfrac{2}{9} - \dfrac{1}{6} + \dfrac{2}{3}$

CONCEPT EXTENSIONS

Solve. For Exercises 83 and 84, see the second Concept Check in this section; for Exercises 85 and 86, see the third Concept Check.

83. Which of the expressions are equivalent to $\dfrac{x}{5-x}$?

a. $\dfrac{-x}{5-x}$

b. $\dfrac{-x}{-5+x}$

c. $\dfrac{x}{x-5}$

d. $\dfrac{-x}{x-5}$

84. Which of the expressions are equivalent to $\dfrac{-2+x}{x}$?

a. $\dfrac{2-x}{-x}$

b. $-\dfrac{2-x}{x}$

c. $\dfrac{x-2}{x}$

d. $\dfrac{x-2}{-x}$

85. Does $\dfrac{x}{x+5}$ simplify to $\dfrac{1}{5}$? Why or why not?

86. Does $\dfrac{x+7}{x}$ simplify to 7? Why or why not?

△ **87.** Find the area of the rectangle.

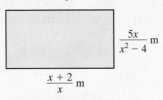

$\dfrac{5x}{x^2-4}$ m

$\dfrac{x+2}{x}$ m

△ **88.** Find the area of the triangle.

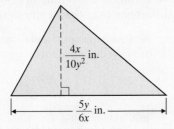

$\dfrac{4x}{10y^2}$ in.

$\dfrac{5y}{6x}$ in.

△ **89.** A parallelogram has an area of $\dfrac{x^2+x-2}{x^3}$ square feet and a height of $\dfrac{x^2}{x-1}$ feet. Express the length of its base as a rational expression in x. (*Hint:* Since $A = b \cdot h$, $b = \dfrac{A}{h}$ or $b = A \div h$.)

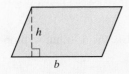

h

b

90. A lottery prize of $\dfrac{15x^3}{y^2}$ dollars is to be divided among $5x$ people. Express the amount of money each person is to receive as a rational expression in x and y.

91. In your own words, explain how to simplify a rational expression.

92. In your own words, explain the difference between multiplying rational expressions and dividing rational expressions.

93. Decide whether each rational expression equals 1, −1, or neither.

a. $\dfrac{x+5}{5+x}$

b. $\dfrac{x-5}{5-x}$

c. $\dfrac{x+5}{x-5}$

d. $\dfrac{-x-5}{x+5}$

e. $\dfrac{x-5}{-x+5}$

f. $\dfrac{-5+x}{x-5}$

94. Find the polynomial in the second numerator such that the following statement is true.

$$\dfrac{x^2-4}{x^2-7x+10} \cdot \dfrac{?}{2x^2+11x+14} = 1$$

95. In our definition of division for

$$\dfrac{P}{Q} \div \dfrac{R}{S}$$

we stated that $Q \neq 0$, $S \neq 0$, and $R \neq 0$. Explain why R cannot equal 0.

96. In your own words, explain how to find the domain of a rational function.

97. Graph a portion of the function $f(x) = \dfrac{20x}{100-x}$. To do so, complete the given table, plot the points, and then connect the plotted points with a smooth curve.

x	0	10	30	50	70	90	95	99
y or $f(x)$								

98. The domain of the function $f(x) = \dfrac{1}{x}$ is all real numbers except 0. This means that the graph of this function will be in two pieces: one piece corresponding to x values less than 0 and one piece corresponding to x values greater than 0. Graph the function by completing the following tables, separately plotting the points, and connecting each set of plotted points with a smooth curve.

x	$\frac{1}{4}$	$\frac{1}{2}$	1	2	4
y or $f(x)$					

x	−4	−2	−1	$-\frac{1}{2}$	$-\frac{1}{4}$
y or $f(x)$					

Simplify. Assume that no denominator is 0.

99. $\dfrac{p^x - 4}{4 - p^x}$

100. $\dfrac{3 + q^n}{q^n + 3}$

101. $\dfrac{x^n + 4}{x^{2n} - 16}$

102. $\dfrac{x^{2k} - 9}{3 + x^k}$

105. $\dfrac{y^{2n} + 9}{10y} \cdot \dfrac{y^n - 3}{y^{4n} - 81}$

106. $\dfrac{y^{4n} - 16}{y^{2n} + 4} \cdot \dfrac{6y}{y^n + 2}$

107. $\dfrac{y^{2n} - y^n - 2}{2y^n - 4} \div \dfrac{y^{2n} - 1}{1 + y^n}$

Perform the indicated operation. Write all answers in lowest terms.

103. $\dfrac{x^{2n} - 4}{7x} \cdot \dfrac{14x^3}{x^n - 2}$

104. $\dfrac{x^{2n} + 4x^n + 4}{4x - 3} \cdot \dfrac{8x^2 - 6x}{x^n + 2}$

108. $\dfrac{y^{2n} + 7y^n + 10}{10} \div \dfrac{y^{2n} + 4y^n + 4}{5y^n + 25}$

6.2 Adding and Subtracting Rational Expressions

OBJECTIVES

1 Add or Subtract Rational Expressions with a Common Denominator.

2 Identify the Least Common Denominator (LCD) of Two or More Rational Expressions.

3 Add or Subtract Rational Expressions with Unlike Denominators.

OBJECTIVE

1 Adding or Subtracting Rational Expressions with a Common Denominator

Rational expressions, like rational numbers, can be added or subtracted. We add or subtract rational expressions in the same way that we add or subtract rational numbers (fractions).

> **Adding or Subtracting Rational Expressions with a Common Denominator**
>
> If $\dfrac{P}{Q}$ and $\dfrac{R}{Q}$ are rational expressions, then
>
> $$\frac{P}{Q} + \frac{R}{Q} = \frac{P + R}{Q} \quad \text{and} \quad \frac{P}{Q} - \frac{R}{Q} = \frac{P - R}{Q}$$

To add or subtract rational expressions with a common denominator, add or subtract the numerators and write the sum or difference over the common denominator.

EXAMPLE 1 Add or subtract.

a. $\dfrac{x}{4} + \dfrac{5x}{4}$ **b.** $\dfrac{5}{7z^2} + \dfrac{x}{7z^2}$ **c.** $\dfrac{x^2}{x + 7} - \dfrac{49}{x + 7}$ **d.** $\dfrac{x}{3y^2} - \dfrac{x + 1}{3y^2}$

Solution The rational expressions have common denominators, so add or subtract their numerators and place the sum or difference over their common denominator.

a. $\dfrac{x}{4} + \dfrac{5x}{4} = \dfrac{x + 5x}{4} = \dfrac{6x}{4} = \dfrac{3x}{2}$ Add the numerators and write the result over the common denominator.

b. $\dfrac{5}{7z^2} + \dfrac{x}{7z^2} = \dfrac{5 + x}{7z^2}$

c. $\dfrac{x^2}{x + 7} - \dfrac{49}{x + 7} = \dfrac{x^2 - 49}{x + 7}$ Subtract the numerators and write the result over the common denominator.

$$= \frac{(x + 7)(x - 7)}{x + 7}$$ Factor the numerator.

$$= x - 7$$ Simplify.

> ▶ **Helpful Hint**
>
> **Very Important:** Be sure to insert parentheses here so that the entire numerator is subtracted.

d. $\dfrac{x}{3y^2} - \dfrac{x + 1}{3y^2} = \dfrac{x - (x + 1)}{3y^2}$ Subtract the numerators.

$$= \frac{x - x - 1}{3y^2}$$ Use the distributive property.

$$= -\frac{1}{3y^2}$$ Simplify.   □

PRACTICE

1   Add or subtract.

a. $\dfrac{9}{11z^2} + \dfrac{x}{11z^2}$ **b.** $\dfrac{x}{8} + \dfrac{5x}{8}$ **c.** $\dfrac{x^2}{x + 4} - \dfrac{16}{x + 4}$ **d.** $\dfrac{z}{2a^2} - \dfrac{z + 3}{2a^2}$

✓CONCEPT CHECK

Find and correct the error.

$$\dfrac{3 + 2y}{y^2 - 1} - \dfrac{y + 3}{y^2 - 1} = \dfrac{3 + 2y - y + 3}{y^2 - 1}$$

$$= \dfrac{y + 6}{y^2 - 1}$$

OBJECTIVE

2   **Identifying the Least Common Denominator (LCD) of Rational Expressions** ▶

To add or subtract rational expressions with unlike denominators, first write the rational expressions as equivalent rational expressions with a common denominator.

The **least common denominator (LCD)** is usually the easiest common denominator to work with. The LCD of a list of rational expressions is a polynomial of least degree whose factors include the denominator factors in the list.

Use the following steps to find the LCD.

> **Finding the Least Common Denominator (LCD)**
>
> **Step 1.**   Factor each denominator completely.
>
> **Step 2.**   The LCD is the product of all unique factors, each raised to a power equal to the greatest number of times that the factor appears in any factored denominator.

EXAMPLE 2   Find the LCD of the rational expressions in each list.

a. $\dfrac{2}{3x^5y^2}, \dfrac{3z}{5xy^3}$ **b.** $\dfrac{7}{z + 1}, \dfrac{z}{z - 1}$

c. $\dfrac{m - 1}{m^2 - 25}, \dfrac{2m}{2m^2 - 9m - 5}, \dfrac{7}{m^2 - 10m + 25}$ **d.** $\dfrac{x}{x^2 - 4}, \dfrac{11}{6 - 3x}$

Solution

a. First we factor each denominator.

$$3x^5y^2 = 3 \cdot x^5 \cdot y^2$$
$$5xy^3 = 5 \cdot x \cdot y^3$$
$$\text{LCD} = 3 \cdot 5 \cdot x^5 \cdot y^3 = 15x^5y^3$$

> ▶ Helpful Hint
>
> The greatest power of x is 5, so we have a factor of x^5. The greatest power of y is 3, so we have a factor of y^3.

b. The denominators $z + 1$ and $z - 1$ do not factor further. Thus,

$$\text{LCD} = (z + 1)(z - 1)$$

c. We first factor each denominator.

$$m^2 - 25 = (m + 5)(m - 5)$$
$$2m^2 - 9m - 5 = (2m + 1)(m - 5)$$
$$m^2 - 10m + 25 = (m - 5)(m - 5)$$
$$\text{LCD} = (m + 5)(2m + 1)(m - 5)^2$$

Answer to Concept Check:

$$\dfrac{3 + 2y}{y^2 - 1} - \dfrac{y + 3}{y^2 - 1}$$

$$= \dfrac{3 + 2y - y - 3}{y^2 - 1} = \dfrac{y}{y^2 - 1}$$

d. Factor each denominator.

$$x^2 - 4 = (x + 2)(x - 2)$$
$$6 - 3x = 3(2 - x) = 3(-1)(x - 2)$$
$$\text{LCD} = 3(-1)(x + 2)(x - 2)$$
$$= -3(x + 2)(x - 2)$$

▶ Helpful Hint

If opposite factors occur, do not use both in the LCD. Instead, factor -1 from one of the opposite factors so that the factors are then identical.

PRACTICE

2   Find the LCD of the rational expression in each list.

a. $\dfrac{7}{6x^3y^5}, \dfrac{2}{9x^2y^4}$

b. $\dfrac{11}{x - 2}, \dfrac{x}{x + 3}$

c. $\dfrac{b + 2}{b^2 - 16}, \dfrac{8}{b^2 - 8b + 16}, \dfrac{5b}{2b^2 - 5b - 12}$

d. $\dfrac{y}{y^2 - 9}, \dfrac{3}{12 - 4y}$

OBJECTIVE

3   **Adding or Subtracting Rational Expressions with Unlike Denominators** ▶

To add or subtract rational expressions with unlike denominators, we write each rational expression as an equivalent rational expression so that their denominators are alike.

Adding or Subtracting Rational Expressions with Unlike Denominators

Step 1.   Find the LCD of the rational expressions.

Step 2.   Write each rational expression as an equivalent rational expression whose denominator is the LCD found in Step 1.

Step 3.   Add or subtract numerators and write the result over the common denominator.

Step 4.   Simplify the resulting rational expression.

EXAMPLE 3   Perform the indicated operation.

a. $\dfrac{2}{x^2y} + \dfrac{5}{3x^3y}$

b. $\dfrac{3}{x + 2} + \dfrac{2x}{x - 2}$

c. $\dfrac{2x - 6}{x - 1} - \dfrac{4}{1 - x}$

Solution

a. The LCD is $3x^3y$. Write each fraction as an equivalent fraction with denominator $3x^3y$. To do this, we multiply both the numerator and denominator of each fraction by the factors needed to obtain the LCD as denominator.

The first fraction is multiplied by $\dfrac{3x}{3x}$ so that the new denominator is the LCD.

$$\frac{2}{x^2y} + \frac{5}{3x^3y} = \frac{2 \cdot 3x}{x^2y \cdot 3x} + \frac{5}{3x^3y} \qquad \text{The second expression already}$$
$$\qquad\qquad\qquad\qquad\qquad\qquad \text{has a denominator of } 3x^3y.$$
$$= \frac{6x}{3x^3y} + \frac{5}{3x^3y}$$
$$= \frac{6x + 5}{3x^3y} \qquad\qquad \text{Add the numerators.}$$

b. The LCD is the product of the two denominators: $(x + 2)(x - 2)$.

$$\frac{3}{x + 2} + \frac{2x}{x - 2} = \frac{3 \cdot (x - 2)}{(x + 2) \cdot (x - 2)} + \frac{2x \cdot (x + 2)}{(x - 2) \cdot (x + 2)} \qquad \text{Write equivalent rational expressions.}$$

$$= \frac{3x - 6}{(x + 2)(x - 2)} + \frac{2x^2 + 4x}{(x + 2)(x - 2)} \qquad \text{Multiply in the numerators.}$$

$$= \frac{3x - 6 + 2x^2 + 4x}{(x + 2)(x - 2)} \qquad \text{Add the numerators.}$$

$$= \frac{2x^2 + 7x - 6}{(x + 2)(x - 2)} \qquad \text{Simplify the numerator.}$$

c. The LCD is either $x - 1$ or $1 - x$. To get a common denominator of $x - 1$, we factor -1 from the denominator of the second rational expression.

$$\frac{2x - 6}{x - 1} - \frac{4}{1 - x} = \frac{2x - 6}{x - 1} - \frac{4}{-1(x - 1)} \qquad \text{Write } 1 - x \text{ as } -1(x - 1).$$

$$= \frac{2x - 6}{x - 1} - \frac{-1 \cdot 4}{x - 1} \qquad \text{Write } \frac{4}{-1(x - 1)} \text{ as } \frac{-1 \cdot 4}{x - 1}.$$

$$= \frac{2x - 6 - (-4)}{x - 1} \qquad \text{Combine the numerators.}$$

$$= \frac{2x - 6 + 4}{x - 1} \qquad \text{Simplify.}$$

$$= \frac{2x - 2}{x - 1}$$

$$= \frac{2(x - 1)}{x - 1} \qquad \text{Factor.}$$

$$= 2 \qquad \text{Simplest form} \qquad \square$$

PRACTICE

3   Perform the indicated operation.

a. $\dfrac{4}{p^3 q} + \dfrac{3}{5p^4 q}$ **b.** $\dfrac{4}{y + 3} + \dfrac{5y}{y - 3}$ **c.** $\dfrac{3z - 18}{z - 5} - \dfrac{3}{5 - z}$

EXAMPLE 4   Subtract $\dfrac{5k}{k^2 - 4} - \dfrac{2}{k^2 + k - 2}$.

Solution $\dfrac{5k}{k^2 - 4} - \dfrac{2}{k^2 + k - 2} = \dfrac{5k}{(k + 2)(k - 2)} - \dfrac{2}{(k + 2)(k - 1)}$ Factor each denominator to find the LCD.

The LCD is $(k + 2)(k - 2)(k - 1)$. We write equivalent rational expressions with the LCD as denominators.

$$\frac{5k}{(k + 2)(k - 2)} - \frac{2}{(k + 2)(k - 1)}$$

$$= \frac{5k \cdot (k - 1)}{(k + 2)(k - 2) \cdot (k - 1)} - \frac{2 \cdot (k - 2)}{(k + 2)(k - 1) \cdot (k - 2)} \qquad \text{Write equivalent rational expressions.}$$

$$= \frac{5k^2 - 5k}{(k + 2)(k - 2)(k - 1)} - \frac{2k - 4}{(k + 2)(k - 2)(k - 1)} \qquad \text{Multiply in the numerators.}$$

$$= \frac{5k^2 - 5k - 2k + 4}{(k + 2)(k - 2)(k - 1)} \qquad \text{Subtract the numerators.}$$

$$= \frac{5k^2 - 7k + 4}{(k + 2)(k - 2)(k - 1)} \qquad \text{Simplify.} \qquad \square$$

▶ **Helpful Hint**
Very Important: Because we are subtracting, notice the sign change on 4.

PRACTICE

4   Subtract $\dfrac{t}{t^2 - 25} - \dfrac{3}{t^2 - 3t - 10}$.

EXAMPLE 5 Add $\dfrac{2x-1}{2x^2-9x-5} + \dfrac{x+3}{6x^2-x-2}$.

Solution

$$\dfrac{2x-1}{2x^2-9x-5} + \dfrac{x+3}{6x^2-x-2} = \dfrac{2x-1}{(2x+1)(x-5)} + \dfrac{x+3}{(2x+1)(3x-2)}$$ Factor the denominators.

The LCD is $(2x+1)(x-5)(3x-2)$.

$$= \dfrac{(2x-1)\cdot(3x-2)}{(2x+1)(x-5)\cdot(3x-2)} + \dfrac{(x+3)\cdot(x-5)}{(2x+1)(3x-2)\cdot(x-5)}$$

$$= \dfrac{6x^2-7x+2}{(2x+1)(x-5)(3x-2)} + \dfrac{x^2-2x-15}{(2x+1)(x-5)(3x-2)}$$ Multiply in the numerators.

$$= \dfrac{6x^2-7x+2+x^2-2x-15}{(2x+1)(x-5)(3x-2)}$$ Add the numerators.

$$= \dfrac{7x^2-9x-13}{(2x+1)(x-5)(3x-2)}$$ Simplify. □

PRACTICE
5 Add $\dfrac{2x+3}{3x^2-5x-2} + \dfrac{x-6}{6x^2-13x-5}$.

EXAMPLE 6 Perform each indicated operation.

$$\dfrac{7}{x-1} + \dfrac{10x}{x^2-1} - \dfrac{5}{x+1}$$

Solution $\dfrac{7}{x-1} + \dfrac{10x}{x^2-1} - \dfrac{5}{x+1} = \dfrac{7}{x-1} + \dfrac{10x}{(x-1)(x+1)} - \dfrac{5}{x+1}$ Factor the denominators.

The LCD is $(x-1)(x+1)$.

$$= \dfrac{7\cdot(x+1)}{(x-1)\cdot(x+1)} + \dfrac{10x}{(x-1)(x+1)} - \dfrac{5\cdot(x-1)}{(x+1)\cdot(x-1)}$$

$$= \dfrac{7x+7}{(x-1)(x+1)} + \dfrac{10x}{(x-1)(x+1)} - \dfrac{5x-5}{(x+1)(x-1)}$$ Multiply in the numerators.

$$= \dfrac{7x+7+10x-5x+5}{(x-1)(x+1)}$$ Add and subtract the numerators.

$$= \dfrac{12x+12}{(x-1)(x+1)}$$ Simplify.

$$= \dfrac{12(x+1)}{(x-1)(x+1)}$$ Factor the numerator.

$$= \dfrac{12}{x-1}$$ Divide out common factors. □

PRACTICE
6 Perform each indicated operation.

$$\dfrac{2}{x-2} + \dfrac{3x}{x^2-x-2} - \dfrac{1}{x+1}$$

Graphing Calculator Explorations

A graphing calculator can be used to support the results of operations on rational expressions. For example, to verify the result of Example 3b, graph

$$Y_1 = \frac{3}{x+2} + \frac{2x}{x-2} \quad \text{and} \quad Y_2 = \frac{2x^2 + 7x - 6}{(x+2)(x-2)}$$

on the same set of axes. The graphs should be the same. Use a Table feature or a Trace feature to see that this is true.

Vocabulary, Readiness & Video Check

Name the operation(s) below that make each statement true.

   a. Addition   **b.** Subtraction   **c.** Multiplication   **d.** Division

1. The denominators must be the same before performing the operation. ____
2. To perform this operation, you multiply the first rational expression by the reciprocal of the second rational expression. ____
3. Numerator times numerator all over denominator times denominator. ____
4. These operations are commutative (order doesn't matter). ____

For the rational expressions $\frac{5}{y}$ and $\frac{7}{y}$, perform each operation mentally.

5. Addition   **6.** Subtraction   **7.** Multiplication   **8.** Division
   ___   ___   ___   ___

Martin-Gay Interactive Videos

See Video 6.2

Watch the section lecture video and answer the following questions.

OBJECTIVE 1
9. In ▣ Example 1, why are we told to be especially careful with subtraction when the second numerator has more than one term?

OBJECTIVE 2
10. In ▣ Example 2, $(a - b)$ appears as a factor three times within the two factored denominators. Why does the LCD only contain two factors of $(a - b)$?

OBJECTIVE 3
11. Based on ▣ Example 3, complete the following statements. To write an equivalent rational expression, you multiply the _____ of the expression by the exact same thing as the denominator. This is the same as multiplying the original rational expression by _____, which doesn't change the _____ of the original expression.

6.2 Exercise Set   MyMathLab®

Be careful when subtracting! For example, $\dfrac{8}{x+1} - \dfrac{x+5}{x+1} = \dfrac{8-(x+5)}{x+1} = \dfrac{3-x}{x+1}$ *or* $\dfrac{-x+3}{x+1}$.

Use this example to help you perform the subtractions.

1. $\dfrac{5}{2x} - \dfrac{x+1}{2x} =$ _____
2. $\dfrac{9}{5x} - \dfrac{6-x}{5x} =$ _____

9. $\dfrac{x^2}{x+2} - \dfrac{4}{x+2}$
10. $\dfrac{x^2}{x+6} - \dfrac{36}{x+6}$

3. $\dfrac{y+11}{y-2} - \dfrac{y-5}{y-2} =$ _____
4. $\dfrac{z-1}{z+6} - \dfrac{z+4}{z+6} =$ _____

11. $\dfrac{2x-6}{x^2+x-6} + \dfrac{3-3x}{x^2+x-6}$
12. $\dfrac{5x+2}{x^2+2x-8} + \dfrac{2-4x}{x^2+2x-8}$

Add or subtract as indicated. Simplify each answer. See Example 1.

▶ **13.** $\dfrac{x-5}{2x} - \dfrac{x+5}{2x}$
14. $\dfrac{x+4}{4x} - \dfrac{x-4}{4x}$

5. $\dfrac{2}{xz^2} - \dfrac{5}{xz^2}$
6. $\dfrac{4}{x^2y} - \dfrac{2}{x^2y}$

7. $\dfrac{2}{x-2} + \dfrac{x}{x-2}$
8. $\dfrac{x}{5-x} + \dfrac{7}{5-x}$

Find the LCD of the rational expressions in each list. See Example 2.

15. $\dfrac{2}{7}, \dfrac{3}{5x}$

16. $\dfrac{4}{5y}, \dfrac{3}{4y^2}$

17. $\dfrac{3}{x}, \dfrac{2}{x+1}$

18. $\dfrac{5}{2x}, \dfrac{7}{2+x}$

19. $\dfrac{12}{x+7}, \dfrac{8}{x-7}$

20. $\dfrac{1}{2x-1}, \dfrac{8}{2x+1}$

21. $\dfrac{5}{3x+6}, \dfrac{2x}{2x-4}$

22. $\dfrac{2}{3a+9}, \dfrac{5}{5a-15}$

▶ 23. $\dfrac{2a}{a^2-b^2}, \dfrac{1}{a^2-2ab+b^2}$

24. $\dfrac{2a}{a^2+8a+16}, \dfrac{7a}{a^2+a-12}$

25. $\dfrac{x}{x^2-9}, \dfrac{5}{x}, \dfrac{7}{12-4x}$

26. $\dfrac{9}{x^2-25}, \dfrac{1}{50-10x}, \dfrac{6}{x}$

Add or subtract as indicated. Simplify each answer. See Examples 3a and 3b.

▶ 27. $\dfrac{4}{3x} + \dfrac{3}{2x}$

28. $\dfrac{10}{7x} + \dfrac{5}{2x}$

29. $\dfrac{5}{2y^2} - \dfrac{2}{7y}$

30. $\dfrac{4}{11x^4} - \dfrac{1}{4x^2}$

▶ 31. $\dfrac{x-3}{x+4} - \dfrac{x+2}{x-4}$

32. $\dfrac{x-1}{x-5} - \dfrac{x+2}{x+5}$

33. $\dfrac{1}{x-5} - \dfrac{19-2x}{(x-5)(x+4)}$

34. $\dfrac{4x-2}{(x-5)(x+4)} - \dfrac{2}{x+4}$

Perform the indicated operation. If possible, simplify your answer. See Example 3c.

35. $\dfrac{1}{a-b} + \dfrac{1}{b-a}$

36. $\dfrac{1}{a-3} - \dfrac{1}{3-a}$

37. $\dfrac{x+1}{1-x} + \dfrac{1}{x-1}$

38. $\dfrac{5}{1-x} - \dfrac{1}{x-1}$

39. $\dfrac{5}{x-2} + \dfrac{x+4}{2-x}$

40. $\dfrac{3}{5-x} + \dfrac{x+2}{x-5}$

Perform each indicated operation. If possible, simplify your answer. See Examples 4 through 6.

▶ 41. $\dfrac{y+1}{y^2-6y+8} - \dfrac{3}{y^2-16}$

42. $\dfrac{x+2}{x^2-36} - \dfrac{x}{x^2+9x+18}$

43. $\dfrac{x+4}{3x^2+11x+6} + \dfrac{x}{2x^2+x-15}$

44. $\dfrac{x+3}{5x^2+12x+4} + \dfrac{6}{x^2-x-6}$

45. $\dfrac{7}{x^2-x-2} - \dfrac{x-1}{x^2+4x+3}$

46. $\dfrac{a}{a^2+10a+25} - \dfrac{4-a}{a^2+6a+5}$

47. $\dfrac{x}{x^2-8x+7} - \dfrac{x+2}{2x^2-9x-35}$

48. $\dfrac{x}{x^2-7x+6} - \dfrac{x+4}{3x^2-2x-1}$

49. $\dfrac{2}{a^2+2a+1} + \dfrac{3}{a^2-1}$

50. $\dfrac{9x+2}{3x^2-2x-8} + \dfrac{7}{3x^2+x-4}$

MIXED PRACTICE

Add or subtract as indicated. If possible, simplify your answer. See Examples 1 through 6.

51. $\dfrac{4}{3x^2y^3} + \dfrac{5}{3x^2y^3}$

52. $\dfrac{7}{2xy^4} + \dfrac{1}{2xy^4}$

53. $\dfrac{13x-5}{2x} - \dfrac{13x+5}{2x}$

54. $\dfrac{17x+4}{4x} - \dfrac{17x-4}{4x}$

55. $\dfrac{3}{2x+10} + \dfrac{8}{3x+15}$

56. $\dfrac{10}{3x-3} + \dfrac{1}{7x-7}$

57. $\dfrac{-2}{x^2-3x} - \dfrac{1}{x^3-3x^2}$

58. $\dfrac{-3}{2a+8} - \dfrac{8}{a^2+4a}$

59. $\dfrac{ab}{a^2-b^2} + \dfrac{b}{a+b}$

60. $\dfrac{x}{25-x^2} + \dfrac{2}{3x-15}$

61. $\dfrac{5}{x^2-4} - \dfrac{3}{x^2+4x+4}$

62. $\dfrac{3z}{z^2-9} - \dfrac{2}{3-z}$

63. $\dfrac{3x}{2x^2-11x+5} + \dfrac{7}{x^2-2x-15}$

64. $\dfrac{2x}{3x^2-13x+4} + \dfrac{5}{x^2-2x-8}$

65. $\dfrac{2}{x+1} - \dfrac{3x}{3x+3} + \dfrac{1}{2x+2}$

66. $\dfrac{5}{3x-6} - \dfrac{x}{x-2} + \dfrac{3+2x}{5x-10}$

67. $\dfrac{3}{x+3} + \dfrac{5}{x^2+6x+9} - \dfrac{x}{x^2-9}$

68. $\dfrac{x+2}{x^2-2x-3} + \dfrac{x}{x-3} - \dfrac{x}{x+1}$

69. $\dfrac{x}{x^2-9} + \dfrac{3}{x^2-6x+9} - \dfrac{1}{x+3}$

70. $\dfrac{3}{x^2 - 9} - \dfrac{x}{x^2 - 6x + 9} + \dfrac{1}{x + 3}$

71. $\left(\dfrac{1}{x} + \dfrac{2}{3}\right) - \left(\dfrac{1}{x} - \dfrac{2}{3}\right)$

72. $\left(\dfrac{1}{2} + \dfrac{2}{x}\right) - \left(\dfrac{1}{2} - \dfrac{1}{x}\right)$

MIXED PRACTICE (SECTIONS 6.1, 6.2)

Perform the indicated operation. If possible, simplify your answer.

73. $\left(\dfrac{2}{3} - \dfrac{1}{x}\right) \cdot \left(\dfrac{3}{x} + \dfrac{1}{2}\right)$

74. $\left(\dfrac{2}{3} - \dfrac{1}{x}\right) \div \left(\dfrac{3}{x} + \dfrac{1}{2}\right)$

75. $\left(\dfrac{2a}{3}\right)^2 \div \left(\dfrac{a^2}{a + 1} - \dfrac{1}{a + 1}\right)$

76. $\left(\dfrac{x + 2}{2x} - \dfrac{x - 2}{2x}\right) \cdot \left(\dfrac{5x}{4}\right)^2$

77. $\left(\dfrac{2x}{3}\right)^2 \div \left(\dfrac{x}{3}\right)^2$ **78.** $\left(\dfrac{2x}{3}\right)^2 \cdot \left(\dfrac{3}{x}\right)^2$

79. $\left(\dfrac{x}{x + 1} - \dfrac{x}{x - 1}\right) \div \dfrac{x}{2x + 2}$

80. $\dfrac{x}{2x + 2} \div \left(\dfrac{x}{x + 1} + \dfrac{x}{x - 1}\right)$

81. $\dfrac{4}{x} \cdot \left(\dfrac{2}{x + 2} - \dfrac{2}{x - 2}\right)$

82. $\dfrac{1}{x + 1} \cdot \left(\dfrac{5}{x} + \dfrac{2}{x - 3}\right)$

REVIEW AND PREVIEW

Use the distributive property to multiply the following. See Section 1.4.

83. $12\left(\dfrac{2}{3} + \dfrac{1}{6}\right)$ **84.** $14\left(\dfrac{1}{7} + \dfrac{3}{14}\right)$

85. $x^2\left(\dfrac{4}{x^2} + 1\right)$ **86.** $5y^2\left(\dfrac{1}{y^2} - \dfrac{1}{5}\right)$

Find each root. See Section 1.3.

87. $\sqrt{100}$ **88.** $\sqrt{25}$

89. $\sqrt[3]{8}$ **90.** $\sqrt[3]{27}$

91. $\sqrt[4]{81}$ **92.** $\sqrt[4]{16}$

Use the Pythagorean theorem to find the unknown length in each right triangle. See Section 5.8.

△ **93.**   △ **94.**

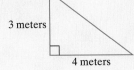

△ **93.** 3 meters, 4 meters

△ **94.** 7 feet, 24 feet

CONCEPT EXTENSIONS

Find and correct each error. See the Concept Check in this section.

95.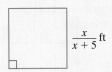

$$\dfrac{2x - 3}{x^2 + 1} - \dfrac{x - 6}{x^2 + 1} = \dfrac{2x - 3 - x - 6}{x^2 + 1}$$

$$= \dfrac{x - 9}{x^2 + 1}$$

96.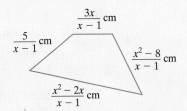

$$\dfrac{7}{x + 7} - \dfrac{x + 3}{x + 7} = \dfrac{7 - x - 3}{(x + 7)^2}$$

$$= \dfrac{-x + 4}{(x + 7)^2}$$

△ **97.** Find the perimeter and the area of the square.

$\dfrac{x}{x + 5}$ ft

△ **98.** Find the perimeter of the quadrilateral.

$\dfrac{3x}{x - 1}$ cm

$\dfrac{5}{x - 1}$ cm

$\dfrac{x^2 - 8}{x - 1}$ cm

$\dfrac{x^2 - 2x}{x - 1}$ cm

99. When is the LCD of two rational expressions equal to the product of their denominators? $\left(\text{Hint: What is the LCD of } \dfrac{1}{x} \text{ and } \dfrac{7}{x + 5}?\right)$

100. When is the LCD of two rational expressions with different denominators equal to one of the denominators? $\left(\text{Hint: What is the LCD of } \dfrac{3x}{x + 2} \text{ and } \dfrac{7x + 1}{(x + 2)^3}?\right)$

101. In your own words, explain how to add rational expressions with different denominators.

102. In your own words, explain how to multiply rational expressions.

103. In your own words, explain how to divide rational expressions.

104. In your own words, explain how to subtract rational expressions with different denominators.

Perform each indicated operation. (Hint: First write each expression with positive exponents.)

105. $x^{-1} + (2x)^{-1}$ **106.** $y^{-1} + (4y)^{-1}$

107. $4x^{-2} - 3x^{-1}$ **108.** $(4x)^{-2} - (3x)^{-1}$

Use a graphing calculator to support the results of each exercise.

109. Exercise 7   **110.** Exercise 8

6.3 | Simplifying Complex Fractions

OBJECTIVES

1 Simplify Complex Fractions by Simplifying the Numerator and Denominator and Then Dividing.

2 Simplify Complex Fractions by Multiplying by a Common Denominator.

3 Simplify Expressions with Negative Exponents.

A rational expression whose numerator, denominator, or both contain one or more rational expressions is called a **complex rational expression** or a **complex fraction**.

$$\textit{Complex Fractions}$$

$$\frac{\dfrac{1}{a}}{\dfrac{b}{2}} \qquad \frac{\dfrac{x}{2y^2}}{\dfrac{6x-2}{9y}} \qquad \frac{x+\dfrac{1}{y}}{y+1}$$

The parts of a complex fraction are

$$\left.\frac{\dfrac{x}{y+2}}{7+\dfrac{1}{y}}\right\} \quad \begin{array}{l}\leftarrow \text{Numerator of complex fraction}\\ \leftarrow \text{Main fraction bar}\\ \leftarrow \text{Denominator of complex fraction}\end{array}$$

Our goal in this section is to simplify complex fractions. A complex fraction is simplified when it is in the form $\dfrac{P}{Q}$, where P and Q are polynomials that have no common factors. Two methods of simplifying complex fractions are introduced. The first method evolves from the definition of a fraction as a quotient.

OBJECTIVE

1 **Simplifying Complex Fractions: Method 1**

> **Simplifying a Complex Fraction: Method I**
>
> **Step 1.** Simplify the numerator and the denominator of the complex fraction so that each is a single fraction.
>
> **Step 2.** Perform the indicated division by multiplying the numerator of the complex fraction by the reciprocal of the denominator of the complex fraction.
>
> **Step 3.** Simplify if possible.

EXAMPLE 1 Simplify each complex fraction.

a. $\dfrac{\dfrac{2x}{27y^2}}{\dfrac{6x^2}{9}}$ **b.** $\dfrac{\dfrac{5x}{x+2}}{\dfrac{10}{x-2}}$ **c.** $\dfrac{\dfrac{x}{y^2}+\dfrac{1}{y}}{\dfrac{y}{x^2}+\dfrac{1}{x}}$

Solution

a. The numerator of the complex fraction is already a single fraction, and so is the denominator. Perform the indicated division by multiplying the numerator, $\dfrac{2x}{27y^2}$, by the reciprocal of the denominator, $\dfrac{6x^2}{9}$. Then simplify.

$$\frac{\dfrac{2x}{27y^2}}{\dfrac{6x^2}{9}} = \frac{2x}{27y^2} \div \frac{6x^2}{9}$$

$$= \frac{2x}{27y^2} \cdot \frac{9}{6x^2} \qquad \text{Multiply by the reciprocal of } \frac{6x^2}{9}.$$

$$= \frac{2x \cdot 9}{27y^2 \cdot 6x^2}$$

$$= \frac{1}{9xy^2}$$

▶ Helpful Hint

Both the numerator and denominator are single fractions, so we perform the indicated division.

b. $\dfrac{\left\{\dfrac{5x}{x+2}\right.}{\left\{\dfrac{10}{x-2}\right.} = \dfrac{5x}{x+2} \div \dfrac{10}{x-2} = \dfrac{5x}{x+2} \cdot \dfrac{x-2}{10}$ Multiply by the reciprocal of $\dfrac{10}{x-2}$.

$= \dfrac{5x(x-2)}{2 \cdot 5(x+2)}$

$= \dfrac{x(x-2)}{2(x+2)}$ Simplify.

c. First simplify the numerator and the denominator of the complex fraction separately so that each is a single fraction. Then perform the indicated division.

$\dfrac{\dfrac{x}{y^2} + \dfrac{1}{y}}{\dfrac{y}{x^2} + \dfrac{1}{x}} = \dfrac{\dfrac{x}{y^2} + \dfrac{1 \cdot y}{y \cdot y}}{\dfrac{y}{x^2} + \dfrac{1 \cdot x}{x \cdot x}}$ Simplify the numerator. The LCD is y^2.

Simplify the denominator. The LCD is x^2.

$= \dfrac{\dfrac{x+y}{y^2}}{\dfrac{y+x}{x^2}}$ Add.

$= \dfrac{x+y}{y^2} \cdot \dfrac{x^2}{y+x}$ Multiply by the reciprocal of $\dfrac{y+x}{x^2}$.

$= \dfrac{x^2(x+y)}{y^2(y+x)}$

$= \dfrac{x^2}{y^2}$ Simplify.   □

PRACTICE

1   Simplify each complex fraction.

a. $\dfrac{\dfrac{5k}{36m}}{\dfrac{15k}{9}}$ **b.** $\dfrac{\dfrac{8x}{x-4}}{\dfrac{3}{x+4}}$ **c.** $\dfrac{\dfrac{5}{a} + \dfrac{b}{a^2}}{\dfrac{5a}{b^2} + \dfrac{1}{b}}$

✓**CONCEPT CHECK**

Which of the following are equivalent to $\dfrac{\dfrac{5}{y}}{\dfrac{2}{z}}$?

a. $\dfrac{5}{y} \div \dfrac{2}{z}$ **b.** $\dfrac{5}{y} \cdot \dfrac{z}{2}$ **c.** $\dfrac{5}{y} \div \dfrac{z}{2}$

Answer to Concept Check:
a and b

OBJECTIVE

2   **Simplifying Complex Fractions: Method 2**

Next we look at another method of simplifying complex fractions. With this method, we multiply the numerator and the denominator of the complex fraction by the LCD of all fractions in the complex fraction.

> **Simplifying a Complex Fraction: Method 2**
>
> **Step 1.** Multiply the numerator and the denominator of the complex fraction by the LCD of the fractions in both the numerator and the denominator.
>
> **Step 2.** Simplify.

EXAMPLE 2 Simplify each complex fraction.

a. $\dfrac{\dfrac{5x}{x+2}}{\dfrac{10}{x-2}}$ **b.** $\dfrac{\dfrac{x}{y^2}+\dfrac{1}{y}}{\dfrac{y}{x^2}+\dfrac{1}{x}}$

Solution

a. Notice we are reworking Example 1b using method 2. The least common denominator of $\dfrac{5x}{x+2}$ and $\dfrac{10}{x-2}$ is $(x+2)(x-2)$. Multiply both the numerator, $\dfrac{5x}{x+2}$, and the denominator, $\dfrac{10}{x-2}$, by the LCD.

$$\dfrac{\dfrac{5x}{x+2}}{\dfrac{10}{x-2}}=\dfrac{\left(\dfrac{5x}{x+2}\right)\cdot(x+2)(x-2)}{\left(\dfrac{10}{x-2}\right)\cdot(x+2)(x-2)}$$ Multiply numerator and denominator by the LCD.

$$=\dfrac{5x\cdot(x-2)}{2\cdot5\cdot(x+2)}$$ Simplify.

$$=\dfrac{x(x-2)}{2(x+2)}$$ Simplify.

b. Here, we are reworking Example 1c using method 2. The least common denominator of $\dfrac{x}{y^2},\dfrac{1}{y},\dfrac{y}{x^2}$, and $\dfrac{1}{x}$ is x^2y^2.

$$\dfrac{\dfrac{x}{y^2}+\dfrac{1}{y}}{\dfrac{y}{x^2}+\dfrac{1}{x}}=\dfrac{\left(\dfrac{x}{y^2}+\dfrac{1}{y}\right)\cdot x^2y^2}{\left(\dfrac{y}{x^2}+\dfrac{1}{x}\right)\cdot x^2y^2}$$ Multiply the numerator and denominator by the LCD.

$$=\dfrac{\dfrac{x}{y^2}\cdot x^2y^2+\dfrac{1}{y}\cdot x^2y^2}{\dfrac{y}{x^2}\cdot x^2y^2+\dfrac{1}{x}\cdot x^2y^2}$$ Use the distributive property.

$$=\dfrac{x^3+x^2y}{y^3+xy^2}$$ Simplify.

$$=\dfrac{x^2(x+y)}{y^2(y+x)}$$ Factor.

$$=\dfrac{x^2}{y^2}$$ Simplify.   □

PRACTICE
2   Use method 2 to simplify:

a. $\dfrac{\dfrac{8x}{x-4}}{\dfrac{3}{x+4}}$ **b.** $\dfrac{\dfrac{b}{a^2}+\dfrac{1}{a}}{\dfrac{a}{b^2}+\dfrac{1}{b}}$

OBJECTIVE

3 **Simplifying Expressions with Negative Exponents**

If an expression contains negative exponents, write the expression as an equivalent expression with positive exponents.

EXAMPLE 3 Simplify.

$$\frac{x^{-1} + 2xy^{-1}}{x^{-2} - x^{-2}y^{-1}}$$

Solution This fraction does not appear to be a complex fraction. If we write it by using only positive exponents, however, we see that it is a complex fraction.

$$\frac{x^{-1} + 2xy^{-1}}{x^{-2} - x^{-2}y^{-1}} = \frac{\dfrac{1}{x} + \dfrac{2x}{y}}{\dfrac{1}{x^2} - \dfrac{1}{x^2y}}$$

The LCD of $\dfrac{1}{x}, \dfrac{2x}{y}, \dfrac{1}{x^2}$, and $\dfrac{1}{x^2y}$ is x^2y. Multiply both the numerator and denominator by x^2y.

$$= \frac{\left(\dfrac{1}{x} + \dfrac{2x}{y}\right) \cdot x^2y}{\left(\dfrac{1}{x^2} - \dfrac{1}{x^2y}\right) \cdot x^2y}$$

$$= \frac{\dfrac{1}{x} \cdot x^2y + \dfrac{2x}{y} \cdot x^2y}{\dfrac{1}{x^2} \cdot x^2y - \dfrac{1}{x^2y} \cdot x^2y} \qquad \text{Apply the distributive property.}$$

$$= \frac{xy + 2x^3}{y - 1} \quad \text{or} \quad \frac{x(y + 2x^2)}{y - 1} \qquad \text{Simplify.} \quad \square$$

PRACTICE

3 Simplify: $\dfrac{3x^{-1} + x^{-2}y^{-1}}{y^{-2} + xy^{-1}}$

⋯⋯⋯⋯⋯⋯⋯⋯⋯⋯⋯⋯⋯⋯⋯⋯⋯⋯⋯⋯⋯⋯⋯⋯⋯⋯⋯⋯⋯⋯⋯⋯ ■

EXAMPLE 4 Simplify: $\dfrac{(2x)^{-1} + 1}{2x^{-1} - 1}$

Solution $\dfrac{(2x)^{-1} + 1}{2x^{-1} - 1} = \dfrac{\dfrac{1}{2x} + 1}{\dfrac{2}{x} - 1}$ Write using positive exponents.

$$= \frac{\left(\dfrac{1}{2x} + 1\right) \cdot 2x}{\left(\dfrac{2}{x} - 1\right) \cdot 2x} \qquad \text{The LDC of } \dfrac{1}{2x} \text{ and } \dfrac{2}{x} \text{ is } 2x.$$

$$= \frac{\dfrac{1}{2x} \cdot 2x + 1 \cdot 2x}{\dfrac{2}{x} \cdot 2x - 1 \cdot 2x} \qquad \text{Use distributive property.}$$

$$= \frac{1 + 2x}{4 - 2x} \quad \text{or} \quad \frac{1 + 2x}{2(2 - x)} \qquad \text{Simplify.} \qquad \square$$

> **Helpful Hint**
>
> Don't forget that $(2x)^{-1} = \dfrac{1}{2x}$, but $2x^{-1} = 2 \cdot \dfrac{1}{x} = \dfrac{2}{x}$.

PRACTICE
4 Simplify: $\dfrac{(3x)^{-1} - 2}{5x^{-1} + 2}$

Vocabulary, Readiness & Video Check

Complete the steps by writing the simplified complex fraction.

1. $\dfrac{\dfrac{7}{x}}{\dfrac{1}{x} + \dfrac{z}{x}} = \dfrac{x\left(\dfrac{7}{x}\right)}{x\left(\dfrac{1}{x}\right) + x\left(\dfrac{z}{x}\right)} = $ _____

2. $\dfrac{\dfrac{x}{4}}{\dfrac{x^2}{2} + \dfrac{1}{4}} = \dfrac{4\left(\dfrac{x}{4}\right)}{4\left(\dfrac{x^2}{2}\right) + 4\left(\dfrac{1}{4}\right)} = $ _____

Write with positive exponents.

3. $x^{-2} = $ _____

4. $y^{-3} = $ _____

5. $2x^{-1} = $ _____

6. $(2x)^{-1} = $ _____

7. $(9y)^{-1} = $ _____

8. $9y^{-2} = $ _____

Martin-Gay Interactive Videos

See Video 6.3

Watch the section lecture video and answer the following questions.

OBJECTIVE 1
9. From Example 2, before you can rewrite the complex fraction as division, describe how it must appear.

OBJECTIVE 2
10. How does finding an LCD in method 2, as in Example 3, differ from finding an LCD in method 1? In your answer, mention the purpose of the LCD in each method.

OBJECTIVE 3
11. Based on Example 4, what connection is there between negative exponents and complex fractions?

6.3 Exercise Set

MyMathLab®

Simplify each complex fraction. See Examples 1 and 2.

1. $\dfrac{\dfrac{10}{3x}}{\dfrac{5}{6x}}$

2. $\dfrac{\dfrac{15}{2x}}{\dfrac{5}{6x}}$

3. $\dfrac{1 + \dfrac{2}{5}}{2 + \dfrac{3}{5}}$

4. $\dfrac{2 + \dfrac{1}{7}}{3 - \dfrac{4}{7}}$

5. $\dfrac{\dfrac{4}{x - 1}}{\dfrac{x}{x - 1}}$

6. $\dfrac{\dfrac{x}{x + 2}}{\dfrac{2}{x + 2}}$

7. $\dfrac{1 - \dfrac{2}{x}}{x + \dfrac{4}{9x}}$

8. $\dfrac{5 - \dfrac{3}{x}}{x + \dfrac{2}{3x}}$

9. $\dfrac{\dfrac{4x^2 - y^2}{xy}}{\dfrac{2}{y} - \dfrac{1}{x}}$

10. $\dfrac{\dfrac{x^2 - 9y^2}{xy}}{\dfrac{1}{y} - \dfrac{3}{x}}$

11. $\dfrac{\dfrac{x + 1}{3}}{\dfrac{2x - 1}{6}}$

12. $\dfrac{\dfrac{x + 3}{12}}{\dfrac{4x - 5}{15}}$

13. $\dfrac{\dfrac{2}{x} + \dfrac{3}{x^2}}{\dfrac{4}{x^2} - \dfrac{9}{x}}$

14. $\dfrac{\dfrac{2}{x^2} + \dfrac{1}{x}}{\dfrac{4}{x^2} - \dfrac{1}{x}}$

15. $\dfrac{\dfrac{1}{x} + \dfrac{2}{x^2}}{x + \dfrac{8}{x^2}}$

16. $\dfrac{\dfrac{1}{y} + \dfrac{3}{y^2}}{y + \dfrac{27}{y^2}}$

17. $\dfrac{\dfrac{4}{5-x}+\dfrac{5}{x-5}}{\dfrac{2}{x}+\dfrac{3}{x-5}}$

18. $\dfrac{\dfrac{3}{x-4}-\dfrac{2}{4-x}}{\dfrac{2}{x-4}-\dfrac{2}{x}}$

▶ **19.** $\dfrac{\dfrac{x+2}{x}-\dfrac{2}{x-1}}{\dfrac{x+1}{x}+\dfrac{x+1}{x-1}}$

20. $\dfrac{\dfrac{5}{a+2}-\dfrac{1}{a-2}}{\dfrac{3}{2+a}+\dfrac{6}{2-a}}$

21. $\dfrac{\dfrac{2}{x}+3}{\dfrac{4}{x^2}-9}$

22. $\dfrac{2+\dfrac{1}{x}}{4x-\dfrac{1}{x}}$

23. $\dfrac{1-\dfrac{x}{y}}{\dfrac{x^2}{y^2}-1}$

24. $\dfrac{1-\dfrac{2}{x}}{x-\dfrac{4}{x}}$

25. $\dfrac{\dfrac{-2x}{x-y}}{\dfrac{y}{x^2}}$

26. $\dfrac{\dfrac{7y}{x^2+xy}}{\dfrac{y^2}{x^2}}$

27. $\dfrac{\dfrac{2}{x}+\dfrac{1}{x^2}}{\dfrac{y}{x^2}}$

28. $\dfrac{\dfrac{5}{x^2}-\dfrac{2}{x}}{\dfrac{1}{x}+2}$

29. $\dfrac{\dfrac{x}{9}-\dfrac{1}{x}}{1+\dfrac{3}{x}}$

30. $\dfrac{\dfrac{x}{4}-\dfrac{4}{x}}{1-\dfrac{4}{x}}$

31. $\dfrac{\dfrac{x-1}{x^2-4}}{1+\dfrac{1}{x-2}}$

32. $\dfrac{\dfrac{x+3}{x^2-9}}{1+\dfrac{1}{x-3}}$

33. $\dfrac{\dfrac{2}{x+5}+\dfrac{4}{x+3}}{\dfrac{3x+13}{x^2+8x+15}}$

34. $\dfrac{\dfrac{2}{x+2}+\dfrac{6}{x+7}}{\dfrac{4x+13}{x^2+9x+14}}$

Simplify. See Examples 3 and 4.

35. $\dfrac{x^{-1}}{x^{-2}+y^{-2}}$

36. $\dfrac{a^{-3}+b^{-1}}{a^{-2}}$

▶ **37.** $\dfrac{2a^{-1}+3b^{-2}}{a^{-1}-b^{-1}}$

38. $\dfrac{x^{-1}+y^{-1}}{3x^{-2}+5y^{-2}}$

39. $\dfrac{1}{x-x^{-1}}$

40. $\dfrac{x^{-2}}{x+3x^{-1}}$

41. $\dfrac{a^{-1}+1}{a^{-1}-1}$

42. $\dfrac{a^{-1}-4}{4+a^{-1}}$

43. $\dfrac{3x^{-1}+(2y)^{-1}}{x^{-2}}$

44. $\dfrac{5x^{-2}-3y^{-1}}{x^{-1}+y^{-1}}$

45. $\dfrac{2a^{-1}+(2a)^{-1}}{a^{-1}+2a^{-2}}$

46. $\dfrac{a^{-1}+2a^{-2}}{2a^{-1}+(2a)^{-1}}$

47. $\dfrac{5x^{-1}+2y^{-1}}{x^{-2}y^{-2}}$

48. $\dfrac{x^{-2}y^{-2}}{5x^{-1}+2y^{-1}}$

49. $\dfrac{5x^{-1}-2y^{-1}}{25x^{-2}-4y^{-2}}$

50. $\dfrac{3x^{-1}+3y^{-1}}{4x^{-2}-9y^{-2}}$

REVIEW AND PREVIEW

Simplify. See Sections 5.1 and 5.2.

51. $\dfrac{3x^3y^2}{12x}$

52. $\dfrac{-36xb^3}{9xb^2}$

53. $\dfrac{144x^5y^5}{-16x^2y}$

54. $\dfrac{48x^3y^2}{-4xy}$

Solve the following. See Section 2.6.

55. $|x-5|=9$

56. $|2y+1|=1$

CONCEPT EXTENSIONS

Solve. See the Concept Check in this section.

57. Which of the following are equivalent to $\dfrac{\dfrac{x+1}{9}}{\dfrac{y-2}{5}}$?

  a. $\dfrac{x+1}{9}\div\dfrac{y-2}{5}$ **b.** $\dfrac{x+1}{9}\cdot\dfrac{y-2}{5}$ **c.** $\dfrac{x+1}{9}\cdot\dfrac{5}{y-2}$

58. Which of the following are equivalent to $\dfrac{\dfrac{a}{7}}{\dfrac{b}{13}}$?

  a. $\dfrac{a}{7}\cdot\dfrac{b}{13}$ **b.** $\dfrac{a}{7}\div\dfrac{b}{13}$ **c.** $\dfrac{a}{7}\div\dfrac{13}{b}$ **d.** $\dfrac{a}{7}\cdot\dfrac{13}{b}$

59. When the source of a sound is traveling toward a listener, the pitch that the listener hears due to the Doppler effect is given by the complex rational compression $\dfrac{a}{1-\dfrac{s}{770}}$, where a is the actual pitch of the sound and s is the speed of the sound source. Simplify this expression.

60. In baseball, the earned run average (ERA) statistic gives the average number of earned runs scored on a pitcher per game. It is computed with the following expression: $\dfrac{E}{\dfrac{I}{9}}$, where E is the number of earned runs scored on a pitcher and I is the total number of innings pitched by the pitcher. Simplify this expression.

61. Which of the following are equivalent to $\dfrac{\frac{1}{x}}{\frac{3}{y}}$?

a. $\dfrac{1}{x} \div \dfrac{3}{y}$ **b.** $\dfrac{1}{x} \cdot \dfrac{y}{3}$ **c.** $\dfrac{1}{x} \div \dfrac{y}{3}$

62. Which of the following are equivalent to $\dfrac{\frac{5}{2}}{a}$?

a. $\dfrac{5}{1} \div \dfrac{2}{a}$ **b.** $\dfrac{1}{5} \div \dfrac{2}{a}$ **c.** $\dfrac{5}{1} \cdot \dfrac{2}{a}$

63. In your own words, explain one method for simplifying a complex fraction.

64. Explain your favorite method for simplifying a complex fraction and why.

Simplify.

65. $\dfrac{1}{1 + (1 + x)^{-1}}$

66. $\dfrac{(x + 2)^{-1} + (x - 2)^{-1}}{(x^2 - 4)^{-1}}$

67. $\dfrac{x}{1 - \dfrac{1}{1 + \dfrac{1}{x}}}$

68. $\dfrac{x}{1 - \dfrac{1}{1 - \dfrac{1}{x}}}$

69. $\dfrac{\dfrac{2}{y^2} - \dfrac{5}{xy} - \dfrac{3}{x^2}}{\dfrac{2}{y^2} + \dfrac{7}{xy} + \dfrac{3}{x^2}}$

70. $\dfrac{\dfrac{2}{x^2} - \dfrac{1}{xy} - \dfrac{1}{y^2}}{\dfrac{1}{x^2} - \dfrac{3}{xy} + \dfrac{2}{y^2}}$

71. $\dfrac{3(a + 1)^{-1} + 4a^{-2}}{(a^3 + a^2)^{-1}}$

72. $\dfrac{9x^{-1} - 5(x - y)^{-1}}{4(x - y)^{-1}}$

In the study of calculus, the difference quotient $\dfrac{f(a + h) - f(a)}{h}$

*is often found and simplified. Find and simplify this quotient for each function f(x) by following steps **a** through **d**.*

a. *Find* $(a + h)$.

b. *Find* $f(a)$.

c. *Use steps **a** and **b** to find* $\dfrac{f(a + h) - f(a)}{h}$

d. *Simplify the result of step **c**.*

73. $f(x) = \dfrac{1}{x}$

74. $f(x) = \dfrac{5}{x}$

75. $\dfrac{3}{x + 1}$

76. $\dfrac{2}{x^2}$

6.4 | Dividing Polynomials: Long Division and Synthetic Division

OBJECTIVES

1 Divide a Polynomial by a Monomial.

2 Divide by a Polynomial.

3 Use Synthetic Division to Divide a Polynomial by a Binomial.

4 Use the Remainder Theorem to Evaluate Polynomials.

OBJECTIVE

1 Dividing a Polynomial by a Monomial

Recall that a rational expression is a quotient of polynomials. An equivalent form of a rational expression can be obtained by performing the indicated division. For example, the rational expression

$$\dfrac{10x^3 - 5x^2 + 20x}{5x}$$

can be thought of as the polynomial $10x^3 - 5x^2 + 20x$ divided by the monomial $5x$. To perform this division of a polynomial by a monomial (which we do on the next page), recall the following addition fact for fractions with a common denominator.

$$\underbrace{\dfrac{a}{c} + \dfrac{b}{c} = \dfrac{a + b}{c}}$$

If a, b, and c are monomials, we might read this equation from right to left and gain insight into dividing a polynomial by a monomial.

> **Dividing a Polynomial by a Monomial**
>
> Divide each term in the polynomial by the monomial.
>
> $$\dfrac{a + b}{c} = \dfrac{a}{c} + \dfrac{b}{c}, \quad \text{where } c \neq 0$$

EXAMPLE 1 Divide $10x^3 - 5x^2 + 20x$ by $5x$.

Solution We divide each term of $10x^3 - 5x^2 + 20x$ by $5x$ and simplify.

$$\frac{10x^3 - 5x^2 + 20x}{5x} = \frac{10x^3}{5x} - \frac{5x^2}{5x} + \frac{20x}{5x} = 2x^2 - x + 4$$

Check: To check, see that (quotient)(divisor) = dividend, or

$$(2x^2 - x + 4)(5x) = 10x^3 - 5x^2 + 20x.$$ □

PRACTICE
1 Divide $18a^3 - 12a^2 + 30a$ by $6a$.

■

EXAMPLE 2 Divide: $\dfrac{3x^5y^2 - 15x^3y - x^2y - 6x}{x^2y}$.

Solution We divide each term in the numerator by x^2y.

$$\frac{3x^5y^2 - 15x^3y - x^2y - 6x}{x^2y} = \frac{3x^5y^2}{x^2y} - \frac{15x^3y}{x^2y} - \frac{x^2y}{x^2y} - \frac{6x}{x^2y}$$

$$= 3x^3y - 15x - 1 - \frac{6}{xy}$$ □

PRACTICE
2 Divide: $\dfrac{5a^3b^4 - 8a^2b^3 + ab^2 - 8b}{ab^2}$.

■

OBJECTIVE
2 Dividing by a Polynomial ▶

To divide a polynomial by a polynomial other than a monomial, we use **long division.** Polynomial long division is similar to long division of real numbers. We review long division of real numbers by dividing 7 into 296.

$$
\begin{array}{r}
42 \\
7{\overline{\smash{)}296}} \\
\underline{-28} \qquad \text{4(7) = 28.} \\
16 \qquad \text{Subtract and bring down the next digit in the dividend.} \\
\underline{-14} \qquad \text{2(7) = 14.} \\
2 \qquad \text{Subtract. The remainder is 2.}
\end{array}
$$

Divisor:

The quotient is $42\dfrac{2 \, (\text{remainder})}{7 \, (\text{divisor})}$.

Check: To check, notice that

$$42(7) + 2 = 296, \text{ the dividend.}$$

This same division process can be applied to polynomials, as shown next.

EXAMPLE 3 Divide $2x^2 - x - 10$ by $x + 2$.

Solution $2x^2 - x - 10$ is the dividend, and $x + 2$ is the divisor.

Step 1. Divide $2x^2$ by x.

$$
\begin{array}{r}
2x \\
x + 2{\overline{\smash{)}2x^2 - x - 10}}
\end{array}
\qquad \frac{2x^2}{x} = 2x, \text{ so } 2x \text{ is the first term of the quotient.}
$$

(Continued on next page)

Step 2. Multiply $2x(x + 2)$.

$$
\begin{array}{r}
2x \\
x + 2 \overline{)\,2x^2 - x - 10} \\
2x^2 - 4x
\end{array}
$$

$2x(x + 2)$
Like terms are lined up vertically.

Step 3. Subtract $(2x^2 + 4x)$ from $(2x^2 - x - 10)$ by changing the signs of $(2x^2 + 4x)$ and adding.

$$
\begin{array}{r}
2x \\
x + 2 \overline{)\,2x^2 - x - 10} \\
{\overline{\mp}2x^2 \,{\overline{\mp}}\, 4x } \\
-5x
\end{array}
$$

Step 4. Bring down the next term, -10, and start the process over.

$$
\begin{array}{r}
2x \\
x + 2 \overline{)\,2x^2 - x - 10} \\
{\overline{\mp}2x^2 \,{\overline{\mp}}\, 4x } \downarrow \\
-5x - 10
\end{array}
$$

Step 5. Divide $-5x$ by x.

$$
\begin{array}{r}
2x - 5 \\
x + 2 \overline{)\,2x^2 - x - 10} \\
{\overline{\mp}2x^2 \,{\overline{\mp}}\, 4x } \\
-5x - 10
\end{array}
$$

$\dfrac{-5x}{x} = -5$, so -5 is the second term of the quotient.

Step 6. Multiply $-5(x + 2)$.

$$
\begin{array}{r}
2x - 5 \\
x + 2 \overline{)\,2x^2 - x - 10} \\
{\overline{\mp}2x^2 \,{\overline{\mp}}\, 4x } \\
-5x - 10 \\
-5x - 10
\end{array}
$$

Multiply: $-5(x + 2)$. Like terms are lined up vertically.

Step 7. Subtract by changing signs of $-5x - 10$ and adding.

$$
\begin{array}{r}
2x - 5 \\
x + 2 \overline{)\,2x^2 - x - 10} \\
{\overline{\mp}2x^2 \,{\overline{\mp}}\, 4x } \\
-5x - 10 \\
{\overline{\mp}5x \,{\overline{\mp}}\, 10} \\
0
\end{array}
$$

Subtract.

Remainder

Then $\dfrac{2x^2 - x - 10}{x + 2} = 2x - 5$. There is no remainder.

Check: Check this result by multiplying $2x - 5$ by $x + 2$. Their product is $(2x - 5)(x + 2) = 2x^2 - x - 10$, the dividend. □

PRACTICE

3 Divide $3x^2 + 7x - 6$ by $x + 3$.

EXAMPLE 4   Divide: $(6x^2 - 19x + 12) \div (3x - 5)$.

Solution

$$
\begin{array}{r}
2x \\
3x - 5 \overline{\smash{)}6x^2 - 19x + 12} \\
\underline{6x^2 - 10x} \downarrow \\
-9x + 12
\end{array}
$$

Divide $\dfrac{6x^2}{3x} = 2x$.

Multiply $2x(3x - 5)$.

Subtract by adding the opposite.
Bring down the next term, $+12$.

$$
\begin{array}{r}
2x - 3 \\
3x - 5 \overline{\smash{)}6x^2 - 19x + 12} \\
\underline{6x^2 - 10x} \\
-9x + 12 \\
\underline{-9x - 15} \\
-3
\end{array}
$$

Divide $\dfrac{-9x}{3x} = -3$.

Multiply $-3(3x - 5)$.

Subtract by adding the opposite.

Check: divisor   ·   quotient   +   remainder

$$(3x - 5) \quad \cdot \quad (2x - 3) \quad + \quad (-3) = 6x^2 - 19x + 15 - 3$$

$$= 6x^2 - 19x + 12 \quad \text{The dividend}$$

The division checks, so

$$\frac{6x^2 - 19x + 12}{3x - 5} = 2x - 3 + \frac{-3}{3x - 5}$$

$$\text{or} \quad 2x - 3 - \frac{3}{3x - 5}$$

> ▶ **Helpful Hint**
> This fraction is the remainder over the divisor.

PRACTICE
4   Divide $(6x^2 - 7x + 8)$ by $(2x - 1)$.

EXAMPLE 5   Divide: $(7x^3 + 16x^2 + 2x - 1) \div (x + 4)$.

Solution

$$
\begin{array}{r}
7x^2 - 12x + 50 \\
x + 4 \overline{\smash{)}7x^3 + 16x^2 + 2x - 1} \\
\underline{7x^3 + 28x^2} \\
-12x^2 + 2x \\
\underline{-12x^2 - 48x} \\
50x - 1 \\
\underline{50x + 200} \\
-201
\end{array}
$$

Divide $\dfrac{7x^3}{x} = 7x^2$.

$7x^2(x + 4)$

Subtract. Bring down $2x$.

$\dfrac{-12x^2}{x} = -12x$, a term of the quotient.

$-12x(x + 4)$ Subtract. Bring down -1.

$\dfrac{50x}{x} = 50$, a term of the quotient.

$50(x + 4)$.   Subtract.

Thus, $\dfrac{7x^3 + 16x^2 + 2x - 1}{x + 4} = 7x^2 - 12x + 50 + \dfrac{-201}{x + 4}$ or

$$7x^2 - 12x + 50 - \frac{201}{x + 4}.$$

PRACTICE
5   Divide $(5x^3 + 9x^2 - 10x + 30) \div (x + 3)$.

EXAMPLE 6 Divide $3x^4 + 2x^3 - 8x + 6$ by $x^2 - 1$.

Solution Before dividing, we represent any "missing powers" by the product of 0 and the variable raised to the missing power. There is no x^2 term in the dividend, so we include $0x^2$ to represent the missing term. Also, there is no x term in the divisor, so we include $0x$ in the divisor.

$$
\begin{array}{r}
3x^2 + 2x + 3 \\
x^2 + 0x - 1\overline{)3x^4 + 2x^3 + 0x^2 - 8x + 6} \\
\underline{3x^4 \not+ 0x^3 \not- 3x^2} \\
2x^3 + 3x^2 - 8x \\
\underline{2x^3 \not- 0x^2 \not+ 2x} \\
3x^2 - 6x + 6 \\
\underline{3x^2 \not+ 0x \not- 3} \\
-6x + 9
\end{array}
$$

$\dfrac{3x^4}{x^2} = 3x^2$
$3x^2(x^2 + 0x - 1)$
Subtract. Bring down $-8x$.

$\dfrac{2x^3}{x^2} = 2x$, a term of the quotient.
$2x(x^2 + 0x - 1)$
Subtract. Bring down 6.

$\dfrac{3x^2}{x^2} = 3$, a term of the quotient.
$3(x^2 + 0x - 1)$
Subtract.

The division process is finished when the degree of the remainder polynomial is less than the degree of the divisor. Thus,

$$\frac{3x^4 + 2x^3 - 8x + 6}{x^2 - 1} = 3x^2 + 2x + 3 + \frac{-6x + 9}{x^2 - 1}$$ □

PRACTICE
6 Divide $2x^4 + 3x^3 - 5x + 2$ by $x^2 + 1$.

EXAMPLE 7 Divide $27x^3 + 8$ by $3x + 2$.

Solution We replace the missing terms in the dividend with $0x^2$ and $0x$.

$$
\begin{array}{r}
9x^2 - 6x + 4 \\
3x + 2\overline{)27x^3 + 0x^2 + 0x + 8} \\
\underline{27x^3 \not+ 18x^2} \\
-18x^2 + 0x \\
\underline{\not+ 18x^2 \not+ 12x} \\
12x + 8 \\
\underline{12x \not+ 8}
\end{array}
$$

$9x^2(3x + 2)$
Subtract. Bring down $0x$.
$-6x(3x + 2)$
Subtract. Bring down 8.
$4(3x + 2)$

Thus, $\dfrac{27x^3 + 8}{3x + 2} = 9x^2 - 6x + 4$. □

PRACTICE
7 Divide $64x^3 - 125$ by $4x - 5$.

✓ CONCEPT CHECK
In a division problem, the divisor is $4x^3 - 5$. The division process can be stopped when which of these possible remainder polynomials is reached?
a. $2x^4 + x^2 - 3$ **b.** $x^3 - 5^2$ **c.** $4x^2 + 25$

OBJECTIVE
3 Using Synthetic Division to Divide a Polynomial by a Binomial

When a polynomial is to be divided by a binomial of the form $x - c$, a shortcut process called **synthetic division** may be used. On the next page, on the left is an example of long division, and on the right, the same example showing the coefficients of the variables only.

$$
\begin{array}{r}
2x^2 + 5x + 2 \\
x - 3\overline{)2x^3 - x^2 - 13x + 1} \\
\underline{2x^3 - 6x^2} \\
5x^2 - 13x \\
\underline{5x^2 - 15x} \\
2x + 1 \\
\underline{2x - 6} \\
7
\end{array}
\qquad
\begin{array}{r}
2 \quad 5 \quad 2 \\
1 - 3\overline{)2 - 1 - 13 + 1} \\
\underline{2 - 6} \\
5 - 13 \\
\underline{5 - 15} \\
2 + 1 \\
\underline{2 - 6} \\
7
\end{array}
$$

Notice that as long as we keep coefficients of powers of x in the same column, we can perform division of polynomials by performing algebraic operations on the coefficients only. This shortcut process of dividing with coefficients only in a special format is called synthetic division. To find $(2x^3 - x^2 - 13x + 1) \div (x - 3)$ by synthetic division, follow the next example.

EXAMPLE 8   Use synthetic division to divide $2x^3 - x^2 - 13x + 1$ by $x - 3$.

Solution   To use synthetic division, the divisor must be in the form $x - c$. Since we are dividing by $x - 3$, c is 3. Write down 3 and the coefficients of the dividend.

c

$\begin{array}{r|rrrr} 3 & 2 & -1 & -13 & 1 \\ & \downarrow \\ \hline & 2 \end{array}$ Next, draw a line and bring down the first coefficient of the dividend.

$\begin{array}{r|rrrr} 3 & 2 & -1 & -13 & 1 \\ & & 6 \\ \hline & 2 \end{array}$ Multiply $3 \cdot 2$ and write down the product, 6.

$\begin{array}{r|rrrr} 3 & 2 & -1 & -13 & 1 \\ & & 6 \\ \hline & 2 & 5 \end{array}$ Add $-1 + 6$. Write down the sum, 5.

$\begin{array}{r|rrrr} 3 & 2 & -1 & -13 & 1 \\ & & 6 & 15 \\ \hline & 2 & 5 & 2 \end{array}$ $3 \cdot 5 = 15.$ $-13 + 15 = 2.$

$\begin{array}{r|rrrr} 3 & 2 & -1 & -13 & 1 \\ & & 6 & 15 & 6 \\ \hline & 2 & 5 & 2 & 7 \end{array}$ $3 \cdot 2 = 6.$ $1 + 6 = 7.$

The quotient is found in the bottom row. The numbers 2, 5, and 2 are the coefficients of the quotient polynomial, and the number 7 is the remainder. The degree of the quotient polynomial is one less than the degree of the dividend. In our example, the degree of the dividend is 3, so the degree of the quotient polynomial is 2. As we found when we performed the long division, the quotient is

$$2x^2 + 5x + 2, \quad \text{remainder } 7$$

or

$$2x^2 + 5x + 2 + \frac{7}{x - 3}$$

PRACTICE 8   Use synthetic division to divide $4x^3 - 3x^2 + 6x + 5$ by $x - 1$.

When using synthetic division, if there are missing powers of the variable, insert 0s as coefficients.

EXAMPLE 9 Use synthetic division to divide $x^4 - 2x^3 - 11x^2 + 34$ by $x + 2$.

Solution The divisor is $x + 2$, which in the form $x - c$ is $x - (-2)$. Thus, c is -2. There is no x-term in the dividend, so we insert coefficient of 0. The dividend coefficients are $1, -2, -11, 0$, and 34.

$$
\begin{array}{r|rrrrr}
-2 & 1 & -2 & -11 & 0 & 34 \\
   &   & -2 & 8 & 6 & -12 \\
\hline
   & 1 & -4 & -3 & 6 & 22
\end{array}
$$

The dividend is a fourth-degree polynomial, so the quotient polynomial is a third-degree polynomial. The quotient is $x^3 - 4x^2 - 3x + 6$ with a remainder of 22. Thus,

$$\frac{x^4 - 2x^3 - 11x^2 + 34}{x + 2} = x^3 - 4x^2 - 3x + 6 + \frac{22}{x + 2}$$

PRACTICE
9 Use synthetic division to divide $x^4 + 3x^3 - 5x^2 + 12$ by $x + 3$.

✓CONCEPT CHECK
Which division problems are candidates for the synthetic division process?

a. $(3x^2 + 5) \div (x + 4)$

b. $(x^3 - x^2 + 2) \div (3x^3 - 2)$

c. $(y^4 + y - 3) \div (x^2 + 1)$

d. $x^5 \div (x - 5)$

> ▶ **Helpful Hint**
> Before dividing by synthetic division, write the dividend in descending order of variable exponents. Any "missing powers" of the variable should be represented by 0 times the variable raised to the missing power.

EXAMPLE 10 If $P(x) = 2x^3 - 4x^2 + 5$,

a. Find $P(2)$ by substitution.

b. Use synthetic division to find the remainder when $P(x)$ is divided by $x - 2$.

Solution

a. $P(x) = 2x^3 - 4x^2 + 5$

$P(2) = 2(2)^3 - 4(2)^2 + 5$

$\qquad = 2(8) - 4(4) + 5 = 16 - 16 + 5 = 5$

Thus, $P(2) = 5$.

b. The coefficients of $P(x)$ are $2, -4, 0$, and 5. The number 0 is the coefficient of the missing power of x^1. The divisor is $x - 2$, so c is 2.

$$
\begin{array}{r|rrrr}
2 & 2 & -4 & 0 & 5 \\
  &   & 4 & 0 & 0 \\
\hline
  & 2 & 0 & 0 & 5 \text{ remainder}
\end{array}
$$

The remainder when $P(x)$ is divided by $x - 2$ is 5.

PRACTICE
10　If $P(x) = x^3 - 5x - 2$,

a. Find $P(2)$ by substitution.

b. Use synthetic division to find the remainder when $P(x)$ is divided by $x - 2$.

OBJECTIVE
4　**Using the Remainder Theorem to Evaluate Polynomials** ▶

Notice in the preceding example that $P(2) = 5$ and that the remainder when $P(x)$ is divided by $x - 2$ is 5. This is no accident. This illustrates the **remainder theorem.**

> **Remainder Theorem**
>
> If a polynomial $P(x)$ is divided by $x - c$, then the remainder is $P(c)$.

EXAMPLE 11　Use the remainder theorem and synthetic division to find $P(4)$ if
$$P(x) = 4x^6 - 25x^5 + 35x^4 + 17x^2.$$

Solution　To find $P(4)$ by the remainder theorem, we divide $P(x)$ by $x - 4$. The coefficients of $P(x)$ are $4, -25, 35, 0, 17, 0$, and 0. Also, c is 4.

$$
\begin{array}{r|rrrrrrr}
c & & & & & & & \\
4 & 4 & -25 & 35 & 0 & 17 & 0 & 0 \\
  &   & 16 & -36 & -4 & -16 & 4 & 16 \\
\hline
  & 4 & -9 & -1 & -4 & 1 & 4 & 16 \quad \text{remainder}
\end{array}
$$

Thus, $P(4) = 16$, the remainder.　□

PRACTICE
11　Use the remainder theorem and synthetic division to find $P(3)$ if $P(x) = 2x^5 - 18x^4 + 90x^2 + 59x$.

Vocabulary, Readiness & Video Check

Martin-Gay Interactive Videos

See Video 6.4 🍐

Watch the section lecture video and answer the following questions.

OBJECTIVE
1　　**1.** In the lecture before ▦ Example 1, dividing a polynomial by a monomial is compared to adding two fractions. What role does the monomial play in the fraction example?

OBJECTIVE
2　　**2.** From ▦ Example 2, how do you know when to stop your long division?

OBJECTIVE
3　　**3.** From ▦ Example 3, once you've completed the synthetic division, what does the bottom row of numbers mean? What is the degree of the quotient?

OBJECTIVE
4　　**4.** From ▦ Example 4, given a polynomial function $P(x)$, under what circumstances might it be easier/faster to use the remainder theorem to find $P(c)$ rather than substituting the value c for x and then simplifying?

6.4 Exercise Set

MyMathLab®

Divide. See Examples 1 and 2.

1. $4a^2 + 8a$ by $2a$

2. $6x^4 - 3x^3$ by $3x^2$

3. $\dfrac{12a^5b^2 + 16a^4b}{4a^4b}$

4. $\dfrac{4x^3y + 12x^2y^2 - 4xy^3}{4xy}$

▶ 5. $\dfrac{4x^2y^2 + 6xy^2 - 4y^2}{2x^2y}$

6. $\dfrac{6x^5y + 75x^4y - 24x^3y^2}{3x^4y}$

Divide. See Examples 3 through 7.

7. $(x^2 + 3x + 2) \div (x + 2)$

8. $(y^2 + 7y + 10) \div (y + 5)$

9. $(2x^2 - 6x - 8) \div (x + 1)$

10. $(3x^2 + 19x + 20) \div (x + 5)$

11. $2x^2 + 3x - 2$ by $2x + 4$

12. $6x^2 - 17x - 3$ by $3x - 9$

13. $(4x^3 + 7x^2 + 8x + 20) \div (2x + 4)$

14. $(8x^3 + 18x^2 + 16x + 24) \div (4x + 8)$

15. $(2x^2 + 6x^3 - 18x - 6) \div (3x + 1)$

16. $(4x - 15x^2 + 10x^3 - 6) \div (2x - 3)$

▶ 17. $(3x^5 - x^3 + 4x^2 - 12x - 8) \div (x^2 - 2)$

18. $(2x^5 - 6x^4 + x^3 - 4x + 3) \div (x^2 - 3)$

19. $\left(2x^4 + \dfrac{1}{2}x^3 + x^2 + x\right) \div (x - 2)$

20. $\left(x^4 - \dfrac{2}{3}x^3 + x\right) \div (x - 3)$

Use synthetic division to divide. See Examples 8 and 9.

21. $\dfrac{x^2 + 3x - 40}{x - 5}$

22. $\dfrac{x^2 - 14x + 24}{x - 2}$

23. $\dfrac{x^2 + 5x - 6}{x + 6}$

24. $\dfrac{x^2 + 12x + 32}{x + 4}$

▶ 25. $\dfrac{x^3 - 7x^2 - 13x + 5}{x - 2}$

26. $\dfrac{x^3 + 6x^2 + 4x - 7}{x + 5}$

27. $\dfrac{4x^2 - 9}{x - 2}$

28. $\dfrac{3x^2 - 4}{x - 1}$

MIXED PRACTICE

Divide. See Examples 1–9.

29. $\dfrac{4x^7y^4 + 8xy^2 + 4xy^3}{4xy^3}$

30. $\dfrac{15x^3y - 5x^2y + 10xy^2}{5x^2y}$

31. $(10x^3 - 5x^2 - 12x + 1) \div (2x - 1)$

32. $(20x^3 - 8x^2 + 5x - 5) \div (5x - 2)$

33. $(2x^3 - 6x^2 - 4) \div (x - 4)$

34. $(3x^3 + 4x - 10) \div (x + 2)$

35. $\dfrac{2x^4 - 13x^3 + 16x^2 - 9x + 20}{x - 5}$

36. $\dfrac{3x^4 + 5x^3 - x^2 + x - 2}{x + 2}$

37. $\dfrac{7x^2 - 4x + 12 + 3x^3}{x + 1}$

38. $\dfrac{4x^3 + x^4 - x^2 - 16x - 4}{x - 2}$

39. $\dfrac{3x^3 + 2x^2 - 4x + 1}{x - \dfrac{1}{3}}$

40. $\dfrac{9y^3 + 9y^2 - y + 2}{y + \dfrac{2}{3}}$

41. $\dfrac{x^3 - 1}{x - 1}$ 42. $\dfrac{y^3 - 8}{y - 2}$

43. $(25xy^2 + 75xyz + 125x^2yz) \div (-5x^2y)$

44. $(x^6y^6 - x^3y^3z + 7x^3y) \div (-7yz^2)$

45. $(9x^5 + 6x^4 - 6x^2 - 4) \div (3x + 2)$

46. $(5x^4 - 5x^2 + 10x^3 - 10x) \div (5x + 10)$

For the given polynomial $P(x)$ and the given c, use the remainder theorem to find $P(c)$. See Examples 10 and 11.

47. $P(x) = x^3 + 3x^2 - 7x + 4;\ 1$

48. $P(x) = x^3 + 5x^2 - 4x - 6;\ 2$

49. $P(x) = 3x^3 - 7x^2 - 2x + 5;\ -3$

50. $P(x) = 4x^3 + 5x^2 - 6x - 4;\ -2$

▶ 51. $P(x) = 4x^4 + x^2 - 2;\ -1$

52. $P(x) = x^4 - 3x^2 - 2x + 5;\ -2$

53. $P(x) = 2x^4 - 3x^2 - 2;\ \dfrac{1}{3}$

54. $P(x) = 4x^4 - 2x^3 + x^2 - x - 4;\ \dfrac{1}{2}$

55. $P(x) = x^5 + x^4 - x^3 + 3;\ \dfrac{1}{2}$

56. $P(x) = x^5 - 2x^3 + 4x^2 - 5x + 6;\ \dfrac{2}{3}$

REVIEW AND PREVIEW

Solve each equation for x. See Sections 2.1 and 5.8.

57. $7x + 2 = x - 3$ **58.** $4 - 2x = 17 - 5x$

59. $x^2 = 4x - 4$ **60.** $5x^2 + 10x = 15$

61. $\dfrac{x}{3} - 5 = 13$ **62.** $\dfrac{2x}{9} + 1 = \dfrac{7}{9}$

Factor the following. See Sections 5.5 and 5.7.

63. $x^3 - 1$

64. $8y^3 + 1$

65. $125z^3 + 8$

66. $a^3 - 27$

67. $xy + 2x + 3y + 6$

68. $x^2 - x + xy - y$

69. $x^3 - 9x$

70. $2x^3 - 32x$

CONCEPT EXTENSIONS

Determine whether each division problem is a candidate for the synthetic division process. See the Concept Checks in this section.

71. $(5x^2 - 3x + 2) \div (x + 2)$

72. $(x^4 - 6) \div (x^3 + 3x - 1)$

73. $(x^7 - 2) \div (x^5 + 1)$

74. $(3x^2 + 7x - 1) \div \left(x - \dfrac{1}{3}\right)$

75. In a long division exercise, if the divisor is $9x^3 - 2x$, the division process can be stopped when the degree of the remainder is

   a. 1   **b.** 3   **c.** 9   **d.** 2

76. In a division exercise, if the divisor is $x - 3$, the division process can be stopped when the degree of the remainder is

   a. 1   **b.** 0   **c.** 2   **d.** 3

△ **77.** A board of length $(3x^4 + 6x^2 - 18)$ meters is to be cut into three pieces of the same length. Find the length of each piece.

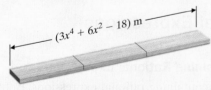

$(3x^4 + 6x^2 - 18)$ m

△ **78.** The perimeter of a regular hexagon is given to be $(12x^5 - 48x^3 + 3)$ miles. Find the length of each side.

△ **79.** If the area of the rectangle is $(15x^2 - 29x - 14)$ square inches, and its length is $(5x + 2)$ inches, find its width.

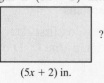

?

$(5x + 2)$ in.

△ **80.** If the area of a parallelogram is $(2x^2 - 17x + 35)$ square centimeters and its base is $(2x - 7)$ centimeters, find its height.

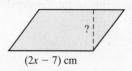

?

$(2x - 7)$ cm

△ **81.** If the area of a parallelogram is $(x^4 - 23x^2 + 9x - 5)$ square centimeters and its base is $(x + 5)$ centimeters, find its height.

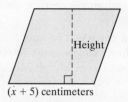

Height

$(x + 5)$ centimeters

△ **82.** If the volume of a box is $(x^4 + 6x^3 - 7x^2)$ cubic meters, its height is x^2 meters, and its length is $(x + 7)$ meters, find its width.

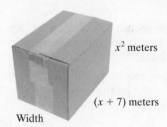

x^2 meters

$(x + 7)$ meters

Width

Divide.

83. $\left(x^4 + \dfrac{2}{3}x^3 + x\right) \div (x - 1)$

84. $\left(2x^3 + \dfrac{9}{2}x^2 - 4x - 10\right) \div (x + 2)$

85. $\left(3x^4 - x - x^3 + \dfrac{1}{2}\right) \div (2x - 1)$

86. $\left(2x^4 + \dfrac{1}{2}x^3 - \dfrac{1}{4}x^2 + x\right) \div (2x + 1)$

87. $(5x^4 - 2x^2 + 10x^3 - 4x) \div (5x + 10)$

88. $(9x^5 + 6x^4 - 6x^2 - 4x) \div (3x + 2)$

For each given f(x) and g(x), find $\dfrac{f(x)}{g(x)}$. Also find any x-values that are not in the domain of $\dfrac{f(x)}{g(x)}$. (Note: Since g(x) is in the denominator, g(x) cannot be 0.)

89. $f(x) = 25x^2 - 5x + 30; g(x) = 5x$

90. $f(x) = 12x^4 - 9x^3 + 3x - 1; g(x) = 3x$

91. $f(x) = 7x^4 - 3x^2 + 2; g(x) = x - 2$

92. $f(x) = 2x^3 - 4x^2 + 1; g(x) = x + 3$

93. Try performing the following division without changing the order of the terms. Describe why this makes the process more complicated. Then perform the division again after putting the terms in the dividend in descending order of exponents.

$$\frac{4x^2 - 12x - 12 + 3x^3}{x - 2}$$

94. Explain how to check polynomial long division.

95. Explain an advantage of using the remainder theorem instead of direct substitution.

96. Explain an advantage of using synthetic division instead of long division.

We say that 2 is a factor of 8 because 2 divides 8 evenly, or with a remainder of 0. In the same manner, the polynomial $x - 2$ is a factor of the polynomial $x^3 - 14x^2 + 24x$ because the remainder is 0 when $x^3 - 14x^2 + 24x$ is divided by $x - 2$. Use this information for Exercises 97 and 98.

97. Use synthetic division to show that $x + 3$ is a factor of $x^3 + 3x^2 + 4x + 12$.

98. Use synthetic division to show that $x - 2$ is a factor of $x^3 - 2x^2 - 3x + 6$.

99. If a polynomial is divided by $x - 5$, the quotient is $2x^2 + 5x - 6$ and the remainder is 3. Find the original polynomial.

100. If a polynomial is divided by $x + 3$, the quotient is $x^2 - x + 10$ and the remainder is -2. Find the original polynomial.

101. eBay is the leading online auction house. eBay's annual net profit can be modeled by the polynomial function $P(x) = 0.48x^3 + 2.06x^2 + 141x + 9.71$, where $P(x)$ is net profit in millions of dollars and x is the number of years since 2000. eBay's annual revenue can be modeled by the function $R(x) = 1011x - 288$, where $R(x)$ is revenue in millions of dollars and x is years after 2000. (*Source:* eBay, Inc., annual reports 2000–2010)

a. Given that

$$\text{Net profit margin} = \frac{\text{net profit}}{\text{revenue}},$$

write a function, $m(x)$, that models eBay's net profit margin.

b. Use part (a) to predict eBay's profit margin in 2015. Round to the nearest hundredth.

102. Kraft Foods is a provider of many of the best-known food brands in our supermarkets. Among their well-known brands are Kraft, Oscar Mayer, Maxwell House, and Oreo. Kraft Foods' annual revenues since 2005 can be modeled by the polynomial function $R(x) = 0.06x^3 + 0.02x^2 + 1.67x + 32.33$, where $R(x)$ is revenue in billions of dollars and x is the number of years since 2005. Kraft Foods' net profit can be modeled by the function $P(x) = 0.07x^3 - 0.42x^2 + 0.7x + 2.63$, where $P(x)$ is the net profit in billions of dollars and x is the number of years since 2005. (*Source:* Based on information from Kraft Foods)

a. Suppose that a market analyst has found the model $P(x)$ and another analyst at the same firm has found the model $R(x)$. The analysts have been asked by their manager to work together to find a model for Kraft Foods' profit margin. The analysts know that a company's profit margin is the ratio of its profit to its revenue. Describe how these two analysts could collaborate to find a function $m(x)$ that models Kraft Foods' net profit margin based on the work they have done independently.

b. Without actually finding $m(x)$, give a general description of what you would expect the answer to be.

103. From the remainder theorem, the polynomial $x - c$ is a factor of a polynomial function $P(x)$ if $P(c)$ is what value?

6.5 Solving Equations Containing Rational Expressions ▶

OBJECTIVE

1 Solve Equations Containing Rational Expressions. ▶

OBJECTIVE

1 Solving Equations Containing Rational Expressions ▶

In this section, we solve equations containing rational expressions. Before beginning this section, make sure that you understand the difference between an *equation* and an *expression*. An **equation** contains an equal sign and an **expression** does not.

Equation	*Expression*
$\dfrac{x}{2} + \dfrac{x}{6} = \dfrac{2}{3}$	$\dfrac{x}{2} + \dfrac{x}{6}$

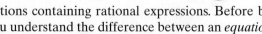

equal sign

▶ **Helpful Hint**

The method described here is for equations only. It may *not* be used for performing operations on expressions.

Solving Equations Containing Rational Expressions

To solve *equations* containing rational expressions, first clear the equation of fractions by multiplying both sides of the equation by the LCD of all rational expressions. Then solve as usual.

✓**CONCEPT CHECK**

True or false? Clearing fractions is valid when solving an equation and when simplifying rational expressions. Explain.

EXAMPLE 1   Solve: $\dfrac{4x}{5} + \dfrac{3}{2} = \dfrac{3x}{10}$.

Solution   The LCD of $\dfrac{4x}{5}, \dfrac{3}{2}$, and $\dfrac{3x}{10}$ is 10. We multiply both sides of the equation by 10.

$$\frac{4x}{5} + \frac{3}{2} = \frac{3x}{10}$$

$$10\left(\frac{4x}{5} + \frac{3}{2}\right) = 10\left(\frac{3x}{10}\right) \qquad \text{Multiply both sides by the LCD.}$$

$$10 \cdot \frac{4x}{5} + 10 \cdot \frac{3}{2} = 10 \cdot \frac{3x}{10} \qquad \text{Use the distributive property.}$$

$$8x + 15 = 3x \qquad \text{Simplify.}$$

$$15 = -5x \qquad \text{Subtract } 8x \text{ from both sides.}$$

$$-3 = x \qquad \text{Solve.}$$

Verify this solution by replacing x with -3 in the original equation.

Check:

$$\frac{4x}{5} + \frac{3}{2} = \frac{3x}{10}$$

$$\frac{4(-3)}{5} + \frac{3}{2} \overset{?}{=} \frac{3(-3)}{10}$$

$$\frac{-12}{5} + \frac{3}{2} \overset{?}{=} \frac{-9}{10}$$

$$-\frac{24}{10} + \frac{15}{10} \overset{?}{=} -\frac{9}{10}$$

$$-\frac{9}{10} = -\frac{9}{10} \qquad \text{True}$$

The solution is -3 or the solution set is $\{-3\}$.   □

PRACTICE
1   Solve: $\dfrac{5x}{4} - \dfrac{3}{2} = \dfrac{7x}{8}$.

The important difference of the equations in this section is that the denominator of a rational expression may contain a variable. Recall that a rational expression is undefined for values of the variable that make the denominator 0. If a proposed solution makes any denominator 0, then it must be rejected as a solution of the original equation. Such proposed solutions are called **extraneous solutions.**

The following steps may be used to solve equations containing rational expressions.

Solving an Equation Containing Rational Expressions

Step 1.   Multiply both sides of the equation by the LCD of all rational expressions in the equation.

Step 2.   Simplify both sides.

Step 3.   Determine whether the equation is linear, quadratic, or higher degree and solve accordingly.

Step 4.   Check the solution in the original equation.

Answer to Concept Check:
false; answers may vary

EXAMPLE 2 Solve: $\dfrac{3}{x} - \dfrac{x+21}{3x} = \dfrac{5}{3}$.

Solution The LCD of the denominators x, $3x$, and 3 is $3x$. We multiply both sides by $3x$.

$$\frac{3}{x} - \frac{x+21}{3x} = \frac{5}{3}$$

$$3x\left(\frac{3}{x} - \frac{x+21}{3x}\right) = 3x\left(\frac{5}{3}\right) \qquad \text{Multiply both sides by the LCD.}$$

$$3x \cdot \frac{3}{x} - 3x \cdot \frac{x+21}{3x} = 3x \cdot \frac{5}{3} \qquad \text{Use the distributive property.}$$

$$9 - (x+21) = 5x \qquad \text{Simplify.}$$

$$9 - x - 21 = 5x$$

$$-12 = 6x$$

$$-2 = x \qquad \text{Solve.}$$

The proposed solution is -2.

Check: Check the proposed solution in the original equation.

$$\frac{3}{x} - \frac{x+21}{3x} = \frac{5}{3}$$

$$\frac{3}{-2} - \frac{-2+21}{3(-2)} \stackrel{?}{=} \frac{5}{3}$$

$$-\frac{9}{6} + \frac{19}{6} \stackrel{?}{=} \frac{5}{3}$$

$$\frac{10}{6} \stackrel{?}{=} \frac{5}{3} \qquad \text{True}$$

The solution is -2 or the solution set is $\{-2\}$.

PRACTICE
2 Solve: $\dfrac{6}{x} - \dfrac{x+9}{5x} = \dfrac{2}{5}$.

Let's talk more about multiplying both sides of an equation by the LCD of the rational expressions in the equation. In Example 3 that follows, the LCD is $x - 2$, so we will first multiply both sides of the equation by $x - 2$. Recall that the multiplication property for equations allows us to multiply both sides of an equation by any *nonzero* number. In other words, for Example 3 below, we may multiply both sides of the equation by $x - 2$ *as long as $x - 2 \neq 0$ or as long as $x \neq 2$*. Keep this in mind when solving these equations.

EXAMPLE 3 Solve: $\dfrac{x+6}{x-2} = \dfrac{2(x+2)}{x-2}$.

Solution First we multiply both sides of the equation by the LCD, $x - 2$. (Remember, we can do this only if $x \neq 2$ so that we are not multiplying by 0.)

$$\frac{x+6}{x-2} = \frac{2(x+2)}{x-2}$$

$$(x-2) \cdot \frac{x+6}{x-2} = (x-2) \cdot \frac{2(x+2)}{x-2} \qquad \text{Multiply both sides by } x - 2.$$

$$x + 6 = 2(x+2) \qquad \text{Simplify.}$$

$$x + 6 = 2x + 4 \qquad \text{Use the distributive property.}$$

$$2 = x \qquad \text{Solve.}$$

From above, we assumed that $x \neq 2$, so this equation has no solution. This will also be the case as we attempt to check this proposed solution.

Check: The proposed solution is 2. Notice that 2 makes a denominator 0 in the original equation. This can also be seen in a check. Check the proposed solution 2 in the original equation.

$$\frac{x+6}{x-2} = \frac{2(x+2)}{x-2}$$

$$\frac{2+6}{2-2} = \frac{2(2+2)}{2-2}$$

$$\frac{8}{0} = \frac{2(4)}{0}$$

The denominators are 0, so 2 is not a solution of the original equation. The solution is $\{\ \}$ or $\varnothing$. □

PRACTICE
3 Solve: $\dfrac{x-5}{x+3} = \dfrac{2(x-1)}{x+3}$.

EXAMPLE 4 Solve: $\dfrac{2x}{2x-1} + \dfrac{1}{x} = \dfrac{1}{2x-1}$.

Solution The LCD is $x(2x-1)$. Multiply both sides by $x(2x-1)$. By the distributive property, this is the same as multiplying each term by $x(2x-1)$.

$$x(2x-1)\cdot\frac{2x}{2x-1} + x(2x-1)\cdot\frac{1}{x} = x(2x-1)\cdot\frac{1}{2x-1}$$

$$x(2x) + (2x-1) = x \quad \text{Simplify.}$$

$$2x^2 + 2x - 1 - x = 0$$

$$2x^2 + x - 1 = 0$$

$$(x+1)(2x-1) = 0$$

$$x+1 = 0 \quad \text{or} \quad 2x-1 = 0$$

$$x = -1 \qquad x = \frac{1}{2}$$

The number $\dfrac{1}{2}$ makes the denominator $2x-1$ equal 0, so it is not a solution. The solution is -1. □

PRACTICE
4 Solve: $\dfrac{5x}{5x-1} + \dfrac{1}{x} = \dfrac{1}{5x-1}$.

EXAMPLE 5 Solve: $\dfrac{2x}{x-3} + \dfrac{6-2x}{x^2-9} = \dfrac{x}{x+3}$.

Solution We factor the second denominator to find that the LCD is $(x+3)(x-3)$. We multiply both sides of the equation by $(x+3)(x-3)$. By the distributive property, this is the same as multiplying each term by $(x+3)(x-3)$.

$$\frac{2x}{x-3} + \frac{6-2x}{x^2-9} = \frac{x}{x+3}$$

$$(x+3)(x-3)\cdot\frac{2x}{x-3} + (x+3)(x-3)\cdot\frac{6-2x}{(x+3)(x-3)}$$

$$= (x+3)(x-3)\left(\frac{x}{x+3}\right)$$

$$2x(x+3) + (6-2x) = x(x-3) \quad \text{Simplify.}$$

$$2x^2 + 6x + 6 - 2x = x^2 - 3x \quad \text{Use the distributive property.}$$

(Continued on next page)

Next we solve this quadratic equation by the factoring method. To do so, we first write the equation so that one side is 0.

$$x^2 + 7x + 6 = 0$$

$$(x + 6)(x + 1) = 0 \qquad \text{Factor.}$$

$$x = -6 \text{ or } x = -1 \qquad \text{Set each factor equal to 0.}$$

Neither -6 nor -1 makes any denominator 0, so they are both solutions. The solutions are -6 and -1. □

PRACTICE
5 Solve: $\dfrac{2}{x - 2} - \dfrac{5 + 2x}{x^2 - 4} = \dfrac{x}{x + 2}$.

EXAMPLE 6 Solve: $\dfrac{z}{2z^2 + 3z - 2} - \dfrac{1}{2z} = \dfrac{3}{z^2 + 2z}$.

Solution Factor the denominators to find that the LCD is $2z(z + 2)(2z - 1)$. Multiply both sides by the LCD. Remember, by using the distributive property, this is the same as multiplying each term by $2z(z + 2)(2z - 1)$.

$$\frac{z}{2z^2 + 3z - 2} - \frac{1}{2z} = \frac{3}{z^2 + 2z}$$

$$\frac{z}{(2z - 1)(z + 2)} - \frac{1}{2z} = \frac{3}{z(z + 2)}$$

$$2z(z + 2)(2z - 1) \cdot \frac{z}{(2z - 1)(z + 2)} - 2z(z + 2)(2z - 1) \cdot \frac{1}{2z}$$

$$= 2z(z + 2)(2z - 1) \cdot \frac{3}{z(z + 2)} \qquad \begin{array}{l}\text{Apply the distributive}\\\text{property.}\end{array}$$

$$2z(z) - (z + 2)(2z - 1) = 3 \cdot 2(2z - 1) \qquad \text{Simplify.}$$

$$2z^2 - (2z^2 + 3z - 2) = 12z - 6$$

$$2z^2 - 2z^2 - 3z + 2 = 12z - 6$$

$$-3z + 2 = 12z - 6$$

$$-15z = -8$$

$$z = \frac{8}{15} \qquad \text{Solve.}$$

The proposed solution $\dfrac{8}{15}$ does not make any denominator 0; the solution is $\dfrac{8}{15}$. □

PRACTICE
6 Solve: $\dfrac{z}{2z^2 - z - 6} - \dfrac{1}{3z} = \dfrac{2}{z^2 - 2z}$.

A graph can be helpful in visualizing solutions of equations. For example, to visualize the solution of the equation $\dfrac{3}{x} - \dfrac{x + 21}{3x} = \dfrac{5}{3}$ in Example 2, the graph of the related rational function $f(x) = \dfrac{3}{x} - \dfrac{x + 21}{3x}$ is shown. A solution of the equation is an x-value that corresponds to a y-value of $\dfrac{5}{3}$.

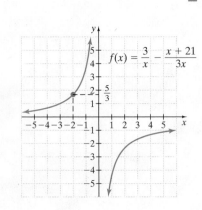

Notice that an x-value of -2 corresponds to a y-value of $\dfrac{5}{3}$. The solution of the equation is indeed -2 as shown in Example 2.

Vocabulary, Readiness & Video Check

Determine whether each is an equation or an expression. Do not solve or simplify.

1. $\dfrac{x}{2} = \dfrac{3x}{5} + \dfrac{x}{6}$

2. $\dfrac{3x}{5} + \dfrac{x}{6}$

3. $\dfrac{x}{x-1} + \dfrac{2x}{x+1}$

4. $\dfrac{x}{x-1} + \dfrac{2x}{x+1} = 5$

5. $\dfrac{y+7}{2} = \dfrac{y+1}{6} + \dfrac{1}{y}$

6. $\dfrac{y+1}{6} + \dfrac{1}{y}$

Choose the least common denominator (LCD) for the rational expressions in each equation. Do not solve.

7. $\dfrac{x}{7} - \dfrac{x}{2} = \dfrac{1}{2}$; LCD = _____

    a. 7   **b.** 2   **c.** 14   **d.** 28

8. $\dfrac{9}{x+1} + \dfrac{5}{(x+1)^2} = \dfrac{x}{x+1}$; LCD = _____

    a. $x+1$ **b.** $(x+1)^2$ **c.** $(x+1)^3$

9. $\dfrac{7}{x-4} = \dfrac{x}{x^2-16} + \dfrac{1}{x+4}$; LCD = _____

    a. $(x+4)(x-4)$ **b.** $x-4$ **c.** $x+4$ **d.** $(x^2-16)(x-4)(x+4)$

10. $3 = \dfrac{1}{x-5} - \dfrac{2}{x^2-5x}$; LCD = _____

    a. $x-5$ **b.** $3(x-5)$ **c.** $3x(x-5)$ **d.** $x(x-5)$

Martin-Gay Interactive Videos

See Video 6.5

Watch the section lecture video and answer the following questions.

OBJECTIVE 1

11. From Examples 2 and 3, why is it important to determine whether you have a linear or a quadratic equation before you finish solving the equation?

OBJECTIVE 1

12. From Examples 2 and 3, what extra check do you make for a proposed solution before you see that it satisfies the original equation?

6.5 Exercise Set   MyMathLab®

Solve each equation. See Examples 1 and 2.

1. $\dfrac{x}{2} - \dfrac{x}{3} = 12$

2. $x = \dfrac{x}{2} - 4$

3. $\dfrac{x}{3} = \dfrac{1}{6} + \dfrac{x}{4}$

4. $\dfrac{x}{2} = \dfrac{21}{10} - \dfrac{x}{5}$

▶ **5.** $\dfrac{2}{x} + \dfrac{1}{2} = \dfrac{5}{x}$

6. $\dfrac{5}{3x} + 1 = \dfrac{7}{6}$

7. $\dfrac{x^2+1}{x} = \dfrac{5}{x}$

8. $\dfrac{x^2-14}{2x} = -\dfrac{5}{2x}$

Solve each equation. See Examples 3 through 6.

9. $\dfrac{x+5}{x+3} = \dfrac{2}{x+3}$

10. $\dfrac{x-7}{x-1} = \dfrac{11}{x-1}$

11. $\dfrac{5}{x-2} - \dfrac{2}{x+4} = -\dfrac{4}{x^2+2x-8}$

12. $\dfrac{1}{x-1} + \dfrac{1}{x+1} = \dfrac{2}{x^2-1}$

13. $\dfrac{1}{x-1} = \dfrac{2}{x+1}$

14. $\dfrac{6}{x+3} = \dfrac{4}{x-3}$

15. $\dfrac{x^2-23}{2x^2-5x-3} + \dfrac{2}{x-3} = \dfrac{-1}{2x+1}$

16. $\dfrac{4x^2-24x}{3x^2-x-2} + \dfrac{3}{3x+2} = \dfrac{-4}{x-1}$

17. $\dfrac{1}{x-4} - \dfrac{3x}{x^2-16} = \dfrac{2}{x+4}$

18. $\dfrac{3}{2x+3} - \dfrac{1}{2x-3} = \dfrac{4}{4x^2-9}$

19. $\dfrac{1}{x-4} = \dfrac{8}{x^2-16}$

20. $\dfrac{2}{x^2-4} = \dfrac{1}{2x-4}$

21. $\dfrac{1}{x-2} - \dfrac{2}{x^2-2x} = 1$

22. $\dfrac{12}{3x^2+12x} = 1 - \dfrac{1}{x+4}$

MIXED PRACTICE

Solve each equation. See Examples 1 through 6.

23. $\dfrac{5}{x} = \dfrac{20}{12}$

24. $\dfrac{2}{x} = \dfrac{10}{5}$

25. $1 - \dfrac{4}{a} = 5$

26. $7 + \dfrac{6}{a} = 5$

27. $\dfrac{x^2+5}{x} - 1 = \dfrac{5(x+1)}{x}$

28. $\dfrac{x^2+6}{x} + 5 = \dfrac{2(x+3)}{x}$

29. $\dfrac{1}{2x} - \dfrac{1}{x+1} = \dfrac{1}{3x^2+3x}$

30. $\dfrac{2}{x-5} + \dfrac{1}{2x} = \dfrac{5}{3x^2-15x}$

31. $\dfrac{1}{x} - \dfrac{x}{25} = 0$

32. $\dfrac{x}{4} + \dfrac{5}{x} = 3$

33. $5 - \dfrac{2}{2y-5} = \dfrac{3}{2y-5}$

34. $1 - \dfrac{5}{y+7} = \dfrac{4}{y+7}$

35. $\dfrac{x-1}{x+2} = \dfrac{2}{3}$

36. $\dfrac{6x+7}{2x+9} = \dfrac{5}{3}$

37. $\dfrac{x+3}{x+2} = \dfrac{1}{x+2}$

▶ 38. $\dfrac{2x+1}{4-x} = \dfrac{9}{4-x}$

39. $\dfrac{1}{a-3} + \dfrac{2}{a+3} = \dfrac{1}{a^2-9}$

40. $\dfrac{12}{9-a^2} + \dfrac{3}{3+a} = \dfrac{2}{3-a}$

41. $\dfrac{64}{x^2-16} + 1 = \dfrac{2x}{x-4}$

42. $2 + \dfrac{3}{x} = \dfrac{2x}{x+3}$

43. $\dfrac{-15}{4y+1} + 4 = y$

44. $\dfrac{36}{x^2-9} + 1 = \dfrac{2x}{x+3}$

45. $\dfrac{28}{x^2-9} + \dfrac{2x}{x-3} + \dfrac{6}{x+3} = 0$

▶ 46. $\dfrac{x^2-20}{x^2-7x+12} = \dfrac{3}{x-3} + \dfrac{5}{x-4}$

47. $\dfrac{x+2}{x^2+7x+10} = \dfrac{1}{3x+6} - \dfrac{1}{x+5}$

48. $\dfrac{3}{2x-5} + \dfrac{2}{2x+3} = 0$

REVIEW AND PREVIEW

TRANSLATING *Write each sentence as an equation and solve. See Section 2.2.*

49. Four more than 3 times a number is 19.

50. The sum of two consecutive integers is 147.

51. The length of a rectangle is 5 inches more than the width. Its perimeter is 50 inches. Find the length and width.

52. The sum of a number and its reciprocal is $\dfrac{5}{2}$.

The following graph is from a recent survey of state and federal prisons. Use this graph to answer Exercises 53 through 58. See Section 2.2.

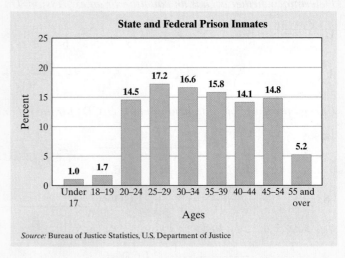

State and Federal Prison Inmates

Source: Bureau of Justice Statistics, U.S. Department of Justice

53. What percent of state and federal prison inmates are age 45 to 54?

54. What percent of state and federal prison inmates are 55 years old or older?

55. What age category shows the highest percent of prison inmates?

56. What percent of state and federal prison inmates are 20 to 34 years old?

57. At the end of 2010, there were 748,728 inmates under the jurisdiction of state and federal correction authorities in the United States. Approximately how many 25- to 29-year-old inmates would you expect to have been held in the United States at the end of 2010? Round to the nearest whole. (*Source:* Bureau of Justice Statistics, U.S. Department of Justice)

58. Use the data from Exercise 57 to answer the following.

   a. Approximate the number of 35- to 39-year-old inmates you might expect to have been held in the United States at the end of 2010. Round to the nearest whole.

   b. Is your answer to part (a) greater than or less than your answer to Exercise 57? Is this reasonable? Why or why not?

CONCEPT EXTENSIONS

59. In your own words, explain the differences between equations and expressions.

60. In your own words, explain why it is necessary to check solutions to equations containing rational expressions.

61. The average cost of producing x game disks for a computer is given by the function $f(x) = 3.3 + \dfrac{5400}{x}$. Find the number of game disks that must be produced for the average cost to be $5.10.

62. The average cost of producing x electric pencil sharpeners is given by the function $f(x) = 20 + \dfrac{4000}{x}$. Find the number of electric pencil sharpeners that must be produced for the average cost to be \$25.

Solve each equation. Begin by writing each equation with positive exponents only.

63. $x^{-2} - 19x^{-1} + 48 = 0$

64. $x^{-2} - 5x^{-1} - 36 = 0$

65. $p^{-2} + 4p^{-1} - 5 = 0$

66. $6p^{-2} - 5p^{-1} + 1 = 0$

Solve each equation. Round solutions to two decimal places.

67. $\dfrac{1.4}{x - 2.6} = \dfrac{-3.5}{x + 7.1}$

68. $\dfrac{-8.5}{x + 1.9} = \dfrac{5.7}{x - 3.6}$

69. $\dfrac{10.6}{y} - 14.7 = \dfrac{9.92}{3.2} + 7.6$

70. $\dfrac{12.2}{x} + 17.3 = \dfrac{9.6}{x} - 14.7$

Solve each equation by substitution.

For example, to solve Exercise 71, first let $u = x - 1$. After substituting, we have $u^2 + 3u + 2 = 0$. Solve for u and then substitute back to solve for x.

71. $(x - 1)^2 + 3(x - 1) + 2 = 0$

72. $(4 - x)^2 - 5(4 - x) + 6 = 0$

73. $\left(\dfrac{3}{x - 1}\right)^2 + 2\left(\dfrac{3}{x - 1}\right) + 1 = 0$

74. $\left(\dfrac{5}{2 + x}\right)^2 + \dfrac{5}{2 + x} - 20 = 0$

Use a graphing calculator to verify the solution of each given exercise.

75. Exercise 23   **76.** Exercise 24

77. Exercise 35   **78.** Exercise 36

Integrated Review) EXPRESSIONS AND EQUATIONS CONTAINING RATIONAL EXPRESSIONS

Sections 6.1–6.5

It is very important that you understand the difference between an expression and an equation containing rational expressions. An equation contains an equal sign; an expression does not.

Expression to be Simplified	**Equation to be Solved**
$\dfrac{x}{2} + \dfrac{x}{6}$	$\dfrac{x}{2} + \dfrac{x}{6} = \dfrac{2}{3}$

Write both rational expressions with the LCD, 6, as the denominator.

$$\dfrac{x}{2} + \dfrac{x}{6} = \dfrac{x \cdot 3}{2 \cdot 3} + \dfrac{x}{6}$$

$$= \dfrac{3x}{6} + \dfrac{x}{6}$$

$$= \dfrac{4x}{6} = \dfrac{2x}{3}$$

Multiply both sides by the LCD, 6.

$$6\left(\dfrac{1}{2} + \dfrac{x}{6}\right) = 6\left(\dfrac{2}{3}\right)$$

$$3 + x = 4$$

$$x = 1$$

Check to see that the solution is 1.

> ▶ Helpful Hint
>
> Remember: Equations can be cleared of fractions; expressions cannot be.

Perform each indicated operation and simplify or solve the equation for the variable.

1. $\dfrac{x}{2} = \dfrac{1}{8} + \dfrac{x}{4}$

2. $\dfrac{x}{4} = \dfrac{3}{2} + \dfrac{x}{10}$

3. $\dfrac{1}{8} + \dfrac{x}{4}$

4. $\dfrac{3}{2} + \dfrac{x}{10}$

5. $\dfrac{4}{x + 2} - \dfrac{2}{x - 1}$

6. $\dfrac{5}{x - 2} - \dfrac{10}{x + 4}$

7. $\dfrac{4}{x+2} = \dfrac{2}{x-1}$

8. $\dfrac{5}{x-2} = \dfrac{10}{x+4}$

9. $\dfrac{2}{x^2-4} = \dfrac{1}{x+2} - \dfrac{3}{x-2}$

10. $\dfrac{3}{x^2-25} = \dfrac{1}{x+5} + \dfrac{2}{x-5}$

11. $\dfrac{5}{x^2-3x} + \dfrac{4}{2x-6}$

12. $\dfrac{5}{x^2-3x} \div \dfrac{4}{2x-6}$

13. $\dfrac{x-1}{x+1} + \dfrac{x+7}{x-1} = \dfrac{4}{x^2-1}$

14. $\left(1 - \dfrac{y}{x}\right) \div \left(1 - \dfrac{x}{y}\right)$

15. $\dfrac{a^2-9}{a-6} \cdot \dfrac{a^2-5a-6}{a^2-a-6}$

16. $\dfrac{2}{a-6} + \dfrac{3a}{a^2-5a-6} - \dfrac{a}{5a+5}$

17. $\dfrac{2x+3}{3x-2} = \dfrac{4x+1}{6x+1}$

18. $\dfrac{5x-3}{2x} = \dfrac{10x+3}{4x+1}$

19. $\dfrac{a}{9a^2-1} + \dfrac{2}{6a-2}$

20. $\dfrac{3}{4a-8} - \dfrac{a+2}{a^2-2a}$

21. $-\dfrac{3}{x^2} - \dfrac{1}{x} + 2 = 0$

22. $\dfrac{x}{2x+6} + \dfrac{5}{x^2-9}$

23. $\dfrac{x-8}{x^2-x-2} + \dfrac{2}{x-2}$

24. $\dfrac{x-8}{x^2-x-2} + \dfrac{2}{x-2} = \dfrac{3}{x+1}$

25. $\dfrac{3}{a} - 5 = \dfrac{7}{a} - 1$

26. $\dfrac{7}{3z-9} + \dfrac{5}{z}$

Use $\dfrac{x}{5} - \dfrac{x}{4} = \dfrac{1}{10}$ *and* $\dfrac{x}{5} - \dfrac{x}{4} + \dfrac{1}{10}$ *for Exercises 27 and 28.*

27. a. Which one above is an expression?

   b. Describe the first step to simplify this expression.

   c. Simplify the expression.

28. a. Which one above is an equation?

   b. Describe the first step to solve this equation.

   c. Solve the equation.

For each exercise, choose the correct statement. [*] *Each figure represents a real number, and no denominators are 0.*

29. a. $\dfrac{\triangle + \square}{\triangle} = \square$ **b.** $\dfrac{\triangle + \square}{\triangle} = 1 + \dfrac{\square}{\triangle}$ **c.** $\dfrac{\triangle + \square}{\triangle} = \dfrac{\square}{\triangle}$ **d.** $\dfrac{\triangle + \square}{\triangle} = 1 + \square$ **e.** $\dfrac{\triangle + \square}{\triangle - \square} = -1$

30. a. $\dfrac{\triangle}{\square} + \dfrac{\square}{\triangle} = \dfrac{\triangle + \square}{\square + \triangle} = 1$ **b.** $\dfrac{\triangle}{\square} + \dfrac{\square}{\triangle} = \dfrac{\triangle + \square}{\triangle\square}$ **c.** $\dfrac{\triangle}{\square} + \dfrac{\square}{\triangle} = \triangle\triangle + \square\square$

   d. $\dfrac{\triangle}{\square} + \dfrac{\square}{\triangle} = \dfrac{\triangle\triangle + \square\square}{\square\triangle}$ **e.** $\dfrac{\triangle}{\square} + \dfrac{\square}{\triangle} = \dfrac{\triangle\square}{\square\triangle} = 1$

31. a. $\dfrac{\triangle}{\square} \cdot \dfrac{\bigcirc}{\square} = \dfrac{\triangle\bigcirc}{\square}$ **b.** $\dfrac{\triangle}{\square} \cdot \dfrac{\bigcirc}{\square} = \triangle\bigcirc$ **c.** $\dfrac{\triangle}{\square} \cdot \dfrac{\bigcirc}{\square} = \dfrac{\triangle + \bigcirc}{\square + \square}$ **d.** $\dfrac{\triangle}{\square} \cdot \dfrac{\bigcirc}{\square} = \dfrac{\triangle\bigcirc}{\square\square}$

32. a. $\dfrac{\triangle}{\square} \div \dfrac{\bigcirc}{\triangle} = \dfrac{\triangle\triangle}{\square\bigcirc}$ **b.** $\dfrac{\triangle}{\square} \div \dfrac{\bigcirc}{\triangle} = \dfrac{\bigcirc\square}{\triangle\triangle}$ **c.** $\dfrac{\triangle}{\square} \div \dfrac{\bigcirc}{\triangle} = \dfrac{\bigcirc}{\square}$ **d.** $\dfrac{\triangle}{\square} \div \dfrac{\bigcirc}{\triangle} = \dfrac{\triangle + \triangle}{\square + \bigcirc}$

33. a. $\dfrac{\dfrac{\triangle + \square}{\bigcirc}}{\dfrac{\triangle}{\bigcirc}} = \square$ **b.** $\dfrac{\dfrac{\triangle + \square}{\bigcirc}}{\dfrac{\triangle}{\bigcirc}} = \dfrac{\triangle\triangle + \triangle\square}{\bigcirc\bigcirc}$ **c.** $\dfrac{\dfrac{\triangle + \square}{\bigcirc}}{\dfrac{\triangle}{\bigcirc}} = 1 + \square$ **d.** $\dfrac{\dfrac{\triangle + \square}{\bigcirc}}{\dfrac{\triangle}{\bigcirc}} = \dfrac{\triangle + \square}{\triangle}$

[*]My thanks to Kelly Champagne for permission to use her Exercises for 29 through 33.

6.6 Rational Equations and Problem Solving ▶

OBJECTIVES

1 Solve an Equation Containing Rational Expressions for a Specified Variable. ▶

2 Solve Number Problems by Writing Equations Containing Rational Expressions. ▶

3 Solve Problems Modeled by Proportions. ▶

4 Solve Problems About Work. ▶

5 Solve Problems About Distance, Rate, and Time. ▶

OBJECTIVE

1 Solving Equations with Rational Expressions for a Specified Variable ▶

In Section 2.3, we solved equations for a specified variable. In this section, we continue practicing this skill by solving equations containing rational expressions for a specified variable. The steps given in Section 2.3 for solving equations for a specified variable are repeated here.

Solving Equations for a Specified Variable

Step 1. Clear the equation of fractions or rational expressions by multiplying each side of the equation by the least common denominator (LCD) of all denominators in the equation.

Step 2. Use the distributive property to remove grouping symbols such as parentheses.

Step 3. Combine like terms on each side of the equation.

Step 4. Use the addition property of equality to rewrite the equation as an equivalent equation with terms containing the specified variable on one side and all other terms on the other side.

Step 5. Use the distributive property and the multiplication property of equality to get the specified variable alone.

EXAMPLE 1 Solve: $\dfrac{1}{x} + \dfrac{1}{y} = \dfrac{1}{z}$ for x.

Solution To clear this equation of fractions, we multiply both sides of the equation by xyz, the LCD of $\dfrac{1}{x}$, $\dfrac{1}{y}$, and $\dfrac{1}{z}$.

$$\frac{1}{x} + \frac{1}{y} = \frac{1}{z}$$

$$xyz\left(\frac{1}{x} + \frac{1}{y}\right) = xyz\left(\frac{1}{z}\right) \quad \text{Multiply both sides by } xyz.$$

$$xyz\left(\frac{1}{x}\right) + xyz\left(\frac{1}{y}\right) = xyz\left(\frac{1}{z}\right) \quad \text{Use the distributive property.}$$

$$yz + xz = xy \quad \text{Simplify.}$$

Notice the two terms that contain the specified variable x.

Next, we subtract xz from both sides so that all terms containing the specified variable x are on one side of the equation and all other terms are on the other side.

$$yz = xy - xz$$

Now we use the distributive property to factor x from $xy - xz$ and then the multiplication property of equality to solve for x.

$$yz = x(y - z)$$

$$\frac{yz}{y - z} = x \quad \text{or} \quad x = \frac{yz}{y - z} \quad \text{Divide both sides by } y - z. \quad \square$$

PRACTICE

1 Solve: $\dfrac{1}{a} - \dfrac{1}{b} = \dfrac{1}{c}$ for a.

2 Solving Number Problems Modeled by Rational Equations

Problem solving sometimes involves modeling a described situation with an equation containing rational expressions. In Examples 2 through 5, we practice solving such problems and use the problem-solving steps first introduced in Section 2.2.

EXAMPLE 2   Finding an Unknown Number

If a certain number is subtracted from the numerator and added to the denominator of $\frac{9}{19}$, the new fraction is equivalent to $\frac{1}{3}$. Find the number.

Solution

1. UNDERSTAND the problem. Read and reread the problem and try guessing the solution. For example, if the unknown number is 3, we have

$$\frac{9 - 3}{19 + 3} = \frac{1}{3}$$

To see if this is a true statement, we simplify the fraction on the left side.

$$\frac{6}{22} = \frac{1}{3} \quad \text{or} \quad \frac{3}{11} = \frac{1}{3} \quad \text{False}$$

Since this is not a true statement, 3 is not the correct number. Remember that the purpose of this step is not to guess the correct solution but to gain an understanding of the problem posed.

   We will let n = the number to be subtracted from the numerator and added to the denominator.

2. TRANSLATE the problem.

In words:

when the number is subtracted from the numerator and added to the denominator of the fraction $\frac{9}{19}$	this is equivalent to	$\frac{1}{3}$
↓	↓	↓

Translate: $\dfrac{9 - n}{19 + n}$ $=$ $\dfrac{1}{3}$

3. SOLVE the equation for n.

$$\frac{9 - n}{19 + n} = \frac{1}{3}$$

To solve for n, we begin by multiplying both sides by the LCD of $3(19 + n)$.

$$3(19 + n) \cdot \frac{9 - n}{19 + n} = 3(19 + n) \cdot \frac{1}{3} \quad \text{Multiply both sides by the LCD.}$$
$$3(9 - n) = 19 + n \quad \text{Simplify.}$$
$$27 - 3n = 19 + n$$
$$8 = 4n$$
$$2 = n \quad \text{Solve.}$$

4. INTERPRET the results.

Check: If we subtract 2 from the numerator and add 2 to the denominator of $\frac{9}{19}$, we have $\dfrac{9 - 2}{19 + 2} = \dfrac{7}{21} = \dfrac{1}{3}$, and the problem checks.

State: The unknown number is 2.

PRACTICE

2   Find a number that when added to the numerator and subtracted from the denominator of $\dfrac{3}{11}$ results in a fraction equivalent to $\dfrac{5}{2}$.

OBJECTIVE

3   Solving Problems Modeled by Proportions

A **ratio** is the quotient of two numbers or two quantities. Since rational expressions are quotients of quantities, rational expressions are ratios also. A **proportion** is a mathematical statement that two ratios are equal.

Let's review two methods for solving a proportion such as $\dfrac{x-3}{10} = \dfrac{7}{15}$. We can multiply both sides of the equation by the LCD, 30.

Multiply both sides by the LCD, 30.

$$30 \cdot \frac{x-3}{10} = 30 \cdot \frac{7}{15}$$

$$3(x-3) = 2 \cdot 7$$

$$3x - 9 = 14$$

$$3x = 23$$

$$x = \frac{23}{3}$$

We can also solve a proportion by setting cross products equal. Here, we are using the fact that if $\dfrac{a}{b} = \dfrac{c}{d}$, then $ad = bc$.

$$\frac{x-3}{10} \diagup\!\!\!\!\diagdown \frac{7}{15}$$

$$15(x-3) = 10 \cdot 7 \qquad \text{Set cross products equal.}$$

$$15x - 45 = 70 \qquad \text{Use the distributive property.}$$

$$15x = 115$$

$$x = \frac{115}{15} \quad \text{or} \quad \frac{23}{3}$$

A ratio of two different quantities is called a **rate**. For example, $\dfrac{3 \text{ miles}}{2 \text{ hours}}$ or 1.5 miles/hour is a rate. The proportions we write to solve problems will sometimes include rates. When this happens, make sure that the rates contain units written in the same order.

EXAMPLE 3   **Calculating Homes Heated by Electricity**

In the United States, 7 out of every 25 homes are heated by electricity. At this rate, how many homes in a community of 36,000 homes would you predict are heated by electricity? (*Source: 2005 American Housing Survey for the United States*)

Solution

1. UNDERSTAND. Read and reread the problem. Try to estimate a reasonable solution. For example, since 7 is less than $\dfrac{1}{3}$ of 25, we might reason that the solution would be less than $\dfrac{1}{3}$ of 36,000 or 12,000.

Let's let $x =$ number of homes in the community heated by electricity.

2. TRANSLATE.

$$\text{homes heated by electricity} \rightarrow \quad \frac{7}{25} = \frac{x}{36{,}000} \quad \leftarrow \text{homes heated by electricity}$$
$$\text{total homes} \rightarrow \qquad\qquad\qquad\qquad\quad \leftarrow \text{total homes}$$

3. SOLVE. To solve this proportion, we can multiply both sides by the LCD, 36,000, or we can set cross products equal. We will set cross products equal.

$$\frac{7}{25} \diagup\!\!\!\!\diagdown \frac{x}{36{,}000}$$

$$25x = 7 \cdot 36{,}000$$

$$x = \frac{252{,}000}{25}$$

$$x = 10{,}080$$

(Continued on next page)

4. INTERPRET.

Check: To check, replace x with 10,080 in the proportion and see that a true statement results. Notice that our answer is reasonable since it is less than 12,000 as we stated in Step 1.

State: We predict that 10,080 homes are heated by electricity. □

PRACTICE
3 In the United States, 1 out of 12 homes is heated by fuel oil. At this rate, how many homes in a community of 36,000 homes are heated by fuel oil? (*Source: American Housing Survey for the United States*)

OBJECTIVE
4 Solving Problems About Work ▶

The following work example leads to an equation containing rational expressions.

> **EXAMPLE 4** Calculating Work Hours

Melissa Scarlatti can clean the house in 4 hours, whereas her husband, Zack, can do the same job in 5 hours. They have agreed to clean together so that they can finish in time to watch a movie on TV that starts in 2 hours. How long will it take them to clean the house together? Can they finish before the movie starts?

Solution

1. UNDERSTAND. Read and reread the problem. The key idea here is the relationship between the *time* (in hours) it takes to complete the job and the *part of the job* completed in 1 unit of time (1 hour). For example, if the *time* it takes Melissa to complete the job is 4 hours, the part of the job she can complete in 1 hour is $\frac{1}{4}$. Similarly, Zack can complete $\frac{1}{5}$ of the job in 1 hour.

We will let $t =$ *the time* in hours it takes Melissa and Zack to clean the house together. Then $\frac{1}{t}$ represents the *part of the job* they complete in 1 hour. We summarize the given information in a chart.

	Hours to Complete the Job	Part of Job Completed in 1 Hour
MELISSA ALONE	4	$\frac{1}{4}$
ZACK ALONE	5	$\frac{1}{5}$
TOGETHER	t	$\frac{1}{t}$

2. TRANSLATE.

In words:	part of job Melissa can complete in 1 hour	added to	part of job Zack can complete in 1 hour	is equal to	part of job they can complete together in 1 hour
	↓	↓	↓	↓	↓
Translate:	$\frac{1}{4}$	$+$	$\frac{1}{5}$	$=$	$\frac{1}{t}$

3. SOLVE.

$$\frac{1}{4} + \frac{1}{5} = \frac{1}{t}$$

$$20t\left(\frac{1}{4} + \frac{1}{5}\right) = 20t\left(\frac{1}{t}\right) \qquad \text{Multiply both sides by the LCD, } 20t.$$

$$5t + 4t = 20$$

$$9t = 20$$

$$t = \frac{20}{9} \quad \text{or} \quad 2\frac{2}{9} \quad \text{Solve.}$$

4. INTERPRET.

Check: The proposed solution is $2\frac{2}{9}$. That is, Melissa and Zack would take $2\frac{2}{9}$ hours to clean the house together. This proposed solution is reasonable since $2\frac{2}{9}$ hours is more than half of Melissa's time and less than half of Zack's time. Check this solution in the originally stated problem.

State: Melissa and Zack can clean the house together in $2\frac{2}{9}$ hours. They cannot complete the job before the movie starts. ☐

PRACTICE

4   Elissa Juarez can clean the animal cages at the animal shelter where she volunteers in 3 hours. Bill Stiles can do the same job in 2 hours. How long would it take them to clean the cages if they work together?

OBJECTIVE

5   Solving Problems About Distance, Rate, and Time

Before we solve Example 5, let's review what we learned in Chapter 2 about the formula

$$d = r \cdot t, \quad \text{or} \quad \text{distance} = \text{rate} \cdot \text{time}$$

For example, if we travel at a rate or speed of 60 mph for a time of 3 hours, the distance we travel is

$$d = 60 \text{ mph} \cdot 3 \text{ hr}$$
$$= 180 \text{ mi}$$

The formula $d = r \cdot t$ is solved for distance, d. We can also solve this formula for rate r or for time t.

Solve $d = r \cdot t$ for r.

$$d = r \cdot t$$

$$\frac{d}{t} = \frac{r \cdot t}{t}$$

$$\frac{d}{t} = r$$

Solve $d = r \cdot t$ for t.

$$d = r \cdot t$$

$$\frac{d}{r} = \frac{r \cdot t}{r}$$

$$\frac{d}{r} = t$$

All three forms of the distance formula are useful, as we shall see.

EXAMPLE 5   **Finding the Speed of a Current**

Steve Deitmer takes $1\frac{1}{2}$ times as long to go 72 miles upstream in his boat as he does to return. If the boat cruises at 30 mph in still water, what is the speed of the current?

Solution

1. UNDERSTAND. Read and reread the problem. Guess a solution. Suppose that the current is 4 mph. The speed of the boat upstream is slowed down by the current:

(Continued on next page)

30 − 4, or 26 mph, and the speed of the boat downstream is speeded up by the current: 30 + 4, or 34 mph. Next let's find out how long it takes to travel 72 miles upstream and 72 miles downstream. To do so, we use the formula $d = rt$, or $\dfrac{d}{r} = t$.

Upstream

$$\frac{d}{r} = t$$

$$\frac{72}{26} = t$$

$$2\frac{10}{13} = t$$

Downstream

$$\frac{d}{r} = t$$

$$\frac{72}{34} = t$$

$$2\frac{2}{17} = t$$

Since the time upstream $\left(2\frac{10}{13} \text{ hours}\right)$ is not $1\frac{1}{2}$ times the time downstream $\left(2\frac{2}{17} \text{ hours}\right)$, our guess is not correct. We do, however, have a better understanding of the problem.

We will let

$$x = \text{the speed of the current}$$
$$30 + x = \text{the speed of the boat downstream}$$
$$30 - x = \text{the speed of the boat upstream}$$

This information is summarized in the following chart, where we use the formula $\dfrac{d}{r} = t$.

	Distance	Rate	Time $\left(\dfrac{d}{r}\right)$
UPSTREAM	72	30 − x	$\dfrac{72}{30 - x}$
DOWNSTREAM	72	30 + x	$\dfrac{72}{30 + x}$

2. **TRANSLATE.** Since the time spent traveling upstream is $1\frac{1}{2}$ times the time spent traveling downstream, we have

In words:

time upstream	is	$1\frac{1}{2}$	times	time downstream
↓	↓	↓	↓	↓

Translate: $\dfrac{72}{30 - x} = \dfrac{3}{2} \cdot \dfrac{72}{30 + x}$

3. **SOLVE.** $\dfrac{72}{30 - x} = \dfrac{3}{2} \cdot \dfrac{72}{30 + x}$

First we multiply both sides by the LCD, $2(30 + x)(30 - x)$.

$$2(30 + x)(30 - x) \cdot \frac{72}{30 - x} = 2(30 + x)(30 - x)\left(\frac{3}{2} \cdot \frac{72}{30 + x}\right)$$

$$72 \cdot 2(30 + x) = 3 \cdot 72 \cdot (30 - x) \quad \text{Simplify.}$$
$$2(30 + x) = 3(30 - x) \quad \text{Divide both sides by 72.}$$
$$60 + 2x = 90 - 3x \quad \text{Use the distributive property.}$$
$$5x = 30$$
$$x = 6 \quad \text{Solve.}$$

4. **INTERPRET.**

Check: Check the proposed solution of 6 mph in the originally stated problem.

State: The current's speed is 6 mph.

PRACTICE
5   A tugboat takes $1\frac{2}{3}$ times as long to go 100 miles upstream along the Mississippi River as it does to return. If the speed of the current of the Mississippi River is 2 miles per hour, find the speed of the tugboat in still water.

Vocabulary, Readiness & Video Check

Martin-Gay Interactive Videos

See Video 6.6

Watch the section lecture video and answer the following questions.

OBJECTIVE 1
1. In Example 1, an equation is solved for a specified variable. What is done differently here to get the specified variable alone on one side of the equation than was done in the past?

OBJECTIVE 2
2. In general, if x represents a nonzero number, how do we represent its reciprocal?

OBJECTIVE 3
3. For Example 3, how are units used to write a correct proportion?

OBJECTIVE 4
4. From Example 4, how can you determine a somewhat reasonable answer to a work application before you even begin to solve it?

OBJECTIVE 5
5. For Example 5, if the boat travels upstream at a rate of $x - y$, what is the boat's rate downstream? Use the letters x and y.

6.6   Exercise Set   MyMathLab®

Solve each equation for the specified variable. See Example 1.

1. $F = \frac{9}{5}C + 32$ for C

△ **2.** $V = \frac{1}{3}\pi r^2 h$ for h

3. $Q = \frac{A - I}{L}$ for I

4. $P = 1 - \frac{C}{S}$ for S

▶ **5.** $\frac{1}{R} = \frac{1}{R_1} + \frac{1}{R_2}$ for R

6. $\frac{1}{R} = \frac{1}{R_1} + \frac{1}{R_2}$ for R_1

7. $S = \frac{n(a + L)}{2}$ for n

8. $S = \frac{n(a + L)}{2}$ for a

△ **9.** $A = \frac{h(a + b)}{2}$ for b

△ **10.** $A = \frac{h(a + b)}{2}$ for h

11. $\frac{P_1 V_1}{T_1} = \frac{P_2 V_2}{T_2}$ for T_2

12. $H = \frac{kA(T_1 - T_2)}{L}$ for T_2

13. $f = \frac{f_1 f_2}{f_1 + f_2}$ for f_2

14. $I = \frac{E}{R + r}$ for r

15. $\lambda = \frac{2L}{n}$ for L

16. $S = \frac{a_1 - a_n r}{1 - r}$ for a_1

17. $\frac{\theta}{\omega} = \frac{2L}{c}$ for c

18. $F = \frac{-GMm}{r^2}$ for M

Solve. For Exercises 19 and 20, the solutions have been started for you. See Example 2.

▶ **19.** The sum of a number and 5 times its reciprocal is 6. Find the number(s).

Start the solution:

1. UNDERSTAND the problem. Reread it as many times as needed. Let's let

x = a number. Then

$\frac{1}{x}$ = its reciprocal.

2. TRANSLATE into an equation. (Fill in the blanks below.)

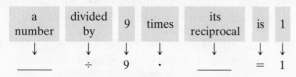

a number	plus	5	times	its reciprocal	is	6
↓	↓	↓	↓	↓	↓	↓
_____	+	5	·	_____	=	6

Finish with:

3. SOLVE and

4. INTERPRET

20. The quotient of a number and 9 times its reciprocal is 1. Find the number(s).

Start the solution:

1. UNDERSTAND the problem. Reread it as many times as needed. Let's let

x = a number. Then

$\dfrac{1}{x}$ = its reciprocal.

2. TRANSLATE into an equation. (Fill in the blanks below.)

a number	divided by	9	times	its reciprocal	is	1
↓	↓	↓	↓	↓	↓	↓
_____	÷	9	·	_____	=	1

Finish with:

3. SOLVE and

4. INTERPRET

21. If a number is added to the numerator of $\dfrac{12}{41}$ and twice the number is added to the denominator of $\dfrac{12}{41}$, the resulting fraction is equivalent to $\dfrac{1}{3}$. Find the number.

22. If a number is subtracted from the numerator of $\dfrac{13}{8}$ and added to the denominator of $\dfrac{13}{8}$, the resulting fraction is equivalent to $\dfrac{2}{5}$. Find the number.

Solve. See Example 3.

23. An Arabian camel can drink 15 gallons of water in 10 minutes. At this rate, how much water can the camel drink in 3 minutes? (*Source:* Grolier, Inc.)

24. An Arabian camel can travel 20 miles in 8 hours, carrying a 300-pound load on its back. At this rate, how far can the camel travel in 10 hours? (*Source:* Grolier, Inc.)

25. In 2010, 13.1 out of every 100 Coast Guard personnel were women. If there were 42,358 total Coast Guard personnel on active duty, estimate the number of women. Round to the nearest whole. (*Source:* Women in Military Service for America Memorial Foundation, Inc.)

26. In 2010, 185 out of every 200 marine personnel were men. If there were 202,441 total marine personnel in 2010, estimate the number of men. Round to the nearest whole. (*Source:* Women in Military Service for America Memorial Foundation, Inc.)

Solve. See Example 4.

27. An experienced roofer can roof a house in 26 hours. A beginning roofer needs 39 hours to complete the same job. Find how long it takes for the two to do the job together.

28. Alan Cantrell can word process a research paper in 6 hours. With Steve Isaac's help, the paper can be processed in 4 hours. Find how long it takes Steve to word process the paper alone.

29. Three postal workers can sort a stack of mail in 20 minutes, 30 minutes, and 60 minutes, respectively. Find how long it takes them to sort the mail if all three work together.

30. A new printing press can print newspapers twice as fast as the old one can. The old one can print the afternoon edition in 4 hours. Find how long it takes to print the afternoon edition if both printers are operating.

Solve. See Example 5.

31. Mattie Evans drove 150 miles in the same amount of time that it took a turbopropeller plane to travel 600 miles. The speed of the plane was 150 mph faster than the speed of the car. Find the speed of the plane.

32. An F-100 plane and a Toyota truck leave the same town at sunrise and head for a town 450 miles away. The speed of the plane is three times the speed of the truck, and the plane arrives 6 hours ahead of the truck. Find the speed of the truck.

33. The speed of Lazy River's current is 5 mph. If a boat travels 20 miles downstream in the same time that it takes to travel 10 miles upstream, find the speed of the boat in still water.

34. The speed of a boat in still water is 24 mph. If the boat travels 54 miles upstream in the same time that it takes to travel 90 miles downstream, find the speed of the current.

MIXED PRACTICE

Solve.

35. The sum of the reciprocals of two consecutive integers is $-\dfrac{15}{56}$. Find the two integers.

36. The sum of the reciprocals of two consecutive odd integers is $\dfrac{20}{99}$. Find the two integers.

37. One hose can fill a goldfish pond in 45 minutes, and two hoses can fill the same pond in 20 minutes. Find how long it takes the second hose alone to fill the pond.

38. If Sarah Clark can do a job in 5 hours and Dick Belli and Sarah working together can do the same job in 2 hours, find how long it takes Dick to do the job alone.

39. Two trains going in opposite directions leave at the same time. One train travels 15 mph faster than the other. In 6 hours, the trains are 630 miles apart. Find the speed of each.

40. The speed of Alberto Contador during the fourth stage of the 2010 Paris–Nice bicycle race was 6 kilometers/hour faster than Jimmy Casper, who came in last in the stage. If Contador traveled 208 kilometers in the time that Casper traveled 181.35 kilometers, find the speed of Casper, rounded to the nearest tenth.

41. A giant tortoise can travel 0.17 miles in 1 hour. At this rate, how long would it take the tortoise to travel 1 mile? Round to the nearest tenth of an hour. (*Source: The World Almanac*)

42. A black mamba snake can travel 88 feet in 3 seconds. At this rate, how long does it take to travel 300 feet (the length of a football field)? Round to the nearest tenth of a second. (*Source: The World Almanac*)

43. A local dairy has three machines to fill half-gallon milk cartons. The machines can fill the daily quota in 5 hours, 6 hours, and 7.5 hours, respectively. Find how long it takes to fill the daily quota if all three machines are running.

44. The inlet pipe of an oil tank can fill the tank in 1 hour, 30 minutes. The outlet pipe can empty the tank in 1 hour. Find how long it takes to empty a full tank if both pipes are open.

45. A plane flies 465 miles with the wind and 345 miles against the wind in the same length of time. If the speed of the wind is 20 mph, find the speed of the plane in still air.

46. Two rockets are launched. The first travels at 9000 mph. Fifteen minutes later, the second is launched at 10,000 mph. Find the distance at which both rockets are an equal distance from Earth.

47. Two joggers, one averaging 8 mph and one averaging 6 mph, start from a designated initial point. The slower jogger arrives at the end of the run a half-hour after the other jogger. Find the distance of the run.

48. A semi truck travels 300 miles through the flatland in the same amount of time that it travels 180 miles through the Great Smoky Mountains. The rate of the truck is 20 miles per hour slower in the mountains than in the flatland. Find both the flatland rate and mountain rate.

49. The denominator of a fraction is 1 more than the numerator. If both the numerator and the denominator are decreased by 3, the resulting fraction is equivalent to $\frac{4}{5}$. Find the fraction.

50. The numerator of a fraction is 4 less than the denominator. If both the numerator and the denominator are increased by 2, the resulting fraction is equivalent to $\frac{2}{3}$. Find the fraction.

51. In 2 minutes, a conveyor belt can move 300 pounds of recyclable aluminum from the delivery truck to a storage area. A smaller belt can move the same quantity of cans the same distance in 6 minutes. If both belts are used, find how long it takes to move the cans to the storage area.

52. Gary Marcus and Tony Alva work at Lombardo's Pipe and Concrete. Mr. Lombardo is preparing an estimate for a customer. He knows that Gary can lay a slab of concrete in 6 hours. Tony can lay the same size slab in 4 hours. If both work on the job and the cost of labor is $45.00 per hour, determine what the labor estimate should be.

53. The world record for the largest white bass caught is held by Ronald Sprouse of Virginia. The bass weighed 6 pounds 13 ounces. If Ronald rows to his favorite fishing spot 9 miles downstream in the same amount of time that he rows 3 miles upstream and if the current is 6 mph, find how long it takes him to cover the 12 miles.

54. An amateur cyclist training for a road race rode the first 20-mile portion of his workout at a constant rate. For the 16-mile cooldown portion of his workout, he reduced his speed by 2 miles per hour. Each portion of the workout took equal time. Find the cyclist's rate during the first portion and his rate during the cooldown portion.

55. Smith Engineering is in the process of reviewing the salaries of their surveyors. During this review, the company found that an experienced surveyor can survey a roadbed in 4 hours. An apprentice surveyor needs 5 hours to survey the same stretch of road. If the two work together, find how long it takes them to complete the job.

56. Mr. Dodson can paint his house by himself in four days. His son will need an additional two days to complete the job if he works by himself. If they work together, find how long it takes to paint the house.

57. An experienced bricklayer can construct a small wall in 3 hours. An apprentice can complete the job in 6 hours. Find how long it takes if they work together.

58. Scanner A can scan a document in 3 hours. Scanner B takes 5 hours to do the same job. If both scanners are used, how long will it take for the document to be scanned?

59. In 2010, 9 out of 20 top grossing movies were rated PG-13. At this rate, how many movies in a year with 599 releases would you predict to be rated PG-13? Round to the nearest whole movie. (*Source:* Motion Picture Association)

60. In 2010, 9 out of 25 U.S./Canada citizens viewed a 3-D movie. At this rate, how many citizens in a town of 46,000 have viewed a 3-D movie? (*Source:* Motion Picture Association)

REVIEW AND PREVIEW

Solve each equation for x. See Section 2.1.

61. $\dfrac{x}{5} = \dfrac{x + 2}{3}$

62. $\dfrac{x}{4} = \dfrac{x + 3}{6}$

63. $\dfrac{x - 3}{2} = \dfrac{x - 5}{6}$

64. $\dfrac{x - 6}{4} = \dfrac{x - 2}{5}$

CONCEPT EXTENSIONS

Calculating body-mass index (BMI) is a way to gauge whether a person should lose weight. Doctors recommend that body-mass index values fall between 19 and 25. The formula for body-mass index B is $B = \dfrac{705w}{h^2}$, where w is weight in pounds and h is height in inches. Use this formula to answer Exercises 65 and 66.

65. A patient is 5 ft 8 in. tall. What should his or her weight be to have a body-mass index of 25? Round to the nearest whole pound.

66. A doctor recorded a body-mass index of 47 on a patient's chart. Later, a nurse notices that the doctor recorded the patient's weight as 240 pounds but neglected to record the patient's height. Explain how the nurse can use the information from the chart to find the patient's height. Then find the height.

In physics, when the source of a sound is traveling toward an observer, the relationship between the actual pitch a of the sound and the pitch h that the observer hears due to the Doppler effect is described by the formula $h = \dfrac{a}{1 - \dfrac{s}{770}}$, *where s is the speed of the sound source in miles per hour. Use this formula to answer Exercise 67 and 68.*

67. An emergency vehicle has a single-tone siren with the pitch of the musical note E. As it approaches an observer standing by the road, the vehicle is traveling 50 mph. Is the pitch that the observer hears due to the Doppler effect lower or higher than the actual pitch? To which musical note is the pitch that the observer hears closest?

Pitch of an Octave of Musical Notes in Hertz (Hz)	
Note	**Pitch**
Middle C	261.63
D	293.66
E	329.63
F	349.23
G	392.00
A	440.00
B	493.88

Note: Greater numbers indicate higher pitches (acoustically).

(*Source:* American Standards Association)

68. Suppose an emergency van has a single-tone siren with the pitch of the musical note G. If the van is traveling at 80 mph approaching a standing observer, name the pitch the observer hears (rounded to the nearest tenth) and the musical note closest to that pitch.

In electronics, the relationship among the resistances R_1 and R_2 of two resistors wired in a parallel circuit and their combined resistance R is described by the formula $\dfrac{1}{R} = \dfrac{1}{R_1} + \dfrac{1}{R_2}$. *Use this formula to solve Exercises 69 through 71.*

69. If the combined resistance is 2 ohms and one of the two resistances is 3 ohms, find the other resistance.

70. Find the combined resistance of two resistors of 12 ohms each when they are wired in a parallel circuit.

71. The relationship among resistance of two resistors wired in a parallel circuit and their combined resistance may be extended to three resistors of resistances R_1, R_2, and R_3. Write an equation you believe may describe the relationship and use it to find the combined resistance if R_1 is 5, R_2 is 6, and R_3 is 2.

72. For the formula $\dfrac{1}{x} = \dfrac{1}{y} + \dfrac{1}{z} - \dfrac{1}{w}$, find x if $y = 2$, $z = 7$, and $w = 6$.

6.7 Variation and Problem Solving

OBJECTIVES

1 Solve Problems Involving Direct Variation.

2 Solve Problems Involving Inverse Variation.

3 Solve Problems Involving Joint Variation.

4 Solve Problems Involving Combined Variation.

OBJECTIVE

1 Solving Problems Involving Direct Variation

A very familiar example of direct variation is the relationship of the circumference C of a circle to its radius r. The formula $C = 2\pi r$ expresses that the circumference is always 2π times the radius. In other words, C is always a constant multiple (2π) of r. Because it is, we say that **C varies directly as r,** that **C varies directly with r,** or that **C is directly proportional to r.**

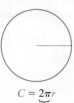

$C = 2\pi r$

constant

> **Direct Variation**
>
> **y varies directly as x**, or **y is directly proportional to x**, if there is a nonzero constant k such that
>
> $$y = kx$$
>
> The number k is called the **constant of variation** or the **constant of proportionality.**

In the above definition, the relationship described between x and y is a linear one. In other words, the graph of $y = kx$ is a line. The slope of the line is k, and the line passes through the origin.

For example, the graph of the direct variation equation $C = 2\pi r$ is shown. The horizontal axis represents the radius r, and the vertical axis is the circumference C. From the graph, we can read that when the radius is 6 units, the circumference is approximately 38 units. Also, when the circumference is 45 units, the radius is between 7 and 8 units. Notice that as the radius increases, the circumference increases.

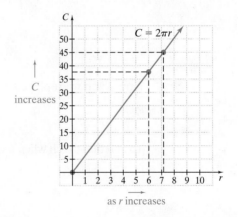

EXAMPLE 1 Suppose that y varies directly as x. If y is 5 when x is 30, find the constant of variation and the direct variation equation.

Solution Since y varies directly as x, we write $y = kx$. If $y = 5$ when $x = 30$, we have that

$$y = kx$$
$$5 = k(30) \quad \text{Replace } y \text{ with 5 and } x \text{ with 30.}$$
$$\frac{1}{6} = k \quad \text{Solve for } k.$$

The constant of variation is $\frac{1}{6}$.

After finding the constant of variation k, the direct variation equation can be written as $y = \frac{1}{6}x$. ☐

PRACTICE

1 Suppose that y varies directly as x. If y is 20 when x is 15, find the constant of variation and the direct variation equation.

EXAMPLE 2 **Using Direct Variation and Hooke's Law**

Hooke's law states that the distance a spring stretches is directly proportional to the weight attached to the spring. If a 40-pound weight attached to the spring stretches the spring 5 inches, find the distance that a 65-pound weight attached to the spring stretches the spring.

(Continued on next page)

Solution

1. **UNDERSTAND.** Read and reread the problem. Notice that we are given that the distance a spring stretches is **directly proportional** to the weight attached. We let

   d = the distance stretched

   w = the weight attached

   The constant of variation is represented by k.

2. **TRANSLATE.** Because d is directly proportional to w, we write

   $$d = kw$$

3. **SOLVE.** When a weight of 40 pounds is attached, the spring stretches 5 inches. That is, when $w = 40$, $d = 5$.

   $$d = kw$$
   $$5 = k(40) \quad \text{Replace } d \text{ with 5 and } w \text{ with 40.}$$
   $$\frac{1}{8} = k \quad \text{Solve for } k.$$

   Now when we replace k with $\frac{1}{8}$ in the equation $d = kw$, we have

   $$d = \frac{1}{8}w$$

   To find the stretch when a weight of 65 pounds is attached, we replace w with 65 to find d.

   $$d = \frac{1}{8}(65)$$
   $$= \frac{65}{8} = 8\frac{1}{8} \quad \text{or} \quad 8.125$$

4. **INTERPRET.**

Check: Check the proposed solution of 8.125 inches in the original problem.

State: The spring stetches 8.125 inches when a 65-pound weight is attached. ☐

PRACTICE
2   Use Hooke's law as stated in Example 2. If a 36-pound weight attached to a spring stretches the spring 9 inches, find the distance that a 75-pound weight attached to the spring stretches the spring.

OBJECTIVE
2   Solving Problems Involving Inverse Variation

When y is proportional to the **reciprocal** of another variable x, we say that **y varies inversely as x**, or that **y is inversely proportional to x**. An example of the inverse variation relationship is the relationship between the pressure that a gas exerts and the volume of its container. As the volume of a container decreases, the pressure of the gas it contains increases.

Inverse Variation

y varies inversely as x, or **y is inversely proportional to x**, if there is a nonzero constant k such that

$$y = \frac{k}{x}$$

The number k is called the **constant of variation** or the **constant of proportionality.**

Notice that $y = \dfrac{k}{x}$ is a rational equation. Its graph for $k > 0$ and $x > 0$ is shown. From the graph, we can see that as x increases, y decreases.

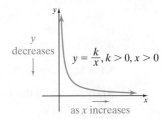

EXAMPLE 3 Suppose that u varies inversely as w. If u is 3 when w is 5, find the constant of variation and the inverse variation equation.

Solution Since u varies inversely as w, we have $u = \dfrac{k}{w}$. We let $u = 3$ and $w = 5$, and we solve for k.

$$u = \frac{k}{w}$$

$$3 = \frac{k}{5} \quad \text{Let } u = 3 \text{ and } w = 5.$$

$$15 = k \quad \text{Multiply both sides by 5.}$$

The constant of variation k is 15. This gives the inverse variation equation

$$u = \frac{15}{w}$$

PRACTICE
3 Suppose that b varies inversely as a. If b is 5 when a is 9, find the constant of variation and the inverse variation equation.

EXAMPLE 4 **Using Inverse Variation and Boyle's Law**

Boyle's law says that if the temperature stays the same, the pressure P of a gas is inversely proportional to the volume V. If a cylinder in a steam engine has a pressure of 960 kilopascals when the volume is 1.4 cubic meters, find the pressure when the volume increases to 2.5 cubic meters.

Solution

1. UNDERSTAND. Read and reread the problem. Notice that we are given that the pressure of a gas is *inversely proportional* to the volume. We will let $P =$ the pressure and $V =$ the volume. The constant of variation is represented by k.

2. TRANSLATE. Because P is inversely proportional to V, we write

$$P = \frac{k}{V}$$

When $P = 960$ kilopascals, the volume $V = 1.4$ cubic meters. We use this information to find k.

$$960 = \frac{k}{1.4} \quad \text{Let } P = 960 \text{ and } V = 1.4.$$

$$1344 = k \quad \text{Multiply both sides by 1.4.}$$

Thus, the value of k is 1344. Replacing k with 1344 in the variation equation, we have

$$P = \frac{1344}{V}$$

Next we find P when V is 2.5 cubic meters.

(Continued on next page)

3. SOLVE.

$$P = \frac{1344}{2.5} \quad \text{Let } V = 2.5.$$

$$= 537.6$$

4. INTERPRET.

Check: Check the proposed solution in the original problem.

State: When the volume is 2.5 cubic meters, the pressure is 537.6 kilopascals.   □

PRACTICE
4   Use Boyle's law as stated in Example 4. When $P = 350$ kilopascals and $V = 2.8$ cubic meters, find the pressure when the volume decreases to 1.5 cubic meters.

■

OBJECTIVE
3   Solving Problems Involving Joint Variation

Sometimes the ratio of a variable to the product of many other variables is constant. For example, the ratio of distance traveled to the product of speed and time traveled is always 1.

$$\frac{d}{rt} = 1 \quad \text{or} \quad d = rt$$

Such a relationship is called **joint variation.**

Joint Variation

If the ratio of a variable y to the product of two or more variables is constant, then **y varies jointly as,** or **is jointly proportional to,** the other variables. If

$$y = kxz$$

then the number k is the **constant of variation** or the **constant of proportionality.**

✓CONCEPT CHECK

Which type of variation is represented by the equation $xy = 8$? Explain.

   a. Direct variation   **b.** Inverse variation   **c.** Joint variation

---

△ **EXAMPLE 5**   **Expressing Surface Area**

The lateral surface area of a cylinder varies jointly as its radius and height. Express this surface area S in terms of radius r and height h.

Solution   Because the surface area varies jointly as the radius r and the height h, we equate S to a constant multiple of r and h.

$$S = krh$$

In the equation, $S = krh$, it can be determined that the constant k is 2π, and we then have the formula $S = 2\pi rh$. (The lateral surface area formula does not include the areas of the two circular bases.)   □

PRACTICE

   5   The area of a regular polygon varies jointly as its apothem and its perimeter. Express the area in terms of the apothem a and the perimeter p.

OBJECTIVE

4 **Solving Problems Involving Combined Variation**

Some examples of variation involve combinations of direct, inverse, and joint variation. We will call these variations **combined variation.**

EXAMPLE 6 Suppose that y varies directly as the square of x. If y is 24 when x is 2, find the constant of variation and the variation equation.

Solution Since y varies directly as the square of x, we have

$$y = kx^2$$

Now let $y = 24$ and $x = 2$ and solve for k.

$$y = kx^2$$
$$24 = k \cdot 2^2$$
$$24 = 4k$$
$$6 = k$$

The constant of variation is 6, so the variation equation is

$$y = 6x^2$$

PRACTICE

   6   Suppose that y varies inversely as the cube of x. If y is $\dfrac{1}{2}$ when x is 2, find the constant of variation and the variation equation.

 EXAMPLE 7 **Finding Column Weight**

The maximum weight that a circular column can support is directly proportional to the fourth power of its diameter and is inversely proportional to the square of its height. A 2-meter-diameter column that is 8 meters in height can support 1 ton. Find the weight that a 1-meter-diameter column that is 4 meters in height can support.

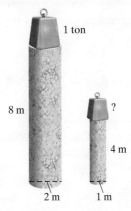

Solution

1. UNDERSTAND. Read and reread the problem. Let w = weight, d = diameter, h = height, and k = the constant of variation.

2. TRANSLATE. Since w is directly proportional to d^4 and inversely proportional to h^2, we have

$$w = \frac{kd^4}{h^2}$$

(Continued on next page)

3. SOLVE. To find k, we are given that a 2-meter-diameter column that is 8 meters in height can support 1 ton. That is, $w = 1$ when $d = 2$ and $h = 8$, or

$$1 = \frac{k \cdot 2^4}{8^2} \quad \text{Let } w = 1, d = 2, \text{ and } h = 8.$$

$$1 = \frac{k \cdot 16}{64}$$

$$4 = k \quad \text{Solve for } k.$$

Now replace k with 4 in the equation $w = \dfrac{kd^4}{h^2}$ and we have

$$w = \frac{4d^4}{h^2}$$

To find weight w for a 1-meter-diameter column that is 4 meters in height, let $d = 1$ and $h = 4$.

$$w = \frac{4 \cdot 1^4}{4^2}$$

$$w = \frac{4}{16} = \frac{1}{4}$$

4. INTERPRET.

Check: Check the proposed solution in the original problem.

State: The 1-meter-diameter column that is 4 meters in height can support $\dfrac{1}{4}$ ton of weight. ☐

PRACTICE

7 Suppose that y varies directly as z and inversely as the cube of x. If y is 15 when $z = 5$ and $x = 3$, find the constant of variation and the variation equation. ∎

Vocabulary, Readiness & Video Check

State whether each equation represents direct, inverse, or joint variation.

1. $y = 5x$

2. $y = \dfrac{700}{x}$

3. $y = 5xz$

4. $y = \dfrac{1}{2}abc$

5. $y = \dfrac{9.1}{x}$

6. $y = 2.3x$

7. $y = \dfrac{2}{3}x$

8. $y = 3.1st$

Martin-Gay Interactive Videos

See Video 6.7 🍐

Watch the section lecture video and answer the following questions.

OBJECTIVE 1
9. Based on the lecture before ▣ Example 1, what kind of equation is a direct variation equation? What does k, the constant of variation, represent in this equation?

OBJECTIVE 2
10. In ▣ Example 3, why is it not necessary to replace the given values of x and y in the inverse variation equation in order to find k?

OBJECTIVE 3
11. Based on ▣ Example 5 and the lecture before, what is the variation equation for "y varies jointly as the square of a and the fifth power of b"?

OBJECTIVE 4
12. From ▣ Example 6, what kind of variation does a combined variation application involve?

6.7   Exercise Set

MyMathLab®

If y varies directly as x, find the constant of variation and the direct variation equation for each situation. See Example 1.

1. $y = 4$ when $x = 20$
2. $y = 5$ when $x = 30$
3. $y = 6$ when $x = 4$
4. $y = 12$ when $x = 8$
5. $y = 7$ when $x = \dfrac{1}{2}$
6. $y = 11$ when $x = \dfrac{1}{3}$
7. $y = 0.2$ when $x = 0.8$
8. $y = 0.4$ when $x = 2.5$

Solve. See Example 2.

9. The weight of a synthetic ball varies directly with the cube of its radius. A ball with a radius of 2 inches weighs 1.20 pounds. Find the weight of a ball of the same material with a 3-inch radius.

10. At sea, the distance to the horizon is directly proportional to the square root of the elevation of the observer. If a person who is 36 feet above the water can see 7.4 miles, find how far a person 64 feet above the water can see. Round to the nearest tenth of a mile.

11. The amount P of pollution varies directly with the population N of people. Kansas City has a population of 460,000 and produces about 270,000 tons of pollutants. Find how many tons of pollution we should expect St. Louis to produce if we know that its population is 319,000. Round to the nearest whole ton. (*Source:* Wikipedia)

12. Charles's law states that if the pressure P stays the same, the volume V of a gas is directly proportional to its temperature T. If a balloon is filled with 20 cubic meters of a gas at a temperature of 300 K, find the new volume if the temperature rises to 360 K while the pressure stays the same.

If y varies inversely as x, find the constant of variation and the inverse variation equation for each situation. See Example 3.

13. $y = 6$ when $x = 5$
14. $y = 20$ when $x = 9$
15. $y = 100$ when $x = 7$
16. $y = 63$ when $x = 3$
17. $y = \dfrac{1}{8}$ when $x = 16$
18. $y = \dfrac{1}{10}$ when $x = 40$
19. $y = 0.2$ when $x = 0.7$
20. $y = 0.6$ when $x = 0.3$

Solve. See Example 4.

21. Pairs of markings a set distance apart are made on highways so that police can detect drivers exceeding the speed limit. Over a fixed distance, the speed R varies inversely with the time T. In one particular pair of markings, R is 45 mph when T is 6 seconds. Find the speed of a car that travels the given distance in 5 seconds.

22. The weight of an object on or above the surface of Earth varies inversely as the square of the distance between the object and Earth's center. If a person weighs 160 pounds on Earth's surface, find the individual's weight if he moves 200 miles above Earth. Round to the nearest whole pound. (Assume that Earth's radius is 4000 miles.)

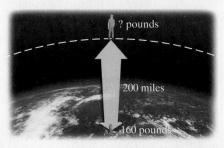

23. If the voltage V in an electric circuit is held constant, the current I is inversely proportional to the resistance R. If the current is 40 amperes when the resistance is 270 ohms, find the current when the resistance is 150 ohms.

24. Because it is more efficient to produce larger numbers of items, the cost of producing a certain computer DVD is inversely proportional to the number produced. If 4000 can be produced at a cost of $1.20 each, find the cost per DVD when 6000 are produced.

25. The intensity I of light varies inversely as the square of the distance d from the light source. If the distance from the light source is doubled (see the figure), determine what happens to the intensity of light at the new location.

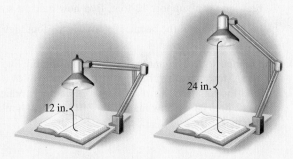

△ **26.** The maximum weight that a circular column can hold is inversely proportional to the square of its height. If an 8-foot column can hold 2 tons, find how much weight a 10-foot column can hold.

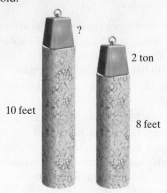

MIXED PRACTICE

Write each statement as an equation. Use k as the constant of variation. See Example 5.

27. *x* varies jointly as *y* and *z*.

28. *P* varies jointly as *R* and the square of *S*.

29. *r* varies jointly as *s* and the cube of *t*.

30. *a* varies jointly as *b* and *c*.

For each statement, find the constant of variation and the variation equation. See Examples 5 and 6.

31. *y* varies directly as the cube of *x*; *y* = 9 when *x* = 3

32. *y* varies directly as the cube of *x*; *y* = 32 when *x* = 4

33. *y* varies directly as the square root of *x*; *y* = 0.4 when *x* = 4

34. *y* varies directly as the square root of *x*; *y* = 2.1 when *x* = 9

35. *y* varies inversely as the square of *x*; *y* = 0.052 when *x* = 5

36. *y* varies inversely as the square of *x*; *y* = 0.011 when *x* = 10

⊙ **37.** *y* varies jointly as *x* and the cube of *z*; *y* = 120 when *x* = 5 and *z* = 2

38. *y* varies jointly as *x* and the square of *z*; *y* = 360 when *x* = 4 and *z* = 3

Solve. See Example 7.

△⊙ **39.** The maximum weight that a rectangular beam can support varies jointly as its width and the square of its height and inversely as its length. If a beam $\frac{1}{2}$ foot wide, $\frac{1}{3}$ foot high, and 10 feet long can support 12 tons, find how much a similar beam can support if the beam is $\frac{2}{3}$ foot wide, $\frac{1}{2}$ foot high, and 16 feet long.

40. The number of cars manufactured on an assembly line at a General Motors plant varies jointly as the number of workers and the time they work. If 200 workers can produce 60 cars in 2 hours, find how many cars 240 workers should be able to make in 3 hours.

△ **41.** The volume of a cone varies jointly as its height and the square of its radius. If the volume of a cone is 32π cubic inches when the radius is 4 inches and the height is 6 inches, find the volume of a cone when the radius is 3 inches and the height is 5 inches.

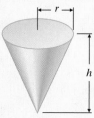

△ **42.** When a wind blows perpendicularly against a flat surface, its force is jointly proportional to the surface area and the speed of the wind. A sail whose surface area is 12 square feet experiences a 20-pound force when the wind speed is 10 miles per hour. Find the force on an 8-square-foot sail if the wind speed is 12 miles per hour.

43. The intensity of light (in foot-candles) varies inversely as the square of *x*, the distance in feet from the light source. The intensity of light 2 feet from the source is 80 foot-candles. How far away is the source if the intensity of light is 5 foot-candles?

44. The horsepower that can be safely transmitted to a shaft varies jointly as the shaft's angular speed of rotation (in revolutions per minute) and the cube of its diameter. A 2-inch shaft making 120 revolutions per minute safely transmits 40 horsepower. Find how much horsepower can be safely transmitted by a 3-inch shaft making 80 revolutions per minute.

MIXED PRACTICE

Write an equation to describe each variation. Use k for the constant of proportionality. See Examples 1 through 7.

45. *y* varies directly as *x*

46. *p* varies directly as *q*

47. *a* varies inversely as *b*

48. *y* varies inversely as *x*

49. *y* varies jointly as *x* and *z*

50. *y* varies jointly as *q*, *r*, and *t*

51. *y* varies inversely as x^3

52. *y* varies inversely as a^4

53. *y* varies directly as *x* and inversely as p^2

54. *y* varies directly as a^5 and inversely as *b*

REVIEW AND PREVIEW

Find the exact circumference and area of each circle. See the inside cover for a list of geometric formulas.

△ **55.**

△ **56.**

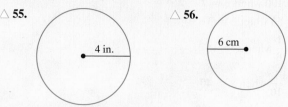

△ **57.**

9 cm

△ **58.**

7 m

Find each square root. See Section 1.3.

59. $\sqrt{81}$ **60.** $\sqrt{36}$

61. $\sqrt{1}$ **62.** $\sqrt{4}$

63. $\sqrt{\dfrac{1}{4}}$ **64.** $\sqrt{\dfrac{1}{25}}$

65. $\sqrt{\dfrac{4}{9}}$ **66.** $\sqrt{\dfrac{25}{121}}$

CONCEPT EXTENSIONS

Solve. See the Concept Check in this section. Choose the type of variation that each equation represents. **a.** *Direct variation* **b.** *Inverse variation* **c.** *Joint variation*

67. $y = \dfrac{2}{3}x$ **68.** $y = \dfrac{0.6}{x}$

69. $y = 9ab$ **70.** $xy = \dfrac{2}{11}$

71. The horsepower to drive a boat varies directly as the cube of the speed of the boat. If the speed of the boat is to double, determine the corresponding increase in horsepower required.

72. The volume of a cylinder varies jointly as the height and the square of the radius. If the height is halved and the radius is doubled, determine what happens to the volume.

73. Suppose that y varies directly as x. If x is doubled, what is the effect on y?

74. Suppose that y varies directly as x^2. If x is doubled, what is the effect on y?

Complete the following table for the inverse variation $y = \dfrac{k}{x}$ over each given value of k. Plot the points on a rectangular coordinate system.

x	$\dfrac{1}{4}$	$\dfrac{1}{2}$	1	2	4
$y = \dfrac{k}{x}$					

75. $k = 3$ **76.** $k = 1$ **77.** $k = \dfrac{1}{2}$ **78.** $k = 5$

Chapter 6   Vocabulary Check

Fill in each blank with one of the words or phrases listed below.

rational expression   equation   complex fraction   opposites   synthetic division

least common denominator   expression   long division   jointly   directly   inversely

1. A rational expression whose numerator, denominator, or both contain one or more rational expressions is called a(n) _____.

2. To divide a polynomial by a polynomial other than a monomial, we use _____.

3. In the equation $y = kx$, y varies _____ as x.

4. In the equation $y = \dfrac{k}{x}$, y varies _____ as x.

5. The _____ of a list of rational expressions is a polynomial of least degree whose factors include the denominator factors in the list.

6. When a polynomial is to be divided by a binomial of the form $x - c$, a shortcut process called _____ may be used.

7. In the equation $y = kxz$, y varies _____ as x and z.

8. The expressions $(x - 5)$ and $(5 - x)$ are called _____.

9. A(n) _____ is an expression that can be written as the quotient $\dfrac{P}{Q}$ of two polynomials P and Q as long as Q is not 0.

10. Which is an expression and which is an equation? An example of an _____ is $\dfrac{2}{x} + \dfrac{2}{x^2} = 7$, and an example of an _____ is $\dfrac{2}{x} + \dfrac{5}{x^2}$.

Chapter 6 Highlights

DEFINITIONS AND CONCEPTS	EXAMPLES

Section 6.1 Rational Functions and Multiplying and Dividing Rational Expressions

A rational expression is the quotient $\dfrac{P}{Q}$ of two polynomials P and Q, as long as Q is not 0.

$$\frac{2x - 6}{7}, \quad \frac{t^2 - 3t + 5}{t - 1}$$

To Simplify a Rational Expression

Step 1. Completely factor the numerator and the denominator.

Step 2. Apply the fundamental principle of rational expressions.

Simplify.

$$\frac{2x^2 + 9x - 5}{x^2 - 25} = \frac{(2x - 1)(x + 5)}{(x - 5)(x + 5)}$$
$$= \frac{2x - 1}{x - 5}$$

To Multiply Rational Expressions

Step 1. Completely factor numerators and denominators.

Step 2. Multiply the numerators and multiply the denominators.

Step 3. Apply the fundamental principle of rational expressions.

Multiply $\dfrac{x^3 + 8}{12x - 18} \cdot \dfrac{14x^2 - 21x}{x^2 + 2x}$.

$$= \frac{(x + 2)(x^2 - 2x + 4)}{6(2x - 3)} \cdot \frac{7x(2x - 3)}{x(x + 2)}$$
$$= \frac{7(x^2 - 2x + 4)}{6}$$

To Divide Rational Expressions

Multiply the first rational expression by the reciprocal of the second rational expression.

Divide $\dfrac{x^2 + 6x + 9}{5xy - 5y} \div \dfrac{x + 3}{10y}$.

$$= \frac{(x + 3)(x + 3)}{5y(x - 1)} \cdot \frac{2 \cdot 5y}{x + 3}$$
$$= \frac{2(x + 3)}{x - 1}$$

A rational function is a function described by a rational expression.

$$f(x) = \frac{2x - 6}{7}, \quad h(t) = \frac{t^2 - 3t + 5}{t - 1}$$

Section 6.2 Adding and Subtracting Rational Expressions

To Add or Subtract Rational Expressions

Step 1. Find the LCD.

Step 2. Write each rational expression as an equivalent rational expression whose denominator is the LCD.

Step 3. Add or subtract numerators and write the result over the common denominator.

Step 4. Simplify the resulting rational expression.

Subtract $\dfrac{3}{x + 2} - \dfrac{x + 1}{x - 3}$.

$$= \frac{3 \cdot (x - 3)}{(x + 2) \cdot (x - 3)} - \frac{(x + 1) \cdot (x + 2)}{(x - 3) \cdot (x + 2)}$$
$$= \frac{3(x - 3) - (x + 1)(x + 2)}{(x + 2)(x - 3)}$$
$$= \frac{3x - 9 - (x^2 + 3x + 2)}{(x + 2)(x - 3)}$$
$$= \frac{3x - 9 - x^2 - 3x - 2}{(x + 2)(x - 3)}$$
$$= \frac{-x^2 - 11}{(x + 2)(x - 3)}$$

DEFINITIONS AND CONCEPTS	EXAMPLES

Section 6.3   Simplifying Complex Fractions

Method 1: Simplify the numerator and the denominator so that each is a single fraction. Then perform the indicated division and simplify if possible.

Simplify $\dfrac{\dfrac{x+2}{x}}{x-\dfrac{4}{x}}$.

Method 1: $\dfrac{\dfrac{x+2}{x}}{\dfrac{x \cdot x}{1 \cdot x}-\dfrac{4}{x}} = \dfrac{\dfrac{x+2}{x}}{\dfrac{x^2-4}{x}}$

$= \dfrac{x+2}{x} \cdot \dfrac{x}{(x+2)(x-2)} = \dfrac{1}{x-2}$

Method 2: Multiply the numerator and the denominator of the complex fraction by the LCD of the fractions in both the numerator and the denominator. Then simplify if possible.

Method 2: $\dfrac{\left(\dfrac{x+2}{x}\right) \cdot x}{\left(x-\dfrac{4}{x}\right) \cdot x} = \dfrac{x+2}{x \cdot x - \dfrac{4}{x} \cdot x}$

$= \dfrac{x+2}{x^2-4} = \dfrac{x+2}{(x+2)(x-2)} = \dfrac{1}{x-2}$

Section 6.4   Dividing Polynomials: Long Division and Synthetic Division

To divide a polynomial by a monomial: Divide each term in the polynomial by the monomial.

Divide $\dfrac{12a^5b^3 - 6a^2b^2 + ab}{6a^2b^2}$

$= \dfrac{12a^5b^3}{6a^2b^2} - \dfrac{6a^2b^2}{6a^2b^2} + \dfrac{ab}{6a^2b^2}$

$= 2a^3b - 1 + \dfrac{1}{6ab}$

To divide a polynomial by a polynomial other than a monomial:

Use **long division.**

Divide $2x^3 - x^2 - 8x - 1$ by $x - 2$.

$$
\begin{array}{r}
2x^2 + 3x - 2 \\
x-2\overline{)2x^3 - x^2 - 8x - 1} \\
\underline{2x^3 - 4x^2} \\
3x^2 - 8x \\
\underline{3x^2 - 6x} \\
-2x - 1 \\
\underline{-2x + 4} \\
-5
\end{array}
$$

The quotient is $2x^2 + 3x - 2 - \dfrac{5}{x-2}$.

A shortcut method called **synthetic division** may be used to divide a polynomial by a binomial of the form $x - c$.

Use synthetic division to divide $2x^3 - x^2 - 8x - 1$ by $x - 2$.

$$
\begin{array}{r|rrrr}
2 & 2 & -1 & -8 & -1 \\
  & \downarrow & 4 & 6 & -4 \\
\hline
  & 2 & 3 & -2 & -5
\end{array}
$$

The quotient is $2x^2 + 3x - 2 - \dfrac{5}{x-2}$.

DEFINITIONS AND CONCEPTS	EXAMPLES

Section 6.5 Solving Equations Containing Rational Expressions

To solve an equation containing rational expressions: Multiply both sides of the equation by the LCD of all rational expressions. Then apply the distributive property and simplify. Solve the resulting equation and then check each proposed solution to see whether it makes a denominator 0. If so, it is an **extraneous solution.**

Solve $x - \dfrac{3}{x} = \dfrac{1}{2}$.

$$2x\left(x - \frac{3}{x}\right) = 2x\left(\frac{1}{2}\right) \quad \text{The LCD is } 2x.$$

$$2x \cdot x - 2x\left(\frac{3}{x}\right) = 2x\left(\frac{1}{2}\right) \quad \text{Distribute.}$$

$$2x^2 - 6 = x$$

$$2x^2 - x - 6 = 0 \quad \text{Subtract } x.$$

$$(2x + 3)(x - 2) = 0 \quad \text{Factor.}$$

$$x = -\frac{3}{2} \quad \text{or} \quad x = 2 \quad \text{Solve.}$$

Both $-\dfrac{3}{2}$ and 2 check. The solutions are 2 and $-\dfrac{3}{2}$.

Section 6.6 Rational Equations and Problem Solving

Solving an Equation for a Specified Variable

Treat the specified variable as the only variable of the equation and solve as usual.

Solve for x.

$$A = \frac{2x + 3y}{5}$$

$$5A = 2x + 3y \quad \text{Multiply both sides by 5.}$$

$$5A - 3y = 2x \quad \text{Subtract } 3y \text{ from both sides.}$$

$$\frac{5A - 3y}{2} = x \quad \text{Divide both sides by 2.}$$

Problem-Solving Steps to Follow

Jeanee and David Dillon volunteer every year to clean a strip of Lake Ponchartrain Beach. Jeanee can clean all the trash in this area of beach in 6 hours; David takes 5 hours. Find how long it will take them to clean the area of beach together.

1. UNDERSTAND.

1. Read and reread the problem.

Let $x =$ time in hours that it takes Jeanee and David to clean the beach together.

	Hours to Complete	**Part Completed in 1 Hour**
Jeanee Alone	6	$\dfrac{1}{6}$
David Alone	5	$\dfrac{1}{5}$
Together	x	$\dfrac{1}{x}$

2. TRANSLATE.

2. In words:

part Jeanee can complete in 1 hour	+	part David can complete in 1 hour	=	part they can complete together in 1 hour
↓		↓		↓

Translate:

$$\frac{1}{6} \quad + \quad \frac{1}{5} \quad = \quad \frac{1}{x}$$

DEFINITIONS AND CONCEPTS	EXAMPLES

Section 6.6 Rational Equations and Problem Solving (continued)

3. SOLVE.

3. $\dfrac{1}{6} + \dfrac{1}{5} = \dfrac{1}{x}$

$5x + 6x = 30$ Multiply both sides by $30x$.

$11x = 30$

$x = \dfrac{30}{11}$ or $2\dfrac{8}{11}$

4. INTERPRET.

4. *Check* and then *state*. Together, they can clean the beach in $2\dfrac{8}{11}$ hours.

Section 6.7 Variation and Problem Solving

y **varies directly as** *x*, or *y* is **directly proportional to** *x*, if there is a nonzero constant *k* such that

$$y = kx$$

y **varies inversely as** *x*, or *y* is **inversely proportional to** *x*, if there is a nonzero constant *k* such that

$$y = \dfrac{k}{x}$$

y **varies jointly as** *x* and *z*, or *y* is **jointly proportional to** *x* and *z*, if there is a nonzero constant *k* such that

$$y = kxz$$

The circumference of a circle *C* varies directly as its radius *r*.

$$C = \underset{k}{2\pi r}$$

Pressure *P* varies inversely with volume *V*.

$$P = \dfrac{k}{V}$$

The lateral surface area *S* of a cylinder varies jointly as its radius *r* and height *h*.

$$S = \underset{k}{2\pi rh}$$

Chapter 6 Review

(6.1) Find the domain for each rational function.

1. $f(x) = \dfrac{3 - 5x}{7}$

2. $g(x) = \dfrac{2x + 4}{11}$

3. $F(x) = \dfrac{-3x^2}{x - 5}$

4. $h(x) = \dfrac{4x}{3x - 12}$

5. $f(x) = \dfrac{x^3 + 2}{x^2 + 8x}$

6. $G(x) = \dfrac{20}{3x^2 - 48}$

Write each rational expression in lowest terms.

7. $\dfrac{x - 12}{12 - x}$

8. $\dfrac{5x - 15}{25x - 75}$

9. $\dfrac{2x}{2x^2 - 2x}$

10. $\dfrac{x + 7}{x^2 - 49}$

11. $\dfrac{2x^2 + 4x - 30}{x^2 + x - 20}$

12. The average cost (per bookcase) of manufacturing *x* bookcases is given by the rational function.

$$C(x) = \dfrac{35x + 4200}{x}$$

a. Find the average cost per bookcase of manufacturing 50 bookcases.

b. Find the average cost per bookcase of manufacturing 100 bookcases.

c. As the number of bookcases increases, does the average cost per bookcase increase or decrease? (See parts (a) and (b).)

Perform each indicated operation. Write your answers in lowest terms.

13. $\dfrac{4 - x}{5} \cdot \dfrac{15}{2x - 8}$

14. $\dfrac{x^2 - 6x + 9}{2x^2 - 18} \cdot \dfrac{4x + 12}{5x - 15}$

15. $\dfrac{a - 4b}{a^2 + ab} \cdot \dfrac{b^2 - a^2}{8b - 2a}$

16. $\dfrac{x^2 - x - 12}{2x^2 - 32} \cdot \dfrac{x^2 + 8x + 16}{3x^2 + 21x + 36}$

17. $\dfrac{4x + 8y}{3} \div \dfrac{5x + 10y}{9}$

18. $\dfrac{x^2 - 25}{3} \div \dfrac{x^2 - 10x + 25}{x^2 - x - 20}$

19. $\dfrac{a - 4b}{a^2 + ab} \div \dfrac{20b - 5a}{b^2 - a^2}$

20. $\dfrac{3x + 3}{x - 1} \div \dfrac{x^2 - 6x - 7}{x^2 - 1}$

21. $\dfrac{2x - x^2}{x^3 - 8} \div \dfrac{x^2}{x^2 + 2x + 4}$

22. $\dfrac{5x - 15}{3 - x} \cdot \dfrac{x + 2}{10x + 20} \cdot \dfrac{x^2 - 9}{x^2 - x - 6}$

(6.2) *Find the LCD of the rational expressions in each list.*

23. $\dfrac{5}{4x^2y^5}, \dfrac{3}{10x^2y^4}, \dfrac{x}{6y^4}$

24. $\dfrac{5}{2x}, \dfrac{7}{x - 2}$

25. $\dfrac{3}{5x}, \dfrac{2}{x - 5}$

26. $\dfrac{1}{5x^3}, \dfrac{4}{x^2 + 3x - 28}, \dfrac{11}{10x^2 - 30x}$

Perform each indicated operation. Write your answers in lowest terms.

27. $\dfrac{4}{x - 4} + \dfrac{x}{x - 4}$

28. $\dfrac{4}{3x^2} + \dfrac{2}{3x^2}$

29. $\dfrac{1}{x - 2} - \dfrac{1}{4 - 2x}$

30. $\dfrac{1}{10 - x} + \dfrac{x - 1}{x - 10}$

31. $\dfrac{x}{9 - x^2} - \dfrac{2}{5x - 15}$

32. $2x + 1 - \dfrac{1}{x - 3}$

33. $\dfrac{2}{a^2 - 2a + 1} + \dfrac{3}{a^2 - 1}$

34. $\dfrac{x}{9x^2 + 12x + 16} - \dfrac{3x + 4}{27x^3 - 64}$

Perform each indicated operation. Write your answers in lowest terms.

35. $\dfrac{2}{x - 1} - \dfrac{3x}{3x - 3} + \dfrac{1}{2x - 2}$

△ **36.** Find the perimeter of the heptagon (a polygon with seven sides).

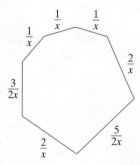

(6.3) *Simplify each complex fraction.*

37. $\dfrac{1 - \dfrac{3x}{4}}{2 + \dfrac{x}{4}}$

38. $\dfrac{\dfrac{x^2}{15}}{\dfrac{x + 1}{5x}}$

39. $\dfrac{2 - \dfrac{3}{2x}}{x - \dfrac{2}{5x}}$

40. $\dfrac{1 + \dfrac{x}{y}}{\dfrac{x^2}{y^2} - 1}$

41. $\dfrac{\dfrac{5}{x} + \dfrac{1}{xy}}{\dfrac{3}{x^2}}$

42. $\dfrac{\dfrac{x}{3} - \dfrac{3}{x}}{1 + \dfrac{3}{x}}$

43. $\dfrac{\dfrac{1}{x - 1} + 1}{\dfrac{1}{x + 1} - 1}$

44. $\dfrac{\dfrac{x - 3}{x + 3} + \dfrac{x + 3}{x - 3}}{\dfrac{x - 3}{x + 3} - \dfrac{x + 3}{x - 3}}$

If $f(x) = \dfrac{3}{x}, x \neq 0$, find each of the following.

45. $f(a + h)$

46. $f(a)$

47. Use Exercises 45 and 46 to find $\dfrac{f(a + h) - f(a)}{h}$.

48. Simplify the results of Exercise 47.

(6.4)

49. $(4xy + 2x^2 - 9) \div 4xy$

50. Divide $12xb^2 + 16xb^4$ by $4xb^3$.

51. $(3x^4 - 25x^2 - 20) \div (x - 3)$

52. $(-x^2 + 2x^4 + 5x - 12) \div (x + 2)$

53. $(2x^3 + 3x^2 - 2x + 2) \div (2x + 3)$

54. $(3x^4 + 5x^3 + 7x^2 + 3x - 2) \div (x^2 + x + 2)$

Use synthetic division to find each quotient.

55. $(3x^3 + 12x - 4) \div (x - 2)$

56. $(x^5 - 1) \div (x + 1)$

57. $(x^3 - 81) \div (x - 3)$

58. $(3x^4 - 2x^2 + 10) \div (x + 2)$

If $P(x) = 3x^5 - 9x + 7$, use the remainder theorem to find the following.

59. $P(4)$

60. $P(-5)$

61. $P\left(-\dfrac{1}{2}\right)$

△ **62.** If the area of the rectangle is $(x^4 - x^3 - 6x^2 - 6x + 18)$ square miles and its width is $(x - 3)$ miles, find the length.

$x^4 - x^3 - 6x^2 - 6x + 18$ square miles $\qquad$ $x - 3$ miles

(6.5) *Solve each equation.*

63. $\dfrac{3}{x} + \dfrac{1}{3} = \dfrac{5}{x}$

64. $\dfrac{2x + 3}{5x - 9} = \dfrac{3}{2}$

65. $\dfrac{1}{x - 2} - \dfrac{3x}{x^2 - 4} = \dfrac{2}{x + 2}$

66. $\dfrac{7}{x} - \dfrac{x}{7} = 0$

Solve each equation or perform each indicated operation. Simplify.

67. $\dfrac{5}{x^2 - 7x} + \dfrac{4}{2x - 14}$

68. $\dfrac{4}{3 - x} - \dfrac{7}{2x - 6} + \dfrac{5}{x}$

69. $3 - \dfrac{5}{x} - \dfrac{2}{x^2} = 0$ **70.** $2 + \dfrac{15}{x^2} = \dfrac{13}{x}$

(6.6) Solve each equation for the specified variable.

△ **71.** $A = \dfrac{h(a + b)}{2}$ for a

72. $\dfrac{1}{R} = \dfrac{1}{R_1} + \dfrac{1}{R_2}$ for R_2

73. $I = \dfrac{E}{R + r}$ for R

74. $A = P + Prt$ for r

75. $\dfrac{1}{x} = \dfrac{1}{y} - \dfrac{1}{z}$ for x

76. $H = \dfrac{kA(T_1 - T_2)}{L}$ for A

Solve.

77. The sum of a number and twice its reciprocal is 3. Find the number(s).

78. If a number is added to the numerator of $\dfrac{3}{7}$, and twice that number is added to the denominator of $\dfrac{3}{7}$, the result is equivalent to $\dfrac{10}{21}$. Find the number.

79. Three boys can paint a fence in 4 hours, 5 hours, and 6 hours, respectively. Find how long it will take all three boys to paint the fence working together.

80. If Sue Katz can type a certain number of mailing labels in 6 hours and Tom Neilson and Sue working together can type the same number of mailing labels in 4 hours, find how long it takes Tom alone to type the mailing labels.

81. The speed of a Ranger boat in still water is 32 mph. If the boat travels 72 miles upstream in the same time that it takes to travel 120 miles downstream, find the current of the stream.

82. The speed of a jogger is 3 mph faster than the speed of a walker. If the jogger travels 14 miles in the same amount of time that the walker travels 8 miles, find the speed of the walker.

(6.7) Solve each variation problem.

83. A is directly proportional to B. If $A = 6$ when $B = 14$, find A when $B = 21$.

84. According to Boyle's law, the pressure exerted by a gas is inversely proportional to the volume as long as the temperature stays the same. If a gas exerts a pressure of 1250 kilopascals when the volume is 2 cubic meters, find the volume when the pressure is 800 kilopascals.

MIXED REVIEW

For expressions, perform the indicated operation and/or simplify. For equations, solve the equation for the unknown variable.

85. $\dfrac{22x + 8}{11x + 4}$

86. $\dfrac{xy - 3x + 2y - 6}{x^2 + 4x + 4}$

87. $\dfrac{2}{5x} \div \dfrac{4 - 18x}{6 - 27x}$

88. $\dfrac{7x + 28}{2x + 4} \div \dfrac{x^2 + 2x - 8}{x^2 - 2x - 8}$

89. $\dfrac{5a^2 - 20}{a^3 + 2a^2 + a + 2} \div \dfrac{7a}{a^3 + a}$

90. $\dfrac{4a + 8}{5a^2 - 20} \cdot \dfrac{3a^2 - 6a}{a + 3} \div \dfrac{2a^2}{5a + 15}$

91. $\dfrac{7}{2x} + \dfrac{5}{6x}$ **92.** $\dfrac{x - 2}{x + 1} - \dfrac{x - 3}{x - 1}$

93. $\dfrac{2x + 1}{x^2 + x - 6} + \dfrac{2 - x}{x^2 + x - 6}$

94. $\dfrac{2}{x^2 - 16} - \dfrac{3x}{x^2 + 8x + 16} + \dfrac{3}{x + 4}$

95. $\dfrac{\dfrac{1}{x} - \dfrac{2}{3x}}{\dfrac{5}{2x} - \dfrac{1}{3}}$ **96.** $\dfrac{2}{1 - \dfrac{2}{x}}$

97. $\dfrac{\dfrac{x^2 + 5x - 6}{4x + 3}}{\dfrac{(x + 6)^2}{8x + 6}}$ **98.** $\dfrac{\dfrac{3}{x - 1} - \dfrac{2}{1 - x}}{\dfrac{2}{x - 1} - \dfrac{2}{x}}$

99. $4 + \dfrac{8}{x} = 8$

100. $\dfrac{x - 2}{x^2 - 7x + 10} = \dfrac{1}{5x - 10} - \dfrac{1}{x - 5}$

101. The denominator of a fraction is 2 more than the numerator. If the numerator is decreased by 3 and the denominator is increased by 5, the resulting fraction is equivalent to $\dfrac{2}{3}$. Find the fraction.

102. The sum of the reciprocals of two consecutive even integers is $-\dfrac{9}{40}$. Find the two integers.

103. The inlet pipe of a water tank can fill the tank in 2 hours and 30 minutes. The outlet pipe can empty the tank in 2 hours. Find how long it takes to empty a full tank if both pipes are open.

104. Timmy Garnica drove 210 miles in the same amount of time that it took a DC-10 jet to travel 1715 miles. The speed of the jet was 430 mph faster than the speed of the car. Find the speed of the jet.

105. Two Amtrak trains traveling on parallel tracks leave Tucson at the same time. In 6 hours, the faster train is 382 miles from Tucson and the trains are 112 miles apart. Find how fast each train is traveling.

△ **106.** The surface area of a sphere varies directly as the square of its radius. If the surface area is 36π square inches when the radius is 3 inches, find the surface area when the radius is 4 inches.

107. In 2013, it is predicted that 3 out of 4 U.S. citizens will own MP3-capable phones. At this rate, how many MP3-capable phones would you predict to be owned in a community with a 2013 population of 43,560?

108. Currently, 8 out of 25 U.S. Internet users purchase digital music. At this rate, how many Internet users purchase digital music in a group of 2000 Internet users?

109. Divide $(x^3 - x^2 + 3x^4 - 2)$ by $(x - 4)$.

110. C is inversely proportional to D. If $C = 12$ when $D = 8$, find C when $D = 24$.

Chapter 6 Test

Find the domain of each rational function.

1. $f(x) = \dfrac{5x^2}{1 - x}$

2. $g(x) = \dfrac{9x^2 - 9}{x^2 + 4x + 3}$

Write each rational expression in lowest terms.

3. $\dfrac{7x - 21}{24 - 8x}$

4. $\dfrac{x^2 - 4x}{x^2 + 5x - 36}$

5. $\dfrac{x^3 - 8}{x - 2}$

Perform the indicated operation. If possible, simplify your answer.

6. $\dfrac{2x^3 + 16}{6x^2 + 12x} \cdot \dfrac{5}{x^2 - 2x + 4}$

7. $\dfrac{5}{4x^3} + \dfrac{7}{4x^3}$

8. $\dfrac{3x^2 - 12}{x^2 + 2x - 8} \div \dfrac{6x + 18}{x + 4}$

9. $\dfrac{4x - 12}{2x - 9} \div \dfrac{3 - x}{4x^2 - 81} \cdot \dfrac{x + 3}{5x + 15}$

10. $\dfrac{3 + 2x}{10 - x} + \dfrac{13 + x}{x - 10}$

11. $\dfrac{2x^2 + 7}{2x^4 - 18x^2} - \dfrac{6x + 7}{2x^4 - 18x^2}$

12. $\dfrac{3}{x^2 - x - 6} + \dfrac{2}{x^2 - 5x + 6}$

13. $\dfrac{5}{x - 7} - \dfrac{2x}{3x - 21} + \dfrac{x}{2x - 14}$

14. $\dfrac{3x}{5} \cdot \left(\dfrac{5}{x} - \dfrac{5}{2x} \right)$

Simplify each complex fraction.

15. $\dfrac{\dfrac{5}{x} - \dfrac{7}{3x}}{\dfrac{9}{8x} - \dfrac{1}{x}}$

16. $\dfrac{\dfrac{x^2 - 5x + 6}{x + 3}}{\dfrac{x^2 - 4x + 4}{x^2 - 9}}$

Divide.

17. $(4x^2y + 9x + 3xz) \div 3xz$

18. $(4x^3 - 5x) \div (2x + 1)$

19. Use synthetic division to divide $(4x^4 - 3x^3 - x - 1)$ by $(x + 3)$.

20. If $P(x) = 4x^4 + 7x^2 - 2x - 5$, use the remainder theorem to find $P(-2)$.

Solve each equation for x.

21. $\dfrac{5x + 3}{3x - 7} = \dfrac{19}{7}$

22. $\dfrac{x}{x - 4} = 3 - \dfrac{4}{x - 4}$

23. $\dfrac{3}{x + 2} - \dfrac{1}{5x} = \dfrac{2}{5x^2 + 10x}$

24. $\dfrac{x^2 + 8}{x} - 1 = \dfrac{2(x + 4)}{x}$

25. Solve for x: $\dfrac{x + b}{a} = \dfrac{4x - 7a}{b}$

26. The product of one more than a number and twice the reciprocal of the number is $\dfrac{12}{5}$. Find the number.

27. If Jan can weed the garden in 2 hours and her husband can weed it in 1 hour and 30 minutes, find how long it takes them to weed the garden together.

28. Suppose that W is inversely proportional to V. If $W = 20$ when $V = 12$, find W when $V = 15$.

29. Suppose that Q is jointly proportional to R and the square of S. If $Q = 24$ when $R = 3$ and $S = 4$, find Q when $R = 2$ and $S = 3$.

30. When an anvil is dropped into a gorge, the speed with which it strikes the ground is directly proportional to the square root of the distance it falls. An anvil that falls 400 feet hits the ground at a speed of 160 feet per second. Find the height of a cliff over the gorge if a dropped anvil hits the ground at a speed of 128 feet per second.

Chapter 6 Cumulative Review

1. Translate each phrase to an algebraic expression. Use the variable x to represent each unknown number.

   a. Eight times a number

   b. Three more than eight times a number

   c. The quotient of a number and -7

   d. One and six-tenths subtracted from twice a number

   e. Six less than a number

   f. Twice the sum of four and a number

2. Translate each phrase to an algebraic expression. Use the variable x to represent each unknown number.

   a. One third subtracted from a number

   b. Six less than five times a number

   c. Three more than eight times a number

   d. The quotient of seven and the difference of two and a number.

3. Solve for y: $\dfrac{y}{3} - \dfrac{y}{4} = \dfrac{1}{6}$

4. Solve $\dfrac{x}{7} + \dfrac{x}{5} = \dfrac{12}{5}$.

5. In the United States, the annual consumption of cigarettes is declining. The consumption c in billions of cigarettes per year since the year 2000 can be approximated by the formula $c = -9.4t + 431$ where t is the number of years after 2000. Use this formula to predict the years that the consumption of cigarettes will be less than 200 billion per year.

6. Olivia has scores of 78, 65, 82, and 79 on her algebra tests. Use an inequality to find the minimum score she can make on her final exam to pass the course with a 78 average or higher, given that the final exam counts as two tests.

7. Solve: $\left| \dfrac{3x + 1}{2} \right| = -2$

8. Solve: $\left| \dfrac{2x - 1}{3} \right| + 6 = 3$

9. Solve for x: $\left| \dfrac{2(x + 1)}{3} \right| \leq 0$

10. Solve for x: $\left| \dfrac{3(x - 1)}{4} \right| \geq 2$

11. Graph the equation $y = -2x + 3$.

12. Graph the equation $y = -x + 3$.

13. Determine whether the following relations are also functions.

   a. $\{(-2, 5), (2, 7), (-3, 5), (9, 9)\}$

   b.

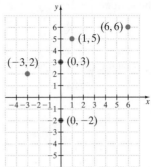

   c.

Input	Correspondence	Output
People in a certain city	Each person's age	The set of nonegative integers

14. If $f(x) = -x^2 + 3x - 2$, find

   a. $f(0)$ **b.** $f(-3)$ **c.** $f\left(\dfrac{1}{3}\right)$

15. Find the intercepts and graph: $3x + 4y = -12$.

16. Find the intercepts and graph: $3x - y = 6$.

17. Find an equation of the line with slope -3 containing the point $(1, -5)$. Write the equation in slope–intercept form $y = mx + b$.

18. Find an equation of the line with slope $\dfrac{1}{2}$ containing the point $(-1, 3)$. Use function notation to write the equation.

19. Graph the intersection of $x \geq 1$ and $y \geq 2x - 1$.

20. Graph the union of $2x + y \leq 4$ or $y > 2$.

21. Use the elimination method to solve the system.
$$\begin{cases} 3x - 2y = 10 \\ 4x - 3y = 15 \end{cases}$$

22. Use the substitution method to solve the system.
$$\begin{cases} -2x + 3y = 6 \\ 3x - y = 5 \end{cases}$$

23. Solve the system. $\begin{cases} 2x - 4y + 8z = 2 \\ -x - 3y + z = 11 \\ x - 2y + 4z = 0 \end{cases}$

24. Solve the system. $\begin{cases} 2x - 2y + 4z = 6 \\ -4x - y + z = -8 \\ 3x - y + z = 6 \end{cases}$

25. The measure of the largest angle of a triangle is $80°$ more than the measure of the smallest angle, and the measure of the remaining angle is $10°$ more than the measure of the smallest angle. Find the measure of each angle.

26. Kernersville office supply sold three reams of paper and two boxes of manila folders for $21.90. Also, five reams of paper and one box of manila folders cost $24.25. Find the price of a ream of paper and a box of manila folders.

27. Use matrices to solve the system. $\begin{cases} x + 2y + z = 2 \\ -2x - y + 2z = 5 \\ x + 3y - 2z = -8 \end{cases}$

28. Use matrices to solve the system. $\begin{cases} x + y + z = 9 \\ 2x - 2y + 3z = 2 \\ -3x + y - z = 1 \end{cases}$

29. Evaluate the following.
   a. 7^0 **b.** -7^0
   c. $(2x + 5)^0$ **d.** $2x^0$

30. Simplify the following. Write answers with positive exponents.
   a. $2^{-2} + 3^{-1}$ **b.** $-6a^0$ **c.** $\dfrac{x^{-5}}{x^{-2}}$

31. Simplify each expression. Assume that a and b are integers and that x and y are not 0.
   a. $x^{-b}(2x^b)^2$ **b.** $\dfrac{(y^{3a})^2}{y^{a-6}}$

32. Simplify each expression. Assume that a and b are integers and that x and y are not 0.
   a. $3x^{4a}(4x^{-a})^2$ **b.** $\dfrac{(y^{4b})^3}{y^{2b-3}}$

33. Find the degree of each term.
   a. $3x^2$ **b.** -2^3x^5 **c.** y
   d. $12x^2yz^3$ **e.** 5.27

34. Subtract $(2x - 7)$ from $2x^2 + 8x - 3$.

35. Multiply $[3 + (2a + b)]^2$.

36. Multiply $[4 + (3x - y)]^2$.

37. Factor $ab - 6a + 2b - 12$.

38. Factor $xy + 2x - 5y - 10$.

39. Factor $2n^2 - 38n + 80$.

40. Factor $6x^2 - x - 35$.

41. Factor $x^2 + 4x + 4 - y^2$.

42. Factor $4x^2 - 4x + 1 - 9y^2$.

43. Solve $(x + 2)(x - 6) = 0$.

44. Solve $2x(3x + 1)(x - 3) = 0$.

45. Simplify each rational expression.
   a. $\dfrac{2x^2}{10x^3 - 2x^2}$
   b. $\dfrac{9x^2 + 13x + 4}{8x^2 + x - 7}$

46. For the graph of $f(x)$, answer the following:

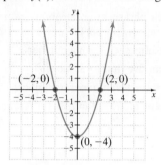

   a. Find the domain and range.
   b. List the x- and y-intercepts.
   c. Find the coordinates of the point with the greatest y-value.
   d. Find the coordinates of the point with the least y-value.
   e. List the x-values whose y-values are equal to 0.
   f. List the x-values whose y-values are less than 0.
   g. Find the solutions of $f(x) = 0$.

47. Subtract $\dfrac{5k}{k^2 - 4} - \dfrac{2}{k^2 + k - 2}$.

48. Subtract $\dfrac{5a}{a^2 - 4} - \dfrac{3}{2 - a}$.

49. Solve $\dfrac{3}{x} - \dfrac{x + 21}{3x} = \dfrac{5}{3}$.

50. Solve $\dfrac{3x - 4}{2x} = -\dfrac{8}{x}$.

Rational Exponents, Radicals, and Complex Numbers

The Google Lunar X PRIZE is an international competition to safely land a robot on the surface of the moon, travel 500 meters over the lunar surface, and send images and data back to Earth. Teams needed to be registered by December 31, 2010. There are multiple prizes and bonuses, but the first team to land on the moon and complete the mission objectives by December 31, 2012, will be awarded $20 million. After this time, the first prize drops to $15 million. The deadline for winning the competition is December 31, 2014, and thus far, 20 teams are competing for the prize.

To reach the moon, these vehicles must first leave the gravity of Earth. In Exercises 115 and 116 of Section 7.1, you will calculate the escape velocity of Earth and the moon, the minimum speed an object must reach to escape the pull of a planet's gravity. (*Source:* X PRIZE Foundation)

In this chapter, radical notation is reviewed, and then rational exponents are introduced. As the name implies, rational exponents are exponents that are rational numbers. We present an interpretation of rational exponents that is consistent with the meaning and rules already established for integer exponents, and we present two forms of notation for roots: radical and exponent. We conclude this chapter with complex numbers, a natural extension of the real number system.

Composition of the Moon

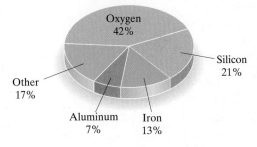

Oxygen 42%

Silicon 21%

Other 17%

Aluminum 7%

Iron 13%

Why the moon? To name a few reasons:

- It is closest to Earth (1.3 seconds for light or radio) so that lunar machines can be directly controlled from Earth.

- It is also the closest source of materials to use for any other space project, and it is 22 times easier to launch from the moon than from Earth.

- The moon is 42% oxygen by weight (see the circle graph to the left) and oxygen is the main ingredient of rocket fuel.

- We can collect energy from the moon's surface and transmit it to Earth.

7.1 | Radicals and Radical Functions

OBJECTIVES

1 Find Square Roots.

2 Approximate Roots.

3 Find Cube Roots.

4 Find nth Roots.

5 Find $\sqrt[n]{a^n}$ Where a Is a Real Number.

6 Graph Square and Cube Root Functions.

OBJECTIVE

1 Finding Square Roots

Recall from Section 1.3 that to find a **square root** of a number a, we find a number that was squared to get a.

Thus, because

$$5^2 = 25 \quad \text{and} \quad (-5)^2 = 25,$$

both 5 and -5 are square roots of 25.

Recall that we denote the **nonnegative**, or **principal, square root** with the **radical sign**.

$$\sqrt{25} = 5$$

We denote the **negative square root** with the **negative radical sign.**

$$-\sqrt{25} = -5$$

An expression containing a radical sign is called a **radical expression.** An expression within, or "under," a radical sign is called a **radicand.**

radical expression: $\overset{\text{radical sign}}{\underset{\text{radicand}}{\sqrt{a}}}$

Principal and Negative Square Roots

If a is a nonnegative number, then

$\sqrt{a}$ is the **principal,** or **nonnegative, square root** of a

$-\sqrt{a}$ is the **negative square root** of a

EXAMPLE 1 Simplify. Assume that all variables represent positive numbers.

a. $\sqrt{36}$ **b.** $\sqrt{0}$ **c.** $\sqrt{\dfrac{4}{49}}$ **d.** $\sqrt{0.25}$

e. $\sqrt{x^6}$ **f.** $\sqrt{9x^{12}}$ **g.** $-\sqrt{81}$ **h.** $\sqrt{-81}$

Solution

a. $\sqrt{36} = 6$ because $6^2 = 36$ and 6 is not negative.

b. $\sqrt{0} = 0$ because $0^2 = 0$ and 0 is not negative.

c. $\sqrt{\dfrac{4}{49}} = \dfrac{2}{7}$ because $\left(\dfrac{2}{7}\right)^2 = \dfrac{4}{49}$ and $\dfrac{2}{7}$ is not negative.

d. $\sqrt{0.25} = 0.5$ because $(0.5)^2 = 0.25$.

e. $\sqrt{x^6} = x^3$ because $(x^3)^2 = x^6$.

f. $\sqrt{9x^{12}} = 3x^6$ because $(3x^6)^2 = 9x^{12}$.

g. $-\sqrt{81} = -9$. The negative in front of the radical indicates the negative square root of 81.

h. $\sqrt{-81}$ is not a real number. ☐

PRACTICE

1 Simplify. Assume that all variables represent positive numbers.

a. $\sqrt{49}$ **b.** $\sqrt{\dfrac{0}{1}}$ **c.** $\sqrt{\dfrac{16}{81}}$ **d.** $\sqrt{0.64}$

e. $\sqrt{z^8}$ **f.** $\sqrt{16b^4}$ **g.** $-\sqrt{36}$ **h.** $\sqrt{-36}$

Recall from Section 1.3 our discussion of the square root of a negative number. For example, can we simplify $\sqrt{-4}$? That is, can we find a real number whose square is -4? No, there is no real number whose square is -4, and we say that $\sqrt{-4}$ is not a real number. In general:

The square root of a negative number is not a real number.

> ▶ Helpful Hint
> - Remember: $\sqrt{0} = 0$.
> - Don't forget that the square root of a negative number is not a real number. For example,
>
> $$\sqrt{-9} \text{ is not a real number}$$
>
> because there is no real number that when multiplied by itself would give a product of -9. In Section 7.7, we will see what kind of a number $\sqrt{-9}$ is.

OBJECTIVE
2 Approximating Roots

Recall that numbers such as 1, 4, 9, and 25 are called **perfect squares,** since $1 = 1^2, 4 = 2^2, 9 = 3^2$, and $25 = 5^2$. Square roots of perfect square radicands simplify to rational numbers. What happens when we try to simplify a root such as $\sqrt{3}$? Since there is no rational number whose square is 3, $\sqrt{3}$ is not a rational number. It is called an **irrational number,** and we can find a decimal **approximation** of it. To find decimal approximations, use a calculator. For example, an approximation for $\sqrt{3}$ is

$$\sqrt{3} \underset{\underset{\text{approximation symbol}}{\uparrow}}{\approx} 1.732$$

To see if the approximation is reasonable, notice that since

$$1 < 3 < 4,$$
$$\sqrt{1} < \sqrt{3} < \sqrt{4}, \text{ or}$$
$$1 < \sqrt{3} < 2.$$

We found $\sqrt{3} \approx 1.732$, a number between 1 and 2, so our result is reasonable.

EXAMPLE 2   Use a calculator to approximate $\sqrt{20}$. Round the approximation to 3 decimal places and check to see that your approximation is reasonable.

$$\sqrt{20} \approx 4.472$$

Solution   Is this reasonable? Since $16 < 20 < 25$, $\sqrt{16} < \sqrt{20} < \sqrt{25}$, or $4 < \sqrt{20} < 5$. The approximation is between 4 and 5 and thus is reasonable.   □

PRACTICE
2   Use a calculator to approximate $\sqrt{45}$. Round the approximation to three decimal places and check to see that your approximation is reasonable.

OBJECTIVE
3 Finding Cube Roots

Finding roots can be extended to other roots such as cube roots. For example, since $2^3 = 8$, we call 2 the **cube root** of 8. In symbols, we write

$$\sqrt[3]{8} = 2$$

> **Cube Root**
> The **cube root** of a real number a is written as $\sqrt[3]{a}$, and
> $$\sqrt[3]{a} = b \text{ only if } b^3 = a$$

From this definition, we have

$$\sqrt[3]{64} = 4 \text{ since } 4^3 = 64$$
$$\sqrt[3]{-27} = -3 \text{ since } (-3)^3 = -27$$
$$\sqrt[3]{x^3} = x \text{ since } x^3 = x^3$$

Notice that, unlike with square roots, *it is possible to have a negative radicand when finding a cube root.* This is so because the *cube* of a negative number is a negative number. Therefore, the *cube root* of a negative number is a negative number.

EXAMPLE 3   Find the cube roots.

a. $\sqrt[3]{1}$ **b.** $\sqrt[3]{-64}$ **c.** $\sqrt[3]{\dfrac{8}{125}}$ **d.** $\sqrt[3]{x^6}$ **e.** $\sqrt[3]{-27x^9}$

Solution

a. $\sqrt[3]{1} = 1$ because $1^3 = 1$.

b. $\sqrt[3]{-64} = -4$ because $(-4)^3 = -64$.

c. $\sqrt[3]{\dfrac{8}{125}} = \dfrac{2}{5}$ because $\left(\dfrac{2}{5}\right)^3 = \dfrac{8}{125}$.

d. $\sqrt[3]{x^6} = x^2$ because $(x^2)^3 = x^6$.

e. $\sqrt[3]{-27x^9} = -3x^3$ because $(-3x^3)^3 = -27x^9$.

PRACTICE

3   Find the cube roots.

a. $\sqrt[3]{-1}$ **b.** $\sqrt[3]{27}$ **c.** $\sqrt[3]{\dfrac{27}{64}}$ **d.** $\sqrt[3]{x^{12}}$ **e.** $\sqrt[3]{-8x^3}$

OBJECTIVE

4   Finding *n*th Roots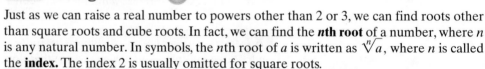

Just as we can raise a real number to powers other than 2 or 3, we can find roots other than square roots and cube roots. In fact, we can find the **nth root** of a number, where n is any natural number. In symbols, the nth root of a is written as $\sqrt[n]{a}$, where n is called the **index.** The index 2 is usually omitted for square roots.

> ▶ **Helpful Hint**
>
> If the index is even, such as $\sqrt{\ }$, $\sqrt[4]{\ }$, $\sqrt[6]{\ }$, and so on, the radicand must be nonnegative for the root to be a real number. For example,
>
> $$\sqrt[4]{16} = 2, \text{ but } \sqrt[4]{-16} \text{ is not a real number.}$$
> $$\sqrt[6]{64} = 2, \text{ but } \sqrt[6]{-64} \text{ is not a real number.}$$
>
> If the index is odd, such as $\sqrt[3]{\ }$, $\sqrt[5]{\ }$, and so on, the radicand may be any real number. For example,
>
> $$\sqrt[3]{64} = 4 \quad \text{and} \quad \sqrt[3]{-64} = -4$$
> $$\sqrt[5]{32} = 2 \quad \text{and} \quad \sqrt[5]{-32} = -2$$

✓ **CONCEPT CHECK**

Which one is not a real number?

a. $\sqrt[3]{-15}$ **b.** $\sqrt[4]{-15}$ **c.** $\sqrt[5]{-15}$ **d.** $\sqrt{(-15)^2}$

Answer to Concept Check: b

EXAMPLE 4 Simplify the following expressions.

a. $\sqrt[4]{81}$ **b.** $\sqrt[5]{-243}$ **c.** $-\sqrt{25}$ **d.** $\sqrt[4]{-81}$ **e.** $\sqrt[3]{64x^3}$

Solution

a. $\sqrt[4]{81} = 3$ because $3^4 = 81$ and 3 is positive.

b. $\sqrt[5]{-243} = -3$ because $(-3)^5 = -243$.

c. $-\sqrt{25} = -5$ because -5 is the opposite of $\sqrt{25}$.

d. $\sqrt[4]{-81}$ is not a real number. There is no real number that, when raised to the fourth power, is -81.

e. $\sqrt[3]{64x^3} = 4x$ because $(4x)^3 = 64x^3$.

PRACTICE

4 Simplify the following expressions.

a. $\sqrt[4]{10,000}$ **b.** $\sqrt[5]{-1}$ **c.** $-\sqrt{81}$ **d.** $\sqrt[4]{-625}$ **e.** $\sqrt[3]{27x^9}$

OBJECTIVE

5 Finding $\sqrt[n]{a^n}$ Where a Is a Real Number ▶

Recall that the notation $\sqrt{a^2}$ indicates the positive square root of a^2 only. For example,

$$\sqrt{(-7)^2} = \sqrt{49} = 7$$

When variables are present in the radicand and it is *unclear whether the variable represents a positive number or a negative number*, absolute value bars are sometimes needed to ensure that the result is a positive number. For example,

$$\sqrt{x^2} = |x|$$

This ensures that the result is positive. This same situation may occur when the index is any *even* positive integer. When the index is any *odd* positive integer, absolute value bars are not necessary.

Finding $\sqrt[n]{a^n}$

If n is an *even* positive integer, then $\sqrt[n]{a^n} = |a|$.

If n is an *odd* positive integer, then $\sqrt[n]{a^n} = a$.

EXAMPLE 5 Simplify.

a. $\sqrt{(-3)^2}$ **b.** $\sqrt{x^2}$ **c.** $\sqrt[4]{(x-2)^4}$ **d.** $\sqrt[3]{(-5)^3}$

e. $\sqrt[5]{(2x-7)^5}$ **f.** $\sqrt{25x^2}$ **g.** $\sqrt{x^2 + 2x + 1}$

Solution

a. $\sqrt{(-3)^2} = |-3| = 3$ When the index is even, the absolute value bars ensure that our result is not negative.

b. $\sqrt{x^2} = |x|$

c. $\sqrt[4]{(x-2)^4} = |x-2|$

d. $\sqrt[3]{(-5)^3} = -5$

e. $\sqrt[5]{(2x-7)^5} = 2x-7$ Absolute value bars are not needed when the index is odd.

f. $\sqrt{25x^2} = 5|x|$

g. $\sqrt{x^2 + 2x + 1} = \sqrt{(x+1)^2} = |x+1|$

PRACTICE

5   Simplify.

a. $\sqrt{(-4)^2}$ **b.** $\sqrt{x^{14}}$ **c.** $\sqrt[4]{(x+7)^4}$ **d.** $\sqrt[3]{(-7)^3}$

e. $\sqrt[5]{(3x-5)^5}$ **f.** $\sqrt{49x^2}$ **g.** $\sqrt{x^2+16x+64}$

OBJECTIVE

6   Graphing Square and Cube Root Functions

Recall that an equation in x and y describes a function if each x-value is paired with exactly one y-value. With this in mind, does the equation

$$y = \sqrt{x}$$

describe a function? First, notice that replacement values for x must be nonnegative real numbers, since $\sqrt{x}$ is not a real number if $x < 0$. The notation $\sqrt{x}$ denotes the principal square root of x, so for every nonnegative number x, there is exactly one number, $\sqrt{x}$. Therefore, $y = \sqrt{x}$ describes a function, and we may write it as

$$f(x) = \sqrt{x}$$

In general, radical functions are functions of the form

$$f(x) = \sqrt[n]{x}.$$

Recall that the domain of a function in x is the set of all possible replacement values of x. This means that if n is even, the domain is the set of all nonnegative numbers, or $\{x \mid x \geq 0\}$ or $[0, \infty)$. If n is odd, the domain is the set of all real numbers, or $(-\infty, \infty)$. Keep this in mind as we find function values.

EXAMPLE 6   If $f(x) = \sqrt{x-4}$ and $g(x) = \sqrt[3]{x+2}$, find each function value.

a. $f(8)$ **b.** $f(6)$ **c.** $g(-1)$ **d.** $g(1)$

Solution

a. $f(8) = \sqrt{8-4} = \sqrt{4} = 2$ **b.** $f(6) = \sqrt{6-4} = \sqrt{2}$

c. $g(-1) = \sqrt[3]{-1+2} = \sqrt[3]{1} = 1$ **d.** $g(1) = \sqrt[3]{1+2} = \sqrt[3]{3}$ □

PRACTICE

6   If $f(x) = \sqrt{x+5}$ and $g(x) = \sqrt[3]{x-3}$, find each function value.

a. $f(11)$ **b.** $f(-1)$ **c.** $g(11)$ **d.** $g(-6)$

▶ **Helpful Hint**

Notice that for the function $f(x) = \sqrt{x-4}$, the domain includes all real numbers that make the radicand ≥ 0. To see what numbers these are, solve $x - 4 \geq 0$ and find that $x \geq 4$. The domain is $\{x \mid x \geq 4\}$, or $[4, \infty)$.

The domain of the cube root function $g(x) = \sqrt[3]{x+2}$ is the set of real numbers, or $(-\infty, \infty)$.

EXAMPLE 7   Graph the square root function $f(x) = \sqrt{x}$.

Solution   To graph, we identify the domain, evaluate the function for several values of x, plot the resulting points, and connect the points with a smooth curve. Since $\sqrt{x}$ represents the nonnegative square root of x, the domain of this function is the set of all nonnegative numbers, $\{x \mid x \geq 0\}$, or $[0, \infty)$. We have approximated $\sqrt{3}$ in the table on the next page to help us locate the point corresponding to $\left(3, \sqrt{3}\right)$.

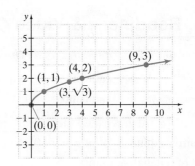

x	$f(x) = \sqrt{x}$
0	0
1	1
3	$\sqrt{3} \approx 1.7$
4	2
9	3

Notice that the graph of this function passes the vertical line test, as expected.   □

PRACTICE
7   Graph the square root function $h(x) = \sqrt{x+2}$.

The equation $f(x) = \sqrt[3]{x}$ also describes a function. Here, x may be any real number, so the domain of this function is the set of all real numbers, or $(-\infty, \infty)$. A few function values are given next.

$$f(0) = \sqrt[3]{0} = 0$$
$$f(1) = \sqrt[3]{1} = 1$$
$$f(-1) = \sqrt[3]{-1} = -1$$
$$\left.\begin{array}{l} f(6) = \sqrt[3]{6} \\ f(-6) = \sqrt[3]{-6} \end{array}\right\}$$ Here, there is no rational number whose cube is 6. Thus, the radicals do not simplify to rational numbers.
$$f(8) = \sqrt[3]{8} = 2$$
$$f(-8) = \sqrt[3]{-8} = -2$$

EXAMPLE 8   Graph the function $f(x) = \sqrt[3]{x}$.

Solution   To graph, we identify the domain, plot points, and connect the points with a smooth curve. The domain of this function is the set of all real numbers. The table comes from the function values obtained earlier. We have approximated $\sqrt[3]{6}$ and $\sqrt[3]{-6}$ for graphing purposes.

x	$f(x) = \sqrt[3]{x}$
0	0
1	1
−1	−1
6	$\sqrt[3]{6} \approx 1.8$
−6	$\sqrt[3]{-6} \approx -1.8$
8	2
−8	−2

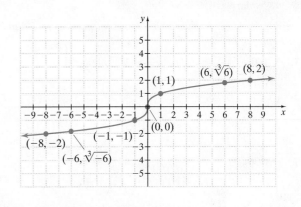

The graph of this function passes the vertical line test, as expected.   □

PRACTICE
8   Graph the function $f(x) = \sqrt[3]{x} - 4$.

Vocabulary, Readiness & Video Check

Use the choices below to fill in each blank. Not all choices will be used.

is	cubes	$-\sqrt{a}$	radical sign	index
is not	squares	$\sqrt{-a}$	radicand	

1. In the expression $\sqrt[n]{a}$, the n is called the _____, the $\sqrt{}$ is called the _____, and a is called the _____.

2. If $\sqrt{a}$ is the positive square root of a, $a \neq 0$, then _____ is the negative square root of a.

3. The square root of a negative number _____ a real number.

4. Numbers such as 1, 4, 9, and 25 are called perfect _____, whereas numbers such as 1, 8, 27, and 125 are called perfect _____.

Fill in the blank.

5. The domain of the function $f(x) = \sqrt{x}$ is _____.

6. The domain of the function $f(x) = \sqrt[3]{x}$ is _____.

7. If $f(16) = 4$, the corresponding ordered pair is _____.

8. If $g(-8) = -2$, the corresponding ordered pair is _____.

Martin-Gay Interactive Videos

See Video 7.1

Watch the section lecture video and answer the following questions.

OBJECTIVE 1
9. From Examples 5 and 6, when simplifying radicals containing variables with exponents, describe a shortcut you can use.

OBJECTIVE 2
10. From Example 9, how can you determine a reasonable approximation for a non-perfect square root without using a calculator?

OBJECTIVE 3
11. From Example 11, what is an important difference between the square root and the cube root of a negative number?

OBJECTIVE 4
12. From Example 12, what conclusion is made about the even root of a negative number?

OBJECTIVE 5
13. From the lecture before Example 17, why do you think no absolute value bars are used when n is odd?

OBJECTIVE 6
14. In Example 19, the domain is found by looking at the graph. How can the domain be found by looking at the function?

7.1 Exercise Set   MyMathLab®

Simplify. Assume that variables represent positive real numbers. See Example 1.

1. $\sqrt{100}$
2. $\sqrt{400}$

3. $\sqrt{\dfrac{1}{4}}$
4. $\sqrt{\dfrac{9}{25}}$

5. $\sqrt{0.0001}$
6. $\sqrt{0.04}$

7. $-\sqrt{36}$
8. $-\sqrt{9}$

9. $\sqrt{x^{10}}$
10. $\sqrt{x^{16}}$

11. $\sqrt{16y^6}$
12. $\sqrt{64y^{20}}$

Use a calculator to approximate each square root to 3 decimal places. Check to see that each approximation is reasonable. See Example 2.

13. $\sqrt{7}$
14. $\sqrt{11}$

▶15. $\sqrt{38}$
16. $\sqrt{56}$

17. $\sqrt{200}$
18. $\sqrt{300}$

Find each cube root. See Example 3.

19. $\sqrt[3]{64}$
20. $\sqrt[3]{27}$

▶21. $\sqrt[3]{\dfrac{1}{8}}$
22. $\sqrt[3]{\dfrac{27}{64}}$

23. $\sqrt[3]{-1}$
24. $\sqrt[3]{-125}$

25. $\sqrt[3]{x^{12}}$
26. $\sqrt[3]{x^{15}}$

▶27. $\sqrt[3]{-27x^9}$
28. $\sqrt[3]{-64x^6}$

Find each root. Assume that all variables represent nonnegative real numbers. See Example 4.

29. $-\sqrt[4]{16}$ **30.** $\sqrt[5]{-243}$

31. $\sqrt[4]{-16}$ **32.** $\sqrt{-16}$

33. $\sqrt[5]{-32}$ **34.** $\sqrt[5]{-1}$

35. $\sqrt[5]{x^{20}}$ **36.** $\sqrt[4]{x^{20}}$

37. $\sqrt[6]{64x^{12}}$ **38.** $\sqrt[5]{-32x^{15}}$

39. $\sqrt{81x^4}$ **40.** $\sqrt[4]{81x^4}$

41. $\sqrt[4]{256x^8}$ **42.** $\sqrt{256x^8}$

Simplify. Assume that the variables represent any real number. See Example 5.

43. $\sqrt{(-8)^2}$ **44.** $\sqrt{(-7)^2}$

45. $\sqrt[3]{(-8)^3}$ **46.** $\sqrt[5]{(-7)^5}$

47. $\sqrt{4x^2}$ **48.** $\sqrt[4]{16x^4}$

49. $\sqrt[3]{x^3}$ **50.** $\sqrt[5]{x^5}$

51. $\sqrt{(x-5)^2}$ **52.** $\sqrt{(y-6)^2}$

53. $\sqrt{x^2+4x+4}$ **54.** $\sqrt{x^2-8x+16}$
(*Hint:* Factor the polynomial first.)   (*Hint:* Factor the polynomial first.)

MIXED PRACTICE

Simplify each radical. Assume that all variables represent positive real numbers.

55. $-\sqrt{121}$ **56.** $-\sqrt[3]{125}$

57. $\sqrt[3]{8x^3}$ **58.** $\sqrt{16x^8}$

59. $\sqrt{y^{12}}$ **60.** $\sqrt[3]{y^{12}}$

61. $\sqrt{25a^2b^{20}}$ **62.** $\sqrt{9x^4y^6}$

63. $\sqrt[3]{-27x^{12}y^9}$ **64.** $\sqrt[3]{-8a^{21}b^6}$

65. $\sqrt[4]{a^{16}b^4}$ **66.** $\sqrt[4]{x^8y^{12}}$

67. $\sqrt[5]{-32x^{10}y^5}$ **68.** $\sqrt[5]{-243x^5z^{15}}$

69. $\sqrt{\dfrac{25}{49}}$ **70.** $\sqrt{\dfrac{4}{81}}$

71. $\sqrt{\dfrac{x^{20}}{4y^2}}$ **72.** $\sqrt{\dfrac{y^{10}}{9x^6}}$

73. $-\sqrt[3]{\dfrac{z^{21}}{27x^3}}$ **74.** $-\sqrt[3]{\dfrac{64a^3}{b^9}}$

75. $\sqrt[4]{\dfrac{x^4}{16}}$ **76.** $\sqrt[4]{\dfrac{y^4}{81x^4}}$

If $f(x)=\sqrt{2x+3}$ and $g(x)=\sqrt[3]{x-8}$, find the following function values. See Example 6.

77. $f(0)$ **78.** $g(0)$

79. $g(7)$ **80.** $f(-1)$

81. $g(-19)$ **82.** $f(3)$

83. $f(2)$ **84.** $g(1)$

Identify the domain and then graph each function. See Example 7.

85. $f(x)=\sqrt{x}+2$ **86.** $f(x)=\sqrt{x}-2$

87. $f(x)=\sqrt{x-3}$; use the following table.

x	$f(x)$
3	
4	
7	
12	

88. $f(x)=\sqrt{x+1}$; use the following table.

x	$f(x)$
-1	
0	
3	
8	

Identify the domain and then graph each function. See Example 8.

89. $f(x)=\sqrt[3]{x}+1$

90. $f(x)=\sqrt[3]{x}-2$

91. $g(x)=\sqrt[3]{x-1}$; use the following table.

x	$g(x)$
1	
2	
0	
9	
-7	

92. $g(x)=\sqrt[3]{x+1}$; use the following table.

x	$g(x)$
-1	
0	
-2	
7	
-9	

REVIEW AND PREVIEW

Simplify each exponential expression. See Sections 5.1 and 5.2.

93. $(-2x^3y^2)^5$

94. $(4y^6z^7)^3$

95. $(-3x^2y^3z^5)(20x^5y^7)$

96. $(-14a^5bc^2)(2abc^4)$

97. $\dfrac{7x^{-1}y}{14(x^5y^2)^{-2}}$

98. $\dfrac{(2a^{-1}b^2)^3}{(8a^2b)^{-2}}$

CONCEPT EXTENSIONS

Determine whether the following are real numbers. See the Concept Check in this section.

99. $\sqrt{-17}$

100. $\sqrt[3]{-17}$

101. $\sqrt[10]{-17}$

102. $\sqrt[15]{-17}$

Choose the correct letter or letters. No pencil is needed, just think your way through these.

103. Which radical is not a real number?

    a. $\sqrt{3}$ **b.** $-\sqrt{11}$ **c.** $\sqrt[3]{-10}$ **d.** $\sqrt{-10}$

104. Which radical(s) simplify to 3?

    a. $\sqrt{9}$ **b.** $\sqrt{-9}$ **c.** $\sqrt[3]{27}$ **d.** $\sqrt[3]{-27}$

105. Which radical(s) simplify to -3?

    a. $\sqrt{9}$ **b.** $\sqrt{-9}$ **c.** $\sqrt[3]{27}$ **d.** $\sqrt[3]{-27}$

106. Which radical does not simplify to a whole number?

    a. $\sqrt{64}$ **b.** $\sqrt[3]{64}$ **c.** $\sqrt{8}$ **d.** $\sqrt[3]{8}$

For Exercises 107 through 110, do not use a calculator.

107. $\sqrt{160}$ is closest to

    a. 10   **b.** 13   **c.** 20   **d.** 40

108. $\sqrt{1000}$ is closest to

    a. 10   **b.** 30   **c.** 100   **d.** 500

△ **109.** The perimeter of the triangle is closest to

    a. 12   **b.** 18
    c. 66   **d.** 132

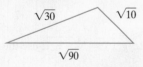

110. The length of the bent wire is closest to

    a. 5   **b.** $\sqrt{28}$
    c. 7   **d.** 14

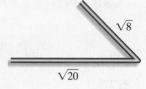

111. Explain why $\sqrt{-64}$ is not a real number.

112. Explain why $\sqrt[3]{-64}$ is a real number.

The Mosteller formula for calculating adult body surface area is $B = \sqrt{\dfrac{hw}{3131}}$, where B is an individual's body surface area in square meters, h is the individual's height in inches, and w is the individual's weight in pounds. Use this information to answer Exercises 113 and 114. Round answers to 2 decimal places.

△ **113.** Find the body surface area of an individual who is 66 inches tall and who weighs 135 pounds.

△ **114.** Find the body surface area of an individual who is 74 inches tall and who weighs 225 pounds.

115. Escape velocity is the minimum speed that an object must reach to escape the pull of a planet's gravity. Escape velocity v is given by the equation $v = \sqrt{\dfrac{2Gm}{r}}$, where m is the mass of the planet, r is its radius, and G is the universal gravitational constant, which has a value of $G = 6.67 \times 10^{-11}$ m³/kg·s². The mass of Earth is 5.97×10^{24} kg, and its radius is 6.37×10^6 m. Use this information to find the escape velocity for Earth in meters per second. Round to the nearest whole number. (*Source:* National Space Science Data Center)

116. Use the formula from Exercise 115 to determine the escape velocity for the moon. The mass of the moon is 7.35×10^{22} kg, and its radius is 1.74×10^6 m. Round to the nearest whole number. (*Source:* National Space Science Data Center)

117. Suppose a classmate tells you that $\sqrt{13} \approx 5.7$. Without a calculator, how can you convince your classmate that he or she must have made an error?

118. Suppose a classmate tells you that $\sqrt[3]{10} \approx 3.2$. Without a calculator, how can you convince your friend that he or she must have made an error?

Use a graphing calculator to verify the domain of each function and its graph.

119. Exercise 85

120. Exercise 86

121. Exercise 89

122. Exercise 90

7.2 | Rational Exponents ▶

OBJECTIVE

1   Understanding the Meaning of $a^{1/n}$ ▶

So far in this text, we have not defined expressions with rational exponents such as $3^{1/2}$, $x^{2/3}$, and $-9^{-1/4}$. We will define these expressions so that the rules for exponents will apply to these rational exponents as well.

Suppose that $x = 5^{1/3}$. Then

$$x^3 = (5^{1/3})^3 = 5^{1/3 \cdot 3} = 5^1 \text{ or } 5$$

using rules
for exponents

Since $x^3 = 5$, x is the number whose cube is 5, or $x = \sqrt[3]{5}$. Notice that we also know that $x = 5^{1/3}$. This means

$$5^{1/3} = \sqrt[3]{5}$$

> **Definition of $a^{1/n}$**
>
> If n is a positive integer greater than 1 and $\sqrt[n]{a}$ is a real number, then
>
> $$a^{1/n} = \sqrt[n]{a}$$

Notice that the denominator of the rational exponent corresponds to the index of the radical.

EXAMPLE 1   Use radical notation to write the following. Simplify if possible.

a. $4^{1/2}$ **b.** $64^{1/3}$ **c.** $x^{1/4}$ **d.** $0^{1/6}$ **e.** $-9^{1/2}$ **f.** $(81x^8)^{1/4}$ **g.** $5y^{1/3}$

Solution

a. $4^{1/2} = \sqrt{4} = 2$ **b.** $64^{1/3} = \sqrt[3]{64} = 4$

c. $x^{1/4} = \sqrt[4]{x}$ **d.** $0^{1/6} = \sqrt[6]{0} = 0$

e. $-9^{1/2} = -\sqrt{9} = -3$ **f.** $(81x^8)^{1/4} = \sqrt[4]{81x^8} = 3x^2$

g. $5y^{1/3} = 5\sqrt[3]{y}$ □

PRACTICE

1   Use radical notation to write the following. Simplify if possible.

a. $36^{1/2}$ **b.** $1000^{1/3}$ **c.** $x^{1/3}$ **d.** $1^{1/4}$

e. $-64^{1/2}$ **f.** $(125x^9)^{1/3}$ **g.** $3x^{1/4}$

OBJECTIVE

2   Understanding the Meaning of $a^{m/n}$ ▶

As we expand our use of exponents to include $\dfrac{m}{n}$, we define their meaning so that rules for exponents still hold true. For example, by properties of exponents,

$$8^{2/3} = (8^{1/3})^2 = (\sqrt[3]{8})^2 \qquad \text{or}$$
$$8^{2/3} = (8^2)^{1/3} = \sqrt[3]{8^2}$$

> **Definition of $a^{m/n}$**
>
> If m and n are positive integers greater than 1 with $\dfrac{m}{n}$ in simplest form, then
>
> $$a^{m/n} = \sqrt[n]{a^m} = (\sqrt[n]{a})^m$$
>
> as long as $\sqrt[n]{a}$ is a real number.

Notice that the denominator n of the rational exponent corresponds to the index of the radical. The numerator m of the rational exponent indicates that the base is to be raised to the mth power. This means

$$8^{2/3} = \sqrt[3]{8^2} = \sqrt[3]{64} = 4 \qquad \text{or}$$

$$8^{2/3} = \left(\sqrt[3]{8}\right)^2 = 2^2 = 4$$

From simplifying $8^{2/3}$, can you see that it doesn't matter whether you raise to a power first and then take the nth root or you take the nth root first and then raise to a power?

> ▶ **Helpful Hint**
> Most of the time, $\left(\sqrt[n]{a}\right)^m$ will be easier to calculate than $\sqrt[n]{a^m}$.

EXAMPLE 2 Use radical notation to write the following. Then simplify if possible.

a. $4^{3/2}$ **b.** $-16^{3/4}$ **c.** $(-27)^{2/3}$

d. $\left(\dfrac{1}{9}\right)^{3/2}$ **e.** $(4x-1)^{3/5}$

Solution

a. $4^{3/2} = \left(\sqrt{4}\right)^3 = 2^3 = 8$ **b.** $-16^{3/4} = -\left(\sqrt[4]{16}\right)^3 = -(2)^3 = -8$

c. $(-27)^{2/3} = \left(\sqrt[3]{-27}\right)^2 = (-3)^2 = 9$ **d.** $\left(\dfrac{1}{9}\right)^{3/2} = \left(\sqrt{\dfrac{1}{9}}\right)^3 = \left(\dfrac{1}{3}\right)^3 = \dfrac{1}{27}$

e. $(4x-1)^{3/5} = \sqrt[5]{(4x-1)^3}$

PRACTICE
2 Use radical notation to write the following. Simplify if possible.

a. $16^{3/2}$ **b.** $-1^{3/5}$ **c.** $-(81)^{3/4}$

d. $\left(\dfrac{1}{25}\right)^{3/2}$ **e.** $(3x+2)^{5/9}$

> ▶ **Helpful Hint**
> The *denominator* of a rational exponent is the index of the corresponding radical. For example, $x^{1/5} = \sqrt[5]{x}$ and $z^{2/3} = \sqrt[3]{z^2}$, or $z^{2/3} = \left(\sqrt[3]{z}\right)^2$.

OBJECTIVE
3 Understanding the Meaning of $a^{-m/n}$

The rational exponents we have given meaning to exclude negative rational numbers. To complete the set of definitions, we define $a^{-m/n}$.

> **Definition of $a^{-m/n}$**
>
> $$a^{-m/n} = \dfrac{1}{a^{m/n}}$$
>
> as long as $a^{m/n}$ is a nonzero real number.

EXAMPLE 3 Write each expression with a positive exponent, and then simplify.

a. $16^{-3/4}$ **b.** $(-27)^{-2/3}$

Solution

a. $16^{-3/4} = \dfrac{1}{16^{3/4}} = \dfrac{1}{\left(\sqrt[4]{16}\right)^3} = \dfrac{1}{2^3} = \dfrac{1}{8}$

b. $(-27)^{-2/3} = \dfrac{1}{(-27)^{2/3}} = \dfrac{1}{\left(\sqrt[3]{-27}\right)^2} = \dfrac{1}{(-3)^2} = \dfrac{1}{9}$

PRACTICE

3   Write each expression with a positive exponent; then simplify.

a. $9^{-3/2}$

b. $(-64)^{-2/3}$

▶ **Helpful Hint**

If an expression contains a negative rational exponent, such as $9^{-3/2}$, you may want to first write the expression with a positive exponent and then interpret the rational exponent. Notice that the sign of the base is not affected by the sign of its exponent. For example,

$$9^{-3/2} = \frac{1}{9^{3/2}} = \frac{1}{(\sqrt{9})^3} = \frac{1}{27}$$

Also,

$$(-27)^{-1/3} = \frac{1}{(-27)^{1/3}} = -\frac{1}{3}$$

✓ **CONCEPT CHECK**
Which one is correct?

a. $-8^{2/3} = \frac{1}{4}$

b. $8^{-2/3} = -\frac{1}{4}$

c. $8^{-2/3} = -4$

d. $-8^{-2/3} = -\frac{1}{4}$

OBJECTIVE

4   Using Rules for Exponents to Simplify Expressions

It can be shown that the properties of integer exponents hold for rational exponents. By using these properties and definitions, we can now simplify expressions that contain rational exponents.

These rules are repeated here for review.

Note: For the remainder of this chapter, we will assume that variables represent positive real numbers. Since this is so, we need not insert absolute value bars when we simplify even roots.

Summary of Exponent Rules

If m and n are rational numbers, and a, b, and c are numbers for which the expressions below exist, then

Product rule for exponents: $a^m \cdot a^n = a^{m+n}$

Power rule for exponents: $(a^m)^n = a^{m \cdot n}$

Power rules for products and quotients: $(ab)^n = a^n b^n$ and

$$\left(\frac{a}{c}\right)^n = \frac{a^n}{c^n}, c \neq 0$$

Quotient rule for exponents: $\dfrac{a^m}{a^n} = a^{m-n}, a \neq 0$

Zero exponent: $a^0 = 1, a \neq 0$

Negative exponent: $a^{-n} = \dfrac{1}{a^n}, a \neq 0$

EXAMPLE 4   Use properties of exponents to simplify. Write results with only positive exponents.

a. $b^{1/3} \cdot b^{5/3}$

b. $x^{1/2} x^{1/3}$

c. $\dfrac{7^{1/3}}{7^{4/3}}$

d. $y^{-4/7} \cdot y^{6/7}$

e. $\dfrac{(2x^{2/5}y^{-1/3})^5}{x^2 y}$

(Continued on next page)

Solution

a. $b^{1/3} \cdot b^{5/3} = b^{(1/3+5/3)} = b^{6/3} = b^2$

b. $x^{1/2}x^{1/3} = x^{(1/2+1/3)} = x^{3/6+2/6} = x^{5/6}$ Use the product rule.

c. $\dfrac{7^{1/3}}{7^{4/3}} = 7^{1/3-4/3} = 7^{-3/3} = 7^{-1} = \dfrac{1}{7}$ Use the quotient rule.

d. $y^{-4/7} \cdot y^{6/7} = y^{-4/7+6/7} = y^{2/7}$ Use the product rule.

e. We begin by using the power rule $(ab)^m = a^m b^m$ to simplify the numerator.

$$\frac{(2x^{2/5}y^{-1/3})^5}{x^2y} = \frac{2^5(x^{2/5})^5(y^{-1/3})^5}{x^2y} = \frac{32x^2y^{-5/3}}{x^2y} \quad \text{Use the power rule and simplify}$$

$$= 32x^{2-2}y^{-5/3-3/3} \qquad \text{Apply the quotient rule.}$$

$$= 32x^0y^{-8/3}$$

$$= \frac{32}{y^{8/3}}$$

PRACTICE
4   Use properties of exponents to simplify.

a. $y^{2/3} \cdot y^{8/3}$ **b.** $x^{3/5} \cdot x^{1/4}$ **c.** $\dfrac{9^{2/7}}{9^{9/7}}$

d. $b^{4/9} \cdot b^{-2/9}$ **e.** $\dfrac{(3x^{1/4}y^{-2/3})^4}{x^4y}$

EXAMPLE 5   Multiply.

a. $z^{2/3}(z^{1/3} - z^5)$ **b.** $(x^{1/3} - 5)(x^{1/3} + 2)$

Solution

a. $z^{2/3}(z^{1/3} - z^5) = z^{2/3}z^{1/3} - z^{2/3}z^5$ Apply the distributive property.

$\qquad = z^{(2/3+1/3)} - z^{(2/3+5)}$ Use the product rule.

$\qquad = z^{3/3} - z^{(2/3+15/3)}$

$\qquad = z - z^{17/3}$

b. $(x^{1/3} - 5)(x^{1/3} + 2) = x^{2/3} + 2x^{1/3} - 5x^{1/3} - 10$ Think of $(x^{1/3} - 5)$ and $(x^{1/3} + 2)$ as 2 binomials, and FOIL.

$\qquad = x^{2/3} - 3x^{1/3} - 10$

PRACTICE
5   Multiply.

a. $x^{3/5}(x^{1/3} - x^2)$ **b.** $(x^{1/2} + 6)(x^{1/2} - 2)$

EXAMPLE 6   Factor $x^{-1/2}$ from the expression $3x^{-1/2} - 7x^{5/2}$. Assume that all variables represent positive numbers.

Solution

$$3x^{-1/2} - 7x^{5/2} = (x^{-1/2})(3) - (x^{-1/2})(7x^{6/2})$$

$$= x^{-1/2}(3 - 7x^3)$$

To check, multiply $x^{-1/2}(3 - 7x^3)$ to see that the product is $3x^{-1/2} - 7x^{5/2}$.

PRACTICE
6   Factor $x^{-1/5}$ from the expression $2x^{-1/5} - 7x^{4/5}$.

OBJECTIVE

5   Using Rational Exponents to Simplify Radical Expressions

Some radical expressions are easier to simplify when we first write them with rational exponents. We can simplify some radical expressions by first writing the expression with rational exponents. Use properties of exponents to simplify, and then convert back to radical notation.

EXAMPLE 7   Use rational exponents to simplify. Assume that variables represent positive numbers.

a. $\sqrt[8]{x^4}$ **b.** $\sqrt[6]{25}$ **c.** $\sqrt[4]{r^2 s^6}$

Solution

a. $\sqrt[8]{x^4} = x^{4/8} = x^{1/2} = \sqrt{x}$

b. $\sqrt[6]{25} = 25^{1/6} = (5^2)^{1/6} = 5^{2/6} = 5^{1/3} = \sqrt[3]{5}$

c. $\sqrt[4]{r^2 s^6} = (r^2 s^6)^{1/4} = r^{2/4} s^{6/4} = r^{1/2} s^{3/2} = (rs^3)^{1/2} = \sqrt{rs^3}$ □

PRACTICE

7   Use rational exponents to simplify. Assume that the variables represent positive numbers.

a. $\sqrt[9]{x^3}$ **b.** $\sqrt[4]{36}$ **c.** $\sqrt[8]{a^4 b^2}$

EXAMPLE 8   Use rational exponents to write as a single radical.

a. $\sqrt{x} \cdot \sqrt[4]{x}$ **b.** $\dfrac{\sqrt{x}}{\sqrt[3]{x}}$ **c.** $\sqrt[3]{3} \cdot \sqrt{2}$

Solution

a. $\sqrt{x} \cdot \sqrt[4]{x} = x^{1/2} \cdot x^{1/4} = x^{1/2 + 1/4}$

$\qquad = x^{3/4} = \sqrt[4]{x^3}$

b. $\dfrac{\sqrt{x}}{\sqrt[3]{x}} = \dfrac{x^{1/2}}{x^{1/3}} = x^{1/2 - 1/3} = x^{3/6 - 2/6}$

$\qquad = x^{1/6} = \sqrt[6]{x}$

c. $\sqrt[3]{3} \cdot \sqrt{2} = 3^{1/3} \cdot 2^{1/2}$ Write with rational exponents.

$\qquad = 3^{2/6} \cdot 2^{3/6}$ Write the exponents so that they have the same denominator.

$\qquad = (3^2 \cdot 2^3)^{1/6}$ Use $a^n b^n = (ab)^n$

$\qquad = \sqrt[6]{3^2 \cdot 2^3}$ Write with radical notation.

$\qquad = \sqrt[6]{72}$ Multiply $3^2 \cdot 2^3$.   □

PRACTICE

8   Use rational expressions to write each of the following as a single radical.

a. $\sqrt[3]{x} \cdot \sqrt[4]{x}$ **b.** $\dfrac{\sqrt[3]{y}}{\sqrt[5]{y}}$ **c.** $\sqrt[3]{5} \cdot \sqrt{3}$

Vocabulary, Readiness & Video Check

Answer each true or false.

1. $9^{-1/2}$ is a positive number. _____

2. $9^{-1/2}$ is a whole number. _____

3. $\dfrac{1}{a^{-m/n}} = a^{m/n}$ (where $a^{m/n}$ is a nonzero real number). _____

Fill in the blank with the correct choice.

4. To simplify $x^{2/3} \cdot x^{1/5}$, _____ the exponents.

    a. add   **b.** subtract   **c.** multiply   **d.** divide

5. To simplify $(x^{2/3})^{1/5}$, _____ the exponents.

    a. add   **b.** subtract   **c.** multiply   **d.** divide

6. To simplify $\dfrac{x^{2/3}}{x^{1/5}}$, _____ the exponents.

    a. add   **b.** subtract   **c.** multiply   **d.** divide

---

Martin-Gay Interactive Videos

See Video 7.2 🍐

Watch the section lecture video and answer the following questions.

OBJECTIVE 1

7. From looking at ▭ Example 2, what is $-(3x)^{1/5}$ in radical notation?

OBJECTIVE 2

8. From ▭ Examples 3 and 4, in a fractional exponent, what do the numerator and denominator each represent in radical form?

OBJECTIVE 3

9. Based on ▭ Example 5, complete the following statements. A negative fractional exponent will move a base from the numerator to the _____ with the fractional exponent becoming _____.

OBJECTIVE 4

10. Based on ▭ Examples 7–9, complete the following statements. Assume you have an expression with fractional exponents. If applying the product rule of exponents, you _____ the exponents. If applying the quotient rule of exponents, you _____ the exponents. If applying the power rule of exponents, you _____ the exponents.

OBJECTIVE 5

11. From ▭ Example 10, describe a way to simplify a radical of a variable raised to a power if the index and the exponent have a common factor.

7.2 Exercise Set   MyMathLab®

Use radical notation to write each expression. Simplify if possible. See Example 1.

▶ **1.** $49^{1/2}$ **2.** $64^{1/3}$

3. $27^{1/3}$ **4.** $8^{1/3}$

5. $\left(\dfrac{1}{16}\right)^{1/4}$ **6.** $\left(\dfrac{1}{64}\right)^{1/2}$

7. $169^{1/2}$ **8.** $81^{1/4}$

▶ **9.** $2m^{1/3}$ **10.** $(2m)^{1/3}$

11. $(9x^4)^{1/2}$ **12.** $(16x^8)^{1/2}$

13. $(-27)^{1/3}$ **14.** $-64^{1/2}$

15. $-16^{1/4}$ **16.** $(-32)^{1/5}$

Use radical notation to write each expression. Simplify if possible. See Example 2.

▶ **17.** $16^{3/4}$ **18.** $4^{5/2}$

▶ **19.** $(-64)^{2/3}$ **20.** $(-8)^{4/3}$

21. $(-16)^{3/4}$ **22.** $(-9)^{3/2}$

23. $(2x)^{3/5}$ **24.** $2x^{3/5}$

25. $(7x+2)^{2/3}$ **26.** $(x-4)^{3/4}$

27. $\left(\dfrac{16}{9}\right)^{3/2}$ **28.** $\left(\dfrac{49}{25}\right)^{3/2}$

Write with positive exponents. Simplify if possible. See Example 3.

▶ **29.** $8^{-4/3}$ **30.** $64^{-2/3}$

31. $(-64)^{-2/3}$ **32.** $(-8)^{-4/3}$

33. $(-4)^{-3/2}$ **34.** $(-16)^{-5/4}$

▶ **35.** $x^{-1/4}$ **36.** $y^{-1/6}$

37. $\dfrac{1}{a^{-2/3}}$ **38.** $\dfrac{1}{n^{-8/9}}$

39. $\dfrac{5}{7x^{-3/4}}$ **40.** $\dfrac{2}{3y^{-5/7}}$

Use the properties of exponents to simplify each expression. Write with positive exponents. See Example 4.

▶ **41.** $a^{2/3}a^{5/3}$ **42.** $b^{9/5}b^{8/5}$

43. $x^{-2/5} \cdot x^{7/5}$ **44.** $y^{4/3} \cdot y^{-1/3}$

45. $3^{1/4} \cdot 3^{3/8}$ **46.** $5^{1/2} \cdot 5^{1/6}$

47. $\dfrac{y^{1/3}}{y^{1/6}}$ **48.** $\dfrac{x^{3/4}}{x^{1/8}}$

49. $(4u^2)^{3/2}$ **50.** $(32^{1/5}x^{2/3})^3$

51. $\dfrac{b^{1/2}b^{3/4}}{-b^{1/4}}$ **52.** $\dfrac{a^{1/4}a^{-1/2}}{a^{2/3}}$

53. $\dfrac{(x^3)^{1/2}}{x^{7/2}}$ **54.** $\dfrac{y^{11/3}}{(y^5)^{1/3}}$

▶ **55.** $\dfrac{(3x^{1/4})^3}{x^{1/12}}$

56. $\dfrac{(2x^{1/5})^4}{x^{3/10}}$

57. $\dfrac{(y^3z)^{1/6}}{y^{-1/2}z^{1/3}}$

58. $\dfrac{(m^2n)^{1/4}}{m^{-1/2}n^{5/8}}$

59. $\dfrac{(x^3y^2)^{1/4}}{(x^{-5}y^{-1})^{-1/2}}$

60. $\dfrac{(a^{-2}b^3)^{1/8}}{(a^{-3}b)^{-1/4}}$

Multiply. See Example 5.

61. $y^{1/2}(y^{1/2} - y^{2/3})$

62. $x^{1/2}(x^{1/2} + x^{3/2})$

▶ **63.** $x^{2/3}(x - 2)$

64. $3x^{1/2}(x + y)$

65. $(2x^{1/3} + 3)(2x^{1/3} - 3)$

66. $(y^{1/2} + 5)(y^{1/2} + 5)$

Factor the given factor from the expression. See Example 6.

67. $x^{8/3}; x^{8/3} + x^{10/3}$

68. $x^{3/2}; x^{5/2} - x^{3/2}$

69. $x^{1/5}; x^{2/5} - 3x^{1/5}$

70. $x^{2/7}; x^{3/7} - 2x^{2/7}$

71. $x^{-1/3}; 5x^{-1/3} + x^{2/3}$

72. $x^{-3/4}; x^{-3/4} + 3x^{1/4}$

Use rational exponents to simplify each radical. Assume that all variables represent positive numbers. See Example 7.

▶ **73.** $\sqrt[6]{x^3}$

74. $\sqrt[9]{a^3}$

75. $\sqrt[6]{4}$

76. $\sqrt[4]{36}$

▶ **77.** $\sqrt[4]{16x^2}$

78. $\sqrt[8]{4y^2}$

79. $\sqrt[8]{x^4y^4}$

80. $\sqrt[9]{y^6z^3}$

81. $\sqrt[12]{a^8b^4}$

82. $\sqrt[10]{a^5b^5}$

83. $\sqrt[4]{(x + 3)^2}$

84. $\sqrt[8]{(y + 1)^4}$

Use rational expressions to write as a single radical expression. See Example 8.

85. $\sqrt[3]{y} \cdot \sqrt[5]{y^2}$

86. $\sqrt[3]{y^2} \cdot \sqrt[6]{y}$

87. $\dfrac{\sqrt[3]{b^2}}{\sqrt[4]{b}}$

88. $\dfrac{\sqrt[4]{a}}{\sqrt[5]{a}}$

89. $\sqrt[3]{x} \cdot \sqrt[4]{x} \cdot \sqrt[8]{x^3}$

90. $\sqrt[6]{y} \cdot \sqrt[3]{y} \cdot \sqrt[5]{y^2}$

91. $\dfrac{\sqrt[3]{a^2}}{\sqrt[6]{a}}$

92. $\dfrac{\sqrt[5]{b^2}}{\sqrt[10]{b^3}}$

93. $\sqrt{3} \cdot \sqrt[3]{4}$

94. $\sqrt[3]{5} \cdot \sqrt{2}$

95. $\sqrt[5]{7} \cdot \sqrt[3]{y}$

96. $\sqrt[4]{5} \cdot \sqrt[3]{x}$

97. $\sqrt{5r} \cdot \sqrt[3]{s}$

98. $\sqrt[3]{b} \cdot \sqrt[5]{4a}$

REVIEW AND PREVIEW

Write each integer as a product of two integers such that one of the factors is a perfect square. For example, write 18 as $9 \cdot 2$ because 9 is a perfect square.

99. 75

100. 20

101. 48

102. 45

Write each integer as a product of two integers such that one of the factors is a perfect cube. For example, write 24 as $8 \cdot 3$ because 8 is a perfect cube.

103. 16

104. 56

105. 54

106. 80

CONCEPT EXTENSIONS

Choose the correct letter for each exercise. Letters will be used more than once. No pencil is needed. Just think about the meaning of each expression.

$A = 2, B = -2, C =$ not a real number

107. $4^{1/2}$ _____

108. $-4^{1/2}$ _____

109. $(-4)^{1/2}$ _____

110. $8^{1/3}$ _____

111. $-8^{1/3}$ _____

112. $(-8)^{1/3}$ _____

Basal metabolic rate (BMR) is the number of calories per day a person needs to maintain life. A person's basal metabolic rate $B(w)$ in calories per day can be estimated with the function $B(w) = 70w^{3/4}$, where w is the person's weight in kilograms. Use this information to answer Exercises 113 and 114.

113. Estimate the BMR for a person who weighs 60 kilograms. Round to the nearest calorie. (*Note:* 60 kilograms is approximately 132 pounds.)

114. Estimate the BMR for a person who weighs 90 kilograms. Round to the nearest calorie. (*Note:* 90 kilograms is approximately 198 pounds.)

The number of cell telephone subscribers in the United States from 1995–2010 can be modeled by $f(x) = 25x^{23/25}$, where $f(x)$ is the number of cellular telephone subscriptions in millions, x years after 1995. (Source: CTIA-Wireless Association, 1995–2010) Use this information to answer Exercises 115 and 116.

115. Use this model to estimate the number of cellular subscriptions in 2010. Round to the nearest tenth of a million.

116. Predict the number of cellular telephone subscriptions in 2015. Round to the nearest tenth of a million.

117. Explain how writing x^{-7} with positive exponents is similar to writing $x^{-1/4}$ with positive exponents.

118. Explain how writing $2x^{-5}$ with positive exponents is similar to writing $2x^{-3/4}$ with positive exponents.

Fill in each box with the correct expression.

119. $\square \cdot a^{2/3} = a^{3/3}$, or a

120. $\square \cdot x^{1/8} = x^{4/8}$, or $x^{1/2}$

121. $\dfrac{\square}{x^{-2/5}} = x^{3/5}$

122. $\dfrac{\square}{y^{-3/4}} = y^{4/4}$, or y

Use a calculator to write a four-decimal-place approximation of each number.

123. $8^{1/4}$

124. $20^{1/5}$

125. $18^{3/5}$

126. $76^{5/7}$

127. In physics, the speed of a wave traveling over a stretched string with tension t and density u is given by the expression $\dfrac{\sqrt{t}}{\sqrt{u}}$. Write this expression with rational exponents.

128. In electronics, the angular frequency of oscillations in a certain type of circuit is given by the expression $(LC)^{-1/2}$. Use radical notation to write this expression.

7.3 Simplifying Radical Expressions

OBJECTIVES

1 Use the Product Rule for Radicals.

2 Use the Quotient Rule for Radicals.

3 Simplify Radicals.

4 Use the Distance and Midpoint Formulas.

OBJECTIVE

1 Using the Product Rule

It is possible to simplify some radicals that do not evaluate to rational numbers. To do so, we use a product rule and a quotient rule for radicals. To discover the product rule, notice the following pattern.

$$\sqrt{9} \cdot \sqrt{4} = 3 \cdot 2 = 6$$
$$\sqrt{9 \cdot 4} = \sqrt{36} = 6$$

Since both expressions simplify to 6, it is true that

$$\sqrt{9} \cdot \sqrt{4} = \sqrt{9 \cdot 4}$$

This pattern suggests the following product rule for radicals.

Product Rule for Radicals

If $\sqrt[n]{a}$ and $\sqrt[n]{b}$ are real numbers, then

$$\sqrt[n]{a} \cdot \sqrt[n]{b} = \sqrt[n]{ab}$$

Notice that the product rule is the relationship $a^{1/n} \cdot b^{1/n} = (ab)^{1/n}$ stated in radical notation.

EXAMPLE 1 Multiply.

a. $\sqrt{3} \cdot \sqrt{5}$ **b.** $\sqrt{21} \cdot \sqrt{x}$ **c.** $\sqrt[3]{4} \cdot \sqrt[3]{2}$

d. $\sqrt[4]{5y^2} \cdot \sqrt[4]{2x^3}$ **e.** $\sqrt{\dfrac{2}{a}} \cdot \sqrt{\dfrac{b}{3}}$

Solution

a. $\sqrt{3} \cdot \sqrt{5} = \sqrt{3 \cdot 5} = \sqrt{15}$

b. $\sqrt{21} \cdot \sqrt{x} = \sqrt{21x}$

c. $\sqrt[3]{4} \cdot \sqrt[3]{2} = \sqrt[3]{4 \cdot 2} = \sqrt[3]{8} = 2$

d. $\sqrt[4]{5y^2} \cdot \sqrt[4]{2x^3} = \sqrt[4]{5y^2 \cdot 2x^3} = \sqrt[4]{10y^2x^3}$

e. $\sqrt{\dfrac{2}{a}} \cdot \sqrt{\dfrac{b}{3}} = \sqrt{\dfrac{2}{a} \cdot \dfrac{b}{3}} = \sqrt{\dfrac{2b}{3a}}$

PRACTICE

1 Multiply.

a. $\sqrt{5} \cdot \sqrt{7}$ **b.** $\sqrt{13} \cdot \sqrt{z}$ **c.** $\sqrt[4]{125} \cdot \sqrt[4]{5}$

d. $\sqrt[3]{5y} \cdot \sqrt[3]{3x^2}$ **e.** $\sqrt{\dfrac{5}{m}} \cdot \sqrt{\dfrac{t}{2}}$

OBJECTIVE

2 Using the Quotient Rule

To discover a quotient rule for radicals, notice the following pattern.

$$\sqrt{\frac{4}{9}} = \frac{2}{3}$$

$$\frac{\sqrt{4}}{\sqrt{9}} = \frac{2}{3}$$

Since both expressions simplify to $\frac{2}{3}$, it is true that

$$\sqrt{\frac{4}{9}} = \frac{\sqrt{4}}{\sqrt{9}}$$

This pattern suggests the following quotient rule for radicals.

> **Quotient Rule for Radicals**
>
> If $\sqrt[n]{a}$ and $\sqrt[n]{b}$ are real numbers and $\sqrt[n]{b}$ is not zero, then
>
> $$\sqrt[n]{\frac{a}{b}} = \frac{\sqrt[n]{a}}{\sqrt[n]{b}}$$

Notice that the quotient rule is the relationship $\left(\dfrac{a}{b}\right)^{1/n} = \dfrac{a^{1/n}}{b^{1/n}}$ stated in radical notation. We can use the quotient rule to simplify radical expressions by reading the rule from left to right or to divide radicals by reading the rule from right to left.

For example,

$$\sqrt{\frac{x}{16}} = \frac{\sqrt{x}}{\sqrt{16}} = \frac{\sqrt{x}}{4} \qquad \text{Using } \sqrt[n]{\frac{a}{b}} = \frac{\sqrt[n]{a}}{\sqrt[n]{b}}$$

$$\frac{\sqrt{75}}{\sqrt{3}} = \sqrt{\frac{75}{3}} = \sqrt{25} = 5 \quad \text{Using } \frac{\sqrt[n]{a}}{\sqrt[n]{b}} = \sqrt[n]{\frac{a}{b}}$$

Note: *Recall that from Section 7.2 on, we assume that variables represent positive real numbers. Since this is so, we need not insert absolute value bars when we simplify even roots.*

EXAMPLE 2 Use the quotient rule to simplify.

a. $\sqrt{\dfrac{25}{49}}$ **b.** $\sqrt{\dfrac{x}{9}}$ **c.** $\sqrt[3]{\dfrac{8}{27}}$ **d.** $\sqrt[4]{\dfrac{3}{16y^4}}$

Solution

a. $\sqrt{\dfrac{25}{49}} = \dfrac{\sqrt{25}}{\sqrt{49}} = \dfrac{5}{7}$ **b.** $\sqrt{\dfrac{x}{9}} = \dfrac{\sqrt{x}}{\sqrt{9}} = \dfrac{\sqrt{x}}{3}$

c. $\sqrt[3]{\dfrac{8}{27}} = \dfrac{\sqrt[3]{8}}{\sqrt[3]{27}} = \dfrac{2}{3}$ **d.** $\sqrt[4]{\dfrac{3}{16y^4}} = \dfrac{\sqrt[4]{3}}{\sqrt[4]{16y^4}} = \dfrac{\sqrt[4]{3}}{2y}$

PRACTICE

2 Use the quotient rule to simplify.

a. $\sqrt{\dfrac{36}{49}}$ **b.** $\sqrt{\dfrac{z}{16}}$ **c.** $\sqrt[3]{\dfrac{125}{8}}$ **d.** $\sqrt[4]{\dfrac{5}{81x^8}}$

OBJECTIVE

3 Simplifying Radicals

Both the product and quotient rules can be used to simplify a radical. If the product rule is read from right to left, we have that

$$\sqrt[n]{ab} = \sqrt[n]{a} \cdot \sqrt[n]{b}.$$

This is used to simplify the following radicals.

EXAMPLE 3 Simplify the following.

a. $\sqrt{50}$ **b.** $\sqrt[3]{24}$ **c.** $\sqrt{26}$ **d.** $\sqrt[4]{32}$

Solution

a. Factor 50 such that one factor is the largest perfect square that divides 50. The largest perfect square factor of 50 is 25, so we write 50 as $25 \cdot 2$ and use the product rule for radicals to simplify.

$$\sqrt{50} = \sqrt{25 \cdot 2} = \sqrt{25} \cdot \sqrt{2} = 5\sqrt{2}$$

         The largest perfect square
         factor of 50

> **Helpful Hint**
> Don't forget that, for example, $5\sqrt{2}$ means $5 \cdot \sqrt{2}$.

b. $\sqrt[3]{24} = \sqrt[3]{8 \cdot 3} = \sqrt[3]{8} \cdot \sqrt[3]{3} = 2\sqrt[3]{3}$

     The largest perfect cube factor of 24

c. $\sqrt{26}$ The largest perfect square factor of 26 is 1, so $\sqrt{26}$ cannot be simplified further.

d. $\sqrt[4]{32} = \sqrt[4]{16 \cdot 2} = \sqrt[4]{16} \cdot \sqrt[4]{2} = 2\sqrt[4]{2}$

     The largest fourth power factor of 32

PRACTICE

3 Simplify the following.

a. $\sqrt{98}$ **b.** $\sqrt[3]{54}$ **c.** $\sqrt{35}$ **d.** $\sqrt[4]{243}$

After simplifying a radical such as a square root, always check the radicand to see that it contains no other perfect square factors. It may, if the largest perfect square factor of the radicand was not originally recognized. For example,

$$\sqrt{200} = \sqrt{4 \cdot 50} = \sqrt{4} \cdot \sqrt{50} = 2\sqrt{50}$$

Notice that the radicand 50 still contains the perfect square factor 25. This is because 4 is not the largest perfect square factor of 200. We continue as follows.

$$2\sqrt{50} = 2\sqrt{25 \cdot 2} = 2 \cdot \sqrt{25} \cdot \sqrt{2} = 2 \cdot 5 \cdot \sqrt{2} = 10\sqrt{2}$$

The radical is now simplified since 2 contains no perfect square factors (other than 1).

> **Helpful Hint**
> To help you recognize largest perfect power factors of a radicand, it will help if you are familiar with some perfect powers. A few are listed below.
>
> | Perfect Squares | 1, | 4, | 9, | 16, | 25, | 36, | 49, | 64, | 81, | 100, | 121, | 144 |
> |---|---|---|---|---|---|---|---|---|---|---|---|---|
> | | 1^2 | 2^2 | 3^2 | 4^2 | 5^2 | 6^2 | 7^2 | 8^2 | 9^2 | 10^2 | 11^2 | 12^2 |
>
> | Perfect Cubes | 1, | 8, | 27, | 64, | 125 |
> |---|---|---|---|---|---|
> | | 1^3 | 2^3 | 3^3 | 4^3 | 5^3 |
>
> | Perfect Fourth Powers | 1, | 16, | 81, | 256 |
> |---|---|---|---|---|
> | | 1^4 | 2^4 | 3^4 | 4^4 |

In general, we say that a radicand of the form $\sqrt[n]{a}$ is simplified when the radicand a contains no factors that are perfect nth powers (other than 1 or −1).

EXAMPLE 4 Use the product rule to simplify.

a. $\sqrt{25x^3}$ b. $\sqrt[3]{54x^6y^8}$ c. $\sqrt[4]{81z^{11}}$

Solution

a. $\sqrt{25x^3} = \sqrt{25x^2 \cdot x}$ Find the largest perfect square factor.

   $= \sqrt{25x^2} \cdot \sqrt{x}$ Apply the product rule.

   $= 5x\sqrt{x}$ Simplify.

b. $\sqrt[3]{54x^6y^8} = \sqrt[3]{27 \cdot 2 \cdot x^6 \cdot y^6 \cdot y^2}$ Factor the radicand and identify perfect cube factors.

   $= \sqrt[3]{27x^6y^6 \cdot 2y^2}$

   $= \sqrt[3]{27x^6y^6} \cdot \sqrt[3]{2y^2}$ Apply the product rule.

   $= 3x^2y^2\sqrt[3]{2y^2}$ Simplify.

c. $\sqrt[4]{81z^{11}} = \sqrt[4]{81 \cdot z^8 \cdot z^3}$ Factor the radicand and identify perfect fourth power factors.

   $= \sqrt[4]{81z^8} \cdot \sqrt[4]{z^3}$ Apply the product rule.

   $= 3z^2\sqrt[4]{z^3}$ Simplify.

PRACTICE
4 Use the product rule to simplify.

a. $\sqrt{36z^7}$ b. $\sqrt[3]{32p^4q^7}$ c. $\sqrt[4]{16x^{15}}$

EXAMPLE 5 Use the quotient rule to divide, and simplify if possible.

a. $\dfrac{\sqrt{20}}{\sqrt{5}}$ b. $\dfrac{\sqrt{50x}}{2\sqrt{2}}$ c. $\dfrac{7\sqrt[3]{48x^4y^8}}{\sqrt[3]{6y^2}}$ d. $\dfrac{2\sqrt[4]{32a^8b^6}}{\sqrt[4]{a^{-1}b^2}}$

Solution

a. $\dfrac{\sqrt{20}}{\sqrt{5}} = \sqrt{\dfrac{20}{5}}$ Apply the quotient rule.

   $= \sqrt{4}$ Simplify.

   $= 2$ Simplify.

b. $\dfrac{\sqrt{50x}}{2\sqrt{2}} = \dfrac{1}{2} \cdot \sqrt{\dfrac{50x}{2}}$ Apply the quotient rule.

   $= \dfrac{1}{2} \cdot \sqrt{25x}$ Simplify.

   $= \dfrac{1}{2} \cdot \sqrt{25} \cdot \sqrt{x}$ Factor 25x.

   $= \dfrac{1}{2} \cdot 5 \cdot \sqrt{x}$ Simplify.

   $= \dfrac{5}{2}\sqrt{x}$

c. $\dfrac{7\sqrt[3]{48x^4y^8}}{\sqrt[3]{6y^2}} = 7 \cdot \sqrt[3]{\dfrac{48x^4y^8}{6y^2}}$ Apply the quotient rule.

   $= 7 \cdot \sqrt[3]{8x^4y^6}$ Simplify.

   $= 7\sqrt[3]{8x^3y^6 \cdot x}$ Factor.

   $= 7 \cdot \sqrt[3]{8x^3y^6} \cdot \sqrt[3]{x}$ Apply the product rule.

(Continued on next page)

$$= 7 \cdot 2xy^2 \cdot \sqrt[3]{x} \quad \text{Simplify.}$$
$$= 14xy^2\sqrt[3]{x}$$

d. $\dfrac{2\sqrt[4]{32a^8b^6}}{\sqrt[4]{a^{-1}b^2}} = 2\sqrt[4]{\dfrac{32a^8b^6}{a^{-1}b^2}} = 2\sqrt[4]{32a^9b^4} = 2\sqrt[4]{16 \cdot a^8 \cdot b^4 \cdot 2 \cdot a}$

$$= 2\sqrt[4]{16a^8b^4} \cdot \sqrt[4]{2a} = 2 \cdot 2a^2b \cdot \sqrt[4]{2a} = 4a^2b\sqrt[4]{2a} \qquad \square$$

PRACTICE
5 Use the quotient rule to divide and simplify.

a. $\dfrac{\sqrt{80}}{\sqrt{5}}$ **b.** $\dfrac{\sqrt{98z}}{3\sqrt{2}}$ **c.** $\dfrac{5\sqrt[3]{40x^5y^7}}{\sqrt[3]{5y}}$ **d.** $\dfrac{3\sqrt[5]{64x^9y^8}}{\sqrt[5]{x^{-1}y^2}}$

✓CONCEPT CHECK
Find and correct the error:

OBJECTIVE
4   Using the Distance and Midpoint Formulas

Now that we know how to simplify radicals, we can derive and use the distance formula. The midpoint formula is often confused with the distance formula, so to clarify both, we will also review the midpoint formula.

The Cartesian coordinate system helps us visualize a distance between points. To find the distance between two points, we use the distance formula, which is derived from the Pythagorean theorem.

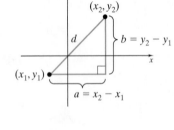

To find the distance d between two points (x_1, y_1) and (x_2, y_2) as shown to the left, notice that the length of leg a is $x_2 - x_1$ and that the length of leg b is $y_2 - y_1$.

Thus, the Pythagorean theorem tells us that

$$d^2 = a^2 + b^2$$

or

$$d^2 = (x_2 - x_1)^2 + (y_2 - y_1)^2$$

or

$$d = \sqrt{(x_2 - x_1)^2 + (y_2 - y_1)^2}$$

This formula gives us the distance between any two points on the real plane.

> **Distance Formula**
>
> The distance d between two points (x_1, y_1) and (x_2, y_2) is given by
> $$d = \sqrt{(x_2 - x_1)^2 + (y_2 - y_1)^2}$$

EXAMPLE 6 Find the distance between $(2, -5)$ and $(1, -4)$. Give an exact distance and a three-decimal-place approximation.

Solution To use the distance formula, it makes no difference which point we call (x_1, y_1) and which point we call (x_2, y_2). We will let $(x_1, y_1) = (2, -5)$ and $(x_2, y_2) = (1, -4)$.

Answer to Concept Check:
$$\dfrac{\sqrt[3]{27}}{\sqrt{9}} = \dfrac{3}{3} = 1$$

$$d = \sqrt{(x_2 - x_1)^2 + (y_2 - y_1)^2}$$
$$= \sqrt{(1 - 2)^2 + [-4 - (-5)]^2}$$
$$= \sqrt{(-1)^2 + (1)^2}$$
$$= \sqrt{1 + 1}$$
$$= \sqrt{2} \approx 1.414$$

The distance between the two points is exactly $\sqrt{2}$ units, or approximately 1.414 units.

□

PRACTICE

6   Find the distance between $(-3, 7)$ and $(-2, 3)$. Give an exact distance and a three-decimal-place approximation.

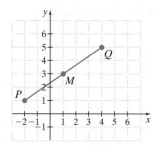

The **midpoint** of a line segment is the **point** located exactly halfway between the two endpoints of the line segment. On the graph to the left, the point M is the midpoint of line segment PQ. Thus, the distance between M and P equals the distance between M and Q.

Note: We usually need no knowledge of roots to calculate the midpoint of a line segment. We review midpoint here only because it is often confused with the distance between two points.

The x-coordinate of M is at half the distance between the x-coordinates of P and Q, and the y-coordinate of M is at half the distance between the y-coordinates of P and Q. That is, the x-coordinate of M is the average of the x-coordinates of P and Q; the y-coordinate of M is the average of the y-coordinates of P and Q.

> **Midpoint Formula**
>
> The midpoint of the line segment whose endpoints are (x_1, y_1) and (x_2, y_2) is the point with coordinates
>
> $$\left(\frac{x_1 + x_2}{2}, \frac{y_1 + y_2}{2} \right)$$

EXAMPLE 7   Find the midpoint of the line segment that joins points $P(-3, 3)$ and $Q(1, 0)$.

Solution   Use the midpoint formula. It makes no difference which point we call (x_1, y_1) or which point we call (x_2, y_2). Let $(x_1, y_1) = (-3, 3)$ and $(x_2, y_2) = (1, 0)$.

$$\text{midpoint} = \left(\frac{x_1 + x_2}{2}, \frac{y_1 + y_2}{2} \right)$$
$$= \left(\frac{-3 + 1}{2}, \frac{3 + 0}{2} \right)$$
$$= \left(\frac{-2}{2}, \frac{3}{2} \right)$$
$$= \left(-1, \frac{3}{2} \right)$$

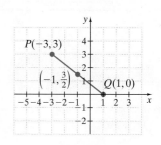

The midpoint of the segment is $\left(-1, \frac{3}{2} \right)$.

□

PRACTICE

7   Find the midpoint of the line segment that joins points $P(5, -2)$ and $Q(8, -6)$.

> ▶ Helpful Hint
> The distance between two points is a distance. The midpoint of a line segment is the point halfway between the endpoints of the segment.
>
> distance—measured in units
>
> midpoint—it is a point

Vocabulary, Readiness & Video Check

Use the choices below to fill in each blank. Some choices may be used more than once.

distance   midpoint   point

1. The _____ of a line segment is a _____ exactly halfway between the two endpoints of the line segment.

2. The _____ between two points is a distance, measured in units.

3. The _____ formula is $d = \sqrt{(x_2 - x_1)^2 + (y_2 - y_1)^2}$.

4. The _____ formula is $\left(\dfrac{x_1 + x_2}{2}, \dfrac{y_1 + y_2}{2}\right)$.

Martin-Gay Interactive Videos

See Video 7.3 🍐

Watch the section lecture video and answer the following questions.

OBJECTIVE 1
5. From ▣ Example 1 and the lecture before, in order to apply the product rule for radicals, what must be true about the indexes of the radicals being multiplied?

OBJECTIVE 2
6. From ▣ Examples 2–6, when might you apply the quotient rule (in either direction) in order to simplify a fractional radical expression?

OBJECTIVE 3
7. From ▣ Example 8, we know that an even power of a variable is a perfect square factor of the variable, leaving no factor in the radicand once simplified. Therefore, what must be true about the power of any variable left in the radicand of a simplified square root? Explain.

OBJECTIVE 4
8. From ▣ Example 10, the formula uses the coordinates of two points similar to the slope formula. What caution should you take when replacing values in the formula?

OBJECTIVE 4
9. Based on ▣ Example 11, complete the following statement. The x-value of the midpoint is the _____ of the x-values of the endpoints and the y-value of the midpoint is the _____ of the y-values of the endpoints.

7.3 Exercise Set

MyMathLab®

Use the product rule to multiply. See Example 1.

1. $\sqrt{7} \cdot \sqrt{2}$

2. $\sqrt{11} \cdot \sqrt{10}$

3. $\sqrt[4]{8} \cdot \sqrt[4]{2}$

4. $\sqrt[4]{27} \cdot \sqrt[4]{3}$

5. $\sqrt[3]{4} \cdot \sqrt[3]{9}$

6. $\sqrt[3]{10} \cdot \sqrt[3]{5}$

▶ 7. $\sqrt{2} \cdot \sqrt{3x}$

8. $\sqrt{3y} \cdot \sqrt{5x}$

9. $\sqrt{\dfrac{7}{x}} \cdot \sqrt{\dfrac{2}{y}}$

10. $\sqrt{\dfrac{6}{m}} \cdot \sqrt{\dfrac{n}{5}}$

11. $\sqrt[4]{4x^3} \cdot \sqrt[4]{5}$

12. $\sqrt[4]{ab^2} \cdot \sqrt[4]{27ab}$

Use the quotient rule to simplify. See Examples 2 and 3.

▶ 13. $\sqrt{\dfrac{6}{49}}$

14. $\sqrt{\dfrac{8}{81}}$

15. $\sqrt{\dfrac{2}{49}}$

16. $\sqrt{\dfrac{5}{121}}$

▶ 17. $\sqrt[4]{\dfrac{x^3}{16}}$

18. $\sqrt[4]{\dfrac{y}{81x^4}}$

19. $\sqrt[3]{\dfrac{4}{27}}$

20. $\sqrt[3]{\dfrac{3}{64}}$

21. $\sqrt[4]{\dfrac{8}{x^8}}$

22. $\sqrt[4]{\dfrac{a^3}{81}}$

23. $\sqrt[3]{\dfrac{2x}{81y^{12}}}$

24. $\sqrt[3]{\dfrac{3}{8x^6}}$

25. $\sqrt{\dfrac{x^2 y}{100}}$

26. $\sqrt{\dfrac{y^2 z}{36}}$

27. $\sqrt{\dfrac{5x^2}{4y^2}}$

28. $\sqrt{\dfrac{y^{10}}{9x^6}}$

29. $-\sqrt[3]{\dfrac{z^7}{27x^3}}$

30. $-\sqrt[3]{\dfrac{64a}{b^9}}$

Simplify. See Examples 3 and 4.

31. $\sqrt{32}$

32. $\sqrt{27}$

33. $\sqrt[3]{192}$

34. $\sqrt[3]{108}$

35. $5\sqrt{75}$

36. $3\sqrt{8}$

37. $\sqrt{24}$

38. $\sqrt{20}$

39. $\sqrt{100x^5}$

40. $\sqrt{64y^9}$

41. $\sqrt[3]{16y^7}$

42. $\sqrt[3]{64y^9}$

43. $\sqrt[4]{a^8 b^7}$

44. $\sqrt[5]{32z^{12}}$

45. $\sqrt{y^5}$

46. $\sqrt[3]{y^5}$

47. $\sqrt{25a^2 b^3}$

48. $\sqrt{9x^5 y^7}$

49. $\sqrt[5]{-32x^{10}y}$

50. $\sqrt[5]{-243z^9}$

51. $\sqrt[3]{50x^{14}}$

52. $\sqrt[3]{40y^{10}}$

53. $-\sqrt{32a^8 b^7}$

54. $-\sqrt{20ab^6}$

55. $\sqrt{9x^7 y^9}$

56. $\sqrt{12r^9 s^{12}}$

57. $\sqrt[3]{125r^9 s^{12}}$

58. $\sqrt[3]{8a^6 b^9}$

59. $\sqrt[4]{32x^{12}y^5}$

60. $\sqrt[4]{162x^7 y^{20}}$

Use the quotient rule to divide. Then simplify if possible. See Example 5.

61. $\dfrac{\sqrt{14}}{\sqrt{7}}$

62. $\dfrac{\sqrt{45}}{\sqrt{9}}$

63. $\dfrac{\sqrt[3]{24}}{\sqrt[3]{3}}$

64. $\dfrac{\sqrt[3]{10}}{\sqrt[3]{2}}$

65. $\dfrac{5\sqrt[4]{48}}{\sqrt[4]{3}}$

66. $\dfrac{7\sqrt[4]{162}}{\sqrt[4]{2}}$

67. $\dfrac{\sqrt{x^5 y^3}}{\sqrt{xy}}$

68. $\dfrac{\sqrt{a^7 b^6}}{\sqrt{a^3 b^2}}$

69. $\dfrac{8\sqrt[3]{54m^7}}{\sqrt[3]{2m}}$

70. $\dfrac{\sqrt[3]{128x^3}}{-3\sqrt[3]{2x}}$

71. $\dfrac{3\sqrt{100x^2}}{2\sqrt{2x^{-1}}}$

72. $\dfrac{\sqrt{270y^2}}{5\sqrt{3y^{-4}}}$

73. $\dfrac{\sqrt[4]{96a^{10}b^3}}{\sqrt[4]{3a^2 b^3}}$

74. $\dfrac{\sqrt[4]{160x^{10}y^5}}{\sqrt[4]{2x^2 y^2}}$

75. $\dfrac{\sqrt[5]{64x^{10}y^3}}{\sqrt[5]{2x^3 y^{-7}}}$

76. $\dfrac{\sqrt[5]{192x^6 y^{12}}}{\sqrt[5]{2x^{-1}y^{-3}}}$

Find the distance between each pair of points. Give an exact distance and a three-decimal-place approximation. See Example 6.

77. $(5, 1)$ and $(8, 5)$

78. $(2, 3)$ and $(14, 8)$

79. $(-3, 2)$ and $(1, -3)$

80. $(3, -2)$ and $(-4, 1)$

81. $(-9, 4)$ and $(-8, 1)$

82. $(-5, -2)$ and $(-6, -6)$

83. $\left(0, -\sqrt{2}\right)$ and $\left(\sqrt{3}, 0\right)$

84. $\left(-\sqrt{5}, 0\right)$ and $\left(0, \sqrt{7}\right)$

85. $(1.7, -3.6)$ and $(-8.6, 5.7)$

86. $(9.6, 2.5)$ and $(-1.9, -3.7)$

Find the midpoint of the line segment whose endpoints are given. See Example 7.

87. $(6, -8), (2, 4)$

88. $(3, 9), (7, 11)$

89. $(-2, -1), (-8, 6)$

90. $(-3, -4), (6, -8)$

91. $(7, 3), (-1, -3)$

92. $(-2, 5), (-1, 6)$

93. $\left(\dfrac{1}{2}, \dfrac{3}{8}\right), \left(-\dfrac{3}{2}, \dfrac{5}{8}\right)$

94. $\left(-\dfrac{2}{5}, \dfrac{7}{15}\right), \left(-\dfrac{2}{5}, -\dfrac{4}{15}\right)$

95. $\left(\sqrt{2}, 3\sqrt{5}\right), \left(\sqrt{2}, -2\sqrt{5}\right)$

96. $\left(\sqrt{8}, -\sqrt{12}\right), \left(3\sqrt{2}, 7\sqrt{3}\right)$

97. $(4.6, -3.5), (7.8, -9.8)$

98. $(-4.6, 2.1), (-6.7, 1.9)$

REVIEW AND PREVIEW

Perform each indicated operation. See Sections 1.4 and 5.4.

99. $6x + 8x$

100. $(6x)(8x)$

101. $(2x + 3)(x - 5)$

102. $(2x + 3) + (x - 5)$

103. $9y^2 - 8y^2$

104. $(9y^2)(-8y^2)$

105. $-3(x + 5)$

106. $-3 + x + 5$

107. $(x - 4)^2$

108. $(2x + 1)^2$

CONCEPT EXTENSIONS

Answer true or false. Assume all radicals represent nonzero real numbers.

109. $\sqrt[n]{a} \cdot \sqrt[n]{b} = \sqrt[n]{ab}$, _____

110. $\sqrt[3]{7} \cdot \sqrt[3]{11} = \sqrt[3]{18}$, _____

111. $\sqrt[3]{7} \cdot \sqrt{11} = \sqrt{77}$, _____

112. $\sqrt{x^7 y^8} = \sqrt{x^7} \cdot \sqrt{y^8}$, _____

113. $\dfrac{\sqrt[n]{a}}{\sqrt[n]{b}} = \sqrt[n]{\dfrac{a}{b}}$, _____

114. $\dfrac{\sqrt[3]{12}}{\sqrt[3]{4}} = \sqrt[3]{8}$, _____

Find and correct the error. See the Concept Check in this section.

115. $\dfrac{\sqrt[3]{64}}{\sqrt{64}} = \sqrt[3]{\dfrac{64}{64}} = \sqrt[3]{1} = 1$ ✗

116. $\dfrac{\sqrt[4]{16}}{\sqrt{4}} = \sqrt[4]{\dfrac{16}{4}} = \sqrt[4]{4}$ ✗

Simplify. See a Concept Check in this section. Assume variables represent positive numbers.

117. $\sqrt[5]{x^{35}}$

118. $\sqrt[6]{y^{48}}$

119. $\sqrt[4]{a^{12}b^4c^{20}}$

120. $\sqrt[3]{a^9b^{21}c^3}$

121. $\sqrt[3]{z^{32}}$

122. $\sqrt[5]{x^{49}}$

123. $\sqrt[4]{q^{17}r^{40}s^7}$

124. $\sqrt[4]{p^{11}q^4r^{45}}$

125. The formula for the radius r of a sphere with surface area A is given by $r = \sqrt{\dfrac{A}{4\pi}}$. Calculate the radius of a standard zorb whose outside surface area is 32.17 sq m. Round to the nearest tenth. (A zorb is a large inflated ball within a ball in which a person, strapped inside, may choose to roll down a hill. *Source:* Zorb, Ltd.)

126. The owner of Knightime Classic Movie Rentals has determined that the demand equation for renting older released DVDs is $F(x) = 0.6\sqrt{49 - x^2}$, where x is the price in dollars per two-day rental and $F(x)$ is the number of times the DVD is demanded per week.

a. Approximate to one decimal place the demand per week of an older released DVD if the rental price is $3 per two-day rental.

b. Approximate to one decimal place the demand per week of an older released DVD if the rental price is $5 per two-day rental.

c. Explain how the owner of the store can use this equation to predict the number of copies of each DVD that should be in stock.

127. The formula for the lateral surface area A of a cone with height h and radius r is given by

$$A = \pi r \sqrt{r^2 + h^2}$$

a. Find the lateral surface area of a cone whose height is 3 centimeters and whose radius is 4 centimeters.

b. Approximate to two decimal places the lateral surface area of a cone whose height is 7.2 feet and whose radius is 6.8 feet.

128. Before Mount Vesuvius, a volcano in Italy, erupted violently in 79 C.E., its height was 4190 feet. Vesuvius was roughly cone-shaped, and its base had a radius of approximately 25,200 feet. Use the formula for the lateral surface area of a cone, given in Exercise 127, to approximate the surface area this volcano had before it erupted. (*Source:* Global Volcanism Network)

4190 ft

25,200 ft

7.4 Adding, Subtracting, and Multiplying Radical Expressions

OBJECTIVES

1 Add or Subtract Radical Expressions.

2 Multiply Radical Expressions.

OBJECTIVE

1 Adding or Subtracting Radical Expressions

We have learned that sums or differences of like terms can be simplified. To simplify these sums or differences, we use the distributive property. For example,

$$2x + 3x = (2 + 3)x = 5x \quad \text{and} \quad 7x^2y - 4x^2y = (7 - 4)x^2y = 3x^2y$$

The distributive property can also be used to add **like radicals.**

> **Like Radicals**
>
> Radicals with the same index and the same radicand are like radicals.

For example, $2\sqrt{7} + 3\sqrt{7} = (2 + 3)\sqrt{7} = 5\sqrt{7}$. Also,

Like radicals

$$5\sqrt{3x} - 7\sqrt{3x} = (5 - 7)\sqrt{3x} = -2\sqrt{3x}$$

The expression $2\sqrt{7} + 2\sqrt[3]{7}$ cannot be simplified further since $2\sqrt{7}$ and $2\sqrt[3]{7}$ are not like radicals.

Unlike radicals

EXAMPLE 1   Add or subtract as indicated. Assume all variables represent positive real numbers.

a. $4\sqrt{11} + 8\sqrt{11}$ **b.** $5\sqrt[3]{3x} - 7\sqrt[3]{3x}$ **c.** $4\sqrt{5} + 4\sqrt[3]{5}$

Solution

a. $4\sqrt{11} + 8\sqrt{11} = (4 + 8)\sqrt{11} = 12\sqrt{11}$

b. $5\sqrt[3]{3x} - 7\sqrt[3]{3x} = (5 - 7)\sqrt[3]{3x} = -2\sqrt[3]{3x}$

c. $4\sqrt{5} + 4\sqrt[3]{5}$

This expression cannot be simplified since $4\sqrt{5}$ and $4\sqrt[3]{5}$ do not contain like radicals.

PRACTICE

1   Add or subtract as indicated.

a. $3\sqrt{17} + 5\sqrt{17}$ **b.** $7\sqrt[3]{5z} - 12\sqrt[3]{5z}$ **c.** $3\sqrt{2} + 5\sqrt[3]{2}$

When adding or subtracting radicals, always check first to see whether any radicals can be simplified.

✓**CONCEPT CHECK**

True or false?

$$\sqrt{a} + \sqrt{b} = \sqrt{a + b}$$

Explain.

EXAMPLE 2   Add or subtract. Assume that variables represent positive real numbers.

a. $\sqrt{20} + 2\sqrt{45}$ **b.** $\sqrt[3]{54} - 5\sqrt[3]{16} + \sqrt[3]{2}$ **c.** $\sqrt{27x} - 2\sqrt{9x} + \sqrt{72x}$

d. $\sqrt[3]{98} + \sqrt{98}$ **e.** $\sqrt[3]{48y^4} + \sqrt[3]{6y^4}$

Solution   First, simplify each radical. Then add or subtract any like radicals.

a. $\sqrt{20} + 2\sqrt{45} = \sqrt{4 \cdot 5} + 2\sqrt{9 \cdot 5}$ Factor 20 and 45.

$\phantom{\sqrt{20} + 2\sqrt{45}} = \sqrt{4} \cdot \sqrt{5} + 2 \cdot \sqrt{9} \cdot \sqrt{5}$ Use the product rule.

$\phantom{\sqrt{20} + 2\sqrt{45}} = 2 \cdot \sqrt{5} + 2 \cdot 3 \cdot \sqrt{5}$ Simplify $\sqrt{4}$ and $\sqrt{9}$.

$\phantom{\sqrt{20} + 2\sqrt{45}} = 2\sqrt{5} + 6\sqrt{5}$

$\phantom{\sqrt{20} + 2\sqrt{45}} = 8\sqrt{5}$ Add like radicals.

b. $\sqrt[3]{54} - 5\sqrt[3]{16} + \sqrt[3]{2}$

$\phantom{\sqrt[3]{54}} = \sqrt[3]{27} \cdot \sqrt[3]{2} - 5 \cdot \sqrt[3]{8} \cdot \sqrt[3]{2} + \sqrt[3]{2}$ Factor and use the product rule.

$\phantom{\sqrt[3]{54}} = 3 \cdot \sqrt[3]{2} - 5 \cdot 2 \cdot \sqrt[3]{2} + \sqrt[3]{2}$ Simplify $\sqrt[3]{27}$ and $\sqrt[3]{8}$.

$\phantom{\sqrt[3]{54}} = 3\sqrt[3]{2} - 10\sqrt[3]{2} + \sqrt[3]{2}$ Write $5 \cdot 2$ as 10.

$\phantom{\sqrt[3]{54}} = -6\sqrt[3]{2}$ Combine like radicals.

(Continued on next page)

Answer to Concept Check:
false; answers may vary

▶ Helpful Hint

None of these terms contain like radicals. We can simplify no further.

c. $\sqrt{27x} - 2\sqrt{9x} + \sqrt{72x}$

$= \sqrt{9} \cdot \sqrt{3x} - 2 \cdot \sqrt{9} \cdot \sqrt{x} + \sqrt{36} \cdot \sqrt{2x}$ Factor and use the product rule.

$= 3 \cdot \sqrt{3x} - 2 \cdot 3 \cdot \sqrt{x} + 6 \cdot \sqrt{2x}$ Simplify $\sqrt{9}$ and $\sqrt{36}$.

$= 3\sqrt{3x} - 6\sqrt{x} + 6\sqrt{2x}$ Write $2 \cdot 3$ as 6.

d. $\sqrt[3]{98} + \sqrt{98} = \sqrt[3]{98} + \sqrt{49} \cdot \sqrt{2}$ Factor and use the product rule.

$= \sqrt[3]{98} + 7\sqrt{2}$ No further simplification is possible.

e. $\sqrt[3]{48y^4} + \sqrt[3]{6y^4} = \sqrt[3]{8y^3} \cdot \sqrt[3]{6y} + \sqrt[3]{y^3} \cdot \sqrt[3]{6y}$ Factor and use the product rule.

$= 2y\sqrt[3]{6y} + y\sqrt[3]{6y}$ Simplify $\sqrt[3]{8y^3}$ and $\sqrt[3]{y^3}$.

$= 3y\sqrt[3]{6y}$ Combine like radicals.   □

PRACTICE
2   Add or subtract.

a. $\sqrt{24} + 3\sqrt{54}$ **b.** $\sqrt[3]{24} - 4\sqrt[3]{81} + \sqrt[3]{3}$ **c.** $\sqrt{75x} - 3\sqrt{27x} + \sqrt{12x}$
d. $\sqrt{40} + \sqrt[3]{40}$ **e.** $\sqrt[3]{81x^4} + \sqrt[3]{3x^4}$

Let's continue to assume that variables represent positive real numbers.

EXAMPLE 3   Add or subtract as indicated.

a. $\dfrac{\sqrt{45}}{4} - \dfrac{\sqrt{5}}{3}$ **b.** $\sqrt[3]{\dfrac{7x}{8}} + 2\sqrt[3]{7x}$

Solution

a. $\dfrac{\sqrt{45}}{4} - \dfrac{\sqrt{5}}{3} = \dfrac{3\sqrt{5}}{4} - \dfrac{\sqrt{5}}{3}$ To subtract, notice that the LCD is 12.

$= \dfrac{3\sqrt{5} \cdot 3}{4 \cdot 3} - \dfrac{\sqrt{5} \cdot 4}{3 \cdot 4}$ Write each expression as an equivalent expression with a denominator of 12.

$= \dfrac{9\sqrt{5}}{12} - \dfrac{4\sqrt{5}}{12}$ Multiply factors in the numerator and the denominator.

$= \dfrac{5\sqrt{5}}{12}$ Subtract.

b. $\sqrt[3]{\dfrac{7x}{8}} + 2\sqrt[3]{7x} = \dfrac{\sqrt[3]{7x}}{\sqrt[3]{8}} + 2\sqrt[3]{7x}$ Apply the quotient rule for radicals.

$= \dfrac{\sqrt[3]{7x}}{2} + 2\sqrt[3]{7x}$ Simplify.

$= \dfrac{\sqrt[3]{7x}}{2} + \dfrac{2\sqrt[3]{7x} \cdot 2}{2}$ Write each expression as an equivalent expression with a denominator of 2.

$= \dfrac{\sqrt[3]{7x}}{2} + \dfrac{4\sqrt[3]{7x}}{2}$

$= \dfrac{5\sqrt[3]{7x}}{2}$ Add.   □

PRACTICE
3   Add or subtract as indicated.

a. $\dfrac{\sqrt{28}}{3} - \dfrac{\sqrt{7}}{4}$ **b.** $\sqrt[3]{\dfrac{6y}{64}} + 3\sqrt[3]{6y}$

OBJECTIVE

2 Multiplying Radical Expressions

We can multiply radical expressions by using many of the same properties used to multiply polynomial expressions. For instance, to multiply $\sqrt{2}(\sqrt{6} - 3\sqrt{2})$, we use the distributive property and multiply $\sqrt{2}$ by each term inside the parentheses.

$$\sqrt{2}(\sqrt{6} - 3\sqrt{2}) = \sqrt{2}(\sqrt{6}) - \sqrt{2}(3\sqrt{2}) \quad \text{Use the distributive property.}$$
$$= \sqrt{2\cdot 6} - 3\sqrt{2\cdot 2}$$
$$= \sqrt{2\cdot 2\cdot 3} - 3\cdot 2 \quad \text{Use the product rule for radicals.}$$
$$= 2\sqrt{3} - 6$$

EXAMPLE 4   Multiply.

a. $\sqrt{3}(5 + \sqrt{30})$ **b.** $(\sqrt{5} - \sqrt{6})(\sqrt{7} + 1)$ **c.** $(7\sqrt{x} + 5)(3\sqrt{x} - \sqrt{5})$

d. $(4\sqrt{3} - 1)^2$ **e.** $(\sqrt{2x} - 5)(\sqrt{2x} + 5)$ **f.** $(\sqrt{x - 3} + 5)^2$

Solution

a. $\sqrt{3}(5 + \sqrt{30}) = \sqrt{3}(5) + \sqrt{3}(\sqrt{30})$
$$= 5\sqrt{3} + \sqrt{3\cdot 30}$$
$$= 5\sqrt{3} + \sqrt{3\cdot 3\cdot 10}$$
$$= 5\sqrt{3} + 3\sqrt{10}$$

b. To multiply, we can use the FOIL method.

$$\overset{\text{First}\quad\text{Outer}\quad\text{Inner}\quad\text{Last}}{(\sqrt{5} - \sqrt{6})(\sqrt{7} + 1) = \sqrt{5}\cdot\sqrt{7} + \sqrt{5}\cdot 1 - \sqrt{6}\cdot\sqrt{7} - \sqrt{6}\cdot 1}$$
$$= \sqrt{35} + \sqrt{5} - \sqrt{42} - \sqrt{6}$$

c. $(7\sqrt{x} + 5)(3\sqrt{x} - \sqrt{5}) = 7\sqrt{x}(3\sqrt{x}) - 7\sqrt{x}(\sqrt{5}) + 5(3\sqrt{x}) - 5(\sqrt{5})$
$$= 21x - 7\sqrt{5x} + 15\sqrt{x} - 5\sqrt{5}$$

d. $(4\sqrt{3} - 1)^2 = (4\sqrt{3} - 1)(4\sqrt{3} - 1)$
$$= 4\sqrt{3}(4\sqrt{3}) - 4\sqrt{3}(1) - 1(4\sqrt{3}) - 1(-1)$$
$$= 16\cdot 3 - 4\sqrt{3} - 4\sqrt{3} + 1$$
$$= 48 - 8\sqrt{3} + 1$$
$$= 49 - 8\sqrt{3}$$

e. $(\sqrt{2x} - 5)(\sqrt{2x} + 5) = \sqrt{2x}\cdot\sqrt{2x} + 5\sqrt{2x} - 5\sqrt{2x} - 5\cdot 5$
$$= 2x - 25$$

f. $(\underbrace{\sqrt{x - 3}}_{a} + \underbrace{5}_{b})^2 = \underbrace{(\sqrt{x - 3})^2}_{a^2} + \underbrace{2\cdot}_{+\,2\cdot} \underbrace{\sqrt{x - 3}}_{a} \cdot \underbrace{5}_{\cdot\,b} + \underbrace{5^2}_{b^2}$

$$= x - 3 + 10\sqrt{x - 3} + 25 \quad \text{Simplify.}$$
$$= x + 22 + 10\sqrt{x - 3} \quad \text{Combine like terms.} \qquad \square$$

PRACTICE

4   Multiply.

a. $\sqrt{5}(2 + \sqrt{15})$ **b.** $(\sqrt{2} - \sqrt{5})(\sqrt{6} + 2)$

c. $(3\sqrt{z} - 4)(2\sqrt{z} + 3)$ **d.** $(\sqrt{6} - 3)^2$

e. $(\sqrt{5x} + 3)(\sqrt{5x} - 3)$ **f.** $(\sqrt{x + 2} + 3)^2$

Vocabulary, Readiness & Video Check

Complete the table with "Like" or "Unlike."

	Terms	Like or Unlike Radical Terms?
1.	$\sqrt{7}, \sqrt[3]{7}$	
2.	$\sqrt[3]{x^2y}, \sqrt[3]{yx^2}$	
3.	$\sqrt[3]{abc}, \sqrt[3]{cba}$	
4.	$2x\sqrt{5}, 2x\sqrt{10}$	

Simplify. Assume that all variables represent positive real numbers.

5. $2\sqrt{3} + 4\sqrt{3} =$ _____

6. $5\sqrt{7} + 3\sqrt{7} =$ _____

7. $8\sqrt{x} - \sqrt{x} =$ _____

8. $3\sqrt{y} - \sqrt{y} =$ _____

9. $7\sqrt[3]{x} + \sqrt[3]{x} =$ _____

10. $8\sqrt[3]{z} + \sqrt[3]{z} =$ _____

Martin-Gay Interactive Videos

See Video 7.4

Watch the section lecture video and answer the following questions.

OBJECTIVE
1

11. From Examples 1 and 2, why should you always check to see if all terms in your expression are simplified before attempting to add or subtract radicals?

OBJECTIVE
2

12. In Example 4, what are you told to remember about the square root of a positive number?

7.4 Exercise Set   MyMathLab®

Add or subtract. See Examples 1 through 3.

1. $\sqrt{8} - \sqrt{32}$

2. $\sqrt{27} - \sqrt{75}$

3. $2\sqrt{2x^3} + 4x\sqrt{8x}$

4. $3\sqrt{45x^3} + x\sqrt{5x}$

▶ **5.** $2\sqrt{50} - 3\sqrt{125} + \sqrt{98}$

6. $4\sqrt{32} - \sqrt{18} + 2\sqrt{128}$

7. $\sqrt[3]{16x} - \sqrt[3]{54x}$

8. $2\sqrt[3]{3a^4} - 3a\sqrt[3]{81a}$

9. $\sqrt{9b^3} - \sqrt{25b^3} + \sqrt{49b^3}$

10. $\sqrt{4x^7} + 9x^2\sqrt{x^3} - 5x\sqrt{x^5}$

11. $\dfrac{5\sqrt{2}}{3} + \dfrac{2\sqrt{2}}{5}$

12. $\dfrac{\sqrt{3}}{2} + \dfrac{4\sqrt{3}}{3}$

▶ **13.** $\sqrt[3]{\dfrac{11}{8}} - \dfrac{\sqrt[3]{11}}{6}$

14. $\dfrac{2\sqrt[3]{4}}{7} - \dfrac{\sqrt[3]{4}}{14}$

15. $\dfrac{\sqrt{20x}}{9} + \sqrt{\dfrac{5x}{9}}$

16. $\dfrac{3x\sqrt{7}}{5} + \sqrt{\dfrac{7x^2}{100}}$

17. $7\sqrt{9} - 7 + \sqrt{3}$

18. $\sqrt{16} - 5\sqrt{10} + 7$

19. $2 + 3\sqrt{y^2} - 6\sqrt{y^2} + 5$

20. $3\sqrt{7} - \sqrt[3]{x} + 4\sqrt{7} - 3\sqrt[3]{x}$

21. $3\sqrt{108} - 2\sqrt{18} - 3\sqrt{48}$

22. $-\sqrt{75} + \sqrt{12} - 3\sqrt{3}$

23. $-5\sqrt[3]{625} + \sqrt[3]{40}$

24. $-2\sqrt[3]{108} - \sqrt[3]{32}$

25. $a^3\sqrt{9ab^3} - \sqrt{25a^7b^3} + \sqrt{16a^7b^3}$

26. $\sqrt{4x^7y^5} + 9x^2\sqrt{x^3y^5} - 5xy\sqrt{x^5y^3}$

27. $5y\sqrt{8y} + 2\sqrt{50y^3}$

28. $3\sqrt{8x^2y^3} - 2x\sqrt{32y^3}$

29. $\sqrt[3]{54xy^3} - 5\sqrt[3]{2xy^3} + y\sqrt[3]{128x}$

30. $2\sqrt[3]{24x^3y^4} + 4x\sqrt[3]{81y^4}$

31. $6\sqrt[3]{11} + 8\sqrt{11} - 12\sqrt{11}$

32. $3\sqrt[3]{5} + 4\sqrt{5} - 8\sqrt{5}$

33. $-2\sqrt[4]{x^7} + 3\sqrt[4]{16x^7} - x\sqrt[4]{x^3}$

34. $6\sqrt[3]{24x^3} - 2\sqrt[3]{81x^3} - x\sqrt[3]{3}$

35. $\dfrac{4\sqrt{3}}{3} - \dfrac{\sqrt{12}}{3}$

36. $\dfrac{\sqrt{45}}{10} + \dfrac{7\sqrt{5}}{10}$

37. $\dfrac{\sqrt[3]{8x^4}}{7} + \dfrac{3x\sqrt[3]{x}}{7}$

38. $\dfrac{\sqrt[4]{48}}{5x} - \dfrac{2\sqrt[4]{3}}{10x}$

39. $\sqrt{\dfrac{28}{x^2}} + \sqrt{\dfrac{7}{4x^2}}$

40. $\dfrac{\sqrt{99}}{5x} - \sqrt{\dfrac{44}{x^2}}$

41. $\sqrt[3]{\dfrac{16}{27}} - \dfrac{\sqrt[3]{54}}{6}$

42. $\dfrac{\sqrt[3]{3}}{10} + \sqrt[3]{\dfrac{24}{125}}$

43. $-\dfrac{\sqrt[3]{2x^4}}{9} + \sqrt[3]{\dfrac{250x^4}{27}}$

44. $\dfrac{\sqrt[3]{y^5}}{8} + \dfrac{5y\sqrt[3]{y^2}}{4}$

△ **45.** Find the perimeter of the trapezoid.

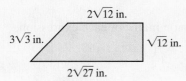

△ **46.** Find the perimeter of the triangle.

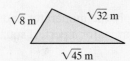

Multiply and then simplify if possible. See Example 4.

▶ **47.** $\sqrt{7}(\sqrt{5} + \sqrt{3})$

48. $\sqrt{5}(\sqrt{15} - \sqrt{35})$

49. $(\sqrt{5} - \sqrt{2})^2$

50. $(3x - \sqrt{2})(3x - \sqrt{2})$

51. $\sqrt{3x}(\sqrt{3} - \sqrt{x})$

52. $\sqrt{5y}(\sqrt{y} + \sqrt{5})$

53. $(2\sqrt{x} - 5)(3\sqrt{x} + 1)$

54. $(8\sqrt{y} + z)(4\sqrt{y} - 1)$

55. $(\sqrt[3]{a} - 4)(\sqrt[3]{a} + 5)$

56. $(\sqrt[3]{a} + 2)(\sqrt[3]{a} + 7)$

57. $6(\sqrt{2} - 2)$

58. $\sqrt{5}(6 - \sqrt{5})$

59. $\sqrt{2}(\sqrt{2} + x\sqrt{6})$

60. $\sqrt{3}(\sqrt{3} - 2\sqrt{5x})$

▶ **61.** $(2\sqrt{7} + 3\sqrt{5})(\sqrt{7} - 2\sqrt{5})$

62. $(\sqrt{6} - 4\sqrt{2})(3\sqrt{6} + \sqrt{2})$

63. $(\sqrt{x} - y)(\sqrt{x} + y)$

64. $(\sqrt{3x} + 2)(\sqrt{3x} - 2)$

65. $(\sqrt{3} + x)^2$

66. $(\sqrt{y} - 3x)^2$

67. $(\sqrt{5x} - 2\sqrt{3x})(\sqrt{5x} - 3\sqrt{3x})$

68. $(5\sqrt{7x} - \sqrt{2x})(4\sqrt{7x} + 6\sqrt{2x})$

69. $(\sqrt[3]{4} + 2)(\sqrt[3]{2} - 1)$

70. $(\sqrt[3]{3} + \sqrt[3]{2})(\sqrt[3]{9} - \sqrt[3]{4})$

71. $(\sqrt[3]{x} + 1)(\sqrt[3]{x^2} - \sqrt[3]{x} + 1)$

72. $(\sqrt[3]{3x} + 2)(\sqrt[3]{9x^2} - 2\sqrt[3]{3x} + 4)$

73. $(\sqrt{x - 1} + 5)^2$

74. $(\sqrt{3x + 1} + 2)^2$

75. $(\sqrt{2x + 5} - 1)^2$

76. $(\sqrt{x - 6} - 7)^2$

REVIEW AND PREVIEW

Factor each numerator and denominator. Then simplify if possible. See Section 6.1.

77. $\dfrac{2x - 14}{2}$

78. $\dfrac{8x - 24y}{4}$

79. $\dfrac{7x - 7y}{x^2 - y^2}$

80. $\dfrac{x^3 - 8}{4x - 8}$

81. $\dfrac{6a^2b - 9ab}{3ab}$

82. $\dfrac{14r - 28r^2s^2}{7rs}$

83. $\dfrac{-4 + 2\sqrt{3}}{6}$

84. $\dfrac{-5 + 10\sqrt{7}}{5}$

CONCEPT EXTENSIONS

△ **85.** Find the perimeter and area of the rectangle.

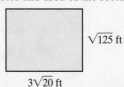

△ **86.** Find the area and perimeter of the trapezoid. (*Hint:* The area of a trapezoid is the product of half the height $6\sqrt{3}$ meters and the sum of the bases $2\sqrt{63}$ and $7\sqrt{7}$ meters.)

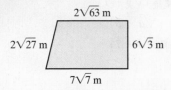

$2\sqrt{63}$ m

$2\sqrt{27}$ m $6\sqrt{3}$ m

$7\sqrt{7}$ m

87. a. Add: $\sqrt{3} + \sqrt{3}$.

   b. Multiply: $\sqrt{3} \cdot \sqrt{3}$.

   c. Describe the differences in parts (a) and (b).

88. a. Add: $2\sqrt{5} + \sqrt{5}$

   b. Multiply: $2\sqrt{5} \cdot \sqrt{5}$

   c. Describe the differences in parts (a) and (b).

89. Multiply: $\left(\sqrt{2} + \sqrt{3} - 1\right)^2$.

90. Multiply: $\left(\sqrt{5} - \sqrt{2} + 1\right)^2$

91. Explain how simplifying $2x + 3x$ is similar to simplifying $2\sqrt{x} + 3\sqrt{x}$.

92. Explain how multiplying $(x - 2)(x + 3)$ is similar to multiplying $\left(\sqrt{x} - \sqrt{2}\right)\left(\sqrt{x} + 3\right)$.

7.5 Rationalizing Denominators and Numerators of Radical Expressions

OBJECTIVES

1 Rationalize Denominators.

2 Rationalize Denominators Having Two Terms.

3 Rationalize Numerators.

OBJECTIVE

1 Rationalizing Denominators of Radical Expressions

Often in mathematics, it is helpful to write a radical expression such as $\dfrac{\sqrt{3}}{\sqrt{2}}$ either without a radical in the denominator or without a radical in the numerator. The process of writing this expression as an equivalent expression but without a radical in the denominator is called **rationalizing the denominator.** To rationalize the denominator of $\dfrac{\sqrt{3}}{\sqrt{2}}$, we use the fundamental principle of fractions and multiply the numerator and the denominator by $\sqrt{2}$. Recall that this is the same as multiplying by $\dfrac{\sqrt{2}}{\sqrt{2}}$, which simplifies to 1.

$$\frac{\sqrt{3}}{\sqrt{2}} = \frac{\sqrt{3} \cdot \sqrt{2}}{\sqrt{2} \cdot \sqrt{2}} = \frac{\sqrt{6}}{\sqrt{4}} = \frac{\sqrt{6}}{2}$$

In this section, we continue to assume that variables represent positive real numbers.

EXAMPLE 1 Rationalize the denominator of each expression.

a. $\dfrac{2}{\sqrt{5}}$ **b.** $\dfrac{2\sqrt{16}}{\sqrt{9x}}$ **c.** $\sqrt[3]{\dfrac{1}{2}}$

Solution

a. To rationalize the denominator, we multiply the numerator and denominator by a factor that makes the radicand in the denominator a perfect square.

$$\frac{2}{\sqrt{5}} = \frac{2 \cdot \sqrt{5}}{\sqrt{5} \cdot \sqrt{5}} = \frac{2\sqrt{5}}{5} \quad \text{The denominator is now rationalized.}$$

b. First, we simplify the radicals and then rationalize the denominator.

$$\frac{2\sqrt{16}}{\sqrt{9x}} = \frac{2(4)}{3\sqrt{x}} = \frac{8}{3\sqrt{x}}$$

To rationalize the denominator, multiply the numerator and denominator by $\sqrt{x}$. Then

$$\frac{8}{3\sqrt{x}} = \frac{8 \cdot \sqrt{x}}{3\sqrt{x} \cdot \sqrt{x}} = \frac{8\sqrt{x}}{3x}$$

c. $\sqrt[3]{\dfrac{1}{2}} = \dfrac{\sqrt[3]{1}}{\sqrt[3]{2}} = \dfrac{1}{\sqrt[3]{2}}$. Now we rationalize the denominator. Since $\sqrt[3]{2}$ is a cube root, we want to multiply by a value that will make the radicand 2 a perfect cube. If we multiply $\sqrt[3]{2}$ by $\sqrt[3]{2^2}$, we get $\sqrt[3]{2^3} = \sqrt[3]{8} = 2$.

$$\dfrac{1 \cdot \sqrt[3]{2^2}}{\sqrt[3]{2} \cdot \sqrt[3]{2^2}} = \dfrac{\sqrt[3]{4}}{\sqrt[3]{2^3}} = \dfrac{\sqrt[3]{4}}{2} \qquad \text{Multiply the numerator and denominator by } \sqrt[3]{2^2} \text{ and then simplify.}$$

PRACTICE
1   Rationalize the denominator of each expression.

a. $\dfrac{5}{\sqrt{3}}$ **b.** $\dfrac{3\sqrt{25}}{\sqrt{4x}}$ **c.** $\sqrt[3]{\dfrac{2}{9}}$

✓CONCEPT CHECK

Determine by which number both the numerator and denominator can be multiplied to rationalize the denominator of the radical expression.

a. $\dfrac{1}{\sqrt[3]{7}}$ **b.** $\dfrac{1}{\sqrt[4]{8}}$

EXAMPLE 2   Rationalize the denominator of $\sqrt{\dfrac{7x}{3y}}$.

Solution $\sqrt{\dfrac{7x}{3y}} = \dfrac{\sqrt{7x}}{\sqrt{3y}}$ Use the quotient rule. No radical may be simplified further.

$\qquad = \dfrac{\sqrt{7x} \cdot \sqrt{3y}}{\sqrt{3y} \cdot \sqrt{3y}}$ Multiply numerator and denominator by $\sqrt{3y}$ so that the radicand in the denominator is a perfect square.

$\qquad = \dfrac{\sqrt{21xy}}{3y}$ Use the product rule in the numerator and denominator. Remember that $\sqrt{3y} \cdot \sqrt{3y} = 3y$.

PRACTICE
2   Rationalize the denominator of $\sqrt{\dfrac{3z}{5y}}$.

EXAMPLE 3   Rationalize the denominator of $\dfrac{\sqrt[4]{x}}{\sqrt[4]{81y^5}}$.

Solution   First, simplify each radical if possible.

$\dfrac{\sqrt[4]{x}}{\sqrt[4]{81y^5}} = \dfrac{\sqrt[4]{x}}{\sqrt[4]{81y^4} \cdot \sqrt[4]{y}}$ Use the product rule in the denominator.

$\qquad = \dfrac{\sqrt[4]{x}}{3y\sqrt[4]{y}}$ Write $\sqrt[4]{81y^4}$ as $3y$.

$\qquad = \dfrac{\sqrt[4]{x} \cdot \sqrt[4]{y^3}}{3y\sqrt[4]{y} \cdot \sqrt[4]{y^3}}$ Multiply numerator and denominator by $\sqrt[4]{y^3}$ so that the radicand in the denominator is a perfect fourth power.

$\qquad = \dfrac{\sqrt[4]{xy^3}}{3y\sqrt[4]{y^4}}$ Use the product rule in the numerator and denominator.

$\qquad = \dfrac{\sqrt[4]{xy^3}}{3y^2}$ In the denominator, $\sqrt[4]{y^4} = y$ and $3y \cdot y = 3y^2$.

Answer to Concept Check:

a. $\sqrt[3]{7^2}$ or $\sqrt[3]{49}$ **b.** $\sqrt[4]{2}$

PRACTICE
3   Rationalize the denominator of $\dfrac{\sqrt[3]{z^2}}{\sqrt[3]{27x^4}}$.

OBJECTIVE

2 Rationalizing Denominators Having Two Terms

Remember the product of the sum and difference of two terms?

$$(a + b)(a - b) = a^2 - b^2$$

These two expressions are called **conjugates** of each other.

To rationalize a numerator or denominator that is a sum or difference of two terms, we use conjugates. To see how and why this works, let's rationalize the denominator of the expression $\dfrac{5}{\sqrt{3} - 2}$. To do so, we multiply both the numerator and the denominator by $\sqrt{3} + 2$, the **conjugate** of the denominator $\sqrt{3} - 2$, and see what happens.

$$\frac{5}{\sqrt{3} - 2} = \frac{5(\sqrt{3} + 2)}{(\sqrt{3} - 2)(\sqrt{3} + 2)}$$

$$= \frac{5(\sqrt{3} + 2)}{(\sqrt{3})^2 - 2^2} \quad \text{Multiply the sum and difference of two terms: } (a + b)(a - b) = a^2 - b^2.$$

$$= \frac{5(\sqrt{3} + 2)}{3 - 4}$$

$$= \frac{5(\sqrt{3} + 2)}{-1}$$

$$= -5(\sqrt{3} + 2) \quad \text{or} \quad -5\sqrt{3} - 10$$

Notice in the denominator that the product of $(\sqrt{3} - 2)$ and its conjugate, $(\sqrt{3} + 2)$, is -1. In general, the product of an expression and its conjugate will contain no radical terms. This is why, when rationalizing a denominator or a numerator containing two terms, we multiply by its conjugate. Examples of conjugates are

$$\sqrt{a} - \sqrt{b} \quad \text{and} \quad \sqrt{a} + \sqrt{b}$$
$$x + \sqrt{y} \quad \text{and} \quad x - \sqrt{y}$$

EXAMPLE 4 Rationalize each denominator.

a. $\dfrac{2}{3\sqrt{2} + 4}$ **b.** $\dfrac{\sqrt{6} + 2}{\sqrt{5} - \sqrt{3}}$ **c.** $\dfrac{2\sqrt{m}}{3\sqrt{x} + \sqrt{m}}$

Solution

a. Multiply the numerator and denominator by the conjugate of the denominator, $3\sqrt{2} + 4$.

$$\frac{2}{3\sqrt{2} + 4} = \frac{2(3\sqrt{2} - 4)}{(3\sqrt{2} + 4)(3\sqrt{2} - 4)}$$

$$= \frac{2(3\sqrt{2} - 4)}{(3\sqrt{2})^2 - 4^2}$$

$$= \frac{2(3\sqrt{2} - 4)}{18 - 16}$$

$$= \frac{2(3\sqrt{2} - 4)}{2}, \quad \text{or} \quad 3\sqrt{2} - 4$$

It is often useful to leave a numerator in factored form to help determine whether the expression can be simplified.

b. Multiply the numerator and denominator by the conjugate of $\sqrt{5} - \sqrt{3}$.

$$\frac{\sqrt{6} + 2}{\sqrt{5} - \sqrt{3}} = \frac{\left(\sqrt{6} + 2\right)\left(\sqrt{5} + \sqrt{3}\right)}{\left(\sqrt{5} - \sqrt{3}\right)\left(\sqrt{5} + \sqrt{3}\right)}$$

$$= \frac{\sqrt{6}\sqrt{5} + \sqrt{6}\sqrt{3} + 2\sqrt{5} + 2\sqrt{3}}{\left(\sqrt{5}\right)^2 - \left(\sqrt{3}\right)^2}$$

$$= \frac{\sqrt{30} + \sqrt{18} + 2\sqrt{5} + 2\sqrt{3}}{5 - 3}$$

$$= \frac{\sqrt{30} + 3\sqrt{2} + 2\sqrt{5} + 2\sqrt{3}}{2}$$

c. Multiply by the conjugate of $3\sqrt{x} + \sqrt{m}$ to eliminate the radicals from the denominator.

$$\frac{2\sqrt{m}}{3\sqrt{x} + \sqrt{m}} = \frac{2\sqrt{m}\left(3\sqrt{x} - \sqrt{m}\right)}{\left(3\sqrt{x} + \sqrt{m}\right)\left(3\sqrt{x} - \sqrt{m}\right)} = \frac{6\sqrt{mx} - 2m}{\left(3\sqrt{x}\right)^2 - \left(\sqrt{m}\right)^2}$$

$$= \frac{6\sqrt{mx} - 2m}{9x - m}$$

PRACTICE
4   Rationalize the denominator.

a. $\dfrac{5}{3\sqrt{5} + 2}$
b. $\dfrac{\sqrt{2} + 5}{\sqrt{3} - \sqrt{5}}$
c. $\dfrac{3\sqrt{x}}{2\sqrt{x} + \sqrt{y}}$

OBJECTIVE
3   Rationalizing Numerators

As mentioned earlier, it is also often helpful to write an expression such as $\dfrac{\sqrt{3}}{\sqrt{2}}$ as an equivalent expression without a radical in the numerator. This process is called **rationalizing the numerator.** To rationalize the numerator of $\dfrac{\sqrt{3}}{\sqrt{2}}$, we multiply the numerator and the denominator by $\sqrt{3}$.

$$\frac{\sqrt{3}}{\sqrt{2}} = \frac{\sqrt{3} \cdot \sqrt{3}}{\sqrt{2} \cdot \sqrt{3}} = \frac{\sqrt{9}}{\sqrt{6}} = \frac{3}{\sqrt{6}}$$

EXAMPLE 5   Rationalize the numerator of $\dfrac{\sqrt{7}}{\sqrt{45}}$.

Solution   First we simplify $\sqrt{45}$.

$$\frac{\sqrt{7}}{\sqrt{45}} = \frac{\sqrt{7}}{\sqrt{9 \cdot 5}} = \frac{\sqrt{7}}{3\sqrt{5}}$$

Next we rationalize the numerator by multiplying the numerator and the denominator by $\sqrt{7}$.

$$\frac{\sqrt{7}}{3\sqrt{5}} = \frac{\sqrt{7} \cdot \sqrt{7}}{3\sqrt{5} \cdot \sqrt{7}} = \frac{7}{3\sqrt{5 \cdot 7}} = \frac{7}{3\sqrt{35}}$$

PRACTICE
5   Rationalize the numerator of $\dfrac{\sqrt{32}}{\sqrt{80}}$.

EXAMPLE 6 Rationalize the numerator of $\dfrac{\sqrt[3]{2x^2}}{\sqrt[3]{5y}}$.

Solution The numerator and the denominator of this expression are already simplified. To rationalize the numerator, $\sqrt[3]{2x^2}$, we multiply the numerator and denominator by a factor that will make the radicand a perfect cube. If we multiply $\sqrt[3]{2x^2}$ by $\sqrt[3]{4x}$, we get $\sqrt[3]{8x^3} = 2x$.

$$\frac{\sqrt[3]{2x^2}}{\sqrt[3]{5y}} = \frac{\sqrt[3]{2x^2} \cdot \sqrt[3]{4x}}{\sqrt[3]{5y} \cdot \sqrt[3]{4x}} = \frac{\sqrt[3]{8x^3}}{\sqrt[3]{20xy}} = \frac{2x}{\sqrt[3]{20xy}}$$

□

PRACTICE
6 Rationalize the numerator of $\dfrac{\sqrt[3]{5b}}{\sqrt[3]{2a}}$.

EXAMPLE 7 Rationalize the numerator of $\dfrac{\sqrt{x} + 2}{5}$.

Solution We multiply the numerator and the denominator by the conjugate of the numerator, $\sqrt{x} + 2$.

$$\frac{\sqrt{x} + 2}{5} = \frac{(\sqrt{x} + 2)(\sqrt{x} - 2)}{5(\sqrt{x} - 2)} \quad \text{Multiply by } \sqrt{x} - 2, \text{ the conjugate of } \sqrt{x} + 2.$$

$$= \frac{(\sqrt{x})^2 - 2^2}{5(\sqrt{x} - 2)} \quad (a + b)(a - b) = a^2 - b^2$$

$$= \frac{x - 4}{5(\sqrt{x} - 2)}$$

□

PRACTICE
7 Rationalize the numerator of $\dfrac{\sqrt{x} - 3}{4}$.

Vocabulary, Readiness & Video Check

Use the choices below to fill in each blank. Not all choices will be used.

rationalizing the numerator   conjugate $\dfrac{\sqrt{3}}{\sqrt{3}}$

rationalizing the denominator $\dfrac{5}{5}$

1. The _____ of $a + b$ is $a - b$.
2. The process of writing an equivalent expression, but without a radical in the denominator is, called

   _____.
3. The process of writing an equivalent expression, but without a radical in the numerator, is called _____.
4. To rationalize the denominator of $\dfrac{5}{\sqrt{3}}$, we multiply by _____.

Martin-Gay Interactive Videos

See Video 7.5

Watch the section lecture video and answer the following questions.

OBJECTIVE 1 5. From Examples 1–3, what is the goal of rationalizing a denominator?

OBJECTIVE 2 6. From Example 4, why will multiplying a denominator by its conjugate always rationalize the denominator?

OBJECTIVE 3 7. From Example 5, is the process of rationalizing a numerator any different from rationalizing a denominator?

7.5   Exercise Set   MyMathLab®

Rationalize each denominator. See Examples 1 through 3.

▶ 1. $\dfrac{\sqrt{2}}{\sqrt{7}}$

2. $\dfrac{\sqrt{3}}{\sqrt{2}}$

3. $\sqrt{\dfrac{1}{5}}$

4. $\sqrt{\dfrac{1}{2}}$

5. $\sqrt{\dfrac{4}{x}}$

6. $\sqrt{\dfrac{25}{y}}$

▶ 7. $\dfrac{4}{\sqrt[3]{3}}$

8. $\dfrac{6}{\sqrt[3]{9}}$

▶ 9. $\dfrac{3}{\sqrt{8x}}$

10. $\dfrac{5}{\sqrt{27a}}$

11. $\dfrac{3}{\sqrt[3]{4x^2}}$

12. $\dfrac{5}{\sqrt[3]{3y}}$

13. $\dfrac{9}{\sqrt{3a}}$

14. $\dfrac{x}{\sqrt{5}}$

15. $\dfrac{3}{\sqrt[3]{2}}$

16. $\dfrac{5}{\sqrt[3]{9}}$

17. $\dfrac{2\sqrt{3}}{\sqrt{7}}$

18. $\dfrac{-5\sqrt{2}}{\sqrt{11}}$

19. $\sqrt{\dfrac{2x}{5y}}$

20. $\sqrt{\dfrac{13a}{2b}}$

21. $\sqrt[3]{\dfrac{3}{5}}$

22. $\sqrt[3]{\dfrac{7}{10}}$

23. $\sqrt{\dfrac{3x}{50}}$

24. $\sqrt{\dfrac{11y}{45}}$

25. $\dfrac{1}{\sqrt{12z}}$

26. $\dfrac{1}{\sqrt{32x}}$

27. $\dfrac{\sqrt[3]{2y^2}}{\sqrt[3]{9x^2}}$

28. $\dfrac{\sqrt[3]{3x}}{\sqrt[3]{4y^4}}$

29. $\sqrt[4]{\dfrac{81}{8}}$

30. $\sqrt[4]{\dfrac{1}{9}}$

31. $\sqrt[4]{\dfrac{16}{9x^7}}$

32. $\sqrt[5]{\dfrac{32}{m^6n^{13}}}$

33. $\dfrac{5a}{\sqrt[5]{8a^9b^{11}}}$

34. $\dfrac{9y}{\sqrt[4]{4y^9}}$

Write the conjugate of each expression.

35. $\sqrt{2} + x$

36. $\sqrt{3} + y$

37. $5 - \sqrt{a}$

38. $6 - \sqrt{b}$

39. $-7\sqrt{5} + 8\sqrt{x}$

40. $-9\sqrt{2} - 6\sqrt{y}$

Rationalize each denominator. See Example 4.

41. $\dfrac{6}{2 - \sqrt{7}}$

42. $\dfrac{3}{\sqrt{7} - 4}$

▶ 43. $\dfrac{-7}{\sqrt{x} - 3}$

44. $\dfrac{-8}{\sqrt{y} + 4}$

45. $\dfrac{\sqrt{2} - \sqrt{3}}{\sqrt{2} + \sqrt{3}}$

46. $\dfrac{\sqrt{3} + \sqrt{4}}{\sqrt{2} - \sqrt{3}}$

47. $\dfrac{\sqrt{a} + 1}{2\sqrt{a} - \sqrt{b}}$

48. $\dfrac{2\sqrt{a} - 3}{2\sqrt{a} + \sqrt{b}}$

49. $\dfrac{8}{1 + \sqrt{10}}$

50. $\dfrac{-3}{\sqrt{6} - 2}$

51. $\dfrac{\sqrt{x}}{\sqrt{x} + \sqrt{y}}$

52. $\dfrac{2\sqrt{a}}{2\sqrt{x} - \sqrt{y}}$

53. $\dfrac{2\sqrt{3} + \sqrt{6}}{4\sqrt{3} - \sqrt{6}}$

54. $\dfrac{4\sqrt{5} + \sqrt{2}}{2\sqrt{5} - \sqrt{2}}$

Rationalize each numerator. See Examples 5 and 6.

55. $\sqrt{\dfrac{5}{3}}$

56. $\sqrt{\dfrac{3}{2}}$

▶ 57. $\sqrt{\dfrac{18}{5}}$

58. $\sqrt{\dfrac{12}{7}}$

59. $\dfrac{\sqrt{4x}}{7}$

60. $\dfrac{\sqrt{3x^5}}{6}$

61. $\dfrac{\sqrt[3]{5y^2}}{\sqrt[3]{4x}}$

62. $\dfrac{\sqrt[3]{4x}}{\sqrt[3]{z^4}}$

63. $\sqrt{\dfrac{2}{5}}$

64. $\sqrt{\dfrac{3}{7}}$

65. $\dfrac{\sqrt{2x}}{11}$

66. $\dfrac{\sqrt{y}}{7}$

67. $\sqrt[3]{\dfrac{7}{8}}$

68. $\sqrt[3]{\dfrac{25}{2}}$

69. $\dfrac{\sqrt[3]{3x^5}}{10}$

70. $\sqrt[3]{\dfrac{9y}{7}}$

71. $\sqrt{\dfrac{18x^4y^6}{3z}}$

72. $\sqrt{\dfrac{8x^5y}{2z}}$

Rationalize each numerator. See Example 7.

73. $\dfrac{2 - \sqrt{11}}{6}$

74. $\dfrac{\sqrt{15} + 1}{2}$

75. $\dfrac{2 - \sqrt{7}}{-5}$

76. $\dfrac{\sqrt{5} + 2}{\sqrt{2}}$

77. $\dfrac{\sqrt{x} + 3}{\sqrt{x}}$

78. $\dfrac{5 + \sqrt{2}}{\sqrt{2x}}$

79. $\dfrac{\sqrt{2} - 1}{\sqrt{2} + 1}$

80. $\dfrac{\sqrt{8} - \sqrt{3}}{\sqrt{2} + \sqrt{3}}$

81. $\dfrac{\sqrt{x} + 1}{\sqrt{x} - 1}$

82. $\dfrac{\sqrt{x} + \sqrt{y}}{\sqrt{x} - \sqrt{y}}$

REVIEW AND PREVIEW

Solve each equation. See Sections 2.1 and 5.8.

83. $2x - 7 = 3(x - 4)$ **84.** $9x - 4 = 7(x - 2)$

85. $(x - 6)(2x + 1) = 0$ **86.** $(y + 2)(5y + 4) = 0$

87. $x^2 - 8x = -12$ **88.** $x^3 = x$

CONCEPT EXTENSIONS

△ **89.** The formula of the radius r of a sphere with surface area A is

$$r = \sqrt{\frac{A}{4\pi}}$$

Rationalize the denominator of the radical expression in this formula.

△ **90.** The formula for the radius r of a cone with height 7 centimeters and volume V is

$$r = \sqrt{\frac{3V}{7\pi}}$$

Rationalize the numerator of the radical expression in this formula.

91. Given $\dfrac{\sqrt{5y^3}}{\sqrt{12x^3}}$, rationalize the denominator by following parts (a) and (b).

  a. Multiply the numerator and denominator by $\sqrt{12x^3}$.

  b. Multiply the numerator and denominator by $\sqrt{3x}$.

  c. What can you conclude from parts (a) and (b)?

92. Given $\dfrac{\sqrt[3]{5y}}{\sqrt[3]{4}}$, rationalize the denominator by following parts (a) and (b).

  a. Multiply the numerator and denominator by $\sqrt[3]{16}$.

  b. Multiply the numerator and denominator by $\sqrt[3]{2}$.

  c. What can you conclude from parts (a) and (b)?

Determine the smallest number both the numerator and denominator should be multiplied by to rationalize the denominator of the radical expression. See the Concept Check in this section.

93. $\dfrac{9}{\sqrt[3]{5}}$ **94.** $\dfrac{5}{\sqrt{27}}$

95. When rationalizing the denominator of $\dfrac{\sqrt{5}}{\sqrt{7}}$, explain why both the numerator and the denominator must be multiplied by $\sqrt{7}$.

96. When rationalizing the numerator of $\dfrac{\sqrt{5}}{\sqrt{7}}$, explain why both the numerator and the denominator must be multiplied by $\sqrt{5}$.

97. Explain why rationalizing the denominator does not change the value of the original expression.

98. Explain why rationalizing the numerator does not change the value of the original expression.

Integrated Review RADICALS AND RATIONAL EXPONENTS

Sections 7.1–7.5

Throughout this review, assume that all variables represent positive real numbers.
Find each root.

1. $\sqrt{81}$ **2.** $\sqrt[3]{-8}$ **3.** $\sqrt[4]{\dfrac{1}{16}}$ **4.** $\sqrt{x^6}$

5. $\sqrt[3]{y^9}$ **6.** $\sqrt{4y^{10}}$ **7.** $\sqrt[5]{-32y^5}$ **8.** $\sqrt[4]{81b^{12}}$

Use radical notation to write each expression. Simplify if possible.

9. $36^{1/2}$ **10.** $(3y)^{1/4}$ **11.** $64^{-2/3}$ **12.** $(x + 1)^{3/5}$

Use the properties of exponents to simplify each expression. Write with positive exponents.

13. $y^{-1/6} \cdot y^{7/6}$ **14.** $\dfrac{(2x^{1/3})^4}{x^{5/6}}$ **15.** $\dfrac{x^{1/4}x^{3/4}}{x^{-1/4}}$ **16.** $4^{1/3} \cdot 4^{2/5}$

Use rational exponents to simplify each radical.

17. $\sqrt[3]{8x^6}$ **18.** $\sqrt[12]{a^9b^6}$

Use rational exponents to write each as a single radical expression.

19. $\sqrt[4]{x} \cdot \sqrt{x}$

20. $\sqrt{5} \cdot \sqrt[3]{2}$

Simplify.

21. $\sqrt{40}$

22. $\sqrt[4]{16x^7 y^{10}}$

23. $\sqrt[3]{54x^4}$

24. $\sqrt[5]{-64b^{10}}$

Multiply or divide. Then simplify if possible.

25. $\sqrt{5} \cdot \sqrt{x}$

26. $\sqrt[3]{8x} \cdot \sqrt[3]{8x^2}$

27. $\dfrac{\sqrt{98y^6}}{\sqrt{2y}}$

28. $\dfrac{\sqrt[4]{48a^9 b^3}}{\sqrt[4]{ab^3}}$

Perform each indicated operation.

29. $\sqrt{20} - \sqrt{75} + 5\sqrt{7}$

30. $\sqrt[3]{54y^4} - y\sqrt[3]{16y}$

31. $\sqrt{3}(\sqrt{5} - \sqrt{2})$

32. $(\sqrt{7} + \sqrt{3})^2$

33. $(2x - \sqrt{5})(2x + \sqrt{5})$

34. $(\sqrt{x+1} - 1)^2$

Rationalize each denominator.

35. $\sqrt{\dfrac{7}{3}}$

36. $\dfrac{5}{\sqrt[3]{2x^2}}$

37. $\dfrac{\sqrt{3} - \sqrt{7}}{2\sqrt{3} + \sqrt{7}}$

Rationalize each numerator.

38. $\sqrt{\dfrac{7}{3}}$

39. $\sqrt[3]{\dfrac{9y}{11}}$

40. $\dfrac{\sqrt{x} - 2}{\sqrt{x}}$

7.6 Radical Equations and Problem Solving

OBJECTIVES

1 Solve Equations That Contain Radical Expressions.

2 Use the Pythagorean Theorem to Model Problems.

OBJECTIVE

1   Solving Equations That Contain Radical Expressions

In this section, we present techniques to solve equations containing radical expressions such as

$$\sqrt{2x - 3} = 9$$

We use the power rule to help us solve these radical equations.

> **Power Rule**
>
> If both sides of an equation are raised to the same power, **all** solutions of the original equation are **among** the solutions of the new equation.

This property *does not* say that raising both sides of an equation to a power yields an equivalent equation. A solution of the new equation *may or may not* be a solution of the original equation. For example, $(-2)^2 = 2^2$, but $-2 \neq 2$. Thus, *each solution of the new equation must be checked* to make sure it is a solution of the original equation. Recall that a proposed solution that is not a solution of the original equation is called an **extraneous solution**.

EXAMPLE 1   Solve: $\sqrt{2x - 3} = 9$.

Solution   We use the power rule to square both sides of the equation to eliminate the radical.

$$\sqrt{2x - 3} = 9$$
$$(\sqrt{2x - 3})^2 = 9^2$$
$$2x - 3 = 81$$
$$2x = 84$$
$$x = 42$$

Now we check the solution in the original equation.

(Continued on next page)

Check:

$$\sqrt{2x - 3} = 9$$
$$\sqrt{2(42) - 3} \stackrel{?}{=} 9 \quad \text{Let } x = 42.$$
$$\sqrt{84 - 3} \stackrel{?}{=} 9$$
$$\sqrt{81} \stackrel{?}{=} 9$$
$$9 = 9 \quad \text{True}$$

The solution checks, so we conclude that the solution is 42, or the solution set is $\{42\}$. □

PRACTICE
1 Solve: $\sqrt{3x - 5} = 7$.

To solve a radical equation, first isolate a radical on one side of the equation.

EXAMPLE 2 Solve: $\sqrt{-10x - 1} + 3x = 0$.

Solution First, isolate the radical on one side of the equation. To do this, we subtract $3x$ from both sides.

$$\sqrt{-10x - 1} + 3x = 0$$
$$\sqrt{-10x - 1} + 3x - 3x = 0 - 3x$$
$$\sqrt{-10x - 1} = -3x$$

Next we use the power rule to eliminate the radical.

$$(\sqrt{-10x - 1})^2 = (-3x)^2$$
$$-10x - 1 = 9x^2$$

Since this is a quadratic equation, we can set the equation equal to 0 and try to solve by factoring.

$$9x^2 + 10x + 1 = 0$$
$$(9x + 1)(x + 1) = 0 \quad \text{Factor.}$$
$$9x + 1 = 0 \quad \text{or} \quad x + 1 = 0 \quad \text{Set each factor equal to 0.}$$
$$x = -\frac{1}{9} \quad \text{or} \quad x = -1$$

Check: Let $x = -\frac{1}{9}$.

$$\sqrt{-10x - 1} + 3x = 0$$
$$\sqrt{-10\left(-\frac{1}{9}\right) - 1} + 3\left(-\frac{1}{9}\right) \stackrel{?}{=} 0$$
$$\sqrt{\frac{10}{9} - \frac{9}{9}} - \frac{3}{9} \stackrel{?}{=} 0$$
$$\sqrt{\frac{1}{9}} - \frac{1}{3} \stackrel{?}{=} 0$$
$$\frac{1}{3} - \frac{1}{3} = 0 \quad \text{True}$$

Let $x = -1$.

$$\sqrt{-10x - 1} + 3x = 0$$
$$\sqrt{-10(-1) - 1} + 3(-1) \stackrel{?}{=} 0$$
$$\sqrt{10 - 1} - 3 \stackrel{?}{=} 0$$
$$\sqrt{9} - 3 \stackrel{?}{=} 0$$
$$3 - 3 = 0 \quad \text{True}$$

Both solutions check. The solutions are $-\frac{1}{9}$ and -1, or the solution set is $\left\{-\frac{1}{9}, -1\right\}$. □

PRACTICE
2 Solve: $\sqrt{16x - 3} - 4x = 0$.

The following steps may be used to solve a radical equation.

> **Solving a Radical Equation**
>
> **Step 1.** Isolate one radical on one side of the equation.
>
> **Step 2.** Raise each side of the equation to a power equal to the index of the radical and simplify.
>
> **Step 3.** If the equation still contains a radical term, repeat Steps 1 and 2. If not, solve the equation.
>
> **Step 4.** Check all proposed solutions in the original equation.

EXAMPLE 3   Solve: $\sqrt[3]{x + 1} + 5 = 3$.

Solution   First we isolate the radical by subtracting 5 from both sides of the equation.

$$\sqrt[3]{x + 1} + 5 = 3$$
$$\sqrt[3]{x + 1} = -2$$

Next we raise both sides of the equation to the third power to eliminate the radical.

$$\left(\sqrt[3]{x + 1}\right)^3 = (-2)^3$$
$$x + 1 = -8$$
$$x = -9$$

The solution checks in the original equation, so the solution is -9.   □

PRACTICE
3   Solve: $\sqrt[3]{x - 2} + 1 = 3$.

··

EXAMPLE 4   Solve: $\sqrt{4 - x} = x - 2$.

Solution

$$\sqrt{4 - x} = x - 2$$
$$\left(\sqrt{4 - x}\right)^2 = (x - 2)^2$$
$$4 - x = x^2 - 4x + 4$$
$$x^2 - 3x = 0 \qquad \text{Write the quadratic equation in standard form.}$$
$$x(x - 3) = 0 \qquad \text{Factor.}$$
$$x = 0 \quad \text{or} \quad x - 3 = 0 \qquad \text{Set each factor equal to 0.}$$
$$x = 3$$

Check:

$$\sqrt{4 - x} = x - 2 \qquad\qquad\qquad \sqrt{4 - x} = x - 2$$
$$\sqrt{4 - 0} \overset{?}{=} 0 - 2 \quad \text{Let } x = 0. \qquad \sqrt{4 - 3} \overset{?}{=} 3 - 2 \quad \text{Let } x = 3.$$
$$2 = -2 \qquad \text{False} \qquad\qquad\qquad 1 = 1 \qquad \text{True}$$

The proposed solution 3 checks, but 0 does not. Since 0 is an extraneous solution, the only solution is 3.   □

PRACTICE
4   Solve: $\sqrt{16 + x} = x - 4$.

··

> ▶ **Helpful Hint**
> In Example 4, notice that $(x - 2)^2 = x^2 - 4x + 4$. Make sure binomials are squared correctly.

✓CONCEPT CHECK
How can you immediately tell that the equation $\sqrt{2y + 3} = -4$ has no real solution?

EXAMPLE 5 Solve: $\sqrt{2x + 5} + \sqrt{2x} = 3$.

Solution We get one radical alone by subtracting $\sqrt{2x}$ from both sides.

$$\sqrt{2x + 5} + \sqrt{2x} = 3$$
$$\sqrt{2x + 5} = 3 - \sqrt{2x}$$

Now we use the power rule to begin eliminating the radicals. First we square both sides.

$$\left(\sqrt{2x + 5}\right)^2 = \left(3 - \sqrt{2x}\right)^2$$
$$2x + 5 = 9 - 6\sqrt{2x} + 2x \quad \text{Multiply } (3 - \sqrt{2x})(3 - \sqrt{2x}).$$

There is still a radical in the equation, so we get a radical alone again. Then we square both sides.

$$2x + 5 = 9 - 6\sqrt{2x} + 2x$$
$$6\sqrt{2x} = 4 \qquad\qquad \text{Get the radical alone.}$$
$$36(2x) = 16 \qquad\qquad \text{Square both sides of the equation to eliminate the radical.}$$
$$72x = 16 \qquad\qquad \text{Multiply.}$$
$$x = \frac{16}{72} \qquad\qquad \text{Solve.}$$
$$x = \frac{2}{9} \qquad\qquad \text{Simplify.}$$

The proposed solution, $\frac{2}{9}$, checks in the original equation. The solution is $\frac{2}{9}$. □

PRACTICE
5 Solve: $\sqrt{8x + 1} + \sqrt{3x} = 2$.

▶ Helpful Hint
Make sure expressions are squared correctly. In Example 5, we squared $(3 - \sqrt{2x})$ as

$$(3 - \sqrt{2x})^2 = (3 - \sqrt{2x})(3 - \sqrt{2x})$$
$$= 3 \cdot 3 - 3\sqrt{2x} - 3\sqrt{2x} + \sqrt{2x} \cdot \sqrt{2x}$$
$$= 9 - 6\sqrt{2x} + 2x$$

✓CONCEPT CHECK
What is wrong with the following solution?

$$\sqrt{2x + 5} + \sqrt{4 - x} = 8$$
$$\left(\sqrt{2x + 5} + \sqrt{4 - x}\right)^2 = 8^2$$
$$(2x + 5) + (4 - x) = 64$$
$$x + 9 = 64$$
$$x = 55$$

Answers to Concept Checks:
answers may vary
$(\sqrt{2x + 5} + \sqrt{4 - x})^2$ is not
$(2x + 5) + (4 - x)$.

OBJECTIVE
2 **Using the Pythagorean Theorem** ▶

Recall that the Pythagorean theorem states that in a right triangle, the length of the hypotenuse squared equals the sum of the lengths of each of the legs squared.

Pythagorean Theorem

If a and b are the lengths of the legs of a right triangle and c is the length of the hypotenuse, then $a^2 + b^2 = c^2$.

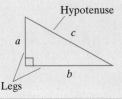

Hypotenuse

c

a

b

Legs

△ **EXAMPLE 6**   Find the length of the unknown leg of the right triangle.

4 m

10 m

b

Solution   In the formula $a^2 + b^2 = c^2$, c is the hypotenuse. Here, $c = 10$, the length of the hypotenuse, and $a = 4$. We solve for b. Then $a^2 + b^2 = c^2$ becomes

$$4^2 + b^2 = 10^2$$
$$16 + b^2 = 100$$
$$b^2 = 84 \quad \text{Subtract 16 from both sides.}$$
$$b = \pm\sqrt{84} = \pm\sqrt{4 \cdot 21} = \pm 2\sqrt{21}$$

Since b is a length and thus is positive, we will use the positive value only. The unknown leg of the triangle is $2\sqrt{21}$ meters long.   ☐

PRACTICE
6   Find the length of the unknown leg of the right triangle.

6 m

12 m

a

△ **EXAMPLE 7**   **Calculating Placement of a Wire**

A 50-foot supporting wire is to be attached to a 75-foot antenna. Because of surrounding buildings, sidewalks, and roadways, the wire must be anchored exactly 20 feet from the base of the antenna.

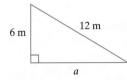

75 ft

50 ft

←— 20 ft —→

a. How high from the base of the antenna is the wire attached?

b. Local regulations require that a supporting wire be attached at a height no less than $\dfrac{3}{5}$ of the total height of the antenna. From part (a), have local regulations been met?

Solution

1. UNDERSTAND. Read and reread the problem. From the diagram, we notice that a right triangle is formed with hypotenuse 50 feet and one leg 20 feet. Let x be the height from the base of the antenna to the attached wire.

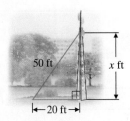

50 ft

x ft

←— 20 ft —→

(Continued on next page)

2. TRANSLATE. Use the Pythagorean theorem.

$$a^2 + b^2 = c^2$$
$$20^2 + x^2 = 50^2 \quad a = 20, c = 50$$

3. SOLVE.

$$20^2 + x^2 = 50^2$$
$$400 + x^2 = 2500$$
$$x^2 = 2100 \qquad \text{Subtract 400 from both sides.}$$
$$x = \pm\sqrt{2100}$$
$$= \pm 10\sqrt{21}$$

4. INTERPRET. *Check* the work and *state* the solution.

Check: We will use only the positive value, $x = 10\sqrt{21}$, because x represents length. The wire is attached exactly $10\sqrt{21}$ feet from the base of the pole, or approximately 45.8 feet.

State: The supporting wire must be attached at a height no less than $\frac{3}{5}$ of the total height of the antenna. This height is $\frac{3}{5}$ (75 feet), or 45 feet. Since we know from part (a) that the wire is to be attached at a height of approximately 45.8 feet, local regulations have been met. □

PRACTICE

7 Keith Robinson bought two Siamese fighting fish, but when he got home, he found he only had one rectangular tank that was 12 in. long, 7 in. wide, and 5 in. deep. Since the fish must be kept separated, he needed to insert a plastic divider in the diagonal of the tank. He already has a piece that is 5 in. in one dimension, but how long must it be to fit corner to corner in the tank?

Graphing Calculator Explorations

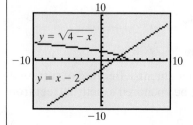

We can use a graphing calculator to solve radical equations. For example, to use a graphing calculator to approximate the solutions of the equation solved in Example 4, we graph the following.

$$Y_1 = \sqrt{4 - x} \qquad \text{and} \qquad Y_2 = x - 2$$

The x-value of the point of intersection is the solution. Use the Intersect feature or the Zoom and Trace features of your graphing calculator to see that the solution is 3.

Use a graphing calculator to solve each radical equation. Round all solutions to the nearest hundredth.

1. $\sqrt{x + 7} = x$ **2.** $\sqrt{3x + 5} = 2x$

3. $\sqrt{2x + 1} = \sqrt{2x + 2}$ **4.** $\sqrt{10x - 1} = \sqrt{-10x + 10} - 1$

5. $1.2x = \sqrt{3.1x + 5}$ **6.** $\sqrt{1.9x^2 - 2.2} = -0.8x + 3$

Vocabulary, Readiness & Video Check

Use the choices below to fill in each blank. Not all choices will be used.

| hypotenuse | right | $x^2 + 25$ | $16 - 8\sqrt{7x} + 7x$ |
| extraneous solution | legs | $x^2 - 10x + 25$ | $16 + 7x$ |

1. A proposed solution that is not a solution of the original equation is called a(n) _____.

2. The Pythagorean theorem states that $a^2 + b^2 = c^2$ where a and b are the lengths of the _____ of a(n) _____ triangle and c is the length of the _____.

3. The square of $x - 5$, or $(x - 5)^2 =$ _____.

4. The square of $4 - \sqrt{7x}$, or $(4 - \sqrt{7x})^2 =$ _____.

Martin-Gay Interactive Videos

See Video 7.6

Watch the section lecture video and answer the following questions.

 OBJECTIVE
1
5. From Examples 1–4, why must you be careful and check your proposed solution(s) in the original equation?

OBJECTIVE
2
6. From ⊟ Example 5, when solving problems using the Pythagorean theorem, what two things must you remember?

OBJECTIVE
2
7. What important reminder is given as the final answer to ⊟ Example 5 is being found?

7.6   Exercise Set   MyMathLab®

Solve. See Examples 1 and 2.

1. $\sqrt{2x} = 4$
2. $\sqrt{3x} = 3$
▶ **3.** $\sqrt{x - 3} = 2$
4. $\sqrt{x + 1} = 5$
5. $\sqrt{2x} = -4$
6. $\sqrt{5x} = -5$
7. $\sqrt{4x - 3} - 5 = 0$
8. $\sqrt{x - 3} - 1 = 0$
9. $\sqrt{2x - 3} - 2 = 1$
10. $\sqrt{3x + 3} - 4 = 8$

Solve. See Example 3.

11. $\sqrt[3]{6x} = -3$
12. $\sqrt[3]{4x} = -2$
13. $\sqrt[3]{x - 2} - 3 = 0$
14. $\sqrt[3]{2x - 6} - 4 = 0$

Solve. See Examples 4 and 5.

15. $\sqrt{13 - x} = x - 1$
16. $\sqrt{2x - 3} = 3 - x$
▶ **17.** $x - \sqrt{4 - 3x} = -8$
18. $2x + \sqrt{x + 1} = 8$
19. $\sqrt{y + 5} = 2 - \sqrt{y - 4}$
20. $\sqrt{x + 3} + \sqrt{x - 5} = 3$
21. $\sqrt{x - 3} + \sqrt{x + 2} = 5$
22. $\sqrt{2x - 4} - \sqrt{3x + 4} = -2$

MIXED PRACTICE

Solve. See Examples 1 through 5.

23. $\sqrt{3x - 2} = 5$
24. $\sqrt{5x - 4} = 9$
25. $-\sqrt{2x} + 4 = -6$
26. $-\sqrt{3x + 9} = -12$
27. $\sqrt{3x + 1} + 2 = 0$
28. $\sqrt{3x + 1} - 2 = 0$
29. $\sqrt[4]{4x + 1} - 2 = 0$
30. $\sqrt[4]{2x - 9} - 3 = 0$
31. $\sqrt{4x - 3} = 7$
32. $\sqrt{3x + 9} = 6$
33. $\sqrt[3]{6x - 3} - 3 = 0$
34. $\sqrt[3]{3x + 4} = 7$
▶ **35.** $\sqrt[3]{2x - 3} - 2 = -5$
36. $\sqrt[3]{x - 4} - 5 = -7$
37. $\sqrt{x + 4} = \sqrt{2x - 5}$
38. $\sqrt{3y + 6} = \sqrt{7y - 6}$
39. $x - \sqrt{1 - x} = -5$
40. $x - \sqrt{x - 2} = 4$
41. $\sqrt[3]{-6x - 1} = \sqrt[3]{-2x - 5}$
42. $\sqrt[3]{-4x - 3} = \sqrt[3]{-x - 15}$
▶ **43.** $\sqrt{5x - 1} - \sqrt{x + 2} = 3$
44. $\sqrt{2x - 1} - 4 = -\sqrt{x - 4}$
45. $\sqrt{2x - 1} = \sqrt{1 - 2x}$
46. $\sqrt{7x - 4} = \sqrt{4 - 7x}$

47. $\sqrt{3x + 4} - 1 = \sqrt{2x + 1}$
48. $\sqrt{x - 2} + 3 = \sqrt{4x + 1}$
49. $\sqrt{y + 3} - \sqrt{y - 3} = 1$
50. $\sqrt{x + 1} - \sqrt{x - 1} = 2$

Find the length of the unknown side of each triangle. See Example 6.

▶ **51.**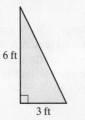

6 ft
3 ft

△ **52.**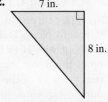

7 in.
8 in.

△ ▶ **53.**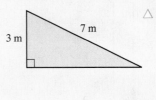

3 m
7 m

△ **54.**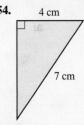

4 cm
7 cm

Find the length of the unknown side of each triangle. Give the exact length and a one-decimal-place approximation. See Example 6.

△ **55.**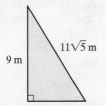

9 m
$11\sqrt{5}$ m

△ **56.**

$5\sqrt{3}$ cm
10 cm

△ **57.**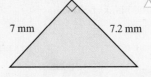

7 mm
7.2 mm

△ **58.**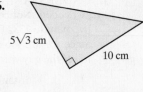

2.7 in.
2.3 in.

Solve. Give exact answers and two-decimal-place approximations where appropriate. For Exercises 59 and 60, the solutions have been started for you. See Example 7.

59. A wire is needed to support a vertical pole 15 feet tall. The cable will be anchored to a stake 8 feet from the base of the pole. How much cable is needed?

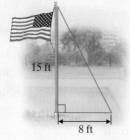

15 ft

8 ft

Start the solution:

1. **UNDERSTAND** the problem. Reread it as many times as needed. Notice that a right triangle is formed with legs of length 8 ft and 15 ft.
   Since we are looking for how much cable is needed, let

   x = amount of cable needed

2. **TRANSLATE** into an equation. We use the Pythagorean theorem. (Fill in the blanks below.)

$$a^2 \quad + \quad b^2 \quad = \quad c^2$$

$$\underline{\quad}^2 \quad + \quad \underline{\quad}^2 \quad = \quad x^2$$

Finish with:

3. **SOLVE** and 4. **INTERPRET**

60. The tallest structure in the United States is a TV tower in Blanchard, North Dakota. Its height is 2063 feet. A 2382-foot length of wire is to be used as a guy wire attached to the top of the tower. Approximate to the nearest foot how far from the base of the tower the guy wire must be anchored. (*Source:* U.S. Geological Survey)

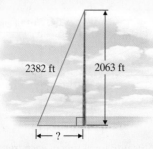

2382 ft   2063 ft

← ? →

Start the solution:

1. **UNDERSTAND** the problem. Reread it as many times as needed. Notice that a right triangle is formed with hypotenuse 2382 ft and one leg 2063 ft.
   Since we are looking for how far from the base of the tower the guy wire is anchored, let

   x = distance from base of tower to where guy wire is anchored.

2. **TRANSLATE** into an equation. We use the Pythagorean theorem. (Fill in the blanks below.)

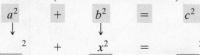

$$a^2 \quad + \quad b^2 \quad = \quad c^2$$

$$\underline{\quad}^2 \quad + \quad x^2 \quad = \quad \underline{\quad}^2$$

Finish with:

3. **SOLVE** and 4. **INTERPRET**

61. A spotlight is mounted on the eaves of a house 12 feet above the ground. A flower bed runs between the house and the sidewalk, so the closest a ladder can be placed to the house is 5 feet. How long of a ladder is needed so that an electrician can reach the place where the light is mounted?

12 ft

5 ft

62. A wire is to be attached to support a telephone pole. Because of surrounding buildings, sidewalks, and roadways, the wire must be anchored exactly 15 feet from the base of the pole. Telephone company workers have only 30 feet of cable, and 2 feet of that must be used to attach the cable to the pole and to the stake on the ground. How high from the base of the pole can the wire be attached?

←15 ft→

63. The radius of the moon is 1080 miles. Use the formula for the radius r of a sphere given its surface area A,

$$r = \sqrt{\frac{A}{4\pi}}$$

to find the surface area of the moon. Round to the nearest square mile. (*Source:* National Space Science Data Center)

64. Police departments find it very useful to be able to approximate the speed of a car when they are given the distance that the car skidded before it came to a stop. If the road surface is wet concrete, the function $S(x) = \sqrt{10.5x}$ is used, where $S(x)$ is the speed of the car in miles per hour and x is the distance skidded in feet. Find how fast a car was moving if it skidded 280 feet on wet concrete.

65. The formula $v = \sqrt{2gh}$ gives the velocity v, in feet per second, of an object when it falls h feet accelerated by gravity g, in feet per second squared. If g is approximately 32 feet per second squared, find how far an object has fallen if its velocity is 80 feet per second.

66. Two tractors are pulling a tree stump from a field. If two forces A and B pull at right angles (90°) to each other, the size of the resulting force R is given by the formula $R = \sqrt{A^2 + B^2}$. If tractor A is exerting 600 pounds of force and the resulting force is 850 pounds, find how much force tractor B is exerting.

600 lb

In psychology, it has been suggested that the number S of nonsense syllables that a person can repeat consecutively depends on his or her IQ score I according to the equation $S = 2\sqrt{I} - 9$.

67. Use this relationship to estimate the IQ of a person who can repeat 11 nonsense syllables consecutively.

68. Use this relationship to estimate the IQ of a person who can repeat 15 nonsense syllables consecutively.

*The **period** of a pendulum is the time it takes for the pendulum to make one full back-and-forth swing. The period of a pendulum depends on the length of the pendulum. The formula for the period P, in seconds, is $P = 2\pi\sqrt{\dfrac{l}{32}}$, where l is the length of the pendulum in feet. Use this formula for Exercises 69 through 74.*

69. Find the period of a pendulum whose length is 2 feet. Give an exact answer and a two-decimal-place approximation.

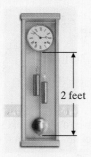

2 feet

70. Klockit sells a 43-inch lyre pendulum. Find the period of this pendulum. Round your answer to 2 decimal places. (*Hint:* First convert inches to feet.)

71. Find the length of a pendulum whose period is 4 seconds. Round your answer to 2 decimal places.

72. Find the length of a pendulum whose period is 3 seconds. Round your answer to 2 decimal places.

73. Study the relationship between period and pendulum length in Exercises 69 through 72 and make a conjecture about this relationship.

74. Galileo experimented with pendulums. He supposedly made conjectures about pendulums of equal length with different bob weights. Try this experiment. Make two pendulums 3 feet long. Attach a heavy weight (lead) to one and a light weight (a cork) to the other. Pull both pendulums back the same angle measure and release. Make a conjecture from your observations.

If the three lengths of the sides of a triangle are known, Heron's formula can be used to find its area. If a, b, and c are the lengths of the three sides, Heron's formula for area is

$$A = \sqrt{s(s - a)(s - b)(s - c)}$$

where s is half the perimeter of the triangle, or $s = \dfrac{1}{2}(a + b + c)$.

Use this formula to find the area of each triangle. Give an exact answer and then a two-decimal-place approximation.

△ **75.**

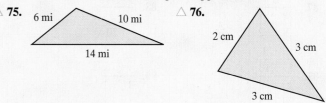

6 mi   10 mi
14 mi

△ **76.**

2 cm   3 cm
3 cm

77. Describe when Heron's formula might be useful.

78. In your own words, explain why you think s in Heron's formula is called the *semiperimeter*.

The maximum distance $D(h)$ in kilometers that a person can see from a height h kilometers above the ground is given by the function $D(h) = 111.7\sqrt{h}$. Use this function for Exercises 79 and 80. Round your answers to two decimal places.

79. Find the height that would allow a person to see 80 kilometers.

80. Find the height that would allow a person to see 40 kilometers.

REVIEW AND PREVIEW

Use the vertical line test to determine whether each graph represents the graph of a function. See Section 3.2.

81.

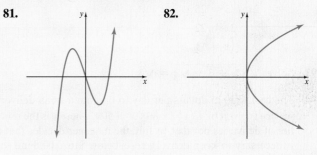

82.

83.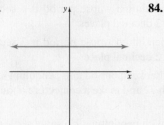

84.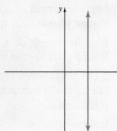

85.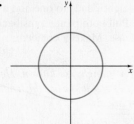

86.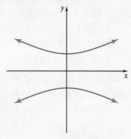

Simplify. See Section 6.3.

87. $\dfrac{\dfrac{x}{6}}{\dfrac{2x}{3} + \dfrac{1}{2}}$

88. $\dfrac{\dfrac{1}{y} + \dfrac{4}{5}}{-\dfrac{3}{20}}$

89. $\dfrac{\dfrac{z}{5} + \dfrac{1}{10}}{\dfrac{z}{20} - \dfrac{z}{5}}$

90. $\dfrac{\dfrac{1}{y} + \dfrac{1}{x}}{\dfrac{1}{y} - \dfrac{1}{x}}$

CONCEPT EXTENSIONS

Find the error in each solution and correct. See the second Concept Check in this section.

91.
$$\sqrt{5x - 1} + 4 = 7$$
$$(\sqrt{5x - 1} + 4)^2 = 7^2$$
$$5x - 1 + 16 = 49$$
$$5x = 34$$
$$x = \frac{34}{5}$$

92.
$$\sqrt{2x + 3} + 4 = 1$$
$$\sqrt{2x + 3} = 5$$
$$(\sqrt{2x + 3})^2 = 5^2$$
$$2x + 3 = 25$$
$$2x = 22$$
$$x = 11$$

93. Solve: $\sqrt{\sqrt{x + 3} + \sqrt{x}} = \sqrt{3}$

94. The cost $C(x)$ in dollars per day to operate a small delivery service is given by $C(x) = 80\sqrt[3]{x} + 500$, where x is the number of deliveries per day. In July, the manager decides that it is necessary to keep delivery costs below \$1620.00. Find the greatest number of deliveries this company can make per day and still keep overhead below \$1620.00.

95. Consider the equations $\sqrt{2x} = 4$ and $\sqrt[3]{2x} = 4$.
   a. Explain the difference in solving these equations.
   b. Explain the similarity in solving these equations.

96. Explain why proposed solutions of radical equations must be checked.

Example

For Exercises 97 through 100, see the example below.

Solve $(t^2 - 3t) - 2\sqrt{t^2 - 3t} = 0$.

Solution

Substitution can be used to make this problem somewhat simpler. Since $t^2 - 3t$ occurs more than once, let $x = t^2 - 3t$.

$$(t^2 - 3t) - 2\sqrt{t^2 - 3t} = 0$$
$$x - 2\sqrt{x} = 0$$
$$x = 2\sqrt{x}$$
$$x^2 = (2\sqrt{x})^2$$
$$x^2 = 4x$$
$$x^2 - 4x = 0$$
$$x(x - 4) = 0$$
$$x = 0 \quad \text{or} \quad x - 4 = 0$$
$$x = 4$$

Now we "undo" the substitution.
$x = 0$ Replace x with $t^2 - 3t$.

$$t^2 - 3t = 0$$
$$t(t - 3) = 0$$
$$t = 0 \quad \text{or} \quad t - 3 = 0$$
$$t = 3$$

$x = 4$ Replace x with $t^2 - 3t$.

$$t^2 - 3t = 4$$
$$t^2 - 3t - 4 = 0$$
$$(t - 4)(t + 1) = 0$$
$$t - 4 = 0 \quad \text{or} \quad t + 1 = 0$$
$$t = 4 \qquad t = -1$$

In this problem, we have four possible solutions: $0, 3, 4,$ and -1. All four solutions check in the original equation, so the solutions are $-1, 0, 3, 4$.

Solve. See the preceding example.

97. $3\sqrt{x^2 - 8x} = x^2 - 8x$

98. $\sqrt{(x^2 - x) + 7} = 2(x^2 - x) - 1$

99. $7 - (x^2 - 3x) = \sqrt{(x^2 - 3x) + 5}$

100. $x^2 + 6x = 4\sqrt{x^2 + 6x}$

7.7 Complex Numbers

OBJECTIVES

1 Write Square Roots of Negative Numbers in the Form bi.

2 Add or Subtract Complex Numbers.

3 Multiply Complex Numbers.

4 Divide Complex Numbers.

5 Raise i to Powers.

OBJECTIVE

1 Writing Numbers in the Form bi

Our work with radical expressions has excluded expressions such as $\sqrt{-16}$ because $\sqrt{-16}$ is not a real number; there is no real number whose square is -16. In this section, we discuss a number system that includes roots of negative numbers. This number system is the **complex number system,** and it includes the set of real numbers as a subset. The complex number system allows us to solve equations such as $x^2 + 1 = 0$ that have no real number solutions. The set of complex numbers includes the **imaginary unit.**

Imaginary Unit

The imaginary unit, written i, is the number whose square is -1. That is,

$$i^2 = -1 \quad \text{and} \quad i = \sqrt{-1}$$

To write the square root of a negative number in terms of i, use the property that if a is a positive number, then

$$\sqrt{-a} = \sqrt{-1} \cdot \sqrt{a}$$
$$= i \cdot \sqrt{a}$$

Using i, we can write $\sqrt{-16}$ as

$$\sqrt{-16} = \sqrt{-1 \cdot 16} = \sqrt{-1} \cdot \sqrt{16} = i \cdot 4, \text{ or } 4i$$

EXAMPLE 1 Write with i notation.

a. $\sqrt{-36}$ **b.** $\sqrt{-5}$ **c.** $-\sqrt{-20}$

Solution

a. $\sqrt{-36} = \sqrt{-1 \cdot 36} = \sqrt{-1} \cdot \sqrt{36} = i \cdot 6, \text{ or } 6i$

b. $\sqrt{-5} = \sqrt{-1(5)} = \sqrt{-1} \cdot \sqrt{5} = i\sqrt{5}$.

c. $-\sqrt{-20} = -\sqrt{-1 \cdot 20} = -\sqrt{-1} \cdot \sqrt{4 \cdot 5} = -i \cdot 2\sqrt{5} = -2i\sqrt{5}$

> ▶ Helpful Hint
> Since $\sqrt{5}i$ can easily be confused with $\sqrt{5i}$, we write $\sqrt{5}i$ as $i\sqrt{5}$.

PRACTICE

1 Write with i notation.

a. $\sqrt{-4}$ **b.** $\sqrt{-7}$ **c.** $-\sqrt{-18}$

The product rule for radicals does not necessarily hold true for imaginary numbers. *To multiply square roots of negative numbers, first we write each number in terms of the imaginary unit i.* For example, to multiply $\sqrt{-4}$ and $\sqrt{-9}$, we first write each number in the form bi.

$$\sqrt{-4}\sqrt{-9} = 2i(3i) = 6i^2 = 6(-1) = -6 \quad \text{Correct}$$

We will also use this method to simplify quotients of square roots of negative numbers. Why? The product rule does not work for this example. In other words,

$$\sqrt{-4} \cdot \sqrt{-9} = \sqrt{(-4)(-9)} = \sqrt{36} = 6 \quad \text{Incorrect}$$

EXAMPLE 2 Multiply or divide as indicated.

a. $\sqrt{-3} \cdot \sqrt{-5}$ **b.** $\sqrt{-36} \cdot \sqrt{-1}$ **c.** $\sqrt{8} \cdot \sqrt{-2}$ **d.** $\dfrac{\sqrt{-125}}{\sqrt{5}}$

Solution

a. $\sqrt{-3} \cdot \sqrt{-5} = i\sqrt{3}(i\sqrt{5}) = i^2\sqrt{15} = -1\sqrt{15} = -\sqrt{15}$

b. $\sqrt{-36} \cdot \sqrt{-1} = 6i(i) = 6i^2 = 6(-1) = -6$

c. $\sqrt{8} \cdot \sqrt{-2} = 2\sqrt{2}(i\sqrt{2}) = 2i(\sqrt{2}\sqrt{2}) = 2i(2) = 4i$

d. $\dfrac{\sqrt{-125}}{\sqrt{5}} = \dfrac{i\sqrt{125}}{\sqrt{5}} = i\sqrt{25} = 5i$ □

PRACTICE
2   Multiply or divide as indicated.

a. $\sqrt{-5} \cdot \sqrt{-6}$ **b.** $\sqrt{-9} \cdot \sqrt{-1}$ **c.** $\sqrt{125} \cdot \sqrt{-5}$ **d.** $\dfrac{\sqrt{-27}}{\sqrt{3}}$

Now that we have practiced working with the imaginary unit, we define complex numbers.

> **Complex Numbers**
>
> A **complex number** is a number that can be written in the form $a + bi$, where a and b are real numbers.

Notice that the set of real numbers is a subset of the complex numbers since any real number can be written in the form of a complex number. For example,

$$16 = 16 + 0i$$

In general, a complex number $a + bi$ is a real number if $b = 0$. Also, a complex number is called a **pure imaginary number** or an imaginary number if $a = 0$ and $b \neq 0$. For example,

$$3i = 0 + 3i \quad \text{and} \quad i\sqrt{7} = 0 + i\sqrt{7}$$

are pure imaginary numbers.

The following diagram shows the relationship between complex numbers and their subsets.

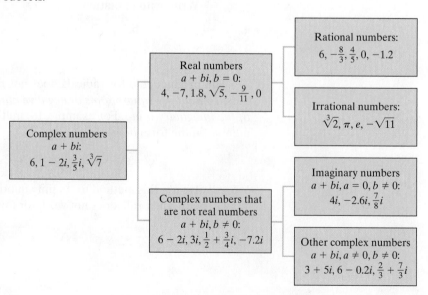

✓CONCEPT CHECK
True or false? Every complex number is also a real number.

OBJECTIVE

2   Adding or Subtracting Complex Numbers

Two complex numbers $a + bi$ and $c + di$ are equal if and only if $a = c$ and $b = d$. Complex numbers can be added or subtracted by adding or subtracting their real parts and then adding or subtracting their imaginary parts.

> **Sum or Difference of Complex Numbers**
>
> If $a + bi$ and $c + di$ are complex numbers, then their sum is
> $$(a + bi) + (c + di) = (a + c) + (b + d)i$$
> Their difference is
> $$(a + bi) - (c + di) = a + bi - c - di = (a - c) + (b - d)i$$

EXAMPLE 3   Add or subtract the complex numbers. Write the sum or difference in the form $a + bi$.

a. $(2 + 3i) + (-3 + 2i)$ **b.** $5i - (1 - i)$ **c.** $(-3 - 7i) - (-6)$

Solution

a. $(2 + 3i) + (-3 + 2i) = (2 - 3) + (3 + 2)i = -1 + 5i$

b. $5i - (1 - i) = 5i - 1 + i$
$$= -1 + (5 + 1)i$$
$$= -1 + 6i$$

c. $(-3 - 7i) - (-6) = -3 - 7i + 6$
$$= (-3 + 6) - 7i$$
$$= 3 - 7i \qquad \square$$

PRACTICE

3   Add or subtract the complex numbers. Write the sum or difference in the form $a + bi$.

a. $(3 - 5i) + (-4 + i)$ **b.** $4i - (3 - i)$ **c.** $(-5 - 2i) - (-8)$

OBJECTIVE

3   Multiplying Complex Numbers

To multiply two complex numbers of the form $a + bi$, we multiply as though they are binomials. Then we use the relationship $i^2 = -1$ to simplify.

EXAMPLE 4   Multiply the complex numbers. Write the product in the form $a + bi$.

a. $-7i \cdot 3i$ **b.** $3i(2 - i)$ **c.** $(2 - 5i)(4 + i)$
d. $(2 - i)^2$ **e.** $(7 + 3i)(7 - 3i)$

Solution

a. $-7i \cdot 3i = -21i^2$
$$= -21(-1) \quad \text{Replace } i^2 \text{ with } -1.$$
$$= 21 + 0i$$

Answer to Concept Check:
false

(Continued on next page)

b. $3i(2 - i) = 3i \cdot 2 - 3i \cdot i$ Use the distributive property.

$\qquad\qquad = 6i - 3i^2$ Multiply.

$\qquad\qquad = 6i - 3(-1)$ Replace i^2 with -1.

$\qquad\qquad = 6i + 3$

$\qquad\qquad = 3 + 6i$

$\qquad\qquad\qquad$ Use the FOIL order below. (First, Outer, Inner, Last)

c. $(2 - 5i)(4 + i) = 2(4) + 2(i) - 5i(4) - 5i(i)$

$\qquad\qquad\qquad\qquad\qquad$ F$\quad$O$\quad$I$\quad$L

$\qquad\qquad\qquad = 8 + 2i - 20i - 5i^2$

$\qquad\qquad\qquad = 8 - 18i - 5(-1)$ $\qquad\qquad\qquad i^2 = -1$

$\qquad\qquad\qquad = 8 - 18i + 5$

$\qquad\qquad\qquad = 13 - 18i$

d. $(2 - i)^2 = (2 - i)(2 - i)$

$\qquad\qquad\quad = 2(2) - 2(i) - 2(i) + i^2$

$\qquad\qquad\quad = 4 - 4i + (-1)$ $\qquad\qquad i^2 = -1$

$\qquad\qquad\quad = 3 - 4i$

e. $(7 + 3i)(7 - 3i) = 7(7) - 7(3i) + 3i(7) - 3i(3i)$

$\qquad\qquad\qquad\qquad = 49 - 21i + 21i - 9i^2$

$\qquad\qquad\qquad\qquad = 49 - 9(-1)$ $\qquad\qquad\qquad i^2 = -1$

$\qquad\qquad\qquad\qquad = 49 + 9$

$\qquad\qquad\qquad\qquad = 58 + 0i$ $\qquad\qquad\qquad\qquad\qquad\qquad$ □

PRACTICE

4   Multiply the complex numbers. Write the product in the form $a + bi$.

a. $-4i \cdot 5i$ $\qquad\qquad$ **b.** $5i(2 + i)$ $\qquad\qquad$ **c.** $(2 + 3i)(6 - i)$

d. $(3 - i)^2$ $\qquad\qquad$ **e.** $(9 + 2i)(9 - 2i)$

Notice that if you add, subtract, or multiply two complex numbers, just like real numbers, the result is a complex number.

OBJECTIVE

4   **Dividing Complex Numbers**

From Example 4e, notice that the product of $7 + 3i$ and $7 - 3i$ is a real number. These two complex numbers are called **complex conjugates** of one another. In general, we have the following definition.

Complex Conjugates

The complex numbers $(a + bi)$ and $(a - bi)$ are called **complex conjugates** of each other, and

$$(a + bi)(a - bi) = a^2 + b^2.$$

To see that the product of a complex number $a + bi$ and its conjugate $a - bi$ is the real number $a^2 + b^2$, we multiply.

$$(a + bi)(a - bi) = a^2 - abi + abi - b^2 i^2$$

$$= a^2 - b^2(-1)$$

$$= a^2 + b^2$$

We use complex conjugates to divide by a complex number.

EXAMPLE 5   Divide. Write in the form $a + bi$.

a. $\dfrac{2 + i}{1 - i}$ **b.** $\dfrac{7}{3i}$

Solution

a. Multiply the numerator and denominator by the complex conjugate of $1 - i$ to eliminate the imaginary number in the denominator.

$$\frac{2 + i}{1 - i} = \frac{(2 + i)(1 + i)}{(1 - i)(1 + i)}$$

$$= \frac{2(1) + 2(i) + 1(i) + i^2}{1^2 - i^2}$$

$$= \frac{2 + 3i - 1}{1 + 1} \qquad \text{Here, } i^2 = -1.$$

$$= \frac{1 + 3i}{2} \quad \text{or} \quad \frac{1}{2} + \frac{3}{2}i$$

b. Multiply the numerator and denominator by the conjugate of $3i$. Note that $3i = 0 + 3i$, so its conjugate is $0 - 3i$ or $-3i$.

$$\frac{7}{3i} = \frac{7(-3i)}{(3i)(-3i)} = \frac{-21i}{-9i^2} = \frac{-21i}{-9(-1)} = \frac{-21i}{9} = \frac{-7i}{3} \quad \text{or} \quad 0 - \frac{7}{3}i \qquad \square$$

PRACTICE
5   Divide. Write in the form $a + bi$.

a. $\dfrac{4 - i}{3 + i}$ **b.** $\dfrac{5}{2i}$

▶ **Helpful Hint**

Recall that division can be checked by multiplication.

To check that $\dfrac{2 + i}{1 - i} = \dfrac{1}{2} + \dfrac{3}{2}i$, in Example 5a, multiply $\left(\dfrac{1}{2} + \dfrac{3}{2}i\right)(1 - i)$ to verify that the product is $2 + i$.

OBJECTIVE
5   **Finding Powers of i** ▶

We can use the fact that $i^2 = -1$ to find higher powers of i. To find i^3, we rewrite it as the product of i^2 and i.

$$i^3 = i^2 \cdot i = (-1)i = -i$$

$$i^4 = i^2 \cdot i^2 = (-1) \cdot (-1) = 1$$

We continue this process and use the fact that $i^4 = 1$ and $i^2 = -1$ to simplify i^5 and i^6.

$$i^5 = i^4 \cdot i = 1 \cdot i = i$$

$$i^6 = i^4 \cdot i^2 = 1 \cdot (-1) = -1$$

If we continue finding powers of i, we generate the following pattern. Notice that the values $i, -1, -i,$ and 1 repeat as i is raised to higher and higher powers.

$i^1 = i$	$i^5 = i$	$i^9 = i$
$i^2 = -1$	$i^6 = -1$	$i^{10} = -1$
$i^3 = -i$	$i^7 = -i$	$i^{11} = -i$
$i^4 = 1$	$i^8 = 1$	$i^{12} = 1$

This pattern allows us to find other powers of i. To do so, we will use the fact that $i^4 = 1$ and rewrite a power of i in terms of i^4. For example,

$$i^{22} = i^{20} \cdot i^2 = (i^4)^5 \cdot i^2 = 1^5 \cdot (-1) = 1 \cdot (-1) = -1.$$

EXAMPLE 6 Find the following powers of i.

a. i^7 **b.** i^{20} **c.** i^{46} **d.** i^{-12}

Solution

a. $i^7 = i^4 \cdot i^3 = 1(-i) = -i$

b. $i^{20} = (i^4)^5 = 1^5 = 1$

c. $i^{46} = i^{44} \cdot i^2 = (i^4)^{11} \cdot i^2 = 1^{11}(-1) = -1$

d. $i^{-12} = \dfrac{1}{i^{12}} = \dfrac{1}{(i^4)^3} = \dfrac{1}{(1)^3} = \dfrac{1}{1} = 1$

PRACTICE
6 Find the following powers of i.

a. i^9 **b.** i^{16} **c.** i^{34} **d.** i^{-24}

Vocabulary, Readiness & Video Check

Use the choices below to fill in each blank. Not all choices will be used.

-1	$\sqrt{-1}$	real	imaginary unit
1	$\sqrt{1}$	complex	pure imaginary

1. A _____ number is one that can be written in the form $a + bi$, where a and b are real numbers.

2. In the complex number system, i denotes the _____.

3. $i^2 =$ _____

4. $i =$ _____

5. A complex number, $a + bi$, is a _____ number if $b = 0$.

6. A complex number, $a + bi$, is a _____ number if $a = 0$ and $b \neq 0$.

Martin-Gay Interactive Videos

See Video 7.7

Watch the section lecture video and answer the following questions.

OBJECTIVE 1
7. From ▭ Example 4, with what rule must you be especially careful when working with imaginary numbers and why?

OBJECTIVE 2
8. In ▭ Examples 5 and 6, what is the process of adding and subtracting complex numbers compared to? What important reminder is given about i?

OBJECTIVE 3
9. In ▭ Examples 7 and 8, what part of the definition of the imaginary unit i may be used during the multiplication of complex numbers to help simplify products?

OBJECTIVE 4
10. In ▭ Example 9, using complex conjugates to divide complex numbers is compared to what process?

OBJECTIVE 5
11. From the lecture before ▭ Example 10, what are the first four powers of i whose values keep repeating?

7.7 Exercise Set MyMathLab®

Simplify. See Example 1.

1. $\sqrt{-81}$ **2.** $\sqrt{-49}$ **3.** $\sqrt{-7}$

4. $\sqrt{-3}$ **5.** $-\sqrt{16}$ **6.** $-\sqrt{4}$

7. $\sqrt{-64}$ **8.** $\sqrt{-100}$

Write in terms of i. See Example 1.

9. $\sqrt{-24}$ **10.** $\sqrt{-32}$

11. $-\sqrt{-36}$ **12.** $-\sqrt{-121}$

13. $8\sqrt{-63}$ **14.** $4\sqrt{-20}$

15. $-\sqrt{54}$ **16.** $\sqrt{-63}$

Multiply or divide. See Example 2.

17. $\sqrt{-2}\cdot\sqrt{-7}$ **18.** $\sqrt{-11}\cdot\sqrt{-3}$

19. $\sqrt{-5}\cdot\sqrt{-10}$ **20.** $\sqrt{-2}\cdot\sqrt{-6}$

21. $\sqrt{16}\cdot\sqrt{-1}$ **22.** $\sqrt{3}\cdot\sqrt{-27}$

23. $\dfrac{\sqrt{-9}}{\sqrt{3}}$ **24.** $\dfrac{\sqrt{49}}{\sqrt{-10}}$

25. $\dfrac{\sqrt{-80}}{\sqrt{-10}}$ **26.** $\dfrac{\sqrt{-40}}{\sqrt{-8}}$

Add or subtract. Write the sum or difference in the form a + bi. See Example 3.

27. $(4-7i)+(2+3i)$ **28.** $(2-4i)-(2-i)$

29. $(6+5i)-(8-i)$ **30.** $(8-3i)+(-8+3i)$

31. $6-(8+4i)$ **32.** $(9-4i)-9$

Multiply. Write the product in the form a + bi. See Example 4.

33. $-10i\cdot-4i$ **34.** $-2i\cdot-11i$

35. $6i(2-3i)$ **36.** $5i(4-7i)$

37. $(\sqrt{3}+2i)(\sqrt{3}-2i)$ **38.** $(\sqrt{5}-5i)(\sqrt{5}+5i)$

39. $(4-2i)^2$ **40.** $(6-3i)^2$

Write each quotient in the form a + bi. See Example 5.

41. $\dfrac{4}{i}$ **42.** $\dfrac{5}{6i}$

43. $\dfrac{7}{4+3i}$ **44.** $\dfrac{9}{1-2i}$

45. $\dfrac{3+5i}{1+i}$ **46.** $\dfrac{6+2i}{4-3i}$

47. $\dfrac{5-i}{3-2i}$ **48.** $\dfrac{6-i}{2+i}$

MIXED PRACTICE

Perform each indicated operation. Write the result in the form a + bi.

49. $(7i)(-9i)$ **50.** $(-6i)(-4i)$

51. $(6-3i)-(4-2i)$ **52.** $(-2-4i)-(6-8i)$

53. $-3i(-1+9i)$ **54.** $-5i(-2+i)$

55. $\dfrac{4-5i}{2i}$ **56.** $\dfrac{6+8i}{3i}$

57. $(4+i)(5+2i)$ **58.** $(3+i)(2+4i)$

59. $(6-2i)(3+i)$ **60.** $(2-4i)(2-i)$

61. $(8-3i)+(2+3i)$ **62.** $(7+4i)+(4-4i)$

63. $(1-i)(1+i)$ **64.** $(6+2i)(6-2i)$

65. $\dfrac{16+15i}{-3i}$ **66.** $\dfrac{2-3i}{-7i}$

67. $(9+8i)^2$ **68.** $(4-7i)^2$

69. $\dfrac{2}{3+i}$ **70.** $\dfrac{5}{3-2i}$

71. $(5-6i)-4i$ **72.** $(6-2i)+7i$

73. $\dfrac{2-3i}{2+i}$ **74.** $\dfrac{6+5i}{6-5i}$

75. $(2+4i)+(6-5i)$ **76.** $(5-3i)+(7-8i)$

77. $(\sqrt{6}+i)(\sqrt{6}-i)$ **78.** $(\sqrt{14}-4i)(\sqrt{14}+4i)$

79. $4(2-i)^2$ **80.** $9(2-i)^2$

Find each power of i. See Example 6.

81. i^8 **82.** i^{10} **83.** i^{21} **84.** i^{15}

85. i^{11} **86.** i^{40} **87.** i^{-6} **88.** i^{-9}

89. $(2i)^6$ **90.** $(5i)^4$ **91.** $(-3i)^5$ **92.** $(-2i)^7$

REVIEW AND PREVIEW

Recall that the sum of the measures of the angles of a triangle is 180°. Find the unknown angle in each triangle.

93.

94.

Use synthetic division to divide the following. See Section 6.4.

95. $(x^3-6x^2+3x-4)\div(x-1)$

96. $(5x^4-3x^2+2)\div(x+2)$

Thirty people were recently polled about the average monthly balance in their checking accounts. The results of this poll are shown in the following histogram. Use this graph to answer Exercises 97 through 102. See Section 1.3.

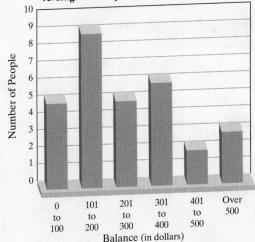

97. How many people polled reported an average checking balance of $201 to $300?

98. How many people polled reported an average checking balance of $0 to $100?

99. How many people polled reported an average checking balance of $200 or less?

100. How many people polled reported an average checking balance of $301 or more?

101. What percent of people polled reported an average checking balance of $201 to $300? Round to the nearest tenth of a percent.

102. What percent of people polled reported an average checking balance of $0 to $100? Round to the nearest tenth of a percent.

CONCEPT EXTENSIONS

Write in the form $a + bi$.

103. $i^3 - i^4$

104. $i^8 - i^7$

105. $i^6 + i^8$

106. $i^4 + i^{12}$

107. $2 + \sqrt{-9}$

108. $5 - \sqrt{-16}$

109. $\dfrac{6 + \sqrt{-18}}{3}$

110. $\dfrac{4 - \sqrt{-8}}{2}$

111. $\dfrac{5 - \sqrt{-75}}{10}$

112. $\dfrac{7 + \sqrt{-98}}{14}$

113. Describe how to find the conjugate of a complex number.

114. Explain why the product of a complex number and its complex conjugate is a real number.

Simplify.

115. $\left(8 - \sqrt{-3}\right) - \left(2 + \sqrt{-12}\right)$

116. $\left(8 - \sqrt{-4}\right) - \left(2 + \sqrt{-16}\right)$

117. Determine whether $2i$ is a solution of $x^2 + 4 = 0$.

118. Determine whether $-1 + i$ is a solution of $x^2 + 2x = -2$.

Chapter 7 Vocabulary Check

Fill in each blank with one of the words or phrases listed below.

| index | rationalizing | conjugate | principal square root | cube root | midpoint |
| complex number | like radicals | radicand | imaginary unit | | distance |

1. The _____ of $\sqrt{3} + 2$ is $\sqrt{3} - 2$.

2. The _____ of a nonnegative number a is written as $\sqrt{a}$.

3. The process of writing a radical expression as an equivalent expression but without a radical in the denominator is called _____ the denominator.

4. The _____, written i, is the number whose square is -1.

5. The _____ of a number is written as $\sqrt[3]{a}$.

6. In the notation $\sqrt[n]{a}$, n is called the _____ and a is called the _____.

7. Radicals with the same index and the same radicand are called _____.

8. A(n) _____ is a number that can be written in the form $a + bi$, where a and b are real numbers.

9. The _____ formula is $d = \sqrt{(x_2 - x_1)^2 + (y_2 - y_1)^2}$.

10. The _____ formula is $\left(\dfrac{x_1 + x_2}{2}, \dfrac{y_1 + y_2}{2}\right)$.

Chapter 7 Highlights

DEFINITIONS AND CONCEPTS	EXAMPLES
Section 7.1 Radicals and Radical Functions	
The **positive**, or **principal**, **square root** of a nonnegative number a is written as $\sqrt{a}$. $\sqrt{a} = b$ only if $b^2 = a$ and $b \geq 0$ The **negative square root of** a is written as $-\sqrt{a}$.	$\sqrt{36} = 6 \qquad \sqrt{\dfrac{9}{100}} = \dfrac{3}{10}$ $-\sqrt{36} = -6 \quad -\sqrt{0.04} = -0.2$

DEFINITIONS AND CONCEPTS	EXAMPLES

Section 7.1 Radicals and Radical Functions (continued)

The **cube root** of a real number a is written as $\sqrt[3]{a}$.

$$\sqrt[3]{a} = b \text{ only if } b^3 = a$$

If n is an even positive integer, then $\sqrt[n]{a^n} = |a|$.

If n is an odd positive integer, then $\sqrt[n]{a^n} = a$.

A **radical function** in x is a function defined by an expression containing a root of x.

$$\sqrt[3]{27} = 3 \qquad \sqrt[3]{-\frac{1}{8}} = -\frac{1}{2}$$

$$\sqrt[3]{y^6} = y^2 \qquad \sqrt[3]{64x^9} = 4x^3$$

$$\sqrt{(-3)^2} = |-3| = 3$$

$$\sqrt[3]{(-7)^3} = -7$$

If $f(x) = \sqrt{x} + 2$,

$$f(1) = \sqrt{(1)} + 2 = 1 + 2 = 3$$

$$f(3) = \sqrt{(3)} + 2 \approx 3.73$$

Section 7.2 Rational Exponents

$a^{1/n} = \sqrt[n]{a}$ if $\sqrt[n]{a}$ is a real number.

If m and n are positive integers greater than 1 with $\dfrac{m}{n}$ in lowest terms and $\sqrt[n]{a}$ is a real number, then

$$a^{m/n} = (a^{1/n})^m = (\sqrt[n]{a})^m$$

$a^{-m/n} = \dfrac{1}{a^{m/n}}$ as long as $a^{m/n}$ is a nonzero number.

Exponent rules are true for rational exponents.

$$81^{1/2} = \sqrt{81} = 9$$

$$(-8x^3)^{1/3} = \sqrt[3]{-8x^3} = -2x$$

$$4^{5/2} = (\sqrt{4})^5 = 2^5 = 32$$

$$27^{2/3} = (\sqrt[3]{27})^2 = 3^2 = 9$$

$$16^{-3/4} = \frac{1}{16^{3/4}} = \frac{1}{(\sqrt[4]{16})^3} = \frac{1}{2^3} = \frac{1}{8}$$

$$x^{2/3} \cdot x^{-5/6} = x^{2/3 - 5/6} = x^{-1/6} = \frac{1}{x^{1/6}}$$

$$(8^4)^{1/2} = 8^2 = 64$$

$$\frac{a^{4/5}}{a^{-2/5}} = a^{4/5 - (-2/5)} = a^{6/5}$$

Section 7.3 Simplifying Radical Expressions

Product and Quotient Rules

If $\sqrt[n]{a}$ and $\sqrt[n]{b}$ are real numbers,

$$\sqrt[n]{a} \cdot \sqrt[n]{b} = \sqrt[n]{a \cdot b}$$

$$\frac{\sqrt[n]{a}}{\sqrt[n]{b}} = \sqrt[n]{\frac{a}{b}}, \text{ provided } \sqrt[n]{b} \neq 0$$

A radical of the form $\sqrt[n]{a}$ is **simplified** when a contains no factors that are perfect nth powers.

Distance Formula

The distance d between two points (x_1, y_1) and (x_2, y_2) is given by

$$d = \sqrt{(x_2 - x_1)^2 + (y_2 - y_1)^2}$$

Multiply or divide as indicated:

$$\sqrt{11} \cdot \sqrt{3} = \sqrt{33}$$

$$\frac{\sqrt[3]{40x}}{\sqrt[3]{5x}} = \sqrt[3]{8} = 2$$

$$\sqrt{40} = \sqrt{4 \cdot 10} = 2\sqrt{10}$$

$$\sqrt{36x^5} = \sqrt{36x^4 \cdot x} = 6x^2\sqrt{x}$$

$$\sqrt[3]{24x^7y^3} = \sqrt[3]{8x^6y^3 \cdot 3x} = 2x^2y\sqrt[3]{3x}$$

Find the distance between points $(-1, 6)$ and $(-2, -4)$. Let $(x_1, y_1) = (-1, 6)$ and $(x_2, y_2) = (-2, -4)$.

$$d = \sqrt{(x_2 - x_1)^2 + (y_2 - y_1)^2}$$

$$= \sqrt{(-2 - (-1))^2 + (-4 - 6)^2}$$

$$= \sqrt{1 + 100} = \sqrt{101}$$

(continued)

DEFINITIONS AND CONCEPTS	EXAMPLES

Section 7.3 Simplifying Radical Expressions (continued)

Midpoint Formula

The midpoint of the line segment whose endpoints are (x_1, y_1) and (x_2, y_2) is the point with coordinates

$$\left(\frac{x_1 + x_2}{2}, \frac{y_1 + y_2}{2} \right)$$

Find the midpoint of the line segment whose endpoints are $(-1, 6)$ and $(-2, -4)$.

$$\left(\frac{-1 + (-2)}{2}, \frac{6 + (-4)}{2} \right)$$

The midpoint is $\left(-\frac{3}{2}, 1 \right)$.

Section 7.4 Adding, Subtracting, and Multiplying Radical Expressions

Radicals with the same index and the same radicand are **like radicals.**

The distributive property can be used to add like radicals.

$$5\sqrt{6} + 2\sqrt{6} = (5 + 2)\sqrt{6} = 7\sqrt{6}$$

$$\sqrt[3]{3x} - 10\sqrt[3]{3x} + 3\sqrt[3]{10x}$$
$$= (-1 - 10)\sqrt[3]{3x} + 3\sqrt[3]{10x}$$
$$= -11\sqrt[3]{3x} + 3\sqrt[3]{10x}$$

Radical expressions are multiplied by using many of the same properties used to multiply polynomials.

Multiply:

$$(\sqrt{5} - \sqrt{2x})(\sqrt{2} + \sqrt{2x})$$
$$= \sqrt{10} + \sqrt{10x} - \sqrt{4x} - 2x$$
$$= \sqrt{10} + \sqrt{10x} - 2\sqrt{x} - 2x$$
$$(2\sqrt{3} - \sqrt{8x})(2\sqrt{3} + \sqrt{8x})$$
$$= 4(3) - 8x = 12 - 8x$$

Section 7.5 Rationalizing Denominators and Numerators of Radical Expressions

The **conjugate** of $a + b$ is $a - b$.

The conjugate of $\sqrt{7} + \sqrt{3}$ is $\sqrt{7} - \sqrt{3}$.

The process of writing the denominator of a radical expression without a radical is called **rationalizing the denominator.**

Rationalize each denominator.

$$\frac{\sqrt{5}}{\sqrt{3}} = \frac{\sqrt{5} \cdot \sqrt{3}}{\sqrt{3} \cdot \sqrt{3}} = \frac{\sqrt{15}}{3}$$

$$\frac{6}{\sqrt{7} + \sqrt{3}} = \frac{6(\sqrt{7} - \sqrt{3})}{(\sqrt{7} + \sqrt{3})(\sqrt{7} - \sqrt{3})}$$

$$= \frac{6(\sqrt{7} - \sqrt{3})}{7 - 3}$$

$$= \frac{6(\sqrt{7} - \sqrt{3})}{4} = \frac{3(\sqrt{7} - \sqrt{3})}{2}$$

DEFINITIONS AND CONCEPTS	EXAMPLES

Section 7.5 Rationalizing Denominators and Numerators of Radical Expressions (continued)

The process of writing the numerator of a radical expression without a radical is called **rationalizing the numerator.**

Rationalize each numerator:

$$\frac{\sqrt[3]{9}}{\sqrt[3]{5}} = \frac{\sqrt[3]{9} \cdot \sqrt[3]{3}}{\sqrt[3]{5} \cdot \sqrt[3]{3}} = \frac{\sqrt[3]{27}}{\sqrt[3]{15}} = \frac{3}{\sqrt[3]{15}}$$

$$\frac{\sqrt{9} + \sqrt{3x}}{12} = \frac{(\sqrt{9} + \sqrt{3x})(\sqrt{9} - \sqrt{3x})}{12(\sqrt{9} - \sqrt{3x})}$$

$$= \frac{9 - 3x}{12(\sqrt{9} - \sqrt{3x})}$$

$$= \frac{3(3 - x)}{3 \cdot 4(3 - \sqrt{3x})} = \frac{3 - x}{4(3 - \sqrt{3x})}$$

Section 7.6 Radical Equations and Problem Solving

To Solve a Radical Equation

Step 1. Write the equation so that one radical is by itself on one side of the equation.

Step 2. Raise each side of the equation to a power equal to the index of the radical and simplify.

Step 3. If the equation still contains a radical, repeat Steps 1 and 2. If not, solve the equation.

Step 4. Check all proposed solutions in the original equation.

Solve: $x = \sqrt{4x + 9} + 3$.

1. $\qquad x - 3 = \sqrt{4x + 9}$

2. $\quad (x - 3)^2 = (\sqrt{4x + 9})^2$
$x^2 - 6x + 9 = 4x + 9$

3. $\quad x^2 - 10x = 0$
$x(x - 10) = 0$
$\qquad x = 0 \quad \text{or} \quad x = 10$

4. The proposed solution 10 checks, but 0 does not. The solution is 10.

Section 7.7 Complex Numbers

$i^2 = -1$ and $i = \sqrt{-1}$

A **complex number** is a number that can be written in the form $a + bi$, where a and b are real numbers.

Simplify: $\sqrt{-9}$.

$$\sqrt{-9} = \sqrt{-1 \cdot 9} = \sqrt{-1} \cdot \sqrt{9} = i \cdot 3 \text{ or } 3i$$

Complex Numbers	**Written in Form $a + bi$**
12	$12 + 0i$
$-5i$	$0 + (-5)i$
$-2 - 3i$	$-2 + (-3)i$

Multiply,

$$\sqrt{-3} \cdot \sqrt{-7} = i\sqrt{3} \cdot i\sqrt{7}$$
$$= i^2\sqrt{21}$$
$$= -\sqrt{21}$$

To add or subtract complex numbers, add or subtract their real parts and then add or subtract their imaginary parts.

To multiply complex numbers, multiply as though they are binomials.

Perform each indicated operation.

$$(-3 + 2i) - (7 - 4i) = -3 + 2i - 7 + 4i$$
$$= -10 + 6i$$
$$(-7 - 2i)(6 + i) = -42 - 7i - 12i - 2i^2$$
$$= -42 - 19i - 2(-1)$$
$$= -42 - 19i + 2$$
$$= -40 - 19i$$

(continued)

DEFINITIONS AND CONCEPTS	EXAMPLES
Section 7.7 Complex Numbers (continued)	

The complex numbers $(a + bi)$ and $(a - bi)$ are called **complex conjugates.**

The complex conjugate of
$$(3 + 6i) \text{ is } (3 - 6i).$$
Their product is a real number:
$$(3 - 6i)(3 + 6i) = 9 - 36i^2$$
$$= 9 - 36(-1) = 9 + 36 = 45$$

To divide complex numbers, multiply the numerator and the denominator by the conjugate of the denominator.

Divide.
$$\frac{4}{2 - i} = \frac{4(2 + i)}{(2 - i)(2 + i)}$$
$$= \frac{4(2 + i)}{4 - i^2}$$
$$= \frac{4(2 + i)}{5}$$
$$= \frac{8 + 4i}{5} = \frac{8}{5} + \frac{4}{5}i$$

Chapter 7 Review

(7.1) *Find the root. Assume that all variables represent positive numbers.*

1. $\sqrt{81}$
2. $\sqrt[4]{81}$
3. $\sqrt[3]{-8}$
4. $\sqrt[4]{-16}$
5. $-\sqrt{\frac{1}{49}}$
6. $\sqrt{x^{64}}$
7. $-\sqrt{36}$
8. $\sqrt[3]{64}$
9. $\sqrt[3]{-a^6b^9}$
10. $\sqrt{16a^4b^{12}}$
11. $\sqrt[5]{32a^5b^{10}}$
12. $\sqrt[5]{-32x^{15}y^{20}}$
13. $\sqrt{\frac{x^{12}}{36y^2}}$
14. $\sqrt[3]{\frac{27y^3}{z^{12}}}$

Simplify. Use absolute value bars when necessary.

15. $\sqrt{(-x)^2}$
16. $\sqrt[4]{(x^2 - 4)^4}$
17. $\sqrt[3]{(-27)^3}$
18. $\sqrt[5]{(-5)^5}$
19. $-\sqrt[5]{x^5}$
20. $-\sqrt[3]{x^3}$
21. $\sqrt[4]{16(2y + z)^4}$
22. $\sqrt{25(x - y)^2}$
23. $\sqrt[5]{y^5}$
24. $\sqrt[6]{x^6}$

25. Let $f(x) = \sqrt{x} + 3$.
    a. Find $f(0)$ and $f(9)$.
    b. Find the domain of $f(x)$.
    c. Graph $f(x)$.

26. Let $g(x) = \sqrt[3]{x} - 3$.
    a. Find $g(11)$ and $g(20)$.
    b. Find the domain $g(x)$.
    c. Graph $g(x)$.

(7.2) *Evaluate.*

27. $\left(\frac{1}{81}\right)^{1/4}$
28. $\left(-\frac{1}{27}\right)^{1/3}$
29. $(-27)^{-1/3}$
30. $(-64)^{-1/3}$
31. $-9^{3/2}$
32. $64^{-1/3}$
33. $(-25)^{5/2}$
34. $\left(\frac{25}{49}\right)^{-3/2}$
35. $\left(\frac{8}{27}\right)^{-2/3}$
36. $\left(-\frac{1}{36}\right)^{-1/4}$

Write with rational exponents.

37. $\sqrt[3]{x^2}$
38. $\sqrt[5]{5x^2y^3}$

Write using radical notation.

39. $y^{4/5}$
40. $5(xy^2z^5)^{1/3}$
41. $(x + 2)^{-1/3}$
42. $(x + 2y)^{-1/2}$

Simplify each expression. Assume that all variables represent positive real numbers. Write with only positive exponents.

43. $a^{1/3}a^{4/3}a^{1/2}$
44. $\frac{b^{1/3}}{b^{4/3}}$
45. $(a^{1/2}a^{-2})^3$
46. $(x^{-3}y^6)^{1/3}$
47. $\left(\frac{b^{3/4}}{a^{-1/2}}\right)^8$
48. $\frac{x^{1/4}x^{-1/2}}{x^{2/3}}$
49. $\left(\frac{49c^{5/3}}{a^{-1/4}b^{5/6}}\right)^{-1}$
50. $a^{-1/4}(a^{5/4} - a^{9/4})$

Use a calculator and write a three-decimal-place approximation of each number.

51. $\sqrt{20}$

52. $\sqrt[3]{-39}$

53. $\sqrt[4]{726}$

54. $56^{1/3}$

55. $-78^{3/4}$

56. $105^{-2/3}$

Use rational exponents to write each as a single radical.

57. $\sqrt[3]{2} \cdot \sqrt{7}$

58. $\sqrt[3]{3} \cdot \sqrt[4]{x}$

(7.3) Perform each indicated operation and then simplify if possible. Assume that all variables represent positive real numbers.

59. $\sqrt{3} \cdot \sqrt{8}$

60. $\sqrt[3]{7y} \cdot \sqrt[3]{x^2z}$

61. $\dfrac{\sqrt{44x^3}}{\sqrt{11x}}$

62. $\dfrac{\sqrt[4]{a^6b^{13}}}{\sqrt[4]{a^2b}}$

Simplify.

63. $\sqrt{60}$

64. $-\sqrt{75}$

65. $\sqrt[3]{162}$

66. $\sqrt[3]{-32}$

67. $\sqrt{36x^7}$

68. $\sqrt[3]{24a^5b^7}$

69. $\sqrt{\dfrac{p^{17}}{121}}$

70. $\sqrt[3]{\dfrac{y^5}{27x^6}}$

71. $\sqrt[4]{\dfrac{xy^6}{81}}$

72. $\sqrt{\dfrac{2x^3}{49y^4}}$

△ The formula for the radius r of a circle of area A is $r = \sqrt{\dfrac{A}{\pi}}$. Use this for Exercises 73 and 74.

73. Find the exact radius of a circle whose area is 25 square meters.

74. Approximate to two decimal places the radius of a circle whose area is 104 square inches.

Find the distance between each pair of points. Give an exact value and a three-decimal-place approximation.

75. $(-6, 3)$ and $(8, 4)$

76. $(-4, -6)$ and $(-1, 5)$

77. $(-1, 5)$ and $(2, -3)$

78. $(-\sqrt{2}, 0)$ and $(0, -4\sqrt{6})$

79. $(-\sqrt{5}, -\sqrt{11})$ and $(-\sqrt{5}, -3\sqrt{11})$

80. $(7.4, -8.6)$ and $(-1.2, 5.6)$

Find the midpoint of each line segment whose endpoints are given.

81. $(2, 6); (-12, 4)$

82. $(-6, -5); (-9, 7)$

83. $(4, -6); (-15, 2)$

84. $\left(0, -\dfrac{3}{8}\right); \left(\dfrac{1}{10}, 0\right)$

85. $\left(\dfrac{3}{4}, -\dfrac{1}{7}\right); \left(-\dfrac{1}{4}, -\dfrac{3}{7}\right)$

86. $(\sqrt{3}, -2\sqrt{6}); (\sqrt{3}, -4\sqrt{6})$

(7.4) Perform each indicated operation. Assume that all variables represent positive real numbers.

87. $\sqrt{20} + \sqrt{45} - 7\sqrt{5}$

88. $x\sqrt{75x} - \sqrt{27x^3}$

89. $\sqrt[3]{128} + \sqrt[3]{250}$

90. $3\sqrt[4]{32a^5} - a\sqrt[4]{162a}$

91. $\dfrac{5}{\sqrt{4}} + \dfrac{\sqrt{3}}{3}$

92. $\sqrt{\dfrac{8}{x^2}} - \sqrt{\dfrac{50}{16x^2}}$

93. $2\sqrt{50} - 3\sqrt{125} + \sqrt{98}$

94. $2a\sqrt[4]{32b^5} - 3b\sqrt[4]{162a^4b} + \sqrt[4]{2a^4b^5}$

Multiply and then simplify if possible. Assume that all variables represent positive real numbers.

95. $\sqrt{3}(\sqrt{27} - \sqrt{3})$

96. $(\sqrt{x} - 3)^2$

97. $(\sqrt{5} - 5)(2\sqrt{5} + 2)$

98. $(2\sqrt{x} - 3\sqrt{y})(2\sqrt{x} + 3\sqrt{y})$

99. $(\sqrt{a} + 3)(\sqrt{a} - 3)$

100. $(\sqrt[3]{a} + 2)^2$

101. $(\sqrt[3]{5x} + 9)(\sqrt[3]{5x} - 9)$

102. $(\sqrt[3]{a} + 4)(\sqrt[3]{a^2} - 4\sqrt[3]{a} + 16)$

(7.5) Rationalize each denominator. Assume that all variables represent positive real numbers.

103. $\dfrac{3}{\sqrt{7}}$

104. $\sqrt{\dfrac{x}{12}}$

105. $\dfrac{5}{\sqrt[3]{4}}$

106. $\sqrt{\dfrac{24x^5}{3y}}$

107. $\sqrt[3]{\dfrac{15x^6y^7}{z^2}}$

108. $\sqrt[4]{\dfrac{81}{8x^{10}}}$

109. $\dfrac{3}{\sqrt{y} - 2}$

110. $\dfrac{\sqrt{2} - \sqrt{3}}{\sqrt{2} + \sqrt{3}}$

Rationalize each numerator. Assume that all variables represent positive real numbers.

111. $\dfrac{\sqrt{11}}{3}$

112. $\sqrt{\dfrac{18}{y}}$

113. $\dfrac{\sqrt[3]{9}}{7}$

114. $\sqrt{\dfrac{24x^5}{3y^2}}$

115. $\sqrt[3]{\dfrac{xy^2}{10z}}$

116. $\dfrac{\sqrt{x} + 5}{-3}$

(7.6) Solve each equation.

117. $\sqrt{y - 7} = 5$

118. $\sqrt{2x} + 10 = 4$

119. $\sqrt[3]{2x - 6} = 4$

120. $\sqrt{x + 6} = \sqrt{x + 2}$

121. $2x - 5\sqrt{x} = 3$

122. $\sqrt{x + 9} = 2 + \sqrt{x - 7}$

Find each unknown length.

△ **123.**

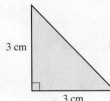

3 cm

3 cm

△ **124.**

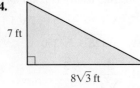

7 ft

$8\sqrt{3}$ ft

△ **125.** Craig and Daniel Cantwell want to determine the distance *x* across a pond on their property. They are able to measure the distances shown on the following diagram. Find how wide the pond is at the crossing point indicated by the triangle to the nearest tenth of a foot.

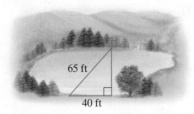

65 ft

40 ft

△ **126.** Andrea Roberts, a pipefitter, needs to connect two underground pipelines that are offset by 3 feet, as pictured in the diagram. Neglecting the joints needed to join the pipes, find the length of the shortest possible connecting pipe rounded to the nearest hundredth of a foot.

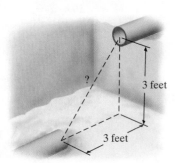

?

3 feet

3 feet

(7.7) *Perform each indicated operation and simplify. Write the results in the form a + bi.*

127. $\sqrt{-8}$

128. $-\sqrt{-6}$

129. $\sqrt{-4} + \sqrt{-16}$

130. $\sqrt{-2} \cdot \sqrt{-5}$

131. $(12 - 6i) + (3 + 2i)$

132. $(-8 - 7i) - (5 - 4i)$

133. $(2i)^6$　　　　　**134.** $(3i)^4$

135. $-3i(6 - 4i)$　　**136.** $(3 + 2i)(1 + i)$

137. $(2 - 3i)^2$　　　**138.** $(\sqrt{6} - 9i)(\sqrt{6} + 9i)$

139. $\dfrac{2 + 3i}{2i}$　　　**140.** $\dfrac{1 + i}{-3i}$

MIXED REVIEW

Simplify. Use absolute value bars when necessary.

141. $\sqrt[3]{x^3}$　　　　　　**142.** $\sqrt{(x + 2)^2}$

Simplify. Assume that all variables represent positive real numbers. If necessary, write answers with positive exponents only.

143. $-\sqrt{100}$　　　　　**144.** $\sqrt[3]{-x^{12}y^3}$

145. $\sqrt[4]{\dfrac{y^{20}}{16x^{12}}}$　　　　**146.** $9^{1/2}$

147. $64^{-1/2}$　　　　　**148.** $\left(\dfrac{27}{64}\right)^{-2/3}$

149. $\dfrac{(x^{2/3}x^{-3})^3}{x^{-1/2}}$　　　**150.** $\sqrt{200x^9}$

151. $\sqrt{\dfrac{3n^3}{121m^{10}}}$

152. $3\sqrt{20} - 7x\sqrt[3]{40} + 3\sqrt[3]{5x^3}$

153. $(2\sqrt{x} - 5)^2$

154. Find the distance between $(-3, 5)$ and $(-8, 9)$.

155. Find the midpoint of the line segment joining $(-3, 8)$ and $(11, 24)$.

Rationalize each denominator.

156. $\dfrac{7}{\sqrt{13}}$　　　　　**157.** $\dfrac{2}{\sqrt{x} + 3}$

Solve.

158. $\sqrt{x + 2} = x$　　　**159.** $\sqrt{2x - 1} + 2 = x$

Chapter 7 Test　MyMathLab®　　**CHAPTER Test Prep VIDEOS**　 You Tube™

Raise to the power or find the root. Assume that all variables represent positive numbers. Write with only positive exponents.

▶ **1.** $\sqrt{216}$

▶ **2.** $-\sqrt[4]{x^{64}}$

▶ **3.** $\left(\dfrac{1}{125}\right)^{1/3}$

▶ **4.** $\left(\dfrac{1}{125}\right)^{-1/3}$

▶ **5.** $\left(\dfrac{8x^3}{27}\right)^{2/3}$

▶ **6.** $\sqrt[3]{-a^{18}b^9}$

▶ **7.** $\left(\dfrac{64c^{4/3}}{a^{-2/3}b^{5/6}}\right)^{1/2}$

▶ **8.** $a^{-2/3}(a^{5/4} - a^3)$

Find the root. Use absolute value bars when necessary.

▶ **9.** $\sqrt[4]{(4xy)^4}$

▶ **10.** $\sqrt[3]{(-27)^3}$

Rationalize the denominator. Assume that all variables represent positive numbers.

▶ **11.** $\sqrt{\dfrac{9}{y}}$　　▶ **12.** $\dfrac{4 - \sqrt{x}}{4 + 2\sqrt{x}}$　　▶ **13.** $\dfrac{\sqrt[3]{ab}}{\sqrt[3]{ab^2}}$

▶ **14.** Rationalize the numerator of $\dfrac{\sqrt{6} + x}{8}$ and simplify.

Perform the indicated operations. Assume that all variables represent positive numbers.

15. $\sqrt{125x^3} - 3\sqrt{20x^3}$

16. $\sqrt{3}(\sqrt{16} - \sqrt{2})$

17. $(\sqrt{x} + 1)^2$

18. $(\sqrt{2} - 4)(\sqrt{3} + 1)$

19. $(\sqrt{5} + 5)(\sqrt{5} - 5)$

Use a calculator to approximate each to three decimal places.

20. $\sqrt{561}$

21. $386^{-2/3}$

Solve.

22. $x = \sqrt{x - 2} + 2$

23. $\sqrt{x^2 - 7} + 3 = 0$

24. $\sqrt[3]{x + 5} = \sqrt[3]{2x - 1}$

Perform the indicated operation and simplify. Write the result in the form $a + bi$.

25. $\sqrt{-2}$

26. $-\sqrt{-8}$

27. $(12 - 6i) - (12 - 3i)$

28. $(6 - 2i)(6 + 2i)$

29. $(4 + 3i)^2$

30. $\dfrac{1 + 4i}{1 - i}$

31. Find x.

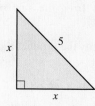

32. Identify the domain of $g(x)$. Then complete the accompanying table and graph $g(x)$.

$$g(x) = \sqrt{x + 2}$$

x	−2	−1	2	7
$g(x)$				

33. Find the distance between the points $(-6, 3)$ and $(-8, -7)$.

34. Find the distance between the points $(-2\sqrt{5}, \sqrt{10})$ and $(-\sqrt{5}, 4\sqrt{10})$.

35. Find the midpoint of the line segment whose endpoints are $(-2, -5)$ and $(-6, 12)$.

36. Find the midpoint of the line segment whose endpoints are $\left(-\dfrac{2}{3}, -\dfrac{1}{5}\right)$ and $\left(-\dfrac{1}{3}, \dfrac{4}{5}\right)$.

Solve.

37. The function $V(r) = \sqrt{2.5r}$ can be used to estimate the maximum safe velocity V in miles per hour at which a car can travel if it is driven along a curved road with a *radius of curvature r* in feet. To the nearest whole number, find the maximum safe speed if a cloverleaf exit on an expressway has a radius of curvature of 300 feet.

38. Use the formula from Exercise 37 to find the radius of curvature if the safe velocity is 30 mph.

Chapter 7 Cumulative Review

1. Simplify each expression.

a. $3xy - 2xy + 5 - 7 + xy$

b. $7x^2 + 3 - 5(x^2 - 4)$

c. $(2.1x - 5.6) - (-x - 5.3)$

d. $\dfrac{1}{2}(4a - 6b) - \dfrac{1}{3}(9a + 12b - 1) + \dfrac{1}{4}$

2. Simplify each expression.

a. $2(x - 3) + (5x + 3)$

b. $4(3x + 2) - 3(5x - 1)$

c. $7x + 2(x - 7) - 3x$

3. Solve for x: $\dfrac{x + 5}{2} + \dfrac{1}{2} = 2x - \dfrac{x - 3}{8}$.

4. Solve: $\dfrac{a - 1}{2} + a = 2 - \dfrac{2a + 7}{8}$.

5. A part-time salesperson earns \$600 per month plus a commission of 20% of sales. Find the minimum amount of sales needed to receive a total income of at least \$1500 per month.

6. The Smith family owns a lake house 121.5 miles from home. If it takes them $4\dfrac{1}{2}$ hours round-trip to drive from their house to their lake house, find their average speed.

7. Solve: $2|x| + 25 = 23$

8. Solve: $|3x - 2| + 5 = 5$

9. Solve: $\left|\dfrac{x}{3} - 1\right| - 7 \geq -5$

10. Solve: $\left|\dfrac{x}{2} - 1\right| \leq 0$.

11. Graph the equation $y = |x|$.

12. Graph $y = |x - 2|$.

13. Determine the domain and range of each relation.

a. $\{(2,3),(2,4),(0,-1),(3,-1)\}$

b.

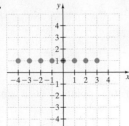

c.

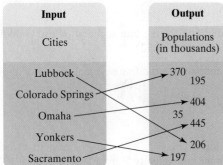

14. Find the domain and the range of each relation. Use the vertical line test to determine whether each graph is the graph of a function.

a.

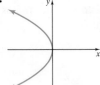

b.

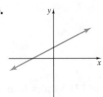

c.

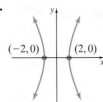

15. Graph $y = -3$.

16. Graph $f(x) = -2$.

17. Find the slope of the line $x = -5$.

18. Find the slope of $y = -3$.

19. Use the substitution method to solve the system.

$$\begin{cases} -\dfrac{x}{6} + \dfrac{y}{2} = \dfrac{1}{2} \\ \dfrac{x}{3} - \dfrac{y}{6} = -\dfrac{3}{4} \end{cases}$$

20. Use the substitution method to solve the system.

$$\begin{cases} \dfrac{x}{6} - \dfrac{y}{2} = 1 \\ \dfrac{x}{3} - \dfrac{y}{4} = 2 \end{cases}$$

21. Use the product rule to simplify.

a. $2^2 \cdot 2^5$

b. $x^7 x^3$

c. $y \cdot y^2 \cdot y^4$

22. At a seasonal clearance sale, Nana Long spent $33.75. She paid $3.50 for tee-shirts and $4.25 for shorts. If she bought 9 items, how many of each item did she buy?

23. Use scientific notation to simplify $\dfrac{2000 \times 0.000021}{700}$.

24. Use scientific notation to simplify and write the answer in scientific notation $\dfrac{0.0000035 \times 4000}{0.28}$.

25. If $P(x) = 3x^2 - 2x - 5$, find the following.

a. $P(1)$

b. $P(-2)$

26. Subtract $(2x - 5)$ from the sum of $(5x^2 - 3x + 6)$ and $(4x^2 + 5x - 3)$.

27. Multiply and simplify the product if possible.

a. $(x + 3)(2x + 5)$

b. $(2x - 3)(5x^2 - 6x + 7)$

28. Multiply and simplify the product if possible.

a. $(y - 2)(3y + 4)$

b. $(3y - 1)(2y^2 + 3y - 1)$

29. Find the GCF of $20x^3y$, $10x^2y^2$, and $35x^3$.

30. Factor $x^3 - x^2 + 4x - 4$.

31. Simplify each rational expression.

a. $\dfrac{x^3 + 8}{2 + x}$

b. $\dfrac{2y^2 + 2}{y^3 - 5y^2 + y - 5}$

32. Simplify each rational expression.

a. $\dfrac{a^3 - 8}{2 - a}$

b. $\dfrac{3a^2 - 3}{a^3 + 5a^2 - a - 5}$

33. Perform the indicated operation.

a. $\dfrac{2}{x^2y} + \dfrac{5}{3x^3y}$

b. $\dfrac{3}{x + 2} + \dfrac{2x}{x - 2}$

c. $\dfrac{2x - 6}{x - 1} - \dfrac{4}{1 - x}$

34. Perform the indicated operations.

a. $\dfrac{3}{xy^2} - \dfrac{2}{3x^2y}$

b. $\dfrac{5x}{x+3} - \dfrac{2x}{x-3}$

c. $\dfrac{x}{x-2} - \dfrac{5}{2-x}$

35. Simplify each complex fraction.

a. $\dfrac{\dfrac{5x}{x+2}}{\dfrac{10}{x-2}}$

b. $\dfrac{\dfrac{x}{y^2} + \dfrac{1}{y}}{\dfrac{y}{x^2} + \dfrac{1}{x}}$

36. Simplify each complex fraction.

a. $\dfrac{\dfrac{y-2}{16}}{\dfrac{2y+3}{12}}$

b. $\dfrac{\dfrac{x}{16} - \dfrac{1}{x}}{1 - \dfrac{4}{x}}$

37. Divide $10x^3 - 5x^2 + 20x$ by $5x$.

38. Divide $x^3 - 2x^2 + 3x - 6$ by $x - 2$.

39. Use synthetic division to divide $2x^3 - x^2 - 13x + 1$ by $x - 3$.

40. Use synthetic division to divide $4y^3 - 12y^2 - y + 12$ by $y - 3$.

41. Solve: $\dfrac{x+6}{x-2} = \dfrac{2(x+2)}{x-2}$

42. Solve: $\dfrac{28}{9-a^2} = \dfrac{2a}{a-3} + \dfrac{6}{a+3}$

43. Solve: $\dfrac{1}{x} + \dfrac{1}{y} = \dfrac{1}{z}$ for x.

44. Solve: $A = \dfrac{h(a+b)}{2}$ for a.

45. Suppose that u varies inversely as w. If u is 3 when w is 5, find the constant of variation and the inverse variation equation.

46. Suppose that y varies directly as x. If $y = 0.51$ when $x = 3$, find the constant of variation and the direct variation equation.

47. Write each expression with a positive exponent and then simplify.

a. $16^{-3/4}$

b. $(-27)^{-2/3}$

48. Write each expression with a positive exponent and then simplify.

a. $(81)^{-3/4}$

b. $(-125)^{-2/3}$

49. Rationalize the numerator of $\dfrac{\sqrt{x}+2}{5}$.

50. Add or subtract.

a. $\sqrt{36a^3} - \sqrt{144a^3} + \sqrt{4a^3}$

b. $\sqrt[3]{128ab^3} - 3\sqrt[3]{2ab^3} + b\sqrt[3]{16a}$

c. $\dfrac{\sqrt[3]{81}}{10} + \sqrt[3]{\dfrac{192}{125}}$

Quadratic Equations and Functions

An important part of the study of algebra is learning to model and solve problems. Often, the model of a problem is a quadratic equation or a function containing a second-degree polynomial. In this chapter, we continue the work begun in Chapter 5, when we solved polynomial equations in one variable by factoring. Two additional methods of solving quadratic equations are analyzed as well as methods of solving nonlinear inequalities in one variable.

Man has always desired to reach the stars, and some buildings seem to be trying to do just that. As populations expand and land becomes scarcer, ever taller and more spectacular buildings are being constructed. As of 2010, the tallest building in the world was the Burj Khalifa, in Dubai. In Exercise 80, Section 8.1, you will explore the height of the Burj Khalifa. (*Source:* Council on Tall Buildings and Urban Habitat, *Fast Company*)

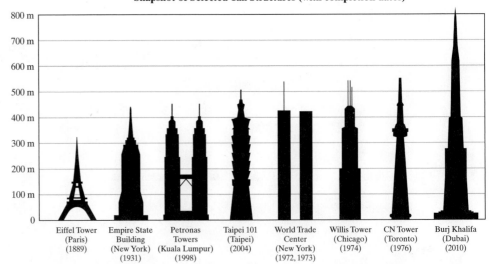

Snapshot of Selected Tall Structures (with completion dates)

Source: Council on Tall Buildings and Urban Habitat, (*Fast Company*)

8.1 Solving Quadratic Equations by Completing the Square

OBJECTIVES

1 Use the Square Root Property to Solve Quadratic Equations.

2 Solve Quadratic Equations by Completing the Square.

3 Use Quadratic Equations to Solve Problems.

OBJECTIVE

1   Using the Square Root Property

In Chapter 5, we solved quadratic equations by factoring. Recall that a **quadratic,** or **second-degree, equation** is an equation that can be written in the form $ax^2 + bx + c = 0$, where a, b, and c are real numbers and a is not 0. To solve a quadratic equation such as $x^2 = 9$ by factoring, we use the zero factor theorem. To use the zero factor theorem, the equation must first be written in standard form, $ax^2 + bx + c = 0$.

$$x^2 = 9$$
$$x^2 - 9 = 0 \quad \text{Subtract 9 from both sides.}$$
$$(x + 3)(x - 3) = 0 \quad \text{Factor.}$$
$$x + 3 = 0 \quad \text{or} \quad x - 3 = 0 \quad \text{Set each factor equal to 0.}$$
$$x = -3 \qquad\qquad x = 3 \quad \text{Solve.}$$

The solution set is $\{-3, 3\}$, the positive and negative square roots of 9. Not all quadratic equations can be solved by factoring, so we need to explore other methods. Notice that the solutions of the equation $x^2 = 9$ are two numbers whose square is 9.

$$3^2 = 9 \quad \text{and} \quad (-3)^2 = 9$$

Thus, we can solve the equation $x^2 = 9$ by taking the square root of both sides. Be sure to include both $\sqrt{9}$ and $-\sqrt{9}$ as solutions since both $\sqrt{9}$ and $-\sqrt{9}$ are numbers whose square is 9.

$$x^2 = 9$$
$$\sqrt{x^2} = \pm\sqrt{9} \quad \text{The notation } \pm\sqrt{9} \text{ (read as "plus or minus } \sqrt{9}\text{")}$$
$$x = \pm 3 \qquad \text{indicates the pair of numbers } +\sqrt{9} \text{ and } -\sqrt{9}.$$

This illustrates the square root property.

Square Root Property

If b is a real number and if $a^2 = b$, then $a = \pm\sqrt{b}$.

▶ Helpful Hint

The notation ± 3, for example, is read as "plus or minus 3." It is a shorthand notation for the pair of numbers $+3$ and -3.

EXAMPLE 1   Use the square root property to solve $x^2 = 50$.

Solution

$$x^2 = 50$$
$$x = \pm\sqrt{50} \quad \text{Use the square root property.}$$
$$x = \pm 5\sqrt{2} \quad \text{Simplify the radical.}$$

Check: Let $x = 5\sqrt{2}$.   Let $x = -5\sqrt{2}$.

$$x^2 = 50 \qquad\qquad\qquad x^2 = 50$$
$$(5\sqrt{2})^2 \stackrel{?}{=} 50 \qquad\qquad (-5\sqrt{2})^2 \stackrel{?}{=} 50$$
$$25 \cdot 2 \stackrel{?}{=} 50 \qquad\qquad\quad 25 \cdot 2 \stackrel{?}{=} 50$$
$$50 = 50 \quad \text{True} \qquad\qquad 50 = 50 \quad \text{True}$$

The solutions are $5\sqrt{2}$ and $-5\sqrt{2}$, or the solution set is $\{-5\sqrt{2}, 5\sqrt{2}\}$.

PRACTICE

1   Use the square root property to solve $x^2 = 32$.

EXAMPLE 2 Use the square root property to solve $2x^2 - 14 = 0$.

Solution First we get the squared variable alone on one side of the equation.

$$2x^2 - 14 = 0$$
$$2x^2 = 14 \qquad \text{Add 14 to both sides.}$$
$$x^2 = 7 \qquad \text{Divide both sides by 2.}$$
$$x = \pm\sqrt{7} \qquad \text{Use the square root property.}$$

Check to see that the solutions are $\sqrt{7}$ and $-\sqrt{7}$, or the solution set is $\{-\sqrt{7}, \sqrt{7}\}$. □

PRACTICE

2 Use the square root property to solve $5x^2 - 50 = 0$.

EXAMPLE 3 Use the square root property to solve $(x + 1)^2 = 12$.

Solution

$$(x + 1)^2 = 12$$
$$x + 1 = \pm\sqrt{12} \qquad \text{Use the square root property.}$$
$$x + 1 = \pm2\sqrt{3} \qquad \text{Simplify the radical.}$$
$$x = \underbrace{-1 \pm 2\sqrt{3}} \qquad \text{Subtract 1 from both sides.}$$

> ▶ **Helpful Hint**
> Don't forget that $-1 \pm 2\sqrt{3}$, for example, means $-1 + 2\sqrt{3}$ and $-1 - 2\sqrt{3}$. In other words, the equation in Example 3 has two solutions.

Check: Below is a check for $-1 + 2\sqrt{3}$. The check for $-1 - 2\sqrt{3}$ is almost the same and is left for you to do on your own.

$$(x + 1)^2 = 12$$
$$\left(-1 + 2\sqrt{3} + 1\right)^2 \overset{?}{=} 12$$
$$\left(2\sqrt{3}\right)^2 \overset{?}{=} 12$$
$$4 \cdot 3 \overset{?}{=} 12$$
$$12 = 12 \quad \text{True}$$

The solutions are $-1 + 2\sqrt{3}$ and $-1 - 2\sqrt{3}$. □

PRACTICE

3 Use the square root property to solve $(x + 3)^2 = 20$.

EXAMPLE 4 Use the square root property to solve $(2x - 5)^2 = -16$.

Solution

$$(2x - 5)^2 = -16$$
$$2x - 5 = \pm\sqrt{-16} \qquad \text{Use the square root property.}$$
$$2x - 5 = \pm4i \qquad \text{Simplify the radical.}$$
$$2x = 5 \pm 4i \qquad \text{Add 5 to both sides.}$$
$$x = \frac{5 \pm 4i}{2} \qquad \text{Divide both sides by 2.}$$

The solutions are $\dfrac{5 + 4i}{2}$ and $\dfrac{5 - 4i}{2}$. □

PRACTICE

4 Use the square root property to solve $(5x - 2)^2 = -9$.

✓ **CONCEPT CHECK**
How do you know just by looking that $(x - 2)^2 = -4$ has complex but not real solutions?

OBJECTIVE

2   Solving by Completing the Square

Notice from Examples 3 and 4 that, if we write a quadratic equation so that one side is the square of a binomial, we can solve by using the square root property. To write the square of a binomial, we write perfect square trinomials. Recall that a perfect square trinomial is a trinomial that can be factored into two identical binomial factors.

Perfect Square Trinomials	*Factored Form*
$x^2 + 8x + 16$	$(x + 4)^2$
$x^2 - 6x + 9$	$(x - 3)^2$
$x^2 + 3x + \dfrac{9}{4}$	$\left(x + \dfrac{3}{2}\right)^2$

Notice that for each perfect square trinomial in x, **the constant term of the trinomial is the square of half the coefficient of the x-term.** For example,

$$x^2 + 8x + 16 \qquad\qquad x^2 - 6x + 9$$

$$\frac{1}{2}(8) = 4 \text{ and } 4^2 = 16 \qquad \frac{1}{2}(-6) = -3 \text{ and } (-3)^2 = 9$$

The process of writing a quadratic equation so that one side is a perfect square trinomial is called **completing the square.**

EXAMPLE 5   Solve $p^2 + 2p = 4$ by completing the square.

Solution   First, add the square of half the coefficient of p to both sides so that the resulting trinomial will be a perfect square trinomial. The coefficient of p is 2.

$$\frac{1}{2}(2) = 1 \quad \text{and} \quad 1^2 = 1$$

Add 1 to both sides of the original equation.

$$p^2 + 2p = 4$$
$$p^2 + 2p + 1 = 4 + 1 \quad \text{Add 1 to both sides.}$$
$$(p + 1)^2 = 5 \qquad \text{Factor the trinomial; simplify the right side.}$$

We may now use the square root property and solve for p.

$$p + 1 = \pm\sqrt{5} \quad \text{Use the square root property.}$$
$$p = -1 \pm \sqrt{5} \quad \text{Subtract 1 from both sides.}$$

Notice that there are two solutions: $-1 + \sqrt{5}$ and $-1 - \sqrt{5}$.   □

PRACTICE

5   Solve $b^2 + 4b = 3$ by completing the square.

· ▪

EXAMPLE 6   Solve $m^2 - 7m - 1 = 0$ for m by completing the square.

Solution   First, add 1 to both sides of the equation so that the left side has no constant term.

$$m^2 - 7m - 1 = 0$$
$$m^2 - 7m = 1$$

Answer to Concept Check:
answers may vary

(Continued on next page)

Now find the constant term that makes the left side a perfect square trinomial by squaring half the coefficient of m. Add this constant to both sides of the equation.

$$\frac{1}{2}(-7) = -\frac{7}{2} \quad \text{and} \quad \left(-\frac{7}{2}\right)^2 = \frac{49}{4}$$

$$m^2 - 7m + \frac{49}{4} = 1 + \frac{49}{4} \qquad \text{Add } \frac{49}{4} \text{ to both sides of the equation.}$$

$$\left(m - \frac{7}{2}\right)^2 = \frac{53}{4} \qquad \text{Factor the perfect square trinomial and simplify the right side.}$$

$$m - \frac{7}{2} = \pm\sqrt{\frac{53}{4}} \qquad \text{Apply the square root property.}$$

$$m = \frac{7}{2} \pm \frac{\sqrt{53}}{2} \qquad \text{Add } \frac{7}{2} \text{ to both sides and simplify } \sqrt{\frac{53}{4}}.$$

$$m = \frac{7 \pm \sqrt{53}}{2} \qquad \text{Simplify.}$$

The solutions are $\dfrac{7 + \sqrt{53}}{2}$ and $\dfrac{7 - \sqrt{53}}{2}$. ☐

PRACTICE

6   Solve $p^2 - 3p + 1 = 0$ by completing the square.

The following steps may be used to solve a quadratic equation such as $ax^2 + bx + c = 0$ by completing the square. This method may be used whether or not the polynomial $ax^2 + bx + c$ is factorable.

> **Solving a Quadratic Equation in x by Completing the Square**
>
> **Step 1.**   If the coefficient of x^2 is 1, go to Step 2. Otherwise, divide both sides of the equation by the coefficient of x^2.
>
> **Step 2.**   Isolate all variable terms on one side of the equation.
>
> **Step 3.**   Complete the square for the resulting binomial by adding the square of half of the coefficient of x to both sides of the equation.
>
> **Step 4.**   Factor the resulting perfect square trinomial and write it as the square of a binomial.
>
> **Step 5.**   Use the square root property to solve for x.

EXAMPLE 7   Solve: $2x^2 - 8x + 3 = 0$.

Solution   Our procedure for finding the constant term to complete the square works only if the coefficient of the squared variable term is 1. Therefore, to solve this equation, the first step is to divide both sides by 2, the coefficient of x^2.

$$2x^2 - 8x + 3 = 0$$

Step 1. $x^2 - 4x + \dfrac{3}{2} = 0$ Divide both sides by 2.

Step 2. $x^2 - 4x = -\dfrac{3}{2}$ Subtract $\dfrac{3}{2}$ from both sides.

Next find the square of half of -4.

$$\frac{1}{2}(-4) = -2 \quad \text{and} \quad (-2)^2 = 4$$

Add 4 to both sides of the equation to complete the square.

Step 3. $x^2 - 4x + 4 = -\dfrac{3}{2} + 4$

Step 4. $(x - 2)^2 = \dfrac{5}{2}$ Factor the perfect square and simplify the right side.

Step 5. $x - 2 = \pm\sqrt{\dfrac{5}{2}}$ Apply the square root property.

$x - 2 = \pm\dfrac{\sqrt{10}}{2}$ Rationalize the denominator.

$x = 2 \pm \dfrac{\sqrt{10}}{2}$ Add 2 to both sides.

$= \dfrac{4}{2} \pm \dfrac{\sqrt{10}}{2}$ Find a common denominator.

$= \dfrac{4 \pm \sqrt{10}}{2}$ Simplify.

The solutions are $\dfrac{4 + \sqrt{10}}{2}$ and $\dfrac{4 - \sqrt{10}}{2}$.   ☐

PRACTICE
7   Solve: $3x^2 - 12x + 1 = 0$.

- ■

EXAMPLE 8   Solve $3x^2 - 9x + 8 = 0$ by completing the square.

Solution $3x^2 - 9x + 8 = 0$

Step 1. $x^2 - 3x + \dfrac{8}{3} = 0$ Divide both sides of the equation by 3.

Step 2. $x^2 - 3x = -\dfrac{8}{3}$ Subtract $\dfrac{8}{3}$ from both sides.

Since $\dfrac{1}{2}(-3) = -\dfrac{3}{2}$ and $\left(-\dfrac{3}{2}\right)^2 = \dfrac{9}{4}$, we add $\dfrac{9}{4}$ to both sides of the equation.

Step 3. $x^2 - 3x + \dfrac{9}{4} = \underbrace{-\dfrac{8}{3} + \dfrac{9}{4}}$

Step 4. $\left(x - \dfrac{3}{2}\right)^2 = -\dfrac{5}{12}$ Factor the perfect square trinomial.

Step 5. $x - \dfrac{3}{2} = \pm\sqrt{-\dfrac{5}{12}}$ Apply the square root property.

$x - \dfrac{3}{2} = \pm\dfrac{i\sqrt{5}}{2\sqrt{3}}$ Simplify the radical.

$x - \dfrac{3}{2} = \pm\dfrac{i\sqrt{15}}{6}$ Rationalize the denominator.

$x = \dfrac{3}{2} \pm \dfrac{i\sqrt{15}}{6}$ Add $\dfrac{3}{2}$ to both sides.

$= \dfrac{9}{6} \pm \dfrac{i\sqrt{15}}{6}$ Find a common denominator.

$= \dfrac{9 \pm i\sqrt{15}}{6}$ Simplify.

The solutions are $\dfrac{9 + i\sqrt{15}}{6}$ and $\dfrac{9 - i\sqrt{15}}{6}$.   ☐

PRACTICE
8   Solve $2x^2 - 5x + 7 = 0$ by completing the square.

- ■

3 Solving Problems Modeled by Quadratic Equations

Recall the **simple interest** formula $I = Prt$, where I is the interest earned, P is the principal, r is the rate of interest, and t is time in years. If $100 is invested at a simple interest rate of 5% annually, at the end of 3 years the total interest I earned is

$$I = P \cdot r \cdot t$$

or

$$I = 100 \cdot 0.05 \cdot 3 = \$15$$

and the new principal is

$$\$100 + \$15 = \$115$$

Most of the time, the interest computed on money borrowed or money deposited is **compound interest.** Compound interest, unlike simple interest, is computed on original principal *and* on interest already earned. To see the difference between simple interest and compound interest, suppose that $100 is invested at a rate of 5% compounded annually. To find the total amount of money at the end of 3 years, we calculate as follows.

$$I = P \cdot r \cdot t$$

First year:	Interest = $100 \cdot 0.05 \cdot 1$ = $5.00
	New principal = $100.00 + $5.00 = $105.00
Second year:	Interest = $105.00 \cdot 0.05 \cdot 1$ = $5.25
	New principal = $105.00 + $5.25 = $110.25
Third year:	Interest = $110.25 \cdot 0.05 \cdot 1 \approx$ $5.51
	New principal = $110.25 + $5.51 = $115.76

At the end of the third year, the total compound interest earned is $15.76, whereas the total simple interest earned is $15.

It is tedious to calculate compound interest as we did above, so we use a compound interest formula. The formula for calculating the total amount of money when interest is compounded annually is

$$A = P(1 + r)^t$$

where P is the original investment, r is the interest rate per compounding period, and t is the number of periods. For example, the amount of money A at the end of 3 years if $100 is invested at 5% compounded annually is

$$A = \$100(1 + 0.05)^3 \approx \$100(1.1576) = \$115.76$$

as we previously calculated.

EXAMPLE 9 Finding Interest Rates

Use the formula $A = P(1 + r)^t$ to find the interest rate r if $2000 compounded annually grows to $2420 in 2 years.

Solution

1. UNDERSTAND the problem. Since the $2000 is compounded annually, we use the compound interest formula. For this example, make sure that you understand the formula for compounding interest annually.

2. TRANSLATE. We substitute the given values into the formula.

$$A = P(1 + r)^t$$

$$2420 = 2000(1 + r)^2 \quad \text{Let } A = 2420, P = 2000, \text{ and } t = 2.$$

3. SOLVE. Solve the equation for r.

$$2420 = 2000(1 + r)^2$$

$$\frac{2420}{2000} = (1 + r)^2 \qquad \text{Divide both sides by 2000.}$$

$$\frac{121}{100} = (1 + r)^2 \qquad \text{Simplify the fraction.}$$

$$\pm\sqrt{\frac{121}{100}} = 1 + r \qquad \text{Use the square root property.}$$

$$\pm\frac{11}{10} = 1 + r \qquad \text{Simplify.}$$

$$-1 \pm \frac{11}{10} = r$$

$$-\frac{10}{10} \pm \frac{11}{10} = r$$

$$\frac{1}{10} = r \quad \text{or} \quad -\frac{21}{10} = r$$

4. INTERPRET. The rate cannot be negative, so we reject $-\dfrac{21}{10}$.

Check: $\dfrac{1}{10} = 0.10 = 10\%$ per year. If we invest \$2000 at 10% compounded annually, in 2 years the amount in the account would be $2000(1 + 0.10)^2 = 2420$ dollars, the desired amount.

State: The interest rate is 10% compounded annually.   □

PRACTICE

9 Use the formula from Example 9 to find the interest rate r if \$5000 compounded annually grows to \$5618 in 2 years.

Graphing Calculator Explorations

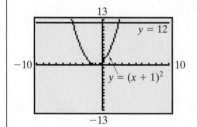

In Section 5.8, we showed how we can use a grapher to approximate real number solutions of a quadratic equation written in standard form. We can also use a grapher to solve a quadratic equation when it is not written in standard form. For example, to solve $(x + 1)^2 = 12$, the quadratic equation in Example 3, we graph the following on the same set of axes. Use Xmin $= -10$, Xmax $= 10$, Ymin $= -13$, and Ymax $= 13$.

$$Y_1 = (x + 1)^2 \quad \text{and} \quad Y_2 = 12$$

Use the Intersect feature or the Zoom and Trace features to locate the points of intersection of the graphs. (See your manuals for specific instructions.) The x-values of these points are the solutions of $(x + 1)^2 = 12$. The solutions, rounded to two decimal places, are 2.46 and -4.46.

Check to see that these numbers are approximations of the exact solutions $-1 \pm 2\sqrt{3}$.

Use a graphing calculator to solve each quadratic equation. Round all solutions to the nearest hundredth.

1. $x(x - 5) = 8$ 　　　　　　　　　　**2.** $x(x + 2) = 5$

3. $x^2 + 0.5x = 0.3x + 1$ 　　　　　　**4.** $x^2 - 2.6x = -2.2x + 3$

5. Use a graphing calculator and solve $(2x - 5)^2 = -16$, Example 4 in this section, using the window

$$\text{Xmin} = -20$$
$$\text{Xmax} = 20$$
$$\text{Xscl} = 1$$
$$\text{Ymin} = -20$$
$$\text{Ymax} = 20$$
$$\text{Yscl} = 1$$

Explain the results. Compare your results with the solution found in Example 4.

6. What are the advantages and disadvantages of using a graphing calculator to solve quadratic equations?

Vocabulary, Readiness & Video Check

Use the choices below to fill in each blank. Not all choices will be used.

binomial	$\sqrt{b}$	$\pm\sqrt{b}$	b^2	9	25	completing the square
quadratic	$-\sqrt{b}$	$\dfrac{b}{2}$	$\left(\dfrac{b}{2}\right)^2$	3	5	

1. By the square root property, if b is a real number, and $a^2 = b$, then $a =$ _____.
2. A _____ equation can be written in the form $ax^2 + bx + c = 0, a \neq 0$.
3. The process of writing a quadratic equation so that one side is a perfect square trinomial is called _____.
4. A perfect square trinomial is one that can be factored as a _____ squared.
5. To solve $x^2 + 6x = 10$ by completing the square, add _____ to both sides.

6. To solve $x^2 + bx = c$ by completing the square, add _____ to both sides.

Martin-Gay Interactive Videos

See Video 8.1

Watch the section lecture video and answer the following questions.

OBJECTIVE 1
7. From ▭ Examples 2 and 3, explain a step you can perform so that you may easily apply the square root property to $2x^2 = 16$. Explain why you perform this step.

OBJECTIVE 2
8. In ▭ Example 5, why is the equation first divided through by 3?

OBJECTIVE 3
9. In ▭ Example 6, why is the negative solution not considered?

8.1 Exercise Set

MyMathLab®

Use the square root property to solve each equation. These equations have real number solutions. See Examples 1 through 3.

▶ 1. $x^2 = 16$
2. $x^2 = 49$
3. $x^2 - 7 = 0$
4. $x^2 - 11 = 0$
5. $x^2 = 18$
6. $y^2 = 20$
7. $3z^2 - 30 = 0$
8. $2x^2 - 4 = 0$
9. $(x + 5)^2 = 9$
10. $(y - 3)^2 = 4$

▶ 11. $(z - 6)^2 = 18$
12. $(y + 4)^2 = 27$
13. $(2x - 3)^2 = 8$
14. $(4x + 9)^2 = 6$

Use the square root property to solve each equation. See Examples 1 through 4.

15. $x^2 + 9 = 0$
16. $x^2 + 4 = 0$
▶ 17. $x^2 - 6 = 0$
18. $y^2 - 10 = 0$
19. $2z^2 + 16 = 0$
20. $3p^2 + 36 = 0$

21. $(3x - 1)^2 = -16$

22. $(4y + 2)^2 = -25$

23. $(z + 7)^2 = 5$

24. $(x + 10)^2 = 11$

25. $(x + 3)^2 + 8 = 0$

26. $(y - 4)^2 + 18 = 0$

Add the proper constant to each binomial so that the resulting trinomial is a perfect square trinomial. Then factor the trinomial.

27. $x^2 + 16x +$ _____   **28.** $y^2 + 2y +$ _____

29. $z^2 - 12z +$ _____   **30.** $x^2 - 8x +$ _____

31. $p^2 + 9p +$ _____   **32.** $n^2 + 5n +$ _____

33. $x^2 + x +$ _____   **34.** $y^2 - y +$ _____

MIXED PRACTICE

Solve each equation by completing the square. These equations have real number solutions. See Examples 5 through 7.

35. $x^2 + 8x = -15$

36. $y^2 + 6y = -8$

▶ **37.** $x^2 + 6x + 2 = 0$

38. $x^2 - 2x - 2 = 0$

39. $x^2 + x - 1 = 0$

40. $x^2 + 3x - 2 = 0$

41. $x^2 + 2x - 5 = 0$

42. $x^2 - 6x + 3 = 0$

43. $y^2 + y - 7 = 0$

44. $x^2 - 7x - 1 = 0$

45. $3p^2 - 12p + 2 = 0$

46. $2x^2 + 14x - 1 = 0$

47. $4y^2 - 2 = 12y$

48. $6x^2 - 3 = 6x$

49. $2x^2 + 7x = 4$

50. $3x^2 - 4x = 4$

51. $x^2 + 8x + 1 = 0$

52. $x^2 - 10x + 2 = 0$

▶ **53.** $3y^2 + 6y - 4 = 0$

54. $2y^2 + 12y + 3 = 0$

55. $2x^2 - 3x - 5 = 0$ **56.** $5x^2 + 3x - 2 = 0$

Solve each equation by completing the square. See Examples 5 through 8.

57. $y^2 + 2y + 2 = 0$

58. $x^2 + 4x + 6 = 0$

59. $y^2 + 6y - 8 = 0$

60. $y^2 + 10y - 26 = 0$

61. $2a^2 + 8a = -12$

62. $3x^2 + 12x = -14$

63. $5x^2 + 15x - 1 = 0$

64. $16y^2 + 16y - 1 = 0$

65. $2x^2 - x + 6 = 0$

66. $4x^2 - 2x + 5 = 0$

67. $x^2 + 10x + 28 = 0$

68. $y^2 + 8y + 18 = 0$

69. $z^2 + 3z - 4 = 0$

70. $y^2 + y - 2 = 0$

71. $2x^2 - 4x = -3$

72. $9x^2 - 36x = -40$

73. $3x^2 + 3x = 5$

74. $10y^2 - 30y = 2$

Use the formula $A = P(1 + r)^t$ to solve Exercises 75 through 78. See Example 9.

▶ **75.** Find the rate r at which $3000 compounded annually grows to $4320 in 2 years.

76. Find the rate r at which $800 compounded annually grows to $882 in 2 years.

77. Find the rate at which $15,000 compounded annually grows to $16,224 in 2 years.

78. Find the rate at which $2000 compounded annually grows to $2880 in 2 years.

Neglecting air resistance, the distance $s(t)$ in feet traveled by a freely falling object is given by the function $s(t) = 16t^2$, where t is time in seconds. Use this formula to solve Exercises 79 through 82. Round answers to two decimal places.

79. The Petronas Towers in Kuala Lumpur, completed in 1998, are the tallest buildings in Malaysia. Each tower is 1483 feet tall. How long would it take an object to fall to the ground from the top of one of the towers? (*Source:* Council on Tall Buildings and Urban Habitat, Lehigh University)

80. The Burj Khalifa, the tallest building in the world, was completed in 2010 in Dubai. It is estimated to be 2717 feet tall. How long would it take an object to fall to the ground from the top of the building? (*Source:* Council on Tall Buildings and Urban Habitat)

81. The Rogun Dam in Tajikistan (part of the former USSR that borders Afghanistan) is the tallest dam in the world at 1100 feet. How long would it take an object to fall from the top to the base of the dam? (*Source:* U.S. Committee on Large Dams of the International Commission on Large Dams)

82. The Hoover Dam, located on the Colorado River on the border of Nevada and Arizona near Las Vegas, is 725 feet tall. How long would it take an object to fall from the top to the base of the dam? (*Source:* U.S. Committee on Large Dams of the International Commission on Large Dams)

Solve.

△ **83.** The area of a square room is 225 square feet. Find the dimensions of the room.

△ **84.** The area of a circle is 36π square inches. Find the radius of the circle.

△ **85.** An isosceles right triangle has legs of equal length. If the hypotenuse is 20 centimeters long, find the length of each leg.

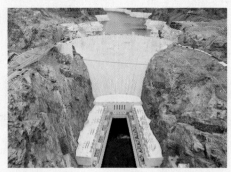

△ **86.** The top of a square coffee table has a diagonal that measures 30 inches. Find the length of each side of the top of the coffee table.

REVIEW AND PREVIEW

Simplify each expression. See Section 7.1

87. $\dfrac{1}{2} - \sqrt{\dfrac{9}{4}}$

88. $\dfrac{9}{10} - \sqrt{\dfrac{49}{100}}$

Simplify each expression. See Section 7.5.

89. $\dfrac{6 + 4\sqrt{5}}{2}$

90. $\dfrac{10 - 20\sqrt{3}}{2}$

91. $\dfrac{3 - 9\sqrt{2}}{6}$

92. $\dfrac{12 - 8\sqrt{7}}{16}$

Evaluate $\sqrt{b^2 - 4ac}$ for each set of values. See Section 7.3.

93. $a = 2, b = 4, c = -1$

94. $a = 1, b = 6, c = 2$

95. $a = 3, b = -1, c = -2$

96. $a = 1, b = -3, c = -1$

CONCEPT EXTENSIONS

Without solving, determine whether the solutions of each equation are real numbers or complex but not real numbers. See the Concept Check in this section.

97. $(x + 1)^2 = -1$

98. $(y - 5)^2 = -9$

99. $3z^2 = 10$

100. $4x^2 = 17$

101. $(2y - 5)^2 + 7 = 3$

102. $(3m + 2)^2 + 4 = 1$

Find two possible missing terms so that each is a perfect square trinomial.

103. $x^2 + \underline{} + 16$

104. $y^2 + \underline{} + 9$

105. $z^2 + \underline{} + \dfrac{25}{4}$

106. $x^2 + \underline{} + \dfrac{1}{4}$

107. In your own words, explain how to calculate the number that will complete the square on an expression such as $x^2 - 5x$.

108. In your own words, what is the difference between simple interest and compound interest?

109. If you are depositing money in an account that pays 4%, would you prefer the interest to be simple or compound? Explain your answer.

110. If you are borrowing money at a rate of 10%, would you prefer the interest to be simple or compound? Explain your answer.

A common equation used in business is a demand equation. It expresses the relationship between the unit price of some commodity and the quantity demanded. For Exercises 111 and 112, p represents the unit price and x represents the quantity demanded in thousands.

111. A manufacturing company has found that the demand equation for a certain type of scissors is given by the equation $p = -x^2 + 47$. Find the demand for the scissors if the price is $11 per pair.

112. Acme, Inc., sells desk lamps and has found that the demand equation for a certain style of desk lamp is given by the equation $p = -x^2 + 15$. Find the demand for the desk lamp if the price is $7 per lamp.

8.2 Solving Quadratic Equations by the Quadratic Formula

OBJECTIVES

1 Solve Quadratic Equations by Using the Quadratic Formula.

2 Determine the Number and Type of Solutions of a Quadratic Equation by Using the Discriminant.

3 Solve Problems Modeled by Quadratic Equations.

OBJECTIVE

1 **Solving Quadratic Equations by Using the Quadratic Formula**

Any quadratic equation can be solved by completing the square. Since the same sequence of steps is repeated each time we complete the square, let's complete the square for a general quadratic equation, $ax^2 + bx + c = 0, a \neq 0$. By doing so, we find a pattern for the solutions of a quadratic equation known as the **quadratic formula.**

Recall that to complete the square for an equation such as $ax^2 + bx + c = 0$, we first divide both sides by the coefficient of x^2.

$$ax^2 + bx + c = 0$$

$$x^2 + \frac{b}{a}x + \frac{c}{a} = 0 \qquad \text{Divide both sides by } a, \text{ the coefficient of } x^2.$$

$$x^2 + \frac{b}{a}x = -\frac{c}{a} \qquad \text{Subtract the constant } \frac{c}{a} \text{ from both sides.}$$

Next, find the square of half $\frac{b}{a}$, the coefficient of x.

$$\frac{1}{2}\left(\frac{b}{a}\right) = \frac{b}{2a} \quad \text{and} \quad \left(\frac{b}{2a}\right)^2 = \frac{b^2}{4a^2}$$

Add this result to both sides of the equation.

$$x^2 + \frac{b}{a}x + \frac{b^2}{4a^2} = -\frac{c}{a} + \frac{b^2}{4a^2} \qquad \text{Add } \frac{b^2}{4a^2} \text{ to both sides.}$$

$$x^2 + \frac{b}{a}x + \frac{b^2}{4a^2} = \frac{-c \cdot 4a}{a \cdot 4a} + \frac{b^2}{4a^2} \qquad \begin{array}{l}\text{Find a common denominator} \\ \text{on the right side.}\end{array}$$

$$x^2 + \frac{b}{a}x + \frac{b^2}{4a^2} = \frac{b^2 - 4ac}{4a^2} \qquad \text{Simplify the right side.}$$

$$\left(x + \frac{b}{2a}\right)^2 = \frac{b^2 - 4ac}{4a^2} \qquad \begin{array}{l}\text{Factor the perfect square} \\ \text{trinomial on the left side.}\end{array}$$

$$x + \frac{b}{2a} = \pm\sqrt{\frac{b^2 - 4ac}{4a^2}} \qquad \text{Apply the square root property.}$$

$$x + \frac{b}{2a} = \pm\frac{\sqrt{b^2 - 4ac}}{2a} \qquad \text{Simplify the radical.}$$

$$x = -\frac{b}{2a} \pm \frac{\sqrt{b^2 - 4ac}}{2a} \qquad \text{Subtract } \frac{b}{2a} \text{ from both sides.}$$

$$x = \frac{-b \pm \sqrt{b^2 - 4ac}}{2a} \qquad \text{Simplify.}$$

This equation identifies the solutions of the general quadratic equation in standard form and is called the quadratic formula. It can be used to solve any equation written in standard form $ax^2 + bx + c = 0$ as long as a is not 0.

Quadratic Formula

A quadratic equation written in the form $ax^2 + bx + c = 0$ has the solutions

$$x = \frac{-b \pm \sqrt{b^2 - 4ac}}{2a}$$

EXAMPLE 1 Solve $3x^2 + 16x + 5 = 0$ for x.

Solution This equation is in standard form, so $a = 3$, $b = 16$, and $c = 5$. Substitute these values into the quadratic formula.

$$x = \frac{-b \pm \sqrt{b^2 - 4ac}}{2a} \qquad \text{Quadratic formula}$$

$$= \frac{-16 \pm \sqrt{16^2 - 4(3)(5)}}{2 \cdot 3} \qquad \text{Use } a = 3, b = 16, \text{ and } c = 5.$$

$$= \frac{-16 \pm \sqrt{256 - 60}}{6}$$

$$= \frac{-16 \pm \sqrt{196}}{6} = \frac{-16 \pm 14}{6}$$

$$x = \frac{-16 + 14}{6} = -\frac{1}{3} \quad \text{or} \quad x = \frac{-16 - 14}{6} = -\frac{30}{6} = -5$$

The solutions are $-\dfrac{1}{3}$ and -5, or the solution set is $\left\{ -\dfrac{1}{3}, -5 \right\}$.   ☐

PRACTICE
1   Solve $3x^2 - 5x - 2 = 0$ for x.

> **▶ Helpful Hint**
> To replace a, b, and c correctly in the quadratic formula, write the quadratic equation in standard form $ax^2 + bx + c = 0$.

EXAMPLE 2 Solve: $2x^2 - 4x = 3$.

Solution First write the equation in standard form by subtracting 3 from both sides.

$$2x^2 - 4x - 3 = 0$$

Now $a = 2$, $b = -4$, and $c = -3$. Substitute these values into the quadratic formula.

$$x = \frac{-b \pm \sqrt{b^2 - 4ac}}{2a}$$

$$= \frac{-(-4) \pm \sqrt{(-4)^2 - 4(2)(-3)}}{2 \cdot 2}$$

$$= \frac{4 \pm \sqrt{16 + 24}}{4}$$

$$= \frac{4 \pm \sqrt{40}}{4} = \frac{4 \pm 2\sqrt{10}}{4}$$

$$= \frac{2(2 \pm \sqrt{10})}{2 \cdot 2} = \frac{2 \pm \sqrt{10}}{2}$$

The solutions are $\dfrac{2 + \sqrt{10}}{2}$ and $\dfrac{2 - \sqrt{10}}{2}$, or the solution set is $\left\{ \dfrac{2 - \sqrt{10}}{2}, \dfrac{2 + \sqrt{10}}{2} \right\}$.   ☐

PRACTICE
2   Solve: $3x^2 - 8x = 2$.

> **▶ Helpful Hint**
>
> To simplify the expression $\dfrac{4 \pm 2\sqrt{10}}{4}$ in the preceding example, note that 2 is factored out of both terms of the numerator *before* simplifying.
>
> $$\frac{4 \pm 2\sqrt{10}}{4} = \frac{2(2 \pm \sqrt{10})}{2 \cdot 2} = \frac{2 \pm \sqrt{10}}{2}$$

✓CONCEPT CHECK

For the quadratic equation $x^2 = 7$, which substitution is correct?

a. $a = 1$, $b = 0$, and $c = -7$
b. $a = 1$, $b = 0$, and $c = 7$
c. $a = 0$, $b = 0$, and $c = 7$
d. $a = 1$, $b = 1$, and $c = -7$

EXAMPLE 3   Solve: $\dfrac{1}{4}m^2 - m + \dfrac{1}{2} = 0$.

Solution   We could use the quadratic formula with $a = \dfrac{1}{4}$, $b = -1$, and $c = \dfrac{1}{2}$. Instead, we find a simpler, equivalent standard form equation whose coefficients are not fractions.

Multiply both sides of the equation by the LCD 4 to clear fractions.

$$4\left(\frac{1}{4}m^2 - m + \frac{1}{2}\right) = 4 \cdot 0$$

$$m^2 - 4m + 2 = 0 \qquad \text{Simplify.}$$

Substitute $a = 1$, $b = -4$, and $c = 2$ into the quadratic formula and simplify.

$$m = \frac{-(-4) \pm \sqrt{(-4)^2 - 4(1)(2)}}{2 \cdot 1} = \frac{4 \pm \sqrt{16 - 8}}{2}$$

$$= \frac{4 \pm \sqrt{8}}{2} = \frac{4 \pm 2\sqrt{2}}{2} = \frac{2(2 \pm \sqrt{2})}{2}$$

$$= 2 \pm \sqrt{2}$$

The solutions are $2 + \sqrt{2}$ and $2 - \sqrt{2}$.   □

PRACTICE
3   Solve: $\dfrac{1}{8}x^2 - \dfrac{1}{4}x - 2 = 0$.

EXAMPLE 4   Solve: $x = -3x^2 - 3$.

Solution   The equation in standard form is $3x^2 + x + 3 = 0$. Thus, let $a = 3$, $b = 1$, and $c = 3$ in the quadratic formula.

$$x = \frac{-1 \pm \sqrt{1^2 - 4(3)(3)}}{2 \cdot 3} = \frac{-1 \pm \sqrt{1 - 36}}{6} = \frac{-1 \pm \sqrt{-35}}{6} = \frac{-1 \pm i\sqrt{35}}{6}$$

The solutions are $\dfrac{-1 + i\sqrt{35}}{6}$ and $\dfrac{-1 - i\sqrt{35}}{6}$.   □

PRACTICE
4   Solve: $x = -2x^2 - 2$.

✓CONCEPT CHECK

What is the first step in solving $-3x^2 = 5x - 4$ using the quadratic formula?

Answer to Concept Checks:
a
Write the equation in standard form.

In Example 1, the equation $3x^2 + 16x + 5 = 0$ had 2 real roots, $-\dfrac{1}{3}$ and -5. In Example 4, the equation $3x^2 + x + 3 = 0$ (written in standard form) had no real roots. How do their related graphs compare? Recall that the x-intercepts of $f(x) = 3x^2 + 16x + 5$ occur where $f(x) = 0$ or where $3x^2 + 16x + 5 = 0$. Since this equation has 2 real roots, the graph has 2 x-intercepts. Similarly, since the equation $3x^2 + x + 3 = 0$ has no real roots, the graph of $f(x) = 3x^2 + x + 3$ has no x-intercepts.

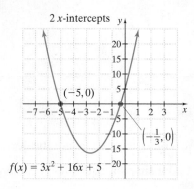

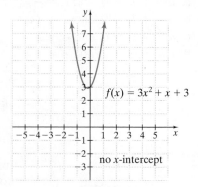

OBJECTIVE

2 Using the Discriminant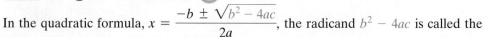

In the quadratic formula, $x = \dfrac{-b \pm \sqrt{b^2 - 4ac}}{2a}$, the radicand $b^2 - 4ac$ is called the **discriminant** because, by knowing its value, we can **discriminate** among the possible number and type of solutions of a quadratic equation. Possible values of the discriminant and their meanings are summarized next.

Discriminant

The following table corresponds the discriminant $b^2 - 4ac$ of a quadratic equation of the form $ax^2 + bx + c = 0$ with the number and type of solutions of the equation.

$b^2 - 4ac$	*Number and Type of Solutions*
Positive	Two real solutions
Zero	One real solution
Negative	Two complex but not real solutions

EXAMPLE 5 Use the discriminant to determine the number and type of solutions of each quadratic equation.

a. $x^2 + 2x + 1 = 0$ **b.** $3x^2 + 2 = 0$ **c.** $2x^2 - 7x - 4 = 0$

Solution

a. In $x^2 + 2x + 1 = 0$, $a = 1$, $b = 2$, and $c = 1$. Thus,

$$b^2 - 4ac = 2^2 - 4(1)(1) = 0$$

Since $b^2 - 4ac = 0$, this quadratic equation has one real solution.

b. In this equation, $a = 3$, $b = 0$, $c = 2$. Then $b^2 - 4ac = 0 - 4(3)(2) = -24$. Since $b^2 - 4ac$ is negative, the quadratic equation has two complex but not real solutions.

c. In this equation, $a = 2$, $b = -7$, and $c = -4$. Then

$$b^2 - 4ac = (-7)^2 - 4(2)(-4) = 81$$

Since $b^2 - 4ac$ is positive, the quadratic equation has two real solutions. □

PRACTICE

5 Use the discriminant to determine the number and type of solutions of each quadratic equation.

a. $x^2 - 6x + 9 = 0$ **b.** $x^2 - 3x - 1 = 0$ **c.** $7x^2 + 11 = 0$

The discriminant helps us determine the number and type of solutions of a quadratic equation, $ax^2 + bx + c = 0$. Recall from Section 5.8 that the solutions of this equation are the same as the x-intercepts of its related graph $f(x) = ax^2 + bx + c$. This means that the discriminant of $ax^2 + bx + c = 0$ also tells us the number of x-intercepts for the graph of $f(x) = ax^2 + bx + c$ or, equivalently, $y = ax^2 + bx + c$.

Graph of $f(x) = ax^2 + bx + c$ or $y = ax^2 + bx + c$

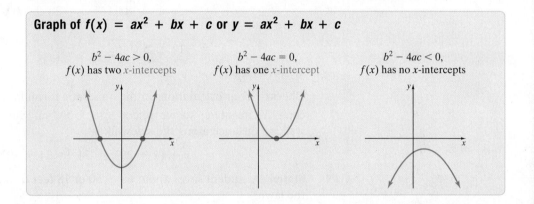

$b^2 - 4ac > 0$,
$f(x)$ has two x-intercepts

$b^2 - 4ac = 0$,
$f(x)$ has one x-intercept

$b^2 - 4ac < 0$,
$f(x)$ has no x-intercepts

OBJECTIVE

3   Solving Problems Modeled by Quadratic Equations

The quadratic formula is useful in solving problems that are modeled by quadratic equations.

⚠ **EXAMPLE 6**   **Calculating Distance Saved**

At a local university, students often leave the sidewalk and cut across the lawn to save walking distance. Given the diagram below of a favorite place to cut across the lawn, approximate how many feet of walking distance a student saves by cutting across the lawn instead of walking on the sidewalk.

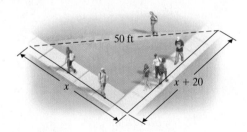

50 ft

x

$x + 20$

Solution

1. UNDERSTAND. Read and reread the problem. In the diagram, notice that a triangle is formed. Since the corner of the block forms a right angle, we use the Pythagorean theorem for right triangles. You may want to review this theorem.

2. TRANSLATE. By the Pythagorean theorem, we have

$$\text{In words: } (\text{leg})^2 + (\text{leg})^2 = (\text{hypotenuse})^2$$
$$\text{Translate: } x^2 + (x + 20)^2 = 50^2$$

3. SOLVE. Use the quadratic formula to solve.

$$x^2 + x^2 + 40x + 400 = 2500 \quad \text{Square } (x + 20) \text{ and 50.}$$
$$2x^2 + 40x - 2100 = 0 \quad \text{Set the equation equal to 0.}$$
$$x^2 + 20x - 1050 = 0 \quad \text{Divide by 2.}$$

Here, $a = 1, b = 20, c = -1050$. By the quadratic formula,

$$x = \frac{-20 \pm \sqrt{20^2 - 4(1)(-1050)}}{2 \cdot 1}$$

$$= \frac{-20 \pm \sqrt{400 + 4200}}{2} = \frac{-20 \pm \sqrt{4600}}{2}$$

$$= \frac{-20 \pm \sqrt{100 \cdot 46}}{2} = \frac{-20 \pm 10\sqrt{46}}{2}$$

$$= -10 \pm 5\sqrt{46} \quad \text{Simplify.}$$

4. INTERPRET

Check: Your calculations in the quadratic formula. The length of a side of a triangle can't be negative, so we reject $-10 - 5\sqrt{46}$. Since $-10 + 5\sqrt{46} \approx 24$ feet, the walking distance along the sidewalk is

$$x + (x + 20) \approx 24 + (24 + 20) = 68 \text{ feet.}$$

State: A student saves about $68 - 50$ or 18 feet of walking distance by cutting across the lawn. □

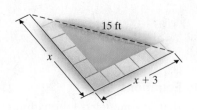

PRACTICE

6   Given the diagram, approximate to the nearest foot how many feet of walking distance a person can save by cutting across the lawn instead of walking on the sidewalk. ■

EXAMPLE 7   **Calculating Landing Time**

An object is thrown upward from the top of a 200-foot cliff with a velocity of 12 feet per second. The height h in feet of the object after t seconds is

$$h = -16t^2 + 12t + 200$$

How long after the object is thrown will it strike the ground? Round to the nearest tenth of a second.

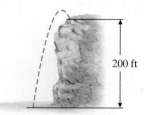

200 ft

Solution

1. UNDERSTAND. Read and reread the problem.

2. TRANSLATE. Since we want to know when the object strikes the ground, we want to know when the height $h = 0$, or

$$0 = -16t^2 + 12t + 200$$

3. SOLVE. First we divide both sides of the equation by -4.

$$0 = 4t^2 - 3t - 50 \quad \text{Divide both sides by } -4.$$

Here, $a = 4, b = -3$, and $c = -50$. By the quadratic formula,

$$t = \frac{-(-3) \pm \sqrt{(-3)^2 - 4(4)(-50)}}{2 \cdot 4}$$

$$= \frac{3 \pm \sqrt{9 + 800}}{8}$$

$$= \frac{3 \pm \sqrt{809}}{8}$$

4. INTERPRET.

Check: We check our calculations from the quadratic formula. Since the time won't be negative, we reject the proposed solution

$$\frac{3 - \sqrt{809}}{8}.$$

State: The time it takes for the object to strike the ground is exactly

$$\frac{3 + \sqrt{809}}{8} \text{ seconds} \approx 3.9 \text{ seconds}.$$

PRACTICE

7   A toy rocket is shot upward from the top of a building, 45 feet high, with an initial velocity of 20 feet per second. The height h in feet of the rocket after t seconds is

$$h = -16t^2 + 20t + 45$$

How long after the rocket is launched will it strike the ground? Round to the nearest tenth of a second.

Vocabulary, Readiness & Video Check

Fill in each blank.

1. The quadratic formula is _____ .

2. For $2x^2 + x + 1 = 0$, if $a = 2$, then $b =$ _____ and $c =$ _____ .

3. For $5x^2 - 5x - 7 = 0$, if $a = 5$, then $b =$ _____ and $c =$ _____ .

4. For $7x^2 - 4 = 0$, if $a = 7$, then $b =$ _____ and $c =$ _____ .

5. For $x^2 + 9 = 0$, if $c = 9$, then $a =$ _____ and $b =$ _____ .

6. The correct simplified form of $\dfrac{5 \pm 10\sqrt{2}}{5}$ is _____ .

 a. $1 \pm 10\sqrt{2}$ **b.** $2\sqrt{2}$ **c.** $1 \pm 2\sqrt{2}$ **d.** $\pm 5\sqrt{2}$

Martin-Gay Interactive Videos

See Video 8.2

Watch the section lecture video and answer the following questions.

OBJECTIVE
1

7. Based on ▭ Examples 1–3, answer the following.
   a. Must a quadratic equation be written in standard form in order to use the quadratic formula? Why or why not?
   b. Must fractions be cleared from an equation before using the quadratic formula? Why or why not?

OBJECTIVE
2

8. Based on ▭ Example 4 and the lecture before, complete the following statements. The discriminant is the _____ in the quadratic formula and can be used to find the number and type of solutions of a quadratic equation without _____ the equation. To use the discriminant, the quadratic equation needs to be written in _____ form.

OBJECTIVE
3

9. In ▭ Example 5, the value of x is found, which is then used to find the dimensions of the triangle. Yet all this work still does solve the problem. Explain.

8.2   Exercise Set   MyMathLab®

Use the quadratic formula to solve each equation. These equations have real number solutions only. See Examples 1 through 3.

1. $m^2 + 5m - 6 = 0$

2. $p^2 + 11p - 12 = 0$

3. $2y = 5y^2 - 3$

4. $5x^2 - 3 = 14x$

5. $x^2 - 6x + 9 = 0$

6. $y^2 + 10y + 25 = 0$

7. $x^2 + 7x + 4 = 0$

8. $y^2 + 5y + 3 = 0$

9. $8m^2 - 2m = 7$

10. $11n^2 - 9n = 1$

11. $3m^2 - 7m = 3$

12. $x^2 - 13 = 5x$

13. $\frac{1}{2}x^2 - x - 1 = 0$

14. $\frac{1}{6}x^2 + x + \frac{1}{3} = 0$

15. $\frac{2}{5}y^2 + \frac{1}{5}y = \frac{3}{5}$

16. $\frac{1}{8}x^2 + x = \frac{5}{2}$

17. $\frac{1}{3}y^2 = y + \frac{1}{6}$

18. $\frac{1}{2}y^2 = y + \frac{1}{2}$

19. $x^2 + 5x = -2$

20. $y^2 - 8 = 4y$

21. $(m + 2)(2m - 6) = 5(m - 1) - 12$

22. $7p(p - 2) + 2(p + 4) = 3$

MIXED PRACTICE

Use the quadratic formula to solve each equation. These equations have real solutions and complex but not real solutions. See Examples 1 through 4.

23. $x^2 + 6x + 13 = 0$

24. $x^2 + 2x + 2 = 0$

25. $(x + 5)(x - 1) = 2$

26. $x(x + 6) = 2$

27. $6 = -4x^2 + 3x$

28. $2 = -9x^2 - x$

29. $\frac{x^2}{3} - x = \frac{5}{3}$

30. $\frac{x^2}{2} - 3 = -\frac{9}{2}x$

31. $10y^2 + 10y + 3 = 0$

32. $3y^2 + 6y + 5 = 0$

33. $x(6x + 2) = 3$

34. $x(7x + 1) = 2$

35. $\frac{2}{5}y^2 + \frac{1}{5}y + \frac{3}{5} = 0$

36. $\frac{1}{8}x^2 + x + \frac{5}{2} = 0$

37. $\frac{1}{2}y^2 = y - \frac{1}{2}$

38. $\frac{2}{3}x^2 - \frac{20}{3}x = -\frac{100}{6}$

39. $(n - 2)^2 = 2n$

40. $\left(p - \frac{1}{2}\right)^2 = \frac{p}{2}$

Use the discriminant to determine the number and types of solutions of each equation. See Example 5.

41. $x^2 - 5 = 0$

42. $x^2 - 7 = 0$

43. $4x^2 + 12x = -9$

44. $9x^2 + 1 = 6x$

45. $3x = -2x^2 + 7$

46. $3x^2 = 5 - 7x$

47. $6 = 4x - 5x^2$

48. $8x = 3 - 9x^2$

49. $9x - 2x^2 + 5 = 0$

50. $5 - 4x + 12x^2 = 0$

Solve. See Examples 7 and 8.

51. Nancy, Thelma, and John Varner live on a corner lot. Often, neighborhood children cut across their lot to save walking distance. Given the diagram below, approximate to the nearest foot how many feet of walking distance is saved by cutting across their property instead of walking around the lot.

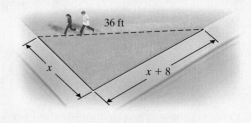

△ **52.** Given the diagram below, approximate to the nearest foot how many feet of walking distance a person saves by cutting across the lawn instead of walking on the sidewalk.

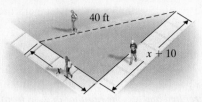

△ **53.** The hypotenuse of an isosceles right triangle is 2 centimeters longer than either of its legs. Find the exact length of each side. (*Hint:* An isosceles right triangle is a right triangle whose legs are the same length.)

△ **54.** The hypotenuse of an isosceles right triangle is one meter longer than either of its legs. Find the length of each side.

△ **55.** Bailey's rectangular dog pen for his Irish setter must have an area of 400 square feet. Also, the length must be 10 feet longer than the width. Find the dimensions of the pen.

△ **56.** An entry in the Peach Festival Poster Contest must be rectangular and have an area of 1200 square inches. Furthermore, its length must be 20 inches longer than its width. Find the dimensions each entry must have.

△ **57.** A holding pen for cattle must be square and have a diagonal length of 100 meters.

  a. Find the length of a side of the pen.

  b. Find the area of the pen.

△ **58.** A rectangle is three times longer than it is wide. It has a diagonal of length 50 centimeters.

  a. Find the dimensions of the rectangle.

  b. Find the perimeter of the rectangle.

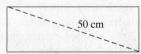

△ **59.** The heaviest reported door in the world is the 708.6 ton radiation shield door in the National Institute for Fusion Science at Toki, Japan. If the height of the door is 1.1 feet longer than its width, and its front area (neglecting depth) is 1439.9 square feet, find its width and height [Interesting note: the door is 6.6 feet thick.] (*Source: Guiness World Records*)

△ **60.** Christi and Robbie Wegmann are constructing a rectangular stained glass window whose length is 7.3 inches longer than its width. If the area of the window is 569.9 square inches, find its width and length.

△ **61.** The base of a triangle is four more than twice its height. If the area of the triangle is 42 square centimeters, find its base and height.

62. If a point *B* divides a line segment such that the smaller portion is to the larger portion as the larger is to the whole, the whole is the length of the *golden ratio*.

$$x \text{ (whole)}$$
$$\overbrace{\quad 1 \quad\quad x - 1 \quad}$$
$$A \qquad B \qquad C$$

The golden ratio was thought by the Greeks to be the most pleasing to the eye, and many of their buildings contained numerous examples of the golden ratio. The value of the golden ratio is the positive solution of

$$\text{(smaller)} \quad \frac{x - 1}{1} = \frac{1}{x} \quad \text{(larger)}$$
$$\text{(larger)} \qquad\qquad\qquad \text{(whole)}$$

Find this value.

The Wollomombi Falls in Australia have a height of 1100 feet. A pebble is thrown upward from the top of the falls with an initial velocity of 20 feet per second. The height of the pebble h after t seconds is given by the equation $h = -16t^2 + 20t + 1100$. Use this equation for Exercises 63 and 64.

63. How long after the pebble is thrown will it hit the ground? Round to the nearest tenth of a second.

64. How long after the pebble is thrown will it be 550 feet from the ground? Round to the nearest tenth of a second.

A ball is thrown downward from the top of a 180-foot building with an initial velocity of 20 feet per second. The height of the ball h after t seconds is given by the equation $h = -16t^2 - 20t + 180$. Use this equation to answer Exercises 65 and 66.

65. How long after the ball is thrown will it strike the ground? Round the result to the nearest tenth of a second.

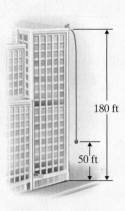

66. How long after the ball is thrown will it be 50 feet from the ground? Round the result to the nearest tenth of a second.

REVIEW AND PREVIEW

Solve each equation. See Sections 6.6 and 7.6.

67. $\sqrt{5x - 2} = 3$

68. $\sqrt{y + 2} + 7 = 12$

69. $\dfrac{1}{x} + \dfrac{2}{5} = \dfrac{7}{x}$

70. $\dfrac{10}{z} = \dfrac{5}{z} - \dfrac{1}{3}$

Factor. See Section 5.7.

71. $x^4 + x^2 - 20$

72. $2y^4 + 11y^2 - 6$

73. $z^4 - 13z^2 + 36$

74. $x^4 - 1$

CONCEPT EXTENSIONS

For each quadratic equation, choose the correct substitution for a, b, and c in the standard form $ax^2 + bx + c = 0$.

75. $x^2 = -10$

  a. $a = 1, b = 0, c = -10$

  b. $a = 1, b = 0, c = 10$

  c. $a = 0, b = 1, c = -10$

  d. $a = 1, b = 1, c = 10$

76. $x^2 + 5 = -x$

  a. $a = 1, b = 5, c = -1$

  b. $a = 1, b = -1, c = 5$

  c. $a = 1, b = 5, c = 1$

  d. $a = 1, b = 1, c = 5$

77. Solve Exercise 1 by factoring. Explain the result.

78. Solve Exercise 2 by factoring. Explain the result.

Use the quadratic formula and a calculator to approximate each solution to the nearest tenth.

79. $2x^2 - 6x + 3 = 0$

80. $3.6x^2 + 1.8x - 4.3 = 0$

The accompanying graph shows the daily low temperatures for one week in New Orleans, Louisiana.

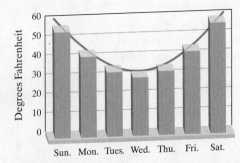

81. Between which days of the week was there the greatest decrease in the low temperature?

82. Between which days of the week was there the greatest increase in the low temperature?

83. Which day of the week had the lowest low temperature?

84. Use the graph to estimate the low temperature on Thursday.

Notice that the shape of the temperature graph is similar to the curve drawn. In fact, this graph can be modeled by the quadratic function $f(x) = 3x^2 - 18x + 56$, where $f(x)$ is the temperature in degrees Fahrenheit and x is the number of days from Sunday. (This graph is shown in blue.) Use this function to answer Exercises 85 and 86.

85. Use the quadratic function given to approximate the temperature on Thursday. Does your answer agree with the graph?

86. Use the function given and the quadratic formula to find when the temperature was 35° F. [*Hint:* Let $f(x) = 35$ and solve for *x*.] Round your answer to one decimal place and interpret your result. Does your answer agree with the graph?

87. The number of college students in the United States can be modeled by the quadratic function $f(x) = 22x^2 + 274x + 15,628$, where $f(x)$ is the number of college students in thousands of students, and x is the number of years after 2000. (*Source:* Based on data from the U.S. Department of Education)

  a. Find the number of college students in the United States in 2010.

  b. If the trend described by this model continues, find the year after 2000 in which the population of American college students reaches 24,500 students.

88. The projected number of Wi-Fi-enabled cell phones in the United States can be modeled by the quadratic function $c(x) = -0.4x^2 + 21x + 35$, where $c(x)$ is the projected number of Wi-Fi-enabled cell phones in millions and x is the number of years after 2009. Round to the nearest million. (*Source:* Techcrunchies.com)

  a. Find the number of Wi-Fi-enabled cell phones in the United States in 2010.

  b. Find the projected number of Wi-Fi-enabled cell phones in the United States in 2012.

  c. If the trend described by this model continues, find the year in which the projected number of Wi-Fi-enabled cell phones in the United States reaches 150 million.

89. The average total daily supply y of motor gasoline (in thousands of barrels per day) in the United States for the period 2000–2008 can be approximated by the equation $y = -10x^2 + 193x + 8464$, where x is the number of years after 2000. (*Source:* Based on data from the Energy Information Administration)

  a. Find the average total daily supply of motor gasoline in 2004.

  b. According to this model, in what year, from 2000 to 2008, was the average total daily supply of gasoline 9325 thousand barrels per day?

  c. According to this model, in what year, from 2009 on, will the average total supply of gasoline be 9325 thousand barrels per day?

90. The relationship between body weight and the Recommended Dietary Allowance (RDA) for vitamin A in children up to age 10 is modeled by the quadratic equation $y = 0.149x^2 - 4.475x + 406.478$, where y is the RDA for vitamin A in micrograms for a child whose weight is x pounds. (*Source:* Based on data from the Food and Nutrition Board, National Academy of Sciences–Institute of Medicine, 1989)

a. Determine the vitamin A requirements of a child who weighs 35 pounds.

b. What is the weight of a child whose RDA of vitamin A is 600 micrograms? Round your answer to the nearest pound.

The solutions of the quadratic equation $ax^2 + bx + c = 0$ are $\dfrac{-b + \sqrt{b^2 - 4ac}}{2a}$ *and* $\dfrac{-b - \sqrt{b^2 - 4ac}}{2a}$.

91. Show that the sum of these solutions is $\dfrac{-b}{a}$.

92. Show that the product of these solutions is $\dfrac{c}{a}$.

Use the quadratic formula to solve each quadratic equation.

93. $3x^2 - \sqrt{12}x + 1 = 0$
(*Hint:* $a = 3, b = -\sqrt{12}, c = 1$)

94. $5x^2 + \sqrt{20}x + 1 = 0$

95. $x^2 + \sqrt{2}x + 1 = 0$

96. $x^2 - \sqrt{2}x + 1 = 0$

97. $2x^2 - \sqrt{3}x - 1 = 0$

98. $7x^2 + \sqrt{7}x - 2 = 0$

99. Use a graphing calculator to solve Exercises 63 and 65.

100. Use a graphing calculator to solve Exercises 64 and 66.

Recall that the discriminant also tells us the number of x-intercepts of the related function.

101. Check the results of Exercise 49 by graphing $y = 9x - 2x^2 + 5$.

102. Check the results of Exercise 50 by graphing $y = 5 - 4x + 12x^2$.

8.3 Solving Equations by Using Quadratic Methods

OBJECTIVES

1 Solve Various Equations That Are Quadratic in Form.

2 Solve Problems That Lead to Quadratic Equations.

OBJECTIVE

1 Solving Equations That Are Quadratic in Form

In this section, we discuss various types of equations that can be solved in part by using the methods for solving quadratic equations.

Once each equation is simplified, you may want to use these steps when deciding which method to use to solve the quadratic equation.

> **Solving a Quadratic Equation**
>
> **Step 1.** If the equation is in the form $(ax + b)^2 = c$, use the square root property and solve. If not, go to Step 2.
>
> **Step 2.** Write the equation in standard form: $ax^2 + bx + c = 0$.
>
> **Step 3.** Try to solve the equation by the factoring method. If not possible, go to Step 4.
>
> **Step 4.** Solve the equation by the quadratic formula.

The first example is a radical equation that becomes a quadratic equation once we square both sides.

EXAMPLE 1   Solve: $x - \sqrt{x} - 6 = 0$.

Solution   Recall that to solve a radical equation, first get the radical alone on one side of the equation. Then square both sides.

$$x - 6 = \sqrt{x} \qquad \text{Add } \sqrt{x} \text{ to both sides.}$$
$$(x - 6)^2 = \left(\sqrt{x}\right)^2 \qquad \text{Square both sides.}$$
$$x^2 - 12x + 36 = x$$
$$x^2 - 13x + 36 = 0 \qquad \text{Set the equation equal to 0.}$$
$$(x - 9)(x - 4) = 0$$
$$x - 9 = 0 \quad \text{or} \quad x - 4 = 0$$
$$x = 9 \qquad\qquad x = 4$$

(Continued on next page)

Check:

Let $x = 9$	Let $x = 4$
$x - \sqrt{x} - 6 = 0$	$x - \sqrt{x} - 6 = 0$
$9 - \sqrt{9} - 6 \stackrel{?}{=} 0$	$4 - \sqrt{4} - 6 \stackrel{?}{=} 0$
$9 - 3 - 6 \stackrel{?}{=} 0$	$4 - 2 - 6 \stackrel{?}{=} 0$
$0 = 0$ True	$-4 = 0$ False

The solution is 9 or the solution set is {9}. ☐

PRACTICE
1   Solve: $x - \sqrt{x + 1} - 5 = 0$.

EXAMPLE 2   Solve: $\dfrac{3x}{x - 2} - \dfrac{x + 1}{x} = \dfrac{6}{x(x - 2)}$.

Solution   In this equation, x cannot be either 2 or 0 because these values cause denominators to equal zero. To solve for x, we first multiply both sides of the equation by $x(x - 2)$ to clear the fractions. By the distributive property, this means that we multiply each term by $x(x - 2)$.

$$x(x - 2)\left(\dfrac{3x}{x - 2}\right) - x(x - 2)\left(\dfrac{x + 1}{x}\right) = x(x - 2)\left[\dfrac{6}{x(x - 2)}\right]$$

$$3x^2 - (x - 2)(x + 1) = 6 \quad \text{Simplify.}$$

$$3x^2 - (x^2 - x - 2) = 6 \quad \text{Multiply.}$$

$$3x^2 - x^2 + x + 2 = 6$$

$$2x^2 + x - 4 = 0 \quad \text{Simplify.}$$

This equation cannot be factored using integers, so we solve by the quadratic formula.

$$x = \dfrac{-1 \pm \sqrt{1^2 - 4(2)(-4)}}{2 \cdot 2} \quad \begin{array}{l}\text{Use } a = 2, b = 1, \text{ and } c = -4 \\ \text{in the quadratic formula.}\end{array}$$

$$= \dfrac{-1 \pm \sqrt{1 + 32}}{4} \quad \text{Simplify.}$$

$$= \dfrac{-1 \pm \sqrt{33}}{4}$$

Neither proposed solution will make the denominators 0.

The solutions are $\dfrac{-1 + \sqrt{33}}{4}$ and $\dfrac{-1 - \sqrt{33}}{4}$ or the solution set is $\left\{\dfrac{-1 + \sqrt{33}}{4}, \dfrac{-1 - \sqrt{33}}{4}\right\}$. ☐

PRACTICE
2   Solve: $\dfrac{5x}{x + 1} - \dfrac{x + 4}{x} = \dfrac{3}{x(x + 1)}$.

EXAMPLE 3   Solve: $p^4 - 3p^2 - 4 = 0$.

Solution   First we factor the trinomial.

$$p^4 - 3p^2 - 4 = 0$$

$$(p^2 - 4)(p^2 + 1) = 0 \quad \text{Factor.}$$

$$(p - 2)(p + 2)(p^2 + 1) = 0 \quad \text{Factor further.}$$

$$p - 2 = 0 \quad \text{or} \quad p + 2 = 0 \quad \text{or} \quad p^2 + 1 = 0 \quad \begin{array}{l}\text{Set each factor equal} \\ \text{to 0 and solve.}\end{array}$$

$$p = 2 \qquad\qquad p = -2 \qquad\qquad p^2 = -1$$

$$p = \pm\sqrt{-1} = \pm i$$

The solutions are 2, −2, i and −i. □

PRACTICE
3   Solve: $p^4 - 7p^2 - 144 = 0$.

> ▶ Helpful Hint
>
> Example 3 can be solved using substitution also. Think of $p^4 - 3p^2 - 4 = 0$ as
>
> $$(p^2)^2 - 3p^2 - 4 = 0 \qquad \text{Then let } x = p^2 \text{ and solve and substitute back.}$$
> $$\text{The solutions will be the same.}$$
> $$x^2 - 3x - 4 = 0$$

✔**CONCEPT CHECK**
   a. True or false? The maximum number of solutions that a quadratic equation can have is 2.
   b. True or false? The maximum number of solutions that an equation in quadratic form can have is 2.

EXAMPLE 4   Solve: $(x - 3)^2 - 3(x - 3) - 4 = 0$.

Solution   Notice that the quantity $(x - 3)$ is repeated in this equation. Sometimes it is helpful to substitute a variable (in this case other than x) for the repeated quantity. We will let $y = x - 3$. Then

$$(x - 3)^2 - 3(x - 3) - 4 = 0$$

becomes

$$y^2 - 3y - 4 = 0 \qquad \text{Let } x - 3 = y.$$
$$(y - 4)(y + 1) = 0 \qquad \text{Factor.}$$

To solve, we use the zero factor property.

$$y - 4 = 0 \quad \text{or} \quad y + 1 = 0 \qquad \text{Set each factor equal to 0.}$$
$$y = 4 \qquad\qquad y = -1 \qquad \text{Solve.}$$

> ▶ Helpful Hint
>
> When using substitution, don't forget to substitute back to the original variable.

To find values of x, we substitute back. That is, we substitute $x - 3$ for y.

$$x - 3 = 4 \quad \text{or} \quad x - 3 = -1$$
$$x = 7 \qquad\qquad x = 2$$

Both 2 and 7 check. The solutions are 2 and 7. □

PRACTICE
4   Solve: $(x + 2)^2 - 2(x + 2) - 3 = 0$.

EXAMPLE 5   Solve: $x^{2/3} - 5x^{1/3} + 6 = 0$.

Solution   The key to solving this equation is recognizing that $x^{2/3} = (x^{1/3})^2$. We replace $x^{1/3}$ with m so that

$$(x^{1/3})^2 - 5x^{1/3} + 6 = 0$$

becomes

$$m^2 - 5m + 6 = 0$$

Now we solve by factoring.

$$m^2 - 5m + 6 = 0$$
$$(m - 3)(m - 2) = 0 \qquad\qquad\qquad \text{Factor.}$$
$$m - 3 = 0 \quad \text{or} \quad m - 2 = 0 \qquad \text{Set each factor equal to 0.}$$
$$m = 3 \qquad\qquad m = 2$$

Answer to Concept Check:
a. true   **b.** false

(Continued on next page)

Since $m = x^{1/3}$, we have

$$x^{1/3} = 3 \qquad \text{or} \quad x^{1/3} = 2$$
$$x = 3^3 = 27 \quad \text{or} \quad x = 2^3 = 8$$

Both 8 and 27 check. The solutions are 8 and 27.

PRACTICE

5   Solve: $x^{2/3} - 5x^{1/3} + 4 = 0$.

OBJECTIVE

2   Solving Problems That Lead to Quadratic Equations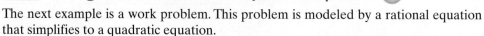

The next example is a work problem. This problem is modeled by a rational equation that simplifies to a quadratic equation.

EXAMPLE 6   Finding Work Time

Together, an experienced word processor and an apprentice word processor can create a word document in 6 hours. Alone, the experienced word processor can create the document 2 hours faster than the apprentice word processor can. Find the time in which each person can create the word document alone.

Solution

1. **UNDERSTAND.** Read and reread the problem. The key idea here is the relationship between the *time* (hours) it takes to complete the job and the *part of the job* completed in one unit of time (hour). For example, because they can complete the job together in 6 hours, the *part of the job* they can complete in 1 hour is $\frac{1}{6}$.

Let

      $x =$ the *time* in hours it takes the apprentice word processor to complete the job alone

  $x - 2 =$ the *time* in hours it takes the experienced word processor to complete the job alone

We can summarize in a chart the information discussed

	Total Hours to Complete Job	*Part of Job Completed in 1 Hour*
Apprentice Word Processor	x	$\frac{1}{x}$
Experienced Word Processor	$x - 2$	$\frac{1}{x - 2}$
Together	6	$\frac{1}{6}$

2. **TRANSLATE.**

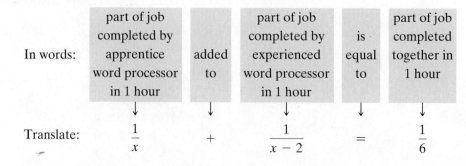

3. SOLVE.

$$\frac{1}{x} + \frac{1}{x-2} = \frac{1}{6}$$

$$6x(x-2)\left(\frac{1}{x} + \frac{1}{x-2}\right) = 6x(x-2) \cdot \frac{1}{6} \qquad \text{Multiply both sides by the LCD } 6x(x-2).$$

$$6x(x-2) \cdot \frac{1}{x} + 6x(x-2) \cdot \frac{1}{x-2} = 6x(x-2) \cdot \frac{1}{6} \qquad \text{Use the distributive property.}$$

$$6(x-2) + 6x = x(x-2)$$

$$6x - 12 + 6x = x^2 - 2x$$

$$0 = x^2 - 14x + 12$$

Now we can substitute $a = 1$, $b = -14$, and $c = 12$ into the quadratic formula and simplify.

$$x = \frac{-(-14) \pm \sqrt{(-14)^2 - 4(1)(12)}}{2 \cdot 1} = \frac{14 \pm \sqrt{148}^*}{2}$$

Using a calculator or a square root table, we see that $\sqrt{148} \approx 12.2$ rounded to one decimal place. Thus,

$$x \approx \frac{14 \pm 12.2}{2}$$

$$x \approx \frac{14 + 12.2}{2} = 13.1 \quad \text{or} \quad x \approx \frac{14 - 12.2}{2} = 0.9$$

4. INTERPRET.

Check: If the apprentice word processor completes the job alone in 0.9 hours, the experienced word processor completes the job alone in $x - 2 = 0.9 - 2 = -1.1$ hours. Since this is not possible, we reject the solution of 0.9. The approximate solution thus is 13.1 hours.

State: The apprentice word processor can complete the job alone in approximately 13.1 hours, and the experienced word processor can complete the job alone in approximately

$$x - 2 = 13.1 - 2 = 11.1 \text{ hours.} \qquad \square$$

PRACTICE

6   Together, Katy and Steve can groom all the dogs at the Barkin' Doggie Day Care in 4 hours. Alone, Katy can groom the dogs 1 hour faster than Steve can groom the dogs alone. Find the time in which each of them can groom the dogs alone.

EXAMPLE 7   **Finding Driving Speeds**

Beach and Fargo are about 400 miles apart. A salesperson travels from Fargo to Beach one day at a certain speed. She returns to Fargo the next day and drives 10 mph faster. Her total travel time was $14\frac{2}{3}$ hours. Find her speed to Beach and the return speed to Fargo.

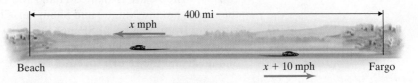

(*This expression can be simplified further, but this will suffice because we are approximating.)

Solution

1. UNDERSTAND. Read and reread the problem. Let

$$x = \text{the speed to Beach, so}$$

$$x + 10 = \text{the return speed to Fargo.}$$

Then organize the given information in a table.

> ▶ **Helpful Hint**
>
> Since $d = rt$, $t = \dfrac{d}{r}$. The time column was completed using $\dfrac{d}{r}$.

	distance	=	rate	·	time	
To Beach	400		x		$\dfrac{400}{x}$	← distance ← rate
Return to Fargo	400		$x + 10$		$\dfrac{400}{x + 10}$	← distance ← rate

2. TRANSLATE.

In words: $\boxed{\begin{array}{c}\text{time to}\\\text{Beach}\end{array}}$ $+$ $\boxed{\begin{array}{c}\text{return}\\\text{time to}\\\text{Fargo}\end{array}}$ $=$ $\boxed{\begin{array}{c}14\frac{2}{3}\\\text{hours}\end{array}}$

Translate: $\dfrac{400}{x}$ $+$ $\dfrac{400}{x + 10}$ $=$ $\dfrac{44}{3}$

3. SOLVE.

$$\frac{400}{x} + \frac{400}{x + 10} = \frac{44}{3}$$

$$\frac{100}{x} + \frac{100}{x + 10} = \frac{11}{3} \qquad \text{Divide both sides by 4.}$$

$$3x(x + 10)\left(\frac{100}{x} + \frac{100}{x + 10}\right) = 3x(x + 10) \cdot \frac{11}{3} \qquad \text{Multiply both sides by the LCD } 3x(x + 10).$$

$$3x(x + 10) \cdot \frac{100}{x} + 3x(x + 10) \cdot \frac{100}{x + 10} = 3x(x + 10) \cdot \frac{11}{3} \qquad \text{Use the distributive property.}$$

$$3(x + 10) \cdot 100 + 3x \cdot 100 = x(x + 10) \cdot 11$$

$$300x + 3000 + 300x = 11x^2 + 110x$$

$$0 = 11x^2 - 490x - 3000 \qquad \text{Set equation equal to 0.}$$

$$0 = (11x + 60)(x - 50) \qquad \text{Factor.}$$

$$11x + 60 = 0 \quad \text{or} \quad x - 50 = 0 \qquad \text{Set each factor equal to 0.}$$

$$x = -\frac{60}{11} \text{ or } -5\frac{5}{11}; \quad x = 50$$

4. INTERPRET.

Check: The speed is not negative, so it's not $-5\dfrac{5}{11}$. The number 50 does check.

State: The speed to Beach was 50 mph, and her return speed to Fargo was 60 mph. □

PRACTICE

7   The 36-km S-shaped Hangzhou Bay Bridge is the longest cross-sea bridge in the world, linking Ningbo and Shanghai, China. A merchant drives over the bridge one morning from Ningbo to Shanghai in very heavy traffic and returns home that night driving 50 km per hour faster. The total travel time was 1.3 hours. Find the speed to Shanghai and the return speed to Ningbo.

Vocabulary, Readiness & Video Check

Martin-Gay Interactive Videos

See Video 8.3 🍐

Watch the section lecture video and answer the following questions.

OBJECTIVE 1

1. From Examples 1 and 2, what's the main thing to remember when using a substitution in order to solve an equation by quadratic methods?

OBJECTIVE 2

2. In Example 4, the translated equation is actually a rational equation. Explain how we end up solving it using quadratic methods.

8.3   Exercise Set   MyMathLab®

Solve. See Example 1.

1. $2x = \sqrt{10 + 3x}$
2. $3x = \sqrt{8x + 1}$
3. $x - 2\sqrt{x} = 8$
4. $x - \sqrt{2x} = 4$
5. $\sqrt{9x} = x + 2$
6. $\sqrt{16x} = x + 3$

Solve. See Example 2.

▶ 7. $\dfrac{2}{x} + \dfrac{3}{x - 1} = 1$

8. $\dfrac{6}{x^2} = \dfrac{3}{x + 1}$

9. $\dfrac{3}{x} + \dfrac{4}{x + 2} = 2$

10. $\dfrac{5}{x - 2} + \dfrac{4}{x + 2} = 1$

11. $\dfrac{7}{x^2 - 5x + 6} = \dfrac{2x}{x - 3} - \dfrac{x}{x - 2}$

12. $\dfrac{11}{2x^2 + x - 15} = \dfrac{5}{2x - 5} - \dfrac{x}{x + 3}$

Solve. See Example 3.

13. $p^4 - 16 = 0$
14. $x^4 + 2x^2 - 3 = 0$
15. $4x^4 + 11x^2 = 3$
16. $z^4 = 81$
17. $z^4 - 13z^2 + 36 = 0$
18. $9x^4 + 5x^2 - 4 = 0$

Solve. See Examples 4 and 5.

▶ 19. $x^{2/3} - 3x^{1/3} - 10 = 0$

20. $x^{2/3} + 2x^{1/3} + 1 = 0$

21. $(5n + 1)^2 + 2(5n + 1) - 3 = 0$

22. $(m - 6)^2 + 5(m - 6) + 4 = 0$

23. $2x^{2/3} - 5x^{1/3} = 3$

24. $3x^{2/3} + 11x^{1/3} = 4$

25. $1 + \dfrac{2}{3t - 2} = \dfrac{8}{(3t - 2)^2}$

26. $2 - \dfrac{7}{x + 6} = \dfrac{15}{(x + 6)^2}$

27. $20x^{2/3} - 6x^{1/3} - 2 = 0$

28. $4x^{2/3} + 16x^{1/3} = -15$

MIXED PRACTICE

Solve. See Examples 1 through 5.

29. $a^4 - 5a^2 + 6 = 0$

30. $x^4 - 12x^2 + 11 = 0$

31. $\dfrac{2x}{x - 2} + \dfrac{x}{x + 3} = -\dfrac{5}{x + 3}$

32. $\dfrac{5}{x - 3} + \dfrac{x}{x + 3} = \dfrac{19}{x^2 - 9}$

▶ 33. $(p + 2)^2 = 9(p + 2) - 20$

34. $2(4m - 3)^2 - 9(4m - 3) = 5$

35. $2x = \sqrt{11x + 3}$

36. $4x = \sqrt{2x + 3}$

37. $x^{2/3} - 8x^{1/3} + 15 = 0$

38. $x^{2/3} - 2x^{1/3} - 8 = 0$

39. $y^3 + 9y - y^2 - 9 = 0$

40. $x^3 + x - 3x^2 - 3 = 0$

41. $2x^{2/3} + 3x^{1/3} - 2 = 0$

42. $6x^{2/3} - 25x^{1/3} - 25 = 0$

43. $x^{-2} - x^{-1} - 6 = 0$

44. $y^{-2} - 8y^{-1} + 7 = 0$

45. $x - \sqrt{x} = 2$

46. $x - \sqrt{3x} = 6$

47. $\dfrac{x}{x - 1} + \dfrac{1}{x + 1} = \dfrac{2}{x^2 - 1}$

48. $\dfrac{x}{x - 5} + \dfrac{5}{x + 5} = -\dfrac{1}{x^2 - 25}$

49. $p^4 - p^2 - 20 = 0$

50. $x^4 - 10x^2 + 9 = 0$

51. $(x + 3)(x^2 - 3x + 9) = 0$

52. $(x - 6)(x^2 + 6x + 36) = 0$

53. $1 = \dfrac{4}{x - 7} + \dfrac{5}{(x - 7)^2}$

54. $3 + \dfrac{1}{2p + 4} = \dfrac{10}{(2p + 4)^2}$

55. $27y^4 + 15y^2 = 2$

56. $8z^4 + 14z^2 = -5$

57. $x - \sqrt{19 - 2x} - 2 = 0$

58. $x - \sqrt{17 - 4x} - 3 = 0$

Solve. For Exercises 59 and 60, the solutions have been started for you. See Examples 6 and 7.

59. Roma Sherry drove 330 miles from her hometown to Tucson. During her return trip, she was able to increase her speed by 11 miles per hour. If her return trip took 1 hour less time, find her original speed and her speed returning home.

Start the solution:

1. UNDERSTAND the problem. Reread it as many times as needed. Let

$$x = \text{original speed}$$
$$x + 11 = \text{return-trip speed}$$

Organize the information in a table.

	distance	=	rate	·	time	
To Tucson	330		x		$\dfrac{330}{x}$	← distance ← rate
Return trip	330		—		$\dfrac{330}{}$	← distance ← rate

2. TRANSLATE into an equation. (Fill in the blanks below.)

Time to Tucson	equals	Return trip time	plus	1 hour
↓	↓	↓	↓	↓
___	=	___	+	1

Finish with:

3. SOLVE and **4.** INTERPRET

60. A salesperson drove to Portland, a distance of 300 miles. During the last 80 miles of his trip, heavy rainfall forced him to decrease his speed by 15 miles per hour. If his total driving time was 6 hours, find his original speed and his speed during the rainfall.

Start the solution:

1. UNDERSTAND the problem. Reread it as many times as needed. Let

$$x = \text{original speed}$$
$$x - 15 = \text{rainfall speed}$$

Organize the information in a table.

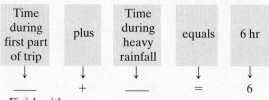

	distance	=	rate	·	time	
First part of trip	300 − 80, or 220		x		$\dfrac{220}{x}$	← distance ← rate
Heavy rainfall part of trip	80		$x - 15$		$\dfrac{80}{}$	← distance ← rate

2. TRANSLATE into an equation. (Fill in the blanks below.)

Time during first part of trip	plus	Time during heavy rainfall	equals	6 hr
↓	↓	↓	↓	↓
___	+	___	=	6

Finish with:

3. SOLVE and **4.** INTERPRET

61. A jogger ran 3 miles, decreased her speed by 1 mile per hour, and then ran another 4 miles. If her total time jogging was $1\dfrac{3}{5}$ hours, find her speed for each part of her run.

62. Mark Keaton's workout consists of jogging for 3 miles and then riding his bike for 5 miles at a speed 4 miles per hour faster than he jogs. If his total workout time is 1 hour, find his jogging speed and his biking speed.

63. A Chinese restaurant in Mandeville, Louisiana, has a large goldfish pond around the restaurant. Suppose that an inlet pipe and a hose together can fill the pond in 8 hours. The inlet pipe alone can complete the job in one hour less time than the hose alone. Find the time that the hose can complete the job alone and the time that the inlet pipe can complete the job alone. Round each to the nearest tenth of an hour.

64. A water tank on a farm in Flatonia, Texas, can be filled with a large inlet pipe and a small inlet pipe in 3 hours. The large inlet pipe alone can fill the tank in 2 hours less time than the small inlet pipe alone. Find the time to the nearest tenth of an hour each pipe can fill the tank alone.

65. Bill Shaughnessy and his son Billy can clean the house together in 4 hours. When the son works alone, it takes him an hour longer to clean than it takes his dad alone. Find how long to the nearest tenth of an hour it takes the son to clean alone.

66. Together, Noodles and Freckles eat a 50-pound bag of dog food in 30 days. Noodles by himself eats a 50-pound bag in 2 weeks less time than Freckles does by himself. How many days to the nearest whole day would a 50-pound bag of dog food last Freckles?

67. The product of a number and 4 less than the number is 96. Find the number.

68. A whole number increased by its square is two more than twice itself. Find the number.

△ **69.** Suppose that an open box is to be made from a square sheet of cardboard by cutting out squares from each corner as shown and then folding along the dotted lines. If the box is to have a volume of 300 cubic inches, find the original dimensions of the sheet of cardboard.

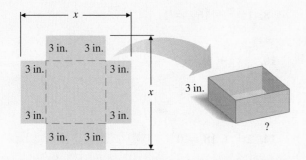

a. The ? in the drawing above will be the length (and the width) of the box as shown. Represent this length in terms of x.

b. Use the formula for volume of a box, $V = l \cdot w \cdot h$, to write an equation in x.

c. Solve the equation for x and give the dimensions of the sheet of cardboard. Check your solution.

△ **70.** Suppose that an open box is to be made from a square sheet of cardboard by cutting out squares from each corner as shown and then folding along the dotted lines. If the box is to have a volume of 128 cubic inches, find the original dimensions of the sheet of cardboard.

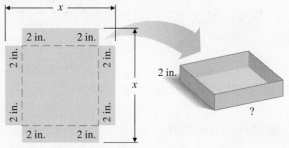

a. The ? in the drawing above will be the length (and the width) of the box as shown. Represent this length in terms of x.

b. Use the formula for volume of a box, $V = l \cdot w \cdot h$, to write an equation in x.

c. Solve the equation for x and give the dimensions of the sheet of cardboard. Check your solution.

△ **71.** A sprinkler that sprays water in a circular pattern is to be used to water a square garden. If the area of the garden is 920 square feet, find the smallest whole number *radius* that the sprinkler can be adjusted to so that the entire garden is watered.

△ **72.** Suppose that a square field has an area of 6270 square feet. See Exercise 71 and find a new sprinkler radius.

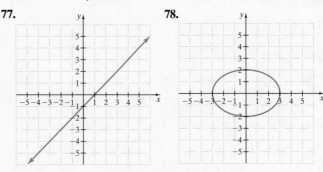

REVIEW AND PREVIEW

Solve each inequality. See Section 2.4.

73. $\dfrac{5x}{3} + 2 \le 7$

74. $\dfrac{2x}{3} + \dfrac{1}{6} \ge 2$

75. $\dfrac{y - 1}{15} > -\dfrac{2}{5}$

76. $\dfrac{z - 2}{12} < \dfrac{1}{4}$

Find the domain and range of each graphed relation. Decide which relations are also functions. See Section 3.2.

77.

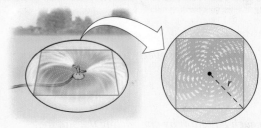

78.

79.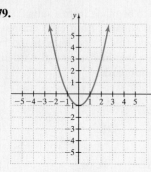

80.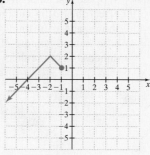

CONCEPT EXTENSIONS

Solve.

81. $5y^3 + 45y - 5y^2 - 45 = 0$

82. $10x^3 + 10x - 30x^2 - 30 = 0$

83. $3x^{-2} - 3x^{-1} - 18 = 0$

84. $2y^{-2} - 16y^{-1} + 14 = 0$

85. $2x^3 = -54$

86. $y^3 - 216 = 0$

87. Write a polynomial equation that has three solutions: 2, 5, and −7.

88. Write a polynomial equation that has three solutions: 0, $2i$, and $-2i$.

89. During the seventh stage of the 2010 Paris–Nice bicycle race, Thomas Voeckler posted the fastest average speed, but Alberto Contador won the race. The seventh stage was 119 kilometers long. Voeckler's average speed was 0.0034 meters per second faster than Contador's. Traveling at these average speeds, Contador took 3 seconds longer than Voeckler to complete the race stage. (*Source:* Based on data from cyclingnews.com)

   a. Find Thomas Voeckler's average speed during the seventh stage of the 2010 Paris–Nice cycle race. Round to three decimal places.

   b. Find Alberto Contador's average speed during the seventh stage of the 2010 Paris–Nice cycle race. Round to three decimal places.

   c. Convert Voeckler's average speed to miles per hour. Round to three decimal places.

 90. Use a graphing calculator to solve Exercise 29. Compare the solution with the solution from Exercise 29. Explain any differences.

Integrated Review   SUMMARY ON SOLVING QUADRATIC EQUATIONS

Sections 8.1–8.3

Use the square root property to solve each equation.

1. $x^2 - 10 = 0$

2. $x^2 - 14 = 0$

3. $(x - 1)^2 = 8$

4. $(x + 5)^2 = 12$

Solve each equation by completing the square.

5. $x^2 + 2x - 12 = 0$

6. $x^2 - 12x + 11 = 0$

7. $3x^2 + 3x = 5$

8. $16y^2 + 16y = 1$

Use the quadratic formula to solve each equation

9. $2x^2 - 4x + 1 = 0$

10. $\frac{1}{2}x^2 + 3x + 2 = 0$

11. $x^2 + 4x = -7$

12. $x^2 + x = -3$

Solve each equation. Use a method of your choice.

13. $x^2 + 3x + 6 = 0$

14. $2x^2 + 18 = 0$

15. $x^2 + 17x = 0$

16. $4x^2 - 2x - 3 = 0$

17. $(x - 2)^2 = 27$

18. $\frac{1}{2}x^2 - 2x + \frac{1}{2} = 0$

19. $3x^2 + 2x = 8$

20. $2x^2 = -5x - 1$

21. $x(x - 2) = 5$

22. $x^2 - 31 = 0$

23. $5x^2 - 55 = 0$

24. $5x^2 + 55 = 0$

25. $x(x + 5) = 66$

26. $5x^2 + 6x - 2 = 0$

27. $2x^2 + 3x = 1$

28. $x - \sqrt{13 - 3x} - 3 = 0$

29. $\dfrac{5x}{x-2} - \dfrac{x+1}{x} = \dfrac{3}{x(x-2)}$

△ **30.** The diagonal of a square room measures 20 feet. Find the exact length of a side of the room. Then approximate the length to the nearest tenth of a foot.

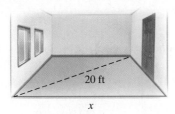

20 ft

x

31. Together, Jack and Lucy Hoag can prepare a crawfish boil for a large party in 4 hours. Lucy alone can complete the job in 2 hours less time than Jack alone. Find the time that each person can prepare the crawfish boil alone. Round each time to the nearest tenth of an hour.

32. Diane Gray exercises at Total Body Gym. On the treadmill, she runs 5 miles, then increases her speed by 1 mile per hour and runs an additional 2 miles. If her total time on the treadmill is $1\frac{1}{3}$ hours, find her speed during each part of her run.

8.4 Nonlinear Inequalities in One Variable

OBJECTIVES

1 Solve Polynomial Inequalities of Degree 2 or Greater.

2 Solve Inequalities That Contain Rational Expressions with Variables in the Denominator.

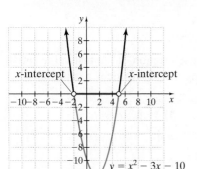

x-values
corresponding to
negative y-values

OBJECTIVE

1 Solving Polynomial Inequalities

Just as we can solve linear inequalities in one variable, so can we also solve quadratic inequalities in one variable. A **quadratic inequality** is an inequality that can be written so that one side is a quadratic expression and the other side is 0. Here are examples of quadratic inequalities in one variable. Each is written in **standard form.**

$$x^2 - 10x + 7 \le 0 \qquad 3x^2 + 2x - 6 > 0$$
$$2x^2 + 9x - 2 < 0 \qquad x^2 - 3x + 11 \ge 0$$

A solution of a quadratic inequality in one variable is a value of the variable that makes the inequality a true statement.

The value of an expression such as $x^2 - 3x - 10$ will sometimes be positive, sometimes negative, and sometimes 0, depending on the value substituted for x. To solve the inequality $x^2 - 3x - 10 < 0$, we are looking for all values of x that make the expression $x^2 - 3x - 10$ **less than 0**, or **negative**. To understand how we find these values, we'll study the graph of the quadratic function $y = x^2 - 3x - 10$.

Notice that the x-values for which y is positive are separated from the x values for which y is negative by the x-intercepts. (Recall that the x-intercepts correspond to values of x for which $y = 0$.) Thus, the solution set of $x^2 - 3x - 10 < 0$ consists of all real numbers from -2 to 5 or, in interval notation, $(-2, 5)$.

It is not necessary to graph $y = x^2 - 3x - 10$ to solve the related inequality $x^2 - 3x - 10 < 0$. Instead, we can draw a number line representing the x-axis and keep the following in mind: *A region on the number line for which the value of $x^2 - 3x - 10$ is positive is separated from a region on the number line for which the value of $x^2 - 3x - 10$ is negative by a value for which the expression is 0.*

Let's find these values for which the expression is 0 by solving the related equation:

$$x^2 - 3x - 10 = 0$$
$$(x - 5)(x + 2) = 0 \qquad \text{Factor.}$$
$$x - 5 = 0 \quad \text{or} \quad x + 2 = 0 \qquad \text{Set each factor equal to 0.}$$
$$x = 5 \qquad\qquad x = -2 \qquad \text{Solve.}$$

These two numbers, -2 and 5, divide the number line into three regions. We will call the regions A, B, and C. These regions are important because, if the value of $x^2 - 3x - 10$ is negative when a number from a region is substituted for x, then $x^2 - 3x - 10$ is negative when any number in that region is substituted for x. The same is true if the value of $x^2 - 3x - 10$ is positive for a particular value of x in a region.

To see whether the inequality $x^2 - 3x - 10 < 0$ is true or false in each region, we choose a test point from each region and substitute its value for x in the inequality $x^2 - 3x - 10 < 0$. If the resulting inequality is true, the region containing the test point is a solution region.

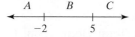

Region	Test Point Value	$(x - 5)(x + 2) < 0$	Result
A	-3	$(-8)(-1) < 0$	False
B	0	$(-5)(2) < 0$	True
C	6	$(1)(8) < 0$	False

The values in region B satisfy the inequality. The numbers -2 and 5 are not included in the solution set since the inequality symbol is $<$. The solution set is $(-2, 5)$, and its graph is shown.

$$A \quad B \quad C$$
$$F\ -2 \quad T \quad 5\ F$$

EXAMPLE 1 Solve: $(x + 3)(x - 3) > 0$.

Solution First we solve the related equation, $(x + 3)(x - 3) = 0$.

$$(x + 3)(x - 3) = 0$$
$$x + 3 = 0 \quad \text{or} \quad x - 3 = 0$$
$$x = -3 \qquad\qquad x = 3$$

The two numbers -3 and 3 separate the number line into three regions, A, B, and C.

$$A \quad B \quad C$$
$$-3 \qquad 3$$

Now we substitute the value of a test point from each region. If the test value satisfies the inequality, every value in the region containing the test value is a solution.

Region	Test Point Value	$(x + 3)(x - 3) > 0$	Result
A	-4	$(-1)(-7) > 0$	True
B	0	$(3)(-3) > 0$	False
C	4	$(7)(1) > 0$	True

The points in regions A and C satisfy the inequality. The numbers -3 and 3 are not included in the solution since the inequality symbol is $>$. The solution set is $(-\infty, -3) \cup (3, \infty)$, and its graph is shown.

$$A \quad B \quad C$$
$$T\ -3 \quad F \quad 3\ T$$

PRACTICE

1 Solve: $(x - 4)(x + 3) > 0$.

The following steps may be used to solve a polynomial inequality.

Solving a Polynomial Inequality

Step 1. Write the inequality in standard form and then solve the related equation.

Step 2. Separate the number line into regions with the solutions from Step 1.

Step 3. For each region, choose a test point and determine whether its value satisfies the *original inequality*.

Step 4. The solution set includes the regions whose test point value is a solution. If the inequality symbol is $\leq$ or $\geq$, the values from Step 1 are solutions; if $<$ or $>$, they are not.

✓**CONCEPT CHECK**
When choosing a test point in Step 3, why would the solutions from Step 1 not make good choices for test points?

EXAMPLE 2   Solve: $x^2 - 4x \leq 0$.

Solution   First we solve the related equation, $x^2 - 4x = 0$.

$$x^2 - 4x = 0$$
$$x(x - 4) = 0$$
$$x = 0 \quad \text{or} \quad x = 4$$

The numbers 0 and 4 separate the number line into three regions, A, B, and C.

We check a test value in each region in the original inequality. Values in region B satisfy the inequality. The numbers 0 and 4 are included in the solution since the inequality symbol is $\leq$. The solution set is $[0, 4]$, and its graph is shown.

PRACTICE
2   Solve: $x^2 - 8x \leq 0$.

EXAMPLE 3   Solve: $(x + 2)(x - 1)(x - 5) \leq 0$.

Solution   First we solve $(x + 2)(x - 1)(x - 5) = 0$. By inspection, we see that the solutions are -2, 1, and 5. They separate the number line into four regions, A, B, C, and D. Next we check test points from each region.

Region	Test Point Value	$(x + 2)(x - 1)(x - 5) \leq 0$	Result
A	-3	$(-1)(-4)(-8) \leq 0$	True
B	0	$(2)(-1)(-5) \leq 0$	False
C	2	$(4)(1)(-3) \leq 0$	True
D	6	$(8)(5)(1) \leq 0$	False

Answer to Concept Check:
The solutions found in Step 1 have a value of 0 in the original inequality.

(Continued on next page)

The solution set is $(-\infty, -2] \cup [1, 5]$, and its graph is shown. We include the numbers -2, 1, and 5 because the inequality symbol is $\leq$.

$$
\begin{array}{ccccc}
 & A & B & C & D \\
\end{array}
$$

$$
\overset{}{\underset{\text{T } -2 \text{ F } 1 \text{ T } 5 \text{ F}}{\longleftrightarrow}}
$$

☐

PRACTICE
3 Solve: $(x + 3)(x - 2)(x + 1) \leq 0$.

■

OBJECTIVE

2 Solving Rational Inequalities

Inequalities containing rational expressions with variables in the denominator are solved by using a similar procedure.

EXAMPLE 4 Solve: $\dfrac{x + 2}{x - 3} \leq 0$.

Solution First we find all values that make the denominator equal to 0. To do this, we solve $x - 3 = 0$ and find that $x = 3$.

Next, we solve the related equation $\dfrac{x + 2}{x - 3} = 0$.

$$\frac{x + 2}{x - 3} = 0$$

$$x + 2 = 0 \qquad \text{Multiply both sides by the LCD, } x - 3.$$

$$x = -2$$

Now we place these numbers on a number line and proceed as before, checking test point values in the original inequality.

$$
\begin{array}{ccc}
A & B & C \\
\end{array}
$$

$$
\overset{}{\underset{-2 \qquad 3}{\longleftrightarrow}}
$$

Choose -3 from region A.

$$\frac{x + 2}{x - 3} \leq 0$$

$$\frac{-3 + 2}{-3 - 3} \leq 0$$

$$\frac{-1}{-6} \leq 0$$

$$\frac{1}{6} \leq 0 \quad \text{False}$$

Choose 0 from region B.

$$\frac{x + 2}{x - 3} \leq 0$$

$$\frac{0 + 2}{0 - 3} \leq 0$$

$$-\frac{2}{3} \leq 0 \quad \text{True}$$

Choose 4 from region C.

$$\frac{x + 2}{x - 3} \leq 0$$

$$\frac{4 + 2}{4 - 3} \leq 0$$

$$6 \leq 0 \quad \text{False}$$

The solution set is $[-2, 3)$. This interval includes -2 because -2 satisfies the original inequality. This interval does not include 3 because 3 would make the denominator 0.

$$
\begin{array}{ccc}
A & B & C \\
\end{array}
$$

$$
\overset{}{\underset{\text{F } -2 \quad \text{T } \quad 3 \text{ F}}{\longleftrightarrow}}
$$

☐

PRACTICE
4 Solve: $\dfrac{x - 5}{x + 4} \leq 0$.

■

The following steps may be used to solve a rational inequality with variables in the denominator.

Solving a Rational Inequality

Step 1. Solve for values that make all denominators 0.

Step 2. Solve the related equation.

Step 3. Separate the number line into regions with the solutions from Steps 1 and 2.

Step 4. For each region, choose a test point and determine whether its value satisfies the *original inequality*.

Step 5. The solution set includes the regions whose test point value is a solution. Check whether to include values from Step 2. Be sure *not* to include values that make any denominator 0.

EXAMPLE 5   Solve: $\dfrac{5}{x+1} < -2$.

Solution   First we find values for x that make the denominator equal to 0.

$$x + 1 = 0$$
$$x = -1$$

Next we solve $\dfrac{5}{x+1} = -2$.

$$(x+1) \cdot \frac{5}{x+1} = (x+1) \cdot -2 \quad \text{Multiply both sides by the LCD, } x+1.$$
$$5 = -2x - 2 \qquad \text{Simplify.}$$
$$7 = -2x$$
$$-\frac{7}{2} = x$$

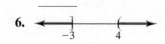

We use these two solutions to divide a number line into three regions and choose test points. Only a test point value from region B satisfies the *original inequality*. The solution set is $\left(-\dfrac{7}{2}, -1 \right)$, and its graph is shown.

PRACTICE
5   Solve: $\dfrac{7}{x+3} < 5$.

Write the graphed solution set in interval notation.

1.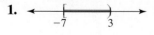
$-7 \qquad 3$

_____

2.
$-1 \qquad 5$

_____

3.
0

_____

4.
-8

_____

5.
$-12 \qquad -10$

_____

6.
$-3 \qquad 4$

_____

Martin-Gay Interactive Videos

See Video 8.4 🍐

Watch the section lecture video and answer the following questions.

OBJECTIVE
1

7. From Examples 1–3, how does solving a related equation help you solve a polynomial inequality? Are the solutions to the related equation ever solutions to the inequality?

OBJECTIVE
2

8. In Example 4, one of the values that separates the number line into regions is 4. The inequality is ≥, so why isn't 4 included in the solution set?

8.4 Exercise Set

MyMathLab®

Solve each polynomial inequality. Write the solution set in interval notation. See Examples 1 through 3.

1. $(x + 1)(x + 5) > 0$

2. $(x + 1)(x + 5) \le 0$

▶ 3. $(x - 3)(x + 4) \le 0$

4. $(x + 4)(x - 1) > 0$

5. $x^2 - 7x + 10 \le 0$

6. $x^2 + 8x + 15 \ge 0$

7. $3x^2 + 16x < -5$

8. $2x^2 - 5x < 7$

9. $(x - 6)(x - 4)(x - 2) > 0$

10. $(x - 6)(x - 4)(x - 2) \le 0$

11. $x(x - 1)(x + 4) \le 0$

12. $x(x - 6)(x + 2) > 0$

13. $(x^2 - 9)(x^2 - 4) > 0$

14. $(x^2 - 16)(x^2 - 1) \le 0$

Solve each inequality. Write the solution set in interval notation. See Example 4.

15. $\dfrac{x + 7}{x - 2} < 0$

16. $\dfrac{x - 5}{x - 6} > 0$

17. $\dfrac{5}{x + 1} > 0$

18. $\dfrac{3}{y - 5} < 0$

▶ 19. $\dfrac{x + 1}{x - 4} \ge 0$

20. $\dfrac{x + 1}{x - 4} \le 0$

Solve each inequality. Write the solution set in interval notation. See Example 5.

21. $\dfrac{3}{x - 2} < 4$

22. $\dfrac{-2}{y + 3} > 2$

23. $\dfrac{x^2 + 6}{5x} \ge 1$

24. $\dfrac{y^2 + 15}{8y} \le 1$

25. $\dfrac{x + 2}{x - 3} < 1$

26. $\dfrac{x - 1}{x + 4} > 2$

MIXED PRACTICE

Solve each inequality. Write the solution set in interval notation.

27. $(2x - 3)(4x + 5) \le 0$

28. $(6x + 7)(7x - 12) > 0$

▶ 29. $x^2 > x$

30. $x^2 < 25$

31. $(2x - 8)(x + 4)(x - 6) \le 0$

32. $(3x - 12)(x + 5)(2x - 3) \ge 0$

33. $6x^2 - 5x \ge 6$

34. $12x^2 + 11x \le 15$

35. $4x^3 + 16x^2 - 9x - 36 > 0$

36. $x^3 + 2x^2 - 4x - 8 < 0$

▶ 37. $x^4 - 26x^2 + 25 \ge 0$

38. $16x^4 - 40x^2 + 9 \le 0$

39. $(2x - 7)(3x + 5) > 0$

40. $(4x - 9)(2x + 5) < 0$

41. $\dfrac{x}{x - 10} < 0$

42. $\dfrac{x + 10}{x - 10} > 0$

43. $\dfrac{x - 5}{x + 4} \geq 0$

44. $\dfrac{x - 3}{x + 2} \leq 0$

45. $\dfrac{x(x + 6)}{(x - 7)(x + 1)} \geq 0$

46. $\dfrac{(x - 2)(x + 2)}{(x + 1)(x - 4)} \leq 0$

47. $\dfrac{-1}{x - 1} > -1$

48. $\dfrac{4}{y + 2} < -2$

49. $\dfrac{x}{x + 4} \leq 2$

50. $\dfrac{4x}{x - 3} \geq 5$

51. $\dfrac{z}{z - 5} \geq 2z$

52. $\dfrac{p}{p + 4} \leq 3p$

53. $\dfrac{(x + 1)^2}{5x} > 0$

54. $\dfrac{(2x - 3)^2}{x} < 0$

REVIEW AND PREVIEW

Recall that the graph of $f(x) + K$ is the same as the graph of $f(x)$ shifted K units upward if $K > 0$ and $|K|$ units downward if $K < 0$. Use the graph of $f(x) = |x|$ below to sketch the graph of each function. See Section 3.6.

55. $g(x) = |x| + 2$

56. $H(x) = |x| - 2$

57. $f(x) = |x| - 1$

58. $h(x) = |x| + 5$

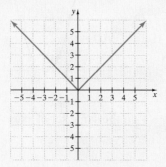

Use the graph of $f(x) = x^2$ below to sketch the graph of each function.

59. $F(x) = x^2 - 3$

60. $h(x) = x^2 - 4$

61. $H(x) = x^2 + 1$

62. $g(x) = x^2 + 3$

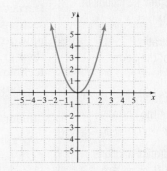

CONCEPT EXTENSIONS

63. Explain why $\dfrac{x + 2}{x - 3} > 0$ and $(x + 2)(x - 3) > 0$ have the same solutions.

64. Explain why $\dfrac{x + 2}{x - 3} \geq 0$ and $(x + 2)(x - 3) \geq 0$ do not have the same solutions.

Find all numbers that satisfy each of the following.

65. A number minus its reciprocal is less than zero. Find the numbers.

66. Twice a number added to its reciprocal is nonnegative. Find the numbers.

67. The total profit function $P(x)$ for a company producing x thousand units is given by

$$P(x) = -2x^2 + 26x - 44$$

Find the values of x for which the company makes a profit. [*Hint:* The company makes a profit when $P(x) > 0$.]

68. A projectile is fired straight up from the ground with an initial velocity of 80 feet per second. Its height $s(t)$ in feet at any time t is given by the function

$$s(t) = -16t^2 + 80t$$

Find the interval of time for which the height of the projectile is greater than 96 feet.

Use a graphing calculator to check each exercise.

69. Exercise 37   **70.** Exercise 38

71. Exercise 39   **72.** Exercise 40

8.5 Quadratic Functions and Their Graphs

OBJECTIVES

1 Graph Quadratic Functions of the Form $f(x) = x^2 + k$.

2 Graph Quadratic Functions of the Form $f(x) = (x - h)^2$.

3 Graph Quadratic Functions of the Form $f(x) = (x - h)^2 + k$.

4 Graph Quadratic Functions of the Form $f(x) = ax^2$.

5 Graph Quadratic Functions of the Form $f(x) = a(x - h)^2 + k$.

OBJECTIVE

1 Graphing $f(x) = x^2 + k$

We first graphed the quadratic equation $y = x^2$ in Section 3.1. In Section 3.2, we learned that this graph defines a function, and we wrote $y = x^2$ as $f(x) = x^2$. In these sections, we discovered that the graph of a quadratic function is a parabola opening upward or downward. In this section, we continue our study of quadratic functions and their graphs. (Much of the contents of this section is a review of shifting and reflecting techniques from Section 3.6, but specific to quadratic functions.)

First, let's recall the definition of a quadratic function.

> **Quadratic Function**
>
> A quadratic function is a function that can be written in the form $f(x) = ax^2 + bx + c$, where a, b, and c are real numbers and $a \neq 0$.

Notice that equations of the form $y = ax^2 + bx + c$, where $a \neq 0$, define quadratic functions, since y is a function of x or $y = f(x)$.

Recall that if $a > 0$, the parabola opens upward and if $a < 0$, the parabola opens downward. Also, the vertex of a parabola is the lowest point if the parabola opens upward and the highest point if the parabola opens downward. The axis of symmetry is the vertical line that passes through the vertex.

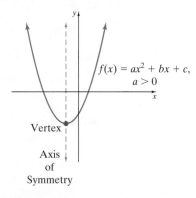

$f(x) = ax^2 + bx + c$, $a > 0$

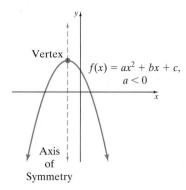

$f(x) = ax^2 + bx + c$, $a < 0$

EXAMPLE 1 Graph $f(x) = x^2$ and $g(x) = x^2 + 6$ on the same set of axes.

Solution First we construct a table of values for $f(x)$ and plot the points. Notice that for each x-value, the corresponding value of $g(x)$ must be 6 more than the corresponding value of $f(x)$ since $f(x) = x^2$ and $g(x) = x^2 + 6$. In other words, the graph of $g(x) = x^2 + 6$ is the same as the graph of $f(x) = x^2$ shifted upward 6 units. The axis of symmetry for both graphs is the y-axis.

x	$f(x) = x^2$	$g(x) = x^2 + 6$
-2	4	10
-1	1	7
0	0	6
1	1	7
2	4	10

Each y-value is increased by 6.

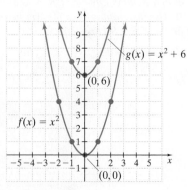

PRACTICE

1 Graph $f(x) = x^2$ and $g(x) = x^2 - 4$ on the same set of axes.

In general, we have the following properties.

Graphing the Parabola Defined by $f(x) = x^2 + k$

If k is positive, the graph of $f(x) = x^2 + k$ is the graph of $y = x^2$ shifted upward k units.

If k is negative, the graph of $f(x) = x^2 + k$ is the graph of $y = x^2$ shifted downward $|k|$ units.

The vertex is $(0, k)$, and the axis of symmetry is the y-axis.

EXAMPLE 2   Graph each function.

a. $F(x) = x^2 + 2$

b. $g(x) = x^2 - 3$

Solution

a. $F(x) = x^2 + 2$

The graph of $F(x) = x^2 + 2$ is obtained by shifting the graph of $y = x^2$ upward 2 units.

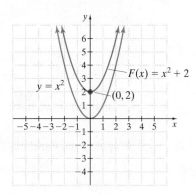

b. $g(x) = x^2 - 3$

The graph of $g(x) = x^2 - 3$ is obtained by shifting the graph of $y = x^2$ downward 3 units.

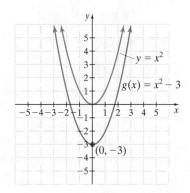

PRACTICE

2   Graph each function.

a. $f(x) = x^2 - 5$

b. $g(x) = x^2 + 3$

OBJECTIVE

2   Graphing $f(x) = (x - h)^2$ ▶

Now we will graph functions of the form $f(x) = (x - h)^2$.

EXAMPLE 3 Graph $f(x) = x^2$ and $g(x) = (x - 2)^2$ on the same set of axes.

Solution By plotting points, we see that for each x-value, the corresponding value of $g(x)$ is the same as the value of $f(x)$ when the x-value is increased by 2. Thus, the graph of $g(x) = (x - 2)^2$ is the graph of $f(x) = x^2$ shifted to the right 2 units. The axis of symmetry for the graph of $g(x) = (x - 2)^2$ is also shifted 2 units to the right and is the line $x = 2$.

x	$f(x) = x^2$	x	$g(x) = (x - 2)^2$
-2	4	0	4
-1	1	1	1
0	0	2	0
1	1	3	1
2	4	4	4

Each x-value increased by 2 corresponds to same y-value.

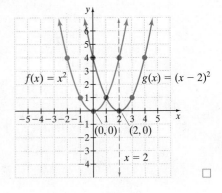

PRACTICE 3 Graph $f(x) = x^2$ and $g(x) = (x + 6)^2$ on the same set of axes.

In general, we have the following properties.

Graphing the Parabola Defined by $f(x) = (x - h)^2$

If h is positive, the graph of $f(x) = (x - h)^2$ is the graph of $y = x^2$ shifted to the right h units.
If h is negative, the graph of $f(x) = (x - h)^2$ is the graph of $y = x^2$ shifted to the left $|h|$ units.
The vertex is $(h, 0)$, and the axis of symmetry is the vertical line $x = h$.

EXAMPLE 4 Graph each function.

a. $G(x) = (x - 3)^2$ **b.** $F(x) = (x + 1)^2$

Solution

a. The graph of $G(x) = (x - 3)^2$ is obtained by shifting the graph of $y = x^2$ to the right 3 units. The graph of $G(x)$ is below on the left.

b. The equation $F(x) = (x + 1)^2$ can be written as $F(x) = [x - (-1)]^2$. The graph of $F(x) = [x - (-1)]^2$ is obtained by shifting the graph of $y = x^2$ to the left 1 unit. The graph of $F(x)$ is below on the right.

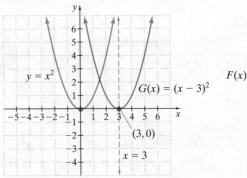
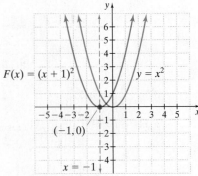

PRACTICE
4   Graph each function.

a. $G(x) = (x + 4)^2$ **b.** $H(x) = (x - 7)^2$

OBJECTIVE

3   Graphing $f(x) = (x - h)^2 + k$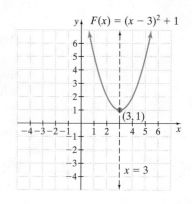

As we will see in graphing functions of the form $f(x) = (x - h)^2 + k$, it is possible to combine vertical and horizontal shifts.

> **Graphing the Parabola Defined by $f(x) = (x - h)^2 + k$**
>
> The parabola has the same shape as $y = x^2$.
> The vertex is (h, k), and the axis of symmetry is the vertical line $x = h$.

EXAMPLE 5   Graph $F(x) = (x - 3)^2 + 1$.

Solution   The graph of $F(x) = (x - 3)^2 + 1$ is the graph of $y = x^2$ shifted 3 units to the right and 1 unit up. The vertex is then $(3, 1)$, and the axis of symmetry is $x = 3$. A few ordered pair solutions are plotted to aid in graphing.

x	$F(x) = (x - 3)^2 + 1$
1	5
2	2
4	2
5	5

PRACTICE
5   Graph $f(x) = (x + 2)^2 + 2$.

OBJECTIVE

4   Graphing $f(x) = ax^2$ ▶

Next, we discover the change in the shape of the graph when the coefficient of x^2 is not 1.

EXAMPLE 6   Graph $f(x) = x^2$, $g(x) = 3x^2$, and $h(x) = \dfrac{1}{2}x^2$ on the same set of axes.

Solution   Comparing the tables of values, we see that for each x-value, the corresponding value of $g(x)$ is triple the corresponding value of $f(x)$. Similarly, the value of $h(x)$ is half the value of $f(x)$.

x	$f(x) = x^2$
-2	4
-1	1
0	0
1	1
2	4

x	$g(x) = 3x^2$
-2	12
-1	3
0	0
1	3
2	12

x	$h(x) = \dfrac{1}{2}x^2$
-2	2
-1	$\dfrac{1}{2}$
0	0
1	$\dfrac{1}{2}$
2	2

(Continued on next page)

The result is that the graph of $g(x) = 3x^2$ is narrower than the graph of $f(x) = x^2$, and the graph of $h(x) = \dfrac{1}{2}x^2$ is wider. The vertex for each graph is $(0, 0)$, and the axis of symmetry is the y-axis.

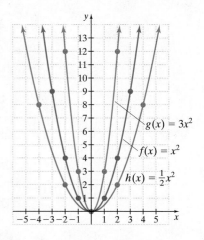

$g(x) = 3x^2$
$f(x) = x^2$
$h(x) = \frac{1}{2}x^2$

PRACTICE
6 Graph $f(x) = x^2$, $g(x) = 4x^2$, and $h(x) = \dfrac{1}{4}x^2$ on the same set of axes.

Graphing the Parabola Defined by $f(x) = ax^2$

If a is positive, the parabola opens upward, and if a is negative, the parabola opens downward.
If $|a| > 1$, the graph of the parabola is narrower than the graph of $y = x^2$.
If $|a| < 1$, the graph of the parabola is wider than the graph of $y = x^2$.

EXAMPLE 7 Graph $f(x) = -2x^2$.

Solution Because $a = -2$, a negative value, this parabola opens downward. Since $|-2| = 2$ and $2 > 1$, the parabola is narrower than the graph of $y = x^2$. The vertex is $(0, 0)$, and the axis of symmetry is the y-axis. We verify this by plotting a few points.

x	$f(x) = -2x^2$
-2	-8
-1	-2
0	0
1	-2
2	-8

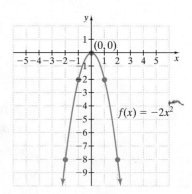

$(0, 0)$
$f(x) = -2x^2$

PRACTICE
7 Graph $f(x) = -\dfrac{1}{2}x^2$.

OBJECTIVE
5 Graphing $f(x) = a(x - h)^2 + k$ ▶

Now we will see the shape of the graph of a quadratic function of the form $f(x) = a(x - h)^2 + k$.

EXAMPLE 8   Graph $g(x) = \frac{1}{2}(x + 2)^2 + 5$. Find the vertex and the axis of symmetry.

Solution The function $g(x) = \frac{1}{2}(x + 2)^2 + 5$ may be written as $g(x) = \frac{1}{2}[x - (-2)]^2 + 5$. Thus, this graph is the same as the graph of $y = x^2$ shifted 2 units to the left and 5 units up, and it is wider because a is $\frac{1}{2}$. The vertex is $(-2, 5)$, and the axis of symmetry is $x = -2$. We plot a few points to verify.

x	$g(x) = \dfrac{1}{2}(x + 2)^2 + 5$
-4	7
-3	$5\dfrac{1}{2}$
-2	5
-1	$5\dfrac{1}{2}$
0	7

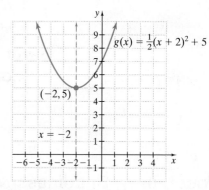

PRACTICE
8   Graph $h(x) = \frac{1}{3}(x - 4)^2 - 3$.

In general, the following holds.

Graph of a Quadratic Function

The graph of a quadratic function written in the form $f(x) = a(x - h)^2 + k$ is a parabola with vertex (h, k).

If $a > 0$, the parabola opens upward.
If $a < 0$, the parabola opens downward.

The axis of symmetry is the line whose equation is $x = h$.

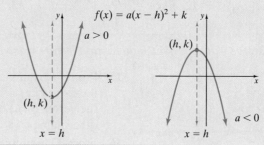

✓CONCEPT CHECK

Which description of the graph of $f(x) = -0.35(x + 3)^2 - 4$ is correct?
a. The graph opens downward and has its vertex at $(-3, 4)$.
b. The graph opens upward and has its vertex at $(-3, 4)$.
c. The graph opens downward and has its vertex at $(-3, -4)$.
d. The graph is narrower than the graph of $y = x^2$.

Answer to Concept Check: c

Graphing Calculator Explorations

Use a graphing calculator to graph the first function of each pair that follows. Then use its graph to predict the graph of the second function. Check your prediction by graphing both on the same set of axes.

1. $F(x) = \sqrt{x}; G(x) = \sqrt{x} + 1$

2. $g(x) = x^3; H(x) = x^3 - 2$

3. $H(x) = |x|; f(x) = |x - 5|$

4. $h(x) = x^3 + 2; g(x) = (x - 3)^3 + 2$

5. $f(x) = |x + 4|; F(x) = |x + 4| + 3$

6. $G(x) = \sqrt{x} - 2; g(x) = \sqrt{x - 4} - 2$

Vocabulary, Readiness & Video Check

Use the choices below to fill in each blank. Some choices will be used more than once.

upward   highest   parabola   downward   lowest   quadratic

1. A(n) _____ function is one that can be written in the form $f(x) = ax^2 + bx + c, a \neq 0$.

2. The graph of a quadratic function is a(n) _____ opening _____ or _____.

3. If $a > 0$, the graph of the quadratic function opens _____.

4. If $a < 0$, the graph of the quadratic function opens _____.

5. The vertex of a parabola is the _____ point if $a > 0$.

6. The vertex of a parabola is the _____ point if $a < 0$.

State the vertex of the graph of each quadratic function.

7. $f(x) = x^2$

8. $f(x) = -5x^2$

9. $g(x) = (x - 2)^2$

10. $g(x) = (x + 5)^2$

11. $f(x) = 2x^2 + 3$

12. $h(x) = x^2 - 1$

13. $g(x) = (x + 1)^2 + 5$

14. $h(x) = (x - 10)^2 - 7$

Martin-Gay Interactive Videos

See Video 8.5

Watch the section lecture video and answer the following questions.

OBJECTIVE 1

15. From ▭ Examples 1 and 2 and the lecture before, how do graphs of the form $f(x) = x^2 + k$ differ from $y = x^2$? Consider the location of the vertex $(0, k)$ on these graphs of the form $f(x) = x^2 + k$—by what other name do we call this point on a graph?

OBJECTIVE 2

16. From ▭ Example 3 and the lecture before, how do graphs of the form $f(x) = (x - h)^2$ differ from $y = x^2$? Consider the location of the vertex $(h, 0)$ on these graphs of the form $f(x) = (x - h)^2$—by what other name do we call this point on a graph?

OBJECTIVE 3

17. From ▭ Example 4 and the lecture before, what general information does the equation $f(x) = (x - h)^2 + k$ tell us about its graph?

OBJECTIVE 4

18. From the lecture before ▭ Example 5, besides the direction a parabola opens, what other graphing information can the value of a tell us?

OBJECTIVE 5

19. In ▭ Examples 6 and 7, what four properties of the graph did we learn from the equation that helped us locate and draw the general shape of the parabola?

8.5   Exercise Set   MyMathLab®

MIXED PRACTICE

Sketch the graph of each quadratic function. Label the vertex and sketch and label the axis of symmetry. See Examples 1 through 5.

1. $f(x) = x^2 - 1$ **2.** $g(x) = x^2 + 3$

3. $h(x) = x^2 + 5$ **4.** $h(x) = x^2 - 4$

5. $g(x) = x^2 + 7$ **6.** $f(x) = x^2 - 2$

7. $f(x) = (x - 5)^2$ **8.** $g(x) = (x + 5)^2$

9. $h(x) = (x + 2)^2$ **10.** $H(x) = (x - 1)^2$

11. $G(x) = (x + 3)^2$ **12.** $f(x) = (x - 6)^2$

13. $f(x) = (x - 2)^2 + 5$ **14.** $g(x) = (x - 6)^2 + 1$

15. $h(x) = (x + 1)^2 + 4$ **16.** $G(x) = (x + 3)^2 + 3$

17. $g(x) = (x + 2)^2 - 5$ **18.** $h(x) = (x + 4)^2 - 6$

Sketch the graph of each quadratic function. Label the vertex, and sketch and label the axis of symmetry. See Examples 6 and 7.

19. $H(x) = 2x^2$ **20.** $f(x) = 5x^2$

21. $h(x) = \dfrac{1}{3}x^2$ **22.** $f(x) = -\dfrac{1}{4}x^2$

23. $g(x) = -x^2$ **24.** $g(x) = -3x^2$

Sketch the graph of each quadratic function. Label the vertex and sketch and label the axis of symmetry. See Example 8.

25. $f(x) = 2(x - 1)^2 + 3$ **26.** $g(x) = 4(x - 4)^2 + 2$

27. $h(x) = -3(x + 3)^2 + 1$ **28.** $f(x) = -(x - 2)^2 - 6$

29. $H(x) = \dfrac{1}{2}(x - 6)^2 - 3$ **30.** $G(x) = \dfrac{1}{5}(x + 4)^2 + 3$

MIXED PRACTICE

Sketch the graph of each quadratic function. Label the vertex and sketch and label the axis of symmetry.

31. $f(x) = -(x - 2)^2$ **32.** $g(x) = -(x + 6)^2$

33. $F(x) = -x^2 + 4$ **34.** $H(x) = -x^2 + 10$

35. $F(x) = 2x^2 - 5$ **36.** $g(x) = \dfrac{1}{2}x^2 - 2$

37. $h(x) = (x - 6)^2 + 4$ **38.** $f(x) = (x - 5)^2 + 2$

39. $F(x) = \left(x + \dfrac{1}{2}\right)^2 - 2$ **40.** $H(x) = \left(x + \dfrac{1}{2}\right)^2 - 3$

41. $F(x) = \dfrac{3}{2}(x + 7)^2 + 1$ **42.** $g(x) = -\dfrac{3}{2}(x - 1)^2 - 5$

43. $f(x) = \dfrac{1}{4}x^2 - 9$ **44.** $H(x) = \dfrac{3}{4}x^2 - 2$

45. $G(x) = 5\left(x + \dfrac{1}{2}\right)^2$ **46.** $F(x) = 3\left(x - \dfrac{3}{2}\right)^2$

47. $h(x) = -(x - 1)^2 - 1$ **48.** $f(x) = -3(x + 2)^2 + 2$

49. $g(x) = \sqrt{3}(x + 5)^2 + \dfrac{3}{4}$ **50.** $G(x) = \sqrt{5}(x - 7)^2 - \dfrac{1}{2}$

51. $h(x) = 10(x + 4)^2 - 6$ **52.** $h(x) = 8(x + 1)^2 + 9$

53. $f(x) = -2(x - 4)^2 + 5$ **54.** $G(x) = -4(x + 9)^2 - 1$

REVIEW AND PREVIEW

Add the proper constant to each binomial so that the resulting trinomial is a perfect square trinomial. See Section 8.1.

55. $x^2 + 8x$ **56.** $y^2 + 4y$

57. $z^2 - 16z$ **58.** $x^2 - 10x$

59. $y^2 + y$ **60.** $z^2 - 3z$

Solve by completing the square. See Section 8.1.

61. $x^2 + 4x = 12$ **62.** $y^2 + 6y = -5$

63. $z^2 + 10z - 1 = 0$ **64.** $x^2 + 14x + 20 = 0$

65. $z^2 - 8z = 2$ **66.** $y^2 - 10y = 3$

CONCEPT EXTENSIONS

Solve. See the Concept Check in this section.

67. Which description of $f(x) = -213(x - 0.1)^2 + 3.6$ is correct?

Graph Opens	Vertex
a. upward	$(0.1, 3.6)$
b. upward	$(-213, 3.6)$
c. downward	$(0.1, 3.6)$
d. downward	$(-0.1, 3.6)$

68. Which description of $f(x) = 5\left(x + \dfrac{1}{2}\right)^2 + \dfrac{1}{2}$ is correct?

Graph Opens	Vertex
a. upward	$\left(\dfrac{1}{2}, \dfrac{1}{2}\right)$
b. upward	$\left(-\dfrac{1}{2}, \dfrac{1}{2}\right)$
c. downward	$\left(\dfrac{1}{2}, -\dfrac{1}{2}\right)$
d. downward	$\left(-\dfrac{1}{2}, -\dfrac{1}{2}\right)$

Write the equation of the parabola that has the same shape as $f(x) = 5x^2$ but with the following vertex.

69. $(2, 3)$ **70.** $(1, 6)$

71. $(-3, 6)$ **72.** $(4, -1)$

The shifting properties covered in this section apply to the graphs of all functions. Given the graph of y = f(x) below, sketch the graph of each of the following.

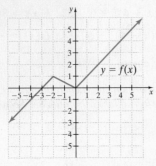

$y = f(x)$

73. $y = f(x) + 1$

74. $y = f(x) - 2$

75. $y = f(x - 3)$

76. $y = f(x + 3)$

77. $y = f(x + 2) + 2$

78. $y = f(x - 1) + 1$

79. The quadratic function $f(x) = 2158x^2 - 10,339x + 6731$ approximates the number of text messages sent in the United States each month between 2000 and 2008, where x is the number of years past 2000 and $f(x)$ is the number of text messages sent in the U.S. each month in millions. (*Source:* cellsigns)

   a. Use this function to find the number of text messages sent in the U.S. each month in 2010.

   b. Use this function to predict the number of text messages sent in the U.S. each month in 2014.

80. Use the function in Exercise 79.

   a. Use this function to predict the number of text messages sent in the U.S. each month in 2018.

   b. Look up the current number of cell phone subscribers in the U.S.

   c. Based on your answers for parts **a.** and **b.**, discuss some possible limitations of using this quadratic function to predict data.

8.6 Further Graphing of Quadratic Functions

OBJECTIVES

1 Write Quadratic Functions in the Form $y = a(x - h)^2 + k$.

2 Derive a Formula for Finding the Vertex of a Parabola.

3 Find the Minimum or Maximum Value of a Quadratic Function.

OBJECTIVE

1   Writing Quadratic Functions in the Form $y = a(x - h)^2 + k$

We know that the graph of a quadratic function is a parabola. If a quadratic function is written in the form

$$f(x) = a(x - h)^2 + k$$

we can easily find the vertex (h, k) and graph the parabola. To write a quadratic function in this form, complete the square. (See Section 8.1 for a review of completing the square.)

EXAMPLE 1   Graph $f(x) = x^2 - 4x - 12$. Find the vertex and any intercepts.

Solution   The graph of this quadratic function is a parabola. To find the vertex of the parabola, we will write the function in the form $y = (x - h)^2 + k$. To do this, we complete the square on the binomial $x^2 - 4x$. To simplify our work, we let $f(x) = y$.

$$y = x^2 - 4x - 12 \quad \text{Let } f(x) = y.$$
$$y + 12 = x^2 - 4x \qquad \text{Add 12 to both sides to get the } x\text{-variable terms alone.}$$

Now we add the square of half of -4 to both sides.

$$\frac{1}{2}(-4) = -2 \quad \text{and} \quad (-2)^2 = 4$$

$$y + 12 + 4 = x^2 - 4x + 4 \qquad \text{Add 4 to both sides.}$$
$$y + 16 = (x - 2)^2 \qquad \text{Factor the trinomial.}$$
$$y = (x - 2)^2 - 16 \quad \text{Subtract 16 from both sides.}$$
$$f(x) = (x - 2)^2 - 16 \quad \text{Replace } y \text{ with } f(x).$$

From this equation, we can see that the vertex of the parabola is $(2, -16)$, a point in quadrant IV, and the axis of symmetry is the line $x = 2$.

Notice that $a = 1$. Since $a > 0$, the parabola opens upward. This parabola opening upward with vertex $(2, -16)$ will have two x-intercepts and one y-intercept. (See the Helpful Hint after this example.)

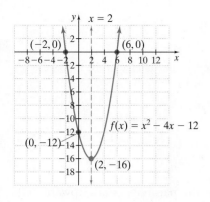

x-intercepts: let y or $f(x) = 0$

$$f(x) = x^2 - 4x - 12$$
$$0 = x^2 - 4x - 12$$
$$0 = (x - 6)(x + 2)$$
$$0 = x - 6 \quad \text{or} \quad 0 = x + 2$$
$$6 = x \qquad\qquad -2 = x$$

y-intercept: let $x = 0$

$$f(x) = x^2 - 4x - 12$$
$$f(0) = 0^2 - 4 \cdot 0 - 12$$
$$= -12$$

The two x-intercepts are $(6, 0)$ and $(-2, 0)$. The y-intercept is $(0, -12)$. The sketch of $f(x) = x^2 - 4x - 12$ is shown.

Notice that the axis of symmetry is always halfway between the x-intercepts. For this example, halfway between -2 and 6 is $\dfrac{-2 + 6}{2} = 2$, and the axis of symmetry is $x = 2$. □

PRACTICE

1   Graph $g(x) = x^2 - 2x - 3$. Find the vertex and any intercepts.

▶ **Helpful Hint**

Parabola Opens Upward
Vertex in I or II: no x-intercept
Vertex in III or IV: 2 x-intercepts

Parabola Opens Downward
Vertex in I or II: 2 x-intercepts
Vertex in III or IV: no x-intercept.

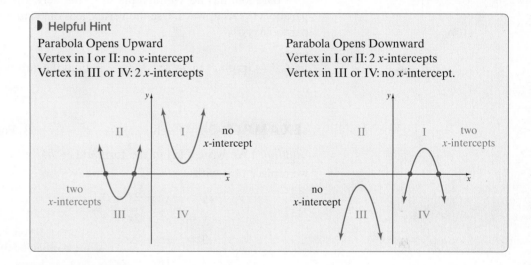

EXAMPLE 2   Graph $f(x) = 3x^2 + 3x + 1$. Find the vertex and any intercepts.

Solution   Replace $f(x)$ with y and complete the square on x to write the equation in the form $y = a(x - h)^2 + k$.

$$y = 3x^2 + 3x + 1 \qquad \text{Replace } f(x) \text{ with } y.$$
$$y - 1 = 3x^2 + 3x \qquad \text{Isolate } x\text{-variable terms.}$$

Factor 3 from the terms $3x^2 + 3x$ so that the coefficient of x^2 is 1.

$$y - 1 = 3(x^2 + x) \qquad \text{Factor out 3.}$$

The coefficient of x in the parentheses above is 1. Then $\dfrac{1}{2}(1) = \dfrac{1}{2}$ and $\left(\dfrac{1}{2}\right)^2 = \dfrac{1}{4}$.

Since we are adding $\dfrac{1}{4}$ inside the parentheses, we are really adding $3\left(\dfrac{1}{4}\right)$, so we *must* add $3\left(\dfrac{1}{4}\right)$ to the left side.

(Continued on next page)

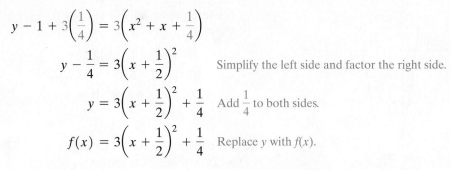

$$y - 1 + 3\left(\frac{1}{4}\right) = 3\left(x^2 + x + \frac{1}{4}\right)$$

$$y - \frac{1}{4} = 3\left(x + \frac{1}{2}\right)^2 \qquad \text{Simplify the left side and factor the right side.}$$

$$y = 3\left(x + \frac{1}{2}\right)^2 + \frac{1}{4} \qquad \text{Add } \frac{1}{4} \text{ to both sides.}$$

$$f(x) = 3\left(x + \frac{1}{2}\right)^2 + \frac{1}{4} \qquad \text{Replace } y \text{ with } f(x).$$

Then $a = 3$, $h = -\dfrac{1}{2}$, and $k = \dfrac{1}{4}$. This means that the parabola opens upward with vertex $\left(-\dfrac{1}{2}, \dfrac{1}{4}\right)$ and that the axis of symmetry is the line $x = -\dfrac{1}{2}$.

To find the y-intercept, let $x = 0$. Then

$$f(0) = 3(0)^2 + 3(0) + 1 = 1$$

Thus the y-intercept is $(0, 1)$.

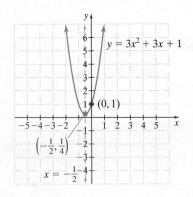

This parabola has no x-intercepts since the vertex is in the second quadrant and the parabola opens upward. Use the vertex, axis of symmetry, and y-intercept to sketch the parabola. □

PRACTICE
2 Graph $g(x) = 4x^2 + 4x + 3$. Find the vertex and any intercepts.

EXAMPLE 3 Graph $f(x) = -x^2 - 2x + 3$. Find the vertex and any intercepts.

Solution We write $f(x)$ in the form $a(x - h)^2 + k$ by completing the square. First we replace $f(x)$ with y.

$$f(x) = -x^2 - 2x + 3$$

$$y = -x^2 - 2x + 3$$

$$y - 3 = -x^2 - 2x \qquad \text{Subtract 3 from both sides to get the } x\text{-variable terms alone.}$$

$$y - 3 = -1(x^2 + 2x) \qquad \text{Factor } -1 \text{ from the terms } -x^2 - 2x.$$

The coefficient of x is 2. Then $\dfrac{1}{2}(2) = 1$ and $1^2 = 1$. We add 1 to the right side inside the parentheses and add $-1(1)$ to the left side.

$$y - 3 - 1(1) = -1(x^2 + 2x + 1)$$

$$y - 4 = -1(x + 1)^2 \qquad \text{Simplify the left side and factor the right side.}$$

$$y = -1(x + 1)^2 + 4 \qquad \text{Add 4 to both sides.}$$

$$f(x) = -1(x + 1)^2 + 4 \qquad \text{Replace } y \text{ with } f(x).$$

> **▶ Helpful Hint**
> This can be written as
> $f(x) = -1[x - (-1)]^2 + 4$.
> Notice that the vertex is $(-1, 4)$.

Since $a = -1$, the parabola opens downward with vertex $(-1, 4)$ and axis of symmetry $x = -1$.

To find the y-intercept, we let $x = 0$ and solve for y. Then

$$f(0) = -0^2 - 2(0) + 3 = 3$$

Thus, $(0, 3)$ is the y-intercept.

To find the x-intercepts, we let y or $f(x) = 0$ and solve for x.

$$f(x) = -x^2 - 2x + 3$$

$$0 = -x^2 - 2x + 3 \qquad \text{Let } f(x) = 0.$$

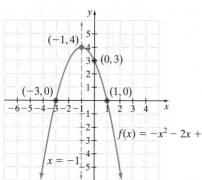

The graph shows points $(-1,4)$, $(0,3)$, $(-3,0)$, $(1,0)$, the line $x = -1$, and the function $f(x) = -x^2 - 2x + 3$.

Now we divide both sides by -1 so that the coefficient of x^2 is 1.

$$\frac{0}{-1} = \frac{-x^2}{-1} - \frac{2x}{-1} + \frac{3}{-1} \qquad \text{Divide both sides by } -1.$$

$$0 = x^2 + 2x - 3 \qquad \text{Simplify.}$$

$$0 = (x + 3)(x - 1) \qquad \text{Factor.}$$

$$x + 3 = 0 \quad \text{or} \quad x - 1 = 0 \qquad \text{Set each factor equal to 0.}$$

$$x = -3 \qquad\qquad x = 1 \qquad \text{Solve.}$$

The x-intercepts are $(-3, 0)$ and $(1, 0)$. Use these points to sketch the parabola. □

PRACTICE

3   Graph $g(x) = -x^2 + 5x + 6$. Find the vertex and any intercepts.

OBJECTIVE

2   Deriving a Formula for Finding the Vertex

There is also a formula that may be used to find the vertex of a parabola. Now that we have practiced completing the square, we will show that the x-coordinate of the vertex of the graph of $f(x)$ or $y = ax^2 + bx + c$ can be found by the formula $x = \dfrac{-b}{2a}$. To do so, we complete the square on x and write the equation in the form $y = a(x - h)^2 + k$.

First, isolate the x-variable terms by subtracting c from both sides.

$$y = ax^2 + bx + c$$

$$y - c = ax^2 + bx$$

Next, factor a from the terms $ax^2 + bx$.

$$y - c = a\left(x^2 + \frac{b}{a}x\right)$$

Next, add the square of half of $\dfrac{b}{a}$, or $\left(\dfrac{b}{2a}\right)^2 = \dfrac{b^2}{4a^2}$, to the right side inside the parentheses. Because of the factor a, what we really added was $a\left(\dfrac{b^2}{4a^2}\right)$, and this must be added to the left side.

$$y - c + a\left(\frac{b^2}{4a^2}\right) = a\left(x^2 + \frac{b}{a}x + \frac{b^2}{4a^2}\right)$$

$$y - c + \frac{b^2}{4a} = a\left(x + \frac{b}{2a}\right)^2 \qquad \begin{array}{l}\text{Simplify the left side and}\\ \text{factor the right side.}\end{array}$$

$$y = a\left(x + \frac{b}{2a}\right)^2 + c - \frac{b^2}{4a} \qquad \begin{array}{l}\text{Add } c \text{ to both sides and subtract } \dfrac{b^2}{4a}\\ \text{from both sides.}\end{array}$$

Compare this form with $f(x)$ or $y = a(x - h)^2 + k$ and see that h is $\dfrac{-b}{2a}$, which means that the x-coordinate of the vertex of the graph of $f(x) = ax^2 + bx + c$ is $\dfrac{-b}{2a}$.

Vertex Formula

The graph of $f(x) = ax^2 + bx + c$, when $a \neq 0$, is a parabola with vertex

$$\left(\frac{-b}{2a}, f\left(\frac{-b}{2a}\right)\right)$$

Let's use this formula to find the vertex of the parabola we graphed in Example 1.

EXAMPLE 4 Find the vertex of the graph of $f(x) = x^2 - 4x - 12$.

Solution In the quadratic function $f(x) = x^2 - 4x - 12$, notice that $a = 1$, $b = -4$, and $c = -12$. Then

$$\frac{-b}{2a} = \frac{-(-4)}{2(1)} = 2$$

The x-value of the vertex is 2. To find the corresponding $f(x)$ or y-value, find $f(2)$. Then

$$f(2) = 2^2 - 4(2) - 12 = 4 - 8 - 12 = -16$$

The vertex is $(2, -16)$. These results agree with our findings in Example 1.   □

PRACTICE

4 Find the vertex of the graph of $g(x) = x^2 - 2x - 3$.

OBJECTIVE

3 Finding Minimum and Maximum Values

The vertex of a parabola gives us some important information about its corresponding quadratic function. The quadratic function whose graph is a parabola that opens upward has a minimum value, and the quadratic function whose graph is a parabola that opens downward has a maximum value. The $f(x)$ or y-value of the vertex is the minimum or maximum value of the function.

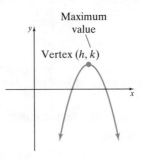
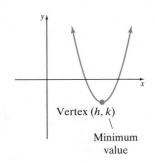

✓CONCEPT CHECK

Without making any calculations, tell whether the graph of $f(x) = 7 - x - 0.3x^2$ has a maximum value or a minimum value. Explain your reasoning.

EXAMPLE 5 Finding Maximum Height

A rock is thrown upward from the ground. Its height in feet above ground after t seconds is given by the function $f(t) = -16t^2 + 20t$. Find the maximum height of the rock and the number of seconds it took for the rock to reach its maximum height.

Solution

1. UNDERSTAND. The maximum height of the rock is the largest value of $f(t)$. Since the function $f(t) = -16t^2 + 20t$ is a quadratic function, its graph is a parabola. It opens downward since $-16 < 0$. Thus, the maximum value of $f(t)$ is the $f(t)$ or y-value of the vertex of its graph.

Answer to Concept Check:
$f(x)$ has a maximum value since it opens downward.

2. TRANSLATE. To find the vertex (h, k), notice that for $f(t) = -16t^2 + 20t$, $a = -16, b = 20$, and $c = 0$. We will use these values and the vertex formula

$$\left(\frac{-b}{2a}, f\left(\frac{-b}{2a}\right)\right)$$

3. SOLVE.

$$h = \frac{-b}{2a} = \frac{-20}{-32} = \frac{5}{8}$$

$$f\left(\frac{5}{8}\right) = -16\left(\frac{5}{8}\right)^2 + 20\left(\frac{5}{8}\right)$$

$$= -16\left(\frac{25}{64}\right) + \frac{25}{2}$$

$$= -\frac{25}{4} + \frac{50}{4} = \frac{25}{4}$$

4. INTERPRET. The graph of $f(t)$ is a parabola opening downward with vertex $\left(\frac{5}{8}, \frac{25}{4}\right)$. This means that the rock's maximum height is $\frac{25}{4}$ feet, or $6\frac{1}{4}$ feet, which was reached in $\frac{5}{8}$ second.   □

PRACTICE

5   A ball is tossed upward from the ground. Its height in feet above ground after t seconds is given by the function $h(t) = -16t^2 + 24t$. Find the maximum height of the ball and the number of seconds it took for the ball to reach the maximum height.

···

Vocabulary, Readiness & Video Check

Fill in each blank.

1. If a quadratic function is in the form $f(x) = a(x - h)^2 + k$, the vertex of its graph is _____.

2. The graph of $f(x) = ax^2 + bx + c, a \neq 0$, is a parabola whose vertex has x-value _____.

Martin-Gay Interactive Videos

See Video 8.6

Watch the section lecture video and answer the following questions.

OBJECTIVE 1

3. From ⊞ Example 1, how does writing a quadratic function in the form $f(x) = a(x - h)^2 + k$ help us graph the function? What procedure can we use to write a quadratic function in this form?

OBJECTIVE 2

4. From ⊞ Example 2, how can locating the vertex and knowing whether parabola opens upward or downward potentially help save unnecessary work? Explain.

OBJECTIVE 3

5. From ⊞ Example 4, when an application involving a quadratic function asks for the maximum or minimum, what part of a parabola should we find?

8.6 Exercise Set   MyMathLab®

Fill in each blank.

	Parabola Opens	Vertex Location	Number of x-intercept(s)	Number of y-intercept(s)
1.	up	Q I		
2.	up	Q III		
3.	down	Q II		
4.	down	Q IV		
5.	up	x-axis		
6.	down	x-axis		
7.		Q III	0	
8.		Q I	2	
9.		Q IV	2	
10.		Q II	0	

Find the vertex of the graph of each quadratic function. See Examples 1 through 4.

11. $f(x) = x^2 + 8x + 7$

12. $f(x) = x^2 + 6x + 5$

13. $f(x) = -x^2 + 10x + 5$

14. $f(x) = -x^2 - 8x + 2$

15. $f(x) = 5x^2 - 10x + 3$

16. $f(x) = -3x^2 + 6x + 4$

17. $f(x) = -x^2 + x + 1$

18. $f(x) = x^2 - 9x + 8$

Match each function with its graph. See Examples 1 through 4.

A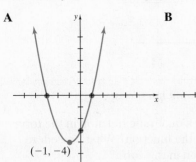
(−1, −4)

B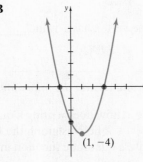
(1, −4)

C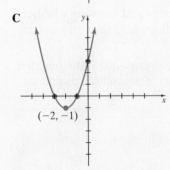
(−2, −1)

D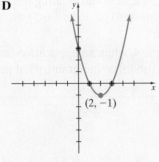
(2, −1)

19. $f(x) = x^2 - 4x + 3$

20. $f(x) = x^2 + 2x - 3$

21. $f(x) = x^2 - 2x - 3$

22. $f(x) = x^2 + 4x + 3$

MIXED PRACTICE

Find the vertex of the graph of each quadratic function. Determine whether the graph opens upward or downward, find any intercepts, and sketch the graph. See Examples 1 through 4.

23. $f(x) = x^2 + 4x - 5$

24. $f(x) = x^2 + 2x - 3$

25. $f(x) = -x^2 + 2x - 1$

26. $f(x) = -x^2 + 4x - 4$

27. $f(x) = x^2 - 4$

28. $f(x) = x^2 - 1$

29. $f(x) = 4x^2 + 4x - 3$

30. $f(x) = 2x^2 - x - 3$

31. $f(x) = \frac{1}{2}x^2 + 4x + \frac{15}{2}$

32. $\frac{1}{5}x^2 + 2x + \frac{9}{5}$

33. $f(x) = x^2 - 6x + 5$

34. $f(x) = x^2 - 4x + 3$

35. $f(x) = x^2 - 4x + 5$

36. $f(x) = x^2 - 6x + 11$

37. $f(x) = 2x^2 + 4x + 5$

38. $f(x) = 3x^2 + 12x + 16$

39. $f(x) = -2x^2 + 12x$

40. $f(x) = -4x^2 + 8x$

41. $f(x) = x^2 + 1$

42. $f(x) = x^2 + 4$

43. $f(x) = x^2 - 2x - 15$

44. $f(x) = x^2 - x - 12$

45. $f(x) = -5x^2 + 5x$

46. $f(x) = 3x^2 - 12x$

47. $f(x) = -x^2 + 2x - 12$

48. $f(x) = -x^2 + 8x - 17$

49. $f(x) = 3x^2 - 12x + 15$

50. $f(x) = 2x^2 - 8x + 11$

51. $f(x) = x^2 + x - 6$

52. $f(x) = x^2 + 3x - 18$

53. $f(x) = -2x^2 - 3x + 35$

54. $f(x) = 3x^2 - 13x - 10$

Solve. See Example 5.

55. If a projectile is fired straight upward from the ground with an initial speed of 96 feet per second, then its height h in feet after t seconds is given by the equation

$$h(t) = -16t^2 + 96t$$

Find the maximum height of the projectile.

56. If Rheam Gaspar throws a ball upward with an initial speed of 32 feet per second, then its height h in feet after t seconds is given by the equation

$$h(t) = -16t^2 + 32t$$

Find the maximum height of the ball.

57. The cost C in dollars of manufacturing x bicycles at Holladay's Production Plant is given by the function

$$C(x) = 2x^2 - 800x + 92{,}000.$$

a. Find the number of bicycles that must be manufactured to minimize the cost.

b. Find the minimum cost.

58. The Utah Ski Club sells calendars to raise money. The profit P, in cents, from selling x calendars is given by the equation $P(x) = 360x - x^2$.

a. Find how many calendars must be sold to maximize profit.

b. Find the maximum profit.

59. Find two numbers whose sum is 60 and whose product is as large as possible. [*Hint:* Let x and $60 - x$ be the two positive numbers. Their product can be described by the function $f(x) = x(60 - x)$.]

60. Find two numbers whose sum is 11 and whose product is as large as possible. (Use the hint for Exercise 59.)

61. Find two numbers whose difference is 10 and whose product is as small as possible. (Use the hint for Exercise 59.)

62. Find two numbers whose difference is 8 and whose product is as small as possible.

△ **63.** The length and width of a rectangle must have a sum of 40. Find the dimensions of the rectangle that will have the maximum area. (Use the hint for Exercise 59.)

△ **64.** The length and width of a rectangle must have a sum of 50. Find the dimensions of the rectangle that will have maximum area.

REVIEW AND PREVIEW

Sketch the graph of each function. See Section 8.5.

65. $f(x) = x^2 + 2$

66. $f(x) = (x - 3)^2$

67. $g(x) = x + 2$

68. $h(x) = x - 3$

69. $f(x) = (x + 5)^2 + 2$

70. $f(x) = 2(x - 3)^2 + 2$

71. $f(x) = 3(x - 4)^2 + 1$

72. $f(x) = (x + 1)^2 + 4$

73. $f(x) = -(x - 4)^2 + \dfrac{3}{2}$

74. $f(x) = -2(x + 7)^2 + \dfrac{1}{2}$

CONCEPT EXTENSIONS

Without calculating, tell whether each graph has a minimum value or a maximum value. See the Concept Check in the section.

75. $f(x) = 2x^2 - 5$

76. $g(x) = -7x^2 + x + 1$

77. $f(x) = 3 - \dfrac{1}{2}x^2$

78. $G(x) = 3 - \dfrac{1}{2}x + 0.8x^2$

Find the vertex of the graph of each quadratic function. Determine whether the graph opens upward or downward, find the y-intercept, approximate the x-intercepts to one decimal place, and sketch the graph.

79. $f(x) = x^2 + 10x + 15$ **80.** $f(x) = x^2 - 6x + 4$

81. $f(x) = 3x^2 - 6x + 7$ **82.** $f(x) = 2x^2 + 4x - 1$

Find the maximum or minimum value of each function. Approximate to two decimal places.

83. $f(x) = 2.3x^2 - 6.1x + 3.2$

84. $f(x) = 7.6x^2 + 9.8x - 2.1$

85. $f(x) = -1.9x^2 + 5.6x - 2.7$

86. $f(x) = -5.2x^2 - 3.8x + 5.1$

87. The projected number of Wi-Fi-enabled cell phones in the United States can be modeled by the quadratic function $c(x) = -0.4x^2 + 21x + 35$, where $c(x)$ is the projected number of Wi-Fi-enabled cell phones in millions and x is the number of years after 2009. (*Source:* Techcrunchies.com)

a. Will this function have a maximum or a minimum? How can you tell?

b. According to this model, in what year will the number of Wi-Fi-enabled cell phones in the United States be at its maximum or minimum?

c. What is the maximum/minimum number of Wi-Fi-enabled cell phones predicted? Round to the nearest whole million.

88. Methane is a gas produced by landfills, natural gas systems, and coal mining that contributes to the greenhouse effect and global warming. Projected methane emissions in the United States can be modeled by the quadratic function

$$f(x) = -0.072x^2 + 1.93x + 173.9$$

where $f(x)$ is the amount of methane produced in million metric tons and x is the number of years after 2000. (*Source:* Based on data from the U.S. Environmental Protection Agency, 2000–2020)

a. According to this model, what will U.S. emissions of methane be in 2018? (Round to 2 decimal places.)

b. Will this function have a maximum or a minimum? How can you tell?

c. In what year will methane emissions in the United States be at their maximum/minimum? Round to the nearest whole year.

d. What is the level of methane emissions for that year? (Use your rounded answer from part (c).) (Round this answer to 2 decimals places.)

Use a graphing calculator to check each exercise.

 89. Exercise 37   **90.** Exercise 38

 91. Exercise 47   **92.** Exercise 48

Chapter 8 Vocabulary Check

Fill in each blank with one of the words or phrases listed below.

quadratic formula	quadratic	discriminant	$\pm\sqrt{b}$
completing the square	quadratic inequality	(h, k)	$(0, k)$
$(h, 0)$	$\dfrac{-b}{2a}$		

1. The _____ helps us find the number and type of solutions of a quadratic equation.

2. If $a^2 = b$, then $a =$ _____.

3. The graph of $f(x) = ax^2 + bx + c$, where a is not 0, is a parabola whose vertex has x-value _____.

4. A _____ is an inequality that can be written so that one side is a quadratic expression and the other side is 0.

5. The process of writing a quadratic equation so that one side is a perfect square trinomial is called _____.

6. The graph of $f(x) = x^2 + k$ has vertex _____.

7. The graph of $f(x) = (x - h)^2$ has vertex _____.

8. The graph of $f(x) = (x - h)^2 + k$ has vertex _____.

9. The formula $x = \dfrac{-b \pm \sqrt{b^2 - 4ac}}{2a}$ is called the _____.

10. A _____ equation is one that can be written in the form $ax^2 + bx + c = 0$ where $a, b,$ and c are real numbers and a is not 0.

Chapter 8 Highlights

DEFINITIONS AND CONCEPTS	**EXAMPLES**
Section 8.1 Solving Quadratic Equations by Completing the Square	

DEFINITIONS AND CONCEPTS	**EXAMPLES**
Square root property If b is a real number and if $a^2 = b$, then $a = \pm\sqrt{b}$.	Solve: $(x + 3)^2 = 14$. $\qquad x + 3 = \pm\sqrt{14}$ $\qquad\qquad x = -3 \pm \sqrt{14}$
To solve a quadratic equation in x by completing the square **Step 1.** If the coefficient of x^2 is not 1, divide both sides of the equation by the coefficient of x^2. **Step 2.** Isolate the variable terms. **Step 3.** Complete the square by adding the square of half of the coefficient of x to both sides. **Step 4.** Write the resulting trinomial as the square of a binomial. **Step 5.** Apply the square root property and solve for x.	Solve: $3x^2 - 12x - 18 = 0$. **1.** $x^2 - 4x - 6 = 0$ **2.** $\qquad x^2 - 4x = 6$ **3.** $\quad \dfrac{1}{2}(-4) = -2$ and $(-2)^2 = 4$ $\qquad x^2 - 4x + 4 = 6 + 4$ **4.** $\qquad (x - 2)^2 = 10$ **5.** $\qquad x - 2 = \pm\sqrt{10}$ $\qquad\qquad x = 2 \pm \sqrt{10}$

DEFINITIONS AND CONCEPTS	EXAMPLES

Section 8.2   Solving Quadratic Equations by the Quadratic Formula

A quadratic equation written in the form $ax^2 + bx + c = 0$ has solutions

$$x = \frac{-b \pm \sqrt{b^2 - 4ac}}{2a}$$

Solve: $x^2 - x - 3 = 0$.
$$a = 1, b = -1, c = -3$$
$$x = \frac{-(-1) \pm \sqrt{(-1)^2 - 4(1)(-3)}}{2 \cdot 1}$$
$$x = \frac{1 \pm \sqrt{13}}{2}$$

Section 8.3   Solving Equations by Using Quadratic Methods

Substitution is often helpful in solving an equation that contains a repeated variable expression.

Solve: $(2x + 1)^2 - 5(2x + 1) + 6 = 0$.
Let $m = 2x + 1$. Then
$$m^2 - 5m + 6 = 0 \qquad \text{Let } m = 2x + 1.$$
$$(m - 3)(m - 2) = 0$$
$$m = 3 \quad \text{or} \quad m = 2$$
$$2x + 1 = 3 \quad \text{or} \quad 2x + 1 = 2 \quad \text{Substitute back.}$$
$$x = 1 \quad \text{or} \quad x = \frac{1}{2}$$

Section 8.4   Nonlinear Inequalities in One Variable

To solve a polynomial inequality

Step 1.   Write the inequality in standard form.

Step 2.   Solve the related equation.

Step 3.   Use solutions from Step 2 to separate the number line into regions.

Step 4.   Use test points to determine whether values in each region satisfy the original inequality.

Step 5.   Write the solution set as the union of regions whose test point value is a solution.

Solve: $x^2 \geq 6x$.

1. $x^2 - 6x \geq 0$

2. $x^2 - 6x = 0$
$$x(x - 6) = 0$$
$$x = 0 \quad \text{or} \quad x = 6$$

3.

4.

Region	Test Point Value	$x^2 \geq 6x$	Result
A	-2	$(-2)^2 \geq 6(-2)$	True
B	1	$1^2 \geq 6(1)$	False
C	7	$7^2 \geq 6(7)$	True

5.

The solution set is $(-\infty, 0] \cup [6, \infty)$.

To solve a rational inequality

Step 1.   Solve for values that make all denominators 0.

Step 2.   Solve the related equation.

Step 3.   Use solutions from Steps 1 and 2 to separate the number line into regions.

Step 4.   Use test points to determine whether values in each region satisfy the original inequality.

Step 5.   Write the solution set as the union of regions whose test point value is a solution.

Solve: $\dfrac{6}{x - 1} < -2$.

1. $x - 1 = 0$ Set denominator equal to 0.
$$x = 1$$

2. $\dfrac{6}{x - 1} = -2$
$$6 = -2(x - 1) \quad \text{Multiply by } (x - 1).$$
$$6 = -2x + 2$$
$$4 = -2x$$
$$-2 = x$$

(continued)

DEFINITIONS AND CONCEPTS	EXAMPLES

Section 8.4 Nonlinear Inequalities in One Variable (continued)

3.
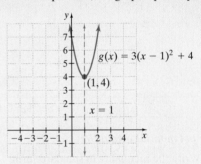

4. Only a test value from region B satisfies the original inequality.

5.

The solution set is $(-2, 1)$.

Section 8.5 Quadratic Functions and Their Graphs

Graph of a quadratic function

The graph of a quadratic function written in the form $f(x) = a(x - h)^2 + k$ is a parabola with vertex (h, k). If $a > 0$, the parabola opens upward; if $a < 0$, the parabola opens downward. The axis of symmetry is the line whose equation is $x = h$.

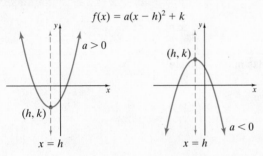

Graph $g(x) = 3(x - 1)^2 + 4$.

The graph is a parabola with vertex $(1, 4)$ and axis of symmetry $x = 1$. Since $a = 3$ is positive, the graph opens upward.

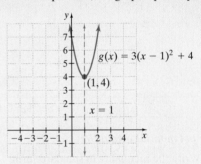

Section 8.6 Further Graphing of Quadratic Functions

The graph of $f(x) = ax^2 + bx + c$, where $a \neq 0$, is a parabola with vertex

$$\left(\frac{-b}{2a}, f\left(\frac{-b}{2a} \right) \right)$$

Graph $f(x) = x^2 - 2x - 8$. Find the vertex and x- and y-intercepts.

$$\frac{-b}{2a} = \frac{-(-2)}{2 \cdot 1} = 1$$

$$f(1) = 1^2 - 2(1) - 8 = -9$$

The vertex is $(1, -9)$.

$$0 = x^2 - 2x - 8$$

$$0 = (x - 4)(x + 2)$$

$$x = 4 \quad \text{or} \quad x = -2$$

The x-intercepts are $(4, 0)$ and $(-2, 0)$.

$$f(0) = 0^2 - 2 \cdot 0 - 8 = -8$$

The y-intercept is $(0, -8)$.

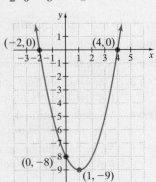

(8.1) Solve by factoring.

1. $x^2 - 15x + 14 = 0$ **2.** $7a^2 = 29a + 30$

Solve by using the square root property.

3. $4m^2 = 196$ **4.** $(5x - 2)^2 = 2$

Solve by completing the square.

5. $z^2 + 3z + 1 = 0$

6. $(2x + 1)^2 = x$

7. If P dollars are originally invested, the formula $A = P(1 + r)^2$ gives the amount A in an account paying interest rate r compounded annually after 2 years. Find the interest rate r such that $2500 increases to $2717 in 2 years. Round the result to the nearest hundredth of a percent.

△ **8.** Two ships leave a port at the same time and travel at the same speed. One ship is traveling due north and the other due east. In a few hours, the ships are 150 miles apart. How many miles has each ship traveled? Give an exact answer and a one-decimal-place approximation.

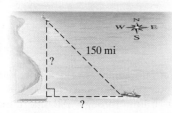

(8.2) If the discriminant of a quadratic equation has the given value, determine the number and type of solutions of the equation.

9. -8 **10.** 48

11. 100   **12.** 0

Solve by using the quadratic formula.

13. $x^2 - 16x + 64 = 0$ **14.** $x^2 + 5x = 0$

15. $2x^2 + 3x = 5$ **16.** $9x^2 + 4 = 2x$

17. $6x^2 + 7 = 5x$ **18.** $(2x - 3)^2 = x$

19. Cadets graduating from military school usually toss their hats high into the air at the end of the ceremony. One cadet threw his hat so that its distance $d(t)$ in feet above the ground t seconds after it was thrown was $d(t) = -16t^2 + 30t + 6$.

   a. Find the distance above the ground of the hat 1 second after it was thrown.

   b. Find the time it takes the hat to hit the ground. Give an exact time and a one-decimal-place approximation.

△ **20.** The hypotenuse of an isosceles right triangle is 6 centimeters longer than either of the legs. Find the length of the legs.

(8.3) Solve each equation for the variable.

21. $x^3 = 27$

22. $y^3 = -64$

23. $\dfrac{5}{x} + \dfrac{6}{x - 2} = 3$

24. $x^4 - 21x^2 - 100 = 0$

25. $x^{2/3} - 6x^{1/3} + 5 = 0$

26. $5(x + 3)^2 - 19(x + 3) = 4$

27. $a^6 - a^2 = a^4 - 1$

28. $y^{-2} + y^{-1} = 20$

29. Two postal workers, Jerome Grant and Tim Bozik, can sort a stack of mail in 5 hours. Working alone, Tim can sort the mail in 1 hour less time than Jerome can. Find the time that each postal worker can sort the mail alone. Round the result to one decimal place.

30. A negative number decreased by its reciprocal is $-\dfrac{24}{5}$. Find the number.

(8.4) Solve each inequality for x. Write each solution set in interval notation.

31. $2x^2 - 50 \le 0$

32. $\dfrac{1}{4}x^2 < \dfrac{1}{16}$

33. $(x^2 - 4)(x^2 - 25) \le 0$

34. $(x^2 - 16)(x^2 - 1) > 0$

35. $\dfrac{x - 5}{x - 6} < 0$

36. $\dfrac{(4x + 3)(x - 5)}{x(x + 6)} > 0$

37. $(x + 5)(x - 6)(x + 2) \le 0$

38. $x^3 + 3x^2 - 25x - 75 > 0$

39. $\dfrac{x^2 + 4}{3x} \le 1$ **40.** $\dfrac{3}{x - 2} > 2$

(8.5) Sketch the graph of each function. Label the vertex and the axis of symmetry.

41. $f(x) = x^2 - 4$

42. $g(x) = x^2 + 7$

43. $H(x) = 2x^2$

44. $h(x) = -\dfrac{1}{3}x^2$

45. $F(x) = (x - 1)^2$

46. $G(x) = (x + 5)^2$

47. $f(x) = (x - 4)^2 - 2$

48. $f(x) = -3(x - 1)^2 + 1$

(8.6) Sketch the graph of each function. Find the vertex and the intercepts.

49. $f(x) = x^2 + 10x + 25$

50. $f(x) = -x^2 + 6x - 9$

51. $f(x) = 4x^2 - 1$

52. $f(x) = -5x^2 + 5$

53. Find the vertex of the graph of $f(x) = -3x^2 - 5x + 4$. Determine whether the graph opens upward or downward, find the y-intercept, approximate the x-intercepts to one decimal place, and sketch the graph.

54. The function $h(t) = -16t^2 + 120t + 300$ gives the height in feet of a projectile fired from the top of a building after t seconds.

   a. When will the object reach a height of 350 feet? Round your answer to one decimal place.

   b. Explain why part (a) has two answers.

55. Find two numbers whose product is as large as possible, given that their sum is 420.

56. Write an equation of a quadratic function whose graph is a parabola that has vertex $(-3, 7)$. Let the value of a be $-\dfrac{7}{9}$.

MIXED REVIEW

Solve each equation or inequality.

57. $x^2 - x - 30 = 0$

58. $10x^2 = 3x + 4$

59. $9y^2 = 36$

60. $(9n + 1)^2 = 9$

61. $x^2 + x + 7 = 0$

62. $(3x - 4)^2 = 10x$

63. $x^2 + 11 = 0$

64. $x^2 + 7 = 0$

65. $(5a - 2)^2 - a = 0$

66. $\dfrac{7}{8} = \dfrac{8}{x^2}$

67. $x^{2/3} - 6x^{1/3} = -8$

68. $(2x - 3)(4x + 5) \geq 0$

69. $\dfrac{x(x + 5)}{4x - 3} \geq 0$

70. $\dfrac{3}{x - 2} > 2$

71. The busiest airport in the world is the Hartsfield International Airport in Atlanta, Georgia. The total amount of passenger traffic through Atlanta during the period 2000 through 2010 can be modeled by the equation $y = -32x^2 + 1733x + 76{,}362$, where y is the number of passengers enplaned and deplaned in thousands, and x is the number of years after 2000. (*Source:* Based on data from Airports Council International)

   a. Estimate the passenger traffic at Atlanta's Hartsfield International Airport in 2015.

   b. According to this model, in what year will the passenger traffic at Atlanta's Hartsfield International Airport first reach 99,000 thousand passengers?

Chapter 8 Test  Test Prep VIDEOS You Tube

Solve each equation.

1. $5x^2 - 2x = 7$

2. $(x + 1)^2 = 10$

3. $m^2 - m + 8 = 0$

4. $u^2 - 6u + 2 = 0$

5. $7x^2 + 8x + 1 = 0$

6. $y^2 - 3y = 5$

7. $\dfrac{4}{x + 2} + \dfrac{2x}{x - 2} = \dfrac{6}{x^2 - 4}$

8. $x^5 + 3x^4 = x + 3$

9. $x^6 + 1 = x^4 + x^2$

10. $(x + 1)^2 - 15(x + 1) + 56 = 0$

Solve by completing the square.

▶ **11.** $x^2 - 6x = -2$

▶ **12.** $2a^2 + 5 = 4a$

Solve each inequality for x. Write the solution set in interval notation.

▶ **13.** $2x^2 - 7x > 15$

▶ **14.** $(x^2 - 16)(x^2 - 25) \geq 0$

▶ **15.** $\dfrac{5}{x + 3} < 1$

▶ **16.** $\dfrac{7x - 14}{x^2 - 9} \leq 0$

Graph each function. Label the vertex.

▶ **17.** $f(x) = 3x^2$

▶ **18.** $G(x) = -2(x - 1)^2 + 5$

Graph each function. Find and label the vertex, y-intercept, and x-intercepts (if any).

▶ **19.** $h(x) = x^2 - 4x + 4$

▶ **20.** $F(x) = 2x^2 - 8x + 9$

▶ **21.** Dave and Sandy Hartranft can paint a room together in 4 hours. Working alone, Dave can paint the room in 2 hours less time than Sandy can. Find how long it takes Sandy to paint the room alone.

▶ **22.** A stone is thrown upward from a bridge. The stone's height in feet, $s(t)$, above the water t seconds after the stone is thrown is a function given by the equation $s(t) = -16t^2 + 32t + 256$.

 a. Find the maximum height of the stone.

 b. Find the time it takes the stone to hit the water. Round the answer to two decimal places.

△ **23.** Given the diagram shown, approximate to the nearest tenth of a foot how many feet of walking distance a person saves by cutting across the lawn instead of walking on the sidewalk.

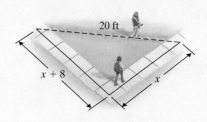

Chapter 8 Cumulative Review

1. Write each sentence using mathematical symbols.

 a. The sum of 5 and y is greater than or equal to 7.

 b. 11 is not equal to z.

 c. 20 is less than the difference of 5 and twice x.

2. Solve $|3x - 2| = -5$.

3. Find the slope of the line containing the points $(0, 3)$ and $(2, 5)$. Graph the line.

4. Use the elimination method to solve the system.
$$\begin{cases} -6x + y = 5 \\ 4x - 2y = 6 \end{cases}$$

5. Use the elimination method to solve the system:
$$\begin{cases} x - 5y = -12 \\ -x + y = 4 \end{cases}$$

6. Simplify. Use positive exponents to write each answer.

 a. $(a^{-2}bc^3)^{-3}$

 b. $\left(\dfrac{a^{-4}b^2}{c^3}\right)^{-2}$

 c. $\left(\dfrac{3a^8b^2}{12a^5b^5}\right)^{-2}$

7. Multiply.

 a. $(2x - 7)(3x - 4)$

 b. $(3x^2 + y)(5x^2 - 2y)$

8. Multiply.

 a. $(4a - 3)(7a - 2)$

 b. $(2a + b)(3a - 5b)$

9. Factor.

 a. $8x^2 + 4$

 b. $5y - 2z^4$

 c. $6x^2 - 3x^3 + 12x^4$

10. Factor.

 a. $9x^3 + 27x^2 - 15x$

 b. $2x(3y - 2) - 5(3y - 2)$

 c. $2xy + 6x - y - 3$

Factor the polynomials in Exercises 11 through 14.

11. $x^2 - 12x + 35$

12. $x^2 - 2x - 48$

13. $12a^2x - 12abx + 3b^2x$

14. Factor. $2ax^2 - 12axy + 18ay^2$

15. Solve $3(x^2 + 4) + 5 = -6(x^2 + 2x) + 13$.

16. Solve $2(a^2 + 2) - 8 = -2a(a - 2) - 5$.

17. Solve $x^3 = 4x$.

18. Find the vertex and any intercepts of $f(x) = x^2 + x - 12$.

19. Simplify $\dfrac{2x^2}{10x^3 - 2x^2}$

20. Simplify $\dfrac{x^2 - 4x + 4}{2 - x}$.

21. Add $\dfrac{2x - 1}{2x^2 - 9x - 5} + \dfrac{x + 3}{6x^2 - x - 2}$.

22. Subtract $\dfrac{a + 1}{a^2 - 6a + 8} - \dfrac{3}{16 - a^2}$

23. Simplify $\dfrac{x^{-1} + 2xy^{-1}}{x^{-2} - x^{-2}y^{-1}}$.

24. Simplify $\dfrac{(2a)^{-1} + b^{-1}}{a^{-1} + (2b)^{-1}}$.

25. Divide $\dfrac{3x^5y^2 - 15x^3y - x^2y - 6x}{x^2y}$.

26. Divide $x^3 - 3x^2 - 10x + 24$ by $x + 3$.

27. If $P(x) = 2x^3 - 4x^2 + 5$

  a. Find $P(2)$ by substitution.

  b. Use synthetic division to find the remainder when $P(x)$ is divided by $x - 2$.

28. If $P(x) = 4x^3 - 2x^2 + 3$,

  a. Find $P(-2)$ by substitution.

  b. Use synthetic division to find the remainder when $P(x)$ is divided by $x + 2$.

29. Solve $\dfrac{4x}{5} + \dfrac{3}{2} = \dfrac{3x}{10}$.

30. Solve $\dfrac{x + 3}{x^2 + 5x + 6} = \dfrac{3}{2x + 4} - \dfrac{1}{x + 3}$.

31. If a certain number is subtracted from the numerator and added to the denominator of $\dfrac{9}{19}$, the new fraction is equivalent to $\dfrac{1}{3}$. Find the number.

32. Mr. Briley can roof his house in 24 hours. His son can roof the same house in 40 hours. If they work together, how long will it take to roof the house?

33. Suppose that y varies directly as x. If y is 5 when x is 30, find the constant of variation and the direct variation equation.

34. Suppose that y varies inversely as x. If y is 8 when x is 14, find the constant of variation and the inverse variation equation.

35. Simplify.

  a. $\sqrt{(-3)^2}$ **b.** $\sqrt{x^2}$

  c. $\sqrt[4]{(x - 2)^4}$ **d.** $\sqrt[3]{(-5)^3}$

  e. $\sqrt[5]{(2x - 7)^5}$ **f.** $\sqrt{25x^2}$

  g. $\sqrt{x^2 + 2x + 1}$

36. Simplify.

  a. $\sqrt{(-2)^2}$ **b.** $\sqrt{y^2}$

  c. $\sqrt[4]{(a - 3)^4}$ **d.** $\sqrt[3]{(-6)^3}$

  e. $\sqrt[5]{(3x - 1)^5}$

37. Use rational exponents to simplify. Assume that variables represent positive numbers.

  a. $\sqrt[8]{x^4}$ **b.** $\sqrt[6]{25}$

  c. $\sqrt[4]{r^2s^6}$

38. Use rational exponents to simplify. Assume that variables represent positive numbers.

  a. $\sqrt[4]{5^2}$ **b.** $\sqrt[12]{x^3}$

  c. $\sqrt[6]{x^2y^4}$

39. Use the product rule to simplify.

  a. $\sqrt{25x^3}$ **b.** $\sqrt[3]{54x^6y^8}$

  c. $\sqrt[4]{81z^{11}}$

40. Use the product rule to simplify. Assume that variables represent positive numbers.

  a. $\sqrt{64a^5}$ **b.** $\sqrt[3]{24a^7b^9}$

  c. $\sqrt[4]{48x^9}$

41. Rationalize the denominator of each expression.

  a. $\dfrac{2}{\sqrt{5}}$ **b.** $\dfrac{2\sqrt{16}}{\sqrt{9x}}$

  c. $\sqrt[3]{\dfrac{1}{2}}$

42. Multiply. Simplify if possible.

  a. $(\sqrt{3} - 4)(2\sqrt{3} + 2)$

  b. $(\sqrt{5} - x)^2$

  c. $(\sqrt{a} + b)(\sqrt{a} - b)$

43. Solve $\sqrt{2x + 5} + \sqrt{2x} = 3$.

44. Solve $\sqrt{x - 2} = \sqrt{4x + 1} - 3$.

45. Divide. Write in the form $a + bi$.

  a. $\dfrac{2 + i}{1 - i}$ **b.** $\dfrac{7}{3i}$

46. Write each product in the form of $a + bi$.

  a. $3i(5 - 2i)$ **b.** $(6 - 5i)^2$

  c. $(\sqrt{3} + 2i)(\sqrt{3} - 2i)$

47. Use the square root property to solve $(x + 1)^2 = 12$.

48. Use the square root property to solve $(y - 1)^2 = 24$.

49. Solve $x - \sqrt{x} - 6 = 0$.

50. Use the quadratic formula to solve $m^2 = 4m + 8$.

A compact fluorescent lamp (or light) (CFL) is a type of fluorescent light that is quickly gaining popularity for many reasons. Compared to an incandescent bulb, CFLs use less power and last between 8 and 15 times as long. Although a CFL has a higher price, the savings per bulb are substantial (possibly $30 per life of bulb). Many CFLs are now manufactured to replace an incandescent bulb and can fit into existing fixtures. It should be noted that since CFLs are a type of fluorescent light, they do contain a small amount of mercury.

Although we have no direct applications in this chapter, it should be noted that the light output of a CFL decays exponentially. By the end of their lives, they produce 70–80% of their original output, with the fastest losses occurring soon after the light is first used. Also, it should be noted that the response of the human eye to light is logarithmic.

In this chapter, we discuss two closely related functions: exponential and logarithmic functions. These functions are vital to applications in economics, finance, engineering, the sciences, education, and other fields. Models of tumor growth and learning curves are two examples of the uses of exponential and logarithmic functions.

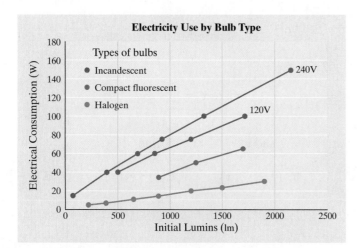

Electricity Use by Bulb Type

Types of bulbs
- Incandescent
- Compact fluorescent
- Halogen

(Note: Lower points correspond to lower energy use.)

9.1 The Algebra of Functions; Composite Functions

OBJECTIVES

1 Add, Subtract, Multiply, and Divide Functions.

2 Construct Composite Functions.

OBJECTIVE

1 Adding, Subtracting, Multiplying, and Dividing Functions

As we have seen in earlier chapters, it is possible to add, subtract, multiply, and divide functions. Although we have not stated it as such, the sums, differences, products, and quotients of functions are themselves functions. For example, if $f(x) = 3x$ and $g(x) = x + 1$, their product, $f(x) \cdot g(x) = 3x(x + 1) = 3x^2 + 3x$, is a new function. We can use the notation $(f \cdot g)(x)$ to denote this new function. Finding the sum, difference, product, and quotient of functions to generate new functions is called the **algebra of functions**.

Algebra of Functions

Let f and g be functions. New functions from f and g are defined as follows.

Sum $\qquad (f + g)(x) = f(x) + g(x)$

Difference $\qquad (f - g)(x) = f(x) - g(x)$

Product $\qquad (f \cdot g)(x) = f(x) \cdot g(x)$

Quotient $\qquad \left(\dfrac{f}{g}\right)(x) = \dfrac{f(x)}{g(x)}, \quad g(x) \neq 0$

EXAMPLE 1 If $f(x) = x - 1$ and $g(x) = 2x - 3$, find

a. $(f + g)(x)$ **b.** $(f - g)(x)$ **c.** $(f \cdot g)(x)$ **d.** $\left(\dfrac{f}{g}\right)(x)$

Solution Use the algebra of functions and replace $f(x)$ by $x - 1$ and $g(x)$ by $2x - 3$. Then we simplify.

a. $(f + g)(x) = f(x) + g(x)$
$\qquad\qquad\quad = (x - 1) + (2x - 3) = 3x - 4$

b. $(f - g)(x) = f(x) - g(x)$
$\qquad\qquad\quad = (x - 1) - (2x - 3)$
$\qquad\qquad\quad = x - 1 - 2x + 3$
$\qquad\qquad\quad = -x + 2$

c. $(f \cdot g)(x) = f(x) \cdot g(x)$
$\qquad\qquad\quad = (x - 1)(2x - 3)$
$\qquad\qquad\quad = 2x^2 - 5x + 3$

d. $\left(\dfrac{f}{g}\right)(x) = \dfrac{f(x)}{g(x)} = \dfrac{x - 1}{2x - 3}$, where $x \neq \dfrac{3}{2}$ $\qquad\qquad\qquad$ □

PRACTICE

1 If $f(x) = x + 2$ and $g(x) = 3x + 5$, find

a. $(f + g)(x)$ **b.** $(f - g)(x)$ **c.** $(f \cdot g)(x)$ **d.** $\left(\dfrac{f}{g}\right)(x)$

There is an interesting but not surprising relationship between the graphs of functions and the graphs of their sum, difference, product, and quotient. For example, the graph of $(f + g)(x)$ can be found by adding the graph of $f(x)$ to the graph of $g(x)$. We add two graphs by adding y-values of corresponding x-values.

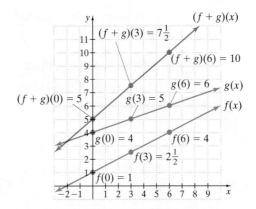

OBJECTIVE

2 **Constructing Composite Functions** ▶

Another way to combine functions is called **function composition.** To understand this new way of combining functions, study the diagrams below. The right diagram shows an illustration by tables, and the left diagram is the same illustration but by thermometers. In both illustrations, we show degrees Celsius $f(x)$ as a function of degrees Fahrenheit x, then Kelvins $g(x)$ as a function of degrees Celsius x. (The Kelvin scale is a temperature scale devised by Lord Kelvin in 1848.) The first function we will call f, and the second function we will call g.

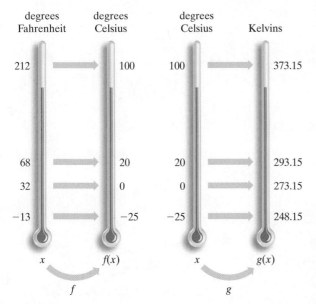

Table Illustration

$x =$ **Degrees Fahrenheit** (**Input**)	-13	32	68	212
$f(x) =$ **Degrees Celsius** (**Output**)	-25	0	20	100

$x =$ **Degrees Celsius** (**Input**)	-25	0	20	100
$g(x) =$ **Kelvins** (**Output**)	248.15	273.15	293.15	373.15

Suppose that we want a function that shows a direct conversion from degrees Fahrenheit to Kelvins. In other words, suppose that a function is needed that shows Kelvins as a function of degrees Fahrenheit. This can easily be done because the output of the first function $f(x)$ is the same as the input of the second function. If we use $g(f(x))$ to represent this, then we get the diagrams below.

$x =$ **Degrees Fahrenheit** (**Input**)	-13	32	68	212
$g(f(x)) =$ **Kelvins** (**Output**)	248.15	273.15	293.15	373.15

For example $g(f(-13)) = 248.15$, and so on.

Since the output of the first function is used as the input of the second function, we write the new function as $g(f(x))$. The new function is formed from the composition of the other two functions. The mathematical symbol for this composition is $(g \circ f)(x)$. Thus, $(g \circ f)(x) = g(f(x))$.

It is possible to find an equation for the composition of the two functions f and g. In other words, we can find a function that converts degrees Fahrenheit directly to Kelvins. The function $f(x) = \dfrac{5}{9}(x - 32)$ converts degrees Fahrenheit to degrees Celsius, and the function $g(x) = x + 273.15$ converts degrees Celsius to Kelvins. Thus,

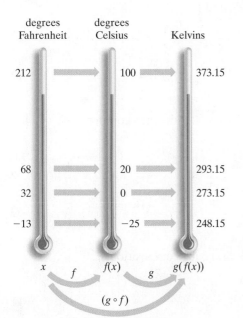

$$(g \circ f)(x) = g(f(x)) = g\left(\frac{5}{9}(x - 32)\right) = \frac{5}{9}(x - 32) + 273.15$$

In general, the notation $g(f(x))$ means "g composed with f" and can be written as $(g \circ f)(x)$. Also $f(g(x))$, or $(f \circ g)(x)$, means "f composed with g."

> **Composition of Functions**
>
> The composition of functions f and g is
>
> $$(f \circ g)(x) = f(g(x))$$

> ▶ **Helpful Hint**
>
> $(f \circ g)(x)$ does not mean the same as $(f \cdot g)(x)$.
>
> $$(f \circ g)(x) = f(g(x)) \text{ while } (f \cdot g)(x) = f(x) \cdot g(x)$$
> $$\qquad\quad \uparrow \qquad\qquad\qquad\qquad\qquad \uparrow$$
> $$\text{Composition of functions} \qquad \text{Multiplication of functions}$$

EXAMPLE 2 If $f(x) = x^2$ and $g(x) = x + 3$, find each composition.

a. $(f \circ g)(2)$ and $(g \circ f)(2)$ **b.** $(f \circ g)(x)$ and $(g \circ f)(x)$

Solution

a. $(f \circ g)(2) = f(g(2))$
$\qquad\qquad\quad = f(5)$ Replace $g(2)$ with 5. [Since $g(x) = x + 3$, then
$\qquad\qquad\quad = 5^2 = 25$ $g(2) = 2 + 3 = 5$.]
$\quad (g \circ f)(2) = g(f(2))$
$\qquad\qquad\quad = g(4)$ Since $f(x) = x^2$, then $f(2) = 2^2 = 4$.
$\qquad\qquad\quad = 4 + 3 = 7$

b. $(f \circ g)(x) = f(g(x))$
$\qquad\qquad\quad = f(x + 3)$ Replace $g(x)$ with $x + 3$.
$\qquad\qquad\quad = (x + 3)^2$ $f(x + 3) = (x + 3)^2$
$\qquad\qquad\quad = x^2 + 6x + 9$ Square $(x + 3)$.
$\quad (g \circ f)(x) = g(f(x))$
$\qquad\qquad\quad = g(x^2)$ Replace $f(x)$ with x^2.
$\qquad\qquad\quad = x^2 + 3$ $g(x^2) = x^2 + 3$ □

PRACTICE
2 If $f(x) = x^2 + 1$ and $g(x) = 3x - 5$, find

a. $(f \circ g)(4)$ **b.** $(f \circ g)(x)$
$\quad (g \circ f)(4)$ $\quad (g \circ f)(x)$

EXAMPLE 3 If $f(x) = |x|$ and $g(x) = x - 2$, find each composition.

a. $(f \circ g)(x)$ **b.** $(g \circ f)(x)$

Solution

a. $(f \circ g)(x) = f(g(x)) = f(x - 2) = |x - 2|$
b. $(g \circ f)(x) = g(f(x)) = g(|x|) = |x| - 2$ □

> ▶ **Helpful Hint**
>
> In Examples 2 and 3, notice that $(g \circ f)(x) \neq (f \circ g)(x)$. In general, $(g \circ f)(x)$ *may or may not* equal $(f \circ g)(x)$.

PRACTICE
3 If $f(x) = x^2 + 5$ and $g(x) = x + 3$, find each composition.

a. $(f \circ g)(x)$ **b.** $(g \circ f)(x)$

EXAMPLE 4 If $f(x) = 5x, g(x) = x - 2$, and $h(x) = \sqrt{x}$, write each function as a composition using two of the given functions.

a. $F(x) = \sqrt{x - 2}$ **b.** $G(x) = 5x - 2$

Solution

a. Notice the order in which the function F operates on an input value x. First, 2 is subtracted from x. This is the function $g(x) = x - 2$. Then the square root *of that result* is taken. The square root function is $h(x) = \sqrt{x}$. This means that $F = h \circ g$. To check, we find $h \circ g$.

$$F(x) = (h \circ g)(x) = h(g(x)) = h(x - 2) = \sqrt{x - 2}$$

b. Notice the order in which the function G operates on an input value x. First, x is multiplied by 5, and then 2 is subtracted from the result. This means that $G = g \circ f$. To check, we find $g \circ f$.

$$G(x) = (g \circ f)(x) = g(f(x)) = g(5x) = 5x - 2 \qquad \square$$

PRACTICE

4 If $f(x) = 3x, g(x) = x - 4$, and $h(x) = |x|$, write each function as a composition using two of the given functions.

a. $F(x) = |x - 4|$ **b.** $G(x) = 3x - 4$

Graphing Calculator Explorations

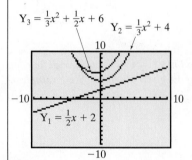

If $f(x) = \dfrac{1}{2}x + 2$ and $g(x) = \dfrac{1}{3}x^2 + 4$, then

$$(f + g)(x) = f(x) + g(x)$$
$$= \left(\frac{1}{2}x + 2\right) + \left(\frac{1}{3}x^2 + 4\right)$$
$$= \frac{1}{3}x^2 + \frac{1}{2}x + 6.$$

To visualize this addition of functions with a graphing calculator, graph

$$Y_1 = \frac{1}{2}x + 2, \qquad Y_2 = \frac{1}{3}x^2 + 4, \qquad Y_3 = \frac{1}{3}x^2 + \frac{1}{2}x + 6$$

Use a TABLE feature to verify that for a given x value, $Y_1 + Y_2 = Y_3$. For example, verify that when $x = 0$, $Y_1 = 2$, $Y_2 = 4$, and $Y_3 = 2 + 4 = 6$.

Vocabulary, Readiness & Video Check

Match each function with its definition.

1. $(f \circ g)(x)$ **4.** $(g \circ f)(x)$ **A.** $g(f(x))$ **D.** $\dfrac{f(x)}{g(x)}, g(x) \neq 0$

2. $(f \cdot g)(x)$ **5.** $\left(\dfrac{f}{g}\right)(x)$ **B.** $f(x) + g(x)$ **E.** $f(x) \cdot g(x)$

3. $(f - g)(x)$ **6.** $(f + g)(x)$ **C.** $f(g(x))$ **F.** $f(x) - g(x)$

Martin-Gay Interactive Videos

See Video 9.1

Watch the section lecture video and answer the following questions.

OBJECTIVE 1

7. From Example 1 and the lecture before, we know that $(f + g)(x) = f(x) + g(x)$. Use this fact to explain two ways you can find $(f + g)(2)$.

OBJECTIVE 2

8. From Example 3, given two functions $f(x)$ and $g(x)$, can $f(g(x))$ ever equal $g(f(x))$?

9.1 Exercise Set   MyMathLab®

For the functions f and g, find ***a.*** *$(f + g)(x)$,* ***b.*** *$(f - g)(x)$,* ***c.*** *$(f \cdot g)(x)$, and* ***d.*** *$\left(\dfrac{f}{g}\right)(x)$. See Example 1.*

1. $f(x) = x - 7, g(x) = 2x + 1$

2. $f(x) = x + 4, g(x) = 5x - 2$

3. $f(x) = x^2 + 1, g(x) = 5x$

4. $f(x) = x^2 - 2, g(x) = 3x$

5. $f(x) = \sqrt[3]{x}, g(x) = x + 5$

6. $f(x) = \sqrt[3]{x}, g(x) = x - 3$

7. $f(x) = -3x, g(x) = 5x^2$

8. $f(x) = 4x^3, g(x) = -6x$

If $f(x) = x^2 - 6x + 2, g(x) = -2x$, and $h(x) = \sqrt{x}$, find each composition. See Example 2.

9. $(f \circ g)(2)$

10. $(h \circ f)(-2)$

11. $(g \circ f)(-1)$

12. $(f \circ h)(1)$

13. $(g \circ h)(0)$

14. $(h \circ g)(0)$

Find $(f \circ g)(x)$ and $(g \circ f)(x)$. See Examples 2 and 3.

15. $f(x) = x^2 + 1, g(x) = 5x$

16. $f(x) = x - 3, g(x) = x^2$

17. $f(x) = 2x - 3, g(x) = x + 7$

18. $f(x) = x + 10, g(x) = 3x + 1$

19. $f(x) = x^3 + x - 2, g(x) = -2x$

20. $f(x) = -4x, g(x) = x^3 + x^2 - 6$

21. $f(x) = |x|; g(x) = 10x - 3$

22. $f(x) = |x|; g(x) = 14x - 8$

23. $f(x) = \sqrt{x}, g(x) = -5x + 2$

24. $f(x) = 7x - 1, g(x) = \sqrt[3]{x}$

If $f(x) = 3x, g(x) = \sqrt{x}$, and $h(x) = x^2 + 2$, write each function as a composition using two of the given functions. See Example 4.

25. $H(x) = \sqrt{x^2 + 2}$

26. $G(x) = \sqrt{3x}$

27. $F(x) = 9x^2 + 2$

28. $H(x) = 3x^2 + 6$

29. $G(x) = 3\sqrt{x}$

30. $F(x) = x + 2$

Find $f(x)$ and $g(x)$ so that the given function $h(x) = (f \circ g)(x)$.

31. $h(x) = (x + 2)^2$

32. $h(x) = |x - 1|$

33. $h(x) = \sqrt{x + 5} + 2$

34. $h(x) = (3x + 4)^2 + 3$

35. $h(x) = \dfrac{1}{2x - 3}$

36. $h(x) = \dfrac{1}{x + 10}$

REVIEW AND PREVIEW

Solve each equation for y. See Section 2.3.

37. $x = y + 2$ **38.** $x = y - 5$

39. $x = 3y$ **40.** $x = -6y$

41. $x = -2y - 7$ **42.** $x = 4y + 7$

CONCEPT EXTENSIONS

Given that $f(-1) = 4$ $g(-1) = -4$

$\qquad f(0) = 5 \qquad g(0) = -3$

$\qquad f(2) = 7 \qquad g(2) = -1$

$\qquad f(7) = 1 \qquad g(7) = 4$

find each function value.

43. $(f + g)(2)$ **44.** $(f - g)(7)$

45. $(f \circ g)(2)$ **46.** $(g \circ f)(2)$

47. $(f \cdot g)(7)$ **48.** $(f \cdot g)(0)$

49. $\left(\dfrac{f}{g}\right)(-1)$ **50.** $\left(\dfrac{g}{f}\right)(-1)$

51. If you are given $f(x)$ and $g(x)$, explain in your own words how to find $(f \circ g)(x)$ and then how to find $(g \circ f)(x)$.

52. Given $f(x)$ and $g(x)$, describe in your own words the difference between $(f \circ g)(x)$ and $(f \cdot g)(x)$.

Solve.

53. Business people are concerned with cost functions, revenue functions, and profit functions. Recall that the profit $P(x)$ obtained from x units of a product is equal to the revenue $R(x)$ from selling the x units minus the cost $C(x)$ of manufacturing the x units. Write an equation expressing this relationship among $C(x)$, $R(x)$, and $P(x)$.

54. Suppose the revenue $R(x)$ for x units of a product can be described by $R(x) = 25x$, and the cost $C(x)$ can be described by $C(x) = 50 + x^2 + 4x$. Find the profit $P(x)$ for x units. (See Exercise 53.)

9.2 | Inverse Functions

OBJECTIVES

1 Determine Whether a Function Is a One-to-One Function.

2 Use the Horizontal Line Test to Decide Whether a Function Is a One-to-One Function.

3 Find the Inverse of a Function.

4 Find the Equation of the Inverse of a Function.

5 Graph Functions and Their Inverses.

6 Determine Whether Two Functions Are Inverses of Each Other.

OBJECTIVE

1 Determining Whether a Function Is One-To-One

In the next three sections, we begin a study of two new functions: exponential and logarithmic functions. As we learn more about these functions, we will discover that they share a special relation to each other: They are inverses of each other.

Before we study these functions, we need to learn about inverses. We begin by defining one-to-one functions.

Study the following table.

Degrees Fahrenheit (Input)	-31	-13	32	68	149	212
Degrees Celsius (Output)	-35	-25	0	20	65	100

Recall that since each degrees Fahrenheit (input) corresponds to exactly one degrees Celsius (output), this pairing of inputs and outputs does describe a function. Also notice that each output corresponds to exactly one input. This type of function is given a special name—a one-to-one function.

Does the set $f = \{(0, 1), (2, 2), (-3, 5), (7, 6)\}$ describe a one-to-one function? It is a function since each x-value corresponds to a unique y-value. For this particular function f, each y-value also corresponds to a unique x-value. Thus, this function is also **a one-to-one function.**

> ### One-to-One Function
>
> For a **one-to-one function**, each x-value (input) corresponds to only one y-value (output), and each y-value (output) corresponds to only one x-value (input).

EXAMPLE 1   Determine whether each function described is one-to-one.

a. $f = \{(6, 2), (5, 4), (-1, 0), (7, 3)\}$

b. $g = \{(3, 9), (-4, 2), (-3, 9), (0, 0)\}$

c. $h = \{(1, 1), (2, 2), (10, 10), (-5, -5)\}$

d.

Mineral (Input)	Talc	Gypsum	Diamond	Topaz	Stibnite
Hardness on the Mohs Scale (Output)	1	2	10	8	2

e.

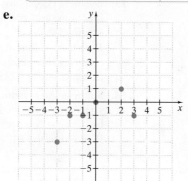

f.

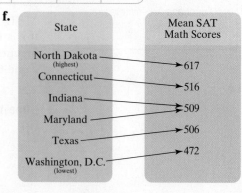

Percent of Cell Phone
Subscribers Who Text

Cities

El Paso (highest) ——————→ 57

Buffalo ——————→ 47

Austin ——————→ 49

Tulsa

Charleston ——————→ 36

Detroit ——————→ 45

Solution

a. *f* is one-to-one since each *y*-value corresponds to only one *x*-value.

b. *g* is not one-to-one because the *y*-value 9 in (3, 9) and (−3, 9) corresponds to different *x*-values.

c. *h* is a one-to-one function since each *y*-value corresponds to only one *x*-value.

d. This table does not describe a one-to-one function since the output 2 corresponds to two inputs, gypsum and stibnite.

e. This graph does not describe a one-to-one function since the *y*-value −1 corresponds to three *x*-values, −2, −1, and 3. (See the graph to the left.)

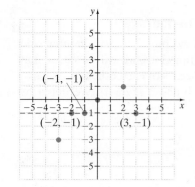

f. The mapping is not one-to-one since 49% corresponds to Austin and Tulsa.   ☐

PRACTICE

1   Determine whether each function described is one-to-one.

a. $f = \{(4, -3), (3, -4), (2, 7), (5, 0)\}$

b. $g = \{(8, 4), (-2, 0), (6, 4), (2, 6)\}$

c. $h = \{(2, 4), (1, 3), (4, 6), (-2, 4)\}$

d.

Year	1950	1963	1968	1975	1997	2008
Federal Minimum Wage	$0.75	$1.25	$1.60	$2.10	$5.15	$6.55

e.

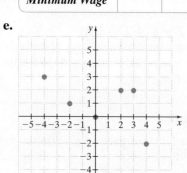

f.

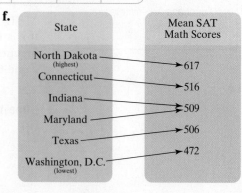

State	Mean SAT Math Scores

North Dakota (highest) ——————→ 617

Connecticut ——————→ 516

Indiana ——————→ 509

Maryland ——————→ 506

Texas ——————→ 472

Washington, D.C. (lowest)

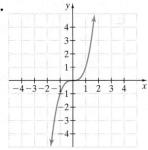

Not a one-to-one function.

OBJECTIVE
2   Using the Horizontal Line Test

Recall that we recognize the graph of a function when it passes the vertical line test. Since every *x*-value of the function corresponds to exactly one *y*-value, each vertical line intersects the function's graph at most once. The graph shown (left), for instance, is the graph of a function.

Is this function a *one-to-one* function? The answer is no. To see why not, notice that the *y*-value of the ordered pair $(-3, 3)$, for example, is the same as the *y*-value of the ordered pair $(3, 3)$. In other words, the *y*-value 3 corresponds to two *x*-values, -3 and 3. This function is therefore not one-to-one.

To test whether a graph is the graph of a one-to-one function, apply the vertical line test to see if it is a function and then apply a similar **horizontal line test** to see if it is a one-to-one function.

> ### Horizontal Line Test
>
> If every horizontal line intersects the graph of a function at most once, then the function is a one-to-one function.

EXAMPLE 2   Determine whether each graph is the graph of a one-to-one function.

a.

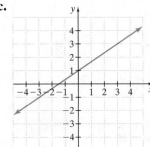

b.

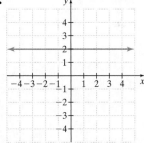

c.

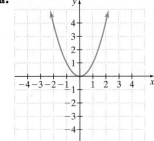

d.

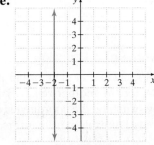

e.

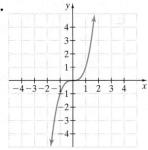

Solution   Graphs **a, b, c,** and **d** all pass the vertical line test, so only these graphs are graphs of functions. But, of these, only **b** and **c** pass the horizontal line test, so only **b** and **c** are graphs of one-to-one functions.   □

Determine whether each graph is the graph of a one-to-one function.

a.

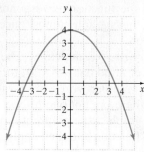

b.

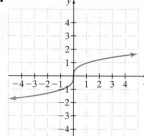

c.

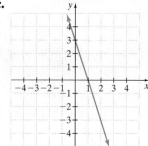

d.

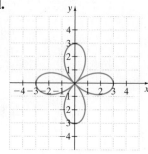

e.

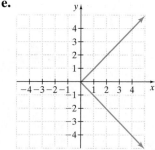

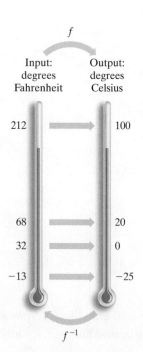

f

Input: degrees Fahrenheit

Output: degrees Celsius

212 → 100

68 → 20

32 → 0

−13 → −25

f^{-1}

▶ **Helpful Hint**

All linear equations are one-to-one functions except those whose graphs are horizontal or vertical lines. A vertical line does not pass the vertical line test and hence is not the graph of a function. A horizontal line is the graph of a function but does not pass the horizontal line test and hence is not the graph of a one-to-one function.

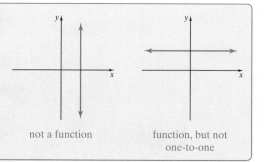

not a function   function, but not one-to-one

OBJECTIVE
3 Finding the Inverse of a Function ▶

One-to-one functions are special in that their graphs pass both the vertical and horizontal line tests. They are special, too, in another sense: For each one-to-one function, we can find its **inverse function** by switching the coordinates of the ordered pairs of the function, or the inputs and the outputs. For example, the inverse of the one-to-one function

Degrees Fahrenheit (Input)	−31	−13	32	68	149	212
Degrees Celsius (Output)	−35	−25	0	20	65	100

is the function

Degrees Celsius (Input)	−35	−25	0	20	65	100
Degrees Fahrenheit (Output)	−31	−13	32	68	149	212

Notice that the ordered pair $(-31, -35)$ of the function, for example, becomes the ordered pair $(-35, -31)$ of its inverse.

Also, the inverse of the one-to-one function $f = \{(2, -3), (5, 10), (9, 1)\}$ is $\{(-3, 2), (10, 5), (1, 9)\}$. For a function f, we use the notation f^{-1}, read "f inverse," to denote its inverse function. Notice that since the coordinates of each ordered pair have been switched, the domain (set of inputs) of f is the range (set of outputs) of f^{-1}, and the range of f is the domain of f^{-1}.

Inverse Function

The inverse of a one-to-one function f is the one-to-one function f^{-1} that consists of the set of all ordered pairs (y, x) where (x, y) belongs to f.

▶ **Helpful Hint**

If a function is not one-to-one, it does not have an inverse function.

EXAMPLE 3   Find the inverse of the one-to-one function.

$$f = \{(0, 1), (-2, 7), (3, -6), (4, 4)\}$$

Solution $f^{-1} = \{(1, 0), (7, -2), (-6, 3), (4, 4)\}$

    ↑   ↑   ↑   ↑ Switch coordinates of each ordered pair.   □

PRACTICE

3   Find the inverse of the one-to-one function.

$$f = \{(3, 4), (-2, 0), (2, 8), (6, 6)\}$$

▶ **Helpful Hint**

The symbol f^{-1} is the single symbol that denotes the inverse of the function f.

It is read as "f inverse." This symbol *does not mean* $\dfrac{1}{f}$.

✓**CONCEPT CHECK**

Suppose that f is a one-to-one function and that $f(1) = 5$.

a. Write the corresponding ordered pair.

b. Write one point that we know must belong to the inverse function f^{-1}.

OBJECTIVE

4   **Finding the Equation of the Inverse of a Function**

If a one-to-one function f is defined as a set of ordered pairs, we can find f^{-1} by interchanging the x- and y-coordinates of the ordered pairs. If a one-to-one function f is given in the form of an equation, we can find f^{-1} by using a similar procedure.

Answer to Concept Check:
a. $(1, 5)$,   **b.** $(5, 1)$

Finding the Inverse of a One-to-One Function $f(x)$

Step 1. Replace $f(x)$ with y.

Step 2. Interchange x and y.

Step 3. Solve the equation for y.

Step 4. Replace y with the notation $f^{-1}(x)$.

EXAMPLE 4   Find an equation of the inverse of $f(x) = x + 3$.

Solution $f(x) = x + 3$

Step 1. $y = x + 3$ Replace $f(x)$ with y.

Step 2. $x = y + 3$ Interchange x and y.

Step 3. $x - 3 = y$ Solve for y.

Step 4. $f^{-1}(x) = x - 3$ Replace y with $f^{-1}(x)$.

The inverse of $f(x) = x + 3$ is $f^{-1}(x) = x - 3$. Notice that, for example,

$$f(1) = 1 + 3 = 4 \quad \text{and} \quad f^{-1}(4) = 4 - 3 = 1$$

Ordered pair: $(1, 4)$ Ordered pair: $(4, 1)$

The coordinates are
switched, as expected.   □

PRACTICE
4   Find the equation of the inverse of $f(x) = 6 - x$.

EXAMPLE 5   Find the equation of the inverse of $f(x) = 3x - 5$. Graph f and f^{-1} on the same set of axes.

Solution $f(x) = 3x - 5$

Step 1. $y = 3x - 5$ Replace $f(x)$ with y.

Step 2. $x = 3y - 5$ Interchange x and y.

Step 3. $3y = x + 5$ Solve for y.

$$y = \frac{x + 5}{3}$$

Step 4. $f^{-1}(x) = \dfrac{x + 5}{3}$ Replace y with $f^{-1}(x)$.

Now we graph $f(x)$ and $f^{-1}(x)$ on the same set of axes. Both $f(x) = 3x - 5$ and $f^{-1}(x) = \dfrac{x + 5}{3}$ are linear functions, so each graph is a line.

$f(x) = 3x - 5$	
x	$y = f(x)$
1	-2
0	-5
$\dfrac{5}{3}$	0

$f^{-1}(x) = \dfrac{x + 5}{3}$	
x	$y = f^{-1}(x)$
-2	1
-5	0
0	$\dfrac{5}{3}$

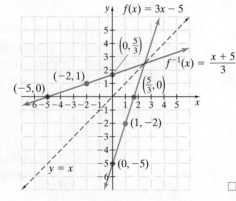

□

5 Find the equation of the inverse of $f(x) = 5x + 2$. Graph f and f^{-1} on the same set of axes.

OBJECTIVE

5 Graphing Inverse Functions

Notice that the graphs of f and f^{-1} in Example 5 are mirror images of each other, and the "mirror" is the dashed line $y = x$. This is true for every function and its inverse. For this reason, we say that *the graphs of f and f^{-1} are symmetric about the line $y = x$.*

To see why this happens, study the graph of a few ordered pairs and their switched coordinates in the diagram below.

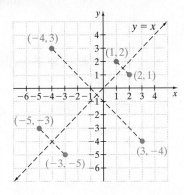

EXAMPLE 6 Graph the inverse of each function.

Solution The function is graphed in blue and the inverse is graphed in red.

a.

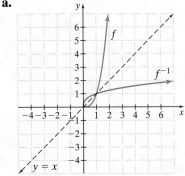

b.

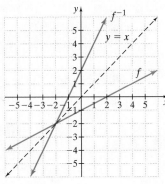

6 Graph the inverse of each function.

a.

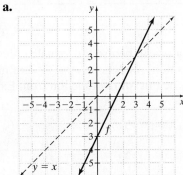

b.

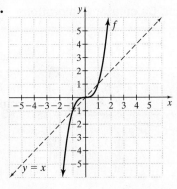

6 Determining Whether Functions Are Inverses of Each Other

Notice in the table of values in Example 5 that $f(0) = -5$ and $f^{-1}(-5) = 0$, as expected. Also, for example, $f(1) = -2$ and $f^{-1}(-2) = 1$. In words, we say that for some input x, the function f^{-1} takes the output of x, called $f(x)$, back to x.

$$x \rightarrow f(x) \quad \text{and} \quad f^{-1}(f(x)) \rightarrow x$$
$$\downarrow \qquad \downarrow \qquad\qquad \downarrow \qquad \downarrow$$
$$f(0) = -5 \quad \text{and} \quad f^{-1}(-5) = 0$$
$$f(1) = -2 \quad \text{and} \quad f^{-1}(-2) = 1$$

In general,

> If f is a one-to-one function, then the inverse of f is the function f^{-1} such that
> $$(f^{-1} \circ f)(x) = x \quad \text{and} \quad (f \circ f^{-1})(x) = x$$

EXAMPLE 7 Show that if $f(x) = 3x + 2$, then $f^{-1}(x) = \dfrac{x - 2}{3}$.

Solution See that $(f^{-1} \circ f)(x) = x$ and $(f \circ f^{-1})(x) = x$.

$$(f^{-1} \circ f)(x) = f^{-1}(f(x))$$
$$= f^{-1}(3x + 2) \qquad \text{Replace } f(x) \text{ with } 3x + 2.$$
$$= \frac{3x + 2 - 2}{3}$$
$$= \frac{3x}{3}$$
$$= x$$

$$(f \circ f^{-1})(x) = f(f^{-1}(x))$$
$$= f\left(\frac{x - 2}{3}\right) \qquad \text{Replace } f^{-1}(x) \text{ with } \frac{x - 2}{3}.$$
$$= 3\left(\frac{x - 2}{3}\right) + 2$$
$$= x - 2 + 2$$
$$= x$$

PRACTICE

7 Show that if $f(x) = 4x - 1$, then $f^{-1}(x) = \dfrac{x + 1}{4}$.

Graphing Calculator Explorations

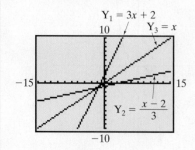

A graphing calculator can be used to visualize the results of Example 7. Recall that the graph of a function f and its inverse f^{-1} are mirror images of each other across the line $y = x$. To see this for the function from Example 7, use a square window and graph

the given function: $Y_1 = 3x + 2$

its inverse: $Y_2 = \dfrac{x - 2}{3}$

and the line: $Y_3 = x$

See Exercises 67–70 in Exercise Set 9.2.

Vocabulary, Readiness & Video Check

Use the choices below to fill in each blank. Some choices will not be used, and some will be used more than once.

vertical	$(3, 7)$	$(11, 2)$	$y = x$	x	true
horizontal	$(7, 3)$	$(2, 11)$	$\dfrac{1}{f}$	the inverse of f	false

1. If $f(2) = 11$, the corresponding ordered pair is _____ .

2. If $(7, 3)$ is an ordered pair solution of $f(x)$, and $f(x)$ has an inverse, then an ordered pair solution of $f^{-1}(x)$ is

_____ .

3. The symbol f^{-1} means _____ .

4. True or false: The function notation $f^{-1}(x)$ means $\dfrac{1}{f(x)}$. _____

5. To tell whether a graph is the graph of a function, use the _____ line test.

6. To tell whether the graph of a function is also a one-to-one function, use the _____ line test.

7. The graphs of f and f^{-1} are symmetric about the line _____ .

8. Two functions are inverses of each other if $(f \circ f^{-1})(x) =$ _____ and $(f^{-1} \circ f)(x) =$ _____ .

Martin-Gay Interactive Videos

See Video 9.2 🍐

Watch the section lecture video and answer the following questions.

OBJECTIVE
1
9. From ▱ Example 1 and the definition before, what makes a one-to-one function different from other types of functions?

OBJECTIVE
2
10. From ▱ Examples 2 and 3, if a graph passes the horizontal line test, but not the vertical line test, is it a one-to-one function? Explain.

OBJECTIVE
3
11. From ▱ Example 4 and the lecture before, if you find the inverse of a one-to-one function, is this inverse function also a one-to-one function? How do you know?

OBJECTIVE
4
12. From ▱ Examples 5 and 6, explain why the interchanging of x and y when finding an inverse equation makes sense given the definition of an inverse function.

OBJECTIVE
5
13. From ▱ Example 7, if you have the equation or graph of a one-to-one function, how can you graph its inverse without finding the inverse's equation?

OBJECTIVE
6
14. Based on ▱ Example 8 and the lecture before, what's wrong with the following statement? "If f is a one-to-one function, you can prove that f and f^{-1} are inverses of each other by showing that $f(f^{-1}(x)) = f^{-1}(f(x))$."

9.2 Exercise Set

MyMathLab®

Determine whether each function is a one-to-one function. If it is one-to-one, list the inverse function by switching coordinates, or inputs and outputs. See Examples 1 and 3.

1. $f = \{(-1, -1), (1, 1), (0, 2), (2, 0)\}$

2. $g = \{(8, 6), (9, 6), (3, 4), (-4, 4)\}$

3. $h = \{(10, 10)\}$

4. $r = \{(1, 2), (3, 4), (5, 6), (6, 7)\}$

▶ **5.** $f = \{(11, 12), (4, 3), (3, 4), (6, 6)\}$

6. $g = \{(0, 3), (3, 7), (6, 7), (-2, -2)\}$

7.

Month of 2009 (Input)	July	August	September	October	November	December
Unemployment Rate in Percent (Output)	9.4	9.7	9.8	10.1	10.0	10.0

(*Source:* U.S. Bureau of Labor Statistics)

8.

State (Input)	Texas	Massachusetts	Nevada	Idaho	Wisconsin
Number of Two-Year Colleges (Output)	70	22	3	3	31

(*Source:* University of Texas at Austin)

9.

State (Input)	California	Alaska	Indiana	Louisiana	New Mexico	Ohio
Rank in Population (Output)	1	47	16	25	36	7

(*Source:* U.S. Bureau of the Census)

△ **10.**

Shape (Input)	Triangle	Pentagon	Quadrilateral	Hexagon	Decagon
Number of Sides (Output)	3	5	4	6	10

Given the one-to-one function $f(x) = x^3 + 2$, find the following.
[Hint: You do not need to find the equation for $f^{-1}(x)$.]

11. a. $f(1)$
   b. $f^{-1}(3)$

12. a. $f(0)$
   b. $f^{-1}(2)$

13. a. $f(-1)$
   b. $f^{-1}(1)$

14. a. $f(-2)$
   b. $f^{-1}(-6)$

Determine whether the graph of each function is the graph of a one-to-one function. See Example 2.

 15.

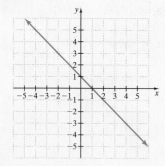

16.

17.

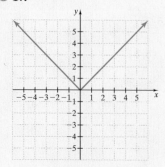

18.

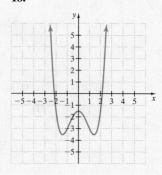

19.

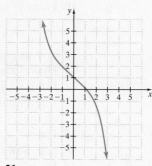

20.

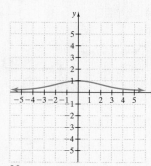

21.

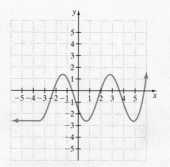

22.

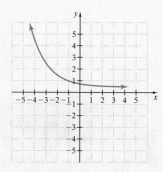

MIXED PRACTICE

Each of the following functions is one-to-one. Find the inverse of each function and graph the function and its inverse on the same set of axes. See Examples 4 and 5.

23. $f(x) = x + 4$

24. $f(x) = x - 5$

25. $f(x) = 2x - 3$

26. $f(x) = 4x + 9$

27. $f(x) = \frac{1}{2}x - 1$

28. $f(x) = -\frac{1}{2}x + 2$

29. $f(x) = x^3$

30. $f(x) = x^3 - 1$

Find the inverse of each one-to-one function. See Examples 4 and 5.

31. $f(x) = 5x + 2$

32. $f(x) = 6x - 1$

33. $f(x) = \frac{x - 2}{5}$

34. $f(x) = \frac{x - 3}{2}$

35. $f(x) = \sqrt[3]{x}$

36. $f(x) = \sqrt[3]{x + 1}$

⊙ **37.** $f(x) = \frac{5}{3x + 1}$

38. $f(x) = \frac{7}{2x + 4}$

39. $f(x) = (x + 2)^3$

40. $f(x) = (x - 5)^3$

Graph the inverse of each function on the same set of axes. See Example 6.

41.

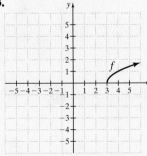

42.

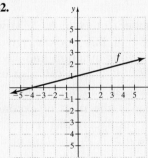

43.

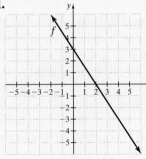

44.

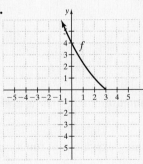

45.

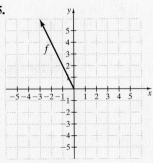

46.

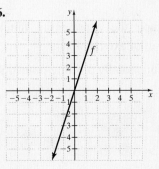

Solve. See Example 7.

⊙ **47.** If $f(x) = 2x + 1$, show that $f^{-1}(x) = \frac{x - 1}{2}$.

48. If $f(x) = 3x - 10$, show that $f^{-1}(x) = \frac{x + 10}{3}$.

49. If $f(x) = x^3 + 6$, show that $f^{-1}(x) = \sqrt[3]{x - 6}$.

50. If $f(x) = x^3 - 5$, show that $f^{-1}(x) = \sqrt[3]{x + 5}$.

REVIEW AND PREVIEW

Evaluate each of the following. See Section 7.2.

51. $25^{1/2}$ **52.** $49^{1/2}$

53. $16^{3/4}$ **54.** $27^{2/3}$

55. $9^{-3/2}$ **56.** $81^{-3/4}$

If $f(x) = 3^x$, find the following. In Exercises 59 and 60, give an exact answer and a two-decimal-place approximation. See Sections 3.2, 5.1, and 7.2.

57. $f(2)$ **58.** $f(0)$

59. $f\left(\frac{1}{2}\right)$ **60.** $f\left(\frac{2}{3}\right)$

CONCEPT EXTENSIONS

Solve. See the Concept Check in this section.

61. Suppose that f is a one-to-one function and that $f(2) = 9$.

   a. Write the corresponding ordered pair.

   b. Name one ordered pair that we know is a solution of the inverse of f, or f^{-1}.

62. Suppose that F is a one-to-one function and that $F\left(\frac{1}{2}\right) = -0.7$.

   a. Write the corresponding ordered pair.

   b. Name one ordered pair that we know is a solution of the inverse of F, or F^{-1}.

For Exercises 63 and 64,

a. *Write the ordered pairs for $f(x)$ whose points are highlighted. (Include the points whose coordinates are given.)*

b. *Write the corresponding ordered pairs for the inverse of f, f^{-1}.*

c. *Graph the ordered pairs for f^{-1} found in part **b**.*

d. *Graph $f^{-1}(x)$ by drawing a smooth curve through the plotted points.*

63.

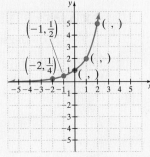

64.

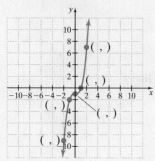

65. If you are given the graph of a function, describe how you can tell from the graph whether a function has an inverse.

66. Describe the appearance of the graphs of a function and its inverse.

Find the inverse of each given one-to-one function. Then use a graphing calculator to graph the function and its inverse on a square window.

67. $f(x) = 3x + 1$

68. $f(x) = -2x - 6$

69. $f(x) = \sqrt[3]{x + 1}$

70. $f(x) = x^3 - 3$

9.3   Exponential Functions

OBJECTIVES

1 Graph Exponential Functions.

2 Solve Equations of the Form $b^x = b^y$.

3 Solve Problems Modeled by Exponential Equations.

OBJECTIVE

1   Graphing Exponential Functions

In earlier chapters, we gave meaning to exponential expressions such as 2^x, where x is a rational number. For example,

$$2^3 = 2 \cdot 2 \cdot 2 \qquad \text{Three factors; each factor is 2}$$

$$2^{3/2} = (2^{1/2})^3 = \sqrt{2} \cdot \sqrt{2} \cdot \sqrt{2} \qquad \text{Three factors; each factor is } \sqrt{2}$$

When x is an irrational number (for example, $\sqrt{3}$), what meaning can we give to $2^{\sqrt{3}}$?

It is beyond the scope of this book to give precise meaning to 2^x if x is irrational. We can confirm your intuition and say that $2^{\sqrt{3}}$ is a real number and, since $1 < \sqrt{3} < 2$, $2^1 < 2^{\sqrt{3}} < 2^2$. We can also use a calculator and approximate $2^{\sqrt{3}}$: $2^{\sqrt{3}} \approx 3.321997$. In fact, as long as the base b is positive, b^x is a real number for all real numbers x. Finally, the rules of exponents apply whether x is rational or irrational as long as b is positive. In this section, we are interested in functions of the form $f(x) = b^x$, where $b > 0$. A function of this form is called an **exponential function.**

> **Exponential Function**
>
> A function of the form
>
> $$f(x) = b^x$$
>
> is called an **exponential function** if $b > 0$, b is not 1, and x is a real number.

Next, we practice graphing exponential functions.

EXAMPLE 1   Graph the exponential functions defined by $f(x) = 2^x$ and $g(x) = 3^x$ on the same set of axes.

Solution   Graph each function by plotting points. Set up a table of values for each of the two functions.

If each set of points is plotted and connected with a smooth curve, the following graphs result.

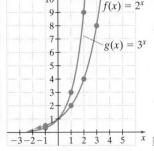

$f(x) = 2^x$	x	0	1	2	3	−1	−2
	$f(x)$	1	2	4	8	$\frac{1}{2}$	$\frac{1}{4}$

$g(x) = 3^x$	x	0	1	2	3	−1	−2
	$g(x)$	1	3	9	27	$\frac{1}{3}$	$\frac{1}{9}$

PRACTICE

1 Graph the exponential functions defined by $f(x) = 2^x$ and $g(x) = 7^x$ on the same set of axes.

A number of things should be noted about the two graphs of exponential functions in Example 1. First, the graphs show that $f(x) = 2^x$ and $g(x) = 3^x$ are one-to-one functions since each graph passes the vertical and horizontal line tests. The y-intercept of each graph is $(0, 1)$, but neither graph has an x-intercept. From the graph, we can also see that the domain of each function is all real numbers and that the range is $(0, \infty)$. We can also see that as x-values are increasing, y-values are increasing also.

EXAMPLE 2 Graph the exponential functions $y = \left(\frac{1}{2}\right)^x$ and $y = \left(\frac{1}{3}\right)^x$ on the same set of axes.

Solution As before, plot points and connect them with a smooth curve.

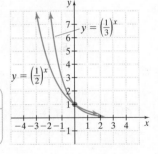

$y = \left(\frac{1}{2}\right)^x$	x	0	1	2	3	−1	−2
	y	1	$\frac{1}{2}$	$\frac{1}{4}$	$\frac{1}{8}$	2	4

$y = \left(\frac{1}{3}\right)^x$	x	0	1	2	3	−1	−2
	y	1	$\frac{1}{3}$	$\frac{1}{9}$	$\frac{1}{27}$	3	9

PRACTICE

2 Graph the exponential functions $f(x) = \left(\frac{1}{3}\right)^x$ and $g(x) = \left(\frac{1}{5}\right)^x$ on the same set of axes.

Each function in Example 2 again is a one-to-one function. The y-intercept of both is $(0, 1)$. The domain is the set of all real numbers, and the range is $(0, \infty)$.

Notice the difference between the graphs of Example 1 and the graphs of Example 2. An exponential function is always increasing if the base is greater than 1.

When the base is between 0 and 1, the graph is always decreasing. The figures on the next page summarize these characteristics of exponential functions.

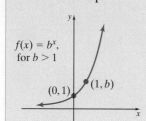
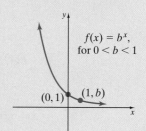

$$f(x) = b^x, \quad b > 0, \quad b \neq 1$$

- one-to-one function
- y-intercept $(0, 1)$
- no x-intercept

- domain: $(-\infty, \infty)$
- range: $(0, \infty)$

EXAMPLE 3 Graph the exponential function $f(x) = 3^{x+2}$.

Solution As before, we find and plot a few ordered pair solutions. Then we connect the points with a smooth curve.

$f(x) = 3^{x+2}$	
x	$f(x)$
0	9
-1	3
-2	1
-3	$\dfrac{1}{3}$
-4	$\dfrac{1}{9}$

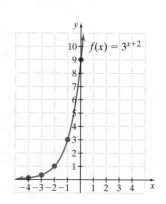

PRACTICE
3 Graph the exponential function $f(x) = 2^{x-3}$.

✓CONCEPT CHECK

Which functions are exponential functions?

a. $f(x) = x^3$

b. $g(x) = \left(\dfrac{2}{3}\right)^x$

c. $h(x) = 5^{x-2}$

d. $w(x) = (2x)^2$

OBJECTIVE
2 Solving Equations of the Form $b^x = b^y$

We have seen that an exponential function $y = b^x$ is a one-to-one function. Another way of stating this fact is a property that we can use to solve exponential equations.

Uniqueness of b^x

Let $b > 0$ and $b \neq 1$. Then $b^x = b^y$ is equivalent to $x = y$.

Thus, one way to solve an exponential equation depends on whether it's possible to write each side of the equation with the same base; that is, $b^x = b^y$. We solve by this method first.

EXAMPLE 4   Solve each equation for x.

a. $2^x = 16$ **b.** $9^x = 27$ **c.** $4^{x+3} = 8^x$

Solution

a. We write 16 as a power of 2 and then use the uniqueness of b^x to solve.

$$2^x = 16$$
$$2^x = 2^4$$

Since the bases are the same and are nonnegative, by the uniqueness of b^x, we then have that the exponents are equal. Thus,

$$x = 4$$

The solution is 4, or the solution set is $\{4\}$.

b. Notice that both 9 and 27 are powers of 3.

$$9^x = 27$$
$$(3^2)^x = 3^3 \quad \text{Write 9 and 27 as powers of 3.}$$
$$3^{2x} = 3^3$$
$$2x = 3 \quad \text{Apply the uniqueness of } b^x.$$
$$x = \frac{3}{2} \quad \text{Divide by 2.}$$

To check, replace x with $\frac{3}{2}$ in the original expression, $9^x = 27$. The solution is $\frac{3}{2}$.

c. Write both 4 and 8 as powers of 2.

$$4^{x+3} = 8^x$$
$$(2^2)^{x+3} = (2^3)^x$$
$$2^{2x+6} = 2^{3x}$$
$$2x + 6 = 3x \quad \text{Apply the uniqueness of } b^x.$$
$$6 = x \quad \text{Subtract } 2x \text{ from both sides.}$$

The solution is 6.   □

PRACTICE
4   Solve each equation for x.

a. $3^x = 9$ **b.** $8^x = 16$ **c.** $125^x = 25^{x-2}$

There is one major problem with the preceding technique. Often the two sides of an equation cannot easily be written as powers of a common base. We explore how to solve an equation such as $4 = 3^x$ with the help of **logarithms** later.

OBJECTIVE
3   Solving Problems Modeled by Exponential Equations

The bar graph on the next page shows the increase in the number of cellular phone users. Notice that the graph of the exponential function $y = 136.76(1.107)^x$ approximates the heights of the bars. This is just one example of how the world abounds with patterns that can be modeled by exponential functions. To make these applications realistic, we use numbers that warrant a calculator.

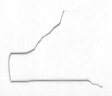

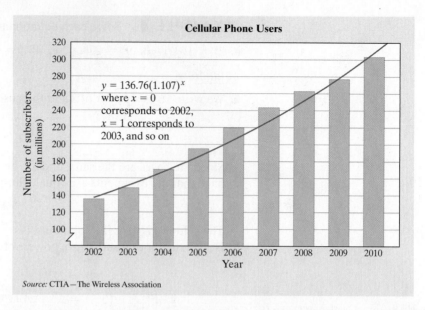

Cellular Phone Users

$y = 136.76(1.107)^x$ where $x = 0$ corresponds to 2002, $x = 1$ corresponds to 2003, and so on

Source: CTIA — The Wireless Association

Another application of an exponential function has to do with interest rates on loans. The exponential function defined by $A = P\left(1 + \dfrac{r}{n}\right)^{nt}$ models the dollars A accrued (or owed) after P dollars are invested (or loaned) at an annual rate of interest r compounded n times each year for t years. This function is known as the compound interest formula.

EXAMPLE 5 Using the Compound Interest Formula

Find the amount owed at the end of 5 years if $1600 is loaned at a rate of 9% compounded monthly.

Solution We use the formula $A = P\left(1 + \dfrac{r}{n}\right)^{nt}$, with the following values.

$P = \$1600$ (the amount of the loan)
$r = 9\% = 0.09$ (the annual rate of interest)
$n = 12$ (the number of times interest is compounded each year)
$t = 5$ (the duration of the loan, in years)

$$A = P\left(1 + \frac{r}{n}\right)^{nt} \qquad \text{Compound interest formula}$$

$$= 1600\left(1 + \frac{0.09}{12}\right)^{12(5)} \qquad \text{Substitute known values.}$$

$$= 1600(1.0075)^{60}$$

To approximate A, use the $\boxed{y^x}$ or $\boxed{\wedge}$ key on your calculator.

$$\boxed{2505.0896}$$

Thus, the amount A owed is approximately $2505.09.   ☐

PRACTICE

5 Find the amount owed at the end of 4 years if $3000 is loaned at a rate of 7% compounded semiannually (twice a year).

EXAMPLE 6 **Estimating Percent of Radioactive Material**

As a result of a nuclear accident, radioactive debris was carried through the atmosphere. One immediate concern was the impact that the debris had on the milk supply. The percent y of radioactive material in raw milk after t days is estimated by $y = 100(2.7)^{-0.1t}$. Estimate the expected percent of radioactive material in the milk after 30 days.

Solution Replace t with 30 in the given equation.

$$y = 100(2.7)^{-0.1t}$$
$$= 100(2.7)^{-0.1(30)} \quad \text{Let } t = 30.$$
$$= 100(2.7)^{-3}$$

To approximate the percent y, the following keystrokes may be used on a scientific calculator.

The display should read

$$\boxed{5.0805263}$$

Thus, approximately 5% of the radioactive material still remained in the milk supply after 30 days. □

PRACTICE

6 The percent p of light that passes through n successive sheets of a particular glass is given approximately by the function $p(n) = 100(2.7)^{-0.05n}$. Estimate the expected percent of light that will pass through the following numbers of sheets of glass. Round each to the nearest hundredth of a percent.

a. 2 sheets of glass   **b.** 10 sheets of glass

Graphing Calculator Explorations

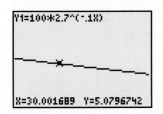

We can use a graphing calculator and its TRACE feature to solve Example 6 graphically.

To estimate the expected percent of radioactive material in the milk after 30 days, enter $Y_1 = 100(2.7)^{-0.1x}$. (The variable t in Example 6 is changed to x here to accommodate our work better on the graphing calculator.) The graph does not appear on a standard viewing window, so we need to determine an appropriate viewing window. Because it doesn't make sense to look at radioactivity *before* the nuclear accident, we use Xmin = 0. We are interested in finding the percent of radioactive material in the milk when $x = 30$, so we choose Xmax = 35 to leave enough space to see the graph at $x = 30$. Because the values of y are percents, it seems appropriate that $0 \leq y \leq 100$. (We also use Xscl = 1 and Yscl = 10.) Now we graph the function.

We can use the TRACE feature to obtain an approximation of the expected percent of radioactive material in the milk when $x = 30$. (A TABLE feature may also be used to approximate the percent.) To obtain a better approximation, let's use the ZOOM feature several times to zoom in near $x = 30$.

The percent of radioactive material in the milk 30 days after the nuclear accident was 5.08%, accurate to two decimal places.

Use a graphing calculator to find each percent. Approximate your solutions so that they are accurate to two decimal places.

1. Estimate the expected percent of radioactive material in the milk 2 days after the nuclear accident.

2. Estimate the expected percent of radioactive material in the milk 10 days after the nuclear accident.

3. Estimate the expected percent of radioactive material in the milk 15 days after the nuclear accident.

4. Estimate the expected percent of radioactive material in the milk 25 days after the nuclear accident.

Vocabulary, Readiness & Video Check

Use the choices to fill in each blank.

1. A function such as $f(x) = 2^x$ is a(n) _____ function.
   A. linear   **B.** quadratic   **C.** exponential

2. If $7^x = 7^y$, then _____.
   A. $x = 7^y$ **B.** $x = y$ **C.** $y = 7^x$ **D.** $7 = 7^y$

Answer the questions about the graph of $y = 2^x$, shown to the right.

3. Is this a function? _____

4. Is this a one-to-one function? _____

5. Is there an x-intercept? _____ If so, name the coordinates. _____

6. Is there a y-intercept? _____ If so, name the coordinates. _____

7. The domain of this function, in interval notation, is _____.

8. The range of this function, in interval notation, is _____.

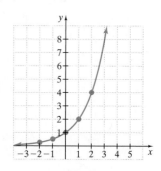

Martin-Gay Interactive Videos

See Video 9.3

Watch the section lecture video and answer the following questions.

OBJECTIVE
1

9. From the lecture before ▣ Example 1, what's the main difference between a polynomial function and an exponential function?

OBJECTIVE
2

10. From ▣ Examples 2 and 3, you can only apply the uniqueness of b^x to solve an exponential equation if you're able to do what?

OBJECTIVE
3

11. For ▣ Example 4, write the equation and find how much uranium will remain after 101 days. Round your answer to the nearest tenth.

9.3 Exercise Set   MyMathLab®

Graph each exponential function. See Examples 1 through 3.

1. $y = 5^x$

2. $y = 4^x$

3. $y = 2^x + 1$

4. $y = 3^x - 1$

5. $y = \left(\dfrac{1}{4}\right)^x$

6. $y = \left(\dfrac{1}{5}\right)^x$

7. $y = \left(\dfrac{1}{2}\right)^x - 2$

8. $y = \left(\dfrac{1}{3}\right)^x + 2$

9. $y = -2^x$

10. $y = -3^x$

11. $y = -\left(\dfrac{1}{4}\right)^x$

12. $y = -\left(\dfrac{1}{5}\right)^x$

13. $f(x) = 2^{x+1}$

14. $f(x) = 3^{x-1}$

15. $f(x) = 4^{x-2}$

16. $f(x) = 2^{x+3}$

Match each exponential equation with its graph below. See Examples 1 through 3.

17. $f(x) = \left(\dfrac{1}{2}\right)^x$

18. $f(x) = \left(\dfrac{1}{4}\right)^x$

19. $f(x) = 2^x$

20. $f(x) = 3^x$

A.

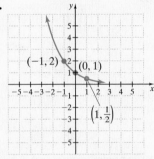

B.

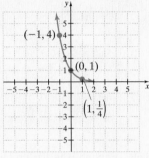

C.

D.

Solve each equation for x. See Example 4.

21. $3^x = 27$

22. $6^x = 36$

▸ **23.** $16^x = 8$

24. $64^x = 16$

25. $32^{2x-3} = 2$

26. $9^{2x+1} = 81$

27. $\dfrac{1}{4} = 2^{3x}$

28. $\dfrac{1}{27} = 3^{2x}$

29. $5^x = 625$

30. $2^x = 64$

31. $4^x = 8$

32. $32^x = 4$

▸ **33.** $27^{x+1} = 9$

34. $125^{x-2} = 25$

35. $81^{x-1} = 27^{2x}$

36. $4^{3x-7} = 32^{2x}$

Solve. Unless otherwise indicated, round results to one decimal place. See Example 6.

▸ **37.** One type of uranium has a radioactive decay rate of 0.4% per day. If 30 pounds of this uranium is available today, how much will still remain after 50 days? Use $y = 30(0.996)^x$ and let x be 50.

38. The nuclear waste from an atomic energy plant decays at a rate of 3% each century. If 150 pounds of nuclear waste is disposed of, how much of it will still remain after 10 centuries? Use $y = 150(0.97)^x$, and let x be 10.

39. Cheese production in the United States is currently growing at a rate of 3% per year. The equation $y = 8.6(1.03)^x$ models the cheese production in the United States from 2003 to 2009. In this equation, y is the amount of cheese produced, in billions of pounds, and x represents the number of years after 2003. Round answers to the nearest tenth of a billion. (*Source*: National Agricultural Statistics Service)

a. Estimate the total cheese production in the United States in 2007.

b. Assuming this equation continues to be valid in the future, use the equation to predict the total amount of cheese produced in the United States in 2015.

40. Retail revenue from shopping on the Internet is currently growing at rate of 26% per year. In 2003, a total of $39 billion in revenue was collected through Internet retail sales. Answer the following questions using $y = 39(1.26)^t$, where y is Internet revenues in billions of dollars and t is the number of years after 2003. Round answers to the nearest tenth of a billion dollars. (*Source*: U.S. Bureau of the Census)

a. According to the model, what level of retail revenues from Internet shopping was expected in 2005?

b. If the given model continues to be valid, predict the level of Internet shopping revenues in 2015.

41. The equation $y = 140{,}242(1.083)^x$ models the number of American college students who studied abroad each year from 2000 through 2009. In the equation, y is the number of American students studying abroad, and x represents the number of years after 2000. Round answers to the nearest whole. (*Source*: Based on data from Institute of International Education, Open Doors)

a. Estimate the number of American students studying abroad in 2004.

b. Assuming this equation continues to be valid in the future, use this equation to predict the number of American students studying abroad in 2015.

42. Carbon dioxide (CO_2) is a greenhouse gas that contributes to global warming. Partially due to the combustion of fossil fuel, the amount of CO_2 in Earth's atmosphere has been increasing by 0.5% annually over the past century. In 2000, the concentration of CO_2 in the atmosphere was 369.4 parts per million by volume. To make the following predictions, use $y = 369.4(1.005)^t$ where y is the concentration of CO_2 in parts per million by volume and t is the number of years after 2000. (*Sources*: Based on data from the United Nations Environment Programme and the Carbon Dioxide Information Analysis Center)

a. Predict the concentration of CO_2 in the atmosphere in the year 2015.

b. Predict the concentration of CO_2 in the atmosphere in the year 2030.

The equation $y = 136.76(1.107)^x$ gives the number of cellular phone users y (in millions) in the United States for the years 2002 through 2010. In this equation, $x = 0$ corresponds to 2002, $x = 1$ corresponds to 2003, and so on. Use this model to solve Exercises 43 and 44. Round answers to the nearest tenth of a million.

43. Predict the number of cell phone users in the year 2012.

44. Predict the number of cell phone users in 2014.

45. An unusually wet spring has caused the size of the Cape Cod mosquito population to increase by 8% each day. If an estimated 200,000 mosquitoes are on Cape Cod on May 12, find how many mosquitoes will inhabit the Cape on May 25. Use $y = 200{,}000(1.08)^x$ where x is number of days since May 12. Round to the nearest thousand.

46. The atmospheric pressure p, in pascals, on a weather balloon decreases with increasing height. This pressure, measured in millimeters of mercury, is related to the number of kilometers h above sea level by the function $p(h) = 760(2.7)^{-0.145h}$. Round to the nearest tenth of a pascal.

  a. Find the atmospheric pressure at a height of 1 kilometer.

  b. Find the atmospheric pressure at a height of 10 kilometers.

Solve. Use $A = P\left(1 + \dfrac{r}{n}\right)^{nt}$. Round answers to two decimal places. See Example 5.

47. Find the amount Erica owes at the end of 3 years if $6000 is loaned to her at a rate of 8% compounded monthly.

48. Find the amount owed at the end of 5 years if $3000 is loaned at a rate of 10% compounded quarterly.

49. Find the total amount Janina has in a college savings account if $2000 was invested and earned 6% compounded semiannually for 12 years.

50. Find the amount accrued if $500 is invested and earns 7% compounded monthly for 4 years.

REVIEW AND PREVIEW

Solve each equation. See Sections 2.1 and 5.8.

51. $5x - 2 = 18$

52. $3x - 7 = 11$

53. $3x - 4 = 3(x + 1)$

54. $2 - 6x = 6(1 - x)$

55. $x^2 + 6 = 5x$

56. $18 = 11x - x^2$

By inspection, find the value for x that makes each statement true. See Section 5.1.

57. $2^x = 8$

58. $3^x = 9$

59. $5^x = \dfrac{1}{5}$

60. $4^x = 1$

CONCEPT EXTENSIONS

Is the given function an exponential function? See the Concept Check in this section.

61. $f(x) = 1.5x^2$

62. $g(x) = 3^x$

63. $h(x) = \left(\dfrac{1}{2}x\right)^2$

64. $F(x) = 0.4^{x+1}$

Match each exponential function with its graph.

65. $f(x) = 2^{-x}$

66. $f(x) = \left(\dfrac{1}{2}\right)^{-x}$

67. $f(x) = 4^{-x}$

68. $f(x) = \left(\dfrac{1}{3}\right)^{-x}$

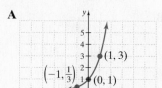

A
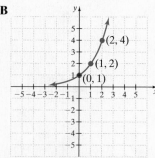

C
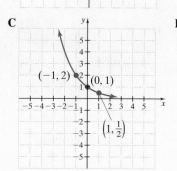

D
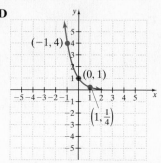

69. Explain why the graph of an exponential function $y = b^x$ contains the point $(1, b)$.

70. Explain why an exponential function $y = b^x$ has a y-intercept of $(0, 1)$.

Graph.

71. $y = |3^x|$

72. $y = \left|\left(\dfrac{1}{3}\right)^x\right|$

73. $y = 3^{|x|}$

74. $y = \left(\dfrac{1}{3}\right)^{|x|}$

75. Graph $y = 2^x$ and $y = \left(\dfrac{1}{2}\right)^{-x}$ on the same set of axes. Describe what you see and why.

76. Graph $y = 2^x$ and $x = 2^y$ on the same set of axes. Describe what you see.

Use a graphing calculator to solve. Estimate your results to two decimal places.

77. Verify the results of Exercise 37.

78. Verify the results of Exercise 38.

79. From Exercise 37, estimate the number of pounds of uranium that will be available after 100 days.

80. From Exercise 37, estimate the number of pounds of uranium that will be available after 120 days.

9.4 | Exponential Growth and Decay Functions

OBJECTIVES

1 Model Exponential Growth.
2 Model Exponential Decay.

Now that we can graph exponential functions, let's learn about exponential growth and exponential decay.

A quantity that grows or decays by the same percent at regular time periods is said to have **exponential growth** or **exponential decay.** There are many real-life examples of exponential growth and decay, such as population, bacteria, viruses, and radioactive substances, just to name a few.

Recall the graphs of exponential functions.

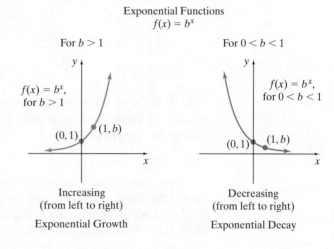

OBJECTIVE

1 Modeling Exponential Growth

We begin with exponential growth, as described below.

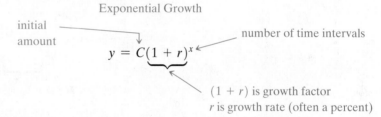

EXAMPLE 1 In 1995, let's suppose a town named Jackson had a population of 15,500 and was consistently increasing by 10% per year. If this yearly increase continues, predict the city's population in 2015. (Round to the nearest whole.)

Solution: Let's begin to understand by calculating the city's population each year:

Time Interval	x = 1	x = 2	3	4	5	and so on ...
Year	1996	1997	1998	1999	2000	
Population	17,050	18,755	20,631	22,694	24,963	

$15,500 + 0.10(15,500)$ $17,050 + 0.10(17,050)$

This is an example of exponential growth, so let's use our formula with

$$C = 15,500; r = 0.10; x = 2015 - 1995 = 20$$

Then,

$$y = C(1 + r)^x$$
$$= 15,500(1 + 0.10)^{20}$$
$$= 15,500(1.1)^{20}$$
$$\approx 104,276$$

(Continued on next page)

In 2015, we predict the population of Jackson to be 104,276.

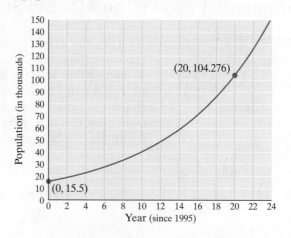

PRACTICE

1 In 2000, the town of Jackson (from Example 1) had a population of 25,000 and started consistently increasing by 12% per year. If this yearly increase continues, predict the city's population in 2015. Round to the nearest whole.

Note: The exponential growth formula, $y = C(1 + r)^x$, should remind you of the compound interest formula from the previous section, $A = P(1 + \frac{r}{n})^{nt}$. In fact, if the number of compoundings per year, n, is 1, the interest formula becomes $A = P(1 + r)^t$, which is the exponential growth formula written with different variables.

OBJECTIVE

2 Modeling Exponential Decay

Now let's study exponential decay.

Exponential Decay

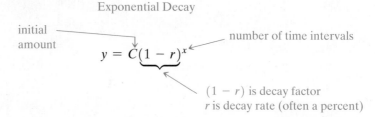

initial amount — $y = C\underbrace{(1 - r)}^x$ — number of time intervals

$(1 - r)$ is decay factor
r is decay rate (often a percent)

EXAMPLE 2 A large golf country club holds a singles tournament each year. At the start of the tournament for a particular year, there are 512 players. After each round, half the players are eliminated. How many players remain after 6 rounds?

Solution: This is an example of exponential decay.

Let's begin to understand by calculating the number of players after a few rounds.

Round (same as interval)	1	2	3	4	and so on …
Players (at end of round)	256	128	64	32	

↑ $\lfloor 512 - 0.50(512) \rfloor$ ↑ $\lfloor 256 - 0.50(256) \rfloor$

Here, $C = 512$; $r = \frac{1}{2}$ or $50\% = 0.50$; $x = 6$

Thus,

$$y = 512(1 - 0.50)^6$$
$$= 512(0.50)^6$$
$$= 8$$

After 6 rounds, there are 8 players remaining.

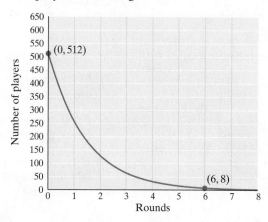

PRACTICE
2   A tournament with 800 persons is played so that after each round, the number of players decreases by 30%. Find the number of players after round 9. Round your answer to the nearest whole.

The **half-life** of a substance is the amount of time it takes for half of the substance to decay.

EXAMPLE 3   A form of DDT pesticide (banned in 1972) has a half-life of approximately 15 years. If a storage unit had 400 pounds of DDT, find how much DDT is remaining after 72 years. Round to the nearest tenth of a pound.

Solution: Here, we need to be careful because each time interval is 15 years, the half-life.

Time Interval	1	2	3	4	5	and so on …
Years Passed	15	$2 \cdot 15 = 30$	45	60	75	
Pounds of DDT	200	100	50	25	12.5	

From the table, we see that after 72 years, between 4 and 5 intervals, there should be between 12.5 and 25 pounds of DDT remaining.

Let's calculate x, the number of time intervals.

$$x = \frac{72 \,(\text{years})}{15 \,(\text{half-life})} = 4.8$$

Now, using our exponential decay formula and the definition of half-life, for each time interval x, the decay rate r is $\frac{1}{2}$ or 50% or 0.50.

$$y = 400(1 - 0.50)^{4.8} \text{—time intervals for 72 years}$$

original amount   decay rate

$$y = 400(0.50)^{4.8}$$
$$y \approx 14.4$$

In 72 years, 14.4 pounds of DDT remain.

PRACTICE
3   Use the information from Example 3 and calculate how much of a 500-gram sample of DDT will remain after 51 years. Round to the nearest tenth of a gram.

Vocabulary, Readiness & Video Check

Martin-Gay Interactive Videos

See Video 9.4

Watch the section lecture video and answer the following questions.

OBJECTIVE 1

1. Example 1 reviews exponential growth. Explain how you find the growth rate and the correct number of time intervals.

OBJECTIVE 2

2. Explain how you know that Example 2 has to do with exponential decay, and not exponential growth.

OBJECTIVE 2

3. For Example 3, which has to do with half-life, explain how to calculate the number of time intervals. Also, what is the decay rate for half-life and why?

9.4 Exercise Set   MyMathLab®

Practice using the exponential growth formula by completing the table below. Round final amounts to the nearest whole. See Example 1.

	Original Amount	Growth Rate per Year	Number of Years, x	Final Amount after x Years of Growth
1.	305	5%	8	
2.	402	7%	5	
3.	2000	11%	41	
4.	1000	47%	19	
5.	17	29%	28	
6.	29	61%	12	

Practice using the exponential decay formula by completing the table below. Round final amounts to the nearest whole. See Example 2.

	Original Amount	Decay Rate per Year	Number of Years, x	Final Amount after x Years of Decay
7.	305	5%	8	
8.	402	7%	5	
9.	10,000	12%	15	
10.	15,000	16%	11	
11.	207,000	32%	25	
12.	325,000	29%	31	

MIXED PRACTICE

Solve. Unless noted otherwise, round answers to the nearest whole. See Examples 1 and 2.

13. Suppose a city with population 500,000 has been growing at a rate of 3% per year. If this rate continues, find the population of this city in 12 years.

14. Suppose a city with population 320,000 has been growing at a rate of 4% per year. If this rate continues, find the population of this city in 20 years.

15. The number of employees for a certain company has been decreasing each year by 5%. If the company currently has 640 employees and this rate continues, find the number of employees in 10 years.

16. The number of students attending summer school at a local community college has been decreasing each year by 7%. If 984 students currently attend summer school and this rate continues, find the number of students attending summer school in 5 years.

17. National Park Service personnel are trying to increase the size of the bison population of Theodore Roosevelt National Park. If 260 bison currently live in the park, and if the population's rate of growth is 2.5% annually, find how many bison there should be in 10 years.

18. The size of the rat population of a wharf area grows at a rate of 8% monthly. If there are 200 rats in January, find how many rats should be expected by next January.

19. A rare isotope of a nuclear material is very unstable, decaying at a rate of 15% each second. Find how much isotope remains 10 seconds after 5 grams of the isotope is created.

20. An accidental spill of 75 grams of radioactive material in a local stream has led to the presence of radioactive debris decaying at a rate of 4% each day. Find how much debris still remains after 14 days.

Practice using the exponential decay formula with half-lives by completing the table below. The first row has been completed for you. See Example 3.

		Half-Life (in years)	Number of Years	Time Intervals, $x = \left(\dfrac{\text{Years}}{\text{Half-Life}}\right)$ Rounded to Tenths if Needed	Final Amount after x Time Intervals (rounded to tenths)	Is Your Final Amount Reasonable?
	Original Amount					
	60	8	10	$\dfrac{10}{8} = 1.25$	25.2	yes
21.	**a.** 40	7	14			
	b. 40	7	11			
22.	**a.** 200	12	36			
	b. 200	12	40			
23.	21	152	500			
24.	35	119	500			

Solve. Round answers to the nearest tenth.

25. A form of nickel has a half-life of 96 years. How much of a 30-gram sample is left after 250 years?

26. A form of uranium has a half-life of 72 years. How much of a 100-gram sample is left after 500 years?

REVIEW AND PREVIEW

By inspection, find the value for x that makes each statement true. See Sections 5.1 and 9.3.

27. $2^x = 8$ **28.** $3^x = 9$ **29.** $5^x = \dfrac{1}{5}$ **30.** $4^x = 1$

CONCEPT EXTENSIONS

31. An item is on sale for 40% off its original price. If it is then marked down an additional 60%, does this mean the item is free? Discuss why or why not.

32. Uranium U-232 has a half-life of 72 years. What eventually happens to a 10 gram sample? Does it ever completely decay and disappear? Discuss why or why not.

9.5 Logarithmic Functions

OBJECTIVES

1 Write Exponential Equations with Logarithmic Notation and Write Logarithmic Equations with Exponential Notation.

2 Solve Logarithmic Equations by Using Exponential Notation.

3 Identify and Graph Logarithmic Functions.

OBJECTIVE

1 Using Logarithmic Notation

Since the exponential function $f(x) = 2^x$ is a one-to-one function, it has an inverse.

We can create a table of values for f^{-1} by switching the coordinates in the accompanying table of values for $f(x) = 2^x$.

x	$y = f(x)$
-3	$\dfrac{1}{8}$
-2	$\dfrac{1}{4}$
-1	$\dfrac{1}{2}$
0	1
1	2
2	4
3	8

x	$y = f^{-1}(x)$
$\dfrac{1}{8}$	-3
$\dfrac{1}{4}$	-2
$\dfrac{1}{2}$	-1
1	0
2	1
4	2
8	3

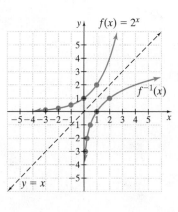

The graphs of $f(x)$ and its inverse are shown on the previous page. Notice that the graphs of f and f^{-1} are symmetric about the line $y = x$, as expected.

Now we would like to be able to write an equation for f^{-1}. To do so, we follow the steps for finding an inverse.

$$f(x) = 2^x$$

Step 1. Replace $f(x)$ by y. $\qquad y = 2^x$

Step 2. Interchange x and y. $\qquad x = 2^y$

Step 3. Solve for y.

At this point, we are stuck. To solve this equation for y, a new notation, the **logarithmic notation,** is needed.

The symbol $\log_b x$ means "the power to which b is raised to produce a result of x." In other words,

$$\log_b x = y \quad \text{means} \quad b^y = x$$

We say that $\log_b x$ is "the logarithm of x to the base b" or "the log of x to the base b."

Logarithmic Definition

If $b > 0$ and $b \neq 1$, then

$$y = \log_b x \text{ means } x = b^y$$

for every $x > 0$ and every real number y.

> **Helpful Hint**
>
> Notice that a *logarithm* is an *exponent*. In other words, $\log_3 9$ is the *power* that we raise 3 in order to get 9.

Before returning to the function $x = 2^y$ and solving it for y in terms of x, let's practice using the new notation $\log_b x$.

It is important to be able to write exponential equations from logarithmic notation and vice versa. The following table shows examples of both forms.

Logarithmic Equation	Corresponding Exponential Equation
$\log_3 9 = 2$	$3^2 = 9$
$\log_6 1 = 0$	$6^0 = 1$
$\log_2 8 = 3$	$2^3 = 8$
$\log_4 \dfrac{1}{16} = -2$	$4^{-2} = \dfrac{1}{16}$
$\log_8 2 = \dfrac{1}{3}$	$8^{1/3} = 2$

EXAMPLE 1 Write each as an exponential equation.

a. $\log_5 25 = 2$ $\qquad$ **b.** $\log_6 \dfrac{1}{6} = -1$ $\qquad$ **c.** $\log_2 \sqrt{2} = \dfrac{1}{2}$ $\qquad$ **d.** $\log_7 x = 5$

Solution

a. $\log_5 25 = 2$ means $5^2 = 25$

b. $\log_6 \dfrac{1}{6} = -1$ means $6^{-1} = \dfrac{1}{6}$

c. $\log_2 \sqrt{2} = \dfrac{1}{2}$ means $2^{1/2} = \sqrt{2}$

d. $\log_7 x = 5$ means $7^5 = x$

PRACTICE
1 Write each as an exponential equation.

a. $\log_3 81 = 4$ **b.** $\log_5 \dfrac{1}{5} = -1$ **c.** $\log_7 \sqrt{7} = \dfrac{1}{2}$ **d.** $\log_{13} y = 4$

EXAMPLE 2 Write each as a logarithmic equation.

a. $9^3 = 729$ **b.** $6^{-2} = \dfrac{1}{36}$ **c.** $5^{1/3} = \sqrt[3]{5}$ **d.** $\pi^4 = x$

Solution

a. $9^3 = 729$ means $\log_9 729 = 3$

b. $6^{-2} = \dfrac{1}{36}$ means $\log_6 \dfrac{1}{36} = -2$

c. $5^{1/3} = \sqrt[3]{5}$ means $\log_5 \sqrt[3]{5} = \dfrac{1}{3}$

d. $\pi^4 = x$ means $\log_\pi x = 4$

PRACTICE
2 Write each as a logarithmic equation.

a. $4^3 = 64$ **b.** $6^{1/3} = \sqrt[3]{6}$ **c.** $5^{-3} = \dfrac{1}{125}$ **d.** $\pi^7 = z$

EXAMPLE 3 Find the value of each logarithmic expression.

a. $\log_4 16$ **b.** $\log_{10} \dfrac{1}{10}$ **c.** $\log_9 3$

Solution

a. $\log_4 16 = 2$ because $4^2 = 16$

b. $\log_{10} \dfrac{1}{10} = -1$ because $10^{-1} = \dfrac{1}{10}$

c. $\log_9 3 = \dfrac{1}{2}$ because $9^{1/2} = \sqrt{9} = 3$

PRACTICE
3 Find the value of each logarithmic expression.

a. $\log_3 9$ **b.** $\log_2 \dfrac{1}{8}$ **c.** $\log_{49} 7$

▶ **Helpful Hint**

Another method for evaluating logarithms such as those in Example 3 is to set the expression equal to x and then write them in exponential form to find x. For example:

a. $\log_4 16 = x$ means $4^x = 16$. Since $4^2 = 16$, $x = 2$ or $\log_4 16 = 2$.

b. $\log_{10} \dfrac{1}{10} = x$ means $10^x = \dfrac{1}{10}$. Since $10^{-1} = \dfrac{1}{10}$, $x = -1$ or $\log_{10} \dfrac{1}{10} = -1$.

c. $\log_9 3 = x$ means $9^x = 3$. Since $9^{1/2} = 3$, $x = \dfrac{1}{2}$ or $\log_9 3 = \dfrac{1}{2}$.

OBJECTIVE

2 Solving Logarithmic Equations

The ability to interchange the logarithmic and exponential forms of a statement is often the key to solving logarithmic equations.

EXAMPLE 4 Solve each equation for x.

a. $\log_4 \frac{1}{4} = x$ **b.** $\log_5 x = 3$ **c.** $\log_x 25 = 2$ **d.** $\log_3 1 = x$ **e.** $\log_b 1 = x$

Solution

a. $\log_4 \frac{1}{4} = x$ means $4^x = \frac{1}{4}$. Solve $4^x = \frac{1}{4}$ for x.

$$4^x = \frac{1}{4}$$
$$4^x = 4^{-1}$$

Since the bases are the same, by the uniqueness of b^x, we have that

$$x = -1$$

The solution is -1 or the solution set is $\{-1\}$. To check, see that $\log_4 \frac{1}{4} = -1$, since $4^{-1} = \frac{1}{4}$.

b. $\log_5 x = 3$

$\quad 5^3 = x$ Write as an exponential equation.

$\quad 125 = x$

The solution is 125.

c. $\log_x 25 = 2$

$\quad x^2 = 25$ Write as an exponential equation. Here $x > 0, x \neq 1$.

$\quad x = 5$

Even though $(-5)^2 = 25$, the base b of a logarithm must be positive. The solution is 5.

d. $\log_3 1 = x$

$\quad 3^x = 1$ Write as an exponential equation.

$\quad 3^x = 3^0$ Write 1 as 3^0.

$\quad x = 0$ Use the uniqueness of b^x.

The solution is 0.

e. $\log_b 1 = x$

$\quad b^x = 1$ Write as an exponential equation. Here, $b > 0$ and $b \neq 1$.

$\quad b^x = b^0$ Write 1 as b^0.

$\quad x = 0$ Apply the uniqueness of b^x.

The solution is 0. ☐

PRACTICE

4 Solve each equation for x.

a. $\log_5 \frac{1}{25} = x$ **b.** $\log_x 8 = 3$ **c.** $\log_6 x = 2$

d. $\log_{13} 1 = x$ **e.** $\log_h 1 = x$

In Example 4e, we proved an important property of logarithms. That is, $\log_b 1$ is always 0. This property as well as two important others are given next.

> **Properties of Logarithms**
>
> If b is a real number, $b > 0$, and $b \neq 1$, then
>
> **1.** $\log_b 1 = 0$
>
> **2.** $\log_b b^x = x$
>
> **3.** $b^{\log_b x} = x$

To see that **2.** $\log_b b^x = x$, change the logarithmic form to exponential form. Then, $\log_b b^x = x$ means $b^x = b^x$. In exponential form, the statement is true, so in logarithmic form, the statement is also true.

To understand **3.** $b^{\log_b x} = x$, write this exponential equation as an equivalent logarithm.

EXAMPLE 5   Simplify.

a. $\log_3 3^2$ **b.** $\log_7 7^{-1}$ **c.** $5^{\log_5 3}$ **d.** $2^{\log_2 6}$

Solution

a. From Property 2, $\log_3 3^2 = 2$.

b. From Property 2, $\log_7 7^{-1} = -1$.

c. From Property 3, $5^{\log_5 3} = 3$.

d. From Property 3, $2^{\log_2 6} = 6$.   ☐

PRACTICE

5   Simplify.

a. $\log_5 5^4$ **b.** $\log_9 9^{-2}$ **c.** $6^{\log_6 5}$ **d.** $7^{\log_7 4}$

OBJECTIVE

3   **Graphing Logarithmic Functions** ▶

Let us now return to the function $f(x) = 2^x$ and write an equation for its inverse, $f^{-1}(x)$. Recall our earlier work.

$$f(x) = 2^x$$

Step 1. Replace $f(x)$ by y. $y = 2^x$

Step 2. Interchange x and y. $x = 2^y$

Having gained proficiency with the notation $\log_b x$, we can now complete the steps for writing the inverse equation by writing $x = 2^y$ as an equivalent logarithm.

Step 3. Solve for y. $y = \log_2 x$

Step 4. Replace y with $f^{-1}(x)$. $f^{-1}(x) = \log_2 x$

Thus, $f^{-1}(x) = \log_2 x$ defines a function that is the inverse function of the function $f(x) = 2^x$. The function $f^{-1}(x)$ or $y = \log_2 x$ is called a **logarithmic function.**

> **Logarithmic Function**
>
> If x is a positive real number, b is a constant positive real number, and b is not 1, then a **logarithmic function** is a function that can be defined by
>
> $$f(x) = \log_b x$$
>
> The domain of f is the set of positive real numbers, and the range of f is the set of real numbers.

✓CONCEPT CHECK

Let $f(x) = \log_3 x$ and $g(x) = 3^x$. These two functions are inverses of each other. Since $(2, 9)$ is an ordered pair solution of $g(x)$ or $g(2) = 9$, what ordered pair do we know to be a solution of $f(x)$? Also, find $f(9)$. Explain why.

We can explore logarithmic functions by graphing them.

EXAMPLE 6 Graph the logarithmic function $y = \log_2 x$.

Solution First we write the equation with exponential notation as $2^y = x$. Then we find some ordered pair solutions that satisfy this equation. Finally, we plot the points and connect them with a smooth curve. The domain of this function is $(0, \infty)$, and the range is all real numbers.

Since $x = 2^y$ is solved for x, we choose y-values and compute corresponding x-values.

If $y = 0, x = 2^0 = 1$
If $y = 1, x = 2^1 = 2$
If $y = 2, x = 2^2 = 4$
If $y = -1, x = 2^{-1} = \dfrac{1}{2}$

$x = 2^y$	y
1	0
2	1
4	2
$\dfrac{1}{2}$	-1

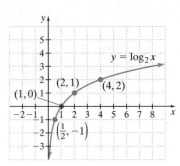

Notice that the x-intercept is $(1, 0)$ and there is no y-intercept.

PRACTICE
6 Graph the logarithmic function $y = \log_9 x$.

EXAMPLE 7 Graph the logarithmic function $f(x) = \log_{1/3} x$.

Solution Replace $f(x)$ with y and write the result with exponential notation.

$$f(x) = \log_{1/3} x$$
$$y = \log_{1/3} x \quad \text{Replace } f(x) \text{ with } y.$$
$$\left(\frac{1}{3}\right)^y = x \quad \text{Write in exponential form.}$$

Now we can find ordered pair solutions that satisfy $\left(\dfrac{1}{3}\right)^y = x$, plot these points, and connect them with a smooth curve.

If $y = 0, x = \left(\dfrac{1}{3}\right)^0 = 1$

If $y = 1, x = \left(\dfrac{1}{3}\right)^1 = \dfrac{1}{3}$

If $y = -1, x = \left(\dfrac{1}{3}\right)^{-1} = 3$

If $y = -2, x = \left(\dfrac{1}{3}\right)^{-2} = 9$

$x = \left(\dfrac{1}{3}\right)^y$	y
1	0
$\dfrac{1}{3}$	1
3	-1
9	-2

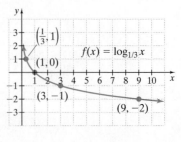

The domain of this function is $(0, \infty)$, and the range is the set of all real numbers. The x-intercept is $(1, 0)$ and there is no y-intercept.

PRACTICE
7 Graph the logarithmic function $y = \log_{1/4} x$.

Answer to Concept Check:
$(9, 2)$; $f(9) = 2$; answers may vary

The following figures summarize characteristics of logarithmic functions.

$$f(x) = \log_b x, b > 0, b \neq 1$$

- one-to-one function
- x-intercept $(1, 0)$
- no y-intercept

- domain: $(0, \infty)$
- range: $(-\infty, \infty)$

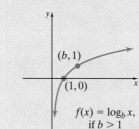

$f(x) = \log_b x,$
if $b > 1$

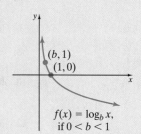

$f(x) = \log_b x,$
if $0 < b < 1$

Vocabulary, Readiness & Video Check

Use the choices to fill in each blank.

1. A function such as $y = \log_2 x$ is a(n) _____ function.

    A. linear   **B.** logarithmic   **C.** quadratic   **D.** exponential

2. If $y = \log_2 x$, then _____ .

    A. $x = y$ **B.** $2^x = y$ **C.** $2^y = x$ **D.** $2y = x$

Answer the questions about the graph of $y = \log_2 x$, shown to the left.

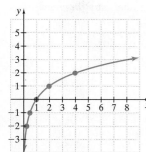

3. Is this a one-to-one function? _____

4. Is there an x-intercept? _____ If so, name the coordinates. _____

5. Is there a y-intercept? _____ If so, name the coordinates. _____

6. The domain of this function, in interval notation, is _____ .

7. The range of this function, in interval notation, is _____ .

Martin-Gay Interactive Videos

See Video 9.5

Watch the section lecture video and answer the following questions.

OBJECTIVE
1

8. Notice from the definition of a logarithm and from ⊞ Examples 1–4 that a logarithmic statement equals the power in the exponent statement, such as $b^y = x$. What conclusion can you make about logarithms and exponents?

OBJECTIVE
2

9. From ⊞ Examples 8 and 9, how do you solve a logarithmic equation?

OBJECTIVE
3

10. In ⊞ Example 12, why is it easier to choose values for y when finding ordered pairs for the graph?

9.5 Exercise Set MyMathLab®

Write each as an exponential equation. See Example 1.

1. $\log_6 36 = 2$

2. $\log_2 32 = 5$

3. $\log_3 \dfrac{1}{27} = -3$

4. $\log_5 \dfrac{1}{25} = -2$

5. $\log_{10} 1000 = 3$

6. $\log_{10} 10 = 1$

7. $\log_9 x = 4$

8. $\log_8 y = 7$

9. $\log_\pi \dfrac{1}{\pi^2} = -2$

10. $\log_e \dfrac{1}{e} = -1$

11. $\log_7 \sqrt{7} = \dfrac{1}{2}$

12. $\log_{11} \sqrt[4]{11} = \dfrac{1}{4}$

13. $\log_{0.7} 0.343 = 3$

14. $\log_{1.2} 1.44 = 2$

15. $\log_3 \dfrac{1}{81} = -4$

16. $\log_{1/4} 16 = -2$

Write each as a logarithmic equation. See Example 2.

17. $2^4 = 16$

18. $5^3 = 125$

19. $10^2 = 100$

20. $10^4 = 10{,}000$

21. $\pi^3 = x$

22. $\pi^5 = y$

23. $10^{-1} = \dfrac{1}{10}$

24. $10^{-2} = \dfrac{1}{100}$

25. $4^{-2} = \dfrac{1}{16}$

26. $3^{-4} = \dfrac{1}{81}$

27. $5^{1/2} = \sqrt{5}$

28. $4^{1/3} = \sqrt[3]{4}$

Find the value of each logarithmic expression. See Examples 3 and 5.

29. $\log_2 8$

30. $\log_3 9$

31. $\log_3 \dfrac{1}{9}$

32. $\log_2 \dfrac{1}{32}$

33. $\log_{25} 5$

34. $\log_8 \dfrac{1}{2}$

35. $\log_{1/2} 2$

36. $\log_{2/3} \dfrac{4}{9}$

37. $\log_6 1$

38. $\log_9 9$

39. $\log_{10} 100$

40. $\log_{10} \dfrac{1}{10}$

41. $\log_3 81$

42. $\log_2 16$

43. $\log_4 \dfrac{1}{64}$

44. $\log_3 \dfrac{1}{9}$

Solve. See Example 4.

45. $\log_3 9 = x$

46. $\log_2 8 = x$

47. $\log_3 x = 4$

48. $\log_2 x = 3$

49. $\log_x 49 = 2$

50. $\log_x 8 = 3$

51. $\log_2 \dfrac{1}{8} = x$

52. $\log_3 \dfrac{1}{81} = x$

53. $\log_3 \dfrac{1}{27} = x$

54. $\log_5 \dfrac{1}{125} = x$

55. $\log_8 x = \dfrac{1}{3}$

56. $\log_9 x = \dfrac{1}{2}$

57. $\log_4 16 = x$

58. $\log_2 16 = x$

59. $\log_{3/4} x = 3$

60. $\log_{2/3} x = 2$

61. $\log_x 100 = 2$

62. $\log_x 27 = 3$

63. $\log_2 2^4 = x$

64. $\log_6 6^{-2} = x$

65. $3^{\log_3 5} = x$

66. $5^{\log_5 7} = x$

67. $\log_x \dfrac{1}{7} = \dfrac{1}{2}$

68. $\log_x 2 = -\dfrac{1}{3}$

Simplify. See Example 5.

69. $\log_5 5^3$

70. $\log_6 6^2$

71. $2^{\log_2 3}$

72. $7^{\log_7 4}$

73. $\log_9 9$

74. $\log_2 2$

75. $\log_8 (8)^{-1}$

76. $\log_{11} (11)^{-1}$

Graph each logarithmic function. Label any intercepts. See Examples 6 and 7.

77. $y = \log_3 x$

78. $y = \log_8 x$

79. $f(x) = \log_{1/4} x$

80. $f(x) = \log_{1/2} x$

81. $f(x) = \log_5 x$

82. $f(x) = \log_6 x$

83. $f(x) = \log_{1/6} x$

84. $f(x) = \log_{1/5} x$

REVIEW AND PREVIEW

Simplify each rational expression. See Section 6.1.

85. $\dfrac{x+3}{3+x}$

86. $\dfrac{x-5}{5-x}$

87. $\dfrac{x^2 - 8x + 16}{2x - 8}$

88. $\dfrac{x^2 - 3x - 10}{2+x}$

Add or subtract as indicated. See Section 6.2.

89. $\dfrac{2}{x} + \dfrac{3}{x^2}$

90. $\dfrac{5}{y+1} - \dfrac{4}{y-1}$

91. $\dfrac{3x}{x+3} + \dfrac{9}{x+3}$

92. $\dfrac{m^2}{m+1} - \dfrac{1}{m+1}$

CONCEPT EXTENSIONS

Solve. See the Concept Check in this section.

93. Let $f(x) = \log_5 x$. Then $g(x) = 5^x$ is the inverse of $f(x)$. The ordered pair $(2, 25)$ is a solution of the function $g(x)$.

   a. Write this solution using function notation.

   b. Write an ordered pair that we know to be a solution of $f(x)$.

   c. Use the answer to part b and write the solution using function notation.

94. Let $f(x) = \log_{0.3} x$. Then $g(x) = 0.3^x$ is the inverse of $f(x)$. The ordered pair $(3, 0.027)$ is a solution of the function $g(x)$.

   a. Write this solution using function notation.

   b. Write an ordered pair that we know to be a solution of $f(x)$.

   c. Use the answer to part b and write the solution using function notation.

95. Explain why negative numbers are not included as logarithmic bases.

96. Explain why 1 is not included as a logarithmic base.

Solve by first writing as an exponent.

97. $\log_7 (5x - 2) = 1$ **98.** $\log_3 (2x + 4) = 2$

99. Simplify: $\log_3(\log_5 125)$

100. Simplify: $\log_7(\log_4(\log_2 16))$

Graph each function and its inverse function on the same set of axes. Label any intercepts.

101. $y = 4^x$; $y = \log_4 x$

102. $y = 3^x$; $y = \log_3 x$

103. $y = \left(\frac{1}{3}\right)^x$; $y = \log_{1/3} x$

104. $y = \left(\frac{1}{2}\right)^x$; $y = \log_{1/2} x$

105. Explain why the graph of the function $y = \log_b x$ contains the point $(1, 0)$ no matter what b is.

106. $\log_3 10$ is between which two integers? Explain your answer.

107. The formula $\log_{10}(1 - k) = \dfrac{-0.3}{H}$ models the relationship between the half-life H of a radioactive material and its rate of decay k. Find the rate of decay of the iodine isotope I-131 if its half-life is 8 days. Round to four decimal places.

108. The formula $\text{pH} = -\log_{10}(\text{H}^+)$ provides the pH for a liquid, where H^+ stands for the concentration of hydronium ions. Find the pH of lemonade, whose concentration of hydronium ions is 0.0050 moles/liter. Round to the nearest tenth.

9.6   Properties of Logarithms

OBJECTIVES

1 Use the Product Property of Logarithms.

2 Use the Quotient Property of Logarithms.

3 Use the Power Property of Logarithms.

4 Use the Properties of Logarithms Together.

In the previous section, we explored some basic properties of logarithms. We now introduce and explore additional properties. Because a logarithm is an exponent, logarithmic properties are just restatements of exponential properties.

OBJECTIVE

1   Using the Product Property

The first of these properties is called the **product property of logarithms** because it deals with the logarithm of a product.

> **Product Property of Logarithms**
>
> If x, y, and b are positive real numbers and $b \neq 1$, then
> $$\log_b xy = \log_b x + \log_b y$$

To prove this, let $\log_b x = M$ and $\log_b y = N$. Now write each logarithm with exponential notation.

$$\log_b x = M \quad \text{is equivalent to} \quad b^M = x$$
$$\log_b y = N \quad \text{is equivalent to} \quad b^N = y$$

When we multiply the left sides and the right sides of the exponential equations, we have that

$$xy = (b^M)(b^N) = b^{M+N}$$

If we write the equation $xy = b^{M+N}$ in equivalent logarithmic form, we have

$$\log_b xy = M + N$$

But since $M = \log_b x$ and $N = \log_b y$, we can write

$$\log_b xy = \log_b x + \log_b y \quad \text{Let } M = \log_b x \text{ and } N = \log_b y.$$

In other words, the logarithm of a product is the sum of the logarithms of the factors. This property is sometimes used to simplify logarithmic expressions.

In the examples that follow, assume that variables represent positive numbers.

EXAMPLE 1 Write each sum as a single logarithm.

a. $\log_{11} 10 + \log_{11} 3$ **b.** $\log_3 \dfrac{1}{2} + \log_3 12$ **c.** $\log_2(x + 2) + \log_2 x$

Solution

In each case, both terms have a common logarithmic base.

a. $\log_{11} 10 + \log_{11} 3 = \log_{11}(10 \cdot 3)$ Apply the product property.

$$= \log_{11} 30$$

b. $\log_3 \dfrac{1}{2} + \log_3 12 = \log_3 \left(\dfrac{1}{2} \cdot 12 \right) = \log_3 6$

c. $\log_2(x + 2) + \log_2 x = \log_2[(x + 2) \cdot x] = \log_2(x^2 + 2x)$ ☐

> ▶ **Helpful Hint**
> Check your logarithm properties. Make sure you understand that $\log_2(x + 2)$ *is not* $\log_2 x + \log_2 2$.

PRACTICE

1 Write each sum as a single logarithm.

a. $\log_8 5 + \log_8 3$

b. $\log_2 \dfrac{1}{3} + \log_2 18$

c. $\log_5(x - 1) + \log_5(x + 1)$

OBJECTIVE

2 Using the Quotient Property

The second property is the **quotient property of logarithms.**

Quotient Property of Logarithms

If x, y, and b are positive real numbers and $b \neq 1$, then

$$\log_b \frac{x}{y} = \log_b x - \log_b y$$

The proof of the quotient property of logarithms is similar to the proof of the product property. Notice that the quotient property says that the logarithm of a quotient is the difference of the logarithms of the dividend and divisor.

✓**CONCEPT CHECK**

Which of the following is the correct way to rewrite $\log_5 \dfrac{7}{2}$?

a. $\log_5 7 - \log_5 2$ **b.** $\log_5(7 - 2)$ **c.** $\dfrac{\log_5 7}{\log_5 2}$ **d.** $\log_5 14$

Answer to Concept Check: **a**

EXAMPLE 2 Write each difference as a single logarithm.

a. $\log_{10} 27 - \log_{10} 3$ **b.** $\log_5 8 - \log_5 x$ **c.** $\log_3(x^2 + 5) - \log_3(x^2 + 1)$

Solution In each case, both terms have a common logarithmic base.

a. $\log_{10} 27 - \log_{10} 3 = \log_{10} \dfrac{27}{3} = \log_{10} 9$

b. $\log_5 8 - \log_5 x = \log_5 \dfrac{8}{x}$

c. $\log_3(x^2 + 5) - \log_3(x^2 + 1) = \log_3 \dfrac{x^2 + 5}{x^2 + 1}$ Apply the quotient property. □

PRACTICE
2 Write each difference as a single logarithm.

a. $\log_5 18 - \log_5 6$ **b.** $\log_6 x - \log_6 3$ **c.** $\log_4(x^2 + 1) - \log_4(x^2 + 3)$

OBJECTIVE
3 Using the Power Property
The third and final property we introduce is the **power property of logarithms.**

Power Property of Logarithms
If x and b are positive real numbers, $b \neq 1$, and r is a real number, then
$$\log_b x^r = r \log_b x$$

EXAMPLE 3 Use the power property to rewrite each expression.

a. $\log_5 x^3$ **b.** $\log_4 \sqrt{2}$

Solution

a. $\log_5 x^3 = 3 \log_5 x$ **b.** $\log_4 \sqrt{2} = \log_4 2^{1/2} = \dfrac{1}{2} \log_4 2$ □

PRACTICE
3 Use the power property to rewrite each expression.

a. $\log_7 x^8$ **b.** $\log_5 \sqrt[4]{7}$

OBJECTIVE
4 Using the Properties Together
Many times, we must use more than one property of logarithms to simplify a logarithmic expression.

EXAMPLE 4 Write as a single logarithm.

a. $2 \log_5 3 + 3 \log_5 2$ **b.** $3 \log_9 x - \log_9(x + 1)$ **c.** $\log_4 25 + \log_4 3 - \log_4 5$

Solution In each case, all terms have a common logarithmic base.

a. $2 \log_5 3 + 3 \log_5 2 = \log_5 3^2 + \log_5 2^3$ Apply the power property.

$\qquad\qquad\qquad\qquad = \log_5 9 + \log_5 8$

$\qquad\qquad\qquad\qquad = \log_5(9 \cdot 8)$ Apply the product property.

$\qquad\qquad\qquad\qquad = \log_5 72$

b. $3 \log_9 x - \log_9(x + 1) = \log_9 x^3 - \log_9(x + 1)$ Apply the power property.

$\qquad\qquad\qquad\qquad\qquad = \log_9 \dfrac{x^3}{x + 1}$ Apply the quotient property.

c. Use both the product and quotient properties.

$$\log_4 25 + \log_4 3 - \log_4 5 = \log_4(25 \cdot 3) - \log_4 5 \quad \text{Apply the product property.}$$
$$= \log_4 75 - \log_4 5 \quad \text{Simplify.}$$
$$= \log_4 \frac{75}{5} \quad \text{Apply the quotient property.}$$
$$= \log_4 15 \quad \text{Simplify.} \qquad \square$$

PRACTICE
4   Write as a single logarithm.

a. $2 \log_5 4 + 5 \log_5 2$ **b.** $2 \log_8 x - \log_8(x + 3)$ **c.** $\log_7 12 + \log_7 5 - \log_7 4$

EXAMPLE 5   Write each expression as sums or differences of multiples of logarithms.

a. $\log_3 \dfrac{5 \cdot 7}{4}$ **b.** $\log_2 \dfrac{x^5}{y^2}$

Solution

a. $\log_3 \dfrac{5 \cdot 7}{4} = \log_3(5 \cdot 7) - \log_3 4 \quad$ Apply the quotient property.

$= \log_3 5 + \log_3 7 - \log_3 4 \quad$ Apply the product property.

b. $\log_2 \dfrac{x^5}{y^2} = \log_2(x^5) - \log_2(y^2) \quad$ Apply the quotient property.

$= 5 \log_2 x - 2 \log_2 y \quad$ Apply the power property. $\square$

PRACTICE
5   Write each expression as sums or differences of multiples of logarithms.

a. $\log_5 \dfrac{4 \cdot 3}{7}$ **b.** $\log_4 \dfrac{a^2}{b^5}$

Answer to Concept Check:
The properties do not give any way to simplify the logarithm of a sum; answers may vary.

▶ **Helpful Hint**

Notice that we are not able to simplify further a logarithmic expression such as $\log_5(2x - 1)$. None of the basic properties gives a way to write the logarithm of a difference (or sum) in some equivalent form.

✓**CONCEPT CHECK**
What is wrong with the following?

$$\log_{10}(x^2 + 5) = \log_{10} x^2 + \log_{10} 5$$
$$= 2 \log_{10} x + \log_{10} 5$$

Use a numerical example to demonstrate that the result is incorrect.

EXAMPLE 6   If $\log_b 2 = 0.43$ and $\log_b 3 = 0.68$, use the properties of logarithms to evaluate.

a. $\log_b 6$ **b.** $\log_b 9$ **c.** $\log_b \sqrt{2}$

Solution

a. $\log_b 6 = \log_b(2 \cdot 3)$ Write 6 as $2 \cdot 3$.

$\qquad = \log_b 2 + \log_b 3$ Apply the product property.

$\qquad = 0.43 + 0.68$ Substitute given values.

$\qquad = 1.11$ Simplify.

b. $\log_b 9 = \log_b 3^2$ Write 9 as 3^2.

$\qquad = 2 \log_b 3$

$\qquad = 2(0.68)$ Substitute 0.68 for $\log_b 3$.

$\qquad = 1.36$ Simplify.

c. First, recall that $\sqrt{2} = 2^{1/2}$. Then

$\log_b \sqrt{2} = \log_b 2^{1/2}$ Write $\sqrt{2}$ as $2^{1/2}$.

$\qquad = \dfrac{1}{2} \log_b 2$ Apply the power property.

$\qquad = \dfrac{1}{2}(0.43)$ Substitute the given value.

$\qquad = 0.215$ Simplify.   □

PRACTICE

6   If $\log_b 5 = 0.83$ and $\log_b 3 = 0.56$, use the properties of logarithms to evaluate.

a. $\log_b 15$ **b.** $\log_b 25$ **c.** $\log_b \sqrt{3}$

A summary of the basic properties of logarithms that we have developed so far is given next.

Properties of Logarithms

If x, y, and b are positive real numbers, $b \neq 1$, and r is a real number, then

1. $\log_b 1 = 0$ 　　　　　　　　　　　**2.** $\log_b b^x = x$

3. $b^{\log_b x} = x$ 　　　　　　　　　　**4.** $\log_b xy = \log_b x + \log_b y$ Product property.

5. $\log_b \dfrac{x}{y} = \log_b x - \log_b y$ Quotient property. **6.** $\log_b x^r = r \log_b x$ Power property.

Vocabulary, Readiness & Video Check

Select the correct choice.

1. $\log_b 12 + \log_b 3 = \log_b$ _____
   a. 36   **b.** 15   **c.** 4   **d.** 9

2. $\log_b 12 - \log_b 3 = \log_b$ _____
   a. 36   **b.** 15   **c.** 4   **d.** 9

3. $7 \log_b 2 =$ _____
   a. $\log_b 14$ **b.** $\log_b 2^7$ **c.** $\log_b 7^2$ **d.** $(\log_b 2)^7$

4. $\log_b 1 =$ _____
   a. b **b.** 1   **c.** 0   **d.** no answer

5. $b^{\log_b x} =$ _____
   a. x **b.** b **c.** 1   **d.** 0

6. $\log_5 5^2 =$ _____
   a. 25   **b.** 2   **c.** 5^{5^2} **d.** 32

See Video 9.6

Watch the section lecture video and answer the following questions.

OBJECTIVE 1

7. Can the product property of logarithms be used again on the bottom line of ⊞ Example 2 to write $\log_{10}(10x^2 + 20)$ as a sum of logarithms, $\log_{10} 10x^2 + \log_{10} 20$? Explain.

OBJECTIVE 2

8. From ⊞ Example 3 and the lecture before, what must be true about bases before you can apply the quotient property of logarithms?

OBJECTIVE 3

9. Based on ⊞ Example 5, explain why $\log_2 \dfrac{1}{x} = -\log_2 x$.

OBJECTIVE 4

10. From the lecture before ⊞ Example 6, where do the logarithmic properties come from?

9.6 Exercise Set

MyMathLab

Write each sum as a single logarithm. Assume that variables represent positive numbers. See Example 1.

1. $\log_5 2 + \log_5 7$
2. $\log_3 8 + \log_3 4$
3. $\log_4 9 + \log_4 x$
4. $\log_2 x + \log_2 y$
5. $\log_6 x + \log_6 (x + 1)$
6. $\log_5 y^3 + \log_5 (y - 7)$
7. $\log_{10} 5 + \log_{10} 2 + \log_{10}(x^2 + 2)$
8. $\log_6 3 + \log_6 (x + 4) + \log_6 5$

Write each difference as a single logarithm. Assume that variables represent positive numbers. See Example 2.

9. $\log_5 12 - \log_5 4$
10. $\log_7 20 - \log_7 4$
11. $\log_3 8 - \log_3 2$
12. $\log_5 12 - \log_5 3$
13. $\log_2 x - \log_2 y$
14. $\log_3 12 - \log_3 z$
15. $\log_2 (x^2 + 6) - \log_2 (x^2 + 1)$
16. $\log_7 (x + 9) - \log_7 (x^2 + 10)$

Use the power property to rewrite each expression. See Example 3.

17. $\log_3 x^2$
18. $\log_2 x^5$
19. $\log_4 5^{-1}$
20. $\log_6 7^{-2}$
21. $\log_5 \sqrt{y}$
22. $\log_5 \sqrt[3]{x}$

MIXED PRACTICE

Write each as a single logarithm. Assume that variables represent positive numbers. See Example 4.

23. $\log_2 5 + \log_2 x^3$
24. $\log_5 2 + \log_5 y^2$
25. $3\log_4 2 + \log_4 6$
26. $2\log_3 5 + \log_3 2$
27. $3\log_5 x + 6\log_5 z$
28. $2\log_7 y + 6\log_7 z$
29. $\log_4 2 + \log_4 10 - \log_4 5$
30. $\log_6 18 + \log_6 2 - \log_6 9$
31. $\log_7 6 + \log_7 3 - \log_7 4$
32. $\log_8 5 + \log_8 15 - \log_8 20$

33. $\log_{10} x - \log_{10} (x + 1) + \log_{10} (x^2 - 2)$
34. $\log_9 (4x) - \log_9 (x - 3) + \log_9 (x^3 + 1)$
35. $3\log_2 x + \dfrac{1}{2}\log_2 x - 2\log_2 (x + 1)$
36. $2\log_5 x + \dfrac{1}{3}\log_5 x - 3\log_5 (x + 5)$
37. $2\log_8 x - \dfrac{2}{3}\log_8 x + 4\log_8 x$
38. $5\log_6 x - \dfrac{3}{4}\log_6 x + 3\log_6 x$

MIXED PRACTICE

Write each expression as a sum or difference of multiples of logarithms. Assume that variables represent positive numbers. See Example 5.

39. $\log_3 \dfrac{4y}{5}$
40. $\log_7 \dfrac{5x}{4}$
41. $\log_4 \dfrac{5}{9z}$
42. $\log_9 \dfrac{7}{8y}$
43. $\log_2 \dfrac{x^3}{y}$
44. $\log_5 \dfrac{x}{y^4}$
45. $\log_b \sqrt{7x}$
46. $\log_b \sqrt{\dfrac{3}{y}}$
47. $\log_6 x^4 y^5$
48. $\log_2 y^3 z$
49. $\log_5 x^3 (x + 1)$
50. $\log_3 x^2 (x - 9)$
51. $\log_6 \dfrac{x^2}{x + 3}$
52. $\log_3 \dfrac{(x + 5)^2}{x}$

If $\log_b 3 = 0.5$ and $\log_b 5 = 0.7$, evaluate each expression. See Example 6.

53. $\log_b 15$
54. $\log_b 25$
55. $\log_b \dfrac{5}{3}$
56. $\log_b \dfrac{3}{5}$
57. $\log_b \sqrt{5}$
58. $\log_b \sqrt[4]{3}$

If $\log_b 2 = 0.43$ and $\log_b 3 = 0.68$, evaluate each expression. See Example 6.

59. $\log_b 8$

60. $\log_b 81$

61. $\log_b \dfrac{3}{9}$

62. $\log_b \dfrac{4}{32}$

63. $\log_b \sqrt{\dfrac{2}{3}}$

64. $\log_b \sqrt{\dfrac{3}{2}}$

REVIEW AND PREVIEW

Graph both functions on the same set of axes. See Sections 9.3 and 9.5.

65. $y = 10^x$

66. $y = \log_{10} x.$

Evaluate each expression. See Section 9.5.

67. $\log_{10} 100$

68. $\log_{10} \dfrac{1}{10}$

69. $\log_7 7^2$

70. $\log_7 \sqrt{7}$

CONCEPT EXTENSIONS

Solve. See the Concept Checks in this section.

71. Which of the following is the correct way to rewrite $\log_3 \dfrac{14}{11}$?

a. $\dfrac{\log_3 14}{\log_3 11}$

b. $\log_3 14 - \log_3 11$

c. $\log_3 (14 - 11)$

d. $\log_3 154$

72. Which of the following is the correct way to rewrite $\log_9 \dfrac{21}{3}$?

a. $\log_9 7$

b. $\log_9(21 - 3)$

c. $\dfrac{\log_9 21}{\log_9 3}$

d. $\log_9 21 - \log_9 3$

Answer the following true or false. Study your logarithm properties carefully before answering.

73. $\log_2 x^3 = 3 \log_2 x$

74. $\log_3(x + y) = \log_3 x + \log_3 y$

75. $\dfrac{\log_7 10}{\log_7 5} = \log_7 2$

76. $\log_7 \dfrac{14}{8} = \log_7 14 - \log_7 8$

77. $\dfrac{\log_7 x}{\log_7 y} = (\log_7 x) - (\log_7 y)$

78. $(\log_3 6) \cdot (\log_3 4) = \log_3 24$

79. It is true that $\log_b 8 = \log_b(8 \cdot 1) = \log_b 8 + \log_b 1$. Explain how $\log_b 8$ can equal $\log_b 8 + \log_b 1$.

80. It is true that $\log_b 7 = \log_b \dfrac{7}{1} = \log_b 7 - \log_b 1$. Explain how $\log_b 7$ can equal $\log_b 7 - \log_b 1$.

Integrated Review FUNCTIONS AND PROPERTIES OF LOGARITHMS

Sections 9.1–9.6

If $f(x) = x - 6$ and $g(x) = x^2 + 1$, find each value.

1. $(f + g)(x)$

2. $(f - g)(x)$

3. $(f \cdot g)(x)$

4. $\left(\dfrac{f}{g}\right)(x)$

If $f(x) = \sqrt{x}$ and $g(x) = 3x - 1$, find each function.

5. $(f \circ g)(x)$

6. $(g \circ f)(x)$

Determine whether each is a one-to-one function. If it is, find its inverse.

7. $f = \{(-2, 6), (4, 8), (2, -6), (3, 3)\}$

8. $g = \{(4, 2), (-1, 3), (5, 3), (7, 1)\}$

Determine from the graph whether each function is one-to-one.

9.

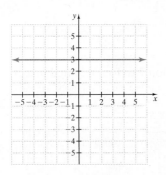

10.

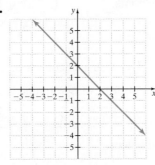

11.

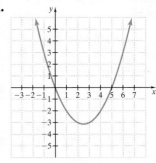

Each function listed is one-to-one. Find the inverse of each function.

12. $f(x) = 3x$

13. $f(x) = x + 4$

14. $f(x) = 5x - 1$

15. $f(x) = 3x + 2$

Graph each function.

16. $y = \left(\dfrac{1}{2}\right)^x$ **17.** $y = 2^x + 1$ **18.** $y = \log_3 x$ **19.** $y = \log_{1/3} x$

Solve.

20. $2^x = 8$ **21.** $9 = 3^{x-5}$ **22.** $4^{x-1} = 8^{x+2}$ **23.** $25^x = 125^{x-1}$

24. $\log_4 16 = x$ **25.** $\log_{49} 7 = x$ **26.** $\log_2 x = 5$ **27.** $\log_x 64 = 3$

28. $\log_x \dfrac{1}{125} = -3$ **29.** $\log_3 x = -2$

Write each as a single logarithm.

30. $5 \log_2 x$ **31.** $x \log_2 5$ **32.** $3 \log_5 x - 5 \log_5 y$ **33.** $9 \log_5 x + 3 \log_5 y$

34. $\log_2 x + \log_2(x - 3) - \log_2(x^2 + 4)$ **35.** $\log_3 y - \log_3(y + 2) + \log_3(y^3 + 11)$

Write each expression as sums or differences of multiples of logarithms.

36. $\log_7 \dfrac{9x^2}{y}$ **37.** $\log_6 \dfrac{5y}{z^2}$

38. An unusually wet spring has caused the size of the mosquito population in a community to increase by 6% each day. If an estimated 100,000 mosquitoes are in the community on April 1, find how many mosquitoes will inhabit the community on April 17. Round to the nearest thousand.

9.7   Common Logarithms, Natural Logarithms, and Change of Base

OBJECTIVES

1. Identify Common Logarithms and Approximate Them by Calculator.

2. Evaluate Common Logarithms of Powers of 10. ▷

3. Identify Natural Logarithms and Approximate Them by Calculator. ▷

4. Evaluate Natural Logarithms of Powers of *e*. ▷

5. Use the Change of Base Formula. ▷

In this section, we look closely at two particular logarithmic bases. These two logarithmic bases are used so frequently that logarithms to their bases are given special names. **Common logarithms** are logarithms to base 10. **Natural logarithms** are logarithms to base *e*, which we introduce in this section. The work in this section is based on the use of a calculator that has both the common "log" ⟨LOG⟩ and the natural "log" ⟨LN⟩ keys.

OBJECTIVE

1   Approximating Common Logarithms ▷

Logarithms to base 10, common logarithms, are used frequently because our number system is a base 10 decimal system. The notation $\log x$ means the same as $\log_{10} x$.

> **Common Logarithms**
>
> $$\log x \text{ means } \log_{10} x$$

EXAMPLE 1   Use a calculator to approximate $\log 7$ to four decimal places.

Solution   Press the following sequence of keys.

$$\boxed{7}\ \boxed{\text{LOG}} \quad \text{or} \quad \boxed{\text{LOG}}\ \boxed{7}\ \boxed{\text{ENTER}}$$

To four decimal places,

$$\log 7 \approx 0.8451$$

PRACTICE

1   Use a calculator to approximate $\log 15$ to four decimal places.

OBJECTIVE

2 Evaluating Common Logarithms of Powers of 10

To evaluate the common log of a power of 10, a calculator is not needed. According to the property of logarithms,

$$\log_b b^x = x$$

It follows that if b is replaced with 10, we have

$$\log 10^x = x$$

▶ Helpful Hint

Remember that the understood base here is 10.

EXAMPLE 2   Find the exact value of each logarithm.

a. $\log 10$ **b.** $\log 1000$ **c.** $\log \dfrac{1}{10}$ **d.** $\log \sqrt{10}$

Solution

a. $\log 10 = \log 10^1 = 1$ **b.** $\log 1000 = \log 10^3 = 3$

c. $\log \dfrac{1}{10} = \log 10^{-1} = -1$ **d.** $\log \sqrt{10} = \log 10^{1/2} = \dfrac{1}{2}$

PRACTICE

2   Find the exact value of each logarithm.

a. $\log \dfrac{1}{100}$ **b.** $\log 100,000$ **c.** $\log \sqrt[5]{10}$ **d.** $\log 0.001$

As we will soon see, equations containing common logarithms are useful models of many natural phenomena.

EXAMPLE 3   Solve $\log x = 1.2$ for x. Give an exact solution and then approximate the solution to four decimal places.

Solution   Remember that the base of a common logarithm is understood to be 10.

$$\log x = 1.2$$

▶ Helpful Hint

The understood base is 10.

$$10^{1.2} = x \qquad \text{Write with exponential notation.}$$

The exact solution is $10^{1.2}$. To four decimal places, $x \approx 15.8489$.

PRACTICE

3   Solve $\log x = 3.4$ for x. Give an exact solution, and then approximate the solution to four decimal places.

The Richter scale measures the intensity, or magnitude, of an earthquake. The formula for the magnitude R of an earthquake is $R = \log\left(\dfrac{a}{T}\right) + B$, where a is the amplitude in micrometers of the vertical motion of the ground at the recording station, T is the number of seconds between successive seismic waves, and B is an adjustment factor that takes into account the weakening of the seismic wave as the distance increases from the epicenter of the earthquake.

EXAMPLE 4   **Finding the Magnitude of an Earthquake**

Find an earthquake's magnitude on the Richter scale if a recording station measures an amplitude of 300 micrometers and 2.5 seconds between waves. Assume that B is 4.2. Approximate the solution to the nearest tenth.

Solution   Substitute the known values into the formula for earthquake intensity.

$$R = \log\left(\frac{a}{T}\right) + B \qquad \text{Richter scale formula}$$

$$= \log\left(\frac{300}{2.5}\right) + 4.2 \quad \text{Let } a = 300, T = 2.5, \text{ and } B = 4.2.$$

$$= \log(120) + 4.2$$

$$\approx 2.1 + 4.2 \qquad \text{Approximate log 120 by 2.1.}$$

$$= 6.3$$

This earthquake had a magnitude of 6.3 on the Richter scale.

PRACTICE

4   Find an earthquake's magnitude on the Richter scale if a recording station measures an amplitude of 450 micrometers and 4.2 seconds between waves with $B = 3.6$. Approximate the solution to the nearest tenth.

OBJECTIVE

3   **Approximating Natural Logarithms**

Natural logarithms are also frequently used, especially to describe natural events hence the label "natural logarithm." Natural logarithms are logarithms to the base e, which is a constant approximately equal to 2.7183. The number e is an irrational number, as is π. The notation $\log_e x$ is usually abbreviated to $\ln x$. (The abbreviation ln is read "el en.")

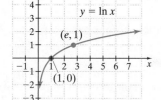

> **Natural Logarithms**
>
> $$\ln x \text{ means } \log_e x$$

The graph of $y = \ln x$ is shown to the left.

EXAMPLE 5   Use a calculator to approximate ln 8 to four decimal places.

Solution   Press the following sequence of keys.

$$\boxed{8}\ \boxed{\text{LN}} \quad \text{or} \quad \boxed{\text{LN}}\ \boxed{8}\ \boxed{\text{ENTER}}$$

To four decimal places,

$$\ln 8 \approx 2.0794$$

PRACTICE

5   Use a calculator to approximate ln 13 to four decimal places.

OBJECTIVE

4   **Evaluating Natural Logarithms of Powers of e.**

As a result of the property $\log_b b^x = x$, we know that $\log_e e^x = x$, or $\boldsymbol{\ln e^x = x.}$
    Since $\ln e^x = x$, $\ln e^5 = 5$, $\ln e^{22} = 22$, and so on. Also,

$$\ln e^1 = 1 \text{ or simply } \ln e = 1.$$

That is why the graph of $y = \ln x$ shown above in the margin passes through $(e, 1)$.
    If $x = e$, then $y = \ln e = 1$, thus the ordered pair is $(e, 1)$.

EXAMPLE 6   Find the exact value of each natural logarithm.

a. $\ln e^3$ 　　　　　**b.** $\ln \sqrt[5]{e}$

Solution

a. $\ln e^3 = 3$ 　　**b.** $\ln \sqrt[5]{e} = \ln e^{1/5} = \dfrac{1}{5}$ 　　□

PRACTICE
6   Find the exact value of each natural logarithm.

a. $\ln e^4$ 　　　　　**b.** $\ln \sqrt[3]{e}$

EXAMPLE 7   Solve $\ln 3x = 5$. Give an exact solution and then approximate the solution to four decimal places.

Solution   Remember that the base of a natural logarithm is understood to be e.

▶ **Helpful Hint**
The understood base is e.

$$\ln 3x = 5$$
$$e^5 = 3x \quad \text{Write with exponential notation.}$$
$$\frac{e^5}{3} = x \quad \text{Solve for } x.$$

The exact solution is $\dfrac{e^5}{3}$. To four decimal places,

$$x \approx 49.4711. \qquad \square$$

PRACTICE
7   Solve $\ln 5x = 8$. Give an exact solution and then approximate the solution to four decimal places.

Recall from Section 9.3 the formula $A = P\left(1 + \dfrac{r}{n}\right)^{nt}$ for compound interest, where n represents the number of compoundings per year. When interest is compounded continuously, the formula $A = Pe^{rt}$ is used, where r is the annual interest rate, and interest is compounded continuously for t years.

EXAMPLE 8   **Finding Final Loan Payment**
Find the amount owed at the end of 5 years if $1600 is loaned at a rate of 9% compounded continuously.

Solution   Use the formula $A = Pe^{rt}$, where

$$P = \$1600 \text{ (the amount of the loan)}$$
$$r = 9\% = 0.09 \text{ (the rate of interest)}$$
$$t = 5 \text{ (the 5-year duration of the loan)}$$
$$A = Pe^{rt}$$
$$= 1600e^{0.09(5)} \quad \text{Substitute in known values.}$$
$$= 1600e^{0.45}$$

Now we can use a calculator to approximate the solution.

$$A \approx 2509.30$$

The total amount of money owed is $2509.30. 　　□

PRACTICE
8   Find the amount owed at the end of 4 years if $2400 is borrowed at a rate of 6% compounded continuously.

OBJECTIVE

5 Using the Change of Base Formula

Calculators are handy tools for approximating natural and common logarithms. Unfortunately, some calculators cannot be used to approximate logarithms to bases other than e or 10—at least not directly. In such cases, we use the change of base formula.

> **Change of Base**
>
> If a, b, and c are positive real numbers and neither b nor c is 1, then
>
> $$\log_b a = \frac{\log_c a}{\log_c b}$$

EXAMPLE 9 Approximate $\log_5 3$ to four decimal places.

Solution Use the change of base property to write $\log_5 3$ as a quotient of logarithms to base 10.

$$\log_5 3 = \frac{\log 3}{\log 5}$$ Use the change of base property. In the change of base property, we let $a = 3$, $b = 5$, and $c = 10$.

$$\approx \frac{0.4771213}{0.69897}$$ Approximate logarithms by calculator.

$$\approx 0.6826063$$ Simplify by calculator.

To four decimal places, $\log_5 3 \approx 0.6826$. □

PRACTICE

9 Approximate $\log_8 5$ to four decimal places.

✓CONCEPT CHECK

If a graphing calculator cannot directly evaluate logarithms to base 5, describe how you could use the graphing calculator to graph the function $f(x) = \log_5 x$.

Vocabulary, Readiness & Video Check

Use the choices to fill in each blank.

1. The base of $\log 7$ is _____.
   a. e **b.** 7   **c.** 10   **d.** no answer

2. The base of $\ln 7$ is ___.
   a. e **b.** 7   **c.** 10   **d.** no answer

3. $\log_{10} 10^7 =$ _____.
   a. e **b.** 7   **c.** 10   **d.** no answer

4. $\log_7 1 =$ _____.
   a. e **b.** 7   **c.** 10   **d.** 0

5. $\log_e e^5 =$ _____.
   a. e **b.** 5   **c.** 0   **d.** 1

6. Study exercise 5 to the left. Then answer: $\ln e^5 =$ _____.
   a. e **b.** 5   **c.** 0   **d.** 1

7. $\log_2 7 =$ _____ (There may be more than one answer.)

   a. $\dfrac{\log 7}{\log 2}$ **b.** $\dfrac{\ln 7}{\ln 2}$ **c.** $\dfrac{\log 2}{\log 7}$ **d.** $\log \dfrac{7}{2}$

Answer to Concept Check:
$$f(x) = \frac{\log x}{\log 5}$$

Martin-Gay Interactive Videos

See Video 9.7

Watch the section lecture video and answer the following questions.

OBJECTIVE
1
8. From ⊞ Example 1 and the lecture before, what is the understood base of a common logarithm?

OBJECTIVE
2
9. From ⊞ Example 2, why can you find exact values of common logarithms of powers of 10?

OBJECTIVE
3
10. From ⊞ Example 4 and the lecture before, what is the understood base of a natural logarithm?

OBJECTIVE
4
11. In ⊞ Examples 5 and 6, consider how the expression is rewritten and the resulting answer. What logarithm property is actually used here?

OBJECTIVE
5
12. From ⊞ Example 8, what two equivalent fractions will give you the exact value of $\log_6 4$?

9.7 Exercise Set MyMathLab®

MIXED PRACTICE

Use a calculator to approximate each logarithm to four decimal places. See Examples 1 and 5.

1. $\log 8$ | **2.** $\log 6$

3. $\log 2.31$ | **4.** $\log 4.86$

5. $\ln 2$ | **6.** $\ln 3$

7. $\ln 0.0716$ | **8.** $\ln 0.0032$

9. $\log 12.6$ | **10.** $\log 25.9$

11. $\ln 5$ | **12.** $\ln 7$

13. $\log 41.5$ | **14.** $\ln 41.5$

MIXED PRACTICE

Find the exact value. See Examples 2 and 6.

15. $\log 100$ | **16.** $\log 10{,}000$

17. $\log \dfrac{1}{1000}$ | **18.** $\log \dfrac{1}{100}$

19. $\ln e^2$ | **20.** $\ln e^4$

21. $\ln \sqrt[4]{e}$ | **22.** $\ln \sqrt[5]{e}$

23. $\log 10^3$ | **24.** $\log 10^7$

25. $\ln e^{-7}$ | **26.** $\ln e^{-5}$

27. $\log 0.0001$ | **28.** $\log 0.001$

29. $\ln \sqrt{e}$ | **30.** $\log \sqrt{10}$

Solve each equation for x. Give an exact solution and a four-decimal-place approximation. See Examples 3 and 7.

31. $\ln 2x = 7$ | **32.** $\ln 5x = 9$

33. $\log x = 1.3$ | **34.** $\log x = 2.1$

35. $\log 2x = 1.1$ | **36.** $\log 3x = 1.3$

37. $\ln x = 1.4$ | **38.** $\ln x = 2.1$

39. $\ln(3x - 4) = 2.3$

40. $\ln(2x + 5) = 3.4$

41. $\log x = 2.3$

42. $\log x = 3.1$

43. $\ln x = -2.3$

44. $\ln x = -3.7$

45. $\log(2x + 1) = -0.5$

46. $\log(3x - 2) = -0.8$

47. $\ln 4x = 0.18$

48. $\ln 3x = 0.76$

Approximate each logarithm to four decimal places. See Example 9.

49. $\log_2 3$ | **50.** $\log_3 2$

51. $\log_{1/2} 5$ | **52.** $\log_{1/3} 2$

53. $\log_4 9$ | **54.** $\log_9 4$

55. $\log_3 \dfrac{1}{6}$ | **56.** $\log_6 \dfrac{2}{3}$

57. $\log_8 6$ | **58.** $\log_6 8$

Use the formula $R = \log\left(\dfrac{a}{T}\right) + B$ to find the intensity R on the Richter scale of the earthquakes that fit the descriptions given. Round answers to one decimal place. See Example 4.

59. Amplitude a is 200 micrometers, time T between waves is 1.6 seconds, and B is 2.1.

60. Amplitude a is 150 micrometers, time T between waves is 3.6 seconds, and B is 1.9.

61. Amplitude a is 400 micrometers, time T between waves is 2.6 seconds, and B is 3.1.

62. Amplitude a is 450 micrometers, time T between waves is 4.2 seconds, and B is 2.7.

Use the formula $A = Pe^{rt}$ to solve. See Example 8.

63. Find how much money Dana Jones has after 12 years if $1400 is invested at 8% interest compounded continuously.

64. Determine the amount in an account in which $3500 earns 6% interest compounded continuously for 1 year.

65. Find the amount of money Barbara Mack owes at the end of 4 years if 6% interest is compounded continuously on her $2000 debt.

66. Find the amount of money for which a $2500 certificate of deposit is redeemable if it has been paying 10% interest compounded continuously for 3 years.

REVIEW AND PREVIEW

Solve each equation for x. See Sections 2.1, 2.3, and 5.8.

67. $6x - 3(2 - 5x) = 6$

68. $2x + 3 = 5 - 2(3x - 1)$

69. $2x + 3y = 6x$

70. $4x - 8y = 10x$

71. $x^2 + 7x = -6$

72. $x^2 + 4x = 12$

Solve each system of equations. See Section 4.1.

73. $\begin{cases} x + 2y = -4 \\ 3x - y = 9 \end{cases}$

74. $\begin{cases} 5x + y = 5 \\ -3x - 2y = -10 \end{cases}$

CONCEPT EXTENSIONS

75. Use a calculator to try to approximate log 0. Describe what happens and explain why.

76. Use a calculator to try to approximate ln 0. Describe what happens and explain why.

77. Without using a calculator, explain which of log 50 or ln 50 must be larger and why.

78. Without using a calculator, explain which of $\log 50^{-1}$ or $\ln 50^{-1}$ must be larger and why.

Graph each function by finding ordered pair solutions, plotting the solutions, and then drawing a smooth curve through the plotted points.

79. $f(x) = e^x$

80. $f(x) = e^{2x}$

81. $f(x) = e^{-3x}$

82. $f(x) = e^{-x}$

83. $f(x) = e^x + 2$

84. $f(x) = e^x - 3$

85. $f(x) = e^{x-1}$

86. $f(x) = e^{x+4}$

87. $f(x) = 3e^x$

88. $f(x) = -2e^x$

89. $f(x) = \ln x$

90. $f(x) = \log x$

91. $f(x) = -2 \log x$

92. $f(x) = 3 \ln x$

93. $f(x) = \log(x + 2)$

94. $f(x) = \log(x - 2)$

95. $f(x) = \ln x - 3$

96. $f(x) = \ln x + 3$

97. Graph $f(x) = e^x$ (Exercise 79), $f(x) = e^x + 2$ (Exercise 83), and $f(x) = e^x - 3$ (Exercise 84) on the same screen. Discuss any trends shown on the graphs.

98. Graph $f(x) = \ln x$ (Exercise 89), $f(x) = \ln x - 3$ (Exercise 95),and $f(x) = \ln x + 3$ (Exercise 96) on the same screen. Discuss any trends shown on the graphs.

9.8 Exponential and Logarithmic Equations and Problem Solving

OBJECTIVES

1 Solve Exponential Equations.

2 Solve Logarithmic Equations.

3 Solve Problems That Can Be Modeled by Exponential and Logarithmic Equations.

OBJECTIVE

1 Solving Exponential Equations

In Section 9.3 we solved exponential equations such as $2^x = 16$ by writing 16 as a power of 2 and applying the uniqueness of b^x.

$$2^x = 16$$
$$2^x = 2^4 \quad \text{Write 16 as } 2^4.$$
$$x = 4 \quad \text{Use the uniqueness of } b^x.$$

Solving the equation in this manner is possible since 16 is a power of 2. If solving an equation such as $2^x = a$ *number,* where the number is not a power of 2, we use logarithms. For example, to solve an equation such as $3^x = 7$, we use the fact that $f(x) = \log_b x$ is a one-to-one function. Another way of stating this fact is as a property of equality.

> **Logarithm Property of Equality**
>
> Let a, b, and c be real numbers such that $\log_b a$ and $\log_b c$ are real numbers and b is not 1. Then
>
> $$\log_b a = \log_b c \text{ is equivalent to } a = c$$

EXAMPLE 1 Solve: $3^x = 7$.

Solution To solve, we use the logarithm property of equality and take the logarithm of both sides. For this example, we use the common logarithm.

$$3^x = 7$$
$$\log 3^x = \log 7 \quad \text{Take the common logarithm of both sides.}$$
$$x \log 3 = \log 7 \quad \text{Apply the power property of logarithms.}$$
$$x = \frac{\log 7}{\log 3} \quad \text{Divide both sides by log 3.}$$

The exact solution is $\dfrac{\log 7}{\log 3}$. If a decimal approximation is preferred,

$$\frac{\log 7}{\log 3} \approx \frac{0.845098}{0.4771213} \approx 1.7712 \text{ to four decimal places.}$$

The solution is $\dfrac{\log 7}{\log 3}$, or *approximately* 1.7712.   □

PRACTICE
1   Solve: $5^x = 9$.

OBJECTIVE
2   **Solving Logarithmic Equations** ▶

By applying the appropriate properties of logarithms, we can solve a broad variety of logarithmic equations.

EXAMPLE 2   Solve: $\log_4(x - 2) = 2$.

Solution   Notice that $x - 2$ must be positive, so x must be greater than 2. With this in mind, we first write the equation with exponential notation.

$$\log_4(x - 2) = 2$$
$$4^2 = x - 2$$
$$16 = x - 2$$
$$18 = x \qquad \text{Add 2 to both sides.}$$

Check: To check, we replace x with 18 in the original equation.

$$\log_4(x - 2) = 2$$
$$\log_4(18 - 2) \stackrel{?}{=} 2 \qquad \text{Let } x = 18.$$
$$\log_4 16 \stackrel{?}{=} 2$$
$$4^2 = 16 \qquad \text{True}$$

The solution is 18.   □

PRACTICE
2   Solve: $\log_2(x - 1) = 5$.

EXAMPLE 3   Solve: $\log_2 x + \log_2(x - 1) = 1$.

Solution   Notice that $x - 1$ must be positive, so x must be greater than 1. We use the product property on the left side of the equation.

$$\log_2 x + \log_2(x - 1) = 1$$
$$\log_2 x(x - 1) = 1 \qquad \text{Apply the product property.}$$
$$\log_2(x^2 - x) = 1$$

Next we write the equation with exponential notation and solve for x.

$$2^1 = x^2 - x$$
$$0 = x^2 - x - 2 \qquad\qquad \text{Subtract 2 from both sides.}$$
$$0 = (x - 2)(x + 1) \qquad\quad \text{Factor.}$$
$$0 = x - 2 \quad \text{or} \quad 0 = x + 1 \quad \text{Set each factor equal to 0.}$$
$$2 = x \qquad\qquad -1 = x$$

Recall that -1 cannot be a solution because x must be greater than 1. If we forgot this, we would still reject -1 after checking. To see this, we replace x with -1 in the original equation.

$$\log_2 x + \log_2(x - 1) = 1$$
$$\log_2(-1) + \log_2(-1 - 1) \stackrel{?}{=} 1 \quad \text{Let } x = -1.$$

(Continued on next page)

Because the logarithm of a negative number is undefined, -1 is rejected. Check to see that the solution is 2. ☐

PRACTICE

3   Solve: $\log_5 x + \log_5(x + 4) = 1$.

EXAMPLE 4   Solve: $\log(x + 2) - \log x = 2$.

We use the quotient property of logarithms on the left side of the equation.

Solution $\log(x + 2) - \log x = 2$

$$\log\frac{x + 2}{x} = 2 \qquad \text{Apply the quotient property.}$$

$$10^2 = \frac{x + 2}{x} \qquad \text{Write using exponential notation.}$$

$$100 = \frac{x + 2}{x} \qquad \text{Simplify.}$$

$$100x = x + 2 \qquad \text{Multiply both sides by } x.$$

$$99x = 2 \qquad \text{Subtract } x \text{ from both sides.}$$

$$x = \frac{2}{99} \qquad \text{Divide both sides by 99.}$$

Verify that the solution is $\frac{2}{99}$. ☐

PRACTICE

4   Solve: $\log(x + 3) - \log x = 1$.

OBJECTIVE

3   Solving Problems Modeled by Exponential and Logarithmic Equations ▶

Logarithmic and exponential functions are used in a variety of scientific, technical, and business settings. A few examples follow.

EXAMPLE 5   Estimating Population Size

The population size y of a community of lemmings varies according to the relationship $y = y_0 e^{0.15t}$. In this formula, t is time in months, and y_0 is the initial population at time 0. Estimate the population after 6 months if there were originally 5000 lemmings.

Solution   We substitute 5000 for y_0 and 6 for t.

$$y = y_0 e^{0.15t}$$

$$= 5000 e^{0.15(6)} \qquad \text{Let } t = 6 \text{ and } y_0 = 5000.$$

$$= 5000 e^{0.9} \qquad \text{Multiply.}$$

Using a calculator, we find that $y \approx 12{,}298.016$. In 6 months, the population will be approximately 12,300 lemmings. ☐

PRACTICE

5   The population size y of a group of rabbits varies according to the relationship $y = y_0 e^{0.916t}$. In this formula, t is time in years and y_0 is the initial population at time $t = 0$. Estimate the population in three years if there were originally 60 rabbits.

EXAMPLE 6   **Doubling an Investment**

How long does it take an investment of $2000 to double if it is invested at 5% interest compounded quarterly? The necessary formula is $A = P\left(1 + \dfrac{r}{n}\right)^{nt}$, where A is the accrued (or owed) amount, P is the principal invested, r is the annual rate of interest, n is the number of compounding periods per year, and t is the number of years.

Solution   We are given that $P = \$2000$ and $r = 5\% = 0.05$. Compounding quarterly means 4 times a year, so $n = 4$. The investment is to double, so A must be $4000. Substitute these values and solve for t.

$$A = P\left(1 + \frac{r}{n}\right)^{nt}$$

$$4000 = 2000\left(1 + \frac{0.05}{4}\right)^{4t} \qquad \text{Substitute in known values.}$$

$$4000 = 2000(1.0125)^{4t} \qquad \text{Simplify } 1 + \frac{0.05}{4}.$$

$$2 = (1.0125)^{4t} \qquad \text{Divide both sides by 2000.}$$

$$\log 2 = \log 1.0125^{4t} \qquad \text{Take the logarithm of both sides.}$$

$$\log 2 = 4t(\log 1.0125) \qquad \text{Apply the power property.}$$

$$\frac{\log 2}{4 \log 1.0125} = t \qquad \text{Divide both sides by } 4 \log 1.0125.$$

$$13.949408 \approx t \qquad \text{Approximate by calculator.}$$

Thus, it takes nearly 14 years for the money to double in value. □

PRACTICE

6   How long does it take for an investment of $3000 to double if it is invested at 7% interest compounded monthly? Round to the nearest year.

Graphing Calculator Explorations

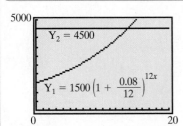

Use a graphing calculator to find how long it takes an investment of $1500 to triple if it is invested at 8% interest compounded monthly.

First, let $P = \$1500$, $r = 0.08$, and $n = 12$ (for 12 months) in the formula

$$A = P\left(1 + \frac{r}{n}\right)^{nt}$$

Notice that when the investment has tripled, the accrued amount A is $4500. Thus,

$$4500 = 1500\left(1 + \frac{0.08}{12}\right)^{12t}$$

Determine an appropriate viewing window and enter and graph the equations

$$Y_1 = 1500\left(1 + \frac{0.08}{12}\right)^{12x}$$

and

$$Y_2 = 4500$$

The point of intersection of the two curves is the solution. The x-coordinate tells how long it takes for the investment to triple.

Use a TRACE feature or an INTERSECT feature to approximate the coordinates of the point of intersection of the two curves. It takes approximately 13.78 years, or 13 years and 9 months, for the investment to triple in value to $4500.

Use this graphical solution method to solve each problem. Round each answer to the nearest hundredth.

1. Find how long it takes an investment of $5000 to grow to $6000 if it is invested at 5% interest compounded quarterly.

2. Find how long it takes an investment of $1000 to double if it is invested at 4.5% interest compounded daily. (Use 365 days in a year.)

3. Find how long it takes an investment of $10,000 to quadruple if it is invested at 6% interest compounded monthly.

4. Find how long it takes $500 to grow to $800 if it is invested at 4% interest compounded semiannually.

Vocabulary, Readiness & Video Check

Martin-Gay Interactive Videos

See Video 9.8

Watch the section lecture video and answer the following questions.

OBJECTIVE 1
1. From the lecture before Example 1, explain why $\ln(4x - 2) = \ln 3$ is equivalent to $4x - 2 = 3$.

OBJECTIVE 2
2. Why is the possible solution of -8 rejected in Example 3?

OBJECTIVE 3
3. For Example 4, write the equation and find the number of years it takes $1000 to double at 7% interest compounded monthly. Explain the similarity to the answer to Example 4. Round your answer to the nearest tenth.

9.8 Exercise Set

MyMathLab®

Solve each equation. Give an exact solution and approximate the solution to four decimal places. See Example 1.

1. $3^x = 6$

2. $4^x = 7$

3. $3^{2x} = 3.8$

4. $5^{3x} = 5.6$

5. $2^{x-3} = 5$

6. $8^{x-2} = 12$

7. $9^x = 5$

8. $3^x = 11$

9. $4^{x+7} = 3$

10. $6^{x+3} = 2$

MIXED PRACTICE

Solve each equation. See Examples 1 through 4.

11. $\log_2(x + 5) = 4$

12. $\log_2(x - 5) = 3$

13. $\log_4 2 + \log_4 x = 0$

14. $\log_3 5 + \log_3 x = 1$

15. $\log_2 6 - \log_2 x = 3$

16. $\log_4 10 - \log_4 x = 2$

17. $\log_6(x^2 - x) = 1$

18. $\log_2(x^2 + x) = 1$

19. $\log_4 x + \log_4(x + 6) = 2$

20. $\log_3 x + \log_3(x + 6) = 3$

21. $\log_5(x + 3) - \log_5 x = 2$

22. $\log_6(x + 2) - \log_6 x = 2$

23. $7^{3x-4} = 11$

24. $5^{2x-6} = 12$

25. $\log_4(x^2 - 3x) = 1$

26. $\log_8(x^2 - 2x) = 1$

27. $e^{6x} = 5$

28. $e^{2x} = 8$

29. $\log_3 x^2 = 4$

30. $\log_2 x^2 = 6$

31. $\ln 5 + \ln x = 0$

32. $\ln 3 + \ln(x - 1) = 0$

33. $3 \log x - \log x^2 = 2$

34. $2 \log x - \log x = 3$

35. $\log_4 x - \log_4(2x - 3) = 3$

36. $\log_2 x - \log_2(3x + 5) = 4$

37. $\log_2 x + \log_2(3x + 1) = 1$

38. $\log_3 x + \log_3(x - 8) = 2$

39. $\log_2 x + \log_2(x + 5) = 1$

40. $\log_4 x + \log_4(x + 7) = 1$

Solve. See Example 5.

41. The size of the wolf population at Isle Royale National Park increases according to the formula $y = y_0 e^{0.043t}$. In this formula, t is time in years and y_0 is the initial population at time 0. If the size of the current population is 83 wolves, find how many there should be in 5 years. Round to the nearest whole number.

42. The number of victims of a flu epidemic is increasing according to the formula $y = y_0 e^{0.075t}$. In this formula, t is time in weeks and y_0 is the given population at time 0. If 20,000 people are currently infected, how many might be infected in 3 weeks? Round to the nearest whole number.

43. The population of the Cook Islands is decreasing according to the formula $y = y_0 e^{-0.0277t}$. In this formula, t is the time in years and y_0 is the initial population at time 0. If the size of the population in 2010 was 11,488, use the formula to predict the population of the Cook Islands in the year 2025. Round to the nearest whole number. (*Source: The World Almanac*)

44. The population of Saint Barthelemy is decreasing according to the formula $y = y_0 e^{-0.0034t}$. In this formula, t is the time in years and y_0 is the initial population at time 0. If the size of the population in 2010 was 6852, use the formula to predict the population of Saint Barthelemy in the year 2025. Round to the nearest whole number. (*Source: The World Almanac*)

Use the formula $A = P\left(1 + \dfrac{r}{n}\right)^{nt}$ to solve these compound interest problems. Round to the nearest tenth. See Example 6.

▶ **45.** Find how long it takes $600 to double if it is invested at 7% interest compounded monthly.

46. Find how long it takes $600 to double if it is invested at 12% interest compounded monthly.

47. Find how long it takes a $1200 investment to earn $200 interest if it is invested at 9% interest compounded quarterly.

48. Find how long it takes a $1500 investment to earn $200 interest if it is invested at 10% compounded semiannually.

49. Find how long it takes $1000 to double if it is invested at 8% interest compounded semiannually.

50. Find how long it takes $1000 to double if it is invested at 8% interest compounded monthly.

The formula $w = 0.00185h^{2.67}$ is used to estimate the normal weight w of a boy h inches tall. Use this formula to solve the height–weight problems. Round to the nearest tenth.

51. Find the expected weight of a boy who is 35 inches tall.

52. Find the expected weight of a boy who is 43 inches tall.

53. Find the expected height of a boy who weighs 85 pounds.

54. Find the expected height of a boy who weighs 140 pounds.

The formula $P = 14.7e^{-0.21x}$ gives the average atmospheric pressure P, in pounds per square inch, at an altitude x, in miles above sea level. Use this formula to solve these pressure problems. Round answers to the nearest tenth.

55. Find the average atmospheric pressure of Denver, which is 1 mile above sea level.

56. Find the average atmospheric pressure of Pikes Peak, which is 2.7 miles above sea level.

57. Find the elevation of a Delta jet if the atmospheric pressure outside the jet is 7.5 lb/sq in.

58. Find the elevation of a remote Himalayan peak if the atmospheric pressure atop the peak is 6.5 lb/sq in.

Psychologists call the graph of the formula $t = \dfrac{1}{c}\ln\left(\dfrac{A}{A - N}\right)$ the learning curve, since the formula relates time t passed, in weeks, to a measure N of learning achieved, to a measure A of maximum learning possible, and to a measure c of an individual's learning style. Round to the nearest week.

59. Norman is learning to type. If he wants to type at a rate of 50 words per minute (N is 50) and his expected maximum rate is 75 words per minute (A is 75), find how many weeks it should take him to achieve his goal. Assume that c is 0.09.

60. An experiment with teaching chimpanzees sign language shows that a typical chimp can master a maximum of 65 signs. Find how many weeks it should take a chimpanzee to master 30 signs if c is 0.03.

61. Janine is working on her dictation skills. She wants to take dictation at a rate of 150 words per minute and believes that the maximum rate she can hope for is 210 words per minute. Find how many weeks it should take her to achieve the 150 words per minute level if c is 0.07.

62. A psychologist is measuring human capability to memorize nonsense syllables. Find how many weeks it should take a subject to learn 15 nonsense syllables if the maximum possible to learn is 24 syllables and c is 0.17.

REVIEW AND PREVIEW

If $x = -2$, $y = 0$, and $z = 3$, find the value of each expression. See Section 1.3.

63. $\dfrac{x^2 - y + 2z}{3x}$

64. $\dfrac{x^3 - 2y + z}{2z}$

65. $\dfrac{3z - 4x + y}{x + 2z}$

66. $\dfrac{4y - 3x + z}{2x + y}$

Find the inverse function of each one-to-one function. See Section 9.2.

67. $f(x) = 5x + 2$

68. $f(x) = \dfrac{x - 3}{4}$

CONCEPT EXTENSIONS

The formula $y = y_0 e^{kt}$ gives the population size y of a population that experiences an annual rate of population growth k (given as a decimal). In this formula, t is time in years and y_0 is the initial population at time 0. Use this formula to solve Exercises 69 and 70.

69. In 2010, the population of Michigan was approximately 9,939,000 and decreasing according to the formula $y = y_0 e^{-0.003t}$. Assume that the population continues to decrease according to the given formula and predict how many years after which the population of Michigan will be 9,500,000. (*Hint:* Let $y_0 = 9,939,000$; $y = 9,500,000$, and solve for t.) (*Source:* U.S. Bureau of the Census)

70. In 2010, the population of Illinois was approximately 12,830,000 and increasing according to the formula $y = y_0 e^{0.005t}$. Assume that the population continues to increase according to the given formula and predict how many years after which the population of Illinois will be 13,500,000. (See the Hint for Exercise 69.) (*Source:* U.S. Bureau of the Census)

71. When solving a logarithmic equation, explain why you must check possible solutions in the original equation.

72. Solve $5^x = 9$ by taking the common logarithm of both sides of the equation. Next, solve this equation by taking the natural logarithm of both sides. Compare your solutions. Are they the same? Why or why not?

Use a graphing calculator to solve each equation. For example, to solve Exercise 73, let $Y_1 = e^{0.3x}$ and $Y_2 = 8$ and graph the equations. The x-value of the point of intersection is the solution. Round all solutions to two decimal places.

73. $e^{0.3x} = 8$

74. $10^{0.5x} = 7$

75. $2 \log(-5.6x + 1.3) + x + 1 = 0$

76. $\ln(1.3x - 2.1) + 3.5x - 5 = 0$

77. Check Exercise 23.

78. Check Exercise 24.

79. Check Exercise 31.

80. Check Exercise 32.

Chapter 9 Vocabulary Check

Fill in each blank with one of the words or phrases listed below. Some words or phrases may be used more than once.

| inverse | common | composition | symmetric | exponential |
| vertical | logarithmic | natural | half-life | horizontal |

1. For a one-to-one function, we can find its _____ function by switching the coordinates of the ordered pairs of the function.

2. The _____ of functions f and g is $(f \circ g)(x) = f(g(x))$.

3. A function of the form $f(x) = b^x$ is called a(n) _____ function if $b > 0$, b is not 1, and x is a real number.

4. The graphs of f and f^{-1} are _____ about the line $y = x$.

5. _____ logarithms are logarithms to base e.

6. _____ logarithms are logarithms to base 10.

7. To see whether a graph is the graph of a one-to-one function, apply the _____ line test to see whether it is a function and then apply the _____ line test to see whether it is a one-to-one function.

8. A(n) _____ function is a function that can be defined by $f(x) = \log_b x$ where x is a positive real number, b is a constant positive real number, and b is not 1.

9. _____ is the amount of time it takes for half of the amount of a substance to decay.

10. A quantity that grows or decays by the same percent at regular time periods is said to have _____ growth or decay.

Chapter 9 Highlights

DEFINITIONS AND CONCEPTS	EXAMPLES

Section 9.1 The Algebra of Functions; Composite Functions

Algebra of Functions

Sum $\quad (f + g)(x) = f(x) + g(x)$

Difference $\quad (f - g)(x) = f(x) - g(x)$

Product $\quad (f \cdot g)(x) = f(x) \cdot g(x)$

Quotient $\quad \left(\dfrac{f}{g}\right)(x) = \dfrac{f(x)}{g(x)}, g(x) \neq 0$

Composite Functions

The notation $(f \circ g)(x)$ means "f composed with g."

$$(f \circ g)(x) = f(g(x))$$
$$(g \circ f)(x) = g(f(x))$$

If $f(x) = 7x$ and $g(x) = x^2 + 1$,

$$(f + g)(x) = f(x) + g(x) = 7x + x^2 + 1$$
$$(f - g)(x) = f(x) - g(x) = 7x - (x^2 + 1)$$
$$= 7x - x^2 - 1$$
$$(f \cdot g)(x) = f(x) \cdot g(x) = 7x(x^2 + 1)$$
$$= 7x^3 + 7x$$
$$\left(\dfrac{f}{g}\right)(x) = \dfrac{f(x)}{g(x)} = \dfrac{7x}{x^2 + 1}$$

If $f(x) = x^2 + 1$ and $g(x) = x - 5$, find $(f \circ g)(x)$.

$$(f \circ g)(x) = f(g(x))$$
$$= f(x - 5)$$
$$= (x - 5)^2 + 1$$
$$= x^2 - 10x + 26$$

Section 9.2 Inverse Functions

If f is a function, then f is a **one-to-one function** only if each y-value (output) corresponds to only one x-value (input).

Horizontal Line Test

If every horizontal line intersects the graph of a function at most once, then the function is a one-to-one function.

Determine whether each graph is a one-to-one function.

A **B**

C

Graphs **A** and **C** pass the vertical line test, so only these are graphs of functions. Of graphs **A** and **C**, only graph **A** passes the horizontal line test, so only graph **A** is the graph of a one-to-one function.

The **inverse** of a one-to-one function f is the one-to-one function f^{-1} that is the set of all ordered pairs (b, a) such that (a, b) belongs to f.

To Find the Inverse of a One-to-One Function $f(x)$

Step 1. Replace $f(x)$ with y.

Step 2. Interchange x and y.

Step 3. Solve for y.

Step 4. Replace y with $f^{-1}(x)$.

Find the inverse of $f(x) = 2x + 7$.

$y = 2x + 7 \quad$ Replace $f(x)$ with y.

$x = 2y + 7 \quad$ Interchange x and y.

$2y = x - 7 \quad$ Solve for y.

$y = \dfrac{x - 7}{2}$

$f^{-1}(x) = \dfrac{x - 7}{2} \quad$ Replace y with $f^{-1}(x)$.

The inverse of $f(x) = 2x + 7$ is $f^{-1}(x) = \dfrac{x - 7}{2}$.

DEFINITIONS AND CONCEPTS	EXAMPLES

Section 9.3   Exponential Functions

A function of the form $f(x) = b^x$ is an **exponential function**, where $b > 0$, $b \neq 1$, and x is a real number.

Graph the exponential function $y = 4^x$.

x	y
-2	$\dfrac{1}{16}$
-1	$\dfrac{1}{4}$
0	1
1	4
2	16

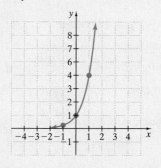

Uniqueness of b^x

If $b > 0$ and $b \neq 1$, then $b^x = b^y$ is equivalent to $x = y$.

Solve $2^{x+5} = 8$.

$2^{x+5} = 2^3$ Write 8 as 2^3.

$x + 5 = 3$ Use the uniqueness of b^x.

$x = -2$ Subtract 5 from both sides.

Section 9.4   Exponential Growth and Decay Functions

A quantity that grows or decays by the same percent at regular time periods is said to have **exponential growth** or **exponential decay.**

Exponential Growth

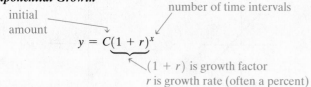

initial amount

number of time intervals

$$y = C(1 + r)^x$$

$(1 + r)$ is growth factor

r is growth rate (often a percent)

Exponential Decay

initial amount

number of time intervals

$$y = C(1 - r)^x$$

$(1 - r)$ is decay factor

r is decay rate (often a percent)

A city has a current population of 37,000 that has been increasing at a rate of 3% per year. At this rate, find the city's population in 20 years.

$$y = C(1 + r)^x$$
$$y = 37,000(1 + 0.03)^{20}$$
$$y \approx 66,826.12$$

In 20 years, the predicted population of the city is 66,826.

A city has a current population of 37,000 that has been decreasing at a rate of 3% per year. At this rate, find the city's population in 20 years.

$$y = C(1 - r)^x$$
$$y = 37,000(1 - 0.03)^{20}$$
$$y \approx 20,120.39$$

In 20 years, predicted population of the city is 20,120.

Section 9.5   Logarithmic Functions

Logarithmic Definition

If $b > 0$ and $b \neq 1$, then

$$y = \log_b x \quad \text{means} \quad x = b^y$$

for any positive number x and real number y.

Properties of Logarithms

If b is a real number, $b > 0$, and $b \neq 1$, then

$$\log_b 1 = 0, \quad \log_b b^x = x, \quad b^{\log_b x} = x$$

Logarithmic Form	Corresponding Exponential Statement
$\log_5 25 = 2$	$5^2 = 25$
$\log_9 3 = \dfrac{1}{2}$	$9^{1/2} = 3$

$$\log_5 1 = 0, \quad \log_7 7^2 = 2, \quad 3^{\log_3 6} = 6$$

DEFINITIONS AND CONCEPTS	EXAMPLES

Section 9.5   Logarithmic Functions (continued)

Logarithmic Function

If $b > 0$ and $b \neq 1$, then a **logarithmic function** is a function that can be defined as

$$f(x) = \log_b x$$

The domain of f is the set of positive real numbers, and the range of f is the set of real numbers.

Graph $y = \log_3 x$.

Write $y = \log_3 x$ as $3^y = x$. Plot the ordered pair solutions listed in the table and connect them with a smooth curve.

x	y
3	1
1	0
$\dfrac{1}{3}$	-1
$\dfrac{1}{9}$	-2

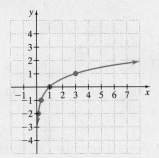

Section 9.6   Properties of Logarithms

Let x, y, and b be positive numbers and $b \neq 1$.

Product Property

$$\log_b xy = \log_b x + \log_b y$$

Quotient Property

$$\log_b \frac{x}{y} = \log_b x - \log_b y$$

Power Property

$$\log_b x^r = r \log_b x$$

Write as a single logarithm.

$2 \log_5 6 + \log_5 x - \log_5 (y + 2)$

$\quad = \log_5 6^2 + \log_5 x - \log_5 (y + 2)$ Power property

$\quad = \log_5 36 \cdot x - \log_5 (y + 2)$ Product property

$\quad = \log_5 \dfrac{36x}{y + 2}$ Quotient property

Section 9.7   Common Logarithms, Natural Logarithms, and Change of Base

Common Logarithms

$$\log x \quad \text{means} \quad \log_{10} x$$

Natural Logarithms

$$\ln x \quad \text{means} \quad \log_e x$$

Continuously Compounded Interest Formula

$$A = Pe^{rt}$$

where r is the annual interest rate for P dollars invested for t years.

$\log 5 = \log_{10} 5 \approx 0.69897$

$\ln 7 = \log_e 7 \approx 1.94591$

Find the amount in an account at the end of 3 years if \$1000 is invested at an interest rate of 4% compounded continuously.

Here, $t = 3$ years, $P = \$1000$, and $r = 0.04$.

$$A = Pe^{rt}$$

$$= 1000e^{0.04(3)}$$

$$\approx \$1127.50$$

Section 9.8   Exponential and Logarithmic Equations and Problem Solving

Logarithm Property of Equality

Let $\log_b a$ and $\log_b c$ be real numbers and $b \neq 1$. Then

$$\log_b a = \log_b c \text{ is equivalent to } a = c$$

Solve $2^x = 5$.

$\log 2^x = \log 5$ Logarithm property of equality

$x \log 2 = \log 5$ Power property

$x = \dfrac{\log 5}{\log 2}$ Divide both sides by log 2.

$x \approx 2.3219$ Use a calculator.

Chapter 9 Review

(9.1) If $f(x) = x - 5$ and $g(x) = 2x + 1$, find

1. $(f + g)(x)$

2. $(f - g)(x)$

3. $(f \cdot g)(x)$

4. $\left(\dfrac{g}{f}\right)(x)$

If $f(x) = x^2 - 2$, $g(x) = x + 1$, and $h(x) = x^3 - x^2$, find each composition.

5. $(f \circ g)(x)$

6. $(g \circ f)(x)$

7. $(h \circ g)(2)$

8. $(f \circ f)(x)$

9. $(f \circ g)(-1)$

10. $(h \circ h)(2)$

(9.2) Determine whether each function is a one-to-one function. If it is one-to-one, list the elements of its inverse.

11. $h = \{(-9, 14), (6, 8), (-11, 12), (15, 15)\}$

12. $f = \{(-5, 5), (0, 4), (13, 5), (11, -6)\}$

13.

U.S. Region (Input)	Northeast	Midwest	South	West
Rank in Housing Starts for 2009 (Output)	4	3	1	2

△ **14.**

Shape (Input)	Square	Triangle	Parallelogram	Rectangle
Number of Sides (Output)	4	3	4	4

Given that $f(x) = \sqrt{x + 2}$ is a one-to-one function, find the following.

15. a. $f(7)$

  b. $f^{-1}(3)$

16. a. $f(-1)$

  b. $f^{-1}(1)$

Determine whether each function is a one-to-one function.

17.

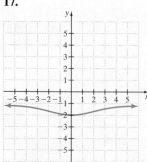

18.

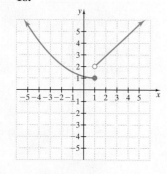

19.

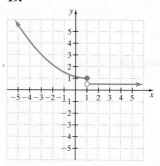

20.

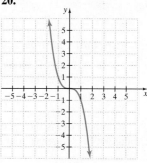

Find an equation defining the inverse function of the given one-to-one function.

21. $f(x) = x - 9$

22. $f(x) = x + 8$

23. $f(x) = 6x + 11$

24. $f(x) = 12x - 1$

25. $f(x) = x^3 - 5$

26. $f(x) = \sqrt[3]{x + 2}$

27. $g(x) = \dfrac{12x - 7}{6}$

28. $r(x) = \dfrac{13x - 5}{2}$

Graph each one-to-one function and its inverse on the same set of axes.

29. $f(x) = -2x + 3$

30. $f(x) = 5x - 5$

(9.3) Solve each equation for x.

31. $4^x = 64$

32. $3^x = \dfrac{1}{9}$

33. $2^{3x} = \dfrac{1}{16}$

34. $5^{2x} = 125$

35. $9^{x+1} = 243$

36. $8^{3x-2} = 4$

Graph each exponential function.

37. $y = 3^x$

38. $y = \left(\dfrac{1}{3}\right)^x$

39. $y = 2^{x-4}$

40. $y = 2^x + 4$

Use the formula $A = P\left(1 + \dfrac{r}{n}\right)^{nt}$ to solve the interest problems. In this formula,

  A = amount accrued (or owed)

  P = principal invested (or loaned)

  r = rate of interest

  n = number of compounding periods per year

  t = time in years

41. Find the amount accrued if \$1600 is invested at 9% interest compounded semiannually for 7 years.

42. A total of \$800 is invested in a 7% certificate of deposit for which interest is compounded quarterly. Find the value that this certificate will have at the end of 5 years.

(9.4) Solve. Round each answer to the nearest whole.

43. The city of Henderson, Nevada, has been growing at a rate of 4.4% per year since the year 2000. If the population of Henderson was 79,087 in 2000 and this rate continues, predict the city's population in 2020.

44. The city of Raleigh, North Carolina, has been growing at a rate of 4.2% per year since the year 2000. If the population of Raleigh was 287,370 in 2000 and this rate continues, predict the city's population in 2018.

45. A summer camp tournament starts with 1024 players. After each round, half the players are eliminated. How many players remain after 7 rounds?

46. The bear population in a certain national park is decreasing by 11% each year. If this rate continues, and there is currently an estimated bear population of 1280, find the bear population in 6 years.

(9.5) Write each equation with logarithmic notation.

47. $49 = 7^2$

48. $2^{-4} = \dfrac{1}{16}$

Write each logarithmic equation with exponential notation.

49. $\log_{1/2} 16 = -4$

50. $\log_{0.4} 0.064 = 3$

Solve for x.

51. $\log_4 x = -3$

52. $\log_3 x = 2$

53. $\log_3 1 = x$

54. $\log_4 64 = x$

55. $\log_4 4^5 = x$

56. $\log_7 7^{-2} = x$

57. $5^{\log_5 4} = x$

58. $2^{\log_2 9} = x$

59. $\log_2(3x - 1) = 4$

60. $\log_3(2x + 5) = 2$

61. $\log_4(x^2 - 3x) = 1$

62. $\log_8(x^2 + 7x) = 1$

Graph each pair of equations on the same coordinate system.

63. $y = 2^x$ and $y = \log_2 x$

64. $y = \left(\dfrac{1}{2}\right)^x$ and $y = \log_{1/2} x$

(9.6) Write each of the following as single logarithms.

65. $\log_3 8 + \log_3 4$

66. $\log_2 6 + \log_2 3$

67. $\log_7 15 - \log_7 20$

68. $\log 18 - \log 12$

69. $\log_{11} 8 + \log_{11} 3 - \log_{11} 6$

70. $\log_5 14 + \log_5 3 - \log_5 21$

71. $2 \log_5 x - 2 \log_5(x + 1) + \log_5 x$

72. $4 \log_3 x - \log_3 x + \log_3(x + 2)$

Use properties of logarithms to write each expression as a sum or difference of multiples of logarithms.

73. $\log_3 \dfrac{x^3}{x + 2}$

74. $\log_4 \dfrac{x + 5}{x^2}$

75. $\log_2 \dfrac{3x^2 y}{z}$

76. $\log_7 \dfrac{yz^3}{x}$

If $\log_b 2 = 0.36$ and $\log_b 5 = 0.83$, find the following.

77. $\log_b 50$

78. $\log_b \dfrac{4}{5}$

(9.7) Use a calculator to approximate the logarithm to four decimal places.

79. $\log 3.6$

80. $\log 0.15$

81. $\ln 1.25$

82. $\ln 4.63$

Find the exact value.

83. $\log 1000$

84. $\log \dfrac{1}{10}$

85. $\ln \dfrac{1}{e}$

86. $\ln e^4$

Solve each equation for x.

87. $\ln(2x) = 2$

88. $\ln(3x) = 1.6$

89. $\ln(2x - 3) = -1$

90. $\ln(3x + 1) = 2$

Use the formula $\ln \dfrac{I}{I_0} = -kx$ to solve the radiation problem in 91 and 92. In this formula,

x = depth in millimeters
I = intensity of radiation
I_0 = initial intensity
k = a constant measure dependent on the material

Round answers to two decimal places.

91. Find the depth at which the intensity of the radiation passing through a lead shield is reduced to 3% of the original intensity if the value of k is 2.1.

92. If k is 3.2, find the depth at which 2% of the original radiation will penetrate.

Approximate the logarithm to four decimal places.

93. $\log_5 1.6$

94. $\log_3 4$

Use the formula $A = Pe^{rt}$ to solve the interest problems in which interest is compounded continuously. In this formula,

A = amount accrued (or owed)
P = principal invested (or loaned)
r = rate of interest
t = time in years

95. Bank of New York offers a 5-year, 3% continuously compounded investment option. Find the amount accrued if $1450 is invested.

96. Find the amount to which a $940 investment grows if it is invested at 4% compounded continuously for 3 years.

(9.8) Solve each exponential equation for x. Give an exact solution and approximate the solution to four decimal places.

97. $3^{2x} = 7$

98. $6^{3x} = 5$

99. $3^{2x+1} = 6$

100. $4^{3x+2} = 9$

101. $5^{3x-5} = 4$

102. $8^{4x-2} = 3$

103. $5^{x-1} = \dfrac{1}{2}$

104. $4^{x+5} = \dfrac{2}{3}$

Solve the equation for x.

105. $\log_5 2 + \log_5 x = 2$

106. $\log_3 x + \log_3 10 = 2$

107. $\log(5x) - \log(x + 1) = 4$

108. $-\log_6(4x + 7) + \log_6 x = 1$

109. $\log_2 x + \log_2 2x - 3 = 1$

110. $\log_3(x^2 - 8x) = 2$

Use the formula $y = y_0 e^{kt}$ to solve the population growth problems. In this formula,

y = size of population

y_0 = initial count of population

k = rate of growth written as a decimal

t = time

Round each answer to the nearest tenth.

111. In 1987, the population of California condors was only 27 birds. They were all brought in from the wild and an intensive breeding program was instituted. If we assume a yearly growth rate of 11.4%, how long will it take the condor population to reach 347 California condors? (*Source:* California Department of Fish and Game)

112. France is experiencing an annual growth rate of 0.4%. In 2010, the population of France was approximately 65,822,000. How long will it take for the population to reach 70,000,000? Round to the nearest tenth. (*Source:* Population Reference Bureau)

113. In 2010, the population of Australia was approximately 22,600,000. How long will it take Australia to double its population if its growth rate is 0.7% annually? Round to the nearest tenth. (*Source:* Population Reference Bureau)

114. Israel's population is increasing in size at a rate of 1.6% per year. How long will it take for its population of 7,746,400 to double in size? Round to the nearest tenth. (*Source:* Population Reference Bureau)

Use the compound interest equation $A = P\left(1 + \dfrac{r}{n}\right)^{nt}$ to solve the following. (See the directions for Exercises 41 and 42 for an explanation of this formula.) Round answers to the nearest tenth.

115. Find how long it will take a $5000 investment to grow to $10,000 if it is invested at 8% interest compounded quarterly.

116. An investment of $6000 has grown to $10,000 while the money was invested at 6% interest compounded monthly. Find how long it was invested.

Use a graphing calculator to solve each equation. Round all solutions to two decimal places.

117. $e^x = 2$ **118.** $10^{0.3x} = 7$

MIXED REVIEW

Solve each equation.

119. $3^x = \dfrac{1}{81}$ **120.** $7^{4x} = 49$ **121.** $8^{3x-2} = 32$

122. $9^{x-2} = 27$ **123.** $\log_4 4 = x$ **124.** $\log_3 x = 4$

125. $\log_5(x^2 - 4x) = 1$ **126.** $\log_4(3x - 1) = 2$

127. $\ln x = -3.2$ **128.** $\log_5 x + \log_5 10 = 2$

129. $\ln x - \ln 2 = 1$ **130.** $\log_6 x - \log_6(4x + 7) = 1$

Chapter 9 **Test**   MyMathLab®   Test Prep VIDEOS   ▶   You Tube™

If $f(x) = x$ and $g(x) = 2x - 3$, find the following.

▶ **1.** $(f \cdot g)(x)$ ▶ **2.** $(f - g)(x)$

If $f(x) = x$, $g(x) = x - 7$, and $h(x) = x^2 - 6x + 5$, find the following.

▶ **3.** $(f \circ h)(0)$ ▶ **4.** $(g \circ f)(x)$

▶ **5.** $(g \circ h)(x)$

On the same set of axes, graph the given one-to-one function and its inverse.

▶ **6.** $f(x) = 7x - 14$

Determine whether the given graph is the graph of a one-to-one function.

▶ **7.**

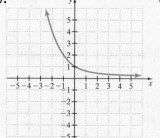

▶ **8.**

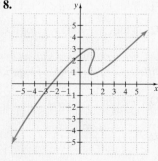

Determine whether each function is one-to-one. If it is one-to-one, find an equation or a set of ordered pairs that defines the inverse function of the given function.

9. $f(x) = 6 - 2x$

10. $f = \{(0,0), (2,3), (-1,5)\}$

11.

Word (Input)	Dog	Cat	House	Desk	Circle
First Letter of Word (Output)	d	c	h	d	c

Use the properties of logarithms to write each expression as a single logarithm.

12. $\log_3 6 + \log_3 4$

13. $\log_5 x + 3 \log_5 x - \log_5(x + 1)$

14. Write the expression $\log_6 \dfrac{2x}{y^3}$ as the sum or difference of multiples of logarithms.

15. If $\log_b 3 = 0.79$ and $\log_b 5 = 1.16$, find the value of $\log_b \dfrac{3}{25}$.

16. Approximate $\log_7 8$ to four decimal places.

17. Solve $8^{x-1} = \dfrac{1}{64}$ for x. Give an exact solution.

18. Solve $3^{2x+5} = 4$ for x. Give an exact solution and approximate the solution to four decimal places.

Solve each logarithmic equation for x. Give an exact solution.

19. $\log_3 x = -2$

20. $\ln \sqrt{e} = x$

21. $\log_8(3x - 2) = 2$

22. $\log_5 x + \log_5 3 = 2$

23. $\log_4(x + 1) - \log_4(x - 2) = 3$

24. Solve $\ln(3x + 7) = 1.31$ accurate to four decimal places.

25. Graph $y = \left(\dfrac{1}{2}\right)^x + 1$.

26. Graph the functions $y = 3^x$ and $y = \log_3 x$ on the same coordinate system.

Use the formula $A = P\left(1 + \dfrac{r}{n}\right)^{nt}$ to solve Exercises 27–29.

27. Find the amount in an account if \$4000 is invested for 3 years at 9% interest compounded monthly.

28. Find how long it will take \$2000 to grow to \$3000 if the money is invested at 7% interest compounded semiannually. Round to the nearest whole.

29. Suppose you have \$3000 to invest. Which investment, rounded to the nearest dollar, yields the greater return over 10 years: 6.5% compounded semiannually or 6% compounded monthly? How much more is yielded by the better investment?

Solve. Round answers to the nearest whole.

30. Suppose a city with population of 150,000 has been decreasing at a rate of 2% per year. If this rate continues, predict the population of the city in 20 years.

31. The prairie dog population of the Grand Forks area now stands at 57,000 animals. If the population is growing at a rate of 2.6% annually, how many prairie dogs will there be in that area 5 years from now?

32. In an attempt to save an endangered species of wood duck, naturalists would like to increase the wood duck population from 400 to 1000 ducks. If the annual population growth rate is 6.2%, how long will it take the naturalists to reach their goal? Round to the nearest whole year.

The reliability of a new model of CD player can be described by the exponential function $R(t) = 2.7^{-(1/3)t}$, where the reliability R is the probability (as a decimal) that the CD player is still working t years after it is manufactured. Round answers to the nearest hundredth. Then write your answers as percents.

33. What is the probability that the CD player will still work half a year after it is manufactured?

34. What is the probability that the CD player will still work 2 years after it is manufactured?

Chapter 9 Cumulative Review

1. Multiply.

a. $(-8)(-1)$ **b.** $(-2)\frac{1}{6}$

c. $-1.2(0.3)$ **d.** $0(-11)$

e. $\left(\frac{1}{5}\right)\left(-\frac{10}{11}\right)$ **f.** $(7)(1)(-2)(-3)$

g. $8(-2)(0)$

2. Solve: $\frac{1}{3}(x-2) = \frac{1}{4}(x+1)$

3. Graph $y = x^2$.

4. Find the equation of a line through $(-2, 6)$ and perpendicular to $f(x) = -3x + 4$. Write the equation using function notation.

5. Solve the system.

$$\begin{cases} x - 5y - 2z = 6 \\ -2x + 10y + 4z = -12 \\ \frac{1}{2}x - \frac{5}{2}y - z = 3 \end{cases}$$

6. Line l and line m are parallel lines cut by transversal t. Find the values of x and y.

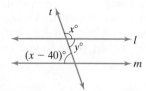

7. Use the quotient rule to simplify.

a. $\frac{x^7}{x^4}$ **b.** $\frac{5^8}{5^2}$

c. $\frac{20x^6}{4x^5}$ **d.** $\frac{12y^{10}z^7}{14y^8z^7}$

8. Use the power rules to simplify the following. Use positive exponents to write all results.

a. $(4a^3)^2$ **b.** $\left(-\frac{2}{3}\right)^3$

c. $\left(\frac{4a^5}{b^3}\right)^3$ **d.** $\left(\frac{3^{-2}}{x}\right)^{-3}$

e. $(a^{-2}b^3c^{-4})^{-2}$

9. For the ICL Production Company, the rational function $C(x) = \dfrac{2.6x + 10,000}{x}$ describes the company's cost per disc for pressing x compact discs. Find the cost per disc for pressing:

a. 100 compact discs

b. 1000 compact discs

10. Multiply.

a. $(3x - 1)^2$

b. $\left(\frac{1}{2}x + 3\right)\left(\frac{1}{2}x - 3\right)$

c. $(2x - 5)(6x + 7)$

11. Add or subtract.

a. $\frac{x}{4} + \frac{5x}{4}$ **b.** $\frac{5}{7z^2} + \frac{x}{7z^2}$

c. $\frac{x^2}{x+7} - \frac{49}{x+7}$ **d.** $\frac{x}{3y^2} - \frac{x+1}{3y^2}$

12. Perform the indicated operations and simplify if possible.

$$\frac{5}{x-2} + \frac{3}{x^2+4x+4} - \frac{6}{x+2}$$

13. Divide $3x^4 + 2x^3 - 8x + 6$ by $x^2 - 1$.

14. Simplify each complex fraction.

a. $\dfrac{\dfrac{a}{5}}{\dfrac{a-1}{10}}$ **b.** $\dfrac{\dfrac{3}{2+a} + \dfrac{6}{2-a}}{\dfrac{5}{a+2} - \dfrac{1}{a-2}}$

c. $\dfrac{x^{-1} + y^{-1}}{xy}$

15. Solve: $\dfrac{2x}{2x-1} + \dfrac{1}{x} = \dfrac{1}{2x-1}$

16. Divide $x^3 - 8$ by $x - 2$.

17. Steve Deitmer takes $1\frac{1}{2}$ times as long to go 72 miles upstream in his boat as he does to return. If the boat cruises at 30 mph in still water, what is the speed of the current?

18. Use synthetic division to divide: $(8x^2 - 12x - 7) \div (x - 2)$

19. Simplify the following expressions.

a. $\sqrt[4]{81}$ **b.** $\sqrt[5]{-243}$

c. $-\sqrt{25}$ **d.** $\sqrt[4]{-81}$

e. $\sqrt[3]{64x^3}$

20. Solve $\dfrac{1}{a+5} = \dfrac{1}{3a+6} - \dfrac{a+2}{a^2+7a+10}$

21. Use rational exponents to write as a single radical.

a. $\sqrt{x} \cdot \sqrt[4]{x}$ **b.** $\dfrac{\sqrt{x}}{\sqrt[3]{x}}$ **c.** $\sqrt[3]{3} \cdot \sqrt{2}$

22. Suppose that y varies directly as x. If $y = \frac{1}{2}$ when $x = 12$, find the constant of variation and the direct variation equation.

23. Multiply.

  a. $\sqrt{3}(5 + \sqrt{30})$

  b. $(\sqrt{5} - \sqrt{6})(\sqrt{7} + 1)$

  c. $(7\sqrt{x} + 5)(3\sqrt{x} - \sqrt{5})$

  d. $(4\sqrt{3} - 1)^2$

  e. $(\sqrt{2x} - 5)(\sqrt{2x} + 5)$

  f. $(\sqrt{x - 3} + 5)^2$

24. Find each root. Assume that all variables represent nonnegative real numbers.

  a. $\sqrt[3]{27}$ **b.** $\sqrt[3]{-27}$ **c.** $\sqrt{\dfrac{9}{64}}$

  d. $\sqrt[4]{x^{12}}$ **e.** $\sqrt[3]{-125y^6}$

25. Rationalize the denominator of $\dfrac{\sqrt[4]{x}}{\sqrt[4]{81y^5}}$.

26. Multiply.

  a. $a^{1/4}(a^{3/4} - a^{7/4})$

  b. $(x^{1/2} - 3)(x^{1/2} + 5)$

27. Solve $\sqrt{4 - x} = x - 2$.

28. Use the quotient rule to divide and simplify if possible. Assume that all variables represent positive numbers.

  a. $\dfrac{\sqrt{54}}{\sqrt{6}}$ **b.** $\dfrac{\sqrt{108a^2}}{3\sqrt{3}}$

  c. $\dfrac{3\sqrt[3]{81a^5b^{10}}}{\sqrt[3]{3b^4}}$

29. Solve $3x^2 - 9x + 8 = 0$ by completing the square.

30. Add or subtract as indicated.

  a. $\dfrac{\sqrt{20}}{3} + \dfrac{\sqrt{5}}{4}$

  b. $\sqrt[3]{\dfrac{24x}{27}} - \dfrac{\sqrt[3]{3x}}{2}$

31. Solve $\dfrac{3x}{x - 2} - \dfrac{x + 1}{x} = \dfrac{6}{x(x - 2)}$.

32. Rationalize the denominator. $\sqrt[3]{\dfrac{27}{m^4n^8}}$

33. Solve $x^2 - 4x \le 0$.

△ **34.** Find the length of the unknown side of the triangle.

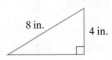

35. Graph $F(x) = (x - 3)^2 + 1$.

36. Find the following powers of i.

  a. i^8 **b.** i^{21}

  c. i^{42} **d.** i^{-13}

37. If $f(x) = x - 1$ and $g(x) = 2x - 3$, find

  a. $(f + g)(x)$ **b.** $(f - g)(x)$

  c. $(f \cdot g)(x)$ **d.** $\left(\dfrac{f}{g}\right)(x)$

38. Solve $4x^2 + 8x - 1 = 0$ by completing the square.

39. Find an equation of the inverse of $f(x) = x + 3$.

40. Solve by using the quadratic formula. $\left(x - \dfrac{1}{2}\right)^2 = \dfrac{x}{2}$

41. Find the value of each logarithmic expression.

  a. $\log_4 16$ **b.** $\log_{10} \dfrac{1}{10}$ **c.** $\log_9 3$

42. Graph $f(x) = -(x + 1)^2 + 1$. Find the vertex and the axis of symmetry.

In Chapter 8, we analyzed some of the important connections between a parabola and its equation. Parabolas are interesting in their own right but are more interesting still because they are part of a collection of curves known as conic sections. This chapter is devoted to quadratic equations in two variables and their conic section graphs: the parabola, circle, ellipse, and hyperbola.

The original Ferris wheel was named after its designer, George Washington Gale Ferris, Jr., a trained engineer who produced the first Ferris wheel for the 1893 World's Columbian Exposition in Chicago. This very first wheel was 264 feet high and was the Columbian Exposition's most noticeable attraction. Since then, Ferris wheels have gotten ever taller, have been built with ever greater capacities, and have changed their designations from Ferris wheels to giant observation wheels because of their closed capsules. In Exercise 92 of Section 10.1, you will explore the dimensions of the Singapore Flyer, the current record-breaking giant observation wheel.

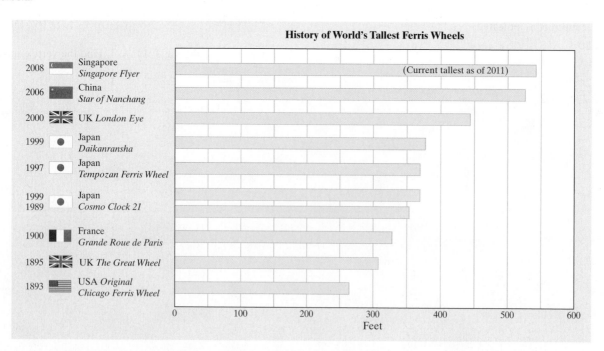

History of World's Tallest Ferris Wheels

2008 Singapore *Singapore Flyer*	(Current tallest as of 2011)
2006 China *Star of Nanchang*	
2000 UK *London Eye*	
1999 Japan *Daikanransha*	
1997 Japan *Tempozan Ferris Wheel*	
1999 1989 Japan *Cosmo Clock 21*	
1900 France *Grande Roue de Paris*	
1895 UK *The Great Wheel*	
1893 USA *Original Chicago Ferris Wheel*	

Feet: 0   100   200   300   400   500   600

10.1 The Parabola and the Circle

OBJECTIVES

1 Graph Parabolas of the Form $x = a(y - k)^2 + h$ and $y = a(x - h)^2 + k$.

2 Graph Circles of the Form $(x - h)^2 + (y - k)^2 = r^2$.

3 Find the Center and the Radius of a Circle, Given Its Equation.

4 Write an Equation of a Circle, Given Its Center and Radius.

Conic sections are named so because each conic section is the intersection of a right circular cone and a plane. The circle, parabola, ellipse, and hyperbola are the conic sections.

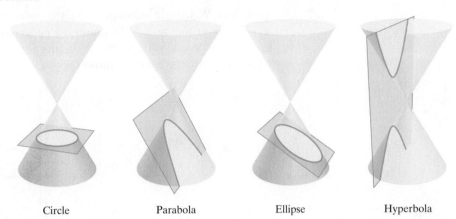

Circle   Parabola   Ellipse   Hyperbola

OBJECTIVE

1 Graphing Parabolas

Thus far, we have seen that $f(x)$ or $y = a(x - h)^2 + k$ is the equation of a parabola that opens upward if $a > 0$ or downward if $a < 0$. Parabolas can also open left or right or even on a slant. Equations of these parabolas are not functions of x, of course, since a parabola opening any way other than upward or downward fails the vertical line test. In this section, we introduce parabolas that open to the left and to the right. Parabolas opening on a slant will not be developed in this book.

Just as $y = a(x - h)^2 + k$ is the equation of a parabola that opens upward or downward, $x = a(y - k)^2 + h$ is the equation of a parabola that opens to the right or to the left. The parabola opens to the right if $a > 0$ and to the left if $a < 0$. The parabola has vertex (h, k), and its axis of symmetry is the line $y = k$.

Parabolas

$$y = a(x - h)^2 + k$$

$a > 0$ (h, k) $x = h$

(h, k) $a < 0$ $x = h$

$$x = a(y - k)^2 + h$$

(h, k) $y = k$ $a > 0$

$a < 0$ $y = k$ (h, k)

The equations $y = a(x - h)^2 + k$ and $x = a(y - k)^2 + h$ are called **standard forms.**

✓CONCEPT CHECK

Does the graph of the parabola given by the equation $x = -3y^2$ open to the left, to the right, upward, or downward?

EXAMPLE 1   Graph the parabola $x = 2y^2$.

Solution   Written in standard form, the equation $x = 2y^2$ is $x = 2(y - 0)^2 + 0$ with $a = 2, k = 0$, and $h = 0$. Its graph is a parabola with vertex $(0, 0)$, and its axis of symmetry is the line $y = 0$. Since $a > 0$, this parabola opens to the right. The table shows a few more ordered pair solutions of $x = 2y^2$. Its graph is also shown.

x	y
8	-2
2	-1
0	0
2	1
8	2

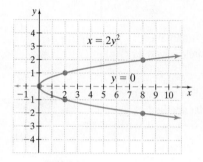

PRACTICE

1   Graph the parabola $x = \dfrac{1}{2}y^2$.

EXAMPLE 2   Graph the parabola $x = -3(y - 1)^2 + 2$.

Solution   The equation $x = -3(y - 1)^2 + 2$ is in the form $x = a(y - k)^2 + h$ with $a = -3, k = 1$, and $h = 2$. Since $a < 0$, the parabola opens to the left. The vertex (h, k) is $(2, 1)$, and the axis of symmetry is the line $y = 1$. When $y = 0, x = -1$, so the x-intercept is $(-1, 0)$. Again, we obtain a few ordered pair solutions and then graph the parabola.

x	y
2	1
-1	0
-1	2
-10	3
-10	-1

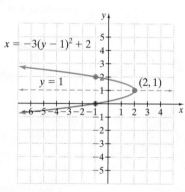

PRACTICE

2   Graph the parabola $x = -2(y + 4)^2 - 1$.

EXAMPLE 3   Graph $y = -x^2 - 2x + 15$.

Solution   Complete the square on x to write the equation in standard form.

$$y - 15 = -x^2 - 2x \qquad \text{Subtract 15 from both sides.}$$
$$y - 15 = -1(x^2 + 2x) \qquad \text{Factor } -1 \text{ from the terms } -x^2 - 2x.$$

The coefficient of x is 2. Find the square of half of 2.

$$\frac{1}{2}(2) = 1 \quad \text{and} \quad 1^2 = 1$$

$$y - 15 - 1(1) = -1(x^2 + 2x + 1) \quad \text{Add } -1(1) \text{ to both sides.}$$

$$y - 16 = -1(x + 1)^2 \quad \text{Simplify the left side and factor the right side.}$$

$$y = -(x + 1)^2 + 16 \quad \text{Add 16 to both sides.}$$

The equation is now in standard form $y = a(x - h)^2 + k$ with $a = -1$, $h = -1$, and $k = 16$.

The vertex is then (h, k), or $(-1, 16)$.

A second method for finding the vertex is by using the formula $\frac{-b}{2a}$.

$$x = \frac{-(-2)}{2(-1)} = \frac{2}{-2} = -1$$

$$y = -(-1)^2 - 2(-1) + 15 = -1 + 2 + 15 = 16$$

Again, we see that the vertex is $(-1, 16)$, and the axis of symmetry is the vertical line $x = -1$. The y-intercept is $(0, 15)$. Now we can use a few more ordered pair solutions to graph the parabola.

x	y
-1	16
0	15
-2	15
1	12
-3	12
3	0
-5	0

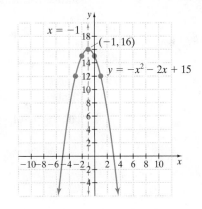

PRACTICE

3   Graph $y = -x^2 + 4x + 6$.

EXAMPLE 4   Graph $x = 2y^2 + 4y + 5$.

Solution   Notice that this equation is quadratic in y, so its graph is a parabola that opens to the left or the right. We can complete the square on y, or we can use the formula $\frac{-b}{2a}$ to find the vertex.

Since the equation is quadratic in y, the formula gives us the y-value of the vertex.

$$y = \frac{-4}{2 \cdot 2} = \frac{-4}{4} = -1$$

$$x = 2(-1)^2 + 4(-1) + 5 = 2 \cdot 1 - 4 + 5 = 3$$

(Continued on next page)

The vertex is $(3, -1)$, and the axis of symmetry is the line $y = -1$. The parabola opens to the right since $a > 0$. The x-intercept is $(5, 0)$.

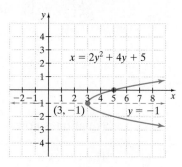

PRACTICE
4 Graph $x = 3y^2 + 6y + 4$.

OBJECTIVE

2 Graphing Circles

Another conic section is the **circle.** A circle is the set of all points in a plane that are the same distance from a fixed point called the **center.** The distance is called the **radius** of the circle. To find a standard equation for a circle, let (h, k) represent the center of the circle and let (x, y) represent any point on the circle. The distance between (h, k) and (x, y) is defined to be the circle's radius, r units. We can find this distance r by using the distance formula.

$$r = \sqrt{(x - h)^2 + (y - k)^2}$$
$$r^2 = (x - h)^2 + (y - k)^2 \qquad \text{Square both sides.}$$

> **Circle**
>
> The graph of $(x - h)^2 + (y - k)^2 = r^2$ is a circle with center (h, k) and radius r.
>
>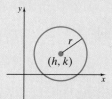
>
> The equation $(x - h)^2 + (y - k)^2 = r^2$ is called **standard form.**

If an equation can be written in the standard form

$$(x - h)^2 + (y - k)^2 = r^2$$

then its graph is a circle, which we can draw by graphing the center (h, k) and using the radius r.

> ▶ **Helpful Hint**
>
> Notice that the radius is the *distance* from the center of the circle to any point of the circle. Also notice that the *midpoint* of a diameter of a circle is the center of the circle.
>
>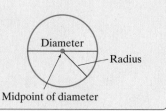

EXAMPLE 5   Graph $x^2 + y^2 = 4$.

Solution   The equation can be written in standard form as

$$(x - 0)^2 + (y - 0)^2 = 2^2$$

The center of the circle is $(0, 0)$, and the radius is 2. Its graph is shown.

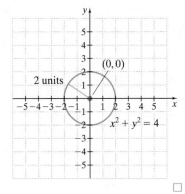

PRACTICE
5   Graph $x^2 + y^2 = 25$.

> ▶ **Helpful Hint**
> Notice the difference between the equation of a circle and the equation of a parabola. The equation of a circle contains both x^2 and y^2 terms on the same side of the equation with equal coefficients. The equation of a parabola has either an x^2 term or a y^2 term but not both.

EXAMPLE 6   Graph $(x + 1)^2 + y^2 = 8$.

Solution   The equation can be written as $(x + 1)^2 + (y - 0)^2 = 8$ with $h = -1$, $k = 0$, and $r = \sqrt{8}$. The center is $(-1, 0)$, and the radius is $\sqrt{8} = 2\sqrt{2} \approx 2.8$.

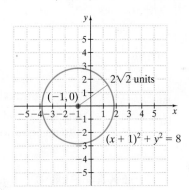

PRACTICE
6   Graph $(x - 3)^2 + (y + 2)^2 = 4$.

✓CONCEPT CHECK
In the graph of the equation $(x - 3)^2 + (y - 2)^2 = 5$, what is the distance between the center of the circle and any point on the circle?

OBJECTIVE
3   Finding the Center and the Radius of a Circle

To find the center and the radius of a circle from its equation, write the equation in standard form. To write the equation of a circle in standard form, we complete the square on both x and y.

Answer to Concept Check:
$\sqrt{5}$ units

EXAMPLE 7 Graph $x^2 + y^2 + 4x - 8y = 16$.

Solution Since this equation contains x^2 and y^2 terms on the same side of the equation with equal coefficients, its graph is a circle. To write the equation in standard form, group the terms involving x and the terms involving y and then complete the square on each variable.

$$(x^2 + 4x) + (y^2 - 8y) = 16$$

Thus, $\frac{1}{2}(4) = 2$ and $2^2 = 4$. Also, $\frac{1}{2}(-8) = -4$ and $(-4)^2 = 16$. Add 4 and then 16 to both sides.

$$(x^2 + 4x + 4) + (y^2 - 8y + 16) = 16 + 4 + 16$$

$$(x + 2)^2 + (y - 4)^2 = 36 \qquad \text{Factor.}$$

This circle the center $(-2, 4)$ and radius 6, as shown.

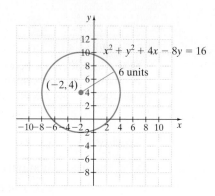

PRACTICE
7   Graph $x^2 + y^2 + 6x - 2y = 6$.

OBJECTIVE
4   Writing Equations of Circles ▶

Since a circle is determined entirely by its center and radius, this information is all we need to write an equation of a circle.

EXAMPLE 8   Find an equation of the circle with center $(-7, 3)$ and radius 10.

Solution   Using the given values $h = -7, k = 3$, and $r = 10$, we write the equation

$$(x - h)^2 + (y - k)^2 = r^2$$

or

$$[x - (-7)]^2 + (y - 3)^2 = 10^2 \quad \text{Substitute the given values.}$$

or

$$(x + 7)^2 + (y - 3)^2 = 100$$

PRACTICE
8   Find an equation of the circle with center $(-2, -5)$ and radius 9.

Graphing Calculator Explorations

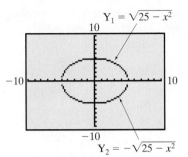

$Y_1 = \sqrt{25 - x^2}$

$Y_2 = -\sqrt{25 - x^2}$

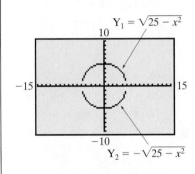

$Y_1 = \sqrt{25 - x^2}$

$Y_2 = -\sqrt{25 - x^2}$

To graph an equation such as $x^2 + y^2 = 25$ with a graphing calculator, we first solve the equation for y.

$$x^2 + y^2 = 25$$
$$y^2 = 25 - x^2$$
$$y = \pm\sqrt{25 - x^2}$$

The graph of $y = \sqrt{25 - x^2}$ will be the top half of the circle, and the graph of $y = -\sqrt{25 - x^2}$ will be the bottom half of the circle.

To graph, press $\boxed{Y=}$ and enter $Y_1 = \sqrt{25 - x^2}$ and $Y_2 = -\sqrt{25 - x^2}$. Insert parentheses around $25 - x^2$ so that $\sqrt{25 - x^2}$ and not $\sqrt{25} - x^2$ is graphed.

The top graph to the left does not appear to be a circle because we are currently using a standard window and the screen is rectangular. This causes the tick marks on the x-axis to be farther apart than the tick marks on the y-axis and, thus, creates the distorted circle. If we want the graph to appear circular, we must define a square window by using a feature of the graphing calculator or by redefining the window to show the x-axis from -15 to 15 and the y-axis from -10 to 10. Using a square window, the graph appears as shown on the bottom to the left.

Use a graphing calculator to graph each circle.

1. $x^2 + y^2 = 55$ **2.** $x^2 + y^2 = 20$
3. $5x^2 + 5y^2 = 50$ **4.** $6x^2 + 6y^2 = 105$
5. $2x^2 + 2y^2 - 34 = 0$ **6.** $4x^2 + 4y^2 - 48 = 0$
7. $7x^2 + 7y^2 - 89 = 0$ **8.** $3x^2 + 3y^2 - 35 = 0$

Vocabulary, Readiness & Video Check

Use the choices below to fill in each blank. Some choices may be used more than once.

radius   center   vertex
diameter   circle   conic sections

1. The circle, parabola, ellipse, and hyperbola are called the _____.
2. For a parabola that opens upward, the lowest point is the _____.
3. A _____ is the set of all points in a plane that are the same distance from a fixed point. The fixed point is called the _____.
4. The midpoint of a diameter of a circle is the _____.
5. The distance from the center of a circle to any point of the circle is called the _____.
6. Twice a circle's radius is its _____.

Martin-Gay Interactive Videos

See Video 10.1

Watch the section lecture video and answer the following questions.

OBJECTIVE 1
7. Based on ▣ Example 1 and the lecture before, would you say that parabolas of the form $x = a(y - k)^2 + h$ are functions? Why or why not?

OBJECTIVE 2
8. Based on the lecture before ▣ Example 2, what would be the standard form of a circle with its center at the origin? Simplify your answer.

OBJECTIVE 3
9. From ▣ Example 3, if you know the center and radius of a circle, how can you write that circle's equation?

OBJECTIVE 4
10. From ▣ Example 4, why do we need to complete the square twice when writing this equation of a circle in standard form?

10.1 Exercise Set MyMathLab®

The graph of each equation is a parabola. Determine whether the parabola opens upward, downward, to the left, or to the right. Do not graph. See Examples 1 through 4.

1. $y = x^2 - 7x + 5$
2. $y = -x^2 + 16$
3. $x = -y^2 - y + 2$
4. $x = 3y^2 + 2y - 5$
5. $y = -x^2 + 2x + 1$
6. $x = -y^2 + 2y - 6$

The graph of each equation is a parabola. Find the vertex of the parabola and then graph it. See Examples 1 through 4.

7. $x = 3y^2$
8. $x = 5y^2$
9. $x = -2y^2$
10. $x = -4y^2$
11. $y = -4x^2$
12. $y = -2x^2$
13. $x = (y - 2)^2 + 3$
14. $x = (y - 4)^2 - 1$
15. $y = -3(x - 1)^2 + 5$
16. $y = -4(x - 2)^2 + 2$
17. $x = y^2 + 6y + 8$
18. $x = y^2 - 6y + 6$
19. $y = x^2 + 10x + 20$
20. $y = x^2 + 4x - 5$
21. $x = -2y^2 + 4y + 6$
22. $x = 3y^2 + 6y + 7$

The graph of each equation is a circle. Find the center and the radius and then graph the circle. See Examples 5 through 7.

23. $x^2 + y^2 = 9$
24. $x^2 + y^2 = 25$
25. $x^2 + (y - 2)^2 = 1$
26. $(x - 3)^2 + y^2 = 9$
27. $(x - 5)^2 + (y + 2)^2 = 1$
28. $(x + 3)^2 + (y + 3)^2 = 4$
29. $x^2 + y^2 + 6y = 0$
30. $x^2 + 10x + y^2 = 0$
31. $x^2 + y^2 + 2x - 4y = 4$
32. $x^2 + y^2 + 6x - 4y = 3$
33. $(x + 2)^2 + (y - 3)^2 = 7$
34. $(x + 1)^2 + (y - 2)^2 = 5$
35. $x^2 + y^2 - 4x - 8y - 2 = 0$
36. $x^2 + y^2 - 2x - 6y - 5 = 0$

Hint: For Exercises 37 through 42, first divide the equation through by the coefficient of x^2 (or y^2).

37. $3x^2 + 3y^2 = 75$
38. $2x^2 + 2y^2 = 18$
39. $6(x - 4)^2 + 6(y - 1)^2 = 24$
40. $7(x - 1)^2 + 7(y - 3)^2 = 63$
41. $4(x + 1)^2 + 4(y - 3)^2 = 12$
42. $5(x - 2)^2 + 5(y + 1) = 50$

Write an equation of the circle with the given center and radius. See Example 8.

43. $(2, 3); 6$
44. $(-7, 6); 2$
45. $(0, 0); \sqrt{3}$
46. $(0, -6); \sqrt{2}$
47. $(-5, 4); 3\sqrt{5}$
48. the origin; $4\sqrt{7}$

MIXED PRACTICE

Sketch the graph of each equation. If the graph is a parabola, find its vertex. If the graph is a circle, find its center and radius.

49. $x = y^2 - 3$
50. $x = y^2 + 2$
51. $y = (x - 2)^2 - 2$
52. $y = (x + 3)^2 + 3$
53. $x^2 + y^2 = 1$
54. $x^2 + y^2 = 49$
55. $x = (y + 3)^2 - 1$
56. $x = (y - 1)^2 + 4$
57. $(x - 2)^2 + (y - 2)^2 = 16$
58. $(x + 3)^2 + (y - 1)^2 = 9$
59. $x = -(y - 1)^2$
60. $x = -2(y + 5)^2$
61. $(x - 4)^2 + y^2 = 7$
62. $x^2 + (y + 5)^2 = 5$
63. $y = 5(x + 5)^2 + 3$
64. $y = 3(x - 4)^2 + 2$
65. $\dfrac{x^2}{8} + \dfrac{y^2}{8} = 2$
66. $2x^2 + 2y^2 = \dfrac{1}{2}$
67. $y = x^2 + 7x + 6$
68. $y = x^2 - 2x - 15$
69. $x^2 + y^2 + 2x + 12y - 12 = 0$
70. $x^2 + y^2 + 6x + 10y - 2 = 0$
71. $x = y^2 + 8y - 4$
72. $x = y^2 + 6y + 2$
73. $x^2 - 10y + y^2 + 4 = 0$
74. $x^2 + y^2 - 8y + 5 = 0$
75. $x = -3y^2 + 30y$
76. $x = -2y^2 - 4y$
77. $5x^2 + 5y^2 = 25$
78. $\dfrac{x^2}{3} + \dfrac{y^2}{3} = 2$
79. $y = 5x^2 - 20x + 16$
80. $y = 4x^2 - 40x + 105$

REVIEW AND PREVIEW

Graph each equation. See Section 3.3.

81. $y = 2x + 5$
82. $y = -3x + 3$
83. $y = 3$
84. $x = -2$

Rationalize each denominator and simplify if possible. See Section 7.5.

85. $\dfrac{1}{\sqrt{3}}$
86. $\dfrac{\sqrt{5}}{\sqrt{8}}$
87. $\dfrac{4\sqrt{7}}{\sqrt{6}}$
88. $\dfrac{10}{\sqrt{5}}$

CONCEPT EXTENSIONS

For Exercises 89 and 90, explain the error in each statement.

89. The graph of $x = 5(y + 5)^2 + 1$ is a parabola with vertex $(-5, 1)$ and opening to the right.

90. The graph of $x^2 + (y + 3)^2 = 10$ is a circle with center $(0, -3)$ and radius 5.

91. **The Sarsen Circle** The first image that comes to mind when one thinks of Stonehenge is the very large sandstone blocks with sandstone lintels across the top. The Sarsen Circle of Stonehenge is the outer circle of the sandstone blocks, each of which weighs up to 50 tons. There were originally 30 of these monolithic blocks, but only 17 remain upright to this day. The "altar stone" lies at the center of this circle, which has a diameter of 33 meters.

   a. What is the radius of the Sarsen Circle?

   b. What is the circumference of the Sarsen Circle? Round your result to 2 decimal places.

c. Since there were originally 30 Sarsen stones located on the circumference, how far apart would the centers of the stones have been? Round to the nearest tenth of a meter.

d. Using the axes in the drawing, what are the coordinates of the center of the circle?

e. Use parts (a) and (d) to write the equation of the Sarsen Circle.

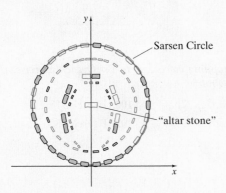

92. Although there are many larger observation wheels on the horizon, as of this writing the largest observation wheel in the world is the Singapore Flyer. From the Flyer, you can see up to 45 km away. Each of the 28 enclosed capsules holds 28 passengers and completes a full rotation every 32 minutes. Its diameter is 150 meters, and the height of this giant wheel is 165 meters. (*Source:* singaporeflyer.com)

a. What is the radius of the Singapore Flyer?

b. How close is the wheel to the ground?

c. How high is the center of the wheel from the ground?

d. Using the axes in the drawing, what are the coordinates of the center of the wheel?

e. Use parts (a) and (d) to write an equation of the Singapore Flyer.

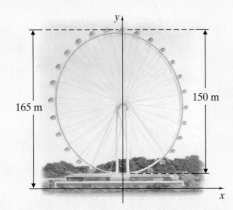

93. In 1893, Pittsburgh bridge builder George Ferris designed and built a gigantic revolving steel wheel whose height was 264 feet and diameter was 250 feet. This Ferris wheel opened at the 1893 exposition in Chicago. It had 36 wooden cars, each capable of holding 60 passengers. (*Source: The Handy Science Answer Book*)

a. What was the radius of this Ferris wheel?

b. How close was the wheel to the ground?

c. How high was the center of the wheel from the ground?

d. Using the axes in the drawing, what are the coordinates of the center of the wheel?

e. Use parts (a) and (d) to write an equation of the wheel.

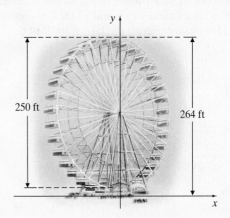

94. The world's largest-diameter Ferris wheel currently operating is the Cosmo Clock 21 at Yokohama City, Japan. It has a 60-armed wheel, its diameter is 100 meters, and it has a height of 105 meters. (*Source: The Handy Science Answer Book*)

a. What is the radius of this Ferris wheel?

b. How close is the wheel to the ground?

c. How high is the center of the wheel from the ground?

d. Using the axes in the drawing, what are the coordinates of the center of the wheel?

e. Use parts (a) and (d) to write an equation of the wheel.

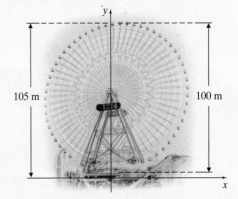

95. If you are given a list of equations of circles and parabolas and none are in standard form, explain how you would determine which is an equation of a circle and which is an equation of a parabola. Explain also how you would distinguish the upward or downward parabolas from the left-opening or right-opening parabolas.

△ **96.** Determine whether the triangle with vertices $(2, 6)$, $(0, -2)$, and $(5, 1)$ is an isosceles triangle.

Solve.

97. Two surveyors need to find the distance across a lake. They place a reference pole at point *A* in the diagram. Point *B* is 3 meters east and 1 meter north of the reference point *A*. Point *C* is 19 meters east and 13 meters north of point *A*. Find the distance across the lake, from *B* to *C*.

98. A bridge constructed over a bayou has a supporting arch in the shape of a parabola. Find an equation of the parabolic arch if the length of the road over the arch is 100 meters and the maximum height of the arch is 40 meters.

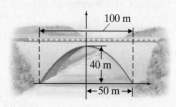

 Use a graphing calculator to verify each exercise. Use a square viewing window.

99. Exercise 77. **100.** Exercise 78.

101. Exercise 79. **102.** Exercise 80.

10.2 The Ellipse and the Hyperbola

OBJECTIVES

1 Define and Graph an Ellipse.

2 Define and Graph a Hyperbola.

OBJECTIVE

1 Graphing Ellipses

An **ellipse** can be thought of as the set of points in a plane such that the sum of the distances of those points from two fixed points is constant. Each of the two fixed points is called a **focus.** (The plural of focus is **foci.**) The point midway between the foci is called the **center.**

An ellipse may be drawn by hand by using two thumbtacks, a piece of string, and a pencil. Secure the two thumbtacks in a piece of cardboard, for example, and tie each end of the string to a tack. Use your pencil to pull the string tight and draw the ellipse. The two thumbtacks are the foci of the drawn ellipse.

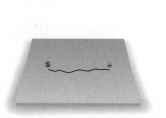

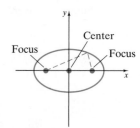

Ellipse with Center (0, 0)

The graph of an equation of the form $\dfrac{x^2}{a^2} + \dfrac{y^2}{b^2} = 1$ is an ellipse with center $(0,0)$.

The *x*-intercepts are $(a, 0)$ and $(-a, 0)$, and the *y*-intercepts are $(0, b)$, and $(0, -b)$.

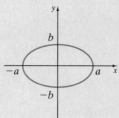

The **standard form** of an ellipse with center $(0, 0)$ is $\dfrac{x^2}{a^2} + \dfrac{y^2}{b^2} = 1$.

EXAMPLE 1   Graph $\dfrac{x^2}{9} + \dfrac{y^2}{16} = 1$.

Solution   The equation is of the form $\dfrac{x^2}{a^2} + \dfrac{y^2}{b^2} = 1$, with $a = 3$ and $b = 4$, so its graph is an ellipse with center $(0, 0)$, x-intercepts $(3, 0)$ and $(-3, 0)$, and y-intercepts $(0, 4)$ and $(0, -4)$.

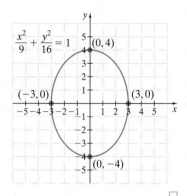

PRACTICE
1   Graph $\dfrac{x^2}{25} + \dfrac{y^2}{4} = 1$.

EXAMPLE 2   Graph $4x^2 + 16y^2 = 64$.

Solution   Although this equation contains a sum of squared terms in x and y on the same side of an equation, this is not the equation of a circle since the coefficients of x^2 and y^2 are not the same. The graph of this equation is an ellipse. Since the standard form of the equation of an ellipse has 1 on one side, divide both sides of this equation by 64.

$$4x^2 + 16y^2 = 64$$

$$\frac{4x^2}{64} + \frac{16y^2}{64} = \frac{64}{64} \qquad \text{Divide both sides by 64.}$$

$$\frac{x^2}{16} + \frac{y^2}{4} = 1 \qquad \text{Simplify.}$$

We now recognize the equation of an ellipse with $a = 4$ and $b = 2$. This ellipse has center $(0, 0)$, x-intercepts $(4, 0)$ and $(-4, 0)$, and y-intercepts $(0, 2)$ and $(0, -2)$.

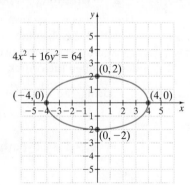

PRACTICE
2   Graph $9x^2 + 4y^2 = 36$.

The center of an ellipse is not always $(0, 0)$, as shown in the next example.

Ellipse with Center (h, k)

The standard form of the equation of an ellipse with center (h, k) is

$$\frac{(x - h)^2}{a^2} + \frac{(y - k)^2}{b^2} = 1$$

EXAMPLE 3 Graph $\dfrac{(x+3)^2}{25} + \dfrac{(y-2)^2}{36} = 1$.

Solution The center of this ellipse is found in a way that is similar to finding the center of a circle. This ellipse has center $(-3, 2)$. Notice that $a = 5$ and $b = 6$. To find four points on the graph of the ellipse, first graph the center, $(-3, 2)$. Since $a = 5$, count 5 units right and then 5 units left of the point with coordinates $(-3, 2)$. Next, since $b = 6$, start at $(-3, 2)$ and count 6 units up and then 6 units down to find two more points on the ellipse.

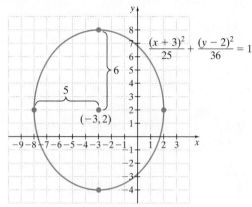

PRACTICE
3 Graph $\dfrac{(x-4)^2}{49} + \dfrac{(y+1)^2}{81} = 1$.

✓**CONCEPT CHECK**

In the graph of the equation $\dfrac{x^2}{64} + \dfrac{y^2}{36} = 1$, which distance is longer: the distance between the x-intercepts or the distance between the y-intercepts? How much longer? Explain.

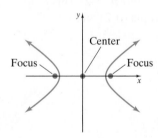

OBJECTIVE
2 Graphing Hyperbolas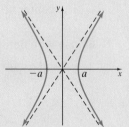

The final conic section is the **hyperbola**. A hyperbola is the set of points in a plane such that the absolute value of the difference of the distances from two fixed points is constant. Each of the two fixed points is called a **focus.** The point midway between the foci is called the **center.**

Using the distance formula, we can show that the graph of $\dfrac{x^2}{a^2} - \dfrac{y^2}{b^2} = 1$ is a hyperbola with center $(0, 0)$ and x-intercepts $(a, 0)$ and $(-a, 0)$. Also, the graph of $\dfrac{y^2}{b^2} - \dfrac{x^2}{a^2} = 1$ is a hyperbola with center $(0, 0)$ and y-intercepts $(0, b)$ and $(0, -b)$.

Hyperbola with Center (0, 0)

The graph of an equation of the form $\dfrac{x^2}{a^2} - \dfrac{y^2}{b^2} = 1$ is a hyperbola with center $(0, 0)$ and x-intercepts $(a, 0)$ and $(-a, 0)$.

Answer to Concept Check:
x-intercepts, by 4 units

The graph of an equation of the form $\dfrac{y^2}{b^2} - \dfrac{x^2}{a^2} = 1$ is a hyperbola with center $(0,0)$ and y-intercepts $(0,b)$ and $(0,-b)$.

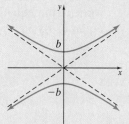

The equations $\dfrac{x^2}{a^2} - \dfrac{y^2}{b^2} = 1$ and $\dfrac{y^2}{b^2} - \dfrac{x^2}{a^2} = 1$ are the **standard forms** for the equation of a hyperbola.

> ▶ **Helpful Hint**
>
> Notice the difference between the equation of an ellipse and a hyperbola. The equation of the ellipse contains x^2 and y^2 terms on the same side of the equation with same-sign coefficients. For a hyperbola, the coefficients on the same side of the equation have different signs.

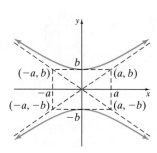

Graphing a hyperbola such as $\dfrac{y^2}{b^2} - \dfrac{x^2}{a^2} = 1$ is made easier by recognizing one of its important characteristics. Examining the figure to the left, notice how the sides of the branches of the hyperbola extend indefinitely and seem to approach the dashed lines in the figure. These dashed lines are called the **asymptotes** of the hyperbola.

To sketch these lines, or asymptotes, draw a rectangle with vertices (a,b), $(-a,b)$, $(a,-b)$ and $(-a,-b)$. The asymptotes of the hyperbola are the extended diagonals of this rectangle.

EXAMPLE 4   Graph $\dfrac{x^2}{16} - \dfrac{y^2}{25} = 1$.

Solution   This equation has the form $\dfrac{x^2}{a^2} - \dfrac{y^2}{b^2} = 1$, with $a = 4$ and $b = 5$. Thus, its graph is a hyperbola that opens to the left and right. It has center $(0,0)$ and x-intercepts $(4, 0)$ and $(-4, 0)$. To aid in graphing the hyperbola, we first sketch its asymptotes. The extended diagonals of the rectangle with corners $(4, 5)$, $(4, -5)$, $(-4, 5)$, and $(-4, -5)$ are the asymptotes of the hyperbola. Then we use the asymptotes to aid in sketching the hyperbola.

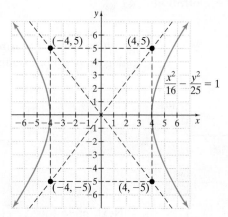

PRACTICE
4   Graph $\dfrac{x^2}{9} - \dfrac{y^2}{16} = 1$.

EXAMPLE 5 Graph $4y^2 - 9x^2 = 36$.

Solution Since this is a difference of squared terms in x and y on the same side of the equation, its graph is a hyperbola as opposed to an ellipse or a circle. The standard form of the equation of a hyperbola has a 1 on one side, so divide both sides of the equation by 36.

$$4y^2 - 9x^2 = 36$$

$$\frac{4y^2}{36} - \frac{9x^2}{36} = \frac{36}{36} \quad \text{Divide both sides by 36.}$$

$$\frac{y^2}{9} - \frac{x^2}{4} = 1 \quad \text{Simplify.}$$

The equation is of the form $\dfrac{y^2}{b^2} - \dfrac{x^2}{a^2} = 1$, with $a = 2$ and $b = 3$, so the hyperbola is centered at $(0, 0)$ with y-intercepts $(0, 3)$ and $(0, -3)$. The sketch of the hyperbola is shown.

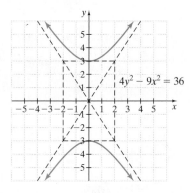

PRACTICE
5 Graph $9y^2 - 25x^2 = 225$.

Although this is beyond the scope of this text, the standard forms of the equations of hyperbolas with center (h, k) are given below. The Concept Extensions section in Exercise Set 10.2 contains some hyperbolas of this form.

> **Hyperbola with Center (h, k)**
>
> Standard forms of the equations of hyperbolas with center (h, k) are:
>
> $$\frac{(x - h)^2}{a^2} - \frac{(y - k)^2}{b^2} = 1 \qquad \frac{(y - k)^2}{b^2} - \frac{(x - h)^2}{a^2} = 1$$

Graphing Calculator Explorations

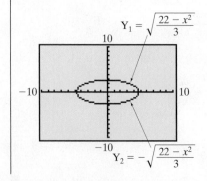

To graph an ellipse by using a graphing calculator, use the same procedure as for graphing a circle. For example, to graph $x^2 + 3y^2 = 22$, first solve for y.

$$3y^2 = 22 - x^2$$

$$y^2 = \frac{22 - x^2}{3}$$

$$y = \pm\sqrt{\frac{22 - x^2}{3}}$$

Next, press the $\boxed{Y =}$ key and enter $Y_1 = \sqrt{\dfrac{22 - x^2}{3}}$ and $Y_2 = -\sqrt{\dfrac{22 - x^2}{3}}$.

(Insert two sets of parentheses in the radicand as $\sqrt{((22 - x^2)/3)}$ so that the desired graph is obtained.) The graph appears as shown to the left.

Use a graphing calculator to graph each ellipse.

1. $10x^2 + y^2 = 32$

2. $x^2 + 6y^2 = 35$

3. $20x^2 + 5y^2 = 100$

4. $4y^2 + 12x^2 = 48$

5. $7.3x^2 + 15.5y^2 = 95.2$

6. $18.8x^2 + 36.1y^2 = 205.8$

Vocabulary, Readiness & Video Check

Use the choices below to fill in each blank. Some choices will be used more than once and some not at all.

ellipse	$(0, 0)$	x	$(a, 0)$ and $(-a, 0)$	$(0, a)$ and $(0, -a)$	focus
hyperbola	center	y	$(b, 0)$ and $(-b, 0)$	$(0, b)$ and $(0, -b)$	

1. A(n) _____ is the set of points in a plane such that the absolute value of the differences of their distances from two fixed points is constant.

2. A(n) _____ is the set of points in a plane such that the sum of their distances from two fixed points is constant.

For exercises 1 and 2 above,

3. The two fixed points are each called a _____.

4. The point midway between the foci is called the _____.

5. The graph of $\dfrac{x^2}{a^2} - \dfrac{y^2}{b^2} = 1$ is a(n) _____ with center _____ and _____-intercepts of _____.

6. The graph of $\dfrac{x^2}{a^2} + \dfrac{y^2}{b^2} = 1$ is a(n) _____ with center _____ and *x*-intercepts of _____.

Martin-Gay Interactive Videos

Watch the section lecture video and answer the following questions.

OBJECTIVE 1

7. From Example 1, what information do the values of *a* and *b* give us about the graph of an ellipse? Answer this same question for Example 2.

OBJECTIVE 2

8. From Example 3, we know the points $(a, b), (a, -b), (-a, b),$ and $(-a, -b)$ are not part of the graph. Explain the role of these points.

See Video 10.2

10.2 Exercise Set   MyMathLab®

Identify the graph of each equation as an ellipse or a hyperbola. Do not graph. See Examples 1 through 5.

1. $\dfrac{x^2}{16} + \dfrac{y^2}{4} = 1$

2. $\dfrac{x^2}{16} - \dfrac{y^2}{4} = 1$

3. $x^2 - 5y^2 = 3$

4. $-x^2 + 5y^2 = 3$

5. $-\dfrac{y^2}{25} + \dfrac{x^2}{36} = 1$

6. $\dfrac{y^2}{25} + \dfrac{x^2}{36} = 1$

Sketch the graph of each equation. See Examples 1 and 2.

7. $\dfrac{x^2}{4} + \dfrac{y^2}{25} = 1$

8. $\dfrac{x^2}{16} + \dfrac{y^2}{9} = 1$

9. $\dfrac{x^2}{9} + y^2 = 1$

10. $x^2 + \dfrac{y^2}{4} = 1$

11. $9x^2 + y^2 = 36$

12. $x^2 + 4y^2 = 16$

13. $4x^2 + 25y^2 = 100$

14. $36x^2 + y^2 = 36$

Sketch the graph of each equation. See Example 3.

15. $\dfrac{(x + 1)^2}{36} + \dfrac{(y - 2)^2}{49} = 1$

16. $\dfrac{(x - 3)^2}{9} + \dfrac{(y + 3)^2}{16} = 1$

17. $\dfrac{(x - 1)^2}{4} + \dfrac{(y - 1)^2}{25} = 1$

18. $\dfrac{(x + 3)^2}{16} + \dfrac{(y + 2)^2}{4} = 1$

Sketch the graph of each equation. See Examples 4 and 5.

19. $\dfrac{x^2}{4} - \dfrac{y^2}{9} = 1$ **20.** $\dfrac{x^2}{36} - \dfrac{y^2}{36} = 1$

21. $\dfrac{y^2}{25} - \dfrac{x^2}{16} = 1$ **22.** $\dfrac{y^2}{25} - \dfrac{x^2}{49} = 1$

23. $x^2 - 4y^2 = 16$ **24.** $4x^2 - y^2 = 36$

25. $16y^2 - x^2 = 16$ **26.** $4y^2 - 25x^2 = 100$

MIXED PRACTICE

Graph each equation. See Examples 1 through 5.

27. $\dfrac{y^2}{36} = 1 - x^2$ **28.** $\dfrac{x^2}{36} = 1 - y^2$

29. $4(x - 1)^2 + 9(y + 2)^2 = 36$

30. $25(x + 3)^2 + 4(y - 3)^2 = 100$

31. $8x^2 + 2y^2 = 32$ **32.** $3x^2 + 12y^2 = 48$

33. $25x^2 - y^2 = 25$ **34.** $x^2 - 9y^2 = 9$

MIXED PRACTICE—SECTIONS 10.1, 10.2

Identify whether each equation, when graphed, will be a parabola, circle, ellipse, or hyperbola. Sketch the graph of each equation.

> *If a parabola, label the vertex.*
> *If a circle, label the center and note the radius.*
> *If an ellipse, label the center.*
> *If a hyperbola, label the x- or y-intercepts.*

35. $(x - 7)^2 + (y - 2)^2 = 4$ **36.** $y = x^2 + 4$

37. $y = x^2 + 12x + 36$ **38.** $\dfrac{x^2}{4} + \dfrac{y^2}{9} = 1$

39. $\dfrac{y^2}{9} - \dfrac{x^2}{9} = 1$ **40.** $\dfrac{x^2}{16} - \dfrac{y^2}{4} = 1$

41. $\dfrac{x^2}{16} + \dfrac{y^2}{4} = 1$ **42.** $x^2 + y^2 = 16$

43. $x = y^2 + 4y - .1$ **44.** $x = -y^2 + 6y$

45. $9x^2 - 4y^2 = 36$ ▶ **46.** $9x^2 + 4y^2 = 36$

47. $\dfrac{(x - 1)^2}{49} + \dfrac{(y + 2)^2}{25} = 1$ ▶ **48.** $y^2 = x^2 + 16$

49. $\left(x + \dfrac{1}{2}\right)^2 + \left(y - \dfrac{1}{2}\right)^2 = 1$ **50.** $y = -2x^2 + 4x - 3$

REVIEW AND PREVIEW

Perform the indicated operations. See Sections 5.1 and 5.3.

51. $(2x^3)(-4x^2)$ **52.** $2x^3 - 4x^3$

53. $-5x^2 + x^2$ **54.** $(-5x^2)(x^2)$

CONCEPT EXTENSIONS

The graph of each equation is an ellipse. Determine which distance is longer, the distance between the x-intercepts or the distance between the y-intercepts. How much longer? See the Concept Check in this section.

55. $\dfrac{x^2}{16} + \dfrac{y^2}{25} = 1$ **56.** $\dfrac{x^2}{100} + \dfrac{y^2}{49} = 1$

57. $4x^2 + y^2 = 16$ **58.** $x^2 + 4y^2 = 36$

59. If you are given a list of equations of circles, parabolas, ellipses, and hyperbolas, explain how you could distinguish the different conic sections from their equations.

60. We know that $x^2 + y^2 = 25$ is the equation of a circle. Rewrite the equation so that the right side is equal to 1. Which type of conic section does this equation form resemble? In fact, the circle is a special case of this type of conic section. Describe the conditions under which this type of conic section is a circle.

The orbits of stars, planets, comets, asteroids, and satellites all have the shape of one of the conic sections. Astronomers use a measure called eccentricity to describe the shape and elongation of an orbital path. For the circle and ellipse, eccentricity e is calculated with the formula $e = \dfrac{c}{d}$, where $c^2 = |a^2 - b^2|$ and d is the larger value of a or b. For a hyperbola, eccentricity e is calculated with the formula $e = \dfrac{c}{d}$, where $c^2 = a^2 + b^2$ and the value of d is equal to a if the hyperbola has x-intercepts or equal to b if the hyperbola has y-intercepts. Use equations A–H to answer Exercises 61–70.

A. $\dfrac{x^2}{36} - \dfrac{y^2}{13} = 1$ **B.** $\dfrac{x^2}{4} + \dfrac{y^2}{4} = 1$ **C.** $\dfrac{x^2}{25} + \dfrac{y^2}{16} = 1$

D. $\dfrac{y^2}{25} - \dfrac{x^2}{39} = 1$ **E.** $\dfrac{x^2}{17} + \dfrac{y^2}{81} = 1$ **F.** $\dfrac{x^2}{36} + \dfrac{y^2}{36} = 1$

G. $\dfrac{x^2}{16} - \dfrac{y^2}{65} = 1$ **H.** $\dfrac{x^2}{144} + \dfrac{y^2}{140} = 1$

61. Identify the type of conic section represented by each of the equations A–H.

62. For each of the equations A–H, identify the values of a^2 and b^2.

63. For each of the equations A–H, calculate the value of c^2 and c.

64. For each of the equations A–H, find the value of d.

65. For each of the equations A–H, calculate the eccentricity e.

66. What do you notice about the values of e for the equations you identified as ellipses?

67. What do you notice about the values of e for the equations you identified as circles?

68. What do you notice about the values of e for the equations you identified as hyperbolas?

69. The eccentricity of a parabola is exactly 1. Use this information and the observations you made in Exercises 66, 67, and 68 to describe a way that could be used to identify the type of conic section based on its eccentricity value.

70. Graph each of the conic sections given in equations A–H. What do you notice about the shape of the ellipses for increasing values of eccentricity? Which is the most elliptical? Which is the least elliptical, that is, the most circular?

71. A planet's orbit about the sun can be described as an ellipse. Consider the sun as the origin of a rectangular coordinate system. Suppose that the x-intercepts of the elliptical path of the planet are $\pm 130{,}000{,}000$ and that the y-intercepts are $\pm 125{,}000{,}000$. Write the equation of the elliptical path of the planet.

72. Comets orbit the sun in elongated ellipses. Consider the sun as the origin of a rectangular coordinate system. Suppose that the equation of the path of the comet is

$$\frac{(x - 1{,}782{,}000{,}000)^2}{3.42 \times 10^{23}} + \frac{(y - 356{,}400{,}000)^2}{1.368 \times 10^{22}} = 1$$

Find the center of the path of the comet.

73. Use a graphing calculator to verify Exercise 46.

74. Use a graphing calculator to verify Exercise 12.

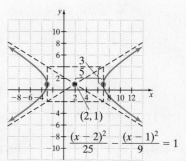

$$\frac{(x - 2)^2}{25} - \frac{(x - 1)^2}{9} = 1$$

For Exercises 75 through 80, see the example below.

Example

Sketch the graph of $\dfrac{(x - 2)^2}{25} - \dfrac{(y - 1)^2}{9} = 1.$

Solution

This hyperbola has center $(2, 1)$. Notice that $a = 5$ and $b = 3$.

Sketch the graph of each equation.

75. $\dfrac{(x - 1)^2}{4} - \dfrac{(y + 1)^2}{25} = 1$ **76.** $\dfrac{(x + 2)^2}{9} - \dfrac{(y - 1)^2}{4} = 1$

77. $\dfrac{y^2}{16} - \dfrac{(x + 3)^2}{9} = 1$ **78.** $\dfrac{(y + 4)^2}{4} - \dfrac{x^2}{25} = 1$

79. $\dfrac{(x + 5)^2}{16} - \dfrac{(y + 2)^2}{25} = 1$ **80.** $\dfrac{(x - 3)^2}{9} - \dfrac{(y - 2)^2}{4} = 1$

Integrated Review   GRAPHING CONIC SECTIONS

Following is a summary of conic sections.

Conic Sections

	Standard Form	Graph
Parabola	$y = a(x - h)^2 + k$	
Parabola	$x = a(y - k)^2 + h$	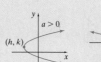
Circle	$(x - h)^2 + (y - k)^2 = r^2$	
Ellipse center $(0, 0)$	$\dfrac{x^2}{a^2} + \dfrac{y^2}{b^2} = 1$	
Hyperbola center $(0, 0)$	$\dfrac{x^2}{a^2} - \dfrac{y^2}{b^2} = 1$	
Hyperbola center $(0, 0)$	$\dfrac{y^2}{b^2} - \dfrac{x^2}{a^2} = 1$	

Identify whether each equation, when graphed, will be a parabola, circle, ellipse, or hyperbola. Then graph each equation.

1. $(x - 7)^2 + (y - 2)^2 = 4$

2. $y = x^2 + 4$

3. $y = x^2 + 12x + 36$

4. $\dfrac{x^2}{4} + \dfrac{y^2}{9} = 1$

5. $\dfrac{y^2}{9} - \dfrac{x^2}{9} = 1$

6. $\dfrac{x^2}{16} - \dfrac{y^2}{4} = 1$

7. $\dfrac{x^2}{16} + \dfrac{y^2}{4} = 1$

8. $x^2 + y^2 = 16$

9. $x = y^2 + 4y - 1$

10. $x = -y^2 + 6y$

11. $9x^2 - 4y^2 = 36$

12. $9x^2 + 4y^2 = 36$

13. $\dfrac{(x - 1)^2}{49} + \dfrac{(y + 2)^2}{25} = 1$

14. $y^2 = x^2 + 16$

15. $\left(x + \dfrac{1}{2}\right)^2 + \left(y - \dfrac{1}{2}\right)^2 = 1$

10.3 Solving Nonlinear Systems of Equations

OBJECTIVES

1 Solve a Nonlinear System by Substitution.

2 Solve a Nonlinear System by Elimination.

In Section 4.1, we used graphing, substitution, and elimination methods to find solutions of systems of linear equations in two variables. We now apply these same methods to nonlinear systems of equations in two variables. A **nonlinear system of equations** is a system of equations at least one of which is not linear. Since we will be graphing the equations in each system, we are interested in real number solutions only.

OBJECTIVE

1 Solving Nonlinear Systems by Substitution

First, nonlinear systems are solved by the substitution method.

EXAMPLE 1 Solve the system

$$\begin{cases} x^2 - 3y = 1 \\ x - y = 1 \end{cases}$$

Solution We can solve this system by substitution if we solve one equation for one of the variables. Solving the first equation for x is not the best choice since doing so introduces a radical. Also, solving for y in the first equation introduces a fraction. We solve the second equation for y.

$$x - y = 1 \quad \text{Second equation}$$
$$x - 1 = y \quad \text{Solve for } y.$$

Replace y with $x - 1$ in the first equation, and then solve for x.

$$x^2 - 3y = 1 \quad \text{First equation}$$
$$x^2 - 3(x - 1) = 1 \quad \text{Replace } y \text{ with } x - 1.$$
$$x^2 - 3x + 3 = 1$$
$$x^2 - 3x + 2 = 0$$
$$(x - 2)(x - 1) = 0$$
$$x = 2 \quad \text{or} \quad x = 1$$

Let $x = 2$ and then let $x = 1$ in the equation $y = x - 1$ to find corresponding y-values.

Let $x = 2$.	Let $x = 1$.
$y = x - 1$	$y = x - 1$
$y = 2 - 1 = 1$	$y = 1 - 1 = 0$

The solutions are $(2, 1)$ and $(1, 0)$, or the solution set is $\{(2, 1), (1, 0)\}$. Check both solutions in both equations. Both solutions satisfy both equations, so both are solutions of the system. The graph of each equation in the system is shown next. Intersections of the graphs are at $(2, 1)$ and $(1, 0)$.

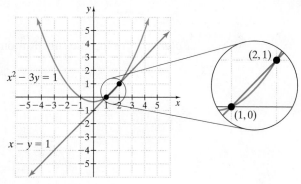

PRACTICE
1 Solve the system $\begin{cases} x^2 - 4y = 4 \\ x + y = -1 \end{cases}$.

EXAMPLE 2 Solve the system

$$\begin{cases} y = \sqrt{x} \\ x^2 + y^2 = 6 \end{cases}$$

Solution This system is ideal for substitution since y is expressed in terms of x in the first equation. Notice that if $y = \sqrt{x}$, then both x and y must be nonnegative if they are real numbers. Substitute $\sqrt{x}$ for y in the second equation, and solve for x.

$$x^2 + y^2 = 6$$
$$x^2 + (\sqrt{x})^2 = 6 \quad \text{Let } y = \sqrt{x}$$
$$x^2 + x = 6$$
$$x^2 + x - 6 = 0$$
$$(x + 3)(x - 2) = 0$$
$$x = -3 \quad \text{or} \quad x = 2$$

The solution -3 is discarded because we have noted that x must be nonnegative. To see this, let $x = -3$ in the first equation. Then let $x = 2$ in the first equation to find a corresponding y-value.

Let $x = -3$.
$$y = \sqrt{x}$$
$$y = \sqrt{-3} \quad \text{Not a real number}$$

Let $x = 2$.
$$y = \sqrt{x}$$
$$y = \sqrt{2}$$

Since we are interested only in real number solutions, the only solution is $(2, \sqrt{2})$. Check to see that this solution satisfies both equations. The graph of each equation in the system is shown to the right.

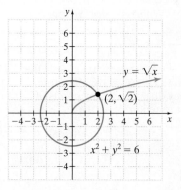

PRACTICE
2 Solve the system $\begin{cases} y = -\sqrt{x} \\ x^2 + y^2 = 20 \end{cases}$.

EXAMPLE 3 Solve the system

$$\begin{cases} x^2 + y^2 = 4 \\ x + y = 3 \end{cases}$$

Solution We use the substitution method and solve the second equation for x.

$$x + y = 3 \qquad \text{Second equation}$$
$$x = 3 - y$$

Now we let $x = 3 - y$ in the first equation.

$$x^2 + y^2 = 4 \qquad \text{First equation}$$
$$(3 - y)^2 + y^2 = 4 \qquad \text{Let } x = 3 - y.$$
$$9 - 6y + y^2 + y^2 = 4$$
$$2y^2 - 6y + 5 = 0$$

By the quadratic formula, where $a = 2$, $b = -6$, and $c = 5$, we have

$$y = \frac{6 \pm \sqrt{(-6)^2 - 4 \cdot 2 \cdot 5}}{2 \cdot 2} = \frac{6 \pm \sqrt{-4}}{4}$$

Since $\sqrt{-4}$ is not a real number, there is no real solution, or $\varnothing$. Graphically, the circle and the line do not intersect, as shown below.

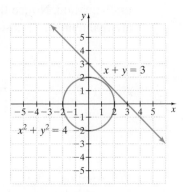

PRACTICE
3   Solve the system $\begin{cases} x^2 + y^2 = 9 \\ x - y = 5 \end{cases}$.

✓**CONCEPT CHECK**
Without solving, how can you tell that $x^2 + y^2 = 9$ and $x^2 + y^2 = 16$ do not have any points of intersection?

Answer to Concept Check:
$x^2 + y^2 = 9$ is a circle inside the circle $x^2 + y^2 = 16$, therefore they do not have any points of intersection.

OBJECTIVE
2   Solving Nonlinear Systems by Elimination
Some nonlinear systems may be solved by the elimination method.

EXAMPLE 4 Solve the system

$$\begin{cases} x^2 + 2y^2 = 10 \\ x^2 - y^2 = 1 \end{cases}$$

Solution We will use the elimination, or addition, method to solve this system. To eliminate x^2 when we add the two equations, multiply both sides of the second equation by -1. Then

$$\begin{cases} x^2 + 2y^2 = 10 \\ (-1)(x^2 - y^2) = -1 \cdot 1 \end{cases}$$ is equivalent to $$\begin{cases} x^2 + 2y^2 = 10 \\ \underline{-x^2 + y^2 = -1} \end{cases}$$ Add.
$$3y^2 = 9$$
$$y^2 = 3 \qquad \text{Divide both}$$
$$y = \pm\sqrt{3} \qquad \text{sides by 3.}$$

To find the corresponding x-values, we let $y = \sqrt{3}$ and $y = -\sqrt{3}$ in either original equation. We choose the second equation.

Let $y = \sqrt{3}$.	Let $y = -\sqrt{3}$.
$x^2 - y^2 = 1$	$x^2 - y^2 = 1$
$x^2 - (\sqrt{3})^2 = 1$	$x^2 - (-\sqrt{3})^2 = 1$
$x^2 - 3 = 1$	$x^2 - 3 = 1$
$x^2 = 4$	$x^2 = 4$
$x = \pm\sqrt{4} = \pm2$	$x = \pm\sqrt{4} = \pm2$

The solutions are $(2, \sqrt{3})$, $(-2, \sqrt{3})$, $(2, -\sqrt{3})$, and $(-2, -\sqrt{3})$. Check all four ordered pairs in both equations of the system. The graph of each equation in this system is shown.

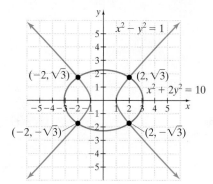

PRACTICE
4 Solve the system $\begin{cases} x^2 + 4y^2 = 16 \\ x^2 - y^2 = 1 \end{cases}$.

Vocabulary, Readiness & Video Check

Martin-Gay Interactive Videos

See Video 10.3

Watch the section lecture video and answer the following questions.

OBJECTIVE 1

1. In ▱ Example 1, why do we choose not to solve either equation for y?

OBJECTIVE 2

2. In ▱ Example 2, what important reminder is made as the second equation is multiplied by a number to get opposite coefficients of x?

10.3 Exercise Set MyMathLab®

MIXED PRACTICE

Solve each nonlinear system of equations for real solutions. See Examples 1 through 4.

1. $\begin{cases} x^2 + y^2 = 25 \\ 4x + 3y = 0 \end{cases}$

2. $\begin{cases} x^2 + y^2 = 25 \\ 3x + 4y = 0 \end{cases}$

3. $\begin{cases} x^2 + 4y^2 = 10 \\ y = x \end{cases}$

4. $\begin{cases} 4x^2 + y^2 = 10 \\ y = x \end{cases}$

5. $\begin{cases} y^2 = 4 - x \\ x - 2y = 4 \end{cases}$

6. $\begin{cases} x^2 + y^2 = 4 \\ x + y = -2 \end{cases}$

7. $\begin{cases} x^2 + y^2 = 9 \\ 16x^2 - 4y^2 = 64 \end{cases}$

8. $\begin{cases} 4x^2 + 3y^2 = 35 \\ 5x^2 + 2y^2 = 42 \end{cases}$

9. $\begin{cases} x^2 + 2y^2 = 2 \\ x - y = 2 \end{cases}$

10. $\begin{cases} x^2 + 2y^2 = 2 \\ x^2 - 2y^2 = 6 \end{cases}$

11. $\begin{cases} y = x^2 - 3 \\ 4x - y = 6 \end{cases}$

12. $\begin{cases} y = x + 1 \\ x^2 - y^2 = 1 \end{cases}$

13. $\begin{cases} y = x^2 \\ 3x + y = 10 \end{cases}$

14. $\begin{cases} 6x - y = 5 \\ xy = 1 \end{cases}$

15. $\begin{cases} y = 2x^2 + 1 \\ x + y = -1 \end{cases}$

16. $\begin{cases} x^2 + y^2 = 9 \\ x + y = 5 \end{cases}$

17. $\begin{cases} y = x^2 - 4 \\ y = x^2 - 4x \end{cases}$

18. $\begin{cases} x = y^2 - 3 \\ x = y^2 - 3y \end{cases}$

19. $\begin{cases} 2x^2 + 3y^2 = 14 \\ -x^2 + y^2 = 3 \end{cases}$

20. $\begin{cases} 4x^2 - 2y^2 = 2 \\ -x^2 + y^2 = 2 \end{cases}$

21. $\begin{cases} x^2 + y^2 = 1 \\ x^2 + (y + 3)^2 = 4 \end{cases}$

22. $\begin{cases} x^2 + 2y^2 = 4 \\ x^2 - y^2 = 4 \end{cases}$

23. $\begin{cases} y = x^2 + 2 \\ y = -x^2 + 4 \end{cases}$

24. $\begin{cases} x = -y^2 - 3 \\ x = y^2 - 5 \end{cases}$

25. $\begin{cases} 3x^2 + y^2 = 9 \\ 3x^2 - y^2 = 9 \end{cases}$

26. $\begin{cases} x^2 + y^2 = 25 \\ x = y^2 - 5 \end{cases}$

27. $\begin{cases} x^2 + 3y^2 = 6 \\ x^2 - 3y^2 = 10 \end{cases}$

28. $\begin{cases} x^2 + y^2 = 1 \\ y = x^2 - 9 \end{cases}$

29. $\begin{cases} x^2 + y^2 = 36 \\ y = \dfrac{1}{6}x^2 - 6 \end{cases}$

30. $\begin{cases} x^2 + y^2 = 16 \\ y = -\dfrac{1}{4}x^2 + 4 \end{cases}$

31. $\begin{cases} y = \sqrt{x} \\ x^2 + y^2 = 12 \end{cases}$ 32. $\begin{cases} y = \sqrt{x} \\ x^2 + y^2 = 20 \end{cases}$

REVIEW AND PREVIEW

Graph each inequality in two variables. See Section 3.7.

33. $x > -3$ 34. $y \le 1$

35. $y < 2x - 1$ 36. $3x - y \le 4$

Find the perimeter of each geometric figure. See Section 5.3.

△ **37.**
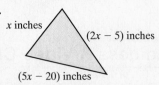
x inches, $(2x - 5)$ inches, $(5x - 20)$ inches

△ **38.**

$(3x + 2)$ centimeters

△ **39.** $(x^2 + 3x + 1)$ meters
x^2 meters

△ **40.**
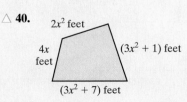
$2x^2$ feet, $4x$ feet, $(3x^2 + 1)$ feet, $(3x^2 + 7)$ feet

CONCEPT EXTENSIONS

For the exercises below, see the Concept Check in this section.

41. Without graphing, how can you tell that the graph of $x^2 + y^2 = 1$ and $x^2 + y^2 = 4$ do not have any points of intersection?

42. Without solving, how can you tell that the graphs of $y = 2x + 3$ and $y = 2x + 7$ do not have any points of intersection?

43. How many real solutions are possible for a system of equations whose graphs are a circle and a parabola? Draw diagrams to illustrate each possibility.

44. How many real solutions are possible for a system of equations whose graphs are an ellipse and a line? Draw diagrams to illustrate each possibility.

Solve.

45. The sum of the squares of two numbers is 130. The difference of the squares of the two numbers is 32. Find the two numbers.

46. The sum of the squares of two numbers is 20. Their product is 8. Find the two numbers.

△ **47.** During the development stage of a new rectangular keypad for a security system, it was decided that the area of the rectangle should be 285 square centimeters and the perimeter should be 68 centimeters. Find the dimensions of the keypad.

△ **48.** A rectangular holding pen for cattle is to be designed so that its perimeter is 92 feet and its area is 525 feet. Find the dimensions of the holding pen.

*Recall that in business, a demand function expresses the quantity of a commodity demanded as a function of the commodity's unit price. A supply function expresses the quantity of a commodity supplied as a function of the commodity's unit price. When the quantity produced and supplied is equal to the quantity demanded, then we have what is called **market equilibrium.***

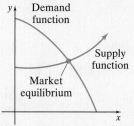

49. The demand function for a certain compact disc is given by the function
$$p = -0.01x^2 - 0.2x + 9$$
and the corresponding supply function is given by
$$p = 0.01x^2 - 0.1x + 3$$
where p is in dollars and x is in thousands of units. Find the equilibrium quantity and the corresponding price by solving the system consisting of the two given equations.

50. The demand function for a certain style of picture frame is given by the function
$$p = -2x^2 + 90$$
and the corresponding supply function is given by
$$p = 9x + 34$$
where p is in dollars and x is in thousands of units. Find the equilibrium quantity and the corresponding price by solving the system consisting of the two given equations.

Use a graphing calculator to verify the results of each exercise.

51. Exercise 3.   **52.** Exercise 4.

53. Exercise 23.   **54.** Exercise 24.

10.4 Nonlinear Inequalities and Systems of Inequalities

OBJECTIVES

1 Graph a Nonlinear Inequality.

2 Graph a System of Nonlinear Inequalities.

OBJECTIVE

1 Graphing Nonlinear Inequalities

We can graph a nonlinear inequality in two variables such as $\dfrac{x^2}{9} + \dfrac{y^2}{16} \le 1$ in a way similar to the way we graphed a linear inequality in two variables in Section 3.7. First, graph the related equation $\dfrac{x^2}{9} + \dfrac{y^2}{16} = 1$. The graph of the equation is our boundary. Then, using test points, we determine and shade the region whose points satisfy the inequality.

EXAMPLE 1   Graph $\dfrac{x^2}{9} + \dfrac{y^2}{16} \le 1$.

Solution   First, graph the equation $\dfrac{x^2}{9} + \dfrac{y^2}{16} = 1$. Sketch a solid curve since the graph of $\dfrac{x^2}{9} + \dfrac{y^2}{16} \le 1$ includes the graph of $\dfrac{x^2}{9} + \dfrac{y^2}{16} = 1$. The graph is an ellipse, and it

(Continued on next page)

divides the plane into two regions, the "inside" and the "outside" of the ellipse. To determine which region contains the solutions, select a test point in either region and determine whether the coordinates of the point satisfy the inequality. We choose $(0, 0)$ as the test point.

$$\frac{x^2}{9} + \frac{y^2}{16} \leq 1$$

$$\frac{0^2}{9} + \frac{0^2}{16} \leq 1 \quad \text{Let } x = 0 \text{ and } y = 0.$$

$$0 \leq 1 \quad \text{True}$$

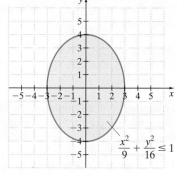

Since this statement is true, the solution set is the region containing $(0, 0)$. The graph of the solution set includes the points on and inside the ellipse, as shaded in the figure.

PRACTICE
1   Graph $\dfrac{x^2}{36} + \dfrac{y^2}{16} \geq 1$.

EXAMPLE 2   Graph $4y^2 > x^2 + 16$.

Solution   The related equation is $4y^2 = x^2 + 16$. Subtract x^2 from both sides and divide both sides by 16, and we have $\dfrac{y^2}{4} - \dfrac{x^2}{16} = 1$, which is a hyperbola. Graph the hyperbola as a dashed curve since the graph of $4y^2 > x^2 + 16$ does _not_ include the graph of $4y^2 = x^2 + 16$. The hyperbola divides the plane into three regions. Select a test point in each region—not on a boundary line—to determine whether that region contains solutions of the inequality.

Test Region A with $(0, 4)$	_Test Region B with_ $(0, 0)$	_Test Region C with_ $(0, -4)$
$4y^2 > x^2 + 16$	$4y^2 > x^2 + 16$	$4y^2 > x^2 + 16$
$4(4)^2 > 0^2 + 16$	$4(0)^2 > 0^2 + 16$	$4(-4)^2 > 0^2 + 16$
$64 > 16$ True	$0 > 16$ False	$64 > 16$ True

The graph of the solution set includes the shaded regions A and C only, not the boundary.

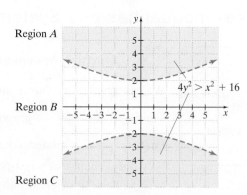

PRACTICE
2   Graph $16y^2 > 9x^2 + 144$.

OBJECTIVE

2   **Graphing Systems of Nonlinear Inequalities** ▶

In Sections 3.7 and 4.5 we graphed systems of linear inequalities. Recall that the graph of a system of inequalities is the intersection of the graphs of the inequalities.

EXAMPLE 3 Graph the system

$$\begin{cases} x \le 1 - 2y \\ y \le x^2 \end{cases}$$

Solution We graph each inequality on the same set of axes. The intersection is shown in the third graph below. It is the darkest shaded (appears purple) region along with its boundary lines. The coordinates of the points of intersection can be found by solving the related system.

$$\begin{cases} x = 1 - 2y \\ y = x^2 \end{cases}$$

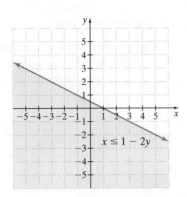 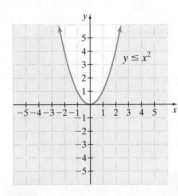 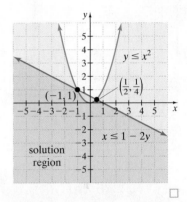

PRACTICE
3 Graph the system $\begin{cases} y \ge x^2 \\ y \le -3x + 2 \end{cases}$.

EXAMPLE 4 Graph the system

$$\begin{cases} x^2 + y^2 < 25 \\ \dfrac{x^2}{9} - \dfrac{y^2}{25} < 1 \\ y < x + 3 \end{cases}$$

Solution We graph each inequality. The graph of $x^2 + y^2 < 25$ contains points "inside" the circle that has center $(0, 0)$ and radius 5. The graph of $\dfrac{x^2}{9} - \dfrac{y^2}{25} < 1$ is the region between the two branches of the hyperbola with x-intercepts -3 and 3 and center $(0, 0)$. The graph of $y < x + 3$ is the region "below" the line with slope 1 and y-intercept $(0, 3)$. The graph of the solution set of the system is the intersection of all the graphs, the darkest shaded region shown. The boundary of this region is not part of the solution.

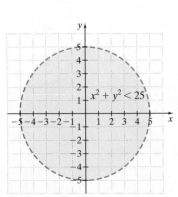 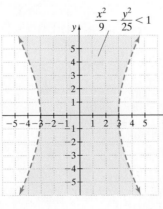 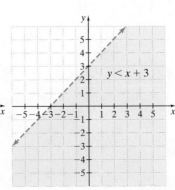 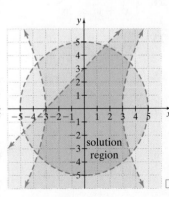

PRACTICE
4 Graph the system $\begin{cases} x^2 + y^2 < 16 \\ \dfrac{x^2}{4} - \dfrac{y^2}{9} < 1 \\ y < x + 3 \end{cases}$.

Vocabulary, Readiness & Video Check

Martin-Gay Interactive Videos

See Video 10.4

Watch the section lecture video and answer the following questions.

OBJECTIVE
1

1. From Example 1, explain the similarities between graphing linear inequalities and graphing nonlinear inequalities.

OBJECTIVE
2

2. From Example 2, describe one possible illustration of graphs of two circle inequalities in which the system has no solution—that is, the graph of the inequalities in the system do not overlap.

10.4 Exercise Set   MyMathLab®

Graph each inequality. See Examples 1 and 2.

1. $y < x^2$

2. $y < -x^2$

3. $x^2 + y^2 \geq 16$

4. $x^2 + y^2 < 36$

5. $\dfrac{x^2}{4} - y^2 < 1$

6. $x^2 - \dfrac{y^2}{9} \geq 1$

7. $y > (x - 1)^2 - 3$

8. $y > (x + 3)^2 + 2$

9. $x^2 + y^2 \leq 9$

10. $x^2 + y^2 > 4$

11. $y > -x^2 + 5$

12. $y < -x^2 + 5$

▶ **13.** $\dfrac{x^2}{4} + \dfrac{y^2}{9} \leq 1$

14. $\dfrac{x^2}{25} + \dfrac{y^2}{4} \geq 1$

15. $\dfrac{y^2}{4} - x^2 \leq 1$

16. $\dfrac{y^2}{16} - \dfrac{x^2}{9} > 1$

17. $y < (x - 2)^2 + 1$

18. $y > (x - 2)^2 + 1$

19. $y \leq x^2 + x - 2$

20. $y > x^2 + x - 2$

Graph each system. See Examples 3 and 4.

21. $\begin{cases} 4x + 3y \geq 12 \\ x^2 + y^2 < 16 \end{cases}$

22. $\begin{cases} 3x - 4y \leq 12 \\ x^2 + y^2 < 16 \end{cases}$

23. $\begin{cases} x^2 + y^2 \leq 9 \\ x^2 + y^2 \geq 1 \end{cases}$

24. $\begin{cases} x^2 + y^2 \geq 9 \\ x^2 + y^2 \geq 16 \end{cases}$

25. $\begin{cases} y > x^2 \\ y \geq 2x + 1 \end{cases}$

26. $\begin{cases} y \leq -x^2 + 3 \\ y \leq 2x - 1 \end{cases}$

▶ **27.** $\begin{cases} x^2 + y^2 > 9 \\ y > x^2 \end{cases}$

28. $\begin{cases} x^2 + y^2 \leq 9 \\ y < x^2 \end{cases}$

29. $\begin{cases} \dfrac{x^2}{4} + \dfrac{y^2}{9} \geq 1 \\ x^2 + y^2 \geq 4 \end{cases}$

30. $\begin{cases} x^2 + (y - 2)^2 \geq 9 \\ \dfrac{x^2}{4} + \dfrac{y^2}{25} < 1 \end{cases}$

31. $\begin{cases} x^2 - y^2 \geq 1 \\ y \geq 0 \end{cases}$

32. $\begin{cases} x^2 - y^2 \geq 1 \\ x \geq 0 \end{cases}$

33. $\begin{cases} x + y \geq 1 \\ 2x + 3y < 1 \\ x > -3 \end{cases}$

34. $\begin{cases} x - y < -1 \\ 4x - 3y > 0 \\ y > 0 \end{cases}$

35. $\begin{cases} x^2 - y^2 < 1 \\ \dfrac{x^2}{16} + y^2 \leq 1 \\ x \geq -2 \end{cases}$

36. $\begin{cases} x^2 - y^2 \geq 1 \\ \dfrac{x^2}{16} + \dfrac{y^2}{4} \leq 1 \\ y \geq 1 \end{cases}$

REVIEW AND PREVIEW

Determine whether each graph is the graph of a function. See Section 3.2.

37.

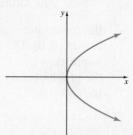

38.

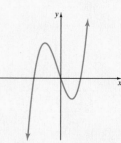

39.

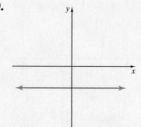

40.

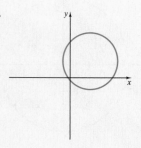

Find each function value if $f(x) = 3x^2 - 2$. See Section 3.2.

41. $f(-1)$ **42.** $f(-3)$

43. $f(a)$ **44.** $f(b)$

47. Graph the system $\begin{cases} y \le x^2 \\ y \ge x + 2 \\ x \ge 0 \\ y \ge 0 \end{cases}$.

CONCEPT EXTENSIONS

45. Discuss how graphing a linear inequality such as $x + y < 9$ is similar to graphing a nonlinear inequality such as $x^2 + y^2 < 9$.

46. Discuss how graphing a linear inequality such as $x + y < 9$ is different from graphing a nonlinear inequality such as $x^2 + y^2 < 9$.

48. Graph the system: $\begin{cases} x \ge 0 \\ y \ge 0 \\ y \ge x^2 + 1 \\ y \le 4 - x \end{cases}$

Chapter 10 Vocabulary Check

Fill in each blank with one of the words or phrases listed below.

circle	ellipse	hyperbola
conic sections	vertex	diameter
center	radius	nonlinear system of equations

1. A(n) _____ is the set of all points in a plane that are the same distance from a fixed point, called the _____.

2. A(n) _____ is a system of equations at least one of which is not linear.

3. A(n) _____ is the set of points in a plane such that the sum of the distances of those points from two fixed points is a constant.

4. In a circle, the distance from the center to a point of the circle is called its _____.

5. A(n) _____ is the set of points in a plane such that the absolute value of the difference of the distance from two fixed points is constant.

6. The circle, parabola, ellipse, and hyperbola are called the _____.

7. For a parabola that opens upward, the lowest point is the _____.

8. Twice a circle's radius is its _____.

Chapter 10 Highlights

DEFINITIONS AND CONCEPTS	EXAMPLES

Section 10.1   The Parabola and the Circle

Parabolas

$$y = a(x - h)^2 + k$$

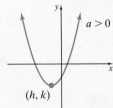

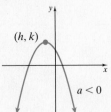

Graph

$$x = 3y^2 - 12y + 13.$$
$$x - 13 = 3y^2 - 12y$$
$$x - 13 + 3(4) = 3(y^2 - 4y + 4) \quad \text{Add } 3(4) \text{ to}$$
$$x = 3(y - 2)^2 + 1 \quad \text{both sides.}$$

Since $a = 3$, this parabola opens to the right with vertex $(1, 2)$. Its axis of symmetry is $y = 2$. The x-intercept is $(13, 0)$.

(continued)

DEFINITIONS AND CONCEPTS	EXAMPLES

Section 10.1   The Parabola and the Circle (continued)

$$x = a(y - k)^2 + h$$

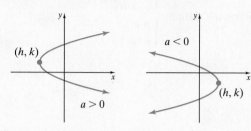

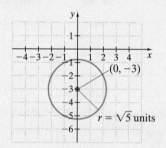

Circle

The graph of $(x - h)^2 + (y - k)^2 = r^2$ is a circle with center (h, k) and radius r.

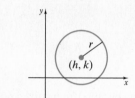

Graph $x^2 + (y + 3)^2 = 5$.

This equation can be written as

$$(x - 0)^2 + (y + 3)^2 = 5 \text{ with } h = 0,$$
$$k = -3, \text{ and } r = \sqrt{5}.$$

The center of this circle is $(0, -3)$, and the radius is $\sqrt{5}$.

Section 10.2   The Ellipse and the Hyperbola

Ellipse with center $(0, 0)$

The graph of an equation of the form $\dfrac{x^2}{a^2} + \dfrac{y^2}{b^2} = 1$ is an ellipse with center $(0, 0)$. The x-intercepts are $(a, 0)$ and $(-a, 0)$, and the y-intercepts are $(0, b)$ and $(0, -b)$.

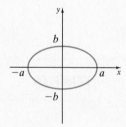

Graph $4x^2 + 9y^2 = 36$.

$$\frac{x^2}{9} + \frac{y^2}{4} = 1 \quad \text{Divide by 36.}$$

$$\frac{x^2}{3^2} + \frac{y^2}{2^2} = 1$$

The ellipse has center $(0, 0)$, x-intercepts $(3, 0)$ and $(-3, 0)$, and y-intercepts $(0, 2)$ and $(0, -2)$.

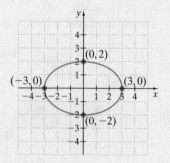

Hyperbola with center $(0, 0)$

The graph of an equation of the form

$\dfrac{x^2}{a^2} - \dfrac{y^2}{b^2} = 1$ is a hyperbola with

center $(0, 0)$ and x-intercepts $(a, 0)$ and $(-a, 0)$.

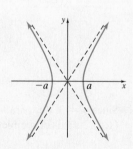

DEFINITIONS AND CONCEPTS	EXAMPLES

Section 10.2   The Ellipse and the Hyperbola (continued)

The graph of an equation of the form $\dfrac{y^2}{b^2} - \dfrac{x^2}{a^2} = 1$ is a hyperbola with center $(0,0)$ and y-intercepts $(0,b)$ and $(0,-b)$.

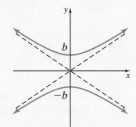

Graph $\dfrac{x^2}{9} - \dfrac{y^2}{4} = 1$. Here $a = 3$ and $b = 2$.

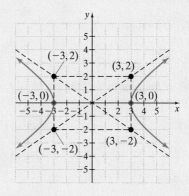

Section 10.3   Solving Nonlinear Systems of Equations

A **nonlinear system of equations** is a system of equations at least one of which is not linear. Both the substitution method and the elimination method may be used to solve a nonlinear system of equations.

Solve the nonlinear system $\begin{cases} y = x + 2 \\ 2x^2 + y^2 = 3 \end{cases}$.

Substitute $x + 2$ for y in the second equation.

$$2x^2 + y^2 = 3$$
$$2x^2 + (x + 2)^2 = 3$$
$$2x^2 + x^2 + 4x + 4 = 3$$
$$3x^2 + 4x + 1 = 0$$
$$(3x + 1)(x + 1) = 0$$
$$x = -\frac{1}{3}, x = -1$$

If $x = -\dfrac{1}{3}, y = x + 2 = -\dfrac{1}{3} + 2 = \dfrac{5}{3}$.

If $x = -1, y = x + 2 = -1 + 2 = 1$.

The solutions are $\left(-\dfrac{1}{3}, \dfrac{5}{3}\right)$ and $(-1, 1)$.

Section 10.4   Nonlinear Inequalities and Systems of Inequalities

The graph of a system of inequalities is the intersection of the graphs of the inequalities.

Graph the system $\begin{cases} x \geq y^2 \\ x + y \leq 4 \end{cases}$.

The graph of the system is the purple shaded region along with its boundary lines.

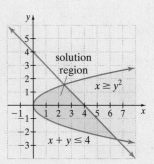

Chapter 10 Review

(10.1) *Write an equation of the circle with the given center and radius.*

1. center $(-4, 4)$, radius 3

2. center $(5, 0)$, radius 5

3. center $(-7, -9)$, radius $\sqrt{11}$

4. center $(0, 0)$, radius $\dfrac{7}{2}$

Sketch the graph of the equation. If the graph is a circle, find its center. If the graph is a parabola, find its vertex.

5. $x^2 + y^2 = 7$

6. $x = 2(y - 5)^2 + 4$

7. $x = -(y + 2)^2 + 3$

8. $(x - 1)^2 + (y - 2)^2 = 4$

9. $y = -x^2 + 4x + 10$

10. $x = -y^2 - 4y + 6$

11. $x = \dfrac{1}{2}y^2 + 2y + 1$

12. $y = -3x^2 + \dfrac{1}{2}x + 4$

13. $x^2 + y^2 + 2x + y = \dfrac{3}{4}$

14. $x^2 + y^2 - 3y = \dfrac{7}{4}$

15. $4x^2 + 4y^2 + 16x + 8y = 1$

16. $3x^2 + 3y^2 + 18x - 12y = -12$

(10.1, 10.2) *Graph each equation.*

17. $x^2 - \dfrac{y^2}{4} = 1$

18. $x^2 + \dfrac{y^2}{4} = 1$

19. $4y^2 + 9x^2 = 36$

20. $-5x^2 + 25y^2 = 125$

21. $x^2 - y^2 = 1$

22. $\dfrac{(x + 3)^2}{9} + \dfrac{(y - 4)^2}{25} = 1$

23. $y = x^2 + 9$

24. $36y^2 - 49x^2 = 1764$

25. $x = 4y^2 - 16$

26. $y = x^2 + 4x + 6$

27. $y^2 + 2(x - 1)^2 = 8$

28. $x - 4y = y^2$

29. $x^2 - 4 = y^2$

30. $x^2 = 4 - y^2$

31. $36y^2 = 576 + 16x^2$

32. $3(x - 7)^2 + 3(y + 4)^2 = 1$

(10.3) *Solve each system of equations.*

33. $\begin{cases} y = 2x - 4 \\ y^2 = 4x \end{cases}$

34. $\begin{cases} x^2 + y^2 = 4 \\ x - y = 4 \end{cases}$

35. $\begin{cases} y = x + 2 \\ y = x^2 \end{cases}$

36. $\begin{cases} 4x - y^2 = 0 \\ 2x^2 + y^2 = 16 \end{cases}$

37. $\begin{cases} x^2 + 4y^2 = 16 \\ x^2 + y^2 = 4 \end{cases}$

38. $\begin{cases} x^2 + 2y = 9 \\ 5x - 2y = 5 \end{cases}$

39. $\begin{cases} y = 3x^2 + 5x - 4 \\ y = 3x^2 - x + 2 \end{cases}$

40. $\begin{cases} x^2 - 3y^2 = 1 \\ 4x^2 + 5y^2 = 21 \end{cases}$

△ **41.** Find the length and the width of a room whose area is 150 square feet and whose perimeter is 50 feet.

42. What is the greatest number of real number solutions possible for a system of two equations whose graphs are an ellipse and a hyperbola?

(10.4) *Graph each inequality or system of inequalities.*

43. $y \le -x^2 + 3$

44. $x < y^2 - 1$

45. $x^2 + y^2 < 9$

46. $\dfrac{x^2}{4} + \dfrac{y^2}{9} \ge 1$

47. $\begin{cases} 3x + 4y \le 12 \\ x - 2y > 6 \end{cases}$

48. $\begin{cases} x^2 + y^2 \le 16 \\ x^2 + y^2 \ge 4 \end{cases}$

49. $\begin{cases} x^2 + y^2 < 4 \\ x^2 - y^2 \le 1 \end{cases}$

50. $\begin{cases} x^2 + y^2 < 4 \\ y \ge x^2 - 1 \\ x \ge 0 \end{cases}$

MIXED REVIEW

51. Write an equation of the circle with center $(-7, 8)$ and radius 5.

Graph each equation.

52. $y = x^2 + 6x + 9$

53. $x = y^2 + 6y + 9$

54. $\dfrac{y^2}{4} - \dfrac{x^2}{16} = 1$

55. $\dfrac{y^2}{4} + \dfrac{x^2}{16} = 1$

56. $\dfrac{(x - 2)^2}{4} + (y - 1)^2 = 1$

57. $y^2 = x^2 + 6$

58. $y^2 + (x - 2)^2 = 10$

59. $3x^2 + 6x + 3y^2 = 9$

60. $x^2 + y^2 - 8y = 0$

61. $6(x - 2)^2 + 9(y + 5)^2 = 36$

62. $\dfrac{x^2}{16} - \dfrac{y^2}{25} = 1$

Solve each system of equations.

63. $\begin{cases} y = x^2 - 5x + 1 \\ y = -x + 6 \end{cases}$

64. $\begin{cases} x^2 + y^2 = 10 \\ 9x^2 + y^2 = 18 \end{cases}$

Graph each inequality or system of inequalities.

65. $x^2 - y^2 < 1$

66. $\begin{cases} y > x^2 \\ x + y \ge 3 \end{cases}$

Chapter 10 Test MyMathLab®  Test Prep VIDEOS ▶ YouTube

Sketch the graph of each equation.

▶ **1.** $x^2 + y^2 = 36$

▶ **2.** $x^2 - y^2 = 36$

▶ **3.** $16x^2 + 9y^2 = 144$

▶ **4.** $y = x^2 - 8x + 16$

▶ **5.** $x^2 + y^2 + 6x = 16$

▶ **6.** $x = y^2 + 8y - 3$

▶ **7.** $\dfrac{(x-4)^2}{16} + \dfrac{(y-3)^2}{9} = 1$

▶ **8.** $y^2 - x^2 = 1$

Solve each system.

▶ **9.** $\begin{cases} x^2 + y^2 = 169 \\ 5x + 12y = 0 \end{cases}$

▶ **10.** $\begin{cases} x^2 + y^2 = 26 \\ x^2 - 2y^2 = 23 \end{cases}$

▶ **11.** $\begin{cases} y = x^2 - 5x + 6 \\ y = 2x \end{cases}$

▶ **12.** $\begin{cases} x^2 + 4y^2 = 5 \\ y = x \end{cases}$

Graph each system.

▶ **13.** $\begin{cases} 2x + 5y \geq 10 \\ y \geq x^2 + 1 \end{cases}$

▶ **14.** $\begin{cases} \dfrac{x^2}{4} + y^2 \leq 1 \\ x + y > 1 \end{cases}$

▶ **15.** $\begin{cases} x^2 + y^2 > 1 \\ \dfrac{x^2}{4} - y^2 \geq 1 \end{cases}$

▶ **16.** $\begin{cases} x^2 + y^2 \geq 4 \\ x^2 + y^2 < 16 \\ y \geq 0 \end{cases}$

▶ **17.** Which graph in the next column best resembles the graph of $x = a(y - k)^2 + h$ if $a > 0, h < 0,$ and $k > 0$?

A.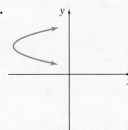

B.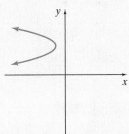

C.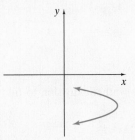

D.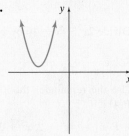

▶ **18.** A bridge has an arch in the shape of half an ellipse. If the equation of the ellipse, measured in feet, is $100x^2 + 225y^2 = 22{,}500$, find the height of the arch from the road and the width of the arch.

Chapter 10 Cumulative Review

1. Use the associative property of multiplication to write an expression equivalent to $4 \cdot (9y)$. Then simplify the equivalent expression.

2. Solve $3x + 4 > 1$ *and* $2x - 5 \leq 9$. Write the solution in interval notation.

3. Graph $x = -2y$ by plotting intercepts.

4. Find the slope of the line that goes through $(3, 2)$ and $(1, -4)$.

5. Use the elimination method to solve the system:

$$\begin{cases} 3x + \dfrac{y}{2} = 2 \\ 6x + y = 5 \end{cases}$$

6. Two planes leave Greensboro, one traveling north and the other south. After 2 hours, they are 650 miles apart. If one plane is flying 25 mph faster than the other, what is the speed of each?

7. Use the power rules to simplify the following. Use positive exponents to write all results.

  a. $(5x^2)^3$

  b. $\left(\dfrac{2}{3}\right)^3$

  c. $\left(\dfrac{3p^4}{q^5}\right)^2$

  d. $\left(\dfrac{2^{-3}}{y}\right)^{-2}$

  e. $(x^{-5}y^2z^{-1})^7$

8. Use the quotient rule to simplify.

  a. $\dfrac{4^8}{4^3}$

  b. $\dfrac{y^{11}}{y^5}$

  c. $\dfrac{32x^7}{4x^6}$

  d. $\dfrac{18a^{12}b^6}{12a^8b^6}$

9. Solve $2x^2 = \dfrac{17}{3}x + 1$.

10. Factor.

  a. $3y^2 + 14y + 15$

  b. $20a^5 + 54a^4 + 10a^3$

  c. $(y - 3)^2 - 2(y - 3) - 8$

11. Perform each indicated operation. $\dfrac{7}{x-1} + \dfrac{10x}{x^2-1} - \dfrac{5}{x+1}$

12. Perform the indicated operation and simplify if possible.
$$\frac{2}{3a - 15} - \frac{a}{25 - a^2}$$

13. Simplify each complex fraction.

a. $\dfrac{\dfrac{2x}{27y^2}}{\dfrac{6x^2}{9}}$
b. $\dfrac{\dfrac{5x}{x + 2}}{\dfrac{10}{x - 2}}$
c. $\dfrac{\dfrac{x}{y^2} + \dfrac{1}{y}}{\dfrac{y}{x^2} + \dfrac{1}{x}}$

14. Simplify each complex fraction.

a. $(a^{-1} - b^{-1})^{-1}$
b. $\dfrac{2 - \dfrac{1}{x}}{4x - \dfrac{1}{x}}$

15. Divide $2x^2 - x - 10$ by $x + 2$.

16. Solve $\dfrac{2}{x + 3} = \dfrac{1}{x^2 - 9} - \dfrac{1}{x - 3}$.

17. Use the remainder theorem and synthetic division to find $P(4)$ if
$$P(x) = 4x^6 - 25x^5 + 35x^4 + 17x^2.$$

18. Suppose that y varies inversely as x. If $y = 3$ when $x = \dfrac{2}{3}$, find the constant of variation and the direct variation equation.

19. Solve: $\dfrac{2x}{x - 3} + \dfrac{6 - 2x}{x^2 - 9} = \dfrac{x}{x + 3}$.

20. Simplify the following expressions. Assume that all variables represent nonnegative real numbers.

a. $\sqrt[5]{-32}$
b. $\sqrt[4]{625}$
c. $-\sqrt{36}$
d. $-\sqrt[3]{-27x^3}$
e. $\sqrt{144y^2}$

21. Melissa Scarlatti can clean the house in 4 hours, whereas her husband, Zack, can do the same job in 5 hours. They have agreed to clean together so that they can finish in time to watch a movie on TV that starts in 2 hours. How long will it take them to clean the house together? Can they finish before the movie starts?

22. Use the quotient rule to simplify.

a. $\dfrac{\sqrt{32}}{\sqrt{4}}$
b. $\dfrac{\sqrt[3]{240y^2}}{5\sqrt[3]{3y^{-4}}}$
c. $\dfrac{\sqrt[5]{64x^9y^2}}{\sqrt[5]{2x^2y^{-8}}}$

23. Find the cube roots.

a. $\sqrt[3]{1}$
b. $\sqrt[3]{-64}$
c. $\sqrt[3]{\dfrac{8}{125}}$
d. $\sqrt[3]{x^6}$
e. $\sqrt[3]{-27x^9}$

24. Multiply and simplify if possible.

a. $\sqrt{5}(2 + \sqrt{15})$
b. $(\sqrt{3} - \sqrt{5})(\sqrt{7} - 1)$
c. $(2\sqrt{5} - 1)^2$
d. $(3\sqrt{2} + 5)(3\sqrt{2} - 5)$

25. Multiply.

a. $z^{2/3}(z^{1/3} - z^5)$
b. $(x^{1/3} - 5)(x^{1/3} + 2)$

26. Rationalize the denominator. $\dfrac{-2}{\sqrt{3} + 3}$

27. Use the quotient rule to divide, and simplify if possible.

a. $\dfrac{\sqrt{20}}{\sqrt{5}}$
b. $\dfrac{\sqrt{50x}}{2\sqrt{2}}$
c. $\dfrac{7\sqrt[3]{48x^4y^8}}{\sqrt[3]{6y^2}}$
d. $\dfrac{2\sqrt[4]{32a^8b^6}}{\sqrt[4]{a^{-1}b^2}}$

28. Solve: $\sqrt{2x - 3} = x - 3$.

29. Add or subtract as indicated.

a. $\dfrac{\sqrt{45}}{4} - \dfrac{\sqrt{5}}{3}$
b. $\sqrt[3]{\dfrac{7x}{8}} + 2\sqrt[3]{7x}$

30. Use the discriminant to determine the number and type of solutions for $9x^2 - 6x = -4$.

31. Rationalize the denominator of $\sqrt{\dfrac{7x}{3y}}$.

32. Solve: $\dfrac{4}{x - 2} - \dfrac{x}{x + 2} = \dfrac{16}{x^2 - 4}$.

33. Solve: $\sqrt{2x - 3} = 9$.

34. Solve: $x^3 + 2x^2 - 4x \geq 8$.

35. Find the following powers of i.

a. i^7
b. i^{20}
c. i^{46}
d. i^{-12}

36. Graph $f(x) = (x + 2)^2 - 1$

37. Solve $p^2 + 2p = 4$ by completing the square.

38. Find the maximum value of $f(x) = -x^2 - 6x + 4$.

39. Solve: $\dfrac{1}{4}m^2 - m + \dfrac{1}{2} = 0$.

40. Find the inverse of $f(x) = \dfrac{x + 1}{2}$.

41. Solve: $p^4 - 3p^2 - 4 = 0$.

42. If $f(x) = x^2 - 3x + 2$ and $g(x) = -3x + 5$, find

a. $(f \circ g)(x)$
b. $(f \circ g)(-2)$
c. $(g \circ f)(x)$
d. $(g \circ f)(5)$

43. Solve: $\dfrac{x + 2}{x - 3} \leq 0$.

44. Graph $4x^2 + 9y^2 = 36$.

45. Graph $g(x) = \dfrac{1}{2}(x + 2)^2 + 5$. Find the vertex and the axis of symmetry.

46. Solve each equation for x.

a. $64^x = 4$
b. $125^{x-3} = 25$
c. $\dfrac{1}{81} = 3^{2x}$

47. Find the vertex of the graph of $f(x) = x^2 - 4x - 12$.

48. Graph the system: $\begin{cases} x + 2y < 8 \\ y \geq x^2 \end{cases}$

49. Find the distance between $(2, -5)$ and $(1, -4)$. Give an exact distance and a three-decimal-place approximation.

50. Solve the system $\begin{cases} x^2 + y^2 = 36 \\ y = x + 6 \end{cases}$

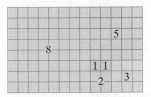

11 Sequences, Series, and the Binomial Theorem

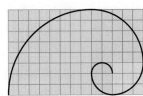

A tiling with squares whose sides are successive Fibonacci numbers in length

A Fibonacci spiral, created by drawing arcs connecting the opposite corners of squares in the Fibonacci tiling

The Fibonacci sequence is a special sequence in which the first two terms are 1 and each term thereafter is the sum of the two previous terms:

$$1, 1, 2, 3, 5, 8, 13, 21, \ldots$$

The Fibonacci numbers are named after Leonardo of Pisa, known as Fibonacci, although there is some evidence that these numbers had been described earlier in India.

There are numerous interesting facts about this sequence, and some are shown on the diagrams on this page. In Section 11.1, Exercise 46, you will have the opportunity to check a formula for this sequence.

Having explored in some depth the concept of function, we turn now in this final chapter to *sequences*. In one sense, a sequence is simply an ordered list of numbers. In another sense, a sequence is itself a function. Phenomena modeled by such functions are everywhere around us. The starting place for all mathematics is the sequence of natural numbers: 1, 2, 3, 4, and so on.

Sequences lead us to *series*, which are a sum of ordered numbers. Through series, we gain new insight, for example about the expansion of a binomial $(a + b)^n$, the concluding topic of this book.

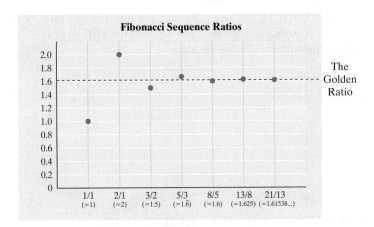

Fibonacci Sequence Ratios

| 1/1 | 2/1 | 3/2 | 5/3 | 8/5 | 13/8 | 21/13 |
| (=1) | (=2) | (=1.5) | (=1.6̄) | (=1.6) | (=1.625) | (=1.61538...) |

The Golden Ratio

The ratio of successive numbers in the Fibonacci sequence approaches a number called the golden ratio or golden number, which is approximately 1.618034.

11.1 Sequences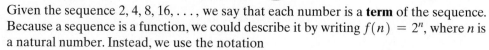

OBJECTIVES

1 Write the Terms of a Sequence Given Its General Term.

2 Find the General Term of a Sequence.

3 Solve Applications That Involve Sequences.

Suppose that a town's present population of 100,000 is growing by 5% each year. After the first year, the town's population will be

$$100{,}000 + 0.05(100{,}000) = 105{,}000$$

After the second year, the town's population will be

$$105{,}000 + 0.05(105{,}000) = 110{,}250$$

After the third year, the town's population will be

$$110{,}250 + 0.05(110{,}250) \approx 115{,}763$$

If we continue to calculate, the town's yearly population can be written as the **infinite sequence** of numbers

$$105{,}000, 110{,}250, 115{,}763, \ldots$$

If we decide to stop calculating after a certain year (say, the fourth year), we obtain the **finite sequence**

$$105{,}000, \ 110{,}250, \ 115{,}763, \ 121{,}551$$

Sequences

An infinite sequence is a function whose domain is the set of natural numbers $\{1, 2, 3, 4, \ldots\}$.

A finite sequence is a function whose domain is the set of natural numbers $\{1, 2, 3, 4, \ldots, n\}$, where n is some natural number.

OBJECTIVE

1   Writing the Terms of a Sequence

Given the sequence 2, 4, 8, 16, . . . , we say that each number is a **term** of the sequence. Because a sequence is a function, we could describe it by writing $f(n) = 2^n$, where n is a natural number. Instead, we use the notation

$$a_n = 2^n$$

Some function values are

$$a_1 = 2^1 = 2 \qquad \text{First term of the sequence}$$
$$a_2 = 2^2 = 4 \qquad \text{Second term}$$
$$a_3 = 2^3 = 8 \qquad \text{Third term}$$
$$a_4 = 2^4 = 16 \qquad \text{Fourth term}$$
$$a_{10} = 2^{10} = 1024 \qquad \text{Tenth term}$$

The nth term of the sequence a_n is called the **general term.**

▶ **Helpful Hint**

If it helps, think of a sequence as simply a list of values in which a position is assigned. For the sequence directly above,

Value: 2,   4,   8,   16, . . . , 1024
         ↑   ↑   ↑   ↑   ↑
Position 1st 2nd 3rd 4th 10th

EXAMPLE 1   Write the first five terms of the sequence whose general term is given by

$$a_n = n^2 - 1$$

Solution   Evaluate a_n, where n is 1, 2, 3, 4, and 5.

$$a_n = n^2 - 1$$
$$a_1 = 1^2 - 1 = 0 \quad \text{Replace } n \text{ with 1.}$$

$$a_2 = 2^2 - 1 = 3 \quad \text{Replace } n \text{ with 2.}$$
$$a_3 = 3^2 - 1 = 8 \quad \text{Replace } n \text{ with 3.}$$
$$a_4 = 4^2 - 1 = 15 \quad \text{Replace } n \text{ with 4.}$$
$$a_5 = 5^2 - 1 = 24 \quad \text{Replace } n \text{ with 5.}$$

Thus, the first five terms of the sequence $a_n = n^2 - 1$ are 0, 3, 8, 15, and 24. ☐

PRACTICE

1 Write the first five terms of the sequence whose general term is given by $a_n = 5 + n^2$.

EXAMPLE 2 If the general term of a sequence is given by $a_n = \dfrac{(-1)^n}{3n}$, find

a. the first term of the sequence   **b.** a_8

c. the one-hundredth term of the sequence   **d.** a_{15}

Solution

a. $a_1 = \dfrac{(-1)^1}{3(1)} = -\dfrac{1}{3}$ $\qquad$ Replace n with 1.

b. $a_8 = \dfrac{(-1)^8}{3(8)} = \dfrac{1}{24}$ $\qquad$ Replace n with 8.

c. $a_{100} = \dfrac{(-1)^{100}}{3(100)} = \dfrac{1}{300}$ $\qquad$ Replace n with 100.

d. $a_{15} = \dfrac{(-1)^{15}}{3(15)} = -\dfrac{1}{45}$ $\qquad$ Replace n with 15. ☐

PRACTICE

2 If the general term of a sequence is given by $a_n = \dfrac{(-1)^n}{5n}$, find

a. the first term of the sequence   **b.** a_4

c. The thirtieth term of the sequence   **d.** a_{19}

OBJECTIVE

2 Finding the General Term of a Sequence

Suppose we know the first few terms of a sequence and want to find a general term that fits the pattern of the first few terms.

EXAMPLE 3 Find a general term a_n of the sequence whose first few terms are given.

a. 1, 4, 9, 16, . . .   **b.** $\dfrac{1}{1}, \dfrac{1}{2}, \dfrac{1}{3}, \dfrac{1}{4}, \dfrac{1}{5}, \ldots$

c. $-3, -6, -9, -12, \ldots$ **d.** $\dfrac{1}{2}, \dfrac{1}{4}, \dfrac{1}{8}, \dfrac{1}{16}, \ldots$

Solution

a. These numbers are the squares of the first four natural numbers, so a general term might be $a_n = n^2$.

b. These numbers are the reciprocals of the first five natural numbers, so a general term might be $a_n = \dfrac{1}{n}$.

c. These numbers are the product of -3 and the first four natural numbers, so a general term might be $a_n = -3n$.

d. Notice that the denominators double each time.

$$\frac{1}{2}, \quad \frac{1}{2 \cdot 2}, \quad \frac{1}{2(2 \cdot 2)}, \quad \frac{1}{2(2 \cdot 2 \cdot 2)}$$

or

$$\frac{1}{2^1}, \quad \frac{1}{2^2}, \quad \frac{1}{2^3}, \quad \frac{1}{2^4}$$

We might then suppose that the general term is $a_n = \dfrac{1}{2^n}$.

PRACTICE
3 Find the general term a_n of the sequence whose first few terms are given.

a. $1, 3, 5, 7, \ldots$

b. $3, 9, 27, 81, \ldots$

c. $\dfrac{1}{2}, \dfrac{2}{3}, \dfrac{3}{4}, \dfrac{4}{5}, \ldots$

d. $-\dfrac{1}{2}, -\dfrac{1}{3}, -\dfrac{1}{4}, -\dfrac{1}{5}, \ldots$

OBJECTIVE
3 Solving Applications Modeled by Sequences

Sequences model many phenomena of the physical world, as illustrated by the following example.

EXAMPLE 4 **Finding a Puppy's Weight Gain**

The amount of weight, in pounds, a puppy gains in each month of its first year is modeled by a sequence whose general term is $a_n = n + 4$, where n is the number of the month. Write the first five terms of the sequence and find how much weight the puppy should gain in its fifth month.

Solution Evaluate $a_n = n + 4$ when n is $1, 2, 3, 4,$ and 5.

$$a_1 = 1 + 4 = 5$$
$$a_2 = 2 + 4 = 6$$
$$a_3 = 3 + 4 = 7$$
$$a_4 = 4 + 4 = 8$$
$$a_5 = 5 + 4 = 9$$

The puppy should gain 9 pounds in its fifth month.

PRACTICE
4 The value v, in dollars, of an office copier depreciates according to the sequence $v_n = 3950(0.8)^n$, where n is the time in years. Find the value of the copier after three years.

Vocabulary, Readiness & Video Check

Use the choices below to fill in each blank.

infinite   finite   general

1. The nth term of the sequence a_n is called the _____ term.

2. A(n) _____ sequence is a function whose domain is $\{1, 2, 3, 4, \ldots, n\}$ where n is some natural number.

3. A(n) _____ sequence is a function whose domain is $\{1, 2, 3, 4, \ldots\}$.

Write the first term of each sequence.

4. $a_n = 7^n; a_1 = $ _____.

5. $a_n = \dfrac{(-1)^n}{n}; a_1 = $ _____.

6. $a_n = (-1)^n \cdot n^4; a_1 = $ _____.

Martin-Gay Interactive Videos

See Video 11.1

Watch the section lecture video and answer the following questions.

OBJECTIVE
1

7. Based on the lecture before Example 1, complete the following statements. A sequence is a _____ whose _____ is the set of natural numbers. We use _____ to mean the general term of a sequence.

OBJECTIVE
2

8. In Example 3, why can't the general term be $a_n = (-2)^n$?

OBJECTIVE
3

9. For Example 4, write the equation for the specific term and find the allowance amount for day 9 of the vacation.

11.1 Exercise Set   MyMathLab®

Write the first five terms of each sequence, whose general term is given. See Example 1.

1. $a_n = n + 4$

2. $a_n = 5 - n$

3. $a_n = (-1)^n$

4. $a_n = (-2)^n$

5. $a_n = \dfrac{1}{n + 3}$

6. $a_n = \dfrac{1}{7 - n}$

7. $a_n = 2n$

8. $a_n = -6n$

9. $a_n = -n^2$

10. $a_n = n^2 + 2$

11. $a_n = 2^n$

12. $a_n = 3^{n-2}$

13. $a_n = 2n + 5$

14. $a_n = 1 - 3n$

15. $a_n = (-1)^n n^2$

16. $a_n = (-1)^{n+1}(n - 1)$

Find the indicated term for each sequence, whose general term is given. See Example 2.

17. $a_n = 3n^2; a_5$

18. $a_n = -n^2; a_{15}$

19. $a_n = 6n - 2; a_{20}$

20. $a_n = 100 - 7n; a_{50}$

21. $a_n = \dfrac{n + 3}{n}; a_{15}$

22. $a_n = \dfrac{n}{n + 4}; a_{24}$

23. $a_n = (-3)^n; a_6$

24. $a_n = 5^{n+1}; a_3$

25. $a_n = \dfrac{n - 2}{n + 1}; a_6$

26. $a_n = \dfrac{n + 3}{n + 4}; a_8$

27. $a_n = \dfrac{(-1)^n}{n}; a_8$

28. $a_n = \dfrac{(-1)^n}{2n}; a_{100}$

29. $a_n = -n^2 + 5; a_{10}$

30. $a_n = 8 - n^2; a_{20}$

31. $a_n = \dfrac{(-1)^n}{n + 6}; a_{19}$

32. $a_n = \dfrac{n - 4}{(-2)^n}; a_6$

Find a general term a_n for each sequence, whose first four terms are given. See Example 3.

33. $3, 7, 11, 15$

34. $2, 7, 12, 17$

35. $-2, -4, -8, -16$

36. $-4, 16, -64, 256$

37. $\dfrac{1}{3}, \dfrac{1}{9}, \dfrac{1}{27}, \dfrac{1}{81}$

38. $\dfrac{2}{5}, \dfrac{2}{25}, \dfrac{2}{125}, \dfrac{2}{625}$

Solve. See Example 4.

39. The distance, in feet, that a Thermos dropped from a cliff falls in each consecutive second is modeled by a sequence whose general term is $a_n = 32n - 16$, where n is the number of seconds. Find the distance the Thermos falls in the second, third, and fourth seconds.

40. The population size of a culture of bacteria triples every hour such that its size is modeled by the sequence $a_n = 50(3)^{n-1}$, where n is the number of the hour just beginning. Find the size of the culture at the beginning of the fourth hour and the size of the culture at the beginning of the first hour.

41. Mrs. Laser agrees to give her son Mark an allowance of $0.10 on the first day of his 14-day vacation, $0.20 on the second day, $0.40 on the third day, and so on. Write an equation of a sequence whose terms correspond to Mark's allowance. Find the allowance Mark will receive on the last day of his vacation.

42. A small theater has 10 rows with 12 seats in the first row, 15 seats in the second row, 18 seats in the third row, and so on. Write an equation of a sequence whose terms correspond to the seats in each row. Find the number of seats in the eighth row.

43. The number of cases of a new infectious disease is doubling every year such that the number of cases is modeled by a sequence whose general term is $a_n = 75(2)^{n-1}$, where n is the number of the year just beginning. Find how many cases there will be at the beginning of the sixth year. Find how many cases there were at the beginning of the first year.

44. A new college had an initial enrollment of 2700 students in 2000, and each year the enrollment increases by 150 students. Find the enrollment for each of 5 years, beginning with 2000.

45. An endangered species of sparrow had an estimated population of 800 in 2000, and scientists predicted that its population would decrease by half each year. Estimate the population in 2004. Estimate the year the sparrow was extinct.

46. A **Fibonacci sequence** is a special type of sequence in which the first two terms are 1, and each term thereafter is the sum of the two previous terms: $1, 1, 2, 3, 5, 8$, etc. The formula for the nth Fibonacci term is $a_n = \dfrac{1}{\sqrt{5}}\left[\left(\dfrac{1 + \sqrt{5}}{2}\right)^n - \left(\dfrac{1 - \sqrt{5}}{2}\right)^n\right]$. Verify that the first two terms of the Fibonacci sequence are each 1.

REVIEW AND PREVIEW

Sketch the graph of each quadratic function. See Section 8.5.

47. $f(x) = (x - 1)^2 + 3$
48. $f(x) = (x - 2)^2 + 1$
49. $f(x) = 2(x + 4)^2 + 2$
50. $f(x) = 3(x - 3)^2 + 4$

Find the distance between each pair of points. See Section 7.3.

51. $(-4, -1)$ and $(-7, -3)$
52. $(-2, -1)$ and $(-1, 5)$

53. $(2, -7)$ and $(-3, -3)$
54. $(10, -14)$ and $(5, -11)$

CONCEPT EXTENSIONS

Find the first five terms of each sequence. Round each term after the first to four decimal places.

55. $a_n = \dfrac{1}{\sqrt{n}}$

56. $\dfrac{\sqrt{n}}{\sqrt{n + 1}}$

57. $a_n = \left(1 + \dfrac{1}{n}\right)^n$

58. $a_n = \left(1 + \dfrac{0.05}{n}\right)^n$

11.2 Arithmetic and Geometric Sequences

OBJECTIVES

1 Identify Arithmetic Sequences and Their Common Differences.

2 Identify Geometric Sequences and Their Common Ratios.

OBJECTIVE

1 Identifying Arithmetic Sequences

Find the first four terms of the sequence whose general term is $a_n = 5 + (n - 1)3$.

$$a_1 = 5 + (1 - 1)3 = 5 \qquad \text{Replace } n \text{ with 1.}$$
$$a_2 = 5 + (2 - 1)3 = 8 \qquad \text{Replace } n \text{ with 2.}$$
$$a_3 = 5 + (3 - 1)3 = 11 \qquad \text{Replace } n \text{ with 3.}$$
$$a_4 = 5 + (4 - 1)3 = 14 \qquad \text{Replace } n \text{ with 4.}$$

The first four terms are $5, 8, 11$, and 14. Notice that the difference of any two successive terms is 3.

$$8 - 5 = 3$$
$$11 - 8 = 3$$
$$14 - 11 = 3$$
$$\vdots$$
$$a_n - a_{n-1} = 3$$
$$\uparrow \qquad \uparrow$$
$$n\text{th} \qquad \text{previous}$$
$$\text{term} \qquad \text{term}$$

Because the difference of any two successive terms is a constant, we call the sequence an **arithmetic sequence,** or an **arithmetic progression.** The constant difference d in successive terms is called the **common difference.** In this example, d is 3.

> **Arithmetic Sequence and Common Difference**
>
> An **arithmetic sequence** is a sequence in which each term (after the first) differs from the preceding term by a constant amount d. The constant d is called the **common difference** of the sequence.

The sequence $2, 6, 10, 14, 18, \ldots$ is an arithmetic sequence. Its common difference is 4. Given the first term a_1 and the common difference d of an arithmetic sequence, we can find any term of the sequence.

EXAMPLE 1   Write the first five terms of the arithmetic sequence whose first term is 7 and whose common difference is 2.

Solution

$$a_1 = 7$$
$$a_2 = 7 + 2 = 9$$
$$a_3 = 9 + 2 = 11$$
$$a_4 = 11 + 2 = 13$$
$$a_5 = 13 + 2 = 15$$

The first five terms are $7, 9, 11, 13, 15$.   □

PRACTICE

1   Write the first five terms of the arithmetic sequence whose first term is 4 and whose common difference is 5.

Notice the general pattern of the terms in Example 1.

$$a_1 = 7$$
$$a_2 = 7 + 2 = 9 \quad \text{or} \quad a_2 = a_1 + d$$
$$a_3 = 9 + 2 = 11 \quad \text{or} \quad a_3 = a_2 + d = (a_1 + d) + d = a_1 + 2d$$
$$a_4 = 11 + 2 = 13 \quad \text{or} \quad a_4 = a_3 + d = (a_1 + 2d) + d = a_1 + 3d$$
$$a_5 = 13 + 2 = 15 \quad \text{or} \quad a_5 = a_4 + d = (a_1 + 3d) + d = a_1 + 4d$$

$\longrightarrow$ (subscript $- 1$) is multiplier $\longrightarrow$

The pattern on the right suggests that the general term a_n of an arithmetic sequence is given by

$$a_n = a_1 + (n - 1)d$$

General Term of an Arithmetic Sequence

The general term a_n of an arithmetic sequence is given by

$$a_n = a_1 + (n - 1)d$$

where a_1 is the first term and d is the common difference.

EXAMPLE 2   Consider the arithmetic sequence whose first term is 3 and whose common difference is -5.

a. Write an expression for the general term a_n.

b. Find the twentieth term of this sequence.

Solution

a. Since this is an arithmetic sequence, the general term a_n is given by $a_n = a_1 + (n - 1)d$. Here, $a_1 = 3$ and $d = -5$, so

$$a_n = 3 + (n - 1)(-5) \quad \text{Let } a_1 = 3 \text{ and } d = -5.$$
$$= 3 - 5n + 5 \qquad \text{Multiply.}$$
$$= 8 - 5n \qquad\qquad \text{Simplify.}$$

b. $a_n = 8 - 5n$
$$a_{20} = 8 - 5 \cdot 20 \qquad \text{Let } n = 20.$$
$$= 8 - 100 = -92 \qquad\qquad\qquad □$$

2 Consider the arithmetic sequence whose first term is 2 and whose common difference is -3.

a. Write an expression for the general term a_n.

b. Find the twelfth term of the sequence.

. . . . ■

EXAMPLE 3 Find the eleventh term of the arithmetic sequence whose first three terms are 2, 9, and 16.

Solution Since the sequence is arithmetic, the eleventh term is

$$a_{11} = a_1 + (11 - 1)d = a_1 + 10d$$

We know a_1 is the first term of the sequence, so $a_1 = 2$. Also, d is the constant difference of terms, so $d = a_2 - a_1 = 9 - 2 = 7$. Thus,

$$a_{11} = a_1 + 10d$$
$$= 2 + 10 \cdot 7 \quad \text{Let } a_1 = 2 \text{ and } d = 7.$$
$$= 72 \qquad\qquad\qquad\qquad\qquad \square$$

PRACTICE
3 Find the ninth term of the arithmetic sequence whose first three terms are 3, 9, and 15.

. . . . ■

EXAMPLE 4 If the third term of an arithmetic sequence is 12 and the eighth term is 27, find the fifth term.

Solution We need to find a_1 and d to write the general term, which then enables us to find a_5, the fifth term. The given facts about terms a_3 and a_8 lead to a system of linear equations.

$$\begin{cases} a_3 = a_1 + (3 - 1)d \\ a_8 = a_1 + (8 - 1)d \end{cases} \text{ or } \begin{cases} 12 = a_1 + 2d \\ 27 = a_1 + 7d \end{cases}$$

Next, we solve the system $\begin{cases} 12 = a_1 + 2d \\ 27 = a_1 + 7d \end{cases}$ by elimination. Multiply both sides of the second equation by -1 so that

$$\begin{cases} 12 = a_1 + 2d \\ -1(27) = -1(a_1 + 7d) \end{cases} \begin{array}{c} \text{simplifies} \\ \text{to} \end{array} \begin{cases} 12 = a_1 + 2d \\ \underline{-27 = -a_1 - 7d} \\ -15 = -5d \qquad \text{Add the equations.} \\ 3 = d \qquad\quad \text{Divide both sides by } -5. \end{cases}$$

To find a_1, let $d = 3$ in $12 = a_1 + 2d$. Then

$$12 = a_1 + 2(3)$$
$$12 = a_1 + 6$$
$$6 = a_1$$

Thus, $a_1 = 6$ and $d = 3$, so

$$a_n = 6 + (n - 1)(3)$$
$$= 6 + 3n - 3$$
$$= 3 + 3n$$

and

$$a_5 = 3 + 3 \cdot 5 = 18 \qquad\qquad\qquad\qquad \square$$

PRACTICE
4 If the third term of an arithmetic sequence is 23 and the eighth term is 63, find the sixth term.

. . . . ■

EXAMPLE 5 **Finding Salary**

Donna Theime has an offer for a job starting at \$40,000 per year and guaranteeing her a raise of \$1600 per year for the next 5 years. Write the general term for the arithmetic sequence that models Donna's potential annual salaries and find her salary for the fourth year.

Solution The first term, a_1, is 40,000, and d is 1600. So

$$a_n = 40,000 + (n - 1)(1600) = 38,400 + 1600n$$
$$a_4 = 38,400 + 1600 \cdot 4 = 44,800$$

Her salary for the fourth year will be \$44,800. □

PRACTICE

5 A starting salary for a consulting company is \$57,000 per year with guaranteed annual increases of \$2200 for the next 4 years. Write the general term for the arithmetic sequence that models the potential annual salaries and find the salary for the third year.

OBJECTIVE

2 **Identifying Geometric Sequences**

We now investigate a **geometric sequence**, also called a **geometric progression.** In the sequence 5, 15, 45, 135, . . . , each term after the first is the *product* of 3 and the preceding term. This pattern of multiplying by a constant to get the next term defines a geometric sequence. The constant is called the **common ratio** because it is the ratio of any term (after the first) to its preceding term.

$$\frac{15}{5} = 3$$

$$\frac{45}{15} = 3$$

$$\frac{135}{45} = 3$$

$$\vdots$$

$$nth\ term \longrightarrow \frac{a_n}{a_{n-1}} = 3$$
$$previous\ term \longrightarrow$$

Geometric Sequence and Common Ratio

A **geometric sequence** is a sequence in which each term (after the first) is obtained by multiplying the preceding term by a constant r. The constant r is called the **common ratio** of the sequence.

The sequence $12, 6, 3, \frac{3}{2}, \ldots$ is geometric since each term after the first is the product of the previous term and $\frac{1}{2}$.

EXAMPLE 6 Write the first five terms of a geometric sequence whose first term is 7 and whose common ratio is 2.

Solution
$$a_1 = 7$$
$$a_2 = 7(2) = 14$$
$$a_3 = 14(2) = 28$$
$$a_4 = 28(2) = 56$$
$$a_5 = 56(2) = 112$$

The first five terms are 7, 14, 28, 56, and 112. □

PRACTICE
6   Write the first four terms of a geometric sequence whose first term is 8 and whose common ratio is -3

Notice the general pattern of the terms in Example 6.

$$a_1 = 7$$
$$a_2 = 7(2) = 14 \quad \text{or} \quad a_2 = a_1(r)$$
$$a_3 = 14(2) = 28 \quad \text{or} \quad a_3 = a_2(r) = (a_1 \cdot r) \cdot r = a_1 r^2$$
$$a_4 = 28(2) = 56 \quad \text{or} \quad a_4 = a_3(r) = (a_1 \cdot r^2) \cdot r = a_1 r^3$$
$$a_5 = 56(2) = 112 \quad \text{or} \quad a_5 = a_4(r) = (a_1 \cdot r^3) \cdot r = a_1 r^4$$

$$\longrightarrow (\text{subscript} - 1) \text{ is power} \longleftarrow$$

The pattern on the right above suggests that the general term of a geometric sequence is given by $a_n = a_1 r^{n-1}$.

General Term of a Geometric Sequence

The general term a_n of a geometric sequence is given by

$$a_n = a_1 r^{n-1}$$

where a_1 is the first term and r is the common ratio.

EXAMPLE 7   Find the eighth term of the geometric sequence whose first term is 12 and whose common ratio is $\frac{1}{2}$.

Solution   Since this is a geometric sequence, the general term a_n is given by

$$a_n = a_1 r^{n-1}$$

Here $a_1 = 12$ and $r = \frac{1}{2}$, so $a_n = 12\left(\frac{1}{2}\right)^{n-1}$. Evaluate a_n for $n = 8$.

$$a_8 = 12\left(\frac{1}{2}\right)^{8-1} = 12\left(\frac{1}{2}\right)^7 = 12\left(\frac{1}{128}\right) = \frac{3}{32}$$

PRACTICE
7   Find the seventh term of the geometric sequence whose first term is 64 and whose common ratio is $\frac{1}{4}$.

EXAMPLE 8   Find the fifth term of the geometric sequence whose first three terms are 2, -6, and 18.

Solution   Since the sequence is geometric and $a_1 = 2$, the fifth term must be $a_1 r^{5-1}$, or $2r^4$. We know that r is the common ratio of terms, so r must be $\frac{-6}{2}$, or -3. Thus,

$$a_5 = 2r^4$$
$$a_5 = 2(-3)^4 = 162$$

PRACTICE
8   Find the seventh term of the geometric sequence whose first three terms are $-3, 6$, and -12.

EXAMPLE 9 If the second term of a geometric sequence is $\frac{5}{4}$ and the third term is $\frac{5}{16}$, find the first term and the common ratio.

Solution Notice that $\frac{5}{16} \div \frac{5}{4} = \frac{1}{4}$, so $r = \frac{1}{4}$. Then

$$a_2 = a_1 \left(\frac{1}{4}\right)^{2-1}$$

$$\frac{5}{4} = a_1 \left(\frac{1}{4}\right)^1, \quad \text{or} \quad a_1 = 5 \quad \text{Replace } a_2 \text{ with } \frac{5}{4}.$$

The first term is 5. ☐

PRACTICE
9 If the second term of a geometric sequence is $\frac{9}{2}$ and the third term is $\frac{27}{4}$, find the first term and the common ratio.

EXAMPLE 10 **Predicting Population of a Bacterial Culture**

The population size of a bacterial culture growing under controlled conditions is doubling each day. Predict how large the culture will be at the beginning of day 7 if it measures 10 units at the beginning of day 1.

Solution Since the culture doubles in size each day, the population sizes are modeled by a geometric sequence. Here $a_1 = 10$ and $r = 2$. Thus,

$$a_n = a_1 r^{n-1} = 10(2)^{n-1} \quad \text{and} \quad a_7 = 10(2)^{7-1} = 640$$

The bacterial culture should measure 640 units at the beginning of day 7. ☐

PRACTICE
10 After applying a test antibiotic, the population of a bacterial culture is reduced by one-half every day. Predict how large the culture will be at the start of day 7 if it measures 4800 units at the beginning of day 1.

Vocabulary, Readiness & Video Check

Use the choices below to fill in each blank. Some choices may be used more than once and some not at all.

| first | arithmetic | difference |
| last | geometric | ratio |

1. A(n) _____ sequence is one in which each term (after the first) is obtained by multiplying the preceding term by a constant r. The constant r is called the common _____.

2. A(n) _____ sequence is one in which each term (after the first) differs from the preceding term by a constant amount d. The constant d is called the common _____.

3. The general term of an arithmetic sequence is $a_n = a_1 + (n-1)d$ where a_1 is the _____ term and d is the common _____.

4. The general term of a geometric sequence is $a_n = a_1 r^{n-1}$ where a_1 is the _____ term and r is the common _____.

Martin-Gay Interactive Videos

See Video 11.2 🍐

Watch the section lecture video and answer the following questions.

OBJECTIVE 1
5. From the lecture before Example 1, what makes a sequence an arithmetic sequence?

OBJECTIVE 2
6. From the lecture before Example 3, what's the difference between an arithmetic and a geometric sequence?

11.2 Exercise Set MyMathLab®

Write the first five terms of the arithmetic or geometric sequence, whose first term, a_1, and common difference, d, or common ratio, r, are given. See Examples 1 and 6.

1. $a_1 = 4; d = 2$

2. $a_1 = 3; d = 10$

▶ **3.** $a_1 = 6; d = -2$

4. $a_1 = -20; d = 3$

5. $a_1 = 1; r = 3$

6. $a_1 = -2; r = 2$

▶ **7.** $a_1 = 48; r = \dfrac{1}{2}$

8. $a_1 = 1; r = \dfrac{1}{3}$

Find the indicated term of each sequence. See Examples 2 and 7.

9. The eighth term of the arithmetic sequence whose first term is 12 and whose common difference is 3

10. The twelfth term of the arithmetic sequence whose first term is 32 and whose common difference is −4

11. The fourth term of the geometric sequence whose first term is 7 and whose common ratio is −5

12. The fifth term of the geometric sequence whose first term is 3 and whose common ratio is 3

13. The fifteenth term of the arithmetic sequence whose first term is −4 and whose common difference is −4

14. The sixth term of the geometric sequence whose first term is 5 and whose common ratio is −4

Find the indicated term of each sequence. See Examples 3 and 8.

15. The ninth term of the arithmetic sequence 0, 12, 24, . . .

16. The thirteenth term of the arithmetic sequence −3, 0, 3, . . .

▶ **17.** The twenty-fifth term of the arithmetic sequence 20, 18, 16, . . .

18. The ninth term of the geometric sequence 5, 10, 20, . . .

19. The fifth term of the geometric sequence 2, −10, 50, . . .

20. The sixth term of the geometric sequence $\dfrac{1}{2}, \dfrac{3}{2}, \dfrac{9}{2}, \ldots$

Find the indicated term of each sequence. See Examples 4 and 9.

21. The eighth term of the arithmetic sequence whose fourth term is 19 and whose fifteenth term is 52

22. If the second term of an arithmetic sequence is 6 and the tenth term is 30, find the twenty-fifth term.

23. If the second term of an arithmetic progression is −1 and the fourth term is 5, find the ninth term.

24. If the second term of a geometric progression is 15 and the third term is 3, find a_1 and r.

25. If the second term of a geometric progression is $-\dfrac{4}{3}$ and the third term is $\dfrac{8}{3}$, find a_1 and r.

26. If the third term of a geometric sequence is 4 and the fourth term is −12, find a_1 and r.

27. Explain why 14, 10, and 6 may be the first three terms of an arithmetic sequence when it appears we are subtracting instead of adding to get the next term.

28. Explain why 80, 20, and 5 may be the first three terms of a geometric sequence when it appears we are dividing instead of multiplying to get the next term.

MIXED PRACTICE

Given are the first three terms of a sequence that is either arithmetic or geometric. If the sequence is arithmetic, find a_1 and d. If a sequence is geometric, find a_1 and r.

29. 2, 4, 6

30. 8, 16, 24

31. 5, 10, 20

32. 2, 6, 18

33. $\dfrac{1}{2}, \dfrac{1}{10}, \dfrac{1}{50}$

34. $\dfrac{2}{3}, \dfrac{4}{3}, 2$

35. $x, 5x, 25x$

36. $y, -3y, 9y$

37. $p, p + 4, p + 8$

38. $t, t - 1, t - 2$

Find the indicated term of each sequence.

39. The twenty-first term of the arithmetic sequence whose first term is 14 and whose common difference is $\dfrac{1}{4}$

40. The fifth term of the geometric sequence whose first term is 8 and whose common ratio is −3

41. The fourth term of the geometric sequence whose first term is 3 and whose common ratio is $-\dfrac{2}{3}$

42. The fourth term of the arithmetic sequence whose first term is 9 and whose common difference is 5

43. The fifteenth term of the arithmetic sequence $\frac{3}{2}, 2, \frac{5}{2}, \ldots$

44. The eleventh term of the arithmetic sequence $2, \frac{5}{3}, \frac{4}{3}, \ldots$

45. The sixth term of the geometric sequence $24, 8, \frac{8}{3}, \ldots$

46. The eighteenth term of the arithmetic sequence $5, 2, -1, \ldots$

47. If the third term of an arithmetic sequence is 2 and the seventeenth term is -40, find the tenth term.

48. If the third term of a geometric sequence is -28 and the fourth term is -56, find a_1 and r.

Solve. See Examples 5 and 10.

49. An auditorium has 54 seats in the first row, 58 seats in the second row, 62 seats in the third row, and so on. Find the general term of this arithmetic sequence and the number of seats in the twentieth row.

50. A triangular display of cans in a grocery store has 20 cans in the first row, 17 cans in the next row, and so on, in an arithmetic sequence. Find the general term and the number of cans in the fifth row. Find how many rows there are in the display and how many cans are in the top row.

51. The initial size of a virus culture is 6 units, and it triples its size every day. Find the general term of the geometric sequence that models the culture's size.

52. A real estate investment broker predicts that a certain property will increase in value 15% each year. Thus, the yearly property values can be modeled by a geometric sequence whose common ratio r is 1.15. If the initial property value was $500,000, write the first four terms of the sequence and predict the value at the end of the third year.

53. A rubber ball is dropped from a height of 486 feet, and it continues to bounce one-third the height from which it last fell. Write out the first five terms of this geometric sequence and find the general term. Find how many bounces it takes for the ball to rebound less than 1 foot.

54. On the first swing, the length of the arc through which a pendulum swings is 50 inches. The length of each successive swing is 80% of the preceding swing. Determine whether this sequence is arithmetic or geometric. Find the length of the fourth swing.

55. Jose takes a job that offers a monthly starting salary of $4000 and guarantees him a monthly raise of $125 during his first year of training. Find the general term of this arithmetic sequence and his monthly salary at the end of his training.

56. At the beginning of Claudia Schaffer's exercise program, she rides 15 minutes on the Lifecycle. Each week, she increases her riding time by 5 minutes. Write the general term of this arithmetic sequence, and find her riding time after 7 weeks. Find how many weeks it takes her to reach a riding time of 1 hour.

57. If a radioactive element has a half-life of 3 hours, then x grams of the element dwindles to $\frac{x}{2}$ grams after 3 hours. If a nuclear reactor has 400 grams of that radioactive element, find the amount of radioactive material after 12 hours.

REVIEW AND PREVIEW

Evaluate. See Section 1.3.

58. $5(1) + 5(2) + 5(3) + 5(4)$

59. $\frac{1}{3(1)} + \frac{1}{3(2)} + \frac{1}{3(3)}$

60. $2(2 - 4) + 3(3 - 4) + 4(4 - 4)$

61. $3^0 + 3^1 + 3^2 + 3^3$

62. $\frac{1}{4(1)} + \frac{1}{4(2)} + \frac{1}{4(3)}$

63. $\frac{8 - 1}{8 + 1} + \frac{8 - 2}{8 + 2} + \frac{8 - 3}{8 + 3}$

CONCEPT EXTENSIONS

Write the first four terms of the arithmetic or geometric sequence, whose first term, a_1, and common difference, d, or common ratio, r, are given.

64. $a_1 = \$3720, d = -\$268.50$

65. $a_1 = \$11{,}782.40, r = 0.5$

66. $a_1 = 26.8, r = 2.5$

67. $a_1 = 19.652; d = -0.034$

68. Describe a situation in your life that can be modeled by a geometric sequence. Write an equation for the sequence.

69. Describe a situation in your life that can be modeled by an arithmetic sequence. Write an equation for the sequence.

11.3 Series

OBJECTIVES

1 Identify Finite and Infinite Series and Use Summation Notation. ▷

2 Find Partial Sums. ▷

OBJECTIVE

1 **Identifying Finite and Infinite Series and Using Summation Notation** ▷

A person who conscientiously saves money by saving first $100 and then saving $10 more each month than he saved the preceding month is saving money according to the arithmetic sequence

$$a_n = 100 + 10(n - 1)$$

Following this sequence, he can predict how much money he should save for any particular month. But if he also wants to know how much money *in total* he has saved, say, by the fifth month, he must find the *sum* of the first five terms of the sequence

$$\underbrace{100}_{a_1} + \underbrace{100 + 10}_{a_2} + \underbrace{100 + 20}_{a_3} + \underbrace{100 + 30}_{a_4} + \underbrace{100 + 40}_{a_5}$$

A sum of the terms of a sequence is called a **series** (the plural is also "series"). As our example here suggests, series are frequently used to model financial and natural phenomena.

A series is a **finite series** if it is the sum of a finite number of terms. A series is an **infinite series** if it is the sum of all the terms of an infinite sequence. For example,

Sequence	*Series*	
$5, 9, 13$	$5 + 9 + 13$	Finite; sum of 3 terms
$5, 9, 13, \ldots$	$5 + 9 + 13 + \cdots$	Infinite
$4, -2, 1, -\dfrac{1}{2}, \dfrac{1}{4}$	$4 + (-2) + 1 + \left(-\dfrac{1}{2}\right) + \left(\dfrac{1}{4}\right)$	Finite; sum of 5 terms
$4, -2, 1, \ldots$	$4 + (-2) + 1 + \cdots$	Infinite
$3, 6, \ldots, 99$	$3 + 6 + \cdots + 99$	Finite; sum of 33 terms

A shorthand notation for denoting a series when the general term of the sequence is known is called **summation notation.** The Greek uppercase letter **sigma,** Σ, is used to mean "sum." The expression $\displaystyle\sum_{n=1}^{5}(3n + 1)$ is read "the sum of $3n + 1$ as n goes from 1 to 5"; this expression means the sum of the first five terms of the sequence whose general term is $a_n = 3n + 1$. Often, the variable i is used instead of n in summation notation: $\displaystyle\sum_{i=1}^{5}(3i + 1)$. Whether we use n, i, k, or some other variable, the variable is called the **index of summation.** The notation $i = 1$ below the symbol Σ indicates the beginning value of i, and the number 5 above the symbol Σ indicates the ending value of i. Thus, the terms of the sequence are found by successively replacing i with the natural numbers $1, 2, 3, 4, 5$. To find the sum, we write out the terms and then add.

$$\sum_{i=1}^{5}(3i + 1) = (3 \cdot 1 + 1) + (3 \cdot 2 + 1) + (3 \cdot 3 + 1)$$
$$+ (3 \cdot 4 + 1) + (3 \cdot 5 + 1)$$
$$= 4 + 7 + 10 + 13 + 16 = 50$$

EXAMPLE 1 Evaluate.

a. $\displaystyle\sum_{i=0}^{6}\dfrac{i - 2}{2}$ **b.** $\displaystyle\sum_{i=3}^{5}2^i$

Solution

a. $\displaystyle\sum_{i=0}^{6}\frac{i-2}{2} = \frac{0-2}{2} + \frac{1-2}{2} + \frac{2-2}{2} + \frac{3-2}{2} + \frac{4-2}{2} + \frac{5-2}{2} + \frac{6-2}{2}$

$$= (-1) + \left(-\frac{1}{2}\right) + 0 + \frac{1}{2} + 1 + \frac{3}{2} + 2$$

$$= \frac{7}{2}, \text{ or } 3\frac{1}{2}$$

b. $\displaystyle\sum_{i=3}^{5}2^{i} = 2^3 + 2^4 + 2^5$

$$= 8 + 16 + 32$$

$$= 56$$

PRACTICE

1   Evaluate.

a. $\displaystyle\sum_{i=0}^{4}\frac{i-3}{4}$ **b.** $\displaystyle\sum_{i=2}^{5}3^{i}$

EXAMPLE 2   Write each series with summation notation.

a. $3 + 6 + 9 + 12 + 15$ **b.** $\dfrac{1}{2} + \dfrac{1}{4} + \dfrac{1}{8} + \dfrac{1}{16}$

Solution

a. Since the *difference* of each term and the preceding term is 3, the terms correspond to the first five terms of the arithmetic sequence $a_n = a_1 + (n-1)d$ with $a_1 = 3$ and $d = 3$. So $a_n = 3 + (n-1)3 = 3n$ when simplified. Thus, in summation notation,

$$3 + 6 + 9 + 12 + 15 = \sum_{i=1}^{5}3i.$$

b. Since each term is the *product* of the preceding term and $\dfrac{1}{2}$, these terms correspond to the first four terms of the geometric sequence $a_n = a_1 r^{n-1}$. Here $a_1 = \dfrac{1}{2}$ and $r = \dfrac{1}{2}$, so $a_n = \left(\dfrac{1}{2}\right)\left(\dfrac{1}{2}\right)^{n-1} = \left(\dfrac{1}{2}\right)^{1+(n-1)} = \left(\dfrac{1}{2}\right)^{n}$. In summation notation,

$$\frac{1}{2} + \frac{1}{4} + \frac{1}{8} + \frac{1}{16} = \sum_{i=1}^{4}\left(\frac{1}{2}\right)^{i}$$

PRACTICE

2   Write each series with summation notation.

a. $5 + 10 + 15 + 20 + 25 + 30$ **b.** $\dfrac{1}{5} + \dfrac{1}{25} + \dfrac{1}{125} + \dfrac{1}{625}$

OBJECTIVE

2   **Finding Partial Sums** ▶

The sum of the first n terms of a sequence is a finite series known as a **partial sum,** S_n. Thus, for the sequence $a_1, a_2, \ldots, a_n$, the first three partial sums are

$$S_1 = a_1$$
$$S_2 = a_1 + a_2$$
$$S_3 = a_1 + a_2 + a_3$$

In general, S_n is the sum of the first n terms of a sequence.

$$S_n = \sum_{i=1}^{n}a_n$$

EXAMPLE 3 Find the sum of the first three terms of the sequence whose general term is $a_n = \dfrac{n+3}{2n}$.

Solution

$$S_3 = \sum_{i=1}^{3} \frac{i+3}{2i} = \frac{1+3}{2 \cdot 1} + \frac{2+3}{2 \cdot 2} + \frac{3+3}{2 \cdot 3}$$

$$= 2 + \frac{5}{4} + 1 = 4\frac{1}{4} \qquad \square$$

PRACTICE

3 Find the sum of the first four terms of the sequence whose general term is $a_n = \dfrac{2+3n}{n^2}$.

The next example illustrates how these sums model real-life phenomena.

EXAMPLE 4 **Number of Baby Gorillas Born**

The number of baby gorillas born at the San Diego Zoo is a sequence defined by $a_n = n(n-1)$, where n is the number of years the zoo has owned gorillas. Find the *total* number of baby gorillas born in the *first 4 years*.

Solution To solve, find the sum

$$S_4 = \sum_{i=1}^{4} i(i-1)$$

$$= 1(1-1) + 2(2-1) + 3(3-1) + 4(4-1)$$

$$= 0 + 2 + 6 + 12 = 20$$

Twenty gorillas were born in the first 4 years. $\qquad \square$

PRACTICE

4 The number of new strawberry plants growing in a garden each year is a sequence defined by $a_n = n(2n-1)$, where n is the number of years after planting a strawberry plant. Find the total number of strawberry plants after 5 years.

Vocabulary, Readiness & Video Check

Use the choices below to fill in each blank. Not all choices may be used.

index of summation	infinite	sigma	1	7
partial sum	finite	summation	5	

1. A series is a(n) _____ series if it is the sum of all the terms of an infinite sequence.

2. A series is a(n) _____ series if it is the sum of a finite number of terms.

3. A shorthand notation for denoting a series when the general term of the sequence is known is called _____ notation.

4. In the notation $\displaystyle\sum_{i=1}^{7}(5i-2)$, the Σ is the Greek uppercase letter _____ and the i is called the _____.

5. The sum of the first n terms of a sequence is a finite series known as a _____.

6. For the notation in Exercise 4 above, the beginning value of i is _____ and the ending value of i is _____.

Watch the section lecture video and answer the following questions.

OBJECTIVE 1

7. From the lecture before Example 1, for the series with the summation notation $\sum_{i=2}^{10} \frac{(-1)^i}{i}$, identify/explain each piece of the notation:

$\Sigma, i, 2, 10, \dfrac{(-1)^i}{i}$.

OBJECTIVE 2

8. From Example 2 and the lecture before, if you're finding the series S_7 of a sequence, what are you actually finding?

11.3   Exercise Set   MyMathLab®

Evaluate. See Example 1.

1. $\sum_{i=1}^{4}(i - 3)$

2. $\sum_{i=1}^{5}(i + 6)$

▶ 3. $\sum_{i=4}^{7}(2i + 4)$

4. $\sum_{i=2}^{3}(5i - 1)$

5. $\sum_{i=2}^{4}(i^2 - 3)$

6. $\sum_{i=3}^{5}i^3$

7. $\sum_{i=1}^{3}\left(\frac{1}{i + 5}\right)$

8. $\sum_{i=2}^{4}\left(\frac{2}{i + 3}\right)$

9. $\sum_{i=1}^{3}\frac{1}{6i}$

10. $\sum_{i=1}^{3}\frac{1}{3i}$

11. $\sum_{i=2}^{6}3i$

12. $\sum_{i=3}^{6}-4i$

13. $\sum_{i=3}^{5}i(i + 2)$

14. $\sum_{i=2}^{4}i(i - 3)$

15. $\sum_{i=1}^{5}2^i$

16. $\sum_{i=1}^{4}3^{i-1}$

17. $\sum_{i=1}^{4}\frac{4i}{i + 3}$

18. $\sum_{i=2}^{5}\frac{6 - i}{6 + i}$

Write each series with summation notation. See Example 2.

19. $1 + 3 + 5 + 7 + 9$

20. $4 + 7 + 10 + 13$

21. $4 + 12 + 36 + 108$

22. $5 + 10 + 20 + 40 + 80 + 160$

23. $12 + 9 + 6 + 3 + 0 + (-3)$

24. $5 + 1 + (-3) + (-7)$

25. $12 + 4 + \dfrac{4}{3} + \dfrac{4}{9}$

26. $80 + 20 + 5 + \dfrac{5}{4} + \dfrac{5}{16}$

27. $1 + 4 + 9 + 16 + 25 + 36 + 49$

28. $1 + (-4) + 9 + (-16)$

Find each partial sum. See Example 3.

29. Find the sum of the first two terms of the sequence whose general term is $a_n = (n + 2)(n - 5)$.

▶ 30. Find the sum of the first two terms of the sequence whose general term is $a_n = n(n - 6)$.

31. Find the sum of the first six terms of the sequence whose general term is $a_n = (-1)^n$.

32. Find the sum of the first seven terms of the sequence whose general term is $a_n = (-1)^{n-1}$.

33. Find the sum of the first four terms of the sequence whose general term is $a_n = (n + 3)(n + 1)$.

34. Find the sum of the first five terms of the sequence whose general term is $a_n = \dfrac{(-1)^n}{2n}$.

35. Find the sum of the first four terms of the sequence whose general term is $a_n = -2n$.

36. Find the sum of the first five terms of the sequence whose general term is $a_n = (n - 1)^2$.

37. Find the sum of the first three terms of the sequence whose general term is $a_n = -\dfrac{n}{3}$.

38. Find the sum of the first three terms of the sequence whose general term is $a_n = (n + 4)^2$.

Solve. See Example 4.

39. A gardener is making a triangular planting with 1 tree in the first row, 2 trees in the second row, 3 trees in the third row, and so on for 10 rows. Write the sequence that describes the number of trees in each row. Find the total number of trees planted.

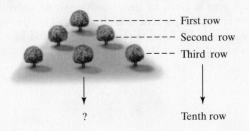

40. Some surfers at the beach form a human pyramid with 2 surfers in the top row, 3 surfers in the second row, 4 surfers in the third row, and so on. If there are 6 rows in the pyramid, write the sequence that describes the number of surfers in each row of the pyramid. Find the total number of surfers.

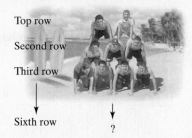

Top row

Second row

Third row

Sixth row

?

41. A culture of fungus starts with 6 units and grows according to the sequence defined by $a_n = 6 \cdot 2^{n-1}$, where n is the number of fungus units at the end of the day. Find the total number of fungus units there will be at the end of the fifth day.

42. A bacterial colony begins with 100 bacteria and grows according to the sequence defined by $a_n = 100 \cdot 2^{n-1}$, where n is the number of 6-hour periods. Find the total number of bacteria there will be after 24 hours.

43. The number of species born each year in a new aquarium forms a sequence whose general term is $a_n = (n + 1)(n + 3)$. Find the number of species born in the fourth year, and find the total number born in the first four years.

44. The number of otters born each year in a new aquarium forms a sequence whose general term is $a_n = (n - 1)(n + 3)$. Find the number of otters born in the third year and find the total number of otters born in the first three years.

45. The number of opossums killed each month on a new highway forms the sequence whose general term is $a_n = (n + 1)(n + 2)$, where n is the number of the month. Find the number of opossums killed in the fourth month and find the total number killed in the first four months.

46. In 2007, the population of the Northern Spotted Owl continued to decline, and the owl remained on the endangered species list as old-growth Northwest forests were logged. The size of the decrease in the population in a given year can be estimated by $200 - 6n$ pairs of birds. Find the decrease in population in 2010 if year 1 is 2007. Find the estimated total decrease in the spotted owl population for the years 2007 through 2010. (*Source: United States Forest Service*)

47. The amount of decay in pounds of a radioactive isotope each year is given by the sequence whose general term is $a_n = 100(0.5)^n$, where n is the number of the year. Find the amount of decay in the fourth year and find the total amount of decay in the first four years.

48. A person has a choice between two job offers. Job A has an annual starting salary of \$30,000 with guaranteed annual raises of \$1200 for the next four years, whereas job B has an annual starting salary of \$28,000 with guaranteed annual raises of \$2500 for the next four years. Compare the fifth partial sums for each sequence to determine which job would pay more money over the next 5 years.

49. A pendulum swings a length of 40 inches on its first swing. Each successive swing is $\frac{4}{5}$ of the preceding swing. Find the length of the fifth swing and the total length swung during the first five swings. (Round to the nearest tenth of an inch.)

50. Explain the difference between a sequence and a series.

REVIEW AND PREVIEW

Evaluate. See Section 1.3.

51. $\dfrac{5}{1 - \dfrac{1}{2}}$

52. $\dfrac{-3}{1 - \dfrac{1}{7}}$

53. $\dfrac{\frac{1}{3}}{1 - \dfrac{1}{10}}$

54. $\dfrac{\frac{6}{11}}{1 - \dfrac{1}{10}}$

55. $\dfrac{3(1 - 2^4)}{1 - 2}$

56. $\dfrac{2(1 - 5^3)}{1 - 5}$

57. $\dfrac{10}{2}(3 + 15)$

58. $\dfrac{12}{2}(2 + 19)$

CONCEPT EXTENSIONS

59. a. Write the sum $\sum\limits_{i=1}^{7} (i + i^2)$ without summation notation.

b. Write the sum $\sum\limits_{i=1}^{7} i + \sum\limits_{i=1}^{7} i^2$ without summation notation.

c. Compare the results of parts (a) and (b).

d. Do you think the following is true or false? Explain your answer.

$$\sum_{i=1}^{n} (a_n + b_n) = \sum_{i=1}^{n} a_n + \sum_{i=1}^{n} b_n$$

60. a. Write the sum $\sum\limits_{i=1}^{6} 5i^3$ without summation notation.

b. Write the expression $5 \cdot \sum\limits_{i=1}^{6} i^3$ without summation notation.

c. Compare the results of parts (a) and (b).

d. Do you think the following is true or false? Explain your answer.

$$\sum_{i=1}^{n} c \cdot a_n = c \cdot \sum_{i=1}^{n} a_n, \text{ where } c \text{ is a constant}$$

Integrated Review SEQUENCES AND SERIES

Write the first five terms of each sequence, whose general term is given.

1. $a_n = n - 3$

2. $a_n = \dfrac{7}{1 + n}$

3. $a_n = 3^{n-1}$

4. $a_n = n^2 - 5$

Find the indicated term for each sequence.

5. $(-2)^n; a_6$

6. $-n^2 + 2; a_4$

7. $\dfrac{(-1)^n}{n}; a_{40}$

8. $\dfrac{(-1)^n}{2n}; a_{41}$

Write the first five terms of the arithmetic or geometric sequence, whose first term is a_1 and whose common difference, d, or common ratio, r, are given.

9. $a_1 = 7; d = -3$

10. $a_1 = -3; r = 5$

11. $a_1 = 45; r = \dfrac{1}{3}$

12. $a_1 = -12; d = 10$

Find the indicated term of each sequence.

13. The tenth term of the arithmetic sequence whose first term is 20 and whose common difference is 9

14. The sixth term of the geometric sequence whose first term is 64 and whose common ratio is $\dfrac{3}{4}$

15. The seventh term of the geometric sequence $6, -12, 24, \ldots$

16. The twentieth term of the arithmetic sequence $-100, -85, -70, \ldots$

17. The fifth term of the arithmetic sequence whose fourth term is -5 and whose tenth term is -35

18. The fifth term of a geometric sequence whose fourth term is 1 and whose seventh term is $\dfrac{1}{8}$

Evaluate.

19. $\displaystyle\sum_{i=1}^{4} 5i$

20. $\displaystyle\sum_{i=1}^{7} (3i + 2)$

21. $\displaystyle\sum_{i=3}^{7} 2^{i-4}$

22. $\displaystyle\sum_{i=2}^{5} \dfrac{i}{i + 1}$

Find each partial sum.

23. Find the sum of the first three terms of the sequence whose general term is $a_n = n(n - 4)$.

24. Find the sum of the first ten terms of the sequence whose general term is $a_n = (-1)^n(n + 1)$.

11.4	**Partial Sums of Arithmetic and Geometric Sequences**

OBJECTIVES

1 Find the Partial Sum of an Arithmetic Sequence.

2 Find the Partial Sum of a Geometric Sequence.

3 Find the Sum of the Terms of an Infinite Geometric Sequence.

OBJECTIVE

1   Finding Partial Sums of Arithmetic Sequences

Partial sums S_n are relatively easy to find when n is small—that is, when the number of terms to add is small. But when n is large, finding S_n can be tedious. For a large n, S_n is still relatively easy to find if the addends are terms of an arithmetic sequence or a geometric sequence.

For an arithmetic sequence, $a_n = a_1 + (n - 1)d$ for some first term a_1 and some common difference d. So S_n, the sum of the first n terms, is

$$S_n = a_1 + (a_1 + d) + (a_1 + 2d) + \cdots + (a_1 + (n - 1)d)$$

We might also find S_n by working backward from the nth term a_n, finding the preceding term a_{n-1}, by subtracting d each time.

$$S_n = a_n + (a_n - d) + (a_n - 2d) + \cdots + (a_n - (n - 1)d)$$

Now add the left sides of these two equations and add the right sides.

$$2S_n = (a_1 + a_n) + (a_1 + a_n) + (a_1 + a_n) + \cdots + (a_1 + a_n)$$

The d terms subtract out, leaving n sums of the first term, a_1, and last term, a_n. Thus, we write

$$2S_n = n(a_1 + a_n)$$

or

$$S_n = \frac{n}{2}(a_1 + a_n)$$

Partial Sum S_n of an Arithmetic Sequence

The partial sum S_n of the first n terms of an arithmetic sequence is given by

$$S_n = \frac{n}{2}(a_1 + a_n)$$

where a_1 is the first term of the sequence and a_n is the nth term.

EXAMPLE 1   Use the partial sum formula to find the sum of the first six terms of the arithmetic sequence $2, 5, 8, 11, 14, 17, \ldots$.

Solution   Use the formula for S_n of an arithmetic sequence, replacing n with 6, a_1 with 2, and a_n with 17.

$$S_n = \frac{n}{2}(a_1 + a_n)$$

$$S_6 = \frac{6}{2}(2 + 17) = 3(19) = 57 \qquad \square$$

PRACTICE

1   Use the partial sum formula to find the sum of the first five terms of the arithmetic sequence $2, 9, 16, 23, 30, \ldots$.

EXAMPLE 2   Find the sum of the first 30 positive integers.

Solution   Because $1, 2, 3, \ldots, 30$ is an arithmetic sequence, use the formula for S_n with $n = 30$, $a_1 = 1$, and $a_n = 30$. Thus,

$$S_n = \frac{n}{2}(a_1 + a_n)$$

$$S_{30} = \frac{30}{2}(1 + 30) = 15(31) = 465 \qquad \square$$

PRACTICE

2   Find the sum of the first 50 positive integers.

EXAMPLE 3   **Stacking Rolls of Carpet**

Rolls of carpet are stacked in 20 rows with 3 rolls in the top row, 4 rolls in the next row, and so on, forming an arithmetic sequence. Find the total number of carpet rolls if there are 22 rolls in the bottom row.

3 rolls
4 rolls
5 rolls

Solution   The list $3, 4, 5, \ldots, 22$ is the first 20 terms of an arithmetic sequence. Use the formula for S_n with $a_1 = 3$, $a_n = 22$, and $n = 20$ terms. Thus,

$$S_{20} = \frac{20}{2}(3 + 22) = 10(25) = 250$$

There are a total of 250 rolls of carpet. $\qquad \square$

PRACTICE

3   An ice sculptor is creating a gigantic castle-facade ice sculpture for First Night festivities in Boston. To get the volume of ice necessary, large blocks of ice were stacked atop each other in 10 rows. The topmost row comprised 6 blocks of ice, the next row 7 blocks of ice, and so on, forming an arithmetic sequence. Find the total number of ice blocks needed if there were 15 blocks in the bottom row.

2   Finding Partial Sums of Geometric Sequences ▷

We can also derive a formula for the partial sum S_n of the first n terms of a geometric series. If $a_n = a_1 r^{n-1}$, then

$$S_n = a_1 + a_1 r + a_1 r^2 + \cdots + a_1 r^{n-1}$$

↑   ↑   ↑   ↑
1st 2nd 3rd   nth
term term term   term

Multiply each side of the equation by $-r$.

$$-rS_n = -a_1 r - a_1 r^2 - a_1 r^3 - \cdots - a_1 r^n$$

Add the two equations.

$$S_n - rS_n = a_1 + (a_1 r - a_1 r) + (a_1 r^2 - a_1 r^2) + (a_1 r^3 - a_1 r^3) + \cdots - a_1 r^n$$

$$S_n - rS_n = a_1 - a_1 r^n$$

Now factor each side.

$$S_n(1 - r) = a_1(1 - r^n)$$

Solve for S_n by dividing both sides by $1 - r$. Thus,

$$S_n = \frac{a_1(1 - r^n)}{1 - r}$$

as long as r is not 1.

Partial Sum S_n of a Geometric Sequence

The partial sum S_n of the first n terms of a geometric sequence is given by

$$S_n = \frac{a_1(1 - r^n)}{1 - r}$$

where a_1 is the first term of the sequence, r is the common ratio, and $r \neq 1$.

EXAMPLE 4   Find the sum of the first six terms of the geometric sequence $5, 10, 20, 40, 80, 160$.

Solution   Use the formula for the partial sum S_n of the terms of a geometric sequence. Here, $n = 6$, the first term $a_1 = 5$, and the common ratio $r = 2$.

$$S_n = \frac{a_1(1 - r^n)}{1 - r}$$

$$S_6 = \frac{5(1 - 2^6)}{1 - 2} = \frac{5(-63)}{-1} = 315 \qquad \square$$

PRACTICE

4   Find the sum of the first five terms of the geometric sequence $32, 8, 2, \frac{1}{2}, \frac{1}{8}$. ▪

EXAMPLE 5   **Finding Amount of Donation**

A grant from an alumnus to a university specified that the university was to receive $800,000 during the first year and 75% of the preceding year's donation during each of the following 5 years. Find the total amount donated during the 6 years.

Solution   The donations are modeled by the first six terms of a geometric sequence. Evaluate S_n when $n = 6$, $a_1 = 800,000$, and $r = 0.75$.

$$S_6 = \frac{800{,}000[1 - (0.75)^6]}{1 - 0.75}$$

$$= \$2{,}630{,}468.75$$

The total amount donated during the 6 years is $2,630,468.75. $\qquad \square$

PRACTICE

5 A new youth center is being established in a downtown urban area. A philanthropic charity has agreed to help it get off the ground. The charity has pledged to donate $250,000 in the first year, with 80% of the preceding year's donation for each of the following 6 years. Find the total amount donated during the 7 years.

OBJECTIVE

3 Finding Sums of Terms of Infinite Geometric Sequences

Is it possible to find the sum of all the terms of an infinite sequence? Examine the partial sums of the geometric sequence $\dfrac{1}{2}, \dfrac{1}{4}, \dfrac{1}{8}, \dots$.

$$S_1 = \frac{1}{2}$$

$$S_2 = \frac{1}{2} + \frac{1}{4} = \frac{3}{4}$$

$$S_3 = \frac{1}{2} + \frac{1}{4} + \frac{1}{8} = \frac{7}{8}$$

$$S_4 = \frac{1}{2} + \frac{1}{4} + \frac{1}{8} + \frac{1}{16} = \frac{15}{16}$$

$$S_5 = \frac{1}{2} + \frac{1}{4} + \frac{1}{8} + \frac{1}{16} + \frac{1}{32} = \frac{31}{32}$$

$$\vdots$$

$$S_{10} = \frac{1}{2} + \frac{1}{4} + \frac{1}{8} + \cdots + \frac{1}{2^{10}} = \frac{1023}{1024}$$

Even though each partial sum is larger than the preceding partial sum, we see that each partial sum is closer to 1 than the preceding partial sum. If n gets larger and larger, then S_n gets closer and closer to 1. We say that 1 is the **limit** of S_n and that 1 is the sum of the terms of this infinite sequence. In general, if $|r| < 1$, the following formula gives the sum of the terms of an infinite geometric sequence.

Sum of the Terms of an Infinite Geometric Sequence

The sum S_∞ of the terms of an infinite geometric sequence is given by

$$S_\infty = \frac{a_1}{1 - r}$$

where a_1 is the first term of the sequence, r is the common ratio, and $|r| < 1$. If $|r| \geq 1$, S_∞ does not exist.

What happens for other values of r? For example, in the following geometric sequence, $r = 3$.

$$6, 18, 54, 162, \dots$$

Here, as n increases, the sum S_n increases also. This time, though, S_n does not get closer and closer to a fixed number but instead increases without bound.

EXAMPLE 6 Find the sum of the terms of the geometric sequence $2, \dfrac{2}{3}, \dfrac{2}{9}, \dfrac{2}{27}, \ldots$.

Solution For this geometric sequence, $r = \dfrac{1}{3}$. Since $|r| < 1$, we may use the formula for S_∞ of a geometric sequence with $a_1 = 2$ and $r = \dfrac{1}{3}$.

$$S_\infty = \frac{a_1}{1 - r} = \frac{2}{1 - \dfrac{1}{3}} = \frac{2}{\dfrac{2}{3}} = 3 \qquad \square$$

PRACTICE
6 Find the sum of the terms of the geometric sequence $7, \dfrac{7}{4}, \dfrac{7}{16}, \dfrac{7}{64}, \ldots$.

The formula for the sum of the terms of an infinite geometric sequence can be used to write a repeating decimal as a fraction. For example,

$$0.33\overline{3} = \frac{3}{10} + \frac{3}{100} + \frac{3}{1000} + \cdots$$

This sum is the sum of the terms of an infinite geometric sequence whose first term a_1 is $\dfrac{3}{10}$ and whose common ratio r is $\dfrac{1}{10}$. Using the formula for S_∞,

$$S_\infty = \frac{a_1}{1 - r} = \frac{\dfrac{3}{10}}{1 - \dfrac{1}{10}} = \frac{1}{3}$$

So, $0.33\overline{3} = \dfrac{1}{3}$.

EXAMPLE 7 **Distance Traveled by a Pendulum**

On its first pass, a pendulum swings through an arc whose length is 24 inches. On each pass thereafter, the arc length is 75% of the arc length on the preceding pass. Find the total distance the pendulum travels before it comes to rest.

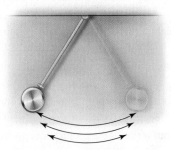

Solution We must find the sum of the terms of an infinite geometric sequence whose first term, a_1, is 24 and whose common ratio, r, is 0.75. Since $|r| < 1$, we may use the formula for S_∞.

$$S_\infty = \frac{a_1}{1 - r} = \frac{24}{1 - 0.75} = \frac{24}{0.25} = 96$$

The pendulum travels a total distance of 96 inches before it comes to rest. $\square$

PRACTICE
7 The manufacturers of the "perpetual bouncing ball" claim that the ball rises to 96% of its dropped height on each bounce of the ball. Find the total distance the ball travels before it comes to rest if it is dropped from a height of 36 inches.

Vocabulary, Readiness & Video Check

Decide whether each sequence is geometric or arithmetic.

1. $5, 10, 15, 20, 25, \dots$; _____

2. $5, 10, 20, 40, 80, \dots$; _____

3. $-1, 3, -9, 27, -81 \dots$; _____

4. $-1, 1, 3, 5, 7, \dots$; _____

5. $-7, 0, 7, 14, 21, \dots$; _____

6. $-7, 7, -7, 7, -7, \dots$; _____

Martin-Gay Interactive Videos

See Video 11.4

Watch the section lecture video and answer the following questions.

OBJECTIVE 1

7. From Example 1, suppose you are asked to find the sum of the first 100 terms of an arithmetic sequence in which you were given only the first few terms. You need the 100th term for the partial sum formula—how can you find this term without actually writing down the first 100 terms?

OBJECTIVE 2

8. From the lecture before Example 2, we know $r \neq 1$ in the partial sum formula because it would make the denominator 0. What would a geometric sequence with $r = 1$ look like? How could the partial sum of such a sequence be found (without the given formula)?

OBJECTIVE 3

9. From the lecture before Example 3, why can't you find S_∞ for the geometric sequence $-1, -3, -9, -27, \dots$?

11.4 Exercise Set MyMathLab®

Use the partial sum formula to find the partial sum of the given arithmetic or geometric sequence. See Examples 1 and 4.

1. Find the sum of the first six terms of the arithmetic sequence $1, 3, 5, 7, \dots$.

2. Find the sum of the first seven terms of the arithmetic sequence $-7, -11, -15, \dots$.

3. Find the sum of the first five terms of the geometric sequence $4, 12, 36, \dots$.

4. Find the sum of the first eight terms of the geometric sequence $-1, 2, -4, \dots$.

5. Find the sum of the first six terms of the arithmetic sequence $3, 6, 9, \dots$.

6. Find the sum of the first four terms of the arithmetic sequence $-4, -8, -12, \dots$.

7. Find the sum of the first four terms of the geometric sequence $2, \dfrac{2}{5}, \dfrac{2}{25}, \dots$.

8. Find the sum of the first five terms of the geometric sequence $\dfrac{1}{3}, -\dfrac{2}{3}, \dfrac{4}{3}, \dots$.

Solve. See Example 2.

9. Find the sum of the first ten positive integers.

10. Find the sum of the first eight negative integers.

11. Find the sum of the first four positive odd integers.

12. Find the sum of the first five negative odd integers.

Find the sum of the terms of each infinite geometric sequence. See Example 6.

13. $12, 6, 3, \dots$

14. $45, 15, 5, \dots$

15. $\dfrac{1}{10}, \dfrac{1}{100}, \dfrac{1}{1000}, \dots$

16. $\dfrac{3}{5}, \dfrac{3}{20}, \dfrac{3}{80}, \dots$

17. $-10, -5, -\dfrac{5}{2}, \dots$

18. $-16, -4, -1, \dots$

19. $2, -\dfrac{1}{4}, \dfrac{1}{32}, \dots$

20. $-3, \dfrac{3}{5}, -\dfrac{3}{25}, \dots$

21. $\dfrac{2}{3}, -\dfrac{1}{3}, \dfrac{1}{6}, \dots$

22. $6, -4, \dfrac{8}{3}, \dots$

MIXED PRACTICE

Solve.

23. Find the sum of the first ten terms of the sequence $-4, 1, 6, \dots, 41$ where 41 is the tenth term.

24. Find the sum of the first twelve terms of the sequence $-3, -13, -23, \dots, -113$ where -113 is the twelfth term.

25. Find the sum of the first seven terms of the sequence $3, \dfrac{3}{2}, \dfrac{3}{4}, \dots$.

26. Find the sum of the first five terms of the sequence $-2, -6, -18, \dots$.

27. Find the sum of the first five terms of the sequence $-12, 6, -3, \dots$.

28. Find the sum of the first four terms of the sequence $-\frac{1}{4}$, $-\frac{3}{4}$, $-\frac{9}{4}$, ...

29. Find the sum of the first twenty terms of the sequence $\frac{1}{2}$, $\frac{1}{4}$, 0, ..., $-\frac{17}{4}$ where $-\frac{17}{4}$ is the twentieth term.

30. Find the sum of the first fifteen terms of the sequence $-5, -9, -13, \ldots, -61$ where -61 is the fifteenth term.

31. If a_1 is 8 and r is $-\frac{2}{3}$, find S_3.

32. If a_1 is 10, a_{18} is $\frac{3}{2}$, and d is $-\frac{1}{2}$, find S_{18}.

Solve. See Example 3.

33. Modern Car Company has come out with a new car model. Market analysts predict that 4000 cars will be sold in the first month and that sales will drop by 50 cars per month after that during the first year. Write out the first five terms of the sequence and find the number of sold cars predicted for the twelfth month. Find the total predicted number of sold cars for the first year.

34. A company that sends faxes charges $3 for the first page sent and $0.10 less than the preceding page for each additional page sent. The cost per page forms an arithmetic sequence. Write the first five terms of this sequence and use a partial sum to find the cost of sending a nine-page document.

35. Sal has two job offers: Firm A starts at $22,000 per year and guarantees raises of $1000 per year, whereas Firm B starts at $20,000 and guarantees raises of $1200 per year. Over a 10-year period, determine the more profitable offer.

36. The game of pool uses 15 balls numbered 1 to 15. In the variety called rotation, a player who sinks a ball receives as many points as the number on the ball. Use an arithmetic series to find the score of a player who sinks all 15 balls.

Solve. See Example 5.

37. A woman made $30,000 during the first year she owned her business and made an additional 10% over the previous year in each subsequent year. Find how much she made during her fourth year of business. Find her total earnings during the first four years.

38. In free fall, a parachutist falls 16 feet during the first second, 48 feet during the second second, 80 feet during the third second, and so on. Find how far she falls during the eighth second. Find the total distance she falls during the first 8 seconds.

39. A trainee in a computer company takes 0.9 times as long to assemble each computer as he took to assemble the preceding computer. If it took him 30 minutes to assemble the first computer, find how long it takes him to assemble the fifth computer. Find the total time he takes to assemble the first five computers (round to the nearest minute).

40. On a gambling trip to Reno, Carol doubled her bet each time she lost. If her first losing bet was $5 and she lost six consecutive bets, find how much she lost on the sixth bet. Find the total amount lost on these six bets.

Solve. See Example 7.

41. A ball is dropped from a height of 20 feet and repeatedly rebounds to a height that is $\frac{4}{5}$ of its previous height. Find the total distance the ball covers before it comes to rest.

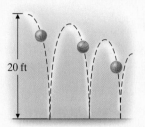

42. A rotating flywheel coming to rest makes 300 revolutions in the first minute and in each minute thereafter makes $\frac{2}{5}$ as many revolutions as in the preceding minute. Find how many revolutions the wheel makes before it comes to rest.

MIXED PRACTICE

Solve.

43. In the pool game of rotation, player A sinks balls numbered 1 to 9, and player B sinks the rest of the balls. Use an arithmetic series to find each player's score (see Exercise 36).

44. A godfather deposited $250 in a savings account on the day his godchild was born. On each subsequent birthday, he deposited $50 more than he deposited the previous year. Find how much money he deposited on his godchild's twenty-first birthday. Find the total amount deposited over the 21 years.

45. During the holiday rush, a business can rent a computer system for $200 the first day, with the rental fee decreasing $5 for each additional day. Find the fee paid for 20 days during the holiday rush.

46. The spraying of a field with insecticide killed 6400 weevils the first day, 1600 the second day, 400 the third day, and so on. Find the total number of weevils killed during the first 5 days.

47. A college student humorously asks his parents to charge him room and board according to this geometric sequence: $0.01 for the first day of the month, $0.02 for the second day, $0.04 for the third day, and so on. Find the total room and board he would pay for 30 days.

48. Following its television advertising campaign, a bank attracted 80 new customers the first day, 120 the second day, 160 the third day, and so on in an arithmetic sequence. Find how many new customers were attracted during the first 5 days following its television campaign.

REVIEW AND PREVIEW

Evaluate. See Section 1.3.

49. $6 \cdot 5 \cdot 4 \cdot 3 \cdot 2 \cdot 1$

50. $8 \cdot 7 \cdot 6 \cdot 5 \cdot 4 \cdot 3 \cdot 2 \cdot 1$

51. $\dfrac{3 \cdot 2 \cdot 1}{2 \cdot 1}$

52. $\dfrac{5 \cdot 4 \cdot 3 \cdot 2 \cdot 1}{3 \cdot 2 \cdot 1}$

Multiply. See Section 5.4.

53. $(x + 5)^2$

54. $(x - 2)^2$

55. $(2x - 1)^3$

56. $(3x + 2)^3$

CONCEPT EXTENSIONS

57. Write $0.88\overline{8}$ as an infinite geometric series and use the formula for S_∞ to write it as a rational number.

58. Write $0.54\overline{54}$ as an infinite geometric series and use the formula S_∞ to write it as a rational number.

59. Explain whether the sequence $5, 5, 5, \dots$ is arithmetic, geometric, neither, or both.

60. Describe a situation in everyday life that can be modeled by an infinite geometric series.

11.5 The Binomial Theorem

OBJECTIVES

1 Use Pascal's Triangle to Expand Binomials.

2 Evaluate Factorials.

3 Use the Binomial Theorem to Expand Binomials.

4 Find the nth Term in the Expansion of a Binomial Raised to a Positive Power.

In this section, we learn how to **expand** binomials of the form $(a + b)^n$ easily. Expanding a binomial such as $(a + b)^n$ means to write the factored form as a sum. First, we review the patterns in the expansions of $(a + b)^n$.

$(a + b)^0 = 1$	1 term
$(a + b)^1 = a + b$	2 terms
$(a + b)^2 = a^2 + 2ab + b^2$	3 terms
$(a + b)^3 = a^3 + 3a^2b + 3ab^2 + b^3$	4 terms
$(a + b)^4 = a^4 + 4a^3b + 6a^2b^2 + 4ab^3 + b^4$	5 terms
$(a + b)^5 = a^5 + 5a^4b + 10a^3b^2 + 10a^2b^3 + 5ab^4 + b^5$	6 terms

Notice the following patterns.

1. The expansion of $(a + b)^n$ contains $n + 1$ terms. For example, for $(a + b)^3$, $n = 3$, and the expansion contains $3 + 1$ terms, or 4 terms.

2. The first term of the expansion of $(a + b)^n$ is a^n, and the last term is b^n.

3. The powers of a decrease by 1 for each term, whereas the powers of b increase by 1 for each term.

4. For each term of the expansion of $(a + b)^n$, the sum of the exponents of a and b is n. (For example, the sum of the exponents of $5a^4b$ is $4 + 1$, or 5, and the sum of the exponents of $10a^3b^2$ is $3 + 2$, or 5.)

OBJECTIVE
1 Using Pascal's Triangle

There are patterns in the coefficients of the terms as well. Written in a triangular array, the coefficients are called **Pascal's triangle.**

$(a + b)^0$:						1						$n = 0$
$(a + b)^1$:					1		1					$n = 1$
$(a + b)^2$:				1		2		1				$n = 2$
$(a + b)^3$:			1		3		3		1			$n = 3$
$(a + b)^4$:		1		4		6		4		1		$n = 4$
$(a + b)^5$:	1		5		10		10		5		1	$n = 5$

Each row in Pascal's triangle begins and ends with 1. Any other number in a row is the sum of the two closest numbers above it. Using this pattern, we can write the next row, for $n = 6$, by first writing the number 1. Then we can add the consecutive numbers in the row for $n = 5$ and write each sum between and below the pair. We complete the row by writing a 1.

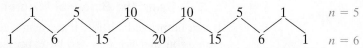

We can use Pascal's triangle and the patterns noted to expand $(a + b)^n$ without actually multiplying any terms.

EXAMPLE 1   Expand $(a + b)^6$.

Solution   Using the $n = 6$ row of Pascal's triangle as the coefficients and following the patterns noted, $(a + b)^6$ can be expanded as

$$a^6 + 6a^5b + 15a^4b^2 + 20a^3b^3 + 15a^2b^4 + 6ab^5 + b^6$$

PRACTICE

1   Expand $(p + r)^7$.

OBJECTIVE

2   **Evaluating Factorials** ▶

For a large n, the use of Pascal's triangle to find coefficients for $(a + b)^n$ can be tedious. An alternative method for determining these coefficients is based on the concept of a **factorial**.

The **factorial of n,** written $n!$ (read "n factorial"), is the product of the first n consecutive natural numbers.

> **Factorial of n: $n!$**
>
> If n is a natural number, then $n! = n(n-1)(n-2)(n-3) \cdots 3 \cdot 2 \cdot 1$. The factorial of 0, written $0!$, is defined to be 1.

For example, $3! = 3 \cdot 2 \cdot 1 = 6, 5! = 5 \cdot 4 \cdot 3 \cdot 2 \cdot 1 = 120$, and $0! = 1$.

EXAMPLE 2   Evaluate each expression.

a. $\dfrac{5!}{6!}$ **b.** $\dfrac{10!}{7!3!}$ **c.** $\dfrac{3!}{2!1!}$ **d.** $\dfrac{7!}{7!0!}$

Solution

a. $\dfrac{5!}{6!} = \dfrac{5 \cdot 4 \cdot 3 \cdot 2 \cdot 1}{6 \cdot 5 \cdot 4 \cdot 3 \cdot 2 \cdot 1} = \dfrac{1}{6}$

b. $\dfrac{10!}{7!3!} = \dfrac{10 \cdot 9 \cdot 8 \cdot 7!}{7! \cdot 3 \cdot 2 \cdot 1} = \dfrac{10 \cdot 9 \cdot 8}{3 \cdot 2 \cdot 1} = 10 \cdot 3 \cdot 4 = 120$

c. $\dfrac{3!}{2!1!} = \dfrac{3 \cdot 2 \cdot 1}{2 \cdot 1 \cdot 1} = 3$

d. $\dfrac{7!}{7!0!} = \dfrac{7!}{7! \cdot 1} = 1$

PRACTICE

2   Evaluate each expression.

a. $\dfrac{6!}{7!}$ **b.** $\dfrac{8!}{4!2!}$ **c.** $\dfrac{5!}{4!1!}$ **d.** $\dfrac{9!}{9!0!}$

> ▶ Helpful Hint
> We can use a calculator with a factorial key to evaluate a factorial. A calculator uses scientific notation for large results.

OBJECTIVE

3 Using the Binomial Theorem

It can be proved, although we won't do so here, that the coefficients of terms in the expansion of $(a + b)^n$ can be expressed in terms of factorials. Following patterns 1 through 4 given earlier and using the factorial expressions of the coefficients, we have what is known as the **binomial theorem.**

Binomial Theorem

If n is a positive integer, then

$$(a + b)^n = a^n + \frac{n}{1!}a^{n-1}b^1 + \frac{n(n-1)}{2!}a^{n-2}b^2$$
$$+ \frac{n(n-1)(n-2)}{3!}a^{n-3}b^3 + \cdots + b^n$$

We call the formula for $(a + b)^n$ given by the binomial theorem the **binomial formula.**

EXAMPLE 3 Use the binomial theorem to expand $(x + y)^{10}$.

Solution Let $a = x, b = y,$ and $n = 10$ in the binomial formula.

$$(x + y)^{10} = x^{10} + \frac{10}{1!}x^9y + \frac{10 \cdot 9}{2!}x^8y^2 + \frac{10 \cdot 9 \cdot 8}{3!}x^7y^3 + \frac{10 \cdot 9 \cdot 8 \cdot 7}{4!}x^6y^4$$
$$+ \frac{10 \cdot 9 \cdot 8 \cdot 7 \cdot 6}{5!}x^5y^5 + \frac{10 \cdot 9 \cdot 8 \cdot 7 \cdot 6 \cdot 5}{6!}x^4y^6$$
$$+ \frac{10 \cdot 9 \cdot 8 \cdot 7 \cdot 6 \cdot 5 \cdot 4}{7!}x^3y^7$$
$$+ \frac{10 \cdot 9 \cdot 8 \cdot 7 \cdot 6 \cdot 5 \cdot 4 \cdot 3}{8!}x^2y^8$$
$$+ \frac{10 \cdot 9 \cdot 8 \cdot 7 \cdot 6 \cdot 5 \cdot 4 \cdot 3 \cdot 2}{9!}xy^9 + y^{10}$$
$$= x^{10} + 10x^9y + 45x^8y^2 + 120x^7y^3 + 210x^6y^4 + 252x^5y^5 + 210x^4y^6$$
$$+ 120x^3y^7 + 45x^2y^8 + 10xy^9 + y^{10} \qquad \square$$

PRACTICE

3 Use the binomial theorem to expand $(a + b)^9$.

EXAMPLE 4 Use the binomial theorem to expand $(x + 2y)^5$.

Solution Let $a = x$ and $b = 2y$ in the binomial formula.

$$(x + 2y)^5 = x^5 + \frac{5}{1!}x^4(2y) + \frac{5 \cdot 4}{2!}x^3(2y)^2 + \frac{5 \cdot 4 \cdot 3}{3!}x^2(2y)^3$$
$$+ \frac{5 \cdot 4 \cdot 3 \cdot 2}{4!}x(2y)^4 + (2y)^5$$
$$= x^5 + 10x^4y + 40x^3y^2 + 80x^2y^3 + 80xy^4 + 32y^5 \qquad \square$$

PRACTICE

4 Use the binomial theorem to expand $(a + 5b)^3$.

EXAMPLE 5   Use the binomial theorem to expand $(3m - n)^4$.

Solution   Let $a = 3m$ and $b = -n$ in the binomial formula.

$$(3m - n)^4 = (3m)^4 + \frac{4}{1!}(3m)^3(-n) + \frac{4 \cdot 3}{2!}(3m)^2(-n)^2$$

$$+ \frac{4 \cdot 3 \cdot 2}{3!}(3m)(-n)^3 + (-n)^4$$

$$= 81m^4 - 108m^3n + 54m^2n^2 - 12mn^3 + n^4 \qquad \square$$

PRACTICE
5   Use the binomial theorem to expand $(3x - 2y)^3$.

OBJECTIVE
4   **Finding the *n*th Term of a Binomial Expansion** ▶

Sometimes it is convenient to find a specific term of a binomial expansion without writing out the entire expansion. By studying the expansion of binomials, a pattern forms for each term. This pattern is most easily stated for the $(r + 1)$st term.

$(r + 1)$st Term in a Binomial Expansion

The $(r + 1)$st term of the expansion of $(a + b)^n$ is $\dfrac{n!}{r!(n - r)!}a^{n-r}b^r$.

EXAMPLE 6   Find the eighth term in the expansion of $(2x - y)^{10}$.

Solution   Use the formula with $n = 10$, $a = 2x$, $b = -y$, and $r + 1 = 8$. Notice that, since $r + 1 = 8$, $r = 7$.

$$\frac{n!}{r!(n - r)!}a^{n-r}b^r = \frac{10!}{7!3!}(2x)^3(-y)^7$$

$$= 120(8x^3)(-y^7)$$

$$= -960x^3y^7 \qquad \square$$

PRACTICE
6   Find the seventh term in the expansion of $(x - 4y)^{11}$.

Vocabulary, Readiness & Video Check

Fill in each blank.

1. $0! = $ _____   **2.** $1! = $ _____   **3.** $4! = $ _____   **4.** $2! = $ _____   **5.** $3!0! = $ _____   **6.** $0!2! = $ _____

Martin-Gay Interactive Videos

See Video 11.5 🍐

Watch the section lecture video and answer the following questions.

OBJECTIVE
1   **7.** From ▥ Example 1 and the lecture before, when expanding a binomial such as $(x + y)^7$, what does Pascal's triangle tell you? What does the power on the binomial tell you?

OBJECTIVE
2   **8.** From ▥ Example 2 and the lecture before, write the definition of 4! and evaluate it. What is the value of 0!?

OBJECTIVE
3   **9.** From ▥ Example 4, what point is made about the terms of a binomial when applying the binomial theorem?

OBJECTIVE
4   **10.** In ▥ Example 5, we are looking for the 4th term, so why do we let $r = 3$?

11.5 Exercise Set

MyMathLab®

Use Pascal's triangle to expand the binomial. See Example 1.

1. $(m + n)^3$

2. $(x + y)^4$

3. $(c + d)^5$

4. $(a + b)^6$

5. $(y - x)^5$

6. $(q - r)^7$

7. Explain how to generate a row of Pascal's triangle.

8. Write the $n = 8$ row of Pascal's triangle.

Evaluate each expression. See Example 2.

9. $\dfrac{8!}{7!}$

10. $\dfrac{6!}{0!}$

11. $\dfrac{7!}{5!}$

12. $\dfrac{8!}{5!}$

13. $\dfrac{10!}{7!2!}$

14. $\dfrac{9!}{5!3!}$

15. $\dfrac{8!}{6!0!}$

16. $\dfrac{10!}{4!6!}$

MIXED PRACTICE

Use the binomial formula to expand each binomial. See Examples 3 through 5.

17. $(a + b)^7$

18. $(x + y)^8$

19. $(a + 2b)^5$

20. $(x + 3y)^6$

21. $(q + r)^9$

22. $(b + c)^6$

23. $(4a + b)^5$

24. $(3m + n)^4$

25. $(5a - 2b)^4$

26. $(m - 4)^6$

27. $(2a + 3b)^3$

28. $(4 - 3x)^5$

29. $(x + 2)^5$

30. $(3 + 2a)^4$

Find the indicated term. See Example 6.

31. The fifth term of the expansion of $(c - d)^5$

32. The fourth term of the expansion of $(x - y)^6$

33. The eighth term of the expansion of $(2c + d)^7$

34. The tenth term of the expansion of $(5x - y)^9$

35. The fourth term of the expansion of $(2r - s)^5$

36. The first term of the expansion of $(3q - 7r)^6$

37. The third term of the expansion of $(x + y)^4$

38. The fourth term of the expansion of $(a + b)^8$

39. The second term of the expansion of $(a + 3b)^{10}$

40. The third term of the expansion of $(m + 5n)^7$

REVIEW AND PREVIEW

Sketch the graph of each function. Decide whether each function is one-to-one. See Sections 3.2 and 9.2.

41. $f(x) = |x|$

42. $g(x) = 3(x - 1)^2$

43. $H(x) = 2x + 3$

44. $F(x) = -2$

45. $f(x) = x^2 + 3$

46. $h(x) = -(x + 1)^2 - 4$

CONCEPT EXTENSIONS

47. Expand the expression $\left(\sqrt{x} + \sqrt{3}\right)^5$.

48. Find the term containing x^2 in the expansion of $\left(\sqrt{x} - \sqrt{5}\right)^6$.

Evaluate the following.

The notation $\dbinom{n}{r}$ means $\dfrac{n!}{r!(n-r)!}$. For example,

$$\binom{5}{3} = \frac{5!}{3!(5-3)!} = \frac{5!}{3!2!} = \frac{5 \cdot 4 \cdot 3 \cdot 2 \cdot 1}{(3 \cdot 2 \cdot 1) \cdot (2 \cdot 1)} = 10.$$

49. $\dbinom{9}{5}$

50. $\dbinom{4}{3}$

51. $\dbinom{8}{2}$

52. $\dbinom{12}{11}$

53. Show that $\dbinom{n}{n} = 1$ for any whole number n.

Chapter 11 Vocabulary Check

Fill in each blank with one of the words or phrases listed below.

general term   common difference   finite sequence   common ratio   Pascal's triangle

infinite sequence   factorial of *n*   arithmetic sequence   geometric sequence   series

1. A(n) _____ is a function whose domain is the set of natural numbers $\{1, 2, 3, \ldots, n\}$, where n is some natural number.

2. The _____, written $n!$, is the product of the first n consecutive natural numbers.

3. A(n) _____ is a function whose domain is the set of natural numbers.

4. A(n) _____ is a sequence in which each term (after the first) is obtained by multiplying the preceding term by a constant amount r. The constant r is called the _____ of the sequence.

5. The sum of the terms of a sequence is called a(n) _____.

6. The nth term of the sequence a_n is called the _____.

7. A(n) _____ is a sequence in which each term (after the first) differs from the preceding term by a constant amount d. The constant d is called the _____ of the sequence.

8. A triangular array of the coefficients of the terms of the expansions of $(a + b)^n$ is called _____.

Chapter 11 Highlights

DEFINITIONS AND CONCEPTS	EXAMPLES
Section 11.1 Sequences	
An **infinite sequence** is a function whose domain is the set of natural numbers $\{1, 2, 3, 4, \ldots \}$. A **finite sequence** is a function whose domain is the set of natural numbers $\{1, 2, 3, 4, \ldots, n\}$, where n is some natural number. The notation a_n, where n is a natural number, denotes a sequence.	***Infinite Sequence*** $$2, 4, 6, 8, 10, \ldots$$ ***Finite Sequence*** $$1, -2, 3, -4, 5, -6$$ Write the first four terms of the sequence whose general term is $a_n = n^2 + 1$. $$a_1 = 1^2 + 1 = 2$$ $$a_2 = 2^2 + 1 = 5$$ $$a_3 = 3^2 + 1 = 10$$ $$a_4 = 4^2 + 1 = 17$$
Section 11.2 Arithmetic and Geometric Sequences	
An **arithmetic sequence** is a sequence in which each term differs from the preceding term by a constant amount d, called the **common difference.** The **general term** a_n of an arithmetic sequence is given by $$a_n = a_1 + (n - 1)d$$ where a_1 is the first term and d is the common difference.	***Arithmetic Sequence*** $$5, 8, 11, 14, 17, 20, \ldots$$ Here, $a_1 = 5$ and $d = 3$. The general term is $$a_n = a_1 + (n - 1)d \text{ or}$$ $$a_n = 5 + (n - 1)3$$

(continued)

DEFINITIONS AND CONCEPTS	EXAMPLES

Section 11.2   Arithmetic and Geometric Sequences (continued)

A **geometric sequence** is a sequence in which each term is obtained by multiplying the preceding term by a constant r, called the **common ratio.**

The **general term** a_n of a geometric sequence is given by

$$a_n = a_1 r^{n-1}$$

where a_1 is the first term and r is the common ratio.

Geometric Sequence

$$12, -6, 3, -\frac{3}{2}, \ldots$$

Here $a_1 = 12$ and $r = -\frac{1}{2}$.

The general term is

$$a_n = a_1 r^{n-1} \text{ or}$$

$$a_n = 12\left(-\frac{1}{2}\right)^{n-1}$$

Section 11.3   Series

A sum of the terms of a sequence is called a **series.**

A shorthand notation for denoting a series is called **summation notation:**

$$\text{index of summation} \rightarrow \sum_{i=1}^{4} \rightarrow \begin{array}{l}\text{Greek letter sigma}\\ \text{used to mean sum}\end{array}$$

Sequence	***Series***	
$3, 7, 11, 15$	$3 + 7 + 11 + 15$	finite
$3, 7, 11, 15, \ldots$	$3 + 7 + 11 + 15 + \cdots$	infinite

$$\sum_{i=1}^{4} 3^i = 3^1 + 3^2 + 3^3 + 3^4$$

$$= 3 + 9 + 27 + 81$$

$$= 120$$

Section 11.4   Partial Sums of Arithmetic and Geometric Sequences

Partial sum, S_n, of the first n terms of an arithmetic sequence:

$$S_n = \frac{n}{2}(a_1 + a_n)$$

where a_1 is the first term and a_n is the nth term.

Partial sum, S_n, of the first n terms of a geometric sequence:

$$S_n = \frac{a_1(1 - r^n)}{1 - r}$$

where a_1 is the first term, r is the common ratio, and $r \neq 1$.

Sum of the terms of an infinite geometric sequence:

$$S_\infty = \frac{a_1}{1 - r}$$

where a_1 is the first term, r is the common ratio, and $|r| < 1$. (If $|r| \geq 1$, S_∞ does not exist.)

The sum of the first five terms of the arithmetic sequence

$$12, 24, 36, 48, 60, \ldots \text{ is}$$

$$S_5 = \frac{5}{2}(12 + 60) = 180$$

The sum of the first five terms of the geometric sequence

$$15, 30, 60, 120, 240, \ldots \text{ is}$$

$$S_5 = \frac{15(1 - 2^5)}{1 - 2} = 465$$

The sum of the terms of the infinite geometric sequence

$$1, \frac{1}{3}, \frac{1}{9}, \frac{1}{27}, \ldots \text{ is}$$

$$S_\infty = \frac{1}{1 - \frac{1}{3}} = \frac{3}{2}$$

Section 11.5   The Binomial Theorem

The **factorial of n**, written $n!$, is the product of the first n consecutive natural numbers.

Binomial Theorem

If n is a positive integer, then

$$(a + b)^n = a^n + \frac{n}{1!}a^{n-1}b^1 + \frac{n(n-1)}{2!}a^{n-2}b^2$$

$$+ \frac{n(n-1)(n-2)}{3!}a^{n-3}b^3 + \cdots + b^n$$

$$5! = 5 \cdot 4 \cdot 3 \cdot 2 \cdot 1 = 120$$

Expand $(3x + y)^4$.

$$(3x + y)^4 = (3x)^4 + \frac{4}{1!}(3x)^3(y)^1$$

$$+ \frac{4 \cdot 3}{2!}(3x)^2(y)^2 + \frac{4 \cdot 3 \cdot 2}{3!}(3x)^1 y^3 + y^4$$

$$= 81x^4 + 108x^3 y + 54x^2 y^2 + 12xy^3 + y^4$$

Chapter 11 **Review**

(11.1) Find the indicated term(s) of the given sequence.

1. The first five terms of the sequence $a_n = -3n^2$

2. The first five terms of the sequence $a_n = n^2 + 2n$

3. The one-hundredth term of the sequence $a_n = \dfrac{(-1)^n}{100}$

4. The fiftieth term of the sequence $a_n = \dfrac{2n}{(-1)^n}$

5. The general term a_n of the sequence $\dfrac{1}{6}, \dfrac{1}{12}, \dfrac{1}{18}, \ldots$

6. The general term a_n of the sequence $-1, 4, -9, 16, \ldots$

Solve the following applications.

7. The distance in feet that an olive falling from rest in a vacuum will travel during each second is given by an arithmetic sequence whose general term is $a_n = 32n - 16$, where n is the number of the second. Find the distance the olive will fall during the fifth, sixth, and seventh seconds.

8. A culture of yeast measures 80 and doubles every day in a geometric progression, where n is the number of the day just ending. Write the measure of the yeast culture for the end of the next 5 days. Find how many days it takes the yeast culture to measure at least 10,000.

9. The Colorado Forest Service reported that western pine beetle infestation, which kills trees, affected approximately 660,000 acres of lodgepole forests in Colorado in 2006. The forest service predicted that during the next 5 years, the beetles would infest twice the number of acres per year as the year before. Write out the first 5 terms of this geometric sequence and find the number of acres of infested trees there were in 2010.

10. The first row of an amphitheater contains 50 seats, and each row thereafter contains 8 additional seats. Write the first ten terms of this arithmetic progression and find the number of seats in the tenth row.

(11.2)

11. Find the first five terms of the geometric sequence whose first term is -2 and whose common ratio is $\dfrac{2}{3}$.

12. Find the first five terms of the arithmetic sequence whose first term is 12 and whose common difference is -1.5.

13. Find the thirtieth term of the arithmetic sequence whose first term is -5 and whose common difference is 4.

14. Find the eleventh term of the arithmetic sequence whose first term is 2 and whose common difference is $\dfrac{3}{4}$.

15. Find the twentieth term of the arithmetic sequence whose first three terms are $12, 7$, and 2.

16. Find the sixth term of the geometric sequence whose first three terms are $4, 6$, and 9.

17. If the fourth term of an arithmetic sequence is 18 and the twentieth term is 98, find the first term and the common difference.

18. If the third term of a geometric sequence is -48 and the fourth term is 192, find the first term and the common ratio.

19. Find the general term of the sequence $\dfrac{3}{10}, \dfrac{3}{100}, \dfrac{3}{1000}, \ldots$

20. Find a general term that satisfies the terms shown for the sequence $50, 58, 66, \ldots$

Determine whether each of the following sequences is arithmetic, geometric, or neither. If a sequence is arithmetic, find a_1 and d. If a sequence is geometric, find a_1 and r.

21. $\dfrac{8}{3}, 4, 6, \ldots$

22. $-10.5, -6.1, -1.7$

23. $7x, -14x, 28x$

24. $3x^2, 9x^4, 81x^8, \ldots$

Solve the following applications.

25. To test the bounce of a racquetball, the ball is dropped from a height of 8 feet. The ball is judged "good" if it rebounds at least 75% of its previous height with each bounce. Write out the first six terms of this geometric sequence (round to the nearest tenth). Determine if a ball is "good" that rebounds to a height of 2.5 feet after the fifth bounce.

26. A display of oil cans in an auto parts store has 25 cans in the bottom row, 21 cans in the next row, and so on, in an arithmetic progression. Find the general term and the number of cans in the top row.

27. Suppose that you save $1 the first day of a month, $2 the second day, $4 the third day, continuing to double your savings each day. Write the general term of this geometric sequence and find the amount you will save on the tenth day. Estimate the amount you will save on the thirtieth day of the month and check your estimate with a calculator.

28. On the first swing, the length of an arc through which a pendulum swings is 30 inches. The length of the arc for each successive swing is 70% of the preceding swing. Find the length of the arc for the fifth swing.

29. Rosa takes a job that has a monthly starting salary of $900 and guarantees her a monthly raise of $150 during her 6-month training period. Find the general term of this sequence and her salary at the end of her training.

30. A sheet of paper is $\dfrac{1}{512}$-inch thick. By folding the sheet in half, the total thickness will be $\dfrac{1}{256}$-inch. A second fold produces a total thickness of $\dfrac{1}{128}$-inch. Estimate the thickness of the stack after 15 folds and then check your estimate with a calculator.

(11.3) Write out the terms and find the sum for each of the following.

31. $\sum_{i=1}^{5} (2i - 1)$

32. $\sum_{i=1}^{5} i(i + 2)$

33. $\sum_{i=2}^{4} \frac{(-1)^i}{2i}$

34. $\sum_{i=3}^{5} 5(-1)^{i-1}$

Write the sum with Σ notation.

35. $1 + 3 + 9 + 27 + 81 + 243$

36. $6 + 2 + (-2) + (-6) + (-10) + (-14) + (-18)$

37. $\frac{1}{4} + \frac{1}{16} + \frac{1}{64} + \frac{1}{256}$

38. $1 + \left(-\frac{3}{2}\right) + \frac{9}{4}$

Solve.

39. A yeast colony begins with 20 yeast and doubles every 8 hours. Write the sequence that describes the growth of the yeast and find the total yeast after 48 hours.

40. The number of cranes born each year in a new aviary forms a sequence whose general term is $a_n = n^2 + 2n - 1$. Find the number of cranes born in the fourth year and the total number of cranes born in the first four years.

41. Harold has a choice between two job offers. Job A has an annual starting salary of $39,500 with guaranteed annual raises of $2200 for the next four years, whereas job B has an annual starting salary of $41,000 with guaranteed annual raises of $1400 for the next four years. Compare the salaries for the fifth year under each job offer.

42. A sample of radioactive waste is decaying such that the amount decaying in kilograms during year n is $a_n = 200(0.5)^n$. Find the amount of decay in the third year and the total amount of decay in the first three years.

(11.4) Find the partial sum of the given sequence.

43. S_4 of the sequence $a_n = (n - 3)(n + 2)$

44. S_6 of the sequence $a_n = n^2$

45. S_5 of the sequence $a_n = -8 + (n - 1)3$

46. S_3 of the sequence $a_n = 5(4)^{n-1}$

47. The sixth partial sum of the sequence $15, 19, 23, \ldots$

48. The ninth partial sum of the sequence $5, -10, 20, \ldots$

49. The sum of the first 30 odd positive integers

50. The sum of the first 20 positive multiples of 7

51. The sum of the first 20 terms of the sequence $8, 5, 2, \ldots$

52. The sum of the first eight terms of the sequence $\frac{3}{4}, \frac{9}{4}, \frac{27}{4}, \ldots$

53. S_4 if $a_1 = 6$ and $r = 5$

54. S_{100} if $a_1 = -3$ and $d = -6$

Find the sum of each infinite geometric sequence.

55. $5, \frac{5}{2}, \frac{5}{4}, \ldots$

56. $18, -2, \frac{2}{9}, \ldots$

57. $-20, -4, -\frac{4}{5}, \ldots$

58. $0.2, 0.02, 0.002, \ldots$

Solve.

59. A frozen yogurt store owner cleared $20,000 the first year he owned his business and made an additional 15% over the previous year in each subsequent year. Find how much he made during his fourth year of business. Find his total earnings during the first 4 years (round to the nearest dollar).

60. On his first morning in a television assembly factory, a trainee takes 0.8 times as long to assemble each television as he took to assemble the one before. If it took him 40 minutes to assemble the first television, find how long it takes him to assemble the fourth television. Find the total time he takes to assemble the first four televisions (round to the nearest minute).

61. During the harvest season, a farmer can rent a combine machine for $100 the first day, with the rental fee decreasing $7 for each additional day. Find how much the farmer pays for the rental on the seventh day. Find how much total rent the farmer pays for 7 days.

62. A rubber ball is dropped from a height of 15 feet and rebounds 80% of its previous height after each bounce. Find the total distance the ball travels before it comes to rest.

63. After a pond was sprayed once with insecticide, 1800 mosquitoes were killed the first day, 600 the second day, 200 the third day, and so on. Find the total number of mosquitoes killed during the first 6 days after the spraying (round to the nearest unit).

64. See Exercise 63. Find the day on which the insecticide is no longer effective and find the total number of mosquitoes killed (round to the nearest mosquito).

65. Use the formula S_∞ to write $0.55\overline{5}$ as a fraction.

66. A movie theater has 27 seats in the first row, 30 seats in the second row, 33 seats in the third row, and so on. Find the total number of seats in the theater if there are 20 rows.

(11.5) Use Pascal's triangle to expand each binomial.

67. $(x + z)^5$

68. $(y - r)^6$

69. $(2x + y)^4$

70. $(3y - z)^4$

Use the binomial formula to expand the following.

71. $(b + c)^8$

72. $(x - w)^7$

73. $(4m - n)^4$

74. $(p - 2r)^5$

Find the indicated term.

75. The fourth term of the expansion of $(a + b)^7$

76. The eleventh term of the expansion of $(y + 2z)^{10}$

77. Evaluate: $\sum_{i=1}^{4} i^2(i + 1)$

78. Find the fifteenth term of the arithmetic sequence whose first three terms are $14, 8,$ and 2.

79. Find the sum of the infinite geometric sequence $27, 9, 3, 1, \ldots$

80. Expand: $(2x - 3)^4$

Chapter 11 Test    MyMathLab®   Test Prep VIDEOS   You Tube

Find the indicated term(s) of the given sequence.

▶ **1.** The first five terms of the sequence $a_n = \dfrac{(-1)^n}{n + 4}$

▶ **2.** The eightieth term of the sequence $a_n = 10 + 3(n - 1)$

▶ **3.** The general term of the sequence $\dfrac{2}{5}, \dfrac{2}{25}, \dfrac{2}{125}, \ldots$

▶ **4.** The general term of the sequence $-9, 18, -27, 36, \ldots$

Find the partial sum of the given sequence.

▶ **5.** S_5 of the sequence $a_n = 5(2)^{n-1}$

▶ **6.** S_{30} of the sequence $a_n = 18 + (n - 1)(-2)$

▶ **7.** S_∞ of the sequence $a_1 = 24$ and $r = \dfrac{1}{6}$

▶ **8.** S_∞ of the sequence $\dfrac{3}{2}, -\dfrac{3}{4}, \dfrac{3}{8}, \ldots$

▶ **9.** $\sum_{i=1}^{4} i(i - 2)$

▶ **10.** $\sum_{i=2}^{4} 5(2)^i(-1)^{i-1}$

Expand each binomial.

▶ **11.** $(a - b)^6$

▶ **12.** $(2x + y)^5$

Solve the following applications.

▶ **13.** The population of a small town is growing yearly according to the sequence defined by $a_n = 250 + 75(n - 1)$, where n is the number of the year just beginning. Predict the population at the beginning of the tenth year. Find the town's initial population.

▶ **14.** A gardener is making a triangular planting with one shrub in the first row, three shrubs in the second row, five shrubs in the third row, and so on, for eight rows. Write the finite series of this sequence and find the total number of shrubs planted.

▶ **15.** A pendulum swings through an arc of length 80 centimeters on its first swing. On each successive swing, the length of the arc is $\dfrac{3}{4}$ the length of the arc on the preceding swing. Find the length of the arc on the fourth swing and find the total arc length for the first four swings.

▶ **16.** See Exercise 15. Find the total arc length before the pendulum comes to rest.

▶ **17.** A parachutist in free-fall falls 16 feet during the first second, 48 feet during the second second, 80 feet during the third second, and so on. Find how far he falls during the tenth second. Find the total distance he falls during the first 10 seconds.

▶ **18.** Use the formula S_∞ to write $0.42\overline{42}$ as a fraction.

Chapter 11 Cumulative Review

1. Divide.

   a. $\dfrac{20}{-4}$ **b.** $\dfrac{-9}{-3}$

   c. $-\dfrac{3}{8} \div 3$ **d.** $\dfrac{-40}{10}$

   e. $\dfrac{-1}{10} \div \dfrac{-2}{5}$ **f.** $\dfrac{8}{0}$

2. Simplify each expression.

   a. $3a - (4a + 3)$

   b. $(5x - 3) + (2x + 6)$

   c. $4(2x - 5) - 3(5x + 1)$

3. Suppose that a computer store just announced an 8% decrease in the price of a particular computer model. If this computer sells for $2162 after the decrease, find the original price of this computer.

4. Sara bought a digital camera for $344.50 including tax. If the tax rate is 6%, what was the price of the camera before taxes?

5. Solve: $3y - 2x = 7$ for y.

6. Find an equation of the line through $(3, -2)$ and parallel to $3x - 2y = 6$. Write the equation using function notation.

7. Use the product rule to simplify.

   a. $(3x^6)(5x)$ **b.** $(-2.4x^3p^2)(4xp^{10})$

8. Solve: $y^3 + 5y^2 - y = 5$

9. Use synthetic division to divide $x^4 - 2x^3 - 11x^2 + 34$ by $x + 2$.

10. Perform the indicated operation and simplify if possible.
$$\frac{5}{3a - 6} - \frac{a}{a - 2} + \frac{3 + 2a}{5a - 10}$$

11. Simplify the following.
   a. $\sqrt{50}$ b. $\sqrt[3]{24}$
   c. $\sqrt{26}$ d. $\sqrt[4]{32}$

12. Solve: $\sqrt{3x + 6} - \sqrt{7x - 6} = 0$

13. Use the formula $A = P(1 + r)^t$ to find the interest rate r if $2000 compounded annually grows to $2420 in 2 years.

14. Rationalize each denominator.
   a. $\sqrt[3]{\dfrac{4}{3x}}$ b. $\dfrac{\sqrt{2} + 1}{\sqrt{2} - 1}$

15. Solve: $(x - 3)^2 - 3(x - 3) - 4 = 0$.

16. Solve: $\dfrac{10}{(2x + 4)^2} - \dfrac{1}{2x + 4} = 3$

17. Solve: $\dfrac{5}{x + 1} < -2$.

18. Graph $f(x) = (x + 2)^2 - 6$. Find the vertex and axis of symmetry.

19. A rock is thrown upward from the ground. Its height in feet above ground after t seconds is given by the function $f(t) = -16t^2 + 20t$. Find the maximum height of the rock and the number of seconds it took for the rock to reach its maximum height.

20. Find the vertex of $f(x) = x^2 + 3x - 18$.

21. If $f(x) = x^2$ and $g(x) = x + 3$, find each composition.
   a. $(f \circ g)(2)$ and $(g \circ f)(2)$
   b. $(f \circ g)(x)$ and $(g \circ f)(x)$

22. Find the inverse of $f(x) = -2x + 3$.

23. Find the inverse of the one-to-one function $f = \{(0, 1), (-2, 7), (3, -6), (4, 4)\}$.

24. If $f(x) = x^2 - 2$ and $g(x) = x + 1$, find each composition.
   a. $(f \circ g)(2)$ and $(g \circ f)(2)$
   b. $(f \circ g)(x)$ and $(g \circ f)(x)$

25. Solve each equation for x.
   a. $2^x = 16$ b. $9^x = 27$ c. $4^{x+3} = 8^x$

26. Solve each equation.
   a. $\log_2 32 = x$ b. $\log_4 \dfrac{1}{64} = x$
   c. $\log_{1/2} x = 5$

27. Simplify.
   a. $\log_3 3^2$ b. $\log_7 7^{-1}$
   c. $5^{\log_5 3}$ d. $2^{\log_2 6}$

28. Solve each equation for x.
   a. $4^x = 64$ b. $8^x = 32$
   c. $9^{x+4} = 243^x$

29. Write each sum as a single logarithm.
   a. $\log_{11} 10 + \log_{11} 3$
   b. $\log_3 \dfrac{1}{2} + \log_3 12$
   c. $\log_2(x + 2) + \log_2 x$

30. Find the exact value.
   a. $\log 100{,}000$ b. $\log 10^{-3}$
   c. $\ln \sqrt[5]{e}$ d. $\ln e^4$

31. Find the amount owed at the end of 5 years if $1600 is loaned at a rate of 9% compounded continuously.

32. Write each expression as a single logarithm.
   a. $\log_6 5 + \log_6 4$
   b. $\log_8 12 - \log_8 4$
   c. $2 \log_2 x + 3 \log_2 x - 2 \log_2(x - 1)$

33. Solve: $3^x = 7$.

34. Using $A = P\left(1 + \dfrac{r}{n}\right)^{nt}$, find how long it takes $5000 to double if it is invested at 2% interest compounded quarterly. Round to the nearest tenth.

35. Solve: $\log_4(x - 2) = 2$.

36. Solve: $\log_4 10 - \log_4 x = 2$.

37. Graph $\dfrac{x^2}{16} - \dfrac{y^2}{25} = 1$.

38. Find the distance between $(8, 5)$ and $(-2, 4)$.

39. Solve the system $\begin{cases} y = \sqrt{x} \\ x^2 + y^2 = 6 \end{cases}$.

40. Solve the system $\begin{cases} x^2 + y^2 = 36 \\ x - y = 6 \end{cases}$.

41. Graph $\dfrac{x^2}{9} + \dfrac{y^2}{16} \leq 1$.

42. Graph $\begin{cases} y \geq x^2 \\ y \leq 4 \end{cases}$.

43. Write the first five terms of the sequence whose general term is given by $a_n = n^2 - 1$.

44. If the general term of a sequence is $a_n = \dfrac{n}{n + 4}$, find a_8.

45. Find the eleventh term of the arithmetic sequence whose first three terms are $2, 9$, and 16.

46. Find the sixth term of the geometric sequence $2, 10, 50, \ldots$.

47. Evaluate.
   a. $\displaystyle\sum_{i=0}^{6} \dfrac{i - 2}{2}$ b. $\displaystyle\sum_{i=3}^{5} 2^i$

48. Evaluate.
   a. $\displaystyle\sum_{i=0}^{4} i(i + 1)$ b. $\displaystyle\sum_{i=0}^{3} 2^i$

49. Find the sum of the first 30 positive integers.

50. Find the third term of the expansion of $(x - y)^6$.

Appendix A
Geometry

A.1 Geometric Formulas

Rectangle

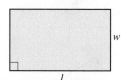

Perimeter: $P = 2l + 2w$
Area: $A = lw$

Square

Perimeter: $P = 4s$
Area: $A = s^2$

Triangle

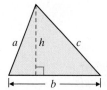

Perimeter: $P = a + b + c$
Area: $A = \frac{1}{2}bh$

Sum of Angles of Triangle

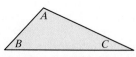

$A + B + C = 180°$
The sum of the measures of the three angles is 180°.

Right Triangles

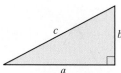

Perimeter: $P = a + b + c$
Area: $A = \frac{1}{2}ab$
One 90° (right) angle

Pythagorean Theorem (for right triangles)

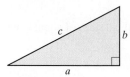

$a^2 + b^2 = c^2$

Isosceles Triangle

Triangle has:
two equal sides and
two equal angles.

Equilateral Triangle

Triangle has:
three equal sides and
three equal angles.
Measure of each angle is 60°.

Trapezoid

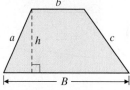

Perimeter: $P = a + b + c + B$
Area: $A = \frac{1}{2}h(B + b)$

Parallelogram

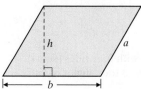

Perimeter: $P = 2a + 2b$
Area: $A = bh$

Circle

Circumference: $C = \pi d$
$C = 2\pi r$
Area: $A = \pi r^2$

Rectangular Solid

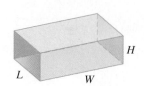

Volume: $V = LWH$
Surface Area:
$S = 2LW + 2HL + 2HW$

Cube

Volume: $V = s^3$
Surface Area: $S = 6s^2$

Cone

Volume: $V = \frac{1}{3}\pi r^2 h$
Lateral Surface Area:
$S = \pi r \sqrt{r^2 + h^2}$

Right Circular Cylinder

Volume: $V = \pi r^2 h$
Surface Area: $S = 2\pi r^2 + 2\pi rh$

Sphere

Volume: $V = \frac{4}{3}\pi r^3$
Surface Area: $S = 4\pi r^2$

Other Formulas

Distance: $d = rt$ ($r =$ rate, $t =$ time)
Percent: $p = br$ ($p =$ percentage, $b =$ base, $r =$ rate)

Compound Interest: $A = P\left(1 + \dfrac{r}{n}\right)^{nt}$

($P =$ principal, $r =$ annual interest rate, $t =$ time in years, $n =$ number of compoundings per year)

Temperature: $F = \dfrac{9}{5}C + 32$ $C = \dfrac{5}{9}(F - 32)$

Simple Interest: $I = Prt$
($P =$ principal, $r =$ annual interest rate, $t =$ time in years)

A.2 Review of Geometric Figures

	Plane figures have length and width but no thickness or depth	
Name	*Description*	*Figure*
Polygon	Union of three or more coplanar line segments that intersect with each other only at each end point, with each end point shared by two segments.	
Triangle	Polygon with three sides (sum of measures of three angles is 180°).	
Scalene Triangle	Triangle with no sides of equal length.	
Isosceles Triangle	Triangle with two sides of equal length.	
Equilateral Triangle	Triangle with all sides of equal length.	
Right Triangle	Triangle that contains a right angle.	hypotenuse, leg, leg
Quadrilateral	Polygon with four sides (sum of measures of four angles is 360°).	
Trapezoid	Quadrilateral with exactly one pair of opposite sides parallel.	base, leg, parallel sides, leg, base
Isosceles Trapezoid	Trapezoid with legs of equal length.	
Parallelogram	Quadrilateral with both pairs of opposite sides parallel and equal in length.	

Plane figures have length and width but no thickness or depth

Name	Description	Figure
Rhombus	Parallelogram with all sides of equal length.	
Rectangle	Parallelogram with four right angles.	
Square	Rectangle with all sides of equal length.	
Circle	All points in a plane the same distance from a fixed point called the **center.**	radius center diameter

Solids have length, width, and depth

Name	Description	Figure
Rectangular Solid	A solid with six sides, all of which are rectangles.	
Cube	A rectangular solid whose six sides are squares.	
Sphere	All points the same distance from a fixed point called the center.	radius center
Right Circular Cylinder	A cylinder with two circular bases that are perpendicular to their altitude.	
Right Circular Cone	A cone with a circular base that is perpendicular to its altitude.	

A.3 Review of Volume and Surface Area

A **convex solid** is a set of points, S, not all in one plane, such that for any two points A and B in S, all points between A and B are also in S. In this appendix, we will find the volume and surface area of special types of solids called polyhedrons. A solid formed by the intersection of a finite number of planes is called a **polyhedron.** The box below is an example of a polyhedron.

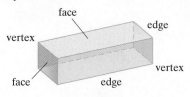

Polyhedron

Each of the plane regions of the polyhedron is called a **face** of the polyhedron. If the intersection of two faces is a line segment, this line segment is an **edge** of the polyhedron. The intersections of the edges are the **vertices** of the polyhedron.

Volume is a measure of the space of a solid. The volume of a box or can, for example, is the amount of space inside. Volume can be used to describe the amount of juice in a pitcher or the amount of concrete needed to pour a foundation for a house.

The volume of a solid is the number of **cubic units** in the solid. A cubic centimeter and a cubic inch are illustrated.

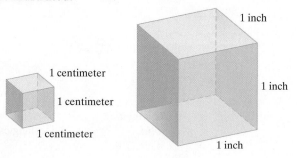

1 cubic centimeter   **1 cubic inch**

The **surface area** of a polyhedron is the sum of the areas of the faces of the polyhedron. For example, each face of the cube to the left above has an area of 1 square centimeter. Since there are 6 faces of the cube, the sum of the areas of the faces is 6 square centimeters. Surface area can be used to describe the amount of material needed to cover or form a solid. Surface area is measured in square units.

Formulas for finding the volumes, V, and surface areas, SA, of some common solids are given next.

Volume and Surface Area Formulas for Common Solids	
Solid	*Formulas*
RECTANGULAR SOLID height width length	$V = lwh$ $SA = 2lh + 2wh + 2lw$ where h = height, w = width, l = length
CUBE side side side	$V = s^3$ $SA = 6s^2$ where s = side

Volume and Surface Area Formulas for Common Solids	
Solid	***Formulas***
SPHERE radius	$V = \dfrac{4}{3}\pi r^3$ $SA = 4\pi r^2$ where r = radius
CIRCULAR CYLINDER height radius	$V = \pi r^2 h$ $SA = 2\pi rh + 2\pi r^2$ where h = height, r = radius
CONE height radius	$V = \dfrac{1}{3}\pi r^2 h$ $SA = \pi r\sqrt{r^2 + h^2} + \pi r^2$ where h = height, r = radius
SQUARE-BASED PYRAMID height slant height side	$V = \dfrac{1}{3}s^2 h$ $SA = B + \dfrac{1}{2}pl$ where B = area of base, p = perimeter of base, h = height, s = side, l = slant height

▶ **Helpful Hint**

Volume is measured in cubic units. Surface area is measured in square units.

EXAMPLE 1   Find the volume and surface area of a rectangular box that is 12 inches long, 6 inches wide, and 3 inches high.

3 in.

6 in.   12 in.

Solution   Let h = 3 in., l = 12 in., and w = 6 in.

$$V = lwh$$
$$V = 12 \text{ inches} \cdot 6 \text{ inches} \cdot 3 \text{ inches} = 216 \text{ cubic inches}$$

The volume of the rectangular box is 216 cubic inches.

$$SA = 2lh + 2wh + 2lw$$
$$= 2(12 \text{ in.})(3 \text{ in.}) + 2(6 \text{ in.})(3 \text{ in.}) + 2(12 \text{ in.})(6 \text{ in.})$$
$$= 72 \text{ sq in.} + 36 \text{ sq in.} + 144 \text{ sq in.}$$
$$= 252 \text{ sq in.}$$

The surface area of the rectangular box is 252 square inches.   □

EXAMPLE 2 Find the volume and surface area of a ball of radius 2 inches. Give the exact volume and surface area and then use the approximation $\frac{22}{7}$ for π.

Solution

2 in.

$$V = \frac{4}{3}\pi r^3 \qquad \text{Formula for volume of a sphere}$$

$$V = \frac{4}{3}\pi(2 \text{ in.})^3 \qquad \text{Let } r = 2 \text{ inches.}$$

$$= \frac{32}{3}\pi \text{ cu in.} \qquad \text{Simplify.}$$

$$\approx \frac{32}{3} \cdot \frac{22}{7} \text{ cu in.} \qquad \text{Approximate } \pi \text{ with } \frac{22}{7}.$$

$$= \frac{704}{21} \text{ or } 33\frac{11}{21} \text{ cu in.}$$

The volume of the sphere is exactly $\frac{32}{3}\pi$ cubic inches or approximately $33\frac{11}{21}$ cubic inches.

$$SA = 4\pi r^2 \qquad \text{Formula for surface area}$$

$$SA = 4\pi(2 \text{ in.})^2 \qquad \text{Let } r = 2 \text{ inches.}$$

$$= 16\pi \text{ sq in.} \qquad \text{Simplify.}$$

$$\approx 16 \cdot \frac{22}{7} \text{ sq in.} \qquad \text{Approximate } \pi \text{ with } \frac{22}{7}.$$

$$= \frac{352}{7} \text{ or } 50\frac{2}{7} \text{ sq in.}$$

The surface area of the sphere is exactly 16π square inches or approximately $50\frac{2}{7}$ square inches. ☐

A.3 Exercise Set   MyMathLab®

Find the volume and surface area of each solid. See Examples 1 and 2. For formulas that contain π, _give an exact answer and then approximate using_ $\frac{22}{7}$ _for_ π.

1.
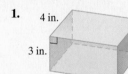
4 in.
3 in.
6 in.

2.

3 mi

3.
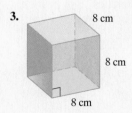
8 cm
8 cm
8 cm

4.
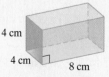
4 cm
4 cm
8 cm

5. (For surface area, use 3.14 to approximate π and round to two decimal places.)
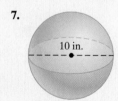
3 yd
2 yd

6.

10 ft
6 ft

7.

10 in.

8. Find the volume only.

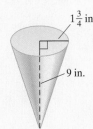

$1\frac{3}{4}$ in

9 in.

9.

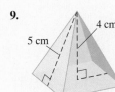

4 cm

5 cm

6 cm

10.

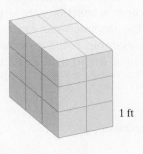

1 ft

Solve.

11. Find the volume of a cube with edges of $1\frac{1}{3}$ inches.

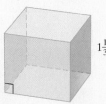

$1\frac{1}{3}$ in.

12. A water storage tank is in the shape of a cone with the pointed end down. If the radius is 14 ft and the depth of the tank is 15 ft, approximate the volume of the tank in cubic feet. Use $\frac{22}{7}$ for π.

14 ft

15 ft

13. Find the surface area of a rectangular box 2 ft by 1.4 ft by 3 ft.

14. Find the surface area of a box in the shape of a cube that is 5 ft on each side.

15. Find the volume of a pyramid with a square base 5 in. on a side and a height of 1.3 in.

16. Approximate to the nearest hundredth the volume of a sphere with a radius of 2 cm. Use 3.14 for π.

17. A paperweight is in the shape of a square-based pyramid 20 cm tall. If an edge of the base is 12 cm, find the volume of the paperweight.

18. A bird bath is made in the shape of a hemisphere (half-sphere). If its radius is 10 in., approximate the volume. Use $\frac{22}{7}$ for π.

10 in.

19. Find the exact surface area of a sphere with a radius of 7 in.

20. A tank is in the shape of a cylinder 8 ft tall and 3 ft in radius. Find the exact surface area of the tank.

21. Find the volume of a rectangular block of ice 2 ft by $2\frac{1}{2}$ ft by $1\frac{1}{2}$ ft.

22. Find the capacity (volume in cubic feet) of a rectangular ice chest with inside measurements of 3 ft by $1\frac{1}{2}$ ft by $1\frac{3}{4}$ ft.

23. An ice cream cone with a 4-cm diameter and 3-cm depth is filled exactly level with the top of the cone. Approximate how much ice cream (in cubic centimeters) is in the cone. Use $\frac{22}{7}$ for π.

24. A child's toy is in the shape of a square-based pyramid 10 in. tall. If an edge of the base is 7 in., find the volume of the toy.

Appendix B

Stretching and Compressing Graphs of Absolute Value Functions

In Section 3.6, we learned to shift and reflect graphs of common functions: $f(x) = x$, $f(x) = x^2$, $f(x) = |x|$, and $f(x) = \sqrt{x}$. Since other common functions are studied throughout this text, in this appendix we concentrate on the absolute value function.

Recall that the graph of $h(x) = -|x - 1| + 2$, for example, is the same as the graph of $f(x) = |x|$ reflected about the x-axis, moved 1 unit to the right and 2 units upward. In other words,

$$h(x) = -|x - 1| + 2$$

opens downward $(1, 2)$ location of vertex of V-shape

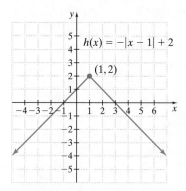

$h(x) = -|x - 1| + 2$

$(1, 2)$

Let's now study the graphs of a few other absolute value functions.

EXAMPLE 1   Graph $h(x) = 2|x|$, and $g(x) = \dfrac{1}{2}|x|$.

Solution   Let's find and plot ordered pair solutions for the functions.

x	$h(x)$	$g(x)$
-2	4	1
-1	2	$\dfrac{1}{2}$
0	0	0
1	2	$\dfrac{1}{2}$
2	4	2

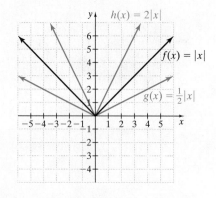

$h(x) = 2|x|$

$f(x) = |x|$

$g(x) = \dfrac{1}{2}|x|$

Notice that the graph of $h(x) = 2|x|$ is narrower than the graph of $f(x) = |x|$ and the graph of $g(x) = \dfrac{1}{2}|x|$ is wider than the graph of $f(x) = |x|$.   □

In general, for the absolute value function, we have the following:

> **The Graph of the Absolute Value Function**
>
> The graph of $f(x) = a|x - h| + k$
> - Has vertex (h, k) and is V-shaped.
> - Opens up if $a > 0$ and down if $a < 0$.
> - If $|a| < 1$, the graph is wider than the graph of $y = |x|$.
> - If $|a| > 1$, the graph is narrower than the graph of $y = |x|$.

EXAMPLE 2   Graph $f(x) = -\dfrac{1}{3}|x + 2| + 4$.

Solution   Let's write this function in the form $f(x) = a|x - h| + k$. For our function, we have $f(x) = -\dfrac{1}{3}|x - (-2)| + 4$. Thus,

- vertex is $(-2, 4)$
- since $a < 0$, V-shape opens down
- since $|a| = \left|-\dfrac{1}{3}\right| = \dfrac{1}{3} < 1$, the graph is wider than $y = |x|$

We will also find and plot ordered pair solutions.

If $x = -5$, $f(-5) = -\dfrac{1}{3}|-5 + 2| + 4$, or 3

If $x = 1$, $f(1) = -\dfrac{1}{3}|1 + 2| + 4$, or 3

If $x = 3$, $f(3) = -\dfrac{1}{3}|3 + 2| + 4$, or $\dfrac{7}{3}$, or $2\dfrac{1}{3}$

x	$f(x)$
-5	3
1	3
3	$2\dfrac{1}{3}$

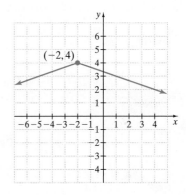

B Exercise Set   MyMathLab®

Sketch the graph of each function. Label the vertex of the V-shape.

1. $f(x) = 3|x|$

2. $f(x) = 5|x|$

3. $f(x) = \dfrac{1}{4}|x|$

4. $f(x) = \dfrac{1}{3}|x|$

5. $g(x) = 2|x| + 3$

6. $g(x) = 3|x| + 2$

7. $h(x) = -\dfrac{1}{2}|x|$

8. $h(x) = -\dfrac{1}{3}|x|$

9. $f(x) = 4|x - 1|$

10. $f(x) = 3|x - 2|$

11. $g(x) = -\dfrac{1}{3}|x| - 2$

12. $g(x) = -\dfrac{1}{2}|x| - 3$

13. $f(x) = -2|x - 3| + 4$

14. $f(x) = -3|x - 1| + 5$

15. $f(x) = \dfrac{2}{3}|x + 2| - 5$

16. $f(x) = \dfrac{3}{4}|x + 1| - 4$

Appendix C

Solving Systems of Equations Using Determinants

OBJECTIVES

1 Define and Evaluate a 2 × 2 Determinant.

2 Use Cramer's Rule to Solve a System of Two Linear Equations in Two Variables.

3 Define and Evaluate a 3 × 3 Determinant.

4 Use Cramer's Rule to Solve a System of Three Linear Equations in Three Variables.

We have solved systems of two linear equations in two variables in four ways: graphically, by substitution, by elimination, and by matrices. Now we analyze another method, called **Cramer's rule.**

OBJECTIVE

1   Evaluating 2 × 2 Determinants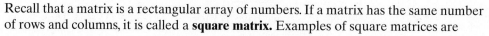

Recall that a matrix is a rectangular array of numbers. If a matrix has the same number of rows and columns, it is called a **square matrix.** Examples of square matrices are

$$\begin{bmatrix} 1 & 6 \\ 5 & 2 \end{bmatrix} \qquad \begin{bmatrix} 2 & 4 & 1 \\ 0 & 5 & 2 \\ 3 & 6 & 9 \end{bmatrix}$$

A **determinant** is a real number associated with a square matrix. The determinant of a square matrix is denoted by placing vertical bars about the array of numbers. Thus,

The determinant of the square matrix $\begin{bmatrix} 1 & 6 \\ 5 & 2 \end{bmatrix}$ is $\begin{vmatrix} 1 & 6 \\ 5 & 2 \end{vmatrix}$.

The determinant of the square matrix $\begin{bmatrix} 2 & 4 & 1 \\ 0 & 5 & 2 \\ 3 & 6 & 9 \end{bmatrix}$ is $\begin{vmatrix} 2 & 4 & 1 \\ 0 & 5 & 2 \\ 3 & 6 & 9 \end{vmatrix}$.

We define the determinant of a 2 × 2 matrix first. (Recall that 2 × 2 is read "two by two." It means that the matrix has 2 rows and 2 columns.)

Determinant of a 2 × 2 Matrix

$$\begin{vmatrix} a & b \\ c & d \end{vmatrix} = ad - bc$$

EXAMPLE 1   Evaluate each determinant

a. $\begin{vmatrix} -1 & 2 \\ 3 & -4 \end{vmatrix}$
  b. $\begin{vmatrix} 2 & 0 \\ 7 & -5 \end{vmatrix}$

Solution   First we identify the values of $a, b, c,$ and $d.$ Then we perform the evaluation.

a. Here $a = -1, b = 2, c = 3,$ and $d = -4.$

$$\begin{vmatrix} -1 & 2 \\ 3 & -4 \end{vmatrix} = ad - bc = (-1)(-4) - (2)(3) = -2$$

b. In this example, $a = 2, b = 0, c = 7,$ and $d = -5.$

$$\begin{vmatrix} 2 & 0 \\ 7 & -5 \end{vmatrix} = ad - bc = 2(-5) - (0)(7) = -10 \qquad \square$$

OBJECTIVE

2   Using Cramer's Rule to Solve a System of Two Linear Equations

To develop Cramer's rule, we solve the system $\begin{cases} ax + by = h \\ cx + dy = k \end{cases}$ using elimination. First, we eliminate y by multiplying both sides of the first equation by d and both sides of the second equation by $-b$ so that the coefficients of y are opposites. The result is that

$$\begin{cases} d(ax + by) = d \cdot h \\ -b(cx + dy) = -b \cdot k \end{cases} \quad \text{simplifies to} \quad \begin{cases} adx + bdy = hd \\ -bcx - bdy = -kb \end{cases}$$

We now add the two equations and solve for x.

$$\begin{aligned} adx + bdy &= hd \\ \underline{-bcx - bdy} &= \underline{-kb} \\ adx - bcx &= hd - kb \quad \text{Add the equations.} \\ (ad - bc)x &= hd - kb \\ x &= \frac{hd - kb}{ad - bc} \quad \text{Solve for } x. \end{aligned}$$

When we replace x with $\dfrac{hd - kb}{ad - bc}$ in the equation $ax + by = h$ and solve for y, we find that $y = \dfrac{ak - ch}{ad - bc}$.

Notice that the numerator of the value of x is the determinant of

$$\begin{vmatrix} h & b \\ k & d \end{vmatrix} = hd - kb$$

Also, the numerator of the value of y is the determinant of

$$\begin{vmatrix} a & h \\ c & k \end{vmatrix} = ak - hc$$

Finally, the denominators of the values of x and y are the same and are the determinant of

$$\begin{vmatrix} a & b \\ c & d \end{vmatrix} = ad - bc$$

This means that the values of x and y can be written in determinant notation:

$$x = \frac{\begin{vmatrix} h & b \\ k & d \end{vmatrix}}{\begin{vmatrix} a & b \\ c & d \end{vmatrix}} \quad \text{and} \quad y = \frac{\begin{vmatrix} a & h \\ c & k \end{vmatrix}}{\begin{vmatrix} a & b \\ c & d \end{vmatrix}}$$

For convenience, we label the determinants D, D_x, and D_y.

x-coefficients

y-coefficients

$$\begin{vmatrix} a & b \\ c & d \end{vmatrix} = D \qquad \begin{vmatrix} h & b \\ k & d \end{vmatrix} = D_x \qquad \begin{vmatrix} a & h \\ c & k \end{vmatrix} = D_y$$

x-column replaced y-column replaced
by constants   by constants

These determinant formulas for the coordinates of the solution of a system are known as **Cramer's rule.**

Cramer's Rule for Two Linear Equations in Two Variables

The solution of the system $\begin{cases} ax + by = h \\ cx + dy = k \end{cases}$ is given by

$$x = \frac{\begin{vmatrix} h & b \\ k & d \end{vmatrix}}{\begin{vmatrix} a & b \\ c & d \end{vmatrix}} = \frac{D_x}{D} \qquad y = \frac{\begin{vmatrix} a & h \\ c & k \end{vmatrix}}{\begin{vmatrix} a & b \\ c & d \end{vmatrix}} = \frac{D_y}{D}$$

as long as $D = ad - bc$ is not 0.

When $D = 0$, the system is either inconsistent or the equations are dependent. When this happens, we need to use another method to see which is the case.

EXAMPLE 2 Use Cramer's rule to solve the system

$$\begin{cases} 3x + 4y = -7 \\ x - 2y = -9 \end{cases}$$

Solution First we find D, D_x, and D_y.

$$\begin{array}{ccc} a & b & h \\ \downarrow & \downarrow & \downarrow \end{array}$$
$$\begin{cases} 3x + 4y = -7 \\ x - 2y = -9 \end{cases}$$
$$\begin{array}{ccc} \uparrow & \uparrow & \uparrow \\ c & d & k \end{array}$$

$$D = \begin{vmatrix} a & b \\ c & d \end{vmatrix} = \begin{vmatrix} 3 & 4 \\ 1 & -2 \end{vmatrix} = 3(-2) - 4(1) = -10$$

$$D_x = \begin{vmatrix} h & b \\ k & d \end{vmatrix} = \begin{vmatrix} -7 & 4 \\ -9 & -2 \end{vmatrix} = (-7)(-2) - 4(-9) = 50$$

$$D_y = \begin{vmatrix} a & h \\ c & d \end{vmatrix} = \begin{vmatrix} 3 & -7 \\ 1 & -9 \end{vmatrix} = 3(-9) - (-7)(1) = -20$$

Then $x = \dfrac{D_x}{D} = \dfrac{50}{-10} = -5$ and $y = \dfrac{D_y}{D} = \dfrac{-20}{-10} = 2$.

The ordered pair solution is $(-5, 2)$.

As always, check the solution in both original equations. □

EXAMPLE 3 Use Cramer's rule to solve the system

$$\begin{cases} 5x + y = 5 \\ -7x - 2y = -7 \end{cases}$$

Solution First we find D, D_x, and D_y.

$$D = \begin{vmatrix} 5 & 1 \\ -7 & -2 \end{vmatrix} = 5(-2) - (-7)(1) = -3$$

$$D_x = \begin{vmatrix} 5 & 1 \\ -7 & -2 \end{vmatrix} = 5(-2) - (-7)(1) = -3$$

$$D_y = \begin{vmatrix} 5 & 5 \\ -7 & -7 \end{vmatrix} = 5(-7) - 5(-7) = 0$$

Then

$$x = \frac{D_x}{D} = \frac{-3}{-3} = 1 \qquad y = \frac{D_y}{D} = \frac{0}{-3} = 0$$

The ordered pair solution is $(1, 0)$.   □

3   Evaluating 3 × 3 Determinants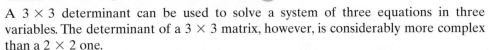

A 3×3 determinant can be used to solve a system of three equations in three variables. The determinant of a 3×3 matrix, however, is considerably more complex than a 2×2 one.

Determinant of a 3 × 3 Matrix

$$\begin{vmatrix} a_1 & b_1 & c_1 \\ a_2 & b_2 & c_2 \\ a_3 & b_3 & c_3 \end{vmatrix} = a_1 \cdot \begin{vmatrix} b_2 & c_2 \\ b_3 & c_3 \end{vmatrix} - a_2 \cdot \begin{vmatrix} b_1 & c_1 \\ b_3 & c_3 \end{vmatrix} + a_3 \cdot \begin{vmatrix} b_1 & c_1 \\ b_2 & c_2 \end{vmatrix}$$

Notice that the determinant of a 3×3 matrix is related to the determinants of three 2×2 matrices. Each determinant of these 2×2 matrices is called a **minor,** and every element of a 3×3 matrix has a minor associated with it. For example, the minor of c_2 is the determinant of the 2×2 matrix found by deleting the row and column containing c_2.

$$\begin{array}{ccc} a_1 & b_1 & c_1 \\ a_2 & b_2 & c_2 \\ a_3 & b_3 & c_3 \end{array} \qquad \text{The minor of } c_2 \text{ is} \qquad \begin{vmatrix} a_1 & b_1 \\ a_3 & b_3 \end{vmatrix}$$

Also, the minor of element a_1 is the determinant of the 2×2 matrix that has no row or column containing a_1.

$$\begin{array}{ccc} a_1 & b_1 & c_1 \\ a_2 & b_2 & c_2 \\ a_3 & b_3 & c_3 \end{array} \qquad \text{The minor of } a_1 \text{ is} \qquad \begin{vmatrix} b_2 & c_2 \\ b_3 & c_3 \end{vmatrix}$$

So the determinant of a 3×3 matrix can be written as

$$a_1 \cdot (\text{minor of } a_1) - a_2 \cdot (\text{minor of } a_2) + a_3 \cdot (\text{minor of } a_3)$$

Finding the determinant by using minors of elements in the first column is called **expanding** by the minors of the first column. *The value of a determinant can be found by expanding by the minors of any row or column.* The following **array of signs** is helpful in determining whether to add or subtract the product of an element and its minor.

$$\begin{array}{ccc} + & - & + \\ - & + & - \\ + & - & + \end{array}$$

If an element is in a position marked $+$, we add. If marked $-$, we subtract.

✓CONCEPT CHECK
Suppose you are interested in finding the determinant of a 4×4 matrix. Study the pattern shown in the array of signs for a 3×3 matrix. Use the pattern to expand the array of signs for use with a 4×4 matrix.

EXAMPLE 4 Evaluate by expanding by the minors of the given row or column.

$$\begin{vmatrix} 0 & 5 & 1 \\ 1 & 3 & -1 \\ -2 & 2 & 4 \end{vmatrix}$$

a. First column **b.** Second row

Solution

a. The elements of the first column are 0, 1, and -2. The first column of the array of signs is $+, -, +$.

$$\begin{vmatrix} 0 & 5 & 1 \\ 1 & 3 & -1 \\ -2 & 2 & 4 \end{vmatrix} = 0 \cdot \begin{vmatrix} 3 & -1 \\ 2 & 4 \end{vmatrix} - 1 \cdot \begin{vmatrix} 5 & 1 \\ 2 & 4 \end{vmatrix} + (-2) \cdot \begin{vmatrix} 5 & 1 \\ 3 & -1 \end{vmatrix}$$

$$= 0(12 - (-2)) - 1(20 - 2) + (-2)(-5 - 3)$$

$$= 0 - 18 + 16 = -2$$

b. The elements of the second row are 1, 3, and -1. This time, the signs begin with $-$ and again alternate.

$$\begin{vmatrix} 0 & 5 & 1 \\ 1 & 3 & -1 \\ -2 & 2 & 4 \end{vmatrix} = -1 \cdot \begin{vmatrix} 5 & 1 \\ 2 & 4 \end{vmatrix} + 3 \cdot \begin{vmatrix} 0 & 1 \\ -2 & 4 \end{vmatrix} - (-1) \cdot \begin{vmatrix} 0 & 5 \\ -2 & 2 \end{vmatrix}$$

$$= -1(20 - 2) + 3(0 - (-2)) - (-1)(0 - (-10))$$

$$= -18 + 6 + 10 = -2$$

Notice that the determinant of the 3×3 matrix is the same regardless of the row or column you select to expand by. □

✓CONCEPT CHECK

Why would expanding by minors of the second row be a good choice for the determinant $\begin{vmatrix} 3 & 4 & -2 \\ 5 & 0 & 0 \\ 6 & -3 & 7 \end{vmatrix}$?

OBJECTIVE

4 Using Cramer's Rule to Solve a System of Three Linear Equations ▶

A system of three equations in three variables may be solved with Cramer's rule also. Using the elimination process to solve a system with unknown constants as coefficients leads to the following.

Cramer's Rule for Three Equations in Three Variables

The solution of the system $\begin{cases} a_1x + b_1y + c_1z = k_1 \\ a_2x + b_2y + c_2z = k_2 \\ a_3x + b_3y + c_3z = k_3 \end{cases}$ is given by

$$x = \frac{D_x}{D} \qquad y = \frac{D_y}{D} \qquad \text{and} \qquad z = \frac{D_z}{D}$$

where

$$D = \begin{vmatrix} a_1 & b_1 & c_1 \\ a_2 & b_2 & c_2 \\ a_3 & b_3 & c_3 \end{vmatrix} \qquad D_x = \begin{vmatrix} k_1 & b_1 & c_1 \\ k_2 & b_2 & c_2 \\ k_3 & b_3 & c_3 \end{vmatrix}$$

$$D_y = \begin{vmatrix} a_1 & k_1 & c_1 \\ a_2 & k_2 & c_2 \\ a_3 & k_3 & c_3 \end{vmatrix} \qquad D_z = \begin{vmatrix} a_1 & b_1 & k_1 \\ a_2 & b_2 & k_2 \\ a_3 & b_3 & k_3 \end{vmatrix}$$

as long as D is not 0.

Answer to Concept Check:
Two elements of the second row are 0, which makes calculations easier.

EXAMPLE 5 Use Cramer's rule to solve the system

$$\begin{cases} x - 2y + z = 4 \\ 3x + y - 2z = 3 \\ 5x + 5y + 3z = -8 \end{cases}$$

Solution First we find $D, D_x, D_y,$ and D_z. Beginning with D, we expand by the minors of the first column.

$$D = \begin{vmatrix} 1 & -2 & 1 \\ 3 & 1 & -2 \\ 5 & 5 & 3 \end{vmatrix} = 1 \cdot \begin{vmatrix} 1 & -2 \\ 5 & 3 \end{vmatrix} - 3 \cdot \begin{vmatrix} -2 & 1 \\ 5 & 3 \end{vmatrix} + 5 \cdot \begin{vmatrix} -2 & 1 \\ 1 & -2 \end{vmatrix}$$

$$= 1(3 - (-10)) - 3(-6 - 5) + 5(4 - 1)$$

$$= 13 + 33 + 15 = 61$$

$$D_x = \begin{vmatrix} 4 & -2 & 1 \\ 3 & 1 & -2 \\ -8 & 5 & 3 \end{vmatrix} = 4 \cdot \begin{vmatrix} 1 & -2 \\ 5 & 3 \end{vmatrix} - 3 \cdot \begin{vmatrix} -2 & 1 \\ 5 & 3 \end{vmatrix} + (-8) \cdot \begin{vmatrix} -2 & 1 \\ 1 & -2 \end{vmatrix}$$

$$= 4(3 - (-10)) - 3(-6 - 5) + (-8)(4 - 1)$$

$$= 52 + 33 - 24 = 61$$

$$D_y = \begin{vmatrix} 1 & 4 & 1 \\ 3 & 3 & -2 \\ 5 & -8 & 3 \end{vmatrix} = 1 \cdot \begin{vmatrix} 3 & -2 \\ -8 & 3 \end{vmatrix} - 3 \cdot \begin{vmatrix} 4 & 1 \\ -8 & 3 \end{vmatrix} + 5 \cdot \begin{vmatrix} 4 & 1 \\ 3 & -2 \end{vmatrix}$$

$$= 1(9 - 16) - 3(12 - (-8)) + 5(-8 - 3)$$

$$= -7 - 60 - 55 = -122$$

$$D_z = \begin{vmatrix} 1 & -2 & 4 \\ 3 & 1 & 3 \\ 5 & 5 & -8 \end{vmatrix} = 1 \cdot \begin{vmatrix} 1 & 3 \\ 5 & -8 \end{vmatrix} - 3 \cdot \begin{vmatrix} -2 & 4 \\ 5 & -8 \end{vmatrix} + 5 \cdot \begin{vmatrix} -2 & 4 \\ 1 & 3 \end{vmatrix}$$

$$= 1(-8 - 15) - 3(16 - 20) + 5(-6 - 4)$$

$$= -23 + 12 - 50 = -61$$

From these determinants, we calculate the solution:

$$x = \frac{D_x}{D} = \frac{61}{61} = 1 \quad y = \frac{D_y}{D} = \frac{-122}{61} = -2 \quad z = \frac{D_z}{D} = \frac{-61}{61} = -1$$

The ordered triple solution is $(1, -2, -1)$. Check this solution by verifying that it satisfies each equation of the system. □

Vocabulary, Readiness & Video Check

Evaluate each determinant mentally.

1. $\begin{vmatrix} 7 & 2 \\ 0 & 8 \end{vmatrix}$

2. $\begin{vmatrix} 6 & 0 \\ 1 & 2 \end{vmatrix}$

3. $\begin{vmatrix} -4 & 2 \\ 0 & 8 \end{vmatrix}$

4. $\begin{vmatrix} 5 & 0 \\ 3 & -5 \end{vmatrix}$

5. $\begin{vmatrix} -2 & 0 \\ 3 & -10 \end{vmatrix}$

6. $\begin{vmatrix} -1 & 4 \\ 0 & -18 \end{vmatrix}$

C Exercise Set   MyMathLab®

Evaluate each determinant. See Example 1.

1. $\begin{vmatrix} 3 & 5 \\ -1 & 7 \end{vmatrix}$

2. $\begin{vmatrix} -5 & 1 \\ 1 & -4 \end{vmatrix}$

3. $\begin{vmatrix} 9 & -2 \\ 4 & -3 \end{vmatrix}$

4. $\begin{vmatrix} 4 & -1 \\ 9 & 8 \end{vmatrix}$

5. $\begin{vmatrix} -2 & 9 \\ 4 & -18 \end{vmatrix}$

6. $\begin{vmatrix} -40 & 8 \\ 70 & -14 \end{vmatrix}$

7. $\begin{vmatrix} \frac{3}{4} & \frac{5}{2} \\ -\frac{1}{6} & \frac{7}{3} \end{vmatrix}$

8. $\begin{vmatrix} \frac{5}{7} & \frac{1}{3} \\ \frac{6}{7} & \frac{2}{3} \end{vmatrix}$

Use Cramer's rule, if possible, to solve each system of linear equations. See Examples 2 and 3.

9. $\begin{cases} 2y - 4 = 0 \\ x + 2y = 5 \end{cases}$

10. $\begin{cases} 4x - y = 5 \\ 3x - 3 = 0 \end{cases}$

11. $\begin{cases} 3x + y = 1 \\ 2y = 2 - 6x \end{cases}$

12. $\begin{cases} y = 2x - 5 \\ 8x - 4y = 20 \end{cases}$

13. $\begin{cases} 5x - 2y = 27 \\ -3x + 5y = 18 \end{cases}$

14. $\begin{cases} 4x - y = 9 \\ 2x + 3y = -27 \end{cases}$

15. $\begin{cases} 2x - 5y = 4 \\ x + 2y = -7 \end{cases}$

16. $\begin{cases} 3x - y = 2 \\ -5x + 2y = 0 \end{cases}$

17. $\begin{cases} \frac{2}{3}x - \frac{3}{4}y = -1 \\ -\frac{1}{6}x + \frac{3}{4}y = \frac{5}{2} \end{cases}$

18. $\begin{cases} \frac{1}{2}x - \frac{1}{3}y = -3 \\ \frac{1}{8}x + \frac{1}{6}y = 0 \end{cases}$

Evaluate. See Example 4.

19. $\begin{vmatrix} 2 & 1 & 0 \\ 0 & 5 & -3 \\ 4 & 0 & 2 \end{vmatrix}$

20. $\begin{vmatrix} -6 & 4 & 2 \\ 1 & 0 & 5 \\ 0 & 3 & 1 \end{vmatrix}$

21. $\begin{vmatrix} 4 & -6 & 0 \\ -2 & 3 & 0 \\ 4 & -6 & 1 \end{vmatrix}$

22. $\begin{vmatrix} 5 & 2 & 1 \\ 3 & -6 & 0 \\ -2 & 8 & 0 \end{vmatrix}$

23. $\begin{vmatrix} 1 & 0 & 4 \\ 1 & -1 & 2 \\ 3 & 2 & 1 \end{vmatrix}$

24. $\begin{vmatrix} 0 & 1 & 2 \\ 3 & -1 & 2 \\ 3 & 2 & -2 \end{vmatrix}$

25. $\begin{vmatrix} 3 & 6 & -3 \\ -1 & -2 & 3 \\ 4 & -1 & 6 \end{vmatrix}$

26. $\begin{vmatrix} 2 & -2 & 1 \\ 4 & 1 & 3 \\ 3 & 1 & 2 \end{vmatrix}$

Use Cramer's rule, if possible, to solve each system of linear equations. See Example 5.

27. $\begin{cases} 3x + z = -1 \\ -x - 3y + z = 7 \\ 3y + z = 5 \end{cases}$

28. $\begin{cases} 4y - 3z = -2 \\ 8x - 4y = 4 \\ -8x + 4y + z = -2 \end{cases}$

29. $\begin{cases} x + y + z = 8 \\ 2x - y - z = 10 \\ x - 2y + 3z = 22 \end{cases}$

30. $\begin{cases} 5x + y + 3z = 1 \\ x - y - 3z = -7 \\ -x + y = 1 \end{cases}$

31. $\begin{cases} 2x + 2y + z = 1 \\ -x + y + 2z = 3 \\ x + 2y + 4z = 0 \end{cases}$

32. $\begin{cases} 2x - 3y + z = 5 \\ x + y + z = 0 \\ 4x + 2y + 4z = 4 \end{cases}$

33. $\begin{cases} x - 2y + z = -5 \\ 3y + 2z = 4 \\ 3x - y = -2 \end{cases}$

34. $\begin{cases} 4x + 5y = 10 \\ 3y + 2z = -6 \\ x + y + z = 3 \end{cases}$

CONCEPT EXTENSIONS

Find the value of x that will make each a true statement.

35. $\begin{vmatrix} 1 & x \\ 2 & 7 \end{vmatrix} = -3$

36. $\begin{vmatrix} 6 & 1 \\ -2 & x \end{vmatrix} = 26$

37. If all the elements in a single row of a determinant are zero, what is the value of the determinant? Explain your answer.

38. If all the elements in a single column of a determinant are 0, what is the value of the determinant? Explain your answer.

Appendix D
An Introduction to Using a Graphing Utility

The Viewing Window and Interpreting Window Settings

In this appendix, we will use the term **graphing utility** to mean a graphing calculator or a computer software graphing package. All graphing utilities graph equations by plotting points on a screen. Although plotting several points can be slow and sometimes tedious, a graphing utility can quickly and accurately plot hundreds of points. How does a graphing utility show plotted points? A computer or calculator screen is made up of a grid of small rectangular areas called **pixels.** If a pixel contains a point to be plotted, the pixel is turned "on"; otherwise, the pixel remains "off." The graph of an equation is then a collection of pixels turned "on." The graph of $y = 3x + 1$ from a graphing calculator is shown in Figure A-1. Notice the irregular shape of the line caused by the rectangular pixels.

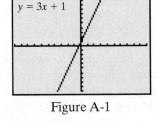

Figure A-1

The portion of the coordinate plane shown on the screen in Figure A-1 is called the **viewing window** or the **viewing rectangle.** Notice the x-axis and the y-axis on the graph. Tick marks are shown on the axes, but they are not labeled. This means that from this screen alone, we do not know how many units each tick mark represents. To see what each tick mark represents and the minimum and maximum values on the axes, check the window setting of the graphing utility. It defines the viewing window. The window of the graph of $y = 3x + 1$ shown in Figure A-1 has the following setting (Figure A-2):

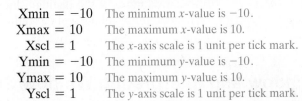

$$\begin{aligned}
\text{Xmin} &= -10 & &\text{The minimum } x\text{-value is } -10. \\
\text{Xmax} &= 10 & &\text{The maximum } x\text{-value is } 10. \\
\text{Xscl} &= 1 & &\text{The } x\text{-axis scale is 1 unit per tick mark.} \\
\text{Ymin} &= -10 & &\text{The minimum } y\text{-value is } -10. \\
\text{Ymax} &= 10 & &\text{The maximum } y\text{-value is } 10. \\
\text{Yscl} &= 1 & &\text{The } y\text{-axis scale is 1 unit per tick mark.}
\end{aligned}$$

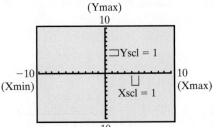

Figure A-2

By knowing the scale, we can find the minimum and the maximum values on the axes simply by counting tick marks. For example, if both the Xscl (x-axis scale) and the Yscl are 1 unit per tick mark on the graph in Figure A-3, we can count the tick marks and find that the minimum x-value is -10 and the maximum x-value is 10. Also, the minimum y-value is -10 and the maximum y-value is 10. If the Xscl (x-axis scale) changes to 2 units per tick mark (shown in Figure A-4), by counting tick marks, we see that the minimum x-value is now -20 and the maximum x-value is now 20.

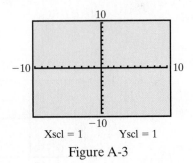

Xscl = 1   Yscl = 1

Figure A-3

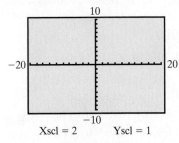

Xscl = 2   Yscl = 1

Figure A-4

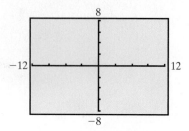

Figure A-5

It is also true that if we know the Xmin and the Xmax values, we can calculate the Xscl by the displayed axes. For example, the Xscl of the graph in Figure A-5 must be 3 units per tick mark for the maximum and minimum x-values to be as shown. Also, the Yscl of that graph must be 2 units per tick mark for the maximum and minimum y-values to be as shown.

We will call the viewing window in Figure A-3 a *standard* viewing window or rectangle. Although a standard viewing window is sufficient for much of this text, special care must be taken to ensure that all key features of a graph are shown. Figures A-6, A-7, and A-8 show the graph of $y = x^2 + 11x - 1$ on three viewing windows. Note that certain viewing windows for this equation are misleading.

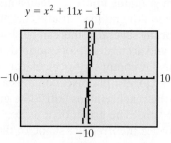

Figure A-6

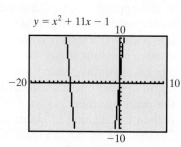

Figure A-7

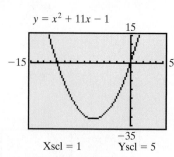

Figure A-8

How do we ensure that all distinguishing features of the graph of an equation are shown? It helps to know about the equation that is being graphed. For example, the equation $y = x^2 + 11x - 1$ is not a linear equation and its graph is not a line. This equation is a quadratic equation and, therefore, its graph is a parabola. By knowing this information, we know that the graph shown in Figure A-6, although correct, is misleading. Of the three viewing rectangles shown, the graph in Figure A-8 is best because it shows more of the distinguishing features of the parabola. Properties of equations needed for graphing will be studied in this text.

Viewing Window and Interpreting Window Settings Exercise Set

In Exercises 1–4, determine whether all ordered pairs listed will lie within a standard viewing rectangle.

1. $(-9, 0), (5, 8), (1, -8)$

2. $(4, 7), (0, 0), (-8, 9)$

3. $(-11, 0), (2, 2), (7, -5)$

4. $(3, 5), (-3, -5), (15, 0)$

In Exercises 5–10, choose an Xmin, Xmax, Ymin, and Ymax so that all ordered pairs listed will lie within the viewing rectangle.

5. $(-90, 0), (55, 80), (0, -80)$

6. $(4, 70), (20, 20), (-18, 90)$

7. $(-11, 0), (2, 2), (7, -5)$

8. $(3, 5), (-3, -5), (15, 0)$

9. $(200, 200), (50, -50), (70, -50)$

10. $(40, 800), (-30, 500), (15, 0)$

Write the window setting for each viewing window shown. Use the following format:

Xmin =	Ymin =
Xmax =	Ymax =
Xscl =	Yscl =

11.

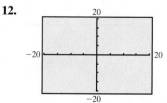

12.

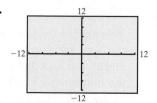

13.

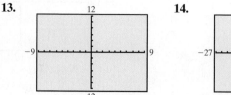

14.

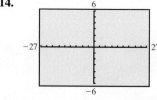

15.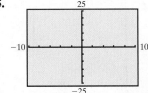

16.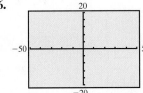

17.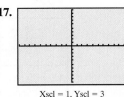
Xscl = 1, Yscl = 3

18.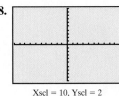
Xscl = 10, Yscl = 2

19.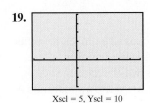
Xscl = 5, Yscl = 10

20.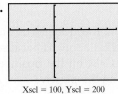
Xscl = 100, Yscl = 200

Graphing Equations and Square Viewing Window

In general, the following steps may be used to graph an equation on a standard viewing window.

> **Graphing an Equation in *X* and *Y* with a Graphing Utility on a Standard Viewing Window**
>
> **Step 1:** Solve the equation for *y*.
> **Step 2:** Using your graphing utility, enter the equation in the form
> Y = *expression involving x*.
> **Step 3:** Activate the graphing utility.

Special care must be taken when entering the *expression involving x* in Step 2. You must be sure that the graphing utility you are using interprets the expression as you want it to. For example, let's graph $3y = 4x$. To do so,

Step 1: Solve the equation for *y*.

$$3y = 4x$$
$$\frac{3y}{3} = \frac{4x}{3}$$
$$y = \frac{4}{3}x$$

Step 2: Using your graphing utility, enter the expression $\frac{4}{3}x$ after the Y = prompt.

For your graphing utility to interpret the expression correctly, you may need to enter $(4/3)x$ or $(4 \div 3)x$.

Step 3: Activate the graphing utility. The graph should appear as in Figure A-9.

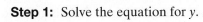

$y = \frac{4}{3}x$

Figure A-9

Distinguishing features of the graph of a line include showing all the intercepts of the line. For example, the window of the graph of the line in Figure A-10 does not show both intercepts of the line, but the window of the graph of the same line in Figure A-11 does show both intercepts. Notice the notation below each graph. This is a short-hand notation of the range setting of the graph. This notation means [Xmin, Xmax] by [Ymin, Ymax].

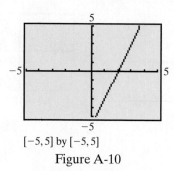

[−5, 5] by [−5, 5]

Figure A-10

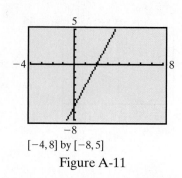

[−4, 8] by [−8, 5]

Figure A-11

On a standard viewing window, the tick marks on the *y*-axis are closer together than the tick marks on the *x*-axis. This happens because the viewing window is a rectangle, so 10 equally spaced tick marks on the positive *y*-axis will be closer together than 10 equally spaced tick marks on the positive *x*-axis. This causes the appearance of graphs to be distorted.

For example, notice the different appearances of the same line graphed using different viewing windows. The line in Figure A-12 is distorted because the tick marks along the *x*-axis are farther apart than the tick marks along the *y*-axis. The graph of the same line in Figure A-13 is not distorted because the viewing rectangle has been selected so that there is equal spacing between tick marks on both axes.

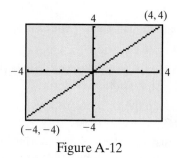

Figure A-12

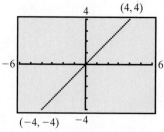

Figure A-13

We say that the line in Figure A-13 is graphed on a *square* setting. Some graphing utilities have a built-in program that, if activated, will automatically provide a square setting. A square setting is especially helpful when we are graphing perpendicular lines, circles, or when a true geometric perspective is desired. Some examples of square screens are shown in Figures A-14 and A-15.

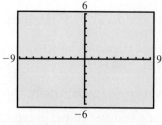

Figure A-14

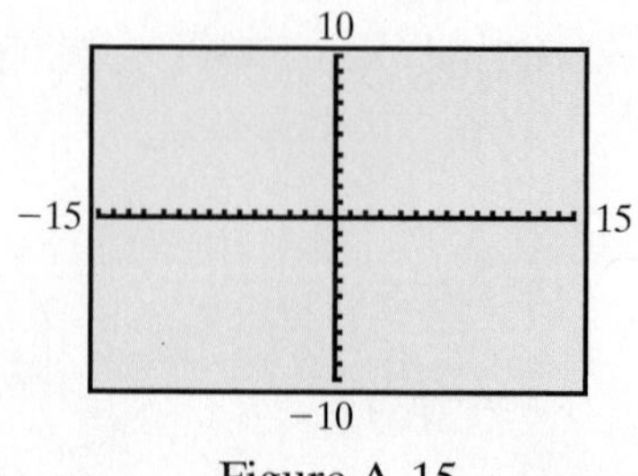

Figure A-15

Other features of a graphing utility such as Trace, Zoom, Intersect, and Table are discussed in appropriate Graphing Calculator Explorations in this text.

Graphing Equations and Square Viewing Window Exercise Set

Graph each linear equation in two variables, using the different range settings given. Determine which setting shows all intercepts of a line.

1. $y = 2x + 12$
   Setting A: $[-10, 10]$ by $[-10, 10]$
   Setting B: $[-10, 10]$ by $[-10, 15]$

2. $y = -3x + 25$
   Setting A: $[-5, 5]$ by $[-30, 10]$
   Setting B: $[-10, 10]$ by $[-10, 30]$

3. $y = -x - 41$
   Setting A: $[-50, 10]$ by $[-10, 10]$
   Setting B: $[-50, 10]$ by $[-50, 15]$

4. $y = 6x - 18$
   Setting A: $[-10, 10]$ by $[-20, 10]$
   Setting B: $[-10, 10]$ by $[-10, 10]$

5. $y = \dfrac{1}{2}x - 15$
   Setting A: $[-10, 10]$ by $[-20, 10]$
   Setting B: $[-10, 35]$ by $[-20, 15]$

6. $y = -\dfrac{2}{3}x - \dfrac{29}{3}$
   Setting A: $[-10, 10]$ by $[-10, 10]$
   Setting B: $[-15, 5]$ by $[-15, 5]$

The graph of each equation is a line. Use a graphing utility and a standard viewing window to graph each equation.

7. $3x = 5y$ **8.** $7y = -3x$ **9.** $9x - 5y = 30$

10. $4x + 6y = 20$ **11.** $y = -7$ **12.** $y = 2$

13. $x + 10y = -5$ **14.** $x - 5y = 9$

Graph the following equations using the square setting given. Some keystrokes that may be helpful are given.

15. $y = \sqrt{x}$ $[-12, 12]$ by $[-8, 8]$
   Suggested keystrokes: $\sqrt{\ }\, x$

16. $y = \sqrt{2x}$ $[-12, 12]$ by $[-8, 8]$
   Suggested keystrokes: $\sqrt{\ }\,(2x)$

17. $y = x^2 + 2x + 1$ $[-15, 15]$ by $[-10, 10]$
   Suggested keystrokes: x^2 + 2x + 1

18. $y = x^2 - 5$ $[-15, 15]$ by $[-10, 10]$
   Suggested keystrokes: x^2 − 5

19. $y = |x|$ $[-9, 9]$ by $[-6, 6]$
   Suggested keystrokes: ABS (x)

20. $y = |x - 2|$ $[-9, 9]$ by $[-6, 6]$
   Suggested keystrokes: ABS$(x - 2)$

Graph each line. Use a standard viewing window; then, if necessary, change the viewing window so that all intercepts of each line show.

21. $x + 2y = 30$ **22.** $1.5x - 3.7y = 40.3$

Contents of Student Resources

Study Skills Builders

Bigger Picture—Study Guide Outline

Practice Final Exam

Answers to Selected Exercises

Student Resources

Study Skills Builders

Attitude and Study Tips

Study Skills Builder 1

Have You Decided to Complete This Course Successfully?

Ask yourself if one of your current goals is to complete this course successfully.

If it is not a goal of yours, ask yourself why. One common reason is fear of failure. Amazingly enough, fear of failure alone can be strong enough to keep many of us from doing our best in any endeavor.

Another common reason is that you simply haven't taken the time to think about or write down your goals for this course. To help accomplish this, answer the exercises below.

Exercises

1. Write down your goal(s) for this course.

2. Now list steps you will take to make sure your goal(s) in Exercise 1 are accomplished.

3. Rate your commitment to this course with a number between 1 and 5. Use the diagram below to help.

High Commitment		Average Commitment		Not Commited at All
5	4	3	2	1

4. If you have rated your personal commitment level (from the exercise above) as a 1, 2, or 3, list the reasons why this is so. Then determine whether it is possible to increase your commitment level to a 4 or 5.

Good luck, and don't forget that a positive attitude will make a big difference.

Study Skills Builder 2

Tips for Studying for an Exam

To prepare for an exam, try the following study techniques:

● Start the study process days before your exam.

● Make sure that you are up to date on your assignments.

● If there is a topic that you are unsure of, use one of the many resources that are available to you. For example,

See your instructor.

View a lecture video on the topic.

Visit a learning resource center on campus.

Read the textbook material and examples on the topic.

● Reread your notes and carefully review the Chapter Highlights at the end of any chapter.

● Work the review exercises at the end of the chapter.

● Find a quiet place to take the Chapter Test found at the end of the chapter. Do not use any resources when taking this sample test. This way, you will have a clear indication of how prepared you are for your exam. Check your answers and use the Chapter Test Prep Videos to make sure that you correct any missed exercises.

Good luck, and keep a positive attitude.

Exercises

Let's see how you did on your last exam.

1. How many days before your last exam did you start studying for that exam?

2. Were you up to date on your assignments at that time, or did you need to catch up on assignments?

3. List the most helpful text supplement (if you used one).

4. List the most helpful campus supplement (if you used one).

5. List your process for preparing for a mathematics test.

6. Was this process helpful? In other words, were you satisfied with your performance on your exam?

7. If not, what changes can you make in your process that will make it more helpful to you?

Study Skills Builder 3

What to Do the Day of an Exam

Your first exam may be soon. On the day of an exam, don't forget to try the following:

- Allow yourself plenty of time to arrive.
- Read the directions on the test carefully.
- Read each problem carefully as you take your test. Make sure that you answer the question asked.
- Watch your time and pace yourself so that you may attempt each problem on your test.
- Check your work and answers.
- ***Do not turn your test in early.*** If you have extra time, spend it double-checking your work.

Good luck!

Exercises

Answer the following questions based on your most recent mathematics exam, whenever that was.

1. How soon before class did you arrive?
2. Did you read the directions on the test carefully?
3. Did you make sure you answered the question asked for each problem on the exam?
4. Were you able to attempt each problem on your exam?
5. If your answer to Exercise 4 is no, list reasons why.
6. Did you have extra time on your exam?
7. If your answer to Exercise 6 is yes, describe how you spent that extra time.

Study Skills Builder 4

Are You Satisfied with Your Performance on a Particular Quiz or Exam?

If not, don't forget to analyze your quiz or exam and look for common errors. Were most of your errors a result of:

- *Carelessness?* Did you turn in your quiz or exam before the allotted time expired? If so, resolve to use any extra time to check your work.
- *Running out of time?* Answer the questions you are sure of first. Then attempt the questions you are unsure of and delay checking your work until all questions have been answered.
- *Not understanding a concept?* If so, review that concept and correct your work so that you make sure you understand it before the next quiz or the final exam.
- *Test conditions?* When studying for a quiz or exam, make sure you place yourself in conditions similar to test conditions. For example, before your next quiz or exam, take a sample test without the aid of your notes or text.

(For a sample test, see your instructor or use the Chapter Test at the end of each chapter.)

Exercises

1. Have you corrected all your previous quizzes and exams?
2. List any errors you have found common to two or more of your graded papers.
3. Is one of your common errors not understanding a concept? If so, are you making sure you understand all the concepts for the next quiz or exam?
4. Is one of your common errors making careless mistakes? If so, are you now taking all the time allotted to check over your work so that you can minimize the number of careless mistakes?
5. Are you satisfied with your grades thus far on quizzes and tests?
6. If your answer to Exercise 5 is no, are there any more suggestions you can make to your instructor or yourself to help? If so, list them here and share these with your instructor.

Study Skills Builder 5
How Are You Doing?

If you haven't done so yet, take a few moments to think about how you are doing in this course. Are you working toward your goal of successfully completing this course? Is your performance on homework, quizzes, and tests satisfactory? If not, you might want to see your instructor to see whether he/she has any suggestions on how you can improve your performance. Reread Section 1.1 for ideas on places to get help with your mathematics course.

Exercises

Answer the following.

1. List any textbook supplements you are using to help you through this course.

2. List any campus resources you are using to help you through this course.

3. Write a short paragraph describing how you are doing in your mathematics course.

4. If improvement is needed, list ways that you can work toward improving your situation as described in Exercise 3.

Study Skills Builder 6
Are You Preparing for Your Final Exam?

To prepare for your final exam, try the following study techniques:

- Review the material that you will be responsible for on your exam. This includes material from your textbook, your notebook, and any handouts from your instructor.

- Review any formulas that you may need to memorize.

- Check to see if your instructor or mathematics department will be conducting a final exam review.

- Check with your instructor to see whether final exams from previous semesters/quarters are available to students for review.

- Use your previously taken exams as a practice final exam. To do so, rewrite the test questions in mixed order on blank sheets of paper. This will help you prepare for exam conditions.

- If you are unsure of a few concepts, see your instructor or visit a learning lab for assistance. Also, view the video segment of any troublesome sections.

- If you need further exercises to work, try the Cumulative Reviews at the end of the chapters.

Once again, good luck! I hope you are enjoying this textbook and your mathematics course.

Organizing Your Work

Study Skills Builder 7
Learning New Terms

Many of the terms used in this text may be new to you. It will be helpful to make a list of new mathematical terms and symbols as you encounter them and to review them frequently. Placing these new terms (including page references) on 3×5 index cards might help you later when you're preparing for a quiz.

Exercises

1. Name one way you might place a word and its definition on a 3×5 card.

2. How do new terms stand out in this text so that they can be found?

Study Skills Builder 8
Are You Organized?

Have you ever had trouble finding a completed assignment? When it's time to study for a test, are your notes neat and organized? Have you ever had trouble reading your own mathematics handwriting? (Be honest—I have.)

When any of these things happens, it's time to get organized. Here are a few suggestions:

- Write your notes and complete your homework assignments in a notebook with pockets (spiral or ring binder).
- Take class notes in this notebook and then follow the notes with your completed homework assignment.
- When you receive graded papers or handouts, place them in the notebook pocket so that you will not lose them.
- Mark (possibly with an exclamation point) any note(s) that seem extra important to you.
- Mark (possibly with a question mark) any notes or homework that you are having trouble with.

- See your instructor or a math tutor to help you with the concepts or exercises that you are having trouble understanding.
- If you are having trouble reading your own handwriting, *slow down* and write your mathematics work clearly!

Exercises

1. Have you been completing your assignments on time?
2. Have you been correcting any exercises you may be having difficulty with?
3. If you are having trouble with a mathematical concept or correcting any homework exercises, have you visited your instructor, a tutor, or your campus math lab?
4. Are you taking lecture notes in your mathematics course? (By the way, these notes should include worked-out examples solved by your instructor.)
5. Is your mathematics course material (handouts, graded papers, lecture notes) organized?
6. If your answer to Exercise 5 is no, take a moment to review your course material. List at least two ways that you might organize it better.

Study Skills Builder 9
Organizing a Notebook

It's never too late to get organized. If you need ideas about organizing a notebook for your mathematics course, try some of these:

- Use a spiral or ring binder notebook with pockets and use it for mathematics only.
- Start each page by writing the book's section number you are working on at the top.
- When your instructor is lecturing, take notes. *Always* include any examples your instructor works for you.
- Place your worked-out homework exercises in your notebook immediately after the lecture notes from that section. This way, a section's worth of material is together.
- Homework exercises: Attempt and check all assigned homework.
- Place graded quizzes in the pockets of your notebook or a special section of your binder.

Exercises

Check your notebook organization by answering the following questions.

1. Do you have a spiral or ring binder notebook for your mathematics course only?
2. Have you ever had to flip through several sheets of notes and work in your mathematics notebook to determine what section's work you are in?
3. Are you now writing the textbook's section number at the top of each notebook page?
4. Have you ever lost or had trouble finding a graded quiz or test?
5. Are you now placing all your graded work in a dedicated place in your notebook?
6. Are you attempting all of your homework and placing all of your work in your notebook?
7. Are you checking and correcting your homework in your notebook? If not, why not?
8. Are you writing in your notebook the examples your instructor works for you in class?

Study Skills Builder 10

How Are Your Homework Assignments Going?

It is very important in mathematics to keep up with homework. Why? Many concepts build on each other. Often your understanding of a day's concepts depends on an understanding of the previous day's material.

Remember that completing your homework assignment involves a lot more than attempting a few of the problems assigned.

To complete a homework assignment, remember these four things:

- Attempt all of it.
- Check it.
- Correct it.
- If needed, ask questions about it.

Exercises

Take a moment to review your completed homework assignments. Answer the questions below based on this review.

1. Approximate the fraction of your homework you have attempted.
2. Approximate the fraction of your homework you have checked (if possible).
3. If you are able to check your homework, have you corrected it when errors have been found?
4. When working homework, if you do not understand a concept, what do you do?

MyMathLab and MathXL

Study Skills Builder 11

Tips for Turning in Your Homework on Time

It is very important to keep up with your mathematics homework assignments. Why? Many concepts in mathematics build upon each other.

Remember these four tips to help ensure that your work is completed on time:

- Know the assignments and due dates set by your instructor.
- Do not wait until the last minute to submit your homework.
- Set a goal to submit your homework 6–8 hours before the scheduled due date in case you have unexpected technology trouble.
- Schedule enough time to complete each assignment.

Following these tips will also help you avoid losing points for late or missed assignments.

Exercises

Take a moment to consider your work on your homework assignments to date and answer the following questions:

1. What percentage of your assignments have you turned in on time?
2. Why might it be a good idea to submit your homework 6–8 hours before the scheduled deadline?
3. If you have missed submitting any homework by the due date, list some of the reasons this occurred.
4. What steps do you plan to take in the future to ensure that your homework is submitted on time?

Study Skills Builder 12
Tips for Doing Your Homework Online

Practice is one of the main keys to success in any mathematics course. Did you know that MyMathLab/MathXL provides you with **immediate feedback** for each exercise? If you are incorrect, you are given hints to work the exercise correctly. You have **unlimited practice opportunities** and can rework any exercises you have trouble with until you master them and submit homework assignments unlimited times before the deadline.

Remember these success tips when doing your homework online:

- Attempt all assigned exercises.
- Write down (neatly) your step-by-step work for each exercise before entering your answer.
- Use the immediate feedback provided by the program to help you check and correct your work for each exercise.
- Rework any exercises you have trouble with until you master them.

- Work through your homework assignment as many times as necessary until you are satisfied.

Exercises

Take a moment to think about your homework assignments to date and answer the following:

1. Have you attempted all assigned exercises?
2. Of the exercises attempted, have you also written out your work before entering your answer so that you can check it?
3. Are you familiar with how to enter answers using the MathXL player so that you avoid answer entry type errors?
4. List some ways the immediate feedback and practice supports have helped you with your homework. If you have not used these supports, how do you plan to use them with the given success tips on your next assignment?

Study Skills Builder 13
Organizing Your Work

Have you ever used any readily available paper (such as the back of a flyer, another course assignment, Post-it notes, etc.) to work out homework exercises before entering the answer in MathXL? To save time, have you ever entered answers directly into MathXL without working the exercises on paper? When it's time to study, have you ever been unable to find your completed work or read and follow your own mathematics handwriting?

When any of these things happen, it's time to get organized. Here are some suggestions:

- Write your step-by-step work for each homework exercise (neatly) on lined, loose-leaf paper and keep this in a 3-ring binder.
- Refer to your step-by-step work when you receive feedback that your answer is incorrect in MathXL. Double-check against the steps and hints provided by the program and correct your work accordingly.
- Keep your written homework with your class notes for that section.

- Identify any exercises you are having trouble with and ask questions about them.
- Keep all graded quizzes and tests in this binder as well to study later.

If you follow these suggestions, you and your instructor or tutor will be able to follow your steps and correct any mistakes. You will have a written copy of your work to refer to later to ask questions and study for tests.

Exercises

1. Why is it important to write out your step-by-step work to homework exercises and keep a hard copy of all work submitted online?
2. If you have gotten an incorrect answer, are you able to follow your steps and find your error?
3. If you were asked today to review your previous homework assignments and first test, could you find them? If not, list some ways you might organize your work better.

Study Skills Builder 14

Getting Help with Your Homework Assignments

Many helpful resources are available to you through MathXL to help you work through any homework exercises you may have trouble with. It is important for you to know what these resources are and when and how to use them.

Let's review these features, found in the homework exercises:

- **Help Me Solve This**—provides step-by-step help for the exercise you are working. You must work an additional exercise of the same type (without this help) before you can get credit for having worked it correctly.

- **View an Example**—allows you to view a correctly worked exercise similar to the one you are having trouble with. You can go back to your original exercise and work it on your own.

- **E-Book**—allows you to read examples from your text and find similar exercises.

- **Video*******—your text author, Elayn Martin-Gay, works an exercise similar to the one you need help with. ***Not all exercises have an accompanying video clip.

- **Ask My Instructor**—allows you to email your instructor for help with an exercise.

Exercises

1. How does the "Help Me Solve This" feature work?

2. If the "View an Example" feature is used, is it necessary to work an additional problem before continuing the assignment?

3. When might be a good time to use the "Video" feature? Do all exercises have an accompanying video clip?

4. Which of the features have you used? List those you found the most helpful to you.

5. If you haven't used the features discussed, list those you plan to try on your next homework assignment.

Study Skills Builder 15

Tips for Preparing for an Exam

Did you know that you can rework your previous homework assignments in MyMathLab and MathXL? This is a great way to prepare for tests. To do this, open a previous homework assignment and click "similar exercise." This will generate new exercises similar to the homework you have submitted. You can then rework the exercises and assignments until you feel confident that you understand them.

To prepare for an exam, follow these tips:

- Review your written work for your previous homework assignments along with your class notes.

- Identify any exercises or topics that you have questions on or have difficulty understanding.

- Rework your previous assignments in MyMathLab and MathXL until you fully understand them and can do them without help.

- Get help for any topics you feel unsure of or for which you have questions.

Exercises

1. Are your current homework assignments up to date and is your written work for them organized in a binder or notebook? If the answer is no, it's time to get organized. For tips on this, see Study Skills Builder 13—Organizing Your Work.

2. How many days in advance of an exam do you usually start studying?

3. List some ways you think that practicing previous homework assignments can help you prepare for your test.

4. List two or three resources you can use to get help for any topics you are unsure of or have questions on.

Good luck!

Study Skills Builder 16

How Well Do You Know the Resources Available to You in MyMathLab?

Many helpful resources are available to you in MyMathLab. Let's take a moment to locate and explore a few of them now. Go into your MyMathLab course and visit the multimedia library, tools for success, and E-book.

Let's see what you found.

Exercises

1. List the resources available to you in the Multimedia Library.

2. List the resources available to you in the Tools for Success folder.

3. Where did you find the English/Spanish Audio Glossary?

4. Can you view videos from the E-book?

5. Did you find any resources you did not know about? If so, which ones?

6. Which resources have you used most often or found most helpful?

Additional Help Inside and Outside Your Textbook

Study Skills Builder 17

How Well Do You Know Your Textbook?

The following questions will help determine whether you are familiar with your textbook. For additional information, see Section 1.1 in this text.

1. What does the ▶ icon mean?

2. What does the ＼ icon mean?

3. What does the △ icon mean?

4. Where can you find a review for each chapter? What answers to this review can be found in the back of your text?

5. Each chapter contains an overview of the chapter along with examples. What is this feature called?

6. Each chapter contains a review of vocabulary. What is this feature called?

7. Practice exercises are contained in this text. What are they and how can they be used?

8. This text contains a student section in the back entitled Student Resources. List the contents of this section and how they might be helpful.

9. What exercise answers are available in this text? Where are they located?

Study Skills Builder 18

Are You Familiar with Your Textbook Supplements?

Below is a review of some of the student supplements available for additional study. Check to see whether you are using the ones most helpful to you.

- Chapter Test Prep Videos. These videos provide video clip solutions to the Chapter Test exercises in this text. You will find this extremely useful when studying for tests or exams.
- Interactive DVD Lecture Series. These are keyed to each section of the text. The material is presented by me, Elayn Martin-Gay, and I have placed a ⊙ by the exercises in the text that I have worked on the video.
- The *Student Solutions Manual.* This contains worked-out solutions to odd-numbered exercises as well as every exercise in the Integrated Reviews, Chapter Reviews, Chapter Tests, and Cumulative Reviews and every Practice exercise.
- Pearson Tutor Center. Mathematics questions may be phoned, faxed, or emailed to this center.

- MyMathLab is a text-specific online course. MathXL is an online homework, tutorial, and assessment system. Take a moment to determine whether these are available to you.

    As usual, your instructor is your best source of information.

Exercises

Let's see how you are doing with textbook supplements.

1. Name one way the Lecture Videos can be helpful to you.
2. Name one way the Chapter Test Prep Video can help you prepare for a chapter test.
3. List any textbook supplements that you have found useful.
4. Have you located and visited a learning resource lab located on your campus?
5. List the textbook supplements that are currently housed in your campus's learning resource lab.

Study Skills Builder 19

Are You Getting All the Mathematics Help That You Need?

Remember that, in addition to your instructor, there are many places to get help with your mathematics course. For example:

- This text has an accompanying video lesson by the author for every section. There are also worked-out video solutions by the author to every Chapter Test exercise.
- The back of the book contains answers to odd-numbered exercises.
- A *Student Solutions Manual* is available that contains worked-out solutions to odd-numbered exercises as well as solutions to every exercise in the Integrated Reviews, Chapter Reviews, Chapter Tests, and Cumulative Reviews and every Practice exercise.

- Don't forget to check with your instructor for other local resources available to you, such as a tutor center.

Exercises

1. List items you find helpful in the text and all student supplements to this text.
2. List all the campus help that is available to you for this course.
3. List any help (besides the textbook) from Exercises 1 and 2 above that you are using.
4. List any help (besides the textbook) that you feel you should try.
5. Write a goal for yourself that includes trying everything you listed in Exercise 4 during the next week.

Bigger Picture–Study Guide Outline

Solving Equations and Inequalities

I. **Equations**

  A. **Linear Equations** (Sec. 2.1)

$$5(x - 2) = \frac{4(2x + 1)}{3}$$

$$3 \cdot 5(x - 2) = \cancel{3} \cdot \frac{4(2x + 1)}{\cancel{3}}$$

$$15x - 30 = 8x + 4$$

$$7x = 34$$

$$x = \frac{34}{7}$$

  B. **Absolute Value Equations** (Sec. 2.6)

$\lvert 3x - 1 \rvert = 8$	$\lvert x - 5 \rvert = \lvert x + 1 \rvert$
$3x - 1 = 8$ or $3x - 1 = -8$	$x - 5 = x + 1$ or $x - 5 = -(x + 1)$
$3x = 9$ or $3x = -7$	$\underbrace{-5 = 1}$ or $x - 5 = -x - 1$
$x = 3$ or $x = -\dfrac{7}{3}$	No solution or $2x = 4$
	$x = 2$

  C. **Quadratic and Higher-Degree Equations** (Secs. 5.8, 8.1, 8.2, 8.3)

$2x^2 - 7x = 9$	$2x^2 + x - 2 = 0$
$2x^2 - 7x - 9 = 0$	$a = 2,\ b = 1,\ c = -2$
$(2x - 9)(x + 1) = 0$	$x = \dfrac{-1 \pm \sqrt{1^2 - 4(2)(-2)}}{2 \cdot 2}$
$2x - 9 = 0$ or $x + 1 = 0$	
$x = \dfrac{9}{2}$ or $x = -1$	$x = \dfrac{-1 \pm \sqrt{17}}{4}$

  D. **Equations with Rational Expressions** (Sec. 6.5)

$$\frac{7}{x - 1} + \frac{3}{x + 1} = \frac{x + 3}{x^2 - 1}$$

$$\cancel{(x - 1)}(x + 1) \cdot \frac{7}{\cancel{x - 1}} + (x - 1)\cancel{(x + 1)} \cdot \frac{3}{\cancel{x + 1}}$$

$$= \cancel{(x - 1)}\cancel{(x + 1)} \cdot \frac{x + 3}{\cancel{(x - 1)}\cancel{(x + 1)}}$$

$$7(x + 1) + 3(x - 1) = x + 3$$

$$7x + 7 + 3x - 3 = x + 3$$

$$9x = -1$$

$$x = -\frac{1}{9}$$

E. Equations with Radicals (Sec. 7.6)

$$\sqrt{5x + 10} - 2 = x$$
$$\sqrt{5x + 10} = x + 2$$
$$(\sqrt{5x + 10})^2 = (x + 2)^2$$
$$5x + 10 = x^2 + 4x + 4$$
$$0 = x^2 - x - 6$$
$$0 = (x - 3)(x + 2)$$
$$x - 3 = 0 \quad \text{or} \quad x + 2 = 0$$
$$x = 3 \quad \text{or} \quad x = -2$$

Both solutions check.

F. Exponential Equations (Secs. 9.3, 9.8)

$9^x = 27^{x+1}$	$5^x = 7$
$(3^2)^x = (3^3)^{x+1}$	$\log 5^x = \log 7$
$3^{2x} = 3^{3x+3}$	$x \log 5 = \log 7$
$2x = 3x + 3$	$x = \dfrac{\log 7}{\log 5}$
$-3 = x$	

G. Logarithmic Equations (Sec. 9.8)

$$\log 7 + \log(x + 3) = \log 5$$
$$\log 7(x + 3) = \log 5$$
$$7(x + 3) = 5$$
$$7x + 21 = 5$$
$$7x = -16$$
$$x = -\frac{16}{7}$$

II. Inequalities

A. Linear Inequalities (Sec. 2.4)

$$-3(x + 2) \geq 6$$
$$-3x - 6 \geq 6$$
$$-3x \geq 12$$
$$\frac{-3x}{-3} \leq \frac{12}{-3}$$
$$x \leq -4 \quad \text{or} \quad (-\infty, -4]$$

B. Compound Inequalities (Sec. 2.5)

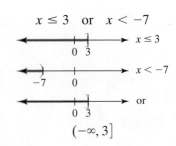

C. Absolute Value Inequalities (Sec. 2.7)

$$|x - 5| - 8 < -2$$
$$|x - 5| < 6$$
$$-6 < x - 5 < 6$$
$$-1 < x < 11$$
$$(-1, 11)$$

$$|2x + 1| \geq 17$$
$$2x + 1 \geq 17 \quad \text{or} \quad 2x + 1 \leq -17$$
$$2x \geq 16 \quad \text{or} \qquad 2x \leq -18$$
$$x \geq 8 \quad \text{or} \qquad x \leq -9$$
$$(-\infty, -9] \cup [8, \infty)$$

D. Nonlinear Inequalities (Sec. 8.4)

$$x^2 - x < 6$$
$$x^2 - x - 6 < 0$$
$$(x - 3)(x + 2) < 0$$

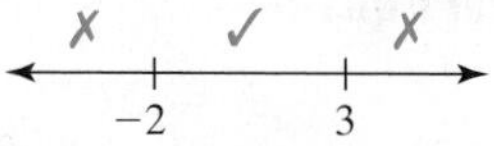

✗ ✓ ✗
$\longleftarrow$|$\quad$|$\longrightarrow$
−2 $\quad$ 3

$$(-2, 3)$$

$$\frac{x - 5}{x + 1} \geq 0$$

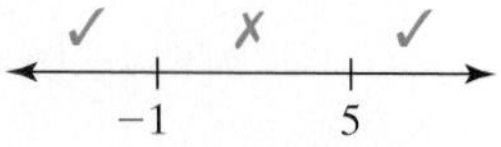

✓ ✗ ✓
$\longleftarrow$|$\quad$|$\longrightarrow$
−1 $\quad$ 5

$$(-\infty, -1) \cup [5, \infty)$$

Simplify. If needed, write answers with positive exponents only.

1. $\sqrt{216}$

2. $\dfrac{(4 - \sqrt{16}) - (-7 - 20)}{-2(1 - 4)^2}$

3. $\left(\dfrac{1}{125}\right)^{-1/3}$

4. $(-9x)^{-2}$

5. $\dfrac{\dfrac{5}{x} - \dfrac{7}{3x}}{\dfrac{9}{8x} - \dfrac{1}{x}}$

6. $\dfrac{6^{-1}a^2b^{-3}}{3^{-2}a^{-5}b^2}$

7. $\left(\dfrac{64c^{4/3}}{a^{-2/3}b^{5/6}}\right)^{1/2}$

Factor completely.

8. $3x^2y - 27y^3$

9. $16y^3 - 2$

10. $x^2y - 9y - 3x^2 + 27$

Perform the indicated operations and simplify if possible.

11. $(4x^3y - 3x - 4) - (9x^3y + 8x + 5)$

12. $(6m + n)^2$

13. $(2x - 1)(x^2 - 6x + 4)$

14. $\dfrac{3x^2 - 12}{x^2 + 2x - 8} \div \dfrac{6x + 18}{x + 4}$

15. $\dfrac{2x^2 + 7}{2x^4 - 18x^2} - \dfrac{6x + 7}{2x^4 - 18x^2}$

16. $\dfrac{3}{x^2 - x - 6} + \dfrac{2}{x^2 - 5x + 6}$

17. $\sqrt{125x^3} - 3\sqrt{20x^3}$

18. $(\sqrt{5} + 5)(\sqrt{5} - 5)$

19. $(4x^3 - 5x) \div (2x + 1)$ [Use long division.]

Solve each equation or inequality. Write inequality solutions using interval notation.

20. $9(x + 2) = 5[11 - 2(2 - x) + 3]$

21. $|6x - 5| - 3 = -2$

22. $3n(7n - 20) = 96$

23. $-3 < 2(x - 3) \le 4$

24. $|3x + 1| > 5$

25. $\dfrac{x^2 + 8}{x} - 1 = \dfrac{2(x + 4)}{x}$

26. $y^2 - 3y = 5$

27. $x = \sqrt{x - 2} + 2$

28. $2x^2 - 7x > 15$

29. Solve the system: $\begin{cases} \dfrac{x}{2} + \dfrac{y}{4} = -\dfrac{3}{4} \\ x + \dfrac{3}{4}y = -4 \end{cases}$

Graph the following.

30. $4x + 6y = 7$

31. $2x - y > 5$

32. $y = -3$

33. $g(x) = -|x + 2| - 1$. Also, find the domain and range of this function.

34. $h(x) = x^2 - 4x + 4$. Label the vertex and any intercepts.

35. $f(x) = \begin{cases} -\dfrac{1}{2}x & \text{if } x \le 0 \\ 2x - 3 & \text{if } x > 0 \end{cases}$. Also, find the domain and range of this function.

Write equations of the following lines. Write each equation using function notation.

36. through $(4, -2)$ and $(6, -3)$

37. through $(-1, 2)$ and perpendicular to $3x - y = 4$

Find the distance or midpoint.

38. Find the distance between the points $(-6, 3)$ and $(-8, -7)$.

39. Find the midpoint of the line segment whose endpoints are $(-2, -5)$ and $(-6, 12)$.

Rationalize each denominator. Assume that variables represent positive numbers.

40. $\sqrt{\dfrac{9}{y}}$

41. $\dfrac{4 - \sqrt{x}}{4 + 2\sqrt{x}}$

Solve.

42. In 2008, the median earnings of young adults with bachelor's degrees were \$46,000. This represents a 28% increase over the median earnings of young adults with associates's degrees. Find the median earnings of young adults with associate's degrees in 2008. Round to the nearest whole dollar. (*Source*: National Center for Educational Statistics)

43. Write the area of the shaded region as a factored polynomial.

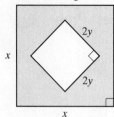

44. The product of one more than a number and twice the reciprocal of the number is $\dfrac{12}{5}$. Find the number.

45. Suppose that W is inversely proportional to V. If $W = 20$ when $V = 12$, find W when $V = 15$.

46. Given the diagram shown, approximate to the nearest tenth of a foot how many feet of walking distance a person saves by cutting across the lawn instead of walking on the sidewalk.

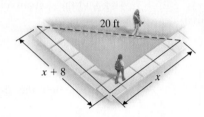

47. A stone is thrown upward from a bridge. The stone's height in feet, $s(t)$, above the water t seconds after the stone is thrown is a function given by the equation

$$s(t) = -16t^2 + 32t + 256$$

a. Find the maximum height of the stone.

b. Find the time it takes the stone to hit the water. Round the answer to two decimal places.

48. The research department of a company that manufactures children's fruit drinks is experimenting with a new flavor. A 17.5% fructose solution is needed, but only 10% and 20% solutions are available. How many gallons of the 10% fructose solution should be mixed with the 20% fructose solution to obtain 20 gallons of a 17.5% fructose solution?

COMPLEX NUMBERS: CHAPTER 7

Perform the indicated operation and simplify. Write the result in the form $a + bi$.

49. $-\sqrt{-8}$

50. $(12 - 6i) - (12 - 3i)$

51. $(4 + 3i)^2$

52. $\dfrac{1 + 4i}{1 - i}$

INVERSE, EXPONENTIAL, AND LOGARITHMIC FUNCTIONS: CHAPTER 9

53. If $g(x) = x - 7$ and $h(x) = x^2 - 6x + 5$, find $(g \circ h)(x)$.

54. Decide whether $f(x) = 6 - 2x$ is a one-to-one function. If it is, find its inverse.

55. Use properties of logarithms to write the expression as a single logarithm.

$$\log_5 x + 3\log_5 x - \log_5(x + 1)$$

Solve. Give exact solutions.

56. $8^{x-1} = \dfrac{1}{64}$

57. $3^{2x+5} = 4$ Give an exact solution and a 4-decimal-place approximation.

58. $\log_8(3x - 2) = 2$

59. $\log_4(x + 1) - \log_4(x - 2) = 3$

60. $\ln\sqrt{e} = x$

61. Graph $y = \left(\dfrac{1}{2}\right)^x + 1$

62. The prairie dog population of the Grand Rapids area now stands at 57,000 animals. If the population is growing at a rate of 2.6% annually, find how many prairie dogs there will be in that area 5 years from now.

CONIC SECTIONS: CHAPTER 10

Sketch the graph of each equation.

63. $x^2 - y^2 = 36$

64. $16x^2 + 9y^2 = 144$

65. $x^2 + y^2 + 6x = 16$

66. Solve the system:
$$\begin{cases} x^2 + y^2 = 26 \\ x^2 - 2y^2 = 23 \end{cases}$$

SEQUENCES, SERIES, AND THE BINOMIAL THEOREM: CHAPTER 11

67. Find the first five terms of the sequence $a_n = \dfrac{(-1)^n}{n + 4}$.

68. Find the partial sum, S_5, of the sequence $a_n = 5(2)^{n-1}$.

69. Find S_∞ of the sequence $\dfrac{3}{2}, -\dfrac{3}{4}, \dfrac{3}{8}, \ldots$

70. Find $\displaystyle\sum_{i=1}^{4} i(i - 2)$

71. Expand: $(2x + y)^5$

Answers to Selected Exercises

CHAPTER 1 REAL NUMBERS AND ALGEBRAIC EXPRESSIONS

Section 1.2
Practice Exercises

1. 14 sq cm   **2.** 31   **3. a.** $\{6, 7, 8, 9\}$ **b.** $\{41, 42, 43, \dots\}$ **4. a.** true   **b.** true   **5. a.** true   **b.** false   **c.** true   **d.** false
6. a. 4   **b.** $\frac{1}{2}$ **c.** 1   **d.** -6.8 **e.** -4 **7. a.** -5.4 **b.** $\frac{3}{5}$ **c.** -18 **8. a.** $3x$ **b.** $2x - 5$ **c.** $3\frac{5}{8} + x$ **d.** $x \div 2$ or $\frac{x}{2}$
e. $x - 14$ **f.** $5(x + 10)$

Vocabulary, Readiness and Video Check 1.2

1. variables   **3.** absolute value   **5.** natural numbers   **7.** integers   **9.** rational number   **11.** This is an application, so the answer needs to be put in context. The decimal actually represents money.   **13.** distance; 0; positive; negative   **15.** Order is important in subtraction so you must read the phrase carefully to determine the order of subtraction in your algebraic expression.

Exercise Set 1.2

1. 35   **3.** 30.38   **5.** $\frac{3}{8}$ **7.** 22   **9.** 2000 mi   **11.** 20.4 sq. ft   **13.** $10,612.80   **15.** $\{1, 2, 3, 4, 5\}$ **17.** $\{11, 12, 13, 14, 15, 16\}$
19. $\{0\}$ **21.** $\{0, 2, 4, 6, 8\}$ **23.**   **25.**   **27.**   **29.**   **31.** $\{3, 0, \sqrt{36}\}$
33. $\{3, \sqrt{36}\}$ **35.** $\{\sqrt{7}\}$ **37.** $\in$ **39.** $\notin$ **41.** $\notin$ **43.** $\notin$ **45.** true   **47.** true   **49.** false   **51.** false   **53.** true
55. false   **57.** -2 **59.** 4   **61.** 0   **63.** -3 **65.** 6.2   **67.** $-\frac{4}{7}$ **69.** $\frac{2}{3}$ **71.** 0   **73.** $2x$ **75.** $2x + 5$ **77.** $x - 10$
79. $x + 2$ **81.** $\frac{x}{11}$ **83.** $12 - 3x$ **85.** $x + 2.3$ or $x + 2\frac{3}{10}$ **87.** $1\frac{1}{3} - x$ **89.** $\frac{5}{4 - x}$ **91.** $2(x + 3)$ **93.** 137   **95.** 69
97. answers may vary   **99.** answers may vary   **101.** answers may vary

Section 1.3
Practice Exercises

1. a. -8 **b.** -3 **c.** 5   **d.** -8.1 **e.** $\frac{1}{15}$ **f.** $-\frac{7}{22}$ **2. a.** -8 **b.** -3 **c.** -12 **d.** 7.7   **e.** $-\frac{22}{21}$ **f.** 1.8   **g.** -7
3. a. 12   **b.** -4 **4. a.** -15 **b.** $\frac{1}{2}$ **c.** -10.2 **d.** 0   **e.** $-\frac{2}{13}$ **f.** 36   **g.** -11.5 **5. a.** -2 **b.** 5   **c.** $-\frac{1}{6}$ **d.** -6
e. $\frac{1}{9}$ **f.** 0   **6. a.** 8   **b.** $\frac{1}{9}$ **c.** -36 **d.** 36   **e.** -64 **f.** -64 **7. a.** 7   **b.** $\frac{1}{4}$ **c.** -8 **d.** not a real number   **e.** 10
8. a. 4   **b.** -1 **c.** 10   **9. a.** 2   **b.** 27   **c.** -29 **10.** 19   **11.** $-\frac{17}{44}$ **12. a.** 67   **b.** -100 **c.** $-\frac{39}{80}$ **13.** 23; 50; 77

Vocabulary, Readiness and Video Check 1.3

1. 0   **3.** reciprocal   **5.** exponent   **7.** square root   **9.** addition   **11.** The parentheses, or lack of them, determine the base of the expression. In Example 7, -7^2, the base is 7 and only 7 is squared.   **13.** It allows each expression to evaluate to a single number.

Exercise Set 1.3

1. 5   **3.** -24 **5.** -11 **7.** -4 **9.** $\frac{4}{3}$ **11.** -2 **13.** $-\frac{1}{2}$ **15.** -6 **17.** -60 **19.** 0   **21.** 0   **23.** -3 **25.** 3   **27.** $-\frac{1}{6}$
29. 0.56   **31.** -7 **33.** -8 **35.** -49 **37.** 36   **39.** -8 **41.** $-\frac{1}{27}$ **43.** 7   **45.** $-\frac{2}{3}$ **47.** 4   **49.** 3   **51.** not a real number
53. 48   **55.** -1 **57.** -9 **59.** 17   **61.** -4 **63.** -2 **65.** 11   **67.** $-\frac{3}{4}$ **69.** 7   **71.** -11 **73.** -2.1 **75.** $-\frac{1}{3}$ **77.** $-\frac{79}{15}$
79. $-\frac{4}{5}$ **81.** -81 **83.** $-\frac{20}{33}$ **85.** 93   **87.** -12 **89.** $-\frac{23}{18}$ **91.** 5   **93.** $-\frac{3}{19}$ **95. a.** 18; 22; 28; 208   **b.** increase; answers may vary
97. a. 600; 150; 105   **b.** decrease; answers may vary   **99.** b, c   **101.** b, d   **103.** b   **105.** $\frac{5}{2}$ **107.** $\frac{13}{35}$ **109.** 4205 m   **111.** $(2 + 7) \cdot (1 + 3)$
113. 20 million   **115.** 70 million   **117.** increasing; answers may vary   **119.** answers may vary   **121.** 3.1623   **123.** 2.8107   **125.** -0.5876

Integrated Review

1. 16   **2.** -16 **3.** 0   **4.** -11 **5.** -5 **6.** $-\frac{1}{60}$ **7.** undefined   **8.** -2.97 **9.** 4   **10.** -50 **11.** 35   **12.** 92
13. $-15 - 2x$ **14.** $3x + 5$ **15.** 0   **16.** true

Section 1.4
Practice Exercises

1. $-4x = 20$ **2.** $3(z - 3) = 9$ **3.** $x + 5 = 2x - 3$ **4.** $y + 2 = 4 + \dfrac{z}{8}$ **5. a.** $<$ **b.** $=$ **c.** $>$ **d.** $<$ **e.** $>$ **f.** $<$

6. a. $x - 3 \le 5$ **b.** $y \ne -4$ **c.** $2 < 4 + \dfrac{1}{2}z$ **7. a.** 7 **b.** -4.7 **c.** $\dfrac{3}{8}$ **8. a.** $-\dfrac{3}{5}$ **b.** $\dfrac{1}{14}$ **c.** $-\dfrac{1}{2}$ **9.** $13x + 8$

10. $(3 \cdot 11)b = 33b$ **11. a.** $4x + 20y$ **b.** $-3 + 2z$ **c.** $0.3xy - 0.9x$ **12. a.** $0.10x$ **b.** $26y$ **c.** $1.75z$ **d.** $0.15t$ **13. a.** $16 - x$
b. $180 - x$ **c.** $x + 2$ **d.** $x + 9$ **14. a.** $5ab$ **b.** $10x - 5$ **c.** $17p - 9$ **15. a.** $-pq + 7$ **b.** $x^2 + 19$ **c.** $5.8x + 3.8$

d. $-c - 8d + \dfrac{1}{4}$

Vocabulary, Readiness and Video Check 1.4

1. $<$ **3.** $\ne$ **5.** $\ge$ **7.** $-a$ **9.** commutative   **11.** distributive   **13.** $=, \ne, <, \le, >, \ge$ **15.** order; grouping   **17.** by combining
like terms; distributive property

Exercise Set 1.4

1. $10 + x = -12$ **3.** $2x + 5 = -14$ **5.** $\dfrac{n}{5} = 4n$ **7.** $z - \dfrac{1}{2} = \dfrac{1}{2}z$ **9.** $7x \le -21$ **11.** $2(x - 6) > \dfrac{1}{11}$ **13.** $2(x - 6) = -27$

15. $>$ **17.** $=$ **19.** $<$ **21.** $<$ **23.** $<$ **25.** $-5; \dfrac{1}{5}$ **27.** $-8; -\dfrac{1}{8}$ **29.** $\dfrac{1}{7}; -7$ **31.** 0; undefined   **33.** $\dfrac{7}{8}; -\dfrac{7}{8}$ **35.** $y + 7x$

37. $w \cdot z$ **39.** $\dfrac{x}{5} \cdot \dfrac{1}{3}$ **41.** $(5 \cdot 7)x$ **43.** $x + (1.2 + y)$ **45.** $14(z \cdot y)$ **47.** $3x + 15$ **49.** $-2a - b$ **51.** $12x + 10y + 4z$

53. $-4x + 8y - 28$ **55.** $3xy - 1.5x$ **57.** $6 + 3x$ **59.** 0 **61.** 7 **63.** $(10 \cdot 2)y$ **65.** $0.1d$ **67.** $112 - x$ **69.** $180 - x$ **71.** $\$6.49x$
73. $x + 2$ **75.** $-6x + 9$ **77.** $2k + 10$ **79.** $-3x + 5$ **81.** $-x^2 + 4xy + 9$ **83.** $4n - 8$ **85.** -24 **87.** $2x + 10$ **89.** $0.8x - 3.6$

91. $\dfrac{11}{12}b - \dfrac{7}{6}$ **93.** $6x + 14$ **95.** $-5x + \dfrac{5}{6}y - 1$ **97.** $3a + \dfrac{3}{35}$ **99.** $3x + 12$ **101.** $(5 \cdot 7)y$ **103.** $a(b + c) = ab + ac$

105. zero; answers may vary   **107.** no; answers may vary   **109.** answers may vary   **111.** $15.4z + 31.11$

Chapter 1 Vocabulary Check

1. algebraic expression   **2.** opposite   **3.** distributive   **4.** absolute value   **5.** exponent   **6.** variable   **7.** inequality   **8.** reciprocals
9. commutative   **10.** associative   **11.** whole   **12.** real

Chapter 1 Review

1. 21 **3.** 4200 **5.** $\{-1, 1, 3\}$ **7.** $\varnothing$ **9.** $\{6, 7, 8, \dots\}$ **11.** true   **13.** true   **15.** false   **17.** false   **19.** true   **21.** false

23. $\left\{5, \dfrac{8}{2}, \sqrt{9}\right\}$ **25.** $\left\{5, -\dfrac{2}{3}, \dfrac{8}{2}, \sqrt{9}, 0.3, 1\dfrac{5}{8}, -1\right\}$ **27.** $\left\{5, -\dfrac{2}{3}, \dfrac{8}{2}, \sqrt{9}, 0.3, \sqrt{7}, 1\dfrac{5}{8}, -1, \pi\right\}$ **29.** $\dfrac{3}{4}$ **31.** 0 **33.** $-\dfrac{4}{3}$ **35.** undefined

37. -4 **39.** -2 **41.** 8 **43.** 0 **45.** undefined   **47.** 4 **49.** $-\dfrac{2}{15}$ **51.** $\dfrac{5}{12}$ **53.** 9 **55.** 3 **57.** $-\dfrac{32}{135}$ **59.** $-\dfrac{5}{4}$ **61.** $\dfrac{5}{8}$ **63.** -1

65. 1 **67.** -4 **69.** $\dfrac{5}{7}$ **71.** $\dfrac{1}{5}$ **73.** -5 **75.** 5 **77.** $6.28; 62.8; 628$ **79.** $-xy + 1$ **81.** $2x^2 - 2$ **83.** $-1.1x - 0.3$ **85.** $12 = -4x$

87. $4(y + 3) = -1$ **89.** $z - 7 = 6$ **91.** $x - 5 \ge 12$ **93.** $\dfrac{2}{3} \ne 2\left(n + \dfrac{1}{4}\right)$ **95.** associative property of addition   **97.** additive inverse

property   **99.** associative and commutative properties of multiplication   **101.** multiplication property of zero   **103.** additive identity property

105. $5(x - 3z)$ **107.** $2 + (-2)$, for example   **109.** $3.4[(0.7)5]$ **111.** $>$ **113.** $<$ **115.** $<$ **117.** $\dfrac{3}{4}; -\dfrac{4}{3}$ **119.** $-10x + 6.1$

121. $-\dfrac{6}{11}$ **123.** $-\dfrac{4}{15}$ **125.** $-x + 3y + 1$ **127.** 2.1 **129.** 2.7 **131.** 0.5

Chapter 1 Test

1. true   **2.** false   **3.** false   **4.** false   **5.** true   **6.** false   **7.** -3 **8.** -56 **9.** -225 **10.** 3 **11.** 1 **12.** $-\dfrac{3}{2}$ **13.** 12

14. 1 **15. a.** $8.75; 26.25; 87.50; 175.00$ **b.** increase   **16.** $2(x + 5) = 30$ **17.** $\dfrac{(6 - y)^2}{7} < -2$ **18.** $\dfrac{9z}{|-12|} \ne 10$ **19.** $3\left(\dfrac{n}{5}\right) = -n$

20. $20 = 2x - 6$ **21.** $-2 = \dfrac{x}{x + 5}$ **22.** distributive property   **23.** associative property of addition   **24.** additive inverse property

25. multiplication property of zero   **26.** $0.05n + 0.1d$ **27.** reciprocal: $-\dfrac{11}{7}$; opposite: $\dfrac{7}{11}$ **28.** $\dfrac{1}{2}a - \dfrac{9}{8}$ **29.** $2y - 10$ **30.** $-1.3x + 1.9$

CHAPTER 2 EQUATIONS, INEQUALITIES, AND PROBLEM SOLVING

Section 2.1
Practice Exercises

1. 5 **2.** 0.2 **3.** -5 **4.** -4 **5.** $\dfrac{5}{6}$ **6.** $\dfrac{5}{4}$ **7.** -3 **8.** $\{\,\}$ or $\varnothing$ **9.** all real numbers or $\{x \mid x \text{ is a real number}\}$

Vocabulary, Readiness & Video Check 2.1

1. equivalent   **3.** addition   **5.** expression   **7.** equation   **9.** both sides; same   **11.** to make the calculations less tedious

Exercise Set 2.1

1. 6   **3.** -22 **5.** 4.7   **7.** 10   **9.** -1.1 **11.** -5 **13.** -2 **15.** 0   **17.** 2   **19.** -9 **21.** $-\dfrac{10}{7}$ **23.** $\dfrac{9}{10}$ **25.** 4   **27.** 1

29. 5   **31.** $\dfrac{40}{3}$ **33.** 17   **35.** all real numbers   **37.** $\varnothing$ **39.** all real numbers   **41.** $\varnothing$ **43.** $\dfrac{1}{8}$ **45.** 0   **47.** all real numbers

49. 4   **51.** $\dfrac{4}{5}$ **53.** 8   **55.** $\varnothing$ **57.** -8 **59.** $-\dfrac{5}{4}$ **61.** -2 **63.** 23   **65.** $-\dfrac{2}{9}$ **67.** $\dfrac{8}{x}$ **69.** $8x$ **71.** $2x - 5$ **73.** subtract 19 instead of adding; -3 **75.** $0.4 - 1.6 = -1.2$, not 1.2; -0.24 **77.** all real numbers   **79.** no solution   **81. a.** $4x + 5$ **b.** -3 **c.** answers may vary   **83.** answers may vary   **85.** $K = -11$ **87.** $K = -23$ **89.** answers may vary   **91.** 1   **93.** 3   **95.** -4.86 **97.** 1.53

Section 2.2
Practice Exercises

1. a. $3x + 6$ **b.** $6x - 1$ **2.** $3x + 18.1$ **3.** 14, 34, 70   **4.** \$450   **5.** width: 32 in.; length: 48 in.   **6.** 25, 27, 29

Vocabulary, Readiness & Video Check 2.2

1. $>$ **3.** $=$ **5.** 31, 32, 33, 34   **7.** 18, 20, 22   **9.** $y, y + 1, y + 2$ **11.** $p, p + 1, p + 2, p + 3$ **13.** distributive property

Exercise Set 2.2

1. $4y$ **3.** $3z + 3$ **5.** $(65x + 30)$ cents   **7.** $10x + 3$ **9.** $2x + 14$ **11.** -5 **13.** 45, 225, 145   **15.** approximately 1612.41 million acres
17. 7747 earthquakes   **19.** 1275 shoppers   **21.** 23%   **23.** 417 employees   **25.** $29°, 35°, 116°$ **27.** 28 m, 36 m, 38 m
29. 18 in., 18 in., 27 in., 36 in.   **31.** 75, 76, 77   **33.** Fallon's ZIP code is 89406; Fernley's ZIP code is 89408; Gardnerville Ranchos' ZIP code is 89410
35. 55 million; 97 million; 138 million   **37.** biomedical engineer: 12 thousand; skin care specialist: 15 thousand; physician assistant: 29 thousand
39. B767-300ER: 207 seats; B737-200: 119 seats; F-100: 87 seats   **41.** \$430.00   **43.** \$446,028   **45.** 1,800,000 people   **47.** $40°, 140°$
49. $64°, 32°, 84°$ **51.** square: 18 cm; triangle: 24 cm   **53.** 76, 78, 80   **55.** Darlington: 61,000; Daytona: 159,000   **57.** Tokyo: 36.67 million; New York: 19.43 million; Mexico City: 19.46 million   **59.** 40.5 ft; 202.5 ft; 240 ft   **61.** incandescent: 1500 bulb hours; fluorescent: 100,000 bulb hours; halogen: 4000 bulb hours   **63.** Milwaukee Brewers: 77 wins; Houston Astros: 76 wins; Chicago Cubs: 75 wins   **65.** Guy's Tower: 469 ft; Queen Mary Hospital: 449 ft; Galter Pavilion: 402 ft   **67.** 208   **69.** -55 **71.** 3195   **73.** yes; answers may vary   **75.** $80°$ **77.** 2022
79. 153   **81.** any three consecutive integers   **83.** 25 skateboards   **85.** company loses money

Section 2.3
Practice Exercises

1. $T = \dfrac{I}{PR}$ **2.** $y = \dfrac{7}{2}x - \dfrac{5}{2}$ **3.** $r = \dfrac{A - P}{Pt}$ **4.** \$10, 134.16   **5.** 25.6 hr; 25 hr 36 min

Vocabulary, Readiness & Video Check 2.3

1. $y = 5 - 2x$ **3.** $a = 5b + 8$ **5.** $k = h - 5j + 6$ **7.** That the specified variable will equal some expression and that this expression should not contain the specified variable.

Exercise Set 2.3

1. $t = \dfrac{d}{r}$ **3.** $R = \dfrac{I}{PT}$ **5.** $y = \dfrac{9x - 16}{4}$ **7.** $W = \dfrac{P - 2L}{2}$ **9.** $A = \dfrac{J + 3}{C}$ **11.** $g = \dfrac{W}{h - 3t^2}$ **13.** $B = \dfrac{T - 2C}{AC}$ **15.** $r = \dfrac{C}{2\pi}$

17. $r = \dfrac{E - IR}{I}$ **19.** $L = \dfrac{2s - na}{n}$ **21.** $v = \dfrac{3st^4 - N}{5s}$ **23.** $H = \dfrac{S - 2LW}{2L + 2W}$ **25.** \$4703.71; \$4713.99; \$4719.22; \$4722.74; \$4724.45
27. a. \$7313.97   **b.** \$7321.14   **c.** \$7325.98   **29.** 3.6 hr or 3 hr 36 min   **31.** $40°C$ **33.** 171 packages   **35.** 9 ft   **37.** 2 gal   **39.** 20 m
41. a. 288π cu mm   **b.** 904.78 cu mm   **43. a.** 1174.86 cu m   **b.** 310.34 cu m   **c.** 1485.20 cu m   **45.** 128.3 mph   **47.** 0.42 ft

49. 41.125π ft ≈ 129.1325 ft   **51.** \$1831.96   **53.** $f = \dfrac{C - 4h - 4p}{9}$ **55.** 178 cal   **57.** 1.5 g   **59.** $-3, -2, -1$ **61.** $-3, -2, -1, 0, 1$

63. answers may vary   **65.** 0.723   **67.** 1.523   **69.** 9.538   **71.** 30.065   **73.** answers may vary   **75.** answers may vary; $g = \dfrac{W}{h - 3t^2}$

77. $\dfrac{1}{8}$ **79.** $\dfrac{1}{8}$ **81.** $\dfrac{1}{2}$ **83.** $\dfrac{3}{8}$ **85.** 0   **87.** 0

Section 2.4
Practice Exercises

1. a. $(-\infty, 3.5)$ **b.** $[-3, \infty)$ **c.** $[-1, 4)$ **2.** $(4, \infty)$

3. $(-\infty, -4]$ **4. a.** $\left[\dfrac{2}{3}, \infty\right)$ **b.** $(-4, \infty)$ **5.** $\left[-\dfrac{3}{2}, \infty\right)$

6. $(-\infty, 13]$ **7.** $(-\infty, \infty)$ **8.** Sales must be $\geq$ \$10,000 per month.   **9.** the entire year 2017 and after

Vocabulary, Readiness & Video Check 2.4

1. d   **3.** b   **5.** $[-0.4, \infty)$ **7.** $(-\infty, -0.4)$ **9.** The graph of Example 1 is shaded from $-\infty$ to, but not including, -3, as indicated by a parenthesis. To write interval notation, write down what is shaded for the inequality from left to right. A parenthesis is always used with $-\infty$, so from the graph, the interval notation is $(-\infty, -3)$.   **11.** same; reverse

Exercise Set 2.4

1. $(-\infty, -3)$ **3.** $[0.3, \infty)$ **5.** $[-7, \infty)$ **7.** $(-2, 5)$

9. $(-1, 5]$ **11.** $[-2, \infty)$ **13.** $(-\infty, 1)$ **15.** $(-\infty, 2]$

17. $[8, \infty)$ **19.** $(-\infty, -4.7)$ **21.** $(-\infty, -3]$ **23.** $(-\infty, -1]$ **25.** $(-\infty, 11]$ **27.** $(0, \infty)$

29. $(-13, \infty)$ **31.** $\left[-\dfrac{79}{3}, \infty\right)$ **33.** $\left(-\infty, -\dfrac{35}{6}\right)$ **35.** $(-\infty, -6)$ **37.** $(4, \infty)$ **39.** $[-0.5, \infty)$ **41.** $(-\infty, 7]$ **43.** $[0, \infty)$ **45.** $(-\infty, -29]$

47. $[3, \infty)$ **49.** $(-\infty, -1]$ **51.** $[-31, \infty)$ **53.** $(-\infty, \infty)$ **55.** $\varnothing$ **57.** $(-\infty, 9)$ **59.** $\left(-\infty, -\dfrac{11}{2}\right]$ **61.** $(-\infty, -2]$ **63.** $(-\infty, -15)$

65. $\left[-\dfrac{37}{3}, \infty\right)$ **67.** $(-\infty, 5)$ **69. a.** $\{x \mid x \geq 81\}$ **b.** A final exam grade of 81 or higher will result in an average of 77 or higher.

71. a. $\{x \mid x \leq 20.5\}$ **b.** She can move at most 20 whole boxes at one time. **73. a.** $\{x \mid x \leq 1040\}$ **b.** The luggage and cargo must weigh 1040 pounds or less. **75. a.** $\{x \mid x > 80\}$ **b.** If you drive more than 80 miles per day, plan A is more economical. **77.** $\{F \mid F \geq 932°\}$

79. a. 2018 **b.** answers may vary **81.** decreasing; answers may vary **83.** 37.55 lb **85.** during 2012 **87.** answers may vary

89. a. $\{t \mid t > 23.3\}$ **b.** 2024 **91.** 2, 3, 4 **93.** 2, 3, 4,... **95.** 5 **97.** $\dfrac{13}{6}$ **99.** $\{x \mid x \geq 2\}; [2, \infty)$ **101.** $(-\infty, 0)$

103. $\{x \mid -2 < x \leq 1.5\};$ **105.** yes **107.** yes **109.** $\{4\}$ **111.** $(-\infty, 4)$ **113.** answers may vary

115. answers may vary

Integrated Review

1. -5 **2.** $(-5, \infty)$ **3.** $\left[\dfrac{8}{3}, \infty\right)$ **4.** $[-1, \infty)$ **5.** 0 **6.** $\left(-\dfrac{1}{10}, \infty\right)$ **7.** $\left(-\infty, -\dfrac{1}{6}\right]$ **8.** 0 **9.** $\varnothing$ **10.** $\left[-\dfrac{3}{5}, \infty\right)$ **11.** 4.2 **12.** 6

13. -8 **14.** $(-\infty, -16)$ **15.** $\dfrac{20}{11}$ **16.** 1 **17.** $(38, \infty)$ **18.** -5.5 **19.** $\dfrac{3}{5}$ **20.** $(-\infty, \infty)$ **21.** 29 **22.** all real numbers **23.** $(-\infty, 1)$

24. $\dfrac{9}{13}$ **25.** $(23, \infty)$ **26.** $(-\infty, 6]$ **27.** $\left(-\infty, \dfrac{3}{5}\right]$ **28.** $\left(-\infty, -\dfrac{19}{32}\right)$

Section 2.5
Practice Exercises

1. $\{1, 3\}$ **2.** $(-\infty, 2)$ **3.** $\{\,\}$ or $\varnothing$ **4.** $(-4, 2)$ **5.** $[-6, 8]$ **6.** $\{1, 2, 3, 4, 5, 6, 7, 9\}$ **7.** $\left(-\infty, \dfrac{3}{8}\right] \cup [3, \infty)$ **8.** $(-\infty, \infty)$

Vocabulary, Readiness & Video Check 2.5

1. compound **3.** or **5.** $\cup$ **7.** and **9.** or

Exercise Set 2.5

1. $\{2, 3, 4, 5, 6, 7\}$ **3.** $\{4, 6\}$ **5.** $\{\ldots, -2, -1, 0, 1, \ldots\}$ **7.** $\{5, 7\}$ **9.** $\{x \mid x \text{ is an odd integer or } x = 2 \text{ or } x = 4\}$ **11.** $\{2, 4\}$

13. $(-3, 1)$ **15.** $\varnothing$ **17.** $(-\infty, -1)$ **19.** $[6, \infty)$ **21.** $(-\infty, -3]$ **23.** $(4, 10)$

25. $(11, 17)$ **27.** $[1, 4]$ **29.** $\left[-3, \dfrac{3}{2}\right]$ **31.** $\left[-\dfrac{7}{3}, 7\right]$ **33.** $(-\infty, 5)$ **35.** $(-\infty, -4] \cup [1, \infty)$

37. $(-\infty, \infty)$ **39.** $[2, \infty)$ **41.** $(-\infty, -4) \cup (-2, \infty)$ **43.** $(-\infty, \infty)$ **45.** $\left(-\dfrac{1}{2}, \dfrac{2}{3}\right)$ **47.** $(-\infty, \infty)$ **49.** $\left[\dfrac{3}{2}, 6\right]$ **51.** $\left(\dfrac{5}{4}, \dfrac{11}{4}\right)$

53. $\varnothing$ **55.** $\left(-\infty, -\dfrac{56}{5}\right) \cup \left(\dfrac{5}{3}, \infty\right)$ **57.** $\left(-5, \dfrac{5}{2}\right)$ **59.** $\left(0, \dfrac{14}{3}\right]$ **61.** $(-\infty, -3]$ **63.** $(-\infty, 1] \cup \left(\dfrac{29}{7}, \infty\right)$ **65.** $\varnothing$ **67.** $\left[-\dfrac{1}{2}, \dfrac{3}{2}\right]$

69. $\left(-\dfrac{4}{3}, \dfrac{7}{3}\right)$ **71.** $(6, 12)$ **73.** -12 **75.** -4 **77.** $-7, 7$ **79.** 0 **81.** 2004, 2005 **83.** answers may vary **85.** $(6, \infty)$ **87.** $[3, 7]$

89. $(-\infty, -1)$ **91.** $-20.2° \leq F \leq 95°$ **93.** $67 \leq \text{final score} \leq 94$

Section 2.6
Practice Exercises

1. $-3, 3$ **2.** $-1, 4$ **3.** $-80, 70$ **4.** $-2, 2$ **5.** 0 **6.** $\{\,\}$ or $\varnothing$ **7.** $\{\,\}$ or $\varnothing$ **8.** $-\dfrac{3}{5}, 5$ **9.** 5

Vocabulary, Readiness & Video Check 2.6

1. C **3.** B **5.** D

Exercise Set 2.6

1. $-3, 3$ **3.** $4.2, -4.2$ **5.** $7, -2$ **7.** $8, 4$ **9.** $5, -5$ **11.** $3, -3$ **13.** 0 **15.** $\varnothing$ **17.** $\dfrac{1}{5}$ **19.** $9, -\dfrac{1}{2}$ **21.** $-\dfrac{5}{2}$ **23.** $4, -4$

25. 0 **27.** $\varnothing$ **29.** $0, \dfrac{14}{3}$ **31.** $2, -2$ **33.** $\varnothing$ **35.** $7, -1$ **37.** $\varnothing$ **39.** $\varnothing$ **41.** $-\dfrac{1}{8}$ **43.** $\dfrac{1}{2}, -\dfrac{5}{6}$ **45.** $2, -\dfrac{12}{5}$ **47.** $3, -2$

49. $-8, \dfrac{2}{3}$ **51.** $\varnothing$ **53.** 4 **55.** $13, -8$ **57.** $3, -3$ **59.** $8, -7$ **61.** $2, 3$ **63.** $2, -\dfrac{10}{3}$ **65.** $\dfrac{3}{2}$ **67.** $\varnothing$ **69.** 31% **71.** 32.4°

73. answers may vary **75.** answers may vary **77.** $\varnothing$ **79.** $|x| = 5$ **81.** answers may vary **83.** $|x-1| = 5$ **85.** answers may vary
87. $|x| = 6$ **89.** $|x - 2| = |3x - 4|$

Section 2.7
Practice Exercises

1. 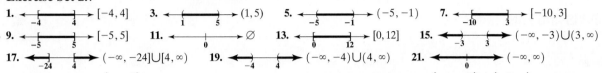 $(-5, 5)$ 2. $(-4, 2)$ 3. $\left[-\frac{2}{3}, 2\right]$ 4. { } or $\varnothing$ 5. {2}

6. $(-\infty, -10] \cup [2, \infty)$ 7. $(-\infty, \infty)$ 8. $(-\infty, 0) \cup (12, \infty)$

Vocabulary, Readiness & Video Check 2.7

1. D   3. C   5. A   7. The solution set involves "or" and "or" means "union."

Exercise Set 2.7

1. $[-4, 4]$ 3. $(1, 5)$ 5. $(-5, -1)$ 7. $[-10, 3]$

9. $[-5, 5]$ 11. $\varnothing$ 13. $[0, 12]$ 15. $(-\infty, -3) \cup (3, \infty)$

17. $(-\infty, -24] \cup [4, \infty)$ 19. $(-\infty, -4) \cup (4, \infty)$ 21. $(-\infty, \infty)$

23. $\left(-\infty, \frac{2}{3}\right) \cup (2, \infty)$ 25. {0}   27. $\left(-\infty, -\frac{3}{8}\right) \cup \left(-\frac{3}{8}, \infty\right)$ 29. $[-2, 2]$

31. $(-\infty, -1) \cup (1, \infty)$ 33. $(-5, 11)$ 35. $(-\infty, 4) \cup (6, \infty)$ 37. $\varnothing$

39. $(-\infty, \infty)$ 41. $[-2, 9]$ 43. $(-\infty, -11] \cup [1, \infty)$ 45. $(-\infty, 0) \cup (0, \infty)$

47. $(-\infty, \infty)$ 49. $\left[-\frac{1}{2}, 1\right]$ 51. $(-\infty, -3) \cup (0, \infty)$ 53. $\varnothing$

55. $\left\{\frac{3}{8}\right\}$ 57. $\left(-\frac{2}{3}, 0\right)$ 59. $(-\infty, -12) \cup (0, \infty)$ 61. $[-1, 8]$

63. $\left[-\frac{23}{8}, \frac{17}{8}\right]$ 65. $(-2, 5)$ 67. $5, -2$ 69. $(-\infty, -7] \cup [17, \infty)$ 71. $-\frac{9}{4}$ 73. $(-2, 1)$ 75. $2, \frac{4}{3}$ 77. $\varnothing$

79. $\frac{19}{2}, -\frac{17}{2}$ 81. $\left(-\infty, -\frac{25}{3}\right) \cup \left(\frac{35}{3}, \infty\right)$ 83. $\frac{1}{6}$ 85. 0   87. $\frac{1}{3}$ 89. -1.5 91. 0   93. $|x| < 7$ 95. $|x| \le 5$ 97. answers may vary   99. $3.45 < x < 3.55$

Chapter 2 Vocabulary Check

1. compound inequality   2. contradiction   3. intersection   4. union   5. identity   6. formula   7. absolute value   8. solution   9. consecutive integers   10. linear inequality in one variable   11. linear equation in one variable

Chapter 2 Review

1. 3   3. $-\frac{45}{14}$ 5. 0   7. 6   9. all real numbers   11. $\varnothing$ 13. -3 15. $\frac{96}{5}$ 17. 32   19. 8   21. $\varnothing$ 23. -7 25. 52

27. width: 40 m; length: 75 m   29. $10, 11, 12, 13$ 31. 258 mi   33. $w = \frac{V}{lh}$ 35. $y = \frac{5x + 12}{4}$ 37. $m = \frac{y - y_1}{x - x_1}$ 39. $r = \frac{E - IR}{I}$

41. $g = \frac{T}{r + vt}$ 43. a. \$3695.27   b. \$3700.81   45. length: 10 in.; width: 8 in.   47. $(3, \infty)$ 49. $(-4, \infty)$ 51. $(-\infty, 7]$ 53. $(-\infty, 1)$

55. more than 35 pounds per week   57. $\left[2, \frac{5}{2}\right]$ 59. $\left(\frac{1}{8}, 2\right)$ 61. $\left(\frac{7}{8}, \frac{27}{20}\right)$ 63. $\left(\frac{11}{3}, \infty\right)$ 65. $\$1750 \le x \le \$3750$ 67. $5, 11$

69. $-1, \frac{11}{3}$ 71. $-\frac{1}{6}$ 73. $\varnothing$ 75. $5, -\frac{1}{3}$ 77. $\left(-\frac{8}{5}, 2\right)$ 79. $(-\infty, -3) \cup (3, \infty)$

81. $\varnothing$ 83. $(-\infty, -27) \cup (-9, \infty)$ 85. 2   87. $B = \frac{2A - hb}{h}$ 89. China: 137 million; United States: 102 million; France: 93 million   91. cylinder holds more ice cream   93. $(2, \infty)$ 95. $(-\infty, \infty)$ 97. $3, -3$ 99. $-10, -\frac{4}{3}$ 101. $\left(-\frac{1}{2}, 2\right)$

Chapter 2 Test

1. 10   2. -32 3. $\varnothing$ 4. all real numbers   5. $-\frac{80}{29}$ 6. $\frac{29}{4}$ 7. $1, \frac{2}{3}$ 8. $\varnothing$ 9. $-4, -\frac{1}{3}$ 10. $\frac{3}{2}$ 11. $y = \frac{3x - 8}{4}$

12. $g = \frac{S}{t^2 + vt}$ 13. $C = \frac{5}{9}(F - 32)$ 14. $(5, \infty)$ 15. $\left(-\infty, -\frac{11}{3}\right]$ 16. $\left(\frac{3}{2}, 5\right]$ 17. $(-\infty, -2) \cup \left(\frac{4}{3}, \infty\right)$ 18. $(3, 7)$ 19. $[5, \infty)$

20. $[4, \infty)$ 21. $\left[1, \frac{11}{2}\right)$ 22. $(-\infty, \infty)$ 23. 9.6   24. 6,720,000 vehicles   25. at most 8 hunting dogs   26. 2,623,000   27. \$3542.27

28. Florida: \$17 billion; California: \$13 billion; New York: \$9 billion

Chapter 2 Cumulative Review

1. a. $\{101, 102, 103, \dots\}$ b. $\{2, 3, 4, 5\}$; Sec. 1.2, Ex. 3   3. a. 3   b. $\frac{1}{7}$ c. -2.7 d. -8 e. 0; Sec. 1.2, Ex. 6   5. a. -14 b. -4

c. 5   d. -10.2 e. $\frac{1}{4}$ f. $-\frac{5}{21}$; Sec. 1.3, Ex. 1   7. a. 3   b. 5   c. $\frac{1}{2}$ d. -6 e. not a real number; Sec. 1.3, Ex. 7   9. a. 33

b. -18 **c.** $\dfrac{1}{12}$; Sec. 1.3, Ex. 12   **11. a.** $x + 5 = 20$ **b.** $2(3 + y) = 4$ **c.** $x - 8 = 2x$ **d.** $\dfrac{z}{9} = 9 + z$; Sec. 1.4, Ex. 1

13. $5 + 7x$; Sec. 1.4, Ex. 9   **15.** 2; Sec. 2.1, Ex. 1   **17.** all real numbers; Sec. 2.1, Ex. 9   **19. a.** $3x + 3$ **b.** $12x - 3$; Sec. 2.2, Ex. 1   **21.** 23, 49, 92;

Sec. 2.2, Ex. 3   **23.** $y = \dfrac{2x + 7}{3}$ or $y = \dfrac{2x}{3} + \dfrac{7}{3}$; Sec. 2.3, Ex. 2   **25.** $b = \dfrac{2A - Bh}{h}$; Sec. 2.3, Ex. 3   **27. a.** ; $[2, \infty)$

b. ; $(-\infty, -1)$ **c.** ; $(0.5, 3]$; Sec. 2.4, Ex. 1   **29.** $\left[\dfrac{5}{2}, \infty\right)$; Sec. 2.4, Ex. 5   **31.** $(-\infty, \infty)$; Sec. 2.4, Ex. 7   **33.** $\{4, 6\}$;

Sec. 2.5, Ex. 1   **35.** $(-\infty, 4)$; Sec. 2.5, Ex. 2   **37.** $\{2, 3, 4, 5, 6, 8\}$; Sec. 2.5, Ex. 6   **39.** $(-\infty, \infty)$; Sec. 2.5, Ex. 8   **41.** $2, -2$; Sec. 2.6, Ex. 1

43. $24, -20$; Sec. 2.6, Ex. 3   **45.** 4; Sec. 2.6, Ex. 9   **47.** $[-3, 3]$; Sec. 2.7, Ex. 1   **49.** $(-\infty, \infty)$; Sec. 2.7, Ex. 7

CHAPTER 3 GRAPHS AND FUNCTIONS

Section 3.1
Practice Exercises

1.    **a.** quadrant IV   **b.** y-axis   **c.** quadrant II   **d.** x-axis   **e.** quadrant III   **f.** quadrant I
2. yes, no, yes   **3. a.** \$4200   **b.** more than \$9000

4. $y = -3x - 2$ **5.**   **6.** $y = 2x^2$ **7.** $y = -|x|$

Graphing Calculator Explorations 3.1

1.   **3.**   **5.**   **7.**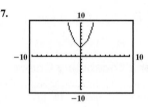

Vocabulary, Readiness & Video Check 3.1

1. origin   **3.** 1   **5.** V-shaped   **7.** origin; left (if negative) or right (if positive); up (if positive) or down (if negative)   **9.** The graph of an equation is a picture of its ordered pair solutions; a third point is found as a check to make sure the points line up and a mistake hasn't been made.

Exercise Set 3.1

1. quadrant I   **3.** quadrant II   **5.** quadrant IV   **7.** y-axis   **9.** quadrant III

11. $(-5, -2)$ **13.** $(-1, 0)$ **15.** $(3, 0)$ **17.** no; yes   **19.** yes; yes   **21.** yes; yes   **23.** yes; no   **25.** yes; yes

27. linear   **29.** linear   **31.** linear   **33.** not linear   **35.** linear   **37.** not linear

39. not linear   **41.** linear   **43.** linear   **45.** not linear   **47.** not linear   **49.** not linear

51. linear   **53.** linear

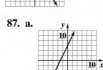

55. −5   **57.** $-\dfrac{1}{10}$ **59.** $(-\infty, -5]$ **61.** $(-\infty, -4)$ **63.** quadrant I   **65.** *y*-axis

67. quadrant III   **69.** quadrant IV   **71.** *x*-axis   **73.** quadrant III   **75.** b   **77.** b   **79.** c

81. 1997   **83.** answers may vary

85.

87. a.

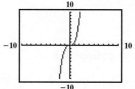

b. 14 in.   **89.** $7000   **91.** $500   **93.** Depreciation is the same from year to year.   **95.** ; answers may vary

97. $y = 3x + 5$ **99.** $y = x^2 + 2$ **101.**   **103.**

Section 3.2
Practice Exercises

1. a. Domain: $\{4, 5\}$; Range: $\{1, -3, -2, 6\}$ **b.** Domain: $\{3\}$; Range: $\{-4, -3, -2, -1, 0, 1, 2, 3, 4\}$ **c.** Domain: {Administrative Secretary, Game Developer, Engineer, Restaurant Manager, Marketing}; Range: $\{27, 50, 73, 35\}$ **2. a.** function   **b.** not a function   **c.** function
3. yes   **4.** yes   **5. a.** function   **b.** function   **c.** not a function   **d.** function   **e.** not a function   **6. a.** Domain: $[-1, 2]$;
Range: $[-2, 9]$; function   **b.** Domain: $[-1, 1]$; Range: $[-4, 4]$; not a function   **c.** Domain: $(-\infty, \infty)$; Range: $(-\infty, 4]$; function   **d.** Domain:
$(-\infty, \infty)$; Range: $(-\infty, \infty)$; function   **7. a.** 1   **b.** 6   **c.** −2   **d.** 15   **8. a.** −3   **b.** −2   **c.** 3   **d.** 1   **e.** −1 and 3   **f.** −3
9. 5.6 million or 5,600,000 students   **10.** 5.8 million or 5,800,000 students

Graphing Calculator Explorations 3.2

1.   **3.**   **5.**

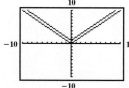

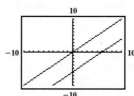

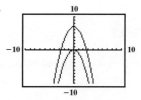

Vocabulary, Readiness & Video Check 3.2

1. relation   **3.** domain   **5.** vertical   **7.** An equation is one way of writing down a correspondence that produces ordered pair solutions; a relation between two sets of coordinates, *x*'s and *y*'s.   **9.** If a vertical line intersects a graph two (or more) times, then there's an *x*-value corresponding to two (or more) different *y*-values and is thus not a function.   **11.** Using function notation, the replacement value for *x* and the resulting $f(x)$ or *y*-value corresponds to an ordered pair (x, y) solution to the function.

Exercise Set 3.2

1. domain: $\{-1, 0, -2, 5\}$; range: $\{7, 6, 2\}$; function   **3.** domain: $\{-2, 6, -7\}$; range: $\{4, -3, -8\}$; not a function   **5.** domain: $\{1\}$;
range: $\{1, 2, 3, 4\}$; not a function   **7.** domain: $\left\{\dfrac{3}{2}, 0\right\}$; range: $\left\{\dfrac{1}{2}, -7, \dfrac{4}{5}\right\}$; not a function   **9.** domain: $\{-3, 0, 3\}$; range: $\{-3, 0, 3\}$; function
11. domain: $\{-1, 1, 2, 3\}$; range: $\{2, 1\}$; function   **13.** domain: {1994, 1998, 2002, 2006, 2010}; range: {6, 9, 10}; function   **15.** domain:
$\{32°, 104°, 212°, 50°\}$; range: $\{0°, 40°, 10°, 100°\}$; function   **17.** domain: $\{0\}$; range: $\{2, -1, 5, 100\}$; not a function   **19.** function
21. not a function   **23.** function   **25.** not a function   **27.** function   **29.** domain: $[0, \infty)$; range: $(-\infty, \infty)$; not a function   **31.** domain: $[-1, 1]$;
range: $(-\infty, \infty)$; not a function   **33.** domain: $(-\infty, \infty)$; range: $(-\infty, -3] \cup [3, \infty)$; not a function   **35.** domain: $[2, 7]$; range $[1, 6]$; not a function
37. domain: $\{-2\}$; range: $(-\infty, \infty)$; not a function   **39.** domain: $(-\infty, \infty)$; range: $(-\infty, 3]$; function   **41.** yes   **43.** no   **45.** yes   **47.** yes
49. yes   **51.** no   **53.** 15   **55.** 38   **57.** 7   **59.** 3   **61. a.** 0   **b.** 1   **c.** −1   **63. a.** 246   **b.** 6   **c.** $\dfrac{9}{2}$ **65. a.** −5   **b.** −5
c. −5   **67. a.** 5.1   **b.** 15.5   **c.** 9.533   **69.** $(1, -10)$ **71.** $(4, 56)$ **73.** $f(-1) = -2$ **75.** $g(2) = 0$ **77.** −4, 0   **79.** 3
81. 28.4; In 2009, about 28.4% of students took at least one online course.   **83.** 54.22; In 2016, we predict that 54.22% of students will take at least one online course.   **85.** answers may vary   **87.** $15.54 billion   **89.** $f(x) = x + 7$ **91.** 25π sq cm   **93.** 2744 cu in.   **95.** 166.38 cm
97. 163.2 mg   **99. a.** 64.14; per capita consumption of beef was 64.14 lb in 2004   **b.** 56.74 lb   **101.** 5, −5, 6
103. $2, \dfrac{8}{7}, \dfrac{12}{7}$ **105.** 0, 0, −6

107. yes; 170 m   **109.** true   **111.** true   **113. a.** 16   **b.** $a^2 + 7$ **115. a.** 0   **b.** $3a - 12$ **c.** $-3x - 12$ **d.** $3x + 3h - 12$
117. infinitely many   **119.** answers may vary   **121.** answers may vary

Section 3.3
Practice Exercises

1. 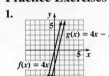   **2.**   **3. a.** $\left(0, -\dfrac{2}{5}\right)$ **b.** $(0, 4.1)$ **4.** 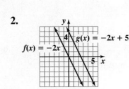   **5.** $y = -3x$

6. $x = -4$ **7.**

Graphing Calculator Explorations 3.3

1. $y = \dfrac{x}{3.5}$ 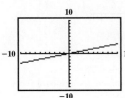   **3.** $y = -\dfrac{5.78}{2.31}x + \dfrac{10.98}{2.31}$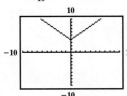

5. $y = |x| + 3.78$ 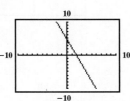   **7.** $y = 5.6x^2 + 7.7x + 1.5$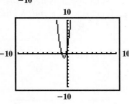

Vocabulary, Readiness & Video Check 3.3

1. linear   **3.** vertical; $(c, 0)$ **5.** $y; f(x); x$ **7.** $f(x) = mx + b$, or slope–intercept form   **9.** For a horizontal line, the coefficient of x is 0 and the coefficient of y is 1; for a vertical line, the coefficient of y is 0 and the coefficient of x is 1.

Exercise Set 3.3

1. 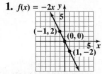   **3.**   **5.** 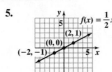   **7.** 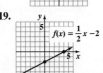   **9.** C   **11.** D   **13.**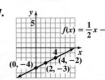

15.   **17.** 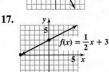   **19.**   **21.**   **23.**   **25.**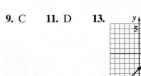

27. C   **29.** A   **31.**   **33.**   **35.**   **37.** 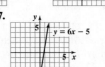   **39.**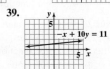

41. 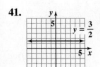   **43.**   **45.**   **47.** 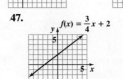   **49.** 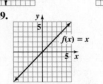   **51.**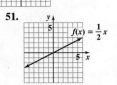

53. 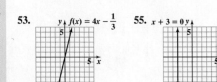   **55.** 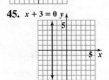   **57.** $9, -3$ **59.** $(-\infty, -4) \cup (-1, \infty)$ **61.** $\left[\dfrac{2}{3}, 2\right]$ **63.** $\dfrac{3}{2}$ **65.** 6   **67.** $-\dfrac{6}{5}$

69. no; answers may vary, 500 chairs can be produced   **71.** yes; answers may vary   **73. a.** $(0, 500)$; if no tables are produced, **b.** $(750, 0)$; if no chairs are produced, 750 tables can be produced   **c.** 466 chairs

75. a. $64   **b.**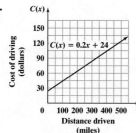

c. The line moves upward from left to right.   **77. a.** $3107   **b.** 2018
c. answers may vary   **79.** answers may vary   **81.** The vertical line $x = 0$ has y-intercepts.   **83.** a line parallel to $y = -4x$ but with y-intercept $(0, 2)$ **85.** d   **87.** c
89. 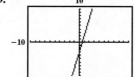   **91.**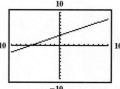

Section 3.4
Practice Exercises

1. $m = -\dfrac{1}{2}$;   **2.** $m = \dfrac{3}{5}$;   **3.** $m = -4$ **4.** $m = \dfrac{2}{3}$; y-intercept: $(0, -3)$ **5. a.** 5,443,000
b. The number of people age 85 or older in the United States increases at a rate of 110,520 per year   **c.** At year $x = 0$, or 1980, there were 2,127,400 people age 85 or older in the United States.   **6.** undefined   **7.** $m = 0$
8. a. perpendicular   **b.** parallel

Graphing Calculator Explorations 3.4

1. 18.4   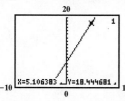   **3.** -1.5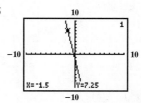

5. $14.0; 4.2, -9.4$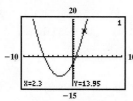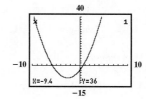

Vocabulary, Readiness & Video Check 3.4

1. slope   **3.** $m; (0, b)$ **5.** horizontal   **7.** -1 **9.** increases; decreases   **11.** Slope–intercept form allows us to see the slope and y-intercept so we can discuss them in the context of the application. Slope is a rate of change and tells us the rate of increase of physician assistants (2.9 thousand per year). The y-intercept tells us the number of physician assistants at "zero" year (75,000 in 2008).   **13.** The slopes of vertical lines are undefined and we can't mathematically work with undefined slopes.

Exercise Set 3.4

1. $\dfrac{9}{5}$ **3.** $-\dfrac{7}{2}$ **5.** $-\dfrac{5}{6}$ **7.** $\dfrac{1}{3}$ **9.** $-\dfrac{4}{3}$ **11.** 0   **13.** undefined   **15.** 2   **17.** -1 **19.** upward   **21.** horizontal   **23.** l_2

25. l_2 **27.** l_2 **29.** $m = 5; (0, -2)$ **31.** $m = -2; (0, 7)$ **33.** $m = \dfrac{2}{3}; \left(0, -\dfrac{10}{3}\right)$ **35.** $m = \dfrac{1}{2}; (0, 0)$ **37.** A   **39.** B

41. undefined   **43.** 0   **45.** undefined   **47.** $m = -1; (0, 5)$ **49.** $m = \dfrac{6}{5}; (0, 6)$ **51.** $m = 3; (0, 9)$ **53.** $m = 0; (0, 4)$

55. $m = 7; (0, 0)$ **57.** $m = 0; (0, -6)$ **59.** slope is undefined, no y-intercept   **61.** parallel   **63.** neither   **65.** perpendicular

67. parallel   **69.** perpendicular   **71.** $\dfrac{3}{2}$ **73.** $-\dfrac{1}{2}$ **75.** $\dfrac{2}{3}$ **77.** approximately -0.12 **79. a.** 76.4 yr   **b.** $m = 0.16$; The life expectancy of a female born in the United States is approximately 0.16 year more than a female born one year before.   **c.** $(0, 71.6)$: At year $= 0$, or 1950, the life expectancy of an American female was 71.6 years.   **81. a.** $m = 2.9; (0, 75)$ **b.** number of people employed as physician assistants increases 2.9 thousand for every 1 year   **c.** 75 thousand physician assistants employed in 2008   **83. a.** 1.42 billion   **b.** in 2017

c. answers may vary   **85.** $y = 5x + 32$ **87.** $y = 2x - 1$ **89.** $m = \dfrac{-14 - 6}{7 - (-2)} = \dfrac{-20}{9}$ or $-\dfrac{20}{9}$ **91.** $m = \dfrac{-10 - (-5)}{-8 - (-11)} = \dfrac{-5}{3}$ or $-\dfrac{5}{3}$

93. $-\dfrac{7}{2}$ **95.** $\dfrac{2}{7}$ **97.** $\dfrac{5}{2}$ **99.** $-\dfrac{2}{5}$ **101.** $l_1: -2, l_2: -1, l_3: -\dfrac{2}{3}$ **103.** $(6, 20)$ **105.** $-\dfrac{7}{4}$ or -1.75 yd per sec   **107.** answers may vary

109. answers may vary   **111.** $\dfrac{1}{48}$ **113.**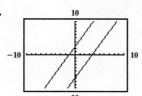

Section 3.5
Practice Exercises

1.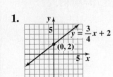

2.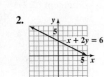

3. $y = -\frac{3}{4}x + 4$ **4.** $y = -4x - 3$ **5.** $f(x) = -\frac{2}{3}x + \frac{4}{3}$ **6.** $2x + 3y = 5$
7. 12,568 house sales **8.** $y = -2$ **9.** $x = 6$ **10.** $3x + 4y = 12$ **11.** $f(x) = \frac{4}{3}x - \frac{41}{3}$

Vocabulary, Readiness & Video Check 3.5

1. $m = -4$, y-intercept: $(0, 12)$ **3.** $m = 5$, y-intercept: $(0, 0)$ **5.** $m = \frac{1}{2}$, y-intercept: $(0, 6)$ **7.** parallel **9.** neither **11.** slope-intercept;

y-intercept; slope **13.** if one of the two points given is the y-intercept **15.** $f(x) = \frac{1}{3}x - \frac{17}{3}$

Exercise Set 3.5

1. 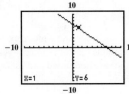 **3.** 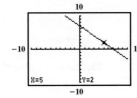 **5.** 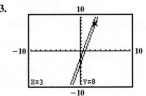 **7.** $y = -x + 1$ **9.** $y = 2x + \frac{3}{4}$ **11.** $y = \frac{2}{7}x$ **13.** $y = 3x - 1$

15. $y = -2x - 1$ **17.** $y = \frac{1}{2}x + 5$ **19.** $y = -\frac{9}{10}x - \frac{27}{10}$ **21.** $f(x) = 3x - 6$ **23.** $f(x) = -2x + 1$ **25.** $f(x) = -\frac{1}{2}x - 5$

27. $f(x) = \frac{1}{3}x - 7$ **29.** $f(x) = -\frac{3}{8}x + \frac{5}{8}$ **31.** $2x + y = 3$ **33.** $2x - 3y = -7$ **35.** -2 **37.** 2 **39.** -2 **41.** $y = -4$

43. $x = 4$ **45.** $y = 5$ **47.** $f(x) = 4x - 4$ **49.** $f(x) = -3x + 1$ **51.** $f(x) = -\frac{3}{2}x - 6$ **53.** $2x - y = -7$ **55.** $f(x) = -x + 7$

57. $x + 2y = 22$ **59.** $2x + 7y = -42$ **61.** $4x + 3y = -20$ **63.** $x = -2$ **65.** $x + 2y = 2$ **67.** $y = 12$ **69.** $8x - y = 47$

71. $x = 5$ **73.** $f(x) = -\frac{3}{8}x - \frac{29}{4}$ **75. a.** $y = 32x$ **b.** 128 ft per sec **77. a.** $y = -250x + 3500$ **b.** 1625 Frisbees

79. a. $y = 58.1x + 2619$ **b.** 2851.4 thousand **81. a.** $y = -4820x + 297,000$ **b.** \$258,440 **83.** $(-\infty, 14]$ **85.** $\left[\frac{7}{2}, \infty\right)$

87. $\left(-\infty, -\frac{1}{4}\right)$ **89.** true **91.** $-4x + y = 4$ **93.** $2x + y = -23$ **95.** $3x - 2y = -13$ **97.** answers may vary

99.   **101.**   **103.**

Integrated Review

1.   **2.**   **3.** 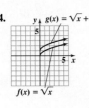 **4.** 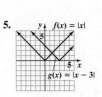 **5.** 0 **6.** $-\frac{3}{5}$ **7.** $m = 3$; $(0, -5)$ **8.** $m = \frac{5}{2}$; $\left(0, -\frac{7}{2}\right)$
9. parallel **10.** perpendicular **11.** $y = -x + 7$
12. $x = -2$ **13.** $y = 0$ **14.** $f(x) = -\frac{1}{2}x - 8$

15. $f(x) = -5x - 6$ **16.** $f(x) = -4x + \frac{1}{3}$ **17.** $f(x) = \frac{1}{2}x - 1$ **18.** $y = 3x - \frac{3}{2}$ **19.** $y = 3x - 2$ **20.** $y = -\frac{5}{4}x + 4$

21. $y = \frac{1}{4}x - \frac{7}{2}$ **22.** $y = -\frac{5}{2}x - \frac{5}{2}$ **23.** $x = -1$ **24.** $y = 3$

Section 3.6
Practice Exercises
1. $f(4) = 5$; $f(-2) = 6$; $f(0) = -2$

2.   **3.**   **4.** 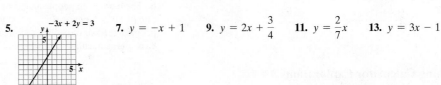 **5.** 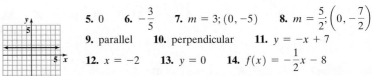 **6.** 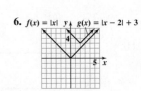 **7.**

Vocabulary, Readiness & Video Check 3.6

1. C   **3.** D   **5.** Although $f(x) = x + 3$ isn't defined for $x = -1$, we need to clearly indicate the point where this piece of the graph ends. Therefore, we find this point and graph it as an open circle.   **7.** *x*-axis

Exercise Set 3.6

1.   **3.**   **5.**   **7.** 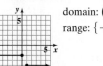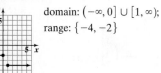   **9.** 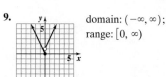 domain: $(-\infty, \infty)$; range: $[0, \infty)$

11. domain: $(-\infty, \infty)$; range: $(-\infty, 5)$ **13.** domain: $(-\infty, \infty)$; range: $(-\infty, 6]$ **15.** domain: $(-\infty, 0] \cup [1, \infty)$; range: $\{-4, -2\}$

17.   **19.**   **21.**   **23.**   **25.**   **27.**

29. 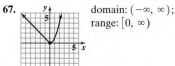   **31.**   **33.** 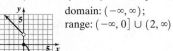   **35.**   **37.**   **39.**

41.   **43.** 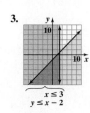   **45.**   **47.**   **49.** A   **51.** D   **53.** answers may vary
55.   **57.** domain: $[2, \infty)$; range: $[3, \infty)$

59. domain: $(-\infty, \infty)$; range: $(-\infty, 3]$ **61.** $[20, \infty)$ **63.** $(-\infty, \infty)$ **65.** $[-103, \infty)$
67. domain: $(-\infty, \infty)$; range: $[0, \infty)$ **69.** domain: $(-\infty, \infty)$; range: $(-\infty, 0] \cup (2, \infty)$

Section 3.7
Practice Exercises

1.   **2.**   **3.** 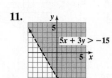   **4.**

Vocabulary, Readiness & Video Check 3.7

1. We find the boundary line equation by replacing the inequality symbol with $=$. The points on this line are solutions (line is solid) if the inequality is $\geq$ or $\leq$; the points on this line are not solutions (line is dashed) if the inequality is $>$ or $<$.

Exercise Set 3.7

1.   **3.**   **5.**   **7.**   **9.**   **11.**

13. **15.** **17.** **19.** **21.** **23.**

25. **27.** **29.** **31.** **33.** **35.**

37. **39.** **41.** **43.**

45. D   **47.** A   **49.** $x \geq 2$ **51.** $y \leq -3$
53. $y > 4$ **55.** $x < 1$ **57.** 8   **59.** -25
61. 16   **63.** $\dfrac{27}{125}$ **65.** domain: $[1, 5]$; range: $[1, 3]$; no
67. with $<$ or $>$ **69.**

Chapter 3 Vocabulary Check

1. relation   **2.** line   **3.** linear inequality   **4.** Standard   **5.** range   **6.** Parallel   **7.** Slope–intercept
8. function   **9.** slope   **10.** perpendicular   **11.** y **12.** domain   **13.** linear function   **14.** x **15.** point–slope

Chapter 3 Review

1. 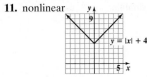 A: quadrant IV;   **3.** no, yes   **5.** yes, yes   **7.** linear   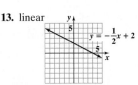   **9.** linear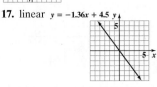
B: quadrant II;
C: y-axis;
D: quadrant III

11. nonlinear   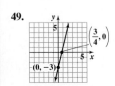   **13.** linear   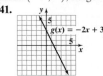   **15.** linear   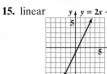   **17.** linear $y = -1.36x + 4.5$

19. domain: $\left\{ -\dfrac{1}{2}, 6, 0, 25 \right\}$; range: $\left\{ \dfrac{3}{4} \text{ (or 0.75)}, -12, 25 \right\}$; function   **21.** domain: $\{2, 4, 6, 8\}$; range: $\{2, 4, 5, 6\}$; not a function
23. domain: $(-\infty, \infty)$; range: $(-\infty, -1] \cup [1, \infty)$; not a function   **25.** domain: $(-\infty, \infty)$; range: $\{4\}$; function   **27.** -3 **29.** 18
31. -3 **33.** 381 lb   **35.** 0   **37.** $-2, 4$ **39.**   **41.**   **43.** C   **45.** B   **47.**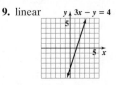

49.   **51.**   **53.** 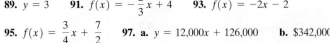   **55.** -3 **57.** $\dfrac{5}{2}$ **59.** $m = -3; \left(0, \dfrac{1}{2} \right)$

61. $m = \dfrac{2}{5}, \left(0, -\dfrac{4}{3} \right)$ **63.** 0   **65.** l_2 **67.** l_2 **69. a.** \$87
b. $m = 0.3$; cost increases by \$0.30 for each additional mile driven
c. (0.42); cost for 0 miles driven is \$42   **71.** neither   **73.** parallel

75. 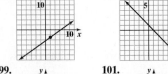   **77.**   **79.**   **81.** $y = -1$ **83.** $x = -4$ **85.** $3x - y = -14$ **87.** $x + 2y = -8$
89. $y = 3$ **91.** $f(x) = -\dfrac{2}{3}x + 4$ **93.** $f(x) = -2x - 2$
95. $f(x) = \dfrac{3}{4}x + \dfrac{7}{2}$ **97. a.** $y = 12{,}000x + 126{,}000$ **b.** \$342,000

99.   **101.**   **103.**   **105.** 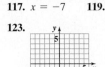   **107.** 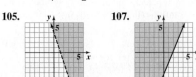   **109.**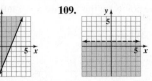

111. 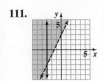   **113.**   **115.**   **117.** $x = -7$ **119.** $y = \dfrac{3}{4}x + 2$ **121.** $y = -\dfrac{3}{2}x - 8$

123. 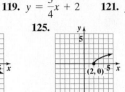   **125.**

Chapter 3 Test

1. A: quadrant IV; B: x-axis; C: quadrant II

2. 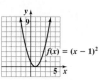 $2x - 3y = -6$

3. 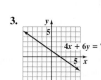 $4x + 6y = 7$

4. $f(x) = \frac{2}{3}x$

5. $y = -3$

6. $-\frac{3}{2}$

7. $m = -\frac{1}{4}, \left(0, \frac{2}{3}\right)$

8. $f(x) = (x-1)^2$

9. $g(x) = |x| + 2$

10. $y = -8$

11. $x = -4$

12. $y = -2$ **13.** $3x + y = 11$ **14.** $5x - y = 2$ **15.** $f(x) = -\frac{1}{2}x$ **16.** $f(x) = -\frac{1}{3}x + \frac{5}{3}$ **17.** $f(x) = -\frac{1}{2}x - \frac{1}{2}$ **18.** neither

19. $x \le -4$

20. $2x - y > 5$

21.

22. domain: $(-\infty, \infty)$; range: $\{5\}$; function
23. domain: $\{-2\}$; range: $(-\infty, \infty)$; not a function
24. domain: $(-\infty, \infty)$; range: $[0, \infty)$; function
25. domain: $(-\infty, \infty)$; range: $(-\infty, \infty)$; function
26. a. 81 games   **b.** 79 games   **c.** \$231 million   **d.** 0.096; Every million dollars spent on payroll increases winnings by 0.096 game.

27. domain: $(-\infty, \infty)$; range: $(-3, \infty)$

28. $(4, 0)$

29. $(-2, -1)$

domain: $(-\infty, \infty)$; range: $(-\infty, -1]$

30. $(0, -1)$

Chapter 3 Cumulative Review

1. 41; Sec. 1.2, Ex. 2   **3. a.** true   **b.** false   **c.** false   **d.** false; Sec. 1.2, Ex. 5   **5. a.** -6 **b.** -7 **c.** -16 **d.** 20.5   **e.** $-\frac{7}{6}$

f. 0.94   **g.** -3; Sec. 1.3, Ex. 2   **7. a.** 9   **b.** $\frac{1}{16}$ **c.** -25 **d.** 25   **e.** -125 **f.** -125; Sec. 1.3, Ex. 6   **9. a.** $>$ **b.** $=$

c. $<$ **d.** $<$ **e.** $>$ **f.** $<$; Sec. 1.4, Ex. 5   **11. a.** $\frac{1}{11}$ **b.** $-\frac{1}{9}$ **c.** $\frac{4}{7}$; Sec. 1.4, Ex. 8   **13.** 0.4; Sec. 2.1, Ex. 2   **15.** $\{\ \}$ or $\varnothing$;

Sec. 2.1, Ex. 8   **17. a.** $3x + 3$ **b.** $12x - 3$; Sec. 2.2, Ex. 1   **19.** 86, 88, and 90; Sec. 2.2, Ex. 6   **21.** $\frac{V}{lw} = h$; Sec. 2.3, Ex. 1   **23.** $\{x \mid x < 7\}$

or $(-\infty, 7)$; 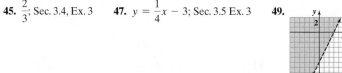 ; Sec. 2.4, Ex. 2   **25.** $\left(-\infty, -\frac{7}{3}\right]$; Sec. 2.4, Ex. 6   **27.** $\varnothing$; Sec. 2.5, Ex. 3   **29.** $\left(-\infty, \frac{13}{5}\right) \cup [4, \infty)$; Sec. 2.5, Ex. 7

31. $-2, \frac{4}{5}$; Sec. 2.6, Ex. 2   **33.** $\frac{3}{4}, 5$; Sec. 2.6, Ex. 8   **35.** $\left[-2, \frac{8}{5}\right]$; Sec. 2.7, Ex. 3   **37.** $(-\infty, -4) \cup (10, \infty)$; Sec. 2.7, Ex. 6

39. solutions: $(0, -12), (2, -6)$; not a solution $(1, 9)$; Sec. 3.1, Ex. 2   **41.** yes; Sec. 3.2, Ex. 3   **43. a.** $\left(0, \frac{3}{7}\right)$ **b.** $(0, -3.2)$; Sec. 3.3, Ex. 3

45. $\frac{2}{3}$; Sec. 3.4, Ex. 3   **47.** $y = \frac{1}{4}x - 3$; Sec. 3.5 Ex. 3   **49.** Sec. 3.7; Ex. 1

CHAPTER 4 SYSTEMS OF EQUATIONS

Section 4.1
Practice Exercises

1. a. yes   **b.** no
2. solution: $(1, 5)$ 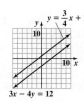 $y = 5x$, $(1, 5)$, $2x + y = 7$

3. no solution ... $y = \frac{3}{4}x + 1$, $3x - 4y = 12$

4. $\{(x, y) \mid 3x - 2y = 4\}$ or $\{(x, y) \mid -9x + 6y = -12\}$ $3x - 2y = 4$, $-9x + 6y = -12$

5. $\left(-\frac{1}{2}, 5\right)$ **6.** $\left(-\frac{9}{5}, -\frac{2}{5}\right)$ **7.** $(2, 1)$ **8.** $(-2, 0)$ **9.** $\varnothing$ or $\{\ \}$ **10.** $\{(x, y) \mid -3x + 2y = -1\}$ or $\{(x, y) \mid 9x - 6y = 3\}$

Graphing Calculator Explorations 4.1

1. $(2.11, 0.17)$ **3.** $(0.57, -1.97)$

Vocabulary, Readiness & Video Check 4.1

1. B **3.** A **5.** The ordered pair must be a solution of *both* equations of the system in order to be a solution of the system. **7.** Solve one equation for a variable. Next be sure to substitute this expression for the variable into the *other* equation.

Exercise Set 4.1

1. yes **3.** no **5.** yes **7.** no **9.** $(2, -1)$ **11.** $(1, 2)$ **13.** $\emptyset$

15. $(2, 8)$ **17.** $(0, -9)$ **19.** $(1, -1)$ **21.** $(-5, 3)$ **23.** $\left(\dfrac{5}{2}, \dfrac{5}{4}\right)$ **25.** $(1, -2)$ **27.** $(8, 2)$ **29.** $(7, 2)$ **31.** $\emptyset$

33. $\{(x, y) \mid 3x + y = 1\}$ **35.** $\left(\dfrac{3}{2}, 1\right)$ **37.** $(2, -1)$ **39.** $(-5, 3)$ **41.** $\{(x, y) \mid 3x + 9y = 12\}$ **43.** $\emptyset$ **45.** $\left(\dfrac{1}{2}, \dfrac{1}{5}\right)$ **47.** $(9, 9)$

49. $\{(x, y) \mid x = 3y + 2\}$ **51.** $\left(-\dfrac{1}{4}, \dfrac{1}{2}\right)$ **53.** $(3, 2)$ **55.** $(7, -3)$ **57.** $\emptyset$ **59.** $(3, 4)$ **61.** $(-2, 1)$ **63.** $(1.2, -3.6)$ **65.** true

67. false **69.** $6y - 4z = 25$ **71.** $x + 10y = 2$ **73.** no solution **75.** infinite number of solutions **77.** no; answers may vary
79. 5000 DVDs; \$21 **81.** supply greater than demand **83.** $(1875, 4687.5)$ **85.** makes money **87.** for x-values greater than 1875

89. answers may vary; One possibility: $\begin{cases} -2x + y = 1 \\ x - 2y = -8 \end{cases}$ **91. a.** Consumption of bottled water is growing faster than that of carbonated diet soft

drinks. **b.** $(3, 14)$ **c.** In the year 1998, per capita bottled water consumption was about the same as per capita carbonated diet soft drink consumption.

93. $\left(\dfrac{1}{4}, 8\right)$ **95.** $\left(\dfrac{1}{3}, \dfrac{1}{2}\right)$ **97.** $\emptyset$ **99.** $\left(-\dfrac{1}{4}, \dfrac{1}{3}\right)$

Section 4.2
Practice Exercises

1. $(-1, 2, 1)$ **2.** $\{\ \}$ or $\emptyset$ **3.** $\left(\dfrac{2}{3}, -\dfrac{1}{2}, 0\right)$ **4.** $\{(x, y, z) \mid 2x + y - 3z = 6\}$ **5.** $(6, 15, -5)$

Vocabulary, Readiness & Video Check 4.2

1. a, b, d **3.** yes; answers may vary **5.** Once we have one equation in two variables, we need to get another equation in the *same* two variables, giving us a system of two equations in two variables. We solve this new system to find the value of two variables. We then substitute these values into an original equation to find the value of the third.

Exercise Set 4.2

1. $(-1, 5, 2)$ **3.** $(-2, 5, 1)$ **5.** $(-2, 3, -1)$ **7.** $\{(x, y, z) \mid x - 2y + z = -5\}$ **9.** $\emptyset$ **11.** $(0, 0, 0)$ **13.** $(-3, -35, -7)$
15. $(6, 22, -20)$ **17.** $\emptyset$ **19.** $(3, 2, 2)$ **21.** $\{(x, y, z) \mid x + 2y - 3z = 4\}$ **23.** $(-3, -4, -5)$ **25.** $\left(0, \dfrac{1}{2}, -4\right)$ **27.** $(12, 6, 4)$
29. 15 and 30 **31.** 5 **33.** $-\dfrac{5}{3}$ **35.** answers may vary **37.** answers may vary **39.** $(1, 1, -1)$ **41.** $(1, 1, 0, 2)$ **43.** $(1, -1, 2, 3)$
45. answers may vary

Section 4.3
Practice Exercises

1. a. 2208 **b.** yes; answers may vary **2.** 17 and 12 **3.** Atlantique: 500 kph; V150: 575 kph **4.** 0.95 liter of water; 0.05 liter of 99% HCL
5. 1500 packages **6.** $40°, 60°, 80°$

Vocabulary, Readiness & Video Check 4.3

1. Up to now we've been choosing one variable/unknown and translating to one equation. To solve by a system of equations, we'll choose two variables to represent two unknowns and translate to two equations. **3.** The ordered triple still needs to be interpreted in the context of the application. Each value actually represents the angle measure of a triangle, in degrees.

Exercise Set 4.3

1. 10 and 8 **3. a.** Enterprise class: 1101 ft; Nimitz class: 1092 ft **b.** 3.67 football fields **5.** plane: 520 mph; wind: 40 mph **7.** 20 qt of 4%;
40 qt of 1% **9.** United Kingdom: 31,342 students; Italy: 27,362 students **11.** 9 large frames; 13 small frames **13.** -10 and -8 **15. a.** 2010
b. answers may vary **17.** tablets: \$0.80; pens: \$0.20 **19.** speed of plane: 630 mph; speed of wind: 90 mph **21. a.** answers may vary
b. 2003 **23.** 28 cm; 28 cm; 37 cm **25.** 600 mi **27.** $x = 75; y = 105$ **29.** 625 units **31.** 3000 units **33.** 1280 units
35. a. $R(x) = 450x$ **b.** $C(x) = 200x + 6000$ **c.** 24 desks **37.** 2 units of Mix A; 3 units of Mix B; 1 unit of Mix C **39.** 5 in.; 7 in.; 7 in.; 10 in.
41. 18, 13, and 9 **43.** free throws: 594; 2-pt field goals: 566; 3-pt fields goals: 145 **45.** $x = 60; y = 55; z = 65$ **47.** $5x + 5z = 10$
49. $-5y + 2z = 2$ **51.** 2007: 780.000; 2010: 1,530,000 **53.** $a = 3, b = 4, c = -1$ **55.** $a = 0.5, b = 24.5, c = 849$; 2015: 1774 thousand
students **57.** $(7, 215)$

Integrated Review

1. C **2.** D **3.** A **4.** B **5.** $(1,3)$ **6.** $\left(\dfrac{4}{3},\dfrac{16}{3}\right)$ **7.** $(2,-1)$ **8.** $(5,2)$ **9.** $\left(\dfrac{3}{2},1\right)$ **10.** $\left(-2,\dfrac{3}{4}\right)$ **11.** $\varnothing$

12. $\{(x,y)\mid 2x-5y=3\}$ **13.** $\left(1,\dfrac{1}{3}\right)$ **14.** $\left(3,\dfrac{3}{4}\right)$ **15.** $(-1,3,2)$ **16.** $(1,-3,0)$ **17.** $\varnothing$ **18.** $\{(x,y,z)\mid x-y+3z=2\}$

19. $\left(2,5,\dfrac{1}{2}\right)$ **20.** $\left(1,1,\dfrac{1}{3}\right)$ **21.** 19 and 27 **22.** 70°; 70°; 100°; 120°

Section 4.4
Practice Exercises
1. $(2,-1)$ **2.** $\varnothing$ **3.** $(-1,1,2)$

Vocabulary, Readiness & Video Check 4.4
1. matrix **3.** row **5.** false **7.** true **9.** Two rows may be interchanged, the elements of any row may be multiplied/divided by the same nonzero number, the elements of any row may be multiplied/divided by the same nonzero number and added to their corresponding elements in any other row; rows were not interchanged in Example 1.

Exercise Set 4.4
1. $(2,-1)$ **3.** $(-4,2)$ **5.** $\varnothing$ **7.** $\{(x,y)\mid 3x-3y=9\}$ **9.** $(-2,5,-2)$ **11.** $(1,-2,3)$ **13.** $(4,-3)$ **15.** $(2,1,-1)$ **17.** $(9,9)$
19. $\varnothing$ **21.** $\varnothing$ **23.** $(1,-4,3)$ **25.** function **27.** not a function **29.** -13 **31.** -36 **33.** 0 **35.** c **37. a.** in 2002 **b.** no; answers may vary **c.** no; it has a positive slope **d.** answers may vary **39.** answers may vary

Section 4.5
Practice Exercises
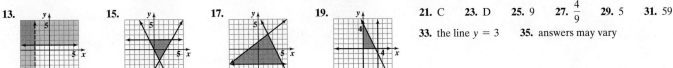

Vocabulary, Readiness & Video Check 4.5
1. system **3.** corner **5.** No; we can choose any test point except a point on the second inequality's own boundary line.

Exercise Set 4.5

21. C **23.** D **25.** 9 **27.** $\dfrac{4}{9}$ **29.** 5 **31.** 59
33. the line $y=3$ **35.** answers may vary

Chapter 4 Vocabulary Check
1. system of equations **2.** solution **3.** consistent **4.** triple **5.** inconsistent **6.** matrix **7.** element **8.** row **9.** column

Chapter 4 Review
1. $(-3,1)$ **3.** $\varnothing$ **5.** $\left(3,\dfrac{8}{3}\right)$ **7.** $(2,0,2)$ **9.** $\left(-\dfrac{1}{2},\dfrac{3}{4},1\right)$ **11.** $\varnothing$ **13.** $(1,1,-2)$

15. 10, 40, and 48 **17.** 58 mph; 65 mph **19.** 20 L of 10% solution; 30 L of 60% solution **21.** 17 pennies; 20 nickels; 16 dimes
23. two sides: 22 cm each; third side; 29 cm **25.** $(-3,1)$ **27.** $\left(-\dfrac{2}{3},3\right)$ **29.** $\left(\dfrac{5}{4},\dfrac{5}{8}\right)$ **31.** $(1,3)$ **33.** $(1,2,3)$ **35.** $(3,-2,5)$
37. $(1,1,-2)$

39. **41.** **43.** **45.** **47.** $\left(\dfrac{7}{3}, -\dfrac{8}{3}\right)$ **49.** $\{(x, y) \mid 5x - 2y = 10\}$

51. $(-1, 3, 5)$ **53.** 28 units, 42 units, 56 units

55. 2000

Chapter 4 Test

1. $(1, 3)$ **2.** $\varnothing$ **3.** $(2, -3)$ **4.** $\{(x, y) \mid 10x + 4y = 10\}$ **5.** $(-1, -2, 4)$ **6.** $\varnothing$ **7.** $\left(\dfrac{7}{2}, -10\right)$

8. $\{(x, y) \mid x - y = -2\}$ **9.** $(5, -3)$ **10.** $(-1, -1, 0)$ **11.** 53 double rooms; 27 single rooms **12.** 5 gal of 10%; 15 gal of 20% **13.** 800 packages **14.** $23°, 45°, 112°$

15.

Chapter 4 Cumulative Review

1. a. true **b.** true; Sec. 1.2, Ex. 4 **3. a.** 6 **b.** -7; Sec. 1.3, Ex. 3 **5. a.** -4 **b.** $-\dfrac{3}{7}$ **c.** 11.2; Sec. 1.4, Ex. 7 **7. a.** $6x + 3y$

b. $-3x + 1$ **c.** $0.7ab - 1.4a$; Sec. 1.4, Ex. 11 **9. a.** $-2x + 4$ **b.** $8yz$ **c.** $4z + 6.1$; Sec. 1.4, Ex. 14 **11.** -4; Sec. 2.1, Ex. 3

13. -4; Sec. 2.1, Ex. 7 **15.** 25 cm, 62 cm, 62 cm; Sec. 2.2, Ex. 5 **17.** $[-10, \infty)$; 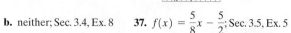 ; Sec. 2.4, Ex. 3 **19.** $(-3, 2)$; Sec. 2.5, Ex. 4

21. $1, -1$; Sec. 2.6, Ex. 4 **23.** $(4, 8)$; Sec. 2.7, Ex. 2 **25. a.** quadrant IV **b.** y-axis **c.** quadrant II **d.** x-axis **e.** quadrant III

f. quadant I;   ; Sec. 3.1, Ex. 1 **27.** no; Sec. 3.2, Ex. 4 **29. a.** 5 **b.** 1 **c.** 35 **d.** -2; Sec. 3.2, Ex. 7

31. 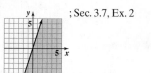 ; Sec. 3.3, Ex. 1 **33.** slope: $\dfrac{3}{4}$; y-intercept: $(0, -1)$; Sec. 3.4, Ex. 4 **35. a.** parallel

b. neither; Sec. 3.4, Ex. 8 **37.** $f(x) = \dfrac{5}{8}x - \dfrac{5}{2}$; Sec. 3.5, Ex. 5 **39.** ; Sec. 3.7, Ex. 2 **41. a.** yes **b.** no; Sec. 4.1, Ex. 1

43. $(-4, 2, -1)$; Sec. 4.2, Ex. 1 **45.** $(-1, 2)$; Sec. 4.4, Ex. 1

CHAPTER 5 EXPONENTS, POLYNOMIALS, AND POLYNOMIAL FUNCTIONS

Section 5.1
Practice Exercises

1. a. 3^6 **b.** x^7 **c.** y^9 **2. a.** $35z^4$ **b.** $-20.5t^6q^8$ **3. a.** 1 **b.** -1 **c.** 1 **d.** 3 **4. a.** z^5 **b.** 3^6 **c.** $9x^4$ **d.** $\dfrac{4}{3}a^7$ or $\dfrac{4a^7}{3}$

5. a. $\dfrac{1}{36}$ **b.** $\dfrac{1}{64}$ **c.** $\dfrac{3}{x^5}$ **d.** $\dfrac{1}{5y}$ **e.** $\dfrac{1}{k^7}$ **f.** $\dfrac{1}{25}$ **g.** $\dfrac{9}{20}$ **h.** z^8 **6. a.** $\dfrac{1}{z^{11}}$ **b.** $7t^8$ **c.** 9 **d.** $\dfrac{b^7}{3a^7}$ **e.** $\dfrac{2}{x^4}$ **7. a.** x^{3a+4}

b. x^{2t+1} **8. a.** 6.5×10^4 **b.** 3.8×10^{-5} **9. a.** 620,000 **b.** 0.03109

Scientific Calculator Explorations 5.1

1. 6×10^{43} **3.** 3.796×10^{28}

Vocabulary, Readiness & Video Check 5.1

1. x **3.** 3   **5.** y^7 **7.** These properties allow us to reorder and regroup factors to put those with the same bases together so that we may apply the product rule.   **9.** Subtract the exponents on like bases when applying the quotient rule.   **11.** When you move the decimal point to the left, the sign of the exponent is positive; when you move the decimal point to the right, the sign of the exponent is negative.

Exercise Set 5.1

1. $\dfrac{5}{xy^2}$ **3.** $\dfrac{a^2}{bc^5}$ **5.** $\dfrac{x^4}{y^2}$ **7.** 4^5 **9.** x^8 **11.** m^{14} **13.** $-20x^2y$ **15.** $-16x^6y^3p^2$ **17.** -1 **19.** 1   **21.** -1 **23.** 9   **25.** a^3

27. $-13z^4$ **29.** x **31.** $\dfrac{4}{3}x^3y^2$ **33.** $-6a^4b^4c^6$ **35.** $\dfrac{1}{16}$ **37.** $-\dfrac{1}{27}$ **39.** $\dfrac{1}{x^8}$ **41.** $\dfrac{5}{a^4}$ **43.** $\dfrac{y^2}{x^7}$ **45.** $\dfrac{1}{x^7}$ **47.** $4r^8$ **49.** 1

51. $\dfrac{b^7}{9a^7}$ **53.** $\dfrac{6x^{16}}{5}$ **55.** $-140x^{12}$ **57.** x^{16} **59.** $10x^{10}$ **61.** 6   **63.** $\dfrac{1}{z^3}$ **65.** -2 **67.** y^4 **69.** $\dfrac{13}{36}$ **71.** $\dfrac{3}{x}$ **73.** r^8 **75.** $\dfrac{1}{x^9y^4}$

77. $24x^7y^6$ **79.** $\dfrac{x}{16}$ **81.** 625   **83.** $\dfrac{1}{8}$ **85.** $\dfrac{a^5}{81}$ **87.** $\dfrac{7}{x^3z^5}$ **89.** x^{7a+5} **91.** x^{2t-1} **93.** y^{11p} **95.** z^{6x-7} **97.** x^{6t-1}

99. 3.125×10^7 **101.** 1.6×10^{-2} **103.** 6.7413×10^4 **105.** 1.25×10^{-2} **107.** 5.3×10^{-5} **109.** 7.783×10^8 **111.** 1.6228×10^{13}
113. 1.24×10^{11} **115.** 1.0×10^{-3} **117.** 0.0000000036   **119.** 93,000,000   **121.** 1,278,000   **123.** 7,350,000,000,000   **125.** 0.000000403

127. 300,000,000   **129.** 0.0095   **131.** 25,500,000,000   **133.** 100   **135.** $\dfrac{27}{64}$ **137.** 64   **139.** answers may vary   **141.** answers may vary

143. a. x^{2a} **b.** $2x^a$ **c.** x^{a-b} **d.** x^{a+b} **e.** $x^a + x^b$ **145.** 7^{13} **147.** 7^{-11}

Section 5.2
Practice Exercises

1. a. z^{15} **b.** 625   **c.** $\dfrac{1}{27}$ **d.** x^{24} **2. a.** $32x^{15}$ **b.** $\dfrac{9}{25}$ **c.** $\dfrac{16a^{20}}{b^{28}}$ **d.** $9x$ **e.** $\dfrac{a^4b^{10}}{c^8}$ **3. a.** $\dfrac{b^{15}}{27a^3}$ **b.** y^{15} **c.** $\dfrac{64}{9}$ **d.** $\dfrac{b^8}{81a^6}$

4. a. $\dfrac{c^3}{125a^{36}b^3}$ **b.** $\dfrac{16x^{16}y^4}{25}$ **5. a.** $27x^a$ **b.** y^{5b+3} **6. a.** 1.7×10^{-2} **b.** 1.4×10^{10} **7.** 4.2×10^{-6}

Vocabulary, Readiness & Video Check 5.2

1. x^{20} **3.** x^9 **5.** y^{42} **7.** z^{36} **9.** z^{18} **11.** The power rule involves a power of a base raised to a power and exponents are multiplied; the product rule involves a product and exponents are added.   **13.** We are asked to write the answer in scientific notation. The first product isn't because the number multiplied by the power of 10 is not between 1 and 10.

Exercise Set 5.2

1. $\dfrac{1}{9}$ **3.** $\dfrac{1}{x^{36}}$ **5.** $9x^4y^6$ **7.** $16x^{20}y^{12}$ **9.** $64\dfrac{c^{18}}{a^{12}b^6}$ **11.** $\dfrac{y^{15}}{x^{35}z^{20}}$ **13.** $-64y^3$ **15.** $\dfrac{1}{a^2}$ **17.** $\dfrac{36}{p^{12}}$ **19.** $-\dfrac{a^6}{512x^3y^9}$ **21.** $\dfrac{64}{27}$

23. $4a^8b^4$ **25.** $\dfrac{x^{14}y^{14}}{a^{21}}$ **27.** $\dfrac{1}{y^{10}}$ **29.** $\dfrac{1}{125}$ **31.** $\dfrac{1}{x^{63}}$ **33.** $\dfrac{1}{x^{20}y^{15}}$ **35.** $16x^4$ **37.** $48x^2y^6$ **39.** $\dfrac{x^9}{8y^3}$ **41.** $\dfrac{x^4}{4z^2}$ **43.** $\dfrac{x^4}{16}$

45. $\dfrac{1}{y^{15}}$ **47.** $\dfrac{16a^2b^9}{9}$ **49.** $\dfrac{3}{8x^8y^7}$ **51.** $\dfrac{1}{x^{30}b^6c^6}$ **53.** $\dfrac{25}{8x^5y^4}$ **55.** $\dfrac{2}{x^4y^{10}}$ **57.** x^{9a+18} **59.** x^{12a+2} **61.** b^{10x-4} **63.** y^{15a+3}
65. $16x^{4t+4}$ **67.** $5x^{-a}y^{-a+2}$ **69.** 1.45×10^9 **71.** 8×10^{15} **73.** 4×10^{-7} **75.** 3×10^{-1} **77.** 2×10^1 **79.** 1×10^1
81. 8×10^{-5} **83.** 1.1×10^7 **85.** 3.5×10^{22} **87.** 2.5808×10^{-5} sq m   **89.** $-2y - 18$ **91.** $-7x + 2$ **93.** $-3z - 14$

95. $\dfrac{8}{x^6y^3}$ cu m   **97.** 2×10^3 lb/ft^3 **99.** yes; $a = \pm 1$ **101.** no; answers may vary   **103.** 87 people per sq mi   **105.** $711 per person

Section 5.3
Practice Exercises

1. a. 5   **b.** 3   **c.** 1   **d.** 11   **e.** 0   **2. a.** degree 4; trinomial   **b.** degree 5; monomial   **c.** degree 5; binomial   **3.** 5
4. a. -15 **b.** -47 **5.** 290 feet; 226 feet   **6. a.** $3x^4 - 5x$ **b.** $7ab - 3b$ **7.** $6x^3 + 6x^2 - 7x - 8$ **8. a.** $3a^4b + 4ab^2 - 5$
b. $2x^5 + y - x - 9$ **9.** $15a^4 - 15a^3 + 3$ **10.** $6x^2y^2 - 4xy^2 - 5y^3$ **11.** C

Graphing Calculator Explorations 5.3

1. $x^3 - 4x^2 + 7x - 8$ **3.** $-2.1x^2 - 3.2x - 1.7$ **5.** $7.69x^2 - 1.26x + 5.3$

Vocabulary, Readiness & Video Check 5.3

1. coefficient   **3.** binomial   **5.** trinomial   **7.** degree   **9.** We are finding the degree of the polynomial, which is the greatest degree of any of its terms, so we find the degree of each term first.   **11.** by combining like terms   **13.** Change the operation from subtraction to addition. Then add the opposite of the polynomial that is being subtracted—that is, change the signs of all terms of the polynomial being subtracted and add.

Exercise Set 5.3

1. 0   **3.** 2   **5.** 3   **7.** 3   **9.** 9   **11.** degree 1; binomial   **13.** degree 2; trinomial   **15.** degree 3; monomial   **17.** degree 4; none of these

19. 57   **21.** 499   **23.** $-\dfrac{11}{16}$ **25.** 1061 ft   **27.** 549 ft   **29.** $-7y^2 + 6y$ **31.** $x^2y - 5x - \dfrac{1}{2}$ **33.** $6x^2 - xy + 16y^2$

35. $18y^2 - 17$ **37.** $3x^2 - 3xy + 6y^2$ **39.** $x^2 - 4x + 8$ **41.** $12x^3y + 8x + 8$ **43.** $4.5x^3 + 0.2x^2 - 3.8x + 9.1$ **45.** $y^2 + 3$
47. $-2x^2 + 5x$ **49.** $7y^2 - 12y - 3$ **51.** $7x^3 + 4x^2 + 8x - 10$ **53.** $-20y^2 + 3yx$ **55.** $-3x^2 + 3$ **57.** $2y^4 - 5y^2 + x^2 + 1$
59. $5x^2 - 9x - 3$ **61.** $3x^2 + x + 18$ **63.** $4x - 13$ **65.** $-x^3 + 8a - 12$ **67.** $14ab + 10a^2b - 18a^2 + 12b^2$ **69.** $5x^2 + 22x + 16$
71. 0 **73.** $8xy^2 + 2x^3 + 3x^2 - 3$ **75.** $3x^2 - 9x + 15$ **77.** $15x^2 + 8x - 6$ **79.** $\frac{1}{3}x^2 - x + 1$ **81.** 202 sq in.   **83. a.** 284 ft   **b.** 536 ft
c. 1400 ft   **d.** 1064 ft   **e.** answers may vary   **f.** 19 sec   **85.** \$80,000   **87.** \$16,500   **89.** A   **91.** D   **93.** $15x - 10$
95. $-2x^2 + 10x - 12$ **97.** a and c   **99.** $(12x - 1.7) - (15x + 6.2) = 12x - 1.7 - 15x - 6.2 = -3x - 7.9$ **101.** answers may vary
103. answers may vary   **105.** $3x^{2a} + 2x^a + 0.7$ **107.** $4x^{2y} + 2x^y - 11$ **109.** $(6x^2 + 14y)$ units   **111.** $4x^2 - 3x + 6$ **113.** $-x^2 - 6x + 10$
115. $3x^2 - 12x + 13$ **117.** $15x^2 + 12x - 9$ **119. a.** $2a - 3$ **b.** $-2x - 3$ **c.** $2x + 2h - 3$ **121. a.** $4a$ **b.** $-4x$ **c.** $4x + 4h$
123. a. $4a - 1$ **b.** $-4x - 1$ **c.** $4x + 4h - 1$ **125. a.** \$474.9 billion   **b.** \$574.5 billion   **c.** \$683.6 billion   **d.** answers may vary

Section 5.4
Practice Exercises

1. a. $6x^6$ **b.** $40m^5n^2p^8$ **2. a.** $21x^2 - 3x$ **b.** $-15a^4 + 30a^3 - 25a^2$ **c.** $-5m^3n^5 - 2m^2n^4 + 5m^2n^3$ **3. a.** $2x^2 + 13x + 15$
b. $3x^3 - 19x^2 + 12x - 2$ **4.** $3x^4 - 12x^3 - 13x^2 - 8x - 10$ **5.** $x^2 - 2x - 15$ **6. a.** $6x^2 - 31x + 35$ **b.** $8x^4 - 10x^2y - 3y^2$
7. a. $x^2 + 12x + 36$ **b.** $x^2 - 4x + 4$ **c.** $9x^2 + 30xy + 25y^2$ **d.** $9x^4 - 48x^2b + 64b^2$ **8. a.** $x^2 - 49$ **b.** $4a^2 - 25$ **c.** $25x^4 - \frac{1}{16}$
d. $a^6 - 16b^4$ **9.** $4 + 12x - 4y + 9x^2 - 6xy + y^2$ **10.** $9x^2 - 6xy + y^2 - 25$ **11.** $x^4 - 32x^2 + 256$ **12.** $h^2 - h + 3$

Graphing Calculator Explorations 5.4

1. $x^2 - 16$ **3.** $9x^2 - 42x + 49$ **5.** $5x^3 - 14x^2 - 13x - 2$

Vocabulary, Readiness & Video Check 5.4

1. $3x^6$; b   **3.** $x^2 - 49$; b   **5.** $(a + 1)^2 + 1$; d   **7.** distributive property, product rule   **9.** FOIL order or distributive property   **11.** Multiply
two at a time and multiply as usual, using patterns if you recognize them. Keep multiplying until you are through.

Exercise Set 5.4

1. $-12x^5$ **3.** $86a^5b^8c^3$ **5.** $12x^2 + 21x$ **7.** $-24x^2y - 6xy^2$ **9.** $-4a^3bx - 4a^3by + 12ab$ **11.** $2x^2 - 2x - 12$
13. $2x^4 + 3x^3 - 2x^2 + x + 6$ **15.** $15x^2 - 7x - 2$ **17.** $15m^3 + 16m^2 - m - 2$ **19.** $-30a^2b^4 + 36a^3b^2 + 36a^2b^3$ **21.** $x^2 + x - 12$
23. $10x^2 - 21xy + 8y^2$ **25.** $16x^2 - \frac{2}{3}x - \frac{1}{6}$ **27.** $5x^4 - 17x^2y^2 + 6y^4$ **29.** $x^2 + 8x + 16$ **31.** $36y^2 - 1$ **33.** $9x^2 - 6xy + y^2$
35. $49a^2b^2 - 9c^2$ **37.** $9x^2 - \frac{1}{4}$ **39.** $16b^2 + 32b + 16$ **41.** $4s^2 - 12s + 8$ **43.** $x^2y^2 - 4xy + 4$ **45.** $x^4 - 2x^2y^2 + y^4$
47. $x^4 - 8x^3 + 24x^2 - 32x + 16$ **49.** $x^4 - 625$ **51.** $-24a^2b^3 + 40a^2b^2 - 160a^2b$ **53.** $36x^2 + 12x + 1$ **55.** $25x^6 - 4y^2$
57. $10x^5 + 8x^4 + 2x^3 + 25x^2 + 20x + 5$ **59.** $9x^4 + 12x^3 - 2x^2 - 4x + 1$ **61.** $3x^2 + 8x - 3$ **63.** $9x^6 + 15x^4 + 3x^2 + 5$
65. $9x^2 + 6x + 1$ **67.** $9b^2 - 36y^2$ **69.** $49x^2 - 9$ **71.** $9x^3 + 30x^2 + 12x - 24$ **73.** $12x^3 - 2x^2 + 13x + 5$ **75.** $x^2y^2 - 4xy + 4$
77. $22a^3 + 11a^2 + 2a + 1$ **79.** $\frac{1}{3}n^2 - 7n + 18$ **81.** $45x^3 + 18x^2y - 5x - 2y$ **83.** $a^2 - 3a$ **85.** $a^2 + 2ah + h^2 - 3a - 3h$
87. $b^2 - 7b + 10$ **89.** -2 **91.** $\frac{3}{5}$ **93.** function   **95.** $7y(3z - 2) + 1 = 21yz - 14y + 1$ **97.** answers may vary
99. a. $a^2 + 2ah + h^2 + 3a + 3h + 2$ **b.** $a^2 + 3a + 2$ **c.** $2ah + h^2 + 3h$ **101.** $30x^2y^{2n+1} - 10x^2y^n$ **103.** $x^{3a} + 5x^{2a} - 3x^a - 15$
105. $\pi(25x^2 - 20x + 4)$ sq km or $(25\pi x^2 - 20\pi x + 4\pi)$ sq km   **107.** $(8x^2 - 12x + 4)$ sq in.   **109. a.** $6x + 12$ **b.** $9x^2 + 36x + 35$; one
operation is addition, the other is multiplication.   **111.** $5x^2 + 25x$ **113.** $x^4 - 4x^2 + 4$ **115.** $x^3 + 5x^2 - 2x - 10$

Section 5.5
Practice Exercises

1. $8y$ **2. a.** $3(2x^2 + 3 + 5x)$ **b.** $3x - 8y^3$ **c.** $2a^3(4a - 1)$ **3.** $8x^3y^2(8x^2 - 1)$ **4.** $-xy^2(9x^3 - 5x - 7)$ **5.** $(3 + 5b)(x + 4)$
6. $(8b - 1)(a^3 + 2y)$ **7.** $(x + 2)(y - 5)$ **8.** $(a + 2)(a^2 + 5)$ **9.** $(x^2 + 3)(y^2 - 5)$ **10.** $(q + 3)(p - 1)$

Vocabulary, Readiness & Video Check 5.5

1. factoring   **3.** least   **5.** false   **7.** false   **9.** The GCF is that common variable raised to the smallest exponent in the list.   **11.** Look for a
GCF other than 1 or -1; if you have a four-term polynomial

Exercise Set 5.5

1. a^3 **3.** y^2z^2 **5.** $3x^2y$ **7.** $5xz^3$ **9.** $6(3x - 2)$ **11.** $4y^2(1 - 4xy)$ **13.** $2x^3(3x^2 - 4x + 1)$ **15.** $4ab(2a^2b^2 - ab + 1 + 4b)$
17. $(x + 3)(6 + 5a)$ **19.** $(z + 7)(2x + 1)$ **21.** $(6x^2 + 5)(3x - 2)$ **23.** no common factor other than 1   **25.** $13x^2y^3(3x - 2)$
27. $(a + 2)(b + 3)$ **29.** $(a - 2)(c + 4)$ **31.** $(x - 2)(2y - 3)$ **33.** $(4x - 1)(3y - 2)$ **35.** $3(2x^3 + 3)$ **37.** $x^2(x + 3)$
39. $4a(2a^2 - 1)$ **41.** no common factors other than 1   **43.** $-4xy(5x - 4y^2)$ or $4xy(-5x + 4y^2)$ **45.** $5ab^2(2ab + 1 - 3b)$
47. $3b(3ac^2 + 2a^2c - 2a + c)$ **49.** $(y - 2)(4x - 3)$ **51.** $(2x + 3)(3y + 5)$ **53.** $(x + 3)(y - 5)$ **55.** $(2a - 3)(3b - 1)$
57. $(6x + 1)(2y + 3)$ **59.** $(n - 8)(2m - 1)$ **61.** $3x^2y^2(5x - 6)$ **63.** $(2x + 3y)(x + 2)$ **65.** $(5x - 3)(x + y)$
67. $(x^2 + 4)(x + 3)$ **69.** $(x^2 - 2)(x - 1)$ **71.** $55x^7$ **73.** $125x^6$ **75.** $x^2 - 3x - 10$ **77.** $x^2 + 5x + 6$ **79.** $y^2 - 4y + 3$
81. d   **83.** $2\pi r(r + h)$ **85.** $A = 5600(1 + rt)$ **87.** answers may vary   **89.** none   **91.** answers may vary   **93.** $I(R_1 + R_2) = E$
95. $x^n(x^{2n} - 2x^n + 5)$ **97.** $2x^{3a}(3x^{5a} - x^{2a} - 2)$ **99. a.** $h(t) = -16t(t - 4)$ **b.** 48 ft   **c.** answers may vary
101. $f(x) = 3(67x^3 - 839x^2 + 2325x + 27,878)$

Section 5.6
Practice Exercises

1. $(x + 3)(x + 2)$ **2.** $(x - 3)(x - 8)$ **3.** $3x(x - 5)(x + 2)$ **4.** $2(b^2 - 9b - 11)$ **5.** $(2x + 1)(x + 6)$ **6.** $(4x - 3)(x + 2)$
7. $3b^2(2b - 5)(3b - 2)$ **8.** $(5x + 2y)^2$ **9.** $(5x + 2)(4x + 3)$ **10.** $(5x + 3)(3x - 1)$ **11.** $(x - 3)(3x + 8)$ **12.** $(3x^2 + 2)(2x^2 - 5)$

Vocabulary, Readiness & Video Check 5.6

1. 5 and 2   **3.** 8 and 3   **5.** Check by multiplying. If you get a middle term of $2x$, not $-2x$, switch the signs of your factors of -24.   **7.** Write down the factors of the first and last terms. Try various combinations of these factors and look at the sum of the outer and inner products to see if you get the middle term of the trinomial. If not, try another combination of factors.   **9.** Our factoring involves a substitution from the original expression, so we need to substitute back in order to get a factorization in terms of the original variable/expression, and simplify if necessary.

Exercise Set 5.6

1. $(x + 3)(x + 6)$ **3.** $(x - 8)(x - 4)$ **5.** $(x + 12)(x - 2)$ **7.** $(x - 6)(x + 4)$ **9.** $3(x - 2)(x - 4)$ **11.** $4z(x + 2)(x + 5)$
13. $2(x^2 - 12x - 32)$ **15.** $(5x + 1)(x + 3)$ **17.** $(2x - 3)(x - 4)$ **19.** prime polynomial   **21.** $(2x - 3)^2$ **23.** $2(3x - 5)(2x + 5)$
25. $y^2(3y + 5)(y - 2)$ **27.** $2x(3x^2 + 4x + 12)$ **29.** $(2x + y)(x - 3y)$ **31.** $2(7y + 2)(2y + 1)$ **33.** $(2x - 3)(x + 9)$
35. $(x^2 + 3)(x^2 - 2)$ **37.** $(5x + 8)(5x + 2)$ **39.** $(x^3 - 4)(x^3 - 3)$ **41.** $(a - 3)(a + 8)$ **43.** $(x - 27)(x + 3)$
45. $(x - 18)(x + 3)$ **47.** $3(x - 1)^2$ **49.** $(3x + 1)(x - 2)$ **51.** $(4x - 3)(2x - 5)$ **53.** $3x^2(2x + 1)(3x + 2)$ **55.** $(x + 7z)(x + z)$
57. $(x - 4)(x + 3)$ **59.** $3(a + 2b)^2$ **61.** prime polynomial   **63.** $(2x + 13)(x + 3)$ **65.** $(3x - 2)(2x - 15)$ **67.** $(x^2 - 6)(x^2 + 1)$
69. $x(3x + 1)(2x - 1)$ **71.** $(4a - 3b)(3a - 5b)$ **73.** $(3x + 5)^2$ **75.** $y(3x - 8)(x - 1)$ **77.** $2(x + 3)(x - 2)$ **79.** $(x + 2)(x - 7)$
81. $(2x^3 - 3)(x^3 + 3)$ **83.** $2x(6y^2 - z)^2$ **85.** $2xy(x + 3)(x - 2)$ **87.** $(x + 5y)(x + y)$ **89.** $x^2 - 9$ **91.** $4x^2 + 4x + 1$
93. $x^3 - 8$ **95.** $\pm5, \pm7$ **97.** $x(x + 4)(x - 2)$ **99. a.** 576 ft; 672 ft; 640 ft; 480 ft   **b.** answers may vary   **c.** $-16(t + 4)(t - 9)$
101. $(x^n + 2)(x^n + 8)$ **103.** $(x^n - 6)(x^n + 3)$ **105.** $(2x^n + 1)(x^n + 5)$ **107.** $(2x^n - 3)^2$ **109.** $x^2(x + 5)(x + 1)$
111. $3x(5x - 1)(2x + 1)$

Section 5.7
Practice Exercises

1. $(b + 8)^2$ **2.** $5b(3x - 1)^2$ **3. a.** $(x + 4)(x - 4)$ **b.** $(5b - 7)(5b + 7)$ **c.** $5(3 - 2x)(3 + 2x)$ **d.** $\left(y - \dfrac{1}{9}\right)\left(y + \dfrac{1}{9}\right)$
4. a. $(x^2 + 100)(x + 10)(x - 10)$ **b.** $(x + 9)(x - 5)$ **5.** $(m + 3 + n)(m + 3 - n)$ **6.** $(x + 4)(x^2 - 4x + 16)$
7. $(a + 2b)(a^2 - 2ab + 4b^2)$ **8.** $(3 - y)(9 + 3y + y^2)$ **9.** $x^2(b - 2)(b^2 + 2b + 4)$

Vocabulary, Readiness & Video Check 5.7

1. $(9y)^2$ **3.** $(8x^3)^2$ **5.** $(2x)^3$ **7.** $(4x^2)^3$ **9.** See if the first term is a square, say a^2, and the last term is a square, say b^2. Check to see if the middle term is $2 \cdot a \cdot b$ or $-2 \cdot a \cdot b$.   **11.** First rewrite the original binomial so that each term is some quantity cubed. Your answers will then vary depending on your interpretation.

Exercise Set 5.7

1. $(x + 3)^2$ **3.** $(2x - 3)^2$ **5.** $(5x + 1)^2$ **7.** $3(x - 4)^2$ **9.** $x^2(3y + 2)^2$ **11.** $(4x - 7y)^2$ **13.** $(x + 5)(x - 5)$
15. $\left(\dfrac{1}{3} + 2z\right)\left(\dfrac{1}{3} - 2z\right)$ **17.** $(y + 9)(y - 5)$ **19.** $4(4x + 5)(4x - 5)$ **21.** $(x + 2y + 3)(x + 2y - 3)$ **23.** $(x + 3 + y)(x + 3 - y)$
25. $(x + 8 + x^2)(x + 8 - x^2)$ **27.** $(x + 3)(x^2 - 3x + 9)$ **29.** $(z - 1)(z^2 + z + 1)$ **31.** $(m + n)(m^2 - mn + n^2)$
33. $y^2(3 - x)(9 + 3x + x^2)$ **35.** $a(2b + 3a)(4b^2 - 6ab + 9a^2)$ **37.** $2(5y - 2x)(25y^2 + 10xy + 4x^2)$ **39.** $(x - 6)^2$
41. $2y(3x + 1)(3x - 1)$ **43.** $(3x + 7)(3x - 7)$ **45.** $(x^2 + 1)(x + 1)(x - 1)$ **47.** $(x^2 - y)(x^4 + x^2y + y^2)$
49. $(2x + 3y)(4x^2 - 6xy + 9y^2)$ **51.** $(2x + 1 + z)(2x + 1 - z)$ **53.** $3y^2(x^2 + 3)(x^4 - 3x^2 + 9)$ **55.** $\left(n - \dfrac{1}{3}\right)\left(n^2 + \dfrac{1}{3}n + \dfrac{1}{9}\right)$
57. $-16(y + 2)(y - 2)$ **59.** $(x - 5 + y)(x - 5 - y)$ **61.** $(ab + 5)(a^2b^2 - 5ab + 25)$ **63.** $\left(\dfrac{x}{5} + \dfrac{y}{3}\right)\left(\dfrac{x}{5} - \dfrac{y}{3}\right)$
65. $(x + y + 5)(x^2 + 2xy + y^2 - 5x - 5y + 25)$ **67.** 5   **69.** $-\dfrac{1}{3}$ **71.** 0   **73.** 5   **75.** no; $x^2 - 4$ can be factored further
77. yes   **79.** $\pi R^2 - \pi r^2 = \pi(R + r)(R - r)$ **81.** $x^3 - y^2x; x(x^3 + y)(x - y)$ **83.** $c = 9$ **85.** $c = 49$ **87.** $c = \pm8$
89. a. $(x + 1)(x^2 - x + 1)(x - 1)(x^2 + x + 1)$ **b.** $(x + 1)(x - 1)(x^4 + x^2 + 1)$ **c.** answers may vary   **91.** $(x^n + 5)(x^n - 5)$
93. $(6x^n + 7)(6x^n - 7)$ **95.** $(x^{2n} + 4)(x^n + 2)(x^n - 2)$

Integrated Review Practice Exercises

1. a. $3xy(4x - 1)$ **b.** $(7x + 2)(7x - 2)$ **c.** $(5x - 3)(x + 1)$ **d.** $(3 + x)(x^2 + 2)$ **e.** $(2x + 5)^2$ **f.** cannot be factored
2. a. $(4x + y)(16x^2 - 4xy + y^2)$ **b.** $7y^2(x - 3y)(x + 3y)$ **c.** $3(x + 2 + b)(x + 2 - b)$ **d.** $x^2y(xy + 3)(x^2y^2 - 3xy + 9)$
e. $(x + 7 + 9y)(x + 7 - 9y)$

Integrated Review

1. $2y^2 + 2y - 11$ **2.** $-2z^4 - 6z^2 + 3z$ **3.** $x^2 - 7x + 7$ **4.** $7x^2 - 4x - 5$ **5.** $25x^2 - 30x + 9$ **6.** $5x^2 - 19x - 4$
7. $2x^3 - 3x + 4$ **8.** $4x^3 - 13x^2 - 5x + 2$ **9.** $(x - 4 + y)(x - 4 - y)$ **10.** $2(3x + 2)(2x - 5)$ **11.** $x(x - 1)(x^2 + x + 1)$
12. $2x(2x - 1)$ **13.** $2xy(7x - 1)$ **14.** $6ab(4b - 1)$ **15.** $4(x + 2)(x - 2)$ **16.** $9(x + 3)(x - 3)$ **17.** $(3x - 11)(x + 1)$
18. $(5x + 3)(x - 1)$ **19.** $4(x + 3)(x - 1)$ **20.** $6(x + 1)(x - 2)$ **21.** $(2x + 9)^2$ **22.** $(5x + 4)^2$ **23.** $(2x + 5y)(4x^2 - 10xy + 25y^2)$
24. $(3x - 4y)(9x^2 + 12xy + 16y^2)$ **25.** $8x^2(2y - 1)(4y^2 + 2y + 1)$ **26.** $27x^2y(xy - 2)(x^2y^2 + 2xy + 4)$
27. $(x + 5 + y)(x^2 + 10x - xy - 5y + y^2 + 25)$ **28.** $(y - 1 + 3x)(y^2 - 2y + 1 - 3xy + 3x + 9x^2)$ **29.** $(5a - 6)^2$ **30.** $(4r + 5)^2$

31. $7x(x - 9)$ **32.** $(4x + 3)(5x + 2)$ **33.** $(a + 7)(b - 6)$ **34.** $20(x - 6)(x - 5)$ **35.** $(x^2 + 1)(x - 1)(x + 1)$ **36.** $5x(3x - 4)$
37. $(5x - 11)(2x + 3)$ **38.** $9m^2n^2(5mn - 3)$ **39.** $5a^3b(b^2 - 10)$ **40.** $x(x + 1)(x^2 - x + 1)$ **41.** prime   **42.** $20(x + y)(x^2 - xy + y^2)$
43. $10x(x - 10)(x - 11)$ **44.** $(3y - 7)^2$ **45.** $a^3b(4b - 3)(16b^2 + 12b + 9)$ **46.** $(y^2 + 4)(y + 2)(y - 2)$ **47.** $2(x - 3)(x^2 + 3x + 9)$
48. $(2s - 1)(r + 5)$ **49.** $(y^4 + 2)(3y - 5)$ **50.** prime   **51.** $100(z + 1)(z^2 - z + 1)$ **52.** $2x(5x - 2)(25x^2 + 10x + 4)$
53. $(2b - 9)^2$ **54.** $(a^4 + 3)(2a - 1)$ **55.** $(y - 4)(y - 5)$ **56.** $(c - 3)(c + 1)$ **57.** $A = 9 - 4x^2 = (3 + 2x)(3 - 2x)$

Section 5.8
Practice Exercises

1. $-8, 5$ **2.** $-4, \dfrac{2}{3}$ **3.** $-4, -\dfrac{2}{3}$ **4.** $-\dfrac{3}{4}$ **5.** $-\dfrac{1}{8}, 2$ **6.** $0, 3, -1$ **7.** $3, -3, -2$ **8.** 6 seconds   **9.** $6, 8, 10$ units
10. $f(x):C; g(x):A; h(x):B$

Graphing Calculator Explorations 5.8

1. $-3.562, 0.562$ **3.** $-0.874, 2.787$ **5.** $-0.465, 1.910$

Vocabulary, Readiness & Video Check 5.8

1. $3, -5$ **3.** $3, -7$ **5.** $0, 9$ **7.** One side of the equation must be a factored polynomial and the other side must be zero.   **9.** Finding the
x-intercepts of any equation in two variables means you let $y = 0$ and solve for x. Doing this with a quadratic equation gives us an equation $= 0$, which
may be solved by factoring.

Exercise Set 5.8

1. $-3, \dfrac{4}{3}$ **3.** $\dfrac{5}{2}, -\dfrac{3}{4}$ **5.** $-3, -8$ **7.** $\dfrac{1}{4}, -\dfrac{2}{3}$ **9.** $1, 9$ **11.** $\dfrac{3}{5}, -1$ **13.** 0 **15.** $6, -3$ **17.** $\dfrac{2}{5}, -\dfrac{1}{2}$ **19.** $\dfrac{3}{4}, -\dfrac{1}{2}$ **21.** $-2, 7, \dfrac{8}{3}$

23. $-1, 1, -\dfrac{1}{6}$ **25.** $0, 3, -3$ **27.** $2, 1, -1$ **29.** $-\dfrac{7}{2}, 10$ **31.** $0, 5$ **33.** $-3, 5$ **35.** $-\dfrac{1}{2}, \dfrac{1}{3}$ **37.** $-4, 9$ **39.** $\dfrac{4}{5}$

41. $-5, 0, 2$ **43.** $-3, 0, \dfrac{4}{5}$ **45.** $\varnothing$ **47.** $-7, 4$ **49.** $4, 6$ **51.** $-\dfrac{1}{2}$ **53.** $-4, -3, 3$ **55.** $-5, 0, 5$ **57.** $-6, 5$ **59.** $-\dfrac{1}{3}, 0, 1$

61. $-\dfrac{1}{3}, 0$ **63.** $-\dfrac{7}{8}$ **65.** $\dfrac{31}{4}$ **67.** 1 **69.** -11 and -6 or 6 and 11 **71.** 75 ft   **73.** 105 units   **75.** 12 cm and 9 cm   **77.** 2 in.

79. 10 sec   **81.** width: $7\dfrac{1}{2}$ ft; length: 12 ft   **83.** 10-in. sq tier   **85.** 9 sec   **87.** E   **89.** F   **91.** B   **93.** $(-3, 0), (0, 2)$; function

95. $(-4, 0), (4, 0), (0, 2), (0, -2)$; not a function   **97.** $x - 5 = 0$ or $x + 2 = 0$ **99.**
$$x = 5 \text{ or } x = -2$$

$$y(y - 5) = -6$$
$$y^2 - 5y + 6 = 0$$
$$(y - 2)(y - 3) = 0$$
$$y - 2 = 0 \text{ or } y - 3 = 0$$
$$y = 2 \text{ or } y = 3$$

101. $-3, -\dfrac{1}{3}, 2, 5$ **103.** $x = \dfrac{4}{3}$ or $x = 7$ **105.** answers may vary   **107.** no; answers may vary   **109.** answers may vary
111. answers may vary   **113.** answers may vary

Chapter 5 Vocabulary Check

1. polynomial   **2.** Factoring   **3.** Exponents   **4.** degree of a term   **5.** monomial   **6.** 1   **7.** trinomial   **8.** quadratic equation
9. scientific notation   **10.** degree of a polynomial   **11.** binomial   **12.** 0

Chapter 5 Review

1. 4   **3.** -4 **5.** 1   **7.** $-\dfrac{1}{16}$ **9.** $-x^2y^7z$ **11.** $\dfrac{1}{a^9}$ **13.** $\dfrac{1}{x^{11}}$ **15.** $\dfrac{1}{y^5}$ **17.** 3.689×10^7 **19.** 0.000001678 **21.** 8^{15} **23.** $27x^3$

25. $\dfrac{36x^2}{25}$ **27.** $\dfrac{9}{16}$ **29.** $\dfrac{1}{4p^4}$ **31.** $x^{25}y^{15}z^{15}$ **33.** $\dfrac{x^2}{625y^4z^4}$ **35.** $27x^{19a}$ **37.** 5.76×10^4 **39.** $\dfrac{3x^7}{2y^2}$ **41.** 5   **43.** $12x - 6x^2 - 6x^2y$
45. $4x^2 + 8y + 6$ **47.** $8x^2 + 2b - 22$ **49.** $-x + 5$ **51.** $x^2 - 6x + 3$ **53.** 290   **55.** 58   **57.** $x^2 + 4x - 6$
59. $(6x^2y - 12x + 12)$ cm   **61.** $-24x^3 + 36x^2 - 6x$ **63.** $2x^2 + x - 36$ **65.** $36x^3 - 11x^2 - 8x - 3$ **67.** $15x^2 + 18xy - 81y^2$
69. $x^4 + 18x^3 + 83x^2 + 18x + 1$ **71.** $9x^2 - 6xy + y^2$ **73.** $x^2 - 9y^2$ **75.** $(9y^2 - 49z^2)$ sq units   **77.** $12a^{2b+2} - 28a^b$
79. $8x^2(2x - 3)$ **81.** $2ab(3b + 4 - 2ab)$ **83.** $(a + 3b)(6a - 5)$ **85.** $(x - 6)(y + 3)$ **87.** $(p - 5)(q - 3)$
89. $x(2y - x)$ **91.** $(x - 18)(x + 4)$ **93.** $2(x - 2)(x - 7)$ **95.** $x(2x - 9)(x + 1)$ **97.** $(6x + 5)(x + 2)$ **99.** $2(2x - 3)(x + 2)$
101. $(x + 6)^2(y - 3)(y + 1)$ **103.** $(x^2 - 8)(x^2 + 2)$ **105.** $(x + 10)(x - 10)$ **107.** $2(x + 4)(x - 4)$ **109.** $(9 + x^2)(3 + x)(3 - x)$
111. $(y + 7)(y - 3)$ **113.** $(x + 6)(x^2 - 6x + 36)$ **115.** $(2 - 3y)(4 + 6y + 9y^2)$ **117.** $6xy(x + 2)(x^2 - 2x + 4)$

119. $(x - 1 - y)(x - 1 + y)$ **121.** $(2x + 3)^2$ **123.** $\pi h(R - r)(R + r)$ cu units   **125.** $\dfrac{1}{3}, -7$ **127.** $0, 4, \dfrac{9}{2}$ **129.** $0, 6$ **131.** $-\dfrac{1}{3}, 2$

133. $-4, 1$ **135.** $0, 6, -3$ **137.** $0, -2, 1$ **139.** $-\dfrac{15}{2}$ or 7   **141.** 5 sec   **143.** $3x^3 + 13x^2 - 9x + 5$ **145.** $8x^2 + 3x + 4.5$ **147.** -24
149. $6y^4(2y - 1)$ **151.** $2(3x + 1)(x - 6)$ **153.** $z^5(2z + 7)(2z - 7)$ **155.** $0, 3$

Chapter 5 Test

1. $\dfrac{1}{81x^2}$ **2.** $-12x^2z$ **3.** $\dfrac{3a^7}{2b^5}$ **4.** $-\dfrac{y^{40}}{z^5}$ **5.** 6.3×10^8 **6.** 1.2×10^{-2} **7.** 0.000005 **8.** 0.0009 **9.** $-5x^3y - 11x - 9$

10. $-12x^2y - 3xy^2$ **11.** $12x^2 - 5x - 28$ **12.** $25a^2 - 4b^2$ **13.** $36m^2 + 12mn + n^2$ **14.** $2x^3 - 13x^2 + 14x - 4$

15. $4x^2y(4x - 3y^3)$ **16.** $(x - 15)(x + 2)$ **17.** $(2y + 5)^2$ **18.** $3(2x + 1)(x - 3)$ **19.** $(2x + 5)(2x - 5)$

20. $(x + 4)(x^2 - 4x + 16)$ **21.** $3y(x + 3y)(x - 3y)$ **22.** $6(x^2 + 4)$ **23.** $2(2y - 1)(4y^2 + 2y + 1)$ **24.** $(x + 3)(x - 3)(y - 3)$

25. $4, -\dfrac{8}{7}$ **26.** $-3, 8$ **27.** $-\dfrac{5}{2}, -2, 2$ **28.** $(x + 2y)(x - 2y)$ **29. a.** 960 ft **b.** 953.44 ft **c.** $-16(t - 11)(t + 5)$ **d.** 11 sec

Chapter 5 Cumulative Review

1. a. 3 **b.** 1 **c.** 2; Sec. 1.3, Ex. 8 **3.** 1; Sec. 2.1, Ex. 4 **5.** $11,607.55; Sec. 2.3, Ex. 4 **7. a.** $\left(-\infty, \dfrac{3}{2}\right]$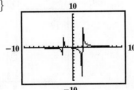

b. $(-3, \infty)$; Sec. 2.4, Ex. 4 **9.** $\left[-9, -\dfrac{9}{2}\right]$; Sec. 2.5, Ex. 5 **11.** 0; Sec. 2.6, Ex. 5 **13.** $\varnothing$; Sec. 2.7, Ex. 4

15. ; Sec. 3.1, Ex. 5 **17.** $(2, -3), (-6, -9), (0, 3)$; Sec. 3.6, Ex. 1 **19.** ; Sec. 3.3, Ex. 6 **21.** 0; Sec. 3.4, Ex. 7

23. $y = 3$; Sec. 3.5, Ex. 8 **25.** ; Sec. 3.7, Ex. 4 **27.** $\left(-4, \dfrac{1}{2}\right)$; Sec. 4.1, Ex. 5 **29.** $\left(\dfrac{1}{2}, 0, \dfrac{3}{4}\right)$; Sec. 4.2, Ex. 3

31. 7 and 11; Sec. 4.3, Ex. 2 **33.** $\varnothing$; Sec. 4.4, Ex. 2 **35.** $30°, 40°, 110°$; Sec. 4.3, Ex. 6 **37. a.** 7.3×10^5 **b.** 1.04×10^{-6}; Sec. 5.1, Ex. 8

39. a. $\dfrac{y^6}{4}$ **b.** x^9 **c.** $\dfrac{49}{4}$ **d.** $\dfrac{y^{16}}{25x^5}$; Sec. 5.2, Ex. 3 **41.** 4; Sec. 5.3, Ex. 3 **43. a.** $10x^9$ **b.** $-7xy^{15}z^9$; Sec. 5.4, Ex. 1

45. $17x^3y^2(1 - 2x)$; Sec. 5.5, Ex. 3 **47.** $(x + 2)(x + 8)$; Sec. 5.6, Ex. 1 **49.** $-5, \dfrac{1}{2}$; Sec. 5.8, Ex. 2

CHAPTER 6 RATIONAL EXPRESSIONS

Section 6.1
Practice Exercises

1. a. $\{x \mid x \text{ is a real number}\}$ **b.** $\{x \mid x \text{ is a real number and } x \neq -3\}$ **c.** $\{x \mid x \text{ is a real number and } x \neq 2, x \neq 3\}$ **2. a.** $\dfrac{1}{2z - 1}$

b. $\dfrac{5x + 3}{6x - 5}$ **3. a.** 1 **b.** -1 **4.** $-\dfrac{5(2 + x)}{x + 3}$ **5. a.** $x^2 - 4x + 16$ **b.** $\dfrac{5}{z - 3}$ **6. a.** $\dfrac{2n + 1}{n(n - 1)}$ **b.** $-x^2$ **7. a.** $-\dfrac{y^3}{21(y + 3)}$

b. $\dfrac{7x + 2}{x + 2}$ **8.** $\dfrac{5}{3(x - 3)}$ **9. a.** $7.20 **b.** $3.60

Graphing Calculator Explorations 6.1

1. $\{x \mid x \text{ is a real number and } x \neq -2, x \neq 2\}$ 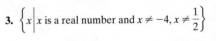 **3.** $\{x \mid x \text{ is a real number and } x \neq -4, x \neq \dfrac{1}{2}\}$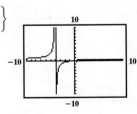

Vocabulary, Readiness & Video Check 6.1

1. rational **3.** domain **5.** 1 **7.** $\dfrac{-a}{b}; \dfrac{a}{-b}$ **9.** Rational expressions are fractions and are therefore undefined if the denominator is zero; the domain of a rational function consists of all real numbers except those for which the rational expression is undefined. **11.** Yes, multiplying and also simplifying rational expressions often require polynomial factoring. Example 6 involves factoring out a GCF and factoring a difference of squares. **13.** Since the domain consists of all allowable x-values and x is defined as the number of years since the book was published, we do not allow x to be negative for the context of this application.

Exercise Set 6.1

1. $\{x \mid x \text{ is a real number}\}$ **3.** $\{t \mid t \text{ is a real number and } t \neq 0\}$ **5.** $\{x \mid x \text{ is a real number and } x \neq 7\}$ **7.** $\{x \mid x \text{ is a real number and } x \neq \dfrac{1}{3}\}$

9. $\{x \mid x \text{ is a real number and } x \neq -2, x \neq 0, x \neq 1\}$ **11.** $\{x \mid x \text{ is a real number and } x \neq 2, x \neq -2\}$ **13.** $1 - 2x$ **15.** $x - 3$ **17.** $\dfrac{9}{7}$

19. $x - 4$ **21.** -1 **23.** $-(x + 7)$ **25.** $\dfrac{2x + 1}{x - 1}$ **27.** $\dfrac{x^2 + 5x + 25}{2}$ **29.** $\dfrac{x - 2}{2x^2 + 1}$ **31.** $\dfrac{1}{3x + 5}$ **33.** $-\dfrac{4}{5}$ **35.** $-\dfrac{6a}{2a + 1}$

37. $\dfrac{3}{2(x - 1)}$ **39.** $\dfrac{x + 2}{x + 3}$ **41.** $\dfrac{3a}{5(a - b)}$ **43.** $\dfrac{1}{6}$ **45.** $\dfrac{x}{3}$ **47.** $\dfrac{4a^2}{a - b}$ **49.** $\dfrac{4}{(x + 2)(x + 3)}$ **51.** $\dfrac{1}{2}$ **53.** -1 **55.** $\dfrac{8(a - 2)}{3(a + 2)}$

57. $\dfrac{(x + 2)(x + 3)}{4}$ **59.** $\dfrac{2(x + 3)(x - 3)}{5(x^2 - 8x - 15)}$ **61.** $r^2 - rs + s^2$ **63.** $\dfrac{8}{x^2 y}$ **65.** $\dfrac{(y + 5)(2x - 1)}{(y + 2)(5x + 1)}$ **67.** $\dfrac{5(3a + 2)}{a}$ **69.** $\dfrac{5x^2 - 2}{(x - 1)^2}$

71. $\dfrac{10}{3}, -8, -\dfrac{7}{3}$ **73.** $-\dfrac{17}{48}, \dfrac{2}{7}, -\dfrac{3}{8}$ **75. a.** \$200 million   **b.** \$500 million   **c.** \$300 million   **d.** $\{x \mid x \text{ is a real number}\}$ **77.** $\dfrac{7}{5}$ **79.** $\dfrac{1}{12}$

81. $\dfrac{11}{16}$ **83.** b and d   **85.** no; answers may vary   **87.** $\dfrac{5}{x - 2}$ sq m   **89.** $\dfrac{(x + 2)(x - 1)^2}{x^5}$ ft   **91.** answers may vary

93. a. 1   **b.** -1 **c.** neither   **d.** -1 **e.** -1 **f.** 1   **95.** answers may vary   **97.** $0, \dfrac{20}{9}, \dfrac{60}{7}, 20, \dfrac{140}{3}, 180, 380, 1980;$

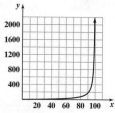

99. -1 **101.** $\dfrac{1}{x^n - 4}$ **103.** $2x^2(x^n + 2)$ **105.** $\dfrac{1}{10y(y^n + 3)}$ **107.** $\dfrac{y^n + 1}{2(y^n - 1)}$

Section 6.2
Practice Exercises

1. a. $\dfrac{9 + x}{11z^2}$ **b.** $\dfrac{3x}{4}$ **c.** $x - 4$ **d.** $\dfrac{-3}{2a^2}$ **2. a.** $18x^3 y^5$ **b.** $(x - 2)(x + 3)$ **c.** $(b - 4)^2(b + 4)(2b + 3)$ **d.** $-4(y - 3)(y + 3)$

3. a. $\dfrac{20p + 3}{5p^4 q}$ **b.** $\dfrac{5y^2 + 19y - 12}{(y + 3)(y - 3)}$ **c.** 3   **4.** $\dfrac{t^2 - t - 15}{(t + 5)(t - 5)(t + 2)}$ **5.** $\dfrac{5x^2 - 12x - 3}{(3x + 1)(x - 2)(2x - 5)}$ **6.** $\dfrac{4}{x - 2}$

Vocabulary, Readiness & Video Check 6.2

1. a, b   **3.** c   **5.** $\dfrac{12}{y}$ **7.** $\dfrac{35}{y^2}$ **9.** We need to be sure we subtract the entire second numerator—that is, make sure we "distribute" the subtraction to each term in the second numerator.   **11.** numerator; one; value

Exercise Set 6.2

1. $\dfrac{-x + 4}{2x}$ **3.** $\dfrac{16}{y - 2}$ **5.** $-\dfrac{3}{xz^2}$ **7.** $\dfrac{x + 2}{x - 2}$ **9.** $x - 2$ **11.** $-\dfrac{1}{x - 2}$ or $\dfrac{1}{2 - x}$ **13.** $-\dfrac{5}{x}$ **15.** $35x$ **17.** $x(x + 1)$

19. $(x + 7)(x - 7)$ **21.** $6(x + 2)(x - 2)$ **23.** $(a + b)(a - b)^2$ **25.** $-4x(x + 3)(x - 3)$ **27.** $\dfrac{17}{6x}$ **29.** $\dfrac{35 - 4y}{14y^2}$

31. $\dfrac{-13x + 4}{(x + 4)(x - 4)}$ **33.** $\dfrac{3}{x + 4}$ **35.** 0   **37.** $-\dfrac{x}{x - 1}$ **39.** $\dfrac{-x + 1}{x - 2}$ **41.** $\dfrac{y^2 + 2y + 10}{(y + 4)(y - 4)(y - 2)}$ **43.** $\dfrac{5(x^2 + x - 4)}{(3x + 2)(x + 3)(2x - 5)}$

45. $\dfrac{-x^2 + 10x + 19}{(x - 2)(x + 1)(x + 3)}$ **47.** $\dfrac{x^2 + 4x + 2}{(2x + 5)(x - 7)(x - 1)}$ **49.** $\dfrac{5a + 1}{(a + 1)^2(a - 1)}$ **51.** $\dfrac{3}{x^2 y^3}$ **53.** $-\dfrac{5}{x}$ **55.** $\dfrac{25}{6(x + 5)}$ **57.** $\dfrac{-2x - 1}{x^2(x - 3)}$

59. $\dfrac{b(2a - b)}{(a + b)(a - b)}$ **61.** $\dfrac{2(x + 8)}{(x + 2)^2(x - 2)}$ **63.** $\dfrac{3x^2 + 23x - 7}{(2x - 1)(x - 5)(x + 3)}$ **65.** $\dfrac{5 - 2x}{2(x + 1)}$ **67.** $\dfrac{2(x^2 + x - 21)}{(x + 3)^2(x - 3)}$ **69.** $\dfrac{6x}{(x + 3)(x - 3)^2}$

71. $\dfrac{4}{3}$ **73.** $\dfrac{2x^2 + 9x - 18}{6x^2}$ or $\dfrac{(x + 6)(2x - 3)}{6x^2}$ **75.** $\dfrac{4a^2}{9(a - 1)}$ **77.** 4   **79.** $-\dfrac{4}{x - 1}$ **81.** $-\dfrac{32}{x(x + 2)(x - 2)}$ **83.** 10   **85.** $4 + x^2$

87. 10   **89.** 2   **91.** 3   **93.** 5 m   **95.** $\dfrac{2x - 3}{x^2 + 1} - \dfrac{x - 6}{x^2 + 1} = \dfrac{2x - 3 - x + 6}{x^2 + 1} = \dfrac{x + 3}{x^2 + 1}$ **97.** $\dfrac{4x}{x + 5}$ ft; $\dfrac{x^2}{(x + 5)^2}$ sq ft   **99.** answers may vary

101. answers may vary   **103.** answers may vary   **105.** $\dfrac{3}{2x}$ **107.** $\dfrac{4 - 3x}{x^2}$ **109.**

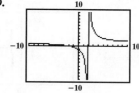

Section 6.3
Practice Exercises

1. a. $\dfrac{1}{12m}$ **b.** $\dfrac{8x(x + 4)}{3(x - 4)}$ **c.** $\dfrac{b^2}{a^2}$ **2. a.** $\dfrac{8x(x + 4)}{3(x - 4)}$ **b.** $\dfrac{b^2}{a^2}$ **3.** $\dfrac{y(3xy + 1)}{x^2(1 + xy)}$ **4.** $\dfrac{1 - 6x}{15 + 6x}$

Vocabulary, Readiness & Video Check 6.3

1. $\dfrac{7}{1 + z}$ **3.** $\dfrac{1}{x^2}$ **5.** $\dfrac{2}{x}$ **7.** $\dfrac{1}{9y}$ **9.** a single fraction in the numerator and in the denominator   **11.** Since a negative exponent moves its base from a numerator to a denominator of the expression only, a rational expression containing negative exponents can become a complex fraction when rewritten with positive exponents.

Exercise Set 6.3

1. 4   **3.** $\dfrac{7}{13}$ **5.** $\dfrac{4}{x}$ **7.** $\dfrac{9(x-2)}{9x^2+4}$ **9.** $2x+y$ **11.** $\dfrac{2(x+1)}{2x-1}$ **13.** $\dfrac{2x+3}{4-9x}$ **15.** $\dfrac{1}{x^2-2x+4}$ **17.** $\dfrac{x}{5(x-2)}$ **19.** $\dfrac{x-2}{2x-1}$

21. $\dfrac{x}{2-3x}$ **23.** $-\dfrac{y}{x+y}$ **25.** $-\dfrac{2x^3}{y(x-y)}$ **27.** $\dfrac{2x+1}{y}$ **29.** $\dfrac{x-3}{9}$ **31.** $\dfrac{1}{x+2}$ **33.** 2   **35.** $\dfrac{xy^2}{x^2+y^2}$ **37.** $\dfrac{2b^2+3a}{b(b-a)}$

39. $\dfrac{x}{(x+1)(x-1)}$ **41.** $\dfrac{1+a}{1-a}$ **43.** $\dfrac{x(x+6y)}{2y}$ **45.** $\dfrac{5a}{2(a+2)}$ **47.** $xy(5y+2x)$ **49.** $\dfrac{xy}{2x+5y}$ **51.** $\dfrac{x^2y^2}{4}$ **53.** $-9x^3y^4$

55. $-4,14$ **57.** a and c   **59.** $\dfrac{770a}{770-s}$ **61.** a, b   **63.** answers may vary   **65.** $\dfrac{1+x}{2+x}$ **67.** $x(x+1)$ **69.** $\dfrac{x-3y}{x+3y}$ **71.** $3a^2+4a+4$

73. a. $\dfrac{1}{a+h}$ **b.** $\dfrac{1}{a}$ **c.** $\dfrac{\dfrac{1}{a+h}-\dfrac{1}{a}}{h}$ **d.** $-\dfrac{1}{a(a+h)}$ **75. a.** $\dfrac{3}{a+h+1}$ **b.** $\dfrac{3}{a+1}$ **c.** $\dfrac{\dfrac{3}{a+h+1}-\dfrac{3}{a+1}}{h}$ **d.** $\dfrac{-3}{(a+h+1)(a+1)}$

Section 6.4
Practice Exercises

1. $3a^2-2a+5$ **2.** $5a^2b^2-8ab+1-\dfrac{8}{ab}$ **3.** $3x-2$ **4.** $3x-2$ **5.** $5x^2-6x+8+\dfrac{6}{x+3}$ **6.** $2x^2+3x-2+\dfrac{-8x+4}{x^2+1}$

7. $16x^2+20x+25$ **8.** $4x^2+x+7+\dfrac{12}{x-1}$ **9.** $x^3-5x+15-\dfrac{33}{x+3}$ **10. a.** -4 **b.** -4 **11.** 15

Vocabulary, Readiness & Video Check 6.4

1. the common denominator.   **3.** The last number is the remainder and the other numbers are the coefficients of the variables in the quotient; the degree of the quotient is one less than the degree of the dividend.

Exercise Set 6.4

1. $2a+4$ **3.** $3ab+4$ **5.** $2y+\dfrac{3y}{x}-\dfrac{2y}{x^2}$ **7.** $x+1$ **9.** $2x-8$ **11.** $x-\dfrac{1}{2}$ **13.** $2x^2-\dfrac{1}{2}x+5$ **15.** $2x^2-6$

17. $3x^3+5x+4-\dfrac{2x}{x^2-2}$ **19.** $2x^3+\dfrac{9}{2}x^2+10x+21+\dfrac{42}{x-2}$ **21.** $x+8$ **23.** $x-1$ **25.** $x^2-5x-23-\dfrac{41}{x-2}$

27. $4x+8+\dfrac{7}{x-2}$ **29.** $x^6y+\dfrac{2}{y}+1$ **31.** $5x^2-6-\dfrac{5}{2x-1}$ **33.** $2x^2+2x+8+\dfrac{28}{x-4}$ **35.** $2x^3-3x^2+x-4$

37. $3x^2+4x-8+\dfrac{20}{x+1}$ **39.** $3x^2+3x-3$ **41.** x^2+x+1 **43.** $-\dfrac{5y}{x}-\dfrac{15z}{x}-25z$ **45.** $3x^4-2x$ **47.** 1   **49.** -133 **51.** 3

53. $-\dfrac{187}{81}$ **55.** $\dfrac{95}{32}$ **57.** $-\dfrac{5}{6}$ **59.** 2   **61.** 54   **63.** $(x-1)(x^2+x+1)$ **65.** $(5z+2)(25z^2-10z+4)$ **67.** $(y+2)(x+3)$

69. $x(x+3)(x-3)$ **71.** yes   **73.** no   **75.** a or d   **77.** (x^4+2x^2-6) m   **79.** $(3x-7)$ in.   **81.** (x^3-5x^2+2x-1) cm

83. $x^3+\dfrac{5}{3}x^2+\dfrac{5}{3}x+\dfrac{8}{3}+\dfrac{8}{3(x-1)}$ **85.** $\dfrac{3}{2}x^3+\dfrac{1}{4}x^2+\dfrac{1}{8}x-\dfrac{7}{16}+\dfrac{1}{16(2x-1)}$ **87.** $x^3-\dfrac{2}{5}x$ **89.** $5x-1+\dfrac{6}{x}; x\neq 0$

91. $7x^3+14x^2+25x+50+\dfrac{102}{x-2}; x\neq 2$ **93.** answers may vary   **95.** answers may vary   **97.** $(x+3)(x^2+4)=x^3+3x^2+4x+12$

99. $2x^3-5x^2-31x+33$ **101. a.** $m(x)=\dfrac{0.48x^3+2.06x^2+141x+9.71}{1011x-288}$ **b.** 0.26   **103.** 0

Section 6.5
Practice Exercises

1. 4   **2.** 7   **3.** $\{\,\}$ or $\varnothing$ **4.** -1 **5.** 1   **6.** $12,-1$

Vocabulary, Readiness & Video Check 6.5

1. equation   **3.** expression   **5.** equation   **7.** $14; c$ **9.** $(x+4)(x-4); a$ **11.** Linear and quadratic equations are solved in very different ways, so you need to determine the next correct move to make.

Exercise Set 6.5

1. 72   **3.** 2   **5.** 6   **7.** $2,-2$ **9.** $\varnothing$ **11.** $-\dfrac{28}{3}$ **13.** 3   **15.** -8 **17.** 3   **19.** $\varnothing$ **21.** 1   **23.** 3   **25.** -1 **27.** 6   **29.** $\dfrac{1}{3}$

31. $-5,5$ **33.** 3   **35.** 7   **37.** $\varnothing$ **39.** $\dfrac{4}{3}$ **41.** -12 **43.** $1,\dfrac{11}{4}$ **45.** $-5,-1$ **47.** $-\dfrac{7}{5}$ **49.** 5   **51.** length, 15 in.; width, 10 in.

53. 14.8%   **55.** 25–29   **57.** 128,781 inmates   **59.** answers may vary   **61.** 3000 game disks   **63.** $\dfrac{1}{16},\dfrac{1}{3}$ **65.** $-\dfrac{1}{5},1$ **67.** -0.17

69. 0.42   **71.** $-1,0$ **73.** -2 **75.**   **77.**

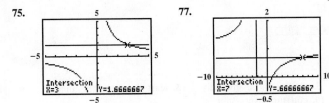

Integrated Review

1. $\dfrac{1}{2}$　**2.** 10　**3.** $\dfrac{1+2x}{8}$　**4.** $\dfrac{15+x}{10}$　**5.** $\dfrac{2(x-4)}{(x+2)(x-1)}$　**6.** $-\dfrac{5(x-8)}{(x-2)(x+4)}$　**7.** 4　**8.** 8　**9.** -5　**10.** $-\dfrac{2}{3}$　**11.** $\dfrac{2x+5}{x(x-3)}$

12. $\dfrac{5}{2x}$　**13.** -2　**14.** $-\dfrac{y}{x}$　**15.** $\dfrac{(a+3)(a+1)}{a+2}$　**16.** $\dfrac{-a^2+31a+10}{5(a-6)(a+1)}$　**17.** $-\dfrac{1}{5}$　**18.** $-\dfrac{3}{13}$　**19.** $\dfrac{4a+1}{(3a+1)(3a-1)}$

20. $\dfrac{-a-8}{4a(a-2)}$ or $-\dfrac{a+8}{4a(a-2)}$　**21.** $-1, \dfrac{3}{2}$　**22.** $\dfrac{x^2-3x+10}{2(x+3)(x-3)}$　**23.** $\dfrac{3}{x+1}$　**24.** $\{x \mid x \text{ is a real number and } x \neq 2, x \neq -1\}$　**25.** -1

26. $\dfrac{22z-45}{3z(z-3)}$　**27. a.** $\dfrac{x}{5} - \dfrac{x}{4} + \dfrac{1}{10}$　**b.** Write each rational expression term so that the denominator is the LCD, 20.　**c.** $\dfrac{-x+2}{20}$

28. a. $\dfrac{x}{5} - \dfrac{x}{4} = \dfrac{1}{10}$　**b.** Clear the equation of fractions by multiplying each term by the LCD, 20.　**c.** -2　**29.** b　**30.** d　**31.** d　**32.** a　**33.** d

Section 6.6
Practice Exercises

1. $a = \dfrac{bc}{b+c}$　**2.** 7　**3.** 3000 homes　**4.** $1\dfrac{1}{5}$ hr　**5.** 8 mph

Vocabulary, Readiness & Video Check 6.6

1. After we multiply through by the LCD, we factor the specified variable out so that we can divide and get the specified variable alone.　**3.** We write the same units in the numerators and the same units in the denominators.　**5.** $x + y$ (or $y + x$)

Exercise Set 6.6

1. $C = \dfrac{5}{9}(F-32)$　**3.** $I = A - QL$　**5.** $R = \dfrac{R_1 R_2}{R_1 + R_2}$　**7.** $n = \dfrac{2S}{a+L}$　**9.** $b = \dfrac{2A - ah}{h}$　**11.** $T_2 = \dfrac{P_2 V_2 T_1}{P_1 V_1}$　**13.** $f_2 = \dfrac{f_1 f}{f_1 - f}$

15. $L = \dfrac{n\lambda}{2}$　**17.** $c = \dfrac{2L\omega}{\theta}$　**19.** 1 or 5　**21.** 5　**23.** 4.5 gal　**25.** 5549 women　**27.** 15.6 hr　**29.** 10 min　**31.** 200 mph

33. 15 mph　**35.** -8 and -7　**37.** 36 min　**39.** 45 mph; 60 mph　**41.** 5.9 hr　**43.** 2 hr　**45.** 135 mph　**47.** 12 mi　**49.** $\dfrac{7}{8}$

51. $1\dfrac{1}{2}$ min　**53.** 1 hr　**55.** $2\dfrac{2}{9}$ hr　**57.** 2 hr　**59.** 270 movies　**61.** -5　**63.** 2　**65.** 164 lb　**67.** higher; F　**69.** 6 ohms

71. $\dfrac{1}{R} = \dfrac{1}{R_1} + \dfrac{1}{R_2} + \dfrac{1}{R_3}$; $R = \dfrac{15}{13}$ ohms

Section 6.7
Practice Exercises

1. $k = \dfrac{4}{3}; y = \dfrac{4}{3}x$　**2.** $18\dfrac{3}{4}$ in.　**3.** $k = 45; b = \dfrac{45}{a}$　**4.** $653\dfrac{1}{3}$ kilopascals　**5.** $A = kpa$　**6.** $k = 4; y = \dfrac{4}{x^3}$　**7.** $k = 81; y = \dfrac{81z}{x^3}$

Vocabulary, Readiness & Video Check 6.7

1. direct　**3.** joint　**5.** inverse　**7.** direct　**9.** linear; slope　**11.** $y = ka^2 b^5$

Exercise Set 6.7

1. $k = \dfrac{1}{5}; y = \dfrac{1}{5}x$　**3.** $k = \dfrac{3}{2}; y = \dfrac{3}{2}x$　**5.** $k = 14; y = 14x$　**7.** $k = 0.25; y = 0.25x$　**9.** 4.05 lb　**11.** 187,239 tons　**13.** $k = 30; y = \dfrac{30}{x}$

15. $k = 700; y = \dfrac{700}{x}$　**17.** $k = 2; y = \dfrac{2}{x}$　**19.** $k = 0.14; y = \dfrac{0.14}{x}$　**21.** 54 mph　**23.** 72 amps　**25.** divided by 4　**27.** $x = kyz$

29. $r = kst^3$　**31.** $k = \dfrac{1}{3}; y = \dfrac{1}{3}x^3$　**33.** $k = 0.2; y = 0.2\sqrt{x}$　**35.** $k = 1.3; y = \dfrac{1.3}{x^2}$　**37.** $k = 3; y = 3xz^3$　**39.** 22.5 tons　**41.** 15π cu in.

43. 8 ft　**45.** $y = kx$　**47.** $a = \dfrac{k}{b}$　**49.** $y = kxz$　**51.** $y = \dfrac{k}{x^3}$　**53.** $y = \dfrac{kx}{p^2}$　**55.** $C = 8\pi$ in.; $A = 16\pi$ sq in.

57. $C = 18\pi$ cm; $A = 81\pi$ sq cm　**59.** 9　**61.** 1　**63.** $\dfrac{1}{2}$　**65.** $\dfrac{2}{3}$　**67.** a　**69.** c　**71.** multiplied by 8　**73.** multiplied by 2

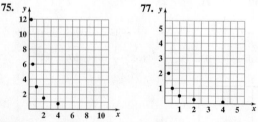

Chapter 6 Vocabulary Check

1. complex fraction　**2.** long division　**3.** directly　**4.** inversely　**5.** least common denominator　**6.** synthetic division　**7.** jointly
8. opposites　**9.** rational expression　**10.** equation, expression

Chapter 6 Review

1. $\{x \mid x \text{ is a real number}\}$ **3.** $\{x \mid x \text{ is a real number and } x \neq 5\}$ **5.** $\{x \mid x \text{ is a real number and } x \neq 0, x \neq -8\}$ **7.** -1 **9.** $\dfrac{1}{x-1}$

11. $\dfrac{2(x-3)}{x-4}$ **13.** $-\dfrac{3}{2}$ **15.** $\dfrac{a-b}{2a}$ **17.** $\dfrac{12}{5}$ **19.** $\dfrac{a-b}{5a}$ **21.** $-\dfrac{1}{x}$ **23.** $60x^2y^5$ **25.** $5x(x-5)$ **27.** $\dfrac{4+x}{x-4}$ **29.** $\dfrac{3}{2(x-2)}$

31. $\dfrac{-7x-6}{5(x-3)(x+3)}$ **33.** $\dfrac{5a-1}{(a-1)^2(a+1)}$ **35.** $\dfrac{5-2x}{2(x-1)}$ **37.** $\dfrac{4-3x}{8+x}$ **39.** $\dfrac{5(4x-3)}{2(5x^2-2)}$ **41.** $\dfrac{x(5y+1)}{3y}$ **43.** $\dfrac{1+x}{1-x}$ **45.** $\dfrac{3}{a+h}$

47. $\dfrac{\dfrac{3}{a+h}-\dfrac{3}{a}}{h}$ **49.** $1+\dfrac{x}{2y}-\dfrac{9}{4xy}$ **51.** $3x^3+9x^2+2x+6-\dfrac{2}{x-3}$ **53.** $x^2-1+\dfrac{5}{2x+3}$ **55.** $3x^2+6x+24+\dfrac{44}{x-2}$

57. $x^2+3x+9-\dfrac{54}{x-3}$ **59.** 3043 **61.** $\dfrac{365}{32}$ **63.** 6 **65.** $\dfrac{3}{2}$ **67.** $\dfrac{2x+5}{x(x-7)}$ **69.** $-\dfrac{1}{3},2$ **71.** $a=\dfrac{2A}{h}-b$ **73.** $R=\dfrac{E}{I}-r$

75. $x=\dfrac{yz}{z-y}$ **77.** $1,2$ **79.** $1\dfrac{23}{37}$ hr **81.** 8 mph **83.** 9 **85.** 2 **87.** $\dfrac{3}{5x}$ **89.** $\dfrac{5(a-2)}{7}$ **91.** $\dfrac{13}{3x}$ **93.** $\dfrac{1}{x-2}$ **95.** $\dfrac{2}{15-2x}$

97. $\dfrac{2(x-1)}{x+6}$ **99.** 2 **101.** $\dfrac{23}{25}$ **103.** 10 hr **105.** $63\dfrac{2}{3}$ mph; 45 mph **107.** $32,670$ **109.** $3x^3+13x^2+51x+204+\dfrac{814}{x-4}$

Chapter 6 Test

1. $\{x \mid x \text{ is a real number and } x \neq 1\}$ **2.** $\{x \mid x \text{ is a real number and } x \neq -3, x \neq -1\}$ **3.** $-\dfrac{7}{8}$ **4.** $\dfrac{x}{x+9}$ **5.** x^2+2x+4 **6.** $\dfrac{5}{3x}$

7. $\dfrac{3}{x^3}$ **8.** $\dfrac{x+2}{2(x+3)}$ **9.** $-\dfrac{4(2x+9)}{5}$ **10.** -1 **11.** $\dfrac{1}{x(x+3)}$ **12.** $\dfrac{5x-2}{(x-3)(x+2)(x-2)}$ **13.** $\dfrac{-x+30}{6(x-7)}$ **14.** $\dfrac{3}{2}$ **15.** $\dfrac{64}{3}$

16. $\dfrac{(x-3)^2}{x-2}$ **17.** $\dfrac{4xy}{3z}+\dfrac{3}{z}+1$ **18.** $2x^2-x-2+\dfrac{2}{2x+1}$ **19.** $4x^3-15x^2+45x-136+\dfrac{407}{x+3}$ **20.** 91 **21.** 7 **22.** 8 **23.** $\dfrac{2}{7}$

24. 3 **25.** $x=\dfrac{7a^2+b^2}{4a-b}$ **26.** 5 **27.** $\dfrac{6}{7}$ hr **28.** 16 **29.** 9 **30.** 256 ft

Chapter 6 Cumulative Review

1. a. $8x$ **b.** $8x+3$ **c.** $x \div -7$ or $\dfrac{x}{-7}$ **d.** $2x-1.6$ **e.** $x-6$ **f.** $2(4+x)$; Sec. 1.2, Ex. 8 **3.** 2; Sec. 2.1, Ex. 5

5. 2025 and after; Sec. 2.4, Ex. 9 **7.** $\varnothing$; Sec. 2.6, Ex. 7 **9.** -1; Sec. 2.7, Ex. 5 **11.** ; Sec. 3.1, Ex. 4

13. a. function **b.** not a function **c.** function; Sec. 3.2, Ex. 2 **15.** ; Sec. 3.3, Ex. 4 **17.** $y=-3x-2$; Sec. 3.5, Ex. 4

19. ; Sec. 3.7, Ex. 3 **21.** $(0,-5)$; Sec. 4.1, Ex. 8 **23.** $\varnothing$; Sec. 4.2, Ex. 2 **25.** $30°, 110°, 40°$; Sec. 4.3, Ex. 6

27. $(1,-1,3)$; Sec. 4.4, Ex. 3 **29. a.** 1 **b.** -1 **c.** 1 **d.** 2; Sec. 5.1, Ex. 3 **31. a.** $4x^b$ **b.** y^{5a+6}; Sec. 5.2, Ex. 5 **33. a.** 2 **b.** 5 **c.** 1 **d.** 6 **e.** 0; Sec. 5.3, Ex. 1 **35.** $9+12a+6b+4a^2+4ab+b^2$; Sec. 5.4, Ex. 9 **37.** $(b-6)(a+2)$; Sec. 5.5, Ex. 7

39. $2(n^2-19n+40)$; Sec. 5.6, Ex. 4 **41.** $(x+2+y)(x+2-y)$; Sec. 5.7, Ex. 5 **43.** $-2,6$; Sec. 5.8, Ex. 1 **45. a.** $\dfrac{1}{5x-1}$

b. $\dfrac{9x+4}{8x-7}$; Sec. 6.1, Ex. 2 **47.** $\dfrac{5k^2-7k+4}{(k+2)(k-2)(k-1)}$; Sec. 6.2, Ex. 4 **49.** -2; Sec. 6.5, Ex. 2

CHAPTER 7 RATIONAL EXPONENTS, RADICALS, AND COMPLEX NUMBERS

Section 7.1
Practice Exercises

1. a. 7 **b.** 0 **c.** $\dfrac{4}{9}$ **d.** 0.8 **e.** z^4 **f.** $4b^2$ **g.** -6 **h.** not a real number **2.** 6.708 **3. a.** -1 **b.** 3 **c.** $\dfrac{3}{4}$ **d.** x^4 **e.** $-2x$

4. a. 10 **b.** -1 **c.** -9 **d.** not a real number **e.** $3x^3$ **5. a.** 4 **b.** $|x^7|$ **c.** $|x+7|$ **d.** -7 **e.** $3x-5$

f. $7|x|$ **g.** $|x+8|$ **6. a.** 4 **b.** 2 **c.** 2 **d.** $\sqrt[3]{-9}$ **7.** **8.**

Vocabulary, Readiness & Video Check 7.1

1. index; radical sign; radicand **3.** is not **5.** $[0,\infty)$ **7.** $(16, 4)$ **9.** Divide the index into each exponent in the radicand. **11.** The square root of a negative number is not a real number, but the cube root of a negative number is a real number. **13.** For odd roots, there's only one root/answer whether the radicand is positive or negative, so absolute value bars aren't needed.

Exercise Set 7.1

1. 10 **3.** $\frac{1}{2}$ **5.** 0.01 **7.** -6 **9.** x^5 **11.** $4y^3$ **13.** 2.646 **15.** 6.164 **17.** 14.142 **19.** 4 **21.** $\frac{1}{2}$ **23.** -1 **25.** x^4
27. $-3x^3$ **29.** -2 **31.** not a real number **33.** -2 **35.** x^4 **37.** $2x^2$ **39.** $9x^2$ **41.** $4x^2$ **43.** 8 **45.** -8 **47.** $2|x|$
49. x **51.** $|x - 5|$ **53.** $|x + 2|$ **55.** -11 **57.** $2x$ **59.** y^6 **61.** $5ab^{10}$ **63.** $-3x^4y^3$ **65.** a^4b **67.** $-2x^2y$ **69.** $\frac{5}{7}$
71. $\frac{x^{10}}{2y}$ **73.** $-\frac{z^7}{3x}$ **75.** $\frac{x}{2}$ **77.** $\sqrt{3}$ **79.** -1 **81.** -3 **83.** $\sqrt{7}$
85. $[0,\infty)$; **87.** $[3,\infty)$; 0, 1, 2, 3 **89.** $(-\infty,\infty)$; **91.** $(-\infty,\infty)$; 0, 1, -1, 2, -2

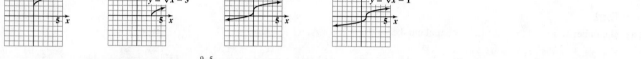

93. $-32x^{15}y^{10}$ **95.** $-60x^7y^{10}z^5$ **97.** $\frac{x^9y^5}{2}$ **99.** not a real number **101.** not a real number **103.** d **105.** d **107.** b **109.** b
111. answers may vary **113.** 1.69 sq m **115.** 11,181 m per sec **117.** answers may vary **119.**
121.

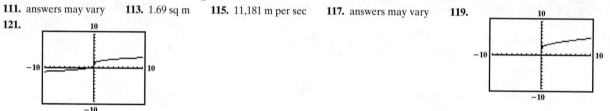

Section 7.2
Practice Exercises

1. a. 6 **b.** 10 **c.** $\sqrt[3]{x}$ **d.** 1 **e.** -8 **f.** $5x^3$ **g.** $3\sqrt[4]{x}$ **2. a.** 64 **b.** -1 **c.** -27 **d.** $\frac{1}{125}$ **e.** $\sqrt[9]{(3x + 2)^5}$
3. a. $\frac{1}{27}$ **b.** $\frac{1}{16}$ **4. a.** $y^{10/3}$ **b.** $x^{17/20}$ **c.** $\frac{1}{9}$ **d.** $b^{2/9}$ **e.** $\frac{81}{x^3y^{11/3}}$ **5. a.** $x^{14/15} - x^{13/5}$ **b.** $x + 4x^{1/2} - 12$ **6.** $x^{-1/5}(2 - 7x)$
7. a. $\sqrt[3]{x}$ **b.** $\sqrt{6}$ **c.** $\sqrt[4]{a^2b}$ **8. a.** $\sqrt[12]{x^7}$ **b.** $\sqrt[15]{y^2}$ **c.** $\sqrt[6]{675}$

Vocabulary, Readiness & Video Check 7.2

1. true **3.** true **5.** multiply, c **7.** $-\sqrt[5]{3x}$ **9.** denominator; positive **11.** Write the radical using an equivalent fractional exponent form, simplify the fraction, then write as a radical again.

Exercise Set 7.2

1. 7 **3.** 3 **5.** $\frac{1}{2}$ **7.** 13 **9.** $2\sqrt[3]{m}$ **11.** $3x^2$ **13.** -3 **15.** -2 **17.** 8 **19.** 16 **21.** not a real number **23.** $\sqrt[5]{(2x)^3}$
25. $\sqrt[3]{(7x + 2)^2}$ **27.** $\frac{64}{27}$ **29.** $\frac{1}{16}$ **31.** $\frac{1}{16}$ **33.** not a real number **35.** $\frac{1}{x^{1/4}}$ **37.** $a^{2/3}$ **39.** $\frac{5x^{3/4}}{7}$ **41.** $a^{7/3}$ **43.** x **45.** $3^{5/8}$
47. $y^{1/6}$ **49.** $8u^3$ **51.** $-b$ **53.** $\frac{1}{x^2}$ **55.** $27x^{2/3}$ **57.** $\frac{y}{z^{1/6}}$ **59.** $\frac{1}{x^{7/4}}$ **61.** $y - y^{7/6}$ **63.** $x^{5/3} - 2x^{2/3}$ **65.** $4x^{2/3} - 9$
67. $x^{8/3}(1 + x^{2/3})$ **69.** $x^{1/5}(x^{1/5} - 3)$ **71.** $x^{-1/3}(5 + x)$ **73.** $\sqrt{x}$ **75.** $\sqrt[3]{2}$ **77.** $2\sqrt{x}$ **79.** $\sqrt{xy}$ **81.** $\sqrt[3]{a^2b}$ **83.** $\sqrt{x + 3}$
85. $\sqrt[15]{y^{11}}$ **87.** $\sqrt[12]{b^5}$ **89.** $\sqrt[24]{x^{23}}$ **91.** $\sqrt{a}$ **93.** $\sqrt[15]{432}$ **95.** $\sqrt[15]{343y^5}$ **97.** $\sqrt[6]{125r^3s^2}$ **99.** $25 \cdot 3$ **101.** $16 \cdot 3$ or $4 \cdot 12$
103. $8 \cdot 2$ **105.** $27 \cdot 2$ **107.** A **109.** C **111.** B **113.** 1509 calories **115.** 302.0 million **117.** answers may vary **119.** $a^{1/3}$
121. $x^{1/5}$ **123.** 1.6818 **125.** 5.6645 **127.** $\frac{t^{1/2}}{u^{1/2}}$

Section 7.3
Practice Exercises

1. a. $\sqrt{35}$ **b.** $\sqrt{13z}$ **c.** 5 **d.** $\sqrt[3]{15x^2y}$ **e.** $\sqrt{\frac{5t}{2m}}$ **2. a.** $\frac{6}{7}$ **b.** $\frac{\sqrt{z}}{4}$ **c.** $\frac{5}{2}$ **d.** $\frac{\sqrt[4]{5}}{3x^2}$ **3. a.** $7\sqrt{2}$ **b.** $3\sqrt[3]{2}$
c. $\sqrt{35}$ **d.** $3\sqrt[4]{3}$ **4. a.** $6z^3\sqrt{z}$ **b.** $2pq^2\sqrt[3]{4pq}$ **c.** $2x^3\sqrt[4]{x^3}$ **5. a.** 4 **b.** $\frac{7}{3}\sqrt{z}$ **c.** $10xy^2\sqrt[3]{x^2}$ **d.** $6x^2y\sqrt[3]{2y}$
6. $\sqrt{17}$ units ≈ 4.123 units **7.** $\left(\frac{13}{2}, -4\right)$

Vocabulary, Readiness & Video Check 7.3

1. midpoint; point **3.** distance **5.** the indexes must be the same **7.** The power must be 1. Any even power is a perfect square and will leave no factor in the radicand; any higher odd power can have an even power factored from it, leaving one factor remaining in the radicand. **9.** average; average

Exercise Set 7.3

1. $\sqrt{14}$ **3.** 2 **5.** $\sqrt[3]{36}$ **7.** $\sqrt{6x}$ **9.** $\sqrt{\dfrac{14}{xy}}$ **11.** $\sqrt[4]{20x^3}$ **13.** $\dfrac{\sqrt{6}}{7}$ **15.** $\dfrac{\sqrt{2}}{7}$ **17.** $\dfrac{\sqrt[4]{x^3}}{2}$ **19.** $\dfrac{\sqrt[3]{4}}{3}$ **21.** $\dfrac{\sqrt[4]{8}}{x^2}$

23. $\dfrac{\sqrt[3]{2x}}{3y^4\sqrt[3]{3}}$ **25.** $\dfrac{x\sqrt{y}}{10}$ **27.** $\dfrac{x\sqrt{5}}{2y}$ **29.** $-\dfrac{z^2\sqrt[3]{z}}{3x}$ **31.** $4\sqrt{2}$ **33.** $4\sqrt[3]{3}$ **35.** $25\sqrt{3}$ **37.** $2\sqrt{6}$ **39.** $10x^2\sqrt{x}$ **41.** $2y^2\sqrt[3]{2y}$

43. $a^2b\sqrt[4]{b^3}$ **45.** $y^2\sqrt{y}$ **47.** $5ab\sqrt{b}$ **49.** $-2x^2\sqrt[5]{y}$ **51.** $x^4\sqrt[3]{50x^2}$ **53.** $-4a^4b^3\sqrt{2b}$ **55.** $3x^3y^4\sqrt{xy}$ **57.** $5r^3s^4$

59. $2x^3y^4\sqrt{2y}$ **61.** $\sqrt{2}$ **63.** 2 **65.** 10 **67.** x^2y **69.** $24m^2$ **71.** $\dfrac{15x\sqrt{2x}}{2}$ or $\dfrac{15x}{2}\sqrt{2x}$ **73.** $2a^2\sqrt[4]{2}$ **75.** $2xy^2\sqrt[5]{x^2}$ **77.** 5 units

79. $\sqrt{41}$ units ≈ 6.403 units **81.** $\sqrt{10}$ units ≈ 3.162 units **83.** $\sqrt{5}$ units ≈ 2.236 units **85.** $\sqrt{192.58}$ units ≈ 13.877 units **87.** $(4,-2)$

89. $\left(-5,\dfrac{5}{2}\right)$ **91.** $(3,0)$ **93.** $\left(-\dfrac{1}{2},\dfrac{1}{2}\right)$ **95.** $\left(\sqrt{2},\dfrac{\sqrt{5}}{2}\right)$ **97.** $(6.2,-6.65)$ **99.** $14x$ **101.** $2x^2-7x-15$ **103.** y^2

105. $-3x-15$ **107.** $x^2-8x+16$ **109.** true **111.** false **113.** true **115.** $\dfrac{\sqrt[3]{64}}{\sqrt{64}}=\dfrac{4}{8}=\dfrac{1}{2}$ **117.** x^7 **119.** a^3bc^5 **121.** $z^{10}\sqrt[3]{z^2}$

123. $q^2r^5s\sqrt[7]{q^3r^5}$ **125.** 1.6 m **127. a.** 20π sq cm **b.** 211.57 sq ft

Section 7.4
Practice Exercises

1. a. $8\sqrt{17}$ **b.** $-5\sqrt[3]{5z}$ **c.** $3\sqrt{2}+5\sqrt[3]{2}$ **2. a.** $11\sqrt{6}$ **b.** $-9\sqrt[3]{3}$ **c.** $-2\sqrt{3x}$ **d.** $2\sqrt{10}+2\sqrt[3]{5}$ **e.** $4x\sqrt[3]{3x}$

3. a. $\dfrac{5\sqrt{7}}{12}$ **b.** $\dfrac{13\sqrt[3]{6y}}{4}$ **4. a.** $2\sqrt{5}+5\sqrt{3}$ **b.** $2\sqrt{3}+2\sqrt{2}-\sqrt{30}-2\sqrt{5}$ **c.** $6z+\sqrt{z}-12$ **d.** $-6\sqrt{6}+15$

e. $5x-9$ **f.** $6\sqrt{x+2}+x+11$

Vocabulary, Readiness & Video Check 7.4

1. Unlike **3.** Like **5.** $6\sqrt{3}$ **7.** $7\sqrt{x}$ **9.** $8\sqrt[3]{x}$ **11.** Sometimes you can't see that there are like radicals until you simplify, so you may incorrectly think you cannot add or subtract if you don't simplify first.

Exercise Set 7.4

1. $-2\sqrt{2}$ **3.** $10x\sqrt{2x}$ **5.** $17\sqrt{2}-15\sqrt{5}$ **7.** $-\sqrt[3]{2x}$ **9.** $5b\sqrt{b}$ **11.** $\dfrac{31\sqrt{2}}{15}$ **13.** $\dfrac{\sqrt[3]{11}}{3}$ **15.** $\dfrac{5\sqrt{5x}}{9}$ **17.** $14+\sqrt{3}$

19. $7-3y$ **21.** $6\sqrt{3}-6\sqrt{2}$ **23.** $-23\sqrt[3]{5}$ **25.** $2b\sqrt{b}$ **27.** $20y\sqrt{2y}$ **29.** $2y\sqrt[3]{2x}$ **31.** $6\sqrt[3]{11}-4\sqrt{11}$ **33.** $3x\sqrt[4]{x^3}$ **35.** $\dfrac{2\sqrt{3}}{3}$

37. $\dfrac{5x\sqrt[3]{x}}{7}$ **39.** $\dfrac{5\sqrt{7}}{2x}$ **41.** $\dfrac{\sqrt[3]{2}}{6}$ **43.** $\dfrac{14x\sqrt[3]{2x}}{9}$ **45.** $15\sqrt{3}$ in. **47.** $\sqrt{35}+\sqrt{21}$ **49.** $7-2\sqrt{10}$ **51.** $3\sqrt{x}-x\sqrt{3}$

53. $6x-13\sqrt{x}-5$ **55.** $\sqrt[3]{a^2}+\sqrt[3]{a}-20$ **57.** $6\sqrt{2}-12$ **59.** $2+2x\sqrt{3}$ **61.** $-16-\sqrt{35}$ **63.** $x-y^2$ **65.** $3+2x\sqrt{3}+x^2$

67. $23x-5x\sqrt{15}$ **69.** $2\sqrt[3]{2}-\sqrt[3]{4}$ **71.** $x+1$ **73.** $x+24+10\sqrt{x-1}$ **75.** $2x+6-2\sqrt{2x+5}$ **77.** $x-7$ **79.** $\dfrac{7}{x+y}$

81. $2a-3$ **83.** $\dfrac{-2+\sqrt{3}}{3}$ **85.** $22\sqrt{5}$ ft; 150 sq ft **87. a.** $2\sqrt{3}$ **b.** 3 **c.** answers may vary **89.** $2\sqrt{6}-2\sqrt{2}-2\sqrt{3}+6$

91. answers may vary

Section 7.5
Practice Exercises

1. a. $\dfrac{5\sqrt{3}}{3}$ **b.** $\dfrac{15\sqrt{x}}{2x}$ **c.** $\dfrac{\sqrt[3]{6}}{3}$ **2.** $\dfrac{\sqrt{15yz}}{5y}$ **3.** $\dfrac{\sqrt[3]{z^2x^2}}{3x^2}$ **4. a.** $\dfrac{5(3\sqrt{5}-2)}{41}$ **b.** $\dfrac{\sqrt{6}+5\sqrt{3}+\sqrt{10}+5\sqrt{5}}{-2}$ **c.** $\dfrac{6x-3\sqrt{xy}}{4x-y}$

5. $\dfrac{2}{\sqrt{10}}$ **6.** $\dfrac{5b}{\sqrt[3]{50ab^2}}$ **7.** $\dfrac{x-9}{4(\sqrt{x}+3)}$

Vocabulary, Readiness & Video Check 7.5

1. conjugate **3.** rationalizing the numerator **5.** To write an equivalent expression without a radical in the denominator. **7.** No, except for the fact you're working with numerators, the process is the same.

Exercise Set 7.5

1. $\dfrac{\sqrt{14}}{7}$ **3.** $\dfrac{\sqrt{5}}{5}$ **5.** $\dfrac{2\sqrt{x}}{x}$ **7.** $\dfrac{4\sqrt[3]{9}}{3}$ **9.** $\dfrac{3\sqrt{2x}}{4x}$ **11.** $\dfrac{3\sqrt[3]{2x}}{2x}$ **13.** $\dfrac{3\sqrt{3a}}{a}$ **15.** $\dfrac{3\sqrt[3]{4}}{2}$ **17.** $\dfrac{2\sqrt{21}}{7}$ **19.** $\dfrac{\sqrt{10xy}}{5y}$ **21.** $\dfrac{\sqrt[3]{75}}{5}$

23. $\dfrac{\sqrt{6x}}{10}$ **25.** $\dfrac{\sqrt{3z}}{6z}$ **27.** $\dfrac{\sqrt[3]{6xy^2}}{3x}$ **29.** $\dfrac{3\sqrt[4]{2}}{2}$ **31.** $\dfrac{2\sqrt[4]{9x}}{3x^2}$ **33.** $\dfrac{5\sqrt[3]{4ab^4}}{2ab^3}$ **35.** $\sqrt{2}-x$ **37.** $5+\sqrt{a}$ **39.** $-7\sqrt{5}-8\sqrt{x}$

41. $-2(2 + \sqrt{7})$ **43.** $\dfrac{7(3 + \sqrt{x})}{9 - x}$ **45.** $-5 + 2\sqrt{6}$ **47.** $\dfrac{2a + 2\sqrt{a} + \sqrt{ab} + \sqrt{b}}{4a - b}$ **49.** $-\dfrac{8(1 - \sqrt{10})}{9}$ **51.** $\dfrac{x - \sqrt{xy}}{x - y}$

53. $\dfrac{5 + 3\sqrt{2}}{7}$ **55.** $\dfrac{5}{\sqrt{15}}$ **57.** $\dfrac{6}{\sqrt{10}}$ **59.** $\dfrac{2x}{7\sqrt{x}}$ **61.** $\dfrac{5y}{\sqrt[3]{100xy}}$ **63.** $\dfrac{2}{\sqrt{10}}$ **65.** $\dfrac{2x}{11\sqrt{2x}}$ **67.** $\dfrac{7}{2\sqrt[3]{49}}$ **69.** $\dfrac{3x^2}{10\sqrt[3]{9x}}$ **71.** $\dfrac{6x^2y^3}{\sqrt{6z}}$

73. $\dfrac{-7}{12 + 6\sqrt{11}}$ **75.** $\dfrac{3}{10 + 5\sqrt{7}}$ **77.** $\dfrac{x - 9}{x - 3\sqrt{x}}$ **79.** $\dfrac{1}{3 + 2\sqrt{2}}$ **81.** $\dfrac{x - 1}{x - 2\sqrt{x} + 1}$ **83.** 5   **85.** $-\dfrac{1}{2}, 6$ **87.** $2, 6$ **89.** $r = \dfrac{\sqrt{A\pi}}{2\pi}$

91. a. $\dfrac{y\sqrt{15xy}}{6x^2}$ **b.** $\dfrac{y\sqrt{15xy}}{6x^2}$ **c.** answers may vary   **93.** $\sqrt[3]{25}$ **95.** answers may vary   **97.** answers may vary

Integrated Review

1. 9   **2.** -2 **3.** $\dfrac{1}{2}$ **4.** x^3 **5.** y^3 **6.** $2y^5$ **7.** $-2y$ **8.** $3b^3$ **9.** 6   **10.** $\sqrt[4]{3y}$ **11.** $\dfrac{1}{16}$ **12.** $\sqrt[5]{(x + 1)^3}$ **13.** y

14. $16x^{1/2}$ **15.** $x^{5/4}$ **16.** $4^{11/15}$ **17.** $2x^2$ **18.** $\sqrt[4]{a^3b^2}$ **19.** $\sqrt[4]{x^3}$ **20.** $\sqrt[6]{500}$ **21.** $2\sqrt{10}$ **22.** $2xy^2\sqrt[4]{x^3y^2}$ **23.** $3x\sqrt[3]{2x}$

24. $-2b^2\sqrt[5]{2}$ **25.** $\sqrt{5x}$ **26.** $4x$ **27.** $7y^2\sqrt{y}$ **28.** $2a^2\sqrt[4]{3}$ **29.** $2\sqrt{5} - 5\sqrt{3} + 5\sqrt{7}$ **30.** $y\sqrt[3]{2y}$ **31.** $\sqrt{15} - \sqrt{6}$ **32.** $10 + 2\sqrt{21}$

33. $4x^2 - 5$ **34.** $x + 2 - 2\sqrt{x + 1}$ **35.** $\dfrac{\sqrt{21}}{3}$ **36.** $\dfrac{5\sqrt[3]{4x}}{2x}$ **37.** $\dfrac{13 - 3\sqrt{21}}{5}$ **38.** $\dfrac{7}{\sqrt{21}}$ **39.** $\dfrac{3y}{\sqrt[3]{33y^2}}$ **40.** $\dfrac{x - 4}{x + 2\sqrt{x}}$

Section 7.6
Practice Exercises

1. 18   **2.** $\dfrac{1}{4}, \dfrac{3}{4}$ **3.** 10   **4.** 9   **5.** $\dfrac{3}{25}$ **6.** $6\sqrt{3}$ meters   **7.** $\sqrt{193}$ in. ≈ 13.89

Graphing Calculator Explorations 7.6

1. 3.19   **3.** $\varnothing$ **5.** 3.23

Vocabulary, Readiness & Video Check 7.6

1. extraneous solution   **3.** $x^2 - 10x + 25$ **5.** Applying the power rule can result in an equation with more solutions than the original equation, so you need to check all proposed solutions in the original equation.   **7.** Our answer is either a positive square root of a value or a negative square root of a value. We're looking for a length, which must be positive, so our answer must be the positive square root.

Exercise Set 7.6

1. 8   **3.** 7   **5.** $\varnothing$ **7.** 7   **9.** 6   **11.** $-\dfrac{9}{2}$ **13.** 29   **15.** 4   **17.** -4 **19.** $\varnothing$ **21.** 7   **23.** 9   **25.** 50   **27.** $\varnothing$

29. $\dfrac{15}{4}$ **31.** 13   **33.** 5   **35.** -12 **37.** 9   **39.** -3 **41.** 1   **43.** 1   **45.** $\dfrac{1}{2}$ **47.** $0, 4$ **49.** $\dfrac{37}{4}$ **51.** $3\sqrt{5}$ ft

53. $2\sqrt{10}$ m   **55.** $2\sqrt{131}$ m ≈ 22.9 m   **57.** $\sqrt{100.84}$ mm ≈ 10.0 mm   **59.** 17 ft   **61.** 13 ft   **63.** 14,657,415 sq mi   **65.** 100 ft

67. 100   **69.** $\dfrac{\pi}{2}$ sec ≈ 1.57 sec   **71.** 12.97 ft   **73.** answers may vary   **75.** $15\sqrt{3}$ sq mi ≈ 25.98 sq mi   **77.** answers may vary

79. 0.51 km   **81.** function   **83.** function   **85.** not a function   **87.** $\dfrac{x}{4x + 3}$ **89.** $-\dfrac{4z + 2}{3z}$

91. $\sqrt{5x - 1} + 4 = 7$
$\sqrt{5x - 1} = 3$
$(\sqrt{5x - 1})^2 = 3^2$
$5x - 1 = 9$
$5x = 10$
$x = 2$
93. 1   **95. a.–b.** answers may vary   **97.** $-1, 0, 8, 9$ **99.** $-1, 4$

Section 7.7
Practice Exercises

1. a. $2i$ **b.** $i\sqrt{7}$ **c.** $-3i\sqrt{2}$ **2. a.** $-\sqrt{30}$ **b.** -3 **c.** $25i$ **d.** $3i$ **3. a.** $-1 - 4i$ **b.** $-3 + 5i$ **c.** $3 - 2i$ **4. a.** 20

b. $-5 + 10i$ **c.** $15 + 16i$ **d.** $8 - 6i$ **e.** 85   **5. a.** $\dfrac{11}{10} - \dfrac{7i}{10}$ **b.** $0 - \dfrac{5i}{2}$ **6. a.** i **b.** 1   **c.** -1 **d.** 1

Vocabulary, Readiness & Video Check 7.7

1. complex   **3.** -1 **5.** real   **7.** The product rule for radicals; you need to first simplify each separate radical and have nonnegative radicands before applying the product rule.   **9.** the fact that $i^2 = -1$ **11.** $i, i^2 = -1, i^3 = -i, i^4 = 1$

Exercise Set 7.7

1. $9i$ **3.** $i\sqrt{7}$ **5.** -4 **7.** $8i$ **9.** $2i\sqrt{6}$ **11.** $-6i$ **13.** $24i\sqrt{7}$ **15.** $-3\sqrt{6} + 0i$ **17.** $-\sqrt{14}$ **19.** $-5\sqrt{2}$ **21.** $4i$ **23.** $i\sqrt{3}$

25. $2\sqrt{2}$ **27.** $6 - 4i$ **29.** $-2 + 6i$ **31.** $-2 - 4i$ **33.** $-40 + 0i$ **35.** $18 + 12i$ **37.** $7 + 0i$ **39.** $12 - 16i$ **41.** $0 - 4i$

43. $\dfrac{28}{25} - \dfrac{21}{25}i$ **45.** $4 + i$ **47.** $\dfrac{17}{13} + \dfrac{7}{13}i$ **49.** $63 + 0i$ **51.** $2 - i$ **53.** $27 + 3i$ **55.** $-\dfrac{5}{2} - 2i$ **57.** $18 + 13i$ **59.** $20 + 0i$

61. $10 + 0i$ **63.** $2 + 0i$ **65.** $-5 + \dfrac{16}{3}i$ **67.** $17 + 144i$ **69.** $\dfrac{3}{5} - \dfrac{1}{5}i$ **71.** $5 - 10i$ **73.** $\dfrac{1}{5} - \dfrac{8}{5}i$ **75.** $8 - i$ **77.** $7 + 0i$

79. $12 - 16i$ **81.** 1 **83.** i **85.** $-i$ **87.** -1 **89.** -64 **91.** $-243i$ **93.** $40°$ **95.** $x^2 - 5x - 2 - \dfrac{6}{x - 1}$ **97.** 5 people

99. 14 people   **101.** 16.7%   **103.** $-1 - i$ **105.** $0 + 0i$ **107.** $2 + 3i$ **109.** $2 + i\sqrt{2}$ **111.** $\dfrac{1}{2} - \dfrac{\sqrt{3}}{2}i$ **113.** answers may vary

115. $6 - 3i\sqrt{3}$ **117.** yes

Chapter 7 Vocabulary Check

1. conjugate   **2.** principal square root   **3.** rationalizing   **4.** imaginary unit   **5.** cube root   **6.** index; radicand   **7.** like radicals
8. complex number   **9.** distance   **10.** midpoint

Chapter 7 Review

1. 9 **3.** -2 **5.** $-\dfrac{1}{7}$ **7.** -6 **9.** $-a^2b^3$ **11.** $2ab^2$ **13.** $\dfrac{x^6}{6y}$ **15.** $|-x|$ **17.** -27 **19.** $-x$ **21.** $2|2y + z|$ **23.** y **25. a.** $3, 6$

b. $[0, \infty)$ **c.**
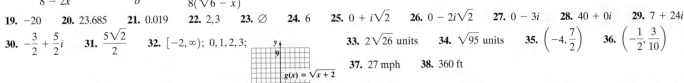

27. $\dfrac{1}{3}$ **29.** $-\dfrac{1}{3}$ **31.** -27 **33.** not a real number   **35.** $\dfrac{9}{4}$ **37.** $x^{2/3}$ **39.** $\sqrt[5]{y^4}$ **41.** $\dfrac{1}{\sqrt[3]{x + 2}}$

43. $a^{13/6}$ **45.** $\dfrac{1}{a^{9/2}}$ **47.** a^4b^6 **49.** $\dfrac{b^{5/6}}{49a^{1/4}c^{5/3}}$ **51.** 4.472 **53.** 5.191 **55.** -26.246 **57.** $\sqrt[6]{1372}$

59. $2\sqrt{6}$ **61.** $2x$ **63.** $2\sqrt{15}$ **65.** $3\sqrt[3]{6}$ **67.** $6x^3\sqrt{x}$ **69.** $\dfrac{p^8\sqrt{p}}{11}$ **71.** $\dfrac{y\sqrt[4]{xy^2}}{3}$

73. $\dfrac{5}{\sqrt{\pi}}$ m or $\dfrac{5\sqrt{\pi}}{\pi}$ m   **75.** $\sqrt{197}$ units ≈ 14.036 units   **77.** $\sqrt{73}$ units ≈ 8.544 units   **79.** $2\sqrt{11}$ units ≈ 6.633 units   **81.** $(-5, 5)$

83. $\left(-\dfrac{11}{2}, -2\right)$ **85.** $\left(\dfrac{1}{4}, \dfrac{2}{7}\right)$ **87.** $-2\sqrt{5}$ **89.** $9\sqrt[3]{2}$ **91.** $\dfrac{15 + 2\sqrt{3}}{6}$ **93.** $17\sqrt{2} - 15\sqrt{5}$ **95.** 6 **97.** $-8\sqrt{5}$

99. $a - 9$ **101.** $\sqrt[3]{25x^2} - 81$ **103.** $\dfrac{3\sqrt{7}}{7}$ **105.** $\dfrac{5\sqrt[3]{2}}{2}$ **107.** $\dfrac{x^2y^2\sqrt[3]{15yz}}{z}$ **109.** $\dfrac{3\sqrt{y} + 6}{y - 4}$ **111.** $\dfrac{11}{3\sqrt{11}}$ **113.** $\dfrac{3}{7\sqrt[3]{3}}$ **115.** $\dfrac{xy}{\sqrt[3]{10x^2yz}}$

117. 32 **119.** 35 **121.** 9 **123.** $3\sqrt{2}$ cm   **125.** 51.2 ft   **127.** $2i\sqrt{2}$ **129.** $6i$ **131.** $15 - 4i$ **133.** -64

135. $-12 - 18i$ **137.** $-5 - 12i$ **139.** $\dfrac{3}{2} - i$ **141.** x **143.** -10 **145.** $\dfrac{y^5}{2x^3}$ **147.** $\dfrac{1}{8}$ **149.** $\dfrac{1}{x^{13/2}}$ **151.** $\dfrac{n\sqrt{3n}}{11m^5}$

153. $4x - 20\sqrt{x} + 25$ **155.** $(4, 16)$ **157.** $\dfrac{2\sqrt{x} - 6}{x - 9}$ **159.** 5

Chapter 7 Test

1. $6\sqrt{6}$ **2.** $-x^{16}$ **3.** $\dfrac{1}{5}$ **4.** 5 **5.** $\dfrac{4x^2}{9}$ **6.** $-a^6b^3$ **7.** $\dfrac{8a^{1/3}c^{2/3}}{b^{5/12}}$ **8.** $a^{7/12} - a^{7/3}$ **9.** $|4xy|$ or $4|xy|$ **10.** -27 **11.** $\dfrac{3\sqrt{y}}{y}$

12. $\dfrac{8 - 6\sqrt{x} + x}{8 - 2x}$ **13.** $\dfrac{\sqrt[3]{b^2}}{b}$ **14.** $\dfrac{6 - x^2}{8(\sqrt{6} - x)}$ **15.** $-x\sqrt{5x}$ **16.** $4\sqrt{3} - \sqrt{6}$ **17.** $x + 2\sqrt{x} + 1$ **18.** $\sqrt{6} - 4\sqrt{3} + \sqrt{2} - 4$

19. -20 **20.** 23.685 **21.** 0.019 **22.** $2, 3$ **23.** $\varnothing$ **24.** 6 **25.** $0 + i\sqrt{2}$ **26.** $0 - 2i\sqrt{2}$ **27.** $0 - 3i$ **28.** $40 + 0i$ **29.** $7 + 24i$

30. $-\dfrac{3}{2} + \dfrac{5}{2}i$ **31.** $\dfrac{5\sqrt{2}}{2}$ **32.** $[-2, \infty); 0, 1, 2, 3;$ **33.** $2\sqrt{26}$ units   **34.** $\sqrt{95}$ units   **35.** $\left(-4, \dfrac{7}{2}\right)$ **36.** $\left(-\dfrac{1}{2}, \dfrac{3}{10}\right)$

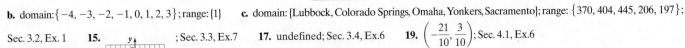

37. 27 mph   **38.** 360 ft

Chapter 7 Cumulative Review

1. a. $2xy - 2$ **b.** $2x^2 + 23$ **c.** $3.1x - 0.3$ **d.** $-a - 7b + \dfrac{7}{12}$; Sec. 1.4 Ex. 15   **3.** $\dfrac{21}{11}$; Sec. 2.1 Ex. 6   **5.** \$4500 per month; Sec. 2.4 Ex. 8

7. $\varnothing$; Sec. 2.6, Ex. 6   **9.** $(-\infty, -3] \cup [9, \infty)$; Sec. 2.7, Ex. 8   **11.** ; Sec. 3.1 Ex.7   **13. a.** domain: $\{2, 0, 3\}$; range: $\{3, 4, -1\}$

b. domain: $\{-4, -3, -2, -1, 0, 1, 2, 3\}$; range: $\{1\}$ **c.** domain: {Lubbock, Colorado Springs, Omaha, Yonkers, Sacramento}; range: $\{370, 404, 445, 206, 197\}$;

Sec. 3.2, Ex. 1   **15.** ; Sec. 3.3, Ex.7   **17.** undefined; Sec. 3.4, Ex.6   **19.** $\left(-\dfrac{21}{10}, \dfrac{3}{10}\right)$; Sec. 4.1, Ex.6

21. a. 2^7 **b.** x^{10} **c.** y^7; Sec. 5.1; Ex. 1   **23.** 6×10^{-5}; Sec. 5.2, Ex. 7   **25. a.** -4 **b.** 11; Sec. 5.3, Ex. 4   **27. a.** $2x^2 + 11x + 15$
b. $10x^3 - 27x^2 + 32x - 21$; Sec. 5.4, Ex. 3   **29.** $5x^2$; Sec. 5.5, Ex. 1   **31. a.** $x^2 - 2x + 4$ **b.** $\dfrac{2}{y-5}$; Sec. 6.1, Ex. 5   **33. a.** $\dfrac{6x+5}{3x^3y}$
b. $\dfrac{2x^2 + 7x - 6}{(x+2)(x-2)}$ **c.** 2; Sec. 6.2, Ex. 3   **35. a.** $\dfrac{x(x-2)}{2(x+2)}$ **b.** $\dfrac{x^2}{y^2}$; Sec. 6.3, Ex. 2   **37.** $2x^2 - x + 4$; Sec. 6.4, Ex. 1
39. $2x^2 + 5x + 2 + \dfrac{7}{x-3}$; Sec. 6.4, Ex. 8   **41.** $\varnothing$; Sec. 6.5, Ex. 3   **43.** $x = \dfrac{yz}{y-z}$; Sec. 6.6, Ex. 1   **45.** constant of variation: 15; $u = \dfrac{15}{w}$;
Sec. 6.7, Ex. 3   **47. a.** $\dfrac{1}{8}$ **b.** $\dfrac{1}{9}$; Sec. 7.2, Ex. 3   **49.** $\dfrac{x-4}{5(\sqrt{x}-2)}$; Sec. 7.5, Ex. 7

CHAPTER 8 QUADRATIC EQUATIONS AND FUNCTIONS

Section 8.1
Practice Exercises

1. $-4\sqrt{2}, 4\sqrt{2}$ **2.** $-\sqrt{10}, \sqrt{10}$ **3.** $-3 - 2\sqrt{5}, -3 + 2\sqrt{5}$ **4.** $\dfrac{2+3i}{5}, \dfrac{2-3i}{5}$ **5.** $-2 - \sqrt{7}, -2 + \sqrt{7}$ **6.** $\dfrac{3-\sqrt{5}}{2}, \dfrac{3+\sqrt{5}}{2}$
7. $\dfrac{6-\sqrt{33}}{3}, \dfrac{6+\sqrt{33}}{3}$ **8.** $\dfrac{5-i\sqrt{31}}{4}, \dfrac{5+i\sqrt{31}}{4}$ **9.** 6%

Graphing Calculator Explorations 8.1

1. $-1.27, 6.27$ 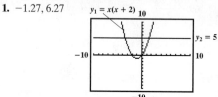   **3.** $-1.10, 0.90$ 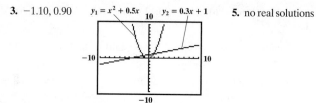   **5.** no real solutions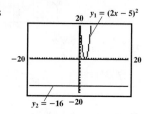

Vocabulary, Readiness & Video Check 8.1

1. $\pm\sqrt{b}$ **3.** completing the square   **5.** 9   **7.** We need a quantity squared by itself on one side of the equation. The only quantity squared is x, so we need to divide both sides by 2 before applying the square root property.   **9.** We're looking for an interest rate so a negative value does not make sense.

Exercise Set 8.1

1. $-4, 4$ **3.** $-\sqrt{7}, \sqrt{7}$ **5.** $-3\sqrt{2}, 3\sqrt{2}$ **7.** $-\sqrt{10}, \sqrt{10}$ **9.** $-8, -2$ **11.** $6 - 3\sqrt{2}, 6 + 3\sqrt{2}$ **13.** $\dfrac{3-2\sqrt{2}}{2}, \dfrac{3+2\sqrt{2}}{2}$
15. $-3i, 3i$ **17.** $-\sqrt{6}, \sqrt{6}$ **19.** $-2i\sqrt{2}, 2i\sqrt{2}$ **21.** $\dfrac{1-4i}{3}, \dfrac{1+4i}{3}$ **23.** $-7 - \sqrt{5}, -7 + \sqrt{5}$ **25.** $-3 - 2i\sqrt{2}, -3 + 2i\sqrt{2}$
27. $x^2 + 16x + 64 = (x+8)^2$ **29.** $z^2 - 12z + 36 = (z-6)^2$ **31.** $p^2 + 9p + \dfrac{81}{4} = \left(p + \dfrac{9}{2}\right)^2$ **33.** $x^2 + x + \dfrac{1}{4} = \left(x + \dfrac{1}{2}\right)^2$
35. $-5, -3$ **37.** $-3 - \sqrt{7}, -3 + \sqrt{7}$ **39.** $\dfrac{-1-\sqrt{5}}{2}, \dfrac{-1+\sqrt{5}}{2}$ **41.** $-1 - \sqrt{6}, -1 + \sqrt{6}$ **43.** $\dfrac{-1-\sqrt{29}}{2}, \dfrac{-1+\sqrt{29}}{2}$
45. $\dfrac{6-\sqrt{30}}{3}, \dfrac{6+\sqrt{30}}{3}$ **47.** $\dfrac{3-\sqrt{11}}{2}, \dfrac{3+\sqrt{11}}{2}$ **49.** $-4, \dfrac{1}{2}$ **51.** $-4 - \sqrt{15}, -4 + \sqrt{15}$ **53.** $\dfrac{-3-\sqrt{21}}{3}, \dfrac{-3+\sqrt{21}}{3}$ **55.** $-1, \dfrac{5}{2}$
57. $-1 - i, -1 + i$ **59.** $3 - \sqrt{17}, 3 + \sqrt{17}$ **61.** $-2 - i\sqrt{2}, -2 + i\sqrt{2}$ **63.** $\dfrac{-15-7\sqrt{5}}{10}, \dfrac{-15+7\sqrt{5}}{10}$ **65.** $\dfrac{1-i\sqrt{47}}{4}, \dfrac{1+i\sqrt{47}}{4}$
67. $-5 - i\sqrt{3}, -5 + i\sqrt{3}$ **69.** $-4, 1$ **71.** $\dfrac{2-i\sqrt{2}}{2}, \dfrac{2+i\sqrt{2}}{2}$ **73.** $\dfrac{-3-\sqrt{69}}{6}, \dfrac{-3+\sqrt{69}}{6}$ **75.** 20%   **77.** 4%   **79.** 9.63 sec
81. 8.29 sec   **83.** 15 ft by 15 ft   **85.** $10\sqrt{2}$ cm   **87.** -1 **89.** $3 + 2\sqrt{5}$ **91.** $\dfrac{1-3\sqrt{2}}{2}$ **93.** $2\sqrt{6}$ **95.** 5   **97.** complex, but not
real numbers   **99.** real numbers   **101.** complex, but not real numbers   **103.** $-8x, 8x$ **105.** $-5z, 5z$ **107.** answers may vary
109. compound; answers may vary

Section 8.2
Practice Exercises

1. $2, -\dfrac{1}{3}$ **2.** $\dfrac{4-\sqrt{22}}{3}, \dfrac{4+\sqrt{22}}{3}$ **3.** $1 - \sqrt{17}, 1 + \sqrt{17}$ **4.** $\dfrac{-1-i\sqrt{15}}{4}, \dfrac{1+i\sqrt{15}}{4}$ **5. a.** one real solution   **b.** two real solutions
c. two complex, but not real solutions   **6.** 6 ft   **7.** 2.4 sec

Vocabulary, Readiness & Video Check 8.2

1. $x = \dfrac{-b \pm \sqrt{b^2 - 4ac}}{2a}$ **3.** $-5; -7$ **5.** $1; 0$ **7. a.** Yes, in order to make sure we have correct values for a, b, and c.   **b.** No; clearing
fractions makes the work less tedious, but it's not a necessary step.   **9.** With applications, we need to make sure we answer the question(s) asked.
Here we're asked how much distance is saved, so once the dimensions of the triangle are known, further calculations are needed to answer this question
and solve the problem.

Exercise Set 8.2

1. $-6, 1$ **3.** $-\dfrac{3}{5}, 1$ **5.** 3 **7.** $\dfrac{-7 - \sqrt{33}}{2}, \dfrac{-7 + \sqrt{33}}{2}$ **9.** $\dfrac{1 - \sqrt{57}}{8}, \dfrac{1 + \sqrt{57}}{8}$ **11.** $\dfrac{7 - \sqrt{85}}{6}, \dfrac{7 + \sqrt{85}}{6}$ **13.** $1 - \sqrt{3}, 1 + \sqrt{3}$

15. $-\dfrac{3}{2}, 1$ **17.** $\dfrac{3 - \sqrt{11}}{2}, \dfrac{3 + \sqrt{11}}{2}$ **19.** $\dfrac{-5 - \sqrt{17}}{2}, \dfrac{-5 + \sqrt{17}}{2}$ **21.** $\dfrac{5}{2}, 1$ **23.** $-3 - 2i, -3 + 2i$ **25.** $-2 - \sqrt{11}, -2 + \sqrt{11}$

27. $\dfrac{3 - i\sqrt{87}}{8}, \dfrac{3 + i\sqrt{87}}{8}$ **29.** $\dfrac{3 - \sqrt{29}}{2}, \dfrac{3 + \sqrt{29}}{2}$ **31.** $\dfrac{-5 - i\sqrt{5}}{10}, \dfrac{-5 + i\sqrt{5}}{10}$ **33.** $\dfrac{-1 - \sqrt{19}}{6}, \dfrac{-1 + \sqrt{19}}{6}$

35. $\dfrac{-1 - i\sqrt{23}}{4}, \dfrac{-1 + i\sqrt{23}}{4}$ **37.** 1 **39.** $3 - \sqrt{5}, 3 + \sqrt{5}$ **41.** two real solutions   **43.** one real solution   **45.** two real solutions

47. two complex but not real solutions   **49.** two real solutions   **51.** 14 ft   **53.** $(2 + 2\sqrt{2})$cm, $(2 + 2\sqrt{2})$cm, $(4 + 2\sqrt{2})$cm

55. width: $(-5 + 5\sqrt{17})$ ft; length: $(5 + 5\sqrt{17})$ ft   **57. a.** $50\sqrt{2}$ m   **b.** 5000 sq m   **59.** 37.4 ft by 38.5 ft   **61.** base, $(2 + 2\sqrt{43})$ cm; height,

$(-1 + \sqrt{43})$ cm   **63.** 8.9 sec   **65.** 2.8 sec   **67.** $\dfrac{11}{5}$ **69.** 15   **71.** $(x^2 + 5)(x + 2)(x - 2)$ **73.** $(z + 3)(z - 3)(z + 2)(z - 2)$

75. b   **77.** answers may vary   **79.** 0.6, 2.4   **81.** Sunday to Monday   **83.** Wednesday   **85.** 32; yes   **87. a.** 20,568 thousand students

b. 2015   **89. a.** 9076 thousand barrels per day   **b.** 2007   **c.** 2012   **91.** answers may vary   **93.** $\dfrac{\sqrt{3}}{3}$ **95.** $\dfrac{-\sqrt{2} - i\sqrt{2}}{2}, \dfrac{-\sqrt{2} + i\sqrt{2}}{2}$

97. $\dfrac{\sqrt{3} - \sqrt{11}}{4}, \dfrac{\sqrt{3} + \sqrt{11}}{4}$ **99.** 8.9 sec: 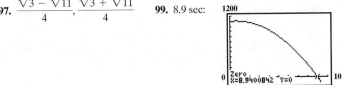   2.8 sec: 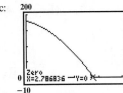   **101.** two real solutions

Section 8.3
Practice Exercises

1. 8   **2.** $\dfrac{5 \pm \sqrt{137}}{8}$ **3.** $4, -4, 3i, -3i$ **4.** $1, -3$ **5.** $1, 64$ **6.** Katy: $\dfrac{7 + \sqrt{65}}{2} \approx 7.5$ hr; Steve: $\dfrac{9 + \sqrt{65}}{2} \approx 8.5$ hr

7. to Shanghai: 40 km/hr; to Ningbo: 90 km/hr

Vocabulary, Readiness & Video Check 8.3

1. The values we get for the substituted variable are *not* our final answers. Remember to always substitute back to the original variable and solve for it if necessary.

Exercise Set 8.3

1. 2 **3.** 16 **5.** $1, 4$ **7.** $3 - \sqrt{7}, 3 + \sqrt{7}$ **9.** $\dfrac{3 - \sqrt{57}}{4}, \dfrac{3 + \sqrt{57}}{4}$ **11.** $\dfrac{1 - \sqrt{29}}{2}, \dfrac{1 + \sqrt{29}}{2}$ **13.** $-2, 2, -2i, 2i$

15. $-\dfrac{1}{2}, \dfrac{1}{2}, -i\sqrt{3}, i\sqrt{3}$ **17.** $-3, 3, -2, 2$ **19.** $125, -8$ **21.** $-\dfrac{4}{5}, 0$ **23.** $-\dfrac{1}{8}, 27$ **25.** $-\dfrac{2}{3}, \dfrac{4}{3}$ **27.** $-\dfrac{1}{125}, \dfrac{1}{8}$ **29.** $-\sqrt{2}, \sqrt{2}, -\sqrt{3}, \sqrt{3}$

31. $\dfrac{-9 - \sqrt{201}}{6}, \dfrac{-9 + \sqrt{201}}{6}$ **33.** $2, 3$ **35.** 3 **37.** $27, 125$ **39.** $1, -3i, 3i$ **41.** $\dfrac{1}{8}, -8$ **43.** $-\dfrac{1}{2}, \dfrac{1}{3}$ **45.** 4 **47.** -3

49. $-\sqrt{5}, \sqrt{5}, -2i, 2i$ **51.** $-3, \dfrac{3 - 3i\sqrt{3}}{2}, \dfrac{3 + 3i\sqrt{3}}{2}$ **53.** $6, 12$ **55.** $-\dfrac{1}{3}, \dfrac{1}{3}, -\dfrac{i\sqrt{6}}{3}, \dfrac{i\sqrt{6}}{3}$ **57.** 5 **59.** 55 mph: 66 mph

61. 5 mph, then 4 mph   **63.** inlet pipe: 15.5 hr; hose: 16.5 hr   **65.** 8.5 hr   **67.** 12 or -8 **69. a.** $(x - 6)$ in.   **b.** $300 = (x - 6) \cdot (x - 6) \cdot 3$
c. 16 in. by 16 in.   **71.** 22 feet   **73.** $(-\infty, 3]$ **75.** $(-5, \infty)$ **77.** domain: $(-\infty, \infty)$; range: $(-\infty, \infty)$; function   **79.** domain: $(-\infty, \infty)$;

range: $[-1, \infty)$; function   **81.** $1, -3i, 3i$ **83.** $-\dfrac{1}{2}, \dfrac{1}{3}$ **85.** $-3, \dfrac{3 - 3i\sqrt{3}}{2}, \dfrac{3 + 3i\sqrt{3}}{2}$ **87.** answers may vary   **89. a.** 11.615 m/sec

b. 11.612 m/sec   **c.** 25.925 mph

Integrated Review

1. $-\sqrt{10}, \sqrt{10}$ **2.** $-\sqrt{14}, \sqrt{14}$ **3.** $1 - 2\sqrt{2}, 1 + 2\sqrt{2}$ **4.** $-5 - 2\sqrt{3}, -5 + 2\sqrt{3}$ **5.** $-1 - \sqrt{13}, -1 + \sqrt{13}$ **6.** $1, 11$
7. $\dfrac{-3 - \sqrt{69}}{6}, \dfrac{-3 + \sqrt{69}}{6}$ **8.** $\dfrac{-2 - \sqrt{5}}{4}, \dfrac{-2 + \sqrt{5}}{4}$ **9.** $\dfrac{2 - \sqrt{2}}{2}, \dfrac{2 + \sqrt{2}}{2}$ **10.** $-3 - \sqrt{5}, -3 + \sqrt{5}$ **11.** $-2 - i\sqrt{3}, -2 + i\sqrt{3}$
12. $\dfrac{-1 - i\sqrt{11}}{2}, \dfrac{-1 + i\sqrt{11}}{2}$ **13.** $\dfrac{-3 - i\sqrt{15}}{2}, \dfrac{-3 + i\sqrt{15}}{2}$ **14.** $-3i, 3i$ **15.** $-17, 0$ **16.** $\dfrac{1 - \sqrt{13}}{4}, \dfrac{1 + \sqrt{13}}{4}$ **17.** $2 - 3\sqrt{3}, 2 + 3\sqrt{3}$

18. $2 - \sqrt{3}, 2 + \sqrt{3}$ **19.** $-2, \dfrac{4}{3}$ **20.** $\dfrac{-5 - \sqrt{17}}{4}, \dfrac{-5 + \sqrt{17}}{4}$ **21.** $1 - \sqrt{6}, 1 + \sqrt{6}$ **22.** $-\sqrt{31}, \sqrt{31}$ **23.** $-\sqrt{11}, \sqrt{11}$

24. $-i\sqrt{11}, i\sqrt{11}$ **25.** $-11, 6$ **26.** $\dfrac{-3 - \sqrt{19}}{5}, \dfrac{-3 + \sqrt{19}}{5}$ **27.** $\dfrac{-3 - \sqrt{17}}{4}, \dfrac{-3 + \sqrt{17}}{4}$ **28.** 4 **29.** $\dfrac{-1 - \sqrt{17}}{8}, \dfrac{-1 + \sqrt{17}}{8}$

30. $10\sqrt{2}$ ft ≈ 14.1 ft   **31.** Jack: 9.1 hr; Lucy: 7.1 hr   **32.** 5 mph during the first part, then 6 mph

Section 8.4
Practice Exercises

1. $(-\infty, -3) \cup (4, \infty)$ **2.** $[0, 8]$ **3.** $(-\infty, -3] \cup [-1, 2]$ **4.** $(-4, 5]$ **5.** $(-\infty, -3) \cup \left(-\dfrac{8}{5}, \infty\right)$

Vocabulary, Readiness & Video Check 8.4

1. $[-7, 3)$ **3.** $(-\infty, 0]$ **5.** $(-\infty, -12) \cup [-10, \infty)$ **7.** We use the solutions to the related equation to divide the number line into regions that either entirely are or entirely are not solution regions; the solutions to the related equation are solutions to the inequality only if the inequality symbol is $\le$ or $\ge$.

Exercise Set 8.4

1. $(-\infty, -5) \cup (-1, \infty)$ **3.** $[-4, 3]$ **5.** $[2, 5]$ **7.** $\left(-5, -\dfrac{1}{3}\right)$ **9.** $(2, 4) \cup (6, \infty)$ **11.** $(-\infty, -4] \cup [0, 1]$

13. $(-\infty, -3) \cup (-2, 2) \cup (3, \infty)$ **15.** $(-7, 2)$ **17.** $(-1, \infty)$ **19.** $(-\infty, -1] \cup (4, \infty)$ **21.** $(-\infty, 2) \cup \left(\dfrac{11}{4}, \infty\right)$ **23.** $(0, 2] \cup [3, \infty)$

25. $(-\infty, 3)$ **27.** $\left[-\dfrac{5}{4}, \dfrac{3}{2}\right]$ **29.** $(-\infty, 0) \cup (1, \infty)$ **31.** $(-\infty, -4] \cup [4, 6]$ **33.** $\left(-\infty, -\dfrac{2}{3}\right] \cup \left[\dfrac{3}{2}, \infty\right)$

35. $\left(-4, -\dfrac{3}{2}\right) \cup \left(\dfrac{3}{2}, \infty\right)$ **37.** $(-\infty, -5] \cup [-1, 1] \cup [5, \infty)$ **39.** $\left(-\infty, -\dfrac{5}{3}\right) \cup \left(\dfrac{7}{2}, \infty\right)$ **41.** $(0, 10)$ **43.** $(-\infty, -4) \cup [5, \infty)$

45. $(-\infty, -6] \cup (-1, 0] \cup (7, \infty)$ **47.** $(-\infty, 1) \cup (2, \infty)$ **49.** $(-\infty, -8] \cup (-4, \infty)$ **51.** $(-\infty, 0] \cup \left(5, \dfrac{11}{2}\right]$ **53.** $(0, \infty)$

55.
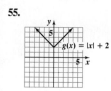

57.
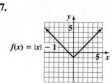

59.

61.

63. answers may vary
65. $(-\infty, -1) \cup (0, 1)$, or any number less than -1 or between 0 and 1

67. x is between 2 and 11   **69.**
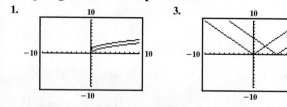
    71.
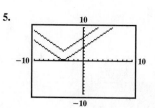

Section 8.5
Practice Exercises

1. $f(x) = x^2$, $g(x) = x^2 - 4$
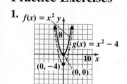
    2. a.

    b.
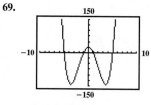
    3. $g(x) = (x + 6)^2$, $f(x) = x^2$
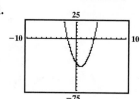
    4. a.
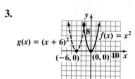
    b.

5.

    6. $f(x) = x^2$, $g(x) = 4x^2$, $h(x) = \dfrac{1}{4}x^2$

    7.

    8.

Graphing Calculator Explorations 8.5

1.

    3.
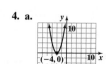
    5.

Vocabulary, Readiness & Video Check 8.5

1. quadratic   **3.** upward   **5.** lowest   **7.** $(0, 0)$ **9.** $(2, 0)$ **11.** $(0, 3)$ **13.** $(-1, 5)$ **15.** Graphs of the form $f(x) = x^2 + k$ shift up or down the y-axis k units from $y = x^2$; the y-intercept.   **17.** The vertex, (h, k) and the axis of symmetry, $x = h$; the basic shape of $y = x^2$ does not change.   **19.** the coordinates of the vertex, whether the graph opens upward or downward, whether the graph is narrower or wider than $y = x^2$ and the graph's axis of symmetry

Exercise Set 8.5

1. $V(0, -1)$, $x = 0$

3. $V(0, 5)$, $x = 0$

5. $V(0, 7)$, $x = 0$

7. $V(5, 0)$, $x = 5$

9. $V(-2, 0)$, $x = -2$

11. $V(-3, 0)$, $x = -3$

13. $V(2, 5)$, $x = 2$

15. 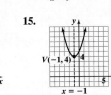 $V(-1, 4)$, $x = -1$

17. $V(-2, -5)$, $x = -2$

19. $V(0, 0)$, $x = 0$

21. $V(0, 0)$, $x = 0$

23. $V(0, 0)$, $x = 0$

25. $V(1, 3)$, $x = 1$

27. $V(-3, 1)$, $x = -3$

29. $V(6, -3)$, $x = 6$

31. $V(2, 0)$, $x = 2$, $y = -(x - 2)^2$

33. 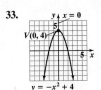 $V(0, 4)$, $x = 0$, $y = -x^2 + 4$

35. 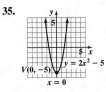 $V(0, -5)$, $x = 0$, $y = 2x^2 - 5$

37. $y = (x - 6)^2 + 4$, $V(6, 4)$, $x = 6$

39. 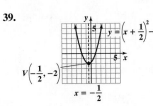 $y = \left(x + \frac{1}{2}\right)^2 - 2$, $V\left(-\frac{1}{2}, -2\right)$, $x = -\frac{1}{2}$

41. 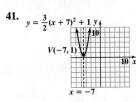 $y = \frac{3}{2}(x + 7)^2 + 1$, $V(-7, 1)$, $x = -7$

43. 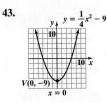 $y = \frac{1}{4}x^2 - 9$, $V(0, -9)$, $x = 0$

45. 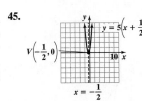 $y = 5\left(x + \frac{1}{2}\right)^2$, $V\left(-\frac{1}{2}, 0\right)$, $x = -\frac{1}{2}$

47. 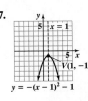 $x = 1$, $V(1, -1)$, $y = -(x - 1)^2 - 1$

49. 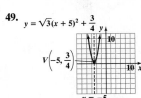 $y = \sqrt{3}(x + 5)^2 + \frac{3}{4}$, $V\left(-5, \frac{3}{4}\right)$, $x = -5$

51. 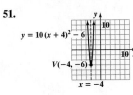 $y = 10(x + 4)^2 - 6$, $V(-4, -6)$, $x = -4$

53. $x = 4$, $V(4, 5)$, $y = -2(x - 4)^2 + 5$

55. $x^2 + 8x + 16$ **57.** $z^2 - 16z + 64$ **59.** $y^2 + y + \frac{1}{4}$ **61.** $-6, 2$ **63.** $-5 - \sqrt{26}, -5 + \sqrt{26}$
65. $4 - 3\sqrt{2}, 4 + 3\sqrt{2}$ **67.** c **69.** $f(x) = 5(x - 2)^2 + 3$ **71.** $f(x) = 5(x + 3)^2 + 6$

73.

75.

77.

79. a. 119,141 million **b.** 284,953 million

Section 8.6
Practice Exercises

1. $(-1, 0)$, $(3, 0)$, $(0, -3)$, $(1, -4)$

2. $\left(-\frac{1}{2}, 2\right)$, $(0, 3)$

3. 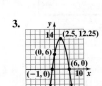 $(2.5, 12.25)$, $(0, 6)$, $(6, 0)$, $(-1, 0)$

4. $(1, -4)$ **5.** Maximum height 9 feet in $\frac{3}{4}$ second

Vocabulary, Readiness & Video Check 8.6

1. (h, k) **3.** We can immediately identify the vertex (h, k), whether the parabola opens upward or downward, and know its axis of symmetry; completing the square. **5.** the vertex

Exercise Set 8.6

1. 0; 1   **3.** 2; 1   **5.** 1; 1   **7.** down   **9.** up   **11.** $(-4, -9)$ **13.** $(5, 30)$ **15.** $(1, -2)$ **17.** $\left(\frac{1}{2}, \frac{5}{4}\right)$ **19.** D   **21.** B

23. 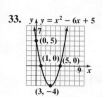   **25.**   **27.**   **29.**   **31.**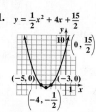

33. 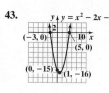   **35.** 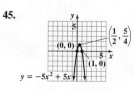   **37.** 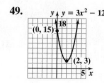   **39.**   **41.**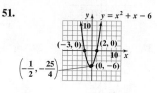

43. 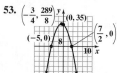   **45.** 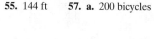   **47.** 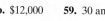   **49.**   **51.**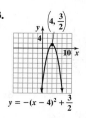

53. $\left(-\frac{3}{4}, \frac{289}{8}\right)$ **55.** 144 ft   **57. a.** 200 bicycles   **b.** \$12,000   **59.** 30 and 30   **61.** 5, -5 **63.** length, 20 units; width, 20 units

65.   **67.** 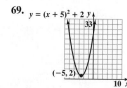   **69.** 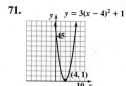   **71.** 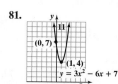   **73.**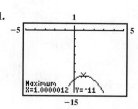

75. minimum value   **77.** maximum value   **79.** 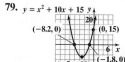   **81.**   **83.** -0.84 **85.** 1.43

87. a. maximum; answers may vary   **b.** 2035   **c.** 311 million   **89.**   **91.**

Chapter 8 Vocabulary Check

1. discriminant   **2.** $\pm\sqrt{b}$ **3.** $\dfrac{-b}{2a}$ **4.** quadratic inequality   **5.** completing the square   **6.** $(0, k)$ **7.** $(h, 0)$ **8.** (h, k)

9. quadratic formula   **10.** quadratic

Chapter 8 Review

1. 14, 1   **3.** $-7, 7$ **5.** $\dfrac{-3 - \sqrt{5}}{2}, \dfrac{-3 + \sqrt{5}}{2}$ **7.** 4.25%   **9.** two complex but not real solutions   **11.** two real solutions   **13.** 8   **15.** $-\dfrac{5}{2}, 1$

17. $\dfrac{5 - i\sqrt{143}}{12}, \dfrac{5 + i\sqrt{143}}{12}$ **19. a.** 20 ft   **b.** $\dfrac{15 + \sqrt{321}}{16}$ sec; 2.1 sec   **21.** 3, $\dfrac{-3 - 3i\sqrt{3}}{2}, \dfrac{-3 + 3i\sqrt{3}}{2}$ **23.** $\dfrac{2}{3}, 5$ **25.** 1, 125

27. $-1, 1, -i, i$ **29.** Jerome: 10.5 hr; Tim: 9.5 hr   **31.** $[-5, 5]$ **33.** $[-5, -2] \cup [2, 5]$ **35.** $(5, 6)$ **37.** $(-, -5] \cup [-2, 6]$ **39.** $(-\infty, 0)$

41.

43.

45.

47.

49.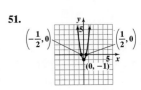

51.

53. $\left(-\dfrac{5}{6}, \dfrac{73}{12}\right)$

55. 210 and 210 **57.** $-5, 6$ **59.** $-2, 2$ **61.** $\dfrac{-1 - 3i\sqrt{3}}{2}, \dfrac{-1 + 3i\sqrt{3}}{2}$ **63.** $-i\sqrt{11}, i\sqrt{11}$

65. $\dfrac{21 - \sqrt{41}}{50}, \dfrac{21 + \sqrt{41}}{50}$ **67.** $8, 64$ **69.** $[-5, 0] \cup \left(\dfrac{3}{4}, \infty\right)$ **71. a.** 95,157 thousand passengers **b.** 2022

Chapter 8 Test

1. $\dfrac{7}{5}, -1$ **2.** $-1 - \sqrt{10}, -1 + \sqrt{10}$ **3.** $\dfrac{1 - i\sqrt{31}}{2}, \dfrac{1 + i\sqrt{31}}{2}$ **4.** $3 - \sqrt{7}, 3 + \sqrt{7}$ **5.** $-\dfrac{1}{7}, -1$ **6.** $\dfrac{3 - \sqrt{29}}{2}, \dfrac{3 + \sqrt{29}}{2}$

7. $-2 - \sqrt{11}, -2 + \sqrt{11}$ **8.** $-1, 1, -i, i, -3$ **9.** $-1, 1, -i, i$ **10.** $6, 7$ **11.** $3 - \sqrt{7}, 3 + \sqrt{7}$ **12.** $\dfrac{2 - i\sqrt{6}}{2}, \dfrac{2 + i\sqrt{6}}{2}$

13. $\left(-\infty, -\dfrac{3}{2}\right) \cup (5, \infty)$ **14.** $(-\infty, -5] \cup [-4, 4] \cup [5, \infty)$ **15.** $(-\infty, -3) \cup (2, \infty)$ **16.** $(-\infty, -3) \cup [2, 3)$

17.

18.

19.

20.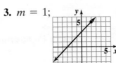

21. $(5 + \sqrt{17})\,\text{hr} \approx 9.12\,\text{hr}$ **22. a.** 272 ft **b.** 5.12 sec **23.** 7.2 ft

Chapter 8 Cumulative Review

1. a. $5 + y \geq 7$ **b.** $11 \neq z$ **c.** $20 < 5 - 2x$; Sec. 1.4, Ex. 6 **3.** $m = 1$;  ; Sec. 3.4, Ex. 1 **5.** $(-2, 2)$; Sec. 4.1, Ex. 7

7. a. $6x^2 - 29x + 28$ **b.** $15x^4 - x^2y - 2y^2$; Sec. 5.4, Ex. 6 **9. a.** $4(2x^2 + 1)$ **b.** prime polynomial **c.** $3x^2(2 - x + 4x^2)$; Sec. 5.5, Ex. 2

11. $(x - 5)(x - 7)$; Sec. 5.6, Ex. 2 **13.** $3x(a - 2b)^2$; Sec. 5.7, Ex. 2 **15.** $-\dfrac{2}{3}$; Sec. 5.8, Ex. 4 **17.** $-2, 0, 2$; Sec. 5.8, Ex. 6 **19.** $\dfrac{1}{5x - 1}$;

Sec. 6.1, Ex. 2a **21.** $\dfrac{7x^2 - 9x - 13}{(2x + 1)(x - 5)(3x - 2)}$; Sec. 6.2, Ex. 5 **23.** $\dfrac{xy + 2x^3}{y - 1}$ or $\dfrac{x(y + 2x^2)}{y - 1}$; Sec. 6.3, Ex. 3 **25.** $3x^3y - 15x - 1 - \dfrac{6}{xy}$; Sec. 6.4, Ex. 2

27. a. 5 **b.** 5; Sec. 6.4, Ex. 10 **29.** -3; Sec. 6.5, Ex. 1 **31.** 2; Sec. 6.6, Ex. 2 **33.** $k = \dfrac{1}{6}; y = \dfrac{1}{6}x$; Sec. 6.7, Ex. 1 **35. a.** 3 **b.** $|x|$

c. $|x - 2|$ **d.** -5 **e.** $2x - 7$ **f.** $5|x|$ **g.** $|x + 1|$; Sec. 7.1, Ex. 5 **37. a.** $\sqrt{x}$ **b.** $\sqrt[3]{5}$ **c.** $\sqrt{rs^3}$; Sec. 7.2, Ex. 7 **39. a.** $5x\sqrt{x}$

b. $3x^2y^2\sqrt[3]{2y^2}$ **c.** $3z^2\sqrt[3]{z^3}$; Sec. 7.3, Ex. 4 **41. a.** $\dfrac{2\sqrt{5}}{5}$ **b.** $\dfrac{8\sqrt{x}}{3x}$ **c.** $\dfrac{\sqrt[3]{4}}{2}$; Sec. 7.5, Ex. 1 **43.** $\dfrac{2}{9}$; Sec. 7.6, Ex. 5 **45. a.** $\dfrac{1}{2} + \dfrac{3}{2}i$

b. $0 - \dfrac{7}{3}i$; Sec. 7.7, Ex. 5 **47.** $-1 + 2\sqrt{3}, -1 - 2\sqrt{3}$; Sec. 8.1, Ex. 3 **49.** 9; Sec. 8.3, Ex. 1

CHAPTER 9 EXPONENTIAL AND LOGARITHMIC FUNCTIONS

Section 9.1
Practice Exercises

1. a. $4x + 7$ **b.** $-2x - 3$ **c.** $3x^2 + 11x + 10$ **d.** $\dfrac{x + 2}{3x + 5}$, where $x \neq -\dfrac{5}{3}$ **2. a.** $50; 46$ **b.** $9x^2 - 30x + 26; 3x^2 - 2$

3. a. $x^2 + 6x + 14$ **b.** $x^2 + 8$ **4. a.** $(h \circ g)(x)$ **b.** $(g \circ f)(x)$

Vocabulary, Readiness & Video Check 9.1
1. C **3.** F **5.** D **7.** You can find $(f + g)(x)$ and then find $(f + g)(2)$ or you can find $f(2)$ and $g(2)$ and then add those results.

Exercise Set 9.1

1. a. $3x - 6$ **b.** $-x - 8$ **c.** $2x^2 - 13x - 7$ **d.** $\dfrac{x - 7}{2x + 1}$, where $x \neq -\dfrac{1}{2}$ **3. a.** $x^2 + 5x + 1$ **b.** $x^2 - 5x + 1$ **c.** $5x^3 + 5x$

d. $\dfrac{x^2 + 1}{5x}$, where $x \neq 0$ **5. a.** $\sqrt{x} + x + 5$ **b.** $\sqrt{x} - x - 5$ **c.** $x\sqrt{x} + 5\sqrt{x}$ **d.** $\dfrac{\sqrt{x}}{x + 5}$, where $x \neq -5$

7. a. $5x^2 - 3x$ **b.** $-5x^2 - 3x$ **c.** $-15x^3$ **d.** $-\dfrac{3}{5x}$, where $x \neq 0$ **9.** 42 **11.** -18 **13.** 0

15. $(f \circ g)(x) = 25x^2 + 1; (g \circ f)(x) = 5x^2 + 5$ **17.** $(f \circ g)(x) = 2x + 11; (g \circ f)(x) = 2x + 4$

19. $(f \circ g)(x) = -8x^3 - 2x - 2; (g \circ f)(x) = -2x^3 - 2x + 4$ **21.** $(f \circ g)(x) = |10x - 3|; (g \circ f)(x) = 10|x| - 3$

23. $(f \circ g)(x) = \sqrt{-5x + 2}; (g \circ f)(x) = -5\sqrt{x} + 2$ **25.** $H(x) = (g \circ h)(x)$ **27.** $F(x) = (h \circ f)(x)$ **29.** $G(x) = (f \circ g)(x)$

31. answers may vary; for example, $g(x) = x + 2$ and $f(x) = x^2$ **33.** answers may vary; for example, $g(x) = x + 5$ and $f(x) = \sqrt{x} + 2$

35. answers may vary; for example, $g(x) = 2x - 3$ and $f(x) = \dfrac{1}{x}$ **37.** $y = x - 2$ **39.** $y = \dfrac{x}{3}$ **41.** $y = -\dfrac{x + 7}{2}$ **43.** 6 **45.** 4 **47.** 4

49. -1 **51.** answers may vary **53.** $P(x) = R(x) - C(x)$

Section 9.2
Practice Exercises

1. a. one-to-one **b.** not one-to-one **c.** not one-to-one **d.** one-to-one **e.** not one-to-one **f.** not one-to-one **2. a.** no, not one-to-one
b. yes **c.** yes **d.** no, not a function **e.** no, not a function **3.** $f^{-1} = \{(4, 3), (0, -2), (8, 2), (6, 6)\}$ **4.** $f^{-1}(x) = 6 - x$

5. **6. a.** **b.**

7. $f(f^{-1}(x)) = f\left(\dfrac{x + 1}{4}\right) = 4\left(\dfrac{x + 1}{4}\right) - 1 = x + 1 - 1 = x$

$f^{-1}(f(x)) = f^{-1}(4x - 1) = \dfrac{(4x - 1) + 1}{4} = \dfrac{4x}{4} = x$

Vocabulary, Readiness & Video Check 9.2

1. $(2, 11)$ **3.** the inverse of f **5.** vertical **7.** $y = x$ **9.** Every function must have each x-value correspond to only one y-value. A one-to-one function must also have each y-value correspond to only one x-value. **11.** Yes; by the definition of an inverse function. **13.** Once you know some points of the original equation or graph, you can switch the x's and y's of these points to find points that satisfy the inverse and then graph it. You can also check that the two graphs (the original and the inverse) are symmetric about the line $y = x$.

Exercise Set 9.2

1. one-to-one; $f^{-1} = \{(-1, -1), (1, 1), (2, 0), (0, 2)\}$ **3.** one-to-one; $h^{-1} = \{(10, 10)\}$ **5.** one-to-one; $f^{-1} = \{(12, 11), (3, 4), (4, 3), (6, 6)\}$
7. not one-to-one **9.** one-to-one;
11. a. 3 **b.** 1 **13. a.** 1 **b.** -1

Rank in Population (Input)	1	47	16	25	36	7
State (Output)	CA	AK	IN	LA	NM	OH

15. one-to-one **17.** not one-to-one
19. one-to-one **21.** not one-to-one

23. $f^{-1}(x) = x - 4$ **25.** $f^{-1}(x) = \dfrac{x + 3}{2}$ **27.** $f^{-1}(x) = 2x + 2$ **29.** $f^{-1}(x) = \sqrt[3]{x}$ **31.** $f^{-1}(x) = \dfrac{x - 2}{5}$ **33.** $f^{-1}(x) = 5x + 2$

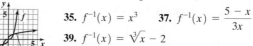

35. $f^{-1}(x) = x^3$ **37.** $f^{-1}(x) = \dfrac{5 - x}{3x}$

39. $f^{-1}(x) = \sqrt[3]{x} - 2$

41. **43.** **45.**

47. $(f \circ f^{-1})(x) = x; (f^{-1} \circ f)(x) = x$ **49.** $(f \circ f^{-1})(x) = x; (f^{-1} \circ f)(x) = x$

51. 5 **53.** 8 **55.** $\dfrac{1}{27}$ **57.** 9 **59.** $3^{1/2} \approx 1.73$ **61. a.** $(2, 9)$ **b.** $(9, 2)$

63. a. $\left(-2, \dfrac{1}{4}\right), \left(-1, \dfrac{1}{2}\right), (0, 1), (1, 2), (2, 5)$ **b.** $\left(\dfrac{1}{4}, -2\right), \left(\dfrac{1}{2}, -1\right), (1, 0), (2, 1), (5, 2)$ **c.** **d.**

65. answers may vary **67.** $f^{-1}(x) = \dfrac{x - 1}{3}$; **69.** $f^{-1}(x) = x^3 - 1$;

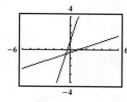 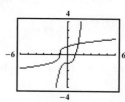

Section 9.3
Practice Exercises

1.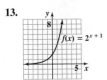

2. $f(x) = \left(\frac{1}{3}\right)^x$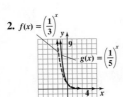

3.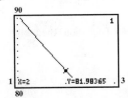

4. a. 2   **b.** $\frac{4}{3}$ **c.** -4 **5.** $3950.43   **6. a.** 90.54%   **b.** 60.86%

Graphing Calculator Explorations 9.3

1. 81.98%; 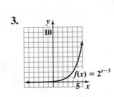   **3.** 22.54%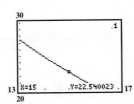

Vocabulary, Readiness & Video Check 9.3

1. exponential   **3.** yes   **5.** no; none   **7.** $(-\infty, \infty)$ **9.** In a polynomial function, the base is the variable and the exponent is the constant; in an exponential function, the base is the constant and the exponent is the variable.   **11.** $y = 30(0.996)^{101} \approx 20.0$ lb

Exercise Set 9.3

1.   **3.**   **5.** 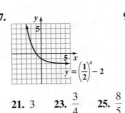   **7.**   **9.**   **11.**

13. 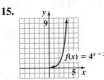   **15.**    **17.** C   **19.** B   **21.** 3   **23.** $\frac{3}{4}$ **25.** $\frac{8}{5}$ **27.** $-\frac{2}{3}$ **29.** 4   **31.** $\frac{3}{2}$

33. $-\frac{1}{3}$ **35.** -2 **37.** 24.6 lb   **39. a.** 9.7 billion lb   **b.** 12.3 billion lb   **41. a.** 192,927 students   **b.** 463,772 students   **43.** 378.0 million

45. 544,000 mosquitoes   **47.** $7621.42   **49.** $4065.59   **51.** 4   **53.** ∅   **55.** 2, 3   **57.** 3   **59.** -1 **61.** no   **63.** no   **65.** C

67. D   **69.** answers may vary   **71.**   **73.**   **75.** The graphs are the same since $\left(\frac{1}{2}\right)^{-x} = 2^x$.

77. 24.55 lb   **79.** 20.09 lb

Section 9.4
Practice Exercises

1. 136,839   **2.** 32   **3.** 47.4 g

Vocabulary, Readiness & Video Check 9.4

1. For Example 1, the growth rate is given as 5% per year. Since this is "per year," the number of time intervals is the "number of years," or 8.

3. time intervals = years/half-life; the decay rate is 50% or $\frac{1}{2}$ because half-life is the amount of time it takes half of a substance to decay

Exercise Set 9.4

1. 451   **3.** 144,302   **5.** 21,231   **7.** 202   **9.** 1470   **11.** 13   **13.** 712,880   **15.** 383   **17.** 333 bison   **19.** 1 g   **21. a.** $\frac{14}{7} = 2$; 10; yes

b. $\frac{11}{7} \approx 1.6$; 13.2; yes   **23.** $\frac{500}{152} \approx 3.3$; 2.1; yes   **25.** 4.9 g   **27.** 3   **29.** -1 **31.** no; answers may vary

Section 9.5
Practice Exercises

1. a. $3^4 = 81$ **b.** $5^{-1} = \dfrac{1}{5}$ **c.** $7^{1/2} = \sqrt{7}$ **d.** $13^4 = y$ **2. a.** $\log_4 64 = 3$ **b.** $\log_6 \sqrt[3]{6} = \dfrac{1}{3}$ **c.** $\log_5 \dfrac{1}{125} = -3$ **d.** $\log_\pi z = 7$

3. a. 2 **b.** -3 **c.** $\dfrac{1}{2}$ **4. a.** -2 **b.** 2 **c.** 36 **d.** 0 **e.** 0 **5. a.** 4 **b.** -2 **c.** 5 **d.** 4

6. 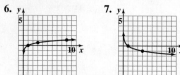 **7.**

Vocabulary, Readiness & Video Check 9.5

1. logarithmic **3.** yes **5.** no; none **7.** $(-\infty, \infty)$ **9.** First write the equation as an equivalent exponential equation. Then solve.

Exercise Set 9.5

1. $6^2 = 36$ **3.** $3^{-3} = \dfrac{1}{27}$ **5.** $10^3 = 1000$ **7.** $9^4 = x$ **9.** $\pi^{-2} = \dfrac{1}{\pi^2}$ **11.** $7^{1/2} = \sqrt{7}$ **13.** $0.7^3 = 0.343$ **15.** $3^{-4} = \dfrac{1}{81}$

17. $\log_2 16 = 4$ **19.** $\log_{10} 100 = 2$ **21.** $\log_\pi x = 3$ **23.** $\log_{10} \dfrac{1}{10} = -1$ **25.** $\log_4 \dfrac{1}{16} = -2$ **27.** $\log_5 \sqrt{5} = \dfrac{1}{2}$

29. 3 **31.** -2 **33.** $\dfrac{1}{2}$ **35.** -1 **37.** 0 **39.** 2 **41.** 4 **43.** -3 **45.** 2 **47.** 81 **49.** 7 **51.** -3

53. -3 **55.** 2 **57.** 2 **59.** $\dfrac{27}{64}$ **61.** 10 **63.** 4 **65.** 5 **67.** $\dfrac{1}{49}$ **69.** 3 **71.** 3 **73.** 1 **75.** -1

77. **79.** **81.** **83.** **85.** 1 **87.** $\dfrac{x-4}{2}$ **89.** $\dfrac{2x+3}{x^2}$ **91.** 3

93. a. $g(2) = 25$ **b.** $(25, 2)$ **c.** $f(25) = 2$ **95.** answers may vary **97.** $\dfrac{9}{5}$ **99.** 1

101. **103.** $y = \left(\dfrac{1}{3}\right)^x$ **105.** answers may vary **107.** 0.0827

Section 9.6
Practice Exercises

1. a. $\log_8 15$ **b.** $\log_2 6$ **c.** $\log_5(x^2 - 1)$ **2. a.** $\log_5 3$ **b.** $\log_6 \dfrac{x}{3}$ **c.** $\log_4 \dfrac{x^2 + 1}{x^2 + 3}$ **3. a.** $8 \log_7 x$ **b.** $\dfrac{1}{4} \log_5 7$

4. a. $\log_5 512$ **b.** $\log_8 \dfrac{x^2}{x+3}$ **c.** $\log_7 15$ **5. a.** $\log_5 4 + \log_5 3 - \log_5 7$ **b.** $2 \log_4 a - 5 \log_4 b$ **6. a.** 1.39 **b.** 1.66 **c.** 0.28

Vocabulary, Readiness & Video Check 9.6

1. 36 **3.** $\log_b 27$ **5.** x **7.** No, the product property says the logarithm of a product can be written as a sum of logarithms—the expression in Example 2 is a logarithm of a sum. **9.** Since $\dfrac{1}{x} = x^{-1}$, this gives us $\log_2 x^{-1}$. Using the power property, we get $-1 \log_2 x$ or $-\log_2 x$.

Exercise Set 9.6

1. $\log_5 14$ **3.** $\log_4 9x$ **5.** $\log_6(x^2 + x)$ **7.** $\log_{10}(10x^2 + 20)$ **9.** $\log_5 3$ **11.** $\log_3 4$ **13.** $\log_2 \dfrac{x}{y}$ **15.** $\log_2 \dfrac{x^2 + 6}{x^2 + 1}$ **17.** $2 \log_3 x$

19. $-1 \log_4 5 = -\log_4 5$ **21.** $\dfrac{1}{2} \log_5 y$ **23.** $\log_2 5x^3$ **25.** $\log_4 48$ **27.** $\log_5 x^3 z^6$ **29.** $\log_4 4$, or 1 **31.** $\log_7 \dfrac{9}{2}$ **33.** $\log_{10} \dfrac{x^3 - 2x}{x+1}$

35. $\log_2 \dfrac{x^{7/2}}{(x+1)^2}$ **37.** $\log_8 x^{16/3}$ **39.** $\log_3 4 + \log_3 y - \log_3 5$ **41.** $\log_4 5 - \log_4 9 - \log_4 z$ **43.** $3 \log_2 x - \log_2 y$ **45.** $\dfrac{1}{2} \log_b 7 + \dfrac{1}{2} \log_b x$

47. $4 \log_6 x + 5 \log_6 y$ **49.** $3 \log_5 x + \log_5(x+1)$ **51.** $2 \log_6 x - \log_6(x+3)$ **53.** 1.2 **55.** 0.2 **57.** 0.35 **59.** 1.29 **61.** -0.68

63. -0.125 **65–66.** **67.** 2 **69.** 2 **71.** b **73.** true **75.** false **77.** false **79.** because $\log_b 1 = 0$

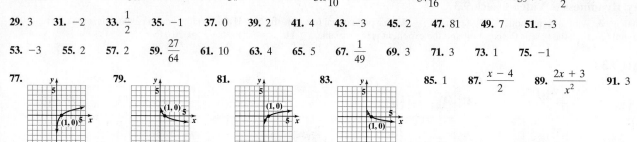

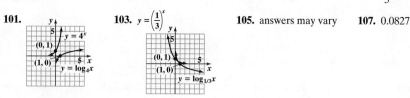

Integrated Review

1. $x^2 + x - 5$ **2.** $-x^2 + x - 7$ **3.** $x^3 - 6x^2 + x - 6$ **4.** $\dfrac{x-6}{x^2+1}$ **5.** $\sqrt{3x-1}$ **6.** $3\sqrt{x} - 1$

7. one-to-one; $\{(6,-2),(8,4),(-6,2),(3,3)\}$ **8.** not one-to-one **9.** not one-to-one **10.** one-to-one

11. not one-to-one **12.** $f^{-1}(x) = \dfrac{x}{3}$ **13.** $f^{-1}(x) = x - 4$ **14.** $f^{-1}(x) = \dfrac{x+1}{5}$ **15.** $f^{-1}(x) = \dfrac{x-2}{3}$

16. $y = \left(\dfrac{1}{2}\right)^x$ **17.** $y = 2^x + 1$ 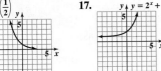 **18.** $y = \log_3 x$ **19.** $y = \log_{1/3} x$ **20.** 3

21. 7 **22.** -8 **23.** 3 **24.** 2 **25.** $\dfrac{1}{2}$ **26.** 32 **27.** 4 **28.** 5 **29.** $\dfrac{1}{9}$ **30.** $\log_2 x^5$ **31.** $\log_2 5^x$ **32.** $\log_5 \dfrac{x^3}{y^5}$ **33.** $\log_5 x^9 y^3$

34. $\log_2 \dfrac{x^2 - 3x}{x^2 + 4}$ **35.** $\log_3 \dfrac{y^4 + 11y}{y + 2}$ **36.** $\log_7 9 + 2\log_7 x - \log_7 y$ **37.** $\log_6 5 + \log_6 y - 2\log_6 z$ **38.** 254,000 mosquitoes

Section 9.7
Practice Exercises

1. 1.1761 **2. a.** -2 **b.** 5 **c.** $\dfrac{1}{5}$ **d.** -3 **3.** $10^{3.4} \approx 2511.8864$ **4.** 5.6 **5.** 2.5649 **6. a.** 4 **b.** $\dfrac{1}{3}$ **7.** $\dfrac{e^8}{5} \approx 596.1916$
8. \$3051.00 **9.** 0.7740

Vocabulary, Readiness & Video Check 9.7

1. 10 **3.** 7 **5.** 5 **7.** $\dfrac{\log 7}{\log 2}$ or $\dfrac{\ln 7}{\ln 2}$ **9.** The understood base of a common logarithm is 10. If you're finding the common logarithm of a known
power of 10, then the common logarithm is the known power of 10. **11.** $\log_b b^x = x$

Exercise Set 9.7

1. 0.9031 **3.** 0.3636 **5.** 0.6931 **7.** -2.6367 **9.** 1.1004 **11.** 1.6094 **13.** 1.6180 **15.** 2 **17.** -3 **19.** 2 **21.** $\dfrac{1}{4}$

23. 3 **25.** -7 **27.** -4 **29.** $\dfrac{1}{2}$ **31.** $\dfrac{e^7}{2} \approx 548.3166$ **33.** $10^{1.3} \approx 19.9526$ **35.** $\dfrac{10^{1.1}}{2} \approx 6.2946$ **37.** $e^{1.4} \approx 4.0552$

39. $\dfrac{4 + e^{2.3}}{3} \approx 4.6581$ **41.** $10^{2.3} \approx 199.5262$ **43.** $e^{-2.3} \approx 0.1003$ **45.** $\dfrac{10^{-0.5} - 1}{2} \approx -0.3419$ **47.** $\dfrac{e^{0.18}}{4} \approx 0.2993$

49. 1.5850 **51.** -2.3219 **53.** 1.5850 **55.** -1.6309 **57.** 0.8617 **59.** 4.2 **61.** 5.3 **63.** \$3656.38 **65.** \$2542.50

67. $\dfrac{4}{7}$ **69.** $x = \dfrac{3y}{4}$ **71.** $-6, -1$ **73.** $(2, -3)$ **75.** answers may vary **77.** ln 50; answers may vary

79. $f(x) = e^x$ **81.** $f(x) = e^{-3x}$ **83.** $f(x) = e^x + 2$ **85.** $f(x) = e^{x-1}$ **87.** $f(x) = 3e^x$ **89.** $f(x) = \ln x$

91. 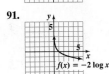 $f(x) = -2\log x$ **93.** $f(x) = \log(x + 2)$ **95.** $f(x) = \ln x - 3$ **97.** answers may vary;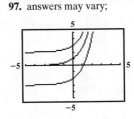

Section 9.8
Practice Exercises

1. $\dfrac{\log 9}{\log 5} \approx 1.3652$ **2.** 33 **3.** 1 **4.** $\dfrac{1}{3}$ **5.** 937 rabbits **6.** 10 yr

Graphing Calculator Explorations 9.8

1. 3.67 years, or 3 years and 8 months **3.** 23.16 years, or 23 years and 2 months

Vocabulary, Readiness & Video Check 9.8

1. $\ln(4x - 2) = \ln 3$ is the same as $\log_e(4x - 2) = \log_e 3$. Therefore, from the logarithm property of equality, we know that $4x - 2 = 3$.

3. $2000 = 1000\left(1 + \dfrac{0.07}{12}\right)^{12 \cdot t} \approx 9.9$ yr; As long as the interest rate and compounding are the same, it takes any amount of money the same time to
double.

Exercise Set 9.8

1. $\dfrac{\log 6}{\log 3}$; 1.6309 **3.** $\dfrac{\log 3.8}{2 \log 3}$; 0.6076 **5.** $\dfrac{3 \log 2 + \log 5}{\log 2}$ or $3 + \dfrac{\log 5}{\log 2}$; 5.3219 **7.** $\dfrac{\log 5}{\log 9}$; 0.7325 **9.** $\dfrac{\log 3 - 7 \log 4}{\log 4}$ or $\dfrac{\log 3}{\log 4} - 7$; −6.2075

11. 11 **13.** $\dfrac{1}{2}$ **15.** $\dfrac{3}{4}$ **17.** −2, 3 **19.** 2 **21.** $\dfrac{1}{8}$ **23.** $\dfrac{4 \log 7 + \log 11}{3 \log 7}$ or $\dfrac{1}{3}\left(4 + \dfrac{\log 11}{\log 7}\right)$; 1.7441 **25.** 4, −1 **27.** $\dfrac{\ln 5}{6}$; 0.2682

29. 9, −9 **31.** $\dfrac{1}{5}$ **33.** 100 **35.** $\dfrac{192}{127}$ **37.** $\dfrac{2}{3}$ **39.** $\dfrac{-5 + \sqrt{33}}{2}$ **41.** 103 wolves **43.** 7582 inhabitants **45.** 9.9 yr **47.** 1.7 yr

49. 8.8 yr **51.** 24.5 lb **53.** 55.7 in. **55.** 11.9 lb/sq in. **57.** 3.2 mi **59.** 12 weeks **61.** 18 weeks **63.** $-\dfrac{5}{3}$ **65.** $\dfrac{17}{4}$

67. $f^{-1}(x) = \dfrac{x - 2}{5}$ **69.** 15 yr **71.** answers may vary

73. 6.93; **75.** −3.68; 0.19 **77.** 1.74 **79.** 0.2

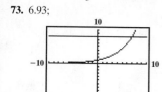 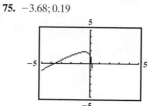 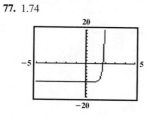 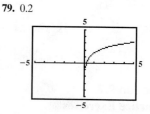

Chapter 9 Vocabulary Check

1. inverse **2.** composition **3.** exponential **4.** symmetric **5.** Natural **6.** Common **7.** vertical; horizontal **8.** logarithmic
9. Half-life **10.** exponential

Chapter 9 Review

1. $3x - 4$ **3.** $2x^2 - 9x - 5$ **5.** $x^2 + 2x - 1$ **7.** 18 **9.** −2 **11.** one-to-one; $h^{-1} = \{(14, -9), (8, 6), (12, -11), (15, 15)\}$
13. one-to-one;

Rank in Housing Starts for 2009 (Input)	4	3	1	2
US Region (Output)	Northeast	Midwest	South	West

15. a. 3 **b.** 7 **17.** not one-to-one **19.** not one-to-one **21.** $f^{-1}(x) = x + 9$ **23.** $f^{-1}(x) = \dfrac{x - 11}{6}$ **25.** $f^{-1}(x) = \sqrt[3]{x + 5}$

27. $g^{-1}(x) = \dfrac{6x + 7}{12}$ **29.** **31.** 3 **33.** $-\dfrac{4}{3}$ **35.** $\dfrac{3}{2}$ **37.** **39.**

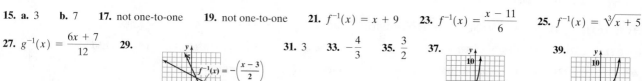 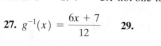 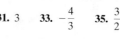 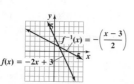

41. \$2963.11 **43.** 187,118 **45.** 8 players **47.** $\log_7 49 = 2$ **49.** $\left(\dfrac{1}{2}\right)^{-4} = 16$ **51.** $\dfrac{1}{64}$ **53.** 0 **55.** 5 **57.** 4 **59.** $\dfrac{17}{3}$ **61.** −1, 4

63. **65.** $\log_3 32$ **67.** $\log_7 \dfrac{3}{4}$ **69.** $\log_{11} 4$ **71.** $\log_5 \dfrac{x^3}{(x + 1)^2}$ **73.** $3 \log_3 x - \log_3 (x + 2)$

75. $\log_2 3 + 2 \log_2 x + \log_2 y - \log_2 z$ **77.** 2.02 **79.** 0.5563 **81.** 0.2231 **83.** 3 **85.** −1 **87.** $\dfrac{e^2}{2}$ **89.** $\dfrac{e^{-1} + 3}{2}$

91. 1.67 mm **93.** 0.2920 **95.** \$1684.66 **97.** $\dfrac{\log 7}{2 \log 3}$; 0.8856 **99.** $\dfrac{\log 6 - \log 3}{2 \log 3}$ or $\dfrac{1}{2}\left(\dfrac{\log 6}{\log 3} - 1\right)$; 0.3155

101. $\dfrac{\log 4 + 5 \log 5}{3 \log 5}$ or $\dfrac{1}{3}\left(\dfrac{\log 4}{\log 5} + 5\right)$; 1.9538 **103.** $\dfrac{\log \frac{1}{2} + \log 5}{\log 5}$ or $-\dfrac{\log 2}{\log 5} + 1$; 0.5693 **105.** $\dfrac{25}{2}$ **107.** $\varnothing$ **109.** $2\sqrt{2}$ **111.** 22.4 yr

113. 99.0 yr **115.** 8.8 yr **117.** 0.69; **119.** −4 **121.** $\dfrac{11}{9}$ **123.** 1 **125.** −1, 5 **127.** $e^{-3.2}$ **129.** $2e$

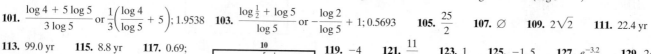

Chapter 9 Test

1. $2x^2 - 3x$ **2.** $3 - x$ **3.** 5 **4.** $x - 7$ **5.** $x^2 - 6x - 2$

6. **7.** one-to-one **8.** not one-to-one **9.** one-to-one; $f^{-1}(x) = \dfrac{-x + 6}{2}$ **10.** one-to-one; $f^{-1} = \{(0, 0), (3, 2), (5, -1)\}$

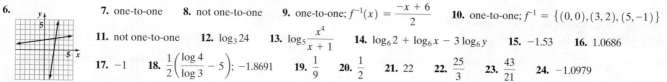

11. not one-to-one **12.** $\log_3 24$ **13.** $\log_5 \dfrac{x^4}{x + 1}$ **14.** $\log_6 2 + \log_6 x - 3 \log_6 y$ **15.** −1.53 **16.** 1.0686

17. −1 **18.** $\dfrac{1}{2}\left(\dfrac{\log 4}{\log 3} - 5\right)$; −1.8691 **19.** $\dfrac{1}{9}$ **20.** $\dfrac{1}{2}$ **21.** 22 **22.** $\dfrac{25}{3}$ **23.** $\dfrac{43}{21}$ **24.** −1.0979

25. **26.** **27.** $5234.58 **28.** 6 yr **29.** 6.5%; $230 **30.** 100,141 **31.** 64,805 prairie dogs
32. 15 yr **33.** 85% **34.** 52%

Chapter 9 Cumulative Review

1. a. 8 **b.** $-\dfrac{1}{3}$ **c.** -0.36 **d.** 0 **e.** $-\dfrac{2}{11}$ **f.** 42 **g.** 0; Sec. 1.3, Ex. 4 **3.** ; Sec. 3.1, Ex. 6

5. $\{(x, y, z)\,|\,x - 5y - 2z = 6\}$; Sec. 4.2, Ex. 4 **7. a.** x^3 **b.** 5^6 **c.** $5x$ **d.** $\dfrac{6y^2}{7}$; Sec. 5.1, Ex. 4 **9. a.** $102.60 **b.** $12.60; Sec. 6.1, Ex. 9

11. a. $\dfrac{3x}{2}$ **b.** $\dfrac{5 + x}{7z^2}$ **c.** $x - 7$ **d.** $-\dfrac{1}{3y^2}$; Sec. 6.2, Ex. 1 **13.** $3x^2 + 2x + 3 + \dfrac{-6x + 9}{x^2 - 1}$; Sec. 6.4, Ex. 6 **15.** -1; Sec. 6.5 Ex. 4

17. 6 mph; Sec. 6.6, Ex. 5 **19. a.** 3 **b.** -3 **c.** -5 **d.** not a real number **e.** $4x$; Sec. 7.1, Ex. 4 **21. a.** $\sqrt[4]{x^3}$ **b.** $\sqrt[6]{x}$

c. $\sqrt[6]{72}$; Sec. 7.2, Ex. 8 **23. a.** $5\sqrt{3} + 3\sqrt{10}$ **b.** $\sqrt{35} + \sqrt{5} - \sqrt{42} - \sqrt{6}$ **c.** $21x - 7\sqrt{5x} + 15\sqrt{x} - 5\sqrt{5}$ **d.** $49 - 8\sqrt{3}$

e. $2x - 25$ **f.** $x + 22 + 10\sqrt{x - 3}$; Sec. 7.4 Ex. 4 **25.** $\dfrac{\sqrt[4]{xy^3}}{3y^2}$; Sec. 7.5, Ex. 3 **27.** 3; Sec. 7.6, Ex. 4 **29.** $\dfrac{9 + i\sqrt{15}}{6}, \dfrac{9 - i\sqrt{15}}{6}$; Sec. 8.1, Ex. 8

31. $\dfrac{-1 + \sqrt{33}}{4}, \dfrac{-1 - \sqrt{33}}{4}$; Sec. 8.3, Ex. 2 **33.** $[0, 4]$; Sec. 8.4, Ex. 2 **35.** 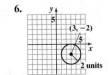 ; Sec. 8.5, Ex. 5 **37. a.** $3x - 4$ **b.** $-x + 2$

$F(x) = (x - 3)^2 + 1$

c. $2x^2 - 5x + 3$ **d.** $\dfrac{x - 1}{2x - 3}$, where $x \neq \dfrac{3}{2}$; Sec. 9.1, Ex. 1 **39.** $f^{-1}(x) = x - 3$; Sec. 9.2, Ex. 4 **41. a.** 2 **b.** -1 **c.** $\dfrac{1}{2}$; Sec. 9.5, Ex. 3

CHAPTER 10 CONIC SECTIONS

Section 10.1
Practice Exercises

1. 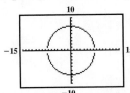 **2.** **3.** **4.** **5.** **6.**

7. **8.** $(x + 2)^2 + (y + 5)^2 = 81$

Graphing Calculator Explorations 10.1

1. **3.** **5.** **7.**

Vocabulary, Readiness & Video Check 10.1

1. conic sections **3.** circle; center **5.** radius **7.** No, their graphs don't pass the vertical line test. **9.** Since the standard form of a circle involves a squared binomial for both x and y, we need to complete the square on both x and y.

Exercise Set 10.1

1. upward **3.** to the left **5.** downward

7. 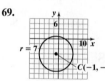 **9.** **11.** **13.**

15. **17.** **19.** **21.** **23.** **25.**

27. **29.** **31.** **33.** (-2, 3) **35.**

43. $(x - 2)^2 + (y - 3)^2 = 36$ **45.** $x^2 + y^2 = 3$
47. $(x + 5)^2 + (y - 4)^2 = 45$

37. **39.** **41.**

49. **51.** **53.** **55.** **57.** **59.**

61. **63.** **65.** **67.** **69.** **71.**

73. **75.** **77.** **79.** **81.** **83.**

85. $\dfrac{\sqrt{3}}{3}$ **87.** $\dfrac{2\sqrt{42}}{3}$ **89.** The vertex is $(1, -5)$. **91. a.** 16.5 m **b.** 103.67 m **c.** 3.5 m **d.** $(0, 16.5)$ **e.** $x^2 + (y - 16.5)^2 = 16.5^2$

93. a. 125 ft **b.** 14 ft **c.** 139 ft **d.** $(0, 139)$ **e.** $x^2 + (y - 139)^2 = 125^2$ **95.** answers may vary **97.** 20 m

99. **101.**

Section 10.2
Practice Exercises

1. **2.** **3.** **4.** **5.**

Graphing Calculator Explorations 10.2

1.
3.
5.

Vocabulary, Readiness & Video Check 10.2

1. hyperbola   **3.** focus   **5.** hyperbola; $(0,0)$; x; $(a,0)$ and $(-a,0)$ **7.** a and b give you the location of 4 points on the ellipse, horizontally and vertically located from the center; these 4 points are intercepts when the center is the origin, $(0,0)$.

Exercise Set 10.2

1. ellipse   **3.** hyperbola   **5.** hyperbola   **7.** 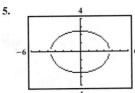   **9.** 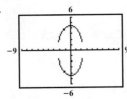   **11.** 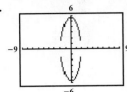   **13.**

15.   **17.**   **19.**   **21.**   **23.**   **25.**

27.   **29.**   **31.**   **33.**   **35.** circle 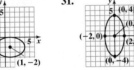   **37.** parabola

39. hyperbola   **41.** ellipse   **43.** parabola   **45.** hyperbola   **47.** ellipse

49. circle 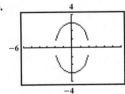   **51.** $-8x^5$ **53.** $-4x^2$ **55.** y-intercepts: 2 units   **57.** y-intercepts: 4 units   **59.** answers may vary
61. ellipses: C, E, H; circles: B, F; hyperbolas: A, D, G   **63.** A: 49, 7; B: 0, 0; C: 9, 3; D: 64, 8; E: 64, 8; F: 0, 0; G: 81, 9; H: 4, 2
65. A: $\frac{7}{6}$; B: 0; C: $\frac{3}{5}$; D: $\frac{8}{5}$; E: $\frac{8}{9}$; F: 0; G: $\frac{9}{4}$; H: $\frac{1}{6}$ **67.** equal to zero   **69.** answers may vary

71. $\dfrac{x^2}{1.69 \times 10^{16}} + \dfrac{y^2}{1.5625 \times 10^{16}} = 1$

73. 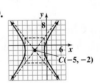   **75.**   **77.** 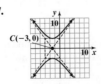   **79.**

Integrated Review

1. circle   **2.** parabola   **3.** parabola   **4.** ellipse   **5.** hyperbola   **6.** hyperbola

7. ellipse

8. circle

9. parabola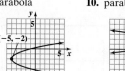

10. parabola

11. hyperbola

12. ellipse

13. ellipse

14. hyperbola

15. circle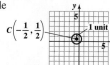

Section 10.3
Practice Exercises

1. $(-4, 3)(0, -1)$ **2.** $(4, -2)$ **3.** $\varnothing$ **4.** $(2, \sqrt{3}); (2, -\sqrt{3}); (-2, \sqrt{3}); (-2, -\sqrt{3})$

Vocabulary, Readiness & Video Check 10.3

1. That would either introduce tedious fractions (2nd equation) or a square root (1st equation) into the calculations.

Exercise Set 10.3

1. $(3, -4), (-3, 4)$ **3.** $(\sqrt{2}, \sqrt{2}), (-\sqrt{2}, -\sqrt{2})$ **5.** $(4, 0), (0, -2)$ **7.** $(-\sqrt{5}, -2), (-\sqrt{5}, 2), (\sqrt{5}, -2), (\sqrt{5}, 2)$ **9.** $\varnothing$
11. $(1, -2), (3, 6)$ **13.** $(2, 4), (-5, 25)$ **15.** $\varnothing$ **17.** $(1, -3)$ **19.** $(-1, -2), (-1, 2), (1, -2), (1, 2)$ **21.** $(0, -1)$
23. $(-1, 3), (1, 3)$ **25.** $(\sqrt{3}, 0), (-\sqrt{3}, 0)$ **27.** $\varnothing$ **29.** $(-6, 0), (6, 0), (0, -6)$ **31.** $(3, \sqrt{3})$

33. 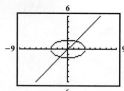 **35.** 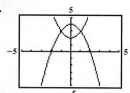 **37.** $(8x - 25)$ in. **39.** $(4x^2 + 6x + 2)$ m **41.** answers may vary **43.** $0, 1, 2, 3,$ or 4; answers may vary **45.** 9 and 7; 9 and -7; -9 and 7; -9 and -7 **47.** 15 cm by 19 cm **49.** 15 thousand compact discs; price: $3.75

51.   **53.**

Section 10.4
Practice Exercises

1.   **2.**   **3.**   **4.**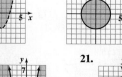

Vocabulary, Readiness & Video Check 10.4

1. For both, you look at the related equation to graph the boundary and graph a solid boundary for $\leq$ and $\geq$; you choose a test point not on the boundary and shade that region if the test point is a solution or shade the other region if not.

Exercise Set 10.4

1.   **3.**   **5.**   **7.**   **9.**   **11.**

13.   **15.** **17.** **19.** **21.** **23.**

25.   **27.**   **29.**   **31.**   **33.**   **35.**

37. not a function **39.** function **41.** 1 **43.** $3a^2 - 2$ **45.** answers may vary **47.**

Chapter 10 Vocabulary Check

1. circle; center **2.** nonlinear system of equations **3.** ellipse **4.** radius **5.** hyperbola **6.** conic sections **7.** vertex **8.** diameter

Chapter 10 Review

1. $(x + 4)^2 + (y - 4)^2 = 9$ **3.** $(x + 7)^2 + (y + 9)^2 = 11$ **5.**   **7.**   **9.**

11. 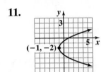 **13.** 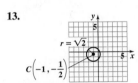 **15.** 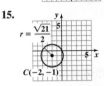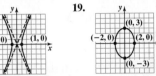 **17.** 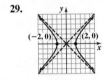 **19.**

21. 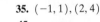 **23.**   **25.**  **27.** 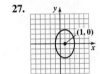 **29.** 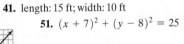 **31.**

33. $(1, -2), (4, 4)$ **35.** $(-1, 1), (2, 4)$ **37.** $(0, 2), (0, -2)$ **39.** $(1, 4)$ **41.** length: 15 ft; width: 10 ft

43.   **45.**   **47.**   **49.** 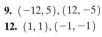 **51.** $(x + 7)^2 + (y - 8)^2 = 25$ **53.**

55.   **57.** 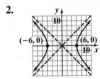 **59.**   **61.**   **63.** $(5, 1), (-1, 7)$ **65.**

Chapter 10 Test

1. 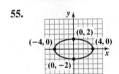 **2.**   **3.**   **4.**   **5.** 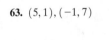 **6.**

7.   **8.** 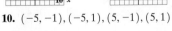 **9.** $(-12, 5), (12, -5)$ **10.** $(-5, -1), (-5, 1), (5, -1), (5, 1)$ **11.** $(6, 12), (1, 2)$ **12.** $(1, 1), (-1, -1)$

13.   **14.**   **15.**   **16.**   **17.** B **18.** height: 10 ft; width: 30 ft

Chapter 10 Cumulative Review

1. $(4 \cdot 9)y = 36y$; Sec. 1.4, Ex. 10 **3.** 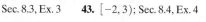 ; Sec. 3.3, Ex. 5 **5.** $\varnothing$; Sec. 4.1, Ex. 9 **7. a.** $125x^6$ **b.** $\dfrac{8}{27}$ **c.** $\dfrac{9p^8}{q^{10}}$ **d.** $64y^2$

e. $\dfrac{y^{14}}{x^{35}z^7}$; Sec. 5.2, Ex. 2 **9.** $-\dfrac{1}{6}, 3$; Sec. 5.8, Ex. 5 **11.** $\dfrac{12}{x-1}$; Sec. 6.2, Ex. 6 **13. a.** $\dfrac{1}{9xy^2}$ **b.** $\dfrac{x(x-2)}{2(x+2)}$ **c.** $\dfrac{x^2}{y^2}$; Sec. 6.3, Ex. 1

15. $2x - 5$; Sec. 6.4, Ex. 3 **17.** 16; Sec. 6.4, Ex. 11 **19.** $-6, -1$; Sec. 6.5, Ex. 5 **21.** $2\dfrac{2}{9}$ hr; no; Sec. 6.6, Ex. 4 **23. a.** 1 **b.** -4 **c.** $\dfrac{2}{5}$

d. x^2 **e.** $-3x^3$; Sec. 7.1, Ex. 3 **25. a.** $z - z^{17/3}$ **b.** $x^{2/3} - 3x^{1/3} - 10$; Sec. 7.2, Ex. 5 **27. a.** 2 **b.** $\dfrac{5}{2}\sqrt{x}$ **c.** $14xy^2\sqrt[3]{x}$

d. $4a^2b\sqrt[4]{2a}$; Sec. 7.3, Ex. 5 **29. a.** $\dfrac{5\sqrt{5}}{12}$ **b.** $\dfrac{5\sqrt[3]{7x}}{2}$; Sec. 7.4, Ex. 3 **31.** $\dfrac{\sqrt{21xy}}{3y}$; Sec. 7.5, Ex. 2 **33.** 42; Sec. 7.6, Ex. 1 **35. a.** $-i$

b. 1 **c.** -1 **d.** 1; Sec. 7.7, Ex. 6 **37.** $-1 + \sqrt{5}, -1 - \sqrt{5}$; Sec. 8.1, Ex. 5 **39.** $2 + \sqrt{2}, 2 - \sqrt{2}$; Sec. 8.2, Ex. 3 **41.** $2, -2, i, -i$;

Sec. 8.3, Ex. 3 **43.** $[-2, 3)$; Sec. 8.4, Ex. 4 **45.** ; Sec. 8.5, Ex. 8 **47.** $(2, -16)$; Sec. 8.6, Ex. 4 **49.** $\sqrt{2} \approx 1.414$; Sec. 7.3, Ex. 6

CHAPTER 11 SEQUENCES, SERIES, AND THE BINOMIAL THEOREM

Section 11.1
Practice Exercises

1. $6, 9, 14, 21, 30$ **2. a.** $-\dfrac{1}{5}$ **b.** $\dfrac{1}{20}$ **c.** $\dfrac{1}{150}$ **d.** $-\dfrac{1}{95}$ **3. a.** $a_n = 2n - 1$ **b.** $a_n = 3^n$ **c.** $a_n = \dfrac{n}{n+1}$ **d.** $a_n = -\dfrac{1}{n+1}$
4. $2022.40

Vocabulary, Readiness & Video Check 11.1

1. general **3.** infinite **5.** -1 **7.** function; domain; a_n **9.** $a_9 = 0.10(2)^{9-1} = 25.60$

Exercise Set 11.1

1. $5, 6, 7, 8, 9$ **3.** $-1, 1, -1, 1, -1$ **5.** $\dfrac{1}{4}, \dfrac{1}{5}, \dfrac{1}{6}, \dfrac{1}{7}, \dfrac{1}{8}$ **7.** $2, 4, 6, 8, 10$ **9.** $-1, -4, -9, -16, -25$ **11.** $2, 4, 8, 16, 32$ **13.** $7, 9, 11, 13, 15$

15. $-1, 4, -9, 16, -25$ **17.** 75 **19.** 118 **21.** $\dfrac{6}{5}$ **23.** 729 **25.** $\dfrac{4}{7}$ **27.** $\dfrac{1}{8}$ **29.** -95 **31.** $-\dfrac{1}{25}$ **33.** $a_n = 4n - 1$ **35.** $a_n = -2^n$

37. $a_n = \dfrac{1}{3^n}$ **39.** 48 ft, 80 ft, and 112 ft **41.** $a_n = 0.10(2)^{n-1}$; $819.20 **43.** 2400 cases; 75 cases **45.** 50 sparrows in 2004; extinct in 2010

47.   **49.** **51.** $\sqrt{13}$ units **53.** $\sqrt{41}$ units **55.** $1, 0.7071, 0.5774, 0.5, 0.4472$ **57.** $2, 2.25, 2.3704, 2.4414, 2.4883$

Section 11.2
Practice Exercises

1. $4, 9, 14, 19, 24$ **2. a.** $a_n = 5 - 3n$ **b.** -31 **3.** 51 **4.** 47 **5.** $a_n = 54{,}800 + 2200n$; $61,400 **6.** $8, -24, 72, -216$ **7.** $\dfrac{1}{64}$
8. -192 **9.** $a_1 = 3; r = \dfrac{3}{2}$ **10.** 75 units

Vocabulary, Readiness & Video Check 11.2

1. geometric; ratio **3.** first; difference **5.** If there is a common difference between each term and its preceding term in a sequence, it's an arithmetic sequence.

Exercise Set 11.2

1. $4, 6, 8, 10, 12$ **3.** $6, 4, 2, 0, -2$ **5.** $1, 3, 9, 27, 81$ **7.** $48, 24, 12, 6, 3$ **9.** 33 **11.** -875 **13.** -60 **15.** 96 **17.** -28 **19.** 1250

21. 31 **23.** 20 **25.** $a_1 = \dfrac{2}{3}; r = -2$ **27.** answers may vary **29.** $a_1 = 2; d = 2$ **31.** $a_1 = 5; r = 2$ **33.** $a_1 = \dfrac{1}{2}; r = \dfrac{1}{5}$

35. $a_1 = x; r = 5$ **37.** $a_1 = p; d = 4$ **39.** 19 **41.** $-\dfrac{8}{9}$ **43.** $\dfrac{17}{2}$ **45.** $\dfrac{8}{81}$ **47.** -19 **49.** $a_n = 4n + 50$; 130 seats **51.** $a_n = 6(3)^{n-1}$

53. 486, 162, 54, 18, 6; $a_n = \dfrac{486}{3^{n-1}}$; 6 bounces   **55.** $a_n = 3875 + 125n$; $5375   **57.** 25 g   **59.** $\dfrac{11}{18}$ **61.** 40   **63.** $\dfrac{907}{495}$

65. $11,782.40, $5891.20, $2945.60, $1472.80   **67.** 19.652, 19.618, 19.584, 19.55   **69.** answers may vary

Section 11.3
Practice Exercises

1. a. $-\dfrac{5}{4}$ **b.** 360   **2. a.** $\displaystyle\sum_{i=1}^{6} 5i$ **b.** $\displaystyle\sum_{i=1}^{4}\left(\dfrac{1}{5}\right)^i$ **3.** $\dfrac{655}{72}$ or $9\dfrac{7}{72}$ **4.** 95 plants

Vocabulary, Readiness & Video Check 11.3

1. infinite   **3.** summation   **5.** partial sum   **7.** sigma/sum, index of summation, beginning value of i, ending value of i, and general term of the sequence

Exercise Set 11.3

1. -2 **3.** 60   **5.** 20   **7.** $\dfrac{73}{168}$ **9.** $\dfrac{11}{36}$ **11.** 60   **13.** 74   **15.** 62   **17.** $\dfrac{241}{35}$ **19.** $\displaystyle\sum_{i=1}^{5}(2i-1)$ **21.** $\displaystyle\sum_{i=1}^{4}4(3)^{i-1}$

23. $\displaystyle\sum_{i=1}^{6}(-3i+15)$ **25.** $\displaystyle\sum_{i=1}^{4}\dfrac{4}{3^{i-2}}$ **27.** $\displaystyle\sum_{i=1}^{7}i^2$ **29.** -24 **31.** 0   **33.** 82   **35.** -20 **37.** -2 **39.** 1, 2, 3, ..., 10; 55 trees

41. 186 units   **43.** 35 species; 82 species   **45.** 30 opossums; 68 opossums   **47.** 6.25 lb; 93.75 lb   **49.** 16.4 in.; 134.5 in.   **51.** 10   **53.** $\dfrac{10}{27}$

55. 45   **57.** 90   **59. a.** $2 + 6 + 12 + 20 + 30 + 42 + 56$ **b.** $1 + 2 + 3 + 4 + 5 + 6 + 7 + 1 + 4 + 9 + 16 + 25 + 36 + 49$

c. answers may vary   **d.** true; answers may vary

Integrated Review

1. $-2, -1, 0, 1, 2$ **2.** $\dfrac{7}{2}, \dfrac{7}{3}, \dfrac{7}{4}, \dfrac{7}{5}, \dfrac{7}{6}$ **3.** 1, 3, 9, 27, 81   **4.** $-4, -1, 4, 11, 20$ **5.** 64   **6.** -14 **7.** $\dfrac{1}{40}$ **8.** $-\dfrac{1}{82}$ **9.** 7, 4, 1, -2, -5

10. $-3, -15, -75, -375, -1875$ **11.** $45, 15, 5, \dfrac{5}{3}, \dfrac{5}{9}$ **12.** $-12, -2, 8, 18, 28$ **13.** 101   **14.** $\dfrac{243}{16}$ **15.** 384   **16.** 185   **17.** -10 **18.** $\dfrac{1}{2}$

19. 50   **20.** 98   **21.** $\dfrac{31}{2}$ **22.** $\dfrac{61}{20}$ **23.** -10 **24.** 5

Section 11.4
Practice Exercises

1. 80   **2.** 1275   **3.** 105 blocks of ice   **4.** $\dfrac{341}{8}$ or $42\dfrac{5}{8}$ **5.** $987,856   **6.** $\dfrac{28}{3}$ or $9\dfrac{1}{3}$ **7.** 900 in.

Vocabulary, Readiness & Video Check 11.4

1. arithmetic   **3.** geometric   **5.** arithmetic   **7.** Use the general term formula from Section 11.2 for the general term of an arithmetic sequence: $a_n = a_1 + (n-1)d$.   **9.** The common ratio r is 3 for this sequence so that $|r| \geq 1$, or $|3| \geq 1$; S_∞ doesn't exist if $|r| \geq 1$.

Exercise Set 11.4

1. 36   **3.** 484   **5.** 63   **7.** $\dfrac{312}{125}$ **9.** 55   **11.** 16   **13.** 24   **15.** $\dfrac{1}{9}$ **17.** -20 **19.** $\dfrac{16}{9}$ **21.** $\dfrac{4}{9}$ **23.** 185   **25.** $\dfrac{381}{64}$

27. $-\dfrac{33}{4}$ or -8.25 **29.** $-\dfrac{75}{2}$ **31.** $\dfrac{56}{9}$ **33.** 4000, 3950, 3900, 3850, 3800; 3450 cars; 44,700 cars   **35.** Firm A (Firm A, $265,000; Firm B, $254,000)

37. $39,930; $139,230   **39.** 20 min; 123 min   **41.** 180 ft   **43.** Player A, 45 points; Player B, 75 points   **45.** $3050   **47.** $10,737,418.23

49. 720   **51.** 3   **53.** $x^2 + 10x + 25$ **55.** $8x^3 - 12x^2 + 6x - 1$ **57.** $\dfrac{8}{10} + \dfrac{8}{100} + \dfrac{8}{1000} + \cdots; \dfrac{8}{9}$ **59.** answers may vary

Section 11.5
Practice Exercises

1. $p^7 + 7p^6r + 21p^5r^2 + 35p^4r^3 + 35p^3r^4 + 21p^2r^5 + 7pr^6 + r^7$ **2. a.** $\dfrac{1}{7}$ **b.** 840   **c.** 5   **d.** 1

3. $a^9 + 9a^8b + 36a^7b^2 + 84a^6b^3 + 126a^5b^4 + 126a^4b^5 + 84a^3b^6 + 36a^2b^7 + 9ab^8 + b^9$ **4.** $a^3 + 15a^2b + 75ab^2 + 125b^3$
5. $27x^3 - 54x^2y + 36xy^2 - 8y^3$ **6.** $1,892,352x^5y^6$

Vocabulary, Readiness & Video Check 11.5

1. 1   **3.** 24   **5.** 6   **7.** Pascal's triangle gives you the coefficients of the terms of the expanded binomial; also, the power tells you how many terms the expansion has (1 more than the power on the binomial).   **9.** The theorem is in terms of $(a+b)^n$, so if your binomial is of the form $(a-b)^n$, then remember to think of it as $(a + (-b))^n$, so your second term is $-b$.

Exercise Set 11.5

1. $m^3 + 3m^2n + 3mn^2 + n^3$ **3.** $c^5 + 5c^4d + 10c^3d^2 + 10c^2d^3 + 5cd^4 + d^5$ **5.** $y^5 - 5y^4x + 10y^3x^2 - 10y^2x^3 + 5yx^4 - x^5$
7. answers may vary   **9.** 8   **11.** 42   **13.** 360   **15.** 56   **17.** $a^7 + 7a^6b + 21a^5b^2 + 35a^4b^3 + 35a^3b^4 + 21a^2b^5 + 7ab^6 + b^7$
19. $a^5 + 10a^4b + 40a^3b^2 + 80a^2b^3 + 80ab^4 + 32b^5$ **21.** $q^9 + 9q^8r + 36q^7r^2 + 84q^6r^3 + 126q^5r^4 + 126q^4r^5 + 84q^3r^6 + 36q^2r^7 + 9qr^8 + r^9$
23. $1024a^5 + 1280a^4b + 640a^3b^2 + 160a^2b^3 + 20ab^4 + b^5$ **25.** $625a^4 - 1000a^3b + 600a^2b^2 - 160ab^3 + 16b^4$ **27.** $8a^3 + 36a^2b + 54ab^2 + 27b^3$

29. $x^5 + 10x^4 + 40x^3 + 80x^2 + 80x + 32$ **31.** $5cd^4$ **33.** d^7 **35.** $-40r^2s^3$ **37.** $6x^2y^2$ **39.** $30a^9b$

41. ; not one-to-one **43.** 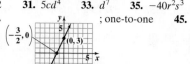 ; one-to-one **45.** ; not one-to-one

47. $x^2\sqrt{x} + 5\sqrt{3}x^2 + 30x\sqrt{x} + 30\sqrt{3}x + 45\sqrt{x} + 9\sqrt{3}$ **49.** 126 **51.** 28 **53.** answers may vary

Chapter 11 Vocabulary Check

1. finite sequence **2.** factorial of n **3.** infinite sequence **4.** geometric sequence, common ratio **5.** series **6.** general term **7.** arithmetic sequence, common difference **8.** Pascal's triangle

Chapter 11 Review

1. $-3, -12, -27, -48, -75$ **3.** $\dfrac{1}{100}$ **5.** $a_n = \dfrac{1}{6n}$ **7.** 144 ft, 176 ft, 208 ft **9.** 660,000; 1,320,000; 2,640,000; 5,280,000; 10,560,000; 2010:

10,560,000 infested acres **11.** $-2, -\dfrac{4}{3}, -\dfrac{8}{9}, -\dfrac{16}{27}, -\dfrac{32}{81}$ **13.** 111 **15.** -83 **17.** $a_1 = 3; d = 5$ **19.** $a_n = \dfrac{3}{10^n}$ **21.** $a_1 = \dfrac{8}{3}, r = \dfrac{3}{2}$

23. $a_1 = 7x, r = -2$ **25.** 8, 6, 4.5, 3.4, 2.5, 1.9; good **27.** $a_n = 2^{n-1}$; 512; $536{,}870{,}912$ **29.** $a_n = 150n + 750$; $1650/month

31. $1 + 3 + 5 + 7 + 9 = 25$ **33.** $\dfrac{1}{4} - \dfrac{1}{6} + \dfrac{1}{8} = \dfrac{5}{24}$ **35.** $\displaystyle\sum_{i=1}^{6} 3^{i-1}$ **37.** $\displaystyle\sum_{i=1}^{4} \dfrac{1}{4^i}$ **39.** $a_n = 20(2)^n$; n represents the number of 8-hour periods;

1280 yeast **41.** Job A, $48,300; Job B, $46,600 **43.** -4 **45.** -10 **47.** 150 **49.** 900 **51.** -410 **53.** 936 **55.** 10

57. -25 **59.** $30,418; $99,868 **61.** $58; $553 **63.** 2696 mosquitoes **65.** $\dfrac{5}{9}$ **67.** $x^5 + 5x^4z + 10x^3z^2 + 10x^2z^3 + 5xz^4 + z^5$

69. $16x^4 + 32x^3y + 24x^2y^2 + 8xy^3 + y^4$ **71.** $b^8 + 8b^7c + 28b^6c^2 + 56b^5c^3 + 70b^4c^4 + 56b^3c^5 + 28b^2c^6 + 8bc^7 + c^8$

73. $256m^4 - 256m^3n + 96m^2n^2 - 16mn^3 + n^4$ **75.** $35a^4b^3$ **77.** 130 **79.** 40.5

Chapter 11 Test

1. $-\dfrac{1}{5}, \dfrac{1}{6}, -\dfrac{1}{7}, \dfrac{1}{8}, -\dfrac{1}{9}$ **2.** 247 **3.** $a_n = \dfrac{2}{5}\left(\dfrac{1}{5}\right)^{n-1}$ **4.** $a_n = (-1)^n 9n$ **5.** 155 **6.** -330 **7.** $\dfrac{144}{5}$ **8.** 1 **9.** 10 **10.** -60

11. $a^6 - 6a^5b + 15a^4b^2 - 20a^3b^3 + 15a^2b^4 - 6ab^5 + b^6$ **12.** $32x^5 + 80x^4y + 80x^3y^2 + 40x^2y^3 + 10xy^4 + y^5$ **13.** 925 people; 250 people initially

14. $1 + 3 + 5 + 7 + 9 + 11 + 13 + 15$; 64 shrubs **15.** 33.75 cm, 218.75 cm **16.** 320 cm **17.** 304 ft; 1600 ft **18.** $\dfrac{14}{33}$

Chapter 11 Cumulative Review

1. a. -5 **b.** 3 **c.** $-\dfrac{1}{8}$ **d.** -4 **e.** $\dfrac{1}{4}$ **f.** undefined; Sec. 1.3, Ex. 5 **3.** $2350; Sec. 2.2, Ex. 4 **5.** $y = \dfrac{2x}{3} + \dfrac{7}{3}$; Sec. 2.3, Ex. 2

7. a. $15x^7$ **b.** $-9.6x^4p^{12}$; Sec. 5.1, Ex. 2 **9.** $x^3 - 4x^2 - 3x + 6 + \dfrac{22}{x+2}$; Sec. 6.4, Ex. 9 **11. a.** $5\sqrt{2}$ **b.** $2\sqrt[3]{3}$ **c.** $\sqrt{26}$

d. $2\sqrt[4]{2}$; Sec. 7.3, Ex. 3 **13.** 10%; Sec. 8.1, Ex. 9 **15.** 2, 7; Sec. 8.3, Ex. 4 **17.** $\left(-\dfrac{7}{2}, -1\right)$; Sec. 8.4, Ex. 5 **19.** $\dfrac{25}{4}$ ft; $\dfrac{5}{8}$ sec; Sec. 8.6, Ex. 5

21. a. 25; 7 **b.** $x^2 + 6x + 9; x^2 + 3$; Sec. 9.1, Ex. 2 **23.** $f^{-1} = \{(1,0), (7,-2), (-6,3), (4,4)\}$; Sec. 9.2, Ex. 3 **25. a.** 4 **b.** $\dfrac{3}{2}$

c. 6; Sec. 9.3, Ex. 4 **27. a.** 2 **b.** -1 **c.** 3 **d.** 6; Sec. 9.5, Ex. 5 **29. a.** $\log_{11} 30$ **b.** $\log_3 6$ **c.** $\log_2(x^2 + 2x)$; Sec. 9.6, Ex. 1

31. $2509.30; Sec. 9.7, Ex. 8 **33.** $\dfrac{\log 7}{\log 3} \approx 1.7712$; Sec. 9.8, Ex. 1 **35.** 18; Sec. 9.8, Ex. 2 **37.** 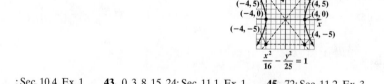 ; Sec. 10.2, Ex. 4

$\dfrac{x^2}{16} - \dfrac{y^2}{25} = 1$

39. $(2, \sqrt{2})$; Sec. 10.3, Ex. 2 **41.** ; Sec. 10.4, Ex. 1 **43.** 0, 3, 8, 15, 24; Sec. 11.1, Ex. 1 **45.** 72; Sec. 11.2, Ex. 3

$\dfrac{x^2}{9} + \dfrac{y^2}{16} \le 1$

47. a. $\dfrac{7}{2}$ **b.** 56; Sec. 11.3, Ex. 1 **49.** 465; Sec. 11.4, Ex. 2

APPENDIX A   GEOMETRY

Exercise Set Appendix A

1. $V = 72$ cu in.; $SA = 108$ sq in.   **3.** $V = 512$ cu cm; $SA = 384$ sq cm   **5.** $V = 4\pi$ cu yd $\approx 12\frac{4}{7}$ cu yd; $SA = \left(2\sqrt{13}\pi + 4\pi\right)$ sq yd ≈ 35.20 sq yd

7. $V = \frac{500}{3}\pi$ cu in. $\approx 523\frac{17}{21}$ cu in.; $SA = 100\pi$ sq in. $\approx 314\frac{2}{7}$ sq in.   **9.** $V = 48$ cu cm; $SA = 96$ sq cm   **11.** $2\frac{10}{27}$ cu in.   **13.** 26 sq ft

15. $10\frac{5}{6}$ cu in.   **17.** 960 cu cm   **19.** 196π sq in.   **21.** $7\frac{1}{2}$ cu ft   **23.** $12\frac{4}{7}$ cu cm

APPENDIX B   STRETCHING AND COMPRESSING GRAPHS OF ABSOLUTE VALUE FUNCTIONS

Exercise Set Appendix B

1.

$f(x) = 3|x|$ $(0,0)$

3.

$f(x) = \frac{1}{4}|x|$ $(0,0)$

5.

$g(x) = 2|x| + 3$ $(0,3)$

7.

$(0,0)$ $h(x) = -\frac{1}{2}|x|$

9.

$f(x) = 4|x - 1|$ $(1,0)$

11.

$(0, -2)$ $g(x) = -\frac{1}{3}|x| - 2$

13.

$(3, 4)$ $f(x) = -2|x - 3| + 4$

15.

$(-2, -5)$ $f(x) = \frac{2}{3}|x + 2| - 5$

APPENDIX C   SOLVING SYSTEMS OF EQUATIONS USING DETERMINANTS

Vocabulary, Readiness & Video Check Appendix C

1. 56   **3.** -32 **5.** 20

Exercise Set Appendix C

1. 26   **3.** -19 **5.** 0   **7.** $\frac{13}{6}$ **9.** $(1, 2)$ **11.** $\{(x, y) \mid 3x + y = 1\}$ **13.** $(9, 9)$ **15.** $(-3, -2)$ **17.** $(3, 4)$ **19.** 8   **21.** 0

23. 15   **25.** 54   **27.** $(-2, 0, 5)$ **29.** $(6, -2, 4)$ **31.** $(-2, 3, -1)$ **33.** $(0, 2, -1)$ **35.** 5   **37.** 0; answers may vary

APPENDIX D AN INTRODUCTION TO USING A GRAPHING UTILITY

Viewing Window and Interpreting Window Settings Exercise Set

1. yes   **3.** no   **5.** answers may vary   **7.** answers may vary   **9.** answers may vary

11. Xmin $= -12$ Ymin $= -12$ **13.** Xmin $= -9$ Ymin $= -12$ **15.** Xmin $= -10$ Ymin $= -25$ **17.** Xmin $= -10$ Ymin $= -30$
Xmax $= 12$ Ymax $= 12$ Xmax $= 9$ Ymax $= 12$ Xmax $= 10$ Ymax $= 25$ Xmax $= 10$ Ymax $= 30$
Xscl $= 3$ Yscl $= 3$ Xscl $= 1$ Yscl $= 2$ Xscl $= 2$ Yscl $= 5$ Xscl $= 1$ Yscl $= 3$

19. Xmin $= -20$ Ymin $= -30$
Xmax $= 30$ Ymax $= 50$
Xscl $= 5$ Yscl $= 10$

Graphing Equations and Square Viewing Window Exercise Set

1. Setting B   **3.** Setting B   **5.** Setting B

7.

9.

11.

13.

15.

17.

19.

21.

PRACTICE FINAL EXAM

1. $6\sqrt{6}$ **2.** $-\dfrac{3}{2}$ **3.** 5 **4.** $\dfrac{1}{81x^2}$ **5.** $\dfrac{64}{3}$ **6.** $\dfrac{3a^7}{2b^5}$ **7.** $\dfrac{8a^{1/3}c^{2/3}}{b^{5/12}}$ **8.** $3y(x+3y)(x-3y)$ **9.** $2(2y-1)(4y^2+2y+1)$

10. $(x+3)(x-3)(y-3)$ **11.** $-5x^3y-11x-9$ **12.** $36m^2+12mn+n^2$ **13.** $2x^3-13x^2+14x-4$ **14.** $\dfrac{x+2}{2(x+3)}$ **15.** $\dfrac{1}{x(x+3)}$

16. $\dfrac{5x-2}{(x-3)(x+2)(x-2)}$ **17.** $-x\sqrt{5x}$ **18.** -20 **19.** $2x^2-x-2+\dfrac{2}{2x+1}$ **20.** -32 **21.** $1,\dfrac{2}{3}$ **22.** $4,-\dfrac{8}{7}$

23. $\left(\dfrac{3}{2},5\right]$ **24.** $(-\infty,-2)\cup\left(\dfrac{4}{3},\infty\right)$ **25.** 3 **26.** $\dfrac{3\pm\sqrt{29}}{2}$ **27.** $2,3$ **28.** $\left(-\infty,-\dfrac{3}{2}\right)\cup(5,\infty)$ **29.** $\left(\dfrac{7}{2},-10\right)$

30.   **31.**   **32.**   **33.** ; domain: $(-\infty,\infty)$; range: $(-\infty,-1]$

34.   **35.** ; domain: $(-\infty,\infty)$; range: $(-3,\infty)$ **36.** $f(x)=-\dfrac{1}{2}x$ **37.** $f(x)=-\dfrac{1}{3}x+\dfrac{5}{3}$ **38.** $2\sqrt{26}$ units

39. $\left(-4,\dfrac{7}{2}\right)$ **40.** $\dfrac{3\sqrt{y}}{y}$ **41.** $\dfrac{8-6\sqrt{x}+x}{8-2x}$ **42.** \$35,938 **43.** $(x+2y)(x-2y)$ **44.** 5 **45.** 16 **46.** 7.2 ft **47. a.** 272 ft

b. 5.12 sec **48.** 5 gal of 10%; 15 gal of 20% **49.** $0-2i\sqrt{2}$ **50.** $0-3i$ **51.** $7+24i$ **52.** $-\dfrac{3}{2}+\dfrac{5}{2}i$ **53.** $(g\circ h)(x)=x^2-6x-2$

54. $f^{-1}(x)=\dfrac{-x+6}{2}$ **55.** $\log_5\dfrac{x^4}{x+1}$ **56.** -1 **57.** $\dfrac{1}{2}\left(\dfrac{\log 4}{\log 3}-5\right)$; -1.8691 **58.** 22 **59.** $\dfrac{43}{21}$ **60.** $\dfrac{1}{2}$

61.   **62.** 64,805 prairie dogs **63.** **64.** **65.**

66. $(-5,-1),(-5,1),(5,-1),(5,1)$ **67.** $-\dfrac{1}{5},\dfrac{1}{6},-\dfrac{1}{7},\dfrac{1}{8},-\dfrac{1}{9}$ **68.** 155 **69.** 1 **70.** 10 **71.** $32x^5+80x^4y+80x^3y^2+40x^2y^3+10xy^4+y^5$

Index

Photo Credits